About the Authors

WILLIAM P. CUNNINGHAM

William P. Cunningham is an emeritus professor at the University of Minnesota. In his 38-year career at the university, he taught a variety of biology courses, including Environmental Science, Conservation Biology, Environmental Health, Environmental Ethics, Plant Physiology, and Cell Biology. He is a member of the Academy of Distinguished Teachers, the highest teaching award granted at the University of Minnesota. He was a member of a number of interdisciplinary programs for international students, teachers, and nontraditional students. He also carried out research or taught in Sweden, Norway, Brazil, New Zealand, China, and Indonesia.

Professor Cunningham has participated in a number of governmental and nongovernmental organizations over the past 40 years. He was chair of the Minnesota chapter of the Sierra Club, a member of the Sierra Club national committee on energy policy, vice president of the Friends of the Boundary Waters Canoe Area, chair of the Minnesota governor's task force on energy policy, and a citizen member of the Minnesota Legislative Commission on Energy.

In addition to environmental science textbooks, Cunningham edited three editions of the *Environmental Encyclopedia,* published by Thompson-Gale Press. He has also authored or coauthored about 50 scientific articles, mostly in the fields of cell biology and conservation biology, as well as several invited chapters or reports in the areas of energy policy and environmental health. His Ph.D. from the University of Texas was in botany.

Professor Cunningham's hobbies include photography birding, hiking, gardening, and traveling. He lives in St. Paul, Minnesota, with his wife, Mary. He has three children (one of whom is coauthor of this book) and seven grandchildren.

Both authors have a long-standing interest in the topic in this book. Nearly half the photos in the book were taken on trips to the places they discuss.

MARY ANN CUNNINGHAM

Mary Ann Cunningham is an associate professor of geography at Vassar College. A biogeographer with interests in landscape ecology, geographic information systems (GIS), and remote sensing, she teaches environmental science, natural resource conservation, and land-use planning, as well as GIS and remote sensing. Field research methods, statistical methods,

and scientific methods in data analysis are regular components of her teaching. As a scientist and an educator, Mary Ann enjoys teaching and conducting research with both science students and nonscience liberal arts students. As a geographer, she likes to engage students with the ways their physical surroundings and social context shape their world experience. In addition to teaching at a liberal arts college, she has taught at community colleges and research universities.

Mary Ann has been writing in environmental science for over a decade, and she has been coauthor of this book since its seventh edition. She is also coauthor of *Principles of Environmental Science* (now in its fifth edition) and an editor of the *Environmental Encyclopedia* (third edition, Thompson-Gale Press). She has published work on pedagogy in cartography, as well as instructional and testing materials in environmental science. With colleagues at Vassar she has published a GIS lab manual, *Exploring Environmental Science with GIS,* designed to provide students with an easy, inexpensive introduction to spatial and environmental analysis with GIS.

In addition to environmental science, Mary Ann's primary research activities focus on land-cover change, habitat fragmentation, and distributions of bird populations. This work allows her to conduct field studies in the grasslands of the Great Plains as well as in the woodlands of the Hudson Valley. In her spare time she loves to travel, hike, and watch birds.

Mary Ann holds a bachelor's degree from Carleton College, a master's degree from the University of Oregon, and a Ph.D. from the University of Minnesota.

Brief Contents

TWELFTH EDITION

Environmental SCIENCE

A Global Concern

William P. Cunningham
University of Minnesota

Mary Ann Cunningham
Vassar College

McGraw Hill

Connect
Learn
Succeed™

ENVIRONMENTAL SCIENCE: A GLOBAL CONCERN, TWELFTH EDITION

Published by McGraw-Hill, a business unit of The McGraw-Hill Companies, Inc., 1221 Avenue of the Americas, New York, NY 10020. Copyright © 2012 by The McGraw-Hill Companies, Inc. All rights reserved. Printed in the United States of America. Previous editions © 2010, 2008, and 2007. No part of this publication may be reproduced or distributed in any form or by any means, or stored in a database or retrieval system, without the prior written consent of The McGraw-Hill Companies, Inc., including, but not limited to, in any network or other electronic storage or transmission, or broadcast for distance learning.

Some ancillaries, including electronic and print components, may not be available to customers outside the United States.

 This book is printed on recycled, acid-free paper containing 10% postconsumer waste.

1 2 3 4 5 6 7 8 9 0 QDB/QDB 1 0 9 8 7 6 5 4 3 2 1

ISBN 978–0–07–131495–4
MHID 0–07–131495–4

All credits appearing on page or at the end of the book are considered to be an extension of the copyright page.

www.mhhe.com

Contents

7 Human Populations   131

8 Environmental Health and Toxicology   153

9 Food and Hunger   177

Preface

ENVIRONMENTAL SCIENCE HAS NEVER BEEN MORE IMPORTANT

A serene tropical coastline, shown on the cover, invokes some of the profoundly important, diverse, and fascinating environmental systems that you can explore in environmental science. Though we live firmly on dry land, our lives are intricately tied to life offshore. Coastal coral reefs, salt marshes, estuaries, mangrove forests, and seagrass beds sustain three-quarters of all commercial fish and shellfish during some part of their life cycles. These species are the main protein sources for at least 1.5 billion people, one-fifth of all humanity, and are important nutritional sources for billions of others. Oceans, which store and distribute heat, strongly shape our climate and ecosystems on land.

These systems are also increasingly vulnerable to our actions. Overfishing and destructive harvesting techniques imperil marine ecosystems. Since 1989, 13 of the 17 major marine fisheries have declined dramatically or become commercially unsustainable. Commercial fisheries settle for smaller and smaller species, as more populations disappear. Pollutants, plastic debris, and nutrients washing off the land surface severely contaminate marine systems. Climate change and warming seas threaten valuable coral reefs, and ocean acidification, resulting from high carbon dioxide emissions, debilitates corals and shellfish. We don't know when the ocean systems we depend on might reach a tipping point and spiral into instability.

What can we do with such challenges? Plenty. A first step is to understand the issues and systems better by studying environmental science, as you are now doing. As we begin to understand environmental systems, we have some hope of working to keep them stable and healthy. As you read this book, you may discover many ways to engage in the issues and ideas involved in environmental science. Whether you are a biologist, a geologist, a chemist, an economist, a political scientist, a writer, or an artist or poet who can capture our imagination, you can find fruitful and interesting ways to engage with the topics in this book.

Another step is to understand how our policies and economic decisions influence the systems on which we depend. We've spent far more money traveling to the moon than we have exploring the ecological treasures on earth and under the sea. We spend more effort debating climate change than it would cost to address it. We often follow shortsighted policies, degrading habitats and biodiversity or exploiting energy resources unsustainably.

At the same time, there is abundant evidence of the progress we can make. Human population growth is slowing almost everywhere, as education for women and economic stability allow for small, well-cared-for families. New energy technologies are proving to be reliable alternatives to fossil fuels in many places. Solar, wind, biomass, geothermal energy, and conservation could supply all the energy we need, if we chose to invest in them. We have also shown that we can dramatically improve water quality and air quality if we put our minds to it.

Governments around the world are acknowledging the costs of environmental degradation and are taking steps to reduce their environmental impacts. China has announced ambitious plans to restore forests, conserve water, reduce air and water pollution, and develop sustainable energy supplies. China has even agreed to reduce greenhouse gas emissions, something it refused to consider when the Kyoto Protocol was signed a decade ago.

In the United States, there has been renewed respect for both science and the environment. Citizens and voters need to remain vigilant to protect the status of science in policy making, but experienced scientists have been appointed to government posts previously given to political appointees. President Obama has involved scientific evidence and analysis in guiding federal policy. He has taken many steps to safeguard our environment and its resources, and public support for these steps has been overwhelmingly enthusiastic. Grants and tax incentives are supporting more sustainable energy and millions of green jobs.

Businesses, too, now recognize the opportunities in conservation, recycling, producing nontoxic products, and reducing their ecological footprints. Many are hiring sustainability experts and beginning to recognize environmental impacts in accounting.

This is a good time to study environmental science. New jobs are being created in environmental fields. Public opinion supports environmental protection because the public sees the importance of environmental health for the economy, society, and quality of life. College and university students are finding new ways to organize, network, and take action to protect the environment they will inherit.

Ecologist Norman Meyers has said, "The present has a unique position in history. Now, as never before, we have technical, political, and economic resources to solve our global environmental crisis. And if we don't do it now, it may be too late for future generations to do so." We hope you'll find ideas in this book to help you do something to make the world a better place.

WHAT SETS THIS BOOK APART?

As practicing scientists and educators, we bring to this book decades of experience in the classroom, in the practice of science, and in civic engagement. This experience can help give students a clear sense of what environmental science is and why it matters.

A positive viewpoint

Our intent with this book is to empower students to make a difference in their communities by becoming informed, critical thinkers with an awareness of environmental issues and the factors that cause them and some ways to resolve them. It's easy to be overwhelmed by the countless environmental problems we face, and certainly it is essential to be aware of these issues and to take them as a wake-up call. It is also essential to see a way forward. Throughout this text we balance evidence of serious environmental challenges with ideas about what we can do to overcome them.

We recognize that many environmental problems remain severe, but also there have been many improvements over past decades, including cleaner water and cleaner air for most Americans, declining hunger rates and birth rates, and increasing access to education. An entire chapter (chapter 13) focuses on ecological restoration, one of the most important aspects of ecology today. Case studies in most chapters show examples of real progress, and "What Can You Do?" lists give students ideas for contributing to solutions.

A balanced presentation encourages critical thinking

Critical thinking is an essential skill, and environmental science provides abundant opportunity to practice critical analysis of contradictory data, conflicting interests, and opposing interpretations of evidence. Among the most important practices a student can learn are to think analytically about evidence, to consider uncertainty, and to skeptically evaluate the sources of information. We give students many opportunities to practice critical thinking in brief "Think About It" questions, in "What Do You Think?" readings, in end-of-chapter Discussion Questions, and throughout the text. We present balanced evidence, and we provide the tools for students to discuss and form their own opinions.

We also devote a special introduction (Learning to Learn) to an explicit examination of how to study, and how to practice critical, analytical, and reflective thinking.

Emphasis on science

Science is critical for understanding environmental change. We emphasize principles and methods of science through the use of quantitative reasoning, statistics, uncertainty and probability. Students can practice these skills in a variety of data analysis graphing exercises. "Exploring Science" readings also show how scientists observe the world and gather data.

An integrated, global perspective

Globalization spotlights the interconnectedness of environmental concerns, as well as economies. To remain competitive in a global economy, it is critical that we understand conditions in other countries and cultures. This book provides case studies and topics from regions around the world, as well as maps and data showing global issues. These examples also show the integration between environmental, social, and economic conditions at home and abroad.

Google Earth™ placemarks

Throughout this book you'll see small globe icons that mark topics particularly suited to exploration in Google Earth. This online program lets you view amazingly detailed satellite images of the earth that will help you understand the geographic context of these places you're studying. We've created placemarks that will help you find the places being discussed, and we've provided brief descriptions and questions to stimulate a thoughtful exploration of each site and its surroundings. This interactive geographical exploration is a wonderful tool to give you an international perspective on environmental issues.

You can download placemarks individually (from www.mhhe.com/cunningham12e) or all at once (from EnvironmentalScience-Cunningham.blogspot.com). You'll also find links there for downloading the free Google Earth program as well as suggestions on how to use it effectively.

Active learning resources

The Google Earth placemarks, questions for Discussion and Critical Thinking, "Think About It" notes, and other resources are designed to be used as starting points for lecture, discussion in class, essays, or other active learning activities. Some data analysis exercises involve simple polls of classes, which can be used for graphing and interpretation. data analysis exercises vary in the kinds of learning and skills involved, and all aim to give students an opportunity to explore data or documents on their own, to conduct their own evaluation and learn about the resources available to them. These activities can serve as starting points for lab exercises as well as independent projects.

WHAT'S NEW IN THIS EDITION?

Of the 25 chapters in this book, 17 have new opening case studies, which introduce new developments, classic cases, and key ideas and problems for a chapter. Discussions of many topics are updated, with the latest available data used throughout the book.

Specific changes to chapters

- **Learning to Learn** has a new boxed essay that explores critical reading of the news ("How Do You Tell the News from the Noise?") as well as revised discussions of critical and analytical thinking strategies.

- **Chapter 1** opens with a new case study about Rizhao, China, where 99 percent of all households use solar collectors for water heating. Rizhao claims to be the first carbon-neutral city in the developing world. Updates are also given on Brazil's reduction in forest destruction by about two-thirds in the past five years, while protected areas have increased nearly fivefold in two decades.

- **Chapter 2** has a new case study demonstrating the design of field experiments to test the effects of climate change on boreal forests. The case study shows how field experiments are similar to and yet different from controlled lab approaches to understanding environmental change.

- **Chapter 3** has a new opening case study on nutrients in the Chesapeake Bay watershed and a data analysis exercise on the Chesapeake environmental report card.

- **Chapter 4** includes two new boxed readings, one on the evolution of influenza strains, which shows evolution in action as well as offering important insights into community health; the other examines the ecological effects of earthworm invasions in northern forests, a surprising and important example of species interactions.

- **Chapter 5** opens with a new case study discussing reforestation by Kenya's Greenbelt movement. An expanded explanation of climate graphs precedes the discussion of biomes.

- **Chapter 6** begins with a new case study on population biology of the overfished bluefin tuna, a subject of ongoing disputes over endangered species listing. A new Exploring Science reading about how we study population viability in fish accompanies the new case study.

- **Chapter 7** includes a new boxed reading on China's highly successful but controversial one-child-per-family policy. This provides an opportunity to discuss family planning, population control, and demographic trends. World demographic data have been updated to the latest available information. A new data analysis feature includes links to and questions about interactive population data with the revolutionary data visualization tools of GapMinder.org.

- **Chapter 8** opens with a new case study about the dangers of bisphenol A (BPA). The heartening story of the control of guinea worms has moved to an Exploring Science reading. Conservation medicine has been enhanced with short case studies about white nose disease in bats and the worldwide *Chytridomycosis* epidemic in amphibians. An Exploring Science box introduces the important topic of epigenetics and the role of environmental factors in a wide variety of chronic diseases.

- **Chapter 9** provides updated data on hunger and obesity, new discussion of why food costs rise despite falling farm income, including factors such as palm oil and ethanol, and climate change, and a section on the economics of food production and agricultural subsidies. Discussions of seafood and other meat protein sources are expanded.

Examination of herbicide tolerance and genetically modified foods has been updated.

- **Chapter 10** has a new boxed reading on the Growing Power urban youth farming program and a new data analysis box on graphing pesticide usage. Updated and revised discussions of pest control, pesticide usage, and organic and sustainable agriculture have been added. New reports on UN studies reporting the importance of sustainable techniques for improving global food production are added.

- **Chapter 11** opens with an updated case study on protecting northern spotted owls. This chapter introduces some novel invasive species, including Asian carp and the emerald ash borer, and a new Exploring Science box considers the role of bison in prairie restoration.

- **Chapter 12** has a new opening case study on an unprecedented partnership between Norway and Indonesia to protect tropical rainforests as part of the UN REDD (reducing emissions from forest destruction and degradation) program. We discuss rapidly increasing palm oil production in Southeast Asia, and the importance of tropical peatland protection in reducing carbon emissions. Efforts to save temperate rainforests in Canada are described in a new Exploring Science box.

- **Chapter 13** the opening case study for this chapter is on restoring Louisiana's coastal wetlands which takes on added importance after the 2010 Gulf oil spill. The history of ecological restoration and the goals and techniques of successful restoration projects are reexamined in light of recent disasters as well as global climate change.

- **Chapter 14** includes a discussion of the 2011 tsunami in Japan and a new opening case study on the 2010 earthquake in Haiti that killed at least 230,000 people and left millions homeless. A new Exploring Science box explains the crisis in high-tech manufacturing in 2011, when China (which currently produces 97 percent of the total world supply) cut its exports of rare earth metals by half to protect domestic production of electronic components. A new section examines dams, water diversion projects, and sedimentation in reservoirs.

- **Chapter 15** has revised discussions of climate circulation, energy in the atmosphere, storms, and climate history, including lessons from the 800,000-year record from EPICA ice core data, which doubles the 400,000-year Vostok core record. We have revised discussions of how climate change works, what greenhouse gases are, and how we know that recent change is anthropogenic. A new section considers some of the reasons we dispute climate change. A new boxed reading details household CO_2 emissions, and an updated discussion of climate solutions ends the chapter.

- **Chapter 16** has a new opening case study on the Great London Smog, which helped to redefine our ideas about air pollution. Discussions of criteria pollutants, CO_2, and halogens are updated, including the relative impact of different halogens. We also present recent findings on economic benefits of the Clean Air Act.

- **Chapter 17** opens with a new case study on declining water levels in Lake Mead. Expanded coverage is given to the importance of freshwater in daily life, causes and effects of shortages around the world, and desalination, a new but expensive freshwater source in coastal areas. A new boxed reading looks at China's current project to channel water 1,600 km north from the Yangtze River to the dry plains around Beijing.

- **Chapter 18** focuses on the origins of the Clean Water Act in its case study. The Exploring Science box on the Gulf dead zone has been updated to include effects of the 2010 Gulf oil spill. Living machines, rain gardens, and other natural systems for treating polluted water are discussed.

- **Chapter 19** has a new opening case study on the causes and effects of the 2010 Gulf oil spill. Ultradeep drilling is explored further in a new boxed reading that explains how wells are being drilled in more than 4,000 m of water and up to 10,000 m beneath the ocean floor. Estimates that we have already passed "peak oil" are discussed, along with alternative ways that we could obtain and use fossil fuels more efficiently, including carbon sequestration. Questions about unconventional sources, such as the very deep and tight Marcellus shale formation in the eastern United States, and Canadian tar sands, are examined, along with ideas about a "nuclear renaissance."

- **Chapter 20** has a new opening case study about Desertech, an ambitious plan to link about 36 large new concentrating solar plants in North Africa and the Middle East with at least 20 offshore wind farms through a vast system of high-voltage direct-current undersea transmission lines to provide most of the electricity used in northern Europe. We examine the latest advances in capturing renewable energy, including wind, solar, geothermal, and biomass, which many analysts say could supply all our energy if we invested in them now.

- **Chapter 21** opens with a new case study on Papahãnau-mokuãkea Marine National Monument, a national treasure that is threatened by plastic marine debris. Discussions of e-waste, municipal solid waste, waste disposal methods, and Superfund and hazardous waste management are updated.

- **Chapter 22** has a new case study on Vauban, a car-free suburb in Germany, and an expanded discussion of mass transit and the growth of private automobiles in both the developed and developing countries.

- **Chapter 23** provides an updated discussion of economics, including expanded discussion of cost externalization. The chapter also has expanded discussions of ecological economics, including ecosystem services and accounting for natural capital.

- **Chapter 24** has a new case study examining the Convention on International Trade in Endangered Species (CITES), as well as a revised discussion of policy formation, including the impact of the Supreme Court's decision in the *Citizens United* case on campaign financing. There are revised discussions of international conventions, enforcement, and the importance of citizen action in policy formation.

- **Chapter 25** opens with a new case study about 350.org, a new global, youth-oriented organization working on climate change. New environmental leaders are featured, including Majora Carter and Van Jones, who combine environmental concerns with social justice.

ACKNOWLEDGMENTS

We owe a great debt to the hardworking, professional team that has made this the best environmental science text possible. We express special thanks for editorial support to Ryan Blankenship, Janice Roerig-Blong, and Wendy Langerud. We are grateful to the excellent production team led by Kelly Heinrichs and marketing leadership by Lisa Nicks. We also thank Wendy Nelson and Cathy Conroy for copyediting. The following individuals helped write and review learning goal-oriented content for **LearnSmart™ for Environmental Science**: William Sylvester Allred, Northern Arizona University; Elaina Cainas, Broward College; Mary Ann Cunningham, Vassar College; Nilo Marin, Broward College; Jessica Miles, Palm Beach State College; Jessica Seares; David Serrano, Broward College; and Gina Seegers Szablewski, University of Wisconsin—Milwaukee. Finally, we thank the many contributions of careful reviewers who shared their ideas with us during revisions.

TWELFTH EDITION REVIEWERS

Elmer Bettis III
University of Iowa

Robert I. Bruck
North Carolina State University

Daniel Cramer
Johnson County Community College

Francette Fey
Macomb Community College

Allan Matthias
University of Arizona

Edward Mondor
Georgia Southern University

Gregory O'Mullan
Queens College, City University of New York

Bruce Olszewski
San Jose State University

Kimberly Schulte
Georgia Perimeter College

Robert Stelzer
University of Wisconsin, Oshkosh

Karen S. Wehn
Buffalo State College and Erie Community College

GUIDED TOUR

A global perspective is vital to learning about environmental science.

Case Studies

At the start of each chapter, case studies utilize stories to portray real-life global issues that affect our food, our quality of life, and our future. Seventeen new case studies have been added to further focus on current events and the success stories of environmental protection progress.

Google Earth™ Placemarks

This feature provides interactive satellite imagery of the earth to give students a geographic context of places and topics in the text. Students can zoom in for detail or they can zoom out for a more global perspective. Placemark links can be found on the website http://www.mhhe.com/cunningham12e.

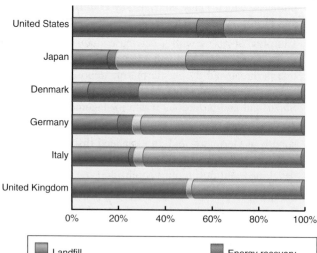

The Latest Global Data

Easy-to-follow graphs, charts, and maps display numerous examples from many regions of the world. Students are exposed to the fact that environmental issues cross borders and oceans.

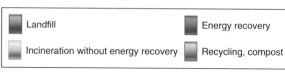

Critical thinking skills support understanding of environmental change.

Exploring Science

Real-life environmental issues drive these readings as students learn about the principles of scientific observation and proper data-gathering techniques.

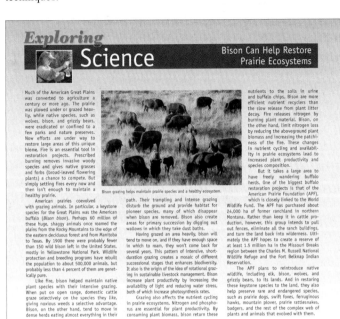

Exploring Science: Bison Can Help Restore Prairie Ecosystems

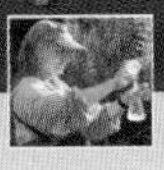

Much of the American Great Plains was converted to agriculture a century or more ago. The prairie was plowed under or grazed heavily, while native species, such as wolves, bison, and grizzly bears, were eradicated or confined to a few parks and nature preserves. Now efforts are under way to restore large areas of this unique biome. Fire is an essential tool in restoration projects. Prescribed burning removes invasive woody species and gives native grasses and forbs (broad-leaved flowering plants) a chance to compete. But simply setting fires every now and then isn't enough to maintain a healthy prairie.

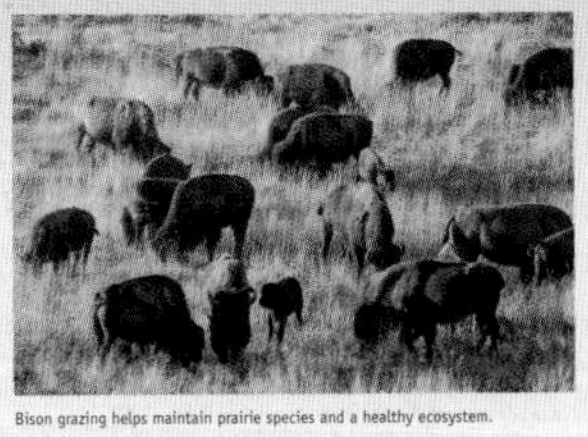

Bison grazing helps maintain prairie species and a healthy ecosystem.

American prairies coevolved with grazing animals. In particular, a keystone species for the Great Plains was the American buffalo (*Bison bison*). Perhaps 60 million of these huge, shaggy animals once roamed the plains from the Rocky Mountains to the edge of the eastern deciduous forest and from Manitoba to Texas. By 1900 there were probably fewer than 150 wild bison left in the United States, mostly in Yellowstone National Park. Wildlife protection and breeding programs have rebuilt the population to about 500,000 animals, but probably less than 4 percent of them are genetically pure.

Like fire, bison helped maintain native plant species with their intensive grazing. When put on open range, domestic cattle graze selectively on the species they like, giving noxious weeds a selective advantage. Bison, on the other hand, tend to move in dense herds eating almost everything in their path. Their trampling and intense grazing disturb the ground and provide habitat for pioneer species, many of which disappear when bison are removed. Bison also create areas for primary succession by digging out wallows in which they take dust baths.

Having grazed an area heavily, bison will tend to move on, and if they have enough space in which to roam, they won't come back for several years. This pattern of intensive, short-duration grazing creates a mosaic of different successional stages that enhances biodiversity. It also is the origin of the idea of rotational grazing in sustainable livestock management. Bison increase plant productivity by increasing the availability of light and reducing water stress, both of which increase photosynthesis rates.

Grazing also affects the nutrient cycling in prairie ecosystems. Nitrogen and phosphorus are essential for plant productivity. By consuming plant biomass, bison return these nutrients to the soils in urine and buffalo chips. Bison are more efficient nutrient recyclers than the slow release from plant litter decay. Fire releases nitrogen by burning plant material. Bison, on the other hand, limit nitrogen loss by reducing the aboveground plant biomass and increasing the patchiness of the fire. These changes in nutrient cycling and availability in prairie ecosystems lead to increased plant productivity and species composition.

But it takes a large area to have freely wandering buffalo herds. One of the biggest buffalo restoration projects is that of the American Prairie Foundation (APF), which is closely linked to the World Wildlife Fund. The APF has purchased about 24,000 ha of former ranchland in northern Montana. Rather than keep it in cattle production, however, this group intends to pull out fences, eliminate all the ranch buildings, and turn the land back into wilderness. Ultimately the APF hopes to create a reserve of at least 1.5 million ha in the Missouri Breaks region between the Charles M. Russell National Wildlife Refuge and the Fort Belknap Indian Reservation.

The APF plans to reintroduce native wildlife, including elk, bison, wolves, and grizzly bears, to its lands. And in restoring these keystone species to the land, they also help preserve rare and endangered species, such as prairie dogs, swift foxes, ferruginous hawks, mountain plover, prairie rattlesnakes, badgers, and the rest of the complex web of plants and animals that evolved with them.

Data Analysis

At the end of every chapter, these exercises ask students to graph and evaluate data while critically analyzing what they observe.

Data Analysis: Examining the IPCC Fourth Assessment Report (AR4)

The Intergovernmental Panel on Climate Change (IPCC) has a rich repository of figures and data, and because these data are likely to influence some policy actions in your future, it's worthwhile taking a few minutes to look at the IPCC reports.

The most brief and to the point is the Summary for Policy Makers (SPM) that accompanies the fourth Assessment Report. You can find the summary at www.ipcc.ch/ipccreports/ar4-syr.htm. If you have time, the full report is also available at this site.

Open the SPM and look at the first page of text, then look at the first figure, SPM1 (reproduced here). Look at this figure carefully and answer the following questions:

1. What is the subject of each graph? Why are all three shown together?
2. Carefully read the caption. What does the area between the blue lines represent? Why are the blue lines shown in this report?
3. The left axis for all three graphs shows the difference between each year's observations and an average value. What values are averaged?
4. What do the blue lines represent? In the third graph, what is the value of the blue line, in million km^2, for the most recent year shown? Approximately what year had the lowest value shown? What does a decline in this graph represent on the ground?
5. Why is the trend in the snow cover graph less steep than the trends in the other two graphs?
6. Nearly every page of the IPCC report has graphs that show quite interesting details when you take the time to look at them. Choose two other graphs in the SPM document and explain the main messages they give. See if you can explain them clearly enough to communicate the main idea to a friend or family member. Have different students select different graphs and explain them to the class.

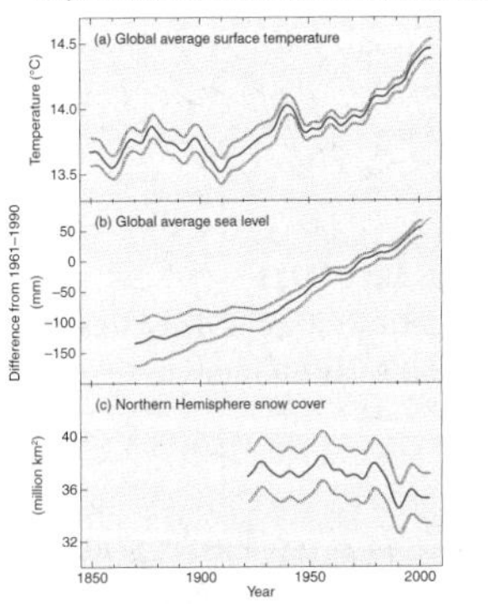
Changes in temperature, sea level and Northern Hemisphere snow cover
(a) Global average surface temperature
Temperature (°C)
14.5
14.0
13.5
(b) Global average sea level
Difference from 1961–1990
(mm)
50
0
−50
−100
−150
(c) Northern Hemisphere snow cover
(million km²)
40
36
32
1850
1900
1950
2000
Year

See the evidence: view the IPCC report at www.ipcc.ch/graphics/graphics/syr/spml.jpg.

For Additional Help in Studying This Chapter, please visit our website at www.mhhe.com/cunningham12e. You will find additional practice quizzes and case studies, flashcards, regional examples, placemarks for Google Earth™ mapping, and an extensive reading list, all of which will help you learn environmental science.

What Do You Think?

Ultradeep Drilling

The Deepwater Horizon, a floating drill rig that sank and spilled about 200 million gallons (nearly 800 million liters) of crude oil into the Gulf of Mexico, wasn't by any means working on the deepest or most remote well in the world. For the Gulf, the record is currently held by the Perdido Spar rig, which is drilling wells in 9,627 ft (nearly 3,000 m) of water about 320 km east of Brownsville, Texas. The Perdido (which means "lost," "missing," or "damned" in Spanish) is a technological marvel. It's a spar platform, meaning that the drill rig sits on top of a huge, hollow cylinder that's nearly as tall as Paris's Eiffel Tower. The spar is tethered to the ocean floor by nine thick cables, and supposedly will withstand hurricane winds, fierce ocean currents, and giant waves. The rig can support 35 wells that go down as much as 6 km below the ocean floor and radiate out horizontally up to 16 km from the well head. It's expected to produce about 130,000 barrels of oil plus 200 million ft^3 (about 6 million m^3) of natural gas per day.

A spar platform is one of several designs for underwater drilling. In shallow water, up to about 500 m, a fixed platform with very long legs that sit on the sea floor can be used. Beyond 500 m, floating or semi-submersible rigs are used. The Deepwater Horizon, which could drill in up to 2,000 m of water, was built on giant pontoons partially filled with seawater for ballast. All the drilling equipment and living quarters for hundreds of workers sat on decks held up by the pontoons. Like spar rigs, floating platforms are anchored for stability. Even greater depths are accessible from dynamically positioned drilling ships. For example, the Discover Clear Leader, owned and operated by Transocean, is capable of drilling in water nearly 4,000 m deep, and then punching down another 10,000 m through the seabed to ultradeep oil deposits.

Conditions at these depths are extreme. The oil can be 200°C, while water temperatures at the seafloor are just above freezing. Temperature shocks can rupture drill pipe. Oil deposits often accumulate beneath splintery shale or thick layers of taffy-like salt. Corrosion from the salt or sulfur in sediments erodes metal while strong ocean currents sweep equipment away. These depths are too great for human divers to do repairs, so drill operators have to depend on remotely operated robots to do all their work. We think of the seafloor at great depths to be a featureless, lifeless, mud-covered abyssal plain, but in fact it's often a jumble of deep canyons, sharp ridges, and huge piles of jumbled rocks with a rich, if largely unknown, community of life. All this makes drilling extremely difficult. Even under normal conditions, operating a drill rig, such as the Perdido, costs about $500,000 per day.

In spite of the disaster at the Deepwater Horizon, many countries are rushing to drill in harsh frontier environments. Before 1995 only about 10 percent of oil from the Gulf of Mexico came from deep water (more than 2,000 m), but now, as the shallow fields are being exhausted, about 70 percent does. The economic rewards of hitting a big find are enormous. The Bureau of Ocean Energy Management, Regulation and Enforcement (the successor to the disgraced Minerals Management Service) estimates that the U.S. outer continental shelf holds about 86 billion barrels of oil and 420 trillion cubic feet (12 trillion m^3) of gas. This represents about 60 percent of the oil and 40 percent of the natural gas resources for the United States. Brazil has recently begun tapping an ultradeep oil field that could hold between 50 to 100 billion barrels of oil about 300 km offshore in the Atlantic Ocean. This find could be worth $10 trillion and make Brazil a major player in international oil.

And even after seeing crude oil hemorrhaging into the Gulf of Mexico, fish and seabirds wallowing in black sludge, and BP responsible for billions in damages, other nations are rushing to do their own ultradeep drilling. Ghana, Nigeria, Angola, Congo, Libya, Egypt, and Australia are among the countries offering deep-sea oil leases in their oceanic territories. If the agencies in the United States that are supposed to regulate offshore drilling are rife with cronyism, corruption, and incompetence, think what the oversight may be in some of these other places. America, Canada, and Russia also are exploring drilling in the Arctic Ocean (remember the Titanic?). All this is fueled, of course, by our insatiable appetite for oil. What do you think? What are the limits to the risks we are willing to take for the oil to which we've become accustomed?

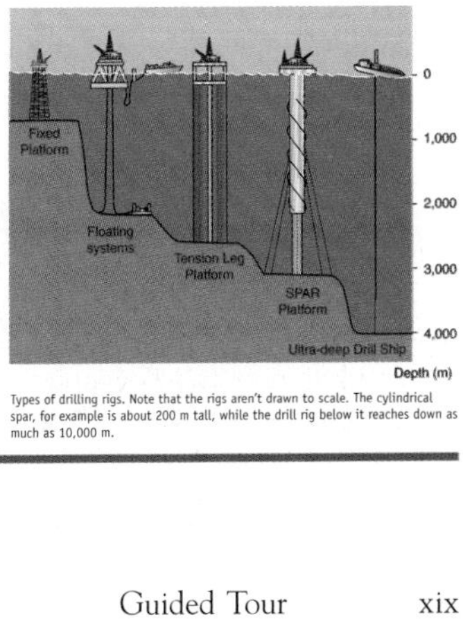
Fixed Platform
Floating systems
Tension Leg Platform
SPAR Platform
Ultra-deep Drill Ship
0
1,000
2,000
3,000
4,000
Depth (m)

Types of drilling rigs. Note that the rigs aren't drawn to scale. The cylindrical spar, for example is about 200 m tall, while the drill rig below it reaches down as much as 10,000 m.

What Do You Think?

This feature provides challenging environmental studies that offer an opportunity for students to consider contradictory data, special interests, and conflicting interpretations within a real scenario.

Sound pedagogy encourages science inquiry and application.

Learning Outcomes

Found at the beginning of each chapter, and organized by major headings, these outcomes give students an overview of the key concepts they will need to understand.

Learning Outcomes

After studying this chapter, you should be able to:

20.1 Describe renewable energy resources.

20.2 Explain how we could tap solar energy.

20.3 Grasp the potential of fuel cells.

20.4 Explain how we get energy from biomass.

20.5 Summarize the prospects for hydropower.

20.6 Report on the applications for wind power.

20.7 Visualize the uses of waves, tides, and geothermal energy.

20.8 Discuss our energy future.

Conclusion

This section summarizes the chapter by highlighting key ideas and relating them to one another.

CONCLUSION

We need materials from the earth to sustain our modern lifestyle, but many of the methods we use to get those materials have severe environmental consequences. Still, there are ways that we can extend resources through recycling and the development of new materials and more efficient ways of using them. We can do much although will nev contami Pollutan sedimen damage.

We also should be aware of geological hazards, such as floods, earthquakes, volcanoes, and landslides. Because these hazards often occur on a geological time scale, residents who haven't experienced one of these catastrophic events assume that they never will. People move into highly risky places without

Reviewing Learning Outcomes

Related to the Learning Outcomes at the beginning of each chapter, this review clearly restates the important concepts associated with each outcome.

REVIEWING LEARNING OUTCOMES

By now you should be able to explain the following points:

18.1 Define *water pollution*.
- Water pollution is anything that degrades water quality.

18.2 Describe the types and effects of water pollutants.
- Infectious agents remain an important threat to human health.
- Bacteria are detected by measuring oxygen levels.
- Nutrient enrichment leads to cultural eutrophication.
- Eutrophication can cause toxic tides and "dead zones."
- Inorganic pollutants include metals, salts, acids, and bases.
- Organic pollutants include drugs, pesticides, and other industrial substances.
- Sediment also degrades water quality.
- Thermal pollution is dangerous for organisms.

18.3 In

- Other countries also have serious water pollution.
- Groundwater is hard to monitor and clean.
- There are few controls on ocean pollution.

18.4 Explain water pollution control.
- Source reduction is often the cheapest and best way to reduce pollution.
- Controlling nonpoint sources requires land management.
- Human waste disposal occurs naturally when concentrations are low.
- Water remediation may involve containment, extraction, or phytoremediation.

18.5 Summarize water pollution legislation.
- The Clean Water Act was ambitious, bipartisan, and largely

Critical Thinking and Discussion Questions

Brief scenarios of everyday occurrences or ideas challenge students to apply what they have learned to their lives.

CRITICAL THINKING AND DISCUSSION QUESTIONS

1. In your opinion, how much environmental protection is too much? Think of a practical example in which some stakeholders may feel oppressed by government regulations. How would you justify or criticize these regulations?

2. Among the steps in the policy cycle, where would you put your efforts if you wanted influence in establishing policy?

3. Do y shou why favo

4. It's sometimes difficult to determine whether a lawsuit is retaliatory or based on valid reason. How would you define a SLAPP suit, and differentiate it from a legitimate case?

5. Create a list of arguments for and against an international body with power to enforce global environmental laws. Can you see a way to create a body that could satisfy both reasons for and against this power?

Practice Quiz

Short-answer questions allow students to check their knowledge of chapter concepts.

PRACTICE QUIZ

1. Describe five ways we could conserve energy individually or collectively.

2. Explain the principle of net energy yield. Give some examples.

3. What is the difference between active and passive solar energy?

4. How do photovoltaic cells generate electricity?

5. What is a fuel cell and how does it work?

6. Why might *Jatropha* be a good source of biodiesel?

7. Why might *Miscanthus* be a good source of ethanol?

8. What are some advantages and disadvantages of large hydroelectric dams?

9. How can geothermal energy be used for home heating?

10. Describe how tidal power or ocean wave power generate electricity.

What Can You Do?

This feature gives students realistic steps for applying their knowledge to make a positive difference in our environment.

What Can You Do?

Saving Energy and Reducing Pollution

- Conserve energy: carpool, bike, walk, use public transport, and buy compact fluorescent bulbs and energy-efficient appliances (see chapter 20 for other suggestions).

- Don't use polluting two-cycle gasoline engines if cleaner four-cycle models are available for lawnmowers, boat motors, etc.

- Buy refrigerators and air conditioners designed for CFC alternatives. If you have old appliances or other CFC sources, dispose of them responsibly.

- Plant a tree and care for it (every year).

- Write to your congressional representatives and support a transition to an energy-efficient economy.

- If green-pricing options are available in your area, buy renewable energy.

Think About It

These boxes provide several opportunities in each chapter for students to review material, practice critical thinking, and apply scientific principles.

Think About It

What barriers do you see to walking, biking, or mass transit in your hometown? How could cities become more friendly to sustainable transportation? Why not write a letter to your city leaders or the editor of your newspaper describing your ideas?

Teaching and Learning Tools

|environmental science

McGraw-Hill Connect® Environmental Science is a Web-based assignment and assessment platform that gives students the means to better connect with their coursework, with their instructors, and with the important concepts that they will need to know for success now and in the future.

With Connect® Environmental Science, instructors can deliver assignments, quizzes, and tests online. Nearly all the questions from the text are presented in an auto-gradable format and tied to the text's learning objectives. Instructors can edit existing questions and author entirely new problems, track individual student performance—by question, assignment or in relation to the class overall—with detailed grade reports, integrate grade reports easily with Learning Management Systems (LMS) such as WebCT and Blackboard®, and much more.

By choosing Connect Environmental Science, instructors are providing their students with a powerful tool for improving academic performance and truly mastering course material. Connect Environmental Science allows students to practice important skills at their own pace and on their own schedule. Importantly, students' assessment results and instructors' feedback are all saved online—so students call continually review their progress and plot their course to success.

Some instructors may also choose ConnectPlus® Environmental Science for their students. Like Connect Environmental Science, ConnectPlus Environmental Science provides students with online assignments and assessments, plus 24/7 online access to an eBook—an online edition of the text—to aid them in successfully completing their work, wherever and whenever they choose.

To learn more, visit
www.mcgrawhillconnect.com

LearnSmart™

McGraw-Hill LearnSmart™ is available as an integrated feature of Connect™ Environmental Science and provides students with a GPS (**G**uided **P**ath to **S**uccess) for your course. Using artificial intelligence, LearnSmart™ intelligently assesses a student's knowledge of course content through a series of adaptive questions. It pinpoints concepts the student does not understand and maps out a personalized study plan for success. This innovative study tool also has features that allow instructors to see exactly what students have accomplished, and a built-in assessment tool for graded assignments.

Visit the site below for a demonstration.
www.mhlearnsmart.com

My Lectures— McGraw-Hill Tegrity™

McGraw-Hill Tegrity records and distributes your class with just a click of a button. Students can view anytime/anywhere via computer, iPod, or mobile device. It indexes as it records your PowerPoint® presentations and anything shown on your computer so students can use keywords to find exactly what they want to study. Tegrity™ is available an integrated feature of Connect Environmental Science or as standalone.

Online Teaching and Study Tools

Text Website: http://www.mhhe.com/cunningham12e

McGraw-Hill offers various tools and technology products to support *Environmental Science: A Global Concern.* Instructors can obtain teaching aids by calling the Customer Service Department at 1-800-334-7344.

Presentation Center (ISBN-13: 978-0-07-733714-8; ISBN-10:0-07-733714-X)

Presentation Center is an online digital library containing assets such as photos, artwork, PowerPoints, animations, and other media types that can be used to create customized lectures, visually enhanced tests and quizzes, compelling course websites, and attractive printed support materials. The following digital assets are grouped by chapter:

- **Color Art** Full-color digital files of illustrations in the text can be readily incorporated into lecture presentations, exams, or custom-made classroom materials. These include all of the 3-D realistic art found in this edition, representing some of the most important concepts in environmental science.

- **Photos** Digital files of photographs from the text can be reproduced for multiple classroom uses.

- **Tables** Every table that appears in the text is provided in electronic format.

- **Videos** This special collection of 69 underwater video clips displays interesting habitats and behaviors of many animals in the ocean.

- **Animations** One hundred full-color animations that illustrate many different concepts covered in the study of environmental science are available for use in creating classroom lectures, testing materials, or online course communication. The visual impact of motion will enhance classroom presentations and increase comprehension.

- **Test Bank** A computerized test bank that uses testing software to quickly create customized exams is available on for this text. The user-friendly program allows instructors to search for questions by topic or format, edit existing questions or add new ones; and scramble questions for multiple versions of the same test. Word files of the test bank questions are provided for those instructors who prefer to work outside the test-generator software.

- **Global Base Maps** Eighty-eight base maps for all world regions and major subregions are offered in four versions: black-and-white and full-color, both with labels and without labels. These choices allow instructors the flexibility to plan class activities, quizzing opportunities, study tools, and PowerPoint enhancements.

- **PowerPoint Lecture Outlines** Ready-made presentations that combine art and photos and lecture notes are provided for each of the 25 chapters of the text. These outlines can be used as they are or tailored to reflect your preferred lecture topics and sequences.

- **PowerPoint Slides** For instructors who prefer to create their lectures from scratch, all illustrations, photos, and tables are preinserted by chapter into blank PowerPoint slides for convenience.

- **Course Delivery Systems** With help from WebCT and Blackboard®, professors can take complete control of their course content. Course cartridges containing website content, online testing, and powerful student tracking features are readily available for use within these platforms.

Electronic Textbook

CourseSmart is a new way for faculty to find and review eTextbooks. It's also a great option for students who are interested in accessing their course materials digitally and saving money. CourseSmart offers thousands of the most commonly adopted textbooks across hundreds of courses from a wide variety of higher education publishers. It is the only place for faculty to review and compare the full text of a textbook online, providing immediate access without the environmental impact of requesting a print exam copy. At CourseSmart, students can save up to 50 percent off the cost of a print book, reduce their impact on the environment, and gain access to powerful web tools for learning including full text search, notes and highlighting, and email tools for sharing notes between classmates. www.CourseSmart.com

Learning Supplements for Students

Website (www.mhhe.com/cunningham12e)

The *Environmental Science: A Global Concern* website provides access to resources such as multiple-choice practice quizzes with immediate feedback and grade, Google Earth links and questions interactive maps, animation quizzes, and a case study library.

Annual Editions: Environment 11/12 by Sharp

(MHID: 0-07-351556-6)

This Twenty-Eighth Edition provides convenient, inexpensive access to current articles selected from some of the most respected magazines, newspapers, and journals published today. Organizational features include: an annotated listing of selected World Wide Web sites; an annotated table of contents; a topic guide; a general introduction; brief overviews for each section; and an instructor's resource guide with testing materials. *Using Annual Editions in the Classroom* is also offered as a practical guide for instructors.

Taking Sides: Clashing Views on Environmental Issues,

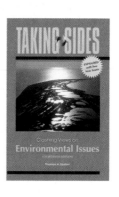

Expanded Fourteenth Edition by Easton (MHID: 0-07-351445-4)

This Expanded Fourteenth Edition of *Taking Sides: Environmental Issues* presents two additional current controversial issues in a debate-style format designed to stimulate student interest and develop critical thinking skills. Each issue is thoughtfully framed with an issue summary, an issue introduction, and a postscript. Taking Sides readers also feature annotated listings of selected World Wide Web sites. An instructor's resource guide with testing material is available for each volume. *Using Taking Sides in the Classroom* is also an excellent instructor resource.

Field & Laboratory Exercises in Environmental Science

Seventh Edition, by Enger and Smith
(ISBN: 978-0-07-290913-5; MHID: 0-07-290913-7)
The major objectives of this manual are to provide students with hands-on experiences that are relevant, easy to understand, applicable to the student's life, and presented in an interesting, informative format. Ranging from field and lab experiments to conducting social and personal assessments of the environmental impact of human activities, the manual presents something for everyone, regardless of the budget or facilities of each class. These labs are grouped by categories that can be used in conjunction with any introductory environmental textbook.

Global Studies: The World at a Glance,

Second Edition, by Tessema
(ISBN: 978-0-07-340408-0;
MHID: 0-07-340408-X)
This book features a compilation of up-to-date data and accurate information on some of the important facts about the world we live in. While it is close to impossible to stay current on every nation's capital, type of government, currency, major languages, population, religions, political structure, climate, economics, and more, this book is intended to help students to understand these essential facts in order to make useful applications.

Sources: Notable Selections in Environmental Studies, Second Edition, by Goldfarb

(ISBN: 978-0-07-303186-6;
MHID: 0-07-303186-0)
This volume brings together primary source selections of enduring intellectual value—classic articles, book excerpts, and research studies—that have shaped environmental studies and our contemporary understanding of it. The book includes carefully edited selections from the works of the most distinguished environmental observers, past and present. Selections are organized topically around the following major areas of study: energy, environmental degradation, population issues and the environment, human health and the environment, and environment and society.

Student Atlas of Environmental Issues

by Allen (ISBN: 978-0-69-736520-0;
MHID: 0-69-736520-4)
This atlas is an invaluable pedagogical tool for exploring the human impact on the air, waters, biosphere, and land in every major world region. This informative resource provides a unique combination of maps and data that help students understand the dimensions of the world's environmental problems and the geographical basis of these problems.

Learning to learn is a lifelong skill.

Learning to Learn

Learning Outcomes

After studying this introduction, you should be able to:

L.1 Form a plan to organize your efforts and become a more effective and efficient student.

L.2 Make an honest assessment of the strengths and weaknesses of your current study skills.

L.3 Assess what you need to do to get the grade you want in this class.

L.4 Set goals, schedule your time, and evaluate your study space.

L.5 Use this textbook effectively, practice active reading, and prepare for exams.

L.6 Be prepared to apply critical and reflective thinking in environmental science.

L.7 Understand the advantages of concept mapping and use it in your studying.

"What kind of world do you want to live in? Demand that your teachers teach you what you need to know to build it."

~Peter Kropotkin

1

Case Study

Why Study Environmental Science?

Welcome to environmental science. We hope you'll enjoy learning about the material presented in this book, and that you'll find it both engaging and useful. There should be something here for just about everyone, whether your interests are in basic ecology, natural resources, or the broader human condition. You'll see, as you go through the book, that it covers a wide range of topics. It defines our environment, not only the natural world, but also the built world of technology, cities, and machines, as well as human social or cultural institutions. All of these interrelated aspects of our life affect us, and, in turn, are affected by what we do.

You'll find that many issues discussed here are part of current news stories on television or in newspapers. Becoming an educated environmental citizen will give you a toolkit of skills and attitudes that will help you understand current events and be a more interesting person. Because this book contains information from so many different disciplines, you will find connections here with many of your other classes. Seeing material in an environmental context may assist you in mastering subject matter in many courses, as well as in life after you leave school.

One of the most useful skills you can learn in any of your classes is critical thinking—a principal topic of this chapter. Much of the most important information in environmental science is highly contested. Facts vary depending on when and by whom they were gathered. For every opinion there is an equal and opposite opinion. How can you make sense out of this welter of ever-changing information? The answer is that you need to develop a capacity to think independently, systematically, and skillfully to form your own opinions (fig. L.1). These qualities and abilities can help you in many aspects of life. Throughout this book you will find "What Do You Think?" boxes that invite you to practice your critical and reflective thinking skills.

There is much to be worried about in our global environment. Evidence is growing relentlessly that we are degrading our environment and consuming resources at unsustainable rates. Biodiversity is disappearing at a pace unequaled since the end of the age of dinosaurs 65 million years ago. Irreplaceable topsoil erodes from farm fields, threatening global food supplies. Ancient forests are being destroyed to make newsprint and toilet paper. Rivers and lakes are polluted with untreated sewage, while soot and smoke obscure our skies. Even our global climate seems to be changing to a new regime that could have catastrophic consequences.

At the same time, we have better tools and knowledge than any previous generation to do something about these crises. Worldwide public awareness of—and support for—environmental protection is at an all-time high. Over the past 50 years, human

FIGURE L.1 What does it all mean? Studying environmental science gives you an opportunity to develop creative, reflective, and critical thinking skills.

ingenuity and enterprise have brought about a breathtaking pace of technological innovations and scientific breakthroughs. We have learned to produce more goods and services with less material. The breathtaking spread of communication technology makes it possible to share information worldwide nearly instantaneously. Since World War II, the average real income in developing countries has doubled; malnutrition has declined by almost one-third; child death rates have been halved; average life expectancy has increased by 30 percent; and the percentage of rural families with access to safe drinking water has risen from less than 10 percent to almost 75 percent.

The world's gross domestic product has increased more than tenfold over the past five decades, but the gap between the rich and poor has grown ever wider. More than a billion people now live in abject poverty without access to adequate food, shelter, medical care, education, and other resources required for a healthy, secure life. The challenge for us is to spread the benefits of our technological and economic progress more equably and to find ways to live sustainably over the long run without diminishing the natural resources and vital ecological services on which all life depends. We've tried to strike a balance in this book between enough doom and gloom to give you a realistic view of our problems, and enough positive examples to give hope that we can discover workable solutions.

What would it mean to become a responsible environmental citizen? What rights and privileges do you enjoy as a member of the global community? What duties and responsibilities earn us the rights and privileges of citizenship? In many chapters of this book you will find practical advice on things you can do to conserve resources and decrease adverse environmental impacts. Ethical perspectives are an important part of our relationship to the environment and the other people with whom we share it. The discussion of ethical principles and worldviews in chapter 2 is a key section of this book. We hope you'll think about the ethics of how we treat our common environment.

Clearly, to become responsible and productive environmental citizens each of us needs a basis in scientific principles, as well as some insights into the social, political, and economic systems that impact our global environment. We hope this book and the class you're taking will give you the information you need to reach those goals. As the noted Senegalese conservationist and educator, Baba Dioum, once said, "in the end, we will conserve only what we love, we will love only what we understand, and we will understand only what we are taught."

L.1 How Can I Get an A in This Class?

"What have I gotten myself into?" you are probably wondering as you began to read this book. "Will environmental science be worth my while? Do I have a chance to get a good grade?" The answers to these questions depend, to a large extent, on you and how you decide to apply yourself. Expecting to be interested and to do either well or poorly in your classes often turns out to be a self-fulfilling prophecy. As Henry Ford once said, "If you think you can do a thing, or think you can't do a thing, you're right." Cultivating good study skills can help you to reach your goals and make your experience in environmental science a satisfying and rewarding one. The purpose of this introduction is to give you some tips to help you get off to a good start in studying. You'll find that many of these techniques are also useful in other courses and after you graduate, as well.

Another thing that will help you do well in this class—and enjoy it—is to understand that science is useful and accessible, if you just take your time with it. You might be someone who loves science, but many people consider science unfamiliar and intimidating. To do well in this class, it will help to identify the ways that science connects with your interests and with the things you like to do. Most environmental scientists are motivated by a love for something: a fishery biologist might love fishing; a plant pathologist might love gardening; an environmental chemist might be motivated by wanting to improve children's health in the city in which she lives. All these people use the tools of science to help them understand something they get excited about. Finding that angle can help you do better in this class, and it can help you be a better and happier member of your community.

Most people think science is the domain of specialists in lab coats. But in fact science is practiced by all kinds of people in all kinds of ways, every day, including you. Basically, science is just about trying to figure out how things work. Understanding some basic ideas in science can be very empowering: learning to look for evidence and to question your assumptions is a life skill; building comfort with thinking about numbers can help you budget your groceries, prioritize your schedule, or plan your vacation. Ideas in this book can help you understand the food you eat, the weather you encounter, the policies you hear about in the news—from energy policy to urban development to economics. A lot of people think science is foreign, but it belongs to all of us, and this book is about helping you see how a better understanding of science can make the world more understandable and interesting for you.

Environmental science, as you can see by skimming through the table of contents of this book, is a complex, transdisciplinary field that draws from many academic specialties. It is loaded with facts, ideas, theories, and confusing data. It is also a dynamic, highly contested subject. Topics such as environmental contributions to cancer rates, potential dangers of pesticides, or when and how much global warming may be caused by human activities are widely disputed. Often you will find distinguished and persuasive experts who take completely opposite positions on any particular question. It will take an active, organized approach on your part to make sense of the arguments and ideas you'll encounter here. And it will take critical, thoughtful reasoning to formulate your own position on the many controversial theories and ideas in environmental science. Learning to learn will help you keep up-to-date on important issues after you leave this course. Becoming educated voters and consumers is essential for a sustainable future.

Develop good study habits

Many students find themselves unprepared for studying in college. In a survey released in 2003 by the Higher Education Research Institute, more than two-thirds of high school seniors nationwide reported studying outside of class less than one hour per day. Nevertheless, because of grade inflation, nearly half those students claim to have an A average. It comes as a rude shock to many to discover that the study habits they developed in high school won't allow them to do as well—or perhaps even to pass their classes—in college. Many will have to triple or even quadruple their study time. In addition, they need urgently to learn to study more efficiently and effectively.

What are your current study skills and habits? Making a frank and honest assessment of your strengths and weaknesses will help you set goals and make plans for achieving them during this class. Answer the questions in table L.1 as a way of assessing where you are as you begin to study environmental science and where you need to work to improve your study habits.

One of the first requirements for success is to set clear, honest, attainable goals for yourself. Are you willing to commit the time and effort necessary to do well in this class? Make goals for yourself in terms that you can measure and in time frames within which you can see progress and adjust your approach if it isn't taking you where you want to go. Be positive but realistic. It's more effective to try to accomplish a positive action than to avoid a negative one. When you set your goals, use proactive language that states what you want rather than negative language about what you're trying to avoid. It's good to be optimistic, but setting impossibly high standards will only lead to disappointment. Be objective about the obstacles you face and be willing to modify your goals if necessary. As you gain more experience and information, you may need to adjust your expectations either up or down. Take stock from time to time to see whether you are on track to accomplish what you expect from your studies. In environmental planning, this is called adaptive management.

One of the most common mistakes many of us make is to procrastinate and waste time. Be honest, are you habitually late for meetings or in getting assignments done? Do you routinely leave your studying until the last minute and then frantically cram the night before your exams? If so, you need to organize your schedule so that you can get your work done and still have a life. Make a study schedule for yourself and stick to it. Allow

Table L.1 Assess Your Study Skills

Rate yourself on each of the following study skills and habits on a scale of 1 (excellent) to 5 (needs improvement). If you rate yourself below 3 on any item, think about an action plan to improve that competence or behavior.

_____ How strong is your commitment to be successful in this class?

_____ How well do you manage your time (e.g., do you always run late or do you complete assignments on time)?

_____ Do you have a regular study environment that reduces distraction and encourages concentration?

_____ How effective are you at reading and note-taking (e.g., do you remember what you've read; can you decipher your notes after you've made them)?

_____ Do you attend class regularly and listen for instructions and important ideas? Do you participate actively in class discussions and ask meaningful questions?

_____ Do you generally read assigned chapters in the textbook before attending class or do you wait until the night before the exam?

_____ Are you usually prepared before class with questions about material that needs clarification or that expresses your own interest in the subject matter?

_____ How do you handle test anxiety (e.g., do you usually feel prepared for exams and quizzes or are you terrified of them)? Do you have techniques to reduce anxiety or turn it into positive energy)?

_____ Do you actively evaluate how you are doing in a course based on feedback from your instructor and then make corrections to improve your effectiveness?

_____ Do you seek out advice and assistance outside of class from your instructors or teaching assistants?

enough time for sleep, regular meals, exercise, and recreation so that you will be rested, healthy, and efficient when you do study. Schedule regular study times between your classes and work. Plan some study times during the day when you are fresh; don't leave all your work until late night hours when you don't get much done. Divide your work into reasonable sized segments that you can accomplish on a daily basis. Plan to have all your reading and assignments completed several days before your exams so you will have adequate time to review and process information. Carry a calendar so you will remember appointments and assignments.

Establish a regular study space in which you can be effective and productive. It might be a desk in your room, a carrel in the library, or some other quiet, private environment. Find a place that works for you and be disciplined about sticking to what you need to do. If you get in the habit of studying in a particular place and time, you will find it easier to get started and to stick to your tasks. Many students make the mistake of thinking that they can study while talking to their friends or watching TV. They may put in many hours but not really accomplish much. On the

other hand, some people think most clearly in the anonymity of a crowd. The famous philosopher, Immanuel Kant, found that he could think best while wandering through the noisy, crowded streets of Königsberg, his home town.

How you behave in class and interact with your instructor can have a big impact on how much you learn and what grade you get. Make an effort to get to know your instructor. She or he is probably not nearly as formidable as you might think. Sit near the front of the room where you can see and be seen. Pay attention and ask questions that show your interest in the subject matter. Practice the skills of good note-taking (table L.2). Attend every class and arrive on time. Don't fold up your papers and prepare to leave until after the class period is over. Arriving late and leaving early says to your instructor that you don't care much about either the class or your grade. If you think of yourself as a good student and act like one, you may well get the benefit of the doubt when your grade is assigned.

Practice active, purposeful learning. It isn't enough to passively absorb knowledge provided by your instructor and this textbook. You need to actively engage the material in order to really understand it. The more you invest yourself in the material, the easier it will be to comprehend and remember. It is very helpful to have a study buddy with whom you can compare notes and try out ideas (fig. L.2). You will get a valuable perspective on whether you're getting the main points and understanding an adequate amount by comparing. It's an old adage that the best way to learn something is to teach it to someone else. Take turns with your study buddy explaining the material you're studying. You may think you've

Table L.2 Learning Skills—Taking Notes

1. Identify the important points in a lecture and organize your notes in an outline form to show main topics and secondary or supporting points. This will help you follow the sense of the lecture.

2. Write down all you can. If you miss something, having part of the notes will help your instructor identify what you've missed.

3. Leave a wide margin in your notes in which you can generate questions to which your notes are the answers. If you can't write a question about the material, you probably don't understand it.

4. Study for your test under test conditions by answering your own questions without looking at your notes. Cover your notes with a sheet of paper on which you write your answers, then slide it to the side to check your accuracy.

5. Go all the way through your notes once in this test mode, then go back to review those questions you missed.

6. Compare your notes and the questions you generated with those of a study buddy. Did you get the same main points from the lecture? Can you answer the questions someone else has written?

7. Review your notes again just before test time, paying special attention to major topics and questions you missed during study time.

Source: Dr. Melvin Northrup, Grand Valley State University.

FIGURE L.2 Cooperative learning, in which you take turns explaining ideas and approaches with a friend, can be one of the best ways to comprehend material.

mastered a topic by quickly skimming the text but you're likely to find that you have to struggle to give a clear description in your own words. Anticipating possible exam questions and taking turns quizzing each other can be a very good way to prepare for tests.

Recognize and hone your learning styles

Each of us has ways that we learn most effectively. Discovering techniques that work for you and fit the material you need to learn is an important step in reaching your goals. Do any of the following fit your preferred ways of learning?

- **Visual Learner:** understands and remembers best by reading, looking at photographs, figures, and diagrams. Good with maps and picture puzzles. Visualizes image or spatial location for recall. Uses flash cards for memorization.

- **Verbal Learner:** understands and remembers best by listening to lectures, reading out loud, and talking things through with a study partner. May like poetry and word games. Memorizes by repeating item verbally.

- **Logical Learner:** understands and remembers best by thinking through a subject and finding reasons that make sense. Good at logical puzzles and mysteries. May prefer to find patterns and logical connections between items rather than memorize.

- **Active Learner:** understands and remembers best those ideas and skills linked to physical activity. Takes notes, makes lists, uses cognitive maps. Good at working with hands and learning by doing. Remembers best by writing, drawing, or physically manipulating items.

The list above represents only a few of the learning styles identified by educational psychologists. How can you determine which approaches are right for you? Think about the one thing in life that you most enjoy and in which you have the greatest skills. What hobbies or special interests do you have? How do you learn new material in that area? Do you read about a procedure in a book and then do it, or do you throw away the manual and use trial and error to figure out how things work? Do you need to see a diagram or a picture before things make sense, or are spoken directions most memorable and meaningful for you? Some people like to learn by themselves in a quiet place where there are no distractions, while others need to discuss ideas with another person to feel really comfortable about what they're learning.

Sometimes you have to adjust your preferred learning style to the specific material you're trying to master. You may be primarily a verbal learner, but if what you need to remember for a particular subject is spatial or structural, you may need to try some visual learning techniques. Memorizing vocabulary items might be best accomplished by oral repetition, while developing your ability to work quantitative problems should be approached by practicing analytical or logical skills.

Use this textbook effectively

An important part of productive learning is to read assigned material in a purposeful, deliberate manner. Ask yourself questions as you read. What is the main point being made here? Does the evidence presented adequately support the assertions being made? What personal experience have you had or what prior knowledge can you bring to bear on this question? Can you suggest alternative explanations for the phenomena being discussed? What additional information would you need in order to make an informed judgment about this subject and how might you go about obtaining that information or making that judgment?

A study technique developed by Frances Robinson and called the SQ3R method (table L.3) can be a valuable aid in

Table L.3 The SQ3R Method for Studying Texts
Survey
Preview the information to be studied before reading.
Question
Ask yourself critical questions about the content of what you are reading.
Read
Conduct the actual reading in small segments.
Recite
Stop periodically to recite to yourself what you have just read.
Review
Once you have completed the section, review the main points to make sure you remember them clearly.

improving your reading comprehension. Start your study session with a *survey* of the entire chapter or section you are about to read so you'll have an idea of how the whole thing fits together. What are the major headings and subdivisions? Notice that there is usually a hierarchical organization that gives you clues about the relationship between the various parts. This survey will help you plan your strategy for approaching the material. Next, *question* what the main points are likely to be in each of the sections. Which parts look most important or interesting? Ask yourself where you should invest the most time and effort. Is one section or topic likely to be more relevant to your particular class? Has your instructor emphasized any of the topics you see? Being alert for important material can help you plan the most efficient way to study.

After developing a general plan, begin active reading of the text. Read in small segments and stop frequently for reflection and to make notes. Don't fall into a trance in which the words swim by without leaving any impression. Highlight or underline the main points but be careful that you don't just paint the whole page yellow. If you highlight too much, nothing will stand out. Try to distinguish what is truly central to the argument being presented. Make brief notes in the margins that identify main points. This can be very helpful in finding important sections or ideas when you are reviewing. Check your comprehension at the end of each major section. Ask yourself: Did I understand what I just read? What are the main points being made here? Does this relate to my own personal experiences or previous knowledge? Are there details or ideas that need clarification or elaboration?

As you read, stop periodically to *recite* the information you've just acquired. Summarize the information in your own words to be sure that you really understand and are not just depending on rote memory. This is a good time to have a study group (fig. L.3). Taking turns to summarize and explain material really helps you internalize it. If you don't have a study group and you feel awkward talking to yourself, you can try writing your summary. Finally, *review* the section. Did you miss any important points? Do you understand things differently the second time through? This is a chance to think critically about the material. Do you agree with the conclusions suggested by the authors? Can you think of alternative explanations for the same evidence? As you review each section, think about how this may be covered on the test. Put yourself in the position of the instructor. What would be some good questions based on this material? Don't try to memorize everything but try to anticipate what might be the most important points.

After class, compare your lecture notes with your study notes. Do they agree? If not, where are the discrepancies? Is it possible that you misunderstood what was said in class, or does your instructor differ with what's printed in the textbook? Are there things that your instructor emphasized in lecture that you missed in your pre-class reading? This is a good time to go back over the readings to reinforce your understanding and memory of the material.

FIGURE L.3 Explaining ideas to your peers is an excellent way to test your knowledge. If you can teach it to someone else, than you probably have a good grasp of the material.

Will this be on the test?

Students often complain that test results don't adequately reflect what they know and how much they've learned in studying. It may well be that test questions won't cover what you think is important or use a style that appeals to you, but you'll probably be more successful if you adapt yourself to the realities of your instructor's test methods rather than trying to force your instructor to accommodate to your preferences. One of your first priorities in studying, therefore, should be to learn your instructor's test style. Are you likely to have short-answer objective questions (multiple choice, true or false, fill in the blank) or does your instructor prefer essay questions? If you have an essay test, will the questions be broad and general or more analytical? You should develop a very different study strategy depending on whether you are expected to remember and choose between a multitude of facts and details, or whether you will be asked to write a paragraph summarizing some broad topic.

Organize the ideas you're reading and hearing in lecture. This course will probably include a great deal of information. Unless you have a photographic memory, you won't be able to remember every detail. What's most important? What's the big picture? If you see how pieces of the course fit together, it will all make more sense and be easier to remember. As you read and review, ask yourself what might be some possible test questions in each section. If you're likely to have factual questions, what are the most significant facts in the material you've read? Memorize some benchmark figures. Just a few will help a lot. Pay special attention to tables, graphs, and diagrams. They were chosen because they illustrate important points.

You probably won't be expected to remember all the specific numbers in this book but you probably should know orders of magnitude. The world population is about 6.5 *billion* people, not thousands, millions, or trillions. Highlight facts and figures in your lecture notes about which your instructor seemed especially interested. There is a good chance you'll see those topics again on a test. It often helps to remember facts and figures if you can relate them to some other familiar example. The United States, for instance, has about 295 million residents. The European Union is slightly larger, India is about three times and China is more than four times as large. Be sure you're familiar with the bold-face key terms in the textbook. Vocabulary terms make good objective questions. The Practice Quiz at the end of each chapter generally covers objective material that makes good short-answer questions.

A number of strategies can help you be successful in test-taking. Look over the whole test at the beginning and answer the questions you know well first, then tackle the harder ones. On multiple choice tests, find out whether there is a penalty for guessing. Use the process of elimination to narrow down the possible choices and improve the odds for guessing. Often you can get hints from the context of the question or from other similar questions. Notice that the longest or most specific answer often is right while those that are vague or general are more likely wrong. Be alert for absolutes (such as always, never, all) which could indicate wrong choices. Qualifiers (such as sometimes, may, or could) on the other hand, often point to correct answers. Exactly opposite answers may indicate that one of them is correct.

If you anticipate essay questions, practice writing one- or two-paragraph summaries of major points in each chapter. Develop your ability to generalize and to make connections between important facts and ideas. Notice that the Critical Thinking and Discussion Questions at the end of each chapter are open-ended topics that can work well either for discussion groups or as questions for an essay test. You'll have a big advantage on a test if you have some carefully thought out arguments for and against the major ideas presented in each chapter. If you don't have any idea what a particular essay question means, you often can make a transition to something you do understand. Look for a handle that links the question to a topic you are prepared to answer. Even if you have no idea what the question means, make an educated guess. You might get some credit. Anything is better than a zero. Sometimes if you explain your answer, you'll get at least some points. "If the question means such and such, then the answer would be _____" may get you partial credit.

Does your instructor like thought questions? Does she/he expect you to be able to interpret graphs or to draw inferences from a data table? Might you be asked to read a paragraph and describe what it means or relate it to other cases you've covered in the class? If so, you should practice these skills. Making up and sharing these types of questions with your study group can greatly increase your understanding of the material as well as improve your performance on exams. Writing a paragraph answer for each of the Critical Thinking and Discussion Questions could be a very good way to study for an essay test.

Concentrate on positive attitudes and building confidence before your tests. If you have fears and test anxiety, practice relaxation techniques and visualize success. Be sure you are rested and well prepared. You certainly won't do well if you're sleep-deprived and a bundle of nerves. Often the worst thing you can do is to stay up all night to cram your brain with a jumble of data. Being able to think clearly and express yourself well may count much more than knowing a pile of unrelated facts. Review your test when it is returned to learn what you did well and where you need to improve. Ask your instructor for pointers on how you might have answered the questions better. Carefully add your score to be sure you got all the points you deserve. Sometimes graders make simple mathematical errors in adding up points.

L.2 THINKING ABOUT THINKING

Perhaps the most valuable skill you can learn in any of your classes is the ability to think clearly, creatively, and purposefully. In a rapidly moving field such as environmental science, facts and explanations change constantly. It's often said that in six years approximately half the information you learn from this class will be obsolete. During your lifetime you will probably change careers four to six times. Unfortunately, we don't know which of the ideas we now hold will be outdated or what qualifications you will need for those future jobs. Developing the ability to learn new skills, examine new facts, evaluate new theories, and formulate your own interpretations is essential to keep up in a changing world. In other words, you need to learn how to learn on your own.

Even in our everyday lives most of us are inundated by a flood of information and misinformation. Competing claims and contradictory ideas battle for our attention. The rapidly growing complexity of our world and our lives intensifies the difficulties in knowing what to believe or how to act. Consider how the communications revolution has brought us computers, e-mail, cell phones, mobile faxes, pagers, the World Wide Web, hundreds of channels of satellite TV, and direct mail or electronic marketing that overwhelm us with conflicting information. We have more choices than we can possibly manage, and know more about the world around us than ever before but, perhaps, understand less. How can we deal with the barrage of often contradictory news and advice that inundates us?

To complicate our difficulty in knowing what to believe, distinguished authorities disagree vehemently about many important topics. A law of environmental science might be that for any expert there is always an equal and opposite expert. How can you decide what is true and meaningful in such a welter of confusing information? Is it simply a matter of what feels good at the moment or supports our preconceived notions? Or are there ways to use logical, orderly, creative thinking procedures to reach decisions?

By now, most of us know not to believe everything we read or hear (fig. L.4). "Tastes great . . . Low, low sale price . . . Vote for me . . . Lose 30 pounds in 3 weeks . . . You may already be a winner . . . Causes no environmental harm . . . I'll never lie to you . . . Two out of three doctors recommend . . ." More and more of the information we used to buy, elect, advise, judge, or heal has been created not to expand our knowledge but to sell a product or advance a cause. It would be unfortunate if we become cynical and apathetic due to information overload. It does make a difference what we think and how we act.

Approaches to truth and knowledge

A number of skills, attitudes, and approaches can help us evaluate information and make decisions. **Analytical thinking** asks, "How can I break this problem down into its constituent parts?" **Creative thinking** asks, "How might I approach this problem in new and inventive ways?" **Logical thinking** asks, "How can orderly, deductive reasoning help me think clearly?" **Critical thinking** asks, "What am I trying to accomplish here and how will I know when I've succeeded?" **Reflective thinking** asks, "What does it all mean?" In this section, we'll look more closely at critical and reflective thinking as a foundation for your study of environmental science. We hope you will apply these ideas consistently as you read this book.

FIGURE L.4 "There is absolutely no cause for alarm at the nuclear plant!"
Source: © Tribune Media Services. Reprinted with permission.

As figure L.5 suggests, critical thinking is central in the constellation of thinking skills. It challenges us to examine theories, facts, and options in a systematic, purposeful, and responsible manner. It shares many methods and approaches with other methods of reasoning but adds some important contextual skills, attitudes, and dispositions. Furthermore, it challenges us to plan methodically and to assess the process of thinking as well as the implications of our decisions. Thinking critically can help us discover hidden ideas and means, develop strategies for evaluating reasons and conclusions in arguments, recognize the differences between facts and values, and avoid jumping to conclusions. Professor Karen J. Warren of Macalester College identifies ten steps in critical thinking (table L.4).

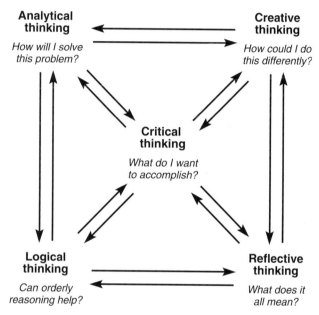

FIGURE L.5 Different approaches to thinking are used to solve different kinds of problems or to study alternate aspects of a single issue.

Table L.4   Steps in Critical Thinking
1. What is the *purpose* of my thinking?
2. What precise *question* am I trying to answer?
3. Within what *point of view* am I thinking?
4. What *information* am I using?
5. How am I *interpreting* that information?
6. What *concepts* or ideas are central to my thinking?
7. What *conclusions* am I aiming toward?
8. What am I taking for granted; what *assumptions* am I making?
9. If I accept the conclusions, what are the *implications*?
10. What would the *consequences* be, if I put my thoughts into action?

Source: Courtesy of Karen Warren, Philosophy Department, Macalester College, St. Paul, MN.

How Do You Tell the News From the Noise?

With the explosion of cable channels, web logs (blogs), social networks, and email access, most of us are interconnected constantly to a degree unique in history. There were at least 150 million blogs on the Web in 2010 and 15,000 new ones are added every day. More than a billion people are linked in social networks. Every day several billion emails, tweets, text messages, online videos, and Facebook postings connect us to one another. Handheld devices make it still easier to surf the Web, watch videos, or link to friends. In 2010, there were 4.6 billion mobile phones in the world, or enough for two-thirds of humanity to have one.

There are many benefits from social networks and rapid communication. They were instrumental in bringing about democratic revolutions in the Middle East. And they help people find others with compatible interests or talents. Whatever you want to discuss or learn about, you can probably find a group on the Internet. You may be the only person in your community fascinated by a particular topic, but elsewhere in the world there are others just like you. Together you make a critical mass that justifies a publication or an affinity group. But there's a darker side of this specialization and narrowing focus. Many people use their amazing degree of interconnection not so much to be educated, or to get new ideas, as to reinforce their existing beliefs. A study on the State of the Media by the Center for Journalistic Excellence at Columbia University concluded that the news is becoming increasingly partisan and ideological. Rumors and outright lies fly through the net at light speed. Conspiracy theorists and political operatives spread sensational accusations that are picked up and amplified in the echo chambers of modern media. Newscasters find they don't have to aim at mass markets any more. With so many channels available, they can cater to a narrow sector of the population and give them just what they want to hear.

One effect of separate conversations for separate communities has been the growth of hyper-partisan news programing, which increasingly involves attack journalism. Commentators often ridicule and demean their opponents rather than weighing ideas or reporting objective facts and sources, because shouting matches are exciting and sell advertising. Most newspapers have laid off almost all their investigative reporters and most television stations have abandoned the traditional written and edited news story. According to the Center for Journalistic Excellence, more than two-thirds of all TV news segments now consist of on-site "stand-up" reports or live interviews in which a single viewpoint is presented as news without any background or perspective. Visual images seem more immediate and believable: after all, pictures don't lie—although they can give a very selective view of the truth. Many topics, such as policy issues, don't make good visuals, and therefore never make it into TV coverage. Crime, accidents, disasters, lifestyle stories, sports, and weather make up more than 90 percent of the coverage on a typical television news program. An entire day of cable TV news would show, on average, only 1 minute each about the environment and health care, 2 minutes each on science and education, and 4 minutes on art and culture. More than 70 percent of

the segments are less than 1 minute long, which allows them to convey lots of emotion but little substance. People who get their news primarily from TV are significantly more fearful and pessimistic than those who get news from print media. And it becomes hard to separate rumor from truth. Evidence and corroboration take a backseat to dogma and passion. As consumers of instantaneous communication, we often don't have time to seek evidence, but depend more on gut instincts, which often means simply our prejudices and preconceived notions.

Partisan journalism has become much more prevalent since the deregulation of public media in 1988. From the birth of the broadcasting industry, the airwaves were regulated as a public trust. Broadcasters, as a condition of their licenses, were required to operate in the "public interest" by covering important policy issues and providing equal time to both sides of contested issues. In 1988, however, the Federal Communications Commission ruled that the proliferation of mass media gives the public adequate access to diverse sources of information. Media outlets no longer are obliged to provide fair and balanced coverage of issues. Presenting a single perspective or even a deceptive version of events is no longer regarded as a betrayal of public trust.

How can you detect bias in a blog or news report? Ask yourself (or your friends) these questions as you practice critical thinking, look for bias, and make sense out of what you see and hear.

1. What political positions are represented? Are they overt or covert?

2. Are speakers discussing facts and rational ideas, or are they resorting to innuendo, name-calling, character assassination, and ad hominem attacks? When people start calling each other Nazi or communist (or both), civil discourse has probably come to an end.

3. What special interests might be involved here? Who stands to gain presenting a particular viewpoint? Who is paying for the message?

4. What sources are used as evidence in this communication? How credible are they?

5. Are facts or statistics cited in the presentation? Are they credible? Are citations provided so you can check the sources?

6. Is the story one-sided, or are alternate viewpoints presented? If it is one-sided, does it represent majority opinion? Does that matter?

7. If the presentation claims to be fair and balanced, are both sides represented by credible spokespersons, or is one simply a foil set up to make the other side look good?

8. Are the arguments presented based on facts and logic, or are they purely emotional appeals?

How many of the critical thinking steps above do you use regularly, as you interpret information from the television or the Internet? How many news sources do you rely on for information? Is it just one, or do you seek out views from multiple sources? What motivates you to do this? What kinds of factors influence the ways you form your opinions on the news?

[1]The State of the News Media 2004 available at http://www.journalism.org.

Notice that many critical thinking processes are self-reflective and self-correcting. This form of thinking is sometimes called "thinking about thinking." It is an attempt to plan rationally how to analyze a problem, to monitor your progress while you are doing it, and to evaluate how your strategy worked and what you have learned when you are finished. It is not critical in the sense of finding fault, but it makes a conscious, active, disciplined effort to be aware of hidden motives and assumptions, to uncover bias, and to recognize the reliability or unreliability of sources (What Do You Think? p. 9).

What do I need to think critically?

Certain attitudes, tendencies, and dispositions are essential for well-reasoned analysis. Professor Karen Warren suggests the following list:

- *Skepticism and independence.* Question authority. Don't believe everything you hear or read—including this book; even experts sometimes are wrong.

- *Open-mindedness and flexibility.* Be willing to consider differing points of view and to entertain alternative explanations. Try arguing from a viewpoint different from your own. It will help you identify weaknesses and limitations in your own position.

- *Accuracy and orderliness.* Strive for as much precision as the subject permits or warrants. Deal systematically with parts of a complex whole. Be disciplined in the standards you apply.

- *Persistence and relevance.* Stick to the main point. Don't allow diversions or personal biases to lead you astray. Information may be interesting or even true, but is it relevant?

- *Contextual sensitivity and empathy.* Consider the total situation, relevant context, feelings, level of knowledge, and sophistication of others as you evaluate information. Imagine being in someone else's place to try to understand how they feel.

- *Decisiveness and courage.* Draw conclusions and take a stand when the evidence warrants doing so. Although we often wish for more definitive information, sometimes a well-reasoned but conditional position has to be the basis for action.

- *Humility.* Realize that you may be wrong and that reconsideration may be called for in the future. Be careful about making absolute declarations; you may need to change your mind someday.

While critical thinking shares many of the orderly, systematic approaches of formal logic, it also invokes traits like empathy, sensitivity, courage, and humility. Formulating intelligent opinions about some of the complex issues you'll encounter in environmental science requires more than simple logic. Developing these attitudes and skills is not easy or simple. It takes practice. You have to develop your mental faculties just as you need to train for a sport. Traits such as intellectual integrity, modesty, fairness, compassion, and fortitude are not things you can use only occasionally. They must be cultivated until they become your normal way of thinking.

Applying critical thinking

We all use critical or reflective thinking at times. Suppose a television commercial tells you that a new breakfast cereal is tasty and good for you. You may be suspicious and ask yourself a few questions. What do they mean by good? Good for whom or what? Does "tasty" simply mean more sugar and salt? Might the sources of this information have other motives in mind besides your health

and happiness? Although you may not have been aware of it, you already have been using some of the techniques of critical analysis. Working to expand these skills helps you recognize the ways information and analysis can be distorted, misleading, prejudiced, superficial, unfair, or otherwise defective.

Here are some steps in critical thinking:

1. *Identify and evaluate premises and conclusions in an argument.* What is the basis for the claims made here? What evidence is presented to support these claims and what conclusions are drawn from this evidence? If the premises and evidence are correct, does it follow that the conclusions are necessarily true?

2. *Acknowledge and clarify uncertainties, vagueness, equivocation, and contradictions.* Do the terms used have more than one meaning? If so, are all participants in the argument using the same meanings? Are ambiguity or equivocation deliberate? Can all the claims be true simultaneously?

3. *Distinguish between facts and values.* Are claims made that can be tested? (If so, these are statements of fact and should be able to be verified by gathering evidence.) Are claims made about the worth or lack of worth of something? (If so, these are value statements or opinions and probably cannot be verified objectively.) For example, claims of what we *ought* to do to be moral or righteous or to respect nature are generally value statements.

4. *Recognize and assess assumptions.* Given the backgrounds and views of the protagonists in this argument, what underlying reasons might there be for the premises, evidence, or conclusions presented? Does anyone have an "axe to grind" or a personal agenda in this issue? What do they think you know, need, want, or believe? Is there a subtext based on race, gender, ethnicity, economics, or some belief system that distorts this discussion?

5. *Distinguish the reliability or unreliability of a source.* What makes the experts qualified in this issue? What special knowledge or information do they have? What evidence do they present? How can we determine whether the information offered is accurate, true, or even plausible?

6. *Recognize and understand conceptual frameworks.* What are the basic beliefs, attitudes, and values that this person, group, or society holds? What dominating philosophy or ethics control their outlook and actions? How do these beliefs and values affect the way people view themselves and the world around them? If there are conflicting or contradictory beliefs and values, how can these differences be resolved?

Some clues for unpacking an argument

In logic, an argument is made up of one or more introductory statements (called **premises**), and a **conclusion** that supposedly follows logically from the premises. Often in ordinary conversation, different kinds of statements are mixed together, so it is difficult to distinguish between them or to decipher hidden or implied meanings. Social theorists call the process of separating and analyzing textual components *unpacking*. Applying this type of analysis to an argument can be useful.

An argument's premises are usually claimed to be based on facts; conclusions are usually opinions and values drawn from, or used to interpret, those facts. Words that often introduce a premise include: *as, because, assume that, given that, since, whereas,* and *we all know that* . . . Words that generally indicate a conclusion or statement of opinion or values include: *and, so, thus, therefore, it follows that, consequently, the evidence shows,* and *we can conclude that.*

For instance, in the example we used earlier, the television ad might have said: "*Since* we all need vitamins, and *since* this cereal contains vitamins, *consequently* the cereal must be good for you." Which are the premises and which is the conclusion? Does one necessarily follow from the other? Remember that even if the facts in a premise are correct, the conclusions drawn from them may not be. Information may be withheld from the argument such as the fact that the cereal is also loaded with unhealthy amounts of sugar.

Avoiding logical errors and fallacies

Formal logic catalogs a large number of fallacies and errors that invalidate arguments. Although we don't have room here to include all of these fallacies and errors, it may be helpful to review a few of the more common ones.

- *Red herring:* Introducing extraneous information to divert attention from the important point.
- *Ad hominem attacks:* Criticizing the opponent rather than the logic of the argument.
- *Hasty generalization:* Drawing conclusions about all members of a group based on evidence that pertains only to a selected sample.
- *False cause:* Drawing a link between premises and conclusions that depends on some imagined causal connection that does not, in fact, exist.

- *Appeal to ignorance:* Because some facts are in doubt, therefore a conclusion is impossible.
- *Appeal to authority:* It's true because _____ says so.
- *Begging the question:* Using some trick to make a premise seem true when it is not.
- *Equivocation:* Using words with double meanings to mislead the listener.
- *Slippery slope:* A claim that some event or action will cause some subsequent action.
- *False dichotomy:* Giving either/or alternatives as if they are the only choices.

Avoiding these fallacies yourself or being aware of them in another's argument can help you be more logical and have more logical and reasonable discussions.

Using critical thinking in environmental science

As you go through this book, you will have many opportunities to practice critical thinking skills. Every chapter includes many facts, figures, opinions, and theories. Are all of them true? No, probably not. They were the best information available when this text was written, but much in environmental science is in a state of flux. Data change constantly as does our interpretation of them. Do the ideas presented here give a complete picture of the state of our environment? Unfortunately, they probably don't. No matter how comprehensive our discussion is of this complex, diverse subject, it can never capture everything worth knowing, nor can it reveal all possible points of view.

When reading this text, try to distinguish between statements of fact and opinion. Ask yourself if the premises support the conclusions drawn from them. Although we have tried to be fair and even-handed in presenting controversies, we, like everyone, have biases and values—some of which we may not even recognize—that affect how we see issues and present arguments. Watch for cases in which you need to think for yourself and utilize your critical and reflective thinking skills to find the truth.

CONCLUSION

Whether you find environmental science interesting and useful depends largely on your own attitudes and efforts. Developing good study habits, setting realistic goals for yourself, taking the initiative to look for interesting topics, finding an appropriate study space, and working with a study partner can both make your study time more efficient and improve your final grade. Each of us has his or her own learning style. You may understand and remember things best if you see them in writing, hear them spoken by someone else, reason them out for yourself, or learn by doing. By determining your preferred style, you can study in the way that is most comfortable and effective for you.

In 2009, China passed Denmark, Germany, and Spain to become the world's largest producer of wind turbines, and in 2010, China also became the leading producer of photovoltaic panels and solar water heaters.

Understanding Our Environment

Learning Outcomes

After studying this chapter, you should be able to:

1.1 Explain what environmental science is, and how it draws on different kinds of knowledge.

1.2 List and describe some current concerns in environmental science.

1.3 Identify some early thinkers on environment and resources, and contrast some of their ideas.

1.4 Appreciate the human dimensions of environmental science, including the connection between poverty and environmental degradation.

1.5 Describe sustainable development and its goals.

1.6 Explain a key point of environmental ethics.

1.7 Identify ways in which faith-based groups share concerns for our environment.

"Today we are faced with a challenge that calls for a shift in our thinking, so that humanity stops threatening its life-support system." ~ *Wangari Maathai*

Winner of 2004 Nobel Peace Prize

Case Study   Renewable Energy in China

From ground level, Rizhao looks like any other midsized Chinese city. Located in Shandong Province about halfway between Beijing and Shanghai, Rizhao sits on the coastal plain with its back to the mountains. Rows of traditional houses alternate with high-rise apartments and office buildings. But from above, Rizhao shows a different face. More than 1 million gleaming solar collectors decorate the rooftops of this city of 2.8 million residents (fig. 1.1). More than 99 percent of all households get hot water and space heating from renewable energy.

In 2008, Rizhao became carbon neutral, one of the first four cities in the world to reach this milestone, a remarkable accomplishment in a developing country. Already, Rizhao has cut its per capita carbon emissions by half, compared to a decade ago, and its energy use by one-third. Generous subsidies for property owners, low-cost loans, and regulations that require renewable energy for all new construction have created mass markets for equipment that brings costs down, cleans the air, saves money, and creates thousands of local jobs. A solar water heater currently costs about (U.S.) $230 in Rizhao—about one-tenth the cost in the United States—and pays for itself in just a few years.

Fortunately, Rizhao isn't an isolated case. In the past few years, China has become the world's leader in clean energy. In 2009, China passed Denmark, Germany, and Spain to become the world's largest producer of wind turbines. And in 2010 China produced about two-thirds of the world's photovoltaic modules as well as about 80 percent of solar water heaters.

China's green technology progress is great news for our global environment. Lower prices for solar, wind, and other sustainable energy sources make it more feasible for people everywhere to wean themselves off of environmentally destructive fossil fuels. And nowhere is this change more important than in China itself. At the same time it has become the leader in solar and wind power, China has also greatly expanded its coal consumption. With its economy expanding at about 9 percent annually, China's energy

FIGURE 1.1 China already has more than 40 million rooftop solar collectors and has recently become the world's leader in renewable energy.

consumption is growing about 3.8 percent per year. Coal currently supplies about 70 percent of China's electricity.

Burning billions of tons of dirty coal every year makes China's air odious and unhealthy. According to the World Bank, 20 of the world's smoggiest cities are in China and acid rain affects at least one-third of the country. More than one million children are born in China each year with birth defects attributed to environmental pollution. In 2006, China passed the U.S. as the largest source of greenhouse gas emissions. For centuries, China has suffered from devastating droughts and floods. The effects of global climate change will very likely exacerbate these tragedies.

Moving to clean energy is a wonderful economic opportunity for this developing country. Already, more than a million Chinese workers are employed in the clean energy sector. With three of the five largest solar producers in the world, China now provides about 40 percent of the solar panels installed in California, the United States' largest market. And a Chinese company using Chinese turbines is building the largest wind farm currently under construction in the U.S. The Chinese government has promised to spend 5 trillion yuan ($736 billion) over the next ten years on clean energy. This is about four times the current level of investment in the United States.

China has several advantages in the race to produce sustainable energy. Around 250 million people have moved from the country to the city since 1990, and an equal number are expected to become urbanized in the next few decades, providing a huge market for new housing, electricity, and technology. To meet growing energy demand in just the next ten years, China will need to add about nine times as much electric generating capacity as the United States. Where utility managers are adding so much new equipment anyway, it isn't hard to make some of it solar or wind. American and European utilities, on the other hand, may have to abandon some existing technology to move in a meaningful way to renewables.

China also benefits from low labor and raw material costs. Already, Chinese companies produce the lowest priced solar panels in the world. Polysilicon, the main ingredient in solar photovoltaics, cost

Case Study   continued

about $400 per kg in 2008. China can now produce it for $45 per kg, and expects to drive prices down even further in coming years. Furthermore, China has a near monopoly on several rare earth elements, such as dysprosium and terbium, essential in green technology (see chapter 14). Solar power stations and wind farms are built with relative ease in China, meeting little of the public resistance that hampers Western developers. And government officials in China can simply order utilities to switch to renewable power.

This case study exemplifies some of the complexities of environmental science. As you'll learn in reading this book, this field incorporates information from many disciplines. Economics, engineering, geography, politics, and social conditions are important in understanding our environment as are biology, chemistry, climatology, or ecology. It's essential to consider many different sources of information to get a comprehensive view of our environmental condition. In this chapter, we'll survey some of the major challenges we face as well as encouraging signs for solutions to these problems. For related resources, including Google Earth™ placemarks that show locations where these issues can be explored, visit http://EnvironmentalScience-Cunningham.blogspot.com.

1.1 WHAT IS ENVIRONMENTAL SCIENCE?

Humans have always inhabited two worlds. One is the natural world of plants, animals, soils, air, and water that preceded us by billions of years and of which we are a part. The other is the world of social institutions and artifacts that we create for ourselves using science, technology, and political organization. Both worlds are essential to our lives, but integrating them successfully causes enduring tensions.

Where earlier people had limited ability to alter their surroundings, we now have power to extract and consume resources, produce wastes, and modify our world in ways that threaten both our continued existence and that of many organisms with which we share the planet. To ensure a sustainable future for ourselves and future generations, we need to understand something about how our world works, what we are doing to it, and what we can do to protect and improve it.

Environment (from the French *environner:* to encircle or surround) can be defined as (1) the circumstances or conditions that surround an organism or group of organisms, or (2) the complex of social or cultural conditions that affect an individual or community. Since humans inhabit the natural world as well as the "built" or technological, social, and cultural world, all constitute important parts of our environment (fig. 1.2).

Environmental science, then, is the systematic study of our environment and our proper place in it. A relatively new field, environmental science is highly interdisciplinary, integrating natural sciences, social sciences, and humanities in a broad, holistic study of the world around us. In contrast to more theoretical disciplines, environmental science is mission-oriented. That is, it seeks new, valid, contextual knowledge about the natural world and our impacts on it, but obtaining this information creates a responsibility to get involved in trying to do something about the problems we have created.

As distinguished economist Barbara Ward pointed out, for an increasing number of environmental issues, the difficulty is not to identify remedies. Remedies are now well understood. The problem is to make them socially, economically, and politically acceptable. Foresters know how to plant trees, but not how to establish conditions under which villagers in developing countries can manage plantations for themselves. Engineers know how to control pollution,

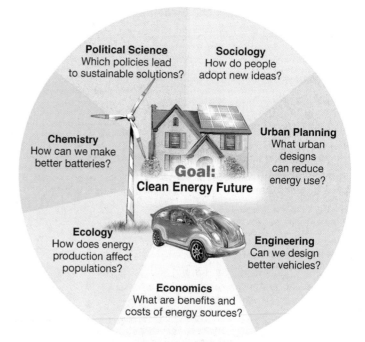

FIGURE 1.2 Many kinds of knowledge contribute to solutions in environmental science. A few examples are shown.

but not how to persuade factories to install the necessary equipment. City planners know how to build housing and design safe drinking water systems, but not how to make them affordable for the poorest members of society. The solutions to these problems increasingly involve human social systems as well as natural science.

As you study environmental science, you should learn the following:

- awareness and appreciation of the natural and built environment;
- knowledge of natural systems and ecological concepts;
- understanding of current environmental issues; and
- the ability to use critical-thinking and problem-solving skills on environmental issues.

For the remainder of this chapter, we'll complete our overview with a short history of environmental ideas and a survey of some important current issues that face us.

1.2 CURRENT CONDITIONS

As you probably already know, many environmental problems now face us. Before surveying them in the following section, we should pause for a moment to consider the extraordinary natural world that we inherited and that we hope to pass on to future generations in as good—perhaps even better—a condition than when we arrived.

We live on a marvelous planet

Imagine that you are an astronaut returning to Earth after a long trip to the moon or Mars. What a relief it would be to come back to this beautiful, bountiful planet after experiencing the hostile, desolate environment of outer space. Although there are dangers and difficulties here, we live in a remarkably prolific and hospitable world that is, as far as we know, unique in the universe. Compared to the conditions on other planets in our solar system, temperatures on the earth are mild and relatively constant. Plentiful supplies of clean air, fresh water, and fertile soil are regenerated endlessly and spontaneously by geological and biological cycles (discussed in chapters 3 and 4).

Perhaps the most amazing feature of our planet is the rich diversity of life that exists here. Millions of beautiful and intriguing species populate the earth and help sustain a habitable environment (fig. 1.3). This vast multitude of life creates complex, interrelated communities where towering trees and huge animals live together with, and depend upon, tiny life-forms such as viruses, bacteria, and fungi. Together all these organisms make up delightfully diverse, self-sustaining communities, including dense, moist forests, vast sunny savannas, and richly colorful coral reefs. From time to time, we should pause to remember that, in spite of the challenges and complications of life on earth, we are incredibly lucky to be here. We should ask ourselves: what is our proper place in nature? What *ought* we do and what *can* we do to protect the irreplaceable habitat that produced and supports us?

FIGURE 1.3 Perhaps the most amazing feature of our planet is its rich diversity of life.

But we also need to get outdoors and appreciate nature. As author Ed Abbey said, "It is not enough to fight for the land; it is even more important to enjoy it. While you can. While it is still there. So get out there and mess around with your friends, ramble out yonder and explore the forests, encounter the grizz, climb the mountains. Run the rivers, breathe deep of that yet sweet and lucid air, sit quietly for a while and contemplate the precious stillness, that lovely, mysterious and awesome space. Enjoy yourselves, keep your brain in your head and your head firmly attached to your body, the body active and alive."

We face many serious environmental problems

It's important for you to be aware of current environmental conditions. We'll cover all these issues in subsequent chapters of this book, but here's an overview to get you started. With more than 7.1 billion humans currently, we're adding about 80 million more to the world every year. While demographers report a transition to slower growth rates in most countries, present trends project a population between

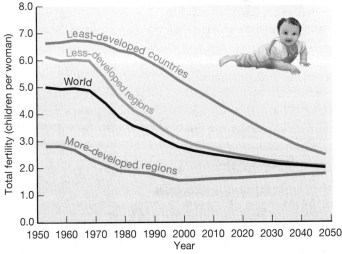

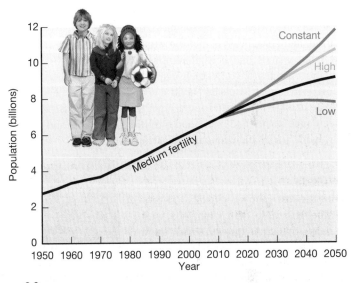

FIGURE 1.4 Bad news and good news: globally, populations continue to rise, but our rate of growth has plummeted. Some countries are below the replacement rate of about two children per woman.

Source: United Nations Population Program, 2007.

8 and 10 billion by 2050 (fig. 1.4). The impacts of that many people on our natural resources and ecological systems is a serious concern.

Climate Change

Burning fossil fuels, making cement, cultivating rice paddies, clearing forests, and other human activities release carbon dioxide and other so-called "greenhouse gases" that trap heat in the atmosphere. Over the past 200 years, atmospheric CO_2 concentrations have increased about 35 percent. By 2100, if current trends continue, climatologists warn that mean global temperatures will probably increase 2° to 6°C compared to 1900 temperatures (3.6° to 12.8°F: fig. 1.5a). Although we can't say whether specific recent storms were influenced by global warming, climate changes caused by greenhouse gases are very likely to result in increasingly severe weather events including droughts, floods, hurricanes, and tornadoes. Melting alpine glaciers and snowfields could threaten water supplies on which millions of people depend.

Already, we are seeing dramatic climate changes in the Antarctic and Arctic where seasons are changing, sea ice is disappearing, and permafrost is melting. Rising sea levels are flooding low-lying islands and coastal regions, while habitat losses and climatic changes are affecting many biological species. Canadian Environment Minister David Anderson has said that global climate change is a greater threat than terrorism because it could threaten the homes and livelihood of billions of people and trigger worldwide social and economic catastrophe.

Hunger

Over the past century, global food production has more than kept pace with human population growth, but there are worries about whether we will be able to maintain this pace (fig. 1.5b). Soil scientists report that about two-thirds of all agricultural lands show signs of degradation. Biotechnology and intensive farming techniques responsible for much of our recent production gains often are too expensive for poor farmers. Can we find ways to produce the food we need without further environmental degradation? And will that food be distributed equitably? In a world of food surpluses, the United Nations estimates that some 925 million people are now chronically undernourished, and at least 60 million face acute food shortages due to natural disasters or conflicts.

Clean Water

Water may well be the most critical resource in the twenty-first century. Already at least 1.1 billion people lack an adequate supply of safe drinking water, and more than twice that many don't have modern sanitation (fig. 1.5c). Polluted water and inadequate sanitation are estimated to contribute to the ill health of more than 1.2 billion people annually, including the death of 15 million children per year. About 40 percent of the world population lives in countries where water demands now exceed supplies, and by 2025 the UN projects that as many as three-fourths of us could live under similar conditions. Water wars may well become the major source of international conflict in coming decades.

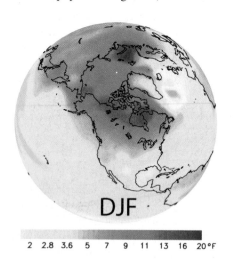

(a) Climate change

(b) Hunger

(c) Water quality

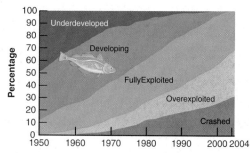

(d) Biodiversity and fisheries

FIGURE 1.5 Major environmental challenges: (a) Climate change is projected to raise temperatures, especially in northern winter months. (b) Nearly a billion people suffered from chronic hunger in 2010. (c) Poor water quality is responsible for 15 million deaths each year. (d) Biodiversity including marine species continues to decline.
Data from United Nations 2010.

Energy

How we obtain and use energy is likely to play a crucial role in our environmental future. Fossil fuels (oil, coal, and natural gas) presently provide around 80 percent of the energy used in industrialized countries. However, problems associated with their acquisition and use—air and water pollution, mining damage, shipping accidents, and geopolitics—may limit what we do with remaining reserves. Cleaner renewable energy resources, such as those now being produced in China including solar power, wind, geothermal, and biomass, together with conservation, could give us cleaner, less destructive options if we invest in appropriate technology.

Biodiversity Loss

Biologists report that habitat destruction, overexploitation, pollution, and introduction of exotic organisms are eliminating species at a rate comparable to the great extinction that marked the end of the age of dinosaurs (fig. 1.5d). The UN Environment Program reports that, over the past century, more than 800 species have disappeared and at least 10,000 species are now considered threatened. This includes about half of all primates and freshwater fish, together with around 10 percent of all plant species. Top predators, including nearly all the big cats in the world, are particularly rare and endangered. A nationwide survey of the United Kingdom in 2004 found that most bird and butterfly populations had declined between 50 and 75 percent over the previous 20 years. At least half of the forests existing before the introduction of agriculture have been cleared, and much of the diverse "old growth" on which many species depend for habitat is rapidly being cut and replaced by secondary growth or monoculture.

Air Pollution

Air quality has worsened dramatically in many areas. Over southern Asia, for example, satellite images recently revealed a 3-km (2-mile)-thick toxic haze of ash, acids, aerosols, dust, and photochemical products regularly covers the entire Indian subcontinent for much of the year. Nobel laureate Paul Crutzen estimates that at least 3 million people die each year from diseases triggered by air pollution. Worldwide, the United Nations estimates that more than 2 billion metric tons of air pollutants (not including carbon dioxide or wind-blown soil) are emitted each year. Air pollution no longer is merely a local problem. Mercury, bisphenol A (BPA), perflurocarbons, and other long-lasting pollutants accumulate in arctic ecosystems and native people after being transported by air currents from industrial regions thousands of kilometers to the south. And during certain days, as much as 75 percent of the smog and particulate pollution recorded on the west coast of North America can be traced to Asia.

Finding solutions to these problems requires good science as well as individual and collective actions. Becoming educated about our global environment is the first step in understanding how to control our impacts on it. We hope this book will help you in that quest.

There are also many signs of hope

Is there hope that we can find solutions to these dilemmas? We think so. As the opening case study for this chapter shows, even developing countries, such as China, are making progress on environmental problems. China now has more than 200,000 wind generators and 10 million biogas generators (the most in the world). Solar collectors on 35 million buildings furnish hot water. China could easily get all its energy from renewable sources, and it may be better able to provide advice and technology to other developing countries than can rich nations.

Population and Pollution

Many cities in Europe and North America are cleaner and much more livable now than they were a century ago. Clean technology, such as the solar panels and wind turbines now being produced in China, help eliminate pollution and save resources. Population has stabilized in most industrialized countries and even in some very poor countries where social security and democracy have been established. Over the last 20 years, the average number of children born per woman worldwide has decreased from 6.1 to 2.7 (see fig. 1.4). By 2050, the UN Population Division predicts that all developed countries and 75 percent of the developing world will experience a below-replacement fertility rate of 2.1 children per woman. This prediction suggests that the world population will stabilize at about 8.9 billion rather than 9.3 billion, as previously estimated.

Health

The incidence of life-threatening infectious diseases has been reduced sharply in most countries during the past century, while life expectancies have nearly doubled on average (fig. 1.6a). Smallpox has been completely eradicated and polio has been vanquished except in a few countries. Since 1990, more than 800 million people have gained access to improved water supplies and modern sanitation. In spite of population growth that added nearly a billion people to the world during the 1990s, the number facing food insecurity and chronic hunger during this period actually declined by about 40 million.

Information and Education

Because so many environmental issues can be fixed by new ideas, technologies, and strategies, expanding access to knowledge is essential to progress. The increased speed at which information now moves around the world offers unprecedented opportunities for sharing ideas. At the same time, literacy and access to education are expanding in most regions of the world (fig. 1.6b). Many developing countries may be able to benefit from the mistakes made by industrialized countries and leapfrog directly to sustainability.

Sustainable Resource Use Around the World

We are finding ways to conserve resources and use them more sustainably. For example, improved monitoring of fisheries and networks of marine protected areas promote species conservation as well as human development (fig. 1.6c).

Habitat Conservation

Brazil, which has the largest area of tropical rainforest in the world, has reduced forest destruction by nearly two-thirds in the past five years. In addition to protecting endangered species, this is great news in the battle to stabilize our global climate. Nature preserves and protected areas have increased nearly five-fold over the past 20 years, from about 2.6 million km^2 to about 12.2 million km^2. This represents only 8.2 percent of all land area—less than the 12 percent thought necessary to protect a viable sample of the world's biodiversity—but is a dramatic expansion nonetheless (fig. 1.6d).

(a) Health care

(c) Sustainable resource use

(b) Education

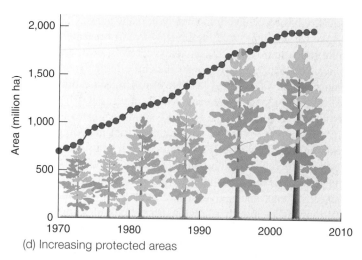

(d) Increasing protected areas

FIGURE 1.6 Conditions are improving in many areas, including access to (a) health care and (b) education. In many areas, (c) sustainable resource use is being improved by expanding (d) networks of protected areas.
Data: IUCN and UNEP, 2010.

What Do You Think?

Calculating Your Ecological Footprint

Can the earth sustain our current lifestyles? Will there be adequate natural resources for future generations? These questions are among the most important in environmental science today. We depend on nature for food, water, energy, oxygen, waste disposal, and other life-support systems. Sustainability implies that we cannot turn our resources into waste faster than nature can recycle that waste and replenish the supplies on which we depend. It also recognizes that degrading ecological systems ultimately threatens everyone's well-being. Although we may be able to overspend nature's budget temporarily, future generations will have to pay the debts we leave them. Living sustainably means meeting our own vital needs without compromising the ability of future generations to meet their own needs.

How can we evaluate and illustrate our ecological impacts? Redefining Progress, a nongovernmental environmental organization, has developed a measure called the **ecological footprint** to compute the demands placed on nature by individuals and nations. A simple questionnaire of 16 items gives a rough estimate of your personal footprint. A more complex assessment of 60 categories including primary commodities (such as milk, wood, or metal ores), as well as the manufactured products derived from them, gives a measure of national consumption patterns.

According to Redefining Progress, the average world citizen has an ecological footprint equivalent to 2.3 hectares (5.6 acres), while the biologically productive land available is only 1.9 hectares (ha) per person. How can this be? The answer is that we're using nonrenewable resources (such as fossil fuels) to support a lifestyle beyond the productive capacity of our environment. It's like living by borrowing on your credit cards. You can do it for a while, but eventually you have to pay off the deficit. The unbalance is far more pronounced in some of the richer countries. The average resident of the United States, for example, lives at a consumption

level that requires 9.7 ha of bioproductive land. A dramatic comparison of consumption levels versus population size is shown in figure 1. If everyone in the world were to adopt a North American lifestyle, we'd need about four more planets to support us all. You can check out your own ecological footprint by going to www.redefiningprogress.org/.

Like any model, an ecological footprint gives a useful description of a system. Also like any model, it is built on a number of assumptions: (1) Various measures of resource consumption and waste flows can be converted into the biologically productive area required to maintain them; (2) different kinds of resource use and dissimilar types of productive land can be standardized into roughly equivalent areas; (3) because these areas stand for mutually exclusive uses, they can be added up to a total—a total representing humanity's demand—that can be compared to the total world area of bioproductive land. The model also implies that our world has a fixed supply of resources that can't be expanded. Part of the power of this metaphor is that we all can visualize a specific area of land and imagine it being divided into smaller and smaller parcels as our demands increase. But this perspective doesn't take into account technological progress. For example, since 1950, world food production has increased about fourfold. Some of this growth has come from expansion of croplands, but most has come from technological advances such as irrigation, fertilizer use, and higher-yielding crop varieties. Whether this level of production is sustainable is another question, but this progress shows that land area isn't always an absolute limit. Similarly, switching to renewable energy sources such as wind and solar power would make a huge impact on estimates of our ecological footprint. Notice that in figure 2 energy consumption makes up about half of the calculated footprint.

What do you think? Does analyzing our ecological footprint inspire you to correct our mistakes, or does it make sustainability seem an impossible goal? If we in the richer nations have the technology and political power to exploit a larger share of resources, do we have a right to do so, or do we have an ethical responsibility to restrain our consumption? And what about future generations? Do we have an obligation to leave resources for them, or can we assume they'll make technological discoveries to solve their own problems if resources become scarce? You'll find that many of the environmental issues we discuss in this book aren't simply a matter of needing more scientific data. Ethical considerations and intergenerational justice often are just as important as having more facts.

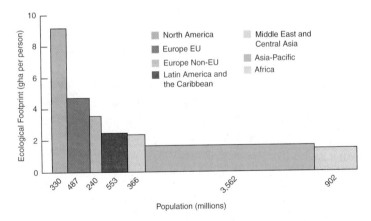

FIGURE 1 Ecological footprint by region, 2005. Bar weight shows footprint per person. Width of bars shows population size. Area of bars shows the region's total ecological footprint.

Source: WWF, 2008.

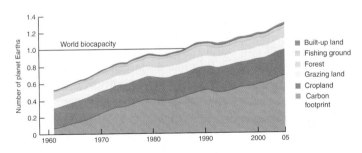

FIGURE 2 Humanity's ecological footprint has nearly tripled since 1961, when we began to collect global environmental data.

Source: WWF, 2008.

Renewable Energy

As the opening case study for this chapter shows, dramatic progress is being made in a transition to renewable energy sources. With mass production in China and progress in thin film technology in the United States, prices for solar panels dropped by 50 percent in 2010 making them much more competitive with fossil fuels. The European Union has pledged to get 20 percent of its energy from renewable sources by 2020. Former British Prime Minister Tony Blair laid out even more ambitious plans to fight global warming by cutting carbon dioxide emissions in his country by 60 percent through energy conservation and a switch to renewables.

Carbon Markets and Standards

Cap and trade programs in which limits are established on greenhouse gas emissions and companies can buy and sell discharge permits have been in place in Europe for several years and are stimulating both conservation measures and technological improvements. In 2010, California, which would be the eighth largest economy in the world, if it were an independent country, established a similar program, the first of its kind in America.

International Cooperation

Currently, more than 500 international environmental protection agreements are now in force. Some, such as the Montreal Protocol on Stratospheric Ozone, have been highly successful. Others, such as the Law of the Sea, lack enforcement powers. Perhaps the most important of all these treaties is the Kyoto Protocol on global climate change, which has been ratified by 191 countries including every industrialized nation except the United States.

1.3 A BRIEF HISTORY OF CONSERVATION AND ENVIRONMENTALISM

Many of our current ideas about our environment and its resources were articulated by writers and thinkers in the past 150 years. Although many earlier societies had negative impacts on their environments, recent technological innovations have greatly increased our impacts. As a consequence of these changes, different approaches have developed for understanding and protecting our environment. We can divide conservation history and environmental activism into at least four distinct stages: (1) pragmatic resource conservation, (2) moral and aesthetic nature preservation, (3) a growing concern about health and ecological damage caused by pollution, and (4) global environmental citizenship. Each era focused on different problems and each suggested a distinctive set of solutions. These stages are not necessarily mutually exclusive, however; parts of each persist today in the environmental movement and one person may embrace them all simultaneously.

Nature protection has historic roots

Recognizing human misuse of nature is not unique to modern times. Plato complained in the fourth century B.C. that Greece once was blessed with fertile soil and clothed with abundant forests of fine trees. After the trees were cut to build houses and ships, however, heavy rains washed the soil into the sea, leaving only a rocky "skeleton of a body wasted by disease" (fig. 1.7). Springs and rivers dried up while farming became all but impossible. Many classical authors regarded Earth as a living being, vulnerable to aging, illness, and even mortality. Periodic threats about the impending death of nature as a result of human misuse have persisted into our own time. Many of these dire warnings have proven to be premature or greatly exaggerated, but others remain relevant to our own times. As Mostafa K. Tolba, former Executive Director of the United Nations Environment Program has said, "The problems that overwhelm us today are precisely those we failed to solve decades ago."

Some of the earliest scientific studies of environmental damage were carried out in the eighteenth century by French and British colonial administrators who often were trained scientists and who observed rapid soil loss and drying wells that resulted from intensive colonial production of sugar and other commodities. Some of these administrators recognized that environmental stewardship was an economic necessity. These early conservationists observed and understood the connection between deforestation, soil erosion, and local climate change. The pioneering British plant physiologist, Stephen Hales, for instance, suggested that conserving green plants preserved rainfall. His ideas were put into practice in 1764 on the Caribbean island of Tobago, where about 20 percent of the land was marked as "reserved in wood for rains."

Pierre Poivre, an early French governor of Mauritius, an island in the Indian Ocean, was appalled at the environmental and

FIGURE 1.7 Nearly 2,500 years ago, Plato lamented land degradation that denuded the hills of Greece. Have we learned from history's lessons?

social devastation caused by destruction of wildlife (such as the flightless dodo) and the felling of ebony forests on the island by early European settlers. In 1769, Poivre ordered that one-quarter of the island was to be preserved in forests, particularly on steep mountain slopes and along waterways. Mauritius remains a model for balancing nature and human needs. Its forest reserves shelter a larger percentage of its original flora and fauna than most other human-occupied islands.

Resource waste inspired pragmatic, utilitarian conservation

Many historians consider the publication of *Man and Nature* in 1864 by geographer George Perkins Marsh as the wellspring of environmental protection in North America. Marsh, who also was a lawyer, politician, and diplomat, traveled widely around the Mediterranean as part of his diplomatic duties in Turkey and Italy. He read widely in the classics (including Plato) and personally observed the damage caused by the excessive grazing by goats and sheep and by the deforesting of steep hillsides. Alarmed by the wanton destruction and profligate waste of resources still occurring on the American frontier in his lifetime, he warned of its ecological consequences. Largely as a result of his book, national forest reserves were established in the United States in 1873 to protect dwindling timber supplies and endangered watersheds.

Among those influenced by Marsh's warnings were President Theodore Roosevelt (fig. 1.8a) and his chief conservation advisor, Gifford Pinchot (fig. 1.8b). In 1905, Roosevelt, who was the leader of the populist, progressive movement, moved the Forest Service out of the corruption-filled Interior Department into the Department of Agriculture. Pinchot, who was the first native-born professional forester in North America, became the founding head of this new agency. He put resource management on an honest, rational, and scientific basis for the first time in our history. Together with naturalists and activists such as John Muir, William Brewster, and George Bird Grinnell, Roosevelt and Pinchot established the framework of our national forest, park, and wildlife refuge systems, passed game protection laws, and tried to stop some of the most flagrant abuses of the public domain.

The basis of Roosevelt's and Pinchot's policies was pragmatic **utilitarian conservation.** They argued that the forests should be saved "not because they are beautiful or because they shelter wild creatures of the wilderness, but only to provide homes and jobs for people." Resources should be used "for the greatest good, for the greatest number for the longest time." "There has been a fundamental misconception," Pinchot said, "that conservation means nothing but husbanding of resources for future generations. Nothing could be further from the truth. The first principle of conservation is development and use of the natural resources now existing on this continent for the benefit of the people who live here now. There may be just as much waste in neglecting the development and use of certain natural resources as there is in their destruction." This pragmatic approach still can be seen today in the multiple use policies of the Forest Service.

(a) President Teddy Roosevelt (b) Gifford Pinchot

(c) John Muir (d) Aldo Leopold

FIGURE 1.8 Some early pioneers of the American conservation movement. President Teddy Roosevelt (a) and his main advisor Gifford Pinchot (b) emphasized pragmatic resource conservation, while John Muir (c) and Aldo Leopold (d) focused on ethical and aesthetic relationships.

Ethical and aesthetic concerns inspired the preservation movement

John Muir (fig. 1.8c), geologist, author, and first president of the Sierra Club, strenuously opposed Pinchot's utilitarian approach. Muir argued that nature deserves to exist for its own sake, regardless of its usefulness to us. Aesthetic and spiritual values formed the core of his philosophy of nature protection. This outlook has been called **biocentric preservation** because it emphasizes the fundamental right of other organisms to exist and to pursue their own interests. Muir wrote: "The world, we are told, was made for man. A presumption that is totally unsupported by the facts. . . . Nature's object in making animals and plants might possibly be first of all the happiness of each one of them. . . . Why ought man to value himself as more than an infinitely small unit of the one great unit of creation?"

Muir, who was an early explorer and interpreter of the Sierra Nevada Mountains in California, fought long and hard for establishment of Yosemite and Kings Canyon National Parks. The National Park Service, established in 1916, was first headed by Muir's disciple, Stephen Mather, and has always been oriented toward preservation of nature in its purest state. It has often been at odds with Pinchot's utilitarian Forest Service.

FIGURE 1.9 Aldo Leopold's Wisconsin shack, the main location for his *Sand County Almanac,* in which he wrote, "A thing is right when it tends to preserve the integrity, stability, and beauty of the biotic community. It is wrong when it tends otherwise." How might you apply this to your life?

In 1935, pioneering wildlife ecologist Aldo Leopold (fig. 1.8*d*) bought a small, worn-out farm in central Wisconsin. A dilapidated chicken shack, the only remaining building, was remodeled into a rustic cabin (fig. 1.9). Working together with his children, Leopold planted thousands of trees in a practical experiment in restoring the health and beauty of the land. Leopold argued for stewardship of the land. He wrote of "the land ethic," by which we should care for the land because it's the right thing to do—as well as the smart thing. "Conservation," he wrote, "is the positive exercise of skill and insight, not merely a negative exercise of abstinence or caution." The shack became a writing refuge and became the main focus of *A Sand County Almanac,* a much beloved collection of essays about our relation with nature. In it, Leopold wrote, "We abuse land because we regard it as a commodity belonging to us. When we see land as a community to which we belong, we may begin to use it with love and respect."

Think About It

Suppose a beautiful grove of trees near your house is scheduled to be cut down for a civic project such as a swimming pool. Would you support this? Why or why not? Which of the philosophies described in this chapter best describes your attitude?

Rising pollution levels led to the modern environmental movement

The undesirable effects of pollution probably have been recognized at least as long as those of forest destruction. In 1273, King Edward I of England threatened to hang anyone burning coal in London because of the acrid smoke it produced. In 1661, the English diarist John Evelyn complained about the noxious air pollution caused by coal fires and factories and suggested that sweet-smelling trees be planted to purify city air. Increasingly dangerous

smog attacks in Britain led, in 1880, to formation of a national Fog and Smoke Committee to combat this problem.

The tremendous industrial expansion during and after the Second World War added a new set of concerns to the environmental agenda. *Silent Spring,* written by Rachel Carson (fig. 1.10*a*) and published in 1962, awakened the public to the threats of pollution and toxic chemicals to humans as well as other species. The movement she engendered might be called **environmentalism** because its concerns are extended to include both environmental resources and pollution. Among the pioneers of this movement were activist David Brower (fig. 1.10*b*) and scientist Barry Commoner (fig. 1.10*c*). Brower, while executive director of the Sierra Club, Friends of the Earth, and the Earth Island Institute, introduced many of the techniques of modern environmentalism, including litigation, intervention in regulatory hearings, book and calendar publishing, and using mass media for publicity campaigns.

Barry Commoner, like Rachel Carson, emphasized the links between science, technology, and society. Trained as a molecular biologist, Commoner was an early example of activist scientists, who speak out about public hazards revealed by their research. Many of today's efforts to curb climate change or reduce biodiversity losses are led by scientists who raise the alarm about environmental problems.

(a) Rachel Carson

(b) David Brower

(c) Barry Commoner

(d) Wangari Maathai

Figure 1.10 Among many distinguished environmental leaders in modern times, Rachel Carson (a), David Brower (b), Barry Commoner (c), and Wangari Maathai (d) stand out for their dedication, innovation, and bravery.

Environmental quality is tied to social progress

Many people today believe that the roots of the environmental movement are elitist—promoting the interests of a wealthy minority, who can afford to vacation in wilderness. In fact, most environmental leaders have seen social justice and environmental equity as closely linked. Gifford Pinchot, Teddy Roosevelt, and John Muir all strove to keep nature and resources accessible to everyone, at a time when public lands, forests, and waterways were increasingly controlled by a few wealthy individuals and private corporations. The idea of national parks, one of our principal strategies for nature conservation, is to provide public access to natural beauty and outdoor recreation. Aldo Leopold, a founder of the Wilderness Society, promoted ideas of land stewardship among farmers, fishers, and hunters. Robert Marshall, also a founder of the Wilderness Society, campaigned all his life for social and economic justice for low-income groups. Both Rachel Carson and Barry Commoner were principally interested in environmental health—an issue that is especially urgent for low-income, minority, and inner-city residents. Many of these individuals grew up in working class families, so their sympathy with social causes is not surprising.

Increasingly, environmental activists are linking environmental quality and social progress on a global scale. One of the core concepts of modern environmental thought is **sustainable development,** the idea that economic improvement for the world's poorest populations is possible without devastating the environment. This idea became widely publicized after the Earth Summit, a United Nations meeting held in Rio de Janeiro, Brazil, in 1992. The Rio meeting was a pivotal event because it brought together many diverse groups. Environmentalists and politicians from wealthy countries, indigenous people and workers struggling for rights and land, and government representatives from developing countries all came together and became more aware of their common needs.

Some of today's leading environmental thinkers come from developing nations, where poverty and environmental degradation together plague hundreds of millions of people. Dr. Wangari Maathai of Kenya is a notable example. In 1977, Dr. Maathai (fig. 1.10d) founded the Green Belt Movement in her native Kenya as a way of both organizing poor rural women and restoring their environment. Beginning at a small, local scale, this organization has grown to more than 600 grassroots networks across Kenya. They have planted more than 30 million trees while mobilizing communities for self-determination, justice, equity, poverty reduction, and environmental conservation. Dr. Maathai was elected to the Kenyan Parliament and served as Assistant Minister for Environment and Natural Resources. Her leadership has helped bring democracy and good government to her country. In 2004, she received the Nobel Peace Prize for her work, the first time a Nobel has been awarded for environmental action. In her acceptance speech, she said, "Working together, we have proven that sustainable development is possible; that reforestation of degraded land is possible; and that exemplary governance is possible when ordinary citizens are informed, sensitized, mobilized and involved in direct action for their environment."

FIGURE 1.11 The life-sustaining ecosystems on which we all depend are unique in the universe, as far as we know.

Under the leadership of a number of other brilliant and dedicated activists and scientists, the environmental agenda was expanded in the 1960s and 1970s to include issues such as human population growth, atomic weapons testing and atomic power, fossil fuel extraction and use, recycling, air and water pollution, wilderness protection, and a host of other pressing problems that are addressed in this textbook. Environmentalism has become well established on the public agenda since the first national Earth Day in 1970. A majority of Americans now consider themselves environmentalists, although there is considerable variation in what that term means.

Photographs of the earth from space (fig. 1.11) provide a powerful icon for the fourth wave of ecological concern that might be called **global environmentalism.** These photos remind us how small, fragile, beautiful, and rare our home planet is. We all share a common environment at this global scale. As our attention shifts from questions of preserving particular landscapes or preventing pollution of a specific watershed or airshed, we begin to worry about the life-support systems of the whole planet.

A growing number of Chinese activists are part of this global environmental movement. In 2006, Yu Xiaogang was awarded the Goldman Prize, the world's top honor for environmental protection. Yu was recognized for his work on Yunan's Lashi Lake where he brought together residents, government officials, and entrepreneurs to protect wetlands, restore fisheries, and improve water quality. He also worked on sustainable development programs, such as women's schools and microcredit loans. His leadership was instrumental in stopping plans for 13 dams on the Nu River (known as the Salween when it crosses into Thailand, and Burma).

Another Goldman Prize winner is Dai Qing, who was jailed for her book that revealed the social and environmental costs of the Three Gorges Dam on the Yangtze River.

Other global environmental leaders who will be discussed later in this book include Professor Muhammad Yunus of Bangladesh, who won the Nobel Peace Prize in 2006 for his microcredit loan program at the Grameen Bank, and former Norwegian Prime Minister Gro Harlem Brundtland, who chaired the World Commission on Environment and Development, which coined the most widely accepted definition of sustainability.

1.4 HUMAN DIMENSIONS OF ENVIRONMENTAL SCIENCE

Because we live in both the natural and social worlds, and because we and our technology have become such dominant forces on the planet, environmental science must take human institutions and the human condition into account. We live in a world of haves and have-nots; a few of us live in increasing luxury, while many others lack the basic necessities for a decent, healthy, productive life. The World Bank estimates that more than 1.4 billion people—about one-fifth of the world's population—live in **extreme poverty** with an income of less than (U.S.)\$1.25 per day (fig. 1.12). These poorest of the poor often lack access to an adequate diet, decent housing, basic sanitation, clean water, education, medical care, and other essentials for a humane existence. Seventy percent of those people are women and children. In fact, four out of five people in the world live in what would be considered poverty in the United States or Canada.

Policymakers are becoming aware that eliminating poverty and protecting our common environment are inextricably interlinked because the world's poorest people are both the victims and the agents of environmental degradation. The poorest people are often forced to meet short-term survival needs at the cost of long-term sustainability. Desperate for croplands to feed themselves and their families, many move into virgin forests or cultivate steep, erosion-prone hillsides, where soil nutrients are exhausted after only a few years. Others migrate to the grimy, crowded slums and ramshackle shantytowns that now surround most major cities in the developing world. With no way to dispose of wastes, the residents often foul their environment further and contaminate the air they breathe and the water on which they depend for washing and drinking.

The cycle of poverty, illness, and limited opportunities can become a self-sustaining process that passes from one generation to another. People who are malnourished and ill can't work productively to obtain food, shelter, or medicine for themselves or their children, who also are malnourished and ill. About 250 million children—mostly in Asia and Africa and some as young as 4 years old—are forced to work under appalling conditions weaving carpets, making ceramics and jewelry, or working in the sex trade. Growing up in these conditions leads to educational, psychological, and developmental deficits that condemn these children to perpetuate this cycle.

FIGURE 1.12 Three-quarters of the world's poorest nations are in Africa. Millions of people lack adequate food, housing medical care, clean water, and safety. The human suffering engendered by this poverty is tragic.

Faced with immediate survival needs and few options, these unfortunate people often have no choice but to overharvest resources; in doing so, however, they diminish not only their own options but also those of future generations. And in an increasingly interconnected world, the environments and resource bases damaged by poverty and ignorance are directly linked to those on which we depend.

The Worldwatch Institute warns that "poverty, disease and environmental decline are the true axis of evil." Terrorist attacks—and the responses they provoke—are the symptoms of the underlying sources of global instability, including the dangerous interplay among poverty, hunger, disease, environmental degradation, and rising resource competition. Failure to deal with these sources of insecurity could plunge the world into a dangerous downward spiral in which instability and radicalization grows. Unless the world takes action to promote sustainability and equity, Worldwatch suggests we will face an uphill battle to deal with the consequences of wars, terrorism, and natural disasters.

We live in an inequitable world

About one-fifth of the world's population lives in the 20 richest countries, where the average per capita income is above (U.S.) \$35,000 per year. Most of these countries are in North America or Western Europe, but Japan, Singapore, and Australia also fall into this group. Almost every country, however, even the richest, such as the United States and Canada, has poor people. No doubt everyone reading this book knows about homeless people or other individuals who lack resources for a safe, productive life. According to the U.S. Census Bureau, 37 million Americans—one-third of them children—live in households below the poverty line.

The other four-fifths of the world's population lives in middle- or low-income countries, where nearly everyone is poor by North American standards. Nearly a billion people live in the poorest nations, where the average per capita income is below (U.S.)$1,000 per year. Among the 41 nations in this category, 33 are in sub-Saharan Africa. All the other lowest-income nations, except Haiti, are in Asia. Although poverty levels in countries such as China and Indonesia have fallen in recent years, most countries in sub-Saharan Africa and much of Latin America have made little progress. The destabilizing and impoverishing effects of earlier colonialism continue to play important roles in the ongoing problems of these unfortunate countries. Meanwhile, the relative gap between rich and poor has increased dramatically.

As table 1.1 shows, the gulf between the richest and poorest nations affects many quality-of-life indicators. The average individual in the highest-income countries has an annual income roughly 40 times that of those in the lowest-income nations. Infant mortality in the least-developed countries is about 25 times as high as in the most-developed countries. Only 23 percent of residents in poorer countries have access to modern sanitation, while this ammenity is essentially universal in richer countries. Carbon dioxide emissions (a measure of both energy use and contributions to global warming) are 150 times greater in rich countries.

The gulf between rich and poor is even greater at the individual level. The richest 200 people in the world have a combined wealth of $1 trillion. This is more than the total owned by the 3 billion people who make up the poorest half of the world's population. According to the United Nations Human Development Program, this inequality is more detrimental to political stability than absolute poverty.

Is there enough for everyone?

Those of us in the richer nations now enjoy a level of affluence and comfort unprecedented in human history. But we consume an inordinate share of the world's resources, and produce an unsustainable amount of pollution to support our lifestyle. What

FIGURE 1.13 "And may we continue to be worthy of consuming a disproportionate share of this planet's resources."
Source: © *The New Yorker Collection*, 1992. Lee Lorenz from cartoonbank.com. All Rights Reserved.

if everyone in the world tried to live at that same level of consumption? The United States, for example, with about 4.6 percent of the world's population, consumes about 25 percent of all oil while producing about 25 percent of all carbon dioxide and 50 percent of all toxic wastes in the world (fig. 1.13). What will the environmental effects be if other nations try to emulate our prosperity?

Take the example of China that we discussed in the opening case study for this chapter. In the early 1960s, it's estimated that 300 million Chinese suffered from chronic hunger, and perhaps 50 million starved to death in the worst famine in world history. Since then, however, China has experienced amazing economic growth. The national GDP has been increasing at about 10 percent per year. If current trends continue, the Chinese economy will surpass the United States and become the world's largest by 2020. This rapid growth has brought many benefits. Hundreds of millions of people have been lifted out of extreme poverty. Chronic hunger has decreased from about 30 percent of the population 40 years ago to less than 10 percent today. Average life expectancy has increased from 42 to 73.5 years. And infant mortality dropped from 150 per 1,000 live births in 1960 to 18 today, while the annual per capita GDP has grown from less than (U.S.)$200 per year to more than $7,250.

Most Chinese continue to live at a low level of material consumption compared to American or European standards. In terms of ecological footprints (What Do You Think? p. 19), it takes about 9.7 global hectares (gha, or hectares-worth of resources) to support the average American each year. By contrast, the average person in China consumes about 2.1 gha per year, close to the global average. Providing the 1.3 billion Chinese with American standards of consumption would require about 10 billion gha, or almost another entire earth's worth of resources.

Many of the environmental problems mentioned in the opening case study for this chapter arise from poverty. China couldn't afford to worry (at least so they thought) about pollution and land degradation in the past. Today, however, the greatest environmental worries are about the effects of rising affluence (fig. 1.14).

Table 1.1 Quality of Life Indicators		
	Least-Developed Countries	Most-Developed Countries
GDP/Person[1]	(U.S.)$1,006	(U.S.)$43,569
Poverty Index[2]	59.7%	~0
Life Expectancy	53.2 years	80.6 years
Adult Literacy	58%	99%
Female Secondary Education	11%	95%
Total Fertility[3]	5.0	1.7
Infant Mortality[4]	99.5	4.1
Improved Sanitation	23%	100%
Improved Water	61%	100%
CO_2/capita[5]	0.1 tons	15.2 tons

[1]Annual gross domestic product
[2]Percent living on less than (U.S.)$1.25/day
[3]Average births/woman
[4]Per 1,000 live births
[5]Metric tons/yr/person

Source: UNDP Human Development Index, 2010.

FIGURE 1.14 A rapidly growing economy has brought increasing affluence to China that has improved standards of living for many Chinese people, but it also brings environmental and social problems associated with Western lifestyles.

In 1985, there were essentially no private automobiles in China. Bicycles and public transportation were how nearly everyone got around. Now, there are about 50 million automobiles in China, and by 2015, if current trends continue, there could be 150 million. Already, Chinese auto efficiency standards are higher than in the United States, but is there enough petroleum in the world to support all these vehicles? China is now the world's largest source of CO_2 (the United States is second). Both China and the United States depend on coal for about 75 percent of their electricity. Both have very large supplies of coal. There are many benefits of expanding China's electrical supply, but if they reach the same level of power consumption—which is now about one-tenth the amount per person as in the United States—by burning coal, the effects on our global climate will be disastrous (fig. 1.15).

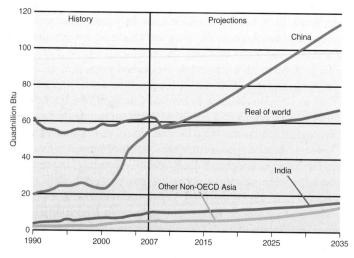

FIGURE 1.15 Coal consumption in China rose sharply in the first decade of the twenty-first century. If it continues on this trajectory, the climate consequences will be disastrous.
Source: U.S. Energy Information Agency, 2010.

On the other hand, as the opening case study for this chapter demonstrates, China is making remarkable strides in developing renewable energy sources. If more countries in both the developed and developing world adopt these environmentally friendly technologies, we could easily have enough resources for everyone.

Recent progress is encouraging

Over the past 50 years, human ingenuity and enterprise have brought about a breathtaking pace of technological innovations and scientific breakthroughs. The world's gross domestic product increased more than tenfold during that period, from $2 trillion to $22 trillion per year. While not all that increased wealth was applied to human development, there has been significant progress in increasing general standard of living nearly everywhere. In 1960, for instance, nearly three-quarters of the world's population lived in abject poverty. Now, less than one-third are still at this low level of development.

Since World War II, average real income in developing countries has doubled; malnutrition declined by almost one-third; child death rates have been reduced by two-thirds; average life expectancy increased by 30 percent. Overall, poverty rates have decreased more in the last 50 years than in the previous 500. Nonetheless, while general welfare has increased, so has the gap between rich and poor worldwide. In 1960, the income ratio between the richest 20 percent of the world and the poorest 20 percent was 30 to 1. In 2000, this ratio was 100 to 1. Because perceptions of poverty are relative, people may feel worse off compared to their rich neighbors than development indices suggest they are.

1.5 SUSTAINABLE DEVELOPMENT

Can we improve the lives of the world's poor without destroying our shared environment? A possible solution to this dilemma is **sustainable development,** a term popularized by *Our Common Future,* the 1987 report of the World Commission on Environment and Development, chaired by Norwegian Prime Minister Gro Harlem Brundtland (and consequently called the Brundtland Commission). In the words of this report, sustainable development means "meeting the needs of the present without compromising the ability of future generations to meet their own needs."

Another way of saying this is that we are dependent on nature for food, water, energy, fiber, waste disposal, and other life-support services. We can't deplete resources or create wastes faster than nature can recycle them if we hope to be here for the long term. Development means improving people's lives. Sustainable development, then, means progress in human well-being that can be extended or prolonged over many generations rather than just a few years. To be truly enduring, the benefits of sustainable development must be available to all humans rather than to just the members of a privileged group.

To many economists, it seems obvious that economic growth is the only way to bring about a long-range transformation to more advanced and productive societies and to provide resources to improve the lot of all people. As former President John F. Kennedy

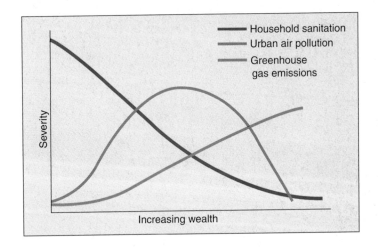

Think About It

Examine figure 1.16. Describe in your own words how increasing wealth affects the three kinds of pollution shown. Why do the trends differ?

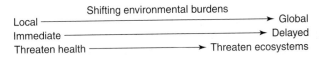

FIGURE 1.16 Environmental indicators show different patterns as incomes rise. Sanitation problems decrease when people can afford septic systems and clean water. Local air pollution, on the other hand, increases as more fuel is burned; eventually, however, development reaches a point at which people can afford both clean air and the benefits of technology. Delayed, distant problems, such as greenhouse gas emissions that lead to global climate change, tend to rise steadily with income because people make decisions based on immediate needs and wants rather than long-term consequences. Thus, we tend to shift environmental burdens from local and immediate to distant and delayed if we can afford to do so.

Source: Graph from World Energy Assessment, UNDP 2000, Figure 3.10, p. 95.

said, "A rising tide lifts all boats." But economic growth is not sufficient in itself to meet all essential needs. As the Brundtland Commission pointed out, political stability, democracy, and equitable economic distribution are needed to ensure that the poor will get a fair share of the benefits of greater wealth in a society. A study released in 2006 by researchers at Yale and Columbia Universities reported a significant correlation between environmental sustainability, open political systems, and good government. Of the 133 countries in this study, New Zealand, Sweden, Finland, Czech Republic, and the United Kingdom held the top five places (in that order). The United States ranked 28th, behind countries such as Japan, and most of Western Europe.

Can development be truly sustainable?

Many ecologists regard "sustainable" growth of any sort as impossible in the long run because of the limits imposed by nonrenewable resources and the capacity of the biosphere to absorb our wastes. Using ever-increasing amounts of goods and services to make human life more comfortable, pleasant, or agreeable must inevitably interfere with the survival of other species and, eventually, of humans themselves in a world of fixed resources. But, supporters of sustainable development assure us, both technology and social organization can be managed in ways that meet essential needs and provide long-term—but not infinite—growth within natural limits, if we use ecological knowledge in our planning.

While economic growth makes possible a more comfortable lifestyle, it doesn't automatically result in a cleaner environment. As figure 1.16 shows, people will purchase clean water and sanitation if they can afford to do so. For low-income people, however, more money tends to result in higher air pollution because they can afford to burn more fuel for transportation and heating. Given enough money, people will be able to afford both convenience *and* clean air. Some environmental problems, such as waste generation and carbon dioxide emissions, continue to rise sharply with increasing wealth because their effects are diffuse and delayed. If we are able to sustain economic growth, we will need to develop personal restraint or social institutions to deal with these problems.

Some projects intended to foster development have been environmental, economic, and social disasters. Large-scale hydropower projects, like that in the James Bay region of Quebec or the Brazilian Amazon that were intended to generate valuable electrical power, also displaced indigenous people, destroyed wildlife, and poisoned local ecosystems with acids from decaying vegetation and heavy metals leached out of flooded soils. Similarly, introduction of "miracle" crop varieties in Asia and huge grazing projects in Africa financed by international lending agencies crowded out wildlife, diminished the diversity of traditional crops, and destroyed markets for small-scale farmers.

Other development projects, however, work more closely with both nature and local social systems. Socially conscious businesses and environmental, nongovernmental organizations sponsor ventures that allow people in developing countries to grow or make high-value products—often using traditional techniques and designs—that can be sold on world markets for good prices (fig. 1.17). Pueblo to People, for example, is a nonprofit organization that buys textiles and crafts directly from producers in Latin America. It sells goods in America, with the profits going to community development projects in Guatemala, El Salvador, and Peru. It also informs customers in wealthy countries about the conditions in the developing world.

As the economist John Stuart Mill wrote in 1857, "It is scarcely necessary to remark that a stationary condition of capital and population implies no stationary state of human improvement. There would be just as much scope as ever for all kinds of mental culture and moral and social progress; as much room for improving the art of living and much more likelihood of its being improved when minds cease to be engrossed by the art of getting on." Somehow, in our rush to exploit nature and consume resources, we have forgotten this sage advice.

FIGURE 1.17 A Mayan woman from Guatemala weaves on a back-strap loom. A member of a women's weaving cooperative, she sells her work to nonprofit organizations in the United States at much higher prices than she would get at the local market.

What is the role of international aid?

Could we eliminate the most acute poverty and ensure basic human needs for everyone in the world? Many experts say this goal is eminently achievable. Economist Jeffery Sachs, director of the UN Millennium Development Project, says we could end extreme poverty worldwide by 2025 if the richer countries would donate just 0.7 percent of their national income for development aid in the poorest nations. These funds could be used for universal childhood vaccination against common infectious diseases, access to primary schools for everyone, family planning services for those who wish them, safe drinking water and sanitation for all, food supplements for the hungry, and strategic microcredit loans for self-employment.

How much would this cost? A rough estimate provided by the United Nations Development Agency is that it would take

FIGURE 1.18 Every year, military spending equals the total income of half the world's people. The cost of a single large air-craft carrier equals 10 years of human development aid given by all the world's industrialized countries.

about (U.S.)$135 billion per year to abolish extreme poverty and the worst infectious diseases over the next 20 years. That's a lot of money—much more than we currently give—but it's not an impossible goal. Annual global military spending is now over $1 trillion (fig. 1.18). If we were to shift one-tenth of that to development aid, we'd not only reduce incalculable suffering but also be safer in the long run, according to many experts.

Although the rich nations have made promises to help aleviate debt and encourage development in poorer countries, the amount actually provided has been far less than is needed. The United States, for example, while the world's largest total donor, sets aside only 0.16 percent of its gross domestic product for development aid. Put another way, the United States currently donates about 18 cents per citizen per day for both private and government aid to foreign nations.

What do you think? Would you be willing to donate an extra dollar per day to reduce suffering and increase political stability? As former Canadian Prime Minister Jean Chrétien says, "Aid to developing countries isn't charity; it's an investment. It will make us safer, and when standards of living increase in those countries, they'll become customers who will buy tons of stuff from us."

Indigenous people are important guardians of nature

Often at the absolute bottom of the social strata, whether in rich or poor countries, are the indigenous or native peoples who are generally the least powerful, most neglected groups in the world. Typically descendants of the original inhabitants of an area taken over by more powerful outsiders, they often are distinct from their country's dominant language, culture, religion, and racial communities. Of the world's nearly 6,000 recognized cultures, 5,000 are indigenous ones that account for only about 10 percent of the total world population. In many countries, these indigenous people are repressed by traditional caste systems, discriminatory laws, economics, or prejudice. Unique cultures are disappearing, along with biological diversity, as natural habitats are destroyed to satisfy industrialized world appetites for resources. Traditional ways of life are disrupted further by dominant Western culture sweeping around the globe.

At least half of the world's 6,000 distinct languages are dying because they are no longer taught to children. When the last few elders who still speak the language die, so will the culture that was its origin. Lost with those cultures will be a rich repertoire of knowledge about nature and a keen understanding about a particular environment and a way of life (fig. 1.19).

Nonetheless, in many places, the 500 million indigenous people who remain in traditional homelands still possess valuable ecological wisdom and remain the guardians of little-disturbed habitats that are the refuge for rare and endangered species and relatively undamaged ecosystems. Author Alan Durning estimates that indigenous homelands harbor more biodiversity than all the world's nature reserves and that greater understanding of nature is encoded in the languages, customs, and practices of native people than is stored in all the libraries of modern science. Interestingly,

FIGURE 1.19 Do indigenous people have unique knowledge about nature and inalienable rights to traditional territories?

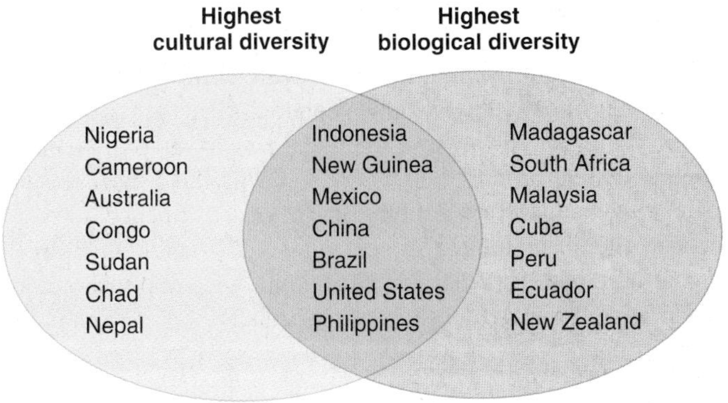

Highest cultural diversity		Highest biological diversity
Nigeria	Indonesia	Madagascar
Cameroon	New Guinea	South Africa
Australia	Mexico	Malaysia
Congo	China	Cuba
Sudan	Brazil	Peru
Chad	United States	Ecuador
Nepal	Philippines	New Zealand

FIGURE 1.20 Cultural diversity and biodiversity often go hand in hand. Seven of the countries with the highest cultural diversity in the world are also on the list of "megadiversity" countries with the highest number of unique biological organisms. (Listed in decreasing order of importance.)
Source: Norman Myers, Conservation International and Cultural Survival Inc., 2002.

just 12 countries account for 60 percent of all human languages (fig. 1.20). Seven of those are also among the "megadiversity" countries that contain more than half of all unique plant and animal species. Conditions that support evolution of many unique species seem to favor development of equally diverse human cultures as well.

Recognizing native land rights and promoting political pluralism is often one of the best ways to safeguard ecological processes and endangered species. As the Kuna Indians of Panama say, "Where there are forests, there are native people, and where there are native people, there are forests." A few countries, such as Papua New Guinea, Fiji, Ecuador, Canada, and Australia acknowledge indigenous title to extensive land areas.

In other countries, unfortunately, the rights of native people are ignored. Indonesia, for instance, claims ownership of nearly three-quarters of its forest lands and all waters and offshore fishing rights, ignoring the interests of indigenous people who have lived in these areas for millennia. Similarly, the Philippine government claims possession of all uncultivated land in its territory, while Cameroon and Tanzania recognize no rights at all for forest-dwelling pygmies who represent one of the world's oldest cultures.

1.6 Environmental Ethics

The ways we interpret environmental issues, or our decisions about what we should or should not do with natural resources, depend partly on our basic worldviews. Perhaps you have a basic ethical assumption that you should be kind to your neighbors, or that you should try to contribute in positive ways to your community. Do you have similar responsibilities to take care of your environment? To conserve energy? To prevent the extinction of rare species? Why? Or why not?

Your position on these questions is partly a matter of **ethics,** or your sense of what is right and wrong. Some of these ideas you learn early in life; some might change over time. Ethical views in society also change over time. In ancient Greece, many philosophers who were concerned with ethics and morality owned slaves; today few societies condone slavery. Most societies now believe it is wrong, or unethical, to treat other humans as property. Often our core beliefs are so deeply held that we have difficulty even identifying them. But they can influence how you act, how you spend money, or how you vote. Try to identify some of your core beliefs. What is a basic thing you simply should or should not do? Where does your understanding come from about those actions?

Ethics also constrain what kinds of questions we are able to ask. Ancient Greeks could not question whether slaves had rights; modern Americans have difficulty asking if it is wrong to consume vastly more energy and goods than other countries do. Many devout religious people find it unconscionable to question basic tenets of their faith. But one of the assumptions of science, including environmental science, is that we should allow ourselves to ask any question, because it is by asking questions that we discover new insights about ourselves and about our world.

We can extend moral value to people and things

One of the reasons we don't accept slavery now, as the ancient Greeks did, is because most societies believe that all humans have basic rights. The Greeks granted **moral value,** or worth, only to adult male citizens within their own community. Women, slaves, and children had few rights and were essentially treated as property. Over time we have gradually extended our sense of moral value to a wider and wider circle, an idea known as **moral extensionism** (fig. 1.21). In most countries, women and minorities have basic civil rights, children cannot be treated as property, even domestic pets have some legal protections against cruel treatment. For many people, moral value also extends to domestic livestock (cattle, hogs, poultry), which makes eating meat a fundamentally wrong thing to do. For others, this moral extension ends with pets, or with humans. Some people extend moral value to include forests, biodiversity, inanimate objects, or the earth as a whole.

These philosophical questions aren't simply academic or historical. In 2004, the journal *Science* caused public uproar by publishing a study demonstrating that fish feel pain. Many recreational anglers had long managed to suppress worries that they were causing pain to fish, and the story was so unsettling that it made national headlines and provoked fresh public debates on the ethics of fishing.

How we treat other people, animals, or things, can also depend on whether we believe they have **inherent value**—an intrinsic right to exist, or **instrumental value** (they have value because they are useful to someone who matters). If I hurt you, I owe you an apology. If I borrow your car and smash it into a tree, I don't owe the car an apology, I owe *you* an apology—or reimbursement.

How does this apply to nonhumans? Domestic animals clearly have an instrumental value because they are useful to their

FIGURE 1.21 Moral extensionism describes an increasing consideration of moral value in other living things—or even nonliving things.

owners. But some philosophers would say they also have inherent values and interests. By living, breathing, struggling to stay alive, the animal carries on its own life independent of its usefulness to someone else.

Some people believe that even nonliving things also have inherent worth. Rocks, rivers, mountains, landscapes, and certainly the earth itself, have value. These things were in existence before we came along, and we couldn't re-create them if they are altered or destroyed. This philosophical debate became a legal dispute in an historic 1969 court case, when the Sierra Club sued the Disney Corporation on behalf of the trees, rocks, and wildlife of Mineral King Valley in the Sierra Nevada Mountains (fig. 1.22) where Disney wanted to build a ski resort. The Sierra Club argued that it represented the interests of beings that could not speak for themselves in court.

A legal brief entitled *Should Trees Have Standing?*, written for this case by Christopher D. Stone, proposed that organisms as well as ecological systems and processes should have standing (or rights) in court. After all, corporations—such as Disney—are treated as persons and given legal rights even though they are really only figments of our imagination. Why shouldn't nature have similar standing? The case went all the way to the Supreme Court but was overturned on a technicality. In the meantime, Disney lost interest in the project and the ski resort was never built. What do you think? Where would you draw the line of what deserves moral considerability? Are there ethical limits on what we can do to nature?

FIGURE 1.22 Mineral King Valley at the southern border of Sequoia National Park was the focus of an important environmental law case in 1969. The Disney Corporation wanted to build a ski resort here, but the Sierra Club sued to protect the valley on behalf of the trees, rocks, and native wildlife.

1.7 Faith, Conservation, and Justice

Ethical and moral values are often rooted in religious traditions, which try to guide us in what is right and wrong to do. With growing public awareness of environmental problems, religious organizations have begun to take stands on environmental concerns. They recognize that some of our most pressing environmental problems don't need technological or scientific solutions; they're not so much a question of what we're able to do, but what we're willing to do. Are we willing to take the steps necessary to stop global climate change? Do our values and ethics require us to do so? In this section, we'll look at some religious perspectives and how they influence our attitudes toward nature.

Environmental scientists have long been concerned about religious perspectives. In 1967, historian Lynn White, Jr. published a widely influential paper, "The Historic Roots of Our Ecological Crisis." He argued that Christian societies have often exploited natural resources carelessly because the Bible says that God commanded Adam and Eve to dominate nature: "Be fruitful, and multiply, and replenish the earth and subdue it: and have dominion over the fish of the sea, and over the fowl of the air, and over every living thing that moveth upon the earth" (Genesis 1:28). Since then, many religious scholars have pointed out that God also commanded Adam and Eve to care for the garden they were given, "to till it and keep it" (Genesis 2:15). Furthermore, Noah was commanded to preserve individuals of all living species, so that they would not perish in the great Flood. Passages such as these inspire many Christians to insist that it is our responsibility to act as stewards of nature, and to care for God's creations.

Calls for both environmental stewardship and anthropocentric domination over nature can be found in the writings of most major faiths. The Koran teaches that "each being exists by virtue of the truth and is also owed its due according to nature," a view that extends moral rights and value to all other creatures. Hinduism and Buddhism teach *ahimsa,* or the practice of not harming other living creatures, because all living beings are divinely connected (fig. 1.23).

Many faiths support environmental conservation

The idea of **stewardship,** or taking care of the resources we are given, inspires many religious leaders to promote conservation. "Creation care" is a term that has become prominent among evangelical Christians in the United States. In 1995, representatives of

FIGURE 1.23 Many religions emphasize the divine relationships among humans and the natural world. The Tibetan Buddhist goddess Tara represents compassion for all beings.

Table 1.2 Principles and Actions in the Ohito Declaration

Spiritual Principles

1. Religious beliefs and traditions call us to care for the earth.
2. For people of faith, maintaining and sustaining environmental life systems is a religious responsibility.
3. Environmental understanding is enhanced when people learn from the example of prophets and of nature itself.
4. People of faith should give more emphasis to a higher quality of life, in preference to a higher standard of living, recognizing that greed and avarice are root causes of environmental degradation and human debasement.
5. People of faith should be involved in the conservation and development process.

Recommended Courses of Action

The Ohito Declaration calls upon religious leaders and communities to

1. emphasize environmental issues within religious teaching: faith should be taught and practiced as if nature mattered.
2. commit themselves to sustainable practices and encourage community use of their land.
3. promote environmental education, especially among youth and children.
4. pursue peacemaking as an essential component of conservation action.
5. take up the challenge of instituting fair trading practices devoid of financial, economic, and political exploitation.

nine major religions met in Ohito, Japan, to discuss views of environmental stewardship in their various traditions. The resulting document, the Ohito Declaration, outlined common beliefs and responsibilities of these different faiths toward protecting the earth and its life (table 1.2). In recent years, religious organizations have played important roles in nature protection. A coalition of evangelical Christians has been instrumental in promoting stewardship of many aspects of our environment, from rare plants and animals to our global climate.

Religious concern extends beyond our treatment of plants and animals. Pope John Paul II and Orthodox Patriarch Bartholomew called on countries bordering the Black Sea to stop pollution, saying that "to commit a crime against nature is a sin." In addition to its campaign to combat global warming described at the beginning of this chapter, the Creation Care Network has also launched initiatives against energy inefficiency, mercury pollution, mountaintop removal mining, and endangered species destruction. For many people, religious beliefs provide the best justification for environmental protection.

Environmental justice combines civil rights and environmental protection

People of color in the United States and around the world are subjected to a disproportionately high level of environmental health risks in their neighborhoods and on their jobs. Minorities, who tend to be poorer and more disadvantaged than other residents, work in the dirtiest jobs where they are exposed to toxic chemicals and other hazards. More often than not they also live in urban ghettos, barrios, reservations, and rural poverty pockets that have shockingly high pollution levels and are increasingly the site of unpopular industrial facilities, such as toxic waste dumps, landfills, smelters, refineries, and incinerators. **Environmental justice** combines civil rights with environmental protection to demand a safe, healthy, life-giving environment for everyone.

Among the evidence of environmental injustice is the fact that three out of five African-Americans and Hispanics, and nearly half of all Native Americans, Asians, and Pacific Islanders live in communities with one or more uncontrolled toxic waste sites, incinerators, or major landfills, while fewer than 10 percent of all whites live in these areas. Using zip codes or census tracts as a unit of measurement, researchers found that minorities make up twice as large a population share in communities with these locally unwanted land uses (**LULUs**) as in communities without them. A recent study using "distance-based" methods found an even greater correlation between race and location of hazardous waste facilities.

Although it is difficult to distinguish between race, class, historical locations of ethnic groups, economic disparities, and other social factors in these disputes, racial origins often seem to play a role in exposure to environmental hazards. Simple correlation doesn't prove causation; still, while poor people in general are more likely to live in polluted neighborhoods than rich people, the discrepancy between the pollution exposure of middle class blacks and middle class whites is even greater than the difference between poorer whites and blacks. Where upper class whites can "vote with their feet" and move out of polluted and dangerous neighborhoods, blacks and other minorities are restricted by color barriers and prejudice (overt or covert) to the less desirable locations (fig. 1.24).

FIGURE 1.24 Poor people and people of color often live in the most dangerous and least desirable places. Here children play next to a chemical refinery in Texas City, Texas.

Environmental racism distributes hazards inequitably

Racial prejudice is a belief that people are inferior merely because of their race. Racism is prejudice with power. **Environmental racism** is inequitable distribution of environmental hazards based on race. Evidence of environmental racism can be seen in lead poisoning in children. The Federal Agency for Toxic Substances and Disease Registry considers lead poisoning to be the number one environmental health problem for children in the United States. Some 4 million children—many of whom are African American, Latino, Native American, or Asian, and most of whom live in inner-city areas—have dangerously high lead levels in their bodies. This lead is absorbed from old lead-based house paint, contaminated drinking water from lead pipes or lead solder, and soil polluted by industrial effluents and automobile exhaust. The evidence of racism is that at every income level, whether rich or poor, black children are two to three times more likely than whites to suffer from lead poisoning.

Because of their quasi-independent status, most Native-American reservations are considered sovereign nations that are not covered by state environmental regulations. Court decisions holding that reservations are specifically exempt from hazardous waste storage and disposal regulations have resulted in a land rush of seductive offers from waste disposal companies to Native-American reservations for onsite waste dumps, incinerators, and landfills. The short-term economic incentives can be overwhelming for communities in which adult unemployment runs between 60 and 80 percent. Uneducated, powerless people often can be tricked or intimidated into signing environmentally and socially disastrous contracts. Nearly every tribe in America has been approached with proposals for some dangerous industry or waste facility.

The practice of targeting poor communities of color in the developing nations for waste disposal and/or experimentation with risky technologies has been described as **toxic colonialism.** Internationally, the trade in toxic waste has mushroomed in recent years as wealthy countries have become aware of the risks of industrial refuse. Poor, minority communities at home and abroad are being increasingly targeted as places to dump unwanted wastes.

FIGURE 1.25 Much of our waste is exported to developing countries where environmental controls are limited. Here workers in a Chinese village sort electronic waste materials. **Source:** Basel Action Network.

Although a treaty regulating international shipping of toxics was signed by 105 nations in 1989, millions of tons of toxic and hazardous materials continue to move—legally or illegally—from the richer countries to the poorer ones every year. This issue is discussed further in chapter 23.

One of the ways we export our pollution is in the form of discarded electrical equipment, such as computers and cell phones. Often these items are broken apart to remove lead, copper, and other components. Conditions for workers can be extremely hazardous (fig. 1.25).

The U.S. Environmental Justice Act was established in 1992 to identify areas threatened by the highest levels of toxic chemicals, assess health effects caused by emissions of those chemicals, and ensure that groups or individuals residing within those areas have opportunities and resources to participate in public discussions concerning siting and cleanup of industrial facilities. Perhaps we need something similar worldwide.

CONCLUSION

We face many environmental dilemmas, but there are also many opportunities for improving lives without damaging our shared environment. China's growth and innovation provide examples of those challenges and opportunities. Both in China and globally, we face air and water pollution, chronic hunger, water shortages, and other problems. On the other hand, we have seen important innovations in transportation, energy sources, food production, and international cooperation for environmental protection. Environmental science is a discipline that draws on many kinds of knowledge to understand these problems and to help find solutions—which can draw on knowledge from technological, biological, economic, political, social, and many other fields of study.

There are deep historic roots to our efforts to protect our environment. Utilitarian conservation has been a common incentive; aesthetic preservation also motivates many people to work for conservation. Social progress, and a concern for making sure that all people have access to a healthy environment, has also important motivating factors in environmental science and in environmental conservation. Inequitable distribution of resources has been a persistent concern. Growing consumption of energy, water, land, and other resources makes many questions in environmental science more urgent.

Sustainable development is the idea that we can improve people's lives without reducing resources and opportunities for future generations. This goal may or may not be achievable, but it is an important ideal that can help us understand and identify appropriate and fair directions for improving people's lives around the world.

Ethics and faith-based perspectives often inspire people to work for resource conservation, because ethical frameworks and religions often promote ideas of fairness and stewardship of the world we have received. One important ethical principle is the notion of moral extensionism. Stewardship, or taking care of our environment, has been a guiding principle for many faith-based groups. Often these groups have led the struggle for environmental justice for minority and low-income communities.

REVIEWING LEARNING OUTCOMES

By now you should be able to explain the following points:

1.1 Explain what environmental science is, and how it draws on different kinds of knowledge.

- Environmental science is the systematic study of our environment and our proper place in it.
- No one discipline has answers to all the environmental challenges we face. It will take integrative, creative, resourceful thinking to find sustainable solutions.

1.2 List and describe some current concerns in environmental science.

- We live on a marvelous planet of rich biodiversity and complex ecological systems.
- We face many serious environmental problems including water supplies, safe drinking water, hunger, land degradation, energy, air quality, and biodiversity losses.
- There are many signs of hope in terms of social progress, environmental protection, energy choices, and the spread of democracy.

1.3 Identify some early thinkers on environment and resources, and contrast some of their ideas.

- Nature protection has historic roots.
- Resource waste inspired pragmatic, utilitarian conservation.
- Ethical and aesthetic concerns inspired the preservation movement.

- Rising pollution levels led to the modern environmental movement.
- Environmental quality is tied to social progress.

1.4 Appreciate the human dimensions of environmental science, including the connection between poverty and environmental degradation.

- We live in an inequitable world.
- Faced with immediate survival needs and few options, poor people often have no choice but to degrade their environment.
- Recent progress in human development is encouraging.

1.5 Describe sustainable development and its goals.

- Can development be truly sustainable?
- What is the role of international aid?
- Indigenous people are important guardians of nature.

1.6 Explain a key point of environmental ethics.

- We can extend moral value to people and things.

1.7 Identify ways in which faith-based groups share concerns for our environment.

- Many faiths support environmental conservation.
- Environmental justice combines civil rights and environmental protection.
- Environmental racism distributes hazards inequitably.

PRACTICE QUIZ

1. Define *environment* and *environmental science*.
2. Describe four stages in conservation history and identify one leader associated with each stage.
3. List six environmental dilemmas that we now face and summarize how each concerns us.
4. Identify some signs of hope for solving environmental problems.
5. What is extreme poverty, and why should we care?
6. How much difference is there in per capita income, infant mortality, and CO_2 production between the poorest and richest countries?
7. Why should we be worried about economic growth in China?
8. Define *sustainable development*.
9. How much would it cost to eliminate acute poverty and ensure basic human needs for everyone?
10. Why are indigenous people important as guardians of nature?

CRITICAL THINKING AND DISCUSSION QUESTIONS

1. Should environmental science include human dimensions? Explain.

2. Overall, do environmental and social conditions in China give you hope or fear about the future?

3. What are the underlying assumptions and values of utilitarian conservation and altruistic preservation? Which do you favor?

4. What resource uses are most strongly represented in the ecological footprint? What are the advantages and disadvantages of using this assessment?

5. Are there enough resources in the world for 8 or 10 billion people to live decent, secure, happy lives? What do these terms mean to you? Try to imagine what they mean to residents of other countries.

6. What would it take for human development to be truly sustainable?

7. Are you optimistic or pessimistic about our chances of achieving sustainability? Why?

 Data Analysis: **Working with Graphs**

Graphs are one of the most common and important ways scientists communicate their results. Learning to understand graphing techniques—the language of graphs—will help you better understand this book.

Graphs are visual presentations of data that help us identify trends and understand relationships. We could present a table of numbers, but most of us have difficulty seeing a pattern in a field of numbers. In a graph, we can quickly and easily see trends and relationships.

Below are two graphs that appeared earlier in this chapter. Often we pass quickly over graphs like these that appear in text, but it's rewarding to investigate them more closely, because their relationships can raise interesting questions. Answer the numbered questions on the next page to make sure you understand the graphs shown.

First let's examine the parts of a graph. Usually there is a horizontal axis (also known as the "X-axis") and a vertical axis (the "Y-axis"). Usually, in the relationship shown in a graph, one variable is thought to explain the other. In figure 1, for example, as *time* passes, the size of our ecological *footprint* grows. In this case, *time* is an **independent variable** that (at least partly) explains changes in the **dependent variable**, *footprint*.

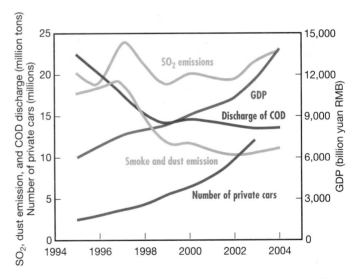

FIGURE 2 Environmental indicators for China, 1994–2005. Income doubled, as measured by gross domestic product (GDP), but the number of cars rose fourfold. Chemical oxygen demand (COD, a measure of water pollution) declined with industrial controls, but sulfur dioxide (SO_2) emissions increased as more coal was burned.
Source: Shao, M., et al., 2006.

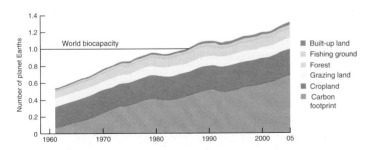

FIGURE 1 Our global ecological footprint has nearly tripled since 1961, when we began to collect global environmental data.
Source: WWF, 2008.

Questions:

1. What units are used for the dependent variable?

2. What is the lowest value on this axis? The highest?

3. What was the approximate size of our global footprint in 1960? In 2005? How many times bigger is the 2005 value? (To answer, divide the bigger number by the smaller number.)

4. This graph has several lines, each contributing part of the total. Which factor had the greatest impact in 1961? The second greatest impact? By 2005, had those values changed greatly or slightly?

5. Which factor had the greatest impact in 2005? What is the proportional increase from 1961 to 2005?

6. Based on this graph, would you say that your ecological footprint is probably greater or less than your parents' footprints when they were your age? What does that mean about the kinds of goods you consume? Are you happier or healthier than your parents were at your age? Why or why not?

7. Examine figure 2, which shows several indicators of China's economy and environment. This is a more complex graph than the first one because it has two Y-axes and more than one value graphed. But it follows the same principles as any other line graph.

8. What is the range of values on the X-axis? What are the values and units on the *right* vertical axis?

9. What is GDP, in general terms?

10. The right axis shows values for only one of the plotted lines. Which line is this?

11. The complex left axis shows the number of cars and how many millions of tons of pollutants are produced. The pollutants shown are SO_2 and dust (sulfur dioxide and airborne dust are important air pollutants) and COD, or chemical oxygen demand (a measure of water contamination). As GDP has risen, have all three pollutants also risen?

12. Based on this graph, would you say that rising GDP *necessarily* causes greater pollution? Why would rising GDP cause more pollution? Why might it not?

13. How does this graph correspond to the theoretical presentation in figure 1.16? Based on theory, which factors would you expect to increase as GDP rises, and which would you expect to fall?

Answers:

1. Units are number of planet earths.

2. The lowest and highest values are 0 and 1.4 planet earths.

3. 1960: about 0.6 earths; 2005: about 1.4 earths. This is an increase of more than twofold.

4. The biggest factors in 1961 were cropland and grazing land. These had changed little by 2005.

5. The biggest factor in 2005 was carbon footprint. This factor rose from about 0.1 to about 0.6 earths, a six fold increase.

6. On average, our ecological footprint has more than doubled compared to a generation ago, mainly through energy use.

For Additional Help in Studying This Chapter, please visit our website at www .mhhe.com/cunningham12e. You will find additional practice quizzes and case studies, flashcards, regional examples, placemarkers for Google Earth™ mapping, and an extensive reading list, all of which will help you learn environmental science.

Researchers measure plant growth in experimental plots in the B4Warmed study in northern Minnesota.

Principles of Science and Systems

Learning Outcomes

After studying this chapter, you should be able to:

2.1 Describe the scientific method and explain how it works.

2.2 Explain systems and how they're useful in science.

2.3 Evaluate the role of scientific consensus and conflict.

"The ultimate test of a moral society is the kind of world that it leaves to its children." ~ *Dietrich Bonhoeffer*

Case Study

Forest Responses to Global Warming

How will forests respond to climate change? This is one of the great unknowns in environmental science today. Will northern regions that now support boreal forest shift to another biome—hardwood forest, open savannah, grassland, or something entirely different? With rising emissions of CO_2 and other greenhouse gases, climate models predict that boreal forests will move north by about 480 km (300 mi) within this century. But there's a great deal of uncertainty in this prediction.

How do environmental scientists approach and analyze such complex questions? One strategy is to grow plants in a greenhouse, and test plant responses to different temperature and moisture levels. By changing just one variable at a time, we can get an approximation of responses to environmental change. But this approach misses the complex species interactions that influence plant growth, a real ecosystem, so an alternative approach is to use field tests in which mixtures of plants are grown in natural settings that include competition for resources, predator/prey interactions, natural climatic variations, and other ecological factors.

Professor Peter Reich, his colleagues and student research assistants are now carrying out such a field study in a patch of boreal (northern) forest in Minnesota. Calling this experiment B4Warmed, which stands for Boreal Forest Warming at an Ecotone in Danger, they are artificially raising ambient temperatures in a series of boreal forest plots, to emulate warming climate conditions.

The group established 96 circular experimental plots, each 3 meters (9.8 ft) in diameter (fig. 2.1). Each plot was planted with a mixture of tree species and annual understory plants. The plots were then randomly assigned to one of four treatments. Half the plots are in mature forest, and half are in forest openings. Half are kept 2°C above ambient temperatures, and half are kept 4°C higher than ambient temperatures, using infrared lamps placed around the plots, as well as buried heat cables (fig 2.2). Control plots (with no temperature manipulations) are also maintained for comparison with treatments.

It's too early to know exactly what the long-term effects of warming will be on the northern forest community. It seems likely that species, such as aspen, spruce, and birch, which are now at the southern edge of their range in the study area won't do as well under a warmer climate as the temperate maple-oak forests now growing further south. However, both northern and temperate species may perform poorly under warmer conditions. If so, neither our current forest trees nor their potential replacements may be well suited to our future climate. This experiment will enable us to assess the potential for climate change to alter future forest composition.

One preliminary result from this study that appears to offer good news is that the CO_2 emissions both from forest plants and from the soil are lower than expected at higher temperatures.

Apparently both standing vegetation and soil microbes alter their metabolic rates to acclimate to ambient environmental conditions. Thus the feedback cycles predicted to exacerbate global warming effects may not be as bad as we feared.

This kind of careful, rational, systematic research is the hallmark of modern science. It has given us powerful insights into how our world works. In this chapter, we'll look at how scientists form and answer other questions about our environment. For related resources, including Google Earth place marks that show locations where these issues can be explored, visit http://EnvironmentalScience-Cunningham.blogspot.com.

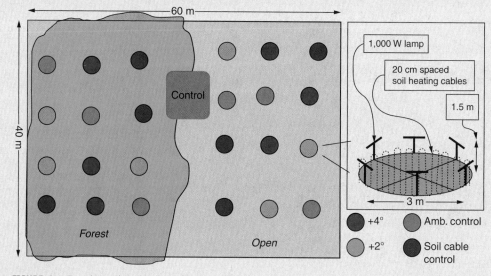

FIGURE 2.1 Experimental design for B4Warmed Study.

FIGURE 2.2 A student researcher adjusts the electrical panel that controls heat lamps and heating cables.

2.1 WHAT IS SCIENCE?

Science is a process for producing knowledge methodically and logically. Derived from *scire*, "to know" in Latin, science depends on making precise observations of natural phenomena. We develop or test theories (proposed explanations of how a process works) using these observations. "Science" also refers to the cumulative body of knowledge produced by many scientists. Science is valuable because it helps us understand the world and meet practical needs, such as new medicines, new energy sources, or new foods. In this section, we'll investigate how and why science follows standard methods.

Science rests on the assumption that the world is knowable and that we can learn about the world by careful observation (table 2.1). For early philosophers of science, this assumption was a radical departure from religious and philosophical approaches. In the Middle Ages, the ultimate sources of knowledge about how crops grow, how diseases spread, or how the stars move, were religious authorities or cultural traditions. While these sources provided many useful insights, there was no way to test their explanations independently and objectively. The benefit of scientific thinking is that it searches for testable evidence. By testing our ideas with observable evidence, we can evaluate whether our explanations are reasonable or not.

Science depends on skepticism and accuracy

Ideally, scientists are skeptical. They are cautious about accepting proposed explanations until there is substantial evidence to support them. Even then, as we saw in the case study about global warming

Table 2.1 Basic Principles of Science

1. *Empiricism:* We can learn about the world by careful observation of empirical (real, observable) phenomena; we can expect to understand fundamental processes and natural laws by observation.

2. *Uniformitarianism:* Basic patterns and processes are uniform across time and space; the forces at work today are the same as those that shaped the world in the past, and they will continue to do so in the future.

3. *Parsimony:* When two plausible explanations are reasonable, the simpler (more parsimonious) one is preferable. This rule is also known as Ockham's razor, after the English philosopher who proposed it.

4. *Uncertainty:* Knowledge changes as new evidence appears, and explanations (theories) change with new evidence. Theories based on current evidence should be tested on additional evidence, with the understanding that new data may disprove the best theories.

5. *Repeatability:* Tests and experiments should be repeatable; if the same results cannot be reproduced, then the conclusions are probably incorrect.

6. *Proof is elusive:* We rarely expect science to provide absolute proof that a theory is correct, because new evidence may always undermine our current understanding.

7. *Testable questions:* To find out whether a theory is correct, it must be tested; we formulate testable statements (hypotheses) to test theories.

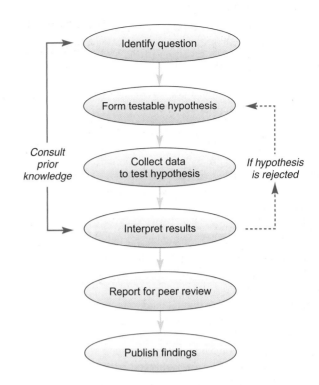

FIGURE 2.3 Ideally, scientific investigation follows a series of logical, orderly steps to formulate and test hypotheses.

that opened this chapter, explanations are considered only provisionally true, because there is always a possibility that some additional evidence may appear to disprove them. Scientists also aim to be methodical and unbiased. Because bias and methodical errors are hard to avoid, scientific tests are subject to review by informed peers, who can evaluate results and conclusions (fig. 2.3). The peer review process is an essential part of ensuring that scientists maintain good standards in study design, data collection, and interpretation of results.

Scientists demand **reproducibility** because they are cautious about accepting conclusions. Making an observation or obtaining a result just once doesn't count for much. You have to produce the same result consistently to be sure that your first outcome wasn't a fluke. Even more important, you must be able to describe the conditions of your study so that someone else can reproduce your findings. Repeating studies or tests is known as **replication.**

Science also relies on accuracy and precision. Accuracy is correctness of measurements. Inaccurate data can produce sloppy and misleading conclusions (fig. 2.4). Precision means repeatability of results and level of detail. The classic analogy for repeatability is throwing darts at a dart board. You might throw ten darts and miss the center every time, but if all the darts hit nearly the same spot, they were very precise. Another way to think of precision is levels of detail. Suppose you want to measure how much snow fell last night, so you take out your ruler, which is marked in centimeters, and you find that the snow is just over 6 cm deep. You cannot tell if it is 6.3 cm or 6.4 cm because the ruler doesn't report that level of detail. If you average several measurements, you might find an average depth of 6.4333 cm. If you report all four decimal places,

FIGURE 2.4 Making careful, accurate measurements and keeping good records are essential in scientific research.

it will imply that you know more than you really do about the snow depth. If you had a ruler marked in millimeters (one-tenth of a centimeter), you could find a depth of 6.4 cm. Here, the one decimal place would be a **significant number,** or a level of detail you actually knew. Reporting 6.4333 cm would be inappropriate because the last three digits are not meaningful.

Deductive and inductive reasoning are both useful

Ideally, scientists deduce conclusions from general laws that they know to be true. For example, if we know that massive objects attract each other (because of gravity), then it follows that an apple will fall to the ground when it releases from the tree. This logical reasoning from general to specific is known as **deductive reasoning.** Often, however, we do not know general laws that guide natural systems. We observe, for example, that birds appear and disappear as a year goes by. Through many repeated observations in different places, we can infer that the birds move from place to place. We can develop a general rule that birds migrate seasonally. Reasoning from many observations to produce a general rule is **inductive reasoning.** Although deductive reasoning is more logically sound than inductive reasoning, it only works when our general laws are correct. We often rely on inductive reasoning to understand the world because we have few immutable laws.

Sometimes it is insight, as much as reasoning, that leads us to an answer. Many people fail to recognize the role that insight, creativity, aesthetics, and luck play in research. Some of our most important discoveries were made not because of superior scientific method and objective detachment, but because the investigators were passionately interested in their topics and pursued hunches that appeared unreasonable to fellow scientists. A good example is Barbara McClintock, the geneticist who discovered that genes in corn can move and recombine spontaneously. Where other corn geneticists saw random patterns of

color and kernel size, McClintock's years of experience in corn breeding, and an uncanny ability to recognize patterns, led her to guess that genes could recombine in ways that no one had yet imagined. Her intuitive understanding led to a theory that took other investigators years to accept.

Testable hypotheses and theories are essential tools

Science also depends on orderly testing of hypotheses, a process known as the scientific method. You may already be using the scientific method without being aware of it. Suppose you have a flashlight that doesn't work. The flashlight has several components (switch, bulb, batteries) that could be faulty. If you change all the components at once, your flashlight might work, but a more methodical series of tests will tell you more about what was wrong with the system—knowledge that may be useful next time you have a faulty flashlight. So you decide to follow the standard scientific steps:

1. *Observe* that your flashlight doesn't light; also, there are three main components of the lighting system (batteries, bulb, and switch).

2. Propose a ***hypothesis,*** a testable explanation: "The flashlight doesn't work because the batteries are dead."

3. Develop a *test* of the hypothesis and *predict* the result that would indicate your hypothesis was correct: "I will replace the batteries; the light should then turn on."

4. Gather *data* from your test: After you replaced the batteries, did the light turn on?

5. *Interpret* your results: If the light works now, then your hypothesis was right; if not, then you should formulate a new hypothesis, perhaps that the bulb is faulty, and develop a new test for that hypothesis.

In systems more complex than a flashlight, it is almost always easier to prove a hypothesis wrong than to prove it unquestionably true. This is because we usually test our hypotheses with observations, but there is no way to make every possible observation. The philosopher Ludwig Wittgenstein illustrated this problem as follows: Suppose you saw hundreds of swans, and all were white. These observations might lead you to hypothesize that all swans were white. You could test your hypothesis by viewing thousands of swans, and each observation might support your hypothesis, but you could never be entirely sure that it was correct. On the other hand, if you saw just one black swan, you would know with certainty that your hypothesis was wrong.

As you'll read in later chapters, the elusiveness of absolute proof is a persistent problem in environmental policy and law. You can never absolutely prove that the toxic waste dump up the street is making you sick. The elusiveness of proof often decides environmental liability lawsuits.

When an explanation has been supported by a large number of tests, and when a majority of experts have reached a general consensus that it is a reliable description or explanation, we call it a **scientific theory.** Note that scientists' use of this term is very different from the way the public uses it. To many people, a theory is speculative and unsupported by facts. To a scientist, it means just the opposite: While all explanations are tentative and open to revision and correction, an explanation that counts as a scientific theory is supported by an overwhelming body of data and experience, and it is generally accepted by the scientific community, at least for the present (fig. 2.5).

Understanding probability helps reduce uncertainty

One strategy to improve confidence in the face of uncertainty is to focus on probability. Probability is a measure of how likely something is to occur. Usually, probability estimates are based on a set of previous observations or on standard statistical measures. Probability does not tell you what *will* happen, but it tells you what *is likely* to happen. If you hear on the news that you have a 20 percent chance of catching a cold this winter, that means that 20 of every 100 people are likely to catch a cold. This doesn't mean that you will catch one. In fact, it's more likely that you won't catch a cold than that you will. If you hear that 80 out of every 100 people will catch a cold, you still don't know whether you'll get sick, but there's a much higher chance that you will.

Science often involves probability, so it is important to be familiar with the idea. Sometimes probability has to do with random chance: If you flip a coin, you have a random chance of getting heads or tails. Every time you flip, you have the same 50 percent probability of getting heads. The chance of getting ten heads in a row is small (in fact, the chance is 1 in 2^{10}, or 1 in 1,024), but on any individual flip, you have exactly the same 50 percent chance, since this is a random test.

Sometimes probability is weighted by circumstances: Suppose that about 10 percent of the students in this class earn an A each semester. Your likelihood of being in that 10 percent depends a great deal on how much time you spend studying, how many questions you ask in class, and other factors. Sometimes there is a combination of chance and circumstances: The probability that you will catch a cold this winter depends partly on whether

you encounter someone who is sick (largely random chance) and whether you take steps to stay healthy (get enough rest, wash your hands frequently, eat a healthy diet, and so on).

Scientists often increase their confidence in a study by comparing results to a random sample or a larger group. Suppose that 40 percent of the students in your class caught a cold last winter. This *seems* like a lot of colds, but is it? One way to decide is to compare to the cold rate in a larger group. You call your state epidemiologist, who took a random sample of the state population last year: She collected 200 names from the telephone book and called each to find out if each got a cold last year. A larger sample, say 2,000 people, would have been more likely to represent the actual statewide cold rate. But a sample of 200 is much better than a sample of 50 or 100. The epidemiologist tells you that in your state as a whole, only 20 percent of people caught a cold.

Now you know that the rate in your class (40 percent) was quite high, and you can investigate possible causes for the difference. Perhaps people in your class got sick because they were short on sleep, because they tended to stay up late studying. You could test whether studying late was a contributing factor by comparing the frequency of colds in two groups: those who study long and late, and those who don't. Suppose it turns out that among the 40 late-night studiers, 30 got colds (a rate of 75 percent). Among the 60 casual studiers, only 10 got colds (17 percent). This difference would give you a good deal of confidence that staying up late contributes to getting sick. (Note, however, that all 40 of the studying group got good grades!)

Statistics can indicate the probability that your results were random

Statistics can help in experimental design as well as in interpreting data (see Exploring Science, pp. 42–43). Many statistical tests focus on calculating the probability that observed results could have occurred by chance. Often, the degree of confidence we can assign to results depends on sample size as well as the amount of variability between groups.

Ecological tests are often considered significant if there is less than 5 percent probability that the results were achieved by random chance. A probability of less than 1 percent gives still greater confidence in the results.

As you read this book, you will encounter many statistics, including many measures of probability. When you see these numbers, stop and think: Is the probability high enough to worry about? How high is it compared to other risks or chances you've read about? What are the conditions that make probability higher or lower? Science involves many other aspects of statistics.

Experimental design can reduce bias

The study of colds and sleep deprivation is an example of an observational experiment, one in which you observe natural events and interpret a causal relationship between the variables. This kind of study is also called a **natural experiment**, one that involves observation of events that have already happened. Many scientists depend on natural experiments: A geologist, for instance, might

FIGURE 2.5 Data collection and repeatable tests support scientific theories. Here students use telemetry to monitor radio-tagged fish.

Exploring Science

What Are Statistics, and Why Are They Important?

Statistics are numbers that let you evaluate and compare things. "Statistics" is also a field of study that has developed meaningful methods of comparing those numbers. By both definitions, statistics are widely used in environmental sciences, partly because they can give us a useful way to assess patterns in a large population, and partly because the numbers can give us a measure of confidence in our research or observations. Understanding the details of statistical tests can take years of study, but a few basic ideas will give you a good start toward interpreting statistics.

1. *Descriptive statistics help you assess the general state of a group.* In many towns and cities, the air contains a dust, or particulate matter, as well as other pollutants. From personal experience you might know your air isn't as clean as you'd like, but you may not know how clean or dirty it is. You could start by collecting daily particulate measurements to find average levels. An averaged value is more useful than a single day's values, because daily values may vary a great deal, but general, long-term conditions affect your general health. Collect a sample every day for a year; then divide the sum by the number of days, to get a **mean** (average) dust

level. Suppose you found a mean particulate level of 30 micrograms per cubic meter ($\mu g/m^3$) of air. Is this level high or low? In 1997 the EPA set a standard of 50 $\mu g/m^3$ as a limit for allowable levels of coarse particulates (2.5–10 micrometers in diameter). Higher levels tend to be associated with elevated rates of asthma and other respiratory diseases. Now you know that your town, with an annual average of 30 $\mu g/m^3$, has relatively clean air, after all.

2. *Statistical samples.* Although your town is clean by EPA standards, how does it compare with the rest of the cities in the country? Testing the air in *every* city is probably not possible. You could compare your town's air quality with a **sample,** or subset of cities, however. A large, random sample of cities should represent the general "population" of cities reasonably well. Taking a large sample reduces the effects of outliers (unusually high or low values) that might be included. A random sample minimizes the chance that you're getting only the worst sites, or only a collection of sites that are close together, which might all have similar conditions. Suppose you get average annual particulate levels from a sample of 50 randomly selected cities.

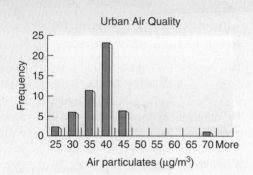

FIGURE 1 Average annual airborne dust levels for 50 cities in 2001.

Source: Data from U.S. Environmental Protection Agency.

You can draw a frequency distribution, or histogram, to display your results (fig. 1). The mean value of this group is 36.8 $\mu g/m^3$, so by comparison your town (at 30 $\mu g/m^3$) is relatively clean.

Many statistical tests assume that the sample has a normal, or Gaussian, frequency distribution, often described as a bell-shaped curve (fig. 2). In this distribution, the mean is near the center of the range of values, and most values are fairly close to the mean. Large and random samples are more likely to fit this shape than are small and nonrandom samples.

---

want to study mountain building, or an ecologist might want to learn about how species coevolve, but neither scientist can spend millions of years watching the process happen. Similarly, a toxicologist cannot give people a disease just to see how lethal it is.

Other scientists can use **manipulative experiments,** such as the B4Warmed experiment in the opening case study for this chapter, in which some conditions are deliberately altered, and all other variables are held constant (fig. 2.6). In one famous manipulative study, ecologists Edward O. Wilson and Robert MacArthur were interested in how quickly species colonize small islands, depending on distance to the mainland. They fumigated several tiny islands in the Florida Keys, killing all resident insects, spiders, and other invertebrates. They then monitored the islands to learn how quickly ants and spiders recolonized them from the mainland or other islands.

Most manipulative experiments are done in the laboratory, where conditions can be carefully controlled. Suppose you were interested in studying whether lawn chemicals contributed to deformities in tadpoles. You might keep two groups of tadpoles in fish tanks, and expose one to chemicals. In the lab, you could ensure that both tanks had identical temperatures, light, food, and

FIGURE 2.6 A researcher gathers data from the B4Warmed field experiment in the boreal forest.

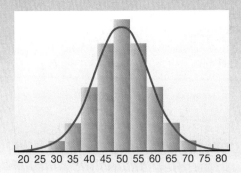

FIGURE 2 A normal distribution.

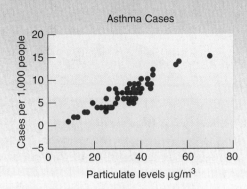

FIGURE 3 A dot plot shows relationships between variables.

3. *Confidence.* How do you know that the 50 cities you sampled really represent all the cities in the country? You can't ever be completely certain, but you can use estimates, such as confidence limits, to express the reliability of your mean statistic. Depending on the size of your sample (not 10, not 100, but 50) and the amount of variability in the sample data, you can calculate a confidence interval that the mean represents the whole population (all cities). Confidence levels, or confidence intervals, represent the likelihood that your statistics represent the entire population correctly. For the mean of your sample, a confidence interval tells you the probability that your sample is similar to other random samples of the population. A common convention is to compare values with a 95 percent confidence level, or a probability of 5 percent or less that your conclusions are misleading. Using statistical software, we can

calculate that, for our 50 cities, the mean is 36.8 $\mu g/m^3$, and the confidence interval is 35.0 to 38.6. This suggests that, if you take 1,000 samples from the entire population of cities, 95 percent of those samples ought to be within 2 $\mu g/m^3$ of your mean. This indicates that your mean is reliable and representative.

4. *Is your group unusual?* Once you have described your group of cities, you can compare it with other groups. For example, you might believe that Canadian cities have cleaner air than U.S. cities. You can compare mean air quality levels for the two groups. Then you can calculate confidence intervals for the difference between the means, to see if the difference is meaningful.

5. *Evaluating relationships between variables.* Are respiratory diseases correlated with air pollution? For each city in your sample, you

could graph pollution and asthma rates (fig. 3). If the graph looks like a loose cloud of dots, there is no clear relationship. A tight, linear pattern of dots trending upward to the right indicates a strong and positive relationship. You can also use a statistical package to calculate an equation to describe the relationship and, again, confidence intervals for the equation. This is known as a regression equation.

6. *Lies, damned lies, and statistics.* Can you trust a number to represent a complex or large phenomenon? One of the devilish details of representing the world with numbers is that those numbers can be tabulated in many ways. If we want to assess the greatest change in air quality statistics, do we report rates of change or the total amount of change? Do we look at change over five years? Twenty-five years? Do we accept numbers selected by the EPA, by the cities themselves, by industries, or by environmental groups? Do we trust that all the data were collected with a level of accuracy and precision that we would accept if we knew the hidden details in the data-gathering process? Like all information, statistics need to be interpreted in terms of who produced them, when, and why. Awareness of some of the standard assumptions behind statistics, such as sampling, confidence, and probability, will help you interpret statistics that you see and hear.

For more discussion of graphs and statistics, see the Data Analysis exercise at the end of this chapter.

---

oxygen. By comparing a treatment (exposed) group and a control (unexposed) group, you have also made this a **controlled study.**

Often, there is a risk of experimenter bias. Suppose the researcher sees a tadpole with a small nub that looks like it might become an extra leg. Whether she calls this nub a deformity might depend on whether she knows that the tadpole is in the treatment group or the control group. To avoid this bias, **blind experiments** are often used, in which the researcher doesn't know which group is treated until after the data have been analyzed. In health studies, such as tests of new drugs, **double-blind experiments** are used, in which neither the subject (who receives a drug or a placebo) nor the researcher knows who is in the treatment group and who is in the control group.

In each of these studies there is one **dependent variable** and one, or perhaps more, **independent variables.** The dependent variable, also known as a response variable, is affected by the independent variables. In a graph, the dependent variable is on the vertical (Y) axis, by convention. Independent variables are rarely really independent (they are affected by the same environmental conditions as the dependent variable, for example). Many people prefer to call them explanatory variables, because we hope they will explain differences in the dependent variable.

Models are an important experimental strategy

Another way to gather information about environmental systems is to use **models.** A model is a simple representation of something. Perhaps you have built a model airplane. The model doesn't have all the elements of a real airplane, but it has the most important ones for your needs. A simple wood or plastic airplane has the proper shape, enough to allow a child to imagine it is flying (fig. 2.7). A more complicated model airplane might have a small gas engine, just enough to let a teenager fly it around for short distances.

Similarly, scientific models vary greatly in complexity, depending on their purposes. Some models are physical models: Engineers test new cars and airplanes in wind tunnels to see how they perform, and biologists often test theories about evolution and genetics using "model organisms" such as fruit flies or rats as a surrogate for humans.

Most models are numeric, though. A model could be a mathematical equation, such as a simple population growth model ($N_t = rN_{(t-1)}$). Here the essential components are number (N) of individuals at time t (N_t), and the model proposes that

FIGURE 2.7 A model uses just the essential elements to represent a complex system.

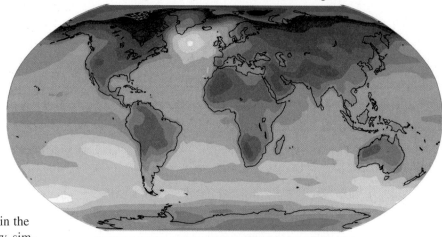

Geographical pattern of surface warming

0 0.5 1 1.5 2 2.5 3 3.5 4 4.5 5 5.5 6 6.5 7 7.5
Temperature change (°C)

FIGURE 2.8 Numerical models, calculated from observed data, can project future scenarios. Here, temperature changes in 2090–2099 are modeled, relative to 1980–1999 temperatures.
Source: IPCC Fourth Assessment Report 2008, model scenario AIB SRES.

N_t is equal to the growth rate (r) times the number in the previous time period ($N_{(t-1)}$). This model is a very simplistic representation of population change, but it is useful because it precisely describes a relationship between population size and growth rate. Also, by converting the symbols to numbers, we can predict populations over time. For example, if last year's rabbit population was 100, and the growth rate is 1.6 per year, then this year's population will be 160. Next year's population will be 160×1.6, or 256. This is a simple model, then, but it can be useful. A more complicated model might account for deaths, immigration, emigration, and other factors.

More complicated mathematical models can be used to describe and calculate more complex processes, such as climate change or economic growth (fig. 2.8). These models are also useful because they allow the researcher to manipulate variables without actually destroying anything. An economist can experiment with different interest rates to see how they affect economic growth. A climatologist can raise CO_2 levels and see how quickly temperatures respond. These models are often called simulation models, because they simulate a complex system. Of course, the results depend on the assumptions built into the models. One model might show temperature rising quickly in response to CO_2; another might show temperature rising more slowly, depending on how evaporation, cloud cover, and other variables are taken into account. Consequently, simulations can produce powerful but controversial results. If multiple models generally agree, though, as in the cases of climate models that agree on generally upward temperature trends, we can have confidence that the overall predictions are reliable. These models are also very useful in laying out and testing our ideas about how a system works.

2.2 SYSTEMS DESCRIBE INTERACTIONS

The forest ecosystem you examined in the opening case study of this chapter is interesting because it is composed of many interdependent parts. By studying those parts, we can understand how similar ecosystems might function, and why. **Systems,** including ecosystems, are a central idea in environmental science. A system is a network of interdependent components and processes, with materials and energy flowing from one component of the system to another. For example, "ecosystem" is probably a familiar term for you. This simple word represents a complex assemblage of animals, plants, and their environment, through which materials and energy move.

The idea of systems is useful because it helps us organize our thoughts about the inconceivably complex phenomena around us. For example, an ecosystem might consist of countless animals, plants, and their physical surroundings. (You are a system consisting of millions of cells, complex organs, and innumerable bits of energy and matter that move through you.) Keeping track of all the elements and their relationships in an ecosystem would probably be an impossible task. But if we step back and think about them in terms of plants, herbivores, carnivores, and decomposers, then we can start to comprehend how it works (fig. 2.9).

We can use some general terms to describe the components of a system. A simple system consists of state variables (also called compartments), which store resources such as energy, matter, or water; and flows, or the pathways by which those resources move from one state variable to another. In figure 2.9, the plant and animals represent state variables. The plant represents many different plant types, all of which are things that store solar energy and create carbohydrates from carbon, water, and sunlight. The rabbit represents many kinds of herbivores, all of which consume plants, then store energy, water, and carbohydrates until they are used, transformed, or consumed by a carnivore. We can describe the flows in terms of herbivory, predation, or photosynthesis, all processes that transfer energy and matter from one state variable to another.

It may seem cold and analytical to describe a rabbit or a flower as a state variable, but it is also helpful to do so. When we

FIGURE 2.9 A system can be described in very simple terms.

start discussing natural complexity in the simple terms of systems, we can identify common characteristics. Understanding these characteristics can help us diagnose disturbances or changes in the system: for example, if rabbits become too numerous, herbivory can become too rapid for plants to sustain. Overgrazing can lead to widespread collapse of this system. Let's examine some of the common characteristics we can find in systems.

Systems can be described in terms of their characteristics

Open systems are those that receive inputs from their surroundings and produce outputs that leave the system. Almost all natural systems are open systems. In principle, a **closed system** exchanges no energy or matter with its surroundings, but these are rare. Often we think of pseudo-closed systems, those that exchange only a little energy but no matter with their surroundings. **Throughput** is a term we can use to describe the energy and matter that flow into, through, and out of a system. Larger throughput might expand the size of state variables. For example, you can consider your household economy in terms of throughput. If you get more income, you have the option of enlarging your state variables (bank account, car, television, . . .). Usually an increase in income is also associated with an increase in outflow (the money spent on that new car and TV). In a grassland, inputs of energy (sunlight) and matter (carbon dioxide and water) are stored in biomass. If there is lots of water, the biomass storage might increase (in the form of trees). If there's little input, biomass might decrease (grass could become short or sparse). Eventually stored matter and energy may be exported (by fire, grazing, land clearing). The exported matter and energy can be thought of as throughput.

A grassland is an *open system:* it exchanges matter and energy with its surroundings (the atmosphere and soil, for example; fig. 2.10). In theory, a closed system would be entirely isolated from its surroundings, but in fact all natural systems are at least partly open. A fish tank is an example of a system that is less open than a grassland, because it can exist with only sunlight and carbon dioxide inputs (fig. 2.11).

Systems also experience positive and negative feedback mechanisms. A **positive feedback** is a self-perpetuating process. In a grassland, a grass plant grows new leaves, and the more leaves it has, the more energy it can capture for producing more leaves. In other words, in a positive feedback mechanism, increases in a state variable (biomass) lead to further increases in that state variable (more biomass). In contrast, a **negative feedback** is a process that supresses change. If grass grows very rapidly, it may produce more leaves than can be supported by available soil moisture. With insufficient moisture, the plant begins to die back.

In climate systems (chapter 15) positive and negative feedbacks are important ideas. For example, as warm summers melt ice in the Arctic, newly exposed water surfaces absorb heat, which leads to further melting, which leads to further heat absorption . . . This is positive feedback. In contrast, clouds can have a negative feedback effect (although there are debates on the net effect of clouds). A warming atmosphere can evaporate more water, producing clouds. Clouds block some solar heat, which reduces the evaporation. Thus clouds can slow the warming process.

Positive and negative feedback mechanisms are also important in understanding population dynamics (chapter 6). For example, more individuals produce more young, which produces more individuals . . . (a positive feedback). But sometimes environmental limits reduce the number of young that survive to reproduce (a negative feedback). Your body is a system with active negative feedback mechanisms: For example, if you exercise, you become hot, and your skin sweats, which cools your body.

FIGURE 2.10 Environmental scientists often study open systems. Here students at Cedar Creek study the climate-vegetation system, gathering plant samples that grew in carbon dioxide-enriched air pumped from the white poles, but other factors (soil, moisture, sunshine, temperature) are not controlled.

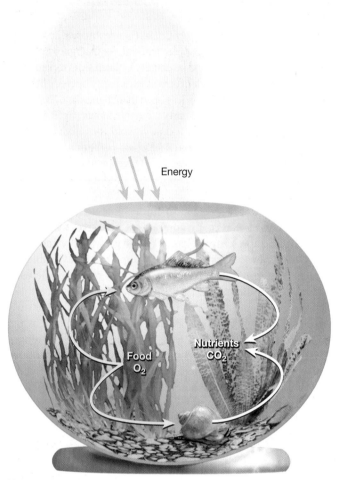

Energy

Nutrients
CO_2

Food
O_2

(a) A simple system

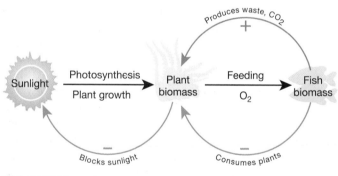

Produces waste, CO_2

Sunlight

Photosynthesis
Plant growth

Blocks sunlight

Plant biomass

Feeding
O_2

Consumes plants

Fish biomass

+

−

−

(b) A model of a system

FIGURE 2.11 Systems consist of compartments (also known as state variables) such as fish and plants, and flows of resources, such as nutrients or O_2 (a). Feedback loops (b) enhance or inhibit flows and the growth of compartments.

Systems may exhibit stability

Negative feedbacks tend to maintain stability in a system. We often think of systems exhibiting **homeostasis,** or a tendency to remain more or less stable and unchanging. Equilibrium is another term for stability in a system. Your body temperature stays remarkably constant despite dramatic changes in your environment and your activity levels. Changing by just a few degrees is extremely unusual—a fever

just 4–6°F above normal is unusual and serious. Natural systems such as grasslands can be fairly stable, too. If the climate isn't too dry or too wet, a grassland tends to remain grassland, although the grass may be dense in some years and sparse in others. Cycles of wet and dry years may be part of the system's normal condition.

Disturbances, events that can destabilize or change the system, might also be normal for the system. There can be many kinds of disturbance in a grassland. Severe drought can set back the community, so that it takes some time to recover. Many grasslands also experience occasional fires, a disturbance that stimulates grass growth (by clearing accumulated litter and recycling nutrients) but destroys trees that might be encroaching on the grassland. Thus disturbances are often a normal part of natural systems. Sometimes we consider this "dynamic equilibrium," or a tendency for a system to change and then return to normal.

Grassland plots show **resilience,** an ability to recover from disturbance. In fact, studies indicate that species-rich plots may show more resilience than species-poor plots. Sometimes severe disturbance can lead to a **state shift,** in which conditions do not return to "normal." For example, a climate shift that drastically reduced rainfall could lead to a transition from grassland to desert. Plowing up grassland to plant crops is basically a state shift from a complex system to a single-species system.

Emergent properties are another interesting aspect of systems. Sometimes a system is more than the sum of its parts. For example, a tree is more than just a mass of stored carbon. It provides structure to a forest, habitat for other organisms, it shades and cools the ground, and it holds soil in place with its roots. An ecosystem can also have beautiful sights and sounds that may be irrelevant to its functioning as a system, but that we appreciate nonetheless (fig. 2.12). In a similar way, you are a system made up of component parts, but you have many emergent properties, including your ability to think, share ideas with people around you, sing, and dance. These are properties that emerge because you function well as a system.

FIGURE 2.12 Emergent properties of systems, including beautiful sights and sounds, make them exciting to study.

2.3 SCIENTIFIC CONSENSUS AND CONFLICT

The scientific method outlined in figure 2.3 is the process used to carry out individual studies. Larger-scale accumulation of scientific knowledge involves cooperation and contributions from countless people. Good science is rarely carried out by a single individual working in isolation. Instead, a community of scientists collaborates in a cumulative, self-correcting process. You often hear about big breakthroughs and dramatic discoveries that change our understanding overnight, but in reality these changes are usually the culmination of the labor of many people, each working on different aspects of a common problem, each adding small insights to solve the problem. Ideas and information are exchanged, debated, tested, and retested to arrive at **scientific consensus,** or general agreement among informed scholars.

The idea of consensus is important. For those not deeply involved in a subject, the multitude of contradictory results can be bewildering: Are shark populations disappearing, and does it matter? Is climate changing, and how much? Among those who have performed and read many studies, there tends to emerge a general agreement about the state of a problem. Scientific consensus now holds that many shark populations are in danger, though opinions vary on how severe the problem is. Consensus is that global climates are changing, though models differ somewhat on how rapidly they will change under different policy scenarios.

Sometimes new ideas emerge that cause major shifts in scientific consensus. Two centuries ago, geologists explained many earth features in terms of Noah's flood. The best scientists held that the flood created beaches well above modern sea level, scattered boulders erratically across the landscape, and gouged enormous valleys where there is little water now (fig. 2.13). Then the Swiss glaciologist Louis Agassiz and others suggested that the earth had once been much colder and that glaciers had covered large areas. Periodic ice ages better explained changing sea levels, boulders transported far from their source rock, and the great, gouged valleys. This new idea completely altered the way geologists explained their subject. Similarly, the idea of tectonic plate movement, in which continents shift slowly around the earth's surface, revolutionized the ways geologists, biogeographers, ecologists, and others explained the development of the earth and its life-forms.

These great changes in explanatory frameworks were termed **paradigm shifts** by Thomas Kuhn, who studied revolutions in scientific thought. According to Kuhn, paradigm shifts occur when a majority of scientists accept that the old explanation no longer explains new observations very well. The shift is often contentious and political, because whole careers and worldviews, based on one sort of research and explanation, can be undermined by a new model. Sometimes a revolution happens rather quickly. Quantum mechanics and Einstein's theory of relativity, for example, overturned classical physics in only about 30 years. Sometimes a whole generation of scholars has to retire before new paradigms can be accepted.

As you study this book, try to identify some of the paradigms that guide our investigations, explanations, and actions today. This

FIGURE 2.13 Paradigm shifts change the ways we explain our world. Geologists now attribute Yosemite's valleys to glaciers, where once they believed events like Noah's flood carved its walls.

is one of the skills involved in critical thinking, discussed in the introductory chapter of this book.

Detecting pseudoscience relies on independent, critical thinking

Ideally, science should serve the needs of society. Deciding what those needs are, however, is often a matter of politics and economics. Should water be taken from a river for irrigation or left in the river for wildlife habitat? Should we force coal-burning power plants to reduce air pollution in order to lower health costs and respiratory illnesses, or are society and our economy better served by having cheap but dirty energy? These thorny questions are decided by a combination of scientific evidence, economic priorities, political positions, and ethical viewpoints.

On the other hand, in every political debate, lawyers and lobbyists can find scientists who will back either side. Politicians hold up favorable studies, proclaiming them "sound science," while they dismiss others as "junk science." Opposing sides dispute the scientific authority of the study they dislike. What is "sound" science, anyway? If science is often embroiled in politics, does this mean that science is always a political process?

If you judge only from reports in newspapers or on television about this issue, you'd probably conclude that scientific opinion is about equally divided on whether global warming is a threat or not. In fact, the vast majority of scientists working on this issue agree with the proposition that the earth's climate is being affected by human activities. Only a handful of maverick scientists disagree. In a study of 928 papers published in refereed scientific journals between 1993 and 2003, not one disagreed with the broad scientific consensus on global warming.

Why, then, is there so much confusion among the public about this issue? Why do politicians continue to assert that the dangers of climate change are uncertain at best, or "the greatest hoax ever perpetrated on the American people," as James Inhofe, former chair of the Senate Committee on Environment and Public Works, claims. A part of the confusion lies in the fact that media often present the debate as if it's evenly balanced. The fact that an overwhelming majority of working scientists are mostly in agreement on this issue doesn't make good drama, so the media give equal time to minority viewpoints just to make an interesting fight.

Perhaps a more important source of misinformation comes from corporate funding for articles and reports denying climate change. The ExxonMobil corporation, for example, has donated at least $20 million over the past decade to more than 100 think tanks, media outlets, and consumer, religious, and civil rights groups that promote skepticism about global warming. Some of these organizations sound like legitimate science or grassroots groups but are really only public relations operations. Others are run by individuals who find it rewarding to offer contrarian views. This tactic of spreading doubt and disbelief through innocuous-sounding organizations or seemingly authentic experts isn't limited to the climate change debate. It was pioneered by the tobacco industry to mislead the public about the dangers of smoking. Interestingly, some of the same individuals, groups, and lobbying firms employed by tobacco companies are now working to spread confusion about climate change.

Given this highly sophisticated battle of "experts," how do you interpret these disputes, and how do you decide whom to trust? The most important strategy is to apply critical thinking as you watch or read the news. What is the position of the person making the report? What is the source of their expertise? What economic or political interests do they serve? Do they appeal to your reason or to your emotions? Do they use inflammatory words (such as "junk"), or do they claim that scientific uncertainty makes their opponents' study meaningless? If they use statistics, what is the context for their numbers?

It helps to seek further information as you answer some of these questions. When you watch or read the news, you can look for places where reporting looks incomplete, you can consider sources and ask yourself what unspoken interests might lie behind the story.

Another strategy for deciphering the rhetoric is to remember that there are established standards of scientific work, and to investigate whether an "expert" follows these standards: Is the

Table 2.2 Questions for Baloney Detection

1. How reliable are the sources of this claim? Is there reason to believe that they might have an agenda to pursue in this case?
2. Have the claims been verified by other sources? What data are presented in support of this opinion?
3. What position does the majority of the scientific community hold in this issue?
4. How does this claim fit with what we know about how the world works? Is this a reasonable assertion or does it contradict established theories?
5. Are the arguments balanced and logical? Have proponents of a particular position considered alternate points of view or only selected supportive evidence for their particular beliefs?
6. What do you know about the sources of funding for a particular position? Are they financed by groups with partisan goals?
7. Where was evidence for competing theories published? Has it undergone impartial peer review or is it only in proprietary publication?

report peer-reviewed? Do a majority of scholars agree? Are the methods used to produce statistics well documented?

Harvard's Edward O. Wilson writes, "We will always have contrarians whose sallies are characterized by willful ignorance, selective quotations, disregard for communications with genuine experts, and destructive campaigns to attract the attention of the media rather than scientists. They are the parasite load on scholars who earn success through the slow process of peer review and approval." How can we identify misinformation and questionable claims? The astronomer Carl Sagan proposed a "Baloney Detection Kit" containing the questions in table 2.2.

Most scientists have an interest in providing knowledge that is useful, and our ideas of what is useful and important depend partly on our worldviews and priorities. Science is not necessarily political, but it is often used for political aims. The main task of educated citizens is to discern where it is being misused or disregarded for purposes that undermine public interests.

CONCLUSION

Science is a process for producing knowledge methodically and logically. Scientists try to understand the world by making observations and trying to discern patterns and rules that explain those observations. Scientists try to remain cautious and skeptical of conclusions, because we understand that any set of observations is only a sample of all possible observations. In order to make sure we follow a careful and methodical approach, we often use the scientific method, which is the step-by-step process of forming a testable question, doing tests, and interpreting results. Scientists use both deductive reasoning (deducing an explanation from general principles) and inductive reasoning (deriving a general rule from observations).

Hypotheses and theories are basic tools of science. A hypothesis is a testable question. A theory is a well-tested explanation that explains observations and that is accepted by the scientific community. Probability is also a key idea: chance is involved in many events, and circumstances can influence probabilities—such as your chances of getting a cold or of getting an A in this class. We often use probability to measure uncertainty when we test our hypotheses.

Models and systems are also central ideas. A system is a network of interdependent components and processes. For example, an ecosystem consists of plants, animals, and other components, and energy and nutrients transfer among those components. Systems have general characteristics we can describe, including throughput, feedbacks, homeostasis, resilience, and emergent properties. Often we use models (simplified representations of systems) to describe or manipulate a system. Models vary in complexity, according to their purposes, from a paper airplane to a global circulation model.

Science aims to foster debate and inquiry, but scientific consensus emerges as most experts come to agree on well-supported theoretical explanations. Sometimes new explanations revolutionize science, but scientific consensus helps us identify which ideas and theories are well supported by evidence, and which are not supported.

REVIEWING LEARNING OUTCOMES

By now you should be able to explain the following points:

2.1 Describe the scientific method and explain how it works.

- Science depends on skepticism and accuracy.
- Deductive and inductive reasoning are both useful.
- Testable hypotheses and theories are essential tools.
- Understanding probability helps reduce uncertainty.
- Statistics can indicate the probability that your results were random.
- Experimental design can reduce bias.
- Models are an important experimental strategy.

2.2 Explain systems and how they're useful in science.

- Systems are composed of processes.
- Disturbances and emergent properties are important characteristics of many systems.

2.2 Evaluate the role of scientific consensus and conflict.

- Detecting pseudoscience relies on independent, critical thinking.
- What's the relation between environmental science and environmentalism?

PRACTICE QUIZ

1. What is science? What are some of its basic principles?

2. Why are widely accepted, well-defended scientific explanations called "theories"?

3. Explain the following terms: probability, dependent variable, independent variable, and model.

4. What are inductive and deductive reasoning? Describe an example in which you have used each.

5. Draw a diagram showing the steps of the scientific method, and explain why each is important.

6. What is scientific consensus and why is it important?

7. What is a positive feedback loop? What is a negative feedback loop? Give an example of each.

8. Explain what a model is. Give an example.

9. Why do we say that proof is elusive in science?

10. What is a manipulative experiment? A natural experiment? A controlled study?

CRITICAL THINKING AND DISCUSSION QUESTIONS

1. Explain why scientific issues are or are not influenced by politics. Can scientific questions ever be entirely free of political interest? If you say no, does that mean that all science is merely politics? Why or why not?

2. Review the questions for "baloney detection" in table 2.2, and apply them to an ad on TV. How many of the critiques in this list are easily detected in the commercial?

3. How important is scientific thinking for you, personally? How important do you think it should be? How important is it for society to have thoughtful scientists? How would your life be different without the scientific method?

4. Many people consider science too remote from everyday life and from nonscientists. Do you feel this way? Are there aspects of scientific methods (such as reasoning from observations) that you use?

5. Many scientific studies rely on models for experiments that cannot be done on real systems, such as climate, human health, or economic systems. If assumptions are built into models, then are model-based studies inherently weak? What would increase your confidence in a model-based study?

Data Analysis: Evaluating Uncertainty

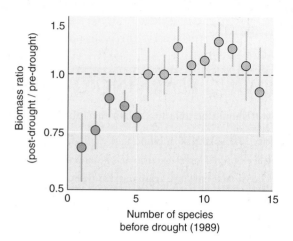

Uncertainty is a key idea in science. We can rarely have absolute proof in experimental results, because our conclusions rest on observations, but we only have a small sample of all possible observations. Because uncertainty is always present, it's useful to describe how much uncertainty you have, relative to what you know. It might seem ironic, but in science, knowing about uncertainty increases our confidence in our conclusions.

The graph at right is from a landmark field study by D. Tilman, et al. It shows change in biomass within, experimental plots containing varying numbers of native prairie plants after a severe drought. Because more than 200 replicate (repeated test) plots were used, this study was able to give an estimate of uncertainty. This uncertainty is shown with error bars. In this graph, dots show means for groups of test plots; the error bars show the range in which that mean could have fallen, if there had been a slightly different set of test plots.

Let's examine the error bars in this graph. To begin, as always, make sure you understand what the axes show. This graph is a relatively complex one, so be patient.

Questions:

1. What variable is shown on the X-axis? What are the lowest and highest values on the axis?

2. Each dot shows the average species count for a set of test plots with a given number of species. About how many species are in the plots represented by the leftmost dot? By the rightmost dot?

3. What is the axis label on the Y-axis? What does a value of 0.75 mean? A value of 1.0?

   (Note: the Y-axis doesn't change at a constant rate. It changes logarithmically. This means values at the low end are more visible.)

4. Each blue dot represents a group of plots with 5 or fewer species; yellow dots represent plots with more than 5 species. Look at the leftmost dot, plots with only 1 species. Was biomass less or more after the drought?

   The error bars show standard error, which you can think of as the range in which the average (the dot) might fall, if you

had a slightly different set of plots. (Standard error is just the standard deviation divided by the square root of the number of observations.) For 1-species plots, there's a small chance that the average could have fallen at the low end of the error bar, or almost as low as about 0.5, or half the pre-drought biomass.

5. How many of the blue error bars overlap the dotted line (no change in biomass)? How many of the yellow error bars overlap the dotted line? Are there any yellow bars entirely above the 0 line?

Where the error bars fall entirely below 1, we can be quite sure that, even if we had had a different set of plots, the after-drought biomass would still have declined. Where the error bars include a value of 1, the averages are not significantly different from 1 (or no change).

The conclusions of this study rest on the fact that the blue bars showed nearly-certain declines in biomass, while the yellow (higher-diversity) bars showed either no change or increases in biomass. Thus the whole paper boiled down to the question of which error bars crossed the dotted line! But the implications of the study are profound: they demonstrate a clear relationship between biodiversity and recovery from drought, at least for this study. One of the exciting things about scientific methods, and of statistics, is that they let us use simple, unambiguous tests to answer important questions.

CHAPTER 3

Chesapeake Bay's ecosystem supports fisheries, recreation, and communities. But the estuary is an ecosystem out of balance.

Matter, Energy, and Life

Learning Outcomes

After studying this chapter, you should be able to:

3.1 Describe matter, atoms, and molecules and give simple examples of the role of four major kinds of organic compounds in living cells.

3.2 Define energy and explain how thermodynamics regulates ecosystems.

3.3 Understand how living organisms capture energy and create organic compounds.

3.4 Define species, populations, communities, and ecosystems, and summarize the ecological significance of trophic levels.

3.5 Compare the ways that water, carbon, nitrogen, sulfur, and phosphorus cycle within ecosystems.

"When one tugs at a single thing in nature, he finds it attached to the rest of the world."

~ *John Muir*

Each year Chesapeake Bay, the largest estuary in the United States, gets a report card, just as you do at the end of a semester. Like your report card, this one summarizes several key performance measures. Unlike your grades, the bay's grades are based on measures such as water clarity, oxygen levels, health of sea grass beds, and the condition of microscopic plankton community. These factors reflect overall stability of fish and shellfish populations, critical to the region's ecosystems and economy. Since record keeping began, the bay's performance has been poor, with scores hovering between 35 and 57 out of 100, and an average grade of low C–. The main reason for the bad grades? Excessive levels of nitrogen and phosphorus, two common life-supporting elements that have destabilized the ecosystem.

Chesapeake Bay's watershed is a vast and complex system, with over 17,600 km (11,000 mi) of tidal shoreline in 6 states, and a population of 20 million people. Approximately 100,000 streams and rivers drain into the bay. All these streams carry runoff from forests, farmlands, cities, and suburbs from as far away as New York (fig. 3.1a).

The system has consistently bad grades, but it's clearly worth saving. Even in its impaired state, the bay provides 240 million kg (500 million lb) of seafood every year. It supports fishing and recreational economies worth $33 billion a year. But this is just a fraction of what it should be. The bay once provided abundant harvests of oysters, blue crabs, rockfish, white perch, shad, sturgeon, flounder, eel, menhaden, alewives, and soft-shell clams. Overharvesting, disease, and declining ecosystem productivity have decimated fisheries. Blue crabs are just above population survival levels. The oyster harvest, which was 15 to 20 million bushels per year in the 1890s, has declined to less than 1 percent of that amount. According to the Environmental Protection Agency (EPA), the bay should support more than twice the fish and shellfish populations that are there today. Human health is also at risk. After heavy rainfall, people are advised to stay out of the water for 48 hours, to avoid contamination from sewer overflows and urban and agricultural runoff.

Among the many challenges for Chesapeake Bay, the principal problem is simply excessive levels of nitrogen and phosphorus. These two elements are essential for life, but the system is overloaded by excess loads from farm fields, livestock manure, urban streets, suburban lawn fertilizer, the legal discharges of over 3,000 sewage treatment plants, and from half a million aging household septic systems. Air pollution from cars, power plants, and factories also introduce nitrogen to the bay (fig. 3.1b). Sediment is also a key issue: it washes in from fields and streets, smothers eelgrass beds, and blocks sunlight that further reduces photosynthesis in the bay.

Just as too many donuts are bad for you, an excessive diet of nutrients is bad for an estuary. Excess nutrients fertilize superabundant growth of algae, which further blocks sunlight and reduces photosynthesis and oxygen levels in the bay. Lifeless, oxygen-depleted areas result, leading to fish die-offs, as well as poor reproduction in oysters, crabs and fish. These algal blooms in nutrient-enriched waters are increasingly common in bays and estuaries worldwide.

Progress has been discouragingly slow for decades, but in 2010 the EPA finally addressed the problem seriously, complying with its charge from Congress (under the Clean Water Act) to protect the bay. Where piecemeal, mostly-voluntary efforts by individual states had long failed to improve the Chesapeake's report cards, the EPA brought all neighboring states to the negotiating table. Total maximum daily loads (TMDLs) for nutrients and sediments were established, and states were given freedom to decide how to meet their share of nitrogen reductions. But the EPA has legal authority from the Clean Water Act to enforce reductions. The aim is to cut nitrogen levels by 25 percent, phosphorus by 24 percent, and sediment by 20 pecent. The nitrogen target of 85 million kg (186 million lb) per year is still 4–5 times

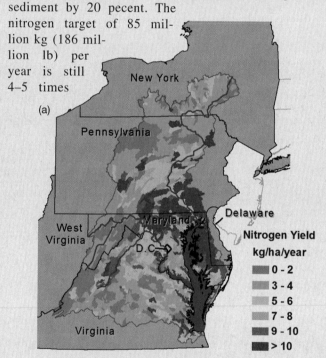

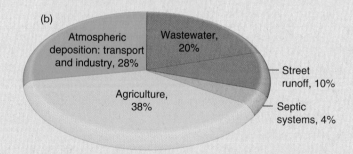

FIGURE 3.1 America's largest and richest estuary, Chesapeake Bay (shown in blue) suffers from pollutants from six states (a), and many sources (b). **Data sources:** USGS, EPA 2010.

Case Study continued

greater than would be released by an undisturbed watershed, but it's a huge improvement.

States from Virginia to New York have chosen their own strategies to meet limits. Maryland plans to capture and sell nitrogen and phosphorus from chicken manure. New York promises better urban wastewater treatment. Pennsylvania is strengthening soil conservation efforts to retain nutrients on farmland. These plans will be implemented gradually, but together, by addressing upstream land uses, they seem likely to turn around this magnificent estuary.

Chesapeake Bay has long been a symbol of the intractable difficulty of managing large, complex systems. Progress has required better understanding of several issues: the integrated functioning of the uplands and the waterways, the interdependence of the diverse human communities and economies that depend on the bay, and the pathways of nitrogen and phosphorus through an ecosystem.

Environmental scientists have led the way to the EPA's solution with years of ecosystem research and data collection. Through their efforts, and with EPA leadership, Chesapeake Bay could become the largest, and perhaps the most broadly beneficial, ecosystem restoration ever attempted in the United States. In this chapter we'll examine how these and other elements, move through systems, and why they are important. Understanding these basic ideas will help you explain functioning of many different systems, including Chesapeake Bay, your local ecosystem, even your own body.

For related resources, including Google Earth™ placemarks that show locations where these issues can be explored via satellite images, visit http:\\EnvironmentalScience-Cunningham.blogspot.com.

3.1 ELEMENTS OF LIFE

The accumulation and transfer of energy and nutrients allows living systems to exist. These processes tie together the parts of an ecosystem—or an organism; you could think of the accumulation and circulation of energy and nutrients as the basis of life. Understanding how nutrients and energy function in a system, and where they come from, and where they go, are essential to understanding **ecology,** the scientific study of relationships between organisms and their environment.

In this chapter we'll introduce a number of concepts that are essential to understanding how living things function in their environment. We review what matter and energy are, then explore the ways organisms acquire and use energy and chemical elements. Then we'll investigate feeding relationships among organisms—the ways that energy and nutrients are passed from one living thing to another—forming the basis of ecosystems. In other words, we'll trace components from atoms to elements to compounds to cells to organisms to ecosystems.

Atoms, elements, and compounds

Everything that takes up space and has mass is **matter.** Matter exists in four distinct states, or phases—solid, liquid, gas, and plasma—which vary in energy intensity and the arrangement of particles that make up the substance. Water, for example, can exist as ice (solid), as liquid water, or as water vapor (gas). The fourth phase, plasma, occurs when matter is heated so intensely that electrons are released, and particles become ionized (electrically charged). We can observe plasma in the sun, lightning, and very hot flames.

Under ordinary circumstances, matter is neither created nor destroyed; rather, it is recycled over and over again. The molecules that make up your body probably contain atoms that once made up the body of a dinosaur. Most certainly you contain atoms that were part of many smaller prehistoric organisms. This is because chemical elements are used and reused by living organisms. Matter is transformed and combined in different ways, but it doesn't disappear; everything goes somewhere. This idea is known as the principle of **conservation of matter.**

How does this principle apply to environmental science? It explains how components of environmental systems are intricately connected. From Chesapeake Bay to your local ecosystem to your own household, all matter comes from somewhere, and all waste goes somewhere. Pause to consider what you have eaten, used, or bought today. Then think of where those materials will go when you are done with them. You are intricately tied to both the sources and the destinations of everything you use. This is a useful idea for us as residents of a finite world. Ultimately when we throw away our disposable goods, they don't really go "away," they just go somewhere else, to stay there for a while and then move on.

Matter consists of **elements,** which are substances that cannot be broken down into simpler forms by ordinary chemical reactions. Each of the 122 known elements (92 natural, plus 30 created under special conditions) has distinct chemical characteristics. Just four elements—oxygen, carbon, hydrogen, and nitrogen—are responsible for more than 96 percent of the mass of most living organisms. See if you can find these four elements in the periodic table of the elements at the end of this book.

Atoms are the smallest particles that exhibit the characteristics of an element. Atoms are composed of positively charged protons, negatively charged electrons, and electrically neutral neutrons. Protons and neutrons, which have approximately the same mass, are clustered in the nucleus in the center of the atom (fig. 3.2). Electrons, which are tiny in comparison to the other particles, orbit the nucleus at the speed of light.

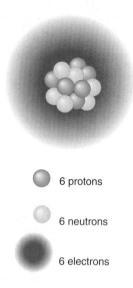

6 protons

6 neutrons

6 electrons

FIGURE 3.2 As difficult as it may be to imagine when you look at a solid object, all matter is composed of tiny, moving particles, separated by space and held together by energy. It is hard to capture these dynamic relationships in a drawing. This model represents carbon-12, with a nucleus of six protons and six neutrons; the six electrons are represented as a fuzzy cloud of potential locations rather than as individual particles.

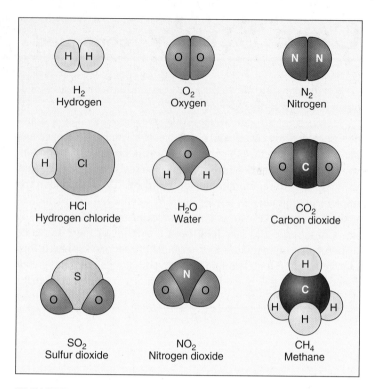

FIGURE 3.3 These common molecules, with atoms held together by covalent bonds, are important components of the atmosphere or important pollutants.

Each element has a characteristic number of protons per atom, called its **atomic number.** Each element also has a characteristic atomic mass, which is the sum of protons and neutrons (each having a mass of about 1). However, the number of neutrons can vary slightly. Forms of an element that differ in atomic mass are called **isotopes.** For example, hydrogen (H) is the lightest element, and normally it has just one proton and one electron (and no neutrons) and an atomic mass of 1. A small percentage of hydrogen atoms also have a neutron in the nucleus, which gives those atoms an atomic mass of 2 (one proton + one neutron). We call this isotope deuterium (^{2}H). An even smaller percentage of natural hydrogen called tritium (^{3}H) has one proton plus two neutrons. Oxygen atoms can also have one or two extra neutrons, making them the isotopes ^{17}O or ^{18}O, instead of the normal ^{16}O.

This difference is interesting to an environmental scientist. Water (H_2O) containing heavy ^{18}O generally evaporates most easily in hot climates, so we can detect ancient climate conditions by examining the abundance of ^{18}O in air bubbles trapped in ancient ice cores (chapter 15). Some isotopes are unstable—that is, they spontaneously emit electromagnetic energy or subatomic particles, or both. Radioactive waste and nuclear energy involve unstable isotopes of elements such as uranium and plutonium (chapters 19, 21).

Chemical bonds hold molecules together

Atoms often join to form **compounds,** or substances composed of different kinds of atoms (fig. 3.3). A pair or group of atoms that can exist as a single unit is known as a **molecule.** Some elements commonly occur as molecules, such as molecular oxygen (O_2) or molecular nitrogen (N_2), and some compounds can exist as molecules, such as glucose ($C_6H_{12}O_6$). In contrast to these molecules, sodium chloride (NaCl, table salt) is a compound that cannot exist as a single pair of atoms. Instead it occurs in a solid mass of Na and Cl atoms or as two ions, Na^+ and Cl^-, dissolved in solution. Most molecules consist of only a few atoms. Others, such as proteins and nucleic acids, discussed below, can include millions or even billions of atoms.

When ions with opposite charges form a compound, the electrical attraction holding them together is an *ionic bond.* Sometimes atoms form bonds by *sharing* electrons. For example, two hydrogen atoms can bond by sharing a pair of electrons—they orbit the two hydrogen nuclei equally and hold the atoms together. Such electron-sharing bonds are known as *covalent bonds.* Carbon (C) can form covalent bonds simultaneously with four other atoms, so carbon can create complex structures such as sugars and proteins. Atoms in covalent bonds do not always share electrons evenly. An important example in environmental science is the covalent bonds in water (H_2O). The oxygen atom attracts the shared electrons more strongly than do the two hydrogen atoms. Consequently, the hydrogen portion of the molecule has a slight positive charge, while the oxygen has a slight negative charge. These charges create a mild attraction between water molecules, so that water tends to be somewhat cohesive. This fact helps explain some of the remarkable properties of water (Exploring Science, p. 55).

When an atom gives up one or more electrons, we say it is *oxidized* (because it is very often oxygen, an abundant and highly reactive element, that takes the electron). When an atom gains electrons, we say it is *reduced.* Oxidation and reduction reactions are necessary for life: Oxidation of sugar and starch molecules, for example, is an important part of how you gain energy from food.

Forming bonds usually releases energy. Breaking bonds generally requires energy. Think of this in burning wood: carbon-rich organic compounds such as cellulose are *broken,* which requires energy; at the same time, oxygen from the air *forms* bonds with carbon from the wood, making CO_2. In a fire, more energy is produced than is consumed, and the net effect is that it feels hot to us.

If travelers from another solar system were to visit our lovely, cool, blue planet, they might call it Aqua rather than Terra because of its outstanding feature: the abundance of streams, rivers, lakes, and oceans of liquid water. Our planet is the only place we know where water exists as a liquid in any appreciable quantity. Water covers nearly three-fourths of the earth's surface and moves around constantly via the hydrologic cycle (discussed in chapter 15) that distributes nutrients, replenishes freshwater supplies, and shapes the land. Water makes up 60 to 70 percent of the weight of most living organisms. It fills cells, giving form and support to tissues. Among water's unique, almost magical qualities, are the following:

1. Water molecules are polar, that is, they have a slight positive charge on one side and a slight negative charge on the other side. Therefore, water readily dissolves polar or ionic substances, including sugars and nutrients, and carries materials to and from cells.

2. Water is the only inorganic liquid that exists under normal conditions at temperatures suitable for life. Most substances exist as either a solid or a gas, with only a very narrow liquid temperature range. Organisms synthesize organic compounds such as oils and alcohols that remain liquid at ambient temperatures and are therefore extremely valuable to life, but the original and predominant liquid in nature is water.

3. Water molecules are cohesive, tending to stick together tenaciously. You have experienced this property if you have ever done a belly flop off a diving board. Water has the highest surface tension of any common, natural liquid. Water also adheres to surfaces. As a result, water is subject to *capillary action:* it can be drawn into small channels. Without capillary action, movement of water and nutrients into groundwater reservoirs and through living organisms might not be possible.

4. Water is unique in that it expands when it crystallizes. Most substances shrink as they change from liquid to solid. Ice floats because it is less dense than liquid water. When temperatures fall below freezing, the surface layers of lakes, rivers, and oceans cool faster and freeze before deeper water. Floating ice then insulates underlying layers, keeping most water bodies liquid (and aquatic organisms alive) throughout the winter in most places. Without this feature, many aquatic systems would freeze solid in winter.

5. Water has a high heat of vaporization, using a great deal of heat to convert from liquid to vapor. Consequently, evaporating water is an effective way for organisms to shed excess heat. Many animals pant or sweat to moisten evaporative cooling surfaces. Why do you feel less comfortable on a hot, humid day than on a hot, dry day? Because the water vapor–laden air inhibits

Surface tension is demonstrated by the resistance of a water surface to penetration, as when it is walked upon by a water strider.

the rate of evaporation from your skin, thereby impairing your ability to shed heat.

6. Water also has a high specific heat; that is, a great deal of heat is absorbed before it changes temperature. The slow response of water to temperature change helps moderate global temperatures, keeping the environment warm in winter and cool in summer. This effect is especially noticeable near the ocean, but it is important globally.

All these properties make water a unique and vitally important component of the ecological cycles that move materials and energy and make life on earth possible.

Generally, some energy input (activation energy) is needed to start these reactions. In your fireplace, a match might provide the needed activation energy. In your car, a spark from the battery provides activation energy to initiate the oxidation (burning) of gasoline.

Ions react and bond to form compounds

Atoms frequently gain or lose electrons, acquiring a negative or positive electrical charge. Charged atoms (or combinations of atoms) are called **ions.** Negatively charged ions (with one or more extra electrons) are *anions.* Positively charged ions are *cations.* A hydrogen (H) atom, for example, can give up its sole electron to become a hydrogen ion (H^+). Chlorine (Cl) readily gains electrons, forming chlorine ions (Cl^-).

Substances that readily give up hydrogen ions in water are known as **acids.** Hydrochloric acid, for example, dissociates in water to form H^+ and Cl^- ions. In later chapters, you may read about acid rain (which has an abundance of H^+ ions), acid mine drainage, and many other environmental problems involving acids. In general, acids cause environmental damage because the H^+ ions react readily with living tissues (such as your skin or tissues of fish larvae) and with nonliving substances (such as the limestone on buildings, which erodes under acid rain).

Substances that readily bond with H^+ ions are called **bases** or alkaline substances. Sodium hydroxide (NaOH), for example, releases hydroxide ions (OH^-) that bond with H^+ ions in water. Bases can be highly reactive, so they also cause significant environmental problems. Acids and bases can also be essential to living

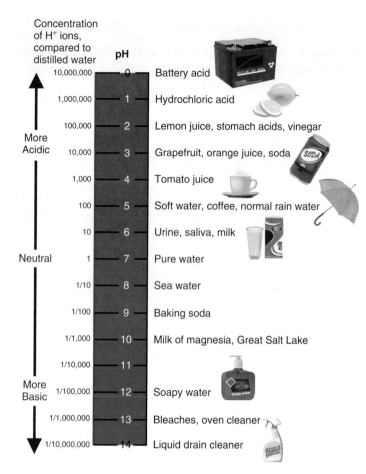

FIGURE 3.4 The pH scale. The numbers represent the negative logarithm of the hydrogen ion concentration in water. Alkaline (basic) solutions have a pH greater than 7. Acids (pH less than 7) have high concentrations of reactive H^+ ions.

things: The acids in your stomach dissolve food, for example, and acids in soil help make nutrients available to growing plants.

We describe the strength of an acid and base by its **pH,** the negative logarithm of its concentration of H^+ ions (fig. 3.4). Acids have a pH below 7; bases have a pH greater than 7. A solution of exactly pH 7 is "neutral." Because the pH scale is logarithmic, pH 6 represents *ten times* more hydrogen ions in solution than pH 7.

A solution can be neutralized by adding buffers—substances that accept or release hydrogen ions. In the environment, for example, alkaline rock can buffer acidic precipitation, decreasing its acidity. Lakes with acidic bedrock, such as granite, are especially vulnerable to acid rain because they have little buffering capacity.

Organic compounds have a carbon backbone

Organisms use some elements in abundance, others in trace amounts, and others not at all. Certain vital substances are concentrated within cells, while others are actively excluded. Carbon is a particularly important element because chains and rings of carbon atoms form the skeletons of **organic compounds,** the material of which biomolecules, and therefore living organisms, are made.

The four major categories of organic compounds in living things ("bio-organic compounds") are lipids, carbohydrates, proteins, and nucleic acids. Lipids (including fats and oils) store energy for cells, and they provide the core of cell membranes and other structures. Lipids do not readily dissolve in water, and their basic structure is a chain of carbon atoms with attached hydrogen atoms. This structure makes them part of the family of hydrocarbons (fig. 3.5a). Carbohydrates (including sugars, starches, and cellulose) also store energy and provide structure to cells. Like lipids, carbohydrates have a basic structure of carbon atoms, but hydroxyl (OH) groups replace half the hydrogen atoms in their basic structure, and they usually consist of long chains of sugars. Glucose (fig. 3.5b) is an example of a very simple sugar.

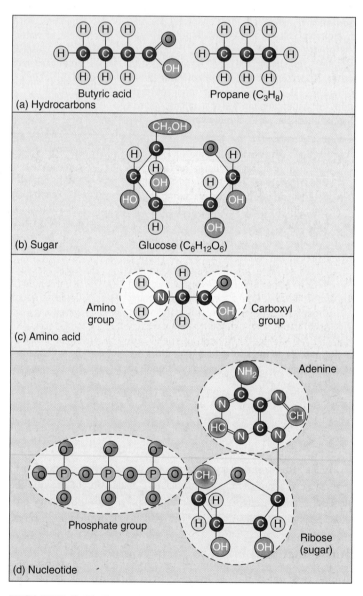

FIGURE 3.5 The four major groups of biologically important organic molecules are based on repeating subunits of these carbon-based structures. Basic structures are shown for (a) butyric acid (a building block of lipids) and a hydrocarbon, (b) a simple carbohydrate, (c) a protein, and (d) a nucleic acid.

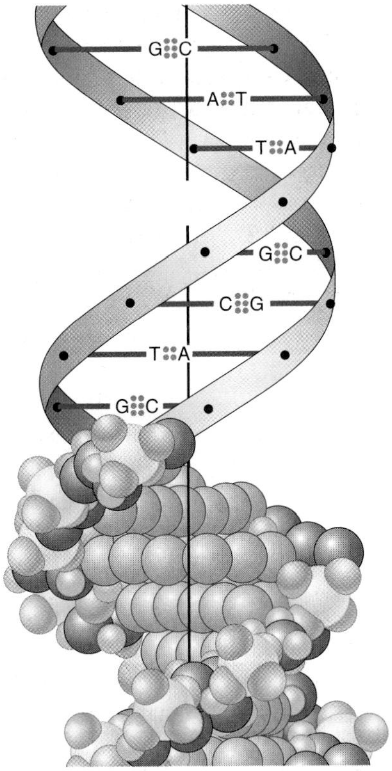

FIGURE 3.6

A composite molecular model of DNA. The lower part shows individual atoms, while the upper part has been simplified to show the strands of the double helix held together by hydrogen bonds (small dots) between matching nucleotides (A, T, G, and C). A complete DNA molecule contains millions of nucleotides and carries genetic information for many specific, inheritable traits.

Proteins are composed of chains of subunits called amino acids (fig. 3.5*c*). Folded into complex three-dimensional shapes, proteins provide structure to cells and are used for countless cell functions. Most enzymes, such as those that release energy from lipids and carbohydrates, are proteins. Proteins also help identify disease-causing microbes, make muscles move, transport oxygen to cells, and regulate cell activity.

Nucleotides are complex molecules made of a five-carbon sugar (ribose or deoxyribose), one or more phosphate groups, and an organic nitrogen-containing base called either a purine or pyrimidine (fig. 3.5*d*). Nucleotides are extremely important as signaling molecules (they carry information between cells, tissues, and organs) and as sources of intracellular energy. They also form long chains called *ribonucleic acid* (RNA) or ***deoxyribonucleic acid* (DNA)** that are essential for storing and expressing genetic information. Only four kinds of nucleotides (adenine, guanine, cytosine, and thyamine) occur in DNA, but there can be billions of these molecules lined up in a very specific sequence. Groups of three nucleotides (called codons) act as the letters in messages that code for the amino acid sequences in proteins. Long chains of DNA bind together to form a stable double helix (fig. 3.6). These chains separate for replication in preparation for cell division or to express their genetic information during protein synthesis. Extracting DNA from cells and reading the nucleotide sequence is widely useful, for medical genetics, agriculture, forensics, taxonomy, and many other fields. Because every individual has a unique set of DNA molecules, sequencing their nucleotide content can provide a distinctive individual identification.

Cells are the fundamental units of life

All living organisms are composed of **cells,** minute compartments within which the processes of life are carried out (fig. 3.7). Microscopic organisms such as bacteria, some algae, and protozoa are composed of single cells. Higher organisms have many cells, usually with many different cell varieties. Your body, for instance, is composed of several trillion cells of about two hundred distinct types. Every cell is

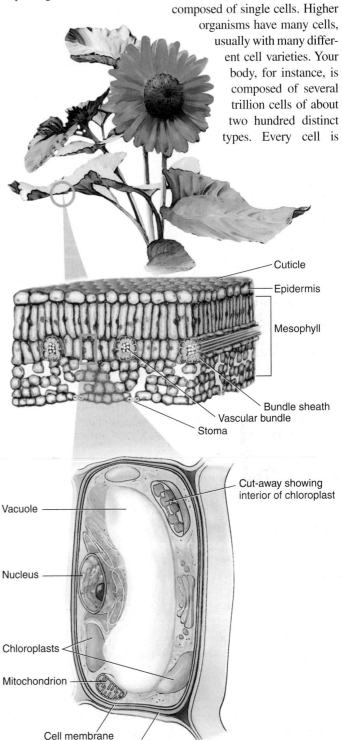

FIGURE 3.7 Plant tissues and a single cell's interior. Cell components include a cellulose cell wall, a nucleus, a large, empty vacuole, and several chloroplasts, which carry out photosynthesis.

surrounded by a thin but dynamic membrane of lipid and protein that receives information about the exterior world and regulates the flow of materials between the cell and its environment. Inside, cells are subdivided into tiny organelles and subcellular particles that provide the machinery for life. Some of these organelles store and release energy. Others manage and distribute information. Still others create the internal structure that gives the cell its shape and allows it to fulfill its role.

A special class of proteins called **enzymes** carry out all the chemical reactions required to create these various structures, provide them with energy and materials to carry out their functions, dispose of wastes, and perform other functions of life at the cellular level. Enzymes are molecular catalysts: they regulate chemical reactions without being used up or inactivated in the process. Like hammers or wrenches, they do their jobs without being consumed or damaged as they work. There are generally thousands of different kinds of enzymes in every cell, all necessary to carry out the many processes on which life depends. Altogether, the multitude of enzymatic reactions performed by an organism is called its **metabolism.**

FIGURE 3.8 Water stored behind this dam represents potential energy. Water flowing over the dam has kinetic energy, some of which is converted to heat.

Potential energy

Kinetic energy

3.2 ENERGY

If matter is the material of which things are made, energy provides the force to hold structures together, tear them apart, and move them from one place to another. In this section we will look at some fundamental characteristics of these components of our world.

Energy occurs in many forms

Energy is the ability to do work such as moving matter over a distance or causing a heat transfer between two objects at different temperatures. Energy can take many different forms. Heat, light, electricity, and chemical energy are examples that we all experience. The energy contained in moving objects is called **kinetic energy.** A rock rolling down a hill, the wind blowing through the trees, water flowing over a dam (fig. 3.8), or electrons speeding around the nucleus of an atom are all examples of kinetic energy. **Potential energy** is stored energy that is latent but available for use. A rock poised at the top of a hill and water stored behind a dam are examples of potential energy. **Chemical energy** stored in the food that you eat and the gasoline that you put into your car are also examples of potential energy that can be released to do useful work. Energy is often measured in units of heat (calories) or work (joules). One joule (J) is the work done when one kilogram is accelerated at one meter per second per second. One calorie is the amount of energy needed to heat one gram of pure water one degree Celsius. A calorie can also be measured as 4.184 J.

Heat describes the energy that can be transferred between objects of different temperature. When a substance absorbs heat, the kinetic energy of its molecules increases, or it may change

state: A solid may become a liquid, or a liquid may become a gas. We sense change in heat content as change in temperature (unless the substance changes state).

An object can have a high heat content but a low temperature, such as a lake that freezes slowly in the fall. Other objects, like a burning match, have a high temperature but little heat content. Heat storage in lakes and oceans is essential to moderating climates and maintaining biological communities. Heat absorbed in changing states is also critical. As you will read in chapter 15, evaporation and condensation of water in the atmosphere helps distribute heat around the globe.

Energy that is diffused, dispersed, and low in temperature is considered low-quality energy because it is difficult to gather and use for productive purposes. The heat stored in the oceans, for instance, is immense but hard to capture and use, so it is low quality. Conversely, energy that is intense, concentrated, and high in temperature is high-quality energy because of its usefulness in carrying out work. The intense flames of a very hot fire or high-voltage electrical energy are examples of high-quality forms that are valuable to humans. Many of our alternative energy sources (such as wind) are diffuse compared to the higher-quality, more concentrated chemical energy in oil, coal, or gas.

Think About It

Can you describe one or two practical examples of the laws of physics and thermodynamics in your own life? Do they help explain why you can recycle cans and bottles but not energy? Which law is responsible for the fact that you get hot and sweaty when you exercise?

Thermodynamics regulates energy transfers

Atoms and molecules cycle endlessly through organisms and their environment, but energy flows in a one-way path. A constant supply of energy—nearly all of it from the sun—is needed to keep biological processes running. Energy can be used repeatedly as it flows through

http://www.mhhe.com/cunningham12e

the system, and it can be stored temporarily in the chemical bonds of organic molecules, but eventually it is released and dissipated.

The study of thermodynamics deals with how energy is transferred in natural processes. More specifically, it deals with the rates of flow and the transformation of energy from one form or quality to another. Thermodynamics is a complex, quantitative discipline, but you don't need a great deal of math to understand some of the broad principles that shape our world and our lives.

The **first law of thermodynamics** states that energy is *conserved;* that is, it is neither created nor destroyed under normal conditions. Energy may be transformed, for example, from the energy in a chemical bond to heat energy, but the total amount does not change.

The second law of thermodynamics states that, with each successive energy transfer or transformation in a system, less energy is available to do work. That is, energy is degraded to lower-quality forms, or it dissipates and is lost, as it is used. When you drive a car, for example, the chemical energy of the gas is degraded to kinetic energy and heat, which dissipates, eventually, to space. The second law recognizes that disorder, or **entropy,** tends to increase in all natural systems. Consequently, there is always less *useful* energy available when you finish a process than there was before you started. Because of this loss, everything in the universe tends to fall apart, slow down, and get more disorganized.

How does the second law of thermodynamics apply to organisms and biological systems? Organisms are highly organized, both structurally and metabolically. Constant care and maintenance is required to keep up this organization, and a constant supply of energy is required to maintain these processes. Every time some energy is used by a cell to do work, some of that energy is dissipated or lost as heat. If cellular energy supplies are interrupted or depleted, the result—sooner or later—is death.

3.3 ENERGY FOR LIFE

Where does the energy for life come from? For nearly all plants and animals on earth, the sun is the ultimate energy source. But some organisms living deep in the earth's crust or at the bottom of the oceans derive energy from chemical reactions, for example between minerals and gases vented from the earth's crust. This energy pathway seems to be more ancient than the light-based pathway we know best. Biologists think that before green plants existed, ancient bacteria-like cells probably lived by processing chemicals in hot springs.

Extremophiles gain energy without sunlight

Until recently, the deep ocean floor was believed to be essentially lifeless. Cold, dark, subject to crushing pressures, and without any known energy supply, it was a place where scientists thought nothing could survive. Undersea explorations in the 1970s, however, revealed dense colonies of animals—blind shrimp, giant tubeworms, strange crabs, and bizarre clams—clustered around vents called black chimneys, where boiling hot, mineral-laden water bubbles up through cracks in the earth's crust. How do these sunless ecosystems get energy? The answer is **chemosynthesis,** the process in which bacteria use chemical bonds between inorganic elements, such as hydrogen sulfide (H_2S) or hydrogen gas (H_2), to provide energy for synthesis of organic molecules.

Discovering organisms living under the severe conditions of deep-sea hydrothermal vents led to exploration of other sites that seem exceptionally harsh to us. A variety of interesting organisms have been discovered in hot springs in thermal areas such as Yellowstone National Park, in intensely salty lakes, and even in deep rock formations (up to 1,500 m, or nearly a mile deep) in Columbia River basalts, for example. Some species are amazingly hardy. The recently described *Pyrolobus fumarii* can withstand temperatures up to 113°C (235°F). Most of these extremophiles are archaea, single-celled organisms that are thought to be the most primitive of all living organisms, and the conditions under which they live are thought to be similar to those in which life first evolved.

Deep-sea explorations of areas without thermal vents also have found abundant life (fig. 3.9). We now know that archaea live in oceanic sediments in astonishing numbers. The deepest of these species (they can be 800 m or more below the ocean floor) make methane from gaseous hydrogen (H_2) and carbon dioxide (CO_2), derived from rocks. Other species oxidize methane using sulfur to create hydrogen sulfide (H_2S), which is consumed by bacteria that serve as a food source for more complex organisms such as tubeworms. Why should we care about this exotic community? It's estimated that the total mass of microbes (microscopic organisms) living beneath the seafloor represents nearly one-third of all the biomass (organic material) on the planet. Furthermore, the vast supply of methane generated by this community could be either a great resource or a terrible threat to us.

The total amount of methane made by these microbes is probably greater than all the known reserves of coal, gas, and oil. If we could safely extract the huge supplies of methane hydrate in ocean sediments, it could supply our energy needs for hundreds of years.

FIGURE 3.9 A colony of tube worms and mussels clusters over a cool, deep-sea methane seep in the Gulf of Mexico.

Of greater immediate importance is that if methane-eating microbes weren't intercepting the methane produced by their neighbors, more than 300 million tons per year of this potent greenhouse gas would probably be bubbling to the surface, and we'd have run-away global warming. Methane-using bacteria can also help clean up pollution. After the Deepwater Horizon oil spill in the Gulf of Mexico in 2010, a deep-sea bloom of methane-metabolizing bacteria apparently consumed most of the methane escaping the spill.

Green plants get energy from the sun

Our sun is a star, a fiery ball of exploding hydrogen gas. Its thermonuclear reactions emit powerful forms of radiation, including potentially deadly ultraviolet and nuclear radiation (fig. 3.10), yet life here is nurtured by, and dependent upon, this searing energy source. Solar energy is essential to life for two main reasons.

First, the sun provides warmth. Most organisms survive within a relatively narrow temperature range. In fact, each species has its own range of temperatures within which it can function normally. At high temperatures (above 40°C), biomolecules begin to break down or become distorted and nonfunctional. At low temperatures (near 0°C), some chemical reactions of metabolism occur too slowly to enable organisms to grow and reproduce. Other planets in our solar system are either too hot or too cold to support life as we know it. The earth's water and atmosphere help to moderate, maintain, and distribute the sun's heat.

Second, nearly all organisms on the earth's surface depend on solar radiation for life-sustaining energy, which is captured by green plants, algae, and some bacteria in a process called **photosynthesis.** Photosynthesis converts radiant energy into useful, high-quality chemical energy in the bonds that hold together organic molecules. Photosynthesis happens on a microscopic scale, but it supports nearly all life on earth. Photosynthetic organisms (plants, algae, and bacteria) capture roughly 105 billion metric tons of carbon into biomass every year. About half of this carbon capture is on land; about half is in the ocean.

This photosynthesis is accomplished using particular wavelengths of solar radiation that pass through our earth's atmosphere and reach the surface. About 45 percent of the radiation at the surface is visible, another 45 percent is infrared, and 10 percent is ultraviolet. Photosynthesis uses mainly the most abundant wavelengths, visible and near infrared. Of light wavelengths, photosynthesis uses mainly red and blue light. Most plants reflect green wavelengths, so that is the color they appear to us. Half of the energy plants absorb is used in evaporating water. In the end, only 1 to 2 percent of the sunlight falling on plants is available for photosynthesis. This small percentage represents the energy base for virtually all life in the biosphere.

Photosynthesis captures energy; respiration releases that energy

Photosynthesis occurs in tiny organelles called chloroplasts that reside within plant cells (see fig. 3.7). The most important key to this process is chlorophyll, a green molecule that can absorb light energy and use the energy to create high-energy chemical in compounds that serve as the fuel for all subsequent cellular metabolism. Chlorophyll doesn't do this important job all alone, however. It is assisted by a large group of other lipid, sugar, protein, and nucleotide molecules. Together these components carry out two interconnected cyclic sets of reactions (fig. 3.11).

Photosynthesis begins with a series of *light-dependent reactions.* These use solar energy directly to split water molecules into oxygen (O_2), which is released to the atmosphere, and hydrogen (H). This is the source of all the oxygen in the atmosphere on which all animals, including you, depend for life. Separating the hydrogen atom from its electron produces H^+ and an electron, both of which are used to form mobile, high-energy molecules called adenosene triphosphate (ATP) and nicotinamide adenine dinucleotide phosphate (NADPH).

Light-independent reactions then use the energy stored in ATP and NADPH molecules to create simple carbohydrates and sugar molecules (glucose, $C_6H_{12}O_6$) from carbon atoms (from CO_2) and water (H_2O). Glucose provides the energy and the building blocks for larger, more complex organic molecules. As ATP and NADPH give up some of their chemical energy, they are transformed to

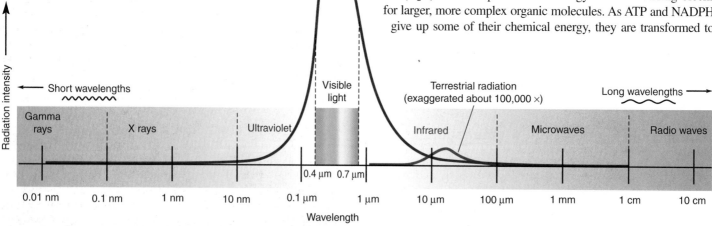

FIGURE 3.10 The electromagnetic spectrum. Our eyes are sensitive to light wavelengths, which make up nearly half the energy that reaches the earth's surface (represented by the area under the curve). Photosynthesizing plants also use the most abundant solar wavelengths. The earth reemits lower-energy, longer wavelengths, mainly the infrared part of the spectrum.

adensosene *di*phosphate (ADP) and NADP. These molecules are then reused in another round of light-dependent reactions. In most temperate-zone plants, photosynthesis can be summarized in the following equation:

$$6H_2O + 6CO_2 + \text{solar energy} \xrightarrow[\text{chlorophyll}]{} C_6H_{12}O_6 \text{ (sugar)} + 6O_2$$

We read this equation as "water plus carbon dioxide plus energy produces sugar plus oxygen." The reason the equation uses six water and six carbon dioxide molecules is that it takes six carbon atoms to make the sugar product.

Note that the CO_2 in the equation above is captured from the air by plant tissues. This means that much of the mass of a plant is made of air, and the rest is largely water. Since you derive carbon from plants you eat, or animals that eat plants, you could say that you are made largely from air, too.

What does the plant do with glucose? Because glucose is an energy-rich compound, it serves as the central, primary fuel for all metabolic processes of cells. The energy in its chemical bonds—created by photosynthesis—can be released by other enzymes and used to make other molecules (lipids, proteins, nucleic acids, or other carbohydrates), or it can drive kinetic processes such as movement of ions across membranes, transmission of messages, changes in cellular shape or structure, or movement of the cell itself in some cases.

This process of releasing chemical energy, called **cellular respiration,** involves splitting carbon and hydrogen atoms from the sugar molecule and recombining them with oxygen to recreate carbon dioxide and water. The net chemical reaction, then, is the reverse of photosynthesis:

$$C_6H_{12}O_6 + 6O_2 \longrightarrow 6H_2O + 6CO_2 + \text{released energy}$$

Note that in photosynthesis, energy is *captured,* while in respiration, energy is *released.* Similarly, photosynthesis *consumes* water and carbon dioxide to *produce* sugar and oxygen, while respiration does just the opposite. In both sets of reactions, chemical bonds are used to capture, store, and deliver energy within a cell. Plants carry out both photosynthesis

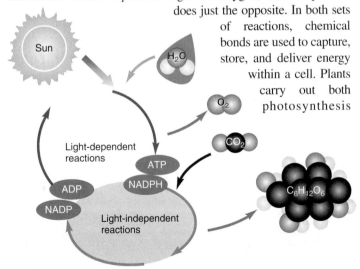

FIGURE 3.11 Photosynthesis involves a series of reactions in which chlorophyll captures light energy and forms high-energy molecules, ATP and NADPH. Light-independent reactions then use energy from ATP and NADPH (converting them to ADP and NADP) to fix carbon from air in organic molecules.

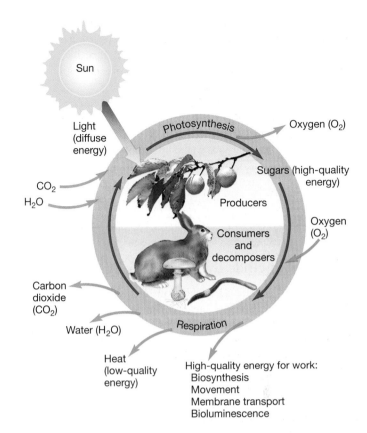

FIGURE 3.12 Energy exchange in ecosystems. Plants use sunlight, water, and carbon dioxide to produce sugars and other organic molecules. Consumers use oxygen and break down sugars during cellular respiration. Plants also carry out respiration, but during the day, if light, water, and CO_2 are available, they have a net production of O_2 and carbohydrates.

and respiration, but during the day, if light, water, and CO_2 are available, they have a net production of O_2 and carbohydrates.

We animals don't have chlorophyll and can't carry out photosynthetic food production. We do perform cellular respiration, however. In fact, this is how we get all our energy for life. We eat plants—or other animals that have eaten plants—and break down the organic molecules in our food through cellular respiration to obtain energy (fig. 3.12). In the process, we also consume oxygen and release carbon dioxide, thus completing the cycle of photosynthesis and respiration.

3.4 FROM SPECIES TO ECOSYSTEMS

When we discuss Chesapeake Bay as a complex system (opening case study) we are concerned with rates of photosynthesis, abundance of photosynthesizing algae, and the ways that changes to the bay's chemistry influence population sizes for different species. Numbers of blue crabs, oysters, menhaden, and other species all contribute to our assessment of the system's stability and health.

Terms like species, population, and community are probably familiar to you, but biologists have particular meanings for these terms. In Latin, *species* literally means *kind*. In biology, **species** generally refers to all organisms of the same kind that are genetically similar enough to breed in nature and produce live, fertile offspring. There are important exceptions to this definition, and increasingly taxonomists rely on genetic differences to define species, but for our purposes this is a useful working definition.

A **population** consists of all the members of a species living in a given area at the same time. All of the populations living and interacting in a particular area make up a **biological community.** What populations make up the biological community of which you are a part? If you consider all the populations of animals, plants, fungi, and microorganisms in your area, your community is probably large and complex. We'll explore the dynamics of populations and communities more in chapters 4 and 6.

Ecosystems include living and nonliving parts

As discussed in chapter 2, systems are networks of interaction among many interdependent factors. Your body, for example, is a very complex, self-regulating system. An ecological system, or **ecosystem,** is composed of a biological community and its physical environment. The environment includes abiotic factors (nonliving components), such as climate, water, minerals, and sunlight, as well as biotic factors, such as organisms, their products (secretions, wastes, and remains), and effects in a given area.

It is useful to think about the biological community and its environment together, because energy and matter flow through both. Understanding how those flows work is a major theme in ecology.

For simplicity, we think of ecosystems as distinct ecological units with fairly clear boundaries. If you look at a patch of woods surrounded by farm fields, for instance, a relatively sharp line might separate the two areas, and conditions such as light levels, wind, moisture, and shelter are quite different in the woods than in the fields around them. Because of these variations, distinct populations of plants and animals live in each place. By studying each of these areas, we can make important and interesting discoveries about who lives where and why and about how conditions are established and maintained there.

The division between the fields and woods is not always clear, however. Air, of course, moves freely from one to another, and the runoff after a rainfall may carry soil, leaf litter, and live organisms between the areas. Birds may feed in the field during the day but roost in the woods at night, giving them roles in both places. Are they members of the woodland community or the field community? Is the edge of the woodland ecosystem where the last tree grows, or does it extend to every place that has an influence on the woods?

As you can see, it may be difficult to draw clear boundaries around communities and ecosystems. To some extent we define these units by what we want to study and how much information we can handle. Thus, an ecosystem might be as large as a whole watershed or as small as a pond or even your own body. The thousands of species of bacteria, fungi, protozoans, and other organisms that live in and on your body make up a complex, interdependent community. You keep the other species warm and fed;

they help you with digestion, nutrition, and other bodily functions. Some members of your community are harmful, but many are beneficial. You couldn't survive easily without them. Interestingly, of the several trillion individual cells that make up your body, only about 10 percent are mammalian. That means that a vast majority (in numerical terms) of cells that make up the ecosystem that is you are nonmammalian.

You, as an ecosystem, have clear boundaries, but you are open in the sense that you take in food, water, energy, and oxygen from your surrounding environment, and you excrete wastes. This is true of most ecosystems, but some are relatively closed: that is, they import and export comparatively little from outside. Others, such as a stream, are in a constant state of flux with materials and even whole organisms coming and going. Because of the second law of thermodynamics, however, every ecosystem must have a constant inflow of energy and a way to dispose of heat. Thus, with regard to energy flow, every ecosystem is open.

Many ecosystems have feedback mechanisms that maintain generally stable structure and functions. A forest tends to remain a forest, for the most part, and to have forest-like conditions if it isn't disturbed by outside forces. Some ecologists suggest that ecosystems—or perhaps all life on the earth—may function as superorganisms because they maintain stable conditions and can be resilient to change.

Food webs link species of different trophic levels

Photosynthesis (and rarely chemosynthesis) is the base of all ecosystems. Organisms that photosynthesize, mainly green plants and algae, are therefore known as **producers.** One of the major properties of an ecosystem is its **productivity,** the amount of **biomass** (biological material) produced in a given area during a given period of time. Photosynthesis is described as *primary productivity* because it is the basis for almost all other growth in an ecosystem. Manufacture of biomass by organisms that eat plants is termed *secondary productivity.* A given ecosystem may have very high total productivity, but if decomposers decompose organic material as rapidly as it is formed, the *net primary productivity* will be low.

Think about what you have eaten today and trace it back to its photosynthetic source. If you have eaten an egg, you can trace it back to a chicken, which probably ate corn. This is an example of a **food chain,** a linked feeding series. Now think about a more complex food chain involving you, a chicken, a corn plant, and a grasshopper. The chicken could eat grasshoppers that had eaten leaves of the corn plant. You also could eat the grasshopper directly—some humans do. Or you could eat corn yourself, making the shortest possible food chain. Humans have several options of where we fit into food chains.

In ecosystems, some consumers feed on a single species, but most consumers have multiple food sources. Similarly, some species are prey to a single kind of predator, but many species in an ecosystem are beset by several types of predators and parasites. In this way, individual food chains become interconnected to form a **food web.** Figure 3.13 shows feeding relationships among some of the larger organisms in a woodland and lake community. If we were to add all the insects, worms, and microscopic organisms

FIGURE 3.13 Each time an organism feeds, it becomes a link in a food chain. In an ecosystem, food chains become interconnected when predators feed on more than one kind of prey, thus forming a food web. The arrows in this diagram and in figure 3.14 indicate the direction in which matter and energy are transferred through feeding relationships. Only a few representative relationships are shown here. What others might you add?

that belong in this picture, however, we would have overwhelming complexity. Perhaps you can imagine the challenge ecologists face in trying to quantify and interpret the precise matter and energy transfers that occur in a natural ecosystem!

An organism's feeding status in an ecosystem can be expressed as its **trophic level** (from the Greek *trophe,* food). In our first example, the corn plant is at the producer level; it transforms solar energy into chemical energy, producing food molecules. Other organisms in the ecosystem are **consumers** of the chemical energy harnessed by the producers. An organism that eats producers is a primary consumer. An organism that eats primary consumers is a secondary consumer, which may, in turn, be eaten by a tertiary consumer, and so on. Most terrestrial food chains are relatively short (seeds → mouse → owl), but aquatic food chains may be quite long (microscopic algae → copepod → minnow → crayfish → bass → osprey). The length of a food chain also may reflect the physical characteristics of a particular ecosystem. A harsh arctic landscape, with relatively low species diversity, can have a much shorter food chain than a temperate or tropical one (fig. 3.14).

Think About It

What would have been the leading primary producers and top consumers in the native ecosystem where you now live? What are they now? Are fewer trophic levels now represented in your ecosystem than in the past?

Organisms can be identified both by the trophic level at which they feed and by the *kinds* of food they eat (fig. 3.15). **Herbivores** are plant eaters, **carnivores** are flesh eaters, and **omnivores** eat both plant and animal matter. What are humans? We are natural omnivores, by history and by habit. Tooth structure is an important clue to understanding animal food preferences, and humans are no exception. Our teeth are suited for an omnivorous diet, with a combination of cutting and crushing surfaces that are not highly adapted for one specific kind of food, as are the teeth of a wolf (carnivore) or a horse (herbivore).

One of the most important trophic levels is occupied by the many kinds of organisms that remove and recycle the dead bodies and waste products of others. **Scavengers** such as crows, jackals, and vultures clean up dead carcasses of larger animals. **Detritivores** such as ants and beetles consume litter, debris, and dung, while **decomposer** organisms such as fungi and bacteria complete the final breakdown and recycling of organic materials. It could be argued that these microorganisms are second in importance only to producers, because without their activity nutrients would remain locked up in the organic compounds of dead organisms and discarded body wastes, rather than being made available to successive generations of organisms.

Ecological pyramids describe trophic levels

If we arrange the organisms according to trophic levels, they generally form a pyramid with a broad base representing primary producers and only a few individuals in the highest trophic levels. This

FIGURE 3.14 Harsh environments tend to have shorter food chains than environments with more favorable physical conditions. Compare the arctic food chains depicted here with the longer food chains in the food web in figure 3.13.

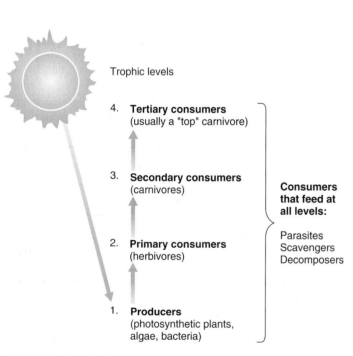

FIGURE 3.15 Organisms in an ecosystem may be identified by how they obtain food for their life processes (producer, herbivore, carnivore, omnivore, scavenger, decomposer, reducer) or by consumer level (producer; primary, secondary, or tertiary consumer) or by trophic level (1st, 2nd, 3rd, 4th).

Trophic levels

4. **Tertiary consumers**
   (usually a "top" carnivore)

3. **Secondary consumers**
   (carnivores)

**Consumers
that feed at
all levels:**

Parasites
Scavengers
Decomposers

2. **Primary consumers**
   (herbivores)

1. **Producers**
   (photosynthetic plants,
   algae, bacteria)

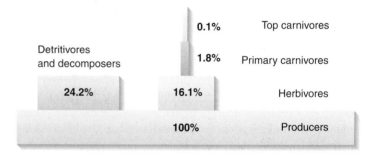

	0.1%	Top carnivores
Detritivores and decomposers	1.8%	Primary carnivores
24.2%	16.1%	Herbivores
100%		Producers

FIGURE 3.16 A classic example of an energy pyramid from Silver Springs, Florida. The numbers in each bar show the percentage of the energy captured in the primary producer level that is incorporated into the biomass of each succeeding level. Detritivores and decomposers feed at every level but are shown attached to the producer bar because this level provides most of their energy.

pyramid arrangement is especially true if we look at the energy content of an ecosystem (fig. 3.16).

Why is there so much less energy in each successive level in figure 3.16? Because of the second law of thermodynamics, which says that energy dissipates and degrades as it is reused. Thus a rabbit consumes a great deal of chemical energy stored in carbohydrates in grass, and much of that energy is transformed to kinetic energy, when the rabbit moves, or to heat, which dissipates to the environment. A fox eats the rabbit, and the same degradation and dissipation happen again. From the fox's point of view,

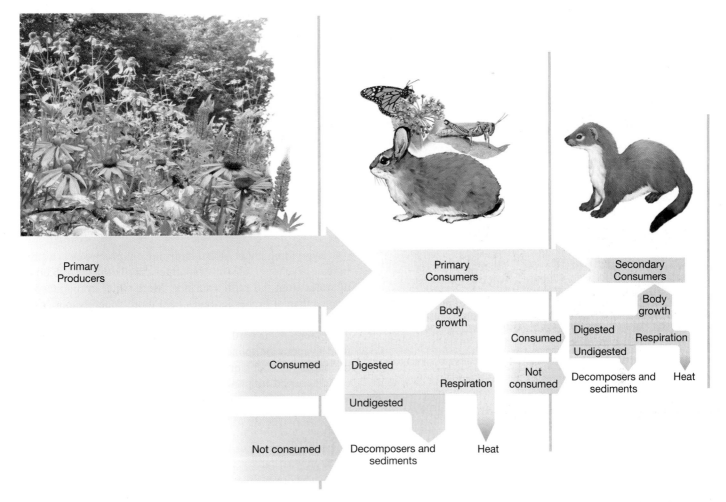

FIGURE 3.17 A biomass pyramid. Like energy, biomass decreases at higher levels. Arrows show how biomass is used and lost.

the lost energy is used in the process of living and growing, and a little of the energy it has eaten is stored in the fox's tissues. From an ecosystem energy perspective, there will always be smaller amounts of energy at successively higher trophic levels. Large top carnivores need a very large pyramid, and a large home range, to support them. A tiger, for example, may require a home range of several hundred square kilometers to survive.

A general rule of thumb is that only about 10 percent of the energy in one consumer level is represented in the next higher level (fig. 3.17). The amount of energy available is often expressed in biomass. For example, it generally takes about 100 kg of clover to make 10 kg of rabbit and 10 kg of rabbit to make 1 kg of fox.

The total number of organisms and the total amount of biomass in each successive trophic level of an ecosystem also may form pyramids (fig. 3.18) similar to those describing energy content. The relationship between biomass and numbers is not as dependable as energy, however. The biomass pyramid, for instance, can be inverted by periodic fluctuations in producer populations (for example, low plant and algal biomass present during winter in temperate aquatic ecosystems). The numbers pyramid also can be inverted. One coyote can support numerous

tapeworms, for example. Numbers inversion also occurs at the lower trophic levels (for example, one large tree can support thousands of caterpillars).

3.5 MATERIAL CYCLES AND LIFE PROCESSES

Earth is the only planet in our solar system that provides a suitable environment for life as we know it. Even our nearest planetary neighbors, Mars and Venus, do not meet these requirements. Maintenance of these conditions requires a constant recycling of materials between the biotic (living) and abiotic (nonliving) components of ecosystems.

The hydrologic cycle redistributes water

The path of water through our environment, known as the **hydrologic cycle,** is perhaps the most familiar material cycle, and it is discussed in greater detail in chapter 17. Most of the earth's water is stored in the oceans, but solar energy continually evaporates

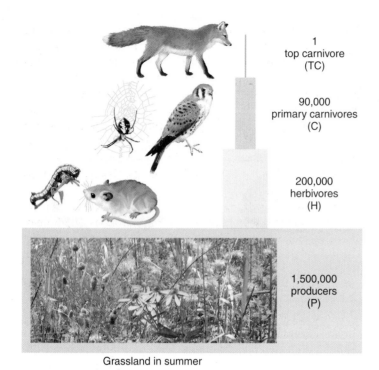

1
top carnivore
(TC)

90,000
primary carnivores
(C)

200,000
herbivores
(H)

1,500,000
producers
(P)

Grassland in summer

FIGURE 3.18 Usually, smaller organisms are eaten by larger organisms and it takes numerous small organisms to feed one large organism. The classic study represented in this pyramid shows numbers of individuals at each trophic level per 1,000 m² of grassland, and reads like this: to support one individual at the top carnivore level, there were 90,000 primary carnivores feeding upon 200,000 herbivores that in turn fed upon 1,500,000 producers.

this water, and winds distribute water vapor around the globe. Water that condenses over land surfaces, in the form of rain, snow, or fog, supports all terrestrial (land-based) ecosystems (fig. 3.19). Living organisms emit the moisture they have consumed through respiration and perspiration. Eventually this moisture reenters the atmosphere or enters lakes and streams, from which it ultimately returns to the ocean again.

As it moves through living things and through the atmosphere, water is responsible for metabolic processes within cells, for maintaining the flows of key nutrients through ecosystems, and for global-scale distribution of heat and energy (chapter 15). Water performs countless services because of its unusual properties. Water is so important that when astronomers look for signs of life on distant planets, traces of water are the key evidence they seek.

Everything about global hydrological processes is awesome in scale. Each year, the sun evaporates approximately 496,000 km³ of water from the earth's surface. More water evaporates in the tropics than at higher latitudes, and more water evaporates over the oceans than over land. Although the oceans cover about 70 percent of the earth's surface, they account for 86 percent of total evaporation. Ninety percent of the water evaporated from the ocean falls back on the ocean as rain. The remaining 10 percent is carried by prevailing winds over the continents where it combines with water evaporated from soil, plant surfaces, lakes, streams, and wetlands to provide a total continental precipitation of about 111,000 km³.

What happens to the surplus water on land—the difference between what falls as precipitation and what evaporates? Some of it is incorporated by plants and animals into biological tissues. A large share of what falls on land seeps into the ground to be stored for a while (from a few days to many thousands of years) as soil moisture or groundwater. Eventually, all the water makes its way back downhill to the oceans. The 40,000 km³ carried back to the ocean each year by surface runoff or underground flow represents the renewable supply available for human uses and sustaining freshwater-dependent ecosystems.

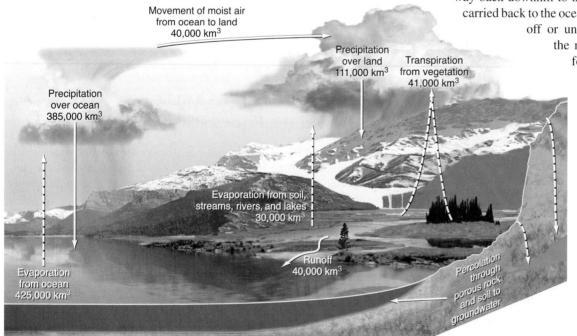

Movement of moist air
from ocean to land
40,000 km³

Precipitation
over land
111,000 km³

Transpiration
from vegetation
41,000 km³

Precipitation
over ocean
385,000 km³

Evaporation from soil,
streams, rivers, and lakes
30,000 km³

Evaporation
from ocean
425,000 km³

Runoff
40,000 km³

Percolation
through
porous rock
and soil to
groundwater

FIGURE 3.19 The hydrologic cycle. Most exchange occurs with evaporation from oceans and precipitation back to oceans. About one-tenth of water evaporated from oceans falls over land, is recycled through terrestrial systems, and eventually drains back to oceans in rivers.

Carbon moves through the carbon cycle

Carbon serves a dual purpose for organisms: (1) it is a structural component of organic molecules, and (2) the energy-holding chemical bonds it forms represent energy "storage." The **carbon cycle** begins with the intake of carbon dioxide (CO_2) by photosynthetic organisms (fig. 3.20). Carbon (and hydrogen and oxygen) atoms are incorporated into sugar molecules during photosynthesis. Carbon dioxide is eventually released during respiration, closing the cycle. The carbon cycle is of special interest because biological accumulation and release of carbon is a major factor in climate regulation (Exploring Science, p. 68).

The path followed by an individual carbon atom in this cycle may be quite direct and rapid, depending on how it is used in an organism's body. Imagine for a moment what happens to a simple sugar molecule you swallow in a glass of fruit juice. The sugar molecule is absorbed into your bloodstream where it is made available to your cells for cellular respiration or for making more complex biomolecules. If it is used in respiration, you may exhale the same carbon atom as CO_2 the same day.

Can you think of examples where carbon may not be recycled for even longer periods of time, if ever? Coal and oil are the compressed, chemically altered remains of plants or microorganisms that lived millions of years ago. Their carbon atoms (and hydrogen, oxygen, nitrogen, sulfur, etc.) are not released until the coal and oil are burned. Enormous amounts of carbon also are locked up as calcium carbonate ($CaCO_3$), used to build shells and skeletons of marine organisms from tiny protozoans to corals. Most of these deposits are at the bottom of the oceans. The world's extensive surface limestone deposits are biologically formed calcium carbonate from ancient oceans, exposed by geological events. The carbon in limestone has been locked away for millennia, which is probably the fate of carbon currently being deposited in ocean sediments. Eventually, even the deep ocean deposits are recycled as they are drawn into deep molten layers and released via volcanic activity. Geologists estimate that every carbon atom on the earth has made about thirty such round trips over the last 4 billion years.

How does tying up so much carbon in the bodies and by-products of organisms affect the biosphere? Favorably. It helps balance CO_2 generation and utilization. Carbon dioxide is one of the so-called greenhouse gases because it absorbs heat radiated from the earth's surface, retaining it instead in the atmosphere (see chapter 15). Photosynthesis, accumulation of organic matter in soils and wetlands, and deposition of $CaCO_3$ remove atmospheric carbon dioxide; therefore, expansive forested areas such as the boreal forests, and the oceans are very important **carbon sinks** (storage deposits). Cellular respiration and combustion both release CO_2, so they are referred to as carbon sources in the cycle.

Presently, combustion of organic fuels (mainly wood, coal, and petroleum products), removal of standing forests, and soil degradation are releasing huge quantities of CO_2 at rates that surpass the pace of CO_2 removal, a problem discussed in chapters 15 and 16.

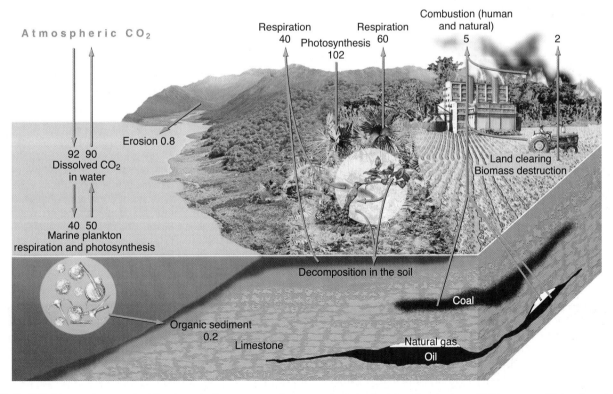

FIGURE 3.20 The carbon cycle. Numbers indicate approximate exchange of carbon in gigatons (Gt) per year. Natural exchanges are balanced, but human sources produce a net increase of CO_2 in the atmosphere.

In Chesapeake Bay, primary productivity is measured using water samples. This method gives precise, accurate information, but it's too labor intensive for larger bodies of water. What if you wanted to know about algal blooms, or biological productivity in all the world's estuaries? Measuring primary productivity essential for understanding ecosystem health; knowing rates of primary productivity is also key to understanding global material questions about material cycles and biological activity:

- Where are oceans affected by nutrient-enriched algal blooms?

- Globally, how much carbon is stored by plants? How does carbon capture differ from the Arctic to the tropics? How does this affect global climates (chapter 15)?

- In global nutrient cycles, how much nitrogen and phosphorus wash offshore, and where?

One of the most important methods of quantifying biological productivity involves remote sensing, or data collected from satellite sensors that observe the energy reflected from the earth's surface.

Green plants appear green to us because chlorophyll *absorbs* red and blue wavelengths better than green, which it *reflects* more. Your eye detects these green wavelengths. Green plants also reflect near-infrared wavelengths, which your eye cannot detect (see fig. 3.10). A white-sand beach, on the other hand, reflects large amounts of all light wavelengths that reach it from the sun, so it looks white (and bright) to your eye. Most surfaces of the earth reflect characteristic wavelengths in this way. Dark-green forests with abundant chlorophyll-rich leaves—and ocean surfaces rich in photosynthetic algae and plants—reflect greens and near-infrared wavelengths. Dry, brown forests with little active chlorophyll reflect more red

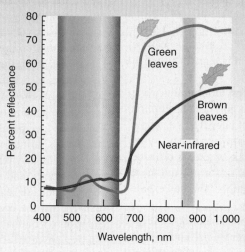

FIGURE 1 Energy wavelengths reflected by green and brown leaves.

and less infrared energy than do dark-green forests (fig. 1).

To detect land-cover patterns on the earth's surface, we can put a sensor on a satellite that orbits the earth. As the satellite travels, the sensor takes "snapshots" and transmits them to earth. One of the best known earth-imaging satellites, *Landsat 7,* produces images that cover an area 185 km (115 mi) wide, and each pixel represents an area of just 30 × 30 m on the ground. *Landsat* orbits approximately from pole to pole, so as the earth spins below the satellite, it captures images of the entire surface every 16 days. Another satellite, *SeaWIFS,* was designed mainly for monitoring biological activity in oceans (fig. 2). *SeaWIFS* follows a path similar to *Landsat's* but it revisits each point on the earth every day and produces images with a pixel resolution of just over 1 km.

Since satellites detect a much greater range of wavelengths than our eyes can, they are able to monitor and map chlorophyll abundance. In

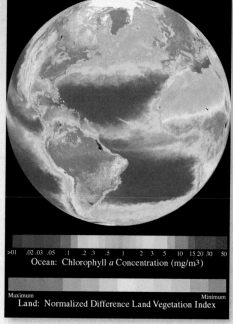

FIGURE 2 *SeaWIFS* image showing chlorophyll abundance in oceans and plant growth on land (normalized difference vegetation index).

oceans and bays, this is an essential indicator of ecosystem health. Primary productivity is also a measure of carbon dioxide uptake. Climatologists are working to estimate the role of ocean ecosystems in moderating climate change: for example, they can estimate the extent of biomass production in the cold, oxygen-rich waters of the North Atlantic (fig. 2). Oceanographers can also detect near-shore areas where nutrients washing off the land surface fertilize marine ecosystems and stimulate high productivity, such as near the mouth of the Amazon or Mississippi Rivers. Monitoring and mapping these patterns helps us estimate human impacts on nutrient flows (figs. 3.21, 3.23) from land to sea.

Nitrogen is not always biologically available

As the opening case study of this chapter shows, nitrogen often is one of the most important limiting factors in ecosystems. The complex interrelationships through which organisms exchange this vital element help shape these biological communities. Organisms cannot exist without amino acids, peptides, nucleic acids, and proteins, all of which are organic molecules containing nitrogen. The nitrogen atoms that form these important molecules are provided by producer organisms. Plants assimilate (take up) inorganic nitrogen from the environment and use it to build their own protein molecules, which are eaten by consumer organisms, digested, and used to build their bodies. However, the most abundant form of nitrogen, N_2 gas (which makes up about 78 percent of the atmosphere), is too stable to be broken up and used by plants.

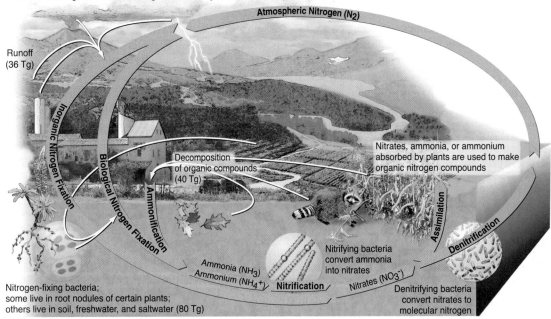

FIGURE 3.21 The nitrogen cycle. Human sources of nitrogen fixation (conversion of molecular nitrogen to ammonia or ammonium) are now about 50 percent greater than natural sources. Bacteria convert ammonia to nitrates, which plants use to create organic nitrogen. Eventually, nitrogen is stored in sediments or converted back to molecular nitrogen (1 Tg = 10^{12} g).

Lightning and volcanoes (10 Tg)
Fossil fuel burning, commercial and agricultural nitrogen fixation (140 Tg)
Atmospheric Nitrogen (N_2)
Runoff (36 Tg)
Inorganic Nitrogen Fixation
Biological Nitrogen Fixation
Ammonification
Decomposition of organic compounds (40 Tg)
Nitrates, ammonia, or ammonium absorbed by plants are used to make organic nitrogen compounds
Assimilation
Denitrification
Ammonia (NH_3)
Ammonium (NH_4^+)
Nitrifying bacteria convert ammonia into nitrates
Nitrification
Nitrates (NO_3^-)
Denitrifying bacteria convert nitrates to molecular nitrogen
Nitrogen-fixing bacteria; some live in root nodules of certain plants; others live in soil, freshwater, and saltwater (80 Tg)

How, then, do green plants get nitrogen? The answer lies in the most complex of the gaseous cycles, the **nitrogen cycle**. Figure 3.21 summarizes the nitrogen cycle. The key natural processes that make nitrogen available are carried out by nitrogen-fixing bacteria (including some blue-green algae or cyanobacteria). These organisms have a highly specialized ability to "fix" nitrogen, meaning they change it to less mobile, more useful forms by combining it with hydrogen to make ammonia (NH_3).

Other bacteria combine the NH_3 with oxygen, forming nitrite (NO_2^-), then nitrate (NO_3^-), which can be absorbed and used by green plants. After nitrates have been absorbed into plant cells, they are reduced to ammonium (NH_4^+), which is used to build amino acids that become the building blocks for peptides and proteins.

Members of the bean family (legumes) and a few other kinds of plants are especially useful in agriculture because they have nitrogen-fixing bacteria actually living *in* their root tissues (fig. 3.22). Legumes and their associated bacteria enrich the soil, so interplanting and rotating legumes with crops such as corn that use but cannot replace soil nitrates are beneficial farming practices that take practical advantage of this relationship.

Nitrogen leaves an organism and reenters the environment in several ways. The most obvious path is through the death of organisms. Their bodies are decomposed by fungi and bacteria, releasing ammonia and ammonium ions, which then are available for nitrate formation. Organisms also release proteins when plants shed their leaves, needles, flowers, fruits, and cones; or when animals shed hair, feathers, skin, exoskeletons, pupal cases, and silk, excrement, or urine, all of which are rich in nitrogen. Urinary wastes are especially high in nitrogen because they contain the detoxified wastes of protein metabolism. All of these by-products of living organisms decompose, replenishing soil fertility.

In oxygen-poor conditions, denitrifying bacteria may convert nitrate (NO_3^-) into N_2 and nitrous oxide (N_2O), both gaseous forms that return to the atmosphere. Denitrification occurs mainly in waterlogged soils that have low oxygen availability and a high amount of decomposable organic matter. Because wetlands lose so much nitrogen to the atmosphere, carnivorous plants often occur in wetlands. These plants acquire nitrogen by capturing and decomposing insects in their leaves.

FIGURE 3.22 The roots of this bean plant are covered with bumps called nodules. Each nodule is a mass of root tissue containing many bacteria that help to convert nitrogen in the soil to a form the bean plants can assimilate and use to manufacture amino acids.

In recent years, humans have profoundly altered the nitrogen cycle. By using synthetic fertilizers, cultivating nitrogen-fixing soybeans and other crops, and burning fossil fuels, we have more than doubled the amount of nitrogen cycled through our global environment. As you are aware, this excess nitrogen input destabilizes rivers, lakes, and estuaries. In terrestrial systems, nitrogen enrichment encourages the spread of weeds into areas such as prairies, where native plants adapted to nitrogen-poor environments compete poorly against quick-responding weeds. In addition, N_2O is an important greenhouse gas.

Phosphorus is an essential nutrient

Minerals become available to organisms after they are released from rocks. Two minerals of particular significance to organisms are phosphorus and sulfur. Phosphorus is a primary ingredient in fertilizers. Why? At the cellular level, energy-rich, phosphorus-containing compounds, such as ATP, are primary participants in energy-transfer reactions. Phosphorus is also a key component of proteins, enzymes, and tissues. The amount of available phosphorus in an environment can, therefore, have a dramatic effect on productivity. Abundant phosphorus stimulates lush plant and algal growth, making it a major contributor to water pollution.

The **phosphorus cycle** (fig. 3.23) is not really a cycle on the time scale of the other cycles discussed here, because phosphorus has no atmospheric form. Instead, phosphorus travels gradually downstream, as it is leached from rocks and minerals, taken up by

the food web, and eventually released into water bodies that deliver it to the ocean. Phosphorus may cycle repeatedly through the food web, as inorganic phosphorus is taken up by primary producers (plants), incorporated into organic molecules, and then passed on to consumers. Eventually, phosphorus washes down river to the ocean. Deep sediments of the oceans are significant phosphorus sinks of extreme longevity. Over geologic time, these deposits may be uplifted into mountains or continents, where they become available to terrestrial life again. Phosphate ores that now are mined to make detergents and inorganic fertilizers represent exposed ocean sediments that are millions of years old. As with nitrogen, we have dramatically accelerated the movement of phosphorus in our environment. Aquatic ecosystems often are dramatically affected, as excess phosphates stimulate explosive growth of algae and photosynthetic bacteria populations, upsetting ecosystem stability.

Sulfur is both a nutrient and an acidic pollutant

Sulfur is a minor but essential component of proteins, so it is important to living organisms. Sulfur compounds are important determinants of the acidity of rainfall, surface water, and soil. Most of the earth's sulfur is tied up underground in rocks and minerals such as iron disulfide (pyrite) or calcium sulfate (gypsum). This inorganic sulfur is released into air and water by weathering, emissions from deep seafloor vents, and by volcanic eruptions (fig. 3.24).

The **sulfur cycle** is complicated by the large number of oxidation states the element can assume, including hydrogen sulfide (H_2S), sulfur dioxide (SO_2), sulfate ion (SO_4^{-2}), and sulfur, among others. Inorganic processes are responsible for many of these transformations, but living organisms, especially bacteria, also sequester sulfur in biogenic deposits or release it into the environment. Which of the several kinds of sulfur bacteria prevail in any given situation depends on oxygen concentrations, pH, and light levels.

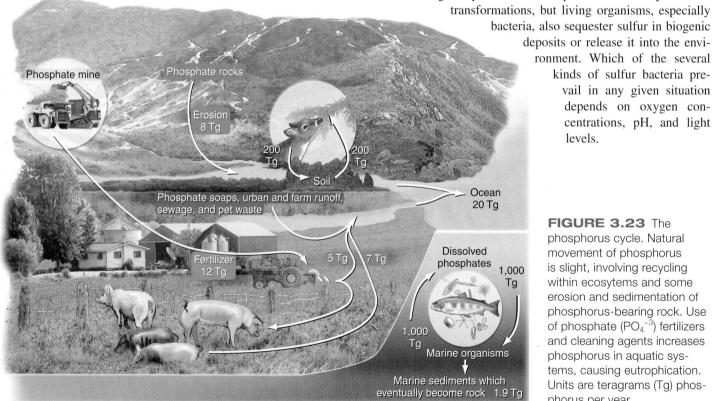

Phosphate mine

Phosphate rocks

Erosion
8 Tg

200 Tg

200 Tg

Soil

Phosphate soaps, urban and farm runoff, sewage, and pet waste

Ocean
20 Tg

Fertilizer
12 Tg

5 Tg   7 Tg

Dissolved phosphates

1,000 Tg

1,000 Tg

Marine organisms

Marine sediments which eventually become rock   1.9 Tg

FIGURE 3.23 The phosphorus cycle. Natural movement of phosphorus is slight, involving recycling within ecosytems and some erosion and sedimentation of phosphorus-bearing rock. Use of phosphate (PO_4^{-3}) fertilizers and cleaning agents increases phosphorus in aquatic systems, causing eutrophication. Units are teragrams (Tg) phosphorus per year.

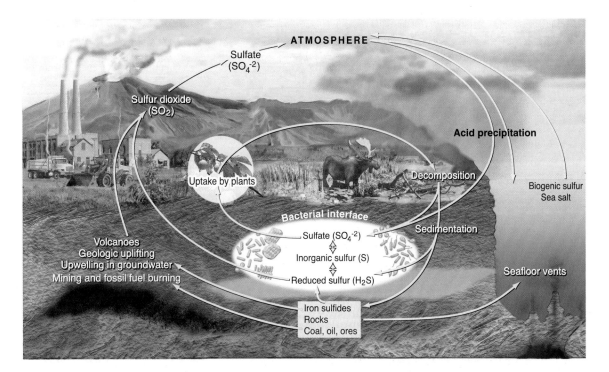

FIGURE 3.24 The sulfur cycle. Sulfur is present mainly in rocks, soil, and water. It cycles through ecosystems when it is taken in by organisms. Combustion of fossil fuels causes increased levels of atmospheric sulfur compounds, which create problems related to acid precipitation.

Human activities also release large quantities of sulfur, primarily through burning fossil fuels. Total yearly anthropogenic sulfur emissions rival those of natural processes, and acid rain caused by sulfuric acid produced as a result of fossil fuel use is a serious problem in many areas (see chapter 16). Sulfur dioxide and sulfate aerosols cause human health problems, damage buildings and vegetation, and reduce visibility. They also absorb UV radiation and create cloud cover that cools cities and may be offsetting greenhouse effects of rising CO_2 concentrations.

CONCLUSION

Matter is conserved as it cycles over and over through ecosystems, but energy is always degraded or dissipated as it is transformed or transferred from one place to another. These laws of physics and thermodynamics mean that elements are continuously recycled, but that living systems need a constant supply of external energy to replace that lost to entropy. Some extremophiles, living in harsh conditions, such as hot springs or the bottom of the ocean, capture energy from chemical reactions. For most organisms, however, the ultimate source of energy is the sun. Plants capture sunlight through the process of photosynthesis, and use the captured energy for metabolic processes and to build biomass (organic material). Herbivores eat plants to obtain energy and nutrients, carnivores eat herbivores or each other, and decomposers eat the waste products of this food web.

This dependence on solar energy is a fundamental limit for most life on earth. It's estimated that humans now dominate roughly 40 percent of the potential terrestrial net productivity. We directly eat only about 10 percent of that total (mainly because of the thermodynamic limits on energy transfers in food webs), but the crops and livestock that feed, clothe, and house us represent the rest of that photosynthetic output. By dominating nature, as we do, we exclude other species.

While energy flows in a complex, but ultimately one-way path through nature, materials are endlessly recycled. Five of the major material cycles (water, carbon, nitrogen, phosphorus, and sulfur) are summarized in this chapter. Each of these materials is critically important to living organisms. As humans interfere with these material cycles, we make it easier for some organisms to survive and more difficult for others. Often, we're intent on manipulating material cycles for our own short-term gain, but we don't think about the consequences for other species or even for ourselves in the long-term. An example of that is the carbon cycle. Our lives are made easier and more comfortable by burning fossil fuels, but in doing so we release carbon dioxide into the atmosphere, causing global warming that could have disastrous results. Clearly, it's important to understand these environmental systems and to take them into account in our public policy.

REVIEWING LEARNING OUTCOMES

By now you should be able to explain the following points:

3.1 Describe matter, atoms, and molecules and give simple examples of the role of four major kinds of organic compounds in living cells.

- Matter is made of atoms, molecules, and compounds.
- Chemical bonds hold molecules together.
- Ions react and bond to form compounds.
- Organic compounds have a carbon backbone.
- Cells are the fundamental units of life.

3.2 Define energy and explain how thermodynamics regulates ecosystems.

- Energy occurs in many forms.
- Thermodynamics regulates energy transfers.

3.3 Understand how living organisms capture energy and create organic compounds.

- Extremophiles gain energy without sunlight.

- Green plants get energy from the sun.
- Photosynthesis captures energy; respiration releases that energy.

3.4 Define species, populations, communities, and ecosystems, and summarize the ecological significance of trophic levels.

- Ecosystems include living and nonliving parts.
- Food webs link species of different trophic levels.
- Ecological pyramids describe trophic levels.

3.5 Compare the ways that water, carbon, nitrogen, sulfur, and phosphorus cycle within ecosystems.

- The hydrologic cycle redistributes water.
- Carbon moves through the carbon cycle.
- Nitrogen is not always biologically available.
- Phosphorus is an essential nutrient.
- Remote sensing helps asssess photosynthesis and material cycles.
- Sulfur is both a nutrient and an acidic pollutant.

PRACTICE QUIZ

1. Define *atom* and *element*. Are these terms interchangeable?
2. Your body contains vast numbers of carbon atoms. How is it possible that some of these carbon atoms may have been part of the body of a prehistoric creature?
3. What are six characteristics of water that make it so valuable for living organisms and their environment?
4. In the biosphere, matter follows a circular pathway while energy follows a linear pathway. Explain.
5. The oceans store a vast amount of heat, but (except for climate moderation) this huge reservoir of energy is of little use to humans. Explain the difference between high-quality and low-quality energy.
6. Ecosystems require energy to function. Where does this energy come from? Where does it go? How does the flow of energy conform to the laws of thermodynamics?

7. Heat is released during metabolism. How is this heat useful to a cell and to a multicellular organism? How might it be detrimental, especially in a large, complex organism?
8. Photosynthesis and cellular respiration are complementary processes. Explain how they exemplify the laws of conservation of matter and thermodynamics.
9. What do we mean by carbon-fixation or nitrogen-fixation? Why is it important to humans that carbon and nitrogen be "fixed"?
10. The population density of large carnivores is always very small compared to the population density of herbivores occupying the same ecosystem. Explain this in relation to the concept of an ecological pyramid.
11. A species is a specific kind of organism. What general characteristics do individuals of a particular species share? Why is it important for ecologists to differentiate among the various species in a biological community?

CRITICAL THINKING AND DISCUSSION QUESTIONS

1. If your dishwasher detergents contained phosphorus, would you change brands? Would you encourage others to change? Why or why not?
2. The laws of thermodynamics are sometimes summarized as "you can't get something for nothing," and "you can't even break even." Explain these ideas.
3. The ecosystem concept revolutionized ecology by introducing holistic systems thinking as opposed to individualistic life history studies. Why was this a conceptual breakthrough?

4. If ecosystems are so difficult to delimit, why is this such a persistent concept? Can you imagine any other ways to define or delimit environmental investigation?
5. Choose one of the material cycles (carbon, nitrogen, phosphorus, or sulfur) and identify the components of the cycle in which you participate. For which of these components would it be easiest to reduce your impacts?

You know that nutrients are an important concern in the Chesapeake Bay watershed in general, but now you can examine the details and see how conditions have changed. Go to www.eco-check.org/reportcard/chesapeake. This site is maintained by the University of Maryland and the National Oceanic and Atmospheric Administration (NOAA), with support of many collaborators and data providers.

Chlorophyll is something you've read about in this chapter. Concentrations of chlorophyll in Chesapeake Bay indicate amounts of tiny floating algae cells—algae nourished by nitrogen and phosphorus from onshore sources. Take a look at chlorophyll-a levels in the bay: roll your mouse over the Indicators and Indices, and click on the "chlorophyll-a" icon (this is one of several kinds of chlorophyll).

Take a few minutes to look at the Threshold map, as well as the definitions to answer the questions below.

1. What is chlorophyll-a used to measure? What factors increase the amount of chlorophyll-a in the water?

2. This map shows areas exceeding healthy levels (thresholds) of chlorophyll. Thresholds differ from fresh to salty parts of the estuary, and by season. Are excessivly high levels detected in much of the bay, or in small areas?

3. How many sampling points were used to produce this map? Were the stations sampled just once? Why or why not?

4. Refer to the map in the opening case study. Which states border the bay? Where is Washington D.C. relative to the chlorophyll measurements on the map?

5. Now look at the Trends Graph tab. Overall would you say that the trend has been an improvement since 1986? Turn on and off the different tributary rivers (different colored boxes below the graph). Which one has had the lowest score in general? Where is the worst one located?

6. Refer to your chapter, and explain what chlorophyll needs to perform photosynthesis. Why are nitrogen and phosphorus needed for plant growth?

7. Choose one other indicator (such as the Biotic Index or Dissolved Oxygen) from the drop-down menu. Explain what that index is, and why it is useful as an indicator of water quality.

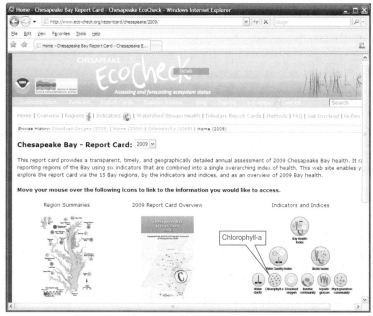

The EcoCheck website provides a wealth of water quality date.

For Additional Help in Studying This Chapter, please visit our website at www.mhhe.com/cunningham12e. You will find additional practice quizzes and case studies, flashcards, regional examples, placemarkers for Google Earth™ mapping, and an extensive reading list, all of which will help you learn environmental science.

The relatively young and barren volcanic islands of the Galápagos isolated from South America by strong, cold currents and high winds, have developed a remarkable community of unique plants and animals.

Evolution, Biological Communities, and Species Interactions

Learning Outcomes

After studying this chapter, you should be able to:

4.1 Describe how evolution produces species diversity.

4.2 Discuss how species interactions shape biological communities.

4.3 Summarize how community properties affect species and populations.

4.4 Explain why communities are dynamic and change over time.

"When I view all beings not as special creations, but as lineal descendents of some few beings which have lived long before the first bed of the Cambrian system was deposited, they seem to me to become ennobled." ~ Charles Darwin

Case Study — Darwin's Voyage of Discovery

Charles Darwin was only 22 years old when he set out in 1831 on his epic five-year, around-the-world voyage aboard the H.M.S. *Beagle* (fig. 4.1). It was to be the adventure of a lifetime, and would lead to insights that would revolutionize the field of biology. Initially an indifferent student, Darwin had found inspiring professors in his last years of college. One of them helped him get a position as an unpaid naturalist on board the *Beagle*. Darwin turned out to be a perceptive observer, an avid collector of specimens, and an extraordinary scientist.

As the *Beagle* sailed slowly along the coast of South America, mapping coastlines and navigational routes, Darwin had time to go ashore on long field trips to explore natural history. He was amazed by the tropical forests of Brazil and the fossils of huge, extinct mammals in Patagonia. He puzzled over the fact that many fossils looked similar, but not quite identical, to contemporary animals. Could species change over time? In Darwin's day, most people believed that everything in the world was exactly as it had been created by God only a few thousand years earlier. But Darwin had read the work of Charles Lyell (1797–1875), who suggested that the world was much older than previously thought, and capable of undergoing gradual, but profound, change over time.

After four years of exploring and mapping, Darwin and the *Beagle* reached the Galápagos Islands, 900 km (540 mi) off the coast of Ecuador. The harsh, volcanic landscape of these remote islands (see page 74) held an extraordinary assemblage of unique plants and animals. Giant land tortoises fed on tree-size cacti. Sea-going iguanas scraped algae off underwater shoals. Sea birds were so unafraid of humans that Darwin could pick them off their nests. The many finches were especially interesting: Every island had its own species, marked by distinct bill shapes, which graded from large and parrot-like to small and warbler-like. Each bird's anatomy and behavior was suited to exploit specific food sources available in its habitat. It seemed obvious that these birds were related, but somehow had been modified to survive under different conditions.

Darwin didn't immediately understand the significance of these observations. Upon returning to England, he began the long process of cataloging and describing the specimens he had collected. Over the next 40 years, he wrote important books on a variety of topics including the formation of oceanic islands from coral reefs, the geology of South America, and the classification and natural history of barnacles. Throughout this time, he puzzled about how organisms might adapt to specific environmental situations.

A key in his understanding was Thomas Malthus's *Essay on the Principle of Population* (1798). From Malthus, Darwin saw that most organisms have the potential to produce far more offspring than can actually survive. Those individuals with superior attributes are more likely to live and reproduce than those less well-endowed. Because the more fit individuals are especially successful in passing along their favorable traits to their offspring, the whole population will gradually change to be better suited for its particular environment. Darwin called this process *natural selection* to distinguish it from the artificial selection that plant and animal breeders used to produce the wide variety of domesticated crops and livestock.

Darwin completed a manuscript outlining his theory of **evolution** (gradual change in species) through natural selection in 1842, but he didn't publish it for another 16 years, perhaps because he was worried about the controversy he knew it would provoke. When his masterpiece, *On the Origin of Species,* was finally made public in 1859, it was both strongly criticized and highly praised. Although Darwin was careful not to question the existence of a Divine Creator, many people interpreted his theory of gradual change in nature as a challenge to their faith. Others took his theory of survival of the fittest much further than Darwin intended, applying it to human societies, economics, and politics.

FIGURE 4.1 Charles Darwin, in a portrait painted shortly after the voyage on the *Beagle*.

One of the greatest difficulties for the theory of evolution was that little was known in Darwin's day of the mechanisms of heredity. No one could explain how genetic variation could arise in a natural population, or how inheritable traits could be sorted and recombined in offspring. It took nearly another century before biologists could use their understanding of molecular genetics to put together a modern synthesis of evolution that clarifies these details.

An overwhelming majority of biologists now consider the theory of evolution through natural selection to be the cornerstone of their science. The theory explains how the characteristics of organisms have arisen from individual molecules, to cellular structures, to tissues and organs, to complex behaviors and population traits. In this chapter, we'll look at the evidence for evolution and how it shapes species and biological communities. We'll examine the ways in which interactions between species and between organisms and their environment allow species to adapt to particular conditions as well as to modify both their habitat and their competitors. For related resources, including Google Earth™ placemarks that show locations where these issues can be explored via satellite images, visit http://EnvironmentalScience-Cunningham.blogspot.com.

For more information, see

Darwin, Charles. *The Voyage of the* Beagle (1837) and *On the Origin of Species* (1859).

Stix, Gary. 2009. Darwin's living legacy. *Scientific American* 300(1): 38–43.

4.1 Evolution Produces Species Diversity

Why do some species live in one place but not another? A more important question to environmental scientists is, what are the mechanisms that promote the great variety of species on earth and that determine which species will survive in one environment but not another? In this section you will come to understand (1) concepts behind the theory of speciation by means of natural selection and adaptation (evolution); (2) the characteristics of species that make some of them weedy and others endangered; and (3) the limitations species face in their environments and implications for their survival. First we'll start with the basics: How do species arise?

Natural selection leads to evolution

How does a polar bear stand the long, sunless, super-cold arctic winter? How does the saguaro cactus survive blistering temperatures and extreme dryness of the desert? We commonly say that each species is *adapted* to the environment where it lives, but what does that mean? **Adaptation,** the acquisition of traits that allow a species to survive in its environment, is one of the most important concepts in biology.

We use the term *adapt* in two ways. An individual organism can respond immediately to a changing environment in a process called acclimation. If you keep a houseplant indoors all winter and then put it out in full sunlight in the spring, the leaves become damaged. If the damage isn't severe, your plant may grow new leaves with thicker cuticles and denser pigments that block the sun's rays. However, the change isn't permanent. After another winter inside, it will still get sun-scald in the following spring. The leaf changes are not permanent and cannot be passed on to offspring, or even carried over from the previous year. Although the capacity to acclimate is inherited, houseplants in each generation must develop their own protective leaf epidermis.

Another type of adaptation affects populations consisting of many individuals. Genetic traits are passed from generation to generation and allow a species to live more successfully in its environment. As the opening case study for this chapter shows, this process of adaptation to environment is explained by the theory of evolution. The basic idea of evolution is that species change over generations because individuals compete for scarce resources. Better competitors in a population survive—they have greater reproductive potential or fitness—and their offspring inherit the beneficial traits. Over generations, those traits become common in a population (fig. 4.2). The process of better-selected individuals passing their traits to the next generation is called **natural selection.** The traits are encoded in a species' DNA, but from where does the original DNA coding come, which then gives some individuals greater fitness? Every organism has a dizzying array of genetic diversity in its DNA. It has been demonstrated in experiments and by observing natural populations that changes to the DNA coding sequence of individuals occurs, and that the changed sequences are inherited by offspring. Exposure to ionizing radiation and toxic materials, and random recombination and mistakes in

FIGURE 4.2 Giraffes don't have long necks because they stretch to reach tree-top leaves, but those giraffes that happened to have longer necks got more food and had more offspring, so the trait became fixed in the population.

replication of DNA strands during reproduction are the main causes of genetic mutations. Sometimes a single mutation has a large effect, but evolutionary change is mostly brought about by many mutations accumulating over time. Only mutations in reproductive cells (gametes) matter; body cell changes—cancers, for example—are not inherited. Most mutations have no effect on fitness, and many actually have a negative effect. During the course of a species' life span—a million or more years—some mutations are thought to have given those individuals an advantage under the **selection pressures** of their environment at that time. The result is a species population that differs from those of numerous preceding generations.

All species live within limits

Environmental factors exert selection pressure and influence the fitness of individuals and their offspring. For this reason, species are limited in where they can live. Limitations include the following: (1) physiological stress due to inappropriate levels of some critical environmental factor, such as moisture, light, temperature, pH, or specific nutrients; (2) competition with other species; (3) predation, including parasitism and disease; and (4) luck. In some cases, the individuals of a population that survive environmental catastrophes or find their way to a new habitat, where they start a new population, may simply be lucky rather than more fit than their contemporaries.

An organism's physiology and behavior allow it to survive only in certain environments. Temperature, moisture level, nutrient supply, soil and water chemistry, living space, and other environmental factors must be at appropriate levels for organisms to persist. In 1840, the chemist Justus von Liebig proposed that the single factor in shortest supply relative to demand is the **critical factor** determining where a species lives. The giant saguaro cactus (*Carnegiea gigantea*), which grows in the dry, hot Sonoran desert of southern Arizona and northern Mexico, offers an example (fig. 4.3). Saguaros are extremely sensitive to freezing temperatures. A single winter night with temperatures below freezing for 12 or more hours kills growing tips on the branches, preventing further development. Thus the northern edge of the saguaro's range corresponds to a zone where freezing temperatures last less than half a day at any time.

FIGURE 4.3 Saguaro cacti, symbolic of the Sonoran desert, are an excellent example of distribution controlled by a critical environmental factor. Extremely sensitive to low temperatures, saguaros are found only where minimum temperatures never dip below freezing for more than a few hours at a time.

Ecologist Victor Shelford (1877–1968) later expanded Liebig's principle by stating that each environmental factor has both minimum and maximum levels, called **tolerance limits,** beyond which a particular species cannot survive or is unable to reproduce (fig. 4.4). The single factor closest to these survival limits, Shelford postulated, is the critical factor that limits where a particular organism can live. At one time, ecologists tried to identify unique factors limiting the growth of every plant and animal population. We now know that several factors working together, even in a clear-cut case like the saguaro, usually determine a species' distribution. If you have ever explored the rocky coasts of

New England or the Pacific Northwest, you have probably noticed that mussels and barnacles grow thickly in the intertidal zone, the place between high and low tides. No one factor decides this pattern. Instead, the distribution of these animals is determined by a combination of temperature extremes, drying time between tides, salt concentrations, competitors, and food availability.

In some species, tolerance limits affect the distribution of young differently than adults. The desert pupfish, for instance, lives in small, isolated populations in warm springs in the northern Sonoran desert. Adult pupfish can survive temperatures between 0° and 42°C (a remarkably high temperature for a fish) and tolerate an equally wide range of salt concentrations. Eggs and juvenile fish, however, can survive only between 20° and 36°C and are killed by high salt levels. Reproduction, therefore, is limited to a small part of the range of the adult fish.

Sometimes the requirements and tolerances of species are useful **indicators** of specific environmental characteristics. The presence or absence of such species indicates something about the community and the ecosystem as a whole. Lichens and eastern white pine, for example, are indicators of air pollution because they are extremely sensitive to sulfur dioxide and ozone, respectively. Bull thistle and many other plant weeds grow on disturbed soil but are not eaten by cattle; therefore, a vigorous population of bull thistle or certain other plants in a pasture indicates it is being overgrazed. Similarly, anglers know that trout species require cool, clean, well-oxygenated water; the presence or absence of trout is used as an indicator of good water quality.

The ecological niche is a species' role and environment

Habitat describes the place or set of environmental conditions in which a particular organism lives. A more functional term, **ecological niche,** describes either the role played by a species in a biological community or the total set of environmental

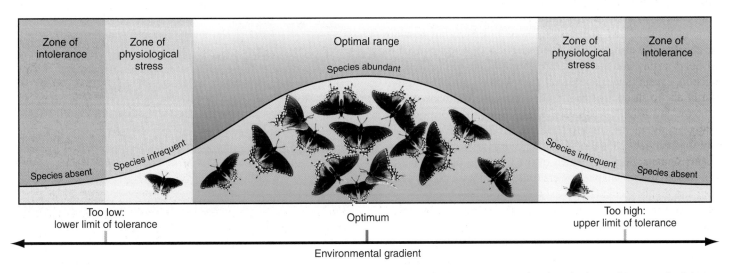

FIGURE 4.4 The principle of tolerance limits states that for every environmental factor, an organism has both maximum and minimum levels beyond which it cannot survive. The greatest abundance of any species along an environmental gradient is around the optimum level of the critical factor most important for that species. Near the tolerance limits, abundance decreases because fewer individuals are able to survive the stresses imposed by limiting factors.

FIGURE 4.5 Each of the species in this African savanna has its own ecological niche that determines where and how it lives.

factors that determine a species distribution. The concept of niche was first defined in 1927 by the British ecologist Charles Elton (1900–1991). To Elton, each species had a role in a community of species, and the niche defined its way of obtaining food, the relationships it had with other species, and the services it provided to its community. Thirty years later, the American limnologist G. E. Hutchinson (1903–1991) proposed a more biophysical definition of niche. Every species, he pointed out, exists within a range of physical and chemical conditions (temperature, light levels, acidity, humidity, salinity, etc.) and also biological interactions (predators and prey present, defenses, nutritional resources available, etc.). The niche is more complex than the idea of a critical factor (fig. 4.5). A graph of a species niche would be multidimensional, with many factors being simultaneously displayed, almost like an electron cloud.

For a generalist, like the brown rat, the ecological niche is broad. In other words, a generalist has a wide range of tolerance for many environmental factors. For others, such as the giant panda (*Ailuropoda melanoleuca*), only a narrow ecological niche exists (fig. 4.6). Bamboo is low in nutrients, but provides 95 percent of a panda's diet, requiring it to spend as much as 16 hours a day eating. There are virtually no competitors for bamboo, except other pandas, yet the species is endangered, primarily due to shrinking habitat. Giant pandas, like many species on earth, are habitat specialists. Specialists have more exacting habitat requirements, tend to have lower reproductive rates, and care for their young longer. They may be less resilient in response to environmental change. Much of the flora and fauna that Darwin studied in the Galápagos were **endemic** (not found anywhere else) and highly specialized to exist in their unique habitat.

Over time, niches change as species develop new strategies to exploit resources. Species of greater intelligence or complex social structures, such as elephants, chimpanzees, and dolphins, learn from their social group how to behave and can invent new ways of doing things when presented with novel opportunities or challenges. In effect, they alter their ecological niche by passing on cultural behavior from one generation to the next. Most organisms, however, are restricted to their niche by their genetically

determined bodies and instinctive behaviors. When two such species compete for limited resources, one eventually gains the larger share, while the other finds different habitat, dies out, or experiences a change in its behavior or physiology so that competition is minimized. The idea that "complete competitors cannot coexist" was proposed by the Russian microbiologist G. F. Gause (1910–1986) to explain why mathematical models of species competition always ended with one species disappearing. The **competitive exclusion principle,** as it is called, states that no two species can occupy the same ecological niche for long. The one that is more efficient in using available resources will exclude the other (see Species Competition at the end of this chapter). We call this process of niche evolution **resource partitioning** (fig. 4.7). Partitioning can allow several species to utilize different parts of the same resource and coexist within a single habitat (fig. 4.8). Species can specialize in time, too. Swallows and insectivorous bats both catch insects, but some insect species are active during the day and others at night, providing noncompetitive feeding opportunities for day-active swallows and night-active bats. The competitive exclusion principle does not explain all situations, however. For example, many similar plant species coexist in some habitats. Do they avoid competition in ways we cannot observe, or are resources so plentiful that no competition need occur?

FIGURE 4.6 The giant panda feeds exclusively on bamboo. Although its teeth and digestive system are those of a carnivore, it is not a good hunter, and has adapted to a vegetarian diet. In the 1970s, huge acreages of bamboo flowered and died, and many pandas starved.

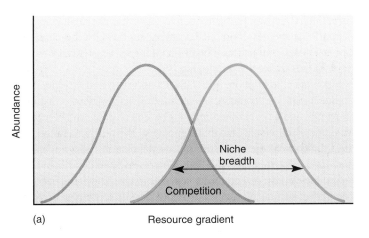

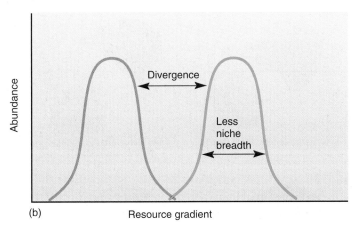

FIGURE 4.7 Competition causes resources partitioning and niche specialization. (a) Where niches of two species overlap along a resource gradient, competition occurs (*shaded area*). Individuals in this part of the niche have less success producing young. (b) Over time the traits of the populations diverge, leading to specialization, narrower niche breadth, and less competition between species.

Speciation maintains species diversity

As an interbreeding species population becomes better adapted to its ecological niche, its genetic heritage (including mutations passed from parents to offspring) gives it the potential to change further as circumstances dictate. In the case of Galápagos finches studied a century and a half ago by Charles Darwin, evidence from body shape, behavior, and genetics leads to the idea that modern Galápagos finches look, behave, and bear DNA related to an original seed-eating finch species that probably blew to the islands from the mainland where a similar species still exists. Today there are 13 distinct species on the islands

FIGURE 4.8 Several species of insect-eating wood warblers occupy the same forests in eastern North America. The competitive exclusion principle predicts that the warblers should partition the resource—insect food—in order to reduce competition. And in fact, the warblers feed in different parts of the forest.
Source: Original observations by R. H. MacArthur (1958).

that differ markedly in appearance, food preferences, and habitat (fig. 4.9). Fruit eaters have thick, parrot-like bills; seed eaters have heavy, crushing bills; insect eaters have thin, probing beaks to catch their prey. One of the most unusual species is the woodpecker finch, which pecks at tree bark for hidden insects. Lacking the woodpecker's long tongue, the finch uses a cactus spine as a tool to extract bugs.

The development of a new species is called **speciation.** Darwin believed that new species arise only very gradually, over immensely long times. In some organisms, however, adaptive changes have occurred fast enough to be observed. Wild European rabbits, for example, were introduced into Australia about 220 years ago. They have changed body size, weight, and ear size as they adapted to the hot, dry Australian climate. Evolutionary scientist Stephen Jay Gould suggested that many species may be relatively stable for long times and then undergo rapid speciation (punctuated equilibrium) in response to environmental change. For further discussion on definitions of species, see chapter 11.

One mechanism of speciation is **geographic isolation.** This is termed **allopatric speciation**—species arise in non-overlapping geographic locations. The original Galápagos finches were separated from the rest of the population on the mainland, could no longer share genetic material, and became reproductively isolated.

The barriers that divide subpopulations are not always physical. For example, two virtually identical tree frogs (*Hyla versicolor, Hyla chrysoscelis*) live in similar habitats of eastern North America but have different mating calls. This is an example of behavioral isolation. It also happens that one species has twice the chromosomes of the other. This example of **sympatric speciation** takes place in the same location as the ancestor species. Fern species and other plants seem prone to sympatric speciation by doubling or quadrupling the chromosome number of their ancestors.

(a) Large ground finch (seeds)

(b) Cactus ground finch (cactus fruits and flowers)

(c) Vegetarian finch (buds)

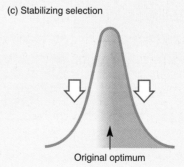

(d) Woodpecker finch (insects)

FIGURE 4.9 Each of the 13 species of Galápagos finches, although originally derived from a common ancestor, has evolved distinctive anatomies and behaviors to exploit different food sources. The woodpecker finch (d) uses cactus thorns to probe for insects under tree bark.

Once isolation is imposed, the two populations begin to diverge in genetics and physical characteristics. Genetic drift ensures that DNA of two formerly joined populations eventually diverges; in several generations, traits are lost from a population during the natural course of reproduction. Under more extreme circumstances, a die-off of most members of an isolated population strips much of the variation in traits from the survivors. The cheetah experienced a genetic bottleneck about 10,000 years ago and exists today as virtually identical individuals.

In isolation, selection pressures shape physical, behavioral, and genetic characteristics of individuals, causing population traits to shift over time (fig. 4.10). From an original range of characteristics, the shift can be toward an extreme of the trait (directional selection), it can narrow the range of a trait (stabilizing selection), or it can cause traits to diverge to the extremes (disruptive selection). Directional selection is implied by increased pesticide resistance in German cockroaches (*Blattella*

FIGURE 4.10 A species trait, such as beak shape, changes in response to selection pressure. (a) The original variation is acted on by selection pressure (*arrows*) that (b) shifts the characteristics of that trait in one direction, or (c) to an intermediate condition. (d) Disruptive selection moves characteristics to the extremes of the trait. Which selection type plausibly resulted in two distinct beak shapes among Galápagos finches—narrow in tree finches versus stout in ground finches?

germanica). Apparently some individuals can make an enzyme that detoxifies pesticides. Individual cockroaches that lack this characteristic are dying out, and as a result, populations of cockroaches with pesticide resistance are developing.

A small population in a new location—island, mountaintop, unique habitat—encounters new environmental conditions that favor some individuals over others (fig. 4.11). The physical and behavioral traits these individuals have are passed to the next generation, and the frequency of the trait shifts in the population. Where a species may have existed but has died out, others arise and contribute to the incredible variety of life-forms seen in nature. The fossil record is one of ever-increasing species diversity, despite several catastrophes, which were recorded in different geological strata and which wiped out a large proportion of the earth's species each time.

Evolution is still at work

You may think that evolution only occurred in the distant past, but it's an ongoing process. Many examples from both laboratory experiments and from nature shows evolution at work (Exploring Science, p. 81). Geneticists have modified many fruit fly properties—including body size, eye color, growth rate, life span, and feeding behavior—using artificial selection. In one experiment, researchers selected fruit flies with many bristles (stiff, hairlike structures) on their abdomen. In each generation, the flies with the most bristles were allowed to mate. After 86 generations, the number of bristles had quadrupled. In a similar experiment with corn, agronomists chose seeds with the

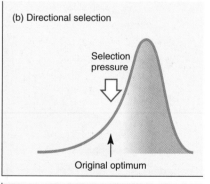

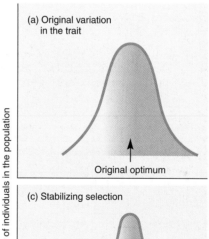

(a) Original variation in the trait

Original optimum

(b) Directional selection

Selection pressure

Original optimum

(c) Stabilizing selection

Original optimum

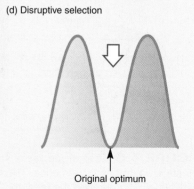

(d) Disruptive selection

Original optimum

Number of individuals in the population

Variation in the trait experiencing natural selection

Why do we need a new flu vaccination every fall? Why can't they make one that lasts for years like the measles/mumps shot that we got as infants? The answer is that the flu virus has an alarming ability to mutate rapidly. Our bodies are constantly trying to identify and build defenses against new viruses, while viruses have evolved methods to evolve rapidly and avoid surveillance by our immune system. Understanding the principles of evolution and genetics has made it possible to defend ourselves from the flu—provided we get the vaccines right each year.

Viruses can't replicate by themselves. They have to invade a cell of a higher organism and hijack the cell's biochemical systems. If multiple viruses infect the same cell, their RNA molecules (genes) can be mixed and recombined to create new virus strains. To invade a cell, the virus binds to a receptor on the cell surface (fig. 1). The binding proteins are called hemagglutin (because they also bind to antibodies in our blood). The viruses also have proteins called neuraminidases on their surface, which play a role in budding of particles from the cell membrane and modifying sugars on the virus exterior. Influenza has 16 groups of H proteins and 9 groups of N proteins. We identify virus strains by code names, such as H5N1, or H3N2, based on their surface proteins.

Every year, new influenza strains sweep across the world, and because they change their surface proteins, our immune system fails to recognize them. The Centers for Disease Control constantly surveys the flu strains occurring elsewhere to try to guess what varieties are most likely to invade the United States.

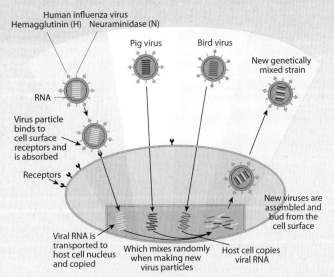

When different strains of the influenza virus infect the same cell, their genetic material can intermix to create a new re-assorted variety.

Vaccines are prepared based on that best guess, but sometimes they're wrong. An unknown variety can suddenly appear against which we have neither residual immunity nor vaccines. The result is a bad flu season.

An example of the surprises caused by rapid flu evolution occurred in 2009. A virus in the H1N1 family emerged in Mexico, where it infected at least 1,000 people and killed around 150. As it spread into the United States, children were particularly susceptible, while adults, particularly those over 60, often had some degree of immunity. While that virus wasn't as lethal as first feared, by November 2009 it had infected about 50 million Americans with 200,000 hospitalizations and 10,000 deaths.

The H1N1 family is notorious as the source of the worst influenza pandemic (worldwide

epidemic) in recorded history. The 1918 Spanish flu killed upward of 50 million people. This family also infects pigs, but it rarely kills them. For years, the swine flu viruses seemed to evolve more slowly than human strains, but this picture is changing. Suddenly, pig viruses have begun to evolve at a much faster rate and move to humans with increasing frequency. Critics of industrial agriculture charge that pigs increasingly are raised in enormous industrial facilities where diseases can quickly sweep through up to a million crowded animals. Many epidemiologists consider the roughly one billion pigs now raised annually to be laboratories for manufacturing new virus strains.

Pigs also serve as a conduit between humans and other animals. That's because they're susceptible to viruses from many sources. And once inside a cell, viral genes can mix freely to create new, more virulent combinations. The 2009 H1N1, for example, was shown to have genes from at least five different strains: a North America swine flu, North American avian flu, human influenza, and two swine viruses typically found in Asia and Europe. It's thought that the recombination of these various strains occurred in pigs, although we don't know when or where that took place.

So for the time being, we must continue to get a new inoculation annually and hope it protects us against the main flu strains we're likely to encounter in the next flu season. Someday, there may be a universal vaccine that will immunize us against all influenza viruses, but for now, that's just a dream.

For more information, see Branswell, H. 2011. Flu factories. *Scientific American* 304(1): 46–51.

highest oil content to plant and mate. After 90 generations, the average oil content had increased 450 percent.

Evolutionary change is also occurring in nature. A classic example is seen in some of the finches on the Galápagos Island of Daphne. Twenty years ago, a large-billed species (*Geospiza magnirostris*) settled on the island, which previously had only a medium-billed species (*Geospiza fortis*). The *G. magnirostris* were better at eating larger seeds and pushed *G. fortis* to depend more and more on smaller seeds. Gradually, birds with smaller bills suited to small seeds became more common in the *G. fortis* population. During a severe drought in 2003–2004, large seeds were scarce, and most

birds with large beaks disappeared. This included almost all of the recently arrived *G. magnirostris* as well as the larger-beaked *G. fortis*. In just two generations, the *G. fortis* population changed to entirely small-beaked individuals. At first, this example of rapid evolution was thought to be a rarity, but subsequent research suggests that it may be more common than previously thought.

Similarly, the widespread application of pesticides in agriculture and urban settings has led to the rapid evolution of resistance in more than 500 insect species. Similarly, the extensive use of antibiotics in human medicine and livestock operations has led to antibiotic resistance in many microbes. The Centers for

1. Single population

2. Geographically isolated populations

FIGURE 4.11 Geographic barriers can result in allopatric speciation. During cool, moist glacial periods, what is now Arizona was forest-covered, and squirrels could travel and interbreed freely. As the climate warmed and dried, desert replaced forest on the plains. Squirrels were confined to cooler mountaintops, which acted as island refugia, where new, reproductively isolated species gradually evolved.

Disease Control estimates that 90,000 Americans die every year from hospital-acquired infections, most of which are resistant to one or more antibiotics. We're engaged in a kind of an arms race with germs. As quickly as new drugs are invented, microbes become impervious to them. Currently, vancomycin is the drug of last resort. When resistance to it becomes widespread, we may have no protection from infections.

Think About It

Try to understand the position of someone who holds an opposite view from your own about evolution. Why would they argue for or against this theory? If you were that person, what evidence would you want to see before you'd change your beliefs?

On the other hand, evolution sometimes works in our favor. We've spread a number of persistent organic pollutants (called POPs), such as pesticides and industrial solvents, throughout

our environment. One of the best ways to get rid of them is with microbes that can destroy or convert them to a nontoxic form. It turns out that the best place to look for these species is in the most contaminated sites. The presence of a new food source has stimulated evolution of organisms that can metabolize it. A little artificial selection and genetic modification in the laboratory can turn these species into very useful bioremediation tools.

Taxonomy describes relationships among species

Taxonomy is the study of types of organisms and their relationships. With it you can trace how organisms have descended from common ancestors. Taxonomic relationships among species are displayed like a family tree. Botanists, ecologists, and other scientists often use the most specific levels of the tree, genus and species, to compose **binomials.** Also called scientific or Latin names, they identify and describe species using Latin, or Latinized nouns and adjectives, or names of people or places. Scientists communicate about species using these scientific names instead of common names (e.g., lion, dandelion, or ant lion), to avoid confusion. A common name can refer to any number of species in different places, and a single species might have many common names. The bionomial *Pinus resinosa,* on the other hand, always is the same tree, whether you call it a red pine, Norway pine, or just pine.

Taxonomy also helps organize specimens and subjects in museum collections and research. You are *Homo sapiens* (human) and eat chips made of *Zea mays* (corn or maize). Both are members of two well-known kingdoms. Scientists, however, recognize six kingdoms (fig. 4.12): animals, plants, fungi (molds and mushrooms), protists (algae, protozoans, slime molds), bacteria (or

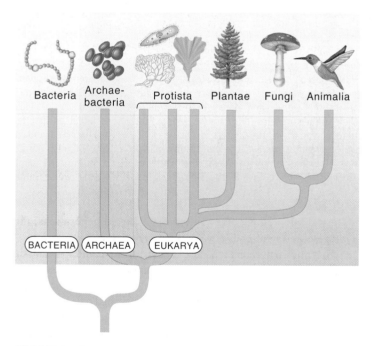

FIGURE 4.12 The six great kingdoms representing all life on earth. The kingdoms are grouped in domains indicating common origins.

eubacteria), and archaebacteria (ancient, single-celled organisms that live in harsh environments, such as hot springs). Within these kingdoms are millions of different species, which you will learn more about in chapters 5 and 11.

4.2 SPECIES INTERACTIONS SHAPE BIOLOGICAL COMMUNITIES

We have learned that adaptation to one's environment, determination of ecological niche, and even speciation is affected not just by bodily limits and behavior, but also by competition and predation. Don't despair. Not all biological interactions are antagonistic, and many, in fact, involve cooperation or at least benign interactions and tolerance. In some cases, different organisms depend on each other to acquire resources. Now we will look at the interactions within and between species that affect their success and shape biological communities.

Competition leads to resource allocation

Competition is a type of antagonistic relationship within a biological community. Organisms compete for resources that are in limited supply: energy and matter in usable forms, living space, and specific sites to carry out life's activities. Plants compete for growing space to develop root and shoot systems so that they can absorb and process sunlight, water, and nutrients (fig. 4.13). Animals compete for living, nesting, and feeding sites, and also for mates. Competition among members of the same species is called **intraspecific competition,** whereas competition between members of different species is called **interspecific competition.** Recall the competitive exclusion principle as it applies to interspecific competition. Competition shapes a species population and biological community by causing individuals and species to shift their focus from one segment of a resource type to another. Thus, warblers all competing with each other for insect food in New England tend to specialize on different areas of the forest's trees, reducing or avoiding competition. Since the 1950s there have been hundreds of interspecific competition studies in natural populations. In general, scientists assume it does occur, but not always, and in some groups—carnivores and plants—it has little effect.

In intraspecific competition, members of the same species compete directly with each other for resources. Several avenues exist to reduce competition in a species population. First, the young of the year disperse. Even plants practice dispersal; seeds are carried by wind, water, and passing animals to less crowded conditions away from the parent plants. Second, by exhibiting strong territoriality, many animals force their offspring or trespassing adults out of their vicinity. In this way territorial species, which include bears, songbirds, ungulates, and even fish, minimize competition between individuals and generations. A third way to reduce intraspecific competition is resource partitioning between generations. The adults and juveniles of these species occupy different ecological niches. For instance, monarch caterpillars munch on milkweed leaves, while metamorphosed butterflies lap nectar. Crabs begin as floating larvae and do not compete with bottom-dwelling adult crabs.

FIGURE 4.13 In this tangled Indonesian rainforest, space and light are at a premium. Plants growing beneath the forest canopy have adaptations that help them secure these limited resources. The ferns and bromeliads seen here are epiphytes; they find space and get closer to the sun by perching on limbs and tree trunks. Strangler figs start out as epiphytes, but send roots down to the forest floor and, once contact is made, put on a growth spurt that kills the supporting tree. These are just some of the adaptations to life in the dark jungle.

We think of competition among animals as a battle for resources—"nature red in tooth and claw" is the phrase. In fact, many animals avoid fighting if possible, or confront one another with noise and predictable movements. Bighorn sheep and many other ungulates, for example, engage in ritualized combat, with the weaker animal knowing instinctively when to back off. It's worse to be injured than to lose. Instead, competition often is simply about getting to food or habitat first, or being able to use it more efficiently. As we discussed, each species has tolerance limits for nonbiological (abiotic) factors. Studies often show that, when two species compete, the one living in the center of its tolerance limits for a range of resources has an advantage and, more often than not, prevails in competition with another species living outside its optimal environmental conditions.

Predation affects species relationships

All organisms need food to live. Producers make their own food, while consumers eat organic matter created by other organisms. As we saw in chapter 3, photosynthetic plants and algae are the producers in most communities. Consumers include herbivores, carnivores, omnivores, scavengers, detritivores, and decomposers. You may think only carnivores are predators, but ecologically a predator is any organism that feeds directly on another living organism, whether or not this kills the prey (fig. 4.14). Herbivores, carnivores, and omnivores, which feed on live prey, are predators, but scavengers, detritivores, and decomposers, which feed on dead

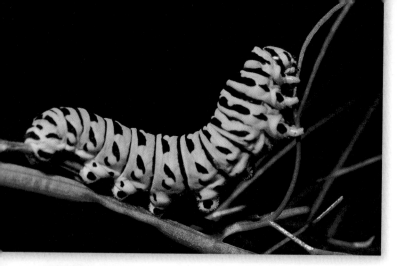

FIGURE 4.14 Insect herbivores are predators as much as are lions and tigers. In fact, insects consume the vast majority of biomass in the world. Complex patterns of predation and defense have often evolved between insect predators and their plant prey.

things, are not. In this sense, parasites (organisms that feed on a host organism or steal resources from it without necessarily killing it) and even pathogens (disease-causing organisms) can be considered predator organisms. Herbivory is the type of predation practiced by grazing and browsing animals on plants.

Predation is a powerful but complex influence on species populations in communities. It affects (1) all stages in the life cycles of predator and prey species; (2) many specialized food-obtaining mechanisms; and (3) the evolutionary adjustments in behavior and body characteristics that help prey escape being eaten, and predators more efficiently catch their prey. Predation also interacts with competition. In **predator-mediated competition,** a superior competitor in a habitat builds up a larger population than its competing species; predators take note and increase their hunting pressure on the superior species, reducing its abundance and allowing the weaker competitor to increase its numbers. To test this idea, scientists remove predators from communities of competing species. Often the superior competitors eliminated other species from the habitat. In a classic example, the ochre starfish (*Pisaster ochraceus*)

FIGURE 4.15 Microscopic plants and animals form the basic levels of many aquatic food chains and account for a large percentage of total world biomass. Many oceanic plankton are larval forms that have habitats and feeding relationships very different from their adult forms.

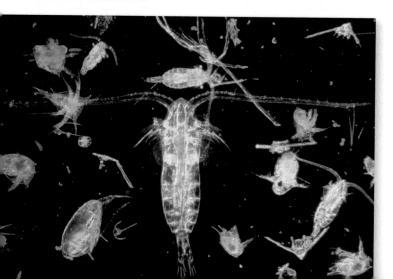

was removed from Pacific tidal zones and its main prey, the common mussel (*Mytilus californicus*), exploded in numbers and crowded out other intertidal species.

Knowing how predators affect prey populations has direct application to human needs, such as pest control in cropland. The cyclamen mite (*Phytonemus pallidus*), for example, is a pest of California strawberry crops. Its damage to strawberry leaves is reduced by predatory mites (*Typhlodromus* and *Neoseiulus*), which arrive naturally or are introduced into fields. Pesticide spraying to control the cyclamen mite can actually increase the infestation because it also kills the beneficial predatory mites.

Predatory relationships may change as the life stage of an organism changes. In marine ecosystems, crustaceans, mollusks, and worms release eggs directly into the water where they and hatchling larvae join the floating plankton community (fig. 4.15). Planktonic animals eat each other and are food for larger carnivores, including fish. As prey species mature, their predators change. Barnacle larvae are planktonic and are eaten by small fish, but as adults their hard shells protect them from fish, but not starfish and predatory snails. Predators often switch prey in the course of their lives. Carnivorous adult frogs usually begin their lives as herbivorous tadpoles. Predators also switch prey when it becomes rare, or something else becomes abundant. Many predators have morphologies and behaviors that make them highly adaptable to a changing prey base, but some, like the polar bear are highly specialized in their prey preferences.

Some adaptations help avoid predation

Predator-prey relationships exert selection pressures that favor evolutionary adaptation. In this world, predators become more efficient at searching and feeding, and prey become more effective at escape and avoidance. Toxic chemicals, body armor, extraordinary speed, and the ability to hide are a few strategies organisms use to protect themselves. Plants have thick bark, spines, thorns, or distasteful and even harmful chemicals in tissues—poison ivy and stinging nettle are examples. Arthropods, amphibians, snakes, and some mammals produce noxious odors or poisonous secretions that cause other species to leave them alone. Animal prey are adept at hiding, fleeing, or fighting back. On the Serengeti Plain of East Africa, the swift Thomson's gazelle and even swifter cheetah are engaged in an arms race of speed, endurance, and quick reactions. The gazelle escapes often because the cheetah lacks stamina, but the cheetah accelerates from 0 to 72 kph in 2 seconds, giving it the edge in a surprise attack. The response of predator to prey and vice versa, over tens of thousands of years, produces physical and behavioral changes in a process known as **coevolution.** Coevolution can be mutually beneficial: many plants and pollinators have forms and behaviors that benefit each other. A classic case is that of fruit bats, which pollinate and disperse seeds of fruit-bearing tropical plants.

Often species with chemical defenses display distinct coloration and patterns to warn away enemies (fig. 4.16). In a neat evolutionary twist, certain species that are harmless resemble poisonous or distasteful ones, gaining protection against predators who remember a bad experience with the actual toxic organism. This is called **Batesian mimicry,** after the English naturalist

FIGURE 4.16 Poison arrow frogs of the family Dendrobatidae display striking patterns and brilliant colors that alert potential predators to the extremely toxic secretions on their skin. Indigenous people in Latin America use the toxin to arm blowgun darts.

H. W. Bates (1825–1892), a traveling companion of Alfred Wallace. Many wasps, for example, have bold patterns of black and yellow stripes to warn potential predators (fig. 4.17a). The much rarer longhorn beetle has no stinger but looks and acts much like a wasp, tricking predators into avoiding it (fig. 4.17b). The distasteful monarch and benign viceroy butterflies are a classic case of Batesian mimicry. Another form of mimicry, **Müllerian mimicry** (after the biologist Fritz Müller) involves two unpalatable or dangerous species who look alike. When predators learn to avoid either species, both benefit. Species also display forms, colors, and patterns that help avoid being discovered. Insects that look like dead leaves or twigs are among the most remarkable examples (fig. 4.18). Unfortunately for prey, predators also often use camouflage to conceal themselves as they lie in wait for their next meal.

Symbiosis involves intimate relations among species

In contrast to predation and competition, some interactions between organisms can be nonantagonistic, even beneficial. In such relationships, called **symbiosis**, two or more species live intimately together, with their fates linked. Symbiotic relationships often enhance the survival of one or both partners. In lichens, a fungus and a photosynthetic partner (either an alga or a cyanobacterium) combine tissues to mutual benefit (fig. 4.19a). This association is called **mutualism.** Some ecologists believe that cooperative, mutualistic relationships may be more important in evolution than commonly thought (fig. 4.19b). Survival of the fittest may also mean survival of organisms that can live together.

Symbiotic relationships often entail some degree of coevolution of the partners, shaping—at least in part—their structural and behavioral characteristics. This mutualistic coadaptation is evident between swollen thorn acacias (*Acacia collinsii*) and the ants (*Pseudomyrmex ferruginea*) that tend them in Central and South America. Acacia ant colonies live inside the swollen thorns on the acacia tree branches. Ants feed on nectar that is produced in glands at the leaf bases and also eat special protein-rich structures that are

(a)

(b)

FIGURE 4.17 An example of Batesian mimicry. The dangerous wasp (a) has bold yellow and black bands to warn predators away. The much rarer longhorn beetle (b) has no poisonous stinger, but looks and acts like a wasp and thus avoids predators as well.

produced on leaflet tips. The acacias thus provide shelter and food for the ants. Although they spend energy to provide these services, the trees are not harmed by the ants. What do the acacias get in return? Ants aggressively defend their territories, driving away herbivorous insects that would feed on the acacias. Ants also trim away vegetation that grows around the tree, reducing competition by other plants for water and nutrients. You can see how mutualism is structuring the biological community in the vicinity of acacias harboring ants, just as competition or predation shapes communities.

Mutualistic relationships can develop quickly. In 2005 the Harvard entomologist E. O. Wilson pieced together evidence to explain a 500-year-old agricultural mystery in the oldest Spanish settlement in the New World, Hispaniola. Using historical accounts and modern research, Dr. Wilson reasoned that mutualism developed between the tropical fire ant (*Solenopsis geminata*),

FIGURE 4.18 This walking stick is highly camouflaged to blend in with the forest floor. Natural selection and evolution have created this remarkable shape and color.

native to the Americas, and a sap-sucking insect that was probably introduced from the Canary Islands in 1516 on a shipment of plantains. The plantains were planted, the sap-suckers were distributed across Hispaniola, and in 1518 a great die-off of crops occurred. Apparently the native fire ants discovered the foreign sap-sucking insects, consumed their excretions of sugar and protein, and protected them from predators, thus allowing the introduced insect population to explode. The Spanish assumed the fire ants caused the agricultural blight, but a little ecological knowledge would have led them to the real culprit.

Commensalism is a type of symbiosis in which one member clearly benefits and the other apparently is neither benefited nor harmed. Many mosses, bromeliads, and other plants growing on trees in the moist tropics are considered commensals (fig. 4.19*c*). These epiphytes are watered by rain and obtain nutrients from leaf litter and falling dust, and often they neither help nor hurt the trees

on which they grow. Robins and sparrows that inhabit suburban yards are commensals with humans. **Parasitism,** a form of predation, may also be considered symbiosis because of the dependency of the parasite on its host.

Keystone species have disproportionate influence

A **keystone species** plays a critical role in a biological community that is out of proportion to its abundance. Originally, keystone species were thought to be top predators—lions, wolves, tigers—which limited herbivore abundance and reduced the herbivory of plants. Scientists now recognize that less-conspicuous species also play keystone roles. Tropical figs, for example, bear fruit year-round at a low but steady rate. If figs are removed from a forest, many fruit-eating animals (frugivores) would starve in the dry season when fruit of other species is scarce. In turn, the disappearance of frugivores would affect plants that depend on them for pollination and seed dispersal. It is clear that the effect of a keystone species on communities often ripples across trophic levels.

Keystone functions have been documented for vegetation-clearing elephants, the predatory ochre sea star, and frog-eating salamanders in coastal North Carolina. Even microorganisms can play keystone roles. In many temperate forest ecosystems, groups of fungi that are associated with tree roots (mycorrhizae) facilitate the uptake of essential minerals. When fungi are absent, trees grow poorly or not at all. Overall, keystone species seem to be more common in aquatic habitats than in terrestrial ones.

The role of keystone species can be difficult to untangle from other species interactions. Off the northern Pacific coast, a giant brown alga (*Macrocystis pyrifera*) forms dense "kelp forests," which shelter fish and shellfish species from predators, allowing them to become established in the community. It turns out, however, that sea otters eat sea urchins living in the kelp forests (fig. 4.20); when sea otters are absent, the urchins graze on and eliminate kelp forests.

(a) Lichen on a rock

(b) Oxpecker and impala

(c) Bromeliad

FIGURE 4.19 Symbiotic relationships. (a) Lichens represent an obligatory mutualism between a fungus and alga or cyanobacterium. (b) Mutualism between a parasite-eating red-billed oxpecker and parasite-infested impala. (c) Commensalism between a tropical tree and free-loading bromeliad.

FIGURE 4.20 Sea otters protect kelp forests in the northern Pacific Ocean by eating sea urchins that would otherwise destroy the kelp. But the otters are being eaten by killer whales. Which is the keystone in this community—or is there a keystone set of organisms?

To complicate things, around 1990, killer whales began preying on otters because of the dwindling stocks of seals and sea lions, thereby creating a cascade of effects. Is the kelp, otter, or orca the keystone here? Whatever the case, keystone species exert their influence by changing competitive relationships. In some communities, perhaps we should call it a "keystone set" of organisms.

4.3 COMMUNITY PROPERTIES AFFECT SPECIES AND POPULATIONS

The processes and principles that we have studied thus far in this chapter—tolerance limits, species interactions, resource partitioning, evolution, and adaptation—play important roles in determining the characteristics of populations and species. In this section we will look at some fundamental properties of biological communities and ecosystems—productivity, diversity, complexity, resilience, stability, and structure—to learn how they are affected by these factors.

Productivity is a measure of biological activity

A community's **primary productivity** is the rate of biomass production, an indication of the rate of solar energy conversion to chemical energy. The energy left after respiration is net primary production. Photosynthetic rates are regulated by light levels, temperature, moisture, and nutrient availability. Figure 4.21 shows approximate productivity levels for some major ecosystems. As you can see, tropical forests, coral reefs, and estuaries (bays or inundated river valleys where rivers meet the ocean) have high levels of productivity because they have abundant supplies of all these resources. In deserts, lack of water limits photosynthesis. On the arctic tundra or in high mountains, low temperatures inhibit plant growth. In the open ocean, a lack of nutrients reduces the ability of algae to make use of plentiful sunshine and water.

Some agricultural crops such as corn (maize) and sugar cane grown under ideal conditions in the tropics approach the productivity levels of tropical forests. Because shallow water ecosystems

such as coral reefs, salt marshes, tidal mud flats, and other highly productive aquatic communities are relatively rare compared to the vast extent of open oceans—which often are effectively biological deserts—marine ecosystems are much less productive on average than terrestrial ecosystems.

Even in the most photosynthetically active ecosystems, only a small percentage of the available sunlight is captured and used to make energy-rich compounds. Between one-quarter and three-quarters of the light reaching plants is reflected by leaf surfaces. Most of the light absorbed by leaves is converted to heat that is either radiated away or dissipated by evaporation of water. Only 0.1 to 0.2 percent of the absorbed energy is used by chloroplasts to synthesize carbohydrates.

In a temperate-climate oak forest, only about half the incident light available on a midsummer day is absorbed by the leaves. Ninety-nine percent of this energy is used to evaporate water. A large oak tree can transpire (evaporate) several thousand liters of water on a warm, dry, sunny day while it makes only a few kilograms of sugars and other energy-rich organic compounds.

What Can You Do?

Working Locally for Ecological Diversity

You might think that diversity and complexity of ecological systems are too large or too abstract for you to have any influence. But you can contribute to a complex, resilient, and interesting ecosystem, whether you live in the inner city, a suburb, or a rural area.

- Keep your cat indoors. Our lovable domestic cats are also very successful predators. Migratory birds, especially those nesting on the ground, have not evolved defenses against these predators.
- Plant a butterfly garden. Use native plants that support a diverse insect population. Native trees with berries or fruit also support birds. (Be sure to avoid non-native invasive species: see chapter 11.) Allow structural diversity (open areas, shrubs, and trees) to support a range of species.
- Join a local environmental organization. Often, the best way to be effective is to concentrate your efforts close to home. City parks and neighborhoods support ecological communities, as do farming and rural areas. Join an organization working to maintain ecosystem health; start by looking for environmental clubs at your school, park organizations, a local Audubon chapter, or a local Nature Conservancy branch.
- Take walks. The best way to learn about ecological systems in your area is to take walks and practice observing your environment. Go with friends and try to identify some of the species and trophic relationships in your area.
- Live in town. Suburban sprawl consumes wildlife habitat and reduces ecosystem complexity by removing many specialized plants and animals. Replacing forests and grasslands with lawns and streets is the surest way to simplify, or eliminate, ecosystems.

Abundance and diversity measure the number and variety of organisms

Abundance is an expression of the total number of organisms in a biological community, while **diversity** is a measure of the number of different species, ecological niches, or genetic variation present. The abundance of a particular species often is inversely related to the total diversity of the community. That is, communities with a very large number of species often have only a few members of any given species in a particular area. As a general rule, diversity decreases but abundance within species increases as we go from the equator toward the poles. The Arctic has vast numbers of insects such as mosquitoes, for example, but only a few species. The tropics, on the other hand, have vast numbers of species—some of which have incredibly bizarre forms and habits—but often only a few individuals of any particular species in a given area.

Consider bird populations. Greenland is home to 56 species of breeding birds, while Colombia, which is only one-fifth the size of Greenland, has 1,395. Why are there so many species in Colombia and so few in Greenland?

Climate and history are important factors. Greenland has such a harsh climate that the need to survive through the winter or escape to milder climates becomes the single most important critical factor that overwhelms all other considerations and severely limits the ability of species to specialize or differentiate into new forms. Furthermore, because Greenland was covered by glaciers until about 10,000 years ago, there has been little time for new species to develop.

Many areas in the tropics, by contrast, have relatively abundant rainfall and warm temperatures year-round so that ecosystems there are highly productive. The year-round dependability of food, moisture, and warmth supports a great exuberance of life and allows a high degree of specialization in physical shape and behavior. Coral reefs are similarly stable, productive, and conducive to proliferation of diverse and amazing life-forms. The enormous abundance of brightly colored and fantastically shaped fish, corals, sponges, and arthropods in the reef community is one of the best examples we have of community diversity.

Productivity is related to abundance and diversity, both of which are dependent on the total resource availability in an ecosystem as well as the reliability of resources, the adaptations of the member species, and the interactions between species. You shouldn't assume that all communities are perfectly adapted to their environment. A relatively new community that hasn't had time for niche specialization, or a disturbed one where roles such as top predators are missing, may not achieve maximum efficiency of resource use or reach its maximum level of either abundance or diversity.

Community structure describes spatial distribution of organisms

Ecological structure refers to patterns of spatial distribution of individuals and populations within a community, as well as the relation of a particular community to its surroundings. At the local level, even in a relatively homogeneous environment, individuals in a single population can be distributed randomly, clumped together, or in highly regular patterns. In randomly arranged populations, individuals live wherever resources are available (fig. 4.22a). Ordered patterns may be determined by the physical environment but are more often the result of biological competition. For example, competition for nesting space in seabird colonies on the Falkland Islands is often fierce.

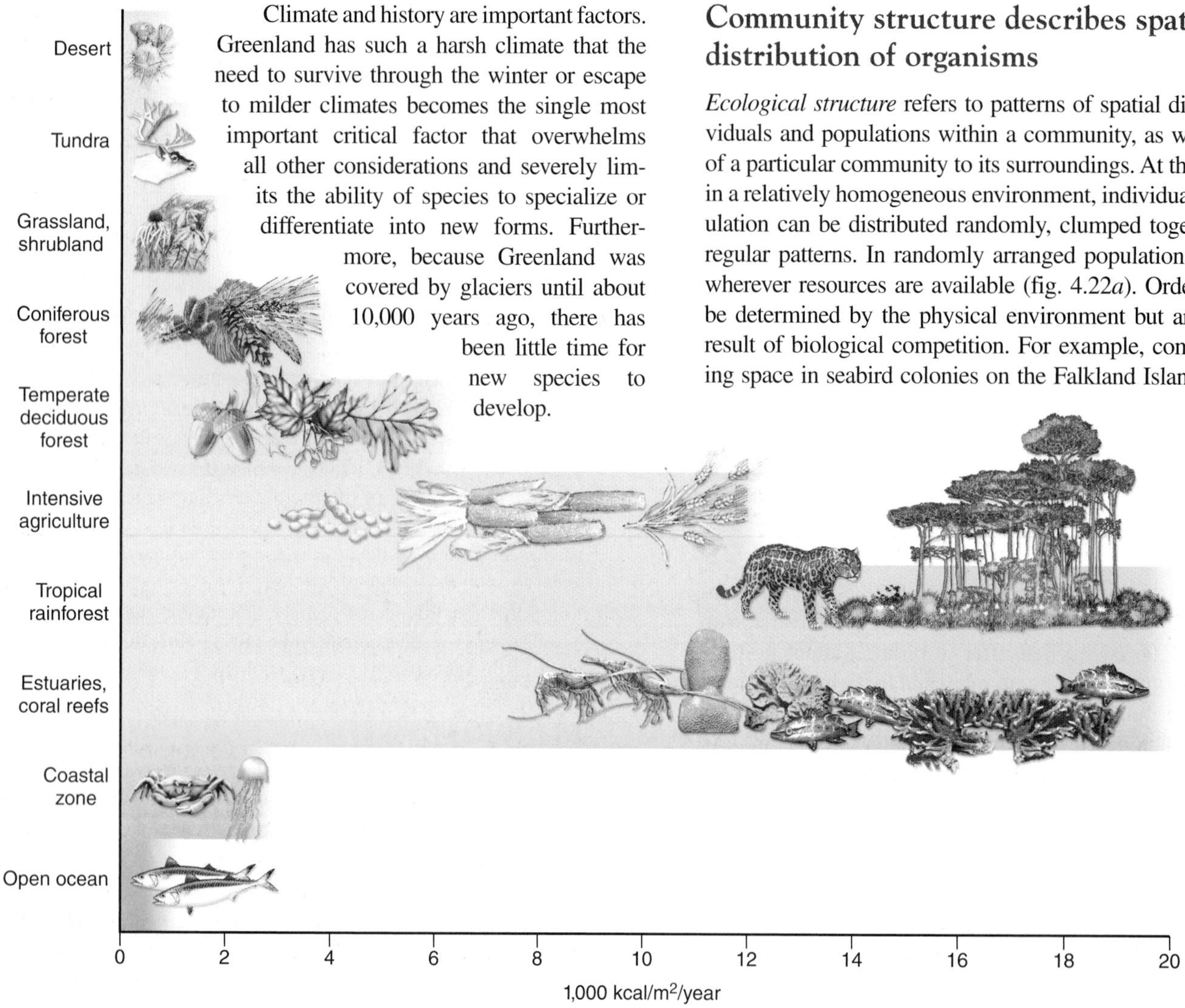

Desert
Tundra
Grassland, shrubland
Coniferous forest
Temperate deciduous forest
Intensive agriculture
Tropical rainforest
Estuaries, coral reefs
Coastal zone
Open ocean

0   2   4   6   8   10   12   14   16   18   20

1,000 kcal/m²/year

FIGURE 4.21 Relative biomass accumulation of major world ecosystems. Only plants and some bacteria capture solar energy. Animals consume biomass to build their own bodies.

(a) Random

(b) Uniform

(c) Clustered

FIGURE 4.22 Distribution of members of a population in a given space can be (a) random, (b) uniform, or (c) clustered. The physical environment and biological interactions determine these patterns. The patterns may produce a graininess or patchiness in community structure.

Each nest tends to be just out of reach of the neighbors sitting on their own nests. Constant squabbling produces a highly regular pattern (fig. 4.22b). Similarly, sagebrush releases toxins from roots and fallen leaves, which inhibit the growth of competitors and create a circle of bare ground around each bush. As neighbors fill in empty spaces up to the limit of this chemical barrier, a regular spacing results.

Some other species cluster together for protection, mutual assistance, reproduction, or access to a particular environmental resource. Dense schools of fish, for instance, cluster closely together in the ocean, increasing their chances of detecting and escaping predators (fig. 4.22c). Similarly, predators, whether sharks, wolves, or humans, often hunt in packs to catch their prey. A flock of blackbirds descending on a cornfield or a troop of baboons traveling across the African savanna band together both to avoid predators and to find food more efficiently.

Plants can cluster for protection, as well. A grove of wind-sheared evergreen trees is often found packed tightly together at the crest of a high mountain or along the seashore. They offer mutual protection from the wind not only to each other but also to other creatures that find shelter in or under their branches.

Most environments are patchy at some scale. Organisms cluster or disperse according to patchy availability of water, nutrients, or other resources. Distribution in a community can be vertical as well as horizontal. The tropical forest, for instance, has many layers, each with different environmental conditions and combinations of species. Distinct communities of smaller plants, animals, and microbes live at different levels. Similarly, aquatic communities are often stratified into layers based on light penetration in the water, temperature, salinity, pressure, or other factors.

Complexity and connectedness are important ecological indicators

Community complexity and connectedness generally are related to diversity and are important because they help us visualize and understand community functions. **Complexity** in ecological terms refers to the number of species at each trophic level and the number of trophic levels in a community. A diverse community may not be very complex if all its species are clustered in only a few trophic levels and form a relatively simple food chain.

By contrast, a complex, highly interconnected community (fig. 4.23) might have many trophic levels, some of which can be compartmentalized into subdivisions. In tropical rainforests, for instance, the herbivores can be grouped into "guilds" based on the specialized ways they feed on plants. There may be fruit eaters, leaf nibblers, root borers, seed gnawers, and sap suckers, each composed of species of very different size, shape, and even biological kingdom, but that feed in related ways. A highly interconnected community such as this can form a very elaborate food web.

Resilience and stability make communities resistant to disturbance

Many biological communities tend to remain relatively stable and constant over time. An oak forest tends to remain an oak forest, for example, because the species that make it up have self-perpetuating mechanisms. We can identify three kinds of stability or resiliency in ecosystems: *constancy* (lack of fluctuations in composition or functions), *inertia* (resistance to perturbations), and *renewal* (ability to repair damage after disturbance).

In 1955, Robert MacArthur, who was then a graduate student at Yale, proposed that the more complex and interconnected a community is, the more stable and resilient it will be in the face of disturbance. If many different species occupy each trophic level, some can fill in if others are stressed or eliminated by external forces, making the whole community resistant to perturbations and able to recover relatively easily from disruptions. This theory has been controversial, however. Some studies support it, while others do not. For example, Minnesota ecologist David Tilman, in studies of native prairie and recovering farm fields, found that plots with high diversity were better able to withstand and recover from drought than those with only a few species.

On the other hand, in a diverse and highly specialized ecosystem, removal of a few keystone members can eliminate many other associated species. Eliminating a major tree species from a tropical forest, for example, may destroy pollinators and fruit distributors as well. We might replant the trees, but could we replace the whole web of relationships on which they depend? In this case, diversity has made the forest less resilient rather than more.

FIGURE 4.23 Tropical rainforests are complex structurally and ecologically. Trees form layers, each with a different amount of light and a unique combination of flora and fauna. Many insects, arthropods, birds, and mammals spend their entire life in the canopy. In Brazil's Atlantic Rainforest, a single hectare had 450 tree species and many times that many insects. With so many species, the ecological relationships are complex and highly interconnected.

Diversity is widely considered important and has received a great deal of attention. In particular, human impacts on diversity are a primary concern of many ecologists (Exploring Science, p. 81).

Edges and boundaries are the interfaces between adjacent communities

An important aspect of community structure is the boundary between one habitat and its neighbors. We call these relationships **edge effects.** Sometimes, the edge of a patch of habitat is relatively sharp and distinct. In moving from a woodland patch into a grassland or cultivated field, you sense a dramatic change from the cool, dark, quiet forest interior to the windy, sunny, warmer, open space of the meadow (fig. 4.24). In other cases, one habitat type intergrades very gradually into another, so there is no distinct border.

Ecologists call the boundaries between adjacent communities **ecotones.** A community that is sharply divided from its neighbors is called a closed community. In contrast, communities with gradual or indistinct boundaries over which many species cross are called open communities. Often this distinction is a matter of degree or perception. As we saw earlier in this chapter, birds might feed in fields or grasslands but nest in the forest. As they fly back and forth, the birds interconnect the ecosystems by moving energy and material from one to the other, making both systems relatively open. Furthermore, the forest edge, while clearly different from the open field, may be sunnier and warmer than the forest interior, and may have a different combination of plant and animal species than either field or forest "core."

Depending on how far edge effects extend from the boundary, differently shaped habitat patches may have very dissimilar amounts of interior area (fig. 4.25). In Douglas fir forests of the Pacific Northwest, for example, increased rates of blowdown, decreased humidity, absence of shade-requiring ground cover, and other edge effects can extend as much as 200 m into a forest. A 40-acre block (about 400 m^2) surrounded by clear-cut would have essentially no true core habitat at all.

Many popular game animals, such as white-tailed deer and pheasants that are adapted to human disturbance, often are most plentiful in boundary zones between different types of habitat. Game managers once were urged to develop as much edge as possible to promote large game populations. Today, however, most wildlife conservationists recognize that the edge effects associated

FIGURE 4.24 Ecotones are edges between ecosystems. For some species, ecotones are barriers to migration, while other species find these edges a particularly hospitable habitat. There are at least two ecotones in this picture: one between the stream and the meadow, and another between the meadow and the forest. As you can see, some edges are sharp boundaries, while others, such as the edge of the forest, are gradual.

What Do You Think?

What's the Harm in Setting Unused Bait Free?

Invasions of landscapes by alien species are one of the most dramatic causes of environmental change today. Effects can range from endangerment or elimination of native species to dramatic changes in whole landscapes. Exotic species often cause huge problems. For example, when brown tree snakes (*Boiga irregularis*) were introduced into Guam in the 1950s, they extirpated most of the native forest bird and amphibian species on the island. In another example, infestation of American lakes and rivers by Zebra mussels (*Driessena polymorpha*) in the 1980s resulted in population explosions of these tiny mollusks that smother fish-spawning beds, destroy native clam species, and clog utility intake pipes. In other cases, the results are subtler and people who release foreign species may think they're doing a good deed. A case in point is the spread of European earthworm species into northern hardwood forests.

For several decades, hikers in the deciduous forests of northern Minnesota, Wisconsin, and Michigan reported that some areas of the forest floor looked strangely denuded of leaf litter and were missing familiar flower species. Ecologists suspected that exotic worm species might be responsible, especially because these denuded areas seemed to be around boat landings and along shorelines where anglers discard unwanted bait. Anglers often think they are being benevolent when they release unwanted worms. They don't realize the ecological effects they may be unleashing.

Northern forests in North America normally lack earthworms because they were removed thousands of years ago when glaciers bulldozed across the landscape. Vegetation has returned since the glaciers retreated, but worms never made it back to these forests. Over the past 10,000 years, or so, local flora and fauna have adapted to the absence of earthworms. For successful growth, seedlings depend on a thick layer of leaf mulch along with associations with fungi and invertebrates that live in the upper soil horizons. Earthworms, which eat up the litter layer,

disrupt nutrient cycling, soil organism populations, and other aspects of the forest floor community.

How do we analyze these changes and show they're really the work of worms? Andy Holdsworth, a graduate student in Conservation Biology at the University of Minnesota, studied the effects of earthworm invasion in the Chequamegon and Chippewa National Forests in Wisconsin and Minnesota. He chose 20 areas in each forest with similar forest type, soils, and management history. Each area bordered on lakes and had no logging activity in the last 40 years. On transects across each area, all vascular plants were identified and recorded. Earthworm populations were sampled both by hand sifting dirt samples and by pouring mustard extraction solutions on plots to drive worms out. Soil samples were taken for pH, texture, and density analysis.

Holdsworth found a mixture of European worm species in most of his sites, reflecting the diversity used as fishing bait. By plotting worm biomass against plant diversity, he showed that worm infestation rates correlated with decreased plant species richness and abundance. Among the species most likely to be missing in worm-invaded plots were wild sarsaparilla (*Aralia nudicaulis*), big-leaved aster (*Aster macrophyllus*), rose twisted stalk (*Streptopus roseus*), hairy Solomon's seal (*Polygonatum pubescens*), and princess pine (*Lycopodium obscurum*).

Perhaps most worrisome were low numbers of some tree species, especially sugar maple (*Acer saccharum*) and basswood (*Tilia americana*), which are among the defining species of these forests. Adult trees don't seem to be adversely affected by the presence of exotic worms, but their seedlings require deep leaf litter to germinate, litter that is consumed by earthworms when infestations are high. This study and others like it suggest that invasions of the lowly worm may lead eventually to dramatic changes in the composition and structure of whole forests.

What do you think? How can we minimize the impacts of well-meaning actions such as setting unused worms free? How can we know when small acts of benevolence to one organism can cause wholesale damage to others? Can you think of other misinformed but well-intentioned actions in your community?

For more information, see A. Holdsworth, L. Frelich, and P. Reich. 2007. *Conservation Biology* 21(4): 997–1008.

No worm invasion.

Heavy worm invasion.

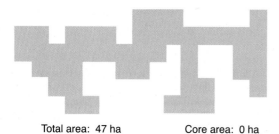

Total area: 47 ha   Core area: 0 ha

Total area: 47 ha   Core area: 20 ha

FIGURE 4.25 Shape can be as important as size in small preserves. While these areas are similar in size, no place in the top figure is far enough from the edge to have characteristics of core habitat, while the bottom patch has a significant core.

with habitat fragmentation are generally detrimental to biodiversity. Preserving large habitat blocks and linking smaller blocks with migration corridors may be the best ways to protect rare and endangered species (chapter 12).

4.4 COMMUNITIES ARE DYNAMIC AND CHANGE OVER TIME

If fire sweeps through a biological community, it's destroyed, right? Not so fast. Fire may be good for that community. Up until now, we've focused on the day-to-day interactions of organisms with their environments, set in a context of adaptation and selection. In this section, we'll step back and look at more dynamic aspects of communities and how they change over time.

The nature of communities is debated

For several decades starting in the early 1900s, ecologists in North America and Europe argued about the basic nature of communities. It doesn't make interesting party conversation, but those discussions affected how we study and understand communities, view the changes taking place within them, and ultimately use them. Both J. E. B. Warming (1841–1924) in Denmark and Henry Chandler Cowles (1869–1939) in the United States came up with the idea that communities develop in a sequence of stages, starting either from bare rock or after a severe disturbance. They worked in sand dunes and watched the changes as plants first took root in bare sand and, with further development, created forest. This example represents constant change, not stability. In sand dunes, the community that developed last and lasted the longest was called the **climax community.**

The idea of climax community was first championed by the biogeographer F. E. Clements (1874–1945). He viewed the process as a relay—species replace each other in predictable groups and in a fixed, regular order. He argued that every landscape has a characteristic climax community, determined mainly by climate. If left undisturbed, this community would mature to a characteristic set of organisms, each performing its optimal functions. A climax community to Clements represented the maximum complexity and stability that was possible. He and others made the analogy that the development of a climax community resembled the maturation of an organism. Both communities and organisms, they argued, began simply and primitively, maturing until a highly integrated, complex community developed.

This organismal theory of community was opposed by Clements' contemporary, H. A. Gleason (1882–1975), who saw community history as an unpredictable process. He argued that species are individualistic, each establishing in an environment according to its ability to colonize, tolerate the environmental conditions, and reproduce there. This idea allows for myriad temporary associations of plants and animals to form, fall apart, and reconstitute in slightly different forms, depending on environmental conditions and the species in the neighborhood. Imagine a time-lapse movie of a busy airport terminal. Passengers come and go; groups form and dissipate. Patterns and assemblages that seem significant may not mean much a year later. Gleason suggested that we think ecosystems are uniform and stable only because our lifetimes are too short and our geographic scope too limited to understand their actual dynamic nature.

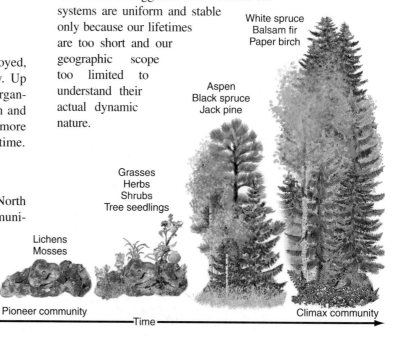

White spruce
Balsam fir
Paper birch

Aspen
Black spruce
Jack pine

Grasses
Herbs
Shrubs
Tree seedlings

Lichens
Mosses

Exposed rocks

Pioneer community

Time

Climax community

FIGURE 4.26 One example of primary succession, shown in five stages (*left to right*). Here, bare rocks are colonized by lichens and mosses, which trap moisture and build soil for grasses, shrubs, and eventually trees.

Ecological succession describes a history of community development

In any landscape, you can read the history of biological communities. That history is revealed by the process of ecological succession. During succession, organisms occupy a site and change the environmental conditions. In **primary succession** land that is bare of soil—a sandbar, mudslide, rock face, volcanic flow—is colonized by living organisms where none lived before (fig. 4.26). When an existing community is disturbed, a new one develops from the biological legacy of the old in a process called **secondary succession.** In both kinds of succession, organisms change the environment by modifying soil, light levels, food supplies, and microclimate. This change permits new species to colonize and eventually replace the previous species, a process known as ecological development or facilitation.

In primary succession on land, the first colonists are hardy **pioneer species,** often microbes, mosses, and lichens that can withstand a harsh environment with few resources. When they die, the bodies of pioneer species create patches of organic matter. Organics and other debris accumulate in pockets and crevices, creating soil where seeds lodge and grow. As succession proceeds, the community becomes more diverse and interspecies competition arises. Pioneers disappear as the environment favors new colonizers that have competitive abilities more suited to the new environment.

You can see secondary succession all around you, in abandoned farm fields, in clear-cut forests, and in disturbed suburbs and lots. Soil and possibly plant roots and seeds are present. Because soil lacks vegetation, plants that live one or two years (annuals and biennials) do well. Their light seeds travel far on the wind, and their seedlings tolerate full sun and extreme heat. When they die, they lay down organic material that improves the soil's fertility and shelters other seedlings. Soon long-lived and deep-rooted perennial grasses, herbs, shrubs, and trees take hold, building up the soil's organic matter and increasing its ability to store moisture. Forest species that cannot survive bare, dry, sunny ground eventually find ample food, a diverse community structure, and shelter from drying winds and low humidity.

Generalists figure prominently in early succession. Over thousands of years, however, competition should decrease as niches proliferate and specialists arise. In theory, long periods of community development lead to greater community complexity, high nutrient conservation and recycling, stable productivity, and great resistance to disturbance—an ideal state to be in when the slings and arrows of misfortune arrive.

Appropriate disturbances can benefit communities

Disturbances are plentiful on earth: landslides, mudslides, hailstorms, earthquakes, hurricanes, tornadoes, tidal waves, wildfires, and volcanoes, to name just the obvious. A **disturbance** is any force that disrupts the established patterns of species diversity and abundance, community structure, or community properties. Animals can cause disturbance. African elephants rip out small trees, trample shrubs, and tear down tree limbs as they forage and move about, opening up forest communities and creating savannas. People also cause disturbances

FIGURE 4.27 These "stump barrens" in Michigan's Upper Peninsula were created over a century ago when clear-cutting of dense white pine forest was followed by repeated burning. The stumps are left from the original forest, which has not grown back in more than 100 years.

with agriculture, forestry, new roads and cities, and construction projects for dams and pipelines. It is customary in ecology to distinguish between natural disturbances and human-caused (or anthropogenic) disturbances, but a subtle point of clarification is needed.

Aboriginal populations have disturbed and continue to disturb communities around the world, setting fire to grasslands and savannas, practicing slash-and-burn agriculture in forests, and so on. Because their populations often are or were relatively small, the disturbances are patchy and limited in scale in forests, or restricted to quickly passing wildfires in grasslands, savannas, or woodland, which are comprised of species already adapted to fire.

The disturbances caused by technologically advanced and numerous people however, may be very different from the disturbances caused by small groups of aborigines. In the Kingston Plains of Michigan's Upper Peninsula, clear-cut logging followed by repeated human-set fires from 1880 to 1900 caused a change in basic ecological conditions such that the white pine forest has never regenerated (fig. 4.27). Given the right combination of disturbances by modern people, or by nature, it may take hundreds of years for a community to return to its predisturbance state.

Ecologists generally find that disturbance benefits most species, much as predation does, because it sets back supreme competitors and allows less-competitive species to persist. In northern temperate forests, maples (especially sugar maple) are more prolific seeders and more shade tolerant at different stages of growth than nearly any other tree species. Given decades of succession, maples outcompete other trees for a place in the forest canopy. Most species of oak, hickory, and other light-requiring trees diminish in abundance, as do species of forest herbs. The dense shade of maples basically starves other species for light. When windstorms, tornadoes, wildfires, or ice storms hit a maple forest, trees are toppled, branches broken, and light again reaches the forest floor and stimulates seedlings of oaks and hickories, as well as forest herbs. Breaking the grip of a supercompetitor is the helpful role disturbances often play.

FIGURE 4.28 This lodgepole pine forest in Yellowstone National Park was once thought to be a climax forest, but we now know that this forest must be constantly renewed by periodic fire. It is an example of an equilibrium, or disclimax, community.

Some landscapes never reach a stable climax in a traditional sense because they are characterized by periodic disturbance and are made up of **disturbance-adapted species** that survive fires underground, or resist the flames, and then reseed quickly after fires. Grasslands, the chaparral scrubland of California and the Mediterranean region, savannas, and some kinds of coniferous forests are shaped and maintained by periodic fires that have long been a part of their history (fig. 4.28). In fact, many of the dominant plant species in these communities need fire to suppress competitors, to prepare the ground for seeds to germinate, or to pop open cones or split thick seed coats and release seeds. Without fire, community structure would be quite different.

People taking an organismal view of such communities believe that disturbance is harmful. In the early 1900s this view merged with the desire to protect timber supplies from ubiquitous wildfires, and to store water behind dams while also controlling floods. Fire suppression and flood control became the central policies in American natural resource management (along with predator control) for most of the twentieth century. Recently, new concepts about natural disturbances are entering land management discussions and bringing change to land management policies. Grasslands and some forests are now considered "fire-adapted" and fires are allowed to burn in them if weather conditions are appropriate. Floods also are seen as crucial for maintaining flood-plain and river health. Policymakers and managers increasingly consider ecological information when deciding on new dams and levee construction projects.

From another view, disturbance resets the successional clock that always operates in every community. Even though all seems chaotic after a disturbance, it may be that preserving species diversity by allowing in natural disturbances (or judiciously applied human disturbances) actually ensures stability over the long run, just as diverse prairies managed with fire recover after drought. In time, community structure and productivity get back to normal, species diversity is preserved, and nature seems to reach its dynamic balance.

Introduced species can cause profound community change

Succession requires the continual introduction of new community members and the disappearance of previously existing species. New species move in as conditions become suitable; others die or move out as the community changes. New species also can be introduced after a stable community already has become established. Some cannot compete with existing species and fail to become established. Others are able to fit into and become part of the community, defining new ecological niches. If, however, an introduced species preys upon or competes more successfully with one or more populations that are native to the community, the entire nature of the community can be altered.

Human introductions of Eurasian plants and animals to non-Eurasian communities often have been disastrous to native

FIGURE 4.29 Mongooses were released in Hawaii in an effort to control rats. The mongooses are active during the day, however, while the rats are night creatures, so they ignored each other. Instead, the mongooses attacked defenseless native birds and became as great a problem as the rats.

species because of competition or overpredation. Oceanic islands offer classic examples of devastation caused by rats, goats, cats, and pigs liberated from sailing ships. All these animals are prolific, quickly developing large populations. Goats are efficient, nonspecific herbivores; they eat nearly everything vegetational, from grasses and herbs to seedlings and shrubs. In addition, their sharp hooves are hard on plants rooted in thin island soils. Rats and pigs are opportunistic omnivores, eating the eggs and nestlings of seabirds that tend to nest in large, densely packed colonies, and digging up sea turtle eggs. Cats prey upon nestlings of both ground- and tree-nesting birds. Native island species are particularly vulnerable because they have not evolved under circumstances that required them to have defensive adaptations to these predators.

Sometimes we introduce new species in an attempt to solve problems created by previous introductions but end up making the situation worse. In Hawaii and on several Caribbean islands, for instance, mongooses were imported to help control rats that had escaped from ships and were destroying indigenous birds and devastating plantations (fig. 4.29). Since the mongooses were diurnal (active in the day), however, and rats are nocturnal, they tended to ignore each other. Instead, the mongooses also killed native birds and further threatened endangered species. Our lessons from this and similar introductions have a new technological twist. Some of the ethical questions currently surrounding the release of genetically engineered organisms are based on concerns that they are novel organisms, and we might not be able to predict how they will interact with other species in natural ecosystems—let alone how they might respond to natural selective forces. It is argued that we can't predict either their behavior or their evolution.

CONCLUSION

Evolution is one of the key organizing principles of biology. It explains how species diversity originates, and how organisms are able to live in highly specialized ecological niches. Natural selection, in which beneficial traits are passed from survivors in one generation to their progeny, is the mechanism by which evolution occurs. Species interactions—competition, predation, symbiosis, and coevolution—are important factors in natural selection. The unique set of organisms and environmental conditions in an ecological community give rise to important properties, such as productivity, abundance, diversity, structure, complexity, connectedness, resilience, and succession. Human introduction of new species as well as removal of existing ones can cause profound changes in biological communities and can compromise the life-supporting ecological services on which we all depend. Understanding these community ecology principles is a vital step in becoming an educated environmental citizen.

REVIEWING LEARNING OUTCOMES

By now you should be able to explain the following points:

4.1 Describe how evolution produces species diversity.
- Natural selection leads to evolution.
- All species live within limits.
- The ecological niche is a species' role and environment.
- Speciation maintains species diversity.
- Evolution is still at work.
- Taxonomy describes relationships among species.

4.2 Discuss how species interactions shape biological communities.
- Competition leads to resource allocation.
- Predation affects species relationships.
- Some adaptations help avoid predation.
- Symbiosis involves intimate relations among species.
- Keystone species have disproportionate influence.

4.3 Summarize how community properties affect species and populations.
- Productivity is a measure of biological activity.
- Abundance and diversity measure the number and variety of organisms.
- Community structure describes spatial distribution of organisms.
- Complexity and connectedness are important ecological indicators.
- Resilience and stability make communities resistant to disturbance.
- Edges and boundaries are the interfaces between adjacent communities.

4.4 Explain why communities are dynamic and change over time.
- The nature of communities is debated.
- Ecological succession describes a history of community development.
- Appropriate disturbances can benefit communities.
- Introduced species can cause profound community change.

PRACTICE QUIZ

1. Explain how tolerance limits to environmental factors determine distribution of a highly specialized species such as the saguaro cactus.

2. Productivity, diversity, complexity, resilience, and structure are exhibited to some extent by all communities and ecosystems. Describe how these characteristics apply to the ecosystem in which you live.

3. Define *selective pressure* and describe one example that has affected species where you live.

4. Define *keystone species* and explain their importance in community structure and function.

5. The most intense interactions often occur between individuals of the same species. What concept discussed in this chapter can be used to explain this phenomenon?

6. Explain how predators affect the adaptations of their prey.

7. Competition for a limited quantity of resources occurs in all ecosystems. This competition can be interspecific or intraspecific. Explain some of the ways an organism might deal with these different types of competition.

8. Describe the process of succession that occurs after a forest fire destroys an existing biological community. Why may periodic fire be beneficial to a community?

9. Which world ecosystems are most productive in terms of biomass (fig. 4.21)? Which are least productive? What units are used in this figure to quantify biomass accumulation?

10. Discuss the dangers posed to existing community members when new species are introduced into ecosystems.

CRITICAL THINKING AND DISCUSSION QUESTIONS

1. The concepts of natural selection and evolution are central to how most biologists understand and interpret the world, and yet the theory of evolution is contrary to the beliefs of many religious groups. Why do you think this theory is so important to science and so strongly opposed by others? What evidence would be required to convince opponents of evolution?

2. What is the difference between saying that a duck has webbed feet because it needs them to swim and saying that a duck is able to swim because it has webbed feet?

3. The concept of keystone species is controversial among ecologists because most organisms are highly interdependent. If each of the trophic levels is dependent on all the others, how can we say one is most important? Choose an ecosystem with which you are familiar and decide whether it has a keystone species or keystone set.

4. Some scientists look at the boundary between two biological communities and see a sharp dividing line. Others looking at the same boundary see a gradual transition with much intermixing of species and many interactions between communities. Why are there such different interpretations of the same landscape?

5. The absence of certain lichens is used as an indicator of air pollution in remote areas such as national parks. How can we be sure that air pollution is really responsible? What evidence would be convincing?

6. We tend to regard generalists or "weedy" species as less interesting and less valuable than rare and highly specialized endemic species. What values or assumptions underlie this attitude?

Data Analysis: Species Competition

In a classic experiment on competition between species for a common food source, the Russian microbiologist G. F. Gause grew populations of different species of ciliated protozoans separately and together in an artificial culture medium. He counted the number of cells of each species and plotted the total volume of each population. The organisms were *Paramecium caudatum* and its close relative, *Paramecium aurelia*. He plotted the aggregate volume of cells rather than the total number in each population because *P. caudatum* is much larger than *P. aurelia* (this

size difference allowed him to distinguish between them in a mixed culture). The graphs in this box show the experimental results. As we mentioned earlier in the text, this was one of the first experimental demonstrations of the principle of competitive exclusion. After studying these graphs, answer the following questions.

1. How do you read these graphs? What is shown in the top and bottom panels?

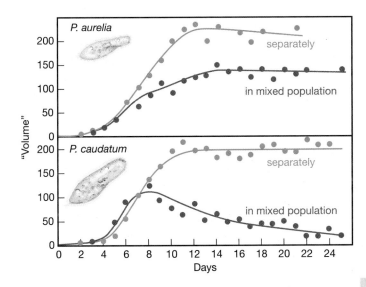

Growth of two *paramecium* species separately and in combination.

Source: Gause, Georgyi Frantsevitch. 1934 *The Struggle for Existence*. Dover Publications, 1971 reprint of original text.

2. How did the total volume of the two species compare after 14 days of separate growth?

3. If *P. caudatum* is roughly twice as large as *P. aurelia*, how did the total number of cells compare after 14 days of separate growth?

4. How did the total volume of the two species compare after 24 days of growth in a mixed population?

5. Which of the two species is the more successful competitor in this experiment?

6. Does the larger species always win in competition for food? Why not?

Kenya's Greenbelt Movement has planted millions of trees and inspired other groups to plant billions more.

Biomes

Global Patterns of Life

Learning Outcomes

After studying this chapter, you should be able to:

5.1 Recognize the characteristics of some major terrestrial biomes as well as the factors that determine their distribution.

5.2 Understand how and why marine environments vary with depth and distance from shore.

5.3 Compare the characteristics and biological importance of major freshwater ecosystems.

5.4 Summarize the overall patterns of human disturbance of world biomes.

"What is the use of a house if you haven't got a tolerable planet to put it on?"

~ Henry David Thoreau

Case Study

Spreading Green Across Kenya

Our environment provides all of us with food, fuel, and shelter, but the world's poorest people often depend most directly on their environment—and they suffer most from a degraded environment. In remote areas of rural Kenya subsistence farmers depend on local forests, soils, and groundwater for fuel, food, and water. What can these villagers do when overuse or careless management degrades these resources? This question challenges rural communities throughout the developing world.

Kenya is a biologically rich country, but millions of people live in severe poverty. Growing populations of farmers and herders depend on dwindling forests and degraded soils. Many forests were cleared decades ago for farming, and remaining woodlands are decimated by people gathering fuel and building materials, as well as by farming and grazing. As the forests disappear, the land becomes dry, soils wash away, and women must travel farther in search of fuelwood. Because women traditionally have responsibility for gathering wood and water, and because they have little economic or political power, women and their children suffer most directly from forest losses. Environmental degradation causes economic instability, and families further exploit remaining forests, causing increasing environmental degradation, in a downward spiral of poverty.

Kenya's story of environmental and social degradation can be found in developing areas worldwide. But Kenyans also have found a strategy to combat the combined problems of social, economic, and environmental devastation. The Greenbelt Movement, initiated by the environmental leader Dr. Wangari Maathai, (fig. 5.1) is working to teach communities to help themselves by growing and planting trees. Starting with the women and expanding to include their families, the movement is mobilizing people to help themselves. In the process, the Greenbelt Movement is teaching peaceful political involvement and local community development.

Dr. Maathai started out working on both environmental issues and women's empowerment. A native of Nyeri, Kenya, she studied in the United States and Germany in the 1960s and 1970s, and earned a PhD from the University of Nairobi. Dr. Maathai taught at the University of Nairobi, worked with the United Nations Environment Program (UNEP), based in Nairobi, and eventually became chair of the National Council of Women of Kenya (NCWK).

According to Dr. Maathai, women in the villages told her they suffered from the loss of trees, so she suggested they plant new trees. But the women said they didn't know how to plant trees.

FIGURE 5.1 Dr. Wangari Maathai has worked to restore trees, communities, and peace.

And so this is where the Greenbelt Movement began, helping villagers create nurseries, grow seedlings, and plant trees. In 1977, Dr. Maathai and her colleagues from the NCWK celebrated World Environment Day by planting trees, the first of what would eventually become an international Greenbelt movement.

Dr. Maathai's experience in multiple fields—like that of many environmental scientists—helped her see the deep ties between the powerlessness of poor women and environmental conditions that made their lives difficult. Dr. Maathai understood that many rural women had to walk miles every day for wood or water, in addition to tending to their farms and families. Making matters worse, many poor women depended on small farm plots with eroding, worn-out soils, to feed their families.

The tree-planting work started small and grew slowly, but it has endured, and community-based reforestation has grown and spread broad roots across Kenya. The Greenbelt Movement has trained people from around the world, and it now has branches in other countries in Africa, Asia, and South America. The program supports community networks that care for over 6,000 tree nurseries. The movement also promotes peace, education, and civic leadership, and recently it has expanded its vision to include climate mitigation through forest conservation. Thousands of community members have planted more than 40 million trees on degraded and eroding lands, in school yards and church yards, on farms, and in cities and villages. Goals of environmental quality and social justice remain a very long way off, but the movement has restored thousands of hectares of land, and it has brought hope to millions. In 2004, Dr. Maathai received the Nobel Prize for Peace for her work on promoting peace through environmental stewardship and social justice.

Tree planting is a powerful act of hope. Planting a tree is an investment in the future, empowering people and showing the world that we care about those who will follow after we are gone. Expanding tree cover in once-forested lands helps nurture soils, biodiversity, and communities. The Greenbelt Movement shows that we have many choices other than simply watching while our environment deteriorates.

Finding ways to live sustainably within the limits of our resource bases, without damaging the life-support systems of ecosystems, is a preeminent challenge of environmental science. Sometimes, as this case study shows, ecological knowledge and local action can lead to positive effects on a global scale. We'll examine these and related issues in this chapter. For related resources, including Google Earth™ placemarks that show locations where these issues can be seen, visit, http://EnvironmentalScience-Cunningham.blogspot.com.

5.1 TERRESTRIAL BIOMES

The Greenbelt movement aims to restore components of an expansive biological community. To understand what that community should be like, it is helpful for us to identify some of the general types of communities, with similar climate conditions, growth patterns, and vegetation types. We call these broad types of biological communities **biomes**. Understanding the global distribution of biomes, and knowing the differences in what grows where and why, is essential to the study of global environmental science. Biological productivity—and ecosystem resilience—varies greatly from one biome to another. Human use of biomes depends largely on those levels of productivity. Our ability to restore ecosystems and nature's ability to restore itself depend largely on biome conditions. Clear-cut forests can regrow relatively quickly in Kenya, but very slowly in Siberia, where logging is currently expanding. Some grasslands rejuvenate rapidly after grazing, and some are very slow to recover. Why these differences? The sections that follow seek to answer this question.

Temperature and precipitation are the most important determinants in biome distribution on land (fig. 5.2). If we know the general temperature range and precipitation level, we can predict what kind of biological community is likely to occur there in the absence of human disturbance.

Because temperatures are cooler at high latitudes (away from the equator), temperature-controlled biomes often occur in latitudinal bands. For example, a band of boreal (northern) forests crosses Canada and Siberia, tropical forests occur near the equator, and expansive grasslands lie near—or just beyond—the tropics (fig. 5.3). Many biomes are even named for their latitudes. Tropical rainforests occur between the Tropic of Cancer (23° north) and the Tropic of Capricorn (23° south); arctic tundra lies near or above the Arctic Circle (66.6° north).

Temperature and precipitation change with elevation as well as with latitude. In mountainous regions, temperatures are cooler and precipitation is usually greater at high elevations. **Vertical zonation** occurs as vegetation types change rapidly from warm and dry to cold and wet as you go up a mountain. A 100-km transect from California's Central Valley up to Mt. Whitney, for example, crosses as many vegetation zones as you would find on a journey from southern California to northern Canada (fig. 5.4).

To compare terrestrial biomes, we often use climate graphs, which show yearly temperature and precipitation. Look carefully at the examples in figure 5.5. Note that months are shown across the bottom, including months above freezing, when primary productivity (plant growth) is active. Temperature and precipitation have different vertical axes and different units (what are the units?). Shading shows when precipitation exceeds evaporation (blue), so that there is plenty of plant growth, and when evaporation exceeds precipitation, and conditions are too dry for abundant vegetation growth. Examine these graphs, and consider the seasonal conditions that control primary productivity as you read about the different biomes.

In this chapter, we'll examine the major terrestrial biomes, then we'll investigate ocean and freshwater communities and environments. Ocean environments are important because they cover two-thirds of the earth's surface, provide food for much of humanity, and help regulate our climate through photosynthesis. Freshwater systems have tremendous influence on environmental health, biodiversity, and water quality. In chapter 12, we'll look at how we use these communities; and in chapter 13, we'll see how we preserve, manage, and restore them when they're degraded.

Tropical moist forests have rain year-round

The humid tropical regions of the world support one of the most complex and biologically rich biome types in the world (fig. 5.6). Although there are several kinds of moist tropical forests, they share common attributes of ample rainfall and uniform temperatures. Cool **cloud forests** are found high in the mountains where fog and mist keep vegetation wet all the time. **Tropical rainforests** occur where rainfall is abundant—more than 200 cm (80 in.) per year—and

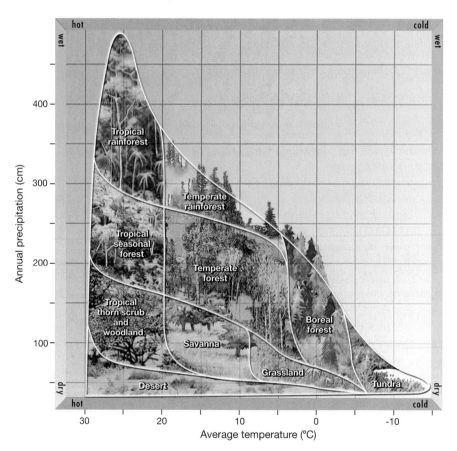

FIGURE 5.2 Biomes most likely to occur in the absence of human disturbance or other disruptions, according to average annual temperature and precipitation.
Note: This diagram does not consider soil type, topography, wind speed, or other important environmental factors. Still, it is a useful general guideline for biome location.

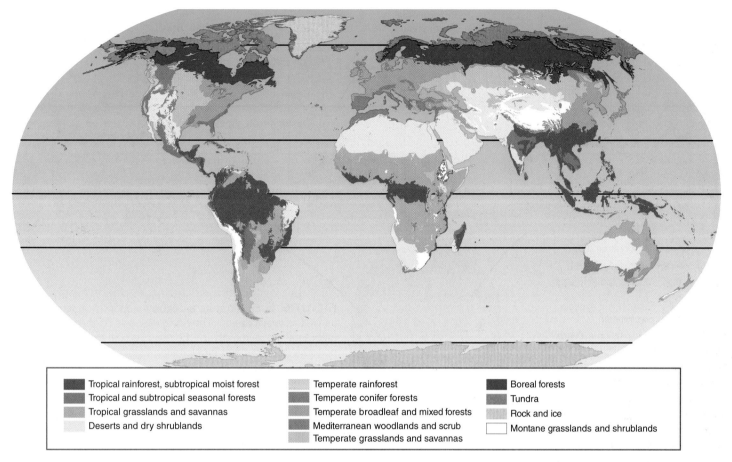

■ Tropical rainforest, subtropical moist forest	■ Temperate rainforest	■ Boreal forests
■ Tropical and subtropical seasonal forests	■ Temperate conifer forests	■ Tundra
■ Tropical grasslands and savannas	■ Temperate broadleaf and mixed forests	■ Rock and ice
■ Deserts and dry shrublands	■ Mediterranean woodlands and scrub	□ Montane grasslands and shrublands
	■ Temperate grasslands and savannas	

FIGURE 5.3 Major world biomes. Compare this map to figure 5.2 for generalized temperature and moisture conditions that control biome distribution. Also compare it to the satellite image of biological productivity (fig. 5.14).
Source: WWF Ecoregions.

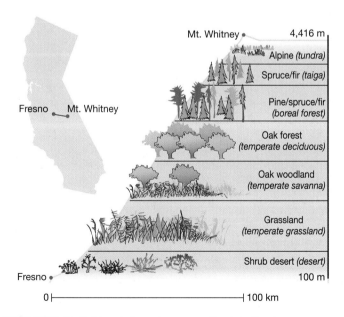

FIGURE 5.4 Vegetation changes with elevation because temperatures are lower and precipitation is greater high on a mountainside. A 100-km transect from Fresno, California, to Mt. Whitney (California's highest point) crosses vegetation zones similar to about seven different biome types.

temperatures are warm to hot year-round. For aid in reading the climate graphs in these figures, see the Data Analysis box at the end of this chapter.

The soil of both these tropical moist forest types tends to be old, thin, acidic, and nutrient-poor, yet the number of species present can be mind-boggling. For example, the number of insect species in the canopy of tropical rainforests has been estimated to be in the millions! It is estimated that one-half to two-thirds of all species of terrestrial plants and insects live in tropical forests.

The nutrient cycles of these forests also are distinctive. Almost all (90 percent) of the nutrients in the system are contained in the bodies of the living organisms. This is a striking contrast to temperate forests, where nutrients are held within the soil and made available for new plant growth. The luxuriant growth in tropical rainforests depends on rapid decomposition and recycling of dead organic material. Leaves and branches that fall to the forest floor decay and are incorporated almost immediately back into living biomass.

When the forest is removed for logging, agriculture, and mineral extraction, the thin soil cannot support continued cropping and cannot resist erosion from the abundant rains. And if the cleared area is too extensive, it cannot be repopulated by the

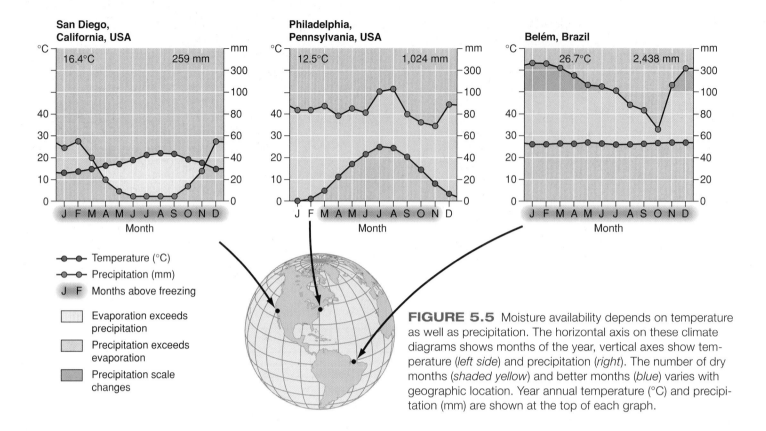

FIGURE 5.5 Moisture availability depends on temperature as well as precipitation. The horizontal axis on these climate diagrams shows months of the year, vertical axes show temperature (*left side*) and precipitation (*right*). The number of dry months (*shaded yellow*) and better months (*blue*) varies with geographic location. Year annual temperature (°C) and precipitation (mm) are shown at the top of each graph.

rainforest community. Rapid deforestation is occurring in many tropical areas as people move into the forests to establish farms and ranches, but the land soon loses its fertility.

Tropical seasonal forests have yearly dry seasons

Many tropical regions are characterized by distinct wet and dry seasons, although temperatures remain hot year-round. These areas support **tropical seasonal forests:** drought-tolerant forests that look brown and dormant in the dry season but burst into vivid green during rainy months. These forests are often called dry tropical forests because they are dry much of the year; however, there must be some periodic rain to support tree growth. Many of the trees and shrubs in a seasonal forest are drought-deciduous: They lose their leaves and cease growing when no water is available. Seasonal forests are often open woodlands that grade into savannas.

Tropical dry forests have typically been more attractive than wet forests for human habitation and have suffered greater degradation. Clearing a dry forest with fire is relatively easy during the dry season. Soils of dry forests often have higher nutrient levels and are more agriculturally productive than those of a rainforest. Finally, having fewer insects, parasites, and fungal diseases than a wet forest makes a dry or seasonal forest a healthier place for humans to live. Consequently, these forests are highly endangered in many places. Less than 1 percent of the dry tropical forests of the Pacific coast of Central America or the Atlantic coast of South America, for instance, remain in an undisturbed state.

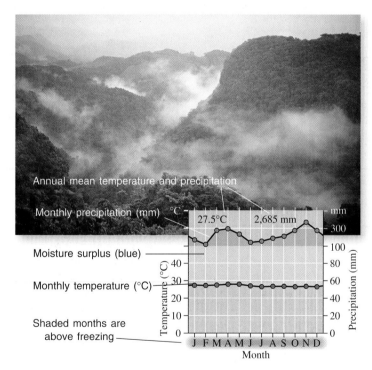

FIGURE 5.6 Tropical rainforests have luxuriant and diverse plant growth. Heavy rainfall in most months, shown in the climate graph, supports this growth.

Tropical savannas, grasslands support few trees

Where there is too little rainfall to support forests, we find open **grasslands** or grasslands with sparse tree cover, which we call **savannas** (fig. 5.7). Like tropical seasonal forests, most tropical savannas and grasslands have a rainy season, but generally the rains are less abundant or less dependable than in a forest. During dry seasons, fires can sweep across a grassland, killing off young trees and keeping the landscape open. Savanna and grassland plants have many adaptations to survive drought, heat, and fires. Many have deep, long-lived roots that seek groundwater and that persist when leaves and stems above the ground die back. After a fire, or after a drought, fresh green shoots grow quickly from the roots. Migratory grazers, such as wildebeest, antelope, or bison thrive on this new growth. Grazing pressure from domestic livestock is an important threat to both the plants and animals of tropical grasslands and savannas.

Deserts are hot or cold, but all are dry

You may think of deserts as barren and biologically impoverished. Their vegetation is sparse, but it can be surprisingly diverse, and most desert plants and animals are highly adapted to survive long droughts, extreme heat, and often extreme cold. **Deserts** occur where precipitation is rare and unpredictable, usually with less than 30 cm of rain per year. Adaptations to these conditions include water-storing leaves and stems, thick epidermal layers to reduce water loss, and salt tolerance. As in other dry environments, many plants are drought-deciduous. Most desert plants also bloom and set seed quickly when a spring rain does fall.

Warm, dry, high-pressure climate conditions (chapter 15) create desert regions at about 30° north and south. Extensive deserts occur in continental interiors (far from oceans, which evaporate the moisture for most precipitation) of North America, Central Asia, Africa, and Australia (fig. 5.8). The rain shadow of the Andes produces the world's driest desert in coastal Chile. Deserts can also be cold. Antarctica is a desert. Some inland valleys apparently get almost no precipitation at all.

Like plants, animals in deserts are specially adapted. Many are nocturnal, spending their days in burrows to avoid the sun's heat and desiccation. Pocket mice, kangaroo rats, and gerbils can get most of their moisture from seeds and plants. Desert rodents also have highly concentrated urine and nearly dry feces that allow them to eliminate body waste without losing precious moisture.

Deserts are more vulnerable than you might imagine. Sparse, slow-growing vegetation is quickly damaged by off-road vehicles. Desert soils recover slowly. Tracks left by army tanks practicing in California deserts during World War II can still be seen today.

Deserts are also vulnerable to overgrazing. In Africa's vast Sahel (the southern edge of the Sahara Desert), livestock are destroying much of the plant cover. Bare, dry soil becomes drifting sand, and restabilization is extremely difficult. Without plant roots and organic matter, the soil loses its ability to retain what rain does fall, and the land becomes progressively drier and more

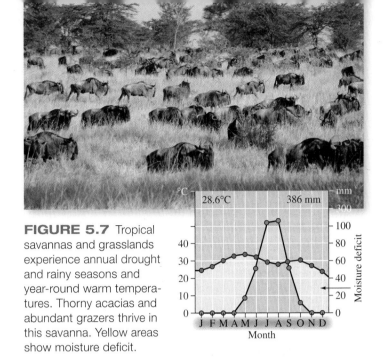

FIGURE 5.7 Tropical savannas and grasslands experience annual drought and rainy seasons and year-round warm temperatures. Thorny acacias and abundant grazers thrive in this savanna. Yellow areas show moisture deficit.

bare. Similar depletion of dryland vegetation is happening in many desert areas, including Central Asia, India, and the American Southwest and Plains states.

Temperate grasslands have rich soils

As in tropical latitudes, temperate (midlatitude) grasslands occur where there is enough rain to support abundant grass but not enough for forests (fig. 5.9). Usually grasslands are a complex, diverse mix of grasses and flowering herbaceous plants, generally known as forbs. Myriad flowering forbs make a grassland colorful and lovely in summer. In dry grasslands, vegetation may be less than a meter tall. In more humid areas, grasses can exceed 2 m. Where scattered trees occur in a grassland, we call it a savanna.

Deep roots help plants in temperate grasslands and savannas survive drought, fire, and extreme heat and cold. These roots, together with an annual winter accumulation of dead leaves on the surface, produce thick, organic-rich soils in temperate grasslands. Because of this rich soil, many grasslands have been converted to farmland. The legendary tallgrass prairies of the central United States and Canada are almost completely replaced by corn, soybeans, wheat, and other crops. Most remaining grasslands in this region are too dry to support agriculture, and their greatest threat is overgrazing. Excessive grazing eventually kills even deep-rooted plants. As ground cover dies off, soil erosion results, and unpalatable weeds, such as cheatgrass or leafy spurge, spread.

Temperate shrublands have summer drought

Often, dry environments support drought-adapted shrubs and trees, as well as grass. These mixed environments can be highly variable. They can also be very rich biologically. Such conditions are often described as Mediterranean (where the hot season coincides with the dry season producing hot, dry summers and cool,

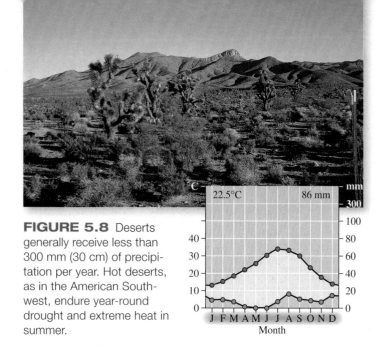

FIGURE 5.8 Deserts generally receive less than 300 mm (30 cm) of precipitation per year. Hot deserts, as in the American Southwest, endure year-round drought and extreme heat in summer.

22.5°C   86 mm

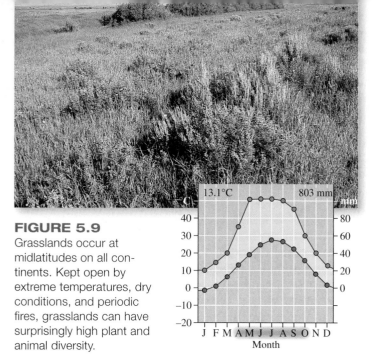

FIGURE 5.9 Grasslands occur at midlatitudes on all continents. Kept open by extreme temperatures, dry conditions, and periodic fires, grasslands can have surprisingly high plant and animal diversity.

13.1°C   803 mm

moist winters). Evergreen shrubs with small, leathery, sclerophyllous (hard, waxy) leaves form dense thickets. Scrub oaks, drought-resistant pines, or other small trees often cluster in sheltered valleys. Periodic fires burn fiercely in this fuel-rich plant assemblage and are a major factor in plant succession. Annual spring flowers often bloom profusely, especially after fires. In California, this landscape is called **chaparral,** Spanish for thicket. Resident animals are drought tolerant such as jackrabbits, kangaroo rats, mule deer, chipmunks, lizards, and many bird species. Very similar landscapes are found along the Mediterranean coast as well as southwestern Australia, central Chile, and South Africa. Although this biome doesn't cover a very large total area, it contains a high number of unique species and is often considered a "hot spot" for biodiversity. It also is highly desired for human habitation, often leading to conflicts with rare and endangered plant and animal species.

Areas that are drier year-round, such as the African Sahel (edge of the Sahara Desert), northern Mexico, or the American Intermountain West (or Great Basin) tend to have a more sparse, open shrubland, characterized by sagebrush (*Artemisia* sp.), chamiso (*Adenostoma* sp.), or saltbush (*Atriplex* sp.). Some typical animals of this biome in America are a wide variety of snakes and lizards, rodents, birds, antelope, and mountain sheep.

Temperate forests can be evergreen or deciduous

Temperate, or midlatitude, forests occupy a wide range of precipitation conditions but occur mainly between about 30 and 55 degrees latitude (see fig. 5.3). In general we can group these forests by tree type, which can be broadleaf **deciduous** (losing leaves seasonally) or evergreen **coniferous** (cone-bearing).

Deciduous Forests

Broadleaf forests occur throughout the world where rainfall is plentiful. In midlatitudes, these forests are deciduous and lose their leaves in winter. The loss of green chlorophyll pigments can

produce brilliant colors in these forests in autumn (fig. 5.10). At lower latitudes, broadleaf forests may be evergreen or drought-deciduous. Southern live oaks, for example, are broadleaf evergreen trees.

Although these forests have a dense canopy in summer, they have a diverse understory that blooms in spring, before the trees leaf out. Spring ephemeral (short-lived) plants produce lovely flowers, and vernal (springtime) pools support amphibians and insects. These forests also shelter a great diversity of songbirds.

FIGURE 5.10 Temperate deciduous forests have year-round precipitation and winters near or below freezing.

12.5°C   1,024 mm

North American deciduous forests once covered most of what is now the eastern half of the United States and southern Canada. Most of western Europe was once deciduous forest but was cleared a thousand years ago. When European settlers first came to North America, they quickly settled and cut most of the eastern deciduous forests for firewood, lumber, and industrial uses, as well as to clear farmland. Many of those regions have now returned to deciduous forest, though the dominant species have changed.

Deciduous forests can regrow quickly because they occupy moist, moderate climates. But most of these forests have been occupied so long that human impacts are extensive, and most native species are at least somewhat threatened. The greatest threat to broadleaf deciduous forests is in eastern Siberia, where deforestation is proceeding rapidly. Siberia may have the highest deforestation rate in the world. As forests disappear, so do Siberian tigers, bears, cranes, and a host of other endangered species.

Coniferous Forests

Coniferous forests grow in a wide range of temperature and moisture conditions. Often they occur where moisture is limited: In cold climates, moisture is unavailable (frozen) in winter; hot climates may have seasonal drought; sandy soils hold little moisture, and they are often occupied by conifers. Thin, waxy leaves (needles) help these trees reduce moisture loss. Coniferous forests provide most wood products in North America. Dominant wood production regions include the southern Atlantic and Gulf coast states, the mountain West, and the Pacific Northwest (northern California to Alaska), but coniferous forests support forestry in many regions.

The coniferous forests of the Pacific coast grow in extremely wet conditions. The wettest coastal forests are known as **temperate rainforest,** a cool, rainy forest often enshrouded in fog (fig. 5.11). Condensation in the canopy (leaf drip) is a major form of precipitation in the understory. Mild year-round temperatures and abundant rainfall, up to 250 cm (100 in.) per year, result in luxuriant plant growth and giant trees such as the California redwoods, the largest trees in the world and the largest above-ground organism ever known to have existed. Redwoods once grew along the Pacific coast from California to Oregon, but logging has reduced them to a few small fragments.

Remaining fragments of ancient temperate rainforests are important areas of biodiversity. Recent battles over old-growth conservation (chapter 12) focus mainly on these areas. As with deciduous forests, Siberian forests are especially vulnerable to old-growth logging. The rate of this clearing, and its environmental effects, remain largely unknown.

Boreal forests occur at high latitudes

Because conifers can survive winter cold, they tend to dominate the **boreal forest,** or northern forests, that lie between about 50° and 60° north (fig. 5.12). Mountainous areas at lower latitudes may also have many characteristics and species of the boreal forest. Dominant trees are pines, hemlocks, spruce, cedar, and fir. Some deciduous trees are also present, such as maples, birch, aspen, and alder. These forests are slow-growing because of the cold temperatures and a short frost-free growing season, but they

FIGURE 5.11
Temperate rainforests have abundant but often seasonal precipitation that supports magnificent trees and luxuriant understory vegetation. Often these forests experience dry summers.

are still an expansive resource. In Siberia, Canada, and the western United States, large regional economies depend on boreal forests.

The extreme, ragged edge of the boreal forest, where forest gradually gives way to open tundra, is known by its Russian name, **taiga.** Here extreme cold and short summer limits the growth rate of trees. A 10 cm diameter tree may be over 200 years old in the far north.

Tundra can freeze in any month

Where temperatures are below freezing most of the year, only small, hardy vegetation can survive. **Tundra,** a treeless landscape that occurs at high latitudes or on mountaintops, has a growing season of only two to three months, and it may have frost any month of the year. Some people consider tundra a variant of grasslands because it has no trees; others consider it a very cold desert because water is unavailable (frozen) most of the year.

Arctic tundra is an expansive biome that has low productivity because it has a short growing season (fig. 5.13). During midsummer, however, 24-hour sunshine supports a burst of plant growth and an explosion of insect life. Tens of millions of waterfowl, shorebirds, terns, and songbirds migrate to the Arctic every year to feast on the abundant invertebrate and plant life and to raise their young on the brief bounty. These birds then migrate to wintering grounds, where they may be eaten by local

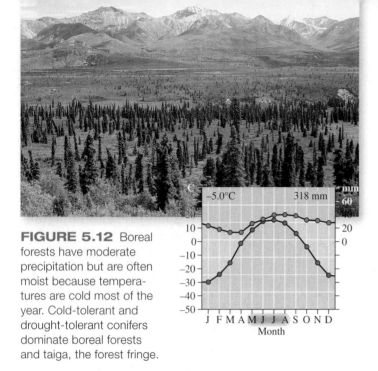

FIGURE 5.12 Boreal forests have moderate precipitation but are often moist because temperatures are cold most of the year. Cold-tolerant and drought-tolerant conifers dominate boreal forests and taiga, the forest fringe.

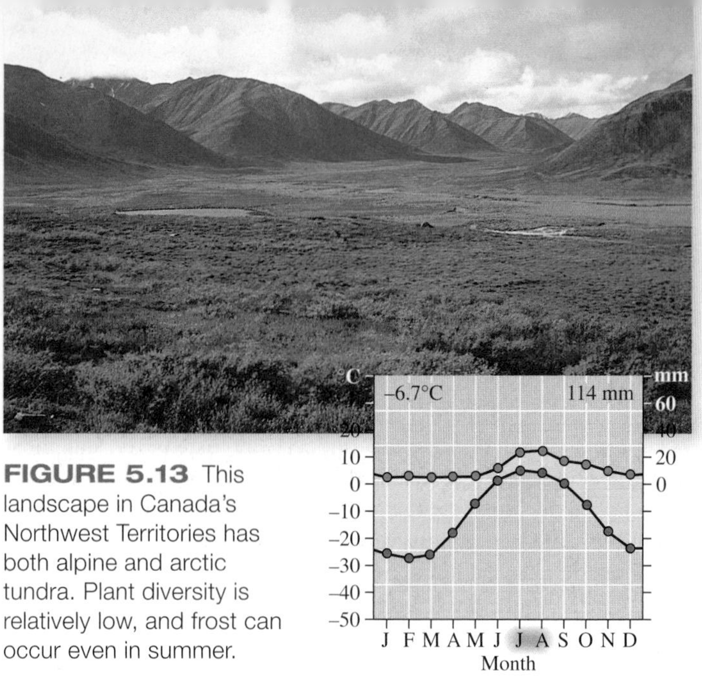

FIGURE 5.13 This landscape in Canada's Northwest Territories has both alpine and arctic tundra. Plant diversity is relatively low, and frost can occur even in summer.

predators—effectively they carry energy and protein from high latitudes to low latitudes. Arctic tundra is essential for global biodiversity, especially for birds.

Alpine tundra, occurring on or near mountaintops, has environmental conditions and vegetation similar to arctic tundra. These areas have a short, intense growing season. Often one sees a splendid profusion of flowers in alpine tundra; everything must flower at once in order to produce seeds in a few weeks before the snow comes again. Many alpine tundra plants also have deep pigmentation and leathery leaves to protect against the strong ultraviolet light in the thin mountain atmosphere.

Compared to other biomes, tundra has relatively low diversity. Dwarf shrubs, such as willows, sedges, grasses, mosses, and lichens tend to dominate the vegetation. Migratory muskox, caribou, or alpine mountain sheep and mountain goats can live on the vegetation because they move frequently to new pastures.

Because these environments are too cold for most human activities, they are not as badly threatened as other biomes. There are important problems, however. Global climate change may be altering the balance of some tundra ecosystems, and air pollution from distant cities tends to accumulate at high latitudes (chapter 15). In eastern Canada, coastal tundra is being badly overgrazed and degraded by overabundant populations of snow geese, whose numbers have exploded due to winter grazing on the rice fields of Arkansas and Louisiana. Oil and gas drilling—and associated truck traffic—threatens tundra in Alaska and Siberia. Clearly, this remote biome is not independent of human activities at lower latitudes.

5.2 MARINE ECOSYSTEMS

The biological communities in oceans and seas are poorly understood, but they are probably as diverse and as complex as terrestrial biomes. In this section, we will explore a few facets of these fascinating environments. Oceans cover nearly three-fourths of the earth's surface, and they contribute in important, although often unrecognized, ways to terrestrial ecosystems. Like land-based systems, most marine communities depend on photosynthetic organisms. Often it is algae or tiny, free-floating photosynthetic plants (**phytoplankton**) that support a marine food web, rather than the trees and grasses we see on land. In oceans, photosynthetic activity tends to be greatest near coastlines, where nitrogen, phosphorus, and other nutrients wash offshore and fertilize primary producers. Ocean currents also contribute to the distribution of biological productivity, as they transport nutrients and phytoplankton far from shore (fig. 5.14).

As plankton, algae, fish, and other organisms die, they sink toward the ocean floor. Deep-ocean ecosystems, consisting of crabs, filter-feeding organisms, strange phosphorescent fish, and many other life-forms, often rely on this "marine snow" as a primary nutrient source. Surface communities also depend on this material. Upwelling currents circulate nutrients from the ocean floor back to the surface. Along the coasts of South America, Africa, and Europe, these currents support rich fisheries.

Vertical stratification is a key feature of aquatic ecosystems, mainly because light decreases rapidly with depth, and communities below the photic zone (light zone, often reaching about 20 m deep) must rely on energy sources other than photosynthesis to persist. Temperature also decreases with depth. Deep-ocean species often grow slowly in part because metabolism is reduced in cold conditions. In contrast, warm, bright, near-surface communities such as coral reefs and estuaries are among the world's most biologically productive environments. Temperature also affects the amount of oxygen and other elements that can be absorbed in water. Cold water holds abundant oxygen, so productivity is often high in cold oceans, as in the North Atlantic, North Pacific, and Antarctic.

Ocean systems can be described by depth and proximity to shore (fig. 5.15). In general, **benthic** communities occur on the bottom, and **pelagic** (from "sea" in Greek) zones are the water column.

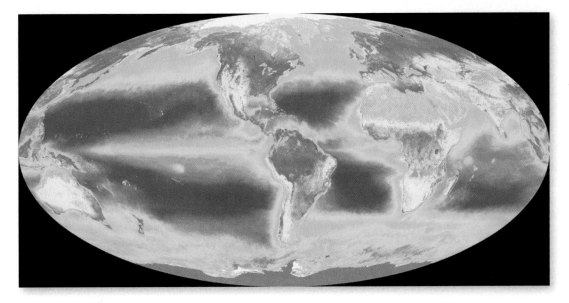

FIGURE 5.14 Satellite measurements of chlorophyll levels in the oceans and on land. Dark green to blue land areas have high biological productivity. Dark blue oceans have little chlorophyll and are biologically impoverished. Light green to yellow ocean zones are biologically rich.

The epipelagic zone (*epi* = on top) has photosynthetic organisms. Below this are the mesopelagic (*meso* = medium), and bathypelagic (*bathos* = deep) zones. The deepest layers are the abyssal zone (to 4,000 m) and hadal zone (deeper than 6,000 m). Shorelines are known as littoral zones, and the area exposed by low tides is known as the intertidal zone. Often there is a broad, relatively shallow region along a continent's coast, which may reach a few kilometers or hundreds of kilometers from shore. This undersea area is the continental shelf.

Open-ocean communities vary from surface to hadal zones

The open ocean is often referred to as a biological desert because it has relatively low productivity. But like terrestrial deserts, the open ocean has areas of rich productivity and diversity. Fish and plankton abound in regions such as the equatorial Pacific and Antarctic oceans, where nutrients are distributed by currents. Another notable exception, the Sargasso Sea in the western Atlantic, is known for its free-floating mats of brown algae. These algae mats support a phenomenal diversity of animals, including sea turtles, fish, and even eels that hatch amid the algae, then eventually migrate up rivers along the Atlantic coasts of North America and Europe.

Deep-sea thermal vent communities are another remarkable type of marine system (fig. 5.16) that was completely unknown until 1977 explorations with the deep-sea submarine *Alvin*. These communities are based on microbes that capture chemical energy, mainly from sulfur compounds released from thermal vents—jets of hot water and minerals on the ocean floor. Magma below the ocean crust heats these vents. Tube worms, mussels, and microbes on these vents are adapted to survive both extreme temperatures, often above 350°C (700°F), and the intense water pressure at depths of 7,000 m (20,000 ft) or more. Oceanographers have discovered thousands of different types of organisms, most of them microscopic, in these communities (chapter 3).

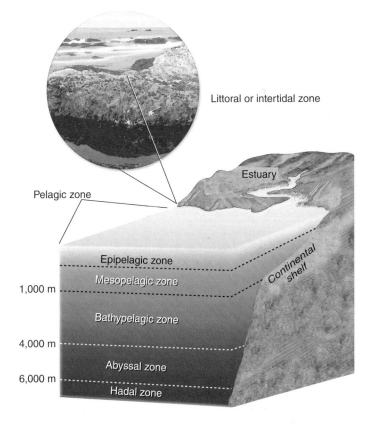

Littoral or intertidal zone

Estuary

Pelagic zone

Continental shelf

Epipelagic zone

Mesopelagic zone

1,000 m

Bathypelagic zone

4,000 m

Abyssal zone

6,000 m

Hadal zone

FIGURE 5.15 Light penetrates only the top 10–20 m of the ocean. Below this level, temperatures drop and pressure increases. Nearshore environments include the intertidal zone and estuaries.

Coastal zones support rich, diverse communities

As in the open ocean, shoreline communities vary with depth, light, nutrient concentrations, and temperature. Some shoreline communities, such as estuaries, have high biological productivity and diversity because they are enriched by nutrients washing from the land. But nutrient loading can be excessive. Around the

FIGURE 5.16 Deep-ocean thermal vent communities have great diversity and are unusual because they rely on chemosynthesis, not photosynthesis, for energy.

world, more than 200 "dead zones" occur in coastal zones where excess nutrients stimulate bacterial growth that consumes almost all oxygen in the water and excludes most other life. We'll discuss this problem further in chapter 18.

Coral reefs are among the best-known marine ecosystems because of their extraordinary biological productivity and their diverse and beautiful organisms. Reefs are aggregations of minute colonial animals (coral polyps) that live symbiotically with photosynthetic algae. Calcium-rich coral skeletons build up to make reefs, atolls, and islands (fig. 5.17a). Reefs protect shorelines and shelter countless species of fish, worms, crustaceans, and other life-forms. Reef-building corals live where water is shallow and clear enough for sunlight to reach the photosynthetic algae. They need warm (but not too warm) water, and can't survive where high nutrient concentrations or runoff from the land create dense layers of algae, fungi, or sediment. Coral reefs also are among the most endangered biomes in the world. As the opening case study for this chapter shows, destructive fishing practices can damage or destroy coral communities. In addition, polluted urban runoff, trash, sewage and industrial effluent, sediment from agriculture, and unsustainable forestry are smothering coral reefs along coastlines that have high human populations. Introduced pathogens and predators also threaten many reefs. Perhaps the greatest threat to reefs is global warming. Elevated water temperatures cause **coral bleaching,** in which corals expel their algal partner and then die. The third UNESCO Conference on Oceans, Coasts, and Islands in 2006 reported that one-third of all coral reefs have been destroyed, and that 60 percent are now degraded and probably will be dead by 2030.

The value of an intact reef in a tourist economy can be upwards of (U.S.)$1 million per square kilometer. The costs of conserving these same reefs in a marine-protected area would be just (U.S.)$775 per square kilometer per year, the UN Environment Program estimates. Of the estimated 30 million small-scale fishers in the developing world, most are dependent to a greater or lesser extent on coral reefs. In the Philippines, the UN estimates that more than 1 million fishers depend directly on coral reefs for their livelihoods. We'll discuss reef restoration efforts further in chapter 13.

Sea-grass beds, or eel-grass beds, often occupy shallow, warm, sandy areas near coral reefs. Like reefs, these communities support a rich diversity of grazers, from snails and worms to turtles and manatees. Also like reefs, these environments are easily smothered by sediment originating from onshore agriculture and development.

Mangroves are trees that grow in salt water. They occur along calm, shallow, tropical coastlines around the world (fig. 5.17b). Mangrove forests or swamps help stabilize shorelines, and they are also critical nurseries for fish, shrimp, and other commercial species. Like coral reefs, mangroves line tropical and subtropical coastlines, where they are vulnerable to development, sedimentation, and overuse. Unlike reefs, mangroves provide commercial timber, and they can be clear-cut to make room for aquaculture (fish farming) and other activities. Ironically, mangroves provide the protected spawning beds for most of the fish and shrimp farmed in these ponds. As mangroves become increasingly threatened in tropical countries, villages relying on fishing for income and sustenance are seeing reduced catches and falling income.

Estuaries are bays where rivers empty into the sea, mixing fresh water with salt water. **Salt marshes,** shallow wetlands flooded regularly or occasionally with seawater, occur on shallow coastlines, including estuaries (fig. 5.17c). Usually calm, warm, and nutrient-rich, estuaries and salt marshes are biologically diverse and productive. Rivers provide nutrients and sediments, and a muddy bottom supports emergent plants (whose leaves emerge above the water surface), as well as the young forms of crustaceans, such as crabs and shrimp, and mollusks, such as clams and oysters. Nearly two-thirds of all marine fish and shellfish rely on estuaries and saline wetlands for spawning and juvenile development.

Estuaries near major American cities once supported an enormous wealth of seafood. Oyster beds and clam banks in the waters adjacent to New York, Boston, and Baltimore provided free and easy food to early residents. Sewage and other contaminants long ago eliminated most of these resources, however. Recently, major efforts have been made to revive Chesapeake Bay, America's largest and most productive estuary. These efforts have shown some success, but many challenges remain (chapter 3).

In contrast to the shallow, calm conditions of estuaries, coral reefs, and mangroves, there are violent, wave-blasted shorelines that support fascinating life-forms in **tide pools.** Tide pools are

(a) Coral reefs

(b) Mangroves

(c) Estuary and salt marsh

(d) Tide pool

FIGURE 5.17 Coastal environments support incredible diversity and help stabilize shorelines. Coral reefs (a), mangroves (b), and estuaries (c) also provide critical nurseries for marine ecosystems. Tide pools (d) also shelter highly specialized organisms.

depressions in a rocky shoreline that are flooded at high tide but retain some water at low tide. These areas remain rocky where wave action prevents most plant growth or sediment (mud) accumulation. Extreme conditions, with frigid flooding at high tide and hot, desiccating sunshine at low tide, make life impossible for most species. But the specialized animals and plants that do occur in this rocky intertidal zone are astonishingly diverse and beautiful (fig. 5.17*d*).

Barrier islands are low, narrow, sandy islands that form parallel to a coastline (fig. 5.18). They occur where the continental shelf is shallow and rivers or coastal currents provide a steady source of sediments. They protect brackish (moderately salty), inshore lagoons and salt marshes from storms, waves, and tides. One of the world's most extensive sets of barrier islands lines the Atlantic coast from New England to Florida, as well as along the Gulf coast of Texas. Composed of sand that is constantly reshaped by wind and waves, these islands can be formed or removed by a single violent storm. Because they are mostly beach, barrier islands are also popular places for real estate development. About

20 percent of the barrier island surface in the United States has been developed. Barrier islands are also critical to preserving coastal shorelines, settlements, estuaries, and wetlands.

Unfortunately, human occupation often destroys the value that attracts us there in the first place. Barrier islands and beaches are dynamic environments, and sand is hard to keep in place. Wind and wave erosion is a constant threat to beach developments. Walking or driving vehicles over dune grass destroys the stabilizing vegetative cover and accelerates, or triggers, erosion. Cutting roads through the dunes further destabilizes these islands, making them increasingly vulnerable to storm damage. When Hurricane Katrina hit the U.S. Gulf coast in 2005, it caused at least $200 billion in property damage and displaced 4 million people. Thousands of homes were destroyed (fig. 5.19), particularly on low-lying barrier islands.

Because of these problems, we spend billions of dollars each year building protective walls and barriers, pumping sand onto beaches from offshore, and moving sand from one beach area

FIGURE 5.18 A barrier island, Assateague, along the Maryland–Virginia coast. Grasses cover and protect dunes, which keep ocean waves from disturbing the bay, salt marshes, and coast at right. Roads cut through the dunes expose them to erosion.

FIGURE 5.19 Winter storms have eroded the beach and undermined the foundations of homes on this barrier island. Breaking through protective dunes to build such houses damages sensitive plant communities and exposes the whole island to storms and erosion. Coastal zone management attempts to limit development on fragile sites.

to another. Much of this expense is borne by the public. Some planners question whether we should allow rebuilding on barrier islands, especially after they've been destroyed multiple times.

5.3 FRESHWATER ECOSYSTEMS

Freshwater environments are far less extensive than marine environments, but they are centers of biodiversity. Most terrestrial communities rely, to some extent, on freshwater environments. In deserts, isolated pools, streams, and even underground

water systems, support astonishing biodiversity as well as provide water to land animals. In Arizona, for example, most birds gather in trees and bushes surrounding the few available rivers and streams.

Lakes have open water

Freshwater lakes, like marine environments, have distinct vertical zones (fig. 5.20). Near the surface a subcommunity of plankton, mainly microscopic plants, animals, and protists (single-celled organisms such as amoebae), float freely in the water column. Insects such as water striders and mosquitoes also live at the air-water interface. Fish move through the water column, sometimes near the surface, and sometimes at depth.

Finally, the bottom, or *benthos,* is occupied by a variety of snails, burrowing worms, fish, and other organisms. These make up the benthic community. Oxygen levels are lowest in the benthic environment, mainly because there is little mixing to introduce oxygen to this zone. Anaerobic bacteria (not using oxygen) may live in low-oxygen sediments. In the littoral zone, emergent plants such as cattails and rushes grow in the bottom sediment. These plants create important functional links between layers of an aquatic ecosystem, and they may provide the greatest primary productivity to the system.

Lakes, unless they are shallow, have a warmer upper layer that is mixed by wind and warmed by the sun. This layer is the *epilimnion.* Below the epilimnion is the hypolimnion (*hypo* = below), a colder, deeper layer that is not mixed. If you have gone swimming in a moderately deep lake, you may have discovered the sharp temperature boundary, known as the **thermocline,** between these layers. Below this boundary, the water is much colder. This boundary is also called the mesolimnion.

Local conditions that affect the characteristics of an aquatic community include (1) nutrient availability (or excess) such as nitrates and phosphates; (2) suspended matter, such as silt, that

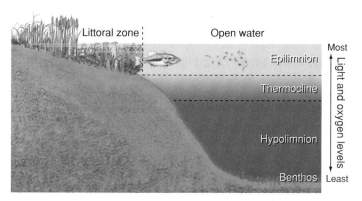

FIGURE 5.20 The layers of a deep lake are determined mainly by gradients of light, oxygen, and temperature. The epilimnion is affected by surface mixing from wind and thermal convections, while mixing between the hypolimnion and epilimnion is inhibited by a sharp temperature and density difference at the thermocline.

FIGURE 5.21 The character of freshwater ecosystems is greatly influenced by the immediately surrounding terrestrial ecosystems, and even by ecosystems far upstream or far uphill from a particular site.

affects light penetration; (3) depth; (4) temperature; (5) currents; (6) bottom characteristics, such as muddy, sandy, or rocky floor; (7) internal currents; and (8) connections to, or isolation from, other aquatic and terrestrial systems (fig. 5.21).

Wetlands are shallow and productive

Wetlands are shallow ecosystems in which the land surface is saturated or submerged at least part of the year. Wetlands have vegetation that is adapted to grow under saturated conditions. These legal definitions are important because although wetlands make up only a small part of most countries, they are disproportionately important in conservation debates and are the focus of continual legal disputes around the world and in North America. Beyond these basic descriptions, defining wetlands is a matter of hot debate. How often must a wetland be saturated, and for how long? How large must it be to deserve legal protection? Answers can vary, depending on political, as well as ecological, concerns.

These relatively small systems support rich biodiversity, and they are essential for both breeding and migrating birds. Although wetlands occupy less than 5 percent of the land in the United States, the Fish and Wildlife Service estimates that one-third of all endangered species spend at least part of their lives in wetlands. Wetlands retain storm water and reduce flooding by slowing the rate at which rainfall reaches river systems. Floodwater storage is worth $3 billion to $4 billion per year in the United States. As water stands in wetlands, it also seeps into the ground, replenishing groundwater supplies. Wetlands filter, and even purify, urban and farm runoff, as bacteria and plants take up nutrients and contaminants in water. They are also in great demand for filling and development. They are often near cities or farms, where land is valuable, and once drained, wetlands are easily converted to more lucrative uses.

Wetlands are described by their vegetation. **Swamps** are wetlands with trees (fig. 5.22*a*). **Marshes** are wetlands without trees (fig. 5.22*b*). **Bogs** are areas of saturated ground, and usually the ground is composed of deep layers of accumulated, undecayed vegetation known as peat. **Fens** are similar to bogs except that they are mainly fed by groundwater, so that they have mineral-rich water and specially adapted plant species. Bogs are fed mainly by precipitation. Swamps and marshes have high biological productivity. Bogs and fens, which are often nutrient-poor, have low biological productivity. They may have unusual and interesting species, though, such as sundews and pitcher plants, which are adapted to capture nutrients from insects rather than from soil.

The water in marshes and swamps usually is shallow enough to allow full penetration of sunlight and seasonal warming (fig. 5.22*c*). These mild conditions favor great photosynthetic activity, resulting in high productivity at all trophic levels. In short, life is abundant and varied. Wetlands are major breeding, nesting, and migration staging areas for waterfowl and shorebirds.

(a) Swamp, or wooded wetland

(b) Marsh

(c) Coastal saltmarsh

FIGURE 5.22 Wetlands provide irreplaceable ecological services, including water filtration, water storage and flood reduction, and habitat. Forested wetlands (a) are often called swamps; marshes (b) have no trees; coastal saltmarshes (c) are tidal and have rich diversity.

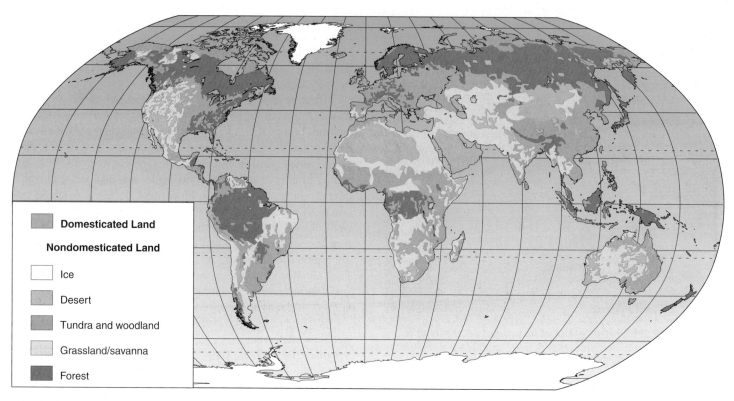

Domesticated Land

Nondomesticated Land

- Ice
- Desert
- Tundra and woodland
- Grassland/savanna
- Forest

FIGURE 5.23 Domesticated land has replaced much of the earth's original land cover.
Source: United Nations Environment Program, *Global Environment Outlook*.

Wetlands may gradually convert to terrestrial communities as they fill with sediment, and as vegetation gradually fills in toward the center. Often this process is accelerated by increased sediment loads from urban development, farms, and roads. Wetland losses are one of the areas of greatest concern among biologists.

5.4 HUMAN DISTURBANCE

Humans have become dominant organisms over most of the earth, damaging or disturbing more than half of the world's terrestrial ecosystems to some extent. By some estimates, humans preempt about 40 percent of the net terrestrial primary productivity of the biosphere either by consuming it directly, by interfering with its production or use, or by altering the species composition or physical processes of human-dominated ecosystems. Conversion of natural habitat to human uses is the largest single cause of biodiversity losses.

Researchers from the environmental group Conservation International have attempted to map the extent of human disturbance of the natural world (fig. 5.23). The greatest impacts have been in Europe, parts of Asia, North and Central America, and islands such as Madagascar, New Zealand, Java, Sumatra, and those in the Caribbean. Data from this study are shown in table 5.1.

Temperate broadleaf forests are the most completely human-dominated of any major biome. The climate and soils that support such forests are especially congenial for human occupation.

In eastern North America or most of Europe, for example, only remnants of the original forest still persist. Regions with a Mediterranean climate generally are highly desired for human habitation. Because these landscapes also have high levels of

Table 5.1 Human Disturbance	
Biome	**% Human Dominated**
Temperate broadleaf forests	81.9
Chaparral	67.8
Temperate grasslands	40.4
Temperate rainforests	46.1
Tropical dry forests	45.9
Mixed mountain systems	25.6
Mixed island systems	41.8
Cold deserts/semideserts	8.5
Warm deserts/semideserts	12.2
Moist tropical forests	24.9
Tropical grasslands	4.7
Temperate coniferous forests	11.8
Tundra and arctic desert	0.3

Note: Where undisturbed and human-dominated areas do not add up to 100 percent, the difference represents partially disturbed lands.

Source: Hannah, Lee, et al., "Human Disturbance and Natural Habitat: A Biome Level Analysis of a Global Data Set," in *Biodiversity and Conservation*, 1995, Vol. 4:128–55.

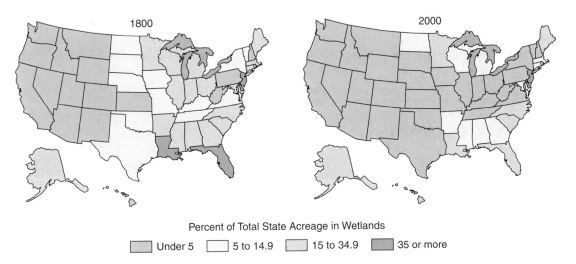

Percent of Total State Acreage in Wetlands

☐ Under 5   ☐ 5 to 14.9   ☐ 15 to 34.9   ☐ 35 or more

FIGURE 5.24 Over the past two centuries, more than half of the original wetlands in the lower 48 states have been drained, filled, polluted, or otherwise degraded. Some of the greatest losses have been in midwestern farming states where up to 99 percent of all wetlands have been lost.

biodiversity, conflicts between human preferences and biological values frequently occur.

Temperate grasslands, temperate rainforests, tropical dry forests, and many islands also have been highly disturbed by human activities. If you have traveled through the American cornbelt states such as Iowa or Illinois, you have seen how thoroughly former prairies have been converted to farmlands. Intensive cultivation of this land exposes the soil to erosion and fertility losses (chapter 9). Islands, because of their isolation, often have high numbers of endemic species. Many islands, such as Madagascar, Haiti, and Java have lost more than 99 percent of their original land cover.

Tundra and arctic deserts are the least disturbed biomes in the world. Harsh climates and unproductive soils make these areas unattractive places to live for most people. Temperate conifer forests also generally are lightly populated and large areas remain in a relatively natural state. However, recent expansion of forest harvesting in Canada and Siberia may threaten the integrity of this biome. Large expanses of tropical moist forests still remain in the Amazon and Congo basins but in other areas of the tropics such as West Africa, Madagascar, Southeast Asia, and the Indo-Malaysian peninsula and archipelago, these forests are disappearing at a rapid rate (chapter 12).

As mentioned earlier, wetlands have suffered severe losses in many parts of the world. About half of all original wetlands in the United States have been drained, filled, polluted, or otherwise degraded over the past 250 years. In the prairie states, small potholes and seasonally flooded marshes have been drained and converted to croplands on a wide scale. Iowa, for example, is estimated to have lost 99 percent of its presettlement wetlands (fig. 5.24). Similarly, California has lost 90 percent of the extensive marshes and deltas that once stretched across its central valley. Wooded swamps and floodplain forests in the southern United States have been widely disrupted by logging and conversion to farmland.

Similar wetland disturbances have occurred in other countries as well. In New Zealand, over 90 percent of natural wetlands have been destroyed since European settlement. In Portugal, some 70 percent of freshwater wetlands and 60 percent of estuarine habitats have been converted to agriculture and industrial areas. In Indonesia, almost all the mangrove swamps that once lined the coasts of Java have been destroyed, while in the Philippines and Thailand, more than two-thirds of coastal mangroves have been cut down for firewood or conversion to shrimp and fish ponds.

Slowing this destruction, or even reversing it, is a challenge that we will discuss in chapter 13.

CONCLUSION

The potential location of biological communities is determined in large part by climate, moisture availability, soil type, geomorphology, and other natural features. Understanding the global distribution of biomes, and knowing the differences in who lives where and why, are essential to the study of global environmental science. Human occupation and use of natural resources is strongly dependent on the biomes found in particular locations. We tend to prefer mild climates and the highly productive biological communities found in temperate zones. These biomes also suffer the highest rates of degradation and overuse.

Being aware of the unique conditions and the characteristics evolved by plants and animals to live in those circumstances can help you appreciate how and why certain species live in particular biomes, such as seasonal tropical forests, alpine tundra, or chaparral shrublands.

Oceans cover over 70 percent of the earth's surface, yet we know relatively little about them. Some marine biomes, such as coral reefs, can be as biologically diverse and productive as any terrestrial biome. People have depended on rich, complex coastal ecosystems for eons, but in recent times rapidly growing human populations, coupled with more powerful ways to harvest resources, have led to damage—and, in some cases, irreversible destruction—of these treasures. Still, there is reason to hope that we'll find ways to protect these living communities. The opening case study of this chapter illustrates how, without expensive technology, people can work to protect and even restore the biological communities on which they depend. This gives us optimism that we'll find similar solutions in other biologically rich but endangered biomes.

REVIEWING LEARNING OUTCOMES

By now you should be able to explain the following points:

5.1 Recognize the characteristics of some major terrestrial biomes as well as the factors that determine their distribution.

- Tropical moist forests have rain year-round.
- Tropical seasonal forests have yearly dry seasons.
- Tropical savannas, grasslands support few trees.
- Deserts are hot or cold, but all are dry.
- Temperate grasslands have rich soils.
- Temperate shrublands have summer drought.
- Temperate forests can be evergreen or deciduous.
- Boreal forests occur at high latitudes.
- Tundra can freeze in any month.

5.2 Understand how and why marine environments vary with depth and distance from shore.

- Open-ocean communities vary from surface to hadal zones.
- Coastal zones support rich, diverse communities.

5.3 Compare the characteristics and biological importance of major freshwater ecosystems.

- Lakes have open water.
- Wetlands are shallow and productive.

5.4 Summarize the overall patterns of human disturbance of world biomes.

- Biomes that humans find comfortable and profitable have high rates of disturbance, while those that are less attractive or have limited resources have large pristine areas.

PRACTICE QUIZ

1. Throughout the central portion of North America is a large biome once dominated by grasses. Describe how physical conditions and other factors control this biome.

2. What is taiga and where is it found? Why might logging in taiga be more disruptive than in southern coniferous forests?

3. Why are tropical moist forests often less suited for agriculture and human occupation than tropical deciduous forests?

4. Find out the annual temperature and precipitation conditions where you live (fig. 5.2). Which biome type do you occupy?

5. Describe four different kinds of wetlands and explain why they are important sites of biodiversity and biological productivity.

6. Forests differ according to both temperature and precipitation. Name and describe a biome that occurs in (a) hot, (b) cold, (c) wet, and (d) dry climates (one biome for each climate).

7. How do physical conditions change with depth in marine environments?

8. Describe four different coastal ecosystems.

CRITICAL THINKING AND DISCUSSION QUESTIONS

1. What physical and biological factors are most important in shaping your biological community? How do the present characteristics of your area differ from those 100 or 1,000 years ago?

2. Forest biomes frequently undergo disturbances such as fire or flooding. As more of us build homes in these areas, what factors should we consider in deciding how to protect people from natural disturbances?

3. Often humans work to preserve biomes that are visually attractive. What biomes might be lost this way? Is this a problem?

4. Disney World in Florida wants to expand onto a wetland. It has offered to buy and preserve a large nature preserve in a different area to make up for the wetland it is destroying. Is that reasonable? What conditions would make it reasonable or unreasonable?

5. Suppose further that the wetland being destroyed in question 4 and its replacement area both contain several endangered species (but different ones). How would you compare different species against each other? How many plant or insect species would one animal species be worth?

6. Historically, barrier islands have been hard to protect because links between them and inshore ecosystems are poorly recognized. What kinds of information would help a community distant from the coast commit to preserving a barrier island?

Data Analysis: Reading Climate Graphs

As you've learned in this chapter, temperature and precipitation are critical factors in determining the distribution of terrestrial biomes. Understanding climate graphs and what they tell us is extremely helpful in making sense of these differences. In the figure below, reproduced from figure 5.5, the graphs show annual patterns in temperature and precipitation (rainfall and snow). They also indicate how much of the year evaporation exceeds precipitation (yellow areas), and when precipitation exceeds evaporation, leaving moisture available for plant growth. Examine these graphs to answer the following questions.

1. What are the maximum and minimum temperatures in each of the three locations shown?
2. What do these temperatures correspond to in Fahrenheit? (*Hint:* look at the conversion table in the back of your book).
3. Which area has the wettest climate; which is driest?
4. How do the maximum and minimum monthly rainfalls in San Diego and Belém compare?
5. Describe these three climates.

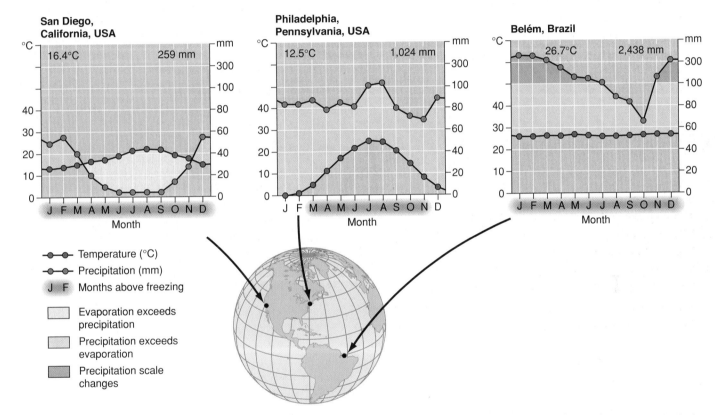

Moisture availability depends on temperature as well as precipitation. The horizontal axis on these climate diagrams shows months of the year; vertical axis show temperature (*left side*) and precipitation (*right*). The number of dry months (*shaded yellow*) and wetter months (*blue*) varies with geographic location. Mean annual temperature (°C) and precipitation (mm) are shown at the top of each graph.

6. What kinds of biomes would you expect to find in these areas?
7. What would a climate graph look like where you live? Try sketching one out, then compare it to a graph for a biome similar to yours in this chapter.
8. Examine fig. 5.3, and identify what kind of biomes exist in Kenya. What sort of tree cover is the Greenbelt movement attempting to restore?

For Additional Help in Studying This Chapter, please visit our website at www.mhhe.com/cunningham12e. You will find additional practice quizzes and case studies, flashcards, regional examples, placemarkers for Google Earth™ mapping, and an extensive reading list, all of which will help you learn environmental science.

A bluefin tina, the largest and most expensive commerically harvested tuna, is disentangled from a net.

Population Biology

Learning Outcomes

After studying this chapter, you should be able to:

6.1 Describe the dynamics of population growth.

6.2 Summarize the BIDE factors that increase or decrease populations.

6.3 Compare and contrast the factors that regulate population growth.

6.4 Identify some applications of population dynamics in conservation biology.

"Nature teaches more than she preaches."

~ *John Burroughs*

Case Study   Fishing to Extinction?

The most expensive tuna ever sold, a 342 kg (754 lb) bluefin tuna, was auctioned in Tokyo in January 2011 for nearly $400,000. This one fish, the auspicious first sale of the new year at the Tsukiji fish market, brought in nearly $1,160 per kg ($527 per lb). The price was extreme because the first fish of the year is thought to bring good luck, but plummeting numbers of bluefins and rising demand for sushi and sashimi also helped to push up the price. The world was watching this sale because bluefin tuna has been the subject of bitter disputes. On the one side, biologists warn that overfishing has cut its population by 70–80 percent and is driving the bluefin toward extinction. On the other side, the fishing industry and traders in Japanese sushi are unwilling to sacrifice the enormous profits it brings. Just months before this sale, under pressure from tuna-fishing nations, Atlantic bluefin tuna were denied international endangered species designation. Then the tuna-fishing industry decided, despite warnings of its own biologists, that reduced catches were unnecessary to protect populations.

Population biology, the science of modeling changes in species abundance, is key to understanding this controversy. The bluefin tuna is a large, long-lived, wide-ranging fish. It can live for at least 20 years, but it matures slowly for a fish—some populations take 8 years or more to reach spawning age. The number of young in a year can be enormous, but that number depends on the number of spawning-age fish and other factors. Biologists use these numbers to calculate the likely rate of decline in the species' numbers, the likely rate of recovery from reduced fishing pressure, or the amount of fishing that the population can safely sustain. In the bluefin's case, spawning-age fish are declining fast, and population models indicate that the species is heading for a crash.

Bluefin tuna are top predators, big and fast enough to eat almost anything they encounter. They can grow to 3 m (over 12 feet) in length and 650 kg (1,430 lbs). Bluefins migrate thousands of km around the world's oceans. Atlantic bluefins spawned in the Mediterranean travel across the Atlantic and to Iceland as they grow. A smaller population spawns on the northern slope of the Gulf of Mexico, then travels up north and mixes with the Mediterranean population in rich foraging areas in the open ocean. Pacific bluefins spawn from the Philippines to Japan and migrate all the way across the Pacific and back again to breed.

This fish had little commercial value until the 1960s, when a market developed for bluefin sushi and sashimi. Its unusually high fat content gives a strong taste when cooked, but its raw flesh is considered especially flavorful. Japan has always been the leader in the raw fish market, consuming 80 percent of the world's bluefin tuna, but other markets have grown recently in China, the United States, and elsewhere.

The International Commission for the Conservation of Atlantic Tunas (ICCAT) is in charge of protecting Atlantic tuna, marlin, swordfish, and other species by setting sustainable catch limits. Ideally, ICCAT uses population models to calculate a sustainable catch rate that maintains a stable spawning-age population. But ICCAT data show that Atlantic spawning stock has dropped to 18–27 percent of pre-1950 levels. Despite this decline, allowable catch limits remain high. A sustainable catch would be 8,500 tons or so of Atlantic bluefin tuna per year, but ICCAT has maintained limits 2–3 times this high. Moreover, ICCAT member states exceed their legal limits every year.

To make matters worse, unreported illegal catches by ICCAT member states are extremely high. Fishing is a notoriously hard industry to monitor. In the free-for-all on the high seas, where enforcement is weak or impossible, where individual nations subsidize fishing fleets, and where so much money is at stake, it's hard to be completely honest—especially if you don't trust the honesty of your competitors. ICCAT estimates that its records represent just half of actual catches in some years (fig. 6.1). According to the United States' National Marine Fisheries Service, comparable problems of overfishing are occurring in nearly all the other

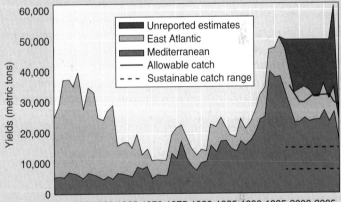

(a)

(b)

FIGURE 6.1 A graph showing the bluefin tuna catch in the Atlantic since 1950. (a) Note the differences between allowable, sustainable, and actual catch estimates. Frozen tuna at auction at Tsukiji market (b).

Data Source: ICCAT.

large-fish fisheries, including marlin, swordfish, and albacore tuna. Some populations are not currently overfished, however, including Atlantic bigeye and yellowfin tuna.

Some tuna-catching nations may see it in their best interest to liquidate the species for short-term profits. Others just want to protect the interests of their own fishing fleets in the face of international competition. Because the species belongs to no individual nation, countries have strong profit incentives to catch the last fish before someone else does. Thus the self-policing ICCAT structure has so far failed to conserve the Atlantic tuna. In 2009, Monaco petitioned for endangered species designation for the east-Atlantic population, but in 2010 that listing was denied, on the grounds that ICCAT was already in charge of conserving the species. Just months later, ICCAT declined to reduce allowable catches sub-stantially, although the organization did promise more thorough monitoring in the future.

Population biology allows us to identify overfishing, to model sustainable catch rates, and to warn about how quickly the species might disappear at current capture rates. In this chapter we'll examine the main concepts of population biology and the uses of these concepts in environmental science.

To find out which fish are best to eat, see the Monterey Bay Aquarium (www.montereybayaquarium.org/cr/seafoodwatch.aspx). You can also see data from ICCAT here: www.iccat.int/Documents/Meetings/Docs/2009-SCRS_ENG.pdf. For related resources, including Google Earth™ placemarks that show locations where these issues can be seen, visit EnvironmentalScience-Cunningham .blogspot.com.

6.1 DYNAMICS OF POPULATION GROWTH

Conserving the bluefin tuna depends on a good understanding of how populations grow and decline. General rules and patterns that describe these changes can greatly improve our understanding of species and their ecosystems.

Population growth can be limited by mortality (as in tuna fishing) or slow reproductive rates. Without these constraints, many organisms can reach unbelievable numbers if environmental conditions are right. Consider the common housefly. Each female fly lays 120 eggs (assume half female) in a generation. In 56 days those eggs become mature adults, able to reproduce. In one year, with seven generations of flies being born and reproducing, that original fly would be the proud parent of 5.6 trillion offspring. If this rate of reproduction continued for ten years, the entire earth would be covered in several meters of housefly bodies. Luckily housefly reproduction, as for most organisms, is constrained in a variety of ways—scarcity of resources, competition, predation, disease, accident. The housefly merely demonstrates the remarkable amplification—the **biotic potential**—of unrestrained biological reproduction (fig. 6.2). Population dynamics describes these changes in the number of organisms in a population over time.

We can describe growth symbolically

Describing the general pattern of population growth is easiest if we can reduce it to a few general factors. Ecologists find it most efficient and simplest to use symbolic terms such as N, r, and t to refer to these factors. At first, this symbolic form might seem hard to interpret, but as you become familiar with the terms, you'll probably find them quicker to follow than longer text would be.

Here are some examples to show how you can describe population change. Figure 6.2 shows a very large population of cockroaches, for example, a species capable of reproducing very rapidly. How rapidly can this population grow? If there are no predators and food is abundant, then that depends mainly on two factors: the number you start with, and the rate of reproduction. Start with 2 cockroaches, one male and one female, and suppose they can lay eggs and increase to about 20 cockroaches in the course of 3 months. You can describe the rate of growth (r) per adult in one 3-month period like this: $r = 20$ per 2 adults, or 10/adult, or "$r = 10$." If nothing limits population growth, numbers will

FIGURE 6.2 Reproduction gives many organisms the potential to expand populations explosively. The cockroaches in this kitchen could have been produced in only a few generations. A single female cockroach can produce up to 80 eggs every six months. This exhibit is in the Smithsonian Institute's National Museum of Natural History.

continue to increase at this rate of $r = 10$ for each 3-month time step. You can call each of these time steps (t). The starting point, before population growth begins, is "time 0" (t_0). The first time step is called t_1, the second time step is t_2, and so on. If $r = 10$, and the population (N) starts at 2 cockroaches, then the numbers will increase like this:

time	N	rate (r)	$r \times N$
t_1	2	10	$10 \times 2 = 20$
t_2	20	10	$10 \times 20 = 200$
t_3	200	10	$10 \times 200 = 2{,}000$
t_4	2,000	10	$10 \times 2{,}000 = 20{,}000$

This is a very rapid rate of increase, from 2 to 20,000 in four time steps (fig. 6.3). It's also a very simplified explanation of growth, but it's fairly easy to follow. This rate is described as a "geometric" rate of increase. Look carefully at the numbers above, and you might notice that the population at t_2 is $2 \times 10 \times 10$, and the population at t_3 is $2 \times 10 \times 10 \times 10$. Another way to say this is that the population at t_2 is 2×10^2, and at t_3 the population is 2×10^3. In fact, the population at any given time is equal to the starting number (2) times the rate (10) raised to the exponent of the number of time steps (10^t). The short way to express the geometric rate of increase is below. Stop here and make sure you understand the terms N, r, and t:

$$N_t = N_0 r^t$$

Exponential growth describes continuous change

The example in the previous section takes growth one time step at a time, but really cockroaches can reproduce continuously if they live in a warm, humid environment. You can describe continuous change using the same terms, r, N, and t, plus the added term delta (d), for *change* (fig. 6.4).

You can read this equation like this: the change in N (dN) per change in time (dt) equals rate of increase (r) times the population size (N). This equation is a model, a very simplified description of the dynamic process of population growth. Models like this are convenient because you can use them to describe many different growth trends, just by changing the "r" term. If $r > 0$, then dN increases over time. If $r < 0$, then dN is negative, and the population is declining. If $r = 0$, then dN is 0 (no change), and the population is stable.

This particular model describes an **exponential growth** rate. An exponential growth rate has a J-shaped curve, as in the upward parts of the curve in figure 6.5. This growth rate describes many species that grow rapidly when food is available, including moose and other prey species.

Exponential growth leads to crashes

A population can only grow at an exponential rate this fast if nothing limits its growth. Usually there are many factors that reduce the rate of increase. Individuals die, they might mature slowly, they may fail to reproduce. But if a population has few or no predators (as in the case of invasive species, see chapter 11), it can grow at an exponential rate, at least for a while.

But all environments have a limited capacity to provide food and other resources for a particular species. **Carrying capacity** is the term for the number or biomass of a species that can be supported in a certain area without depleting resources. Eventually, a rapidly growing population reaches and overshoots this carrying capacity (fig. 6.5). Shortages of food or other resources eventually lead to a **population crash,** or rapid dieback. Once below the carrying capacity, the population may rise again, leading to boom and bust cycles. These oscillations can eventually lower the environmental carrying capacity for an entire food web.

In the case of the bluefin tuna (opening case study), we might say that the population of tuna fishers grew too fast and overshot the carrying capacity of the bluefin resource. The subsequent collapse would appear inevitable to a population biologist.

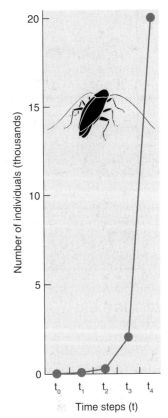

FIGURE 6.3 Population increase with a constant growth rate.

Logistic growth slows with population increase

Sometimes growth rates slow down as the population approaches carrying capacity—as resources become scarce, for example. In symbolic terms, the rate of change (dN/dt) depends on how close population size (N) is to the carrying capacity (K).

For example, suppose you have an area that can support 100 wolves. Let's say that 20 years ago, there were only 50 wolves, so there was abundant space and prey. The 50 wolves were healthy, many pups survived each year, and the population grew rapidly. Now the population has risen to 90. This number is close to the maximum 100 that the environment can support before the wolves begin to deplete their prey. Now, with less food per wolf, fewer cubs are surviving to adulthood, and the rate of increase has slowed. This slowing rate of growth makes an S-shaped curve, or a "sigmoidal" curve (fig. 6.6). This S-shaped growth pattern is also called **logistic growth** because the curve is shaped like a logistic function used in math.

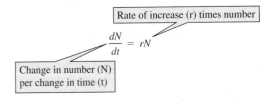

FIGURE 6.4 Exponential growth.

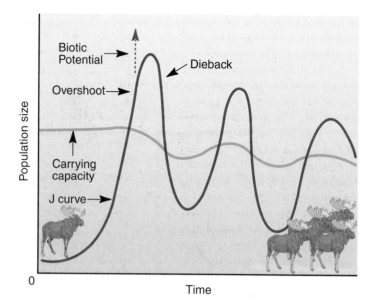

FIGURE 6.5 J curve, or exponential growth curve, with overshoot of carrying capacity. Exponential growth in an unrestrained population (*left side of curve*) leads to a population crash and oscillations below former levels. After the overshoot, carrying capacity may be reduced because of damage to the resources of the habitat. Moose on Isle Royale in Lake Superior may be exhibiting this growth pattern in response to their changing environment.

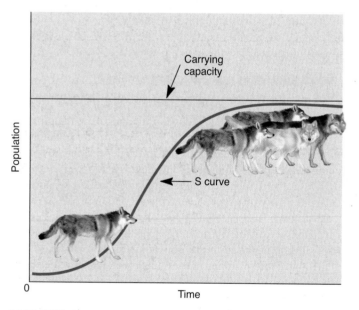

FIGURE 6.6 S curve, or logistic growth curve, describes a population's changing number over time in response to feedback from the environment or its own population density. Over the long run, a conservative and predictable population dynamic may win the race over an exponential population dynamic. Species with this growth pattern tend to be *K*-selected.

You can describe the general case of this growth by modifying the basic exponential equation with a feedback term—a term that can dampen the exponential growth of *N* (fig. 6.7).

If you are patient, you can see interesting patterns in this equation. Look first at the $\frac{N}{K}$ part. For the wolf example, *K* is

$$\frac{dN}{dt} = rN\left(1 - \frac{N}{K}\right)$$

> Population size as a proportion of carrying capacity

FIGURE 6.7 Logistic growth.

100 wolves, the maximum that can be supported. If *N* is 100, then $\frac{N}{K} = \frac{100}{100}$, which is 1. So $1 - \frac{N}{K} = 1 - 1$, which is 0. As a consequence, the right side of the equation is equal to 0 ($rN \times 0 = 0$), so $\frac{dN}{dt} = 0$. So there is no change in *N* if *N* is equal to the carrying capacity. Try working out the following examples on paper as you read, so you can see how *N* changes the equation.

What if *N* is only 50? Then $\frac{N}{K} = \frac{50}{100} = \frac{1}{2}$. So $1 - \frac{N}{K} = 1 - \frac{1}{2} = \frac{1}{2}$. In this case, the rate of increase is $\frac{1}{2}rN$, or half of the maximum possible reproductive rate. If $N = 10$, then $1 - \frac{N}{K} = 1 - \frac{100}{100}$, which is 0.90. So $\frac{dN}{dt}$ is increasing at a rate 90 percent as fast as the maximum possible reproductive rate for that species.

What if the population grows to 120? Overpopulation will likely lead to starvation or low birth rates, and the population will decline to something below 100 again. In terms of the model, now $1 - \frac{N}{K} = 1 - \frac{100}{100} = -0.2$. Now the rate of change, $\frac{dN}{dt}$, is *declining* at a rate of $-0.2rN$.

Logistic growth is **density dependent,** meaning that the growth rate depends on population density. Many density-dependent factors can influence population: overcrowding can increase disease rates, stress, and predation, for example. These factors can lead to smaller body size and lower fertility rates. Crowding stress alone can affect birth rates. In a study of overcrowded house mice ($> 1,600/m^3$), the average litter size was only 5.1 mice per litter, compared to 6.2 per litter in less crowded conditions ($< 34/m^3$). **Density-independent** factors also affect populations. Usually these are abiotic (nonliving) disturbances, such as drought or fire or habitat destruction, which disrupt an ecosystem.

A population can lose a portion of its numbers every year, but that portion depends on *r*, *N*, and *K*, among other factors. A sustainable harvest is possible, as in a tuna fishery, if the number caught is within that sustainable proportion. A "maximum sustained yield" is the highest number that can be regularly captured.

Population biologists have often been very successful in using growth rates to identify a sustainable yield. In North America, game laws restrict hunting of ducks, deer, fish, and other game species, and most hunters and fishers now understand and defend those limits. Acceptable harvest levels are set by population biologists, who have studied reproductive rates, carrying capacities, and population size of each species, in order to determine a sustainable yield. (In some cases, *r* is now too rapid for *K*: see What Do You Think? p. 122.) Similarly, the Pacific salmon fishery, from California to Alaska, is carefully monitored and is considered a healthy and sustainable fishery.

Species respond to limits differently: *r*- and *K*-selected species

Which is more successful for increasing a population, rapid reproduction or long survival within the carrying capacity? Different species place their bets on different strategies. Some organisms, such as dandelions and barnacles, depend on a high rate of reproduction and growth (*rN*) to secure a place in the environment. These organisms are called ***r*-selected species** because they have a high reproductive rate (*r*) but give little or no care to offspring, which have high mortality. Seeds or larvae are cast far and wide, and there is always a chance that some will survive and prosper. If there are no predators or diseases to control their population, these abundantly-reproducing species can overshoot carrying capacity and experience population crashes, but as long as vast quantities of young are produced, a few will survive. Other organisms reproduce more conservatively—with longer generation times, late sexual maturity, and fewer young. These are referred to as ***K*-selected species,** because their growth slows as the carrying capacity (*K*) of their environment is approached.

Many species blend exponential (*r*-selected) and logistic (*K*-selected) growth characteristics. Still, it's useful to contrast the advantages and disadvantages of organisms at the extremes of the continuum. It also helps if we view differences in terms of "strategies" of adaptation and the "logic" of different reproductive modes (table 6.1).

Organisms with *r*-selected, or exponential, growth patterns tend to occupy low trophic levels in their ecosystems (see chapter 3) or they are successional pioneers. These species, which generally have wide tolerance limits for environmental factors, and thus can occupy many different niches and habitats, are the ones we often describe as "weedy." They tend to occupy disturbed or new environments, grow rapidly, mature early, and produce many offspring with excellent dispersal abilities. As individual parents, they do little to care for their offspring or protect them from predation. They invest their energy in producing huge numbers of young and count on some surviving to adulthood.

A female clam, for example, can release up to 1 million eggs in her lifetime. The vast majority of young clams die before reaching maturity, but a few survive, and the species persists. Many marine invertebrates, parasites, insects, rodents, and annual plants follow

this reproductive strategy. Also included in this group are most invasive and pioneer organisms, weeds, and pests.

So-called *K*-selected organisms are usually larger, live long lives, mature slowly, produce few offspring in each generation, and have few natural predators. Elephants, for example, are not reproductively mature until they are 18 to 20 years old. In youth and adolescence, a young elephant belongs to an extended family that cares for it, protects it, and teaches it how to behave. A female elephant normally conceives only once every 4 or 5 years. The gestation period is about 18 months; thus, an elephant herd doesn't produce many babies in any year. Since elephants have few enemies and live a long life (60 or 70 years), this low reproductive rate produces enough elephants to keep the population stable, given good environmental conditions and no poachers.

When you consider the species you recognize from around the world, can you pigeonhole them into categories of *r*- or *K*-selected species? What strategies seem to be operating for ants, bald eagles, cheetahs, clams, dandelions, giraffes, or sharks?

Think About It

Which of the following strategies do humans follow: Do we more closely resemble wolves and elephants in our population growth, or does our population growth pattern more closely resemble that of moose and rabbits? Will we overshoot our environment's carrying capacity (or are we already doing so), or will our population growth come into balance with our resources?

6.2 COMPLICATING THE STORY: *r* = BIDE

By adding carrying capacity, we complicated our first simple population model, and we made it more realistic. To complicate it still further, we can consider the four factors that contribute to *r*, or rate of growth. These factors are **B**irths, **I**mmigration from other areas, **D**eaths, and **E**migration to other areas. More specifically, rate of growth is equal to Births + Immigration – Deaths – Emigration. In a detailed population model, populations receive immigrants and lose individuals to emigration. Number of births might rise more rapidly than number of deaths. Models of human populations (see chapter 7), as well as animal populations involve detailed calculations of the four BIDE factors.

The two terms that make population grow, births and immigration, should be relatively easy to imagine. Birth rates are different for different species (house flies vs. elephants, for example), and birth rate can decline if there are food shortages or if crowding leads to stress, as noted earlier. Of the two negative terms, deaths and emigration, the emigration idea simply means that sometimes individuals leave the population. Deaths, on the other hand, can have some interesting patterns.

Mortality, or death rate, is the portion of the population that dies in any given time period. Some of mortality is determined by environmental factors, and some of it is determined by an organism's physiology, or its natural life span. Life spans vary

Table 6.1 Reproductive Strategies	
r-Selected Species	*K*-Selected Species
1. Short life	1. Long life
2. Rapid growth	2. Slower growth
3. Early maturity	3. Late maturity
4. Many small offspring	4. Few, large offspring
5. Little parental care or protection	5. High parental care or protection
6. Little investment in individual offspring	6. High investment in individual offspring
7. Adapted to unstable environment	7. Adapted to stable environment
8. Pioneers, colonizers	8. Later stages of succession
9. Niche generalists	9. Niche specialists
10. Prey	10. Predators
11. Regulated mainly by extrinsic factors	11. Regulated mainly by intrinsic factors
12. Low trophic level	12. High trophic level

What Do You Think?

Too Many Deer?

A century ago, few Americans had ever seen a wild deer. Uncontrolled hunting and habitat destruction had reduced the deer population to about 500,000 animals nationwide. Some states had no deer at all. To protect the remaining deer, laws were passed in the 1920s and 1930s to restrict hunting, and the main deer predators—wolves and mountain lions—were exterminated throughout most of their former range.

As Americans have moved from rural areas to urban centers, forests have regrown, and deer populations have undergone explosive growth. Maturing at age two, a female deer can give birth to twin fawns every year for a decade or more. Increasing more than 20 percent annually, a deer population can double in just three years, an excellent example of irruptive, exponential growth.

Wildlife biologists estimate that the contiguous 48 states now have a population of more than 30 million white-tailed deer (*Odocoileus virginianus*), probably triple the number present in pre-Columbian times. Some

White-tailed deer (*Odocoileus virginianus*), can become emaciated and sick when they exceed their environment's carrying capacity.

areas have as many as 200 deer per square mile (77/km²). At this density, woodland plant diversity is generally reduced to a few species that deer won't eat. Most deer, in such conditions, suffer from malnourishment, and many die every year of disease and starvation. Other species are diminished as well. Many small mammals and ground-dwelling birds begin to disappear when deer populations reach 25 animals per square mile. At 50 deer per square mile, most ecosystems are seriously impoverished.

The social costs of large deer populations are high. In Pennsylvania alone, where deer numbers are now about 500 times greater than a century ago, deer destroy about $70 million worth of crops and $75 million worth of trees annually. Every year some 40,000 collisions with motor vehicles cause $80 million in property damage. Deer help spread Lyme disease, and, in many states, chronic wasting disease is found in wild deer herds. Some of the most heated criticisms of current deer management policies are in the suburbs. Deer love to browse on the flowers, young trees, and ornamental bushes in suburban yards. Heated disputes often arise between those who love to watch deer and their neighbors who want to exterminate them all.

In remote forest areas, many states have extended hunting seasons, increased the bag limit to four or more animals, and encouraged hunters to shoot does (females) as well as bucks (males). Some hunters criticize these changes because they believe that fewer deer will make it harder to hunt successfully and less likely that they'll find a trophy buck. Others, however, argue that a healthier herd and a more diverse ecosystem is better for all concerned.

In urban areas, increased sport hunting usually isn't acceptable. Wildlife biologists argue that the only practical way to reduce deer herds is culling by professional sharpshooters. Animal rights activists protest lethal control methods as cruel and inhumane. They call instead for fertility controls, reintroduction of predators, such as wolves and mountain lions, or trap and transfer programs. Birth control works in captive populations but is expensive and impractical with wild animals. Trapping, also, is expensive, and there's rarely anyplace willing to take surplus animals, which often die after relocation.

This case shows that carrying capacity can be more complex than simply the maximum number of organisms an ecosystem can support. While it may be possible for 200 deer to survive in a square mile, the ecological carrying—the amount that can be sustained without damage to the ecosystem and to other species—is usually considerably less. There's also an ethical carrying capacity if we don't want to see animals suffer from malnutrition, disease, or starvation. There may also be a cultural carrying capacity if we consider the tolerable rate of depredation on crops and lawns or an acceptable number of motor vehicle collisions.

If you were a wildlife biologist charged with managing the deer herd in your state, how would you reconcile the different interests in this issue? What sources of information or ideas shape views for and against population control in deer? What methods would you suggest to reach the optimal population size? What social or ecological indicators would you look for to gauge whether deer populations are excessive or have reached an appropriate level?

---

enormously. Some microorganisms live whole life cycles in a few hours or even minutes. Bristlecone pine trees in the mountains of California, on the other hand, have life spans up to 4,600 years.

Different rates of growth, maturity, and survival over time can be graphed to compare life histories of different organisms (fig. 6.8). Several general patterns of survivorship can be seen in this idealized figure. Curve (*a*) shows a simplified, general trend for organisms that

have high juvenile survival, high survival in reproductive ages, and a tendency for most individuals to reach old age. Survivorship declines sharply in the older, postreproductive phase, but some persist to near the maximum possible age. Many larger mammals follow this pattern, for example, whales, bears, and elephants (and many human populations). Juvenile survival tends to be fairly high, in part because parents invest considerable energy in tending to one or two young at

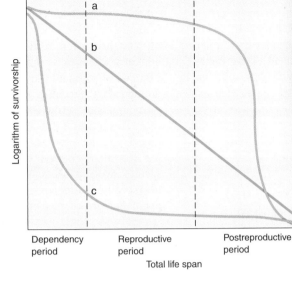

(a) Survive to old age

(c) Long adult life span

(b) Die randomly

FIGURE 6.8 Three basic types of survivorship curves for organisms with different life histories. Curve (a) represents organisms such as humans or elephants, which tend to live out the full physiological life span if they survive early growth. Curve (b) represents organisms such as sea gulls, which have a fairly constant mortality at all age levels. Curve (c) represents such organisms as clams and redwood trees, which have a high mortality rate early in life but live a full life if they reach adulthood.

Within the graph: Logarithm of survivorship — a, b, c — Dependency period | Reproductive period | Postreproductive period — Total life span

a time. Adult mortality is fairly low because these organisms have few predators. There are also very small organisms, including predatory protozoa, that have similar survivorship curves, with a large proportion surviving to a mature age—for a microorganism.

Curve (*b*) shows survivorship for organisms for which the probability of death is unrelated to age, once infancy is past. Sea gulls, mice, rabbits, and other organisms face risks that affect all ages, such as predation, disease, or accidents. Mortality rates can be more or less constant with age, and their survivorship curve can be described as a straight line.

Curve (*c*) is typical of organisms at the base of a food chain or those especially susceptible to mortality early in life. Many tree species, fish, clams, crabs, and other invertebrate species produce a very large number of highly vulnerable offspring. Just a few survive to maturity. Those that do survive to adulthood, however, have a very high chance of living nearly the maximum life span for the species.

Think About It

Which of these survivorship patterns best describes humans? Are we more like elephants or deer? Do wealth and modernity have something to do with it? Might people in Bangladesh have different survivorship prospects than you do?

6.3 FACTORS THAT REGULATE POPULATION GROWTH

So far, we have seen that differing patterns of natality, mortality, life span, and longevity can produce quite different rates of population growth. The patterns of survivorship and age structure created by these interacting factors not only show us how a population is growing but also can indicate what general role that species plays in its ecosystem. They also reveal a good deal about how that species is likely to respond to disasters or resource bonanzas in its environment. But what factors *regulate* natality, mortality, and the other components of population growth? In this section, we will look at some of the mechanisms that determine how a population grows.

Various factors regulate population growth, primarily by affecting natality or mortality, and can be classified in different ways. They can be *intrinsic* (operating within individual organisms or between organisms in the same species) or *extrinsic* (imposed from outside the population). Factors can also be either **biotic** (caused by living organisms) or **abiotic** (caused by nonliving components of the environment). Finally, the regulatory factors can act in a *density-dependent* manner (effects are stronger or a higher proportion of the population is affected as population density increases) or *density-independent* manner (the effect is the same or a constant proportion of the population is affected regardless of population density).

In general, biotic regulatory factors tend to be density-dependent, while abiotic factors tend to be density-independent. There has been much discussion about which of these factors is most important in regulating population dynamics. In fact, it probably depends on the particular species involved, its tolerance levels, the stage of growth and development of the organisms involved, the specific ecosystem in which they live, and the way combinations of factors interact. In most cases, density-dependent and density-independent factors probably exert simultaneous influences. Depending on whether regulatory factors are regular and predictable or irregular and unpredictable, species will develop different strategies for coping with them.

Some population factors are density-independent; others are density-dependent

In general, the factors that affect natality or mortality independently of population density tend to be abiotic components of the ecosystem. Often weather (conditions at a particular time) or climate (average weather conditions over a longer period) are among the most important of these factors. Extreme cold or even moderate cold at the wrong time of year, high heat, drought, excess rain, severe storms, and geologic hazards—such as volcanic eruptions, landslides, and floods—can have devastating impacts on particular populations.

Abiotic factors can have beneficial effects as well, as anyone who has seen the desert bloom after a rainfall can attest. Fire is a powerful shaper of many biomes. Grasslands, savannas, and some montane and boreal forests often are dominated—even created—by periodic fires. Some species, such as jack pine and Kirtland's warblers, are so adapted to periodic disturbances in the environment that they cannot survive without them.

In a sense, these density-independent factors don't necessarily regulate population *per se,* since regulation implies a homeostatic feedback that increases or decreases as density fluctuates. By definition, these factors operate without regard to the number of organisms involved. They may have such a strong impact on a population, however, that they completely overwhelm the influence of any other factor and determine how many individuals make up a particular population at any given time.

Density-dependent mechanisms tend to reduce population size by decreasing natality or increasing mortality as the population size increases. Most of them are the results of interactions *between* populations of a community (especially predation), but some of them are based on interactions *within* a population.

Interspecific interactions occur between species

As we discussed in chapter 4, a predator feeds on—and usually kills—its prey species. While the relationship is one-sided with respect to a particular pair of organisms, the prey species as a whole may benefit from the predation. For instance, the moose that gets eaten by wolves doesn't benefit individually, but the moose *population* is strengthened because the wolves tend to kill old or sick members of the herd. Their predation helps prevent population overshoot, so the remaining moose are stronger and healthier.

Sometimes predator and prey populations oscillate in a sort of synchrony with each other as is shown in figure 6.9, which shows the number of furs brought into Hudson Bay Company trading posts in Canada between 1840 and 1930. As you can see, the numbers of Canada lynx fluctuate on about a ten-year cycle that is similar to, but slightly out of phase with, the population peaks of snowshoe hares. Although there are some doubts now about how and where these data were collected, this remains a classic example of population dynamics. When prey populations (hares) are abundant, predators (lynx) reproduce more successfully and their population grows. When hare populations crash, so do the lynx. This predator-prey oscillation is known as the Lotka-Volterra model after the scientists who first described it mathematically.

Not all interspecific interactions are harmful to one of the species involved. Mutualism and commensalism, for instance, are interspecific interactions that are beneficial or neutral in terms of population growth (see chapter 4).

Intraspecific interactions occur within a species

Individuals within a population also compete for resources. When population density is low, resources are likely to be plentiful and the population growth rate will approach the maximum possible for the species, assuming that individuals are not so dispersed that they cannot find mates. As population density approaches the carrying capacity of the environment, however, one or more of the vital resources becomes limiting. The stronger, quicker, more aggressive, more clever, or luckier members get a larger share, while others get less and then are unable to reproduce successfully or survive.

Territoriality is one principal way many animal species control access to environmental resources. The individual, pair, or group that holds the territory will drive off rivals if possible, either

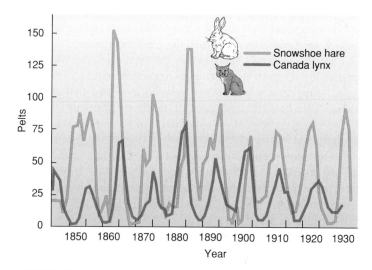

FIGURE 6.9 Ten-year oscillations in the populations of snowshoe hare and lynx in Canada suggest a close linkage of predator and prey, but may not tell the whole story. These data are based on the number of pelts received by the Hudson Bay Company each year, meaning fur-traders were unwitting accomplices in later scientific research.

Source: Data from D. A. MacLulich, *Fluctuations in the Numbers of the Varying Hare* (Lepus americus). Toronto: University of Toronto Press, 1937, reprinted 1974.

by threats, displays of superior features (colors, size, dancing ability), or fighting equipment (teeth, claws, horns, antlers). Members of the opposite sex are attracted to individuals that are able to seize and defend the largest share of the resources. From a selective point of view, these successful individuals presumably represent superior members of the population and the ones best able to produce offspring that will survive.

Stress and crowding can affect reproduction

Stress and crowding also are density-dependent population control factors. When population densities get very high, organisms often exhibit symptoms of what is called stress shock or **stress-related diseases.** These terms describe a loose set of physical, psychological, and/or behavioral changes that are thought to result from the stress of too much competition and too close proximity to other members of the same species. There is a considerable controversy about what causes such changes and how important they are in regulating natural populations. The strange behavior and high mortality of arctic lemmings or hares during periods of high population density may be a manifestation of stress shock (fig. 6.10). On the other hand, they could simply be the result of malnutrition, infectious disease, or some other more mundane mechanism at work.

Some of the best evidence for the existence of stress-related disease comes from experiments in which laboratory animals, usually rats or mice, are grown in very high densities with plenty of food and water but very little living space. A variety of symptoms are reported, including reduced fertility, low resistance to infectious diseases, and pathological behavior. Dominant animals seem to be affected least by crowding, while subordinate animals—the ones presumably subjected to the most stress in intraspecific interactions—seem to be the most severely affected.

Density-dependent effects can be dramatic

The desert locust, *Schistocerca gregarius,* has been called the world's most destructive insect. Throughout recorded human history, locust plagues have periodically swarmed out of deserts and into settled areas. Their impact on human lives has often been so disruptive that records of plagues have taken on religious significance and made their way into sacred and historical texts.

Locusts usually are solitary creatures resembling ordinary grasshoppers. Every few decades, however, when rain comes to the desert and vegetation flourishes, locusts reproduce rapidly until the ground seems to be crawling with bugs. High population densities and stress bring ominous changes in these normally innocuous insects. They stop reproducing, grow longer wings, group together in enormous swarms, and begin to move across the desert. Dense clouds of insects darken the sky, moving as much as 100 km per day. Locusts may be small, but they can eat their own body weight of vegetation every day. A single swarm can cover 1,200 km^2 and contain 50 to 100 billion individuals. The swarm can strip pastures, denude trees, and destroy crops in a matter of hours, consuming as much food in a day as 500,000 people would need for a year. Eventually, having exhausted their food supply and migrated far from the desert where conditions favor reproduction, the locusts die and aren't seen again for decades.

FIGURE 6.10 Animals often battle over resources. This conflict can induce stress and affect reproductive success.

Huge areas of crops and rangeland in northern Africa, the Middle East, and Asia are within the reach of the desert locust. This small insect, with its voracious appetite, can affect the livelihood of at least one-tenth of the world's population. During quiet periods, called recessions, African locusts are confined to the Sahara Desert, but when conditions are right, swarms invade countries as far away as Spain, Russia, and India. Swarms are even reported to have crossed the Atlantic Ocean from Africa to the Caribbean.

Unusually heavy rains in the Sahara in 2004 created the conditions for a locust explosion. Four generations bred in rapid succession, and swarms of insects moved out of the desert. Twenty-eight countries in Africa and the Mediterranean area were afflicted. Crop losses reached 100 percent in some places, and food supplies for millions of people were threatened. Officials at the United Nations warned that we could be headed toward another great plague. Hundreds of thousands of hectares of land were treated with pesticides, but millions of dollars of crop damage were reported anyway.

This case study illustrates the power of exponential growth and the disruptive potential of a boom-and-bust life cycle. Stress, population density, migration, and intraspecific interactions all play a role in this story. Although desert conditions usually keep locust numbers under control, their biotic potential for reproduction is a serious worry for residents of many countries.

6.4 CONSERVATION BIOLOGY

Small, isolated populations can undergo catastrophic declines due to environmental change, genetic problems, or stochastic (random or unpredictable) events. A critical question in conservation biology is the minimum population size of a rare and endangered species required for long-term viability. While much is known about species survival, much also remains to be discovered (see Exploring Science, p. 127). In this section, we'll look at some factors that influence the long-term likelihood of sustaining biodiversity and species.

Island biogeography describes isolated populations

In a classic 1967 study, R. H. MacArthur and E. O. Wilson asked why it is that small islands far from the mainland generally have far fewer species than larger or nearer islands. They proposed the equilibrium theory of **island biogeography,** the idea that diversity in isolated habitats depends on rates of colonization and extinction, which depend on the size or isolation of an island. Colonization by new species tends to be rare on remote islands, which are hard to reach (fig. 6.11). At the same time, small islands have smaller habitats than larger islands, and support fewer individuals of any given species. These small populations are more likely to go extinct due to natural disasters, diseases, or demographic factors such as imbalance between sexes in a particular generation, compared to larger populations. Larger islands are more likely to sustain populations, and islands close to the mainland are readily colonized by new species. Thus they tend to have greater diversity than smaller, more remote places.

Island biogeographical patterns have been observed in many places. In the Caribbean, for instance, Cuba is 100 times as large as Monserrat and has about 10 times as many amphibian species. Similarly, in a study of bird species on the California Channel Islands, Jared Diamond observed that on islands with fewer than 10 breeding pairs, 39 percent of the populations went extinct over an 80-year period. In contrast, only 10 percent of populations numbering between 10 and 100 pairs went extinct, and no species with more than 1,000 pairs disappeared over this time (fig. 6.12). This theory of a balance between colonization and extinction, and the observation that small populations are especially

likely to disappear, has been applied to explain species dynamics in many small, isolated habitat fragments whether on islands or not.

Conservation genetics helps predict survival of endangered species

Genetics plays an important role in the survival or extinction of small, isolated populations. In large populations, genetic variation tends to persist in what is called a Hardy-Weinberg equilibrium, named after the scientists who first described why this occurs. If mating is random, no mutations (changes in genetic material) occur, and there is no gene in-flow or selective pressure for or against particular traits, random distribution of gene types will occur from sexual reproduction. That is, different gene types will be distributed in the offspring in the same ratio they occur in the parents, and genetic diversity is preserved.

In a large population, these conditions for maintaining genetic equilibrium are generally operative. The addition or loss of a few individuals or appearance of new genotypes makes little difference in the total gene pool, and genetic diversity is relatively constant. In small, isolated populations, however, immigration, mortality, mutations, or chance mating events involving only a few individuals can greatly alter the genetic makeup of the whole population. We call the gradual changes in gene frequencies due to random events **genetic drift.**

For many species, loss of genetic diversity causes a number of harmful effects that limit adaptability, reproduction, and species survival. A **founder effect** or **demographic bottleneck** occurs when just a few members of a species survive a catastrophic event or colonize new habitat geographically isolated from other members of the same species. Any deleterious genes present in the founders will be overrepresented in subsequent generations (fig. 6.13). Inbreeding, mating of closely related individuals, also makes expression of rare or recessive genes more likely.

Some species seem not to be harmed by inbreeding or lack of genetic diversity. The northern elephant seal, for example, was reduced by overharvesting a century ago, to fewer than 100 individuals. Today there are more than 150,000 of these enormous animals along the Pacific coast of Mexico and California. No

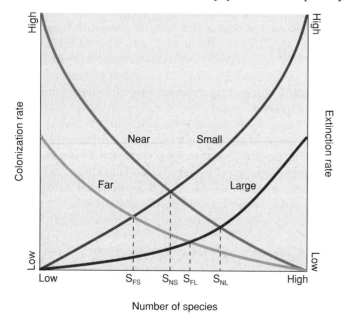

FIGURE 6.11 Predicted species richness on an island resulting from a balance between colonization (immigration) and extinction by natural causes. This island biogeography theory of MacArthur and Wilson (1967) is used to explain why large islands near a mainland (S_{NL}) tend to have more species than small, far islands (S_{FS}).
Source: Based on MacArthur and Wilson, *The Theory of Island Biogeography,* 1967, Princeton University Press.

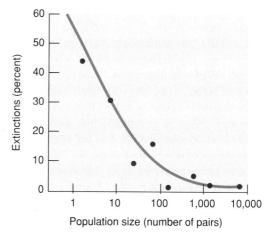

FIGURE 6.12 Extinction rates of bird species on the California Channel Islands as a function of population size over 80 years.
Source: H. L. Jones and J. Diamond, "Short-term-base studies of turnover in breeding bird populations on the California coast island," in *Condor,* vol. 78:526–49, 1976.

Population data are necessary for understanding population stability. Collecting these data is hard in any population, but it's especially difficult when the species live far out in the open ocean, migrate widely, and are rarely seen except by fishing boats that catch them.

Most ocean fish data come from fishing records. A decline in average size, in number of adults, or in size at spawning age indicates that a population is being overfished. The bluefin tuna has shown evidence of all these effects. At Tokyo's Tsukiji fish market, the average weight of a bluefin has fallen since the 1980s from 100–160 kg to just 50 kg today. The proportion of fish younger than 1 year has increased, and the proportion of larger, older fish has fallen sharply.

Another, newer approach is to attach satellite tracking tags, small, plastic-coated rods inserted into the fish's side just below a fin.

An observer measures a big-eye tuna.

Electronic tags can record factors such as the location from satellite readings or water pressure (a measure of depth in the water). Tags can be designed to float to the surface if they are released from the fish, and to transmit recorded data to a satellite and then to the researcher's computer, or tags can be returned if fish are caught.

Tagging studies have revealed the astonishing distances tuna travel, where they go, their preferred feeding grounds, and their fidelity to their home spawning grounds. Ideally, this information can be used to designate no-fishing sanctuaries in spawning grounds, as well as provide information on basic population biology. Mapping tag locations also showed that the western Atlantic and Mediterranean bluefin populations were distinct and that they prefer different types of spawning conditions.

Information such as this gives conservation science a new tool for trying to save a species and the ecosystem that depends on it.

For more information, see Barbara A. Block, et al. 2005. Electronic tagging and population structure of Atlantic bluefin tuna. *Nature* 434: 1121–1127.

---

marine mammal is known to have come closer to extinction and then made such a remarkable recovery. All northern elephant seals today appear to be essentially genetically identical and yet they seem to have no apparent problems. Although interpretations of their situation are controversial, in highly selected populations, where only the most fit individuals reproduce, or in which there are few deleterious genes, inbreeding and a high degree of genetic identity may not be such a negative factor.

Cheetahs, also, appear to have undergone a demographic bottleneck sometime in the not-too-distant past. All the male cheetahs alive today appear to be nearly genetically identical, suggesting that they all share a single male ancestor (fig. 6.14). This lack of diversity is thought to be responsible for an extremely low fertility rate, a high abundance of abnormal sperm, and low survival rate for offspring, all of which threatens the survival of the species.

Population viability analysis calculates chances of survival

Conservation biologists use the concepts of island biogeography, genetic drift, and founder effects to determine **minimum viable population size,** or number of individuals needed for long-term survival of rare and endangered species. A classic example is that

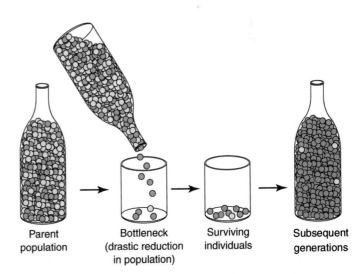

| Parent population | Bottleneck (drastic reduction in population) | Surviving individuals | Subsequent generations |

FIGURE 6.13 Genetic drift: the bottleneck effect. The parent population contains roughly equal numbers of blue and yellow individuals. By chance, the few remaining individuals that comprise the next generation are mostly blue. The bottleneck occurs because so few individuals form the next generation, as might happen after an epidemic or catastrophic storm.

FIGURE 6.14 Sometime in the past, cheetahs underwent a severe population crash. Now all male cheetahs alive today are nearly genetically identical, and deformed sperm, low fertility levels, and low infant survival are common in the species.

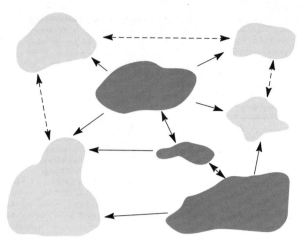

FIGURE 6.15 A metapopulation is composed of several local populations linked by regular (*solid arrows*) or occasional (*dashed lines*) gene flows. Source populations (*dark*) provide excess individuals, which emigrate to and colonize sink habitats (*light*).

of the grizzly bear (*Ursus arctos horribilis*) in North America. Before European settlement, grizzlies roamed from the Great Plains west to California and north to Alaska. Hunting and habitat destruction reduced the number of grizzlies from an estimated 100,000 in 1800 to less than 1,200 animals in six separate subpopulations that now occupy less than 1 percent of the historic range. Recovery target sizes—based on estimated environmental carrying capacities—call for fewer than 100 bears for some subpopulations. Conservation genetics predicts that a completely isolated population of 100 bears cannot be maintained for more than a few generations. Even the 600 bears now in Yellowstone National Park will be susceptible to genetic problems if completely isolated. Interestingly, computer models suggest that translocating only two unrelated bears into small populations every generation (about ten years) could greatly increase population viability.

For mobile organisms, separated populations can have gene exchange if suitable corridors or migration routes exist. A **metapopulation** is a collection of populations that have regular or intermittent gene flow between geographically separate units (fig. 6.15). For example, the Bay checkerspot butterfly (*Euphydrays editha bayensis*) in California exists in several distinct habitat patches.

Individuals occasionally move among these patches, mating with existing animals or recolonizing empty habitats. Thus, the apparently separate groups form a functional metapopulation.

A "source" habitat, where birth rates are higher than death rates, produces surplus individuals that can migrate to new locations within a metapopulation. "Sink" habitats, on the other hand, are places where mortality exceeds birth rates. Sinks may be spatially larger than sources but because of unfavorable conditions, the species would disappear in the sink habitat if it were not periodically replenished from a source population. In general, the larger a reserve is, the better it is for endangered species. Sometimes, however, adding to a reserve can be negative if the extra area is largely sink habitat. Individuals dispersing within the reserve may settle in unproductive areas if better habitat is hard to find. Recent studies using a metapopulation model for spotted owls predict just such a problem for this species in the Pacific Northwest. These endangered owls may end up in habitat patches where they cannot survive, as the once-continuous forest of the Pacific Northwest is reduced to smaller fragments. In this case and many others, conservation biology and population biology are helping to inform and shape policy for conserving species.

CONCLUSION

Given optimum conditions, populations of many organisms can grow exponentially; that is, they can expand at a constant rate per unit of time. This biotic potential can produce enormous populations that far surpass the carrying capacity of the environment if left unchecked. Obviously, no population grows at this rate forever. Sooner or later, predation, disease, starvation, or some other factor will cause the population to crash. Not all species follow this boom-and-bust pattern, however. Most top predators have intrinsic factors that limit their reproduction and prevent overpopulation.

Overharvesting of species, habitat destruction, predator elimination, introduction of exotic species, and other forms of human disruption can also drive populations to boom and/or crash. Population dynamics are an important part of conservation biology. Principles, such as island biogeography, genetic drift, demographic bottlenecks, and metapopulation interactions are critical in endangered species protection.

REVIEWING LEARNING OUTCOMES

By now you should be able to explain the following points:

6.1 Describe the dynamics of population growth.
- We can describe growth symbolically.
- Exponential growth describes continuous change.
- Exponential growth leads to crashes.
- Logistic growth slows with population increase.
- Species respond to limits differently: *r*- and *K*-selected species.

6.2 Summarize the BIDE factors that increase or decrease populations.
- Natality, fecundity, and fertility are measures of birth rates.
- Immigration adds to populations.
- Emigration removes members of a population.

6.3 Compare and contrast the factors that regulate population growth.
- Population factors can be density-independent.
- Population factors also can be density-dependent.
- Density-dependent effects can be dramatic.

6.4 Identify some applications of population dynamics in conservation biology.
- Island biogeography describes isolated populations.
- Conservation genetics helps predict survival of endangered species.
- Population viability analysis calculates chances of survival.
- Metapopulations are connected.

PRACTICE QUIZ

1. What factors caused the collapse of bluefin tuna populations?
2. Define *exponential growth* and *logistic growth*.
3. Explain these terms: *r*, *N*, *t*, *dN/dt*.
4. What is environmental resistance? How does it affect populations?
5. List five or six ways *r*-selected species tend to differ from *K*-selected species.
6. Describe three major types of survivorship patterns and explain what they show about a species' role in its ecosystem.
7. What are the main interspecific population regulatory interactions? How do they work?
8. What is island biogeography and why is it important in conservation biology?
9. Why does genetic diversity tend to persist in large populations, but gradually drift or shift in small populations?
10. Explain the following: *metapopulation, genetic drift, demographic bottleneck*.

CRITICAL THINKING AND DISCUSSION QUESTIONS

1. Compare the advantages and disadvantages to a species that result from exponential or logistic growth. Why do you think hares have evolved to reproduce as rapidly as possible, while lynx appear to have intrinsic or social growth limits?
2. Are humans subject to environmental resistance in the same sense that other organisms are? How would you decide whether a particular factor that limits human population growth is ecological or social?
3. What are advantages and disadvantages in living longer or reproducing more quickly? Why hasn't evolution selected for the most advantageous combination of characteristics so that all organisms would be more or less alike?
4. Why do abiotic factors that influence population growth tend to be density-independent, while biotic factors that regulate population growth tend to be density-dependent?
5. Some people consider stress and crowding studies of laboratory animals highly applicable in understanding human behavior. Other people question the cross-species transferability of these results. What considerations would be important in interpreting these experiments?
6. What implications (if any) for human population control might we draw from our knowledge of basic biological population dynamics?

Data Analysis: Comparing Exponential to Logistic Population Growth

Exponential growth occurs in a series of time steps—days, months, years, or generations. Imagine cockroaches in a room multiplying (or some other species, if you must). Picture a population of ten cockroaches that together produce enough young to increase at a rate of 150 percent per month. What is r for this population?

To find out how this population grows, fill out the table shown. (*Hint:* r remains constant.) Remember, for time step 0 (the first month), you begin with ten roaches, and end (N_e) with a larger number that depends on r, the intrinsic rate of growth. The beginning of the second time step (1) starts with the number at the end of step 0. Round N to the nearest whole number. When you are done, graph the results. At the end of 7 months, how large did this population become? What is the shape of the growth curve?

Time Step (t)	Begin Step (N_b)	Intrinsic Growth Rate (r)	End Step (N_e)
0	10		15
1			
2			
3			
4			
5			
6			
7			

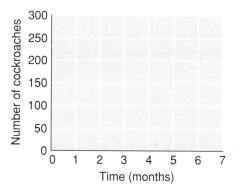

Data Analysis: Experimenting with Population Growth

The previous data analysis lets you work through an example of population growth by hand, which is an important strategy for understanding the equations you've seen in this chapter.

Now try experimenting with more growth rates in an Excel "model." What value of r makes the graph extremely steep? What value makes it flat? Can you model a declining population?

Go to **www.mhhe.com/cunningham12e**, and find the Data Analysis option for this chapter. There you can download an Excel workbook and experiment with different growth rates.

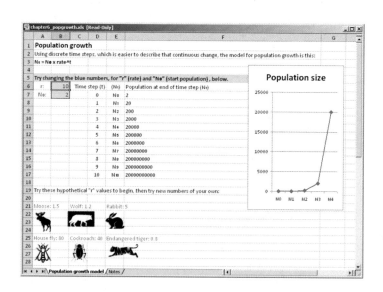

Thailand's highly successful family planning program combines humor and education with economic development.

Human Populations

Learning Outcomes

After studying this chapter, you should be able to:

7.1 Trace the history of human population growth.

7.2 Summarize different perspectives on population growth.

7.3 Analyze some of the factors that determine population growth.

7.4 Explain how ideal family size is culturally and economically dependent.

7.5 Describe how a demographic transition can lead to stable population size.

7.6 Relate how family planning gives us choices.

7.7 Reflect on what kind of future we are creating.

"For every complex problem there is an answer that is clear, simple, and wrong."

~ *H. L. Mencken*

Family Planning in Thailand: A Success Story

Down a narrow lane off Bangkok's busy Sukhumvit Road, is a most unusual café. Called Cabbages and Condoms, it's not only highly rated for its spicy Thai food, but it's also the only restaurant in the world dedicated to birth control. In an adjoining gift shop, baskets of condoms stand next to decorative handicrafts of the northern hill tribes. Piles of T-shirts carry messages, such as, "A condom a day keeps the doctor away," and "Our food is guaranteed not to cause pregnancy." Both businesses are run by the Population and Community Development Association (PDA), Thailand's largest and most influential nongovernmental organization.

The PDA was founded in 1974 by Mechai Viravaidya, a genial and fun-loving former Thai Minister of Health, who is a genius at public relations and human motivation (fig. 7.1). While traveling around Thailand in the early 1970s, Mechai recognized that rapid population growth—particularly in poor rural areas—was an obstacle to community development. Rather than lecture people about their behavior, Mechai decided to use humor to promote family planning. PDA workers handed out condoms at theaters and traffic jams, anywhere a crowd gathered. They challenged governmental officials to condom balloon-blowing contests, and taught youngsters Mechai's condom song: "Too Many Children Make You Poor." The PDA even pays farmers to paint birth control ads on the sides of their water buffalo.

This campaign has been extremely successful at making birth control and family planning, which once had been taboo topics in polite society, into something familiar and unembarrassing. Although condoms—now commonly called "mechais" in Thailand—are the trademark of PDA, other contraceptives, such as pills, spermicidal foam, and IUDs, are promoted as well. Thailand was one of the first countries to allow the use of the injectable contraceptive DMPA, and remains a major user. Free non-scalpel vasectomies are available on the king's birthday. Sterilization has become the most widely used form of contraception

in the country. The campaign to encourage condom use has also been helpful in combating AIDS.

In 1974, when PDA started, Thailand's growth rate was 3.2 percent per year. In just fifteen years, contraceptive use among married couples increased from 15 to 70 percent, and the growth rate had dropped to 1.6 percent, one of the most dramatic birth rate declines ever recorded. Now Thailand's growth rate is 0.7 percent (lower than that of the United States). The fertility rate (or average number of children per woman) decreased from 7.0 in 1979 to 1.64 in 2009. The PDA is credited with the fact that Thailand's population is 20 million less than it would have been if it had followed its former trajectory.

In addition to Mechai's creative genius and flair for showmanship, there are several reasons for this success story. Thai people love humor and are more egalitarian than most developing countries. Thai spouses share in decisions regarding children, family life, and contraception. The government recognizes the need for family planning and is willing to work with volunteer organizations, such as the PDA. And Buddhism, the religion of 95 percent of Thais, promotes family planning.

The PDA hasn't limited itself to family planning and condom distribution. It has expanded into a variety of economic development projects. Microlending provides money for a couple of pigs, or a bicycle, or a small supply of goods to sell at the market. Thousands of water-storage jars and cement rainwater-catchment basins have been distributed. Larger scale community development grants include road building, rural electrification, and irrigation projects. Mechai believes that human development and economic security are keys to successful population programs.

This case study introduces several important themes of this chapter. What might be the effects of exponential growth in human populations? How might we manage fertility and population growth? And what are the links between poverty, birth rates, and our common environment? In this chapter, we'll examine

FIGURE 7.1 Mechai Viravaidya (*right*) is joined by Peter Piot, Executive Director of UNAIDS, in passing out free condoms on family planning and AIDS awareness day in Bangkok.

how scientists form and answer questions such as these about our world. For related resources, including Google Earth™ placemarks that show locations where these issues can be explored, visit EnvironmentalScience-Cunningham.blogspot.com.

7.1 POPULATION GROWTH

Every second, on average, four or five children are born somewhere on the earth. In that same second, two other people die. This difference between births and deaths means a net gain of roughly 2.5 more humans per second in the world's population. In 2011, the total world population passed 7 billion people and was growing at 1.13 percent per year. This means we are now adding nearly 80 million more people per year, and if this rate persists, our global population will double in about 62 years. Humans are now probably the most numerous vertebrate species on the earth. We also are more widely distributed and manifestly have a greater global environmental impact than any other species. For the families to whom these children are born, this may well be a joyous and long-awaited event (fig. 7.2). But is a continuing increase in humans good for the planet in the long run?

Many people worry that overpopulation will cause—or perhaps already is causing—resource depletion and environmental degradation that threaten the ecological life-support systems on which we all depend. These fears often lead to demands for immediate, worldwide birth control programs to reduce fertility rates and to eventually stabilize or even shrink the total number of humans.

Others believe that human ingenuity, technology, and enterprise can extend the world carrying capacity and allow us to overcome any problems we encounter. From this perspective, more people may be beneficial rather than disastrous. A larger population means a larger workforce, more geniuses, more ideas about what to do. Along with every new mouth comes a pair of hands. Proponents of this worldview—many of whom happen to be economists—argue that continued economic and technological growth can both feed the world's billions and enrich everyone enough to end the population explosion voluntarily. Not so, counter many ecologists. Growth is the problem; we must stop both population and economic growth.

Yet another perspective on this subject derives from social justice concerns. In this worldview, there are sufficient resources for everyone. Current shortages are only signs of greed, waste, and oppression. The root cause of environmental degradation, in this view, is inequitable distribution of wealth and power rather than population size. Fostering democracy, empowering women and minorities, and improving the standard of living of the world's poorest people are what are really needed. A narrow focus on population growth only fosters racism and an attitude that blames the poor for their problems while ignoring the deeper social and economic forces at work.

Whether human populations will continue to grow at present rates and what that growth would imply for environmental quality and human life are among the most central and pressing questions in environmental science. In this chapter, we will look at some causes of population growth as well as how populations are measured and described. Family planning and birth control are essential for stabilizing populations. The number of children a couple decides to have and the methods they use to regulate fertility, however, are strongly influenced by culture, religion, politics, and economics, as well as basic biological and medical considerations. We will examine how some of these factors influence human demographics.

Human populations grew slowly until relatively recently

For most of our history, humans have not been very numerous compared to other species. Studies of hunting and gathering societies suggest that the total world population was probably only a few million people before the invention of agriculture and the domestication of animals around 10,000 years ago. The larger and more secure food supply made available by the agricultural revolution allowed the human population to grow, reaching perhaps 50 million people by 5000 B.C. For thousands of years, the number of humans increased very slowly. Archaeological evidence and historical descriptions suggest that only about 300 million people were living at the time of Christ (table 7.1).

Until the Middle Ages, human populations were held in check by diseases, famines, and wars that made life short and uncertain for most people (fig. 7.3). Furthermore, there is evidence that many early societies regulated their population size through cultural taboos and practices such as abstinence and infanticide. Among the most destructive of natural population controls were bubonic plagues (or Black Death) that periodically swept across Europe between 1348 and 1650. During the worst plague years (between 1348 and 1350), it is estimated that at least one-third of the European population perished. Notice, however, that this didn't retard population growth for very long. In 1650, at the end of the last great plague, there were about 600 million people in the world.

FIGURE 7.2 A Mayan family in Guatemala with four of their six living children. Decisions on how many children to have are influenced by many factors, including culture, religion, need for old age security for parents, immediate family finances, household help, child survival rates, and power relationships within the family. Having many children may not be in the best interest of society at large, but may be the only rational choice for individual families.

Table 7.1	World Population Growth and Doubling Times	
Date	Population	Doubling Time
5000 B.C.	50 million	?
800 B.C.	100 million	4,200 years
200 B.C.	200 million	600 years
A.D. 1200	400 million	1,400 years
A.D. 1700	800 million	500 years
A.D. 1900	1,600 million	200 years
A.D. 1965	3,200 million	65 years
A.D. 2000	6,100 million	51 years
A.D. 2050 (estimate)	8,920 million	215 years

Source: United Nations Population Division.

As you can see in figure 7.3, human populations began to increase rapidly after A.D. 1600. Many factors contributed to this rapid growth. Increased sailing and navigating skills stimulated commerce and communication between nations. Agricultural developments, better sources of power, and better health care and hygiene also played a role. We are now in an exponential or J curve pattern of growth.

It took all of human history to reach 1 billion people in 1804, but little more than 150 years to reach 3 billion in 1960. To go from 5 to 6 billion took only 12 years. Another way to look at population growth is that the number of humans tripled during the twentieth century. Will it do so again in the twenty-first century? If it does, will we overshoot the carrying capacity of our environment and experience a catastrophic dieback similar to those described in chapter 6? As you will see later in this chapter, there is evidence that population growth already is slowing, but whether we will reach equilibrium soon enough and at a size that can be sustained over the long term remains a difficult but important question.

7.2 PERSPECTIVES ON POPULATION GROWTH

As with many topics in environmental science, people have widely differing opinions about population and resources. Some believe that population growth is the ultimate cause of poverty and environmental degradation. Others argue that poverty, environmental degradation, and overpopulation are all merely symptoms of deeper social and political factors. The worldview we choose to believe will profoundly affect our approach to population issues. In this section, we will examine some of the major figures and their arguments in this debate.

Does environment or culture control human populations?

Since the time of the Industrial Revolution, when the world population began growing rapidly, individuals have argued about the causes and consequences of population growth. In 1798 Thomas Malthus (1766–1834) wrote *An Essay on the Principle of Population,* changing the way European leaders thought about population

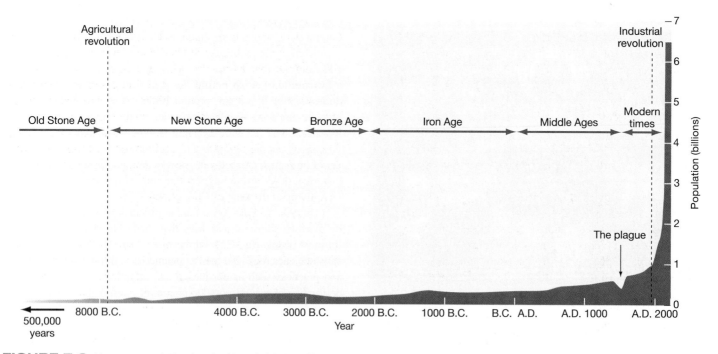

FIGURE 7.3 Human population levels through history. Since about A.D. 1000, our population curve has assumed a J shape. Are we on the upward slope of a population overshoot? Will we be able to adjust our population growth to an S curve? Or can we just continue the present trend indefinitely?

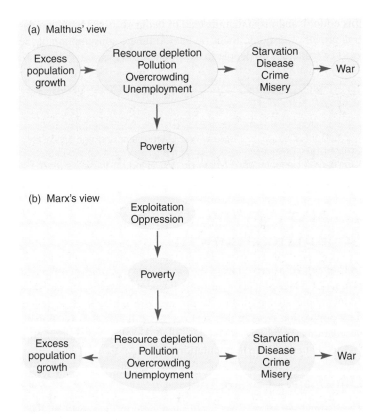

(a) Malthus' view

Excess population growth → Resource depletion Pollution Overcrowding Unemployment → Starvation Disease Crime Misery → War

↓

Poverty

(b) Marx's view

Exploitation Oppression

↓

Poverty

↓

Excess population growth ← Resource depletion Pollution Overcrowding Unemployment → Starvation Disease Crime Misery → War

FIGURE 7.4 (a) Thomas Malthus argued that excess population growth is the ultimate cause of many other social and environmental problems. (b) Karl Marx argued that oppression and exploitation are the real causes of poverty and environmental degradation. Population growth in this view is a symptom or result of other problems, not the source.

growth. Malthus marshaled evidence to show that populations tended to increase at an exponential, or compound, rate while food production either remained stable or increased only slowly. Eventually human populations would outstrip their food supply and collapse into starvation, crime, and misery (fig. 7.4a). He converted most economists of the day from believing that high fertility increased gross domestic output to believing that per capita output actually fell with rapidly rising population.

In Malthusian terms, growing human populations stop growing when disease or famine kills many, or when constraining social conditions compel others to reduce their birth rates—late marriage, insufficient resources, celibacy, and "moral restraint." Several decades later, the economist Karl Marx (1818–1883) presented an opposing view, that population growth resulted from poverty, resource depletion, pollution, and other social ills. Slowing population growth, said Marx, required that people be treated justly, and that exploitation and oppression be eliminated from social arrangements (fig. 7.4b).

Both Marx and Malthus developed their theories about human population growth when understanding of the world, technology, and society were much different than they are today. But these different views of human population growth still inform competing approaches to family planning today. On the one hand, some believe

FIGURE 7.5 Is the world overcrowded already, or are people a resource? In large part, the answer depends on the kinds of resources we use and how we use them. It also depends on democracy, equity, and justice in our social systems.

that we are approaching, or may have surpassed, the earth's carrying capacity. Joel Cohen, a mathematical biologist at Rockefeller University, reviewed published estimates of the maximum human population size the planet can sustain. The estimates, spanning 300 years of thinking, converged on a median value of 10–12 billion. We are at 7 billion strong today, and growing, an alarming prospect for some (fig. 7.5). Cornell University entomologist David Pimental, for example, has said: "By 2100, if current trends continue, twelve billion miserable humans will suffer a difficult life on Earth." In this view, birth control should be our top priority.

On the other hand, many scholars agree with Marx that improved social conditions and educational levels can stabilize populations humanely. In this perspective, the earth is bountiful in its resource base, but poverty and high birth rates result from oppressive social relationships that unevenly distribute wealth and resources. Consequently, this position believes, technological development, education, and just social conditions are the means of achieving population control. Mohandas Gandhi stated it succinctly: "There is enough for everyone's need, but not enough for anyone's greed."

Technology can increase carrying capacity for humans

Optimists argue that Malthus was wrong in his predictions of famine and disaster 200 years ago because he failed to account for scientific and technical progress. In fact, food supplies have increased faster than population growth since Malthus' time. For example, according to the UN FAO Statistics Division, each person on the planet averaged 2,435 calories of food per day in 1970, while in 2000 the caloric intake reached 2,807 calories. Even poorer, developing countries saw a rise, from an average of 2,135 calories per day in 1970 to 2,679 in 2000. In that same period the world population went from 3.7 to more than 6 billion people. Certainly terrible famines have stricken different locations in the past 200 years, but they were caused more by politics and economics than by lack of resources or population size. Whether the world can continue to feed its growing population remains to be seen, but technological advances have vastly increased human carrying capacity so far.

The burst of world population growth that began 200 years ago was stimulated by scientific and industrial revolutions. Progress in agricultural productivity, engineering, information technology, commerce, medicine, sanitation, and other achievements of modern life have made it possible to support thousands of times as many people per unit area as was possible 10,000 years ago. Economist Stephen Moore of the Cato Institute in Washington, D.C., regards this achievement as "a real tribute to human ingenuity and our ability to innovate." There is no reason, he argues, to think that our ability to find technological solutions to our problems will diminish in the future.

Much of our growth and rising standard of living in the past 200 years, however, has been based on easily acquired natural resources, especially cheap, abundant fossil fuels (see chapter 19). Whether rising prices of fossil fuels will constrain that production and result in a crisis in food production and distribution, or in some other critical factor in human society, concerns many people.

However, technology can be a double-edged sword. Our environmental effects aren't just a matter of sheer population size; they also depend on what kinds of resources we use and how we use them. This concept is summarized as the **I = PAT** formula. It says that our environmental impacts (I) are the product of our population size (P) times affluence (A) and the technology (T) used to produce the goods and services we consume. A single American living an affluent lifestyle that depends on high levels of energy and material consumption, and that produces excessive amounts of pollution, probably has a greater environmental impact than a whole village of Asian or African farmers. Ideally, Americans will begin to use nonpolluting, renewable energy and material sources. Better yet, Americans will extend the benefits of environmentally friendly technology to those villages of Asians and Africans so everyone can enjoy the benefits of a better standard of living without degrading their environment.

Population growth could bring benefits

Think of the gigantic economic engine that China has become as it industrializes and its population becomes more affluent. More people mean larger markets, more workers, and efficiencies of scale in mass production of goods. Moreover, adding people boosts human ingenuity and intelligence that will create new resources by finding new materials and discovering new ways of doing things. Economist Julian Simon (1932–1998), a champion of this rosy view of human history, believed that people are the "ultimate resource" and that no evidence suggests that pollution, crime, unemployment, crowding, the loss of species, or any other resource limitations will worsen with population growth. In a famous bet in 1980, Simon challenged Paul Ehrlich, author of *The Population Bomb,* to pick five commodities that would become more expensive by the end of the decade. Ehrlich chose metals that actually became cheaper, and he lost the bet. Leaders of many developing countries share

this outlook and insist that, instead of being obsessed with population growth, we should focus on the inordinate consumption of the world's resources by people in richer countries (see fig. 7.18).

7.3 MANY FACTORS DETERMINE POPULATION GROWTH

Demography is derived from the Greek words *demos* (people) and *graphos* (to write or to measure). It encompasses vital statistics about people, such as births, deaths, and where they live, as well as total population size. In this section, we will survey ways human populations are measured and described, and discuss demographic factors that contribute to population growth.

How many of us are there?

The estimate of more than 7 billion people in the world in 2011 quoted earlier in this chapter is only an educated guess. Even in this age of information technology and communication, counting the number of people in the world is like shooting at a moving target. People continue to be born and die. Furthermore, some countries have never even taken a census, and those that have been done may not be accurate. Governments may overstate or understate their populations to make their countries appear larger and more important or smaller and more stable than they really are. Individuals, especially if they are homeless, refugees, or illegal aliens, may not want to be counted or identified.

FIGURE 7.6 We live in two demographic worlds. One is rich, technologically advanced, and has an elderly population that is growing slowly, if at all. The other is poor, crowded, underdeveloped, and growing rapidly.

We really live in two very different demographic worlds. One is old, rich, and relatively stable. The other is young, poor, and growing rapidly. Most people in Asia, Africa, and Latin America inhabit the latter demographic world (fig. 7.6). These countries represent 80 percent of the world population but more than 90 percent of all projected growth (fig. 7.7)

The highest population growth rates occur in a few "hot spots," such as sub-Saharan Africa and the Middle East, where economics, politics, religion, and civil unrest keep birth rates high and contraceptive use low. In Niger, Yemen, and Palestine, for example, annual population growth is above 3.2 percent. Less than 10 percent of all couples use any form of birth control, women average more than seven children each, and nearly half the population is less than 15 years old. The world's highest current growth rate is in the United Arab Emirates, where births plus immigration are producing an annual increase of 6.8 percent (the highest immigration rate in the world is responsible for 80 percent of that growth). This means that the UAE is doubling its population size approximately every decade. Obviously, a small country with limited resources (except oil) and almost no fresh water or agriculture, can't sustain that high growth rate indefinitely.

Some countries in the developing world have experienced amazing growth rates and are expected to reach extraordinary population sizes by the middle of the twenty-first century. Table 7.2 shows the ten largest countries in the world, arranged by their estimated size in 2011 and projected size in 2050. Note that, while China was the most populous country throughout the twentieth century, India is expected to pass China in the twenty-first century. Nigeria, which had only 33 million residents in 1950, is forecast to have nearly 300 million in 2050. Ethiopia, with about 18 million people 50 years ago, is likely to grow nearly eightfold over a century. In many of these countries, rapid population growth is a serious problem. Bangladesh, about the size of Iowa, is already overcrowded at 150 million people. If rising sea levels flood one-third of the country by 2050, as some climatologists predict, adding another 80 million people will be disastrous.

Table 7.2	The World's Largest Countries		
2011		**2050***	
Country	Population (millions)	Country	Population (millions)
China	1,342	India	1,628
India	1,193	China	1,437
United States	312	United States	420
Indonesia	238	Nigeria	299
Brazil	191	Pakistan	295
Pakistan	172	Indonesia	285
Nigeria	158	Brazil	260
Bangladesh	150	Bangladesh	231
Russia	142	Dem. Rep. of Congo	183
Japan	127	Ethiopia	145

*Estimate.

Source: U.N Population Division 2011.

The other demographic world is made up of the richer countries of North America, western Europe, Japan, Australia, and New Zealand. This world is wealthy, old, and mostly shrinking. Italy, Germany, Hungary, and Japan, for example, all have negative growth rates. The average age in these countries is now 40, and life expectancy of their residents is expected to exceed 90 by 2050. With many couples choosing to have either one or no children, the populations of these countries are expected to decline significantly over the next century. Japan, which has 127 million residents now, is expected to shrink to about 100 million by 2050. Europe, which now makes up about 12 percent of the world population, will constitute less than 7 percent in 50 years, if current trends continue. Even the United States and Canada would have nearly stable populations if immigration were stopped.

It isn't only wealthy countries that have declining populations. Russia, for instance, is now declining by nearly 1 million people per year as death rates have soared and birth rates have plummeted. A collapsing economy, hyperinflation, crime, corruption, and despair have demoralized the population. Horrific pollution levels left from the Soviet era, coupled with poor nutrition and health care, have resulted in high levels of genetic abnormalities, infertility, and infant mortality. Abortions are twice as common as live births, and the average number of children per woman is now 1.4, among the lowest in the world. Death rates, especially among adult men, have risen dramatically. Male life expectancy dropped from 68 years in 1990 to 59 years in 2006. Life expectancy rates have risen since then, but births still lag behind deaths. Russia, which is the world's largest country geographically, could decline from 142 million people currently to below 100 million in 2050. It will then have a smaller population than Vietnam, Egypt, or Uganda. Other former Soviet states are experiencing similar declines. Estonia, Bulgaria, Georgia, and Ukraine, for example, now have negative growth rates and are expected to lose about 40 percent of their population in the next 50 years.

The situation is even worse in many African countries, where AIDS and other communicable diseases are killing people at a terrible

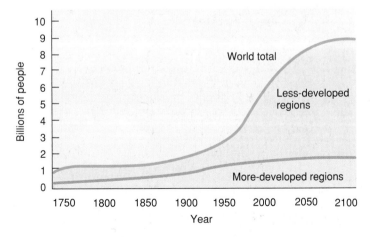

FIGURE 7.7 Estimated human population growth, 1750–2100. In less-developed and more-developed regions. Almost all growth projected for the twenty-first century is in the less-developed countries.
Source: UN Population Division, 2005.

rate. In Zimbabwe, Botswana, Zambia, and Namibia, for example, up to 39 percent of the adult population have AIDS or are HIV positive. Health officials predict that more than two-thirds of the 15-year-olds now living in Botswana will die of AIDS before age 50. Without AIDS, the average life expectancy would be 69.7 years. Now, with

AIDS, Botswana's life expectancy has dropped to only 31.6 years. The populations of many African countries are now falling because of this terrible disease (fig. 7.8). Altogether, Africa's population is expected to be nearly 200 million lower in 2050 than it would have been without AIDS.

AIDS is now spreading in Asia. Because of the large population there, Asia is expected to pass Africa in 2020 in total number of deaths. Although a terrible human tragedy, this probably won't affect total world population very much. Remember that the Black Death killed many people in the fourteenth century but had only a transitory effect on demography.

Figure 7.9 shows human population distribution around the world. Notice the high densities supported by fertile river valleys of the Nile, Ganges, Yellow, Yangtze, and Rhine Rivers and the well-watered coastal plains of India, China, and Europe. Historic factors, such as technology diffusion and geopolitical power, also play a role in geographic distribution.

Fertility measures the number of children born to each woman

As we pointed out in chapter 6, fecundity is the physical ability to reproduce, while fertility describes the actual production of offspring. Those without children may be fecund but not fertile. The most accessible demographic statistic of fertility is usually the **crude birth rate,** the number of births in a year per thousand persons. It is statistically "crude" in the sense that it is not adjusted for population characteristics such as the number of women in reproductive age.

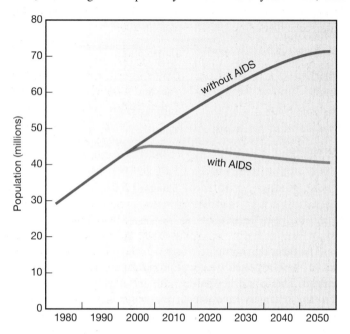

FIGURE 7.8 Projected population of South Africa with and without AIDS.
Data Source: UN Population Division, 2006.

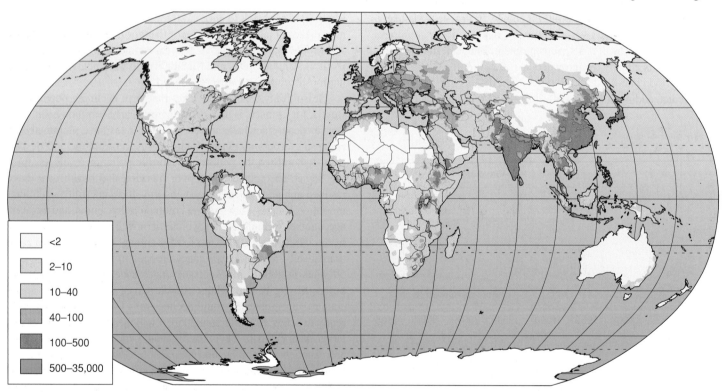

FIGURE 7.9 Population density in persons per square kilometer.
Source: World Bank, 2000.

The **total fertility rate** is the number of children born to an average woman in a population during her entire reproductive life. Upper-class women in seventeenth- and eighteenth-century England, whose babies were given to wet nurses immediately after birth and who were expected to produce as many children as possible, often had 25 or 30 pregnancies. The highest recorded total fertility rates for working-class people is among some Anabaptist agricultural groups in North America who have averaged up to 12 children per woman. In most tribal or traditional societies, food shortages, health problems, and cultural practices limit total fertility to about six or seven children per woman even without modern methods of birth control.

Zero population growth (ZPG) occurs when births plus immigration in a population just equal deaths plus emigration. It takes several generations of replacement level fertility (where people just replace themselves) to reach ZPG. Where infant mortality rates are high, the replacement level may be five or more children per couple. In the more highly developed countries, however, this rate is usually about 2.1 children per couple because some people are infertile, have children who do not survive, or choose not to have children.

Fertility rates have declined dramatically in every region of the world except Africa over the past 50 years (fig. 7.10). Only a few decades ago, total fertility rates above six were common in many countries. The average family in Mexico in 1975, for instance, had seven children. By 2009, however, the average Mexican woman had only 2.37 children. According to the World Health Organization, 100 out of the world's 220 countries are now at or below a replacement rate of 2.1 children per couple, and by 2050, all but a few of the least-developed countries are expected to have reached that milestone. The greatest fertility reduction has been in

Southeast Asia, where rates have fallen by more than half. Most of this decrease has occurred in just the past few decades and, contrary to what many demographers expected, some of the poorest countries in the world have been remarkably successful in lowering growth rates. As the opening case study for this chapter shows, Thailand reduced its total fertility rate from 7.0 in 1979 to 1.64 (lower than that in the United States) in 2009.

China's one-child-per-family policy decreased the fertility rate from 6 in 1970 to 1.54 in 2010 (What Do You Think? p. 140). But as a result of selective abortions for girls, China now reports that 119 boys are being born for every 100 girls. Normal ratios would be about 105 boys to 100 girls. If this imbalance persists, there will be a shortage of brides in another generation. Interestingly, Macao, which is culturally similar to mainland China but hasn't shared its one-child policy, now has a total average fertility rate of only 0.9 and the lowest birth rate in the world.

Although the world as a whole still has an average fertility rate of 2.6, growth rates are now lower than at any time since World War II. If fertility declines like those in Thailand and China were to occur everywhere in the world, global population could begin to decline by 2050, and might be below 6 billion by 2150. Most of Eastern Europe now has fertility levels of 1.2 children per woman. Interestingly, Spain and Italy, although predominately Roman Catholic, have similar fertility rates. Several Indian states have reached zero population growth, but their means of doing so have been very different (What Do You Think? p. 140).

Mortality is the other half of the population equation

A traveler to a foreign country once asked a local resident, "What's the death rate around here?" "Oh, the same as anywhere," was the reply, "about one per person." In demographics, however, **crude death rates** (or crude mortality rates) are expressed in terms of the number of deaths per thousand persons in any given year. Countries in Africa where health care and sanitation are limited may have mortality rates of 20 or more per 1,000 people. Wealthier countries generally have mortality rates around 10 per 1,000. The number of deaths in a population is sensitive to the age structure of the population. Rapidly growing, developing countries such as Libya or Costa Rica have lower crude death rates (4 per 1,000) than do the more-developed, slowly growing countries, such as Denmark (12 per 1,000). This is because there are proportionately more children and fewer elderly people in a rapidly growing country than in a more slowly growing one.

Crude death rate subtracted from crude birth rate gives the **natural increase** of a population. We distinguish natural increase from the **total growth rate,** which includes immigration and emigration, as well as births and deaths. Both of these growth rates are usually expressed as a percent (number per hundred people) rather than per thousand. A useful rule of thumb is that if you divide 70 by the annual percentage growth, you will get the approximate doubling time in years. Niger, for example, which is growing 3.4 percent per year, is doubling its population every 20 years. The United States, which has a natural increase rate of 0.6 percent per

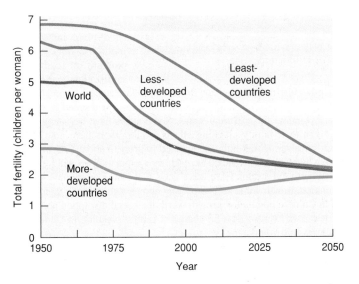

FIGURE 7.10 Average total fertility rates for less-developed countries fell by more than half over the past 50 years. Much of this dramatic change was due to China's one-child policy. Progress has been slower in the least-developed countries, but by 2050, they should be approaching the replacement rate of 2.1 children per woman of reproductive age.

What Do You Think?

China's One-Child Policy

When the People's Republic of China was founded in 1949, it had about 540 million residents, and official government policy encouraged large families. The Republic's First Chairman, Mao Zedong, proclaimed, "Of all things in the world, people are the most precious." He thought that more workers would mean greater output, increasing national wealth, and higher prestige for the country. This optimistic outlook was challenged, however, in the 1960s, when a series of disastrous government policies triggered massive famines and resulted in at least 30 million deaths.

When Deng Xiaoping became Chairman in 1978, he reversed many of Mao's policies including decollectivizing farms, encouraging private enterprise, and discouraging large families. Deng recognized that with an annual growth rate of 2.5 percent, China's population, which had already reached 975 million, would double in only 28 years. China might have nearly 2 billion residents now if that growth had continued. Feeding, housing, educating, and employing all those people would put a severe strain on China's already limited resources.

Deng introduced a highly successful—but controversial—one-child-per-family policy. Rural families and ethnic minorities were supposedly exempt from this rule, but local authorities often were capricious and tyrannical in applying sanctions. Ordinary families were punished harshly for having unauthorized children, while government officials and other powerful individuals could have as many as they wanted. There were many reports of bribery, forced abortions, coerced sterilizations, and even infanticide as a result of this policy.

Critics claim that other approaches to family planning could have reduced population growth while also preserving human rights. The shift to an urban, industrialized society, they argue, might have reduced family size without such draconian intervention. Some point to the successful use of humor, education, and economic development in Thailand (see opening case study for this chapter) as a model for reducing births in a more humane fashion. On the other hand, China is a far larger country than Thailand and has much different social, political, and economic forces at play.

Another result of China's one-child policy is called the 4:2:1 problem. That is, there are now often four grandparents and two parents doting on a single child. Social scientists often refer to this highly spoiled generation as "little emperors." Difficulties occur as those parents and grandparents age. With only one adult child to support and care for elderly relatives, many seniors have to work far beyond normal retirement age because their only child can't provide for them all.

The Chinese government, also, is beginning to worry about a "birth dearth." Will there be enough workers, soldiers, farmers, scientists, inventors, and other productive individuals to keep society functioning in the future? The one-child policy has been eased recently. Couples with no siblings to help care for elderly relatives are being allowed to have two or more children.

The Chinese experiment in population control has been very effective. China's population in 2010 was about 1.35 billion, or several hundred million less than it might have been given its trajectory in 1978. Its annual growth rate is now 0.65 percent, or about one-third less than the 0.98 percent annual growth in the United States. China is already the largest contributor to global warming and is driving up world prices for many commodities with its rapidly growing middle class. Think what the effects would be if there were another half a billion Chinese today.

China has also been far more successful in controlling population growth than India. At about the same time that Deng introduced his one-child plan, India, under Indira Gandhi, started a program of compulsory sterilization in an effort to reduce population growth. This draconian policy caused so much public outrage and opposition that the federal government decided to delegate family planning to individual states. Some states have been highly successful in their family planning efforts, while others have not. The net effect, however, is that India is expected to grow to about 1.65 billion by 2050, while China is expected to reach zero population growth by 2030.

What do you think? Are there ways that China might have reduced population growth while still respecting human rights? Is the rapid reduction in Chinese population growth worth the social disruption and abuses that it caused? If you were in charge of family planning in China, what policies would you pursue?

China's one-child-per-family policy, promoted in this billboard, has been remarkably successful in reducing birth rates, but has had some controversial social effects.

---

year, would double, without immigration, in 116.7 years. Belgium and Sweden, with natural increase rates of 0.1 percent, are doubling in about 700 years. Ukraine, on the other hand, with a growth rate of −0.8 percent, will lose about 40 percent of its population in the next 50 years. The world growth rate is now 1.14 percent, which means that the population will double in about 61 years if this rate persists.

Life span and life expectancy describe our potential longevity

Life span is the oldest age to which a species is known to survive. Although there are many claims in ancient literature of kings living a millennium or more, the oldest age that can be certified by written records was that of Jeanne Louise Calment of Arles, France, who was 122 years old at her death in 1997. The aging process is still a medical mystery, but it appears that cells in our bodies have a limited ability to repair damage and produce new components. At some point they simply wear out, and we fall victim to disease, degeneration, accidents, or senility.

Life expectancy is the average age that a newborn infant can expect to attain in any given society. It is another way of expressing the average age at death. For most of human history, we believe that average life expectancy in most societies has been about 30 years. This doesn't mean that no one lived past age 40, but rather that so many deaths at earlier ages (mostly early childhood) balanced out those who managed to live longer.

Declining mortality, not rising fertility, is the primary cause of most population growth in the past 300 years. Crude death rates began falling in western Europe during the late 1700s. Most of this advance in survivorship came long before the advent of modern medicine and is due primarily to better food and better sanitation.

The twentieth century has seen a global transformation in human health unmatched in history. In 1900 the world average life expectancy was only about 30 years, which was not much higher than the average life span in the Roman Empire 2,000 years earlier. By 2011, the average was 68.9 years (fig. 7.11). Improved nutrition, sanitation, and medical care were responsible for most of that increase. Demographers wonder how much more life expectancies can increase. Notice the great discrepancy in life expectancies between rich and poor countries. Currently, microstates Andorra, San Marino, and Singapore have the world's highest life expectancies (83.5, 82.1, and 81.6 years, respectively). Japan is nearly as high with a countrywide average of 81.5 years.

The lowest national life expectancies are in Africa, where diseases, warfare, poverty, and famine cause many early deaths. In Swaziland, Botswana, and Lesotho, for example, the average person lives only 32.6, 33.7, and 34.4 years, respectively. In many African countries AIDS has reduced life expectancies by about 25 percent in the past two decades. This can be seen in the lag in progress in life expectancies between 1980 and 2000 in these countries in figure 7.11.

Large discrepancies also exist in the United States. While the nationwide average life expectancy is 77.5 years, Asian American women in Bergen County, New Jersey, live 91 years on average, while Native American men on the Pine Ridge Reservation in South Dakota are reported to typically live only 48 years. Two-thirds of African countries have life expectancies greater than Pine Ridge. Women almost always have higher life expectancies than men. Worldwide, the average difference between sexes is three years, but in Russia the difference between men and women is 13 years. Is this because women are biologically superior to men, and thus live longer? Or is it simply that men are generally employed in more hazardous occupations and often engage in more dangerous behaviors (drinking, smoking, reckless driving)?

As figure 7.12 shows, there is a good correlation between annual income and life expectancy up to about (U.S.) $4,000 per person. Beyond that level—which is generally enough for adequate food, shelter, and sanitation for most people—life expectancies level out at about 75 years for men and 85 for women.

Some demographers believe that life expectancy is approaching a plateau, while others predict that advances in biology and medicine might make it possible to live 150 years or more. If our average age at death approaches 100 years, as some expect, society will be profoundly affected. In 1970 the median age in the United States was 30. By 2100 the median age could be over 60. If workers continue to retire at 65, half of the population could be unemployed, and retirees might be facing 35 or 40 years of retirement. We may need to find new ways to structure and finance our lives.

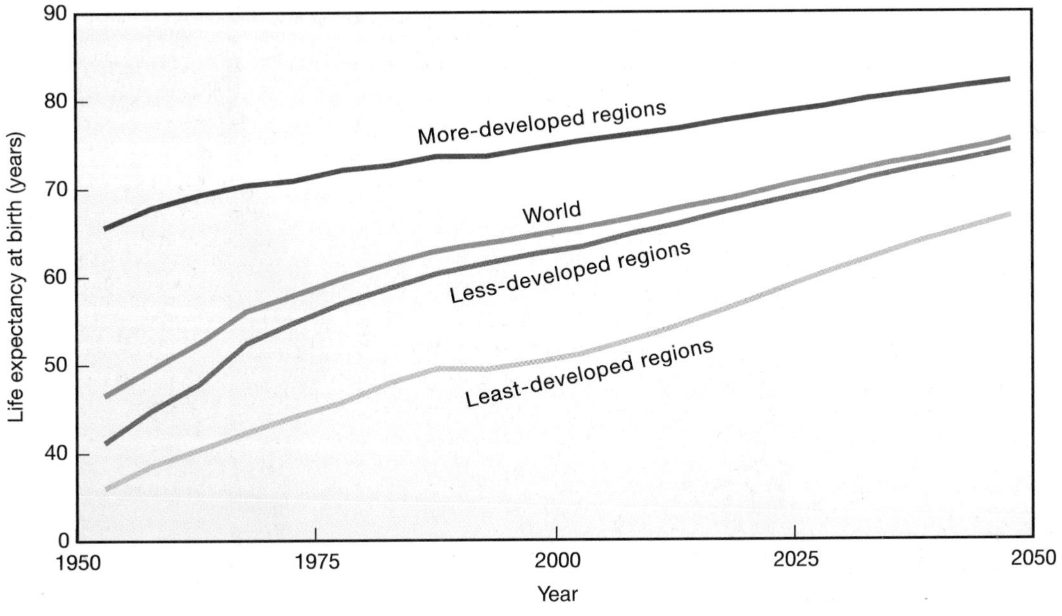

Living longer has demographic implications

A population that is growing rapidly by natural increase has more young people than does a stationary population. One way to show these differences is to graph age classes in a histogram as shown in figure 7.13. In Niger, which is growing at a rate of

FIGURE 7.11 Life expectancy has increased nearly everywhere in the world, but the increase has lagged in the least-developed countries.
Source: Data from the Population Division of the United Nations, 2006.

3.4 percent per year, 49 percent of the population is in the pre-reproductive category (below age 15). Even if total fertility rates were to fall abruptly, the total number of births, and population size, would continue to grow for some years as these young people enter reproductive age. This phenomenon is called population momentum.

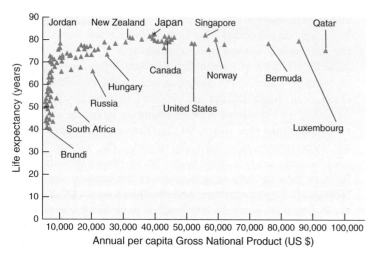

FIGURE 7.12 As incomes rise, so does life expectancy up to about (U.S.) $4,000. Above that amount the curve levels off. Some countries, such as South Africa and Russia, have far lower life expectancies than their GDP would suggest. Jordan, on the other hand, which has only one-tenth the per capita GDP of the United States, actually has a higher life expectancy.
Source: CIA Factbook, 2009.

By contrast, a country with a stable population, like Sweden, has nearly the same number in each age cohort. A population that has recently entered a lower growth rate pattern, such as Singapore, has a bulge in the age classes for the last high-birth-rate generation. Notice that there are more females than males in the older age group in Sweden because of differences in longevity between the sexes.

Both rapidly growing countries and slowly growing countries can have a problem with their **dependency ratio,** or the number of nonworking compared to working individuals in a population. In Mexico, for example, each working person supports a high number of children. In the United States, by contrast, a declining working population is now supporting an ever larger number of retired persons and there are dire predictions that the social security system will soon be bankrupt. This changing age structure and shifting dependency ratio are occurring worldwide (fig. 7.14). By 2050 the UN predicts there will be two older persons for every child in the world. Many countries are rethinking their population policies and beginning to offer incentives for marriage and child-rearing.

Emigration and immigration are important demographic factors

Humans are highly mobile, so emigration and immigration play a larger role in human population dynamics than they do in those of many species. Currently, about 800,000 people immigrate legally to the United States each year, but many more enter illegally. Western Europe receives about 1 million applications each year for asylum from economic chaos and wars in former socialist

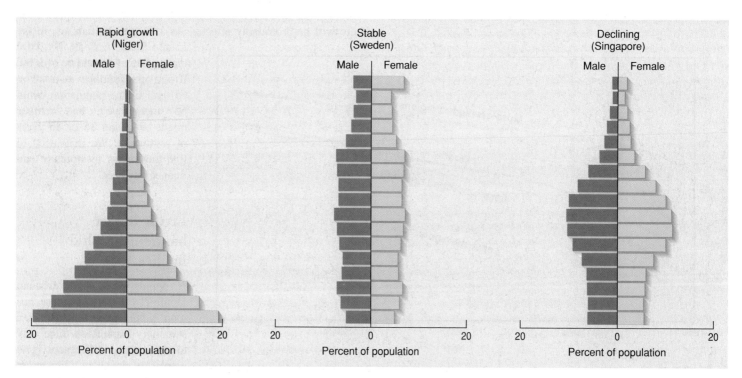

FIGURE 7.13 Age structure graphs for rapidly growing, stable, and declining populations.
Source: U.S. Census Bureau, 2006.

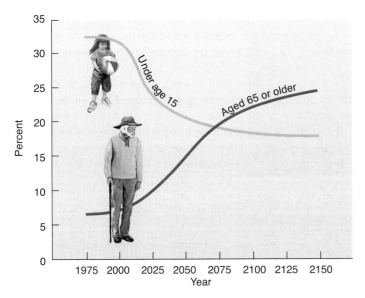

FIGURE 7.14 By the mid-twenty-first century, children under age 15 will make up a smaller percentage of world population, while people over age 65 will contribute a larger and larger share of the population.

states and the Middle East. The United Nations High Commission on Refugees reported that in 2006 there were 20.8 million refugees who had left their countries for political or economic reasons, while about 25 million more were displaced persons in their own countries, and 175 million migrants had left their homes to look for work, greater freedom, or better opportunities.

The number of refugees and migrants has fallen significantly since 2006, but the more-developed regions are expected to gain about 2 million immigrants per year for the next 50 years. Without migration, the population of the wealthiest countries would already be declining and would be more than 126 million less than the current 1.2 billion by 2050. In 2008, nearly 45.5 million U.S. residents (15.2 percent of the total population) classified themselves as Hispanic or Latino. They now constitute the largest U.S. minority.

Immigration is a controversial issue in many countries. "Guest workers" often perform heavy, dangerous, or disagreeable work that citizens are unwilling to do. Many migrants and alien workers are of a different racial or ethnic background than the majority in their new home. They generally are paid low wages and given substandard housing, poor working conditions, and few rights. Local residents often complain, however, that immigrants take away jobs, overload social services, and ignore established rules of behavior or social values. Anti-immigrant groups are springing up in many rich countries.

Some nations encourage, or even force, internal mass migrations as part of a geopolitical demographic policy. In the 1970s, Indonesia embarked on an ambitious "transmigration" plan to move 65 million people from the overcrowded islands of Java and Bali to relatively unpopulated regions of Sumatra, Borneo, and New Guinea. Attempts to turn rainforest into farmland had disastrous environmental and social effects, however, and this plan was

greatly scaled back. China has announced a plan to move up to 100 million people to a sparsely populated region along the Amur River in Heilongjiang. By some estimates, more than 250 million internal migrants in China have moved from rural areas to the cities to look for work.

7.4 IDEAL FAMILY SIZE IS CULTURALLY AND ECONOMICALLY DEPENDENT

A number of social and economic pressures affect decisions about family size, which in turn affects the population at large. In this section we will examine both positive and negative pressures on reproduction.

Many factors increase our desire for children

Factors that increase people's desires to have babies are called **pronatalist pressures.** Raising a family may be the most enjoyable and rewarding part of many people's lives. Children can be a source of pleasure, pride, and comfort. They may be the only source of support for elderly parents in countries without a social security system. Where infant mortality rates are high, couples may need to have many children to be sure that at least a few will survive to take care of them when they are old. Where there is little opportunity for upward mobility, children give status in society, express parental creativity, and provide a sense of continuity and accomplishment otherwise missing from life. Often children are valuable to the family not only for future income, but even more as a source of current income and help with household chores. In much of the developing world, children as young as 6 years old tend domestic animals and younger siblings, fetch water, gather firewood, and help grow crops or sell things in the marketplace (fig. 7.15). Parental desire for children rather than an unmet need for contraceptives may be the most important factor in population growth in many cases.

Society also has a need to replace members who die or become incapacitated. This need often is codified in cultural or religious values that encourage bearing and raising children. In some societies, families with few or no children are looked upon with pity or contempt. The idea of deliberately controlling fertility may be shocking, even taboo. Women who are pregnant or have small children are given special status and protection. Boys frequently are more valued than girls because they carry on the family name and are expected to support their parents in old age. Couples may have more children than they really want in an attempt to produce a son.

Male pride often is linked to having as many children as possible. In Niger and Cameroon, for example, men, on average, want 12.6 and 11.2 children, respectively. Women in these countries consider the ideal family size to be only about one-half that desired by their husbands. Even though a woman might desire fewer children, however, she may have few choices and little control over her own fertility. In many societies, a woman has no status outside of her role as wife and mother. Without children, she has no source of support.

(a)   (b)

FIGURE 7.15 In rural areas with little mechanized agriculture (a) children are needed to tend livestock, care for younger children, and help parents with household chores. Where agriculture is mechanized (b) rural families view children just as urban families do—helpful, but not critical to survival. This affects the decision about how many children to have.

Other factors discourage reproduction

In more highly developed countries, many pressures tend to reduce fertility. Higher education and personal freedom for women often result in decisions to limit childbearing. The desire to have children is offset by a desire for other goods and activities that compete with childbearing and child rearing for time and money. When women have opportunities to earn a salary, they are less likely to stay home and have many children. Not only are the challenge and variety of a career attractive to many women, but the money that they can earn outside the home becomes an important part of the family budget. Thus, education and socioeconomic status are usually inversely related to fertility in richer countries. In developing countries, however, fertility may rise, at least temporarily, as educational levels and socioeconomic status rise. With higher income, families are better able to afford the children they want; more money means that women are likely to be healthier,

and therefore better able to conceive and carry a child to term.

In less-developed countries where feeding and clothing children can be a minimal expense, adding one more child to a family usually doesn't cost much. By contrast, raising a child in the United States can cost hundreds of thousands of dollars by the time the child is through school and is independent. Under these circumstances, parents are more likely to choose to have one or two children on whom they can concentrate their time, energy, and financial resources.

Figure 7.16 shows U.S. birth rates between 1910 and 2010. As you can see, birth rates have fallen and risen in a complex pattern. The period between 1910 and 1930 was a time of industrialization and urbanization. Women were getting more education than ever before and entering the workforce. The Great Depression in the 1930s made it economically difficult for families to have children, and birth rates were low. The birth rate increased at the beginning of World War II (as it often does in wartime). For reasons that are unclear, a higher percentage of boys are usually born during war years.

At the end of the war, there was a "baby boom" as couples were reunited and new families started. This high birth rate persisted through the times of prosperity and optimism of the 1950s, but began to fall in the 1960s. Part of this decline was caused by the small number of babies born in the 1930s. This meant fewer young adults to give birth in the 1960s. Part was due to changed perceptions of the ideal family size. Whereas in the 1950s women typically wanted four children or more, in the 1970s the norm dropped to one or two (or no) children. A small "echo boom" occurred in the 1980s as people born in the 1960s began to have babies, but changing economics and attitudes seem to have permanently altered our view of ideal family size in the United States.

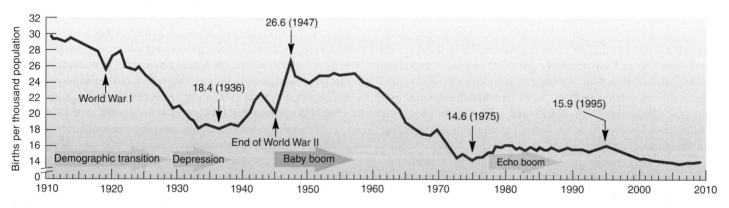

FIGURE 7.16 Birth rates in the United States, 1910–2000. The falling birth rate from 1910 to 1929 represents a demographic transition from an agricultural to an industrial society. The baby boom following World War II lasted from 1945 to 1965. A much smaller "echo boom" occurred around 1980 when the baby boomers started to reproduce.

Sources: Data from Population Reference Bureau and U.S. Bureau of the Census.

Could we have a birth dearth?

Most European countries now have birth rates below replacement rates, and Italy, Russia, Austria, Germany, Greece, and Spain are experiencing negative rates of natural population increase. Asia, Japan, Singapore, and Taiwan are also facing a "child shock" as fertility rates have fallen well below the replacement level of 2.1 children per couple. There are concerns in all these countries about falling military strength (lack of soldiers), economic power (lack of workers), and declining social systems (not enough workers and taxpayers) if low birth rates persist or are not balanced by immigration. In a sense, the United States is fortunate to have a high influx of immigrants that provides youth and energy to its population.

Economist Ben Wattenberg warns that this "birth dearth" might seriously erode the powers of Western democracies in world affairs. He points out that Europe and North America accounted for 22 percent of the world's population in 1950. By the 1980s, this number had fallen to 15 percent, and by the year 2030, Europe and North America probably will make up only 9 percent of the world's population. Germany, Hungary, Denmark, and Russia now offer incentives to encourage women to bear children. Japan offers financial support to new parents, and Singapore provides a dating service to encourage marriages among the upper classes as a way of increasing population.

On the other hand, since Europeans and North Americans consume so many more resources per capita than most other people in the world, a reduction in the population of these countries will do more to spare the environment than would a reduction in population almost anywhere else.

One reason that birth rates have been falling in many industrialized countries may be that toxins and endocrine hormone disrupters in our environment interfere with sperm production. Sperm numbers and quality (fertilization ability) appear to have fallen by about half over the past 50 years in a number of countries. Widespread chemicals, such as phthalates—common ingredients in plastics—that disrupt sperm production may be responsible for this decline. We'll discuss this further in chapter 8.

7.5 A DEMOGRAPHIC TRANSITION CAN LEAD TO STABLE POPULATION SIZE

In 1945, demographer Frank Notestein pointed out that a typical pattern of falling death rates and birth rates due to improved living conditions usually accompanies economic development. He called this pattern the **demographic transition** from high birth and death rates to lower birth and death rates. Figure 7.17 shows an idealized model

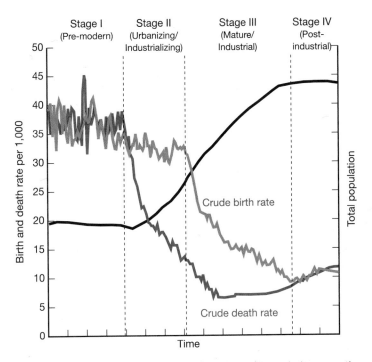

FIGURE 7.17 Theoretical birth, death, and population growth rates in a demographic transition accompanying economic and social development. In a predevelopment society, birth and death rates are both high, and total population remains relatively stable. During development, death rates tend to fall first, followed in a generation or two by falling birth rates. Total population grows rapidly until both birth and death rates stabilize in a fully developed society.

of a demographic transition. This model is often used to explain connections between population growth and economic development.

Economic and social development influence birth and death rates

Stage I in figure 7.17 represents the conditions in a premodern society. Food shortages, malnutrition, lack of sanitation and medicine, accidents, and other hazards generally keep death rates in such a society around 35 per 1,000 people. Birth rates are correspondingly high to keep population densities relatively constant. As economic development brings better jobs, medical care, sanitation, and a generally improved standard of living in Stage II, death rates often fall very rapidly. Birth rates may actually rise at first as more money and better nutrition allow people to have the children they always wanted. Eventually, in a mature industrial economy (Stage III), birth rates fall as people see that all their children are more likely to survive and that the whole family benefits from concentrating more resources on fewer children. Note that population continues to grow rapidly during this stage because of population momentum (baby boomers reaching reproductive age). Depending on how long it takes to complete the transition, the population may go through one or more rounds of doubling before coming into balance again.

Stage IV in figure 7.17 represents conditions in developed countries, where the transition is complete and both birth rates and death rates are low, often, a third or less than those in the predevelopment era. The population comes into a new equilibrium in this phase, but at a much larger size than before. Most of the countries of northern and western Europe went through a demographic transition in the nineteenth or early twentieth century similar to the curves shown in this figure.

Many of the most rapidly growing countries in the world, such as Kenya, Yemen, Libya, and Jordan, now are in the Stage I of this demographic transition. Their death rates have fallen close to the rates of the fully developed countries, but birth rates have not fallen correspondingly. In fact, both their birth rates and total population are higher than those in most European countries when industrialization began 300 years ago. The large disparity between birth and death rates means that many developing countries now are growing at 3 to 4 percent per year. Such high growth rates in developing countries could boost total world population to 9 billion or more before the end of the twenty-first century. This raises what may be the two most important questions in this entire chapter: Why are birth rates not yet falling in these countries, and what can be done about it?

There are reasons to be optimistic about population

Four conditions are necessary for a demographic transition to occur: (1) improved standard of living, (2) increased confidence that children will survive to maturity, (3) improved social status of women, and (4) increased availability and use of birth control. As the example of Thailand in the opening case study for this chapter shows, these conditions can be met, even in relatively poor countries.

Some demographers claim that a demographic transition already is in progress in most developing nations. Problems in taking censuses and a normal lag between falling death and birth rates may hide this for a time, but the world population should stabilize sometime in the next century. Some evidence supports this view. As we mentioned earlier in this chapter, fertility rates have fallen dramatically nearly everywhere in the world over the past half century.

Some countries have had remarkable success in population control. In Thailand, Indonesia, Colombia, and Iran, for instance, total fertility dropped by more than half in 20 years. Morocco, Dominican Republic, Jamaica, Peru, and Mexico all have seen fertility rates fall between 30 percent and 40 percent in a single generation. The following factors could contribute to stabilizing populations:

- Growing prosperity and social reforms that accompany development reduce the need and desire for large families in most countries.

- Technology is available to bring advances to the developing world much more rapidly than was the case a century ago, and the rate of technology transfer is much faster than it was when Europe and North America were developing.

- Less-developed countries have historic patterns to follow. They can benefit from our mistakes and chart a course to stability more quickly than they might otherwise do.

- Modern communications (especially television) have caused a revolution of rising expectations that act as a stimulus to spur change and development.

Many people remain pessimistic about population growth

Economist Lester Brown takes a more pessimistic view. He warns that many of the poorer countries of the world appear to be caught in a "demographic trap" that prevents them from escaping from the middle phase of the demographic transition. Their populations are now growing so rapidly that human demands exceed the sustainable yield of local forests, grasslands, croplands, or water resources. The resulting resource shortages, environmental deterioration, economic decline, and political instability may prevent these countries from ever completing modernization. Their populations may continue to grow until catastrophe intervenes.

Many people argue that the only way to break out of the demographic trap is to immediately and drastically reduce population growth by whatever means are necessary. They argue strongly for birth control education and bold national policies to encourage lower birth rates. Some agree with Malthus that helping the poor will simply increase their reproductive success and further threaten the resources on which we all depend. Author Garret Hardin described this view as lifeboat ethics. "Each rich nation," he wrote, "amounts to a lifeboat full of comparatively rich people. The poor of the world are in other much more crowded lifeboats. Continuously, so to speak, the poor fall out of their lifeboats and swim for a while, hoping to be admitted to a rich lifeboat, or in some other way to benefit from the goodies on board. . . . We cannot risk the safety of all the passengers by helping others in need. What happens if you share space in a lifeboat? The boat is swamped and everyone drowns. Complete justice, complete catastrophe." How would you respond to Professor Hardin?

Social justice is an important consideration

A third view is that **social justice** (a fair share of social benefits for everyone) is the real key to successful demographic transitions. The world has enough resources for everyone, but inequitable social and economic systems cause maldistributions of those resources. Hunger, poverty, violence, environmental degradation, and overpopulation are symptoms of a lack of social justice rather than a lack of resources. Although overpopulation exacerbates other problems, a narrow focus on this factor alone encourages racism and hatred of the poor. A solution for all these problems is to establish fair systems, not to blame the victims. Small nations and minorities often regard calls for population control as a form

FIGURE 7.18 Controlling our population and resources—there may be more than one side to the issue.

of genocide. Figure 7.18 expresses the opinion of many people in less-developed countries about the relationship between resources and population.

An important part of this view is that many of the rich countries are, or were, colonial powers, while the poor, rapidly growing countries were colonies. The wealth that paid for progress and security for developed countries was often extracted from colonies, which now suffer from exhausted resources, exploding populations, and chaotic political systems. Some of the world's poorest countries such as India, Ethiopia, Mozambique, and Haiti had rich resources and adequate food supplies before they were impoverished by colonialism. Those of us who now enjoy abundance may need to help the poorer countries not only as a matter of justice but because we all share the same environment.

In addition to considering the rights of fellow humans, we should also consider those of other species. Rather than ask what is the maximum number of humans that the world can possibly support, perhaps we should think about the needs of other creatures. As we convert natural landscapes into agricultural or industrial areas, species are crowded out that may have just as much right to exist as we do. What do you think would be the optimum number of people to provide a fair and decent life for all humans while causing the minimum impact on nonhuman neighbors?

Women's rights affect fertility

Opportunities for education and paying jobs are critical factors in fertility rates (fig. 7.19). Child survival also is crucial in stabilizing population. When infant and child mortality rates are high, as they are in much of the developing world, parents tend to have high numbers of children to ensure that some will survive to adulthood. There has never been a sustained drop in birth rates that was not first preceded by a sustained drop in infant and child mortality. One of the most important distinctions in our demographically divided world is the high infant mortality rates in the less-developed countries. Better nutrition, improved health care, simple oral rehydration therapy, and immunization against infectious diseases (see chapter 8) have brought about dramatic reductions in child mortality rates, which have been accompanied in most regions by falling birth rates. It has been estimated that saving 5 million children each year from easily preventable communicable diseases would avoid 20 or 30 million extra births.

Increasing family income does not always translate into better welfare for children since men in many cultures control most financial assets. Often the best way to improve child survival is to ensure the rights of mothers. Land reform, political rights, opportunities to earn an independent income, and improved health status

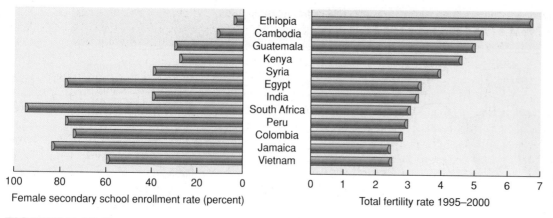

FIGURE 7.19 Total fertility declines as women's education increases.
Source: Worldwatch Institute, 2003.

of women often are better indicators of total fertility and family welfare than rising GNP.

7.6 FAMILY PLANNING GIVES US CHOICES

Family planning allows couples to determine the number and spacing of their children. It doesn't necessarily mean fewer children—people may use family planning to have the maximum number of children possible—but it does imply that the parents will control their reproductive lives and make rational, conscious decisions about how many children they will have and when those children will be born, rather than leaving it to chance. As the desire for smaller families becomes more common, birth control becomes an essential part of family planning in most cases. In this context, **birth control** usually means any method used to reduce births, including abstinence, delayed marriage, contraception, methods that prevent implantation of embryos, and induced abortions. As the opening case study in this chapter shows, there are many ways to encourage family planning.

Fertility control has existed throughout history

Evidence suggests that people in every culture and every historic period have used a variety of techniques to control population size. Studies of hunting and gathering people, such as the !Kung or San of the Kalahari Desert in southwest Africa, indicate that our early ancestors had stable population densities, not because they killed each other or starved to death regularly, but because they controlled fertility.

For instance, San women breast-feed children for three or four years. When calories are limited, lactation depletes body fat stores and suppresses ovulation. Coupled with taboos against intercourse while breast-feeding, this is an effective way of spacing children. Other ancient techniques to control population size include abstinence, folk medicines, abortion, and infanticide. We may find some or all of these techniques unpleasant or morally unacceptable, but we shouldn't assume that other people are too ignorant or too primitive to make decisions about fertility.

Today there are many options

Modern medicine gives us many more options for controlling fertility than were available to our ancestors. The major categories of birth control techniques include (1) avoidance of sex during fertile periods (for example, celibacy or the use of changes in body temperature or cervical mucus to judge when ovulation will occur), (2) mechanical barriers that prevent contact between sperm and egg (for example, condoms, spermicides, diaphragms, cervical caps, and vaginal sponges), (3) surgical methods that prevent release of sperm or egg (for example, tubal ligations in females and vasectomies in males), (4) hormone-like chemicals that prevent maturation or release of sperm or eggs or that prevent embryo implantation in the uterus (for example, estrogen plus progesterone, or progesterone alone, for females: gossypol for males), (5) physical barriers to implantation (for example, intrauterine devices), and (6) abortion.

Not surprisingly, the most effective birth control methods are also the ones most commonly used (table 7.3). In the United States, the majority of women younger than 30 who eventually want to become pregnant use the Pill. Most women over 35, with their child-bearing years behind them, choose sterilization. Male condom use is more effective than the remaining techniques in the table, and increases in effectiveness when used with a spermicide. Only two to six women in a hundred become pregnant in a year using this combination method. Condoms have the added advantage of protecting partners against sexually transmitted diseases, including

Table 7.3 Some Birth Control Methods and Pregnancy Prevention Rates

Method	Number of Women in 100 Who Become Pregnant
Sterilization (male, female)	<1
IUD	<1
Oral contraceptive (the Pill)	1–2
Hormones (implant, patch, injection, etc.)	1–2
Male condom	11
Sponge and spermicide	14–28
Female condom (e.g., cervical cap)	15–23
Diaphragm together with spermicide	17
Abstinence during fertile periods	20
Morning-after-pill (e.g., Preven)	20
Spermicide alone	20–50
Actively seeking pregnancy	85

Source: U.S. Food and Drug Administration, *Birth Control Guide*, 2003 Revision.

AIDS, if they are made of latex and used correctly. That may partly explain why their use in the United States went from 3.5 million users in 1980 to 8 million in 2000. Condoms are an ancient birth control method; the Egyptians used them some 3,000 years ago.

More than 100 new contraceptive methods are now being studied, and some appear to have great promise. Nearly all are biologically based (e.g., hormonal), rather than mechanical (e.g., condom, IUD). Recently, the U.S. Food and Drug Administration approved five new birth control products. Four of these use various methods to administer female hormones that prevent pregnancy. Other methods are years away from use, but take a new direction entirely. Vaccines for women are being developed that will prepare the immune system to reject the hormone choriononic gonadotropin, which maintains the uterine lining and allows egg implant, or that will cause an immune reaction against sperm. Injections for men are focused on reducing sperm production, and have proven effective in mice. Without a doubt, the contemporary couple has access to many more birth control options than their grandparents had.

7.7 WHAT KIND OF FUTURE ARE WE CREATING?

How many people will be in the world a century from now? Most demographers believe that world population will stabilize sometime during the next century. The total number of humans, when we reach that equilibrium, is likely to be somewhere around 8 to 10 billion people, depending on the success of family planning programs and the multitude of other factors affecting human populations. Figure 7.20 shows four scenarios projected by the UN Population Division in its 2004 revision. The optimistic (low) projection shows that world population might reach about 7 billion in 2050, and then fall back below 6 billion by 2150. The medium projection suggests that growth might continue to around 8.9 billion in 2050, and then stabilize. The most pessimistic projection assumes a constant rate of growth (no change from present) to 25 billion people by 2150.

Which of these scenarios will we follow? As you have seen in this chapter, population growth is a complex subject. To accomplish a stabilization or reduction of human populations will require substantial changes from business as usual.

An encouraging sign is that worldwide contraceptive use has increased sharply in recent years. About half of the world's married couples used some family planning techniques in 2000, compared to only 10 percent 30 years earlier, but another 100 million couples say they want, but do not have access to, family planning. Contraceptive use varies widely by region. More than 70 percent of women in Latin America use some form of birth control, compared to 51 percent in Asia (excluding China), and only 21 percent in Africa.

Figure 7.21 shows the unmet need for family planning among married women in some representative countries. When people in developing countries are asked what they want most, men say they want better jobs, but the first choice for a vast majority of women is family planning assistance. In general, a 15 percent increase in contraceptive use equates to about one fewer birth per woman per lifetime. In Mali, for example, where only 8 percent

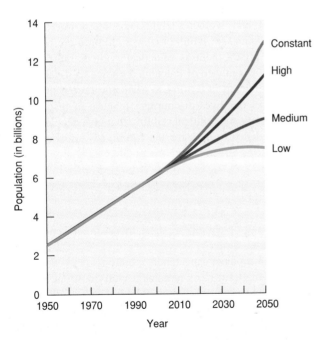

FIGURE 7.20 Population projections for different growth scenarios. Recent progress in family planning and economic development have led to significantly reduced estimates compared to a few years ago. The medium projection is 8.9 billion in 2050, compared to previous estimates of over 10 billion for that date. **Source:** UN Population Division, 2004.

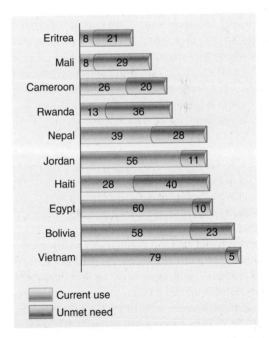

FIGURE 7.21 Unmet need for family planning in selected countries. Globally, more than 100 million women in developing countries would prefer to avoid pregnancy but do not have access to family planning. **Source:** U.S. AID, 2007.

of all women use contraceptives, the average fertility is 7.34 children per woman. In Vietnam, by contrast, where 79 percent of the women who would prefer not to be pregnant use contraceptives, the average fertility is 1.86.

Religion and politics complicate family planning

In 1994, the United Nations convened an historic meeting in Cairo, Egypt, to discuss women's rights and population. The United States played a lead role in the International Conference on Population and Development (ICPD), which identified links between population growth, economic development, environmental degradation, and the social status of women and girls. To address these issues, 179 countries, including the United States, endorsed the goal of universally available reproductive health services, including family planning, by 2015.

During the G. W. Bush administration, however, the United States refused to reaffirm the ICPD because it maintained that the document could be interpreted as promoting abortion—even though the ICPD clearly states, "In no case should abortion be promoted as a method of family planning."

In particular, the United States withheld funds from the United Nations Population Fund (UNFPA) due to claims that, by working in China, the fund tacitly supported the forced abortions reported to be part of that country's one-child policy. A fact-finding team sent to China in 2002 found "no evidence of UNFPA knowledge of or support for such measures," but funding was still halted.

Officials at the UNFPA estimated that the funds withheld by the United States could have prevented 2 million unwanted pregnancies, 800,000 abortions, 4,700 maternal deaths, 60,000 cases of serious maternal illness, and more than 77,000 infant and child deaths. In 2009, President Obama promised to restore funding to the UNFPA.

Many Muslim countries encourage couples to have as many children as possible. Access to birth control is difficult or forbidden outright. Still, some Islamic governments recognize the need for family planning. Iran, for example, decided, in the 1990s, to promote smaller families. It succeeded in cutting birth rates by more than half in ten years.

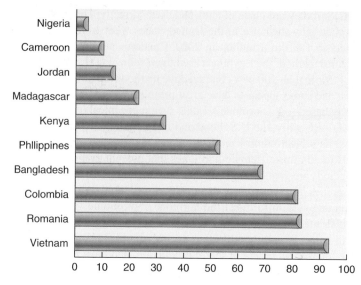

FIGURE 7.22 Percent of married reproductive-age women with two living children who do not want another child. **Source:** Data from UN Population Division, 2006.

The World Health Organization estimates that nearly 1 million conceptions occur daily around the world as a result of some 100 million sex acts. At least half of those conceptions are unplanned or unwanted. But there are still places where people desire large families (fig. 7.22).

Deep societal changes are often required to make family planning programs successful. Among the most important of these are (1) improved social, educational, and economic status for women (birth control and women's rights are often interdependent); (2) improved status for children (fewer children are born as parents come to regard them as valued individuals rather than possessions); (3) acceptance of calculated choice as a valid element in life in general and in fertility in particular (belief that we have no control over our lives discourages a sense of responsibility); (4) social security and political stability that give people the means and the confidence to plan for the future; (5) knowledge, availability, and use of effective and acceptable means of birth control.

CONCLUSION

A few decades ago, we were warned that a human population explosion was about to engulf the world. Exponential population growth was seen as a cause or corollary to nearly every important environmental problem. Some people still warn that the total number of humans might grow to 30 or 40 billion by the end of this century. Birth rates have fallen, however, almost everywhere, and most demographers now believe that we will reach an equilibrium around 9 billion people in about 2050. Some claim that if we promote equality, democracy, human development, and modern family planning techniques, population might even decline to below its current level of 7 billion in the next 50 years. How we should carry out family planning and birth control remains a controversial issue. Should we focus on political and economic reforms, and hope that a demographic transition will

naturally follow; or should we take more direct action (or any action) to reduce births?

Whether our planet can support 9 billion—or even 6 billion—people on a long-term basis remains a vital question. If all those people try to live at a level of material comfort and affluence now enjoyed by residents of the wealthiest nations, using the old, polluting, inefficient technology that we now employ, the answer is almost certain that even 6 billion people is too many in the long run. If we find more sustainable ways to live, however, it may be that 9 billion people could live happy, comfortable, productive lives. If we don't find new ways to live, we probably face a crisis no matter what happens to our population size. We'll discuss pollution problems, energy sources, and sustainability in subsequent chapters of this book.

REVIEWING LEARNING OUTCOMES

By now you should be able to explain the following points:

7.1 Trace the history of human population growth.
- Human populations grew slowly until relatively recently.

7.2 Summarize different perspectives on population growth.
- Does environment or culture control human populations?
- Technology can increase carrying capacity for humans.
- Population growth could bring benefits.

7.3 Analyze some of the factors that determine population growth.
- How many of us are there?
- Fertility measures the number of children born to each woman.
- Mortality is the other half of the population equation.
- Life span and life expectancy describe our potential longevity.
- Living longer has demographic implications.
- Emigration and immigration are important demographic factors.

7.4 Explain how ideal family size is culturally and economically dependent.
- Many factors increase our desire for children.

- Other factors discourage reproduction.
- Could we have a birth dearth?

7.5 Describe how a demographic transition can lead to stable population size.
- Economic and social development influence birth and death rates.
- There are reasons to be optimistic about population.
- Many people remain pessimistic about population growth.
- Social justice is an important consideration.
- Women's rights affect fertility.

7.6 Relate how family planning gives us choices.
- Fertility control has existed throughout history.
- Today there are many options.

7.7 Reflect on what kind of future we are creating.
- Religion and politics complicate family planning.

PRACTICE QUIZ

1. At what point in history did the world population pass its *first* billion? What factors restricted population before that time, and what factors contributed to growth after that point?

2. How might growing populations be beneficial in solving development problems?

3. Why do some economists consider human resources more important than natural resources in determining the future of a country?

4. Where will most population growth occur in the next century? What conditions contribute to rapid population growth in some countries?

5. Define *crude birth rate, total fertility rate, crude death rate,* and *zero population growth.*

6. What is the difference between life expectancy and life span?

7. What is dependency ratio, and how might it affect the United States in the future?

8. What pressures or interests make people want or not want to have babies?

9. Describe the conditions that lead to a demographic transition.

10. Describe some choices in modern birth control.

CRITICAL THINKING AND DISCUSSION QUESTIONS

1. What do you think is the optimum human population? The maximum human population? Are the numbers different? If so, why?

2. Some people argue that technology can provide solutions for environmental problems; others believe that a "technological fix" will make our problems worse. What personal experiences or worldviews do you think might underlie these positions?

3. Karl Marx called Thomas Malthus a "shameless sycophant of the ruling classes." Why would the landed gentry of the eighteenth century be concerned about population growth of the lower classes? Are there comparable class struggles today?

4. Try to imagine yourself in the position of a person your age in a developing world country. What family planning choices and pressures would you face? How would you choose among your options?

5. Some demographers claim that population growth has already begun to slow; others dispute this claim. How would you evaluate the competing claims of these two camps? Is this an issue of uncertain facts or differing beliefs? What sources of evidence would you accept as valid?

6. What role do race, ethnicity, and culture play in our immigration and population policies? How can we distinguish between prejudice and selfishness on one hand and valid concerns about limits to growth on the other?

Data Analysis: Fun with Numbers

Graphs offer us an easy, intuitive way to understand data. In a glance we can see relationships, connections, and trends in a graph that most of us can't discern in a table of numbers. Among the many sources of data on human populations, one of the most entertaining is the Gapminder Foundation. Founded in Stockholm, Sweden, by Ola Rosling, Anna Rosling Ronnlunnd, and Hans Rosling, Gapminder has created a wonderful interactive graphing program to explore international statistics on health, economics, and other social indicators. To learn about their work, go to www.gapminder.org.

Try exploring the data yourself: Go to www.gapminder.org/world, which provides a graph like the one below.

Click on "Play" in the lower left corner to watch how the global life expectancy and income have changed over the past 200 years. Professor Rosling describes it as a race toward higher incomes and longer lives. Notice that it isn't a simple uniform process. Individual nations shoot ahead and then fall back. You can identify the nations by moving your cursor over the bubbles, or show names using the blue check boxes to the right of the graph. You can even turn on "trails" (lower right corner) to watch progress for selected countries. Notice when you roll over the bubbles, the life expectancy and income values show for that country and year. You can also jump to particular years using the slider bar.

While you're exploring this chart, answer the following questions:

1. In 2009, which country had the highest per capita GNP?
2. What was the highest life expectancy in 2009?

3. What's the overall relationship between these two factors?
4. How many countries have a lower per capita income than the United States but a higher life expectancy? Note that the downward-point arrow in the lower left corner of the chart allows you to enlarge a specific area.
5. How does sub-Saharan Africa rank in these indicators? Note the map in the upper right corner that color-codes geographical regions. Switching to the map view also helps you identify locations.
6. What's the lowest life expectancy for any nation in this 200-year span? (*Hint:* scroll slowly through the years and watch individual countries bounce up and down.)
7. What was happening in Russia in 1933? (*Hint:* try Googling "Russia 1933.")
8. What happened to life expectancy and income in China between 1850 and 1870?
9. What explains these Chinese data?
10. What was life expectancy in the United States in 1812? How does that compare to the situation in other nations in 2009?

Now that you're becoming familiar with the graph, click on the bottom axis and change it to "Children per woman (total fertility)." Click "Play." What trends do you see? Are there particular dates of sudden change? Try a comparison between child mortality versus women's education, or child mortality versus total fertility, for example.

To see several remarkable events in the data, click on the "Open graph menu" button at the top left (outside the chart area). Try the "Bangladesh Miracle," for example. Set the bottom axis to GDP/capita. You'll see with stunning clarity how total fertility has fallen by two-thirds, even though per capita income has barely budged over the past 40 years.

While you have the Gapminder World open, look at some of the excellent videos Hans Rosling has made. "200 Countries, 200 Years, 4 Minutes" is wonderfully entertaining. "Asia's Rise, How and When" is also enlightening. Most of all, watch "Population Growth Explained with IKEA Boxes." It's an excellent summary of everything in this chapter.

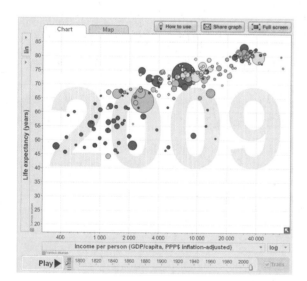

Source: www.gapminder.org/world

Should we be worried about what we're eating? In recent decades, thousands of new, synthetic chemicals have been introduced into our diets and our lives. How dangerous are they? This is an important question in environmental health.

Environmental Health and Toxicology

Learning Outcomes

After studying this chapter, you should be able to:

8.1 Describe health and disease and how global disease burden is now changing.

8.2 Summarize the principles of toxicology.

8.3 Discuss the movement, distribution, and fate of toxins in the environment.

8.4 Characterize mechanisms for minimizing toxic effects.

8.5 Explain ways we measure and describe toxicity.

8.6 Evaluate risk assessment and acceptance.

8.7 Relate how we establish health policy.

"To wish to become well is a part of becoming well."

~ Seneca

153

Case Study

How dangerous is BPA?

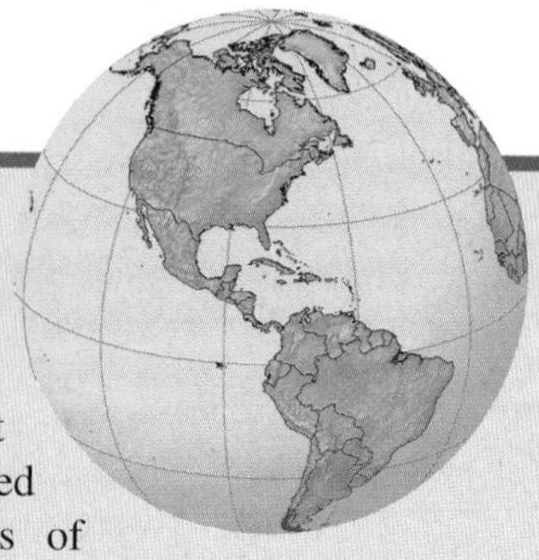

Bisphenol A (BPA), a key ingredient of both polycarbonate plastics and epoxy resins, is one of the world's most widely used chemical compounds. In 2011, total global production was about 3 million metric tons, and the chemical industry expects use to double by 2015 as China and other developing countries manufacture increasing amounts of plastic or plastic-coated wares. BPA is used in items ranging from baby bottles, automobile headlights, eyeglass lenses, CDs, DVDs, water pipes, the linings of cans and bottles, and tooth-protecting sealants.

Traces of BPA are found in humans nearly everywhere. In one study of several thousand normal adult Americans, 95 percent had measurable amounts of this chemical in their bodies. The most likely source of contamination is from food and beverage containers. During plastic polymerization, not all BPA gets locked up into chemical bonds. Unbound molecules can leach out, especially when plastic is heated, washed with harsh detergents, scratched, or exposed to acidic compounds, such as tomato juice, vinegar, or soft drinks. In one study of canned food from major supermarket chains, half the samples had BPA higher than government-recommended dietary levels.

How dangerous is BPA? In recent years dozens of scientists around the globe have linked BPA to myriad health effects in rodents, including mammary and prostate cancer, genital defects in males, early onset of puberty in females, obesity, and even behavior problems such as attention-deficit hyperactivity disorder. Furthermore, epidemiological studies in humans show a correlation between urine concentrations of BPA and cardiovascular disease, type 2 diabetes, and liver-enzyme abnormalities. Scientists find that BPA, phthalates, dioxins, and PCBs act as endocrine hormone disrupters. That is, they upset the normal function of your body's own hormones. Interestingly, the first use for BPA after it was synthesized in 1891 was as a synthetic estrogen.

FIGURE 8.1 What's in our food? How safe are we really?

But rodents, especially those raised in laboratory conditions, may not be accurate models for how humans react in the real world. We have very different genetics, diet, and physiology. And cross-sectional or retrospective studies of human populations show only correlations, not causality. It could be just a coincidence that people exposed to BPA develop common chronic diseases, such as cardiovascular disease or diabetes. Furthermore, as you'll learn in this chapter, detectable levels aren't always dangerous. New technology allows us to measure tiny amounts of chemicals that may or may not be deleterious. Risk assessment is a complex and difficult task.

Industry-funded scientists point to contradictions and unexplained uncertainties in published studies of BPA toxicity. Some investigators find deleterious effects at low BPA levels; others say they can't reproduce those results. Some of this variability may be linked to funding. In one examination of 115 peer-reviewed, published studies of BPA, 94 of those supported by government agencies found adverse health effects from BPA exposure, whereas none of the 15 financed by industry sources found any problem.

Current federal guidelines put the daily upper limit of safe exposure at 50 micrograms of BPA per kilogram of body weight. But that level is based on a small number of high-dose experiments done years ago, rather than on the hundreds of more recent animal and laboratory studies indicating that serious health risks could result from much lower doses of BPA. Several animal studies show adverse effects, such as abnormal reproductive development, at exposures of 2.4 micrograms of BPA per kilogram of body weight per day, a dose that could be reached by a child eating one or two servings daily of certain canned foods.

In response to studies suggesting health risks, particularly for young children, from BPA exposure, Japan, Canada, and most European countries have restricted use of this chemical in consumer applications, especially in food and beverage containers. The United States is still debating this topic. Panels convened by the Food and Drug Administration have come up with conflicting recommendations. Not surprisingly, industry representatives emphasize uncertainty and the need for further research, while most scientists and consumer groups demand action now.

This case study introduces a number of important themes for this chapter. How dangerous are low-level but widespread exposures to a variety of environmental toxins? What are the effects of disruption of endocrine systems by synthetic (or natural) compounds? And how should we test and evaluate toxic substances? For related resources, including Google Earth™ placemarks that show locations where these issues can be explored, visit http://EnvironmentalScience-Cunningham.blogspot.com.

8.1 ENVIRONMENTAL HEALTH

What is health? The WHO defines **health** as a state of complete physical, mental, and social well-being, not merely the absence of disease or infirmity. By that definition, we all are ill to some extent. Likewise, we all can improve our health to live happier, longer, more productive, and more satisfying lives if we think about what we do.

What is disease? A **disease** is an abnormal change in the body's condition that impairs important physical or psychological functions. Diet and nutrition, infectious agents, toxic substances, genetics, trauma, and stress all play roles in **morbidity** (illness) and **mortality** (death). **Environmental health** focuses on external factors that cause disease, including elements of the natural, social, cultural, and technological worlds in which we live. Figure 8.2 shows some major environmental disease agents as well as the media through which we encounter them. Ever since the publication of Rachel Carson's *Silent Spring* in 1962, the discharge, movement, fate, and effects of synthetic chemical toxins have been a special focus of environmental health. Later in this chapter we'll study these topics in detail. First, however, let's look at some of the major causes of illness worldwide.

The global disease burden is changing

World health programs have made tremendous progress in eradicating or greatly reducing many terrible diseases. Smallpox was completely wiped out in 1977. Polio has been eliminated everywhere in the world except for a few remote villages in northern Nigeria.

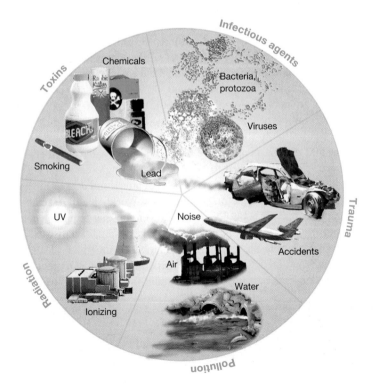

FIGURE 8.2 Major sources of environmental health risks.

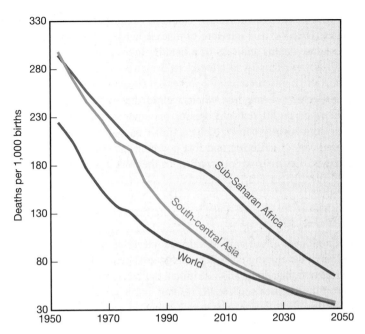

FIGURE 8.3 Child mortality has fallen dramatically over the past 50 years and is expected to continue this decline in the future. Note that South-central Asia saw a significant improvement in child survival in the 1970s and 1980s, while sub-Saharan Africa lagged behind—mainly because of wars and AIDS.

Source: Data from the UN Population Division, 2006.

Guinea worms, river blindness, and yaws appear to be on their way to elimination. Epidemics of typhoid fever, cholera, and yellow fever that regularly killed thousands of people in North America a century ago are now rarely encountered. AIDS, which once was an immediate death sentence, has become a highly treatable disease. The average HIV-positive person in the United States now lives 24 years after diagnosis if treated faithfully with modern medicines.

A way of demonstrating these advances is to look at how much longer we're living. During the twentieth century, world average life expectancies more than doubled, from 30 to 64.3 years. For richer countries, residents can expect to live, on average, about three times as long as their great-grandparents did a century earlier (see chapter 7, Data Analysis, p. 152).

A vital component of rising life expectancies is declining child mortality. In 1950 almost one-quarter (224 per 1,000) of all children born worldwide didn't live to their fifth birthday (fig. 8.3). By 2010 this rate had fallen to about 60 per 1,000. By 2050 it's expected to be nearly 90 percent less than a century earlier. This is essential if we are to reach zero population growth. Sub-Saharan Africa and South-central Asia had much higher child mortality than the world average in 1950. Asia has caught up with the rest of the world, but Africa, for a variety of reasons, lags behind.

In the past, health organizations have focused on the leading causes of death as the best summary of world health. Mortality data, however, fail to capture the impacts of nonfatal outcomes of disease and injury, such as dementia or blindness, on human well-being. When people are ill, work isn't done, crops aren't planted or harvested, meals aren't cooked, and children can't study and learn. Health agencies now calculate **disability-adjusted life years (DALYs)** as a measure of disease burden. DALYs combine premature deaths and loss of a healthy life resulting from illness or disability. This is an attempt to evaluate the total cost of disease, not simply how many people die. Clearly, many more years of expected life are lost when a child dies of neonatal tetanus than when an 80-year-old dies of pneumonia. Similarly, a teenager permanently paralyzed by a traffic accident will have many more years of suffering and lost potential than will a senior citizen who has a stroke. According to the WHO, chronic diseases now account for nearly 60 percent of the 56.5 million total deaths worldwide each year and about half of the global disease burden.

The world is now undergoing a dramatic epidemiological transition. Chronic conditions, such as cardiovascular disease and cancer, no longer afflict only wealthy people. Although the traditional killers in developing countries—infections, maternal and perinatal (birth) complications, and nutritional deficiencies—still take a terrible toll, diseases such as depression and heart attacks that once were thought to occur only in rich countries are rapidly becoming the leading causes of disability and premature death everywhere.

The WHO predicts that in 2020, heart disease, which was fifth in the list of causes of global disease burden a decade ago, will be the leading source of disability and deaths worldwide (table 8.1). Most of that increase will be in the poorer parts of the world where people are rapidly adopting the lifestyles and diet of the richer countries. Similarly, global cancer rates will increase by 50 percent. It's expected that by 2020, 15 million people will have cancer and 9 million will die from it.

A silent epidemic of diabetes is now sweeping through our population. It's estimated that one-third of all children born today in North America will develop this disease in their lifetime. Obesity, diets high in sugar and fat, lack of exercise, and poverty (which encourages fast-food intake and makes health food unavailable) all play important roles in this disease. Blindness, circulatory problems, and kidney failure are common results of severe, uncontrolled diabetes. Seventy percent of all lower limb amputations are diabetes-related. In some Native American groups, more than half of all adults have this disease. It used to be thought that diabetes affected only affluent people, but obesity and diabetes are spreading around the world along with fast-food diets and sedentary lifestyles.

Taking disability as well as death into account in our assessment of disease burden reveals the increasing role of mental health as a worldwide problem. WHO projections suggest that psychiatric and neurological conditions could increase their share of the global burden from the current 10 percent to 15 percent of the total load by 2020. Again, this isn't just a problem of the developed world. Depression is expected to be the second largest cause of all years lived with disability worldwide, as well as the cause of 1.4 percent of all deaths. For women in both developing and developed regions, depression is the leading cause of disease burden, while suicide, which often is the result of untreated depression, is the fourth largest cause of female deaths.

Notice in table 8.1 that diarrhea, which was the second leading cause of disease burden in 1990, is expected to be ninth on the list in 2020, while measles and malaria are expected to drop out of the top 15 causes of disability. Tuberculosis, which is becoming resistant to antibiotics and is spreading rapidly in many areas (especially in the former Soviet Union), is the only infectious disease whose ranking is not expected to change over the next 20 years. Traffic accidents are now soaring as more people drive. War, violence, and self-inflicted injuries similarly are becoming much more important health risks than ever before.

Chronic obstructive lung diseases (e.g., emphysema, asthma, and lung cancer) are expected to increase from eleventh to fifth in disease burden by 2020. A large part of the increase is due to rising use of tobacco in developing countries, sometimes called "the tobacco epidemic." Every day about 100,000 young people—most of them in poorer countries—become addicted to tobacco. At least 1.1 billion people now smoke, and this number is expected to increase at least 50 percent by 2020. If current patterns persist, about 500 million people alive today will eventually be killed by tobacco. This is expected to be the biggest single cause of death worldwide (because illnesses such as heart attack and depression are triggered by multiple factors). In 2003 the World Health Assembly adopted an historic tobacco-control convention that requires countries to impose restrictions on tobacco advertising, establish clean indoor air controls, and clamp down on tobacco smuggling. Dr. Gro Harlem Brundtland, former director-general of the WHO, predicted that the convention, if implemented, could save billions of lives.

Table 8.1 Leading Causes of Global Disease Burden

Rank	1990	Rank	2020
1	Pneumonia	1	Heart disease
2	Diarrhea	2	Depression
3	Perinatal conditions	3	Traffic accidents
4	Depression	4	Stroke
5	Heart disease	5	Chronic lung disease
6	Stroke	6	Pneumonia
7	Tuberculosis	7	Tuberculosis
8	Measles	8	War
9	Traffic accidents	9	Diarrhea
10	Birth defects	10	HIV/AIDS
11	Chronic lung disease	11	Perinatal conditions
12	Malaria	12	Violence
13	Falls	13	Birth defects
14	Iron anemia	14	Self-inflicted injuries
15	Malnutrition	15	Respiratory cancer

Source: World Health Organization, 2002.

Think About It

What changes could you make in your lifestyle to lessen your risks from the diseases in table 8.1? What would have the greatest impact on your future well-being?

FIGURE 8.4 At least 3 million children die every year from easily preventable diseases. This billboard in Guatemala encourages parents to have their children vaccinated against polio, diphtheria, TB, tetanus, pertussis (whooping cough), and scarlet fever.

Infectious and emergent diseases still kill millions of people

Although the ills of modern life have become the leading killers almost everywhere in the world, communicable diseases still are responsible for about one-third of all disease-related mortality. Diarrhea, acute respiratory illnesses, malaria, measles, tetanus, and a few other infectious diseases kill about 11 million children under age five every year in the developing world. Better nutrition, clean water, improved sanitation, and inexpensive inoculations could eliminate most of those deaths (fig. 8.4).

Humans are afflicted by a wide variety of pathogens (disease-causing organisms), including viruses, bacteria, protozoans (single-celled animals), parasitic worms, and flukes (fig. 8.5). Pandemics (worldwide epidemics) have changed the course of history. In the mid-fourteenth century, the "Black Death" (bubonic plague) swept out of Asia and may have killed half the people in Europe. When European explorers and colonists reached the Americas in the late fifteenth and early sixteenth centuries, they brought with them diseases, such as smallpox, measles, cholera, and yellow fever, that killed up to 90 percent of the native population in many areas. The largest loss of life in a pandemic in the past century was in the great influenza pandemic of 1918. Epidemiologists now estimate that at least one-third of all humans living at the time were infected, and that 50 to 100 million died. Businesses, schools, churches, and sports and entertainment events were shut down for months.

We haven't had a pandemic as deadly since the 1918 flu, but epidemiologists warn that new contagious diseases test our defenses every day. Two recent examples are the H5N1 bird flu, or SARS (severe acute respiratory syndrome), that swept around the world in 2003. It's thought to have originated in poultry and wildlife markets in China. A doctor carrying the disease went to Hong Kong, where he passed the infection to other international travelers. Within six months SARS had spread to 31 countries around the globe. The rapid transmission of this disease shows how interconnected we all are. It's thought that a single flight attendant infected 160 people in seven countries in just a few days.

In 2009 another flu pandemic swept around the world. This one, caused by an H1N1 virus related to the 1918 influenza, bore genes from varieties that infect pigs, birds, and humans. Although it wasn't nearly as lethal as some influenza viruses (fig. 8.5*a*), this strain infected an estimated 50 million Americans and killed at least 10,000. The virus's ability to jump between species and recombine genes worries many health experts. Every year the Centers for Disease Control and Prevention in Atlanta surveys the

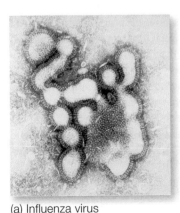

(a) Influenza virus

(b) Pathogenic bacteria

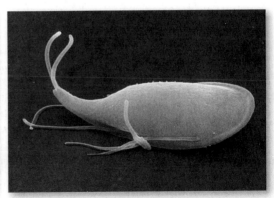

(c) *Giardia*

FIGURE 8.5 (a) A group of influenza viruses magnified about 300,000 times. (b) Pathogenic bacteria magnified about 50,000 times. (c) *Giardia,* a parasitic intestinal protozoan, magnified about 10,000 times.

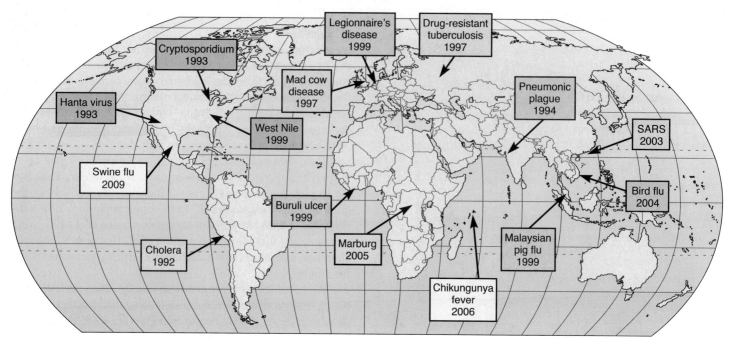

FIGURE 8.6 Some recent outbreaks of highly lethal infectious diseases. Why are supercontagious organisms emerging in so many different places?

Source: Data from U.S. Centers for Disease Control and Prevention.

flu varieties in circulation around the world and tries to guess what strains should be represented in vaccines for the next season. The viruses evolve so rapidly, however, that vaccine manufacturers often guess wrong.

Every year there are 76 million cases of foodborne illnesses in the United States, resulting in 300,000 hospitalizations and 5,000 deaths. Both bacteria and intestinal protozoa cause these illnesses (fig. 8.5b,c). They are spread from feces through food and water. In 2010 nearly 6 million pounds (about 2.7 million kg) of beef was recalled in the United States, mostly due to contamination by *E. coli* O157:H7. At any given time, around 2 billion people—one-third of the world population—suffer from worms, flukes, and other internal parasites. Although such parasites rarely kill people, they can be extremely debilitating and can cause poverty that leads to other, more deadly, diseases.

Malaria is one of the most prevalent remaining infectious diseases. Every year about 500 million new cases of this disease occur, and about 2 million people die from it. The territory infected by this disease is expanding as global climate change allows mosquito vectors to move into new territory. Simply providing insecticide-treated bed nets and a few dollars' worth of antiparasite drugs could prevent tens of millions of cases of this debilitating disease every year. Tragically, some of the countries where malaria is most widespread tax both bed nets and medicine as luxuries, placing them out of reach for ordinary people.

Emergent diseases are those not previously known or that have been absent for at least 20 years. The new strain of swine flu now spreading around the world is a good example. There have been at least 39 outbreaks of emergent diseases over the past

two decades, including the extremely deadly Ebola and Marburg fevers, which have afflicted Central Africa in at least six different locations in the past decade. Similarly, cholera, which had been absent from South America for more than a century, reemerged in Peru in 1992 (fig. 8.6). Some other examples include a new drug-resistant form of tuberculosis, now spreading in Russia; dengue fever, which is spreading through Southeast Asia and the Caribbean; malaria, which is reappearing in places where it had been nearly eradicated; and a new human lymphotropic virus (HTLV), which is thought to have jumped from monkeys into people in Cameroon who handled or ate bushmeat. These HTLV strains are now thought to infect 25 million people.

The largest recent death toll from an emergent disease is HIV/AIDS. Although virtually unknown 15 years ago, acquired immune-deficiency syndrome has now become the fifth greatest cause of contagious deaths. The WHO estimates that 60 million people are now infected with the human immune-deficiency virus, and that 3 million die every year from AIDS complications. Although two-thirds of all current HIV infections are now in sub-Saharan Africa, the disease is spreading rapidly in South and East Asia. Over the next 20 years, there could be an additional 65 million AIDS deaths. In Swaziland, health officials estimate that about one-third of all adults are HIV positive and that two-thirds of all current 15-year-olds will die of AIDS before age 50. As chapter 7 points out, without AIDS the life expectancy in Swaziland would be expected to be about 65 years. With AIDS, the average life expectancy is now only about 33 years. Worldwide, more than 14 million children—the equivalent of all children under age five in America—have lost one or both parents to

AIDS. The economic costs of treating patients and lost productivity from premature deaths resulting from this disease are estimated to be at least (U.S.) $35 billion per year or about one-tenth of the total GDP of sub-Saharan Africa.

Conservation medicine combines ecology and health care

Humans aren't the only ones to suffer from new and devastating diseases. Domestic animals and wildlife also experience sudden and widespread epidemics, which are sometimes called **ecological diseases.** Ebola hemorrhagic fever is one of the most virulent viruses ever seen, killing up to 90 percent of its victims. In 2002 an outbreak of Ebola fever began killing humans along the Gabon-Congo border. A few months later, researchers found that 221 of the 235 western lowland gorillas they had been studying in this area disappeared in just a few months. Many chimpanzees also died. Although the study team could find only a few of the dead gorillas, 75 percent of those tested positive for Ebola. Altogether, researchers estimate that 5,000 gorillas died in this small area of the Congo. Extrapolating to all of central Africa it's possible that Ebola has killed one-quarter of all the gorillas in the world. It's thought that the spread of this disease in humans resulted from the practice of hunting and eating primates.

In 2006, people living near a cave west of Albany, New York, reported something peculiar: little brown bats (*Myotis lucifugus*) were flying outside during daylight in the middle of the winter. Inspection of the cave by the Department of Conservation found numerous dead bats near the cave mouth. Most had white fuzz on their faces and wings, a condition now known as white-nose syndrome (WNS). Little brown bats are tiny creatures, about the size of your thumb. They depend on about 2 grams of stored fat to get them through the winter. Hibernation is essential to making their energy resources last. Being awakened just once can cost a bat a month's worth of fat.

The white fuzz has now been identified as filamentous fungus (*Geomyces destructans*), which thrives in the cool, moist conditions where bats hibernate. We don't know where the fungus came from, but, true to its name, the pathogen has spread like wildfire through half a dozen bat species in 14 states along the Appalachian Mountains and in two Canadian provinces. Biologists estimate that over a million bats already have died from this disease. It isn't known how the pathogen spreads. Perhaps it moves from animal to animal through physical contact. It's also possible that humans introduce fungal spores on their shoes and clothing when they go from one cave to another.

One mammalogist calls WNS "the chestnut blight of bats." So far, six species of bat are known to be susceptible to this plague. In infected colonies, mortality can be 90 percent or more. Some researchers fear that bats could be extinct in 20 years in the eastern United States. Losing these important species would have devastating ecological consequences.

An even more widespread and lethal epidemic is currently sweeping through amphibians worldwide. A disease called Chytridomycosis is causing dramatic losses or even extinctions

FIGURE 8.7 Frogs and toads throughout the world are succumbing to a deadly disease called Chytridomycosis. Is this a newly virulent fungal disease, or are amphibians more susceptible because of other environmental stresses?

of frogs and toads throughout the world (fig. 8.7). A fungus called *Batrachochytrium dendrobatidis* causes the disease. It was first recognized in 1993 in dead and dying frogs in Queensland, Australia and now seems to be spreading rapidly, perhaps because the fungus has become more virulent or amphibians are more susceptible due to environmental change. Most of the world's approximately 6,000 amphibian species appear to be susceptible to the disease, and around 2,000 species have declined or even become extinct in their native habitats as part of this global epidemic. African clawed frogs (*Xenopus* sp), which are resistant to fungal infections and thus may be carriers of the disease, are a possible vector for this disease. Widespread use of these frogs for research and pregnancy testing may have contributed to the rapid spread of the pathogen.

Temperatures above 28°C (82°F) kill the fungus, and treating frogs with warm water can cure the disease in some species. Topical application of the drug chloramphenicol also has successfully cured some frogs. And certain skin bacteria seem to confer immunity to fungal infections. In some places, refuges have been established in which frogs can be maintained under antiseptic conditions until a cure is found. It's hoped that survivors can eventually be reintroduced back to their native habitat and species will be preserved.

Climate change also facilitates expansion of parasites and diseases into new territories. Tropical diseases, such as malaria, cholera, yellow fever, and dengue fever, have been moving into areas from which they were formerly absent as mosquitoes, rodents, and other vectors expand into new habitat. This affects other species besides humans. A disease called Dermo is spreading northward through oyster populations along the Atlantic coast of North America. This disease is caused by a protozoan parasite (*Perkinsus marinus*) that was first recognized in the Gulf of Mexico about 70 years ago. In the 1950s the disease was found in Chesapeake Bay. Since then the parasite has been moving northward, probably assisted by higher sea temperatures caused by global warming. It is now found as far north as Maine. This disease doesn't appear to be harmful to humans, but it is devastating oyster populations.

One thing that emergent diseases in humans and ecological diseases in natural communities have in common is environmental change that stresses biological systems and upsets normal ecological relationships. We cut down forests and drain wetlands, destroying habitat for native species. Invasive organisms and diseases are accidentally or intentionally introduced into new areas where they can grow explosively. Increasing incursion into former wilderness is spurred by human population growth and ecotourism. In 1950 only about 3 million people a year flew on commercial jets; by 2000, more than 300 million did. Diseases can spread around the globe in mere days as people pass through international travel hubs.

We are coming to recognize that the delicate ecological balances that we value so highly—and disrupt so frequently—are important to our own health. **Conservation medicine** is an emerging discipline that attempts to understand how our environmental changes threaten our own health as well as that of the natural communities on which we depend for ecological services. Although it is still small, this new field is gaining recognition from mainstream funding sources such as the World Bank, the World Health Organization, and the U.S. National Institutes of Health.

Resistance to drugs, antibiotics, and pesticides is increasing

Malaria, the most deadly of all insect-borne illnesses, is an example of the return of a disease that once was thought to be nearly vanquished. Malaria now claims about a million lives every year—90 percent are in Africa, and most of them children. With the advent of modern medicines and pesticides, malaria had nearly been wiped out in many places, but recently it has come roaring back. The protozoan parasite that causes the disease is now resistant to most drugs, while the mosquitoes that transmit it have developed resistance to many insecticides. Spraying of DDT in India and Sri Lanka, for instance, reduced malaria from millions of infections per year to only a few thousand in the 1950s and 1960s. Now South Asia is back to its pre-DDT level of about half a million new cases of malaria every year. Other places that never had cases of malaria now have them as a result of climate change and habitat alteration.

In recent years, health workers have become increasingly alarmed about the rapid spread of methicillin-resistant *Staphylococcus aureus* (MRSA). Staphylococcus (or Staph) is very common. Most people have at least some of these bacteria. They are a common cause of sore throats and skin infections, but are usually easily controlled. This new strain is resistant to penicillin and related antibiotics and can cause deadly infections, especially in people with weak immune systems. MRSA is most frequent in hospitals, nursing homes, correctional facilities, and other places where people are in close contact. It's generally spread through direct skin contact. School locker rooms, gymnasiums,

and contact sports also are sources of infections. Several states have closed schools as a result of MRSA contamination. It's estimated that at least 100,000 MRSA infections in the United States resulted in about 19,000 deaths. A much worse situation is reported in China, where about half of the 5 million annual Staph infections are thought to be methicillin-resistant.

Why have vectors such as mosquitoes and pathogens such as Staphylococcus become resistant to pesticides and drugs? Part of the answer is natural selection and the ability of many organisms to evolve rapidly. Another factor is the human tendency to use control measures carelessly. When we discovered that DDT and other insecticides could control mosquito populations, we spread them everywhere. This not only harmed wildlife and beneficial insects, but it created selective pressures that lead to evolution. Many pests and pathogens are exposed to low, chronic doses, allowing those with natural resistance to survive and spread their genes through the population (fig. 8.8). After repeated cycles of exposure and selection, many microorganisms and their vectors are insensitive to almost all our weapons against them.

As chapter 9 discusses, raising huge numbers of cattle, hogs, and poultry in densely packed barns and feedlots helps spread antibiotic resistance in pathogens. Confined animals are dosed constantly with antibiotics and steroid hormones to keep them disease-free and to make them gain weight faster. More than half of all antibiotics used in the United States each year is fed to livestock. A significant amount of these antibiotics and hormones are excreted in urine and feces, which are spread, untreated, on the land or discharged into surface water where they contribute further to the evolution of supervirulent pathogens.

At least half of the 100 million antibiotic doses prescribed for humans every year in the United States are unnecessary or are the wrong ones. Furthermore, many people who start a course

(a) Mutation and selection create drug-resistant strains

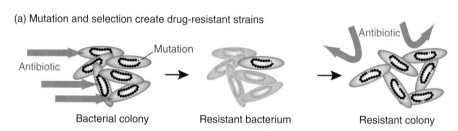

(b) Conjugation transfers drug resistance from one strain to another

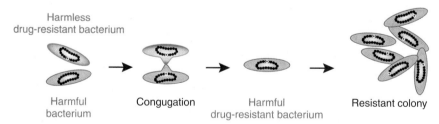

FIGURE 8.8 How microbes acquire antibiotic resistance. (a) Random mutations make a few cells resistant. When challenged by antibiotics, only those cells survive to give rise to a resistant colony. (b) Sexual reproduction (conjugation) or plasmid transfer moves genes from one strain or species to another.

of antibiotic treatment fail to carry it out for the time prescribed. For your own health and that of the people around you, if you are

---

Think About It

If you were making a case for greater U.S. funding for international health care, what points would you stress? Do we have a moral obligation to help others?

---

taking an antibiotic, follow your doctor's orders and don't stop taking the medicine as soon as you start feeling better.

Who should pay for health care?

The heaviest burden of illness is borne by the poorest people, who can afford neither a healthy environment nor adequate health care. Women in sub-Saharan Africa, for example, suffer six times the disease burden per 1,000 population as do women in most European countries. The WHO estimates that 90 percent of all disease burden occurs in developing countries where less than one-tenth of all health care dollars is spent. The group Médecins Sans Frontières (MSF, or Doctors Without Borders) calls this the 10/90 gap. While wealthy nations pursue drugs to treat baldness and obesity, depression in dogs, and erectile dysfunction, billions of people are sick or dying from treatable infections and parasitic diseases to which little attention is paid. Worldwide, only 2 percent of the people with AIDS have access to modern medicines. Every year some 600,000 infants acquire HIV—almost all of them through mother-to-child transmission during birth or breast-feeding. Antiretroviral therapy costing only a few dollars can prevent most of this transmission. The Bill and Melinda Gates Foundation has pledged $200 million for medical aid to developing countries to help fight AIDS, TB, and malaria.

Dr. Jeffrey Sachs of the Columbia University Earth Institute says that disease is as much a cause as a consequence of poverty and political unrest, yet the world's richest countries now spend just $1 per person per year on global health. He predicts that raising our commitment to about $25 billion annually (about 0.1 percent of the annual GDP of the 20 richest countries) not only would save about 8 million lives each year, but would boost the world economy by billions of dollars. There also would be huge social benefits for the rich countries in not living in a world endangered by mass social instability, the spread of pathogens across borders, and the spread of other ills such as terrorism and drug trafficking caused by social problems. Sachs also argues that reducing disease burden would help reduce population growth. When parents believe their offspring will survive, they have fewer children and invest more in food, health, and education for smaller families.

The United States is the least generous of the world's rich countries, donating only about 12 cents per $100 of GDP to international development aid. Could the United States do better? During a time of fear of terrorism and rising anti-American feelings around the globe, it's difficult to interest legislators in international

aid, and yet, helping to reduce disease might win the United States more friends and make the nation safer than buying more bombs and bullets. Improved health care in poorer countries may also help prevent the spread of emergent diseases in a globally interconnected world.

8.2 TOXICOLOGY

Toxicology is the study of **toxins** (poisons) and their effects, particularly on living systems. Because many substances are known to be poisonous to life (whether plant, animal, or microbial), toxicology is a broad field, drawing from biochemistry, histology, pharmacology, pathology, and many other disciplines. Toxins damage or kill living organisms because they react with cellular components to disrupt metabolic functions. Because of this reactivity, toxins often are harmful even in extremely dilute concentrations. In some cases billionths, or even trillionths, of a gram can cause irreversible damage.

All toxins are hazardous, but not all hazardous materials are toxic. Some substances, for example, are dangerous because they're flammable, explosive, acidic, caustic, irritants, or sensitizers. Many of these materials must be handled carefully in large doses or high concentrations, but can be rendered relatively innocuous by dilution, neutralization, or other physical treatment. They don't react with cellular components in ways that make them poisonous at low concentrations.

Environmental toxicology, or ecotoxicology, specifically deals with the interactions, transformation, fate, and effects of natural and synthetic chemicals in the biosphere, including individual organisms, populations, and whole ecosystems. In aquatic systems the fate of the pollutants is primarily studied in relation to mechanisms and processes at interfaces of the ecosystem components. Special attention is devoted to the sediment/water, water/organisms, and water/air interfaces. In terrestrial environments, the emphasis tends to be on effects of metals on soil fauna community and population characteristics.

The U.S. Environmental Protection Agency is responsible for monitoring 275 substances regulated by the Comprehensive Environmental Response, Compensation, and Liability Act (CERCLA), commonly known as the Superfund Act. In 2011 there were 1,280 Superfund sites on the National Priorities List in the United States, and 62 additional sites have been proposed for cleanup. More than 11 million people live within 1 mi (1.2 km) of these sites.

How do toxins affect us?

Allergens are substances that activate the immune system. Some allergens act directly as **antigens;** that is, they are recognized as foreign by white blood cells and stimulate the production of specific antibodies (proteins that recognize and bind to foreign cells or chemicals). Other allergens act indirectly by binding to and changing the chemistry of foreign materials so they become antigenic and cause an immune response.

Formaldehyde is a good example of a widely used chemical that is a powerful sensitizer of the immune system. It is directly allergenic and can also trigger reactions to other substances. Widely used in plastics, wood products, insulation, glue, and fabrics, formaldehyde concentrations in indoor air can be thousands of times higher than in normal outdoor air. Some people suffer from what is called **sick building syndrome:** headaches, allergies, chronic fatigue, and other symptoms caused by poorly vented indoor air contaminated by mold spores, carbon monoxide, nitrogen oxides, formaldehyde, and other toxins released from carpets, insulation, plastics, building materials, and other sources (fig. 8.9). The Environmental Protection Agency estimates that poor indoor air quality may cost the United States $60 billion a year in absenteeism and reduced productivity.

Immune system depressants are pollutants that suppress the immune system rather than activate it. Little is known about how this occurs or which chemicals are responsible. Immune system failure is thought to have played a role, however, in widespread deaths of seals in the North Atlantic and of dolphins in the Mediterranean. These dead animals generally contain high levels of pesticide residues, polychlorinated biphenyls (PCBs), and other contaminants that are suspected of damaging the immune system and making it susceptible to a variety of opportunistic infections.

Endocrine disrupters are chemicals that disrupt normal hormone functions. Hormones are chemicals released into the bloodstream by cells in one part of the body to regulate development and function of tissues and organs elsewhere in the body (fig. 8.10). You undoubtedly have heard about sex hormones and their powerful effects on how we look and behave, but these are only one example of the many regulatory hormones that rule our lives. Some other powerful hormones include thyroxin, insulin, adrenalin, and endorphins.

We now know that some of the most insidious effects of persistent chemicals such as BPA, dioxins, and PCBs are that they interfere with normal growth, development, and physiology of a variety of animals—including humans—at very low

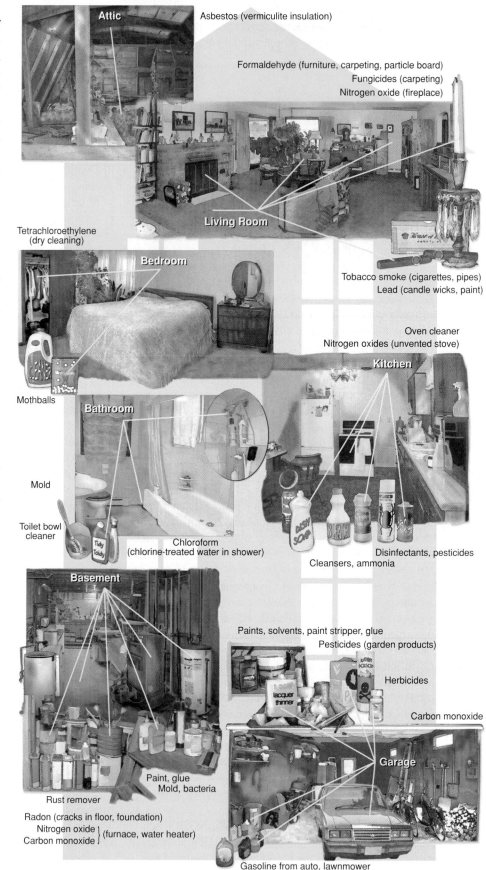

FIGURE 8.9 Some sources of toxic and hazardous substances in a typical home.

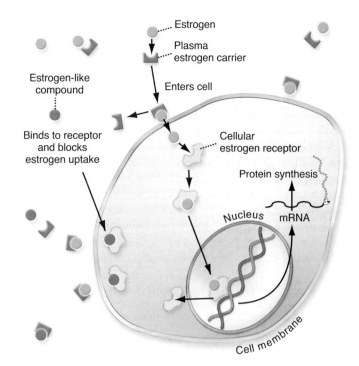

FIGURE 8.10 Steroid hormone action. Plasma hormone carriers deliver regulatory molecules to the cell surface, where they cross the cell membrane. Intracellular carriers deliver hormones to the nucleus, where they bind to and regulate expression of DNA. Estrogen-like compounds bind to receptors and either block uptake of endogenous hormone or act as a substitute hormone to disrupt gene expression.

doses. In some cases, picogram concentrations (trillionths of a gram per liter) may be enough to cause developmental abnormalities in sensitive organisms. Because these chemicals often cause sexual dysfunction (reproductive health problems in females or feminization of males, for example), these chemicals are sometimes called environmental estrogens or androgens. They are just as likely, however, to disrupt functions of other important regulatory molecules as they are to obstruct sex hormones.

Neurotoxins are a special class of metabolic poisons that specifically attack nerve cells (neurons). The nervous system is so important in regulating body activities that disruption of its activities is especially fast-acting and devastating. Different types of neurotoxins act in different ways. Heavy metals such as lead and mercury kill nerve cells and cause permanent neurological damage. Anesthetics (ether, chloroform, halothane, etc.) and chlorinated hydrocarbons (DDT, Dieldrin, Aldrin) disrupt nerve cell membranes necessary for nerve action. Organophosphates (Malathion, Parathion) and carbamates (carbaryl, zeneb, maneb) inhibit acetylcholinesterase, an enzyme that regulates signal transmission between nerve cells and the tissues or organs they innervate (for example, muscle). Most neurotoxins are both extremely toxic and fast-acting.

Mutagens are agents, such as chemicals and radiation, that damage or alter genetic material (DNA) in cells. This damage can lead to birth defects if it occurs during embryonic or fetal growth. Later in life, genetic damage may trigger neoplastic (tumor) growth. When damage occurs in reproductive cells, the results can be passed on to future generations. Cells have repair mechanisms to detect and restore damaged genetic material, but some changes may be hidden, and the repair process itself can be flawed. It is generally accepted that there is no "safe" threshold for exposure to mutagens. Any exposure has some possibility of causing damage.

Teratogens are chemicals or other factors that specifically cause abnormalities during embryonic growth and development. Some compounds that are not otherwise harmful can cause tragic problems in these sensitive stages of life. Perhaps the most prevalent teratogen in the world is alcohol. Drinking during pregnancy can lead to **fetal alcohol syndrome**—a cluster of symptoms including craniofacial abnormalities, developmental delays, behavioral problems, and mental defects that last throughout a child's life. Even one alcoholic drink a day during pregnancy has been associated with decreased birth weight.

By some estimates, 300,000 to 600,000 children born every year in the United States are exposed in the womb to unsafe levels of mercury. The effects are subtle, but include reduced intelligence, attention deficit, and behavioral problems. The total cost of these effects is estimated to be $8.7 billion per year.

Carcinogens are substances that cause **cancer,** invasive, out-of-control cell growth that results in malignant tumors. Cancer rates rose in most industrialized countries during the twentieth century, and cancer is now the second leading cause of death in the United States, killing more than half a million people in 2002. According to the American Cancer Society, 1 in 2 males and 1 in 3 females in the United States will have some form of cancer in their lifetime. Some authors blame this cancer increase on toxic synthetic chemicals in our environment and diet. Others argue that it is attributable mainly to lifestyle (smoking, sunbathing, alcohol) or simply living longer. The U.S. EPA estimates that 200 million U.S. residents live in areas where the combined lifetime cancer risk from environmental carcinogens exceeds 1 in 100,000, or ten times the risk normally considered acceptable.

How does diet influence health?

Diet also has an important effect on health. For instance, there is a strong correlation between cardiovascular disease and the amount of salt and animal fat in one's diet.

Fruits, vegetables, whole grains, complex carbohydrates, and dietary fiber (plant cell walls) often have beneficial health effects. Certain dietary components seem to have anticancer effects—these components include pectins; vitamins A, C, and E; substances produced in cruciferous vegetables (cabbage, broccoli, cauliflower, brussels sprouts); and selenium, which we get from plants.

Eating too much food is a significant dietary health factor in developed countries and among the well-to-do everywhere. Sixty percent of all U.S. adults are now considered overweight, and the worldwide total of obese or overweight people is estimated to be over 1 billion. Every year in the United States, 300,000 deaths are linked to obesity.

The U.S. Centers for Disease Control and Prevention in Atlanta warn that one in three U.S. children will become diabetic unless many more people start eating less and exercising more. The odds are worse for Black and Hispanic children: nearly half of them are likely to develop the disease. And among the Pima tribe of Arizona, nearly 80 percent of all adults are diabetic. More information about food and its health effects is available in chapter 9.

8.3 MOVEMENT, DISTRIBUTION, AND FATE OF TOXINS

There are many sources of toxic and hazardous chemicals in the environment and many factors related to each chemical itself, its route or method of exposure, and its persistence in the environment, as well as characteristics of the target organism (table 8.2), that determine the danger of the chemical. We can think of both individuals and an ecosystem as sets of interacting compartments between which chemicals move, based on molecular size, solubility, stability, and reactivity (fig. 8.11). The dose (amount), route

Table 8.2 Factors in Environmental Toxicity
Factors Related to the Toxic Agent
1. Chemical composition and reactivity
2. Physical characteristics (such as solubility, state)
3. Presence of impurities or contaminants
4. Stability and storage characteristics of toxic agent
5. Availability of vehicle (such as solvent) to carry agent
6. Movement of agent through environment and into cells
Factors Related to Exposure
1. Dose (concentration and volume of exposure)
2. Route, rate, and site of exposure
3. Duration and frequency of exposure
4. Time of exposure (time of day, season, year)
Factors Related to Organism
1. Resistance to uptake, storage, or cell permeability of agent
2. Ability to metabolize, inactivate, sequester, or eliminate agent
3. Tendency to activate or alter nontoxic substances so they become toxic
4. Concurrent infections or physical or chemical stress
5. Species and genetic characteristics of organism
6. Nutritional status of subject
7. Age, sex, body weight, immunological status, and maturity

Source: U. S. Department of Health and Human Services, 1995.

of entry, timing of exposure, and sensitivity of the organism all play important roles in determining toxicity. In this section, we will consider some of these characteristics and how they affect environmental health.

Solubility and mobility determine where and when chemicals move

Solubility is one of the most important characteristics in determining how, where, and when a toxic material will move through the environment or through the body to its site of action. Chemicals can be divided into two major groups: those that dissolve more readily in water and those that dissolve more readily in oil. Water-soluble compounds move rapidly and widely through the environment because water is ubiquitous. They also tend to have ready access to most cells in the body, because aqueous solutions bathe all our cells. Molecules that are oil-or fat-soluble (usually organic molecules) generally need a carrier to move through the environment and into, and within, the body. Once inside the body, however, oil-soluble toxins penetrate readily into tissues and cells, because the membranes that enclose cells are themselves made of similar oil-soluble chemicals. Once they get inside cells, oil-soluble materials are likely to be accumulated and stored in lipid deposits, where they may be protected from metabolic breakdown and persist for many years.

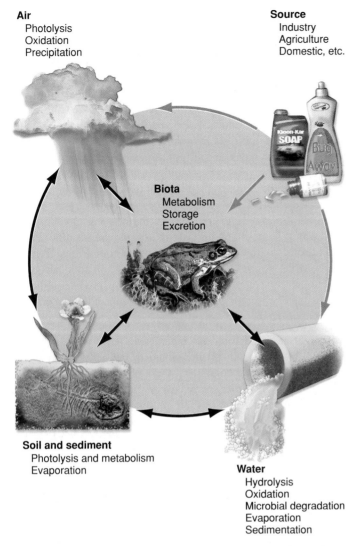

Air
Photolysis
Oxidation
Precipitation

Source
Industry
Agriculture
Domestic, etc.

Biota
Metabolism
Storage
Excretion

Soil and sediment
Photolysis and metabolism
Evaporation

Water
Hydrolysis
Oxidation
Microbial degradation
Evaporation
Sedimentation

FIGURE 8.11 Movement and fate of chemicals in the environment. Toxins also move directly from a source to soil and sediment.

Exposure and susceptibility determine how we respond

Just as there are many sources of toxins in our environment, there are many routes for entry of dangerous substances into our bodies (fig. 8.12). Airborne toxins generally cause more ill health than any other exposure source. We breathe far more air every day than the volume of food we eat or water we drink. Furthermore, the lining of our lungs not only is designed to exchange gases very efficiently but also absorbs toxins very well. Epidemiologists estimate that some 3 million people—two-thirds of them children—die each year from diseases caused or exacerbated by air pollution. Still, food, water, and skin contact also can expose us to a wide variety of toxins. The largest exposures for many toxins are found in industrial settings, where workers may encounter doses thousands of times higher than would be found anywhere else. The European Agency for Safety and Health at Work warns that 32 million people (20 percent of all employees) in the European Union are exposed to unacceptable levels of carcinogens and other toxins in their workplace.

Condition of the organism and timing of exposure also have strong influences on toxicity. Healthy adults, for example, may be relatively insensitive to doses that would be very dangerous for young children or for someone already weakened by disease. Similarly, exposure to a toxin may be very dangerous at certain stages of developmental or metabolic cycles, but may be innocuous at other times. A single dose of the notorious teratogen thalidomide, for example, taken in the third week of pregnancy (a time when many women aren't aware they're pregnant) can cause severe abnormalities in fetal limb development. A complication in measuring toxicity is that great differences in sensitivity exist between species. Thalidomide was tested on a number of laboratory animals without any deleterious effects. Unfortunately, however, it is a powerful teratogen in humans.

Bioaccumulation and biomagnification increase concentrations of chemicals

Cells have mechanisms for **bioaccumulation,** the selective absorption and storage of a great variety of molecules. This allows them to accumulate nutrients and essential minerals, but at the same time they also may absorb and store harmful substances through these same mechanisms. Toxins that are rather dilute in the environment can reach dangerous levels inside cells and tissues through this process of bioaccumulation.

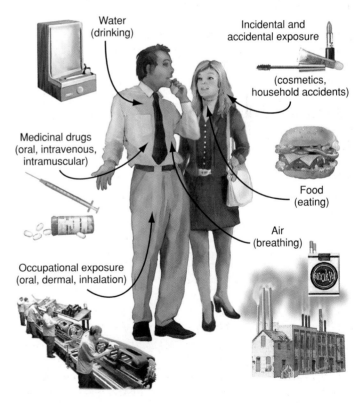

FIGURE 8.12 Routes of exposure to toxic and hazardous environmental factors.

The effects of toxins also are magnified in the environment through food webs. **Biomagnification** occurs when the toxic burden of a large number of organisms at a lower trophic level is accumulated and concentrated by a predator in a higher trophic level. Phytoplankton and bacteria in aquatic ecosystems, for instance, take up heavy metals or toxic organic molecules from water or sediments (fig. 8.13). Their predators—zooplankton and small fish—collect and retain the toxins from many prey organisms, building up higher concentrations of toxins. The top carnivores in the food chain—game fish, fish-eating birds, and humans—can accumulate such high toxin levels that they suffer adverse health effects (see chapter 18). One of the first known examples of bioaccumulation and biomagnification was DDT, which accumulated through food chains so that by the 1960s it was shown to be interfering with reproduction of peregrine falcons, brown pelicans, and other predatory birds at the top of their food chains (see chapter 10).

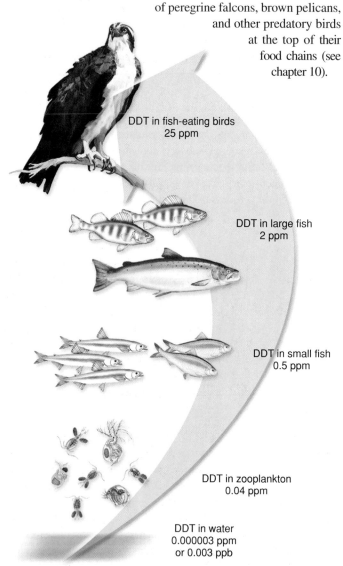

DDT in fish-eating birds
25 ppm

DDT in large fish
2 ppm

DDT in small fish
0.5 ppm

DDT in zooplankton
0.04 ppm

DDT in water
0.000003 ppm
or 0.003 ppb

FIGURE 8.13 Bioaccumulation and biomagnification. Organisms lower on the food chain take up and store toxins from the environment. They are eaten by larger predators, who are eaten, in turn, by even larger predators. The highest members of the food chain can accumulate very high levels of the toxin.

Persistence makes some materials a greater threat

Some chemical compounds are very unstable and degrade rapidly under most environmental conditions, so that their concentrations decline quickly after release. Most modern herbicides and pesticides, for instance, quickly lose their toxicity. Other substances are more persistent and last for years or even centuries in the environment. Metals—such as lead—PVC plastics, chlorinated hydrocarbon pesticides, and asbestos are valuable because they are resistant to degradation. This stability, however, also causes problems because these materials persist in the environment and have unexpected effects far from the sites of their original use.

In addition to BPA, described in the opening case study for this chapter, some other **persistent organic pollutants (POPs)** have become extremely widespread, being found now from the tropics to the Arctic. They often accumulate in food webs and reach toxic concentrations in long-living top predators such as humans, sharks, raptors, swordfish, and bears. POPs of greatest current concern include:

- Polybrominated diphenyl ethers (PBDEs). Widely used as flame retardants in textiles, foam in upholstery, and plastic in appliances and computers, these chemicals are now found in humans and other species everywhere in the world. Nearly 150 million metric tons (330 million lbs) of PBDEs are used every year worldwide. The toxicity and environmental persistence of PBDEs is much like that of PCBs, to which they are closely related chemically. Relatively low exposures in the womb or shortly after birth can irreparably harm children's reproductive and nervous systems. The European Union has already banned PBDEs.

- Perfluorooctane sulfonate (PFOS) and perfluorooctanoic acid (PFOA, also known as C8) are members of a chemical family used to make nonstick, waterproof, and stain-resistant products such as Teflon, Gortex, Scotchguard, and Stainmaster. Industry makes use of their slippery, heat-stable properties to manufacture everything from airplanes and computers to cosmetics and household cleaners. Now these chemicals—which are reported to be infinitely persistent in the environment—are found throughout the world, even the most remote and seemingly pristine sites. Almost all Americans have one or more perfluorinated compounds in their blood. Heating some nonstick cooking pans above 500°F (260°C) can release enough PFOA to kill pet birds. This chemical family has been shown to cause liver damage as well as various cancers and reproductive and developmental problems in rats. Exposure may be especially dangerous to women and girls, who may be 100 times more sensitive than men to these chemicals. In 2005 the EPA announced the start of a study of human health effects of these chemicals.

- Phthalates (pronounced *thalates*) are found in cosmetics, deodorants, and many plastics (such as soft polyvinyl chloride, PVC) used for food packaging, children's toys, and medical devices. Some members of this chemical family are known to be toxic to laboratory animals, causing kidney and liver damage and possibly some cancers. In addition, many phthalates act as endocrine hormone disrupters and have been linked to reproductive abnormalities and decreased fertility in humans. A correlation has been found between phthalate levels in urine and low sperm numbers and decreased sperm motility in men. Nearly everyone in the United States has phthalates in their body at levels reported to cause these problems. While not yet conclusive, these results could help explain a 50-year decline in semen quality in most industrialized countries.

- Perchlorate is a waterborne contaminant left over from propellants and rocket fuels. About 12,000 sites in the United States were used by the military for live munition testing and are contaminated with perchlorate. Polluted water used to irrigate crops such as alfalfa and lettuce has introduced the chemical into the human food chain. Tests of cow's milk and human breast milk detected perchlorate in nearly every sample from throughout the United States. Perchlorate can interfere with iodine uptake in the thyroid gland, disrupting adult metabolism and childhood development.

- Atrazine is the most widely used herbicide in America. More than 60 million pounds of this compound are applied per year, mainly on corn and cereal grains, but also on golf courses, sugarcane, and Christmas trees. It has long been known to disrupt endocrine hormone functions in mammals, resulting in spontaneous abortions, low birth weights, and neurological disorders. Studies of families in corn-producing areas in the American Midwest have found higher rates of developmental defects among infants and certain cancers in families with elevated atrazine levels in their drinking water. University of California professor Tyrone Hayes has shown that atrazine levels as low as 0.1 ppb (30 times less than the EPA maximum contaminant level) caused severe reproductive effects in amphibians, including abnormal gonadal development and hermaphroditism. Atrazine now is found in rain and surface waters nearly everywhere in the United States at levels that could cause abnormal development in frogs. In 2003 the European Union withdrew regulatory approval for this herbicide, and several countries banned its use altogether. Some toxicologists have suggested a similar rule in the United States.

Everyone of us has dozens, if not hundreds, of persistent toxins in our body. This accumulation is called our **body burden.** We acquire it from our air, water, diet, and surroundings. Many of these toxins are present in parts per billion, or even parts per trillion. We don't know how dangerous this persistent burden is, but its presence is a matter of concern. If we're anything like the frogs that Tyrone Hayes studies, this accumulated dose of toxins may be a serious problem. Further discussion of POPs can be found in chapter 10.

Chemical interactions can increase toxicity

Some materials produce *antagonistic* reactions. That is, they interfere with the effects, or stimulate the breakdown of, other chemicals. For instance, vitamins E and A can reduce the response to some carcinogens. Other materials are *additive* when they occur together

in exposures. Rats exposed to both lead and arsenic show twice the toxicity of only one of these elements. Perhaps the greatest concern is synergistic effects. **Synergism** is an interaction in which one substance exacerbates the effects of another. For example, occupational asbestos exposure increases lung cancer rates 20-fold. Smoking increases lung cancer rates by the same amount. Asbestos workers who also smoke, however, have a 400-fold increase in cancer rates. How many other toxic chemicals are we exposed to that are below threshold limits individually but combine to give toxic results?

8.4 MECHANISMS FOR MINIMIZING TOXIC EFFECTS

A fundamental concept in toxicology is that every material can be poisonous under some conditions but most chemicals have some safe level or threshold below which their effects are undetectable or insignificant. Each of us consumes lethal doses of many chemicals over the course of a lifetime. One hundred cups of strong coffee, for instance, contain a lethal dose of caffeine. Similarly, one

---

Think About It

Some of the mechanisms that help repair wounds or fight off infections can result in cancer when we're old. Why hasn't evolution eliminated these processes?
Hint: Do conditions of postreproductive age affect natural selection?

---

hundred aspirin tablets, or 10 kilograms (22 lbs) of spinach or rhubarb, or a liter of alcohol would be deadly if consumed all at once. Taken in small doses, however, most toxins can be broken down or excreted before they do much harm. Furthermore, damage they cause can be repaired. Sometimes, however, mechanisms that protect us from one type of toxin or at one stage in the life cycle become deleterious with another substance or in another stage of development. Let's look at how these processes help protect us from harmful substances as well as how they can go awry.

Metabolic degradation and excretion eliminate toxins

Most organisms have enzymes that process waste products and environmental poisons to reduce their toxicity. In mammals, most of these enzymes are located in the liver, the primary site of detoxification of both natural wastes and introduced poisons. Sometimes, however, these reactions work to our disadvantage. Compounds, such as benzopyrene, for example, that are not toxic in their original form are processed by these same enzymes into cancer-causing carcinogens. Why would we have a system that makes a chemical more dangerous? Evolution and natural selection are expressed through reproductive success or failure. Defense mechanisms that protect us from toxins and hazards early in life are "selected for" by evolution. Factors or conditions that affect postreproductive ages (like cancer or premature senility) usually don't affect reproductive success or exert "selective pressure."

We also reduce the effects of waste products and environmental toxins by eliminating them from our body through excretion. Volatile molecules, such as carbon dioxide, hydrogen cyanide, and ketones, are excreted via breathing. Some excess salts and other substances are excreted in sweat. Primarily, however, excretion is a function of the kidneys, which can eliminate significant amounts of soluble materials through urine formation. Accumulation of toxins in the urine can damage this vital system, however, and the kidneys and bladder often are subjected to harmful levels of toxic compounds. In the same way, the stomach, intestine, and colon often suffer damage from materials concentrated in the digestive system and may be afflicted by diseases and tumors.

Repair mechanisms mend damage

In the same way that individual cells have enzymes to repair damage to DNA and protein at the molecular level, tissues and organs that are exposed regularly to physical wear and tear or to toxic or hazardous materials often have mechanisms for damage repair. Our skin and the epithelial linings of the gastrointestinal tract, blood vessels, lungs, and urogenital system have high cellular reproduction rates to replace injured cells. With each reproduction cycle, however, there is a chance that some cells will lose normal growth controls and run amok, creating a tumor. Thus any agent, such as smoking or drinking, that irritates tissues is likely to be carcinogenic. And tissues with high cell-replacement rates are among the most likely to develop cancers.

8.5 MEASURING TOXICITY

Almost 500 years ago the Swiss scientist Paracelsus said "the dose makes the poison," by which he meant that almost everything is toxic at some level. This remains the most basic principle of toxicology. Sodium chloride (table salt), for instance, is essential for human life in small doses. If you were forced to eat a kilogram of salt all at once, however, it would make you very sick. A similar amount injected into your bloodstream would be lethal. How a material is delivered—at what rate, through which route of entry, and in what medium—plays a vitally important role in determining toxicity.

This does not mean that all toxins are identical, however. Some are so poisonous that a single drop on your skin can kill you. Others require massive amounts injected directly into the blood to be lethal. Measuring and comparing the toxicity of various materials is difficult, because not only do species differ in sensitivity, but individuals within a species respond differently to a given exposure. In this section, we will look at methods of toxicity testing and at how results are analyzed and reported.

We usually test toxins on lab animals

The most commonly used and widely accepted toxicity test is to expose a population of laboratory animals to measured doses of a specific substance under controlled conditions. This procedure is expensive, time-consuming, and often painful and debilitating to the animals being tested. It commonly takes hundreds—or even

thousands—of animals, several years of hard work, and hundreds of thousands of dollars to thoroughly test the effects of a toxin at very low doses. More humane toxicity tests using computer simulation of model reactions, cell cultures, and other substitutes for whole living animals are being developed. However, conventional large-scale animal testing is the method in which we have the most confidence and on which most public policies about pollution and environmental or occupational health hazards are based.

In addition to humanitarian concerns, there are several problems in laboratory animal testing that trouble both toxicologists and policymakers. One problem is differences in sensitivity to a toxin of the members of a specific population. Figure 8.14 shows a typical dose/response curve for exposure to a hypothetical toxin. Some individuals are very sensitive to the toxin, while others are insensitive. Most, however, fall in a middle category forming a bell-shaped curve. The question for regulators and politicians is whether we should set pollution levels that will protect everyone, including the most sensitive people, or only aim to protect the average person. It might cost billions of extra dollars to protect a very small number of individuals at the extreme end of the curve. Is that a good use of resources?

Dose/response curves are not always symmetrical, making it difficult to compare toxicity of unlike chemicals or different species of organisms. A convenient way to describe toxicity of a chemical is to determine the dose to which 50 percent of the test population is sensitive. In the case of a lethal dose (LD), this is called the **LD50** (fig. 8.15).

Unrelated species can react very differently to the same toxin, not only because body sizes vary but also because of differences in physiology and metabolism. Even closely related species can have very dissimilar reactions to a particular toxin. Hamsters, for instance, are nearly 5,000 times less sensitive to some dioxins than are guinea pigs. Of 226 chemicals found to be carcinogenic in either rats or mice, 95 caused cancer in one species but not the other. These variations make it difficult to estimate the risks for

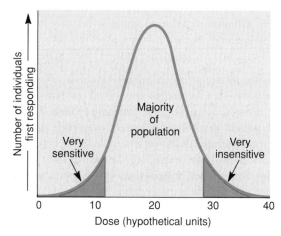

FIGURE 8.14 Probable variations in sensitivity to a toxin within a population. Some members of a population may be very sensitive to a given toxin, while others are much less sensitive. The majority of the population falls somewhere between the two extremes.

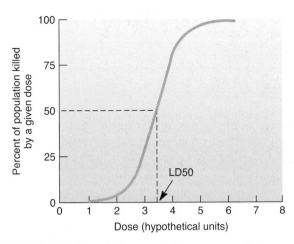

FIGURE 8.15 Cumulative population response to increasing doses of a toxin. The LD50 is the dose that is lethal to half the population.

humans, because we don't consider it ethical to perform controlled experiments in which we deliberately expose people to toxins.

Even within a single species there can be variations in responses between different genetic lines. A current controversy in determining the toxicity of bisphenol A (BPA) concerns the type of rats used for toxicology studies. Standard toxicology protocols call for a sturdy strain called the Sprague-Dawley rat. It turns out, however, that these animals, which were bred to grow fast and breed prolifically in lab conditions, are thousands of times less sensitive to endocrine disrupters than ordinary rats. Industry reports that declare BPA to be harmless based on Sprague-Dawley rats are highly suspect.

There is a wide range of toxicity

It is useful to group materials according to their relative toxicity. A moderate toxin takes about one gram per kilogram of body weight (about two ounces for an average human) to make a lethal dose. Very toxic materials take about one-tenth that amount, while extremely toxic substances take one-hundredth as much (only a few drops) to kill most people. Supertoxic chemicals are extremely potent; for some, a few micrograms (millionths of a gram—an amount invisible to the naked eye) make a lethal dose. These materials are not all synthetic. One of the most toxic chemicals known, for instance, is ricin, a protein found in castor bean seeds. It is so toxic that 0.3 billionths of a gram given intravenously will generally kill a mouse. If aspirin were this toxic, a single tablet, divided evenly, could kill 1 million people.

Many carcinogens, mutagens, and teratogens are dangerous at levels far below their direct toxic effect because abnormal cell growth exerts a kind of biological amplification. A single cell, perhaps altered by a single molecular event, can multiply into millions of tumor cells or an entire organism. Just as there are different levels of direct toxicity, however, there are different degrees of carcinogenicity, mutagenicity, and teratogenicity. Methanesulfonic acid, for instance, is highly carcinogenic, while the sweetener saccharin is a suspected carcinogen whose effects may be vanishingly small.

Acute and chronic doses and effects differ

Most of the toxic effects that we have discussed so far have been **acute effects.** That is, they are caused by a single exposure to the toxin and result in an immediate health crisis of some sort. Often, if the individual experiencing an acute reaction survives this immediate crisis, the effects are reversible. **Chronic effects,** on the other hand, are long-lasting, perhaps even permanent. A chronic effect can result from a single dose of a very toxic substance, or it can be the result of a continuous or repeated sublethal exposure.

We also describe long-lasting *exposures* as chronic, although their effects may or may not persist after the toxin is removed. It usually is difficult to assess the specific health risks of chronic exposures because other factors, such as aging or normal diseases, act simultaneously with the factor under study. Often very large populations of experimental animals are required, to obtain statistically significant results for low-level chronic exposures. Toxicologists talk about "megarat" experiments in which it might take a million rats to determine the health risks of some supertoxic chemicals at very low doses. Such an experiment would be terribly expensive for even a single chemical, let alone for the thousands of chemicals and factors suspected of being dangerous.

An alternative to enormous studies involving millions of animals is to give massive amounts—usually the maximum tolerable dose—of a toxin being studied to a smaller number of individuals and then to extrapolate what the effects of lower doses might have been. This is a controversial approach because it is not clear that responses to toxins are linear or uniform across a wide range of doses.

Figure 8.16 shows three possible results from low doses of a toxin. Curve (*a*) shows a baseline level of response in the population, even at zero dose of the toxin. This suggests that some other factor in the environment also causes this response. Curve (*b*) shows a straight-line relationship from the highest doses to zero exposure.

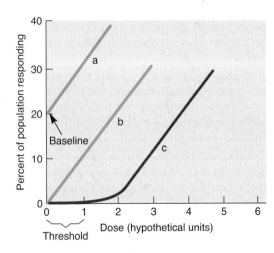

FIGURE 8.16 Three possible dose-response curves at low doses. (a) Some individuals respond, even at zero dose, indicating that some other factor must be involved. (b) Response is linear down to the lowest possible dose. (c) Threshold must be passed before any response is seen.

Many carcinogens and mutagens show this kind of response. Any exposure to such agents, no matter how small, carries some risks. Curve (c) shows a threshold for the response where some minimal dose is necessary before any effect can be observed. This generally suggests the presence of some defense mechanism that either prevents the toxin from reaching its target in an active form or repairs the damage that it causes. Low levels of exposure to the toxin in question may have no deleterious effects, and it might not be necessary to try to keep exposures to zero.

Which, if any, environmental health hazards have thresholds is an important but difficult question. The 1958 Delaney Clause to the U.S. Food and Drug Act forbids the addition of *any* amount of known carcinogens to food and drugs, based on the assumption that any exposure to these substances represents unacceptable risks. This standard was replaced in 1996 by a *no reasonable harm* requirement, defined as less than one cancer for every million people exposed over a lifetime. This change was supported by a report from the National Academy of Sciences concluding that synthetic chemicals in our diet are unlikely to represent an appreciable cancer risk. We will discuss risk analysis in the next section.

Detectable levels aren't always dangerous

You may have seen or heard dire warnings about toxic materials detected in samples of air, water, or food. A typical headline announced recently that 23 pesticides were found in 16 food samples. What does that mean? The implication seems to be that any amount of dangerous materials is unacceptable and that counting the numbers of compounds detected is a reliable way to establish danger. We have seen, however, that the dose makes the poison. It matters not only what is there, but how much, where it is located, how accessible it is, and who is exposed. At some level, the mere presence of a substance is insignificant.

Toxins and pollutants may seem to be more widespread now than in the past, and this is surely a valid perception for many substances. The daily reports we hear of new materials found in new places, however, are also due in part to our more sensitive measuring techniques. Twenty years ago, parts per million was generally the limit of detection for most chemicals. Anything below that amount was often reported as zero or absent rather than more accurately as undetected. A decade ago, new machines and techniques were developed to measure parts per billion. Suddenly chemicals were found where none had been suspected. Now we can detect parts per trillion or even parts per quadrillion in some cases. Increasingly sophisticated measuring capabilities may lead us to believe that toxic materials have become more prevalent. In fact, our environment may be no more dangerous; we are just better at finding trace amounts.

Low doses can have variable effects

A complication in assessing risk is that the effects of low doses of some toxics and health hazards can be nonlinear. They may be either more or less dangerous than would be predicted from exposure to higher doses. For example, low doses of BPA, discussed in the opening case study for this chapter, can have devastating health effects,

Think About It

Why might you and your mother rank some risks differently? List some activities on which the two of you might disagree.

whereas higher doses may shut down the response system and have little noticeable effect. On the other hand, very low amounts of radiation seem to be protective against certain cancers while higher doses are carcinogenic. It's thought now that very low radiation exposure may stimulate DNA repair along with enzymes that destroy free radicals (atoms with unpaired, reactive electrons in their outer shells). Activating these repair mechanisms may defend us from other, unrelated hazards. These nonlinear effects are called hormesis.

Another complication is that some substances can have long-lasting effects on genetic expression. For example, researchers found that exposure of pregnant rats to certain chemicals can have effects, not only on the exposed rats, but on their daughters and granddaughters. A single dose given on a specific day in pregnancy can be expressed several generations later, even if those offspring have never been exposed to the chemical (Exploring Science, p. 171).

8.6 RISK ASSESSMENT AND ACCEPTANCE

Risk is the possibility of suffering harm or loss. **Risk assessment** is the scientific process of estimating the threat that particular hazards pose to human health. This process includes risk identification, dose response assessment, exposure appraisal, and risk characterization. In hazard identification, scientists evaluate all available information about the effects of a toxin to estimate the likelihood that a chemical will cause a certain effect in humans. The best evidence comes from human studies, such as physician case reports. Animal studies are also used to assess health risks. Risk assessment for identified toxicity hazards (for example, lead) includes collection and analysis of site data, development of exposure and risk calculations, and preparation of human health and ecological impact reports. Exposure assessment is the estimation or determination of the magnitude, frequency, duration, and route of exposure to a possible toxin. Toxicity assessment weighs all available evidence and estimates the potential for adverse health effects to occur.

Risk perception isn't always rational

A number of factors influence how we perceive relative risks associated with different situations.

- People with social, political, or economic interests—including environmentalists—tend to downplay certain risks and emphasize others that suit their own agendas. We also tend to tolerate risks that we choose—such as driving, smoking, or overeating—while objecting to risks we cannot control—such as potential exposure to slight amounts of toxic substances.

Could your diet, behavior, or environment affect the lives of your children or grandchildren? For a century or more, scientists assumed that the genes you receive from your parents irreversibly fix your destiny and that factors such as stress, habits, exposure to toxins, or parenting should have no effect on future generations.

Now, however, some startling discoveries are making us reexamine those assumptions. Scientists are finding that a complex set of chemical markers and genetic switches—called the epigenome—consisting of DNA and its associated proteins and other small molecules, regulates gene function in ways that can affect multiple generations. Epi- means "above," and the epigenome is above ordinary genes in the same sense that managers are "above" production workers in a factory. Understanding how this system works helps us see how many environmental factors affect health, and may become useful in treating a variety of diseases.

One of the most striking experiments on the epigenome was carried out a decade ago by researchers at Duke University. They were studying the affects of diet on a strain of mice carrying an agouti gene that makes them obese, yellow, and prone to cancer and diabetes. Starting just before conception, mother agouti mice were fed a diet rich in B vitamins (folic acid and B_{12}). Amazingly, this simple dietary change resulted in baby mice that were sleek, brown, and healthy. Eating a special diet had somehow turned off the agouti gene in the offspring.

We know now that B vitamins as well as some vegetables, such as onions, garlic, and beets, are methyl donors—that is, they can add a carbon atom and three hydrogens to proteins and nucleic acids. Attaching an extra methyl group can switch genes either on or off by changing the way proteins and nucleic acids read and translate the DNA. Similarly, acetylating DNA (addition of an acetyl group: CH_3CO) can also either stimulate or inhibit gene expression.

These reactions work not only directly on genes themselves, but also on a huge set of what we once thought was useless, or junk, DNA in chromosomes as well as a large amount of protein that once seemed to be merely packing material. We now know that both this extra DNA and the protein around which genes are wrapped play vital roles in controlling gene expression. And methylating or acetylating these proteins or nucleic acids also can have profound effects on heredity.

More remarkable is that changes in the epigenome can carry through multiple generations. In 2004 Michael Skinner, a geneticist at Washington State University, was studying the effects of exposure to a commonly used fungicide on rats. He found that male rats exposed in utero had lower sperm counts later in life. It took only a single exposure to cause this effect. Amazingly, the effect lasted for at least four generations, even though those subsequent offspring were never exposed to the fungicide.

The way a mother rodent nurtures her young also can cause changes in methylation patterns in their babies' brains that are quite similar to the prenatal vitamins and nutrients that affected the agouti gene. It's thought that licking and grooming

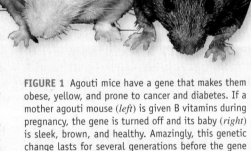

FIGURE 1 Agouti mice have a gene that makes them obese, yellow, and prone to cancer and diabetes. If a mother agouti mouse (*left*) is given B vitamins during pregnancy, the gene is turned off and its baby (*right*) is sleek, brown, and healthy. Amazingly, this genetic change lasts for several generations before the gene resumes its deleterious effects.

activates serotonin receptors that turn on genes to reduce stress responses, resulting in profound brain changes. In another study, rats given extra attention, diet, and mental stimulation (toys) did better at memory tests than did environmentally deprived controls. Altered methylation patterns in the hippocampus—the part of the brain that controls memory—were detected in both these cases. And subsequent generations maintained this methylation.

Epigenetic effects have also been found in humans. One of the most compelling studies involved comparison of two centuries of health records, climate, and food supply in a remote village in northern Sweden. The village of Overkalix was so

isolated that when bad weather caused crop failures, famine struck everyone. In good years, on the other hand, there was plenty of food and people stuffed themselves. A remarkable pattern emerged. When other social factors were factored in, grandfathers who were preteens during lean years had grandsons who lived an amazing 32 years longer than those whose grandfathers were able to gorge themselves as preteens. Similarly, women whose mothers had access to a rich diet while they were pregnant were much more likely to have daughters and granddaughters with health problems and shortened lives.

In an another surprising human health study, researchers found in a long-term study of couples in Bristol, England, that fathers who started smoking before they were 11 years old (just as they were starting puberty and sperm formation was beginning) were much more likely than nonsmokers to have sons and grandsons who were overweight and lived significantly shortened lives. Both these results are attributed to epigenetic effects.

A wide variety of factors can cause epigenetic changes. Smoking, for example, leaves a host of persistent methylation markers in your DNA. So does exposure to a number of pesticides, toxics, drugs, and stressors. At the same time, it is possible to help prevent deleterious methylations by getting plenty of polyphenols in green tea and deeply colored fruit, B vitamins, and garlic, onions, turmeric, and other healthy foods. Not surprisingly, epigenetic changes are implicated in many cancers, including colon, prostate, breast, and blood cancers. This may explain many confusing cases in which our environment seems to have long-lasting effects on health and development that can't be explained by ordinary metabolic effects.

Unlike mutations, epigenetic changes aren't permanent. Eventually the epigenome returns to normal if the exposure wasn't repeated. This makes epigenetic changes candidates for drugs. Currently the Food and Drug Administration has approved two drugs, Vidaza and Dacogen, that inhibit methylation and are used to treat a precursor to leukemia. Another drug, Zolinza, which enhances acetylation, is approved to treat another form of leukemia. Dozens of other drugs that may treat a variety of diseases, including rheumatoid arthritis, neurodegenerative diseases, and diabetes, are under development.

So your diet, behavior, and environment can have a much stronger impact on both your health and that of your descendents than we previously understood. What you ate, drank, smoked, or did last night may have profound effects on future generations.

- Most people have difficulty understanding and believing probabilities. We feel that there must be patterns and connections in events, even though statistical theory says otherwise. If the coin turned up heads last time, we feel certain that it will turn up tails next time. In the same way, it is difficult to understand the meaning of a 1-in-10,000 risk of being poisoned by a chemical.

- Our personal experiences often are misleading. When we have not personally experienced a bad outcome, we feel it is more rare and unlikely to occur than it actually may be. Furthermore, the anxieties generated by life's gambles make us want to deny uncertainty and to misjudge many risks (fig. 8.17).

- We have an exaggerated view of our own abilities to control our fate. We generally consider ourselves above-average drivers, safer than most when using appliances or power tools, and less likely than others to suffer medical problems, such as heart attacks. People often feel they can avoid hazards because they are wiser or luckier than others.

- News media give us a biased perspective on the frequency of certain kinds of health hazards, overreporting some accidents or diseases while downplaying or underreporting others. Sensational, gory, or especially frightful causes of death, like murders, plane crashes, fires, or terrible accidents, occupy a disproportionate amount of attention in the public media. Heart diseases, cancer, and stroke kill nearly 15 times as many people in the United States as do accidents and 75 times as many people as do homicides, but the emphasis placed by the media on accidents and homicides is nearly inversely proportional to their relative frequency compared to either cardiovascular disease or cancer. This gives us an inaccurate picture of the real risks to which we are exposed.

- We tend to have an irrational fear or distrust of certain technologies or activities that leads us to overestimate their dangers. Nuclear power, for instance, is viewed as very risky, while coal-burning power plants seem to be familiar and relatively benign; in fact, coal mining, shipping, and combustion cause an estimated 10,000 deaths each year in the United States. An old, familiar technology seems safer and more acceptable than does a new, unknown one (Data Analysis, p. 176).

FIGURE 8.17 How dangerous are motorcycles? Many parents regard them as extremely risky, while many students—especially males—believe the risks (which are about the same as your chances of dying from surgery or other medical care) are acceptable. Perhaps the more important question is whether the benefits outweigh the risks.

Risk acceptance depends on many factors

How much risk is acceptable? How much is it worth to minimize and avoid exposure to certain risks? Most people will tolerate a higher probability of occurrence of an event if the harm caused by that event is low. Conversely, harm of greater severity is acceptable only at low levels of frequency. A 1-in-10,000 chance of being killed might be of more concern to you than a 1-in-100 chance of being injured. For most people, a 1-in-100,000 chance of dying from some event or some factor is a threshold for changing what we do. That is, if the chance of death is less than 1 in 100,000, we are not likely to be worried enough to change our ways. If the risk is greater, we will probably do something about it. The Environmental Protection Agency generally assumes that a risk of 1 in 1 million is acceptable for most environmental hazards. Critics of this policy ask, acceptable to whom?

For activities that we enjoy or find profitable, we are often willing to accept far greater risks than this general threshold. Conversely, for risks that benefit someone else we demand far higher protection. For instance, your chance of dying in a motor vehicle accident in any given year is about 1 in 5,000, but that doesn't deter many people from riding in automobiles. Your lifetime chance of dying from lung cancer if you smoke one pack of cigarettes per day is about 1 in 4. By comparison, the risk from drinking water with the EPA limit of trichloroethylene is about 1 in 10 million. Strangely, many people demand water with zero levels of trichloroethylene, while continuing to smoke cigarettes.

Table 8.3 lists lifetime odds of dying from a few leading diseases and accidents. These are statistical averages, of course, and there clearly are differences in where one lives or how one behaves that affect the danger level of these activities. Although the average lifetime chance of dying in an automobile accident is 1 in 100, there clearly are things you can do—like wearing a seat belt, following safety rules, and avoiding risky situations—that improve your odds. Still, it is interesting how we readily accept some risks while shunning others.

Our perception of relative risks is strongly affected by whether risks are known or unknown, whether we feel in control of the outcome, and how dreadful the results are. Risks that are unknown or unpredictable and results that are particularly gruesome or disgusting seem far worse than those that are familiar and socially acceptable.

Studies of public risk perception show that most people react more to emotion than to statistics. We go to great lengths to avoid some dangers while gladly accepting others. Factors that are involuntary, unfamiliar, undetectable to those exposed, or catastrophic, or that have delayed effects or are a threat to future generations, are especially feared, whereas those that are voluntary, familiar, detectable, or immediate cause less anxiety. Even though the actual number of deaths from automobile accidents, smoking, or alcohol, for instance, are thousands of times greater than those from pesticides, nuclear energy, or genetic engineering, the latter preoccupy us far more than the former.

Table 8.3	Lifetime Chances of Dying in the United States
Source	**Odds (1 in x)**
Heart disease	2
Cancer	3
Smoking	4
Lung disease	15
Pneumonia	30
Automobile accident	100
Suicide	100
Falls	200
Firearms	200
Fires	1,000
Airplane accident	5,000
Jumping from high places	6,000
Drowning	10,000
Lightning	56,000
Hornets, wasps, bees	76,000
Dog bite	230,000
Poisonous snakes, spiders	700,000
Botulism	1 million
Falling space debris	5 million
Drinking water with EPA limit of trichloroethylene	10 million

Source: U.S. National Safety Council, 2003.

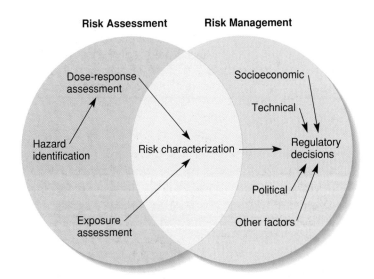

FIGURE 8.18 Risk assessment organizes and analyzes data to determine relative risk. Risk management sets priorities and evaluates relevant factors to make regulatory decisions.

8.7 ESTABLISHING HEALTH POLICY

Risk management combines principles of environmental health and toxicology together with regulatory decisions based on socioeconomic, technical, and political considerations (fig. 8.18). The biggest problem in making regulatory decisions is that we are usually dealing with many sources of harm to which we are exposed, often without being aware of them. It is difficult to separate the effects of all these different hazards and to evaluate their risks accurately, especially when the exposures are near the threshold of measurement and response. In spite of often vague and contradictory data, public policymakers must make decisions.

A current, highly contentious debate surrounds the Endocrine Disrupter Screening Program. In 1996, Congress ordered the U.S. EPA to start testing 87,000 chemicals for their ability to disrupt endocrine hormone functions that regulate almost every aspect of reproduction, growth, development, and functioning of our bodies. After 13 years of study and a $76 million budget, not a single chemical has been declared an endocrine disrupter. Some toxicologists argue that the exposure times may be too short, or the test animals may have been exposed to other chemicals besides the ones being studied. One particularly controversial issue is that an albino rat breed called the Sprague-Dawley is stipulated in these tests. These rats are unusually hardy and fertile. In fact, they were originally bred to be resistant to arsenic trioxide pesticides, and they may be unnaturally resistant to endocrine disrupters as well. Meanwhile, the chemical industry hotly disputes the need for any endocrine testing at all. As you can see, establishing public policy isn't simple.

In setting standards for environmental toxins, we need to consider (1) combined effects of exposure to many different sources of damage, (2) different sensitivities of members of the population, and (3) effects of chronic as well as acute exposures. Some people argue that pollution levels should be set at the highest amount that does *not* cause measurable effects. Others demand that pollution be reduced to zero if possible, or as low as is technologically feasible. It may not be reasonable to demand that we be protected from every potentially harmful contaminant in our environment, no matter how small the risk. As we have seen, our bodies have mechanisms that enable us to avoid or repair many kinds of damage so that most of us can withstand some minimal level of exposure without harm (fig. 8.19).

On the other hand, each challenge to our cells by toxic substances represents stress on our bodies. Although each individual stress may not be life-threatening, the cumulative effects of all the environmental stresses, both natural and human-caused, to which we are exposed may seriously shorten or restrict our lives. Furthermore, some individuals in any population are more susceptible to those stresses than others. Should we set pollution standards so that no one is adversely affected, even the most sensitive individuals, or should the acceptable level of risk be based on the average member of the population?

Finally, policy decisions about hazardous and toxic materials also need to be based on information about how such materials affect the plants, animals, and other organisms that define and maintain our environment. In some cases, pollution can harm or

FIGURE 8.19 "Do you want to stop reading those ingredients while we're trying to eat?"

Table 8.4 Relative Risks to Human Welfare
Relatively High-Risk Problems
Habitat alteration and destruction
Species extinction and loss of biological diversity
Stratospheric ozone depletion
Global climate change
Relatively Medium-Risk Problems
Herbicides/pesticides
Toxics and pollutants in surface waters
Acid deposition
Airborne toxics
Relatively Low-Risk Problems
Oil spills
Groundwater pollution
Radionuclides
Thermal pollution

Source: Environmental Protection Agency.

destroy whole ecosystems with devastating effects on the life-supporting cycles on which we depend. In other cases, only the most sensitive species are threatened. Table 8.4 shows the Environmental Protection Agency's assessment of relative risks to human welfare. This ranking reflects a concern that our exclusive focus on reducing pollution to protect human health has neglected risks to natural ecological systems. While there have been many benefits from a case-by-case approach in which we evaluate the health risks of individual chemicals, we have often missed broader ecological problems that may be of greater ultimate importance.

CONCLUSION

We have made marvelous progress in reducing some of the worst diseases that have long plagued humans. Smallpox is the first major disease to have been completely eliminated. Guinea worms and polio are nearly eradicated worldwide; typhoid fever, cholera, yellow fever, tuberculosis, mumps, and other highly communicable diseases are rarely encountered in advanced countries. Childhood mortality has decreased 90 percent globally, and people almost everywhere are living twice as long, on average, as they did a century ago.

But the technological innovations and affluence that have diminished many terrible diseases have also introduced new risks. Chronic conditions, such as cardiovascular disease, cancer, depression, dementia, diabetes, and traffic accidents, that once were confined to richer countries have now become leading health problems nearly everywhere. Part of this change is that we no longer die at an early age of infectious disease, so we live long enough to develop the infirmities of old age. Another factor is that affluent lifestyles, lack of exercise, and unhealthy diets aggravate these chronic conditions.

New emergent diseases are appearing at an increasing rate. With increased international travel, diseases can spread around the globe in a few days. Epidemiologists warn that the next deadly epidemic may be only a plane ride away. In addition, modern industry is introducing thousands of new chemical substances every year, most of which aren't studied thoroughly for health effects. Endocrine disrupters, neurotoxins, carcinogens, mutagens, teratogens, and other toxins, sometimes even at very low levels, can have tragic outcomes. BPA's role in a wide variety of chronic health effects is an example of how materials we have introduced can have unintended consequences. Many other industrial chemicals could be having similar harmful effects.

Determining what levels of environmental health risk are acceptable is difficult. We are exposed to many different health threats simultaneously. Furthermore, people consider some dangers tolerable, but dread others—especially those that are new, involuntary, and difficult to detect and whose effects are unknown to science. The situation is complicated by the fact that news media gives us a biased perspective on some hazards, while our personal experiences and our sense of our own abilities are often misleading.

There are many steps that each of us can take to protect our health. Eating a healthy diet, exercising regularly, drinking in moderation, driving prudently, and practicing safe sex are among the most important.

REVIEWING LEARNING OUTCOMES

By now you should be able to explain the following points:

8.1 Describe health and disease and how the global disease burden is changing.

- The leading causes of global disease are changing from infectious diseases to chronic conditions associated with an affluent lifestyle.
- However, infectious and emergent diseases still kill millions of people.
- Conservation medicine combines ecology and health care.
- Resistance to drugs, antibiotics, and pesticides is increasing.
- Who should pay for health care?

8.2 Summarize the principles of toxicology.

- How do toxins affect us?
- How does diet influence health?

8.3 Discuss the movement, distribution, and fate of toxins in the environment.

- Solubility and mobility determine where and when chemicals move.
- Exposure and susceptibility determine how we respond.

- Bioaccumulation and biomagnification increase the concentrations of chemicals.
- Persistence makes some materials a greater threat.
- Chemical interactions can increase toxicity.

8.4 Characterize mechanisms for minimizing toxic effects.

- Metabolic degradation and excretion eliminate toxins.
- Repair mechanisms mend damage.

8.5 Explain ways we measure and describe toxicity.

- We usually test toxins on lab animals.
- There is a wide range of toxicity.
- Acute and chronic doses and effects differ.
- Detectable levels aren't always dangerous.

8.6 Evaluate risk assessment and acceptance.

- Risk perception isn't always rational.
- Risk acceptance depends on many factors.

8.7 Relate how we establish health policy.

PRACTICE QUIZ

1. What is BPA and how might you be exposed to it?
2. Define the terms *disease* and *health*.
3. What were some of the most serious diseases in the world in 1990? How is this list expected to change in the next 20 years?
4. What are emergent diseases? Give a few examples, and describe their cause and effects.
5. How do bacteria acquire antibiotic resistance? How might we prevent this?
6. What is the difference between toxic and hazardous? Give some examples of materials in each category.
7. How do the physical and chemical characteristics of materials affect their movement, persistence, distribution, and fate in the environment?
8. What is the difference between acute and chronic toxicity?
9. Define *carcinogenic, mutagenic, teratogenic,* and *neurotoxic.*
10. What are the relative risks of smoking, driving a car, and drinking water with the maximum permissible levels of trichloroethylene? Are these relatively equal risks?

CRITICAL THINKING AND DISCUSSION QUESTIONS

1. What consequences (positive or negative) do you think might result from defining health as a state of complete physical, mental, and social well-being? Who might favor or oppose such a definition?
2. Do rich countries bear any responsibilities if the developing world adopts unhealthy lifestyles or diets? What could (or should) we do about it?
3. Why do we spend more money on heart or cancer research than childhood illnesses?
4. What are the premises in the discussion of assessing risk? Could conflicting conclusions be drawn from the facts presented in this section? What is your perception of risk from your environment?
5. Should pollution levels be set to protect the average person in the population or the most sensitive? Why not have zero exposure to all hazards?
6. What level of risk is acceptable to you? Are there some things for which you would accept more risk than others?

Is it possible to show relationships between two dependent variables on the same graph? Sometimes that's desirable when you want to make comparisons between them. The graph shown here does just that. It's a description of how people perceive different risks. We judge the severity of risks based on how familiar they are and how much control we have over our exposure. The Y-axis represents how mysterious, unknown, or delayed the risk seems to be. Things that are unobservable, unknown to those exposed, delayed in their effects, and unfamiliar or unknown to science tend to be more greatly feared than those that are observable, known, immediate, familiar, and known to science. The X-axis represents a measure of dread, which combines how much control we feel we have over the risk, how terrible the results could potentially be, and how equitably the risks are distributed. The size of the symbol for each risk indicates the combined effect of these two variables. Notice that things such as DNA technology or nuclear waste, which have high levels of both mystery and dread, tend to be regarded with the greatest fear, while familiar, voluntary, personally rewarding behaviors, such as riding in automobiles or on bicycles, or drinking alcohol, are thought to be relatively minor risks. Actuarial experts (statisticians who gather mortality data) would tell you that automobiles, bicycles, and alcohol have killed far more people (so far) than DNA technology or radioactive waste. But this isn't just a question of data. It's a reflection of how much we fear various risks. Notice that this is a kind of scatter plot, mapping categories of data that have no temporal sequence. Still, you can draw some useful inferences from this sort of graphic presentation.

Compare the graph with table 8.3.

1. What is the highest risk factor in table 8.4 that also appears on the perception of risk graph?

2. How do the lifetime risks for smoking and auto accidents calculated by the National Safety Council correspond to the perception of risk in this graph?

3. How do airplane accidents compare in these two assessments?

4. Why is it that DNA technology and radioactive waste appear to generate so much dread, and yet they don't even appear in table 8.3 (lifetime chances of dying)?

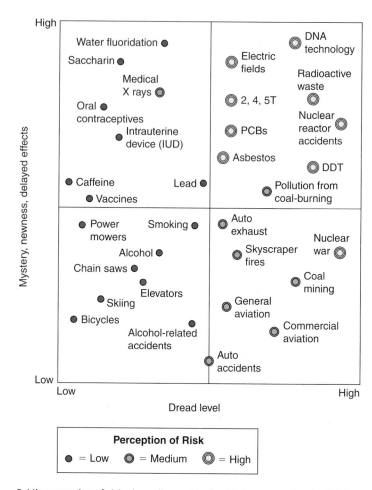

Public perception of risk, depending on the familiarity, apparent potential for harm, and personal control over the risk.

Source: Data from Slovic, Paul, 1987. Perception of Risk, *Science*, 236 (4799): 286–90.

5. Thousands of people die every year from alcohol-related accidents, yet alcohol ranks very low in the two scales shown in this graph. On the other hand, no one has ever died—as far as we know—from DNA technology. Why the great discrepancy in rankings?

Fresh, local, and organic foods can be hard for college students to find on campus.

Food and Hunger

Learning Outcomes

After studying this chapter, you should be able to:

9.1 Describe patterns of world hunger and nutritional requirements.

9.2 Identify key food sources, including protein-rich foods.

9.3 Discuss how policy can affect food resources.

9.4 Explain new crops and genetic engineering.

"It ain't the things we know that cause all the trouble; it's the things we think we know that ain't so."

~ *Will Rogers*

Case Study

Becoming a Locavore in the Dining Hall

Many people care about good food, and sometimes a factor that makes food satisfying is knowing that it's supporting the local farm economy or that it's sustainably grown. But if you're a university or college student, and if you eat most of your meals in a cafeteria, how much control can you have over where your food comes from? Most cafeteria food seems to originate in a large freezer truck at the loading dock behind the dining hall. Any farm behind those boxes of frozen fries and hamburgers is far away and hard to imagine. If you eat from a cafeteria, even if you have time to care about sustainable or healthy foods, you might feel you have very little power to do anything about it.

At a growing number of colleges and universities, students are speaking up about becoming local eaters—one term for the idea is "**locavore**." One of these schools is Vassar College, in New York's Hudson River Valley. Vassar students have started asking the dining service to provide local foods because they're concerned about the local farm economy, because they want pesticide-free and hormone-free foods, and because they're worried about the environmental costs of foods that travel thousands of miles from the farm to the table.

As more students have pushed for local meals, they have empowered food service managers to take the time to work with local producers. Maureen King, the director of campus dining, and her colleague Ken Oldehoff have found local sources of tomato sauce, salsa, squash, fruit, milk, yogurt, fresh produce, juice, desserts, soups, and other foods. Local cider and milk are available in the cafeteria line along with soft drinks. Fresh apples and other fruit are promoted in the fall. King and Oldehoff have initiated pie bake-offs, pumpkin carving events, sauce challenges, even a Local Foods Week that allows students to eat entirely locally all week.

The menu includes more squash, tomatoes, and beets than the standard cafeteria food service truck delivers, but college chefs enjoy using fresh foods, and students keep asking for more local products. There are still plenty of nonlocal alternatives, such as fresh strawberries and melons in midwinter. And of course coffee (from Mexico). But for several years now this coffee has been shade-grown, organic, fair-trade coffee—and it doesn't cost much more at the checkout line than the previous nonorganic coffees did.

Local sourcing isn't always easy. Vassar is located in a region of rapidly growing suburbs, with a struggling farm economy, and Oldehoff and King work constantly on finding new relationships with growers. Sometimes their suppliers give up and go out of business; other times new opportunities suddenly appear. Ordering everything from a single national distributor is usually easier and cheaper than ordering from a miscellaneous group of local growers. But every purchase makes a difference to farmers, and sometimes the college's commitment helps to keep local businesses solvent.

FIGURE 9.1 Ken Oldehoff and his colleagues have been creative and persistent in bringing local foods to college students.

What inspires Maureen King and Ken Oldehoff to take this extra effort? They are concerned about serving healthy food, including lots of vegetables and organic food when possible. They like knowing their suppliers and knowing how food was handled. They also live in the community, and they are happy to put dollars into local pockets, and to help protect the Hudson River Valley's historic agrarian landscape.

King and Oldehoff also worry about the carbon footprint of the college, so they want to minimize the amount of food they buy from the far side of the continent. If a portion of the college's chicken or tomato sauce can be grown just down the road, then why not try to buy them locally?

These are good justifications, but local purchasing wouldn't go anywhere without student interest. It's not necessary for every student to be an organic vegan locavore, but if a few are willing to stand up and show they care, their voices can shift college policy. Many other institutions have exclusive contracts with national suppliers, which restrict purchasing from local suppliers. Partly because of student action, and partly because of the persistence of King and Oldehoff, Vassar's contracts allow for local purchases. The college has also discovered that good environmental citizenship makes good press and generates good feelings about the institution. As Ken Oldehoff notes, committing to local food is one way student activists can make real change.

Local eating is not just a local concern. There are important connections between local consumption and global patterns of food availability, hunger, and nutrition. We can learn a great deal about global food issues by thinking more carefully about what it would take to become a locavore. In this chapter we'll think about those connections. We'll look at some of the different kinds of foods we eat; at how much we eat in wealthier countries, compared to poorer ones; and at how food production differs from one area to another. As you read, think about how these global issues help explain why Ken Oldehoff and Maureen King work as hard as they do to procure local foods, and why the students keep asking for them.

To learn more, go to http://www.foodroutes.org/farmtocollege.jsp. For related resources, including Google Earth™ placemarks that show locations discussed in this chapter, visit http://EnvironmentalScience-Cunningham.blogspot.com.

9.1 WORLD FOOD AND NUTRITION

Despite repeated predictions that runaway population growth would lead to terrible famines (chapter 7), world food supplies have more than kept up with increasing human numbers over the past two centuries. Although population growth slowed to an average 1.7 percent per year during that time, world food production increased an average of 2.2 percent. Increased use of irrigation and fertilizers, improved crop varieties, and better distribution systems have improved the nutrition of billions of people. Globally we consume an average of 3,000 kcal per day, well above the 2,200 kcal considered necessary for a healthy and productive life. Residents of industrialized countries consume an average of 4,000 kcal every day.

The UN Food and Agriculture Organization (FAO) expects world food supplies to continue to increase faster than population growth. In some countries, such as the United States, the problem has long been what to do with surplus food. High production leads to low prices, which make farm profits chronically low. Farmers in these countries are paid billions of dollars per year to keep land out of production. Studies indicate that we still have room to expand farmland and increase production to feed 9 billion people in a few decades, although this may require sacrificing forests and water resources. The main questions may not be whether we can produce enough food, but rather how do we improve access to food, and what are the environmental costs of our food production systems?

Millions of people are still chronically hungry

Despite bountiful production, more than 850 million people in the world today are considered **chronically undernourished**, getting less than the minimum 2,200 kcal on average. This number represents a tragic and persistent problem. Moreover, the number has risen in recent years, after steady decline in the 1980s and early 1990s (fig. 9.2*a*). Much of this increase has occurred in sub-Saharan Africa

and South Asia—which includes India, the second most populous country on earth. Over 95 percent of malnourished people live in these and other developing countries (fig. 9.3). In some regions, on the other hand, progress has been substantial. China has reduced its number of undernourished people by over 75 million in the past decade. Indonesia, Vietnam, Thailand, Nigeria, Ghana, and Peru each reduced chronic hunger by about 3 million people.

Because the global population is growing overall, the *proportion* of people who are hungry is declining. In 1960 nearly 60 percent of people in developing countries were chronically undernourished. Today that proportion has fallen to just 16 percent. For all countries together, the world's undernourished population has declined slightly in number, but the proportion has fallen from 37 percent to 13 percent (fig. 9.2*b*).

Still, poverty threatens **food security**, or the ability to obtain sufficient food on a day-to-day basis. The 1.5 billion people in the world who live on less than $1 per day often can't afford to buy the food they need and lack resources to grow it for themselves. Food security is a concern at multiple scales. In the poorest countries, hunger may affect nearly everyone. In other countries, although the average food availability may be satisfactory, some individual communities or families may not have enough to eat. And within families, males often get both the largest share and the most nutritious food, while women and children—who need food most—all too often get the poorest diet. At least 6 million children under 5 years old die every year (one every 5 seconds) from hunger and malnutrition. Providing a healthy diet might eliminate as much as 60 percent of all premature deaths worldwide.

Hungry people can't work their way out of poverty. Nobel Prize–winning economist Robert Fogel estimates that in 1790 about 20 percent of the population of England and France was effectively excluded from the labor force because they were too weak and hungry to work. Improved nutrition, he calculates, accounted for about half of all European economic growth during the nineteenth century. Because many developing countries are as poor now (in relative terms) as Britain and France were in 1790, his

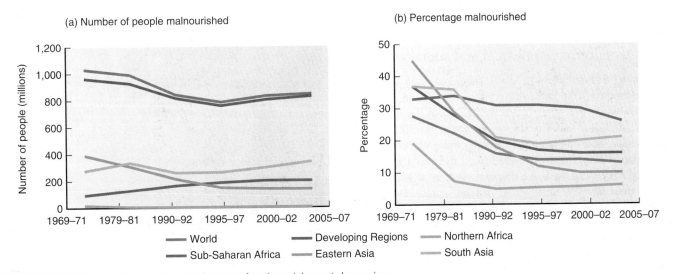

FIGURE 9.2 Changes in numbers and rates of malnourishment, by region.
Source: Data from the UN Food and Agriculture Organization, 2011.

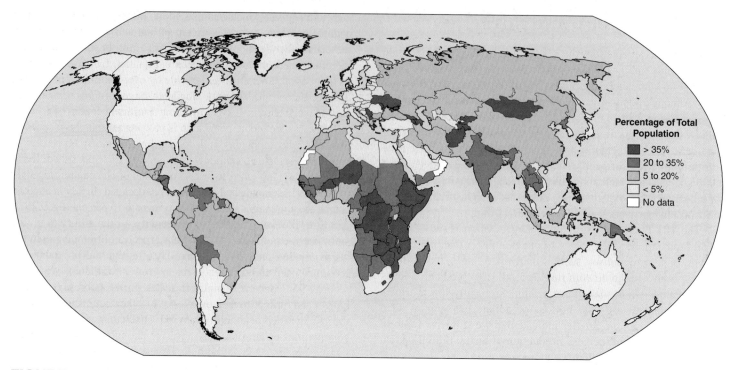

FIGURE 9.3 Hunger around the world. In 2008 the United Nations reported that about 850 million people—830 million of them in developing countries—suffered from chronic hunger and malnutrition. Africa has the largest number of countries with food shortages.
Source: Hunger map from Food and Agriculture Organization of the United Nations website. Used by permission.

analysis suggests that reducing hunger could yield more than (U.S.) $120 billion in economic growth produced by longer, healthier, more productive lives for several hundred million people.

Recognizing the role of women in food production is an important step toward food security for all. Throughout the developing world, women do 50 to 70 percent of all farmwork but control only a tiny fraction of the land and rarely have access to capital or developmental aid. In Nigeria, for example, home gardens occupy only 2 percent of all cropland but provide half the food families eat. Making land, credit, education, and access to markets available to women could contribute greatly to family nutrition.

Famines usually have political and social causes

Chronic hunger and malnutrition can be silent and often invisible, affecting individuals, families, and communities on an ongoing basis. **Famines**, on the other hand, are characterized by large-scale food shortages, massive starvation, social disruption, and economic chaos. Starving people, especially those uprooted from their farms and villages, may be forced to eat their seed grain and slaughter their breeding stock in a desperate attempt to keep themselves and their families alive. Even if better conditions return, they often have sacrificed their productive capacity or lost their land, and recovery will be slow and difficult. Famines are characterized by mass migrations as starving people travel to refugee camps in search of food and medical care (fig. 9.4).

In 2006 the FAO reported that 58 million people in 36 countries (two-thirds of them in sub-Saharan Africa) needed emergency food aid. What causes these emergencies? Droughts, earthquakes, severe storms, and other natural disasters are often the immediate trigger, but politics and economics are often equally important. Bad weather, insect outbreaks, and other environmental factors cause crop failures and create food shortages. But the Nobel Prize-winning work of Harvard economist Amartya K. Sen shows that these factors have often been around for a long time, and local people usually have

FIGURE 9.4 Children wait for their daily ration of porridge at a feeding station in Somalia. When people are driven from their homes by hunger or war, social systems collapse, diseases spread rapidly, and the situation quickly becomes desperate.

adaptations to get through hard times if they aren't thwarted by inept or corrupt governments and greedy elites. National politics, however, together with commodity hoarding, price gouging, poverty, wars, landlessness, and other external factors often make it impossible for poor people to grow their own food or find jobs to earn money to buy the food they need. Professor Sen points out that armed conflict and political oppression almost always are at the root of famine. No democratic country with a relatively free press, he says, has ever had a major famine.

The aid policies of rich countries often don't help as much as we hope. Despite our best intentions, aid often serves as a way to get rid of surplus commodities rather than to stabilize local food production in recipient countries. Even emergency food aid has ambiguous effects. Herding people into feeding camps can badly destabilize communities, and crowding and lack of sanitation in the camps exposes people to epidemic diseases. There are no jobs in the refugee camps, so people can't support themselves. Corruption and violence can occur at food dispensing centers, where aid recipients are highly vulnerable. Having left their land and tools behind, people may have difficulty returning to their farms when conditions return to normal.

Overeating is a growing world problem

Although hunger persists, world food supplies are increasing. This is good news, but the downside is increasing overweight and obese populations. In the United States, and increasingly in developing countries, highly processed foods rich in sugars and fats have become a large part of daily diets. Some 64 percent of adult Americans are overweight, up from 40 percent only a decade ago. About one-third of us are seriously overweight, or **obese**. Obesity is quantified in terms of the body mass index (BMI), calculated as weight/height^2. For example, a person weighing 100 kg and 2 m tall (220 lb and 6 ft 6 inches) would have a body mass of ($100 \text{ kg}/4\text{m}^2$) or 25 kg/m^2. Health officials consider a BMI greater than 25 kg/m^2 overweight; over 30 kg/m^2 is considered obese.

Globally, nearly 2 billion adults (15 and older) are overweight, according to a 2011 Worldwatch study. This number represents 38 percent of the world's adult population. More than twice as many people are overweight than underweight (850 million). About 10 percent of adults are obese (BMI greater than 30 kg/m^2). This trend is no longer limited to richer countries. Obesity is spreading around the world as Western diets and lifestyles are increasingly adopted in the developing world (fig. 9.5).

Being overweight substantially increases risk of hypertension, diabetes, heart attacks, stroke, gallbladder disease, osteoarthritis, respiratory problems, and some cancers. In the United States about 400,000 people die from illnesses related to obesity every year. This number is approaching the number of deaths related to smoking (435,000 annually). Weight-related illnesses and disabilities are now a serious strain on healthcare systems and healthcare budgets worldwide.

Growing rates of obesity result partly from increased consumption of oily and sugary foods and soft drinks, and partly from lifestyles that involve less walking, less physical work, and more leisure than previous generations had. Changing these factors can be hard. Just walking to work regularly can be enough to keep weight down, but many of our daily routines are built around sitting still, at a desk or in a car. Many of our social activities, and our traditional holiday meals, focus on rich foods with gravies and sauces, or sweets. We are probably biologically adapted to prize these energy-rich foods, which were rare and valuable for our ancestors. Today it can take special effort to cut back on them.

Another cause is the economic necessity for food producers to increase profits. When we already have plenty to eat, and when food prices are low, food processors struggle constantly to ensure continuous growth in production and profits. Manufacturers can achieve better profits with "value added" products: Instead of selling plain oatmeal at, say, 50 cents a pound, a manufacturer can convert oats into flavored, sweetened, instant microwavable oatmeal for $2.50 a pound. Better yet, oats processed into sweetened, toasted oat flakes can bring $5 a pound. Increases in sugar and fat content, as well as constant exposure to advertising, encourages us to consume more than we might really need.

Paradoxically, food insecurity and poverty can also contribute to obesity. In one study, more than half the women who reported not having enough to eat were overweight, compared with one-third of the food-secure women. Lack of good-quality food may contribute to a craving for carbohydrates in people with a poor diet. A lack of time for cooking, and limited access to healthy food choices along with ready availability of fast-food snacks and calorie-laden soft drinks, also lead to dangerous dietary imbalances.

Michael Pollan, who writes about food issues at the University of California, Berkeley, says that plain, simple food is what our bodies are adapted to. Products made of manufactured foodlike

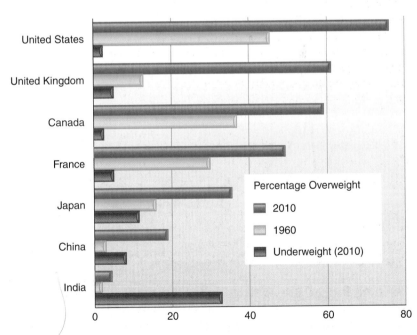

FIGURE 9.5 While nearly a billion people are chronically undernourished, people in wealthier countries are at risk from eating too much.
Source: Worldwatch Institute, 2011.

substances that your grandmother wouldn't recognize probably are not good for you. Pollan sums up the answer to health and obesity problems this way: "Eat food. Not too much. Mostly plants."

High prices remain a widespread threat

Despite surplus production and low prices for farmers (fig. 9.6), food prices are frequently in the news, and food costs threaten struggling families. Nonindustrialized farming economies such as India have also seen long-term price declines, yet impoverished populations still suffer acutely with shorter-term increases in prices for cooking oil, wheat, or other staples. Why do food prices rise despite global abundance?

Floods, droughts, and storms often trigger spikes in food prices, and critical shortages can occur. And droughts and weather extremes are expected to increase with climate change (chapter 15). But because food is now a global commodity, larger market forces also drive prices. Traders in Chicago, London, and Tokyo purchase volumes of grain, sugar, coffee, or other commodities simply to make a profit on the trade. Often this means speculation and trading in futures: I might promise to pay you $4 a bushel for next summer's corn crop, even though the planting season hasn't even started yet, just so I can reserve the crop and settle the price now. But if there's drought in the spring and the year's production looks poor, someone who really needs the corn might pay me $5 a bushel for the same crop that's not yet in the ground. I just made a 25 percent profit on a future corn crop, and my shareholders are delighted. Consumers somewhere else will cover the higher costs. Trading in commodities and futures, then, can drive food prices, even though the exchanges are far removed from the actual food that a farmer plants and a consumer eats. And expected future shortages can drive up prices today.

To complicate matters further, food prices are driven by nonfood demands for crops. In 2007–2008, United States corn prices jumped from around $2 a bushel to over $5 a bushel when the U.S. Congress promised to subsidize corn-based ethanol fuel and to

require that ethanol be sold at gas stations nationwide. In that year, futures speculation for ethanol drove up corn prices, and wheat and other grains followed in the excitement. Because of the ethanol boom, many small bakers and pasta makers couldn't afford wheat and were driven out of business, and U.S. consumers were pinched as food prices rose throughout the grocery store.

The same process occurred in 2008–2010 after the European Union passed new rules requiring biofuel use, with the idea that these fuels would be sustainable and climate neutral. Europe's biofuels are produced largely from palm oil, a tropical oily fruit grown mainly in Malaysia and Indonesia. European biofuel rules produced a boom in global palm oil demand. Unfortunately, palm oil is also a cooking staple for poor families across Asia, for whom a doubling of oil prices can be devastating. In developing countries across the globe, riots broke out over rising cooking oil prices, which were driven by well-meaning European legislation for Malaysian biofuel.

The palm oil boom is also driving accelerated deforestation and wetland drainage across Malaysia, Indonesia, Ecuador, Colombia, and other palm-oil-producing regions, leading to further social and environmental repercussions (chapter 12).

Price changes are merely an inconvenience for food-secure populations, but for impoverished families, and for farmers whose income depends on crop prices, price volatility can trigger disaster. What other factors might drive up prices? (Hint: think about fuel, water, labor, war, and other factors.) If you were a policymaker, which of the issues above would be easiest to modify?

We need the right kinds of food

Generally, eating a good variety of foods provides the range of nutrients you need. In general it's best to have whole grains and vegetables, with only sparing servings of meat, dairy, fats, and sweets. Based on observations of health effects of Mediterranean diets as well as a long-term study of 140,000 U.S. health professionals, Dr. Walter Willett and Dr. Meir Stampfer of Harvard University have recommended a dietary pyramid that minimizes red meat and starchy food such as white rice, white bread, potatoes, and pasta (fig. 9.7). Nuts, legumes (beans, peas, and lentils), fruits, vegetables, and whole grain foods form the basis of this diet. The base of this Harvard pyramid is regular, moderate exercise.

Food-insecure people often can't afford the protein, fruits, and vegetables that would ensure a balanced diet. Starchy foods like maize (corn), polished rice, and manioc (tapioca) form the bulk of the diet for poor populations, especially in developing countries. Even if they get enough calories, they may lack sufficient protein, vitamins, and trace minerals. **Malnourishment** is a term for nutritional imbalance caused by a lack of specific dietary components or an inability to absorb or utilize essential nutrients.

The FAO estimates that perhaps 3 billion people (nearly half the world population) suffer from vitamin, mineral, or protein deficiencies. Effects can include devastating illnesses and deaths as well as slowed mental and physical development. These problems bring an incalculable loss of human potential.

Anemia (low hemoglobin levels in the blood, usually caused by dietary iron deficiency) is the most common nutritional problem

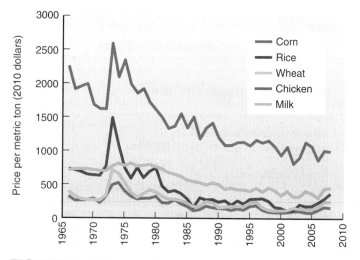

FIGURE 9.6 Prices paid to producers in the United States have declined steadily, yet recent increases in food prices are stressful for consumers. How do we reconcile these two problems? (Data source: UN FAO)

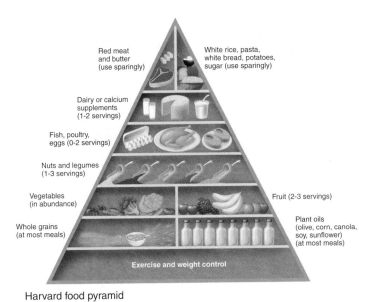

Harvard food pyramid

FIGURE 9.7 The Harvard food pyramid emphasizes fruits, vegetables, and whole grains as the basis of a healthy diet. Red meat, white rice, pasta, and potatoes should be used sparingly. **Source:** Data from Willett and Stampfer, 2002.

in the world. More than 2 billion people suffer from iron deficiencies, especially women and children. The problem is most severe in India, where 80 percent of all pregnant women may be anemic. Anemia increases the risk of maternal deaths from hemorrhage in childbirth and affects childhood development. Red meat, eggs, legumes, and green vegetables all are good sources of dietary iron.

Iodine is essential for synthesis of thyroxin, an endocrine hormone that regulates metabolism and brain development. Chronic iodine deficiency causes goiter (a swollen thyroid gland, fig. 9.8*b*), stunted growth, and mental impairment. The FAO estimates that 740 million people—mainly in South and Southeast Asia—suffer from iodine deficiency and that 177 million children have stunted growth and development. Adding a few pennies' worth of iodine to table salt has nearly eliminated this problem in developed countries.

Vitamin A deficiencies affect 100–140 million children at any given time. At least 350,000 go blind every year from the effects of this vitamin shortage. Folic acid, found in dark green, leafy vegetables, is essential for early fetal development. Ensuring access to leafy greens can be one of the cheapest ways of providing essential vitamins.

Protein deficiency can cause conditions such as kwashiorkor and marasmus. **Kwashiorkor** is a West African word meaning "displaced child." (A young child is displaced—and deprived of nutritious breast milk—when a new baby is born.) This condition most often occurs in young children who subsist mainly on cheap starchy foods. Children with kwashiorkor often have puffy, discolored skin and a bloated belly. **Marasmus** (from the Greek "to waste away") is caused by shortages of both calories and protein. A child suffering from severe marasmus is generally thin and shriveled (fig. 9.8*a*). Children with these deficiencies have low resistance to infections and may suffer lifelong impacts on mental and physical development.

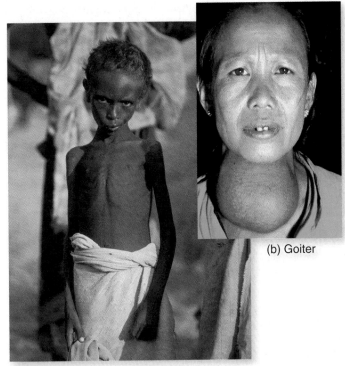

(a) Marasmus

(b) Goiter

FIGURE 9.8 Dietary deficiencies can cause serious illness. *(a)* Marasmus results from protein and calorie deficiency and gives children a wizened look and dry, flaky skin. *(b)* Goiter, a swelling of the thyroid gland, results from an iodine deficiency.

9.2 Key Food Sources

Of the thousands of edible plants and animals in the world, only about a dozen types of seeds and grains, three root crops, twenty or so common fruits and vegetables, six mammals, two domestic fowl, and a few fish and other forms of marine life make up almost all of the food humans eat (table 9.1). In this section, we will highlight the characteristics of some important food sources.

A few major crops supply most of our food

The three crops on which humanity depends for the majority of its nutrients and calories are wheat, rice, and maize (called corn in the United States). Together, some 2 billion metric tons of these three grains are grown each year. Wheat and rice are especially important, as the staple foods for most of the 5.5 billion people in the developing countries of the world. These two grass species supply around 60 percent of the calories consumed directly by humans.

Potatoes, barley, oats, and rye are staples in mountainous regions and high latitudes (northern Europe, north Asia) because they grow well in cool, moist climates. Cassava, sweet potatoes, and other roots and tubers grow well in warm, wet areas and are staples in Amazonian, Africa, Melanesia, and the South Pacific. Barley, oats, and rye can grow in cool, short-season climates. Sorghum and millet are drought-resistant and are staples in the dry regions of Africa.

Table 9.1	Some Important Food Sources
Crop	**2007 Yield (Million Metric Tons)**
Wheat	607
Rice (paddy)	652
Maize (corn)	785
Potatoes	322
Coarse grains*	1,083
Soybeans	216
Cassava and sweet potato	550
Sugar (cane and beet)	150
Pulses (beans, peas)	61
Oil seeds	397
Vegetables and fruits	1,493
Meat and milk	957
Fish and seafood	150

*Barley, oats, sorghum, rye, millet.

Source: Food and Agriculture Organization (FAO), 2009.

Fruits, vegetables, and vegetable oils are usually the most important sources of vitamins, minerals, dietary fiber, and complex carbohydrates. In the United States, however, grains make up a far larger part of our diet. Corn is by far the most abundant crop, followed by soybeans and wheat (fig. 9.9). Of these three, only wheat is primarily consumed directly by humans. Corn and soy are processed into products such as high-fructose corn syrup or fed to livestock.

Rising meat production has costs and benefits

Dramatic increases in corn and soy production have led to rising meat consumption worldwide. In developing countries, meat consumption has risen from just 10 kg per person per year in the 1960s to over 26 kg today (fig. 9.10). In the same interval, meat consumption in the United States has risen from 90 kg to 136 kg

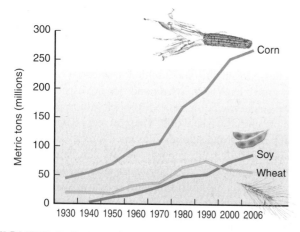

FIGURE 9.9 United States production of the three dominant crops, corn, soybeans, and wheat.
Source: Data from USDA and UN FAO, 2008.

FIGURE 9.10 Meat and dairy consumption have quadrupled in the past 40 years, and China represents about 40 percent of that increased demand.

per person per year. Meat is a concentrated, high-value source of protein, iron, fats, and other nutrients that give us the energy to lead productive lives. Dairy products are also a key protein source: globally we consume more than twice as much dairy as meat. But dairy production per capita has declined slightly while global meat production has doubled in the past 45 years.

Meat is a good indicator of wealth because it is expensive to produce, in terms of the resources needed to grow an animal (fig. 9.11). As discussed in chapter 3, herbivores use most of the energy they consume for moving and growing; only a portion of energy consumed is stored for consumption by carnivores. A beef steer consumes over 8 kg of grain to produce just 1 kg of beef. Pigs, being smaller, are more efficient. Just 3 kg of pig feed are needed to produce 1 kg of pork. Chickens and herbivorous fish (such as catfish) are still more efficient.

Globally, over one-third of cereals (some 660 million metric tons) are used as livestock feed each year. We could feed about eight times as many people by eating that cereal directly, rather than converting it to meat. What differences do you suppose it would make if we did so?

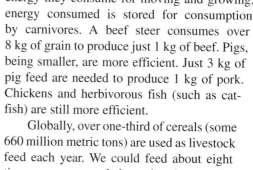

FIGURE 9.11 Number of kilograms of grain needed to produce 1 kg of bread or 1 kg live weight gain.

A number of technological and breeding innovations have made this increased production possible. One of the most important is the **confined animal feeding operation (CAFO)**, where animals are housed and fed—mainly on soy and corn—for rapid growth (fig. 9.12). These operations dominate livestock raising in the United States, Europe, and increasingly in China. Animals are housed in giant enclosures, with up to 10,000 hogs or a million chickens in an enormous barn complex, or 100,000 cattle in a feedlot. These systems require specially prepared mixes of corn, soy, and animal protein that maximizes animals' growth rate. New breeds of livestock have been developed that produce meat rapidly, rather than simply getting fat. The turnaround time is getting shorter, too. A U.S. chicken producer can turn baby chicks into chicken nuggets after just eight weeks of growth. Steers reach full size by just 18 months of age.

Constant use of antibiotics, which are mixed in daily feed, is also necessary for growing animals in such high densities and with unnaturally rich diets. Over 11 million kg of antibiotics are added to animal feed annually in the United States, about eight times as much as is used in human therapy. Nearly 90 percent of U.S. hogs receive antibiotics in their feed.

Because modern meat production is based on energy-intensive farming practices (see chapter 10), meat is also an energy-intensive product. It takes about 16 times as much fossil fuel energy to produce a kilogram of beef as it takes to produce a kilogram of vegetables or rice. The UN Food and Agriculture Organization estimates that livestock produce 20 percent of the world's greenhouse gases, more than is produced by transportation. In fact, by some estimates Americans could cut energy consumption more if we gave up just one-fifth of our meat consumption than if all of us were to drive a hybrid-electric Prius.

Seafood is a key protein source

We currently harvest about 95 million metric tons of wild fish and seafood every year, but we directly eat only about two-thirds of that amount. One-third is used as feed for fish farms, to raise species such as salmon or bluefin tuna, which bring high prices.

Seafood is the main animal protein source for about 1.5 billion people in developing countries, although most of those people eat mainly locally caught fish. In wealthier countries, industrial-scale fishing provides most seafood. Development of freezer technology on oceangoing factory ships since the 1950s allowed annual catches of ocean fish to rise by about 4 percent annually between 1950 and 1988. Since then, all our major marine fisheries have declined dramatically, and most have become commercially unsustainable. An international team of marine biologists warns that if current trends continue, all the world's major fisheries will be exhausted by 2050.

Fish are the only wild-caught meat source still sold commercially on a global scale. Because wild fish belong to nobody in particular, the global competition to catch them is steep. Rising numbers of boats, with increasingly efficient technology, exploit the dwindling resource. Boats as big as ocean liners travel thousands of kilometers and drag nets large enough to scoop up a dozen jumbo jets, sweeping

FIGURE 9.12 Concentrated feeding operations fatten animals quickly and efficiently, but create enormous amounts of waste and expose livestock to unhealthy living conditions.

a large patch of ocean clean of fish in a few hours. Longline fishing boats set cables up to 10 km long with hooks every 2 meters that catch birds, turtles, and other unwanted "by-catch" along with targeted species. Trawlers drag heavy nets across the bottom, reducing broad swaths of spawning habitat to rubble. Most countries subsidize their fishing fleets to preserve jobs and to ensure access to fisheries. The FAO estimates that operating costs for the 4 million boats now harvesting wild fish exceed sales by (U.S.) $50 billion per year.

Aquaculture (growing aquatic species in net pens or tanks) provides about half of the seafood we eat. In addition, about one-third of wild-caught fish is used as food for fish in these operations. Because farmed carnivorous species such as salmon, sea bass, and tuna consume so much wild-caught fish, they also threaten wild fish populations and the seabirds and the organisms that depend on them. Net pens are anchored in near-shore areas and also allow spread of diseases, escape of exotic species, and release of feces, uneaten food, antibiotics, and other pollutants into surrounding ecosystems (fig. 9.13). New designs are being developed, including fully enclosed pens anchored farther from shore, which may mitigate these environmental costs.

Farmed shrimp and many fish are grown in ponds built on former mangrove forests and wetlands, which are also nurseries for marine species.

FIGURE 9.13 Pens for fish-rearing in Thailand.

FIGURE 9.14 This state-of-the-art lagoon is built to store manure from a hog farm. Odors and overflow after storms are risks of open lagoons, but more thorough waste treatment is expensive.

Aquaculture in land-based ponds or warehouses can eliminate many of these problems, especially when raising herbivorous fish, such as catfish, carp, or tilapia, which also consume less feed per pound of meat than do carnivorous species. In China, for example, most fish are raised in ponds or rice paddies. One ecologically balanced system uses four carp species that feed at different levels of the food chain. The grass carp, as its name implies, feeds largely on vegetation, while the common carp is a bottom feeder, living on detritus that settles to the bottom. Silver carp and bighead carp are filter feeders that consume phytoplankton and zooplankton, respectively. Agricultural wastes such as manure, dead silkworms, and rice straw fertilize ponds and encourage phytoplankton growth. These integrated polyculture systems typically boost fish yields per hectare by 50 percent or more compared with monoculture farming.

Antibiotics are needed for intensive production

Intensive food production can have profound environmental effects. Converting land to soy and corn fields raises the rate of soil erosion (chapter 10). Bacteria in the manure in the feedlots, or liquid wastes in manure storage lagoons (holding tanks) around hog farms, can escape into the environment—from airborne dust around feedlots or from breaches in the walls of a manure tank (fig. 9.14). When Hurricane Floyd hit North Carolina's coastal hog production region in 1999, an estimated 10 million m^3 of hog and poultry waste overflowed into local rivers, creating a dead zone in Pamlico Sound.

Constant use of antibiotics raises the very real risk of antibiotic-resistant diseases. Massive and constant exposure produces antibiotic-resistant pathogens, strains that have adapted to survive antibiotics. This process is slowly rendering our standard antibiotics useless for human health care. Next time you are prescribed an antibiotic by your doctor, you might ask whether she or he worries about antibiotic resistance, and you might think about how you would feel if your prescription were ineffectual against your illness.

Although the public is increasingly aware of the environmental and health risks of concentrated meat production, we seem to be willing to accept these risks because this production system has made our favorite foods cheaper, bigger, and more available. A fast-food hamburger today is more than twice the size it was in 1960, especially if you buy the kind with multiple patties and special sauce. At the same time, this larger burger costs less per pound, in constant dollars, than it did in 1960. This helps explain why we now consume more protein and calories than we really need.

As environmental scientists, we are faced with a conundrum, then. Improved efficiency has great environmental costs; it has also given us the abundant, inexpensive foods that we love. We have more protein, but also more obesity, heart disease, and diabetes than ever before. What do you think? Do the environmental risks balance a globally improved quality of life, or should we consider reducing our consumption to reduce environmental costs? How might we go about making changes, if you think any are needed?

9.3 FOOD PRODUCTION POLICIES

The FAO predicts that 70 percent of future world production growth will come from higher yields and new crop varieties, because expanding arable lands is not a reasonable option in many areas. Development of more intensive farming methods, therefore, is a matter of global interest. In this section we'll focus on our dominant strategies for intensification of food production: green revolution hybrids and genetically modified crops.

In addition to these dominant forms, there is also growing interest in alternative agriculture that can reduce our dependence on oil, antibiotics, and other environmental costs of food production. Like the students described in the opening case study, many people are interested in supporting sustainable food production (What Do You Think? p. 187). Organic and sustainable foods are not just vegetables and fruits: meat, eggs, and dairy can be produced sustainably, too. Grass-fed beef, for example, can be an efficient way to convert solar energy into protein. Rotational grazing, using small, easily moved electric fences to concentrate grazing in one area of a field at a time, can invigorate pasture, distribute manure, and keep livestock healthy (fig. 9.15).

FIGURE 9.15 Rotational grazing is one strategy for meat production with less reliance on energy, water, and other resources. Here an electric fence contains cattle in one part of a pasture while another part recovers for several weeks.

What Do You Think?

Shade-Grown Coffee and Cocoa

Has it ever occurred to you that your purchases of coffee and chocolate may be contributing to the protection or destruction of tropical forests? Coffee and cocoa are examples of food products grown exclusively in developing countries but consumed almost entirely in the wealthy nations (vanilla and bananas are some other examples). Coffee grows in cool, mountain areas of the tropics, while cocoa is native to the warm, moist lowlands. Both are small trees of the forest understory, adapted to low light levels.

Until a few decades ago, most of the world's coffee and cocoa were grown under a canopy of large forest trees. Recently, however, new varieties of both crops have been developed that can be grown in full sun. Yields for sun-grown crops are higher because more coffee or cocoa trees can be crowded into these fields, and they get more solar energy than in a shaded plantation.

There are costs, however, in this new technology. Sun-grown trees die earlier from the stress and diseases common in these fields. Furthermore, ornithologists have found that the number of bird species can be cut in half in full-sun plantations, and the number of individual birds may be reduced by 90 percent. Shade-grown coffee and cocoa generally require fewer pesticides (or sometimes none) because the birds and insects residing in the forest canopy eat many of the pests. Shade-grown plantations also need less chemical fertilizer because many of the plants in these complex forests add nutrients to the soil. In addition, shade-grown crops rarely need to be irrigated because heavy leaf fall protects the soil, while forest cover reduces evaporation.

Currently about 40 percent of the world's coffee and cocoa plantations have been converted to full-sun varieties and another 25 percent are in the process of converting. Traditional techniques for coffee and cocoa production are worth preserving. Thirteen

Cocoa pods grow directly on the trunk and large branches of cocoa trees.

of the world's 25 biodiversity hot spots occur in coffee or cocoa regions. If all 20 million ha (49 million) acres of coffee and cocoa plantations in these areas are converted to monocultures, an incalculable number of species will be lost.

The Brazilian state of Bahia is a good example of both the ecological importance of these crops and how they might help preserve forest species. At one time Brazil produced much of the world's cocoa, but in the early 1900s the crop was introduced into West Africa. Now Côte d'Ivoire alone grows more than 40 percent of the world total, and the value of Brazil's harvest has dropped by 90 percent. Côte d'Ivoire is aided in this competition by a labor system that reportedly includes widespread child slavery. Even adult workers in Côte d'Ivoire get only about $165 per year (if they get paid at all) compared to a minimum wage of $850 per year in Brazil. As African cocoa production ratchets up, Brazilian landowners are converting their plantations to pastures or other crops.

The area of Bahia where cocoa was once king is part of Brazil's Atlantic forest, one of the most threatened forest biomes in the world. Only 8 percent of this forest remains undisturbed. Although cocoa plantations don't represent the full diversity of intact forests, they protect a surprisingly large sample of what once was there. And shade-grown cocoa can provide an economic rationale for preserving that biodiversity. Brazilian cocoa will probably never compete with that from other areas for lowest cost. There is room in the market, however, for specialty products. If consumers were willing to pay a small premium for organic, fair-trade, shade-grown chocolate and coffee, this might provide the incentive needed to preserve biodiversity. Wouldn't you like to know that your chocolate or coffee wasn't grown with child slavery, and is helping protect plants and animal species that might otherwise go extinct?

Do sustainable and organic farming offer meaningful contributions to feeding a hungry world, in comparison to the large-scale methods of conventional farming? Opinions vary strongly on this question. To some extent it remains hard to say, because sustainable techniques have received relatively little research and development effort. Most agricultural research has focused on improving inputs (fertilizer, pesticides, seeds, fuel, and irrigation) to intensify production of cereals (mainly corn, rice, wheat, and soy). This strategy has multiplied food production and given us low food prices, especially in wealthier countries. Studies by the FAO and the UN Environment Programme, however, have found that these strategies are expensive for poor farmers and that alternative methods, such as enriching soil with nitrogen-fixing plants, rotating crops, and interplanting crops to reduce pest dispersal, provide greater food security in poor regions. Organic and sustainable farming are discussed further in chapter 10.

Food policy is economic policy

Much of the increase in food production over the past 50 years has been fueled by government support for agricultural education, research, and development projects that support irrigation systems, transportation networks, crop insurance, and direct subsidies. The World Bank estimates that rich countries pay their own farmers $350 billion per year, or nearly six times as much as all developmental aid to poor countries. A typical cow in Europe enjoys annual subsidies three times the average yearly income for most African farmers.

Agricultural subsidies can make a critical difference for farmers, but they are a concern globally. Subsidies allow American farmers to sell their products overseas at as much as 20 percent below the actual cost of production. These cheap commodities, as well as free food aid, frequently flood markets in developing countries, driving local farmers out of business and destabilizing food production. The

FAO argues that ending distorting financial support in the richer countries would have far more positive impact on local food supplies and livelihoods in the developing world than any aid program.

Powerful political and economic interests protect agricultural assistance in many countries. Over the past decade, the United States, for example, has spent $143 billion in farm support. This aid is distributed unevenly. According to the Environmental Working Group, 72 percent of all aid goes to the top 10 percent of recipients. One giant rice-farming operation in Arkansas, for example, received $38 million over a five-year period. Aid also is concentrated geographically. Just 5 percent (22) of the nation's 435 congressional districts collect more than 50 percent of all agricultural payments. Most of this aid is direct payments for each bushel of targeted commodities, mainly corn, wheat, soybeans, rice, and cotton, as well as special subsidies for milk, sugar, and peanuts. Proponents insist that crop supports preserve family farms, but critics claim that the biggest recipients are corporations that don't really need the aid. There have been repeated efforts to roll back agricultural payments, but Congress has found it easier to cut conservation funds and food assistance programs rather than reduce payments to agribusiness interests.

An additional effect of these market interventions is to encourage the oil- and sugar-rich diets that lead to the spreading obesity epidemic. Subsidies help ensure that these processed foods are cheaper and more readily available than fresh fruits, vegetables, and whole grains. Many food policy analysts argue that we should support more vegetables and nutrient-rich foods and fewer commodity crops. Public attention to farm policy could help move us toward such policies.

Farm policies can also protect the land

Every year millions of tons of topsoil and agricultural chemicals wash from U.S. farm fields into rivers, lakes, and, eventually, the ocean. Farmers know that erosion both impoverishes their land and pollutes water, but they're caught in a bind. For every $1 the U.S. government pays farmers to conserve soil and manage nutrients, it pays $7 to support row-crop commodities that require intensive cultivation and promote soil loss and chemical runoff. The USDA estimates that if federal subsidies didn't promote these commodities, farmers would shift 2.5 million ha (6 million acres) of row crops into pasture, hay, and other crops that minimize erosion.

The United States tries to reduce soil erosion and overproduction of crops with the Conservation Reserve Program (CRP), which pays farmers to keep roughly 12 million ha (30 million acres) of highly erodible land out of production. The USDA reports that CRP lands prevent the annual loss of 450 million tons of soil every year, protect 270,000 km (170,000 miles) of streams, and store 48 million tons of carbon per year. Keeping land enrolled in this soil conservation program is vulnerable to political and economic shifts, however, and the amount of land enrolled changes every year.

Land enrolled in the CRP has been declining, as farmers find it more profitable to plant corn for ethanol, and as farm policy commits less money to supporting land retirement. But many agronomists say we should have more CRP land, not less.

The United States could gradually shift payments from production subsidies to conservation programs that would truly support family farms while also protecting the environment.

9.4 THE GREEN REVOLUTION AND GENETIC ENGINEERING

Although at least 3,000 species of plants have been used for food at one time or another, most of the world's food now comes from only 16 species. There is considerable interest in expanding this number and developing new varieties. One of the plants being investigated is the winged bean (fig. 9.16), a perennial plant that grows well in hot climates. The entire plant is edible (pods, mature seeds, shoots, flowers, leaves, and tuberous roots), it is resistant to diseases, and it enriches the soil. Another promising crop is tricale, a hybrid between wheat (*Triticum*) and rye (*Secale*) that grows in light, sandy, infertile soil. It is drought-resistant, has nutritious seeds, and is being tested for salt tolerance for growth in saline soils or irrigation with seawater. Some traditional crop varieties grown by Native Americans, such as tepary beans, amaranth, and Sonoran panicgrass, are being collected by seed conservator Gary Nabhan both as a form of cultural revival for native people and as a possible food crop for harsh environments.

FIGURE 9.16 Winged beans bear fruit year-round in tropical climates and are resistant to many diseases that prohibit growing other bean species. Whole pods can be eaten when they are green, or dried beans can be stored for later use. It is a good protein source in a vegetarian diet.

These innovations are exciting, but our main improvements in farm production have come from breeding hybrid varieties of a few well-known species. Yield increases often have been spectacular. A century ago, when all corn in the United States was open-pollinated, average yields were about 25 bushels per acre. In 2010, average yields were more than 160 bushels per acre, and peak yields were over 350 bushels per acre. It is these kinds of increases that have allowed us to send food aid overseas and to use most of our corn crop for livestock feed, high-fructose corn syrup, ethanol, and other products at home.

Green revolution crops emphasize high yields

Starting in the 1950s and 1960s, agricultural research stations began to breed tropical wheat and rice varieties that would provide food for growing populations in developing countries. Cross-breeding plants with desired traits created new, highly productive hybrids known as "miracle" varieties. The first of these was a dwarf, high-yielding wheat developed by Norman Borlaug, who received a Nobel Peace Prize for his work, at a research center in Mexico (fig. 9.17). At about the same time, the International Rice Research Institute in the Philippines developed dwarf rice strains with three or four times the production of varieties in use at the time.

The dramatic increases obtained as these new varieties spread around the world has been called the **green revolution**. The success of these methods is one of the main reasons that world food supplies have more than kept pace with the growing human population over the past few decades. Miracle varieties were spread around the world as U.S. and European aid programs helped developing countries adopt new methods and seeds. The green revolution replaced traditional crop varieties and growing methods throughout the developed world, and nearly half of all farmers in the developing world were using green revolution seeds, fertilizers, and pesticides by the 1990s.

Most green revolution breeds really are "high responders," meaning that they yield more than other varieties if given optimal levels of fertilizer, water, and pest control (fig. 9.18). Without irrigation and fertilizer, on the other hand, high responders may not produce as well as traditional varieties. New methods and inputs are also expensive. Poor farmers who can't afford hybrid seeds, fertilizer, machinery, fuel, and irrigation are put at a disadvantage, compared to wealthier farmers who can afford these inputs. Thus the green revolution is credited with feeding the world, but it is also often accused of driving poorer farmers off their land, as rising land values and falling commodity prices squeeze them from both sides.

Genetic engineering moves DNA among species

Genetic engineering involves removing genetic material from one organism and splicing it into the chromosomes of another (fig. 9.19). This technology introduces entirely new traits, at a much faster rate compared to cross-breeding methods. It is now possible to build entirely new genes by borrowing bits of DNA from completely unrelated species, or even synthesizing artificial DNA sequences to create desired characteristics in **genetically modified organisms (GMOs)**.

FIGURE 9.17 Semi-dwarf wheat (*right*), bred by Norman Borlaug, has shorter, stiffer stems and is less likely to lodge (fall over) when wet than its conventional cousin (*left*). This "miracle" wheat responds better to water and fertilizer, and has played a vital role in feeding a growing human population.

Genetically modified (GM) crops offer dramatic benefits. Research is under way to improve yields and create crops that resist drought, frost, or diseases. Other strains are being developed to tolerate salty, waterlogged, or nutrient-poor soils. These would allow degraded or marginal farmland to become productive. All of these could be important for reducing hunger in developing countries. Plants that produce their own pesticides might reduce the need for toxic chemicals, and engineering for improved protein or vitamin content could make our food more nutritious. Attempts to remove specific toxins or allergens from crops also could make our food safer. Crops such as bananas and potatoes have been altered to contain oral vaccines that can be grown in developing countries where refrigeration and sterile needles are unavailable. Plants have been engineered to make industrial oils and plastics. Animals, too, are being genetically modified to grow faster, gain

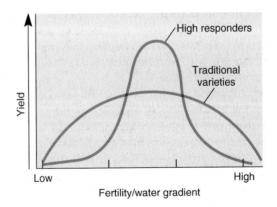

FIGURE 9.18 Green revolution miracle crops are really high responders, meaning that they have excellent yields under optimum conditions. For poor farmers who can't afford the fertilizer and water needed by high responders, traditional varieties may produce better yields.

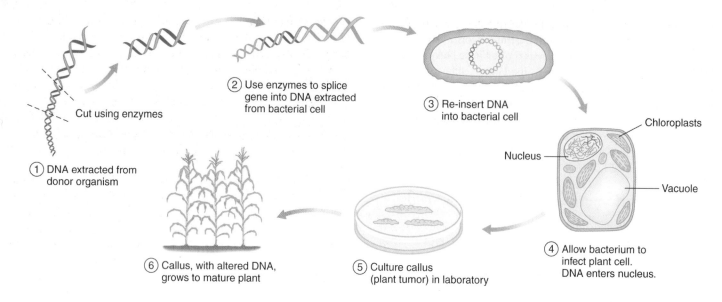

① DNA extracted from donor organism

Cut using enzymes

② Use enzymes to splice gene into DNA extracted from bacterial cell

③ Re-insert DNA into bacterial cell

Chloroplasts

Nucleus

Vacuole

④ Allow bacterium to infect plant cell. DNA enters nucleus.

⑤ Culture callus (plant tumor) in laboratory

⑥ Callus, with altered DNA, grows to mature plant

FIGURE 9.19 One method of gene transfer, using an infectious, tumor-forming bacterium such as agrobacterium. Genes with desired characteristics are cut out of donor DNA and spliced into bacterial DNA using special enzymes. The bacteria then infect plant cells and carry altered DNA into cells' nuclei. The cells multiply, forming a tumor, or callus, which can grow into a mature plant.

weight on less food, and produce pharmaceuticals such as insulin in their milk. It may soon be possible to create animals with human cell-recognition factors that could serve as organ donors.

Most GMOs have been engineered for pest resistance or weed control

The most common gene transfers involve pest resistance or pesticide tolerance. A naturally occurring insecticide from the bacterium *Bacillus thuringiensis* (Bt) has been implanted into a wide variety of crops. The Bt gene produces toxins lethal to Lepidoptera (butterfly family) and Coleoptera (beetle family). The genes for some of these toxins have been transferred into crops such as maize (to protect against European cut worms), potatoes (to fight potato beetles), and cotton (for protection against boll weevils). This allows farmers to reduce insecticide spraying. Arizona cotton farmers, for example, report reducing their use of chemical insecticides by 75 percent. Cotton farms in India report an 80 percent yield increase with Bt cotton compared to neighboring plots growing conventional cotton.

Entomologists worry that because Bt plants produce toxin throughout the growing season, regardless of the level of infestation, they create perfect conditions for selection of Bt resistance in pests. The effectiveness of this natural pesticide—one of the few available to organic growers—is likely to be destroyed within a few years. One solution is to plant at least a part of every field in non-Bt crops that will act as a refuge for nonresistant pests. The hope is that interbreeding between these "wild-type" bugs and those exposed to Bt will dilute out recessive resistance genes. Deliberately harboring pests and letting them munch freely on crops is something that many farmers find hard to do. In addition, devoting a significant part of their land to nonproductive crops lowers the total yield and counteracts the profitability of engineered seed.

There also is a concern about the effects on nontarget species. In laboratory tests, about half of a group of monarch butterfly caterpillars died after being fed on plants dusted with pollen from Bt corn. Under field conditions, however, it has been difficult to demonstrate harm to butterflies.

The other major transgenic crops are engineered to tolerate herbicides. These crops are unaffected when fields are sprayed to kill weeds. These crops make up about three-quarters of all genetically engineered acreage. The two main products in this category are Monsanto's "Roundup Ready" crops—so-called because they tolerate Monsanto's best-selling herbicide, Roundup (glyphosate)—and AgrEvo's "Liberty Link" crops, which resist that company's Liberty (glufosinate) herbicide. Because crops with these genes can grow in spite of high herbicide doses, farmers can spray fields heavily to exterminate weeds. This practice allows for conservation tillage and leaving more crop residue on fields to protect topsoil from erosion, both good ideas, but it may also mean an increase in herbicide use.

GM crops have been introduced to the world's farmers even more rapidly than green revolution crops were in the 1960s and 1970s. A decade after their introduction in 1996, GM varieties were planted on 400 million hectares (1 billion acres) of farmland. Three years later that number had doubled to 800 million ha (2 billion ac). This represents just over half of the world's 1.5 billion ha of cultivated land. The United States accounted for 56 percent of that acreage, followed by Argentina with 19 percent. In 2005 China approved GM rice for commercial production. This was the first GM cereal grain approved for direct human consumption and could move China into the forefront of GM crop production. The first GM animals developed for human consumption are GM Atlantic salmon, which grow much faster than normal because they contain growth hormone genes from an oceanic pout. The "enviropig," meanwhile, is being engineered to produce low-phosphorus manure, which should reduce impacts of concentrated hog operations on water quality.

Although many consumers are wary of GM foods, you have almost certainly eaten them. Over 70 percent of U.S. corn has Bt traits or herbicide tolerance, or both. Nearly 95 percent of soybeans and 80 percent of cotton are modified for herbicide tolerance (fig. 9.20). It has been estimated that over 60 percent of all processed food in America contains GM ingredients.

Is genetic engineering safe?

Opponents of genetically modified crops worry that moving genes willy-nilly could create new pests and health risks. GMOs might interbreed with wild relatives, creating new superweeds, or GM varieties might themselves become pests. Abundant use of herbicides has already produced a variety of herbicide-tolerant weeds, forcing farmers to use increasingly complex cocktails of mixed herbicides to keep weeds down. This heavy use of pesticides can also leave toxic residues in soil and on food.

Like green revolution varieties, GM crops are accused of serving mainly resource-rich farmers or regions. Low-income farmers and farmers in poor countries may be unable to afford these crops and the extra pesticides or fertilizers they require. Corporations producing GM varieties, meanwhile gain new advantages because they own the patents to both seeds and pesticides, which farmers must use in order to be competitive. The concern—in North America as well as in developing areas—is that new varieties make smaller farms uncompetitive and drive developing regions even further into poverty.

Can GM traits spread from fields? In a 10-year study of genetically modified crops, a group from Imperial College, London, concluded that GM crops tested did not survive well in the wild and were no more likely to invade other habitats than other weeds. Other scientists counter that some GM crops already have been shown to spread their genes to nearby fields. One of the greatest concerns is that GM traits will spread to wild relatives of common crops. Normal rapeseed (canola oil) varieties in Canada have been found to contain genes from genetically modified varieties in nearby fields. Monsanto, the owner of the Roundup-ready canola, has successfully sued neighboring farmers for having patented genes in their fields.

Genetically modified animals also raise concerns. GM Atlantic salmon grow seven times faster and are more attractive to the opposite sex than a normal salmon. If they escape from captivity, they may outcompete already endangered wild relatives for food, mates, and habitat. Fish farmers say they will grow only sterile females and will keep them in secure net pens. Opponents point out that salmon frequently escape from aquaculture operations and that wild stocks are already diminishing in areas where salmon farming is common.

Consumer groups worry about unforeseen consequences arising from novel combinations of genetic material, which they sometimes call "Frankenfoods." Industry groups accuse their critics of blindly opposing new technology. Often the unease with genetic modification is a feeling that it simply isn't right to mess with nature. Putting novel genes into the food we eat makes many people uncomfortable. Is this merely a fear of science, or is it a valid ethical issue? Most European nations have bans on genetically engineered crops, on the grounds that their effects are poorly understood. The United States has filed a suit at the World Trade Organization claiming that European policies constitute an unwarranted restraint on trade.

The U.S. Food and Drug Administration, meanwhile, has declined to require labeling of foods containing GMOs, saying that these new varieties are "substantially equivalent" to related varieties bred via traditional practices. After all, proponents say, we have been moving genes around for centuries through plant and animal breeding. Genetic modification just accelerates and expands the modifications we've always done.

Will GM crops feed the world, or will they lead to a greater consolidation of corporate wealth and economic inequality? Can higher yields allow poor farmers in developing countries to stop using marginal land and avoid cutting down forests to expand farmland? Would it be more effective and sustainable to develop fishponds or regenerative farming techniques than to sell them new patented seeds? These are the unresolved and hotly debated issues to consider as we aim to reduce malnutrition and feed 9 billion people in coming decades. We may need all

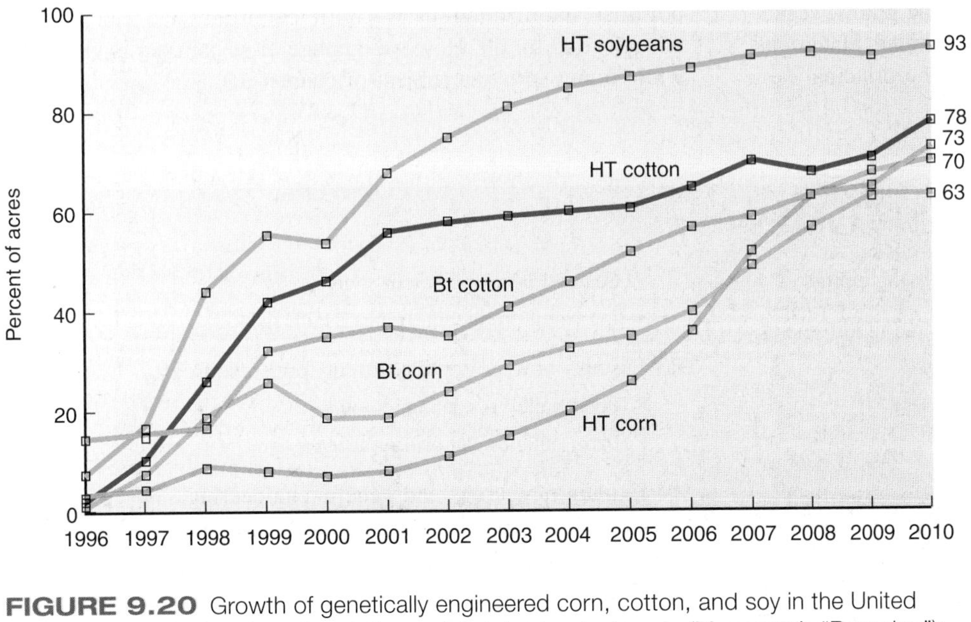

FIGURE 9.20 Growth of genetically engineered corn, cotton, and soy in the United States. HT (herbicide tolerant) varieties mainly tolerate glyphosate (Monsanto's "Roundup"); Bt varieties contain bacterial (Bacillus thuringiensis) proteins that kill insects.
Source: USDA Economic Research Service, 2011.

the tools we can get, including GM foods, less meat-intensive diets, more land conversion, and other approaches. Many people argue that we should take a better-safe-than-sorry "precautionary approach," and err on the side of safety. Our assessment of GM varieties may also depend on whether we are primarily concerned about human health, environmental health, economic stability of farm economies, or other factors. Debates on all these strategies seem likely to continue for years to come.

Think About It

Suppose your grandmother asks you, "What's all this controversy about GMOs? Are they safe or not?" Could you summarize the arguments for and against genetic engineering in a few sentences? Which are the most important issues in this debate, in your opinion?

CONCLUSION

World food supplies have increased dramatically over the past half century. Despite the fact that human population has nearly tripled in that time, food production has increased even faster, and we now grow more than enough food for everyone. Because of uneven distribution of food resources, however, there are still more than 850 million people who don't have enough to eat on a daily basis, and hunger-related diseases remain widespread. Severe famines continue to occur, although most result more from political and social causes (or a combination of political and environmental conditions) than from environmental causes alone.

While hunger persists in many areas, over a billion people consume more food than is healthy on a daily basis. Epidemics of weight-related illnesses are spreading to developing countries, as they adopt diets and lifestyles of wealthier nations. Obesity is a health risk because it can cause or complicate heart conditions, diabetes, hypertension, and other diseases. In the United States, the death rate from illnesses related to obesity is approaching the death rate associated with smoking. Getting the right nutrients is also important. Many preventable diseases are caused by vitamin deficiencies.

Our primary food sources worldwide include grains, vegetables, wheat, rice, corn, and potatoes. In the United States, just three crops—corn, soybeans, and wheat—are the principal farm commodities. Corn and soybeans are mostly fed to livestock, not to people directly. Increasing use of these crops in confined feeding operations has dramatically increased meat production. For this and other reasons, global consumption of protein-rich meat and dairy products has climbed in the past 40 years. Protein gives us the energy to work and study, but raising animals takes a great deal of energy and food, so meat production can be environmentally expensive. However, there are sustainable food alternatives, such as rotational grazing, moderating meat consumption, and eating locally grown foods.

Most increases in food production in recent generations result from "green revolution" varieties of grains, which grow rapidly in response to fertilizer use and irrigation. More recent innovations have focused on genetically modified varieties. Some of these are being developed for improved characteristics, such as vitamin production or tolerance of salty soils. The majority of genetically modified crops are designed to tolerate herbicides, in order to improve competition with weeds.

Meeting the needs of the world's growing population will require a combination of strategies, from new crop varieties to political stabilization in war-torn countries. We can produce enough food for all. How we damage or sustain our environment while doing so is the subject of chapter 10.

REVIEWING LEARNING OUTCOMES

By now you should be able to explain the following points:

9.1 Describe patterns of world hunger and nutritional requirements.

- Millions of people are still chronically hungry.
- Famines usually have political and social causes.
- Overeating is a growing world problem.
- High prices remain a widespread threat.
- We need the right kinds of food.

9.2 Identify key food sources, including protein-rich foods.

- A few major crops supply most of our food.
- A boom in meat production brings costs and benefits.
- Seafood is a key protein source.
- Increased production brings health risks.

9.3 Discuss how policy can affect food resources.

- Food policy is economic policy.
- Farm policies can also protect the land.

9.4 Explain new crops and genetic engineering.

- Green revolution crops emphasize high yields.
- Genetic engineering uses molecular techniques to produce new crop varieties.
- Most GMOs are engineered for pest resistance or weed control.
- Is genetic engineering safe?

PRACTICE QUIZ

1. How many people in the world are chronically undernourished? How many children die each year from starvation and nutrition-related diseases?

2. Which regions of the world face the highest rates of chronic hunger? List at least five African countries with high rates of hunger (fig. 9.3). Use a world map if necessary.

3. What are some of the health risks of overeating? What percentage of adults are overweight in the United States?

4. Explain the relationship between poverty and food security.

5. Why is women's access to food important in food security?

6. According to figure 9.7, what types of food should be most abundant in your diet?

7. List any five of the most abundant food sources produced worldwide. What three food sources are most abundant in the United States?

8. What are some of the environmental risks associated with confined animal feeding operations?

9. What is rotational grazing? What are its benefits?

10. What is the "green revolution," and why was it important?

11. What are genetically modified organisms, and how do they differ from new varieties in the green revolution of the 1960s?

CRITICAL THINKING AND DISCUSSION QUESTIONS

1. Do people around you worry about hunger? Do you think they should? Why or why not? What factors influence the degree to which people worry about hunger in the world?

2. Global issues such as hunger and food production often seem far too large to think about solving, but it may be that many strategies can help us address chronic hunger. Consider your own skills and interests. Think of at least one skill that could be applied (if you had the time and resources) to helping reduce hunger in your community or elsewhere.

3. Suppose you are a farmer who wants to start a confined animal feeding operation. What conditions make this a good strategy for you, and what factors would you consider in weighing its costs and benefits? What would you say to neighbors who wish to impose restrictions on how you run the operation?

4. Debate the claim that famines are caused more by human actions (or inactions) than by environmental forces. What kinds of evidence would be needed to resolve this debate?

5. Outline arguments you would make to your family and friends for why they should buy shade-grown, fair-trade coffee and cocoa. How much of a premium would you pay for these products? What factors would influence how much you would pay?

6. Given what you know about GMO crops, identify some of the costs and benefits associated with them. Which of the costs and benefits do you find most important? Why?

7. Corn is by far the dominant crop in the United States. In what ways is this a good thing for Americans? How is it a problem? Who are the main beneficiaries of this system?

 **Data Analysis:** Using Relative Values

There are many ways to describe trends in an important subject such as world hunger. Figure 9.2 shows two views of this problem: total number and proportion of the population. Another approach is to compare values to a standard value. For example, you could compare all years to 1969, to see how hunger has changed since 1969, when reliable statistics were first gathered by the UN Food and Agriculture Organization (FAO).

These adjusted numbers are **index values,** or values adjusted to be on the same scale or magnitude. In figure 1, index values

were created by dividing all values for a region by the 1969 value. The 1969 value (divided by itself) becomes 1. All other values are either larger or smaller than 1.

Why would you want to adjust values to the same magnitude, rather than show original numbers? One reason might be that values vary a great deal among regions, and it's hard to compare trends on the same graph. Another reason is that you might be more interested in the amount of change than in the absolute numbers. That is, you know there are a lot of undernourished people in

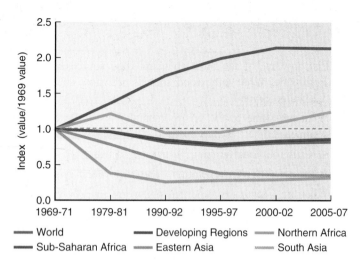

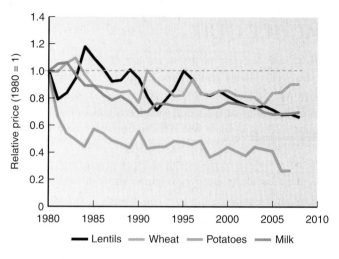

FIGURE 1 Number of people chronically hungry. Index values show change relative to a baseline value (the 1969–71 average).
Source: UN FAO, 2011.

FIGURE 2 Food prices in India, relative to 1980. Index values show prices paid to Indian farmers for products (adjusted for inflation).

sub-Saharan Africa, but you might want to know if the situation is getting worse or better compared to some baseline condition.

Look at figure 1 above carefully, and compare it to figure 9.2 as you answer the following questions.

1. In most of the regions shown, has the total number of undernourished people declined or increased over time?

2. Which region has had the most relative decline? Which region has increased most? If each point on a line shows how many people were hungry relative to the original point (1969–1971), then what does a value of 0.8 represent, in terms of percentage? A value of 1.6? A value of 1.0?

3. Fill in the following:

   Northern Africa had about _____% as many hungry in 2002 as in 1969.

   Developing regions had about _____% as many hungry in 2002 as in 1969.

   Sub-Saharan Africa had about _____% as many hungry in 2002 as in 1969.

4. In fig. 9.2a, the line for Northern Africa is near the bottom of the graph. What does this tell you about the population size in Northern Africa? Why can that population size help explain the next trends shown in the next graph (fig. 9.2b)?

5. Percentage values (figure 9.2b) can be considered another kind of index value. What are all the data divided by in order to make them fit the same scale in this graph?

6. Figure 2 on this page shows prices paid to farmers in India for four main foods. In general, have food prices increased or declined? What other factors might influence affordability of food?

7. What can you infer from this graph about the stability or growth of India's farm economy? Compare to figure 9.6: how would you describe the similarity or difference in these trends?

Enormous farms have been carved out of Brazil's Cerrado (savanna), which once was the most biodiverse grassland and open tropical forest complex in the world.

Farming
Conventional and Sustainable Practices

Learning Outcomes

After studying this chapter, you should be able to:

10.1 Describe the components of soils.

10.2 Explain the ways we use and abuse soils.

10.3 Outline some of the other key resources for agriculture.

10.4 Discuss our principal pests and pesticides.

10.5 List and discuss the environmental effects of pesticides.

10.6 Describe the methods of organic and sustainable agriculture.

10.7 Explain several strategies for soil conservation.

"We abuse the land because we regard it as a commodity belonging to us. When we see land as a community to which we belong, we may begin to use it with love and respect."

~ Aldo Leopold

Case Study   Farming the Cerrado

A soybean boom is sweeping across South America. Inexpensive land, the development of new crop varieties, and government policies that favor agricultural expansion have made South America the fastest-growing agricultural area in the world. The center of this rapid expansion is the Cerrado, a vast area of grassland and tropical forest stretching from Bolivia and Paraguay across the center of Brazil almost to the Atlantic Ocean (fig. 10.1). Biologically, this rolling expanse of grasslands and tropical woodland is the richest savanna in the world, with at least 130,000 plant and animal species, many of which are threatened by agricultural expansion.

Until recently the Cerrado, which is roughly equal in size to the American Midwest, was thought to be unsuitable for cultivation. Its red iron-rich soils are highly acidic and poor in essential plant nutrients. Furthermore, the warm, humid climate harbors many destructive pests and pathogens. For hundreds of years the Cerrado was primarily cattle country with poor-quality pastures producing low livestock yields.

In the past few decades, however, Brazil has developed more than 40 varieties of soybeans specially adapted for the soils and climate of the Cerrado. Most were developed through conventional breeding, but some are genetically modified for pesticide tolerance and other traits. With applications of lime and phosphorus, new varieties can quadruple yields of soybeans, maize, cotton, and other crops.

Until about 40 years ago, soybeans were a minor crop in Brazil. Since 1975, however, the total area planted with soy has doubled about every four years, reaching more than 25 million ha (60 million acres) in 2010. Although that's a large area, it represents only one-eighth of the Cerrado, more than half of which is still pasture.

Brazil is now the world's top soy exporter, shipping some 50 million metric tons per year, or about 10 percent more than the United States. With two crops per year, cheap land, low labor costs, favorable tax rates, and yields per hectare equal to those in the American Midwest, Brazilian farmers can produce soybeans for less than half the cost in America. Agricultural economists predict that by 2020 the global soy crop will be double the current 160 million metric tons per year, and that South America could be responsible for most of that growth. In addition to soy, Brazil now leads the world in exports of beef, maize, oranges, and coffee. This dramatic increase in South American agriculture helps answer the question of how the world's growing human population can be fed.

But it's not people who eat most of the soybeans; rather, it's livestock. A major factor in Brazil's current soy expansion is rising income in China. With more money to spend, the Chinese can afford to feed soy to pigs, chickens, cows, or fish. Meat consumption has grown rapidly, although it's still a fraction of what Americans eat. China now imports about 30 million tons of soy annually. About half of that comes from Brazil, which passed the United States in 2007 as the world's leading soy exporter. In 1997, Brazil shipped only 2 million tons of soy. A decade later, exports reached 28 million tons.

Concerns about mad cow disease (bovine spongiform encephalopathy, or BSE) in Europe, Canada, and Japan fueled increased worldwide demand for Brazilian beef. With 175 million free-range, grass-fed (and presumably BSE-free) cattle, Brazil has become the world's largest beef exporter.

Global demand creates conflicts over land in Brazil. The clearing of pasture and cropland is the leading cause of deforestation and habitat loss, most of which is occurring in the "arc of destruction" between the Cerrado and the Amazon. Small family farms are being gobbled up; many farmworkers, displaced by mechanization, have migrated either to the big cities or to frontier forest areas. Ongoing conflicts between poor farmers and big landowners have led to violent confrontations. The Landless Workers Movement claims that 1,237 rural workers died in Brazil between 1985 and 2000 as a result of assassinations and clashes over land rights.

FIGURE 10.1 Brazil's Cerrado, 2 million ha of savanna (grassland) and open woodland, is the site of the world's fastest growing soybean production. Cattle ranchers and agricultural workers, displaced by mechanized crop production, are moving northward into the "arc of destruction" at the edge of the Amazon rainforest, where the continent's highest rate of forest clearing is occurring.

In 2005 a 74-year-old Catholic nun, Sister Dorothy Stang, was shot by gunmen hired by ranchers who resented her advocacy for native people, workers, and environmental protection. Brazil claims that over the past 20 years it has resettled 600,000 families from the Cerrado. Still, tens of thousands of landless farmworkers and displaced families live in unauthorized squatter camps and shanty-towns across the country, awaiting relocation.

As you can see, rapid growth of beef and soy production in Brazil have both positive and negative aspects. On one hand, more high-quality food is now available to feed the world. The 2 million km² of the Cerrado represents one of the world's last opportunities to open a large area of new, highly productive cropland. On the other hand, the rapid expansion of agriculture in Brazil is destroying biodiversity and creating social conflict. The issues raised in this case study illustrate many of the major themes in this chapter. What factors limit farm production? What are the environmental and social consequences of producing our food? What sustainable approaches are available to help negotiate environmental and social priorities?

For related resources, including Google Earth™ place-marks that show locations discussed in this chapter, visit EnvironmentalScience-Cunningham.blogspot.com.

10.1 RESOURCES FOR AGRICULTURE

Agriculture has dramatically changed our environment, altering patterns of vegetation, soils, and water resources worldwide. The story of Brazil's Cerrado involves the conversion of millions of hectares of tropical savanna and rainforest to crop fields and pasture. This is one recent example of agricultural land conversion, but humans have been converting land to agriculture for thousands of years. Some of these agricultural landscapes are ecologically sustainable and have lasted for centuries or millennia. Others have depleted soil and water resources in just a few decades. What are the differences between farming practices that are sustainable and those that are unsustainable? What aspects of our current farming practices degrade the resources we depend on, and in what ways can farming help to restore and rebuild environmental quality? In this chapter we will examine some of the primary resources we use in farm production, how we use and abuse those resources, and some of the environmental consequences of the ways we cultivate the land.

As you have read in the opening case study, farm expansion has changed the landscape, the environment, and the economy of central Brazil. These changes are driven by financial investments and technological innovations from North American and European corporations. They are supported by rapidly expanding markets in Asia and Europe. But another essential factor has been the development of new ways to modify the region's nutrient-poor, acidic tropical soils. We will begin this chapter by exploring what soils are made of and how they differ from one place to another.

Soils are complex ecosystems

Is soil a renewable resource, or is it a finite resource that we are depleting? It's both. Over time, soil is renewable because it develops gradually through weathering of bedrock and through the accumulation of organic matter, such as decayed leaves and plant roots. But these processes are extremely slow. Building a few millimeters of soil can take anything from a few years (in a healthy grassland) to a few thousand years (in a desert or tundra). Under the best circumstances, topsoil accumulates at about 1 mm per year. With careful management that prevents erosion and adds organic material, soil can be replenished and renewed indefinitely. But most farming techniques deplete soil. Plowing exposes bare soil to erosion by wind or water, and annual harvests remove organic material such as leaves and roots. Severe erosion can carry away 25 mm or more of soil per year, far more than the 1 mm that can accumulate under the best of conditions.

Soil is a marvelous, complex substance. It is a combination of weathered rocks, plant debris, living fungi, and bacteria, an entire ecosystem that is hidden to most of us. In general, soil has six components:

1. *Sand and gravel* (mineral particles from bedrock, either in place or moved from elsewhere, as in windblown sand)

2. *Silts and clays* (extremely small mineral particles; clays are sticky and hold water because of their flat surfaces and ionic charges)

3. *Dead organic material* (decaying plant matter that stores nutrients and gives soils a black or brown color)

4. *Soil fauna and flora* (living organisms, including soil bacteria, worms, fungi, roots of plants, and insects, that recycle organic compounds and nutrients)

5. *Water* (moisture from rainfall or groundwater, essential for soil fauna and plants)

6. *Air* (tiny pockets of air help soil bacteria and other organisms survive)

Variations in these components produce almost infinite variety in the world's soils. Abundant clays make soil sticky and wet. Abundant organic material and sand make the soil soft and easy to dig. Sandy soils drain quickly, often depriving plants of moisture. Silt particles are larger than clays and smaller than sand (fig. 10.2),

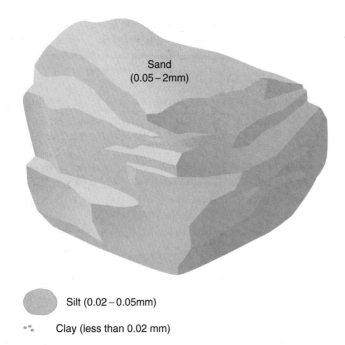

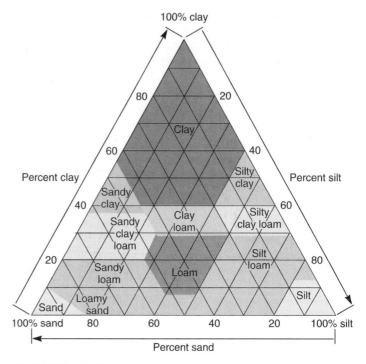

FIGURE 10.2 Relative sizes of soil particles magnified about 100-fold.

FIGURE 10.3 A soil's texture depends on its proportions of sand, clay, and silt particles. Read the graph by following lines across, up, or down from the axes. For example, "loam" has about 50–75 percent sand, 8–30 percent clay, and 18–50 percent silt. Loamy soils have the best texture for most crops, with enough sand to be loose and workable, yet enough silt and clay to retain water and nutrients.

so they aren't sticky and soggy, and they don't drain too quickly. Thus silty soils are ideal for growing crops, but they are also light and blow away easily when exposed to wind. Soils with abundant soil fauna quickly decay dead leaves and roots, making nutrients available for new plant growth. Compacted soils have few air spaces, making soil fauna and plants grow poorly.

You can see some of these differences just by looking at soil. Reddish soils, including most tropical soils, are colored by iron-rich, rust-colored clays, which store few nutrients for plants. Deep black soils, on the other hand, are rich in organic material, and thus rich in nutrients.

Soil texture—the amount of sand, silt, and clay in the soil—is one of the most important characteristics of soils. Texture helps determine whether rainfall drains away quickly or ponds up and drowns plants. Loam soils are usually considered best for farming because they have a mixture of clay, silt, and sand (fig. 10.3).

Most Brazilian tropical soils are deeply weathered, red clays. With frequent rainfall and year-round warm weather, organic material decays quickly and is taken up by living plants or washed away with rainfall. Red, iron-rich, clay soils result. These reddish clays hold few nutrients and little moisture for growing fields of soybeans. In contrast, the rich, black soils of the Corn Belt of the central United States have abundant organic matter and a good mix of sand, silt, and clay. These soils tend to hold enough moisture for crops without becoming waterlogged, and they tend to be rich in nutrients (fig. 10.4). Acidic tropical Brazilian soils can be improved by adding lime (calcium carbonate, as in limestone), which improves the soil's ability to retain nutrients applied in fertilizer. Liming vast areas was not economical until recently, but expanding markets for soybeans and beef in Asia and Europe now make it economical for Brazilian farmers to apply lime to their fields. This is one of the innovations that has allowed recent expansion of Brazilian soy production.

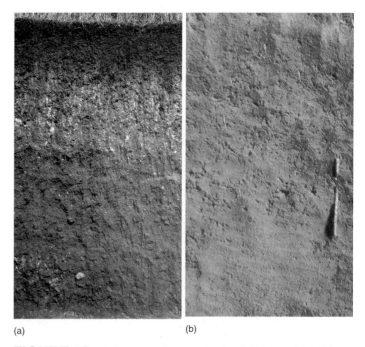

FIGURE 10.4 A temperate grassland soil (a) has a thick, black organic layer. Tropical rainforest soils (b) have little organic matter and are composed mostly of nutrient-poor, deeply weathered iron-rich clays. Each of these profiles is about 1 m deep.

Healthy soil fauna can determine soil fertility

Soil bacteria, algae, and fungi decompose and recycle leaf litter, making nutrients available to plants. These microscopic lifeforms also help to give soils structure and loose texture (fig. 10.5). The abundance of these organisms can be astonishing. One gram of soil can contain hundreds of soil bacteria and 20 m of tiny strands of fungal material. A cubic meter of soil can contain more than 10 kg of bacteria and fungal biomass. Tiny worms and nematodes process organic matter and create air spaces as they burrow through soil. Slightly larger insects, mites, spiders, and earthworms further loosen and aerate the soil. The sweet aroma of freshly turned soil is caused by actinomycetes, bacteria that grow in fungus like strands and give us the antibiotics streptomycin and tetracycline. These organisms mostly stay near the surface, often within the top few centimeters. The roots of plants can reach deeper, however, allowing moisture, nutrients, and organic acids to help break down rocks farther down, and to begin forming new soil.

Many plant species grow best with the help of particular species of soil fungi, in relationships called **mycorrhizal symbiosis**. In this relationship, the mycorrhizal fungus (a fungus growing on and around plant roots) provides water and nutrients to the plant, while the plant provides organic compounds to the fungus. Plants growing with their fungal partners often grow better than those growing alone.

The health of the soil ecosystem depends on environmental conditions, including climate, topography, and parent material (the mineral grains or bedrock on which soil is built), and frequency of disturbance. Too much rain washes away nutrients and organic matter, but soil fauna cannot survive with too little rain. In extreme cold, soil fauna recycle nutrients extremely slowly; in extreme heat they may work so fast that leaf litter on the forest floor is taken up by plants in just weeks or months—so that the soil retains little organic matter. Frequent disturbance prevents the development of a healthy soil ecosystem, as does steep topography that allows rain to wash away soils. In the United States, the best farming soils tend to occur where the climate is not too wet or dry, on glacial silt deposits, such as those in the upper Midwest, and on silt- and clay-rich flood deposits, like those along the Mississippi River.

Most soil fauna occur in the uppermost layers of a soil, where they consume leaf litter. This layer is known as the "O" (organic) horizon. Just below the O horizon is a layer of mixed organic and mineral soil material, the "A" horizon (fig. 10.6), or **surface soil**.

FIGURE 10.5 Soil ecosystems include numerous consumer organisms, as depicted here: (1) snail, (2) termite, (3) nematodes and nematode-killing constricting fungus, (4) earthworm, (5) wood roach, (6) centipede, (7) carabid (ground) beetle, (8) slug, (9) soil fungus, (10) wireworm (click beetle larva), (11) soil protozoan, (12) sow bug, (13) ant, (14) mite, (15) springtail, (16) pseudoscorpion, and (17) cicada nymph.

SOIL HORIZONS

O Surface litter

A Topsoil
organic matter (humus), living organisms, inorganic minerals

E Zone of leaching
dissolved or suspended materials move downward

B Subsoil
accumulation of iron, aluminum, humic compounds, and clay leached down from the A and E horizons

C Regolith
partially broken down inorganic minerals

Bedrock

FIGURE 10.6 Soil profile showing possible soil horizons. The actual number, composition, and thickness of these layers varies in different soil types.

The B horizon, or **subsoil**, tends to be richer in clays than the A; the B horizon is below most organic activity. The B layer accumulates clays that seep downward from the A horizon with rainwater that percolates through the soil. If you dig a hole, you may be able to tell where the B horizon begins, because there the soil tends to become slightly more cohesive. If you squeeze a handful of B-horizon soil, it should hold its shape better than a handful of A-horizon soil.

Sometimes an E (eluviated, or washed-out) layer lies between the A and B horizons. The E layer is loose and light-colored because most of its clays and organic material have been washed down to the B horizon. The C horizon, below the subsoil, is mainly decomposed rock fragments. Parent materials underlie the C layer. Parent material is the sand, windblown silt, bedrock, or other mineral material on which the soil is built. About 70 percent of the parent material in the United States was transported to its present site by glaciers, wind, and water, and is not related to the bedrock formations below it.

FIGURE 10.7 In many areas, soil or climate constraints limit agricultur[al] production. These hungry goats in Sudan feed on a solitary *Acacia* shrub.

Your food comes mostly from the A horizon

Ideal farming soils have a thick, organic-rich A horizon. The soils that support the Corn Belt farm states of the Midwest have a rich, black A horizon that can be more than 2 meters thick (although a century of farming has washed much of this soil away and down the Mississippi River). The A horizon in most soils is less than half a meter thick. Desert soils, with slow rates of organic activity, might have almost no O or A horizons (fig. 10.7).

Because topsoil is so important to our survival, we identify soils largely in terms of the thickness and composition of their upper layers. Soils vary endlessly in depth, color, and composition, but for simplicity we can describe a few general groups. The U.S. Department of Agriculture classifies the soils into 11 soil orders. These soils are described on the USDA website (soils.usda.gov/ technical/classification/orders/). In the Farm Belt of the United States, the dominant soils are mollisols (*mollic* = soft, *sol* = soil). These soils have a thick, organic-rich A horizon that developed from the deep, dense roots of prairie grasses that covered the region until about 150 years ago (see fig. 10.4). Another group that is important for farming is alfisols (*alfa* = first). Alfisols have a slightly thinner A horizon than mollisols do, and slightly less organic matter. Alfisols develop in deciduous forests, where leaf litter is abundant. In contrast, the aridisols (*arid* = dry) of the desert Southwest have little organic matter, and they often contain accumulations of mineral salts. Mollisols and alfisols dominate most of the farming regions of the United States.

10.2 WAYS WE USE AND ABUSE SOILS

Only about 11 percent of the earth's land area (1.5 billion ha out of 13.4 billion ha of land area) is currently in crop production. In theory, up to four times as much land could potentially be converted to cropland, but much of the remaining land is too steep, soggy, salty, cold, or dry for farming. In many developing countries, land continues to be cheaper than other resources, and forests and grasslands are still being converted to farmland. Brazil's expansion of soy farming into the Cerrado (opening case study) is one of the most rapid cases of land conversion; but ancient forests and grasslands are also turning into farmland in many parts of the developing world. The ecological costs of these land conversions are hard to calculate. Farmers can easily count the cash income from the farm products they sell, but it is never easy to calculate the value of biodiversity, clean water, and other ecological services of a forest or grassland, compared to the value of crops.

Arable land is unevenly distributed

The best agricultural lands occur where the climate is moderate—not too cold or too dry—and where thick, fertile soils are found. Take a look at the global map in the back of your book. What regions do you think of as the best agricultural areas? the poorest? Much of the United States, Europe, and Canada are fortunate to have temperate climates, abundant water, and high soil fertility. These produce good crop yields that contribute to high standards of living. Other parts of the world, although rich in land area, lack suitable soil, level land, or climates to sustain good agricultural productivity.

In developed countries, 95 percent of recent agricultural growth in the past century has come from a combination of improved crop varieties and increased use of fertilizers, pesticides, and irrigation. Conversion of new land to crop fields has contributed relatively little to increased production. In fact, less land is being cultivated now than 100 years ago in North America or 600 years ago in Europe. Productivity per unit of land has increased, and some marginal land has been retired. Careful management is important for preserving the remaining farmland.

Soil losses reduce farm productivity

Agriculture both causes and suffers from soil degradation (fig. 10.8). Every year about 3 million hectares of cropland are made unusable by erosion worldwide, and another 4 million hectares are converted to nonagricultural uses, such as urban land, highways, factories, or reservoirs, according to the International Soil Reference and Information Center (ISRIC). In the United States alone we've lost about 140 million hectares of farmland in the past 30 years to urbanization, soil degradation, and other factors (fig. 10.9).

Land degradation is usually slow and incremental. The land doesn't suddenly become useless, but it gradually becomes less fertile, as soil washes and blows away, salts accumulate, and organic matter is lost. About 20 percent of vegetated land in Africa and Asia is degraded enough to reduce productivity; 25 percent of lands in Central America and Mexico are degraded. Wind and water erosion

FIGURE 10.8 More than 43.7 million ha (108 million acres) in the United States are subjected to excess erosion by wind (red) or water (blue) each year. Each dot represents 200,000 tons of average annual soil loss.
Source: USDA Natural Resource Conservation Service.

(a)

(b)

(c)

FIGURE 10.9 Disastrous erosion during the Dust Bowl years (a) led to national erosion control efforts that have reduced, but not eliminated soil loss (b). Nationally, wind and water erosion have declined but continue to degrade farmland (c). **Source:** Natural Resource Conservation Service.

are the primary causes of degradation. Additional causes of degradation include chemical deterioration (mainly salt accumulation from salt-laden irrigation water) and physical deterioration (such as compaction by heavy machinery or waterlogging; fig. 10.10).

As a consequence of soil loss, as well as growth in population, the amount of arable land per person worldwide has shrunk from about 0.38 ha in 1970 to 0.21 ha in 2010. Consider that a hectare is an area 100 m × 100 m, or roughly the size of two football fields. On average, about five people are supported by that land area. By 2050 the arable land per person will decline to 0.15 ha. In the United States, farmland has fallen from 0.7 to 0.45 ha per person in the past 30 years, according to USDA data. To feed a growing population on declining land area, we are likely to need improvements in production methods, reduced consumption of protein (chapter 9), and improved soil management.

Wind and water move most soil

A thin layer taken off the land surface is called **sheet erosion**. When little rivulets of running water gather together and cut small channels in the soil, the process is called **rill erosion** (fig. 10.11*a*). When rills enlarge to form bigger channels or ravines that are too large to be removed by normal tillage operations, we call the process **gully erosion** (fig. 10.11*b*). Streambank erosion refers to the washing away of soil from the banks of established streams, creeks, or rivers, often as a result of removing trees and brush along streambanks and by cattle damage to the banks.

Most soil loss on agricultural land is sheet or rill erosion. Large amounts of soil can be transported a little bit at a time without being very noticeable. A farm field can lose 20 metric tons of soil per hectare during winter and spring runoff in rills so small that they are erased by the first spring cultivation. That represents a loss of only a few millimeters of soil over the whole surface of the field, hardly apparent to any but the most discerning eye. But it doesn't take much mathematical skill to see that if you lose soil twice as fast as it is being replaced, eventually it will run out.

Wind can equal or exceed water in erosive force, especially in a dry climate and on relatively flat land. When plant cover and surface litter are removed from the land by agriculture or grazing, wind lifts loose soil particles and sweeps them away. In extreme conditions, windblown dunes encroach on useful land and cover roads and buildings (fig. 10.11*c*). Over the past 30 years, China has lost 93,000 km² (about the size of Indiana) to **desertification**, or conversion of productive land to desert. Advancing dunes from the Gobi desert are now only 160 km (100 mi) from Beijing. Every year more than 1 million tons of sand and dust blow from Chinese drylands, often traveling across the Pacific Ocean to the west coast of North America.

Some of the highest erosion rates in the world occur in the United States and Canada. The U.S. Department of Agriculture reports that 69 million hectares (170 million acres) of U.S. farmland and range are eroding at rates that reduce long-term productivity. Five tons per acre (11 metric tons per hectare, or

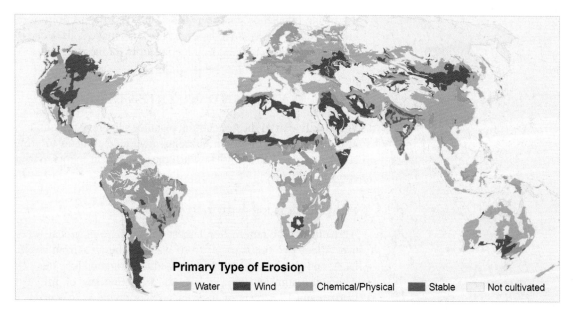

Primary Type of Erosion

☐ Water ■ Wind ☐ Chemical/Physical ■ Stable ☐ Not cultivated

FIGURE 10.10 Global causes of soil erosion and degradation. Globally, 62 percent of eroded land is mainly affected by water; 20 percent is mainly affected by wind.
Source: ISRIC Global Assessment of Human-Induced Soil Degradation, 2008.

about 1 mm depth) is generally considered the maximum tolerable rate of soil loss, because that is generally the highest rate at which soil forms under optimum conditions. Some farms lose soil at more than twice that rate.

Intensive farming practices are largely responsible for this situation. Row crops, such as corn and soybeans, leave soil exposed for much of the growing season (fig. 10.12). Deep plowing and heavy herbicide applications create weed-free fields that look neat but are subject to erosion. Because big machines cannot easily follow contours, they often go straight up and down the hills, creating ready-made gullies for water to follow. Farmers sometimes plow through grass-lined watercourses and have

pulled out windbreaks and fencerows to accommodate the large machines and to get every last square meter into production. Consequently, wind and water carry away the topsoil.

Pressed by economic conditions, many farmers have abandoned traditional crop rotation patterns and the custom of resting land as pasture or fallow every few years. Continuous monoculture cropping can increase soil loss tenfold over other farming patterns. A soil study in Iowa showed that a three-year rotation of corn, wheat, and clover lost an average of about 6 metric tons per hectare. By comparison, continuous wheat production on the same land caused nearly four times as much erosion, and continuous corn cropping resulted in seven times as much soil loss as the rotation with wheat and clover. The Mississippi River carries enough topsoil and fertilizer every year to create a "dead zone" in the Gulf of Mexico that can be as large as 57,000 km². Algal growth stimulated by high nitrogen in runoff from farms and cities depletes oxygen within this zone to levels that are lethal for most marine life. A task force recommended a 20 to 30 percent decrease in nitrogen loading to reduce the size and effects of this zone. Similar hypoxic zones occur near the mouths of many other rivers that drain agricultural areas (chapter 18).

(a) Sheet and rill erosion

(b) Gullying

(c) Wind erosion and desertification

FIGURE 10.11 Land degradation affects more than 1 billion ha yearly, or about two-thirds of all global cropland. Water erosion (a) and Gullying (b) accounts for about half that total. Wind erosion affects a nearly equal area (c).

FIGURE 10.12 Annual row crops leave soil bare and exposed to erosion for most of the year, especially when fields are plowed immediately after harvest, as this one always is.

Deserts are spreading around the world

According to the United Nations, about one-third of the earth's surface and the livelihoods of at least one billion people are threatened by desertification (conversion of productive lands to desert), which contributes to food insecurity, famine, and poverty. Former UN secretary general Kofi Annan called this a "creeping catastrophe" that creates millions of environmental refugees every year. Forced by economic circumstances to overcultivate and overgraze their land, poor people often are both the agents and the victims of desertification.

Rangelands and pastures, which generally are too dry for cultivation, are highly susceptible to desertification. According to the UN, 80 percent of the world's grasslands are suffering from overgrazing and soil degradation, and three-quarters of that area has undergone some degree of desertification. The world's 3 billion domestic grazing animals provide livelihood and food for many people, but can have severe environmental effects.

Two areas of particular concern are Africa and China. Arid lands, where rains are sporadic and infrequent and the economy is based mainly on crop and livestock raising, make up about two-thirds of the African continent. Nearly 400 million people live around the edges of these deserts. Rapid population growth and poverty create unsustainable pressures on the fragile soils of these areas. Stripping trees and land cover for fodder and firewood exposes the soil to erosion and triggers climate changes that spread desertification, which now affects nearly three-quarters of the arable land in Africa. The fringes of the two great African deserts, the Sahara and the Kalahari, are particularly vulnerable. About one-third of the 60 million people who required food aid in 2005 were victims of drought and desertification. Much of northern China, similarly, has little rainfall,

a growing population, and increasing land degradation from overgrazing. Finding ways to reduce pressure and rebuild soils is one of the important tasks in stabilizing food security for these regions.

10.3 WATER AND NUTRIENTS

Soil is only part of the agricultural resource picture. Agriculture is also dependent upon water, nutrients, favorable climates to grow crops, productive crop varieties, and the mechanical energy to tend and harvest them.

All plants need water to grow

Agriculture accounts for the largest single share of global water use. About two-thirds of all fresh water withdrawn from rivers, lakes, and groundwater supplies is used for irrigation (chapter 17). Although estimates vary widely (as do definitions of irrigated land), about 15 percent of all cropland, worldwide, is irrigated.

Some countries are water rich and can readily afford to irrigate farmland, while other countries are water poor and must use water very carefully. The efficiency of irrigation water use is rather low in most countries. High evaporative and seepage losses from unlined and uncovered canals often mean that in some places up to 80 percent of water withdrawn for irrigation never reaches its intended destination (chapter 17). Farmers often tend to overirrigate because water prices are relatively low and because they lack the technology to meter water and distribute just the amount needed. In the United States and Canada, many farmers are adopting water-saving technologies such as drip irrigation or downward-facing sprinklers (fig. 10.13).

Excessive use not only wastes water; it often results in **waterlogging**. Waterlogged soil is saturated with water, and plant roots die from lack of oxygen. **Salinization**, in which mineral salts accumulate in the soil, occurs particularly when soils in dry climates are irrigated with saline water. As the water evaporates, it leaves behind a salty crust on the soil surface that is lethal to most plants. Flushing with excess water can wash away this salt accumulation, but the result is even more saline water for downstream users.

Plants need nutrients, but not too much

In addition to water, sunshine, and carbon dioxide, plants need small amounts of inorganic nutrients for growth. The major elements required by most plants are nitrogen, potassium, phosphorus, calcium, magnesium, and sulfur. Calcium and magnesium often are limited in areas of high rainfall and must be supplied in the form of lime. Lack of nitrogen, potassium, and phosphorus even more often limits plant growth. Adding these elements in fertilizer usually stimulates growth and greatly increases crop yields. A good deal of the doubling in worldwide crop production since 1950 has come from increased use of inorganic fertilizer. In 1950 the average amount of fertilizer used was 20 kg per hectare. By 1990 this had increased to an average of 91 kg per hectare worldwide.

Some farmers overfertilize because they are unaware of the specific nutrient content of their soils or the needs of their crops. European farmers use more than twice as much fertilizer per hectare

FIGURE 10.13 Downward-facing sprinklers on this center-pivot irrigation system deliver water more efficiently than upwardfacing sprinklers.

as do North American farmers, but their yields are not proportionally higher. Phosphates and nitrates from farm fields and cattle feedlots are a major cause of aquatic ecosystem pollution. Nitrate levels in groundwater have risen to dangerous levels in many areas where intensive farming is practiced. Young children are especially sensitive to the presence of nitrates. Using nitrate-contaminated water to mix infant formula can be fatal for newborns.

What are some alternative ways to fertilize crops? Manure and green manure (crops grown specifically to add nutrients to the soil) are important natural sources of soil nutrients. Nitrogen-fixing bacteria living symbiotically in root nodules of legumes are valuable for making nitrogen available as a plant nutrient (chapter 3). Interplanting or rotating beans or some other leguminous crop with such crops as corn and wheat are traditional ways of increasing nitrogen availability.

There is considerable potential for increasing world food supply by increasing fertilizer use in low-production countries if ways can be found to apply fertilizer more effectively and reduce pollution. Africa, for instance, uses an average of only 19 kg of fertilizer per hectare, or about one-fourth of the world average. It has been estimated that the developing world could at least triple its crop production by raising fertilizer use to the world average.

Farming is energy-intensive

Farming as it is generally practiced in the industrialized countries is highly energy-intensive. Reliance on fossil fuels began in the 1920s with the adoption of tractors, and energy use increased sharply after World War II when nitrogen fertilizer made from natural gas became available. Reliance on diesel and gasoline to run tractors, combines, and other machinery has continued to grow in recent decades. Agricultural economist David Pimentel of Cornell University has calculated the many energy inputs, from fertilizer and pesticides to transportation and irrigation. His estimate amounts to an equivalent of 800 liters of oil (5 barrels of oil) per hectare of corn produced in the United States. A third of this energy is used in producing the nitrogen fertilizer applied to fields.

Inputs for machinery and fuel make up another third; herbicides, irrigation, and other fertilizers make up the rest.

After crops leave the farm, additional energy is used in food processing, distribution, storage, and cooking. It has been estimated that the average food item in the American diet travels 2,400 km between the farm that grew it and the person who consumes it. The energy required for this complex processing and distribution system may be five times as much as is used directly in farming. Altogether the food system in the United States consumes about 16 percent of the total energy we use. Most of our foods require more energy to produce, process, and get to market than they yield when we eat them. A British study concluded that eating locally grown food has less environmental impact—even if produced with conventional farming—than organic food from far away.

10.4 PESTS AND PESTICIDES

Every ecosystem has producers and consumers, but in an agricultural system we do our best to simplify the ecosystem to just one type of producer (the crop plant, usually corn or soybeans in the United States) and one type of consumer (humans). This means that other consumers, such as crop-eating insects or fungi, need to be controlled. Although deer are the single largest cause of crop damage in the United States, we spend most of our attention on controlling smaller crop pests, especially insects that attack crops.

Pesticide is a general term for a chemical that kills pests, usually a toxic chemical, but sometimes we also consider chemicals that drive pests away to be pesticides. Some pest-control compounds kill a wide range of living things and are called **biocides** (fig. 10.14). Chemicals such as ethylene dibromide that are used to protect stored grain, or to sterilize soils before planting strawberries, are biocides. In addition, there are chemicals aimed at particular groups of pests. **Herbicides** are chemicals that kill plants; **insecticides** kill insects; and **fungicides** kill fungi.

FIGURE 10.14 Broad-spectrum toxins can eliminate pests quickly and efficiently, but what are the long-term costs to us and to our environment?

Synthetic (artificially made) chemical pesticides have been one of the dominant innovations of modern agricultural production. Our use of pesticides has increased dramatically in recent years (see Data Analysis at the end of this chapter), although pesticides receive relatively little public attention in most areas. Our current food system relies heavily on synthetic chemicals to control pests. These compounds have brought many benefits, but they also bring environmental problems (chapter 8). In this section we will review some of the main types of pesticides, how they work, and some alternative strategies.

People have always used pest controls

Using chemicals to control pests may well have been among our earliest forms of technology. People in every culture have known that salt, smoke, and insect-repelling plants can keep away bothersome organisms and preserve food. The Sumerians controlled insects and mites with sulfur 5,000 years ago. Chinese texts 2,500 years old describe mercury and arsenic compounds used to control body lice and other pests. Greeks and Romans used oil sprays, ash and sulfur ointments, lime, and other natural materials to protect themselves, their livestock, and their crops from a variety of pests.

In addition to these metals and inorganic chemicals, people have used organic compounds, biological controls, and cultural practices for a long time. Alcohol from fermentation and acids in pickling solutions prevent growth of organisms that would otherwise ruin food. Spices were valued both for their flavors and because they deterred spoilage and pest infestations. Romans burned fields and rotated crops to reduce crop diseases. They also employed cover crops to reduce weeds. The Chinese developed plant-derived insecticides and introduced predatory ants in orchards to control caterpillars 1,200 years ago.

Modern pesticides provide benefits but also create problems

The era of synthetic organic pesticides began in 1939 when Swiss chemist Paul Müller discovered the powerful insecticidal properties of dichloro-diphenyl-trichloroethane (DDT). Inexpensive, stable, easily applied, and highly effective, this compound seemed ideal for crop protection and disease prevention. DDT is remarkably lethal to a wide variety of insects but relatively nontoxic to mammals. Mass production of DDT started during World War II, when Allied armies used it to protect troops from insect-borne diseases. In less than a decade, manufacture of the compound soared from a few kilograms to thousands of metric tons per year. It was sprayed on crops and houses, dusted on people and livestock, and used to combat insects nearly everywhere (fig. 10.15).

By the 1960s, however, evidence began to accumulate that indiscriminate use of DDT and other long-lasting industrial toxins was having unexpected effects on wildlife. Peregrine falcons, bald eagles, brown pelicans, and other carnivorous birds were disappearing from former territories in eastern North America. Studies revealed that eggshells were thinning in these species as DDT and its breakdown products were concentrated through food chains

FIGURE 10.15 Before we realized the toxicity of DDT, it was sprayed freely on people to control insects as shown here at Jones Beach, New York, in 1948.

until it reached endocrine hormone-disrupting levels in top predators (see fig. 10.14). In 1962 biologist Rachel Carson published *Silent Spring,* warning that persistent organic pollutants, such as DDT, pose a threat to wildlife and perhaps to humans. DDT was banned for most uses in developed countries in the late 1960s, but it continues to be used in developing countries and remains the most prevalent contaminant on food imported to the United States.

Since the 1940s many new synthetic pesticides have been invented. Many of them, like DDT, have proven to have unintended consequences on nontarget species. Assessing the relative costs and benefits of using these compounds continues to be a contentious topic, especially when unexpected complications arise, such as increasing pest resistance or damage to beneficial insects.

According to the EPA, total pesticide use in the United States amounts to about 5.3 billion pounds (2.4 million metric tons) per year. Roughly half of that amount is chlorine and hypochlorites (bleach) used for water purification (fig. 10.16). Eliminating

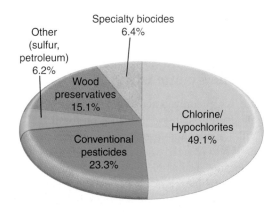

FIGURE 10.16 Of the 5.3 billion pounds of pesticides used in the United States each year is chlorine/hypochlorite disinfectant. Specialty biocides include other antiseptics and sanitizers.
Source: U.S. EPA 2000.

pathogens from drinking water prevents a huge number of infections and deaths, but there's concern that using so much chlorine and hypochlorite to do so may be creating other chronic health risks. The next largest category is conventional pesticides: primarily insecticides, herbicides, and fungicides.

Specialty biocides, such as preservatives used in adhesives and sealants, paints and coatings, leather, petroleum products, and plastics as well as recreational and industrial water treatment amount to some 300 million pounds per year, although they represent only about 6 percent of total pesticide use. The "other" category in figure 10.3 includes sulfur, oil, and chemicals used for insect repellents (such as DEET) and moth control. Wood preservatives represent just 15 percent of total pesticide consumption, but can be especially dangerous to our health because they tend to be both highly toxic and very long-lasting.

Information on pesticide use is often poorly reported, but the U.S. EPA estimates that world usage of conventional pesticides amounts to some 5.7 billion pounds (2.6 million metric tons) per year of active ingredients. In addition, "inert" ingredients are added to pesticides as carriers, stabilizers, emulsifiers, and so on.

Roughly 80 percent of all conventional pesticides applied in the United States are used in agriculture or food storage and shipping. Some 90 million ha of crops in the United States—including 96 percent of all corn and about 90 percent of soybeans—are treated with herbicides every year. In addition, 25 million ha of agricultural fields and 7 million ha of parks, lawns, golf courses, and other lands are treated with insecticides and fungicides. By some accounts, cotton has the highest rate of insecticide application of any crop, while golf courses often have higher rates of application of all conventional pesticides per unit area than any farm fields.

Household uses in homes and gardens account for the fastest-rising sector, about 14 percent of total use, according to the most recent available EPA estimates (from 2001). Three-quarters of all American homes use some type of pesticide, amounting to 20 million applications per year. Often people use much larger quantities of chemicals in their homes, yards, or gardens than farmers would use to eradicate the same pests in their fields. Storage and accessibility of toxins in homes also can be a problem. Children's exposure to toxins in their home may be of greater concern than pesticide residues in food. Health effects of these compounds are discussed in chapter 8.

Global use of pesticides is also hard to evaluate, but the UN Food and Agriculture Organization reports international expenditures on exports and imports. These measures have risen about 60-fold since data collection began in 1962 (fig. 10.17). Approximately 20 percent of global pesticide use is in the United States, according to the U.S. EPA.

There are many types of pesticides

One way to classify pesticides is by their chemical structure and main components. Some are organic (carbon-based) compounds; others are toxic metals (such as arsenic) or halogens (such as bromine).

Organophosphates are among the most abundantly used synthetic pesticides. Glyphosate, the single most heavily used herbicide in the United States, is also known by the trade name

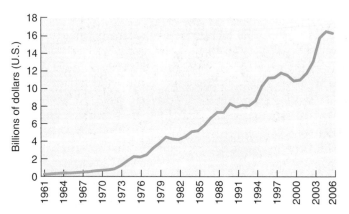

FIGURE 10.17 Value of global trade in pesticides (imports).
Source: Data from the UN Food and Agriculture Organization, 2009.

Roundup. Glyphosate is applied to 90 percent of U.S. soybeans, as well as to other crops. "Roundup-ready" soybeans and corn— varieties genetically modified to tolerate glyphosate while other plants in the field are destroyed—are the most commonly planted genetically modified crops (chapter 9), and these tolerant varieties are one of the factors that make expanding soy production cost-effective in Brazil (opening case study). These "Roundup-ready" varieties have helped glyphosate surpass atrazine (an herbicide used mainly on corn) as the most-used herbicide (fig. 10.18).

Other organophosphates attack the nervous systems of animals and can be dangerous to humans, as well. Parathion, malathion, dichlorvos, and other organophosphates were developed as an outgrowth of nerve gas research during World War II. These compounds can be extremely lethal. Because they break down quickly, usually in just a few days, they are less persistent in the environment than other pesticides. These compounds are very dangerous

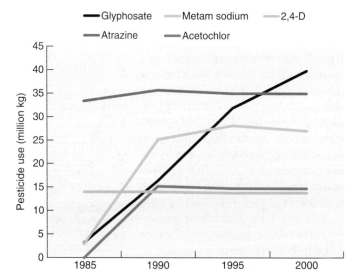

FIGURE 10.18 Usage of the top five pesticides in the United States. All are herbicides applied to soy, corn, or wheat, or to lawns, except metam sodium, a soil fumigant used mainly on ground crops such as carrots, potatoes, peppers, and strawberries.
Source: USDA, 2009.

FIGURE 10.19 The United Farm Workers of America claims that 300,000 farmworkers in the United States suffer from pesticide-related illnesses each year. Worldwide, the WHO estimates that 25 million people suffer from pesticide poisoning and 20,000 die each year from improper use or storage of pesticides.

for workers, however, who are often sent into fields too soon after they have been sprayed (fig. 10.19).

Chlorinated hydrocarbons, also called organochlorines, are persistent and highly toxic to sensitive organisms. Atrazine was the most heavily used herbicide in the United States until the recent increase in glyphosate use. Atrazine is applied to 96 percent of the corn crop in the United States to control weeds in cornfields (fig. 10.20). The widespread use has resulted in concerns about contamination of water supplies. One study of Midwestern Corn Belt states found atrazine in 30 percent of community wells and 60 percent of private wells sampled. This is a worry because atrazine has been linked to sexual abnormalities and population crashes in frogs. Because of its persistence and uncertain health effects, atrazine was banned in Europe in 2003. Among the hundreds of other organochlorines are DDT, chlordane, aldrin, dieldrin, toxaphene, and paradichlorobenzene (mothballs). This group also includes the herbicide 2,4-D, a widely used lawn chemical that selectively suppresses broad-leaf flowering plants, such as dandelions.

Chlorinated hydrocarbons can persist in the soil for decades, and they are stored in fatty tissues of organisms, so they become concentrated through food chains. DDT, which was inexpensive and widely used in the 1950s, has been banned in most developed countries, but it is still produced in the United States and it is used in many developing

countries. Toxaphene is extremely toxic for fish and can kill goldfish at five parts per billion (5 µg/liter).

Fumigants are generally small molecules, such as carbon tetrachloride, ethylene dibromide, and methylene bromide, which can be delivered in the form of a gas so that they readily penetrate soil and other materials. Fumigants are used to control fungus in strawberry fields and other low-growing crops, as well as to prevent decay or rodent and insect infestations in stored grain. Because these compounds are extremely dangerous for workers who apply them, many have been restricted or banned altogether in some areas.

Inorganic pesticides include compounds of toxic elements such as arsenic, sulfur, copper, and mercury. These broad-spectrum poisons are generally highly toxic and indestructible, remaining in the environment forever. They generally act as nerve toxins. Historically, arsenic powder was a primary pesticide applied to apples and other orchard crops, and traces remain in soil and groundwater in many agricultural areas.

Natural organic pesticides, or "botanicals," generally are extracted from plants. Some important examples are nicotine and nicotinoid alkaloids from tobacco, and pyrethrum, a complex of chemicals extracted from the daisy-like *Chrysanthemum cinerariaefolium* (fig. 10.21). These compounds also include turpentine, phenols, and other aromatic oils from conifers. All are toxic to insects, and many prevent wood decay.

Microbial agents and **biological controls** are living organisms or toxins derived from them that are used in place of pesticides. A natural soil bacterium, *Bacillus thuringiensis* is one of the chief pest-control agents allowed in organic farming. This bacterium kills caterpillars and beetles by producing a toxin that ruptures the digestive tract lining when eaten. Parasitic wasps such

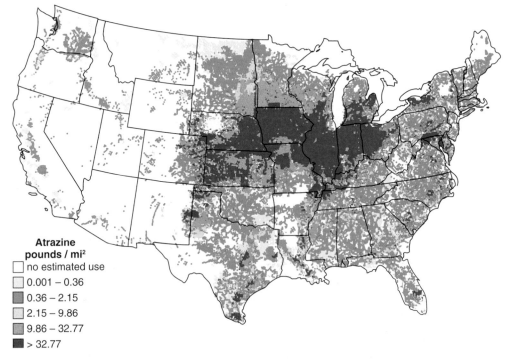

Atrazine
pounds / mi²
☐ no estimated use
☐ 0.001 – 0.36
■ 0.36 – 2.15
☐ 2.15 – 9.86
▨ 9.86 – 32.77
■ > 32.77

FIGURE 10.20 Atrazine herbicide use, average pounds per square mile of farmland.
Source: USDA.

FIGURE 10.21 Chrysanthemum flowers are a source of pyrethrum, a natural insecticide.

FIGURE 10.22 This machine sprays insecticide on orchard trees—and everything nearby. Up to 90 percent of pesticides applied in this fashion never reach target organisms.

What Do You Think?

Organic Farming in the City

Farming is remote for most of us, with some 85 percent of Americans living in cities. We eat foods grown far away, processed in anonymous factories, delivered by national grocery store chains. But a growing movement has been reclaiming food production for city-dwellers. Urban farming, urban gardening, and community gardens are just a few of the names and strategies people in cities are taking to bring some of their food closer to home.

One of the leading examples is Growing Power, an organization formed in Milwaukee, Wisconsin by the former basketball player Will Allen, who received a MacArthur "genius" award for his work. Like many older industrial cities, Milwaukee has seen declining population, housing values, incomes, and economic opportunity for decades. Low-income or unemployed minority groups make up an increasing proportion of the city. Young people have few jobs, few training resources, and little food security.

Difficult conditions can make a fertile ground for a movement promoting self-sustaining food production. Will Allen brought together unemployed teenagers and other community members, starting on just a 2-acre (0.8 hectare) plot of farmland. Allen's organization teaches kids and their parents to improve the soil with compost and mulch, to grow vegetables, tilapia, chickens and other foods, to manage a business and sell food. Growing Power serves kids by teaching them skills and providing internships and paid employment. The organization serves the community by providing a positive focus that brings people together. Who doesn't like fresh food grown by friends? The organization also serves the city by providing wholesome food resources that supports the health and food security of low-income neighborhoods.

One of the first steps Growing Power took was to become a land trust, an organization that could take long-term control of the land they work. This stability allows them to invest in the soil and in green-houses, in projects and plans. Another step they have taken is to provide workshops that spread their philosophy and techniques nationwide. Growing power gives people access to fresh food, teaches kids about nurturing the land, and most important, it invests in the next generation of citizens of Milwaukee and other cities.

Urban farming and gardening movements are growing rapidly and have rich potential. One study found that East Lansing, Michigan could produce 75 percent of its own vegetables on 4,800 acres (2,000 ha) of unbuilt land. Motivations for urban farming include, but are not limited to, issues of food security, community stability, youth employment, improving environmental quality for kids, and fun. Creative and enthusiastic projects abound, from Brooklyn to Detroit to Portland Oregon, and many places between. Do you think community gardens or urban farming would be useful in areas where you live? What do you think it would take to support these efforts in your area?

Urban farming helps young people and their communities grow stronger. These girls are selling produce from Capuchin Soup Kitchen/Earthworks Urban Farm in Detroit, MI.

as the tiny *Trichogramma* genus attack moth caterpillars and eggs, while lacewings and ladybugs are predators that control aphids.

10.5 ENVIRONMENTAL EFFECTS OF PESTICIDES

Although we depend on pesticides for most of our food production, and for other purposes such as biofuel production, widespread use of these compounds brings a number of environmental and health risks. The most common risk is exposure of nontarget organisms. Many pesticides are sprayed broadly and destroy populations of beneficial insects as well as pests (fig. 10.22). The loss of insect diversity has been a growing problem in agricultural regions: at least a third of the crops we eat rely on pollinators, such as bees and other invertebrates, to reproduce.

The disappearance of honeybees has received particular attention in recent years. Many crops, including squash, tomatoes, peppers, apples, and other fruit, rely on bees for pollination, and it is estimated that the economic value of bees for pollination is 100 times the value of their honey. The California almond crop, for example, is worth $1.6 billion annually and is entirely dependent on bees for pollination. For unknown reasons, honeybee hives have been dying, and while there are many possible explanations, pesticide spraying is one of the chief suspects. Other crops, including blueberries and alfalfa, have been devastated by the loss of wild pollinators.

Pest resurgence, or the rebound of resistant populations, is another important problem in overuse of pesticides. This process occurs when a few resistant individuals survive pesticide treatments, and those resistant individuals propagate a new pesticide-resistant population. The Worldwatch Institute reports that at least 1,000 insect pest species and another 550 or so weeds and plant pathogens worldwide have developed chemical resistance. Of the 25 most serious insect pests in California, three-quarters or more are resistant to one or more insecticides. Cornell University entomologist David Pimentel reports that a larger percentage of crops are lost now to insects, diseases, and weeds than in 1944, despite the continuing increase in the use of pest controls (fig. 10.23).

As resistant pests evolve, there is an ever-increasing need for newer, better pesticides—this is called the **pesticide treadmill**. Glyphosate (Roundup), the dominant herbicide used in the United States and one of the primary herbicides in Brazil, Australia, and elsewhere, is no longer effective against a variety of superweeds. Increasingly farmers are advised to mix tanks of various pesticides—metachlor, Flexstar, Gramoxone, diuron, and other combinations are recommended to keep down increasingly aggressive weeds such as pigweed and rye grass. At the same time, ever-larger amounts of glyphosate are needed to combat resistant weeds. In 2010 the U.S. Supreme Court reversed a ban on genetically modified glyphosate-tolerant alfalfa, a decision that crop scientists expect will increase pesticide-tolerant weeds on the 22 million acres of alfalfa grown in the United States. Increasing reliance on glyphosate and other herbicides is sure to increase environmental exposure, with uncertain effects on human health and ecosystems.

Think About It

Pesticide residues in food are a major concern for many people, but most of us also use toxic chemicals in other aspects of our life as well. Look around your home, how many different toxic products can you find? Are they all necessary? Would you have alternatives to these products?

POPs accumulate in remote places

Many pesticides break down to less-harmful components several days or weeks after application. Certain compounds, such as DDT and other chlorinated hydrocarbons, are both effective and dangerous because they don't break down easily. **Persistent organic pollutants (POPs)** is a collective term for these chemicals, which are stable, easily absorbed into fatty tissues, and highly toxic.

Because they persist for years, even decades in some cases, and move freely through air, water, and soil, they often show up far from the point of original application. Some of these compounds have been discovered far from any possible source and long after they most likely were used. Because they have an affinity for fat, many chlorinated hydrocarbons are bio-concentrated and stored in the bodies of predators—such as porpoises, whales, polar bears, trout, eagles, ospreys, and humans—that feed at the top of food webs. In a study of human pesticide uptake and storage, Canadian researchers found that the level of chlorinated hydrocarbons in the breast milk of Inuit mothers living in remote arctic villages was five times that of women from Canada's industrial region some 2,500 km (1,600 mi) to the south. Inuit people have the

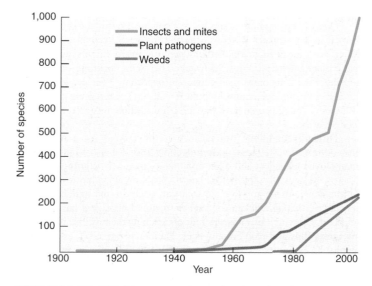

FIGURE 10.23 Many pests have developed resistance to pesticides. Because insecticides were the first class of pesticide to be used widely, selection pressures led insects to show resistance early. More recently, plant pathogens and weeds are also becoming insensitive to pesticides.
Source: Worldwatch Institute, 2003.

highest levels of these persistent pollutants of any human population except those contaminated by industrial accidents.

These compounds accumulate in polar regions by what has been called the "grasshopper effect," in which contaminants evaporate from water and soil in warm areas and then condense and precipitate in colder regions. In a series of long-distance grasshopperlike jumps they eventually collect in polar regions, where they accumulate in top predators. Polar bears, for instance, have been shown to have concentrations of certain chlorinated compounds 3 billion times greater than in the seawater around them. In Canada's St. Lawrence estuary, beluga (white whales), which suffer from a wide range of infectious diseases and tumors thought to be related to environmental toxins, have such high levels of chlorinated hydrocarbons that their carcasses must be treated as toxic waste.

Because POPs are so long-lasting and so dangerous, 127 countries agreed in 2001 to a global ban on the worst of them, including aldrin, chlordane, dieldrin, DDT, endrin, hexachlorobenzene, neptachlor, mirex, toxaphene, polychlorinated biphenyls (PCBs), dioxins, and furans. Most of this "dirty dozen" had been banned or severely restricted in developed countries for years. However, their production has continued. Between 1994 and 1996, U.S. ports shipped more than 100,000 tons of POPs each year. Most of this was sent to developing countries where regulations were lax. Ironically, many of these pesticides returned to the United States on bananas and other imported crops. According to the 2001 POPs treaty, eight of the dirty dozen were banned immediately; PCBs, dioxins, and furans are being phased out; and use of DDT, still allowed for limited uses such as controlling malaria, must be publicly registered in order to permit monitoring. The POPs treaty has been hailed as a triumph for environmental health and international cooperation. Unfortunately other compounds—perhaps just as toxic—have been introduced to replace POPs.

Pesticides cause a variety of health problems

Pesticide effects on human health can be divided into two categories: (1) acute effects, including poisoning and illnesses caused by relatively high doses and accidental exposures, and (2) chronic effects suspected to include cancer, birth defects, immunological problems, endometriosis, neurological problems, Parkinson's disease, and other chronic degenerative diseases.

The World Health Organization (WHO) estimates that 25 million people suffer pesticide poisoning and at least 20,000 die each year (fig. 10.24). At least two-thirds of this illness and death results from occupational exposures in developing countries where people use pesticides without proper warnings or protective clothing. A tragic example of occupational pesticide exposure is found among workers in the Latin American flower industry. Fueled by the year-round demand in North America for fresh vegetables, fruits, and flowers, a booming export trade has developed in countries such as Guatemala, Colombia, Chile, and Ecuador. To meet demands in North American markets for perfect flowers, table grapes, and other produce, growers use high levels of pesticides, often spraying daily with fungicides, insecticides, nematicides, and herbicides. Working in warm, poorly ventilated greenhouses with little protective clothing, the

FIGURE 10.24 Handling pesticides requires protective clothing and an effective respirator. Pesticide applicators in tropical countries, however, often can't afford these safeguards or can't bear to wear them because of the heat.

workers—70 to 80 percent of whom are women—find it hard to avoid pesticide contact. Almost two-thirds of nearly 9,000 workers surveyed in Colombia experienced blurred vision, nausea, headaches, conjunctivitis, rashes, and asthma. Although harder to document, they also reported serious chronic effects such as stillbirths, miscarriages, and neurological problems.

Pesticide use can expose consumers to agricultural chemicals. In studies of a wide range of foods collected by the USDA, the State of California, and the Consumers Union between 1994 and 2000, 73 percent of conventionally grown food had residue from at least one pesticide and were six times as likely as organic foods to contain multiple pesticide residues. Only 23 percent of the organic samples of the same groups had any residues. Using these data, the Environmental Working Group has assembled a list of the fruits and vegetables most commonly contaminated with pesticides (table 10.1).

Table 10.1	The Twelve Most Contaminated Foods
Rank	**Food**
1.	Strawberries
2.	Bell peppers
3.	Spinach
4.	Cherries (U.S.)
5.	Peaches
6.	Cantaloupe (Mexican)
7.	Celery
8.	Apples
9.	Apricots
10.	Green beans
11.	Grapes (Chilean)
12.	Cucumbers

Source: Environmental Working Group, 2002.

10.6 Organic and Sustainable Agriculture

Many farmers and consumers are turning to organic agriculture as a way to reduce pesticide exposure. Sustainable farming can include a multitude of strategies, such as planting nitrogen-fixing plants to avoid fertilizers, using crop rotation to minimize pesticides, strategic water management, mixed cropping, use of perennial or tree crops, and many others. In general soils stay healthier with these strategies than with chemical-intensive monoculture cropping. A Swiss study spanning two decades found that average yields on organic plots were 20 percent less than on adjacent fields farmed by conventional methods, but costs also were lower and prices paid for organic produce were higher, so that net returns were actually higher with organic crops. Energy use was 56 percent less per unit of yield in organic farming than for conventional approaches. In addition, beneficial root fungi were 40 percent higher, earthworms were three times as abundant, and spiders and other pest-eating predators were doubled in the organic plots. The organic farmers and their families reported better health and greater satisfaction than did their neighbors who used conventional farming methods (fig. 10.25). A study of food quality in Sweden reported that organic food contained more cancer-fighting polyphenolics and antioxidants than did pesticide-treated produce. Moreover, farms using sustainable techniques can have up to 400 times less erosion after heavy rains than monoculture row crops.

Can sustainable practices feed the world's growing population? This question is hotly debated. Proponents of conventional agriculture charge that sustainable methods are a boutique strategy incapable of feeding large or poor populations. Proponents of sustainable agriculture say sustainable methods are more productive over time, and that conventional practices degrade the health of soils, waterways, ecosystems, farmers, and consumers. Some proponents of organic production argue that the rapid spread of industrial farming serves mainly the multinational agrochemical corporations, which need to expand markets. Most agricultural research institutions, on the other hand, argue that without innovations in high-responding varieties and pesticides, we would have seen mass starvation in the past 40 years.

In 2011 the UN Commission on Human Rights and the UN Food and Agriculture Organization (FAO) both weighed in on the matter. Their studies indicate that if the aim is to provide food for impoverished regions, then states should promote innovations in sustainable soil-building and water-preserving methods. Data indicate that costs of irrigation, pesticides, fuel, and newly developed seed varieties have risen at least three times as fast as farm income. Resulting high debt and widespread farm failures have forced farmers off their land and into the already-overcrowded cities of the developing world. Studies by the FAO show that areas that have invested in conservation-based farming innovations have increased yields at a rate similar to that of green revolution or genetically modified (GM) crops, while offering more sustainable food security in low-income regions.

Currently less than 1 percent of all American farmland is devoted to organic growing, but the market for organic products may stimulate more conversion to this approach in the future. Organic food is much more popular in Europe than in North America. Tiny Liechtenstein is probably the leader among industrialized nations with 18 percent of its land in certified organic agriculture. Sweden is second with 11 percent of its land in organic production. Much of the developing world is effectively organic, where people can afford few fertilizers or pesticides.

Are organic methods pie in the sky or a necessary strategy? The answer might depend on whether you live in Africa or North America, what evidence you have seen, your beliefs about how much meat the world needs. how you feel about using farmland to produce ethanol and other biofuels, and many other issues.

What does "organic" mean?

In general, organic food is grown without artificial pesticides and with only natural fertilizers, such as manure. Legal definitions of the term, though, are more exact and often more controversial. According to U.S. Department of Agriculture rules, products labeled "100 percent organic" must be produced without hormones, antibiotics, pesticides, synthetic fertilizers, or genetic modification. "Organic" means that at least 95 percent of the ingredients must be organic. "Made with organic ingredients" must contain at least 70 percent organic contents. Products containing less than 70 percent organic ingredients can list them individually. Organic animals must be raised on organic feed, given access to the outdoors, given no steroidal growth hormones, and treated with antibiotics only to treat diseases.

Wal-Mart has become the top seller of organic products in the United States, a step that has done much to move organic products into the mainstream. However, much of the organic food, cotton, and other products we buy from nonlocal producers now comes from overseas, where oversight can be even more difficult than it

FIGURE 10.25 These strawberries were grown organically, but the USDA finds more pesticides in commercial strawberries than in any other fruit.

FIGURE 10.26 Your local farmers' market is a good source of locally grown and organic produce.

is within the United States. More than 2,000 farms in China and India are certified "organic," but how can we be sure what that means? With the market for organic food generating $11 billion per year, it's likely that some farmers and marketers try to pass off foods grown with pesticides as more valuable organic produce. Industrial-scale organic agriculture can also be hard on soils: it often depends on frequent cultivation for weed control, and the constant mechanical disturbance can destroy soil texture and soil microbial communities.

Many who endorse the concept of organic food are disappointed that legal definitions in the United States allow for partial organics and for nonsustainable production methods. The term *organic* is also hard to evaluate clearly when you can buy organic intercontinental grapes in which thousands of calories of jet and diesel fuel were consumed to transport every calorie of food energy from Chile to your supermarket, or when you can buy processed snack foods labeled "organic." Many farmers have declined to pay for organic certification because they regard the term as too broad to be meaningful. Alternative descriptions such as "sustainable" or "natural" are often used, but these terms are also vague. Often the key is to pay attention to how or where our foods are produced. Seeking out local foods is another way to ensure that food is produced in socially and environmentally benign ways (fig. 10.26). Supporting local producers and farmers' markets also benefits the local community and economy. (See What Can You Do?, p. 215.)

Strategic management can reduce pests

Organic farming and sustainable farming use a multitude of practices to control pests. In many cases, improved management programs can cut pesticide use by 50 to 90 percent without reducing crop production or creating new diseases. Some of these techniques are relatively simple and save money while maintaining disease control and yielding crops with just as high quality and quantity as we get with current methods. In this section, we will examine crop management, biological controls, and integrated pest management systems that could substitute for current pest-control methods.

Crop rotation involves growing a different crop in a field each year in a two- to six-year cycle. Most pests are specific to one crop, so rotation keeps pest populations from increasing from year to year. For instance, a three-year soybean/corn/hay rotation is effective and economical protection against white-fringed weevils. Mechanical cultivation keeps weeds down, but it also increases erosion. Flooding fields before planting or burning crop residues and replanting with a cover crop can suppress both weeds and insect pests. Habitat diversification, such as restoring windbreaks, hedgerows, and ground cover on watercourses, provides habitat for insect predators, such as birds, and also reduces erosion. Adjusting the timing of planting or cultivation can help avoid pest outbreaks. Switching from vast monoculture fields to mixed polyculture (many crops grown together) makes it more difficult for pests to find the crops they like.

Useful organisms can help us control pests

Biological controls such as predators (wasps, ladybugs, praying mantises; fig. 10.27) or pathogens (viruses, bacteria, fungi) can control many pests more cheaply and safely than broad-spectrum, synthetic chemicals. *Bacillus thuringiensis* or Bt, for example, is a naturally occurring bacterium that kills the larvae of lepidopteran (butterfly and moth) species but is generally harmless to mammals. A number of important insect pests such as tomato hornworm, corn rootworm, cabbage loopers, and others can be controlled by spraying bacteria on crops. Larger species are effective as well. Ducks, chickens, and geese, among other species, are used to rid fields of both insect pests and weeds. These biological organisms are self-reproducing and often have wide prey tolerance. A few mantises or ladybugs released in your garden in the spring will keep producing offspring and protect your fruits and vegetables against a multitude of pests for the whole growing season.

Herbivorous insects have been used to control weeds. For example, the prickly pear cactus was introduced to Australia about 150 years ago as an ornamental plant. This hardy cactus escaped from gardens and found an ideal home in the dry soils of the outback.

FIGURE 10.27 The praying mantis looks ferocious and is an effective predator against garden pests, but it is harmless to humans. They can even make interesting and useful pets.

FIGURE 10.28 A Nigerian woman examines a neem tree, the leaves, seeds, and bark of which provide a natural insecticide.

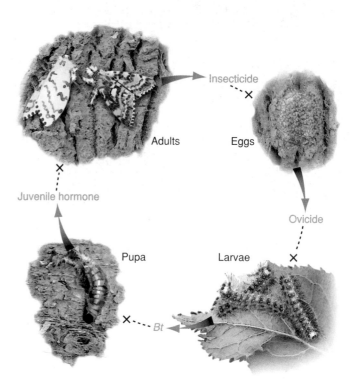

FIGURE 10.29 Different strategies can be used to control pests at various stages of their life cycles. *Bacillus thuringiensis* (*Bt*) kills caterpillars when they eat leaves with these bacteria on the surface. Releasing juvenile hormone in the environment prevents maturation of pupae. Predators attack at all stages.

It quickly established huge, dense stands that dominated 25 million ha (more than 60 million acres) of grazing land. A natural predator from South America, the cactoblastis moth, was introduced into Australia in 1935 to combat the prickly pear. Within a few years, cactoblastis larvae had eaten so much prickly pear that the cactus has become rare and is no longer economically significant.

Some plants make natural pesticides and insect repellents. The neem tree (*Azadirachta indica*) is native to India but is now grown in many tropical countries (fig. 10.28). The leaves, bark, roots, and flowers all contain compounds that repel insects and can be used to combat a number of crop pests and diseases. Another approach is to use hormones that upset development or sex attractants to bait traps containing toxic pesticides. Many municipalities control mosquitoes with these techniques rather than aerial spraying of insecticides because of worries about effects on human health. Briquettes saturated with insect juvenile hormone are scattered in wetlands where mosquitoes breed. The presence of even minute amounts of this hormone prevent larvae from ever turning into biting adults (fig. 10.29).

Genetics and bioengineering can also help in our war against pests. Traditional farmers have long known to save seeds of disease-resistant crop plants or to breed livestock that tolerate pests well. Modern genetic methods have enhanced this process, especially by transferring Bt bacterial genes to corn, soy, and other crops. Heavy reliance on the Bt gene may dilute its effectiveness, however (chapter 9).

IPM uses a combination of techniques

Integrated pest management (IPM) is a flexible, ecologically based strategy that is applied at specific times and aimed at specific crops and pests. It often uses mechanical cultivation and techniques such as vacuuming bugs off crops as an alternative to chemical application (fig. 10.30). IPM doesn't give up chemical pest controls entirely but instead tries to use minimal amounts, only as a last resort, and avoids broad-spectrum, ecologically disruptive products. IPM relies on preventive practices that encourage growth and diversity of beneficial organisms and enhance plant defenses and vigor. Successful IPM requires careful monitoring of pest populations to determine **economic thresholds,** the point at which potential economic damage justifies pest-control expenditures, and the precise time, type, and method of pesticide application.

Trap crops, small areas planted a week or two earlier than the main crop, are also useful. These plots mature before the rest of the field and attract pests away from other plants. The trap crop then is sprayed heavily with pesticides so that no pests are likely to escape. The trap crop is then destroyed, and the rest of the field should be mostly free of both pests and pesticides.

IPM programs are used on a variety of crops. Massachusetts apple growers who use IPM have cut pesticide use by 43 percent in the past ten years while maintaining per-acre yields of marketable fruit equal to that of farmers who use conventional techniques.

FIGURE 10.30 This machine, nicknamed the "salad vac," vacuums bugs off crops as an alternative to treating them with toxic chemicals.

Some of the most dramatic IPM success stories come from the developing world. In Brazil, pesticide use on soybeans has been reduced up to 90 percent with IPM. In Costa Rica, use of IPM on banana plantations has eliminated pesticides altogether in one region. In Africa, mealybugs were destroying up to 60 percent of the cassava crop (the staple food for 200 million people) before IPM was introduced in 1982. A tiny wasp that destroys mealybug eggs was discovered and now controls this pest in over 65 million ha (160 million acres) in 13 countries.

In Indonesia, rice farmers offer a successful IPM model for staple crops. There, brown planthoppers had developed resistance to virtually every insecticide and threatened the country's hardwon self-sufficiency in rice. Researchers found that farmers were spraying their fields habitually—sometimes up to three times a week—regardless of whether fields were infested. In 1986 President Suharto banned 56 of 57 pesticides previously used in Indonesia and declared a crash program to educate farmers about IPM and the dangers of pesticide use. By allowing natural predators to combat pests and spraying only when absolutely necessary with chemicals specific for planthoppers, Indonesian farmers using IPM raised yields and cut pesticide costs by 75 percent. In 1988, only two years after its initiation, the program was declared a success. It has been extended throughout the whole country. Because nearly half the people in the world depend on rice as their staple crop, this example could have important implications elsewhere (fig. 10.31).

Although IPM can be a good alternative to chemical pesticides, it also presents environmental risks in the form of exotic organisms. Wildlife biologist George Boettner of the University of Massachusetts reported in 2000 that biological controls of gypsy moths, which attack fruit trees and ornamental plants, have also decimated populations of native North American moths. *Compsilura* flies, introduced in 1905 to control the gypsy moths, have a voracious appetite for other moth caterpillars as well. One of the largest North American moths, the Cecropia moth (*Hyalophora cecropia*), with a 15 cm wingspan, was once ubiquitous in the eastern United States, but it is now rare in regions where *Compsilura* flies were released.

10.7 SOIL CONSERVATION

With careful husbandry, soil is a renewable resource that can be replenished and renewed indefinitely. Many sustainable farming practices focus on building soil nutrients. Because agriculture is the area in which soil is most essential and also most often lost through erosion, agriculture offers the greatest potential for soil conservation and rebuilding. Some rice paddies in Southeast Asia, for instance, have been farmed continuously for a thousand years without any apparent loss of fertility. The rice-growing cultures that depend on these fields have developed management practices that return organic material to the paddy and carefully nurture the soil (see also Exploring Science, p. 218).

American agriculture causes far more erosion than is sustainable. But conditions were still worse a few generations ago, before USDA soil conservation programs were established. In a study of one Wisconsin watershed, erosion rates were 90 percent less in 1975–1993

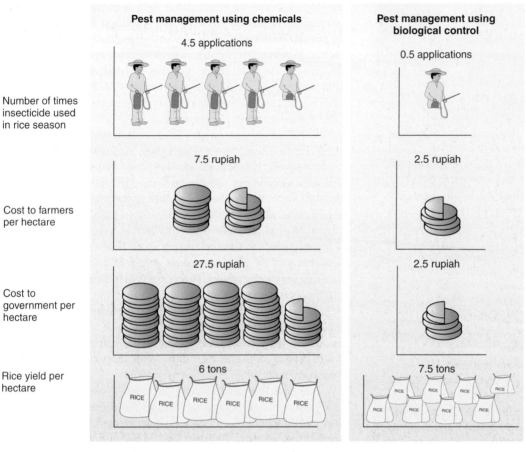

Pest management using chemicals

Number of times insecticide used in rice season — 4.5 applications

Cost to farmers per hectare — 7.5 rupiah

Cost to government per hectare — 27.5 rupiah

Rice yield per hectare — 6 tons

Pest management using biological control

0.5 applications

2.5 rupiah

2.5 rupiah

7.5 tons

FIGURE 10.31 Indonesia has one of the world's most successful integrated pest management (IPM) programs. Switching from toxic chemicals to natural pest predators has saved money while also increasing rice production.
Source: Tolba, et al., *World Environment,* 1972–1992, p. 307, Chapman & Hall, 1992 United Nations Environment Programme.

than they were in the 1930s. Ground cover, irrigation, and tillage systems are the most important elements in soil conservation.

Contours and ground cover reduce runoff

Water runs downhill. The faster it runs, the more soil it carries off the fields. A bare field with a 5 percent slope loses eight times as much soil to erosion as a field with a 1 percent slope. **Contour plowing**—plowing across the hill rather than up and down—is one of the main strategies for controlling soil loss and water runoff. Contour plowing is often combined with **strip farming,** the planting of different kinds of crops in alternating strips along the land contours (fig. 10.32*a*). The ridges created by cultivation also trap water and allow it to seep into the soil.

Terracing involves shaping the land to create level shelves of earth to hold water and soil. The edges of the terrace are planted with soil-anchoring plant species. This is an expensive procedure, requiring either much hand labor or expensive machinery, but makes it possible to farm very steep hillsides. Rice terraces in Asia create beautiful landscapes as well as highly productive

and sustainable agroecosystems (fig. 10.32*b*).

Annual row crops such as corn or beans generally cause the highest erosion rates because they leave soil bare for much of the year (table 10.2). On many steep lands or loose, highly erodible soils, the best way to keep soil in place is to plant **perennial species** (plants that grow for more than two years). Establishing forests, orchards, grassland, or crops such as tea or coffee can minimize the need for regular cultivation. **Cover crops** like rye, alfalfa, or clover can also be planted after harvest to hold and protect the soil. These cover crops can be plowed under at planting time to provide green manure. Many also fix nitrogen and enrich the soil while the land is idle.

In some cases, interplanting of two different crops in the same field not only protects the soil but also is a more efficient use of the land, providing double harvests. Native Americans and pioneer farmers planted beans or pumpkins between the corn rows. The beans provided nitrogen needed by the corn, pumpkins crowded out weeds, and both crops provided foods that nutritionally balance corn. Traditional swidden (slash-and-burn) cultivators in Africa and South America often plant as many as 20 different crops together in small plots. The crops mature at different times so that there is always something to eat, and the soil is never exposed to erosion for very long.

Mulch is a general term for a protective ground cover that can include manure, wood chips, straw, seaweed, leaves, and other natural products. For some high-value crops, such as tomatoes, pineapples, and cucumbers, it is cost-effective to cover the ground with heavy paper or plastic sheets to protect the soil, save water, and prevent weed growth. Israel uses millions of square meters of plastic mulch to grow crops in the Negev desert.

Reduced tillage leaves crop residue

Often the easiest way to provide cover that protects soil from erosion is to leave crop residues on the land after harvest. Residue covers the surface and breaks the erosive power of wind and water; it also reduces evaporation and soil temperature in hot climates and protects soil organisms that help aerate and rebuild

(a) Contour plowing

(b) Terracing

FIGURE 10.32 Contour plowing (a) and terracing, as in these Balinese rice paddies (b), are both strategies to control erosion on farmed hillsides.

soil. In some experiments, 1 ton of crop residue per acre (0.4 ha) increased water infiltration by 99 percent, reduced runoff by 99 percent, and reduced erosion by 98 percent.

Leaving crop residue has been a challenge for many farmers. Since the 1800s farmers have used moldboard plows to keep fields completely "clean" of plant litter. A traditional moldboard plow digs a deep trench and turns the topsoil upside down. Keeping a clean field helps control weeds and pests, but it also exposes soil to erosion and destroys its internal structure, which is important for aeration, moisture, and nutrient retention. Farmers are increasingly finding ways to cultivate less often in order to preserve soil, water, and fuel.

Table 10.2 Soil Cover and Soil Erosion

Cropping System	Average Annual Soil Loss (Tons/Hectare)	Percent Rainfall Runoff
Bare soil (no crop)	41.0	30
Continuous corn	19.7	29
Continuous wheat	10.1	23
Rotation: corn, wheat, clover	2.7	14
Continuous bluegrass	0.3	12

Source: Based on 14 years' data from Missouri Experiment Station, Columbia, MO.

There are several major **reduced tillage systems.** *Minimum till* involves less frequent plowing and cultivating. A chisel plow, with a row of curved chisel-like blades, is often used. A chisel plow doesn't turn over the soil but creates ridges on which seeds can be planted. It leaves up to 75 percent of plant debris on the surface between the rows, preventing erosion. *Conservtill* farming uses a coulter, a sharp disc like a pizza cutter, which slices through the soil, opening up a furrow or slot just wide enough to insert seeds. This disturbs the soil very little and leaves almost all plant debris on the surface. *No-till* planting is accomplished by drilling seeds into the ground directly through mulch and ground cover. This allows a cover crop to be interseeded with a subsequent crop (fig. 10.33).

Farmers who use these conservation tillage techniques often depend on pesticides (insecticides, fungicides, and herbicides) to control insects and weeds. Increased use of toxic agricultural chemicals is a matter of great concern. Massive use of pesticides is not, however, a necessary corollary of soil conservation. It is possible to combat pests and diseases with integrated pest management that combines crop rotation, trap crops, natural repellents, and biological controls.

Low-input agriculture aids farmers and their land

In contrast to the trend toward industrialization and dependence on chemical fertilizers, pesticides, antibiotics, and artificial growth factors common in conventional agriculture, some farmers

FIGURE 10.33 No-till planting involves drilling seeds through debris from last year's crops. Here soybeans grow through corn mulch. Debris keeps weeds down, reduces wind and water erosion, and keeps moisture in the soil.

Although it's ecologically rich, the Amazon rainforest is largely unsuitable for agriculture because of its red, acidic, nutrient-poor soils. But in many parts of the Amazon there are patches of dark, moist, nutrient-rich soils. These patches have long puzzled scientists. Locally known as terra preta de Indio, or "dark earth of the Indians," these patches of soil aren't associated with any particular environmental conditions or vegetation. Instead, the presence of bone fragments and pottery pieces hint that they may have a human origin.

Remote sensing surveys show that these dark earth patches, while usually rather small individually, collectively occupy somewhere between 1 and 10 percent of the Amazon. At the upper estimate, this would be about twice the size of Britain. Archeologists now believe that these fertile soils once supported an extensive civilization of farms, fields, and even large cities in the Amazon basin for 1,000 years or more. After Europeans arrived in the sixteenth century, diseases decimated the indigenous population and cities were abandoned, but in many places the terra preta remains highly fertile 500 years later.

It's now believed the dark soils were created by native people who deliberately worked charcoal, human and animal manure, food waste, and plant debris into their gardens and fields. In some areas these black soils, laced with bits of pottery, reach two meters (6 feet)

Soils enriched by charcoal centuries ago *(left)* still remain darker and more fertile than the usual weathered, red Amazonian soils *(right)*.

in depth. Much of the dark color seems to come from charcoal that has been added to the soil. Charcoal also improves the retention of nutrients, water, and other organic matter. Contrary to what scientists expected, the charcoal also seems to be beneficial for the soil-building activities of microorganisms, fungi, and other soil organisms. In short, what seems like a fairly simple practice of soil husbandry has turned extremely poor soils into highly productive gardens. Crops such as bananas, papaya, and mango are as much as three times more productive in terra preta than on nearby fields. And although most Amazonian soils need to be fallow for eight to ten years to rebuild nutrients after being farmed, these dark soils can recover after only six months or so.

Native people probably produced charcoal by burning biomass in low-temperature fires, in which fuel is allowed to smolder slowly in an oxygen-poor environment. Modern charcoal makers do this in an enclosed kiln. Some soil scientists are now advocating the use of charcoal, which they call "biochar," to help promote growth. But it turns out that charcoal can have another important benefit. When organic material is burned in an open fire or simply allowed to decompose in the open air, the carbon it contains is converted to CO_2 that contributes to global warming. Charcoal that is turned into the soil, on the other hand, can sequester carbon in the soil for centuries. Some of the Amazonian terra preta has five to ten times as much carbon as nearby soils. There's now an international movement to encourage biochar production and use, both to increase food production and to store carbon.

The use of charcoal as a soil amendment wasn't limited to the Amazon. Other places in South America, Africa, and Asia also have had similar soil management traditions, although soil scientists have only recently come to appreciate the benefits of this practice. At recent UN conventions on world food supplies, desertification, and global climate change, there have been discussions of global programs to make and distribute charcoal as a way to combat a whole series of environmental problems. It seems that the rediscovery of ancient methods may improve our soil management today.

---

are going back to a more natural, agroecological farming style. Finding that they can't—or don't want to—compete with factory farms, these producers are making money and staying in farming by returning to smallscale, low-input agriculture. The Minar family, for instance, operates a highly successful 150-cow dairy operation on 97 ha (240 acres) near New Prague, Minnesota. No synthetic chemicals are used on their farm. Cows are rotated every day between 45 pastures or paddocks to reduce erosion and maintain healthy grass. Even in the winter, livestock remain

outdoors to avoid the spread of diseases common in confinement (fig. 10.34). Antibiotics are used only to fight diseases. Milk and meat from this operation are marketed through co-ops and a community-supported agriculture (CSA) program. Sand Creek, which flows across the Minar land, has been shown to be cleaner when it leaves the farm than when it enters. Research at Iowa State University has shown that raising animals on pasture grass rather than grain reduces nitrogen runoff by two-thirds while cutting erosion by more than half.

FIGURE 10.34 On the Minar family's 230-acre dairy farm near New Prague, Minnesota, cows and calves spend the winter outdoors in the snow, bedding down on hay. Dave Minar is part of a growing counterculture that is seeking to keep farmers on the land and bring prosperity to rural areas.

Similarly, the Franzens, who raise livestock on their organic farm near Alta Vista, Iowa, allow their pigs to roam in lush pastures where they can supplement their diet of corn and soybeans with grasses and legumes. Housing for these happy hogs is in spacious, open-ended hoop structures. As fresh layers of straw are added to the bedding, layers of manure beneath are composted, breaking down into odorless organic fertilizer.

Low-input farms such as these typically don't turn out the quantity of meat or milk that their intensive-agriculture neighbors do, but their production costs are lower and they get higher prices for their crops, so the all-important net gain is often higher. The Franzens, for example, calculate that they pay 30 percent less for animal feed, 70 percent less for veterinary bills, and half as much for buildings and equipment than neighboring confinement operations. And on the Minar's farm, erosion after an especially heavy rain was measured to be 400 times lower than on a conventional farm nearby.

Consumers' choices play an important role

Preserving small-scale, family farms also helps preserve rural culture. As Marty Strange of the Center for Rural Affairs in Nebraska asks, "Which is better for the enrollment in rural schools, the membership of rural churches, and the fellowship of rural communities— two farms milking 1,000 cows each or twenty farms milking 100 cows each?" Family farms help keep rural towns alive by purchasing machinery at the local implement dealer, gasoline at the neighborhood filling station, and groceries at the community grocery store.

These are the arguments that lead many people to shift at least part of their diets to local foods. **Locavores** (people who eat locally grown, seasonal food) can help sustain local businesses while they eat. Most profits from conventional foods, in contrast, go to a tiny number of giant food corporations: the top three or four corporations in each commodity group typically control 60 to 80 percent of the U.S. market. Where conventional foods were shipped an average of 2,400 km (1,500 mi) to markets, the average food item at a farmers' market traveled only 72 km (45 miles). Food from local, small-scale farms also often involves less energy for fertilizer, fuel for shipping, and plastic food packaging.

Many co-ops carry food that is locally grown and processed. An even better way to know where your food comes from and how it's produced is to join a **community-supported agriculture (CSA)** farm. In return for an annual contribution to a local CSA farm, you'll receive a weekly "share" of whatever the farm produces. CSA farms generally practice organic or low-input agriculture, and many of them invite members to visit and learn how their food is grown. Much of America's most fertile land is around major cities, and CSAs and farmers' markets are one way to help preserve these landscapes around metropolitan areas.

CONCLUSION

Agriculture leads to some of our most dramatic environmental changes, and agriculture is therefore an area in which improved methods can hold potential for dramatic progress. Soils are complex systems that include biological and mineral components, and soils can be enriched and built up through careful management. Soils can also be eroded and degraded rapidly and irrevocably. Water and wind erosion are the mechanisms damaging most of the world's farming soils. Soil degradation is causing the continuing loss of farmland, even while populations dependent on that farmland grow.

Water for irrigation and energy are two other key resources for agriculture. Irrigation is often necessary, but it can cause salt accumulation or waterlogging in soils. Energy use, in fertilizing, cultivating, harvesting, irrigating, and other activities, continues to grow on farms in the developed world.

Pesticides are an important part of production on modern farms, and their use is increasing dramatically. They bring many benefits but have environmental costs as well. In particular, nontarget organisms are often harmed by pesticides, and extensive use often causes resurgence of pest populations as pests develop immunity to chemicals. Our most abundantly used agricultural chemicals are organophosphates, including glyphosate, and organochlorines, including atrazine. Glyphosate and atrazine are applied to more than 90 percent of soy and corn produced in the United States. Global consumption of these and similar agricultural chemicals continues to grow, but household use is the fastest-growing sector of pesticide use and now makes up about 14 percent of total use.

Alternative strategies for pest control include crop rotation, biological controls, mechanical cultivation, and other methods. Integrated pest management is a flexible, ecologically based approach that involves monitoring pest populations and using small, targeted applications of pesticides. This approach can dramatically reduce pesticide use.

Other sustainable agriculture practices include soil conservation by terracing, by leaving crop residue on the soil, and by reduced frequency of tilling. These practices are still unconventional, but they can save money for farmers and improve the fertility of their land. As a consumer, you can help support environmentally sustainable farming practices in a number of ways: you can buy sustainably or organically produced food, you can buy from local growers, and you can shop at farmers' markets.

REVIEWING LEARNING OUTCOMES

By now you should be able to explain the following points:

10.1 Describe the components of soils.
- Soils are complex ecosystems.
- Healthy soil fauna can determine soil fertility.
- Your food comes mostly from the A horizon.

10.2 Explain the ways we use and abuse soils.
- Arable land is unevenly distributed.
- Soil losses cut farm production.
- Wind and water move most soil.
- Deserts are spreading around the world.

10.3 Outline some of the other key resources for agriculture.
- All plants need water to grow.
- Plants need fertilizer, but not too much.
- Farming is energy-intensive.

10.4 Discuss our principal pests and pesticides.
- People have always used pest controls.

- Modern pesticides provide benefits, but also create problems.
- There are many types of pesticides.

10.5 List and discuss the environmental effects of pesticides.
- POPs accumulate in remote places.
- Pesticides cause a variety of human health problems.

10.6 Describe the methods of organic and sustainable agriculture.
- What does *organic* mean?
- Careful management can reduce pests.
- Useful organisms can help us control pests.
- IPM uses a combination of techniques.

10.7 Explain several strategies for soil conservation.
- Contours and ground cover reduce runoff.
- Reduced tillage leaves crop residue.
- Low-input agriculture can be good for farmers and their land.
- Consumers' choices play an important role.

PRACTICE QUIZ

1. What is the composition of soil? Why are soil organisms important?

2. What are four kinds of erosion? Why is erosion a problem?

3. What is a pest, and what are pesticides? What is the difference between biocides, herbicides, insecticides, and fungicides?

4. What is DDT, and why was it considered a "magic bullet"? Why was it listed among the "dirty dozen" persistent organic pollutants (POPs)?

5. What are endocrine disrupters, and why are they dangerous?

6. Identify three major categories of alternatives to synthetic pesticides.

7. What is IPM, and how is it used in pest control?

8. What is sustainable agriculture?

9. What are some strategies for reducing soil erosion?

10. What is a locavore, and why do some consumers consider them important? In what ways can local food be better or worse than organic food?

CRITICAL THINKING AND DISCUSSION QUESTIONS

1. As you consider the expansion of soybean farming and grazing in Brazil, what are the costs and what are the benefits of these changes? How would you weigh these costs and benefits for Brazilians? If you were a U.S. ambassador to Brazil, how would you advise Brazilians on their farming and grazing policies, and what factors would shape your advice?

2. The discoverer of DDT, Paul Müller, received a Nobel Prize for his work. Would you have given him this prize?

3. Are there steps you could take to minimize your exposure to pesticides, either in things you buy or in your household? What would influence your decision to use household pesticides or not to use them?

4. What criteria should be used to determine whether farmers should use ecologically sound techniques? How would your response differ if you were a farmer, a farmer's neighbor, someone downstream of a farm, or someone far from farming regions?

5. Should we try to increase food production on existing farmland, or should we sacrifice other lands to increase farming areas? Why?

6. Some rice paddies in Southeast Asia have been cultivated continuously for a thousand years or more without losing fertility. Could we, and should we, adapt these techniques to our own country? Why or why not?

7. Terra preta soils were a conundrum for soil scientists for decades. What expectations about tropical soils did these black soils violate? What would it take to make similar investments in soils today?

Data Analysis: Mapping and Graphing Pesticide Use

The National Agricultural Statistics Service (NASS) keeps records of pesticide use in the United States, and you can access those records by going to www.pestmanagement.info/nass/app_usage.cfm. This data source is incomplete and not up to date, but it is the only public monitoring source for chemicals whose use is rapidly increasing, expanding, and diversifying worldwide and in the United States. Both environmental and economic impacts of these uses are substantial. Visit the NASS site, and observe how many pesticides are listed. Monitoring the environmental and health effects of this many compounds is clearly a challenge, but this diversity helps growers respond to the "pesticide treadmill" effect. Refer to your readings to recall what the term *pesticide treadmill* means.

Then look at some of the crops on which growers use the dominant pesticides—glyphosate, atrazine, alachlor, or 2,4-D (for reference, see fig. 10.18). You can experiment with graphing and mapping, as well as tabular reports on uses of these pesticides.

You can download and analyze these data yourself, but to make it easier we have provided an Excel file with a set of this data that is organized for easy graphing (below graph). Acquire this file by going to www.mhhe.com/cunningham12e. Find material for Chapter 10 to locate and download the Data Analysis Excel file. The file contains directions for graphing different crops on which Glyphosate ("Roundup"), the most abundantly used herbicide in the United States, is applied.

Graph data for the different crops, as described in the file, and answer the questions below.

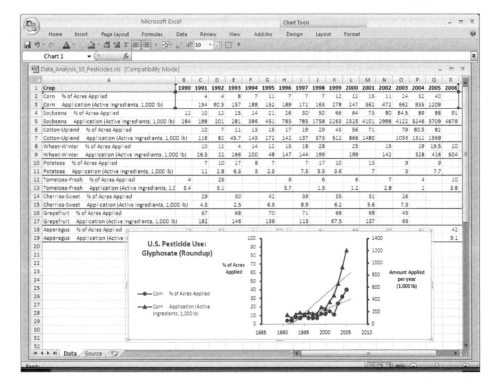

1. What types of information do the two vertical axes represent?

2. For Corn, what is the general trend from 1990 to 2006? Have both variables followed the same trend? (Note that dotted lines indicate the general trend for blue points and for red points.)

3. Roughly what is the amount (blue) applied in 1990? in 2006?

4. Roughly what is the percentage of acres (red) on which glyphosate was used in 1990? in 2006?

5. For Soybeans, answer questions 3 and 4.

6. Crops are sorted roughly according to the amount produced in the United States each year. In general, is more glyphosate used on the most abundant crops or the least abundant crops?

7. Are trends up for all crops? Are trends up for the most abundant crops (corn, soybeans, cotton, wheat, and potatoes)?

Graph trends in pesticide use yourself, using data provided in the excel file at www.mhhe.com/Cunningham12e.

Data source: USDA National Agricultural Statistics Service (NASS).

Habitat degradation is a leading cause of biodiversity loss. Forest fragmentation destroys the old-growth characteristics on which many species, such as the northern spotted owl, depend.

Biodiversity
Preserving Species

Learning Outcomes

After studying this chapter, you should be able to:

11.1 Discuss biodiversity and the species concept.

11.2 Summarize some of the ways we benefit from biodiversity.

11.3 Characterize the threats to biodiversity.

11.4 Evaluate endangered species management.

11.5 Scrutinize captive breeding and species survival plans.

"The first rule of intelligent tinkering is to save all the pieces."

~ *Aldo Leopold*

Case Study

How Can We Save Spotted Owls?

What's the most controversial bird in the world? If you count the number of scientists, lawyers, journalists, and activists who have debated its protection, as well as the amount of money, time, and effort spent on research and recovery, the answer must be the northern spotted owl (*Strix occidentalis caurina*). This brown, medium-size owl (fig. 11.1) lives in the complex old-growth forests of North America's Pacific Northwest. It's thought that before European settlement, northern spotted owls occurred throughout the coastal ranges, including the Cascade Range, from southern British Columbia almost to San Francisco Bay.

Spotted owls nest in cavities in the huge old-growth trees of the ancient forest. They depend on flying squirrels and wood rats as their primary prey, but they'll also eat voles, mice, gophers, hares, birds, and occasionally insects. With 90 percent of their preferred habitat destroyed or degraded, northern spotted owl populations are declining throughout their former range. When the U.S. Congress established the Endangered Species Act (ESA), in 1973, the northern spotted owl was identified as potentially endangered. In 1990, after decades of study—but little action to protect them—northern spotted owls were listed as threatened by the U.S. Fish and Wildlife Service. At that time, estimates placed the population at 5,431 breeding pairs or resident single owls.

Several environmental organizations sued the federal government for its failure to do more to protect the owls. In 1991 a federal district judge agreed that the government wasn't following the requirements of the ESA and temporarily shut down all logging in old-growth habitat in the Pacific Northwest. Timber sales dropped precipitously, and thousands of loggers and mill workers lost their jobs. Although mechanization and export of whole logs to foreign countries accounted for much of this job loss, many people blamed the owls for the economic woes across the region. Fierce debates broke out between loggers, who hung owls in effigy, and conservationists, who regarded them as protectors of the forest as well as the whole biological community that lives there.

FIGURE 11.1 Only about 1,000 pairs of northern spotted owls remain in the old-growth forests of the Pacific Northwest. Cutting old-growth forests threatens the endangered species, but reduced logging threatens the jobs of many timber workers.

In an effort to protect the remaining old-growth forest while still providing timber jobs, President Clinton started a broad planning process for the whole area. After a great deal of study and consultation, a comprehensive Northwest Forest Plan was adopted in 1994 as a management guide for about 9.9 million hectares (24.5 million acres) of federal lands in Oregon, Washington, and northern California. The plan was based on the latest science of ecosystem management and represented compromises on all sides. Nevertheless, loggers complained that this plan locked up forests on which their jobs depended, while environmentalists lamented the fact that millions of hectares of old-growth forest would still be vulnerable to logging (for further discussion, see chapter 12).

In spite of the habitat protection provided by the forest plan, northern spotted owl populations continued to decline. By 2004, researchers could find only 1,044 breeding pairs. They reported that 80 percent of the nesting areas occupied two decades earlier no longer had spotted owls, and that 9 of the 13 geographic populations were declining. The courts ordered the Fish and Wildlife Service to establish a recovery plan as required by the ESA. After four more years of study and deliberation, a recovery plan was published in 2008. The plan identified 133 owl conservation areas encompassing 2.6 million hectares (6.4 million acres) of federal lands that should be managed to protect old-growth habitat and stabilize owl populations. But the Obama administration tossed out this plan, citing political meddling and insufficient protection for old-growth forest habitat. In December 2010 a new draft was released. Many scientists liked it better than the Bush plan, but they said the plan still overemphasizes logging to prevent fire, neglects the impacts of forest thinning, and doesn't protect enough old-growth habitat.

As you can see, protecting rare and endangered species is difficult and controversial. In this chapter we'll look at some of the threats to rare and endangered species as well as the reasons we may want to protect biodiversity and habitat. We'll also discuss the politics of endangered species protection and the difficulty in carrying out recovery projects. For related resources, including Google Earth™ placemarks that show locations where these issues can be explored via satellite images, visit EnvironmentalScience-Cunningham.blogspot.com.

11.1 Biodiversity and the Species Concept

From the driest desert to the dripping rainforests, from the highest mountain peaks to the deepest ocean trenches, life on earth occurs in a marvelous spectrum of sizes, colors, shapes, life cycles, and interrelationships. Think for a moment how remarkable, varied, abundant, and important are the other living creatures with whom we share this planet (fig. 11.2). How will our lives be impoverished if this biological diversity diminishes?

What is biodiversity?

Previous chapters of this book have described some of the fascinating varieties of organisms and complex ecological relationships that give the biosphere its unique, productive characteristics. Three kinds of **biodiversity** are essential to preserve these ecological systems: (1) *genetic diversity* is a measure of the variety of different versions of the same genes within individual species; (2) *species diversity* describes the number of different kinds of organisms within individual communities or ecosystems; and (3) *ecological diversity* assesses the richness and complexity of a biological community, including the number of niches, trophic levels, and ecological processes that capture energy, sustain food webs, and recycle materials within this system.

Within species diversity, we can distinguish between *species richness* (the total number of species in a community) and *species evenness* (the relative abundance of individuals within each species). To illustrate this difference, imagine two ecological communities, each with ten species and 100 individual plants or animals. Suppose that one community has 82 individuals of one species and 2 each of nine other species. In the other community, all ten species are equally abundant, meaning they have 10 individuals each. The species richness is the same, but if you were to walk through these communities, you'd have the impression that the second is much more diverse because you'd be much more likely to encounter a greater variety of organisms.

What are species?

As you can see, the concept of species is fundamental in defining biodiversity, but what exactly do we mean by the term? When Carolus Linnaeus, the great Swedish taxonomist, began our system of scientific nomenclature in the eighteenth century, classification was based entirely on the physical appearance of adult organisms. In recent years taxonomists have introduced other characteristics as means of differentiating species. In chapter 3 we defined species in terms of *reproductive isolation;* that is, all the organisms potentially able to breed in nature and produce fertile offspring. As we pointed out, this definition has some serious problems, especially among plants and protists, many of which either reproduce asexually or regularly make fertile hybrids.

Another definition favored by many evolutionary biologists is the *phylogenetic species concept* (PSC), which emphasizes the branching (or cladistic) relationships among species or higher taxa, regardless of whether organisms can breed successfully.

FIGURE 11.2 This coral reef has both high abundance of some species and high diversity of different genera. What will be lost if this biologically rich community is destroyed?

A third definition, favored by some conservation biologists, is the *evolutionary species concept* (ESC), which defines species in evolutionary and historic terms rather than reproductive potential. The advantage of this definition is that it recognizes that there can be several "evolutionarily significant" populations within a genetically related group of organisms. Unfortunately, we rarely have enough information about a population to judge what its evolutionary importance or fate may be. Paul Ehrlich and Gretchen Daily calculate that, on average, there are 220 evolutionarily significant populations per species. This calculation could mean that there are up to 10 billion different populations in total. Deciding which ones we should protect becomes an even more daunting prospect.

Molecular techniques are revolutionizing taxonomy

Increasingly, DNA sequencing and other molecular techniques are giving us insights into taxonomic and evolutionary relationships. As we described in chapter 3, each individual has a unique hereditary complement called the *genome*. The genome is made up of the millions or billions of nucleotides in DNA arranged in a very specific sequence that spells out the structure of all the proteins that make up the cellular structure and machinery of every organism. As you know from modern court cases and paternity suits, we can use that DNA sequence to identify individuals with a very high degree of certainty. Now this very precise technology is being applied to identify species in nature.

Because only a small amount of tissue is needed for DNA analysis, species classification—or even the identity of individual animals—can be made on samples such as feathers, fur, or feces when it's impossible to capture living creatures. For example, DNA analysis showed that whale meat for sale in Japanese markets was from protected species. Sampling of hair from scratching pads has allowed genetic analysis of lynx and bears in North America without causing them the trauma of being captured. Similarly, a new tiger subspecies (*Tigris panthera jacksoni*) was detected in

Southeast Asia based on blood, skin, and fur samples from zoo and museum specimens (fig. 11.3).

This new technology can help resolve taxonomic uncertainties in conservation. In some cases an apparently widespread and low-risk species may, in reality, comprise a complex of distinct species, some rare or endangered. Such is the case for a unique New Zealand reptile, the tuatara. Genetic marker studies revealed two distinct species, one of which needed additional protection. Similar studies have shown that the northern spotted owl (*Strix occidentalis caurina*) is a genetically distinct subspecies from its close relatives, the California spotted owl (*S. occidentalis occidentalis*) and the Mexican spotted owl (*S. occidentalis lucida*), and therefore deserves continued protection.

On the other hand, in some cases genetic analysis shows that a protected population is closely related to another much more abundant one. For example, the colonial pocket gopher from Georgia is genetically identical to the common pocket gopher and probably doesn't deserve endangered status. The California gnatcatcher (*Polioptila californica californica*), which lives in the coastal sage scrub between Los Angeles and the Mexican border, was listed as a threatened species in 1993, and thousands of hectares of land worth billions of dollars were put off-limits for development. Genetic studies showed, however, that this population is indistinguishable from the black-tailed flycatcher (*Polioptila californica pontilis*), which is abundant in adjacent areas of Mexico.

In some cases molecular taxonomy is causing a revision of the basic phylogenetic ideas of how we think evolution proceeded. Studies of corals and other cnidarians (jellyfish and sea anemones), for example, show that they share more genes with primates than do worms and insects. This evidence suggests a branching of the family tree very early in evolution rather than a single sequence from lower to higher animals.

How many species are there?

At the end of the great exploration era of the nineteenth century, some scientists confidently declared that every important kind of living thing on earth would soon be found and named. Most of those explorations focused on charismatic species such as birds and mammals. Recent studies of less conspicuous organisms such as insects and fungi suggest that millions of new species and varieties remain to be studied scientifically.

Think About It

Compare the estimates of known and threatened species in table 11.1. Are some groups overrepresented? Are we simply more interested in some organisms, or are we really a greater threat to some species?

The 1.7 million species presently known (table 11.1) probably represent only a small fraction of the total number that exist. Based on the rate of new discoveries by research expeditions—especially in the tropics—taxonomists estimate that there may be somewhere between 3 million and 50 million different species alive today.

FIGURE 11.3 DNA analysis revealed a new tiger subspecies (*T. panthera jacksoni*) in Malaysia. This technology has become essential in conservation biology.

In fact, some taxonomists estimate that there are 30 million species of tropical insects alone. The upper limits for these estimates assume a high degree of ecological specialization among tropical insects. A recent study in New Guinea, however, found that 51 plant species were host to 900 species of herbivorous insects. This evidence would suggest no more than 4 to 6 million insect species worldwide.

About 65 percent of all known species are invertebrates (animals without backbones, such as insects, sponges, clams, and worms). This group probably makes up the vast majority of organisms yet to be discovered and may constitute 95 percent of all species. What constitutes a species in bacteria and viruses is even less certain than for other organisms, but there are large numbers of physiologically or genetically distinct varieties of these organisms.

The numbers of endangered species shown in table 11.1 are those officially listed by the International Union for Conservation of Nature and Natural Resources (IUCN). This represents only a small fraction of those actually at risk. It's estimated that one-third of all amphibians, for example, are declining and threatened with extinction. We'll discuss this issue later in this chapter.

Hot spots have exceptionally high biodiversity

Of all the world's currently identified species, only 10 to 15 percent live in North America and Europe. The greatest concentration of different organisms tends to be in the tropics, especially in tropical rainforests and coral reefs. Norman Myers, Russell Mittermeier, and others have identified **biodiversity hot spots** that contain at least 1,500 endemics (species that occur nowhere else) and have lost at least 70 percent of their habitat owing to, for example, deforestation or invasive species. Using plants and land-based vertebrates as indicators, they have proposed 34 hot

Table 11.1 Current Estimates of Known and Threatened Living Species by Taxonomic Group

	Known	Endangered
Mammals	5,491	1,131
Birds	9,990	1,240
Reptiles	8,734	594
Amphibians	6,347	1,898
Fishes	30,700	1,275
Insects	1,000,000	733
Molluscs	85,000	1,288
Crustaceans	47,000	596
Other animals	173,250	253
Mosses	16,236	36
Ferns and allies	12,000	204
Gymnosperms	1,052	178
Flowering plants	268,000	8,296
Green & Red Algae	10,356	10
Lichens	17,000	2
Mushrooms	31,496	1
Brown Algae	3,127	6
Total	**1,727,708**	**17,741**

Source: Data from IUCN Red List, 2011.

spots that are a high priority for conservation because they have both high biodiversity and a high risk of disruption by human activities (fig. 11.4). Although these hot spots occupy only 1.4 percent of the world's land area, they house three-quarters of the world's most threatened mammals, birds, and amphibians. The hot spots also account for about half of all known higher plant species and 42 percent of all terrestrial vertebrate species. The hottest of these hot spots tend to be tropical islands, such as Madagascar, Indonesia, and the Philippines, where geographic isolation has resulted in large numbers of unique plants and animals. Special climatic conditions, such as those found in South Africa, California, and the Mediterranean Basin, also produce highly distinctive flora and fauna.

Some areas with high biodiversity—such as Amazonia, New Guinea, and the Congo basin—aren't included in this hot spot map because most of their land area is relatively undisturbed. Other groups prefer different criteria for identifying important conservation areas. Aquatic biologists, for example, point out that coral reefs, estuaries, and marine shoals host some of the most diverse wildlife communities in the world, and warn that freshwater species are more highly endangered than terrestrial ones. Other scientists worry that the hot spot approach neglects many rare species and major groups that live in less biologically rich areas (cold spots). Nearly half of all terrestrial vertebrates, after all, aren't represented in Myers's hot spots. Focusing on a few hot spots also doesn't recognize the importance of certain species and ecosystems to human beings. Wetlands, for instance, may contain just a few, common plant species, but they perform valuable ecological services, such as filtering water, regulating floods,

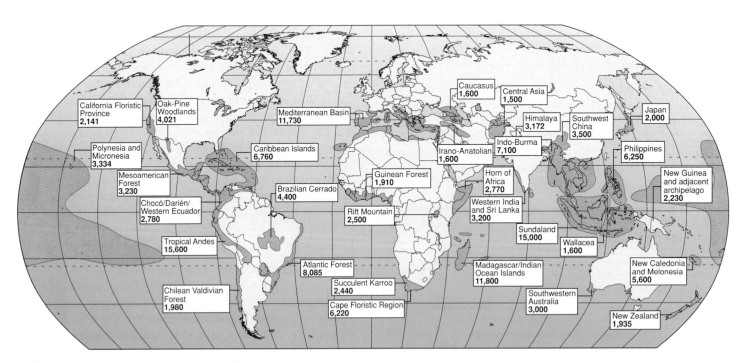

FIGURE 11.4 Biodiversity "hot spots," identified by Conservation International, tend to be in tropical or Mediterranean climates and on islands, coastlines, or mountains where many habitats exist and physical barriers encourage speciation. Numbers indicate endemic species.
Source: Conservation International, 2005.

and serving as nurseries for fish. Some conservationists argue that we should concentrate on saving important biological communities or landscapes rather than rare species.

Anthropologists point out that many of the regions with high biodiversity are also home to high cultural diversity as well (see fig. 1.20). It isn't a precise correlation; some countries, like Madagascar, New Zealand, and Cuba, with a high percentage of endemic species, have only a few cultural groups. Often, however, the varied habitat and high biological productivity of places like Indonesia, New Guinea, and the Philippines that allow extensive species specialization also have fostered great cultural variety. By preserving some of the 7,200 recognized language groups in the world—more than half of which are projected to disappear in this century—we might also protect some of the natural setting in which those cultures evolved.

11.2 HOW DO WE BENEFIT FROM BIODIVERSITY?

We benefit from other organisms in many ways, some of which we don't appreciate until a particular species or community disappears. Even seemingly obscure and insignificant organisms can play irreplaceable roles in ecological systems or be the source of genes or drugs that someday may be indispensable.

All of our food comes from other organisms

Many wild plant species could make important contributions to human food supplies, either as new crops or as a source of genetic material to provide disease resistance or other desirable traits to current domestic crops. Norman Myers estimates that as many as 80,000 edible wild plant species could be utilized by humans. Villagers in Indonesia, for instance, are thought to use some 4,000 native plant and animal species for food, medicine, and other valuable products. Few of these species have been explored for possible domestication or more widespread cultivation. A 1975 study by the National Academy of Science (U.S.) found that Indonesia has 250 edible fruits, only 43 of which have been cultivated widely (fig. 11.5).

Living organisms provide us with many useful drugs and medicines

More than half of all modern medicines are either derived from or modeled on natural compounds from wild species (table 11.2). The United Nations Development Programme estimates the value of pharmaceutical products derived from developing world plants, animals, and microbes to be more than $30 billion per year. Indigenous communities that have protected and nurtured the biodiversity on which these products are based are rarely acknowledged—much less compensated—for the resources extracted from them. Many consider this expropriation "biopiracy" and call for royalties to be paid for folk knowledge and natural assets.

Consider the success story of vinblastine and vincristine. These anticancer alkaloids are derived from the Madagascar periwinkle (*Catharanthus roseus*) (fig. 11.6). They inhibit the growth of cancer

FIGURE 11.5 Mangosteens from Indonesia have been called the world's best-tasting fruit, but they are practically unknown beyond the tropical countries where they grow naturally. There may be thousands of other traditional crops and wild food resources that could be equally valuable but are threatened by extinction.

cells and are very effective in treating certain kinds of cancer. Before these drugs were introduced, childhood leukemias were invariably fatal. Now the remission rate for some childhood leukemias is 99 percent. Hodgkin's disease was 98 percent fatal a few years ago, but is now only 40 percent fatal, thanks to these compounds. The total value of the periwinkle crop is roughly $150 million to $300 million per year, although Madagascar gets little of those profits.

Pharmaceutical companies are actively prospecting for useful products in many tropical countries. Merck, the world's largest biomedical company, paid (U.S.) $1.4 million to the

Table 11.2 Some Natural Medicinal Products		
Product	**Source**	**Use**
Penicillin	Fungus	Antibiotic
Bacitracin	Bacterium	Antibiotic
Tetracycline	Bacterium	Antibiotic
Erythromycin	Bacterium	Antibiotic
Digitalis	Foxglove	Heart stimulant
Quinine	Chincona bark	Malaria treatment
Diosgenin	Mexican yam	Birth-control drug
Cortisone	Mexican yam	Anti-inflammation treatment
Cytarabine	Sponge	Leukemia cure
Vinblastine, vincristine	Periwinkle plant	Anticancer drugs
Reserpine	Rauwolfia	Hypertension drug
Bee venom	Bee	Arthritis relief
Allantoin	Blowfly larva	Wound healer
Morphine	Poppy	Analgesic

Instituto Nacional de Biodiversidad (INBIO) of Costa Rica for plant, insect, and microbe samples to be screened for medicinal applications. INBIO, a public/private collaboration, trained native people as practical "parataxonimists" to locate and catalog all the native flora and fauna—between 500,000 and 1 million species—in Costa Rica. This effort may be a good model both for scientific information gathering and as a way for developing countries to share in the profits from their native resources.

The UN Convention on Biodiversity calls for a more equitable sharing of the gains from exploiting nature between rich and poor nations. Bioprospectors who discover useful genes or biomolecules in native species will be required to share profits with the countries where those species originate. This is not only a question of fairness; it also provides an incentive to poor nations to protect their natural heritage.

Biodiversity provides ecological services

Human life is inextricably linked to ecological services provided by other organisms. Soil formation, waste disposal, air and water purification, nutrient cycling, solar energy absorption, and management of biogeochemical and hydrological cycles all depend on the biodiversity of life (chapter 3). Total value of these ecological services is at least $33 trillion per year, or about half the total world GNP.

There has been a great deal of controversy about the role of biodiversity in ecosystem stability. It seems intuitively obvious that having more kinds of organisms would make a community better able to withstand or recover from disturbance, but few empirical studies show an unequivocal relationship. The opening case study for this chapter describes one of the most famous studies of the stability/diversity relationship.

Because we don't fully understand the complex interrelationships between organisms, we often are surprised and dismayed at the effects of removing seemingly insignificant members of biological communities. For instance, wild insects provide a valuable

FIGURE 11.6 The rosy periwinkle from Madagascar provides anticancer drugs that now make childhood leukemias and Hodgkin's disease highly remissible.

FIGURE 11.7 Birdwatching and other wildlife observation contribute more than $29 million each year to the U.S. economy.

but often unrecognized service in suppressing pests and disease-carrying organisms. It is estimated that 95 percent of the potential pests and disease-carrying organisms in the world are controlled by other species that prey upon them or compete with them in some way. Many unsuccessful efforts to control pests with synthetic chemicals (chapter 9) have shown that biodiversity provides essential pest-control services.

Biodiversity also brings us many aesthetic and cultural benefits

Millions of people enjoy hunting, fishing, camping, hiking, wildlife watching, and other outdoor activities based on nature. These activities keep us healthy by providing invigorating physical exercise. Contact with nature also can be psychologically and emotionally restorative. In some cultures, nature carries spiritual connotations, and a particular species or landscape may be inextricably linked to a sense of identity and meaning. Many moral philosophies and religious traditions hold that we have an ethical responsibility to care for creation and to save "all the pieces" as far as we are able (chapter 2).

Nature appreciation is economically important. The U.S. Fish and Wildlife Service estimates that Americans spend $104 billion every year on wildlife-related recreation (fig. 11.7). This compares to $81 billion spent each year on new automobiles. Forty percent of all adults enjoy wildlife, including 39 million who hunt or fish and 76 million who watch, feed, or photograph wildlife. Ecotourism can be a good form of sustainable economic development, although we have to be careful that we don't abuse the places and cultures we visit.

For many people the value of wildlife goes beyond the opportunity to shoot or photograph, or even see, a particular species. They argue that **existence value,** based on simply knowing that a species exists, is reason enough to protect and preserve it. We contribute to programs to save bald eagles, redwood trees, whooping cranes, whales, and a host of other rare and endangered organisms because we like to know they still exist somewhere, even if we may never have an opportunity to see them.

11.3 What Threatens Biodiversity?

Extinction, the elimination of a species, is a normal process of the natural world. Species die out and are replaced by others, often their own descendants, as part of evolutionary change. In undisturbed ecosystems, the rate of extinction appears to be about one species lost every decade. In this century, however, human impacts on populations and ecosystems have accelerated that rate, causing hundreds or perhaps even thousands of species, subspecies, and varieties to become extinct every year. If present trends continue, we may destroy *millions* of kinds of plants, animals, and microbes in the next few decades. In this section we will look at some ways we threaten biodiversity.

Extinction is a natural process

Studies of the fossil record suggest that more than 99 percent of all species that ever existed are now extinct. Most of those species were gone long before humans came on the scene. Species arise through processes of mutation and natural selection, and they disappear the same way (chapter 4). Often new forms replace their own parents. The tiny *Hypohippus,* for instance, has been replaced by the much larger modern horse, but most of its genes probably still survive in its distant offspring.

Periodically, mass extinctions have wiped out vast numbers of species and even whole families (table 11.3). The best studied of these events occurred at the end of the Cretaceous period, when dinosaurs disappeared along with at least 50 percent of existing genera and 15 percent of marine animal families. An even greater disaster occurred at the end of the Permian period about 250 million years ago, when 95 percent of all marine species and nearly half of all plant and animal families died out over a period of about 10,000 years—a short time by geological standards. Current theories suggest that these catastrophes were caused by climate changes, perhaps triggered when large asteroids struck the earth.

Table 11.3	Mass Extinctions	
Historic Period	**Time (Before Present)**	**Percent of Species Extinct**
Ordovician	444 million	85
Devonian	370 million	83
Permian	250 million	95
Triassic	210 million	80
Cretaceous	65 million	76
Quaternary	Present	33–66

Source: W. W. Gibbs, 2001.

Many ecologists worry that global climate change caused by our release of "greenhouse" gases in the atmosphere could have similarly catastrophic effects (chapter 15).

We are accelerating extinction rates

The rate at which species are disappearing appears to have increased dramatically over the last 150 years. It appears that between A.D. 1600 and 1850, human activities were responsible for the extermination of two or three species per decade. By some estimates, we are now losing species at hundreds or even thousands of times natural rates. If present trends continue, the United Nations Environment Programme warns, half of all primates and one-quarter of all bird species could be extinct in the next 50 years. The eminent biologist E. O. Wilson says the impending biodiversity crash could be more abrupt than any previous mass extinction. Some biologists call this the sixth mass extinction, but note that this time it's not asteroids or volcanoes, but human impacts, that are responsible.

Accurate predictions of biodiversity losses are difficult when many species probably haven't yet been identified. Most predictions of anthropogenic mass extinction are based on an assumption that habitat area and species abundance are tightly correlated. E. O. Wilson calculates, for example, that if you cut down 90 percent of a forest, you'll eliminate at least half of the species originally present. In some of the best-studied biological communities, however, this seems not to be true. More than 90 percent of Costa Rica's dry seasonal forest, for instance, has been converted to pasture land, yet entomologist Dan Janzen reports that no more than 10 percent of the original flora and fauna appear to have been permanently lost. Wilson and others respond that remnants of the native species may be hanging on temporarily, but that in the long run they're doomed without sufficient habitat.

Still, it's clear that habitat is being destroyed in many places, and that numerous species are less abundant than they once were. Shouldn't we try to protect and preserve as much as we can? E. O. Wilson summarizes human threats to biodiversity with the acronym **HIPPO,** which stands for Habitat destruction, Invasive species, Pollution, Population (human), and Overharvesting. Let's look in more detail at each of these issues.

Habitat Destruction

The most important extinction threat for most species—especially terrestrial ones—is habitat loss. Perhaps the most obvious example of habitat destruction is clear-cutting of forests and conversion of grasslands to crop fields. As the opening case study shows, the greatest threat to northern spotted owls is the loss of the old-growth forests on which they depend. Figure. 11.8 shows some of the owl management areas identified by the Fish and Wildlife Service in western Oregon. Before European settlement, almost all of this area would have been dense, structurally complex forest ideal for spotted owls. Although patches of habitat remain, many have been so degraded by human activities (yellow areas) that they will no longer support 20 pairs of breeding owls.

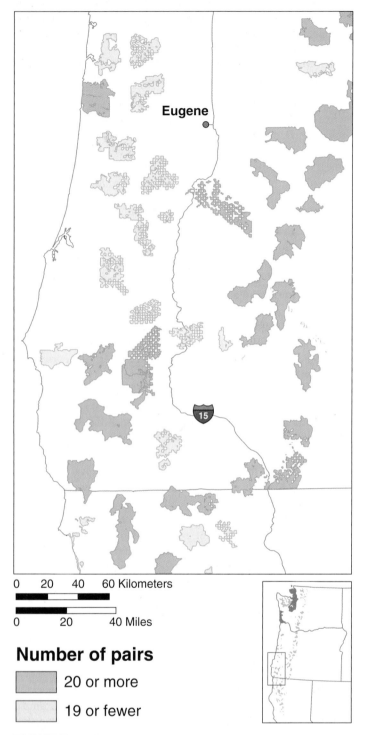

Number of pairs

20 or more

19 or fewer

FIGURE 11.8 A portion of western Oregon and northern California shows some of the 133 owl management areas identified by the U.S. Fish and Wildlife Service. Notice that while most of the habitat along the west side of the Cascade Mountains can support 20 or more pairs of breeding owls, most of the areas in the Coastal Range are already too degraded to support that many. East of the Cascades, the forests are too fire-prone to reliably provide spotted owl habitat.
Source: U.S. Fish and Wildlife Service, 2008.

Notice the "checkerboarding" of many of these areas. To encourage railroad construction in the nineteenth century, the U.S. government gave the Northern Pacific Railroad 40 million acres (16 million ha) of public land to help finance laying track. The railroad was allowed to trade land in the Great Plains that had little perceived value for rich timberlands in the Pacific Northwest. By choosing alternating sections (a section is one square mile or 640 acres or 260 hectares), the companies were able to gain control of an even larger area because no one could cross their land to harvest timber on the enclosed public property.

Fragmentation by clear-cutting (see the opening photo of this chapter) results in a loss of the deep-forest characteristics required by species such as spotted owls. Although as much as half of the forest may remain uncut in many logging operations, most of what's left becomes forest edge (see fig. 4.25).

Sometimes we destroy habitat as side effects of resource extraction, such as mining, dam building, and indiscriminate fishing methods. Surface mining, for example, strips off the land covering along with everything growing on it. Waste from mining operations can bury valleys and streams with toxic material (see mountaintop removal, chapter 14). The building of dams floods vital stream habitat under deep reservoirs and eliminates food sources and breeding habitat for some aquatic species. Our current fishing methods are highly unsustainable. One of the most destructive fishing techniques is bottom trawling, in which heavy nets are dragged across the ocean floor, scooping up every living thing and crushing the bottom structure to lifeless rubble (chapter 9).

Preserving small, scattered areas of habitat often isn't sufficient to maintain a complete species collection. Large mammals, like tigers or wolves, need large expanses of contiguous range relatively free of human incursion. Even species that occupy less space individually suffer when habitat is fragmented into small, isolated pieces. If the intervening areas create a barrier to migration, isolated populations become susceptible to environmental catastrophes such as bad weather or disease epidemics. They also can become inbred and vulnerable to genetic flaws (chapter 6).

Invasive Species

A major threat to native biodiversity in many places is from accidentally or deliberately introduced species. Called a variety of names—*alien, exotic, non-native, non-indigenous, unwanted, disruptive,* or *invaders*—**invasive species** are organisms that move into new territory. These migrants often flourish where they are free of predators, diseases, or resource limitations that may have controlled their population in their native habitat. Although humans have probably transported organisms into new habitats for thousands of years, the rate of movement has increased sharply in recent years with the huge increase in speed and volume of travel by air, water, and land. We move species around the world in a variety of ways. Some are deliberately released because people believe they will be aesthetically pleasing or economically beneficial. Others hitch a ride in ship ballast water, in the wood of packing crates, inside suitcases or shipping containers, in the soil of potted plants, even on people's shoes.

Over the past 300 years, approximately 50,000 non-native species have become established in the United States. Many of these introductions, such as corn, wheat, rice, soybeans, cattle, poultry, and honeybees, have proved to be both socially and economically beneficial. At least 4,500 of these species have established free-living populations, of which 15 percent cause environmental or economic damage (fig. 11.9). Invasive species are estimated to cost the United States some $138 billion annually and are forever changing many ecosystems.

Following are a few important examples of invasive species:

- A major threat to northern spotted owls is invasion of their habitat by the barred owl (*Strix varia*). Originally an eastern species, these larger cousins of the spotted owl have been moving westward, reaching the West Coast toward the end of the twentieth century. They now occur throughout the range of the spotted owl. Barred owls are larger, more aggressive, and more versatile in terms of habitat and diet than spotted owls. When barred owls move in, spotted owls tend to move out. Interbreeding of the species further threatens spotted owls. In an experimental project, removing barred owls resulted in recolonization by spotted owls. This raises the question of whether it's ethical to kill one owl species to protect another.

- Eurasian milfoil (*Myriophyllum spicatum* L.) is an exotic aquatic plant native to Europe, Asia, and Africa. Scientists believe that milfoil arrived in North America during the late nineteenth century in shipping ballast. It grows rapidly and tends to form a dense canopy on the water surface, which displaces native vegetation, inhibits water flow, and obstructs boating, swimming, and fishing. Humans spread the plant between water body systems from boats and boat trailers carrying the plant fragments. Herbicides and mechanical harvesting are effective in milfoil control but can be expensive (up to $5,000 per hectare per year). There is also concern that the methods may harm nontarget organisms. A native milfoil weevil, *Euhrychiopsis lecontei,* is being studied as an agent for milfoil biocontrol.

- Kudzu vine (*Pueraria lobata*) has blanketed large areas of the southeastern United States. Long cultivated in Japan for edible roots, medicines, and fibrous leaves and stems used for paper production, kudzu was introduced by the U.S. Soil Conservation Service in the 1930s to control erosion. Unfortunately it succeeded too well. In the ideal conditions of its new home, kudzu can grow 18 to 30 m in a single season. Smothering everything in its path, it kills trees, pulls down utility lines, and causes millions of dollars in damage every year.

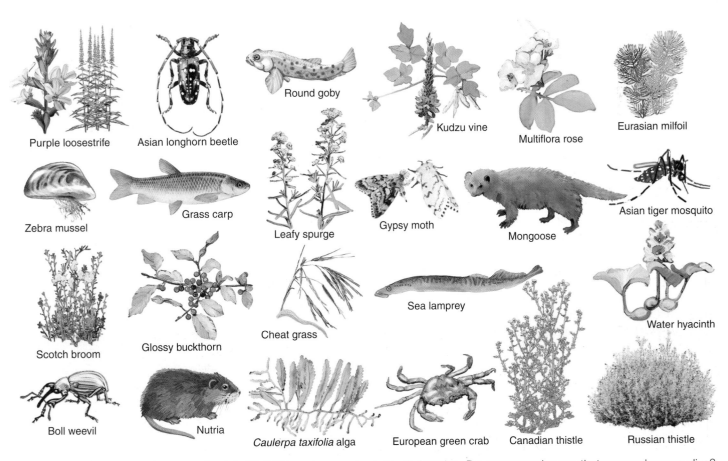

FIGURE 11.9 A few of the approximately 50,000 invasive species in North America. Do you recognize any that occur where you live? What others can you think of?

- The emerald ash borer (*Agrilus planipennis*) is an invasive wood-boring beetle from Siberia and northern China. It was first identified in North America in the summer of 2002 in southeast Michigan and in Windsor, Ontario. It's believed to have been introduced into North America in shipping pallets and wooden containers from Asia. In just eight years the beetle spread into 13 states from West Virginia to Minnesota. Adult emerald ash borers have golden or reddish-green bodies with dark metallic emerald green wing covers. More than 40 million ash trees have died or are dying from emerald ash borer attack in the United States, and more than 7.5 billion trees are at risk.

- In the 1970s several carp species, including bighead carp (*Hypophthalmichthys nobilis*), grass carp (*Ctenopharyngodon idella*), and silver carp (*Hypophthalmichthys molitrix*), were imported from China to control algae in aquaculture ponds. Unfortunately they escaped from captivity and have become established—often in very dense populations—throughout the Mississippi River Basin. Silver carp can grow to 100 pounds (45 kg). They are notorious for being easily frightened by boats and personal watercraft, which causes them to leap as much as 8–10 feet (2.5–3 m) into the air. Getting hit in the face by a large carp when you're traveling 48 km/hr (30 mph) can be life threatening. Large amounts of money have been spent trying to prevent Asian carp from spreading into the Great Lakes, but some carp already have been found in every Great Lake except Lake Superior.

Disease organisms, or pathogens, could also be considered predators. To be successful over the long term, a pathogen must establish a balance in which it is vigorous enough to reproduce, but not so lethal that it completely destroys its host. When a disease is introduced into a new environment, however, this balance may be lacking and an epidemic may sweep through the area.

The American chestnut was once the heart of many Eastern hardwood forests. In the Appalachian Mountains, at least one of every four trees was a chestnut. Often over 45 m (150 ft) tall, 3 m (10 ft) in diameter, fast growing, and able to sprout quickly from a cut stump, it was a forester's dream. Its nutritious nuts were important for birds (like the passenger pigeon), forest mammals, and humans. The wood was straight-grained, light, rot-resistant, and used for everything from fence posts to fine furniture, and its bark was used to tan leather. In 1904 a shipment of nursery stock from China brought a fungal blight to the United States, and within 40 years the American chestnut had all but disappeared from its native range. Efforts are now under way to transfer blight-resistant genes into the few remaining American chestnuts that weren't reached by the fungus or to find biological controls for the fungus that causes the disease.

Of course, the most ubiquitous, ecosystem-changing invasive species is us. We and our domesticated companions have occupied and altered the whole planet. One study calculated that the familiar and generally docile cow (*Bos tarus*), through grazing and trampling, endangers three times as many rare plant and animal species as any nondomesticated invader.

Island ecosystems are particularly susceptible to invasive species

New Zealand is a prime example of the damage that can be done by invasive species in island ecosystems. Having evolved for thousands of years without predators, New Zealand's flora and fauna are particularly susceptible to the introduction of alien organisms. Originally home to more than 3,000 endemic species, including flightless birds such as the kiwi and giant moas, New Zealand has lost at least 40 percent of its native flora and fauna since humans first landed there 1,000 years ago. More than 20,000 plant species have been introduced to New Zealand, and at least 200 of these have become pests that can create major ecological and economic problems. Many animal introductions (both intentional and accidental) also have become major threats to native species. Cats, rats, mice, deer, dogs, goats, pigs, and cattle accompanying human settlers consume native vegetation and eat or displace native wildlife.

Think About It

Domestic and feral house cats are estimated to kill 1 billion birds and small mammals in the United States annually. In 2005 a bill was introduced in the Wisconsin legislature to declare an open hunting season year-round on cats that roam out of their owner's yard. Would you support such a measure? Why or why not? What other measures (if any) would you propose to control feline predation?

One of the most notorious invasive species is the Australian brush-tailed possum, *Trichosurus vulpecula*. This small, furry marsupial was introduced to New Zealand in 1837 to establish a fur trade. In Australia, where their population is held in check by dingoes, fires, diseases, and inhospitable vegetation, possums are rare and endangered. Freed from these constraints in New Zealand, however, possum populations exploded. Now at least 70 million possums chomp their way through about 7 million tons of vegetation per year in their new home. They destroy habitat needed by indigenous New Zealand species, and also eat eggs, nestlings, and even adult birds of species that lack instincts to avoid predators.

Several dozen of New Zealand's offshore islands have been declared nature sanctuaries. Efforts are being made to eliminate invasive pests and to restore endangered species and native ecosystems. One of the most successful examples is Kapiti Island, off the southwest coast of the North Island. In the 1980s the Department of Conservation eradicated 22,500 brush-tailed possums—along with all the feral cats, ferrets, stoats, weasels, dogs, pigs, goats, cattle, and rats—on the 10 km long by 2 km wide island. The ecological benefits were immediately apparent. Native vegetation reappeared as seeds left in the soil sprouted and germinated without being eaten by foreign herbivores. Many native birds, such as the little brown kiwi, saddleback, stitchbird, kokako, and takahe, that are rare and endangered on the main islands now breed successfully in the predator-free environment.

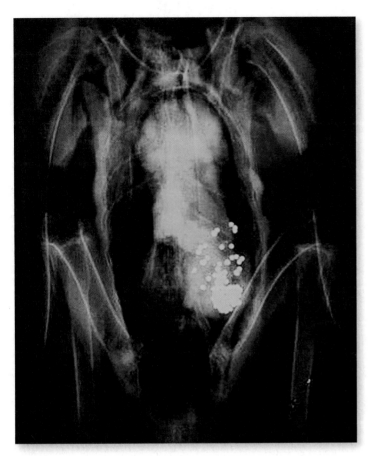

FIGURE 11.10 A bald eagle's stomach contents includes lead shot, which was consumed along with its prey. Fishing weights and shot remain a major cause of lead poisoning in aquatic and fish-eating birds.

Pollution

We have known for a long time that toxic pollutants can have disastrous effects on local populations of organisms. Pesticide-linked declines of top predators, such as eagles, osprey, falcons, and pelicans, were well documented in the 1970s. Declining populations of marine mammals, alligators, fish, and other wildlife alert us to the connection between pollution and health. This connection has led to a new discipline of conservation medicine (chapter 8). Mysterious, widespread deaths of thousands of seals on both sides of the Atlantic in recent years are thought to be linked to an accumulation of persistent chlorinated hydrocarbons, such as DDT, PCBs, and dioxins, in fat, causing weakened immune systems that make animals vulnerable to infections. Similarly, mortality of Pacific sea lions, beluga whales in the St. Lawrence estuary, and striped dolphins in the Mediterranean are thought to be caused by accumulation of toxic pollutants.

Lead poisoning is another major cause of mortality for many species of wildlife. Bottom-feeding waterfowl, such as ducks, swans, and cranes, ingest spent shotgun pellets that fall into lakes and marshes (fig. 11.10). They store the pellets, instead of stones, in their gizzards, and the lead slowly accumulates in their blood and other tissues. The U.S. Fish and Wildlife Service estimates that 3,000 metric tons of lead shot are deposited annually in wetlands and that between 2 and 3 million waterfowl die each year from lead poisoning.

Population

Human population growth represents a threat to biodiversity in several ways. If our consumption patterns remain constant, with more people, we will need to harvest more timber, catch more fish, plow more land for agriculture, dig up more fossil fuels and minerals, build more houses, and use more water. All of these demands impact wild species. Unless we find ways to dramatically increase the crop yield per unit area, it will take much more land than is currently domesticated to feed everyone, if our population grows to 8 to 10 billion as current projections predict. This will be especially true if we abandon intensive (but highly productive) agriculture and introduce more sustainable practices. The human population growth curve is leveling off (chapter 7), but it remains unclear whether we can reduce global inequality and provide a tolerable life for all humans while also preserving healthy natural ecosystems and a high level of biodiversity.

Overharvesting

Overharvesting is responsible for depletion or extinction of many species. A classic example is the extermination of the American passenger pigeon (*Ectopistes migratorius*). Even though it inhabited only eastern North America, 200 years ago this was probably the world's most abundant bird, with a population of 3 to 5 billion animals (fig. 11.11). It once accounted for about one-quarter of all

FIGURE 11.11 A pair of stuffed passenger pigeons (*Ectopistes migratorius*). The last member of this species died in the Cincinnati Zoo in 1914.

birds in North America. In 1830 John James Audubon saw a single flock of birds estimated to be ten miles wide, hundreds of miles long, and thought to contain perhaps a billion birds. In spite of this vast abundance, market hunting and habitat destruction caused the entire population to crash in only about 20 years, between 1870 and 1890. The last known wild passenger pigeon was shot in 1900, and the last existing passenger pigeon, a female named Martha, died in 1914 in the Cincinnati Zoo.

At about the same time that passenger pigeons were being extirpated, the American bison or buffalo (*Bison bison*) was being hunted to near extinction on the Great Plains. In 1850 some 60 million bison roamed the western plains. Many were killed only for their hides or tongues, leaving millions of carcasses to rot. Some of the bison's destruction was carried out by the U.S. Army so that native peoples who depended on bison for food, clothing, and shelter would be bereft of this resource and could then be forced onto reservations. By 1900 there were only about 150 wild bison left and another 250 in captivity.

Fish stocks have been seriously depleted by overharvesting in many parts of the world. A huge increase in fishing fleet size and efficiency in recent years has led to a crash of many oceanic populations. Worldwide, 13 of 17 principal fishing zones are now reported to be commercially exhausted or in steep decline. At least three-quarters of all commercial oceanic species are overharvested. Canadian fisheries biologists estimate that only 10 percent of the top predators, such as swordfish, marlin, tuna, and shark, remain in the Atlantic Ocean. If current trends continue, researchers warn, all major fish stocks could be in collapse—defined as 90 percent depleted—within 50 years (fig. 11.12). You can avoid adding to this overharvest by eating only abundant, sustainably harvested varieties (What Can You Do? p. 236). Facebook currently has a campaign to ban longline fishing that threatens sea birds, turtles, and marine mammals.

Perhaps the most destructive example of harvesting terrestrial wild animal species today is the African bushmeat trade. Wildlife biologists estimate that 1 million tons of bushmeat, including antelope, elephants, primates, and other animals, are sold in African markets every year (fig. 11.13a). For many poor Africans this is the only source of animal protein in their diet. If we hope to protect the animals targeted by bushmeat hunters, we will need to help their hunters and consumers find alternative livelihoods and replacement sources of high-quality protein. The emergence of SARS in 2003 resulted from the wild-food trade in China and Southeast Asia, where millions of civets, monkeys, snakes, turtles, and other animals are consumed each year as luxury foods.

Commercial Products and Live Specimens

In addition to harvesting wild species for food, we also obtain a variety of valuable commercial products from nature. Much of this represents sustainable harvest, but some forms of commercial exploitation are highly destructive and a serious threat to certain rare species. Despite international bans on trade in products from endangered species, the smuggling of furs, hides, horns, live specimens, and folk medicines amounts to millions of dollars each year.

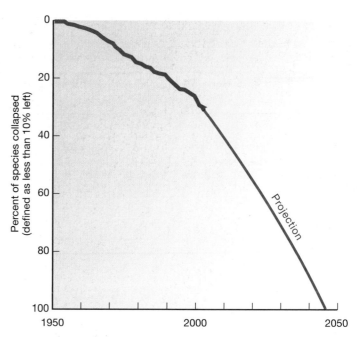

FIGURE 11.12 About one-third of all marine fish species are already in a state of population collapse. If current trends continue, all saltwater fish may reach this state by 2050.
Source: SeaWeb.

Developing countries in Asia, Africa, and Latin America with the richest biodiversity in the world are the main sources of wild animals and animal products, while Europe, North America, and some of the wealthy Asian countries are the principal importers. Japan, Taiwan, and Hong Kong buy three-quarters of all cat and snake skins, for instance, while European countries buy a similar percentage of live birds (fig. 11.13b). The United States imports 99 percent of all live cacti and 75 percent of all orchids sold each year.

The profits to be made in wildlife smuggling are enormous. Tiger or leopard fur coats can bring $100,000 in Japan or Europe. The population of African black rhinos dropped from approximately 100,000 in the 1960s to about 3,000 in the 1980s because of a demand for their horns. In Asia, where it is prized for its supposed medicinal properties, powdered rhino horn fetches (U.S.) $28,000 per kilogram.

Plants also are threatened by overharvesting. Wild ginseng has been nearly eliminated in many areas because of the Asian demand for the roots, which are used as an aphrodisiac and folk medicine. Cactus "rustlers" steal cacti by the ton from the American Southwest and Mexico. With prices ranging as high as $1,000 for rare specimens, it's not surprising that many cacti are now endangered.

The trade in wild species for pets is an enormous business. Worldwide some 5 million live birds are sold each year for pets. This trade endangers many rare species. It also is highly wasteful. Up to 60 percent of the birds die before reaching market. After the United States banned the sale of wild birds in 1992, imports declined 88 percent. Still, pet traders import (often illegally) some 2 million reptiles, 1 million amphibians and mammals, and 128 million tropical fish into the United States each year. About

(a) Bush meat market

(b) Hyacinth macaws

(c) Cyanide fishing

FIGURE 11.13 Threats to wildlife. (a) More than 1 million tons of wild animals are sold each year for human consumption. (b) Wild birds, like these Brazilian hyacinth macaws, are endangered by the pet trade. (c) Cyanide fishing not only kills fish, it also destroys the entire reef community.

75 percent of all saltwater tropical aquarium fish sold come from coral reefs of the Philippines and Indonesia.

Many of these fish are caught by divers using plastic squeeze bottles of cyanide to stun their prey (fig. 11.13c). Far more fish die with this technique than are caught. Worst of all, it kills the coral animals that create the reef. A single diver can destroy all of the life on 200 m² of reef in a day. Altogether, thousands of divers currently destroy about 50 km² of reefs each year. Net fishing would prevent this destruction, and it could be enforced if pet owners would insist on net-caught fish. More than half the world's coral reefs are potentially threatened by human activities; up to 80 percent are at risk in the most populated areas.

11.4 ENDANGERED SPECIES MANAGEMENT

Over the years we have gradually become aware of the harm we have done—and continue to do—to wildlife and biological resources. Slowly we are adopting national legislation and international treaties to protect these irreplaceable assets. Parks, wildlife refuges, nature preserves, zoos, and restoration programs have been established to protect nature and rebuild depleted populations. There has been encouraging progress in this area, but much remains to be done. While most people favor pollution control or protection of favored species such as whales or gorillas, surveys show that few understand what biological diversity is or why it is important.

Hunting and fishing laws have been effective

In 1874 a bill was introduced in the United States Congress to protect the American bison, whose numbers were already falling to dangerous levels. This initiative failed, however, because most legislators believed that all wildlife—and nature in general—was so abundant and prolific that it could never be depleted by human activity.

By the 1890s most states had enacted some hunting and fishing restrictions. The general idea behind these laws was to conserve the

resource for future human use rather than to preserve wildlife for its own sake. The wildlife regulations and refuges established since that time have been remarkably successful for many species. At the turn of the century there were an estimated half million white-tailed deer in the United States; now there are some 14 million—more in some places than the environment can support. Wild turkeys and wood ducks were nearly all gone 50 years ago. Restoring habitat, planting food crops, transplanting breeding stock, building shelters or houses, protecting these birds during breeding season, and other conservation measures have restored populations of these beautiful and interesting birds to several million each. Snowy egrets, which were almost wiped out by plume hunters 80 years ago, are now common again.

Legislation is key to biodiversity protection

The U.S. Endangered Species Act (ESA) and the Canadian Species at Risk law are powerful tools for wildlife protection. Where earlier regulations had been focused almost exclusively on "game" animals, these programs seek to identify all endangered species and populations and to save as much biodiversity as possible, regardless of its usefulness to humans. As defined by the ESA, **endangered species** are those considered to be in imminent danger of extinction, while **threatened species** are those that are likely to become endangered—at least locally—in the foreseeable future. Bald eagles, gray wolves, brown (or grizzly) bears, sea otters, and a number of native orchids and other rare plants are a few of the species considered to be locally threatened even though they remain abundant in other parts of their former range. Polar bears were listed as threatened in 2008 because the sea ice on which they depend for hunting is melting rapidly (fig. 11.14). **Vulnerable species** are naturally rare or have been locally depleted by human activities to a level that puts them at risk. Many of these are candidates for future listing. For vertebrates, protected categories include species, subspecies, and local races or ecotypes.

The ESA regulates a wide range of activities involving endangered species, including "taking" (harassing, harming, pursuing, hunting, shooting, trapping, killing, capturing, or collecting) either

What Can You Do?

Don't Buy Endangered Species Products

You probably are not shopping for a fur coat from an endangered tiger, but there might be other ways you are supporting unsustainable harvest and trade in wildlife species. To be a sustainable consumer, you need to learn about the source of what you buy. Often plant and animal products are farm-raised, not taken from wild populations. But some commercial products are harvested in unsustainable ways. Here are a few products about which you should inquire before you buy:

Seafood includes many top predators that grow slowly and reproduce only when many years old. Despite efforts to manage many fisheries, the following have been severely, sometimes tragically, depleted:

- Top predators: swordfish, marlin, shark, bluefin tuna, albacore ("white") tuna.
- Groundfish and deepwater fish: orange roughy, Atlantic cod, haddock, pollack (source of most fish sticks, artificial crab, generic fish products), yellowtail flounder, monkfish.
- Other species, especially shrimp, yellowfin tuna, and wild sea scallops, are often harvested with methods that destroy other species or habitats.
- Farm-raised species such as shrimp and salmon can be contaminated with PCBs, pesticides, and antibiotics used in their rearing. In addition, aquaculture operations often destroy coastal habitat, pollute surface waters, and deplete wild fish stocks to stock ponds and provide fish meal.

Pets and plants are often collected from wild populations, some sustainably and others not:

- Aquarium fish (often harvested by stunning with dynamite and squirts of cyanide, which destroy tropical reefs and many fish).
- Reptiles: snakes and turtles, especially, are often collected in the wild.
- Plants: orchids and cacti are the best-known, but not the only, group collected in the wild.

Herbal products such as wild ginseng and wild echinacea (purple coneflower) should be investigated before purchasing.

Do buy some of these sustainably harvested products:

- Shade-grown (or organic) coffee, nuts, and other sustainably harvested forest products.
- Pets from the Humane Society, which works to protect stray animals.
- Organic cotton, linen, and other fabrics.
- Fish products that have relatively little environmental impact or fairly stable populations: farm-raised catfish or tilapia, wild-caught salmon, mackerel, Pacific pollack, dolphinfish (mahimahi), squids, crabs, and crayfish.
- Wild freshwater fish like bass, sunfish, pike, catfish, and carp, which are usually better managed than most ocean fish.

---

accidentally or on purpose; importing into or exporting out of the United States; possessing, selling, transporting, or shipping; and selling or offering for sale any endangered species. Prohibitions apply to live organisms, body parts, and products made from endangered species. Violators of the ESA are subject to fines up to $50,000 and one year imprisonment. Vehicles and equipment used in violations may be subject to forfeiture. In 1995 the Supreme Court ruled that critical habitat—habitat essential for a species's survival—must be protected, whether on public or private land.

Currently the United States has 1,372 species on its endangered and threatened species lists and some 386 candidate species waiting to be considered. The number of listed species in different taxonomic groups reflects much more about the kinds of organisms that humans consider interesting and desirable than the actual number in each group. In the United States, invertebrates make up about three-quarters of all known species but only 9 percent of those deemed worthy of protection. Worldwide, the International Union for Conservation of Nature and Natural Resources (IUCN) lists a total of 17,741 endangered and threatened species (table 11.1).

Listing of endangered species is highly selective. We tend to be concerned about the species that we find interesting or useful rather than strive for equal representation from every phylum.

Notice, for instance, that 20 percent of all known mammals on the IUCN red list are described as threatened or endangered, but only 0.06 percent of insects are listed as threatened. This is inequitable in two ways. First, there are probably far more endangered insect species than this, even among those we have identified. Furthermore, it's extremely rare to find a new mammal species, whereas the million known insect species may represent only one-thirtieth of the total insect species on earth.

Listing of new species in the United States has been very slow, generally taking several years from the first petition to final determination. Limited funding, political pressures, listing moratoria, and changing administrative policies have created long delays. At least 18 species have gone extinct since being nominated for protection.

When Congress passed the original ESA, it probably intended to protect only a few charismatic species, such as raptors and big game animals. Sheltering obscure species such as the Delhi Sands flower-loving fly, the Coachella Valley fringe-toed lizard, Mrs. Furbisher's lousewort, or the orange-footed pimple-back mussel most likely never occurred to those who voted for the bill. This raises some interesting ethical questions about the rights and values of seemingly minor species. Although uncelebrated, these species may play important ecological roles. Protecting them usually preserves habitat and a host of unlisted species.

FIGURE 11.14 In 2008, U.S. Interior Secretary Dirk Kemp-thorne listed polar bears as threatened because the arctic sea ice on which they depend is melting rapidly. Nevertheless, Kemp-thorne claimed it would be "inappropriate" to use protection of the bear to reduce greenhouse gases or to address climate change.

Conservatives have tried repeatedly to weaken or eliminate the ESA. President George W. Bush listed only 59 species as endangered or threatened in his two terms in office. By contrast, President Bill Clinton listed 527 species in an equal time. In the Bush administration, political appointees regularly ignored scientific recommendations and obstructed listing or protection of endangered species. Shortly before leaving office, Bush removed the requirement in the Northern Forest Plan that agencies must survey land for vulnerable species, such as northern spotted owls, before starting logging, road building, or other harmful projects. President Obama reversed this order, but with conservative control of the House of Representatives, we may see more attempts to reduce or eliminate the ESA.

Recovery plans rebuild populations of endangered species

Once a species is officially listed as endangered, the Fish and Wildlife Service is required to prepare a recovery plan detailing how populations will be rebuilt to sustainable levels. It usually takes years to reach agreement on specific recovery plans. Among the difficulties are costs, politics, interference by local economic interests, and the fact that once a species is endangered, much of its habitat and its ability to survive are likely already compromised. The total cost of recovery plans for all currently listed species is estimated to be nearly $5 billion.

The recovery plan for the northern spotted owl is expected to cost $489 million over the next 30 years. This includes both management expenses and losses from setting aside 6.4 million acres (2.5 million ha) of old-growth forest needed to maintain 1,600 to 2,400 owls. Timber companies claim that costs will be even higher and that thousands of jobs will be lost. Conservationists counter that protecting owls will benefit watersheds and preserve many other organisms. Geneticists warn that northern spotted owls are undergoing a population bottleneck (chapter 6). There is so little genetic diversity within the population that they are susceptible to diseases and may have little environmental resilience.

The United States currently spends about $150 million per year on endangered species protection and recovery. About half that amount is spent on a dozen charismatic species like the California condor, and the Florida panther and grizzly bear, which receive around $13 million per year. By contrast, the 137 endangered invertebrates and 532 endangered plants get less than $5 million per year altogether. Our funding priorities often are based more on emotion and politics than biology. A variety of terms are used for rare or endangered species thought to merit special attention:

- *Keystone species* are those with major effects on ecological functions and whose elimination would affect many other members of the biological community; examples are prairie dogs (*Cynomys ludovicianus*) or bison (*Bison bison*).

- *Indicator species* are those tied to specific biotic communities or successional stages or a set of environmental conditions. They can be reliably found under certain conditions but not others; an example is brook trout (*Salvelinus fontinalis*).

- *Umbrella species* require large blocks of relatively undisturbed habitat to maintain viable populations. Saving this habitat also benefits other species. Examples of umbrella species are the northern spotted owl (*Strix occidentalis caurina*) and bighorn sheep (*Ovis canadensis*) (fig 11.15).

- *Flagship species* are especially interesting or attractive organisms to which people react emotionally. These species can motivate the public to preserve biodiversity and contribute to conservation; an example is the giant panda (*Ailuropoda melanoleuca*).

FIGURE 11.15 The Endangered Species Act seeks to restore population of species such as the bighorn sheep, which has been listed as endangered over much of its range. Charismatic species are easier to list than obscure ones.

FIGURE 11.16 Bald eagles, and other bird species at the top of the food chain, were decimated by DDT in the 1960s. Many such species have recovered since DDT was banned in the United States and because of protection under the Endangered Species Act.

Some recovery plans have been gratifyingly successful. The American alligator was listed as endangered in 1967 because hunting (for meat, skins, and sport) and habitat destruction had reduced populations to precarious levels. Protection has been so effective that the species is now plentiful throughout its entire southern range. Florida alone estimates that it has at least 1 million alligators.

Sometimes restoring a single species can bring benefits to an entire ecosystem, especially when that species plays a keystone role in the community. Alligators, for example, dig out swimming holes, or wallows, that become dry-season refuges for fish and other aquatic species. American bison are being used in prairie restoration projects to reestablish health and diversity of grassland ecosystems (Exploring Science, p. 239). (See additional discussion in chapter 13.)

Some other successful recovery programs involve bald eagles, peregrine falcons, and whooping cranes (fig. 11.16). Forty years ago, due mainly to DDT poisoning, only 417 nesting pairs of bald eagles (*Haliaeetus leucocephalis*) remained in the contiguous United States. By 2007 the population had rebounded to more than 9,800 nesting pairs, and the birds were removed from the endangered species list. This doesn't mean that eagles are unprotected. Killing, selling, or otherwise harming eagles, their nests, or their eggs is still prohibited. In addition to eagles and falcons, 29 other species, including mammals, fish, reptiles, birds, plants, and even one insect (the Tinian monarch), have been removed or downgraded from the endangered species list.

Opponents of the ESA have repeatedly tried to require that economic costs and benefits be incorporated into endangered species planning. An important test of the ESA occurred in 1978 in Tennessee where construction of the Tellico Dam threatened a tiny fish called the snail darter. As a result of this case, a federal committee (the so-called "God Squad") was given power to override the ESA for economic reasons.

An even more costly recovery program may be required for Columbia River salmon and steelhead endangered by dams that block their migration to the sea. Opening floodgates to allow young fish to run downriver and adults to return to spawning grounds would have high economic costs to barge traffic, farmers, and electric rate payers who have come to depend on abundant water and cheap electricity. On the other hand, commercial and sport fishing for salmon was once worth $1 billion per year and employed about 60,000 people directly or indirectly.

Private land is vital in endangered species protection

Eighty percent of the habitat for more than half of all listed species is on nonpublic property. The Supreme Court has ruled that destroying habitat is as harmful to endangered species as directly taking (killing) them. Many people, however, resist restrictions on how they use their own property to protect what they perceive as insignificant or worthless organisms. This is especially true when the land has potential for economic development. If property is worth millions of dollars as the site of a housing development or shopping center, most owners don't want to be told they have to leave it undisturbed to protect some rare organism. Landowners may be tempted to "shoot, shovel, and shut up," if they discover endangered species on their property. Many feel they should be compensated for lost value caused by ESA regulations.

Recently, to avoid crises like the northern spotted owl, the Fish and Wildlife Service has been negotiating agreements called **habitat conservation plans (HCP)** with private landowners. Under these plans, landowners are allowed to harvest resources or build on part of their land as long as the species benefits overall. In return for improving habitat in some areas, funding conservation research, removing predators and competitors, or taking other steps that benefit the endangered species, developers are allowed to destroy habitat or even "take" endangered organisms.

Scientists and environmentalists often are critical of HCPs, claiming these plans often are based more on politics than biology, and that the potential benefits are frequently overstated. Defenders argue that by making the ESA more landowner-friendly, HCPs benefit wildlife in the long run.

Among the more controversial proposals for HCPs are the so-called Safe Harbor and No-Surprises policies. Under the Safe Harbor clause, any increase in an animal's population resulting from a property owner's voluntary good stewardship would not increase their responsibility or affect future land-use decisions. As long as the property owner complies with the terms of the agreement, he or she can make any use of the property. The No-Surprises provision says that the property owner won't be faced with new requirements or regulations after entering into an HCP. Scientists warn that change, uncertainty, dynamics, and flux are characteristic of all ecosystems. We can't say that natural catastrophes or environmental events won't make it necessary to modify conservation plans in the future.

Bison Can Help Restore Prairie Ecosystems

Much of the American Great Plains was converted to agriculture a century or more ago. The prairie was plowed under or grazed heavily, while native species, such as wolves, bison, and grizzly bears, were eradicated or confined to a few parks and nature preserves. Now efforts are under way to restore large areas of this unique biome. Fire is an essential tool in restoration projects. Prescribed burning removes invasive woody species and gives native grasses and forbs (broad-leaved flowering plants) a chance to compete. But simply setting fires every now and then isn't enough to maintain a healthy prairie.

American prairies coevolved with grazing animals. In particular, a keystone species for the Great Plains was the American buffalo (*Bison bison*). Perhaps 60 million of these huge, shaggy animals once roamed the plains from the Rocky Mountains to the edge of the eastern deciduous forest and from Manitoba to Texas. By 1900 there were probably fewer than 150 wild bison left in the United States, mostly in Yellowstone National Park. Wildlife protection and breeding programs have rebuilt the population to about 500,000 animals, but probably less than 4 percent of them are genetically pure.

Like fire, bison helped maintain native plant species with their intensive grazing. When put on open range, domestic cattle graze selectively on the species they like, giving noxious weeds a selective advantage. Bison, on the other hand, tend to move in dense herds eating almost everything in their

Bison grazing helps maintain prairie species and a healthy ecosystem.

path. Their trampling and intense grazing disturb the ground and provide habitat for pioneer species, many of which disappear when bison are removed. Bison also create areas for primary succession by digging out wallows in which they take dust baths.

Having grazed an area heavily, bison will tend to move on, and if they have enough space in which to roam, they won't come back for several years. This pattern of intensive, short-duration grazing creates a mosaic of different successional stages that enhances biodiversity. It also is the origin of the idea of rotational grazing in sustainable livestock management. Bison increase plant productivity by increasing the availability of light and reducing water stress, both of which increase photosynthesis rates.

Grazing also affects the nutrient cycling in prairie ecosystems. Nitrogen and phosphorus are essential for plant productivity. By consuming plant biomass, bison return these nutrients to the soils in urine and buffalo chips. Bison are more efficient nutrient recyclers than the slow release from plant litter decay. Fire releases nitrogen by burning plant material. Bison, on the other hand, limit nitrogen loss by reducing the aboveground plant biomass and increasing the patchiness of the fire. These changes in nutrient cycling and availability in prairie ecosystems lead to increased plant productivity and species composition.

But it takes a large area to have freely wandering buffalo herds. One of the biggest buffalo restoration projects is that of the American Prairie Foundation (APF), which is closely linked to the World Wildlife Fund. The APF has purchased about 24,000 ha of former ranchland in northern Montana. Rather than keep it in cattle production, however, this group intends to pull out fences, eliminate all the ranch buildings, and turn the land back into wilderness. Ultimately the APF hopes to create a reserve of at least 1.5 million ha in the Missouri Breaks region between the Charles M. Russell National Wildlife Refuge and the Fort Belknap Indian Reservation.

The APF plans to reintroduce native wildlife, including elk, bison, wolves, and grizzly bears, to its lands. And in restoring these keystone species to the land, they also help preserve rare and endangered species, such as prairie dogs, swift foxes, ferruginous hawks, mountain plover, prairie rattlesnakes, badgers, and the rest of the complex web of plants and animals that evolved with them.

Endangered species protection is controversial

The U.S. ESA officially expired in 1992. Since then Congress has debated many alternative proposals, ranging from outright elimination to substantial strengthening of the act. Perhaps no other environmental issue divides Americans more strongly than the ESA. In the western United States, where traditions of individual liberty and freedom are strong and the federal government is viewed with considerable suspicion and hostility, the ESA seems to many to be a diabolical plot to take away private property and trample on individual rights (fig. 11.17). Many people believe that the law puts the welfare of plants and animals above that of humans. Farmers, loggers, miners, ranchers, developers, and other ESA opponents repeatedly have tried to scuttle the law or greatly reduce its power. Environmentalists, on the other hand, see the ESA as essential to protecting nature and maintaining the viability of the planet. They regard it as the single most effective law in their arsenal and want it enhanced and improved.

Critical habitat protection is especially onerous to local residents because it often involves protecting lands that don't now have the endangered species. Conservationists view this as absolutely necessary. How can we hope to restore species if there's no place for them to live? Locals, on the other hand, resent having to curtail their activities for some animal they don't want living in their neighborhood, and that isn't there anyway.

Conservationists, too, have criticisms of our current endangered species protection. Perhaps chief of these is the focus on individual organisms. As we pointed out earlier in this chapter, protecting a keystone or umbrella species, such as wolves or elephants, can benefit entire ecological communities, but often we spend millions of dollars attempting to save a single kind of organism when those funds might have done more good, ecologically, by protecting a functional—if less unique—community. Perhaps it would be better to try to preserve representative samples of many different kinds of biological communities and ecological services (even if those communities are missing a few of their historic members) than to save a few rare species that may be at the end of their evolutionary life cycle anyway.

HERBLOCK'S CARTOON JUN 26 1990

FIGURE 11.17 Endangered species often serve as a barometer for the health of an entire ecosystem and as surrogate protector for a myriad of less well-known creatures.

Source: Copyright 1990 by Herb Block in *The Washington Post.* Reprinted by permission of The Herb Block Foundation.

What Can You Do?

You Can Help Preserve Biodiversity

If you live in an urban area, as most Americans do, you may not think that you have much influence on wildlife, but there are important ways that you can help conserve biodiversity.

- Protect or restore native biomes. If you inquire about environmental organizations or nature preserves near where you live, you'll find opportunities to remove invasive species, gather native seeds, replant native vegetation, or find other ways to preserve or improve habitat.

- Plant local, native species in your garden. Exotic nursery plants often escape and threaten native ecosystems.

- Don't transport firewood from one region to another. It may carry diseases and insects.

- Follow legislation and management plans for natural areas you value. Lobby or write letters supporting funding and biodiversity-friendly policies.

- Help control invasive species. Never release non-native animals (fish, leaches, turtles, etc.) or vegetation into waterways or sewers. If you boat, wash your boat and trailer when moving from one lake or river to another.

- Don't discard worms in the woods. You probably think of earthworms as beneficial for soils—and they are, in the proper place—but many northern deciduous biomes evolved without them. Worms discarded by anglers are now causing severe habitat destruction in many places.

- Keep your cat indoors. House cats are major predators of woodland birds and other native animals. It's estimated that house cats in the United States kill 1 billion birds and small mammals every year.

Large-scale, regional planning is needed

Over the past decade, growing numbers of scientists, land managers, policymakers, and developers have been making the case that it is time to focus on a rational, continent-wide preservation of ecosystems that support maximum biological diversity rather than a species-by-species battle for the rarest or most popular organisms. By focusing on populations already reduced to only a few individuals, we spend most of our conservation funds on species that may be genetically doomed no matter what we do. Furthermore, by concentrating on individual species we spend millions of dollars to breed plants or animals in captivity that have no natural habitat where they can be released. While flagship species such as mountain gorillas or Indian tigers are reproducing well in zoos and wild animal parks, the ecosystems that they formerly inhabited have largely disappeared.

A leader of this new form of conservation is J. Michael Scott, who was project leader of the California condor recovery program in the mid-1980s and had previously spent ten years working on endangered species in Hawaii. In making maps of endangered species, Scott discovered that even Hawaii, where more than 50 percent of the land is federally owned, has many vegetation

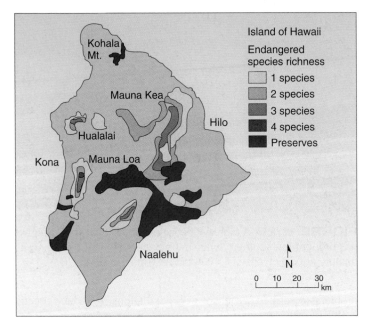

FIGURE 11.18 An example of the biodiversity maps produced by J. Michael Scott and the U.S. Fish and Wildlife Service. Notice that few of the areas of endangered species richness are protected in preserves, which were selected more for scenery or recreation than for biology.

types completely outside of natural preserves (fig. 11.18). The gaps between protected areas may contain more endangered species than are preserved within them.

This observation has led to an approach called **gap analysis** in which conservationists and wildlife managers look for unprotected landscapes that are rich in species. Computers and geographical information systems (GIS) make it possible to store, manage, retrieve, and analyze vast amounts of data and create detailed, high-resolution maps relatively easily. This broad-scale, holistic approach seems likely to save more species than a piecemeal approach.

Conservation biologist R. E. Grumbine suggests four remanagement principles for protecting biodiversity in a large-scale, long-range approach:

1. Protect enough habitat for viable populations of all native species in a given region.

2. Manage at regional scales large enough to accommodate natural disturbances (fire, wind, climate change, and so on).

3. Plan over a period of centuries so that species and ecosystems may continue to evolve.

4. Allow for human use and occupancy at levels that do not result in significant ecological degradation.

International wildlife treaties are important

The 1975 Convention on International Trade in Endangered Species (CITES) was a significant step toward worldwide protection of endangered flora and fauna. It regulated trade in living specimens and products derived from listed species, but it has not been foolproof. Species are smuggled out of countries where they are threatened or endangered, and documents are falsified to make it appear they have come from areas where the species are still common. Investigations and enforcement are especially difficult in developing countries where wildlife is disappearing most rapidly. Still, eliminating markets for endangered wildlife is an effective way to stop poaching. Appendix I of CITES lists 700 species threatened with extinction by international trade.

11.5 CAPTIVE BREEDING AND SPECIES SURVIVAL PLANS

Breeding programs in zoos and botanical gardens are one way to attempt to save severely threatened species. Institutions like the Missouri Botanical Garden and the Bronx Zoo's Wildlife Conservation Society sponsor conservation and research programs. Botanical gardens, such as Kew Gardens in England, and research stations, such as the International Rice Institute in the Philippines, are repositories for rare and endangered plant species, some of which have ceased to exist in the wild. Valuable genetic traits are preserved in these collections, and in some cases, plants with

FIGURE 11.19 Nearly extirpated in the 1950s, the land-dwelling nene of Hawaii has been successfully restored by captive breeding programs. From fewer than 30 birds half a century ago, the wild population has grown to more than 500 birds.

FIGURE 11.20 A highly successful captive breeding program has brought the southern white rhino back from near extinction a century ago to at least 17,480 animals today.

unique cultural or ecological significance may be reintroduced into native habitats after being cultivated for decades or even centuries in these gardens and seed banks.

Zoos can help preserve wildlife

Until fairly recently zoos depended on primarily wild-caught animals for most of their collections. This was a serious drain on wild populations, because up to 80 percent of the animals caught died from the trauma of capture and shipping. With better understanding of reproductive biology and better breeding facilities, most mammals in North American zoos now are produced by captive breeding programs.

Some zoos now participate in programs that reintroduce endangered species to the wild. The California condor is one of the best-known cases of successful captive breeding. In 1986 only nine of these birds existed in their native habitat. Fearing the loss of these last condors, biologists captured them and brought them to the San Diego and Los Angeles zoos, which had begun breeding programs in the 1970s. By 2010 the population had reached 381 birds, including 192 reintroduced to the wild.

The endemic nene of Hawaii (*Nesochen sandvicensis*) also has been successfully bred in captivity and reintroduced into the wild. When Captain Cook arrived in the Hawaiian Islands in 1778, there were probably 25,000 of these land-dwelling geese. By the 1950s, however, habitat destruction and invasive predators had reduced the population to fewer than 30 birds. Today there are about 500 wild nene, and more fledglings are introduced every year (fig. 11.19).

One of the most successful captive breeding programs is that of the white rhino (*Ceratotherium simum simum*) in southern Africa. Although they once ranged widely across southern Africa, these huge animals were considered extinct until a remnant herd was found in Natal, South Africa, in 1895. Today there are an estimated 17,480 southern white rhinos, mainly in national parks and private game ranches (fig. 11.20). The fact that hunters will pay tens of thousands of dollars to shoot one is largely responsible for their preservation. Such breeding programs have limitations, however. In 2007, Canadian officials captured the last 16 wild northern spotted owls in British Columbia and moved them to zoos for captive breeding. Will this help if there isn't habitat to reintroduce them into? Or if barred owls invade former spotted owl breeding sites, can they be displaced?

Moreover bats, whales, and many reptiles rarely reproduce in captivity and still come mainly from the wild. We will never be able to protect the complete spectrum of biological variety in zoos. According to one estimate, if all the space in U.S. zoos were used for captive breeding, only about 100 species of large mammals could be maintained on a long-term basis.

These limitations lead to what is sometimes called the "Noah question": how many species can or should we save? How much are we willing to invest to protect the slimy, smelly, crawly things? Would you favor preserving disease organisms, parasites, and vermin, or should we use our limited resources to protect only beautiful, interesting, or seemingly useful organisms?

Even given adequate area and habitat conditions to perpetuate a given species, continued inbreeding of a small population in captivity can lead to the same kinds of fertility and infant survival problems described earlier for wild populations. To reduce genetic problems, zoos often exchange animals or ship individuals long distances to be bred. It sometimes turns out, however, that zoos far distant from each other unknowingly obtained their animals from the same source. Computer databases operated by the International Species Information System, located at the Minnesota Zoo, now keep track of the genealogy of many species. This system can tell the complete reproductive history of every animal in every zoo in the world for some species. Comprehensive species survival plans based on this genealogy help match breeding pairs and project resource needs.

The ultimate problem with captive breeding, however, is that natural habitat may disappear while we are busy conserving the species itself. Large species such as tigers or apes are sometimes

FIGURE 11.21 The *KM Minnesota* anchored in Tamanjaya Bay in west Java. Funds raised by the Minnesota Zoo paid for local construction of this boat, which allows wardens to patrol Ujung Kulon National Park and protect rare Javanese rhinos from poachers.

called "umbrella species." As long as they persist in their native habitat, many other species survive as well.

We need to save rare species in the wild

Renowned zoologist George Schaller says that ultimately "zoos need to get out of their own walls and put more effort into saving the animals in the wild." An interesting application of this principle is a partnership between the Minnesota Zoo and the Ujung Kulon National Park in Indonesia, home to the world's few remaining Javanese rhinos. Rather than try to capture rhinos and move them to Minnesota, the zoo is helping to protect them in their native habitat by providing patrol boats, radios, housing, training, and salaries for Indonesian guards (fig. 11.21). There are no plans to bring any rhinos to Minnesota, and chances are very slight that any of us will ever see one, but we can gain satisfaction in knowing that, at least for now, a few Javanese rhinos still exist in the wild.

CONCLUSION

Biodiversity provides food, fiber, medicines, clean water, and many other products and services we depend upon every day. Yet nearly one-third of native species in the United States are at risk of disappearing. The Endangered Species Act has proven to be one of the most powerful tools we have for environmental protection. Because of its effectiveness, the act itself is endangered; opponents have succeeded in limiting its scope, and have threatened to eliminate it altogether. Still, the act remains a cornerstone of our most basic environmental protections. It has given new hope for survival to numerous species that were on the brink of extinction—less than 1 percent of species listed under the ESA have gone extinct since 1973, whereas 10 percent of candidate species still waiting to be listed have suffered that fate.

For some species, such as the northern spotted owl, protection and recovery programs are difficult when the critical habitat on which they depend has largely been degraded or destroyed.

Biodiversity protection has gone far beyond the intent of the original framers of this act 30 years ago. In light of the serious threats facing our environment today—including pollution, habitat destruction, invasive species, and global climate change—we probably need to reevaluate which species we will protect, and how we will protect them. It's clear that we need to be concerned about the other organisms on which we depend for a host of ecological services, and with which we share this planet. In the next two chapters, we'll look at programs that work to protect and restore whole communities and landscapes.

REVIEWING LEARNING OUTCOMES

By now you should be able to explain the following points:

11.1 Discuss biodiversity and the species concept.
- What is biodiversity?
- What are species?
- Molecular techniques are revolutionizing taxonomy.
- How many species are there?
- Hot spots have exceptionally high biodiversity.

11.2 Summarize some of the ways we benefit from biodiversity.
- All of our food comes from other organisms.
- Living organisms provide us with many useful drugs and medicines.
- Biodiversity provides ecological services.
- Biodiversity also brings us many aesthetic and cultural benefits.

11.3 Characterize the threats to biodiversity.
- Extinction is a natural process.

- We are accelerating extinction rates.
- Island ecosystems are particularly susceptible to invasive species.

11.4 Evaluate endangered species management.
- Hunting and fishing laws have been effective.
- Legislation is key to biodiversity protection.
- Recovery plans rebuild populations of endangered species.
- Private land is vital in endangered species protection.
- Endangered species protection is controversial.
- Large-scale, regional planning is needed.
- International wildlife treaties are important.

11.5 Scrutinize captive breeding and species survival plans.
- Zoos can help preserve wildlife.
- We need to save rare species in the wild.

PRACTICE QUIZ

1. What is the range of estimates of the total number of species on the earth? Why is the range so great?

2. What group of organisms has the largest number of species?

3. Define *extinction*. What is the natural rate of extinction in an undisturbed ecosystem?

4. What are rosy periwinkles and what products do we derive from them?

5. Describe some foods we obtain from wild plant species.

6. Define *HIPPO* and describe what it means for biodiversity conservation.

7. What is the current rate of extinction and how does this compare to historic rates?

8. Why are barred owls a threat to spotted owls?

9. Define *endangered* and *threatened*. Give an example of each.

10. What is gap analysis and how is it related to ecosystem management and design of nature preserves?

CRITICAL THINKING AND DISCUSSION QUESTIONS

1. Many ecologists would like to move away from protecting individual endangered species to concentrate on protecting whole communities or ecosystems. Others fear that the public will respond to and support only glamorous "flagship" species such as gorillas, tigers, or otters. If you were designing conservation strategy, where would you put your emphasis?

2. Put yourself in the place of a fishing industry worker. If you continue to catch many species, they will quickly become economically extinct if not completely exterminated. On the other hand, there are few jobs in your village and welfare will barely keep you alive. What would you do?

3. Only a few hundred grizzly bears remain in the contiguous United States, but populations are healthy in Canada and Alaska. Should we spend millions of dollars for grizzly recovery and management programs in Yellowstone National Park and adjacent wilderness areas?

4. How could people have believed a century ago that nature is so vast and fertile that human actions could never have a lasting impact on wildlife populations? Are there similar examples of denial or misjudgment occurring now?

5. In the past, mass extinction has allowed for new growth, including the evolution of our own species. Should we assume that another mass extinction would be a bad thing? Could it possibly be beneficial to us? to the world?

6. Some captive breeding programs in zoos are so successful that they often produce surplus animals that cannot be released into the wild because no native habitat remains. Plans to euthanize surplus animals raise storms of protests from animal lovers. What would you do if you were in charge of the zoo?

Data Analysis: Confidence Limits in the Breeding Bird Survey

If you read scientific literature, you often will see graphs with vertical lines on each point. What do those lines mean? They represent standard error, a measure of how much variation there is in a group of observations. This is one way scientists show uncertainty, or their level of confidence in their results.

A central principle of science is the recognition that all knowledge involves uncertainty. No study can observe every possible event in the universe, so there is always missing information. Scientists try to define the limits of their uncertainty, in order to allow a realistic assessment of their results. A corollary of this principle is that the more data we have, the less uncertainty we have. More data increase our confidence that our observations represent the range of possible observations.

One of the most detailed records of wildlife population trends in North America is the Breeding Bird Survey (BBS). Every June more than a thousand volunteers drive established routes and count every bird they see or hear. The accumulated data from thousands of routes, over many years, indicates population *trends,* telling which populations are increasing, decreasing, or expanding into new territory.

Because many scientists use BBS data, it is essential to communicate how much confidence there is in the data. The online BBS database reports measures of data quality, including:

- **N:** the number of survey routes from which population trends are calculated.

- **Confidence limits:** because the reported trend is an average of a small *sample* of year-to-year changes on routes, confidence limits tell us how close the sample's average probably is to the average for the *entire population* of that species. Statistically, 95 percent of all samples should fall in between the confidence limits. In effect, we can be 95 percent sure that the entire population's actual trend falls between the upper and lower confidence limits.

1. Examine the following table, which shows 10 species taken from the online BBS database. How many species have a positive population trend (0)? If a species had a trend of 0, how much would it change from year to year?

2. Which species has the greatest decline per year? For every 100 birds this year, how many fewer will there be next year?

3. If the distance between upper and lower confidence limits (the confidence interval) is narrow, then we can be reasonably sure the trend in our *sample* is close to the trend for the *total population* of that species. What is the reported trend for ring-necked pheasant? What is the number of routes (N) on which this trend is based? What is the range in which the pheasant's true population trend probably falls? Is there a reasonable chance that the pheasant population's average annual change is 0? That the population trend is actually 25?

   Now look at the ruffed grouse, on either the table or the graph. Does the trend show that the population is increasing or decreasing? Can you be certain that the actual trend is not 7, or 17? On how many routes is this trend based?

4. In general, confidence limits depend on the number of observations (N), and how much all the observed values (trends on routes, in this case) differ from the average value. If the values vary greatly, the confidence interval will be wide. Examine the table and graph. Does a large N tend to widen or narrow the confidence interval?

5. A trend of 0 would mean no change at all. When 0 falls within the confidence interval, we have little certainty that the trend is not 0. In this case, we say that the trend is not significant. How many species have trends that are not significant (at 95 percent certainty)?

   Can we be certain that the sharp-tailed grouse and greater prairie-chicken are changing at different rates? How about the sharp-tailed grouse and wild turkey?

6. Does uncertainty in the data mean results are useless? Does reporting of confidence limits increase or decrease your confidence in the results?

For further information on the Breeding Bird Survey, see www .mbr-pwrc.usgs.gov/bbs/.

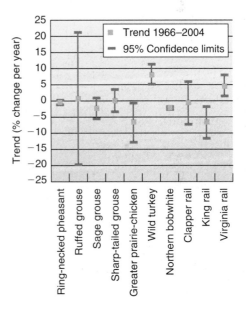

Species	Trend 1966–2004	N	Lower limit	Upper limit
Ring-necked pheasant	−0.5	397	−1.2	0.2
Ruffed grouse	0.7	26	−19.7	21.1
Sage grouse	−2.4	23	−5.6	0.9
Sharp-tailed grouse	0	104	−3.4	3.5
Greater prairie-chicken	−6.8	41	−12.8	−0.8
Wild turkey	8.2	308	5.1	11.2
Northern bobwhite	−2.2	541	−2.7	−1.8
Clapper rail	−0.7	8	−7.3	5.8
King rail	−6.8	22	−11.7	−1.8
Virginia rail	4.7	21	1.4	7.9

Approximately 1.7 billion metric tons of carbon are released annually due to land use change, mainly from tropical deforestation—more than all global transportation emissions combined.

Biodiversity
Preserving Landscapes

Learning Outcomes

After studying this chapter, you should be able to:

12.1 Discuss the types and uses of world forests.
12.2 Describe the location and state of grazing lands around the world.
12.3 Summarize the types and locations of nature preserves.

"If we destroy the land, God may forgive us, but our children will not." ~ *Togiak Elder*

246

In 2010 Norway signed an agreement to support Indonesia's efforts to reduce greenhouse gas emissions from deforestation and forest degradation. Based on Indonesia's performance over the next eight years, Norway will provide up to (U.S.) $1 billion to support this partnership. Indonesia has the third largest area of tropical rainforest in the world (after Brazil and the Democratic Republic of Congo), and because it's an archipelago of more than 16,000 islands, many of which have unique assemblages of plants and animals, Indonesia has some of the highest biological diversity in the world.

Indonesia is an excellent example of the benefits of forest protection. Deforestation, land-use change, and the drying, decomposition, and burning of peatlands cause about 80 percent of the country's current greenhouse gas emissions. This means that Indonesia can make deeper cuts in CO_2 emissions and do it more quickly than most other countries. Reducing deforestation will help preserve biodiversity and protect indigenous forest people. And according to government estimates, up to 80 percent of Indonesia's logging (fig. 12.1) is illegal, so bringing it under control also will increase national revenue and help build civic institutions.

Indonesia recognizes that climate change is one of the greatest challenges facing the world today. In 2009, President Susilo Bambang Yudhoyono committed to reducing Indonesia's CO_2 emissions 26 percent by 2020 compared to a business-as-usual trajectory. This is the largest absolute reduction pledge made by any developing country and could exceed reductions by most industrialized countries as well.

The partnership between Norway and Indonesia is the largest example so far of a new, UN-sponsored program called REDD (Reducing Emissions from Deforestation and Forest Degradation), which aims to slow climate change by paying developing countries to stop cutting down their forests. One of the few positive steps agreed on at the 2010 UN climate conference in Cancun, REDD could result in a major transfer of money from rich countries to poor. It's estimated that it will take about (U.S.) $30 billion per year to fund this program. But it offers a chance to save one of the world's most precious ecosystems. Forests would no longer be viewed merely as timber waiting to be harvested or land awaiting clearance for agriculture.

FIGURE 12.1 Logging valuable hardwoods is generally the first step in tropical forest destruction. Although loggers may take only one or two large trees per hectare, the damage caused by extracting logs exposes the forest to invasive species, poachers, and fires.

Many problems need to be solved for the Norway/Indonesia partnership to work. For one thing, it will be necessary to calculate how much carbon is stored in a particular forest as well as how much carbon could be saved by halting or slowing deforestation. Historical forest data, on which these predictions often are based, is often unreliable or nonexistent in tropical countries. Satellite imaging and computer modeling can give answers to these questions, but technology is expensive. In the first phase of funding, Norway will support political and institutional reform along with infrastructure and capacity building.

Like other donor nations, Norway is also concerned about how permanent the protections will be. What happens if they pay to protect a forest but a future administration decides to log it? Furthermore, loggers are notoriously mobile and adept at circumventing rules by bribing local authorities, if necessary. What's to prevent them from simply moving to new areas to cut trees? If you avoid deforestation in one place but then cut an equal number of trees somewhere else (sometimes known as "leakage"), carbon emissions won't have gone down at all. Similarly, there's concern that a reduction in logging in one country could lead to pressure on other countries to cut down their forests to meet demand. And there would be a financial incentive to do so if reductions in logging pushed up the price of timber.

Will this partnership protect indigenous people's rights? In theory, yes. Indonesia has more than 500 ethnic groups, and many forest communities lack secure land tenure. Large mining, logging, and palm oil operations often push local people off traditional lands with little or no compensation. Indonesia has promised a two-year suspension on new projects to convert natural forests. They also have promised to recognize the rights of native people and local communities.

Could having such a sudden influx of money cause corruption? Yes, that's possible. But Indonesia has a good track record of managing foreign donor funds under President Yudhoyono. The Aceh and Nias Rehabilitation and Reconstruction Agency (BRR), established after the 2004 tsunami, managed around (U.S.) $7 billion of donations in line with the best international standards. Indonesia has promised that the same governance principles will be used to manage REDD funds.

In this chapter, we'll look at other examples of how we protect biodiversity and preserve landscapes. For Google Earth™ placemarks that will help you explore these landscapes via satellite images, visit EnvironmentalScience-Cunningham.blogspot.com.

12.1 World Forests

Forests and woodlands occupy some 4 billion hectares (roughly 15 million mi^2), or about 30 percent of the world's land surface (fig. 12.2). Grasslands (pastures and rangelands) cover about the same percentage. Together these ecosystems supply many essential resources, such as lumber, paper pulp, and grazing lands for livestock. They also provide vital ecological services, including regulating climate, controlling water runoff, providing wildlife habitat, and purifying air and water. Forests and grasslands also have scenic, cultural, and historic values that deserve protection. These biomes are also among the most heavily disturbed (chapter 5) because they're places that people prefer to live and work.

As the opening case study for this chapter shows, these competing land uses and needs often are incompatible. Yet we need wild places as well as the resources they produce. Many conservation debates have concerned protection or use of forests, prairies, and rangelands. This chapter examines the ways we use and abuse these biological communities, as well as some of the ways we can protect them and conserve their resources. We discuss forests first, followed by grasslands and then strategies for conservation and preservation. Chapter 13 focuses on restoration of damaged or degraded ecosystems.

Boreal and tropical forests are most abundant

Forests are widely distributed, but the largest remaining areas are in the humid equatorial regions and the cold boreal forests of high latitudes (fig. 12.3). Five countries—Russia, Brazil, Canada, the United States, and China—together have more than half of the world's forests. The UN Food and Agriculture Organization (FAO) defines **forest** as any area where trees cover more than 10 percent of the land. This definition includes a variety of forest types, ranging from open **savannas,** where trees cover less than 20 percent of the ground, to **closed-canopy forests,** in which tree crowns overlap to cover most of the ground.

The largest tropical forests are in South America, which has about 22 percent of the world's forest area and by far the most extensive area of undisturbed tropical rainforest. Africa and Southeast Asia also have large areas of tropical forest that are highly important biologically, but both continents are suffering

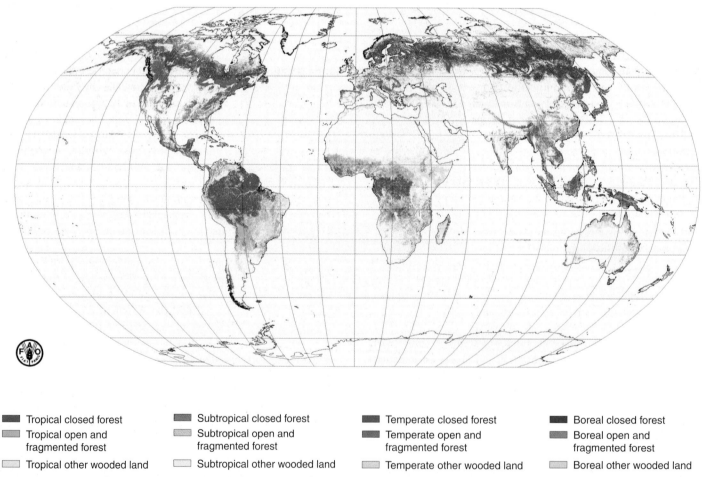

■ Tropical closed forest	■ Subtropical closed forest	■ Temperate closed forest	■ Boreal closed forest
▨ Tropical open and fragmented forest	▨ Subtropical open and fragmented forest	▨ Temperate open and fragmented forest	▨ Boreal open and fragmented forest
□ Tropical other wooded land	□ Subtropical other wooded land	□ Temperate other wooded land	□ Boreal other wooded land

FIGURE 12.2 Major forest types. Note that some of these forests are dense; others may have only 10–20 percent actual tree cover.
Source: UN Food and Agriculture Organization, 2002.

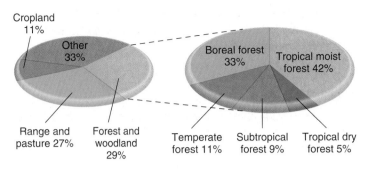

FIGURE 12.3 World land use and forest types. The "other" category includes tundra, desert, wetlands, and urban areas.
Source: UN Food and Agriculture Organization (FAO).

from rapid deforestation. North America and Eurasia have vast areas of relatively unaltered boreal forest. Although many of these forests are harvested regularly, both continents have a net increase in forest area and biomass because of replanting and natural regeneration.

Among the forests of greatest ecological importance are the primeval forests that are home to much of the world's biodiversity, ecological services, and indigenous human cultures. Sometimes called frontier, old-growth, or virgin forests, these are areas large enough and free enough from human modification that native species can live out a natural life cycle, and ecological relationships play out in a relatively normal fashion. The FAO defines **primary forests** as those "composed primarily of native species in which there are no clearly visible indications of human activity and ecological processes are not significantly disturbed."

This doesn't mean that all trees in a primary forest need be enormous or thousands of years old (fig. 12.4). In some biomes, most trees live only a century or so before being killed by disease or some natural disturbance. The successional processes (chapter 4) as trees die and are replaced create structural complexity and a diversity of sizes and ages important for specialists, such as the northern spotted owl (chapter 11). Nor does it mean that humans have never been present. Where human occupation entails relatively little impact, a forest may be inhabited for millennia while still retaining its primary characteristics. Even forests that have been logged or converted to cropland often can revert to natural conditions if left alone long enough.

Globally, about one-third of all forests are categorized as primary forests. Unfortunately, an estimated 6 million ha (15 million acres) of these irreplaceable forests are cleared or heavily damaged every year. According to the FAO, nine of the ten countries that are home to more than 80 percent of the world's primary forest are suffering from unsustainable logging rates.

Forests provide many valuable products

Wood plays a part in more activities of the modern economy than does any other commodity. There is hardly any industry that does not use wood or wood products somewhere in its manufacturing

and marketing processes. Think about the amount of junk mail, newspapers, photocopies, and other paper products that each of us in developed countries handles, stores, and disposes of in a single day. Total annual world wood consumption is about 4 billion m^3. This is more than steel and plastic consumption combined. International trade in wood and wood products amounts to more than (U.S.) $100 billion each year. Developed countries produce less than half of all industrial wood but account for about 80 percent of its consumption. Less-developed countries, mainly in the tropics, produce more than half of all industrial wood but use only 20 percent.

Paper pulp, the fastest-growing type of forest product, accounts for nearly a fifth of all wood consumption. Most of the world's paper is used in the wealthier countries of North America, Europe, and Asia. Global demand for paper is increasing rapidly, however, as other countries develop. The United States, Russia, and Canada are the largest producers of both paper pulp and industrial wood (lumber and panels). Much industrial logging in Europe and North America occurs on managed plantations, rather than in untouched old-growth forest. However, paper production is increasingly blamed for deforestation in Southeast Asia, West Africa, and other regions.

Fuelwood accounts for nearly half of global wood use. Roughly one-third the world's population depends on firewood or charcoal as their principal source of heating and cooking fuel (fig. 12.5). The average amount of fuelwood used in less-developed countries is about 1 m^3 per person per year, roughly equal to the amount that each American consumes each year as paper products alone. Demand for fuelwood, which is increasing at slightly less than the global population growth rate, is causing severe fuelwood shortages and depleting forests in some developing areas, especially around growing cities. Many people have less fuelwood than they need, and experts expect shortages to worsen as poor urban areas grow. Because fuelwood is rarely taken from closed-canopy forest, however, it does not appear to be a major cause of deforestation. Some analysts argue that fuelwood could be produced sustainably in most developing countries, with careful management.

FIGURE 12.4 A tropical rainforest in Queensland, Australia. Primary, or old-growth, forests, such as this, aren't necessarily composed entirely of huge, old trees. Instead, they have trees of many sizes and species that contribute to complex ecological cycles and relationships.

FIGURE 12.5 Firewood accounts for almost half of all wood harvested worldwide and is the main energy source for one-third of all humans.

> **Think About It**
>
> How could modern technology be used to reduce people's dependence on firewood? How might we distribute that technology to the people who need it?

Approximately one-quarter of the world's forests are managed for wood production. Ideally, forest management involves scientific planning for sustainable harvests, with particular attention paid to forest regeneration. In temperate regions, according to the UN Food and Agriculture Organization, more land is being replanted or allowed to regenerate naturally than is being permanently deforested. Much of this reforestation, however, is in large plantations of single-species, single-use, intensive cropping called **monoculture forestry.** Although this produces rapid growth and easier harvesting than a more diverse forest, a dense, single-species stand often supports little biodiversity and does poorly in providing the ecological services, such as soil erosion control and clean water production, that may be the greatest value of native forests (fig. 12.6).

Some of the countries with the most successful reforestation programs are in Asia. China, for instance, cut down most of its

FIGURE 12.6 Monoculture forestry, such as this Wisconsin tree farm, produces valuable timber and pulpwood, but has little biodiversity.

forests 1,000 years ago and has suffered centuries of erosion and terrible floods as a consequence. Recently, however, timber cutting in the headwaters of major rivers has been outlawed, and a massive reforestation project has begun. Since 1990, China has planted 50 billion trees, mainly in Xinjiang Province, to stop the spread of deserts. Korea and Japan also have had very successful forest restoration programs. After being almost totally denuded during World War II, both countries are now about 70 percent forested.

Tropical forests are especially threatened

Tropical forests are among the richest and most diverse terrestrial systems. This is especially true of the moist forests (rainforests) of the Amazon and Congo River basins and Southeast Asia. Although they now occupy less than 10 percent of the earth's land surface, these ecosystems are thought to contain more than two-thirds of all higher plant biomass and at least half of all the plant, animal, and microbial species in the world.

A century ago, an estimated 12.5 million km^2 (nearly 5 million mi^2) of tropical lands were covered by primary forest. This was an area larger than the entire United States. At least half that forest has already been cleared or degraded. Every day logging and burning destroy about 30,000 hectares of tropical forest, while farming, grazing, or conversion to monoculture plantations degrades a roughly equal area. This amounts to an area about the size of the United Kingdom each year. It also represents a serious reduction of nature's capacity to store carbon we release by other means. In addition to their value in carbon storage, forests provide important ecological services, such as generating oxygen, storing water, and protecting biodiversity.

If current rates of destruction continue, no primary forest will be left in many countries outside of parks and nature reserves by the end of this century. In a 2007 survey the FAO reported that 83 countries lost forest area, while 53 reported a net

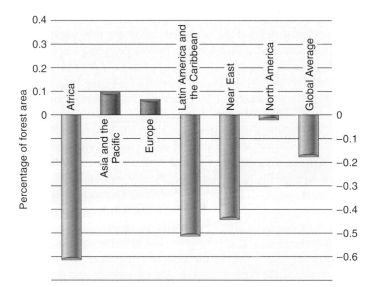

FIGURE 12.7 Annual net change in forest area, 2000–2005. The largest annual net deforestation rate in the world is in Africa. Largely because China has planted 50 billion trees in the past decade, Asia has a net increase in forest area. Europe, also, is gaining forest.
Source: Data from FAO, 2008.

gain (fig. 12.7). The world's highest current rate of forest loss is in Burundi, which is losing 9 percent of its forest annually.

Not only is tropical deforestation a tragic loss of biological diversity, it also represents a declining livelihood for the millions of people who depend on forests for part or all of their sustenance. Furthermore, approximately 1.7 billion metric tons of carbon are released annually due to deforestation and land use changes, mostly in the tropics. This amounts to about 20 percent of all anthropogenic carbon emissions, or more than the emissions from all forms of transportation combined. Halting or substantially decreasing forest destruction and soil degradation would help significantly to avoid global climate change (see chapter 15). But climate change also threatens forests. During severe droughts in 2005 and 2010, the Amazon rainforest lost billions of trees. If severe droughts continue, the forest, which now absorbs about one-quarter of all anthropogenic CO_2, could become a carbon source rather than a sink.

However, as the opening case study for this chapter shows, there have been some encouraging success stories in forest protection. The UN-sponsored REDD program offers hope that we may reduce deforestation. In Brazil, forest losses have decreased more than 60 percent since 2004, when 11,681 sq miles (18,800 sq km) were destroyed. Part of this is probably due to the global recession, which reduced demand for forest products as well as beef and soy grown on deforested land. However, stepped-up law enforcement efforts, which have netted hundreds of illegal loggers and corrupt officials while generating some $1.7 billion in fines, also helped. The government has also dramatically expanded the extent and number of protected areas, setting aside more than 100 million hectares of the Amazon basin from development since 2002.

Causes for Deforestation

There are many causes for deforestation. In Africa, forest clearing by subsistence farmers is responsible for about two-thirds of the forest destruction, but large-scale commercial logging also takes a toll. In Latin America the largest single cause of deforestation is expansion of soy farming and cattle ranching. Loggers start the process by cutting roads into the forest (fig. 12.8) to harvest valuable hardwoods, such as mahogany or cocobolo. This allows subsistence farmers to move into the forest, but they are bought out—or driven out—after a few years by wealthy ranchers.

Fires destroy about 350 million ha (1,350 mi^2) of forest every year (p. 246). Some fires are set by humans to cover up illegal

1975   1989   2001

FIGURE 12.8 Forest destruction in Rondonia, Brazil, between 1975 and 2001. Construction of logging roads creates a feather-like pattern that opens forests to settlement by farmers.

logging or land clearing. Others are started by natural causes. The greatest fire hazard in the world, according to the FAO, is in sub-Saharan Africa, which accounts for about half the global total. Uncontrolled fires tend to be worst in countries with corrupt or ineffective governments and high levels of poverty, civil unrest, and internal refugees. As global climate change brings drought and insect infestations to many parts of the world, there's a worry that forest fires may increase catastrophically.

One of the largest causes of deforestation in Southeast Asia is the demand for palm oil. This oil is widely used for cooking and industrial processes, but the main factor driving plantation expansion is the demand for biodiesel, especially in Europe. Millions of hectares of primary forest have been cleared in the past decade to create palm plantations. Critics of REDD say the original definition didn't distinguish between natural forests and plantations. It also credited replanting trees (even monocultures) on already degraded areas as eligible for carbon credits. Thus a logging company could cut down a native forest and make a fortune selling the valuable hardwoods. It then could claim carbon offset payments by replanting the area with oil palm trees, and in a few years it could begin selling the palm oil at a premium price as biodiesel. The resulting monoculture, though better than bare ground, has little of the biological diversity or ecological richness of the original forest. The version of REDD passed at the Cancun climate convention (called REDD+) expanded the definition of forest destruction to include forest soil degradation from conversion of previously cleared land to agriculture, pulp, or palm oil plantations, or other uses that decrease soil carbon.

In Indonesia, peatlands are a particular concern. Millions of hectares of waterlogged swamp forests of Borneo, Sumatra, and New Guinea have acidic groundwater that preserves dead plant material. Peat (partially decayed wood and plant litter) can be 50 ft (15 m) deep in these forests. Altogether these peatlands contain an estimated 132 gigatons of CO_2. By comparison, the entire Amazon rainforest has about 168 gigatons of CO_2. When peatlands are logged and burned, the soil dries out and oxidizes, releasing its stored carbon. In one case, the logging company Asia Pacific Resources International hopes to receive REDD payments for preserving a one-million acre (400,000 ha) peat bog on the Kampar Peninsula on Sumatra in exchange for concessions to plant a ring of acacia plantations around the core swamp. They say their plantations will protect the peatlands from illegal logging and wildlife poaching. Critics—including local indigenous groups—say this is just a land grab to exclude them from traditional hunting and gathering areas. How this plays out depends to a large degree on whether REDD allows or prohibits conversion of natural forests into industrial tree plantations.

Forest Protection

What can be done to encourage forest protection? While much of news is discouraging, the REDD program offers hope for tropical forest conservation. Many countries now recognize that forests are valuable resources. Investigations are under way to identify the best remaining natural areas (Exploring Science, p. 253).

Nearly 12 percent of all world forests are now in some form of protected status, but the effectiveness of that protection varies greatly. Nominally, Africa has the largest percentage of area in conservation reserves (fig. 12.9). Many of those parks and reserves have little practical protection, however. Park rangers are often outmanned and outgunned by poachers, drug runners, invading militias, and others who threaten the forest and its inhabitants.

Costa Rica has one of the best plans in the world for forest guardianship. Costa Rica is attempting not only to rehabilitate the land (make an area useful to humans) but also to restore the ecosystems to naturally occurring associations. One of the best known of these projects is Dan Janzen's work in Guanacaste National Park. Like many dry tropical forests, the northwestern part of Costa Rica had been almost completely converted to ranchland. By controlling fires, however, Janzen and his coworkers are bringing back the forest. One of the keys to this success is involving local people in the project. Janzen also encourages grazing in the park. The original forest evolved, he reasons, together with ancient grazing animals that are now extinct. Ranching can be a force for forest destruction, but cattle can play a valuable role as seed dispersers (fig. 12.10).

Brazil, also, is a leader in establishing forest reserves. It now recognizes the right of traditional people—Indians, descendants of runaway slaves, traditional fishermen, peasants, and communities engaged in nondestructive extractive activities (such as rubber tapping or nut collecting)—to live in the forest. At the same time, however, Brazil is pressing ahead with building and paving a network of roads to connect the western Amazon with all-weather, high-speed roads to the Pacific. Critics warn that these projects

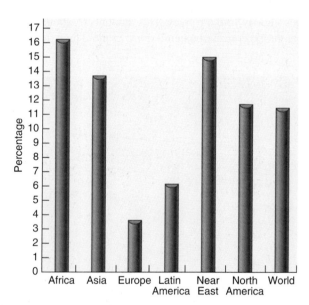

FIGURE 12.9 Percentage of forest area designated for conservation. About 12 percent of all world forests are now in some form of protected status, but the effectiveness of that protection varies greatly.
Source: Data from FAO, 2007.

Exploring Science

Protecting areas in remote places, such as Central Africa, is hard because information is so elusive. Deep, remote, swampy, tropical jungles are difficult to enter, map, and assess for their ecological value. Yet without information about their ecological importance, most people have little reason to care about these remote, trackless forests. How can you conserve ecosystems if you don't know what's there?

For most of history, understanding the extent and conditions of a remote area required an arduous trek to see the place in person. Even on publicly owned lands, only those who could afford the time, or who could afford to pay surveyors, might understand the resources. Over time, maps improved, but most maps show only a few features, such as roads, rivers, and some boundaries.

In recent years, details about public lands and resources have suddenly burst into public view through the use of geographic information systems **(GIS).** A GIS consists of spatial data, such as boundaries or road networks, and software to display and analyze the data. Spatial data can include variables that are hard to see on the ground—watershed boundaries, annual rainfall, landownership, or historical land use. Data can also represent phenomena much larger than we can readily see—land surface slopes and elevation, forested regions, river networks, and so on. By overlaying these layers, GIS analysts can investigate completely new questions about conservation, planning, and restoration.

You have probably used a GIS. Online mapping programs such as MapQuest or Google Earth™ organize and display spatial data. They let you turn layers on and off, or zoom in and out to display different scales. You can also use an online mapping program to calculate distances and driving directions between places. An ecologist, meanwhile, might use a GIS to calculate the extent of habitat areas, to monitor changes in area, to calculate the size of habitat fragments, or to calculate the length of waterways in a watershed.

Identifying Priority Areas

Recently a joint effort of several conservation organizations used GIS and spatial data to identify priority areas for conservation in Central Africa. The project was initiated because new data, including emerging GIS data, were showing dramatic increases in planned logging, in a region that contains the

world's second greatest extent of tropical forest (fig. 1).

Researchers from the Wildlife Conservation Society, Worldwide Fund for Nature, World Resources Institute, USGS, and other agencies and groups, began collecting GIS data on a variety of variables. They identified the range of great apes and other rare or threatened species. They identified areas of extreme plant diversity. They calculated

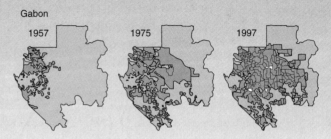

Gabon

1957   1975   1997

FIGURE 1 Gabon, Central Africa, has seen a steady increase in logging concessions.
Source: Wildlife Conservation Society.

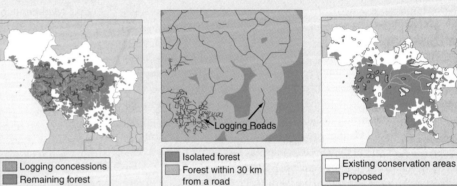

| Logging concessions
| Remaining forest

| Isolated forest
| Forest within 30 km from a road

| Existing conservation areas
| Proposed

Logging Roads

FIGURE 2 A few GIS layers used to identify priority conservation areas.
Source: Wildlife Conservation Society.

the sizes of forest fragments to identify concentrations of intact, ancient forests. Using maps of logging roads, they calculated the area within a 30 km "buffer" around roads, because loggers, settlers, and hunters usually threaten biodiversity near roads. They also mapped existing and planned conservation areas (fig. 2).

By overlaying these and other layers, analysts identified priority conservation areas of extensive original forest, which have high biodiversity and rare species. Overlaying these priority areas with a map of protected lands and a map of timber concessions, they identified *threatened* priority areas (fig. 3).

Most of the unprotected priority areas may never be protected, but having this map provides two important guides for future conservation. First, it assesses the state of the problem. With this map, we know that most of the forest is unprotected but also that the region's primary forest is extensive. Second, this map provides priorities for conservation planning. In addition, maps are very effective tools for publicizing an issue. When a map like this is

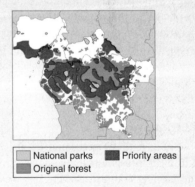

| National parks | | Priority areas
| Original forest

FIGURE 3 Priority areas outside of national parks.
Source: Wildlife Conservation Society.

published, more people become enthusiastic about joining the conservation effort.

GIS has become an essential tool for conserving forests, grasslands, ecosystems, and nature preserves. GIS has revolutionized the science of planning and conservation—examining problems using quantitative data—just as it may have revolutionized the way you plan a driving trip.

FIGURE 12.10 Cattle ranching can increase pressure for forest destruction, but, in the proper setting, cattle also can assist forest regeneration by dispersing seeds.

will accelerate land invasions and will result in displacement of native people and wildlife throughout the forest.

People also are working on the grassroots level to protect and restore forests in other countries. India, for instance, has a long history of nonviolent, passive resistance movements—called *satyagrahas*—to protest unfair government policies. These protests go back to the beginning of Indian culture and often have been associated with forest preservation. Gandhi drew on this tradition in his protests of British colonial rule in the 1930s and 1940s. During the 1970s, commercial loggers began large-scale tree felling in the Garhwal region in the state of Uttar Pradesh in northern India. Landslides and floods resulted from stripping the forest cover from the hills. The firewood on which local people depended was destroyed, and the way of life of the traditional forest culture was threatened. In a remarkable display of courage and determination, the village women wrapped their arms around the trees to protect them, sparking the Chipko Andolan movement (literally, movement to hug trees). They prevented logging on 12,000 km^2 of sensitive watersheds in the Alakanada basin. Today the Chipko Andolan movement has grown to more than 4,000 groups working to save India's forests.

Debt-for-Nature Swaps

Those of us in developed countries also can contribute toward saving tropical forests. Financing nature protection is often a problem in developing countries, where the need is greatest. One promising approach is called **debt-for-nature swaps.** Banks, governments, and lending institutions now hold nearly $1 trillion in loans to developing countries. There is little prospect of ever collecting much of this debt, and banks are often willing to sell bonds at a steep discount—perhaps as little as 10 cents on the dollar. Conservation organizations buy debt obligations on the secondary market at a discount and then offer to cancel the debt if the debtor country agrees to protect or restore an area of biological importance.

There have been many such swaps. Conservation International, for instance, bought $650,000 of Bolivia's debt for $100,000—an 85 percent discount. In exchange for canceling this debt, Bolivia agreed to protect nearly 1 million ha (2.47 million acres) around the Beni Biosphere Reserve in the Andean foothills. Ecuador and Costa Rica have had a different kind of debt-for-nature swap. They have exchanged debt for local currency bonds that fund activities of local private conservation organizations in the country. This has the dual advantage of building and supporting indigenous environmental groups while protecting the land. Critics, however, charge that these swaps compromise national sovereignty and do little to reduce the developing world's debt or to change the situations that led to environmental destruction in the first place.

Temperate forests also are threatened

Tropical countries aren't unique in damaging and degrading their forests. Asia and the Pacific currently have had a net forest increase thanks to an ambitious reforestation effort in China (fig. 12.7). Europe also has increased its forest area with replanting projects and forest regrowth on abandoned fields and previously harvested areas. Although the total forest area in North America has remained nearly constant in recent years, forest management policies in the United States and Canada continue to be controversial.

FIGURE 12.11 The huge old trees of the old-growth temperate rainforest accumulate more total biomass in standing vegetation per unit area than any other ecosystem on earth. They provide habitat to many rare and endangered species, but they are also coveted by loggers who can sell a single tree for thousands of dollars.

As the opening case studies for this chapter and chapter 11 show, large areas of the temperate rainforest of the Pacific Northwest have been set aside to protect endangered species. These forests have more standing biomass per square kilometer than any other ecosystem on earth (fig. 12.11). Because they're so wet, these forests rarely burn, and trees often live to be a thousand years old and many meters in diameter. A unique biological community has evolved in these dense, misty forests. Dozens of species of plants and animals spend their whole lives in the forest canopy, almost never descending to ground level.

In 1994 the U.S. government adopted the Northwest Forest Plan to regulate harvesting on about 9.9 million ha (24.5 million acres) of federal lands in Oregon, Washington, and northern California. This plan was an admirable example of using good science for natural resource planning. Teams of researchers identified specific areas of ancient forest essential for sustaining viable populations of endangered species, such as the northern spotted owl and the marbled murrelet (chapter 11). The plan prohibited most clear-cut logging, especially on steep hillsides and in riparian (streamside) areas where erosion threatens water quality and salmon spawning (fig. 12.12).

Still, logging has been allowed on the "matrix" lands surrounding these islands of ancient, old-growth forests. Conservationists lament the fact that fragmentation reduces the ecological value of the remaining forest, and they claim that many of the areas now lacking old-growth status could achieve the levels of structural complexity and age required for this classification if left uncut for a few more decades.

One of the most controversial aspects of forest management in the United States in recent years has been road building in de facto wilderness areas. Roads fragment forests, provide a route of entry for hunters and invasive species, often result in erosion, and destroy wilderness qualities. Shortly after passage of the Wilderness Act by the U.S. Congress in 1964, the Forest Service began a review of existing roadless (de facto wilderness) lands. Called the Roadless Area Review and Evaluation (RARE), this effort culminated in 1972 with the identification of 56 million acres (230,000 km^2) suitable for wilderness protection. Some of these lands were subsequently included in individual state wilderness bills, but most remained vulnerable to logging, mining, and other extractive activities.

In 2001, during the last days of the Clinton administration, a national guideline called the **Roadless Rule** was established. This rule ended virtually all logging, road building, and development on virtually all the lands identified as deserving of protection in the 1972 RARE assessment. Despite repeated attempts by George W. Bush to either overturn the Roadless Rule or simply not defend it in lawsuits from timber companies, this rule continues to protect de facto wilderness in 38 states. So far the Obama administration has maintained the rule on a year-by-year basis.

A much greater threat to temperate forests may be posed by climate change, insect infestations, and wildfires, all of which are interconnected. Over the past few decades the average temperature over much of North America has risen by more than 1°F (0.5°C). This may not sound like much, but it has caused the worst drought in 500 years. Hot, dry weather weakens trees and makes them

FIGURE 12.12 Clear-cuts, such as this one, threaten species dependent on old-growth forest and exposes steep slopes to soil erosion.

more vulnerable to both insect attacks and fires. In 2009 a research team led by ecologist Jerry Franklin released results showing that tree mortality among a wide variety of species has increased dramatically across a wide area over the past few decades. Infestations by beetles in particular have killed millions of hectares of conifers throughout western North America. This includes pinyon pine forests in the southwest, lodgepole pines throughout the Rocky Mountains, and huge swaths of spruce forests in Canada and Alaska. The billions of dead and dying trees are a huge fire danger, especially where people have built homes in remote areas.

For 70 years the U.S. Forest Service has had a policy of aggressive fire control. The aim has been to extinguish every fire on public land before 10 A.M. Smokey Bear was adopted as the forest mascot and warned us that "only you can prevent forest fires." Recent studies, however, of fire's ecological role suggest that our attempts to suppress all fires may have been misguided. Many biological communities are fire-adapted and require periodic burning for regeneration. Eliminating fire from these ecosystems has allowed woody debris to accumulate, greatly increasing chances for a very big fire (fig. 12.13).

Forests that once were characterized by 50 to 100 mature, fire-resistant trees per hectare and an open understory now have a thick tangle of up to 2,000 small, spindly, mostly dead saplings in the same area. The U.S. Forest Service estimates that 33 million ha (73 million acres), or about 40 percent of all federal forestlands, are at risk of severe fires. To make matters worse, Americans increasingly live in remote areas where wildfires are highly likely. Because there haven't been fires in many of these places in living memory, many people assume there is no danger, but by some estimates 40 million U.S. residents now live in areas with high wildfire risk.

Much of the federal and state firefighting efforts are controlled, in effect, by these homeowners who build in fire-prone areas. A government audit found that 90 percent of the Forest Service firefighting outlays go to save private property. If people who build in forested areas would take some reasonable precautions to protect themselves, we could let fires play their normal ecological role in forests in many cases. For example, you

FIGURE 12.13 By suppressing fires and allowing fuel to accumulate, we make major fires such as this more likely. The safest and most ecologically sound management policy for some forests may be to allow natural or prescribed fires, that don't threaten property or human life, to burn periodically.

shouldn't build a log cabin with a wood shake roof surrounded by dense forest at the end of a long, narrow, winding drive that a fire truck can't safely navigate. If you're going to have a home in the forest, you should use fireproof materials, such as a metal roof and rock or brick walls, and clear all trees and brush from at least 60 m (200 ft) around any buildings.

A recent prolonged drought in the western United States has heightened fire danger there, and in 2006 more than 96,000 wildfires burned 4 million ha (10 million acres) of forests and grasslands in the United States. Federal agencies spent about $1.6 billion to fight these fires, nearly four times the previous ten-year average.

The dilemma is how to undo years of fire suppression and fuel buildup. Fire ecologists favor small, prescribed burns to clean out debris. Loggers decry this approach as a waste of valuable timber, and local residents of fire-prone areas fear that prescribed fires will escape and threaten them. What do you think? What's the best way to restore forest health while also protecting property values and local jobs?

Ecosystem Management

In the 1990s the U.S. Forest Service began to shift its policies from a timber production focus to **ecosystem management,** which attempts to integrate sustainable ecological, economic, and social goals in a unified, systems approach. Some of the principles of this new philosophy include

- Managing across whole landscapes, watersheds, or regions over ecological time scales
- Considering human needs and promoting sustainable economic development and communities
- Maintaining biological diversity and essential ecosystem processes
- Utilizing cooperative institutional arrangements
- Generating meaningful stakeholder and public involvement and facilitating collective decision making

- Adapting management over time, based on conscious experimentation and routine monitoring

Some critics argue that we don't understand ecosystems well enough to make practical decisions in forest management on this basis. They argue we should simply set aside large blocks of untrammeled nature to allow for chaotic, catastrophic, and unpredictable events. Others see this new approach as a threat to industry and customary ways of doing things. Still, elements of ecosystem management appear in the *National Report on Sustainable Forests* prepared by the U.S. Forest Service. Based on the Montreal Working Group criteria and indicators for forest health, this report suggests goals for sustainable forest management (table 12.1).

Science

The wild, rugged coast of British Columbia is home to one of the world's most productive natural communities: the temperate rainforest. Nurtured by abundant rainfall and mild year-round temperatures, forests in the deep, misty fjords shelter giant cedar, spruce, and fir trees. Because this cool, moist forest rarely burns, trees often live for 1,000 years or more, and can be 5 m (16 ft) in diameter and 70 m tall. In addition to huge, moss-draped trees, the forest is home to an abundance of wildlife. One animal in particular has come to symbolize this beautiful landscape: it's a rare, white- or cream-colored black bear. Called a Kermode bear by scientists, these animals are more popularly known as "spirit bears," the name given to them by native Gitga'at people.

The wetlands and adjacent coastal areas also are biologically rich. Whales and dolphins feed in the sheltered fjords and inter-island channels. Sea otters float on the rich offshore kelp forests. It's estimated that 20 percent of the world's remaining wild salmon migrate up the wild rivers of this coastline.

In 2006, officials from the provincial government, native First Nations, logging companies, and environmental groups announced an historic agreement for managing the world's largest remaining intact temperate coastal rainforest. This Great Bear Rainforest encompasses about 6 million ha (15.5 million acres) or about the size of Switzerland. One-third of the area will be entirely protected from logging. In the rest of the land, only selective, sustainable logging will be allowed rather than the more destructive clear-cutting that has devastated surrounding forests. At least $120 million will be provided for conservation projects and ecologically sustainable business ventures, such as eco-tourism lodges and an oyster farm.

A series of factors contributed to preserving this unique area. The largest environmental

British Columbia's Great Bear Rainforest will preserve the home for rare, white-phase black (or spirit) bears along with salmon streams, misty fjords, rich tidal estuaries, and the largest remaining area of old-growth, coastal, temperate rainforest in the world.

protest in Canadian history took place at Clayoquot Sound on nearby Vancouver Island in the 1980s, when logging companies attempted to clear-cut land claimed by First Nations people. This alerted the public to the values of and threats to the coastal temperate rainforest. As a result of the lawsuits and publicity generated by this controversy, most of the largest logging companies have agreed to stop clear-cutting in the remaining virgin forest. The rarity of the spirit bears also caught the public imagination. Tens of thousands of schoolchildren across Canada wrote to the provincial government begging them to set aside a sanctuary for this unique animal. And a growing recognition of the rights of native people also helped convince public officials that traditional lands and ways of living need to be preserved.

More than 60 percent of the world's temperate rainforest has already been logged or

developed. The Great Bear Rainforest contains one-quarter of what's left. It also contains about half the estuaries, coastal wetlands, and healthy, salmon-bearing streams in British Columbia.

How did planners choose the areas to be within the protected area? One of the first steps was a biological survey. Where were the biggest and oldest trees? Which areas are especially valuable for wildlife? Protecting water quality in streams and coastal regions was also a high priority. Keeping logging and roads out of riparian habitats is particularly important. Interestingly, native knowledge of the area was also consulted in drawing boundaries. Which places are mentioned in oral histories? What are the traditional uses of the forest? Although commercial logging is prohibited in the protected areas, First Nations people will be allowed to continue their customary harvest of selected logs for totem poles, longhouses, and canoes. They also will be allowed to harvest berries, catch fish, and hunt wildlife for their own consumption.

Because it's so remote, few people will ever visit the Great Bear Rainforest, yet many of us like knowing that some special places like this continue to exist. Although we depend on wildlands for many products and services, perhaps we don't need to exploit every place on the planet. Which areas we choose to set aside, and how we protect and manage those special places, says a lot about who we are.

12.2 GRASSLANDS

After forests, grasslands are among the biomes most heavily used by humans. Prairies, savannas, steppes, open woodlands, and other grasslands occupy about one-quarter of the world's land surface. Much of the U.S. Great Plains and the Prairie Provinces of Canada fall in this category (fig. 12.14). The 3.8 billion ha (12 million mi^2) of pastures and grazing lands in this biome make up about twice the area of all agricultural crops. When you add to this about 4 billion ha of other lands (forest, desert, tundra, marsh, and thorn scrub) used for raising livestock, more than half of all land is used at least occasionally for grazing. More than 3 billion cattle, sheep, goats, camels, buffalo, and other domestic animals on these lands make a valuable contribution to human nutrition. Sustainable

Table 12.1 Draft Criteria for Sustainable Forestry

1. Conservation of biological diversity
2. Maintenance of productive capacity of forest ecosystems
3. Maintenance of forest ecosystem health and vitality
4. Maintenance of soil and water resources
5. Maintenance of forest contribution to global carbon cycles
6. Maintenance and enhancement of long-term socioeconomic benefits to meet the needs of legal, institutional, and economic framework for forest conservation and sustainable management

Source: Data from USFS, 2002.

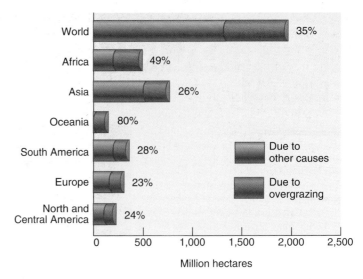

FIGURE 12.15 Rangeland soil degradation due to overgrazing and other causes. Notice that in Europe, Asia, and the Americas, farming, logging, mining, urbanization, etc., are responsible for about three-quarters of all soil degradation. In Africa and Oceania, where more grazing occurs and desert or semiarid scrub make up much of the range, grazing damage is higher. **Source:** World Resources Institute, 2004.

pastoralism can increase productivity while maintaining biodiversity in a grassland ecosystem.

Because grasslands, chaparral, and open woodlands are attractive for human occupation, they frequently are converted to cropland, urban areas, or other human-dominated landscapes. Worldwide the rate of grassland disturbance each year is three times that of tropical forest. Although they may appear to be uniform and monotonous to the untrained eye, native prairies can be highly productive and species-rich. According to the U.S. Department of Agriculture, more threatened plant species occur in rangelands than in any other major American biome.

Grazing can be sustainable or damaging

By carefully monitoring the numbers of animals and the condition of the range, ranchers and **pastoralists** (people who live by herding animals) can adjust to variations in rainfall, seasonal plant conditions, and the nutritional quality of forage to keep livestock healthy and avoid overusing any particular area. Conscientious management can actually improve the quality of the range.

When grazing lands are abused by overgrazing—especially in arid areas—rain runs off quickly before it can soak into the soil to nourish plants or replenish groundwater. Springs and wells dry up. Seeds can't germinate in the dry, overheated soil. The barren ground reflects more of the sun's heat, changing wind patterns, driving away moisture-laden clouds, and leading to further desiccation. This process of conversion of once-fertile land to desert is called **desertification.**

This process is ancient, but in recent years it has been accelerated by expanding populations and the political conditions that force people to overuse fragile lands. According to the International Soil Reference and Information Centre in the Netherlands, nearly three-quarters of all rangelands in the world show signs of either degraded vegetation or soil erosion. Overgrazing is responsible for about one-third of that degradation (fig. 12.15). The highest percentage of moderate, severe, and extreme land degradation is in Mexico and Central America, while the largest total area is in Asia, where the world's most extensive grasslands occur. Can we reverse this process? In some places, people are reclaiming deserts and repairing the effects of neglect and misuse.

Overgrazing threatens many U.S. rangelands

As is the case in many countries, the health of most public grazing lands in the United States is not good. Political and economic pressures encourage managers to increase grazing allotments beyond

FIGURE 12.14 Grasslands are expansive, open environments that can support surprising biodiversity.

the carrying capacity of the range. Lack of enforcement of existing regulations and limited funds for range improvement have resulted in **overgrazing,** damage to vegetation and soil including loss of native forage species and erosion. The Natural Resources Defense Council claims that only 30 percent of public rangelands are in fair condition, and 55 percent are poor or very poor (fig. 12.16).

Overgrazing has allowed populations of unpalatable or inedible species, such as sage, mesquite, cheatgrass, and cactus, to build up on both public and private rangelands. Wildlife conservation groups regard cattle grazing as the most ubiquitous form of ecosystem degradation and the greatest threat to endangered species in the southwestern United States. They call for a ban on cattle and sheep grazing on all public lands, noting that those lands provide only 2 percent of the total forage consumed by beef cattle and supports only 2 percent of all livestock producers.

Like federal timber management policy, grazing fees charged for use of public lands often are far below market value and are an enormous hidden subsidy to western ranchers. Holders of grazing permits generally pay the government less than 25 percent of what it would cost to lease comparable private land. The 31,000 permits on federal range bring in only $11 million in grazing fees but cost $47 million per year for administration and maintenance. The $36 million difference amounts to a massive "cow welfare" system of which few people are aware.

On the other hand, ranchers defend their way of life as an important part of western culture and history. Although few cattle go directly to market from their ranches, they produce almost all the beef calves subsequently shipped to feedlots. And without a viable ranch economy, they claim, even more of the western landscape would be subdivided into small ranchettes to the detriment of both wildlife and environmental quality. Many conservation groups are recognizing that preserving ranches may be the best

FIGURE 12.17 Intensive, rotational grazing encloses livestock in a small area for a short time (often only one day) within a movable electric fence to force them to eat vegetation evenly and fertilize the area heavily.

way to protect wildlife habitat. What do you think? How can we best protect traditional lifestyles and rural communities while also preserving natural resources?

Ranchers are experimenting with new methods

Where a small number of livestock are free to roam a large area, they generally eat the tender, best-tasting grasses and forbs first, leaving the tough, unpalatable species to flourish and gradually dominate the vegetation. In some places farmers and ranchers find that short-term, intensive grazing helps maintain forage quality. As South African range specialist Allan Savory observed, wild ungulates (hoofed animals), such as gnus or zebras in Africa or bison (buffalo) in America, often tend to form dense herds that graze briefly but intensively in a particular location before moving on to the next area. Rest alone doesn't necessarily improve pastures and rangelands. Short-duration, **rotational grazing**—confining animals to a small area for a short time (often only a day or two) before shifting them to a new location—simulates the effects of wild herds (fig. 12.17). Forcing livestock to eat everything equally, to trample the ground thoroughly, and to fertilize heavily with manure before moving on helps keep weeds in check and encourages the growth of more desirable forage species. This approach doesn't work everywhere, however. Many plant communities in the U.S. desert Southwest, for example, apparently evolved in the absence of large, hoofed animals and can't withstand intensive grazing.

Restoring fire and managing grasslands as regional units can have many benefits for both ranchers and wildlife. The Nature Conservancy has participated with private landowners in a number of innovative experiments in range restoration.

Another approach to ranching in some areas is to raise wild species, such as red deer, impala, wildebeest, or oryx (fig. 12.18). These animals forage more efficiently, resist harsh climates, often are more pest- and disease-resistant, and fend off predators better

FIGURE 12.16 More than half of all publicly owned grazing land in the United States is in poor or very poor condition. Overgrazing and invasive weeds are the biggest problems.

FIGURE 12.18 Red deer (*Cervus elaphus*) are raised in New Zealand for antlers and venison.

than usual domestic livestock. Native species also may have different feeding preferences and needs for water and shelter than cows, goats, or sheep. The African Sahel, for instance, can provide only enough grass to raise about 20 to 30 kg (44 to 66 lbs) of beef per hectare. Ranchers can produce three times as much meat with wild native species in the same area because these animals browse on a wider variety of plant materials.

In the United States, ranchers find that elk, American bison, and a variety of African species take less care and supplemental feeding than cattle or sheep and result in a better financial return because their lean meat can bring a better market price than beef or mutton. Media mogul Ted Turner has become both the biggest private landholder in the United States and the owner of more American bison than anyone other than the government.

12.3 PARKS AND PRESERVES

Although most forests and grasslands serve utilitarian purposes, many nations have set aside some natural areas for ecological, cultural, or recreational purposes. Some of these preserves have existed for thousands of years. Ancient Greeks and Druids, for example, protected sacred groves for religious purposes. Royal hunting grounds preserved wild areas in many countries. Although these areas were usually reserved for elite classes in society, they maintained biodiversity and natural landscapes in regions where most lands were heavily used.

The first public parks open to ordinary citizens may have been the tree-sheltered agoras in planned Greek cities. But the idea of providing natural space for recreation, or to preserve natural environments, has really developed in the past half century (fig. 12.19). Currently nearly 12 percent of the land area of the earth is protected in some sort of park, preserve, or wildlife management area. This represents about 19.6 million ha (7.6 million mi^2) in 107,000 different preserves.

Many countries have created nature preserves

Different levels of protection are found in nature preserves. The World Conservation Union divides protected areas into five categories, depending on the intended level of allowed human use (table 12.2). In the most stringent category (ecological reserves and wilderness areas), little or no human impact is allowed. In some strict nature preserves, where particularly sensitive wildlife or natural features are located, human entry may be limited to scientific research groups that visit on rare occasions. In some wildlife sanctuaries, for example, only a few people per year are allowed to visit, to avoid introducing invasive species or disrupting native species. In the least restrictive categories (national forests and other natural resource management areas), on the other hand, there may be a high level of human use.

Venezuela claims to have the highest proportion of its land area protected (70 percent) of any country in the world. About half this land is designated as preserves for indigenous people or for sustainable resource harvesting. With little formal management, there is minimal protection from poaching by hunters, loggers, and illegal gold hunters. Unfortunately it's not uncommon in the developing world to have "paper parks" that exist only as a line drawn on a map with no budget for staff, management, or infrastructure. The United States, by contrast, has only about 15.8 percent of its land area in protected status, and less than one-third of that amount is in IUCN categories I or II (nature reserves, wilderness areas, national parks). The rest is in national forests or wildlife management zones that are designated for sustainable use. With hundreds of thousands of state and federal employees, billions of dollars in public funding, and a high level of public interest and visibility, U.S. public lands are generally well managed.

Brazil, with more than one-quarter of all the world's tropical rainforest, is especially important in biodiversity protection. Currently Brazil has the largest total area in protected status of any country. More than 1.6 million km^2 or 18.7 percent of the nation's

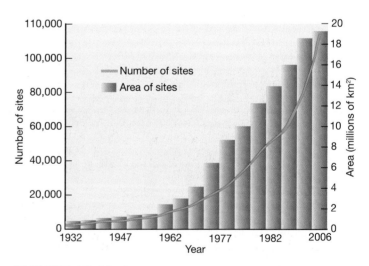

FIGURE 12.19 Growth of protected areas worldwide, 1932–2003.
Source: UN World Commission on Protected Areas.

Table 12.2	IUCN Categories of Protected Areas
Category	Allowed Human Impact or Intervention
1. Ecological reserves and wilderness areas	Little or none
2. National parks	Low
3. Natural monuments and archaeological sites	Low to medium
4. Habitat and wildlife management areas	Medium
5. Cultural or scenic landscapes, recreation areas	Medium to high
6. Managed resource area	High

Source: Data from World Conservation Union, 1990.

FIGURE 12.20 Canada's Quttinirpaaq National Park at the north end of Ellesmere Island offers plenty of solitude and pristine landscapes, but little biodiversity.

land—mostly in the Amazon basin—is in some protected status. In 2006 the northern Brazilian state of Para, in collaboration with Conservation International (CI) and other nongovernmental organizations, announced the establishment of nine new protected areas along the border with Suriname and Guyana. These new areas, about half of which will be strictly protected nature preserves, will link together several existing indigenous areas and nature preserves to create the largest tropical forest reserve in the world. More than 90 percent of the new 15 million ha (58,000 mi^2, or about the size of Illinois) Guyana Shield Corridor is in pristine natural state. CI president Russ Mittermeir says, "If any tropical rainforest on earth remains intact a century from now, it will be this portion of northern Amazonia." In contrast to this dramatic success, the Pantanal, the world's largest wetland/savanna complex, which lies in southern Brazil and is richer in some biodiversity categories than the Amazon, is almost entirely privately owned. There are efforts to set aside some of this important wetland, but so far little is in protected status.

Some other countries with very large reserved areas include Greenland (with a 972,000 km^2 national park that covers most of the northern part of the island) and Saudi Arabia (with a 640,000 km^2 wildlife management area in its Empty Quarter). These areas are relatively easy to set aside, however, being mostly ice covered (Greenland) or desert (Saudi Arabia). Canada's Quttinirpaaq National Park on Ellesmere Island is an example of a preserve with high wilderness values but little biodiversity. Only 800 km (500 miles) from the North Pole, this remote park gets fewer than 100 human visitors per year during its brief, three-week summer season (fig. 12.20). With little evidence of human occupation, it has abundant solitude and stark beauty, but very little wildlife and almost no vegetation. By contrast, the Great Bear Rainforest management area described in Exploring Science has a rich diversity of both marine and terrestrial life, but the valuable timber, mineral, and wildlife resources in the area make protecting it expensive and controversial.

Collectively, according to the World Commission on Protected Areas, Central America has 22.5 percent of its land area in some protected status (table 12.3). The Pacific region, at 1.9 percent in nature reserves, has both the lowest percentage and the lowest total area. With land scarce on small islands, it's hard to find space to set aside for nature sanctuaries. Some biomes are well represented in nature preserves, while others are relatively underprotected. Figure 12.21 shows the percentage of each major biome in protected status. Not surprisingly, there's an inverse relationship between the percentage converted to human use (and where people live) and the percentage protected. Temperate

Table 12.3	World Protected Areas		
Region	Total Area Protected (km^2)	Protected Percent	Number of Areas
North America	4,459,305	16.2%	13,447
South America	1,955,420	19.3%	1,456
North Eurasia	1,816,987	7.7%	17,724
East Asia	1,764,648	14.0%	3,265
Eastern and Southern Africa	1,696,304	14.1%	4,060
Brazil	1,638,867	18.7%	1,287
Australia/New Zealand	1,511,992	16.9%	9,549
North Africa and Middle East	1,320,411	9.8%	1,325
Western and Central Africa	1,131,153	8.7%	2,604
Southeast Asia	867,186	9.6%	2,689
Europe	785,012	12.4%	46,194
South Asia	320,635	6.5%	1,216
Central America	158,193	22.5%	781
Antarctic	70,323	0.5%	122
Pacific	67,502	1.9%	430
Caribbean	66,210	8.2%	958
Total	**19,630,149**	**11.6%**	**107,107**

Source: World Commission on Protected Areas, 2007.

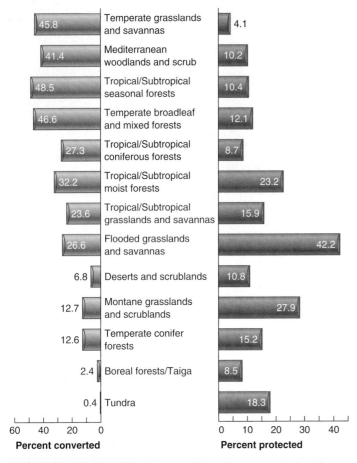

Percent converted	Biome	Percent protected
45.8	Temperate grasslands and savannas	4.1
41.4	Mediterranean woodlands and scrub	10.2
48.5	Tropical/Subtropical seasonal forests	10.4
46.6	Temperate broadleaf and mixed forests	12.1
27.3	Tropical/Subtropical coniferous forests	8.7
32.2	Tropical/Subtropical moist forests	23.2
23.6	Tropical/Subtropical grasslands and savannas	15.9
26.6	Flooded grasslands and savannas	42.2
6.8	Deserts and scrublands	10.8
12.7	Montane grasslands and scrublands	27.9
12.6	Temperate conifer forests	15.2
2.4	Boreal forests/Taiga	8.5
0.4	Tundra	18.3

FIGURE 12.21 With few exceptions, the percent of each biome converted to human use is roughly inverse to the percent protected in parks and preserves. Rock and ice, lakes, and Antarctic ecoregions are excluded.
Source: World Database on Protected Areas, 2009.

In Palau, coral reefs identified as a potential biosphere reserve are damaged by dynamite fishing, while on some beaches in Indonesia every egg laid by endangered sea turtles is taken by egg hunters. These are just a few of the many problems faced by parks and preserves around the world. Often countries with the most important biomes lack funds, trained personnel, and experience to manage the areas under their control.

Even in rich countries, such as the United States, some of the "crown jewels" of the National Park System suffer from overuse and degradation. Yellowstone and Grand Canyon national parks, for example, have large budgets and are highly regulated, but are being "loved to death" because they are so popular. When the U.S. National Park Service was established in 1916, Stephen Mather, the first director, reasoned that he needed to make the parks comfortable and entertaining for tourists as a way of building public support. He created an extensive network of roads in the largest parks so that visitors could view famous sights from the windows of their automobiles, and he encouraged construction of grand lodges in which guests could stay in luxury.

His plan was successful; the National Park System is cherished and supported by many American citizens. But sometimes entertainment seems to have trumped nature protection. Visitors were allowed—in some cases even encouraged—to feed wildlife. Bears lost their fear of humans and became dependent on an unhealthy diet of garbage and handouts (fig. 12.22). In Yellowstone and Grand Teton national parks, the elk herd was allowed

grasslands and savannas (such as the American Midwest) and Mediterranean woodlands and scrub (such as the French Riviera) are highly domesticated and therefore expensive to set aside in large areas. Temperate conifer forests (think of Siberia, or Canada's vast expanse of boreal forest) are relatively uninhabited, and therefore easy to put into some protected category.

Not all preserves are preserved

Even parks and preserves designated with a high level of protection aren't always safe from exploitation or changes in political priorities. Serious problems threaten natural resources and environmental quality in many countries. In Greece, the Pindus National Park is threatened by plans to build a hydroelectric dam in the center of the park. Furthermore, excessive stock grazing and forestry exploitation in the peripheral zone are causing erosion and loss of wildlife habitat. In Colombia, dam building also threatens the Paramillo National Park. Ecuador's largest park, Yasuni National Park, which contains one of the world's most megadiverse regions of lowland Amazonian forest, has been opened to oil drilling, while miners and loggers in Peru have invaded portions of Huascaran National Park.

FIGURE 12.22 Wild animals have always been one of the main attractions in national parks. Many people lose all common sense when interacting with big, dangerous animals. This is not a petting zoo.

Think About It

If you were superintendent of a major national park, how would you reconcile the demand for comfort and recreation with the need to protect nature? If no one comes to your park, you will probably lose public support. But if the landscape is trashed, what's the purpose of having a park?

FIGURE 12.24 Off-road vehicles cause severe, long-lasting environmental damage when driven through wetlands.

to grow to 25,000 animals, or about twice the carrying capacity of the habitat. The excess population overgrazed the vegetation to the detriment of many smaller species and the biological community in general. As we discussed earlier in this chapter, 70 years of fire suppression resulted in changes of forest composition and fuel buildup that made huge fires all but inevitable. In Yosemite, you can stay in a world-class hotel, buy a pizza, play video games, do laundry, play golf or tennis, and shop for curios, but you may find it difficult to experience the solitude or enjoy the natural beauty extolled by John Muir as a prime reason for creating the park.

In many of the most famous parks, traffic congestion and crowds of people stress park resources and detract from the experience of unspoiled nature (fig. 12.23). Some parks, such as Yosemite, and Zion National Park, have banned private automobiles from the most congested areas. Visitors must park in remote lots and ride to popular sites in clean, quiet buses that run on electricity or natural gas. Other parks are considering limits on the number of visitors admitted each day. How would you feel about a lottery system that might allow you to visit some famous parks only once in your lifetime, but to have an uncrowded, peaceful experience on your one allowed visit? Or would you prefer to be able to visit whenever you wish even if it means fighting crowds and congestion?

Originally the great wilderness parks of Canada and the United States were distant from development and isolated from most human impacts. This has changed in many cases. Forests are clear-cut right up to some park boundaries. Mine drainage contaminates streams and groundwater. At least 13 U.S. national monuments are open to oil and gas drilling, including Texas's Padre Island, the only breeding ground for endangered Kemps Ridley sea turtles. Even in the dry desert air of the Grand Canyon, where visibility was once up to 150 km, it's often too smoggy now to see across the canyon due to air pollution from power plants just outside the park. Snowmobiles and off-road vehicles (ORVs) create pollution and noise and cause erosion while disrupting wildlife in many parks (fig. 12.24).

Chronically underfunded, the U.S. National Park System now has a maintenance backlog estimated to be at least $5 billion. Politicians from both major political parties vow during election campaigns to repair park facilities, but then find other uses for public funds once in office. Ironically, a recent study found that, on average, parks generate $4 in user fees for every $1 they receive in federal subsidies. In other words, they more than pay their own way, and should have a healthy surplus if they were allowed to retain all the money they generate.

In recent years the U.S. National Park System has begun to emphasize nature protection and environmental education over entertainment. This new agenda is being adopted by other countries as well. The IUCN has developed a **world conservation strategy** for protecting natural resources that includes the following three objectives: (1) to maintain essential ecological processes and life-support systems (such as soil regeneration and protection, nutrient recycling, and water purification) on which human survival and development depend; (2) to preserve genetic diversity essential for breeding programs to improve cultivated plants and domestic animals; and (3) to ensure that any utilization of wild species and ecosystems is sustainable.

FIGURE 12.23 Thousands of people wait for an eruption of Old Faithful geyser in Yellowstone National Park. Can you find the ranger who's giving a geology lecture?

Marine ecosystems need greater protection

As ocean fish stocks become increasingly depleted globally (chapter 11), biologists are calling for protected areas where marine organisms are sheltered from destructive harvest methods. Research has shown that limiting the amount and kind of fishing in marine reserves can quickly replenish fish stocks in surrounding areas. In a study of 100 marine refuges around the world, researchers found that, on average, the number of organisms inside no-take preserves was twice as high as in surrounding areas where fishing was allowed. In addition, the biomass of organisms was three times as great and individual animals were, on average, 30 percent larger inside the refuge compared to outside. Recent research has shown that closing reserves to fishing even for a few months can have beneficial results in restoring marine populations. The size necessary for a safe haven to protect flora and fauna depends on the species involved, but some marine biologists call on nations to protect at least 20 percent of their nearshore territory as marine refuges.

Coral reefs are among the most threatened marine ecosystems in the world. Remote sensing surveys show that living coral covers only about 285,000 km^2 (110,000 mi^2), or an area about the size of Nevada. This is less than half of previous estimates, and 90 percent of all reefs face threats from rising sea temperatures, destructive fishing methods, coral mining, sediment runoff, and other human disturbance. In many ways coral reefs are the old-growth rainforests of the ocean. Biologically rich, these sensitive communities can take a century or more to recover from damage. Some researchers predict that if current trends continue, in 50 years there will be no viable coral reefs anywhere in the world.

What can be done to reverse this trend? Some countries are establishing large marine reserves specifically to protect coral reefs. In 2007 the United States declared three new marine national monuments in the Pacific Ocean, around three uninhabited islands in the Northern Marianas, Rose Atoll in American Samoa, and seven small islands strung along the equator in the central Pacific. Together these marine reserves total about 195, 000 mi^2 (more than 500,000 km^2), which will be protected from oil and gas extraction and commercial fishing. Australia protects nearly as much area (344,000 km^2) in its Great Barrier Reef (fig. 12.25). Altogether, however, aquatic reserves make up less than 10 percent of all the world's protected areas, despite the fact that 70 percent of the earth's surface is water. A survey of marine biological resources identified the ten richest and most threatened "hot spots," including the Philippines, the Gulf of Guinea and Cape Verde Islands (off the west coast of Africa), Indonesia's Sunda Islands, the Mascarene Islands in the Indian Ocean, South Africa's coast, southern Japan and the east China Sea, the western Caribbean, and the Red Sea and Gulf of Aden. We urgently need more no-take preserves to protect marine resources.

Conservation and economic development can work together

Many of the most biologically rich communities in the world are in developing countries, especially in the tropics. These countries are the guardians of biological resources important to

FIGURE 12.25 Australia's Great Barrier Reef is the world's largest marine reserve. Stretching for nearly 2,000 km (1,200 mi) along Australia's northeast coast, this reef complex is one of the biological wonders of the world.

all of us. Unfortunately, where political and economic systems fail to provide residents with land, jobs, food, and other necessities of life, people do whatever is necessary to meet their own needs. Immediate survival takes precedence over long-term environmental goals. Clearly the struggle to save species and ecosystems can't be divorced from the broader struggle to meet human needs.

As the opening case study for this chapter shows, residents of some developing countries are beginning to realize that their biological resources may be their most valuable assets, and that preserving those resources is vital for sustainable development. **Ecotourism** (tourism that is ecologically and socially sustainable) can be more beneficial in many places over the long term than extractive industries, such as logging and mining. What Can You Do? (p. 265) suggests some ways to ensure that your vacations are ecologically responsible.

Native people can play important roles in nature protection

The American ideal of wilderness parks untouched by humans is unrealistic in many parts of the world. As we mentioned earlier, some biological communities are so fragile that human intrusions have to be strictly limited to protect delicate natural features or particularly sensitive wildlife. In many important biomes, however, indigenous people have been present for thousands of years and have a legitimate right to pursue traditional ways of life. Furthermore, many of the approximately 5,000 indigenous or native peoples that remain today possess ecological knowledge about their ancestral homelands that can be valuable in ecosystem management. According to author Alan Durning, "encoded in indigenous languages, customs, and practices may be as much understanding of nature as is stored in the libraries of modern science."

Some countries have adopted draconian policies to remove native people from parks (fig. 12.26). In South Africa's Kruger

National Park, for example, heavily armed soldiers keep intruders out with orders to shoot to kill. This is very effective in protecting wildlife. In all fairness, before this policy was instituted there was a great deal of poaching by mercenaries armed with automatic weapons. But it also means that people who were forcibly displaced from the park could be killed on sight merely for returning to their former homes to collect firewood or to hunt for rabbits and other small game. Similarly, in 2006 thousands of peasant farmers on the edge of the vast Mau Forest in Kenya's Rift Valley were forced from their homes at gunpoint by police who claimed that the land needed to be cleared to protect the country's natural resources. Critics claimed that the forced removal amounted to "ethnic cleansing" and was based on tribal politics rather than nature protection.

Other countries recognize that finding ways to integrate local human needs with the needs of nature is essential for successful conservation. In 1986 UNESCO (United Nations Educational, Scientific, and Cultural Organization) initiated its **Man and Biosphere (MAB) program,** which encourages the designation of **biosphere reserves,** protected areas divided into zones with different purposes. Critical ecosystem functions and endangered

FIGURE 12.26 Some parks take draconian measures to expel residents and prohibit trespassing. How can we reconcile the rights of local or indigenous people with the need to protect nature?

wildlife are protected in a central core region, where limited scientific study is the only human access allowed. Ecotourism and research facilities are located in a relatively pristine buffer zone around the core, while sustainable resource harvesting and permanent habitation are allowed in multiple-use peripheral regions (fig. 12.27).

Although it has not yet been given a formal MAB designation, the Great Bear Rainforest described in Exploring Science (p. 257) is organized along this general plan. A well-established example of a biosphere reserve is Mexico's 545,000 ha (2,100 mi^2) Sian Ka'an Reserve on the Tulum Coast of the Yucatán. The core area includes 528,000 ha (1.3 million acres) of coral reefs, bays, wetlands, and lowland tropical forest. More than 335 bird species have been observed within the reserve, along with endangered manatees, five types of jungle cats, spider and howler monkeys, and four species of increasingly rare sea turtles. Approximately 25,000 people (about the same number who live in the Great Bear Rainforest) reside in communities and the countryside around Sian Ka'an. In addition to tourism, the economic base of the area includes lobster fishing, small-scale farming, and coconut cultivation.

The Amigos de Sian Ka'an, a local community organization, played a central role in establishing the reserve and is working to protect natural resources while also improving living standards for local people. New intensive farming techniques and sustainable harvesting of forest products enable residents to make a living without harming their ecological base. Better lobster-harvesting techniques developed at the reserve have improved the catch without depleting native stocks. Local people now see the reserve as a benefit rather than an imposition from the outside. Similar success stories from many parts of the world show how we can support local people and recognize indigenous rights while still protecting important environmental features.

What Can You Do?

Being a Responsible Ecotourist

1. *Pretrip preparation.* Learn about the history, geography, ecology, and culture of the area you will visit. Understand the do's and don'ts that will keep you from violating local customs and sensibilities.

2. *Environmental impact.* Stay on designated trails and camp in established sites if available. Take only photographs and memories, and leave only goodwill wherever you go.

3. *Resource impact.* Minimize your use of scarce fuels, food, and water resources. Do you know where your wastes and garbage go?

4. *Cultural impact.* Respect the privacy and dignity of those you meet, and try to understand how you would feel in their place. Don't take photos without asking first. Be considerate of religious and cultural sites and practices. Be as aware of cultural pollution as you are of environmental pollution.

5. *Wildlife impact.* Don't harass wildlife or disturb plant life. Modern cameras make it possible to get good photos from a respectful, safe distance. Don't buy ivory, tortoise shell, animal skins, feathers, or other products taken from endangered species.

6. *Environmental benefit.* Is your trip strictly for pleasure, or will it contribute to protecting the local environment? Can you combine ecotourism with work on cleanup campaigns or delivery of educational materials or equipment to local schools or nature clubs?

7. *Advocacy and education.* Get involved in letter writing, lobbying, or educational campaigns to help protect the lands and cultures you have visited. Give talks at schools or to local clubs after you get home, to inform your friends and neighbors about what you have learned.

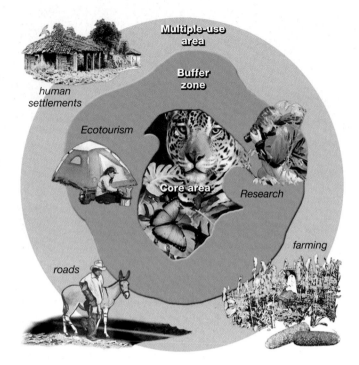

FIGURE 12.27 A model biosphere reserve. Traditional parks and wildlife refuges have well-defined boundaries to keep wildlife in and people out. Biosphere reserves, by contrast, recognize the need for people to have access to resources. Critical ecosystem is preserved in the core. Research and tourism are allowed in the buffer zone, while sustainable resource harvesting and permanent habitations are situated in the multiple-use area around the perimeter.

Species survival can depend on preserve size and shape

Many natural parks and preserves are increasingly isolated, remnant fragments of ecosystems that once extended over large areas. As park ecosystems are shrinking, however, they are also becoming more and more important for maintaining biological diversity. Principles of landscape design and landscape structure become important in managing and restoring these shrinking islands of habitat.

For years, conservation biologists have disputed whether it is better to have a *single large or several small* reserves (the SLOSS debate). Ideally a reserve should be large enough to support viable populations of endangered species, keep ecosystems intact, and isolate critical core areas from damaging external forces. For some species with small territories, several small, isolated refuges can support viable populations, and having several small reserves provides insurance against disease, habitat destruction, or other calamities that might wipe out a single population. But small preserves can't support species such as elephants or tigers, which need large amounts of space. Given human needs and pressures, however, big preserves aren't always possible. One proposed solution has been to create **corridors** of natural habitat that can connect to smaller habitat areas (fig. 12.28). Corridors could effectively create a large preserve from several small ones.

Corridors could also allow populations to maintain genetic diversity or expand into new breeding territory. The effectiveness of corridors probably depends on how long and wide they are, and on how readily a species will use them.

Perhaps the most ambitious corridor project in the world today is the proposed Yellowstone to Yukon (Y2Y) corridor. Linking more than two dozen existing parks, preserves, and wilderness areas, this corridor would stretch 3,200 km (2,000 km) from the Wind River Range in Wyoming to northern Alaska. More than half this corridor is already forested. Some 31 different First Nations and American Indian tribes occupy parts of this land, and are being consulted in ecosystem management.

One of the reasons large preserves are considered better than small preserves is that they have more **core habitat**—areas deep in the interior of a habitat area that have better conditions for specialized species than do edges. **Edge effects** is a term generally used to describe habitat edges: for example, a forest edge is usually more open, bright, and windy than a forest interior, and temperatures and humidity are more varied. For a grassland, on the other hand, edges may be wooded, with more shade, and perhaps more predators, than in the core of the grassland area. As human

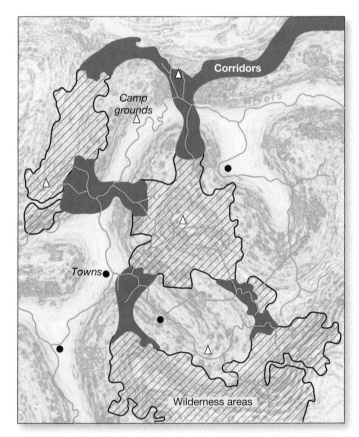

FIGURE 12.28 Corridors serve as routes of migration, linking isolated populations of plants and animals in scattered nature preserves. Although individual preserves may be too small to sustain viable populations, connecting them through river valleys and coastal corridors can facilitate interbreeding and provide an escape route if local conditions become unfavorable.

(a) Patch size and shape: these influence core/edge ratio

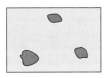

(b) Number of habitat patches, and patch proximity and connectivity

(c) Contrast between habitat patches and surrounding landscape

(d) Complexity of the landscape mosaic

FIGURE 12.29 Some spatial variables examined in landscape ecology. Size, arrangement, context, and other factors often vary simultaneously. Together, they can strongly influence the ability of a wildlife preserve to support rare species.

FIGURE 12.30 How small can a nature preserve be? In an ambitious research project, scientists in the Brazilian rainforest are carefully tracking wildlife in plots of various sizes, either connected to existing forests or surrounded by clear-cuts. As you might expect, the largest and most highly specialized species are the first to disappear.

disturbance fragments an ecosystem, habitat is broken into increasingly isolated islands, with less core and more edge. Small, isolated fragments of habitat often support fewer species, especially fewer rare species, than do extensive, uninterrupted ecosystems. The size and isolation of a wildlife preserve, then, may be critical to the survival of rare species.

Landscape ecology is a science that examines the relationship between these spatial patterns and ecological processes, such as species movement or survival. Landscape ecologists measure variables like habitat size, shape, and the relative amount of core habitat and edge (fig. 12.29). Landscape ecologists also examine the kind of land cover that surrounds habitat areas—is it a city, farm fields, or clear-cut forest? And they include utilitarian landscapes, such as farm fields, in their analysis, as well as pristine wilderness. By quantifying factors such as habitat shape and landscape complexity, and by monitoring the number of species or the size of populations, landscape ecologists try to guide more effective design of nature preserves and parks.

A dramatic experiment in reserve size, shape, and isolation is being carried out in the Brazilian rainforest. In a project funded by the World Wildlife Fund and the Smithsonian Institution, loggers left 23 test sites when they clear-cut a forest. Test sites range from 1 ha (2.47 acres) to 10,000 ha. Clear-cuts surround some, and newly created pasture surrounds others (fig. 12.30); others remain connected to the surrounding forest. Selected species are regularly inventoried to monitor their survival after disturbance. As expected, some species disappear very quickly, especially from small areas. Sun-loving species flourish in the newly created forest edges, but deep-forest, shade-loving species disappear, particularly when the size or shape of a reserve reduces availability of core habitat. This experiment demonstrates the importance of maintaining core habitat in preserves.

CONCLUSION

Forests and grasslands cover nearly 60 percent of global land area. The vast majority of humans live in these biomes, and we obtain many valuable materials from them. And yet these biomes also are the source of much of the world's biodiversity on which we depend for life-supporting ecological services. How we can live sustainably on our natural resources while also preserving enough nature so those resources can be replenished is one of the most important questions in environmental science.

There is some good news in our search for a balance between exploitation and preservation. Although deforestation and land degradation are continuing in some developing countries, many places are more thickly forested now than they were two centuries ago. Protection of Indonesia's rainforest and Australia's Great Barrier Reef shows that we can choose to protect some biodiverse areas in spite of forces that want to exploit them. Overall, nearly 12 percent of the earth's land area is now in some sort of protected status. Although the level of protection in these preserves varies, the rapid recent increase in number and area in protected status exceeds the goals of the United Nations Millennium Project.

We haven't settled the debate between focusing on individual endangered species versus setting aside representative samples of habitat, but pursuing both strategies seem to be working. Protecting charismatic umbrella organisms, such as the "spirit bears" of the Great Bear Rainforest, can result in preservation of innumerable unseen species. At the same time, protecting whole landscapes for aesthetic or recreational purposes can also achieve the same end.

REVIEWING LEARNING OUTCOMES

By now you should be able to explain the following points:

12.1 Discuss the types and uses of world forests.

- Boreal and tropical forests are most abundant.
- Forests provide many valuable products.
- Tropical forests are being cleared rapidly.
- Temperate forests are also threatened.

12.2 Describe the location and state of grazing lands around the world.

- Grazing can be sustainable or damaging.
- Overgrazing threatens many rangelands.
- Ranchers are experimenting with new methods.

12.3 Summarize the types and locations of nature preserves.

- Many countries have created nature preserves.
- Not all preserves are preserved.
- Marine ecosystems need greater protection.
- Conservation and economic development can work together.
- Native people can play important roles in nature protection.
- Species survival can depend on preserve size and shape.

PRACTICE QUIZ

1. What do we mean by *closed-canopy* forest and *primary* forest?

2. Which commodity is used most heavily in industrial economies: steel, plastic, or wood? What portion of the world's population depends on wood or charcoal as the main energy supply?

3. What is a *debt-for-nature swap*?

4. Why is fire suppression a controversial strategy? Why are forest thinning and salvage logging controversial?

5. Are pastures and rangelands always damaged by grazing animals? What are some results of overgrazing?

6. What is *rotational grazing,* and how does it mimic natural processes?

7. What was the first national park in the world, and when was it established? How have the purposes of this park and others changed?

8. How do the size and design of nature preserves influence their effectiveness? What do landscape ecologists mean by *interior habitat* and *edge effects*?

9. What is *ecotourism,* and why is it important?

10. What is a *biosphere reserve,* and how does it differ from a wilderness area or wildlife preserve?

CRITICAL THINKING AND DISCUSSION QUESTIONS

1. Paper and pulp are the fastest-growing sector of the wood products market, as emerging economies of China and India catch up with the growing consumption rates of North America, Europe, and Japan, What should be done to reduce paper use?

2. Conservationists argue that watershed protection and other ecological functions of forests are more economically valuable than timber. Timber companies argue that continued production supports stable jobs and local economies. If you were a judge attempting to decide which group is right, what evidence would you need on both sides? How would you gather this evidence?

3. Divide your class into a ranching group, a conservation group, and a suburban home-builders group, and debate the protection of working ranches versus the establishment of nature preserves. What is the best use of the land? What landscapes are most desirable? Why? How do you propose to maintain these landscapes?

4. Calculating forest area and forest losses is complicated by the difficulty of defining exactly what constitutes a forest. Outline a definition for what counts as forest in your area, in terms of size, density, height, or other characteristics. Compare your definition to those of your colleagues. Is it easy to agree? Would your definition change if you lived in a different region?

5. Why do you suppose dry tropical forest and tundra are well represented in protected areas, while grasslands and wetlands are protected relatively rarely? Consider social, cultural, geographic, and economic reasons in your answer.

6. Oil and gas companies want to drill in several parks, monuments, and wildlife refuges. Do you think this should be allowed? Why or why not? Under what conditions would drilling be allowable?

Data Analysis: Detecting Edge Effects

Edge effects are a fundamental consideration in nature preserves. We usually expect to find dramatic edge effects in pristine habitat with many specialized species. But you may be able to find interior-edge differences on your own college campus, or in a park or other unbuilt area near you. Here are three testable questions you can examine using your own local patch of habitat: (1) Can an edge effect be detected or not? (2) Which species will indicate the difference between edge and interior conditions? (3) At what distance can you detect a difference between edge and interior conditions? To answer these questions, you can form a hypothesis and test it as follows:

1. Choose a study area. Find a distinct patch of habitat, such as woods, unmowed grass, or marshy but walkable wetland, about 50 m wide or larger. With other students, list some local, familiar plant species that you would expect to find in your study area. If possible, make this list on a visit to your site.

2. Form a hypothesis. Examine your list, and predict which species will occur most on edges and (if you can) which you think will occur more in the interior. Form a hypothesis, or a testable statement, based on one of the three questions above. For example, "I will be able to detect an edge effect in my patch," or "I think an edge effect will be indicated by these species: _____," or "I think changes in species abundance will indicate an edge-interior change at _____ m from the edge of the patch."

3. Gather data. Get a meter tape and lay it along the ground from the edge of your habitat patch toward the interior. (You can also use a string and pace distances; treat one pace as a meter.) This line is your transect. At the edge end of the tape (or string), count the number of different species you can see within 1 m² on either side of your line. Repeat this count at each 5 m interval, up to 25 m. Thus you will create a list of species at 0, 5, 10, 15, 20, and 25 m in from the edge.

4. Examine your lists, and determine whether your hypothesis was correct. Can you see a change in species presence/absence from 0 to 25 m? Were you correct in your prediction of which species disappeared with distance from the edge? If you can identify an edge effect, at what distance did it occur?

5. Consider ways your test could be improved. Should you take more frequent samples? larger samples? Should you compare abundance rather than presence/absence? Might you have gotten different results if you had chosen a different study site? or if your class had examined many sites and averaged the results? How else might you modify your test to improve the quality of your results?

For Additional Help in Studying This Chapter, please visit our website at www.mhhe.com/cunningham12e. You will find additional practice quizzes and case studies, flashcards, regional examples, placemarkers for Google Earth™ mapping, and an extensive reading list, all of which will help you learn environmental science.

Student researchers sample aquatic fauna in a Louisiana coastal wetland. Ecological restoration requires careful monitoring such as this.

Restoration Ecology

Learning Outcomes

After studying this chapter, you should be able to:

13.1 Illustrate ways that we can help nature heal.

13.2 Show how nature is resilient.

13.3 Explain how restoring forests has benefits.

13.4 Summarize plans to restore prairies.

13.5 Compare approaches to restoring wetlands and streams.

"When we heal the earth, we heal ourselves." ~ *David Orr*

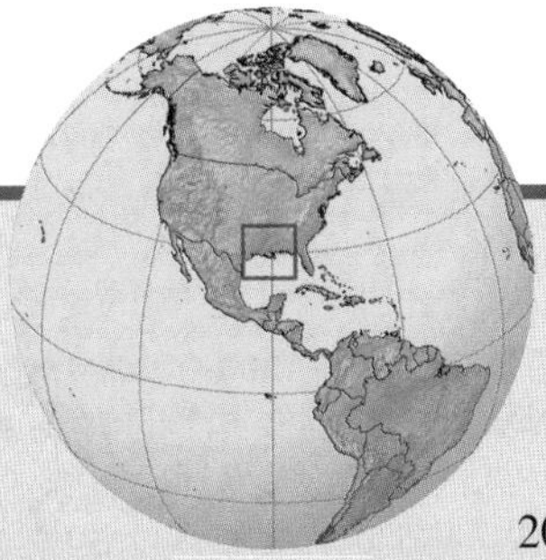

Restoring Louisiana's Coastal Defenses

Wetland restoration on the Louisiana coast has been in the news ever since Hurricane Katrina devastated New Orleans in September 2005. Most of the city flooded, and at least 1,500 people died in the worst natural catastrophe to hit the United States in a generation. Many failures were blamed for the disaster, including weak flood walls and inaction by disaster relief agencies. But one factor that caught the public eye was the erosion of the vast coastal marshes that once protected the coast from storm surges on the Gulf of Mexico.

Historically, vast coastal wetlands protected New Orleans and the region's other towns, farms, and forests from storm surges in the Gulf of Mexico. In the past 60 years these wetlands have shrunk by 4,000 km^2 (photo opposite page). By some estimates, each kilometer of coastal marsh reduces the height of storm surges by 5 cm. Wetland losses have increased storm surges in some parts of the state by up to 3 m. New Orleans, which is mostly below sea level, is highly vulnerable to Gulf Coast storms. Restoring these coastal marshes has become a priority in efforts to defend New Orleans and other coastal cities from future hurricanes.

Why are these wetlands shrinking and dying? Sediment loss, salinization, and physical degradation are three main factors that have deteriorated this dynamic system. The whole northern Gulf Coast is built of sediment dumped by the Mississippi River, which is thick with mud and clay drained from half the continental United States. Over thousands of years this meandering, sediment-rich river has deposited deltas—expanses of sand and silt—all along the Gulf Coast. Salt-marsh grasses root in these shifting, soggy sediments, creating root mats that stabilize the ground and provide nurseries for fish, birds, and the shrimp that make Cajun cuisine legendary. As wetlands expanded outward, the inland areas became less and less salty—saturated increasingly by rainwater and less by seawater. These freshwater marshes support even more plant growth and biodiversity than do the coastal salt marshes.

This coastal wetland system has been radically altered since the 1930s. First the Army Corps of Engineers straightened and contained the Mississippi outlet. Levees now guide the river 50 km out into the Gulf. Mud and silt that once built the coastal wetlands is now dumped off the coastal shelf. Furthermore, dredging and filling has destroyed about 400,000 ha (1 million acres) per year. Oil and gas companies have dug canals deep into the wetlands, allowing boat access to oil and gas wells. Thousands of kilometers of access canals now perforate the region's wetlands, and each canal allows salt water and storm surges, both lethal to the freshwater marshes, deep into the interior of the coastal marsh system. Freshwater marshes now cover less than one-fourth as much area as 60 years ago. This loss represents 80 percent of all coastal wetland loss in the United States and causes an estimated one-half billion dollars per year in economic losses.

The damage caused by the Gulf oil spill in 2010 added urgency to efforts to reverse, or at least slow, coastal wetland losses. A first step in restoration is to reduce the cause of the problem. For example, the Mississippi River Gulf Outlet (MRGO), which helped guide storm surges into New Orleans, has been closed. This canal destroyed thousands of hectares of valuable cypress swamp by flooding them with salt water. A more difficult proposal has been to close and restore some of the web of oil and gas access canals. Who should pay for this is disputed: companies propose that the public is responsible for restoration, but some taxpayers think the oil and gas corporations should pay to clean up the damage.

An additional step in restoration would be to return some of the Mississippi River sediment and fresh water to the marshes. Gaps are being cut in the levees below New Orleans, allowing fresh, muddy river water to re-enter the wetlands. A portion of the sediment would return, but just as important, fresh water could help the wetlands rebuild themselves. In a few experimental restoration areas, such as the Caernarvon Diversion east of New Orleans, simply replenishing freshwater flow has already helped expand the area of wetland vegetation.

FIGURE 13.1 Historic and projected loss of coastal marshes in the Mississippi delta 1932–2050, given a "business-as-usual" scenario.
Source: USGS.

In addition to these strategies, revegetation is going on in some areas, with volunteers replanting wetland grasses by hand to speed the restoration of critical areas, or monitoring the vigor of plants in remaining wetlands (fig 13.1).

Restoration is a new, exciting, and experimental field that applies ecological principles to healing nature. Full restoration of a vast ecosystem is a staggering task, but even small steps can make a difference. In coastal Louisiana and elsewhere, ecological restoration can be essential to human populations and economies, not just natural systems. In this chapter, we'll examine these and other aspects of restoration ecology.

For Google Earth™ placemarks that will help you explore these landscapes via satellite images, visit EnvironmentalScience-Cunningham.-blogspot.com.

13.1 HELPING NATURE HEAL

Humans have disturbed nature as long as we've existed. With the availability of industrial technology, however, our impacts have increased dramatically in both scope and severity. We've chopped down forests, plowed the prairies, slaughtered wildlife, filled in wetlands, and polluted air and water. Our greatly enhanced power also makes it possible, however, to repair some of the damage we've caused. The relatively new field of **ecological restoration** attempts to do this based on both good science and pragmatic approaches. Some see this as a new era in conservation history. As we discussed in chapter 1, the earliest phases in conservation and nature protection consisted in efforts to use resources sustainably and to protect special places from degradation.

Now, thousands of projects are under way to restore or rehabilitate nature. These range from individual efforts to plant native vegetation on small urban yards to huge efforts to restore millions of hectares of prairie or continent-size forests. Examples we'll see in this chapter include proposals to recreate a vast Buffalo Commons something like what Lewis and Clark saw when they crossed the North American Great Plains in 1804. Undoubtedly the biggest reforestation project in history is the Chinese effort to create a "green wall" to hold back the encroaching Gobi and Taklamakan Deserts. More than a billion people are reported to have planted 50 billion trees in China over the past 30 years.

The success of these projects varies. In some cases the land and biota are so degraded that restoration is impossible. In other situations some sort of ecosystem can be recreated, but the available soil, water, nutrients, topography, or genetic diversity limit what can be done on a particular site. Figure 13.2 presents a schematic overview of restoration options. Note that restoration to an original pristine condition is rarely possible. Most often the best choice is a compromise between ideal goals and pragmatic, achievable goals. Choosing these goals is one of the central questions in restoration ecology.

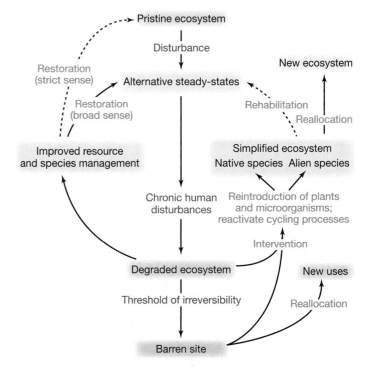

FIGURE 13.2 A model of ecosystem degradation and potential management options.
Source: Data from Walker and Moral, 2003.

Restoration projects range from modest to ambitious

Management goals are described in a variety of ways. Table 13.1 summarizes some of the most common terms employed in ecological restoration. A strict definition of restoration would be to return a biological community as nearly as possible to a pristine, predisturbance condition. Often this isn't possible. A broader definition of restoration has a more pragmatic goal simply to develop a self-sustaining, useful ecosystem with as many of its original

Table 13.1 A Restoration Glossary

Some commonly used terms in restoration ecology:

- **Restoration** (strict sense) to return a biological community to its predisturbance structure and function.
- **Restoration** (broad sense) to reverse degradation and reestablish some aspects of an ecosystem that previously existed on a site.
- **Rehabilitation** to rebuild a community to a useful, functioning state but not necessarily its original condition.
- **Intervention** to apply techniques to discourage or reduce undesired organisms and favor or promote desired species.
- **Reallocation** to use a site (and its resources) to create a new and different kind of biological community rather than the existing one.
- **Remediation** to clean chemical contaminants from a polluted area using relatively mild or nondestructive methods.
- **Reclamation** to use powerful chemical or physical methods to clean and repair severely degraded or even barren sites.
- **Re-creation** to construct an entirely new ecosystem on a severely degraded site.
- **Mitigation** to replace a degraded site with one of more or less equal ecological value somewhere else.

elements as possible. *Rehabilitation* may seek only to repair ecosystem functions. This may take the form of a community generally similar to the original one on a site, or it may aim for an entirely different community that can carry out the desired functions. Sometimes it's enough to leave nature alone to heal itself, but often we need to intervene in some way to remove or discourage unwanted organisms while also promoting the growth of more desirable species. *Reintroduction* generally implies transplanting organisms from some external source (often a nursery or hatchery where native species are grown under controlled conditions) to a site where they have been reduced or eliminated.

Remediation uses chemical, physical, or biological methods to remove pollution, generally with the intention of causing as little disruption as possible. *Reclamation* employs stronger, more extreme techniques to clean up severe pollution or create a newly functioning ecosystem on a seriously degraded or barren site. *Mitigation* implies compensation for destroying a site by purchasing or creating one of more or less equal ecological value somewhere else.

Restoration ecologists tend to be idealistic but pragmatic

Restoration ecologists work in the real world, and although they may dream of returning a disturbed site to its untouched state, they have to deal with the constraints of a specific place, usually faced with multiple obstacles and limited budgets. Restoration ecologists often find it useful to express their goals in broad terms of ecosystem health or integrity. Often the target of restoration is a

matter of debate. In the Nature Conservancy's Hassayampa River Preserve near Phoenix, for instance, pollen grains preserved in sediments reveal that 1,000 years ago the area was a grassy marsh that was unique in the surrounding desert landscape. Some ecologists would try to rebuild a similar marsh. Corn pollen in the same sediments, however, show that 500 years ago Native Americans began farming the marsh. Which is more important, the natural or early agricultural landscape?

Unfortunately, it may be impossible to return to conditions of either 500 or 1,000 years ago, because climate change may have made prehistoric communities incompatible with current conditions. In this case, should restoration ecologists attempt to restore areas to what they used to be, or to create a community that will be more compatible with future conditions? Restoration is fraught with such philosophical questions, as well as pragmatic ones.

13.2 COMPONENTS OF RESTORATION

Because restoration is an experimental and applied science, it often involves technical details of particular problems. General laws and principles in restoration are drawn from a variety of sciences: an understanding of the importance of biodiversity comes from ecology; principles of groundwater movement come from hydrology; soil science provides insights into soil health; and so on.

All restoration projects involve some common activities

Different types of restoration involve separate challenges. Wetland restoration, for example, involves a variety of efforts at reestablishing hydrologic connections, plant and animal diversity, weed removal, and sometimes salinity control. Forest restoration, on the other hand, requires a different set of steps that may include controlled burning, selective cutting, weed removal, control of deer and other herbivores, and so on. Even so, there are at least five main components of restoration, and most efforts share a majority of these activities.

1. *Removing physical stressors.* The first step in most restoration efforts is to remove the cause of degradation or habitat loss. Physical stressors such as pollutants, vehicle traffic, or inadequate water supply may need to be corrected. In wetland restoration, water flow and storage usually must be restored before other steps, such as replanting, can proceed. In forest restoration, clear-cut practices might be replaced with selective logging. In prairie restoration, cultivation might be ended so that land can be replanted with grassland plants, which then provide habitat for grassland butterflies, beetles, and birds.

2. *Controlling invasive species.* Often a few aggressive, "weedy" species suppress the growth of other plants or animals. These invasives may be considered biotic stressors. Removing invasive species can be extremely difficult, but without removal, subsequent steps often fail. Some invasives can be controlled by introducing pests that eat only those species: for example,

purple loosestrife (*Lythrum salicaria*) is a wetland invader that has succeeded in North America partly because it lacks predators. Loosestrife beetles (*Galerucella* sp.), introduced after careful testing to ensure that their introduction wouldn't lead to further invasions, have successfully set back loosestrife populations in many areas. Leafy spurge (*Euphorbia esula*), similarly, has invaded and blanketed vast areas of the Great Plains. Insects, including flea beetles (*Aphthona* sp.), have been a successful strategy to reduce the spread of spurge in many areas, allowing native grassland plants to compete again.

3. *Replanting.* Restoring a site or ecosystem usually involves some replanting of native plant species. Often restorationists try to collect seeds from nearby sources (so that the plants will be genetically similar to the original plants of the area), then grow them in a greenhouse before transplanting to the restoration site. In some cases restoration ecologists can encourage existing plants to grow by removing other plants that outcompete the target plants for space, nutrients, sunshine, or moisture.

4. *Captive breeding and reestablishing fauna.* In some cases restoration involves reintroducing animals. Peregrine falcon restoration involved releasing captive-bred birds, which then managed to survive and nest on their own. In some situations invertebrates, such as butterflies or beetles, may be released. Sometimes a top predator is reintroduced, or allowed to reinvade. Yellowstone National Park has had a 20-year experimental restoration of wolves. Evidence indicates that these predators are reducing excessive deer and elk populations, thus helping to restore vegetation, as well as reducing mid-level carnivores such as coyotes.

5. *Monitoring.* Without before-and-after monitoring, restoration ecologists cannot know if their efforts are working as hoped. Therefore, a central aspect of this science is planned, detailed, ongoing studies of key factors. Repeated counts of species diversity and abundance can tell whether biodiversity is improving. Repeated measures of water quality, salinity, temperature, or other factors can indicate whether suitable conditions have been established for target species.

13.3 ORIGINS OF RESTORATION

As settlers spread across North America in the eighteenth and nineteenth centuries, the woods, prairies, and wildlife populations seemed vast, much too large to be affected by anything humans could do. As we discussed in chapter 1, however, a few pioneers recognized that the rapid destruction of natural communities was unsustainable.

The most influential American forester was Gifford Pinchot. His first job after graduating from college was to manage the wealthy Vanderbilt family's Biltmore estate in North Carolina. With the leisure and money of working on this private estate, Pinchot introduced a system of selective harvest and replanting of choice tree species that increased the value of the forest while also producing a sustainable harvest. Pinchot went on to become the first head of the U.S. Forest Service, where he became the first in U.S. history to promote resource management, including replanting of forests, based on principles of long-term use and scientific research.

Before Pinchot, companies had practiced "cut and run" logging—devastating a region, then moving on to another without any reclamation or replanting, leaving barren expanses of stumps where once had been primeval forests. Pinchot's concepts of science-based management were a vast improvement over these practices. Subsequently critics have complained that his policies have led to replanting of only commercially valuable trees, rather than ecologically important ones, and to a disregard for other ecological functions of a forest, such as protecting water resources and habitat. But Gifford Pinchot's ideas led to the first widespread restoration practices in American forests.

The person most often recognized as the pioneer of restoration ecology was Aldo Leopold. Born at the end of the nineteenth century, Leopold grew up hunting and fishing. He became one of the first generation of professional wildlife managers, in a time when wildlife habitat was disappearing rapidly across North America. Leopold believed that these changes should be turned around. In 1935, Leopold bought a small, worn-out farm on the banks of the Wisconsin River not far from his home in Madison. Originally intended to be merely a hunting camp, it quickly became a year-round retreat from the city, as well as a laboratory in which Leopold could test this theories about conservation, game management, and land restoration (fig 13.3).

FIGURE 13.3 Aldo Leopold's Sand County farm in central Wisconsin served as a refuge from the city and as a laboratory to test theories about land conservation, environmental ethics, and ecologically based land management.

The whole Leopold family participated in planting as many as 6,000 trees each spring. "I have read many definitions of what is a conservationist, and written not a few myself," Leopold wrote, "but I suspect that the best one is written not with a pen, but with an axe. It is a matter of what a man thinks about while chopping, or while deciding what to chop. A conservationist is one who is humbly aware that with each stroke he is writing his signature on the face of his land ... A land ethic then, reflects the existence of an ecological conscience, and this in turn reflects a conviction of individual responsibility for the health of the land ... Health is the capacity of the land for self-renewal. Conservation is our effort to understand and preserve this capacity."

Many modern ecologists now regard goals of restoring the health, beauty, stability, or integrity of nature as unscientific, but they generally view Aldo Leopold as a visionary pioneer and an important figure in conservation history.

Sometimes we can simply let nature heal itself

As is the case in rebuilding Gulf Coast marshes, the first step in conservation and ecological restoration is generally to stop whatever is causing damage. Sometimes this is all that's necessary. Nature has amazing regenerative power. If the damage hasn't passed a threshold of irreversibility, natural successional processes can often rebuild a diverse, stable, interconnected biological community, given enough time.

The first official wildlife refuge established in the United States was Pelican Island, a small sand spit in the Indian River estuary not far from present-day Cape Canaveral. The island was recognized by ornithologists in the mid-1800s as being especially rich in bird life. Pelicans, terns, egrets, and other wading birds nested in huge, noisy colonies. But other people discovered the abundant birds as well. Boatloads of tourists slaughtered birds just for fun, while professional hunters shot thousands of adults during the breeding season, when the birds' plumage was most beautiful, leaving fledglings to starve to death. In 1900 the American Ornithological Union, worried about the wanton destruction of colonies, raised private funds to hire wardens to protect the birds during the breeding season. And in 1903, President Theodore Roosevelt signed an executive order establishing Pelican Island as America's first National Bird Reservation. This set an important precedent that the government could set aside public land for conservation. Pelican Island also became the first of 51 wildlife refuges created by President Roosevelt that may have saved species, such as roseate spoonbills and snowy egrets, from extinction.

Many of the forests and nature preserves described in chapter 12 suffered some degree of degradation before being granted protected status. Often simply prohibiting logging, mining, or excessive burning is enough to allow nature to heal itself. Consider the forests of New England, for example. When the first Europeans arrived in America, New England was mostly densely forested. As settlers spread across the land, they felled the forest to create pastures and farm fields. In 1811 sheep were introduced in New England, and sheep farming expanded

FIGURE 13.4 A mosaic of cropland, pasture, and sugar bush clothes the hills of Vermont. Two hundred years ago, this area was 80 percent cropland and only 20 percent forest. Today that ratio is reversed as the forests have invaded abandoned fields. Many of these forests are reaching late successional stages and are reestablishing ecological associations characteristic of old-growth forests.

rapidly to provide wool to the mills in New Hampshire and Massachusetts. Just 30 years later, in 1840, Vermont had nearly 2 million sheep, and 80 percent of its forests were gone. But competition soon ended this boom in wool.

In 1825 the Erie Canal opened access to western farmlands, which could raise both crops and sheep better than the cold Vermont hills. Eli Whitney invented the cotton gin, and cotton became a cheap and abundant (and more comfortable) alternative to woolen cloth. Within a few decades, most Vermont farmers had abandoned large-scale sheep farming, and abandoned pastures reverted to forest. Today 80 percent of the land in Vermont is once again forested, and less than 20 percent is farmland (fig 13.4).

After a century of natural succession, much of this forest has reacquired many characteristics of old-growth forest: a mixture of native species of different sizes and ages gives the forest diversity and complexity; many of the original animal species—moose, bear, bobcats, pine martins—have become reestablished. There are reports of lynx, mountain lions, and even wolves migrating south from Canada. Now, before woodlot owners can log their land, Vermont law requires that they consult with a professional forester to develop a plan to sustain the biodiversity and quality of their forest.

Native species often need help to become reestablished

Sometimes rebuilding populations of native plants and animals is a simple process of restocking breeding individuals. In other cases, however, it's more difficult. Recovery of a unique indigenous seabird in Bermuda is an inspiring conservation story that gives us hope for other threatened and endangered species.

When Spanish explorers discovered Bermuda early in the fifteenth century, they were frightened by the eerie nocturnal screeching they thought came from ghosts or devils. In fact the cries were those of extremely abundant ground-nesting seabirds now known as the hook-billed petrel, or Bermuda cahow (*Pterodroma cahow*), endemic to the island archipelago (fig. 13.5*a*). It's thought that there may originally have been half a million of these small, agile, gadfly petrels. Colonists soon found that the birds were easy to catch and good to eat. Those overlooked by humans were quickly devoured by the hogs, rats, and cats that accompanied settlers. Although Bermuda holds the distinction of having passed the first conservation laws in the New World, protecting native birds as early as 1616, the cahow was thought to be extinct by the mid-1600s.

In 1951, however, scientists found 18 nesting pairs of cahows on several small islands in Bermuda's main harbor. A protection and recovery program was begun immediately, including establishment of a sanctuary on 6-hectare (15-acre) Nonsuch Island, which has become an excellent example of ecological restoration.

Nonsuch was a near desert after centuries of abuse, neglect, and habitat destruction. All the native flora and fauna were gone, along with most of the soil in which the cahows once dug nesting burrows. This was a case of re-creating nature rather than merely protecting what was left. Sanctuary superintendent David Wingate, who devoted his entire professional career to this project, brought about a remarkable transformation of the island (fig. 13.5*b*).

The first step in restoration was to remove invasive species and reintroduce native vegetation. Wingate and many volunteers trapped and poisoned pigs and rats and other predators that threatened both wildlife and native vegetation. They uprooted millions of exotic plants and replanted native species, including mangroves and Bermuda cedars (*Juniperus bermudiana*). Initial progress was slow as trees struggled to get a foothold; once the forest knit itself into a dense thicket that deflected salt spray and ocean winds, however, the natural community began to reestablish itself. As was the case in New Zealand's Kapiti Island Nature Reserve (chapter 11), native plants that hadn't been seen for decades began to reappear. Once the rats and pigs that ate seedlings were removed, and competition from weedy invasives was eliminated, native seeds that had lain dormant in the soil began to germinate again.

Still, there wasn't enough soil for cahows to dig the underground burrows they need for nesting. Wingate's crews built artificial cement burrows for the birds. Each pair of cahows lays only one egg per year, and only about half survive under ideal conditions. It takes eight to ten years for fledglings to mature, giving the species a low reproductive potential. They also compete poorly against the more common long-tailed tropic birds that steal nesting sites and destroy cahow eggs and fledglings. Wingate designed wooden baffles for the burrow entrances, with holes just large enough for cahows but too small for the larger tropic birds (fig. 13.5*c*). The round cement cap at the back of the burrow can be removed to monitor the nesting cahows.

It takes constant surveillance to eradicate exotic plant species that continue to invade the sanctuary. Rats, cats, and toxic toads also swim from the mainland and must be removed regularly. By 2002, however, the cahow population had rebounded to about 200 individuals with 60 breeding pairs. Hurricane Fabian destroyed many nesting burrows on smaller islets in 2003. Fortunately the restored native forest on Nonsuch Island withstood the winds and preserved the rebuilt soil. The larger island is now being repopulated with chicks, their translocation timed so they will imprint on their new home. Reestablishing a viable population of cahows (which are now Bermuda's national bird) has had the added benefit of rebuilding an entire biological community (fig. 13.6).

It's too early to know if the population is large enough to be stable over the long term, but the progress to date is encouraging. Perhaps more important than rebuilding this single species is that Nonsuch has become a living museum of precolonial Bermuda

(A) Bermuda cahow

(b) David Wingate examines a cahow nest

(c) Long-tall excluder at mouth of burrow

FIGURE 13.5 For more than three centuries, the Bermuda cahow, or hook-billed petrel (a), was thought to be extinct until a few birds were discovered nesting on small islets in the Bermuda harbor. David Wingate (b) devoted his entire career to restoring cahows and their habitat. A key step in this project was to build artificial burrows (c). A long-tail excluder device (a board with a hole just the size of the cahow) keeps other birds out of the burrow. A round cement lump at the back end of the burrow can be removed to view the nesting cahows.

FIGURE 13.6 This brackish pond on Nonsuch Island is a reconstructed wetland. Note the Bermuda cedars on the shore and the mangroves planted in the center of the pond by David Wingate.

that benefits many species besides its most famous resident. It is a heartening example of what can be done with vision, patience, and a great deal of hard work.

There are many other notable reintroduction programs. Once-rare peregrine falcons have been reestablished in the eastern United States, even in the "canyons" of Chicago and New York. California condors have been reintroduced and have begun to breed in the American West. Both of these species were locally extinct 30 years ago, but captive breeding programs produced enough birds to repopulate much of their former range. Similarly, Arabian Oryx have been successfully reestablished in the deserts of Saudi Arabia, and in Hawaii the endemic nene geese were raised in captivity and reintroduced to Volcano National Park, where they exist in a small but self-reproducing population (see fig. 11.18).

13.4 RESTORATION IS GOOD FOR HUMAN ECONOMIES AND CULTURES

Restoration has become a cornerstone of managing economic resources and a source of cultural pride, not just an altruistic ecological activity. In the United States, the largest restoration projects ever attempted have been reforestation of cut-over or degraded forest lands. Building on the policies of Gifford Pinchot, lumber companies routinely replant forests they have harvested to prepare a future crop. Seedlings are grown in huge nurseries, and tractor-drawn planters allow a team of workers to plant thousands of new trees per day (fig 13.7). Usually this mechanical reforestation results in a monoculture of uniformly spaced trees.

These plantings are designed to produce wood quickly, but they have little resemblance to diverse, complex native forests (fig 13.8). Still, these commercial forests supply ground cover, provide habitat for some wildlife species, and grow valuable lumber or paper pulp. As we saw in chapter 12, a recent United Nations survey of world forests found that many countries are more thickly forested now

than they were 200 years ago. Both the total biomass and the quality of the forests in most of these countries have increased as forests have been protected and replanted over the past two centuries.

Japan, for example, was almost completely deforested at the end of World War II. In the 60 years since then, Japan has carried out a massive reforestation program. Now more than 60 percent of the country is forest-covered. Tight restrictions on logging help preserve this forest, which has great cultural value for the Japanese people. Rather than cut their own forest, Japan buys wood from its neighbors (a policy that has drawn some criticism from ecologists and human rights groups).

In 2001, at the Ninth UN Forum on Forests, the African country of Rwanda, torn by civil war and genocide in the 1990s, announced an ambitious program for country-wide restoration of its degraded forest, soil, water, and wildlife resources over the next 25 years. Poor forest management, damaging land-use practices, and war have caused the country's forests to shrink rapidly. Despite brisk economic growth in the past five years, 85 percent of the population still makes a living from subsistence farming on degraded lands. Among the new Forest Landscape Restoration Initiative priorities will be safeguarding the nation's rich wildlife, such as the critically endangered mountain gorilla, which is an important tourist attraction. For help in this project, Rwanda has turned to the International Union for Conservation of Nature (IUCN), the world's oldest and largest global environmental network. "This really is a good news story," said Stewart Maginnis, Director of Environment and Development of the IUCN. "For the first time, we're actually seeing a country recognize that part and parcel of its economic development trajectory has to be rooted in natural resources." If successful, this will be the first nationwide conservation effort in Africa, and one of very few in the world.

FIGURE 13.7 Mechanical planters can plant thousands of trees per day.

FIGURE 13.8 Monoculture forests, such as this tree plantation in Austria, often have far less biodiversity than natural forests.

Tree planting can improve our quality of life

Planting trees within cities can be effective in improving air quality, providing shade, and making urban environments more pleasant. Figure 13.9 shows student volunteers planting native trees and bushes in a project called Greening the Great River. Over the past decade this nonprofit organization has mobilized more than 10,000 volunteers to plant 160,000 native trees and shrubs on vacant land within the Mississippi River corridor as it winds through Minneapolis and St. Paul, Minnesota. This project helps beautify the cities, reduces global warming, and provides habitat for wildlife. Deer, fox, raccoons, coyotes, bobcats, and even an occasional wolf or cougar are now seen in the newly revegetated area.

FIGURE 13.9 Student volunteers plant native trees and shrubs to create an urban forest in the Mississippi River corridor within Minneapolis and St. Paul, Minnesota. This provides wildlife habitat as well as beautifying the urban landscape.

In 2007 the United Nations announced a "billion tree campaign" in the hope of gathering pledges to plant one billion new trees around the world. Everyone can participate in this global reforestation effort. If you look around, there's sure to be some space in your yard (or that of your friends or relatives) or an abandoned piece of land in your neighborhood that could house a tree. Most states have tree farms that will provide seedlings at little or no cost. The American Forestry Association has done an extensive remote sensing survey of the United States. They estimate that the national urban tree deficit now stands at more than 634 million trees. They suggest that everyone has a duty to plant at least one tree every year to help restore our environment (What Can You Do? p. 279).

The billion tree campaign is inspired by the work of Nobel Peace Prize laureate Wangari Maathai. As chapter 1 notes, Dr. Maathai founded the Kenyan Green Belt Movement in 1976 as a way of controlling erosion, providing fodder and food, and empowering women. This network of more than 600 local women's groups from throughout Kenya has planted more than 30 million trees while mobilizing communities for self-determination, justice, equality, poverty reduction, and environmental conservation.

Fire is often an important restoration tool

Oak savannas once covered a broad band at the border between the prairies of the American Great Plains and the eastern deciduous forest. Millions of hectares in what is now Minnesota, Wisconsin, Iowa, Illinois, Michigan, Indiana, Ohio, Missouri, Arkansas, Oklahoma, and Texas had parklike savannas, oak openings, or oak barrens. Although definitions vary, an oak savanna is a forest with scattered "open-grown" trees where the canopy covers 10 to 50 percent of the area and the dappled sunlight reaching the ground supports a variety of grasses and flowering plants (fig. 13.10). The most common tree species is bur oak (*Quercus macrocarpa*), but other species, including white and red oak, also occur in savannas. Due to reduced sunlight, however, some dominant prairie species, such as big and little bluestem grasses, most goldenrods, and asters, are generally missing from savannas.

Because people found them attractive places to live, most oak savannas were converted to agriculture or degraded by logging, overgrazing, and fire suppression. Wisconsin, for example, which probably had the greatest amount of savanna of any Midwestern state (at least 2 million ha) before European colonization, now has less than 200 ha (less than 0.01 percent of its original area) of high-quality savanna. Throughout North America, oak savanna rivals the tallgrass prairie as one of the rarest and most endangered plant communities.

It's difficult to restore, or even maintain, authentic oak savannas. Fire was historically important in controlling vegetation. Before settlement, periodic fires swept in from the prairies and removed shrubs and most trees. Mature oaks, however, have thick bark that allows them to survive low-intensity fires. Grazing by bison and elk may also have helped keep the savanna open. When settlers eliminated fire and grazing by native animals, savannas were invaded by a jumble of shrub and tree growth. Unfortunately, simply resuming occasional burning doesn't result in a high-quality

What Can You Do?

Ecological Restoration in Your Own Neighborhood

Everyone can participate in restoring and improving the quality of their local environment.

1. *Pick up litter*

   - With your friends or fellow students, designate one day to carry a bag and pick up litter as you go. A small amount of collective effort will make your surroundings cleaner and more attractive, but also will draw attention to the local environment. Try establishing a competition to see who can pick up the most. You may be surprised how a simple act can influence others. Working together can also be fun!

   - Join a group in your community or on your campus that conducts cleanup projects. If you look around, you'll likely find groups doing stream cleanups, park cleanups, and other group projects. If you can't find such a group, you can volunteer to organize an event for a local park board, campus organization, fraternity, or other group.

2. *Remove invasive species*

   - Educate yourself on what exotic species are a problem in your area: In your environmental science class, assign an invasive species to each student and do short class presentations to educate each other on why these species are a problem and how they can be controlled.

   - Do your own invasive species removal: In your yard, your parents' yard, or on campus, you can do restoration by pulling weeds. (Make sure you know what you're pulling!) Find a local group that organizes invasive species removal projects. Park boards, wildlife refuges, nature preserves, and organizations such as the Nature Conservancy frequently have volunteer opportunities to help eradicate invasive species.

3. *Replant native species*

   - Once the invasives are eliminated, volunteers are needed to replant native species. Many parks and clubs use volunteers to gather seeds from existing native prairie or wetlands, and to help plant seeds or seedlings. Many of the parks and natural areas you enjoy have benefited from such volunteer labor in the past.

   - Create your own native prairie, wetland, or forest restoration project in your own yard, if you have access to available space. You can learn a great deal and have fun with such a project.

4. *Plant a tree or a garden*

   - Everyone ought to plant at least one tree in their life. It's both repaying a debt to nature and a gratifying experience to see a seedling you've planted grow to a mature tree. If you don't have a yard of your own, ask your neighbors and friends if they would like a tree planted.

   - Plant a garden, or join a community garden. Gardening is good for your mind, your body, and your environment. Flowering or fruiting plants provide habitat for butterflies and pollinating insects. Especially if you live in a city, small fragments of garden can provide critical patches of green space and habitat—even a windowsill garden or potted tomatoes are a good start. You might find your neighbors appreciate the effort, too!

FIGURE 13.10 Oak savannas are parklike forests where the tree canopy covers 10 to 50 percent of the area, and the ground is carpeted with prairie grasses and flowers. This biome once covered a broad swath between the open prairies of the Great Plains and the dense deciduous forest of eastern North America.

savanna. If fires burn too frequently, they kill oak saplings—which are much more vulnerable than mature trees—and prevent forest regeneration. Excessive grazing can do the same thing. If fires are too infrequent, on the other hand, shrubs, dense tree cover, and dead wood accumulate so that when fire does occur, it will be so hot that it kills mature oaks as well as invading species. It's often necessary to clear most of the accumulated vegetation and fuel before starting fires if you want to maintain a savanna. Herbicide treatment may be necessary to prevent regrowth of invasive species until native vegetation can become established.

The Somme Prairie Grove in Cook County, Illinois, is one of the largest and oldest efforts to restore a native oak savanna. The land was purchased by the City of Chicago in the 1920s, and restoration activities began about 50 years later. A complex of wetland, prairie, and forest, much of the upland area was an oak savanna. Although the area was used as pasture for more than a century, most of it was never plowed. Limited grazing probably helped keep brush at bay and preserved the parklike character of the remaining forest. Spring burning was started in 1981 in an effort to preserve and restore the savanna. Burning alone, it was discovered, wasn't enough to eliminate aggressive invasive species, such as glossy buckthorn (*Rhamnus frangula*).

After several years of burning, the ground was mostly bare except for weedy species. In 1985 a more intensive management program was begun. Invasive trees were removed. Seeds of native species were collected from nearby areas and broadcast under the oaks. Weeds thought to be a threat to the restoration were reduced by pulling and scything; these included garlic mustard (*Alliaria officinalis*), burdock (*Arctium minus*), briers (*Rubus* sp.), and tall goldenrod (*Solidago altissima*). A research program monitored the results of this restoration effort. Four transects were established and repeatedly sampled for a wide variety of biota (trees, shrubs, herbs, cryptogams, invertebrates, birds, and small mammals) over a six-year period. Results from this survey are shown in fig. 13.11.

As you can see in the below graph, the biodiversity of the forest increased significantly over the six years of sampling. Native species increased more than total biodiversity. NARI, the Natural Area Rating Index, is a measure of the relative abundance of native species characteristic of high-quality natural communities. Notice that this index shows the greatest increase of any of the measurements in the study. It also continues to increase after 1988, when native species began to replace invasive pioneer species. This suggests that the restoration efforts have been successful, resulting in an increasingly high-quality oak savanna.

Fire is now recognized as a key factor in preserving oak savannas and many other forest types. The Superior National Forest, which manages the Boundary Waters Canoe Area Wilderness in northern Minnesota, for example, has started an ambitious program of prescribed fires to maintain the complex mosaic of mixed conifers and hardwoods in the forest. Pioneering work in the 1960s by forest ecologist M. L. Heinselman showed that the mixture of ages and species in this forest was maintained primarily by burning. Any given area in the forest experiences a low-intensity ground fire about once

FIGURE 13.12 The conifer-hardwood mixture and complex matrix of ages and species in the Great Lakes Forest is dependent on regular but random fire. Maintaining this forest requires prescribed burning.

per decade, on average, and an intense crown fire about once per century. As these fires burned randomly across the forest, they produced the complex forest we see today (fig. 13.12). But to maintain this forest, we need to allow natural fires to burn once again and to reintroduce prescribed fires where necessary to control fuel buildup.

Similarly, many national parks now recognize the necessity for fire to maintain forests. In Sequoia National Park, for example, rangers recognized that giant sequoias have survived for thousands of years because their thick bark is highly fire-resistant, and they shed lower branches so that fire can't get up into their crown. Seventy years of fire suppression, however, allowed a dense undergrowth to crowd around the base of the sequoias. These smaller trees provide a "ladder" for fire to climb up into the sequoia and kill them. Fuel removal and periodic prescribed fires are now a regular management tool for protecting the giant trees.

13.5 RESTORING PRAIRIES

Before European settlement, prairies covered most of the middle third of what is now the United States (fig. 13.13). The eastern edge of the Great Plains was covered by tallgrass prairies where big bluestem (*Andropogon gerardii*) reached heights of 2 m (6 ft). Their roots could extend more than 4 m into the soil and formed a dense, carbon-rich sod. This prairie has almost entirely disappeared, having been plowed and converted to corn and soybean fields. Less than 2 percent of the original 1 million km^2 (400,000 mi^2) of tallgrass prairie remains in its original condition. In Iowa, for example, which once was almost entirely covered by this biome, the largest sample of unplowed tallgrass prairie is only about 80 ha (200 acres), about the extent of many college campuses. The middle of the Great Plains contained a

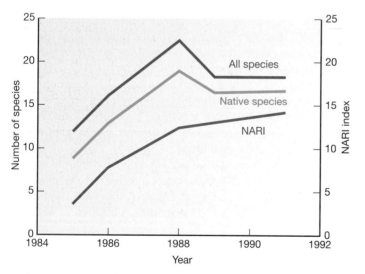

FIGURE 13.11 Biodiversity survey of the Somme Savanna restoration program. NARI, the Natural Area Rating Index, measures the frequency of native species associated with a high-quality community.
Source: Data from S. Packard and J. Balaban, 1994.

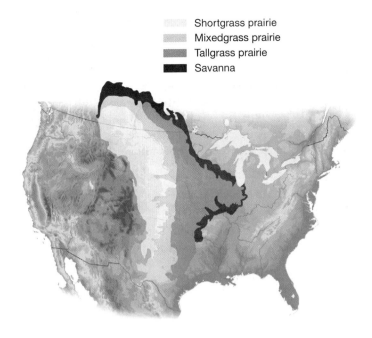

Shortgrass prairie
Mixedgrass prairie
Tallgrass prairie
Savanna

FIGURE 13.13 The eastern edge of the Great Plains was covered by tallgrass prairie, where some grasses reached heights of 2 m (6 ft) and had roots more than 4 m long that formed a dense, carbon-rich sod. Less than 2 percent of the original 10 million km² (400,000 mi²) of tallgrass prairie remains in its original condition. The middle of the Great Plains contained a mixed prairie with both bunch and sod-forming grasses. Few grasses in this region grew to heights of more than 1 m. The westernmost region of the Great Plains, in the rain shadow of the Rocky Mountains, had a shortgrass prairie where sparse bunch grasses rarely grew to more than 30 or 40 cm tall.

mixed prairie with both bunch and sod-forming grasses. The drier climate west of the 100th meridian meant that few grasses grew to heights of more than 1 m. The westernmost band of the Great Plains, in the rain shadow of the Rocky Mountains, received only 10 to 12 inches (25 to 30 cm) of rain per year. In this shortgrass region, the sparse bunch grasses rarely grew to more than 30 or 40 cm tall.

Like the oak savanna immediately to the east, these prairies were maintained by frequent fires and grazing by bison, elk, and other native wildlife. Native American people understood the role of fire in regenerating the prairie. They frequently set fires to provide good hunting grounds and to ease travel.

Fire is also crucial for prairie restoration

One of the earliest attempts to restore native prairie occurred at the University of Wisconsin. Starting in 1934, Aldo Leopold and others worked to re-create a tallgrass prairie on an abandoned farm field at the University Arboretum in Madison. Student volunteers and workers from the Civilian Conservation Corps gathered seed from remnant prairies along railroad rights-of-way and in pioneer cemeteries, and then hand-planted and cultivated them (fig. 13.14a). Prairie plants initially had difficulty getting established and competing against exotics, until it was recognized that fire is an essential part of this ecosystem. Fire not only kills many weedy species, but it also removes nutrients (especially nitrogen). This gives native species, which are adapted to low-nitrogen soils, an advantage. The Curtis Prairie is now an outstanding example of the tallgrass prairie community and a valuable research site (fig. 13.14b).

The Nature Conservancy (TNC) has established many preserves throughout the eastern Great Plains to protect fragments of tallgrass prairie. The biggest of these (and the largest remaining fragment of this once widespread biome) is in Oklahoma, where TNC has purchased or obtained conservation easements on about 18,000 ha (45,000 acres) of land just northwest of Tulsa that was grazed but never plowed. In cooperation with the University of Tulsa, TNC has established the Tallgrass Prairie Ecological Research Station, which is carrying out a number of experimental ecological restoration projects on both public and

(a)

(b)

FIGURE 13.14 In 1934, workers from the Civilian Conservation Corps dug up old farm fields and planted native prairie seeds for the University of Wisconsin's Curtis Prairie (a). The restored prairie (b) has taught ecologists much about ecological succession and restoration.

FIGURE 13.15 A burn-crew technician sets a back fire to control the prescribed fire that will restore a native prairie.

private land. Approximately three dozen prescribed burns are conducted each year, totaling 15,000 to 20,000 acres (fig. 13.15). Since 1991 over 350 randomly selected prescribed burns have been conducted, totaling 210,000 acres. About one-third of each pasture is burned each year. About half of the burns are done in spring and the rest in late summer.

In addition, bison were reintroduced to the preserve in the early 1990s. By summer 2005 the herd numbered about 2,400 head. The long-term goal is to have 2,700 bison on about 23,000 acres. Patch burning and bison grazing create a habitat that can support the diverse group of plants and animals that make up the tallgrass prairie ecosystem. So far, more than three dozen research projects are active on the preserve, and 78 reports from these studies have been published in scientific journals.

The second largest example of this biome in the United States is the Tallgrass Prairie National Preserve in the Flint Hills region of Kansas. Because the land in this area is too rocky for agriculture, the land was grazed but never plowed. Most of this preserve was purchased originally by the Nature Conservancy, but it is now being managed by the National Park Service. Like its neighbor to the south, a part of this preserve—known as the Konza Prairie—is reserved for scientific research. A long-term ecological research (LTER) program, funded by the National Science Foundation and managed by Kansas State University, sponsors a wide variety of basic scientific research and applied ecological restoration experiments. Bison reintroduction and varied fire regimes are being studied at this prairie. The role of grasslands in carbon sequestration and responses of this biological community to climate change are of special interest.

Huge areas of shortgrass prairie are being preserved

Much of the middle, or mixed grass, section of the Great Plains has been converted to crop fields irrigated by water from the Ogallala Aquifer (chapter 17). As this fossil water is being used

up, it may become impossible to continue farming over much of the area using current techniques. A great deal of worry and debate exist about the future of this region of the country. Interestingly, John Wesley Powell, the Colorado River explorer and first head of the U.S. Geological Survey, warned in 1878 that there wasn't enough water to settle this region. He opposed opening the Great Plains to homesteaders, and said, "I tell you gentlemen, you are piling up a heritage of conflict and litigation of water rights, for there is not sufficient water to supply the land."

Powell's predictions have been true, but other economic factors, such as job shortages, high fuel costs, and low farm income, have produced the main exodus from the Great Plains. In the 1990s more than half the counties between the 100th meridian and the Rocky Mountains experienced declining populations (fig 13.16). As farms and ranches fade away, the small towns that once supplied them also dry up. Of the people who remain, the percentage who are above 65 years of age is twice the national average. Ironically, several hundred thousand square miles now have fewer people than they did in 1890, the year Frederick Jackson Turner famously declared the frontier "closed." Large areas have fewer than 2 persons per square mile.

A number of people are now suggesting that we should have followed Powell's recommendations and never tried to settle the arid regions of the American West. Some believe that the best use of much of the empty land is to return it to a **buffalo commons** where bison and other native wildlife could roam freely. This idea was publicized in 1987 when Drs. Frank and Debora Popper, geographers from New Jersey, proposed in an article in the journal *Planning* a grand ecological and social reorganization of the Great Plains. The Poppers were fiercely criticized by many residents of Plains states, but today steps are being taken to accomplish something much like their plan.

FIGURE 13.16 Much of the Great Plains is being depopulated as farms and ranches are abandoned and small towns disappear. This provides an opportunity to restore a buffalo commons much like that discovered two centuries ago by Lewis and Clark.

FIGURE 13.17 Millions of hectares of shortgrass prairie are being converted to nature preserves. Some may be populated by bison and other wildlife rather than cattle, as envisioned by artist George Catlin in 1832.

The idea of a huge buffalo preserve long predates the Poppers. In 1832 the artist George Catlin, who was deeply concerned by the disappearance of both bison and the Native American cultures that depended on them, proposed that most of the Great Plains should be set aside as a national park populated by great herds of buffalo and other wildlife and home to native people living a traditional lifestyle.

There are two competing approaches to saving shortgrass prairie (fig. 13.17) with people, or without them. Working for conservation that includes people, the Nature Conservancy is cooperating with ranchers on joint conservation and restoration programs. On the 24,000 ha (60,000 acre) Matador ranch in Montana, which TNC bought in 2000, 13 neighboring ranchers graze cattle in exchange for specific conservation measures on their home range. The ranchers have agreed, for instance, to protect about 900 ha of prairie dog colonies and sage grouse leks (dancing grounds). All the ranchers have also agreed to control weeds, resulting in almost 120,000 ha of weed-free range. Together with the Conservancy, ranchers are experimenting with fire to improve the prairie, and they plan and manage a grass-banking arrangement in which ranchers access more Conservancy land in drought emergencies. In one case TNC deeded one of its ranches to a young couple who promised to manage it sustainably and to protect some rare wetlands on the property. The Conservancy believes that keeping ranch families on the land is the best way to preserve both the social fabric and the biological resources of the Plains.

Near the Matador land, another group is pursuing a very different strategy for preserving the buffalo commons. The American Prairie Foundation (APF), which is closely linked to the World Wildlife Fund, has also bought about 24,000 ha of former ranchland. Rather than keep it in cattle production, however, this group intends to pull out fences, eliminate all the ranch buildings, and turn the land back into wilderness. Ultimately the APF hopes to create a reserve of at least 1.5 million ha in the Missouri Breaks region between the Charles M. Russell National Wildlife Refuge and the Fort Belknap Indian Reservation (fig. 13.18). The APF plans to reintroduce native wildlife, including elk, bison, wolves, and grizzly bears, to its lands. Neighboring ranchers don't mind having elk or bison nearby, but they object to reintroducing wolves and grizzlies, predators their parents and grandparents exterminated a century ago. Ranchers also bristle at the funding for the APF project. Many of the project's large donations come from Wall Street or California's Silicon Valley. Wealthy individuals, such as media mogul Ted Turner, are using their money to make striking changes in how western land is used. Often locals, who struggle just to stay on the land, resent outsiders coming in and competing to buy the range.

With donations of more than $12 million so far, and fundraising goals of at least $100 million, the APF points out that they aren't forcing anyone from the land. They're only buying from ranchers who want to sell. Locals worry, nonetheless, that the land will be restricted for hunting and other uses. The APF says it will allow tourism, bird-watching, and hunting on nearly all its land. Small towns also are anxious about how it will affect the local economy to take land out of production. The APF points out that without federal subsidies, the average return on ranchland in the Missouri Breaks is less than $5 per acre. Tourism and hunting already bring in more income per acre than does raising cows.

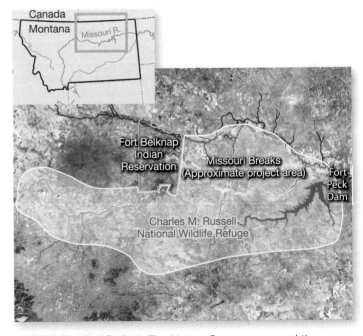

FIGURE 13.18 Both The Nature Conservancy and the American Prairie Foundation have bought large tracts of land in the Missouri Breaks between the Charles M. Russell National Wildlife Refuge and the Fort Belknap Indian Reservation. Ultimately, conservationists hope to protect and restore as much as 2 million ha of shortgrass prairie.

FIGURE 13.19 Bison can be an important tool in prairie restoration. Their trampling and intense grazing disturb the ground and provide an opening for pioneer species. The buffalo chips they leave behind fertilize the soil and help the successional process. Where there were only about two dozen wild bison a century ago, there are now more than 400,000 on ranches and preserves across the Great Plains.

Whether conservation should include human residents and economic activities has often been a deep divide among conservationists, who otherwise agree on the need to preserve biodiversity and ecosystems. Should people expect to maintain an economy in a preserve? How much should those economies be subsidized? Should the ecosystem be restored, predators and all? Or is it complete enough without predators? How would you decide these questions, if you were in charge?

Bison help maintain prairies

As the Exploring Science box on p. 245 shows, bison grazing can help maintain the health and diversity of prairie ecosystems. So far the APF has only 19 bison on its lands in Montana, but it intends to eventually expand its herd to several thousand animals. Others also are returning bison to the shortgrass prairie (fig. 13.19). The biggest of TNC's herds are in southern plains, but they have herds of several hundred animals in the Niobara River Valley in Nebraska and on the Ordway Prairie in north-central South Dakota. The Fort Belknap Indian reservation currently has a herd of 700 animals on a 5,000 ha tribal buffalo reserve. They also have a large prairie dog town that is a refuge for the highly endangered black-footed ferret. Although the native people don't live the lifestyle envisioned by George Catlin, they do welcome tourists for bird-watching, cultural tours, picnicking, sightseeing, wildlife viewing, and other forms of ecotourism.

Many ranchers are coming to recognize the benefits of raising native animals. Where cows require shelter from harsh weather and lots of water to drink, bison are well adapted to the harsh conditions of the prairie. Bison meat is lean and flavorful. It brings a higher price than beef in the marketplace. It's estimated that there are now about 400,000 bison on ranches and farms in America. This is probably less than 1 percent of the original population, but it's 10,000 times more than the tiny remnant left after the wanton slaughter in the nineteenth century.

13.6 RESTORING WETLANDS AND STREAMS

Wetlands and streams provide important ecological services. They play irreplaceable roles in the hydrologic cycle. They also are often highly productive, and provide food and habitat for a wide variety of species. As the opening case study of this chapter points out, Louisiana's freshwater swamps and coastal marshes (fig. 13.20) play important roles in both biological and human communities. Although wetlands currently occupy less than 5 percent of the land in the United States, the Fish and Wildlife Service estimates that one-third of all endangered species spend at least part of their lives in wetlands. As we also saw in the opening case study, coastal wetlands are vital for absorbing storm surges. Storage of floodwaters in wetlands is worth an estimated $3 billion to $4 billion per year. Wetlands also improve water quality by acting as natural water purification systems, removing silt and absorbing nutrients and toxins.

Recognition of these services has come only recently, though. For many years wetlands were considered disagreeable, dangerous, and useless. This attitude was reflected in public policies, such as the U.S. Swamp Lands Act of 1850, which allowed individuals to buy swamps and marshes for as little as 10 cents per acre. Until recently, federal, state, and local governments encouraged wetland drainage and filling to create land for development. In an effort to boost crop production, the government paid farmers to ditch and tile millions of acres of wet meadows and potholes. Cutting channels for oil and gas exploration and shortcuts for navigation through Gulf Coast marshes (fig. 13.21) is one of the main reasons for destruction of the barrier that once protected New Orleans from hurricanes.

FIGURE 13.20 A Louisiana cypress swamp of the sort being threatened by coastal erosion and salt infiltration. Wetlands provide habitat for a wide variety of species, and play irreplaceable ecological roles.

FIGURE 13.21 Louisiana has nearly 40 percent of the remaining coastal wetlands in the United States, but diversion of river sediments that once replenished these marshes and swamps, together with channels and boat wakes, are causing the Gulf shoreline to retreat about 4 m (13.8 ft) per year.

The result of these policies was that from the 1950s to the mid-1970s, the United States lost about 200,000 ha (nearly 500,000 acres) of wetlands per year. The 1972 Clean Water Act began protecting streams and wetlands by requiring discharge permits for dumping waste into surface waters. In 1977, federal courts interpreted this rule to prohibit both pollution and filling of wetlands (but not drainage). The 1985 Farm Bill went farther with a "swamp buster" provision that blocked agricultural subsidies to farmers who drain, fill, or damage wetlands. Many states now have "no net loss" wetlands policies. Between 1998 and 2004, America had a net gain of about 80,000 ha (197,000 acres) of wetlands. This total area concealed an imbalance, however. Continued losses of 210,000 ha of swamp and marsh wetlands were offset by a net gain of 290,000 ha of small ponds and shallow-water wetlands (which are easy to construct and good for duck production). And since 2004, another 700,000 ha have been added to the national total. This is a good start, but the United States needs much more wetland restoration to undo decades of damage. In this section we'll look at some efforts to accomplish this goal.

Restoring water supplies helps wetlands heal

As is the case with other biological communities, sometimes all that's needed is to stop the destructive forces. For wetlands, this often means simply to restore water supplies that have been diverted elsewhere.

For millennia the delta where the Tigris and Euphrates empty into the Persian Gulf created a vast wetland with unique biological and human importance. Covering an area the size of Wales, these marshes provided a resting spot for millions of wildfowl migrating between Eurasia and Africa. The marshes also were the home of a unique group of people, the Marsh Arabs, who built their homes on floating platforms and depended on the wetland for most of their sustenance. Some people believed the verdant marshes to be the biblical Garden of Eden.

Saddam Hussein accused the Marsh Arabs of supporting his enemies during the 1980–1988 war with Iran. In retaliation he ordered the marshes dammed and drained. About 90 percent of the wetlands were destroyed, and more than 140,000 people were forced from their homes. As the marsh dried out, much of it also was burned. After Saddam was toppled in 2003, the Marsh Arabs cut through Saddam's dikes and plugged diversion canals, allowing water to flow once again into natural channels. Aided by a UN-led restoration project, the wetlands have rebounded to about half their former size. Flora and fauna have repopulated the area and people also have begun to return. Still, the marshes haven't returned to their original state. Political instability in Iraq makes it unclear how much restoration will be accomplished, but the situation is much better than it was a few decades ago.

Another simple physical solution to wetland degradation can be seen in the American Midwest. When the U.S. Army Corps of Engineers built 26 locks and dams on the upper Mississippi River to facilitate barge traffic in the 1950s, it created a series of large impoundments. Sediment from the surrounding farm fields began to fill these pools. This might have created a valuable network of wetlands, but waves and currents created by wind, floods, and river traffic keep the sediment constantly roiled up so that wetland plants couldn't take root. The result is a series of wide, shallow, semisolid mud puddles that are too thick for fish but too liquid for vegetation. To remedy this situation, the Army Corps is experimenting with wing dams (see thin white lines connecting islands in fig. 13.22) that separate the backwaters from the main river channel. It's hoped that these dams will allow the mud to solidify and turn into marshlands.

Much of the restoration of the Louisiana wetlands described in the chapter-opening case study depends on an assumption that controlling water flow and sediment deposition will result

FIGURE 13.22 Dams on the upper Mississippi River have created a series of large lakes. Currents created by wind, floods, and river traffic keep the soft sediment stirred up so that wetland plants can't take root. To restore these backwaters, the Army Corps is experimenting with wing dams (see thin white lines between islands) that will allow the mud to solidify and turn into marshlands.

FIGURE 13.23 The Florida Everglades, often described as a "river of grass," is threatened by water pollution and diversion projects.

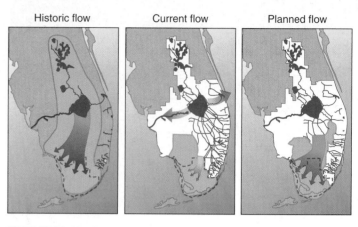

FIGURE 13.24 Planned results of the Comprehensive Everglades Restoration Plan. Red dashed outline shows the national park boundary.

in restoration of a healthy biological community. This may turn out to be true, but monitoring is needed to evaluate results. How do we know whether restoration is working or not? (Exploring Science p. 287)

Replumbing the Everglades is one of the costliest restoration efforts ever

A huge, expensive restoration is now under way in the Florida Everglades. Famously described as a "river of grass," the Everglades are created by a broad, shallow sheet of water that starts in springs near Orlando, in the center of the state, then moves through Lake Okeechobee and flows southward to the Gulf. Spreading over most of the southern tip of the peninsula, and moving imperceptibly across the flat landscape, the fresh water nourishes a vast marshland (fig. 13.23) that supports myriad fish, invertebrates, birds, alligators, and the rare Florida panther.

Farmers found the rich, black muckland of the Everglades could grow fantastic crops if it was drained. Ditching and diverting the water started more than a century ago. A series of floods that threatened the wealthy coastal cities also triggered a demand for more water management. Once-meandering rivers were straightened to shunt surplus water out to sea. Altogether, the Army Corps of Engineers has built more than 1,600 km of canals, 1,000 km of levees, and 200 water-control structures to intercept normal water flow, drain farmlands, and divert floodwater (fig. 13.24). Ironically, many of the cities that demanded the water be diverted are experiencing water shortages during the dry season. Water that might have been stored in the natural wetlands is no longer available. Nature, also, is suffering from water shortages. The Everglades National Park has lost 90 percent of its wading birds, and there are worries that the entire aquatic ecosystem may be collapsing.

After years of debate and acrimony, the various stakeholders have finally agreed on a massive reengineering of the south Florida water system. The plan aims to return some water to the Everglades, yet retain control to prevent flooding. More than 400 km of levees and canals will be removed. New reservoirs will store water currently lost to the ocean, and 500 million liters of water per day will be pumped into underground aquifers for later release. Rivers are being dechannelized to restore natural meanders that store storm water and provide wildlife habitat (fig. 13.25). It's hoped that simply restoring some of the former flow to the Everglades will allow the biological community to recover, although whether it will be that simple remains to be seen. Altogether this project is expected to cost at least $8 billion, making it one of the most costly restoration projects ever undertaken.

FIGURE 13.25 The naturally meandering Kissimmee River (*right channel*) was straightened by the Army Corps of Engineers (*left*) for flood control 30 years ago. Now the Corps is attempting to reverse its actions and restore the Kissimmee and its associated wetlands to their original state.

The science of restoration is complex, with few simple answers. Restoration is also highly optimistic, because it often addresses environmental problems that are huge and persistent. But increasingly we recognize that ecological restoration is a necessity if we are to preserve economies, cultures, and ways of life.

As the opening case study of this chapter reveals, restoring Louisiana's coastal wetlands has been discussed for half a century, but the projects have gained a new urgency since 2005, when Hurricane Katrina flooded New Orleans. Restoring this vast system is almost an inconceivably large project, but without restoration Louisiana will continue to lose communities, roads, and economic activity. Hundreds of small projects have been planned, and some are in progress. A prominent project is the Caernarvon Diversion, a series of structures built on the south bank of the Mississippi east of New Orleans. At a cost of $4.5 million, the project is using culverts (1.25 m diameter pipes) to divert water from the Mississippi, together with fill to plug or block abandoned gas-field canals. These plugs slow the escape of river water.

Engineering this project involved a decade of planning, but monitoring is the most extensive part of the project. Monitoring is a key aspect of all restoration: otherwise how do we know if the project has been successful—and if it's worth doing again?

In addition to the project area, two "reference areas" are also being monitored. Ideally the project area will improve considerably over the reference areas, and over the baseline conditions before the project started. The project plan outlines a 20-year monitoring, with three central concerns:

1. *Ratio of land to open water*. Land and water are being mapped using aerial

Student volunteers sample plant biomass and species composition to evaluate the health of a coastal wetland.

photos. Using a GIS, the monitoring team can calculate the amount of land and water in 2000, then again in 2006 and in 2018. In theory, the amount of land will increase from one time period to the next. With a GIS they can also overlay one year's map on top of another year's map. By subtracting one layer from the other, analysts identify not only the amount of change but which areas have changed.

2. *Plant species composition and relative abundance*. Before water diversions began, ecologists designated a series of square plots of wetland, 2 m on a side. These 4 m^2 plots were placed in reference areas, as well as in the project area. Plant ecologists then visited each plot to list all species in each one. They also estimated the relative abundance of the different species. By leaving permanent markers at the plots, they can revisit the same location years later to monitor change. By sampling a number of replicate plots, they can get a sense of aggregate change in the area.

3. *Salinity*. Salt concentrations will be continuously recorded with a salinity sensor—an electrode that measures the concentration of salt ions in water—which is hooked up to a small computer that automatically records data once an hour. By monitoring hourly, hydrologists and ecologists can observe changes in salinity during storms, spring floods, and other events that might cause rapid variations.

To more clearly organize and structure the data collection and interpretation, scientists framed specific hypotheses (testable statements). For example, one goal is to increase the abundance and diversity of plants. A testable hypothesis is: "After the project implementation, diversity will be significantly greater than before project implementation." By gathering data before and after, the restoration team can test this hypothesis by comparing the average number of plant species, or the average abundance of each species, in the different plots. A simple yes or no answer will diagnose whether the project is working as planned.

Similarly, reducing salinity is a goal. A testable hypothesis is: "After project implementation, salinity will be significantly less than before project implementation." Again, this hypothesis can be tested by comparing before-project samples and after-project samples for mean salinity value and variation from the average.

Restoration often relies on inputs from a variety of sciences, including hydrology, ecology, geology, and other fields. Restoration is also experimental, partly because it is a new science, and much remains uncertain in restoration projects. But restoration is an extremely important—and exciting—field within environmental science.

FIGURE 13.26 Volunteers plant dune grasses as part of Chesapeake Bay restoration. The large cylinder is a "bio log" made of biodegradable material that will help stabilize the shore.

Although announced with great fanfare in 2000, the restoration project is behind schedule, over budget, and at risk of losing congressional support. A blunt internal memo written by a top manager in the Army Corps of Engineers was leaked to the press in 2005. It said, "We haven't built a single project during the first five years, and we've missed almost every milestone." The memo echoed concerns of conservationists, the Miccosukee Tribe, and others who have been complaining about the lack of progress for years. The state of Florida has put up more than $1 billion, nearly five times more than federal agencies to date. These problems show some of the difficulty in carrying out such a huge, complex, and highly political project.

Restoration of the Chesapeake Bay is another long, expensive, and contentious project (see chapter 3). Volunteers are replanting 200,000 ha of coastline, salt marsh, and shallow water areas (fig 13.26).

Wetland mitigation is challenging

Working in large, complex, highly political ecosystems that have been damaged by a large variety of human actions is especially challenging. Smaller ecosystems can be much easier to restore or replace. An encouraging example is the re-creation of prairie potholes in the Great Plains. Before European settlement, millions of these shallow ponds and grassy wetlands once spread across the prairies from Alberta, Saskatchewan, and Manitoba, to northeastern Kansas and western Iowa (fig. 13.27). They produced more than half of all North American migratory waterfowl and mediated flooding by storing enormous amounts of water. As agriculture expanded across the Plains, however, at least half of all prairie wetlands were drained and filled. In the 1900s, Canada's Prairie Provinces had about 10 million potholes; by 1964 an estimated 7 million were gone. The United States had similar losses.

Passage of the Migratory Bird Hunting Stamp Act (popularly known as the Duck Stamp Act) in 1934 marked a turning point in wetland conservation. In the past 70 years this program has collected $700 million that has been used to acquire more than 5.2 million acres of habitat for the National Wildlife Refuge System. In 1934, when the Duck Stamp Act was passed, drought, predators, and habitat destruction had reduced migratory duck populations in the United States to about 26 million birds. John Phillips and Frederick Lincoln, in a 1930 report on the state of American waterfowl, wrote, "We believe it soon will be too late to save [wild-fowl] in numbers sufficient to be of any real importance for recreation in the future." By 2006, however, conservation efforts, a wetter climate, and habitat restoration had increased duck production to about 44 million birds. Duck stamps now provide about $25 million annually for wildlife conservation.

While it's relatively easy to dig new ponds, it's much more difficult to replace other wetland types. As we mentioned earlier in this chapter, between 1998 and 2004 the United States lost 210,000 ha of swamp and marshes. This was offset by repair or construction of 290,000 ha of small ponds and shallow-water wetlands, such as restored prairie potholes. Replacing a damaged wetland with a substitute is called **wetland mitigation.** It's required whenever development destroys a natural wetland. In the past, this process often didn't necessarily replace the native species and ecological functions represented by the original biological community. Figure 13.28, for example, shows a replacement wetland created by a housing developer in Minnesota. In building a housing project, the developer destroyed about 10 ha of a complex, native wetland

FIGURE 13.27 Millions of prairie potholes once covered the Great Plains from Iowa and South Dakota, north to Canada's Prairie Provinces. More than half of these shallow ponds have been drained for agricultural crop production, but now many are being replaced.

FIGURE 13.28 In the past, developers weren't required to plant native vegetation in wetland mitigation. They were allowed to simply dig a hole and wait for it to fill with rainwater and invasive species. New rules now require a more ecological approach.

that contained rare native orchids and several scarce sedge species. To compensate for this loss, the developer simply dug a hole and waited for it to fill with rainwater. He wasn't required to replant wetland species. The law assumes that natural succession will revegetate the disturbed area. In fact, it was soon revegetated, but entirely with exotic invasive species.

Constructed wetlands can filter water

Many cities are finding that artificial wetlands provide a low-cost way to filter and treat sewage effluent. Arcata, California, for instance, needed an expensive sewer plant upgrade. Instead the city transformed a 65 ha garbage dump into a series of ponds and marshes that serve as a simple, low-cost waste treatment facility. Arcata saved millions of dollars and improved its environment simultaneously. The marsh is a haven for wildlife and has become a prized recreation area for the city. Eventually the purified water flows into Humbolt Bay, where marine life flourishes. Similarly, small pools and rain gardens—shallow pits lined with porous surface material and planted with water-tolerant vegetation—are used to collect storm runoff and allow it to seep into the ground rather than run into rivers or lakes. And constructed marshes allow industrial cooling water to equilibrate before entering streams or other surface water bodies. All these created wetlands can be useful to both humans and wildlife. For more on this topic, see chapter 18.

Many streams need rebuilding

Pollution, pathogens and diseases, industrial toxins, invasive organisms, erosion, and a host of other factors degrade streams and rivers. The United States has more than 5.6 million km of rivers and streams. In a 1994 EPA survey of nearly 1 million km of rivers and streams, only 56 percent fully supported multiple uses, including drinking-water supply, fish and wildlife habitat, recreation, and agriculture, as well as flood prevention and erosion control. Sedimentation and excess nutrients were the most significant causes of degradation in the remaining 44 percent.

In 1994, of the nearly 1 million km of rivers and streams that were monitored by the EPA, only 56 percent fully supported multiple uses, including drinking-water supply, fish and wildlife habitat, recreation, and agriculture, as well as flood prevention and erosion control. Sedimentation and excess nutrients were the most significant causes of degradation in the remaining 44 percent. Presumably these results could be extrapolated to the rest of the nation's waterways. Given these statistics, the need for stream restoration is obvious.

One response to erosion and flooding in urban streams has been to turn them into cement channels that rush rainwater off into some larger body of water (fig. 13.29) or to bury them in underground culverts. The result is an artificial system with little resemblance to the living biological community that once made up the stream. Some cities have come to recognize, however, that natural streams can increase property values and improve the livability of the urban environment. Buried streams are being "day-lighted," and channelized ditches are being turned back into living biological communities.

A variety of restoration techniques have been developed for streams. This field has become an important source of jobs for environmental science majors. A simple approach in which everyone can participate is to reduce sediment influx by planting ground cover on uplands and filling gullies with rocks or brush. For small streams, sometimes the quickest way to rebuild a channel is to use heavy earthmoving equipment to simply dig a new one. This can be very disruptive, however, stirring up sediment that can be harmful for fish and other aquatic organisms. Alternative, less intrusive stream improvement methods are available. Most of these methods involve placing barriers (weirs, vanes, dams, log barriers, brush bundles, root wads, or other obstructions) in streams to deflect current away from the banks or trap sediment. Often these barriers will cause currents to scour out deep pools in the stream bottom that provide places for fish to hide and rest.

Other techniques also create fish habitat. Logs, root wads, brush bundles, and boulders can shelter fish. An expensive but effective way to create fish hiding places is the so-called "lunker"

FIGURE 13.29 Many former streams have been turned into concrete-lined ditches to control erosion and speed runoff. The result is an artificial system with little resemblance to the living biological community that once made up the stream.

FIGURE 13.30 A lunker structure is a multilevel wooden framework that rests on the stream bottom and can be anchored to the shore. The top of the box is covered with rock, soil, and vegetation. Openings in the structure provide hiding places for fish.

structure. This is a wood framework that rests on the stream bottom and is anchored securely to the shore. The top of the box is covered with rock, soil, and vegetation. Openings in the structure provide hiding places for fish (fig. 13.30).

Sometimes what's needed is to speed up the current rather than slow it down. Figure 13.31a shows a spring-fed Minnesota trout stream that was degraded by crop production in its uplands, and grazing that broke down the banks and filled the stream with sediment. The stream, which had been about 1 m wide and 1 m deep, spread out to be 5 m wide and only about 10 cm deep. The formerly

cold, swiftly moving water was warmed by the sun as it passed through these shallows so that it became uninhabitable for trout. Furthermore, there was no place to hide from predators. Because this was the last remaining trout stream in the Minneapolis/St. Paul metropolitan area, a decision was made to restore it. Two very different restoration approaches are shown in figure 13.31b and c. Trout Unlimited, an angler's group, offered to bring in a backhoe and completely rebuild the stream. A demonstration section they reconstructed is shown in figure 13.31b. They narrowed the channel with large stone blocks, which also provide a good surface for anglers to walk along the stream. There isn't much biological production in this stream model, but that isn't important to many anglers. They expected the Department of Natural Resources to stock the stream with hatchery-raised fish, which usually are caught before they have time to learn to forage for natural food.

Other groups involved in stream restoration objected to the artificial nature and lack of a native biological community in this design. Figure 13.31c shows an alternative approach that was ultimately adopted for most of the stream. This is the same section shown in figure 13.31a. Straw bales were placed in the deepest part of the stream. This narrowed the stream and increased the speed of the current, which then cut down into the soft sediment. Within three months the stream had deepened about 50 cm. The shallow area between the straw bales and the original shore quickly filled with sediment and became a cattail marsh. The bales were anchored in place with green willow stakes, which rooted in the moist soil and sprouted to make saplings that arched over the stream and shaded it from the sun.

Seeds of native wetland vegetation were scattered on top of the straw bales. They sprouted, and after the first summer the stream was surrounded by a dense growth of vegetation that keeps the water cool, provides fish shelter, and supports a rich community of

(a)

(b)

(c)

FIGURE 13.31 Different visions for restoring a trout stream. (a) Degraded by a century of agriculture and grazing, the stream had become too wide, shallow, and warm for native trout. (b) Trout Unlimited rebuilt a section of the stream to show their preferred option, which featured banks made of large stone blocks. (c) Other environmental organizations preferred a more organic approach. They used straw bales to narrow the stream and increase current flow. This washed away the soft sediment and re-created the deep, narrow, cool, deeply shaded channel that favored native trout.

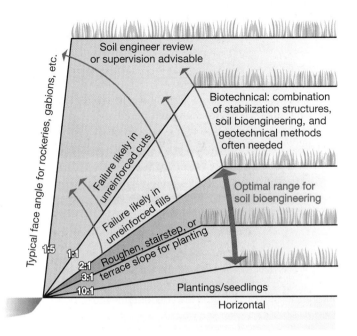

FIGURE 13.32 Steep streambanks need to be reinforced or recontoured to avoid erosion. Shallow slopes can be stabilized with vegetation or mulch. Steeper slopes need stabilization structures or reinforcement to hold the soil.

invertebrates that feed the native trout. An aquatic invertebrate survey found that while the stream was wide and shallow, 58 percent of the aquatic species were snails and copepods, whereas stone flies, caddis flies, and other preferred trout food made up only 20 percent of all invertebrates. After the straw bale restoration, snails and copepods made up only 15 percent of the invertebrate population, and stone flies and caddis flies had increased to 47 percent.

Stabilizing banks is an important step in stream restoration. Where banks have been undercut by erosion and stream action, they will continue to be unstable and cave into the stream. Ideally the bank should be recontoured to a slope of no more than 45 degrees (fig. 13.32). Soil can then be held in place by rocks, planted vegetation, or other ground cover. It may be necessary to install erosion control fabric or mulch to hold soil until vegetation is established. If space isn't available for recontouring, steep banks may have to be supported by rock walls, riprap, or embedded tree trunks.

Severely degraded or polluted sites can be repaired or reconstructed

In a relatively small area—say an old industrial site—it may be economical to simply excavate and replace contaminated soil. If the pollutants are organic, it may be possible to pass soil through an incinerator to eliminate contaminants. After this treatment, the soil won't be worth much for growing vegetation, however.

For polluted surface or groundwater, bacteria can remove organic compounds, such as oil, and other contaminants. Naturally occurring bacteria in groundwater, when provided with oxygen and nutrients, can decontaminate many kinds of toxins. Experiments have shown that pumping air into aquifers can be more effective than pumping water out for treatment. For hostile environments or exotic, human-made chemicals that can't be metabolized by normal organisms, it's sometimes possible to genetically engineer new varieties of bacteria that can survive in extreme conditions and consume materials that would kill ordinary species.

Bioremediation is a growing strategy that uses living things, especially plants or bacteria, to selectively eliminate toxins from the soil. A number of plant species can selectively eliminate toxins from the soil. Some types of mustard, for example, can extract lead, arsenic, zinc, and other metals from contaminated soil. Radioactive strontium and cesium have been removed from soil near the Chernobyl nuclear power plant using common sunflowers. And poplar trees can absorb and break down toxic organic chemicals (fig. 13.33).

In some cases, bioremediation could have multiple benefits. A weedy species called field pennycress or stinkweed (*Thlaspi arvense*), for example, grows well on degraded, polluted soil, absorbing metals as it grows. Its seeds contain high concentrations of oils that make it competitive with canola as an oil seed crop. The oils aren't edible but can be easily converted into biodiesel fuel or can be used for bioplastic production. Furthermore, glucosinolates in the oil can be converted into a soil fumigant that could be an eco-friendly alternative to methyl bromide. Combining all these features, pennycress could be grown on contaminated inner-city lands, producing either fuel or chemical feedstocks while also decontaminating the soil.

Many cities are finding that decontaminating urban "brown fields" (abandoned, contaminated industrial sites) can turn unusable inner-city property into valuable assets. This is a good way to control urban sprawl and make use of existing infrastructure. Cleaning up hazardous and toxic wastes is now a big business in America, and probably will continue to be so for a long time in the future. This is a growth industry in places where most other industry is disappearing.

Reclamation implies using intense physical or chemical methods to clean and repair severely degraded or even totally barren sites. Historically, reclamation meant irrigation projects that brought wetlands and deserts (considered useless wastelands) into

FIGURE 13.33 These poplar trees absorb toxins and nutrients from sewage in India.

agricultural production. In the early part of this century the Bureau of Reclamation and the Army Corps of Engineers dredged, diked, drained, and provided irrigation water to convert millions of acres of wild lands into farm fields. Many of those projects were highly destructive to natural ecosystems. Ironically, we are now using ecological restoration to restore some of these "reclaimed" lands to a more natural state.

Today, reclamation means the repairing of human-damaged land. The Surface Mining Control and Reclamation Act (SMCRA), for example, requires mine operators to restore the shape of the land to its original contour and revegetate it to minimize impacts on local surface water and groundwater. According to the U.S. Office of Surface Mining, more than 8,000 km^2 (3,000 mi^2) of former strip mines have been reclaimed and returned to beneficial uses, such as recreation areas, farming and rangeland, wetlands, wildlife refuges, and sites for facilities such as hospitals, shopping centers, schools, and office and industrial parks.

Ideally, if topsoil is set aside during surface mining, overburden and tailings (waste rock discarded during mining and ore enrichment) could be returned to the pit, smoothed out, and covered with good soil that will support healthy vegetation. Unfortunately, topsoil is often buried deeply during mining, and what ends up on the surface is crushed rock that won't revegetate very well without a great deal of fertilizer and water.

The largest mine pits will never be returned to their original contour. Figure 13.34 shows a view of the Berkely mine pit in Butte, Montana. From the 1860s to the 1980s, this area was one of the world's richest sources of metals, including copper, silver, lead, zinc, manganese, and gold. In the early days all mining was in deep shafts. In 1955 the Anaconda Mining Company switched to open-pit mining and dug the hole you see now. After mining ended in 1981, groundwater, previously controlled with pumps, began to fill the pit. It has now accumulated to make a lake 1.6 km wide and 300 m deep.

FIGURE 13.34 The Berkely mine pit in Butte, Montana, may be the most toxic water body in the United States. Water entering the pit is now being treated and eventually there may be an effort to pump water out of the pit and decontaminate it, but the pit itself will probably never be filled in.

The water has a pH of 2.5 (about the same as vinegar) and is laden with heavy metals and toxic chemicals, such as arsenic, cadmium, zinc, and sulfuric acid. In 1995 a whole flock of migrating snow geese were found dead after landing by mistake in the pit. The pit is now a Superfund site. A water treatment plant has been built on the far shore of the pit (you can see a thin, white stream of treated water from the plant cascading into the lake). By 2018, or whenever the water level hits the critical elevation of 1,649 m above sea level (the height at which it will threaten groundwater used by the city of Butte), the plant will begin to treat the contents of the pit. This is considered a reclamation project, but it's highly unlikely that the pit will ever be filled in.

CONCLUSION

Humans have caused massive damage and degradation to a wide variety of biological communities, but there are many ways to repair this damage and to restore or rehabilitate nature. Ideally, we might prefer to return a site to its pristine, predisturbance condition, but that often isn't possible. A more pragmatic goal is simply to develop a useful, stable, self-sustaining ecosystem with as many of its original ecological elements as possible. Sometimes it's enough to leave nature alone to heal itself, but often we need to intervene in some way to remove or discourage unwanted organisms while also promoting the growth of more desirable species.

Restoration pioneer Aldo Leopold wrote, "A thing is right when it tends to preserve the integrity, stability, and beauty of the biotic community. It is wrong when it tends otherwise." Some modern ecologists object that ecosystems are highly dynamic and

species appear and disappear stochastically and individually. Characteristics such as integrity, health, stability, beauty, and moral responsibility tend to be human interpretations rather than scientific facts. Still, we need to have goals for restoration that the public can understand and accept.

Many ecological restoration projects are now under way. Some are huge efforts, such as rehabilitating Louisiana coastal wetlands, Florida's everglades, or the Chesapeake Bay, and returning a huge swath of shortgrass prairie to a buffalo commons. Others are much more modest: building a rain garden to trap polluted storm runoff, planting trees on empty urban land, or turning a part of your yard into a native prairie or woodland. Within this wide range, there are lots of opportunities for all of us to get involved.

REVIEWING LEARNING OUTCOMES

By now you should be able to explain the following points:

13.1 Illustrate ways that we can help nature heal.
- Restoration projects range from modest to ambitious.
- Restoration ecologists tend to be idealistic but pragmatic.

13.2 Describe the common elements of restoration projects.
- All restoration projects involve some common activities.

13.3 Explain the origins of restoration.
- Sometimes we can simply let nature heal itself.
- Native species often need help to become reestablished.

13.4 Show how restoration can be good for human economies and cultures.
- Tree planting can improve our quality of life.
- Fire is often an important restoration tool.

13.4 Summarize techniques for restoring prairies.
- Fire is also crucial for prairie restoration.
- Huge areas of shortgrass prairie are being preserved.
- Bison help maintain prairies.

13.5 Compare approaches to restoring wetlands, streams, and beaches.
- Reinstating water supplies helps wetlands heal.
- Replumbing the Everglades is one of the costliest restoration efforts ever.
- The Chesapeake Bay is being rehabilitated.
- Beach replenishment is often a losing battle.
- Wetland mitigation is challenging.
- Constructed wetlands can filter water.
- Many streams need rebuilding.
- Severely degraded or polluted sites can be repaired or reconstructed.

PRACTICE QUIZ

1. Why have levees on the lower Mississippi River starved coastal wetlands of sediments?
2. Why does coastal wetland loss matter to New Orleans?
3. Define *ecological restoration.*
4. What's the difference between rehabilitation and remediation?
5. Give an example of letting nature heal itself.
6. Why is restoring savannas difficult?
7. Why are fires essential for prairies?
8. What is the buffalo commons?
9. Why do the Everglades need restoration?
10. What is wetland mitigation?

CRITICAL THINKING AND DISCUSSION QUESTIONS

1. Should we be trying to restore biological communities to what they were in the past, or modify them to be more compatible with anticipated future conditions? What future conditions would you consider most likely to be problematic?

2. How would you balance human preferences (aesthetics, utility, cost) with biological considerations (biodiversity, ecological authenticity, evolutionary potential)?

3. The Nature Conservancy's Hassayampa River Preserve near Phoenix illustrates a situation in which there may be more than one historic condition to which we may wish to restore a landscape. How would you reconcile these different values and goals?

4. Restoring savannas often requires the use of herbicides to remove invasive species. Some people regard this as dangerous and unnatural; how would you respond?

5. The Nature Conservancy believes it's essential to keep productive ranches on the land, both to sustain rural society and for effective protection of the range. The American Prairie Foundation (APF) is buying up ranches and converting them to wilderness where wild animals can roam freely. Which of these approaches would you favor?

6. Sometimes the quickest and easiest way to restore a stream is simply to reconstruct it with heavy equipment. You might even create something more interesting and useful (at least from a human perspective) than the original. Is it okay to replace real nature with something synthetic?

Figure 13.32 on p. 291 is a form of graphic representation we haven't used very often in this book. It's a concept map, or a two-dimensional representation of the relationship between key ideas. It could also be considered a decision flowchart because it's an organized presentation of different policy options. This kind of chart shows how we might think about a situation, and suggests affinities and associations that might not otherwise be obvious. You might like to look at the introductory chapter of this book for more information about concept maps.

1. Using one of the examples presented in this chapter—or another familiar example—replace the descriptions in the colored boxes with brief descriptions of an actual ecosystem. Does this help you see the relationship between different states for the community you've chosen?

2. Now replace the terms associated with the arrows between boxes with actions that cause changes in your particular biological community as well as restoration treatments that could accomplish the proposed restoration outcomes.

3. What do you suppose the authors meant by a "threshold of irreversibility"? If the system is irreversible, why are there arrows for reallocation or intervention?

4. The box for simplified ecosystems has two subcategories, native species and alien species. What does this mean? What would be some examples for the ecological community you've chosen?

5. There are two arrows labeled "reallocation" in this diagram. One leads to "new uses," while the other leads to "new ecosystem." What's the difference? Why are the two boxes separated in the diagram?

6. The arrow labeled "restoration (strict sense)" is a dotted line. What do you think the authors meant by this detail?

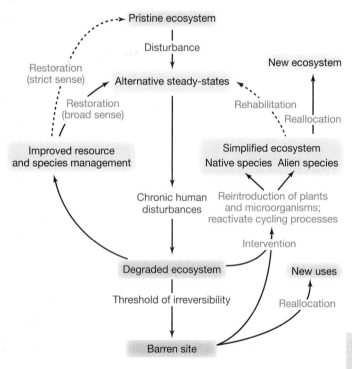

A model of ecosystem degradation and potential management options.

Source: Data from Walker and Moral, 2003.

For Additional Help in Studying This Chapter, please visit our website at www.mhhe.com/cunningham12e. You will find additional practice quizzes and case studies, flashcards, regional examples, placemarkers for Google Earth™ mapping, and an extensive reading list, all of which will help you learn environmental science.

CHAPTER 14

Hundreds of thousands of buildings collapsed and millions of people were left homeless when a magnitude 7.0 earthquake struck the Caribbean island of Haiti in 2010.

Geology and Earth Resources

"Learn geology or die."

~ **Louis Agassiz**

Learning Outcomes

After studying this chapter, you should be able to:

14.1 Summarize the processes that shape the earth and its resources.

14.2 Explain how rocks and minerals are formed.

14.3 Think critically about economic geology and mineralogy.

14.4 Critique the environmental effects of resource extraction.

14.5 Discuss ways we could conserve geological resources.

14.6 Describe geological hazards.

Case Study Earthquake!

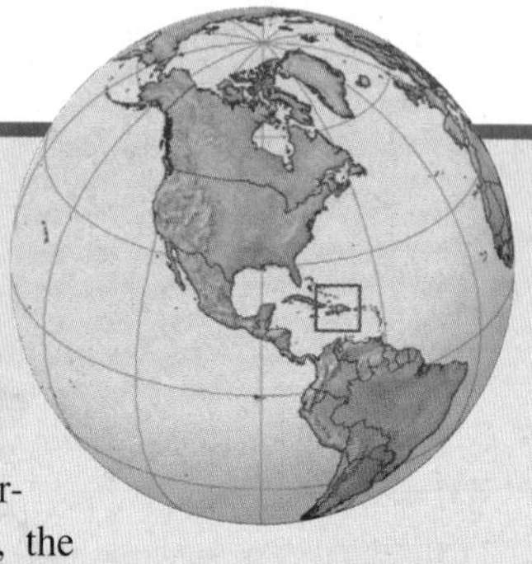

Shortly before 5 p.m. on January 12, 2010, a massive earthquake measuring 7.0 on the Richter scale struck the Caribbean island of Haiti. The earthquake's epicenter was only about 16 km (10 mi) southwest of the capital, Port-au-Prince. It was the worst earthquake in the region in more than 200 years. Huge swaths of Port-au-Prince lay in ruins. Schools, hospitals, commercial buildings, and even the Presidential Palace collapsed. It's estimated that 230,00 people were killed and 300,000 injured. At least 1 million people were left homeless, and more than 3 million suffered contaminated water supplies, food shortages, lost jobs, or missing family members. The Inter-American Development Bank estimated the economic losses could be (U.S.) $15 billion.

Port-au-Prince sits on the coastline where two huge geological features—the Caribbean tectonic plate and the Gonave microplate—slide slowly past each other (fig. 14.1). As the plates grind along what's called a strike-slip fault, strain builds up over centuries until the plates suddenly jerk forward to trigger seismic activity. Two fault systems intersect under the island of Hispaniola, the Caribbean Island Haiti shares with the Dominican Republic. The 2010 quake occurred along the Eniquillo-Plantain Garden Fault, an east–west crack in the earth's crust that runs from Hispaniola through Jamaica and the Cayman Islands.

These faults trace their origins to a broader interaction between the North American plate and the Caribbean plate. The North American plate is diving beneath the Caribbean plate, but one piece of the North American plate, called the Bahamas Platform, is too buoyant to make the plunge easily. The resulting collision deforms and shakes Hispaniola.

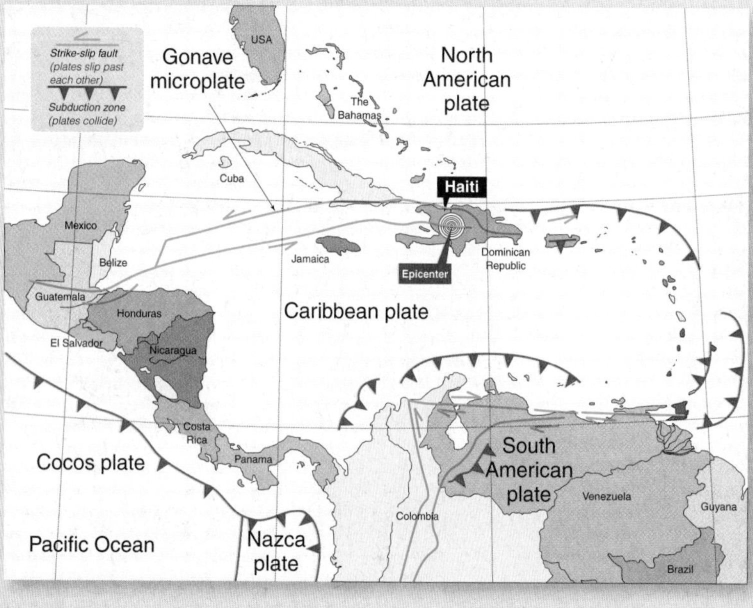

Nearly every island in the Caribbean has experienced earthquakes. Although major quakes occur only every few centuries, they can be extremely catastrophic. In 1692 a 7.5 magnitude megaquake hit the town of Port Royal, Jamaica, which sits on the same fault as Port-au-Prince. Much of the town, which was unusually rich with pirate plunder, sank below the sea with a great loss of life.

The damage in Haiti in 2010 was especially severe because the quake was close to the city center and shallow (only 10 to 15 kilometers below the surface), and more importantly because many homes and buildings in the economically depressed country weren't built to withstand seismic forces. Building codes in Haiti are poorly enforced, and building supplies are expensive, so most concrete there is made with too much sand, too little cement, and not enough reinforcing metal. Furthermore, after the catastrophe occurred, the dysfunctional government was unprepared to offer much assistance to victims. Public services in Haiti are minimal even in the best of times. Port-au-Prince may be the largest city in the world without a public sewer system. A year after the quake, having suffered a major cholera outbreak that sickened more than 100,000 people and killed at least 2,000, as well as torrential rains from Hurricane Tomas that flooded ragged tent cities and added more misery to the grim situation, more than a million suffering Haitians remain homeless.

By contrast, a much larger earthquake hit Chile just six weeks after the one in Haiti. With a magnitude of 8.8 on the Richter scale, the Chile quake was 500 times larger than the one in the Caribbean. But its epicenter was 35 km (21.7 mi) deep, offshore along a relatively remote area of the country, and 105 km (65 mi) from Concepcion, the largest city in the region. Because Chile experiences frequent earthquakes, building codes are far more advanced and more rigorously enforced than in many other countries. Only about 700 people died in Chile compared to about 300 times as many in Haiti.

Geological hazards, such as earthquakes, volcanic eruptions, tsunamis, floods, and landslides, are major threats. Devastating events have altered human history many times in the past, sending geopolitical, economic, genetic, and even artistic repercussions around the planet. In this chapter we'll look at the processes that shape the earth and how rocks and minerals are formed, as well as what we might do to reduce our geological risks and our impacts on our environment as we extract resources.

For related resources, including Google Earth™ placemarks that show locations discussed in this chapter, visit Environmental-Science-Cunningham.blogspot.com.

FIGURE 14.1 The Gonave microplate is squeezed between its larger neighbors as the North American tectonic plate crashes into and dives under the Caribbean plate.

14.1 EARTH PROCESSES SHAPE OUR RESOURCES

Many people are exposed to geological hazards of one type or another, but all of us benefit from the earth's geological resources. Right now you are probably wearing several geological products; plastics, including glasses and synthetic fabric, are made from oil; iron, copper, and aluminum mines produced your snaps and zippers and the screws in your glasses; silver, gold, and diamond mines may have produced your jewelry. All of us also share responsibility for the environmental and social devastation that often results from mining and drilling. Fortunately there are many promising solutions to reduce these costs, including recycling and alternative materials. The question is whether voters will demand that we use these technologies—and whether consumers will share the costs of responsible production.

Why are these resources distributed as they are? To understand how and where earth resources are created, we must examine the earth's structure and the processes that shape it.

Earth is a dynamic planet

Although we think of the ground under our feet as solid and stable, our planet is a dynamic and constantly changing structure. Titanic forces inside the earth cause continents to split, move apart, and then crash into each other in slow but inexorable collisions.

The earth is a layered sphere. The **core**, or interior, is composed of a dense, intensely hot mass of metal—mostly iron—thousands of kilometers in diameter (fig. 14.2). Solid in the center but more fluid in the outer core, this immense mass generates the magnetic field that envelops the earth.

Surrounding the molten outer core is a hot, pliable layer of rock called the **mantle.** The mantle is much less dense than the core because it contains a high concentration of lighter elements, such as oxygen, silicon, and magnesium.

The outermost layer of the earth is the cool, lightweight, brittle rock **crust.** The crust below oceans is relatively thin (8–15 km), dense, and young (less than 200 million years old) because of constant recycling. Crust under continents is relatively thick (25–75 km) and light, and as old as 3.8 billion years, with new material being added continually. It also is predominantly granitic, while oceanic crust is mainly dense basaltic rock. Table 14.1 compares the composition of the whole earth (dominated by the dense core) and the crust.

Tectonic processes reshape continents and cause earthquakes

The huge convection currents in the mantle are thought to break the overlying crust into a mosaic of huge blocks called **tectonic plates** (fig. 14.3). These plates slide slowly across the earth's surface like wind-driven ice sheets on water, in some places breaking up into smaller pieces, in other places crashing ponderously into each other to create new, larger landmasses. Ocean basins form where

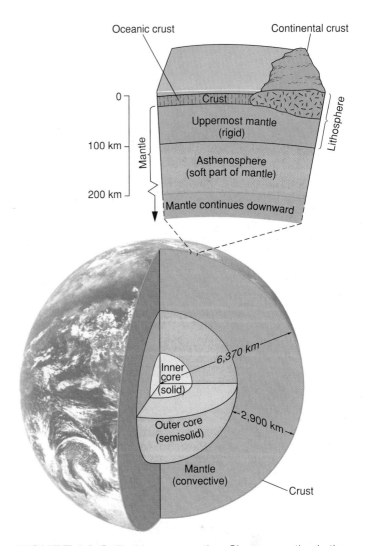

FIGURE 14.2 Earth's cross-section. Slow convection in the mantle causes the thin, brittle crust to move.

Table 14.1	Eight Most Common Chemical Elements (Percent)		
Whole Earth		**Crust**	
Iron	33.3	Oxygen	45.2
Oxygen	29.8	Silicon	27.2
Silicon	15.6	Aluminum	8.2
Magnesium	13.9	Iron	5.8
Nickel	2.0	Calcium	5.1
Calcium	1.8	Magnesium	2.8
Aluminum	1.5	Sodium	2.3
Sodium	0.2	Potassium	1.7

continents crack and pull apart. The Atlantic Ocean, for example, is growing slowly as Europe and Africa move away from the Americas. **Magma** (molten rock) forced up through the cracks forms new oceanic crust that piles up underwater in **mid-ocean ridges.**

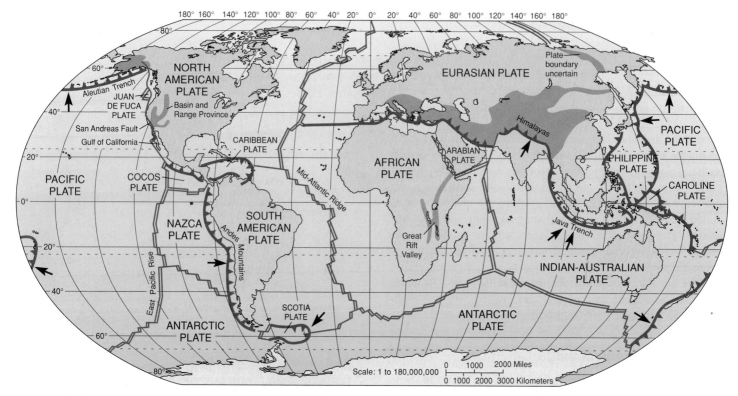

FIGURE 14.3 Map of tectonic plates. Plate boundaries are dynamic zones, characterized by earthquakes and volcanism and the formation of great rifts and mountain ranges. Arrows indicate direction of subduction where one plate is diving beneath another. These zones are sites of deep trenches in the ocean floor and high levels of seismic and volcanic activity.

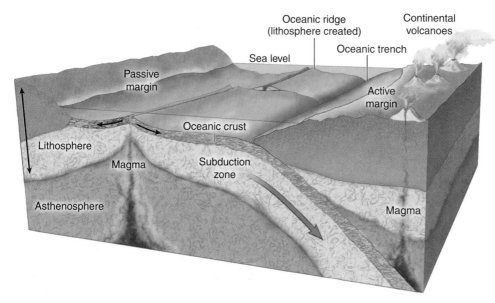

FIGURE 14.4 Tectonic plate movement. Where thin, oceanic plates diverge, upwelling magma forms mid-ocean ridges. A chain of volcanoes, like the Hawaiian Islands, may form as plates pass over a "hot spot." Where plates converge, melting can cause volcanoes, such as the Cascades.

mountains. Slowly spreading from these fracture zones, ocean plates push against continental plates.

Earthquakes are caused by grinding and jerking as plates slide past each other. Mountain ranges like those on the west coast of North America and in Japan are pushed up at the margins of colliding continental plates. The Himalayas are still rising as the Indian subcontinent grinds slowly into Asia. Southern California is slowly sailing north toward Alaska. In about 30 million years, Los Angeles will pass San Francisco, if both still exist by then.

When an oceanic plate collides with a continental landmass, the continental plate usually rides up over the seafloor, while the oceanic plate is **subducted,** or pushed down into the mantle, where it melts and rises back to the surface as magma (fig. 14.4). Deep ocean trenches mark these subduction zones, and volcanoes form where the magma erupts through vents and fissures in the overlying crust. Trenches and volcanic mountains ring the Pacific Ocean rim from Indonesia to Japan to

Creating the largest mountain range in the world, these ridges wind around the earth for 74,000 km (46,000 mi) (fig. 14.3). Although concealed from our view, this jagged range boasts higher peaks, deeper canyons, and sheerer cliffs than any continental

Alaska and down the west coast of the Americas, forming a so-called ring of fire where oceanic plates are being subducted under the continental plates. This ring is the source of more earthquakes and volcanic activity than any other region on the earth.

Over millions of years, continents can drift long distances. Antarctica and Australia once were connected to Africa, for instance, somewhere near the equator, and supported luxuriant forests. Geologists suggest that several times in the earth's history most or all of the continents have gathered to form supercontinents, which have ruptured and re-formed over hundreds

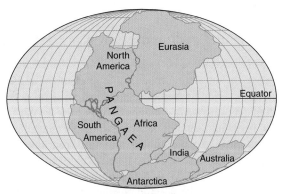

Pangaea, 250 MYA

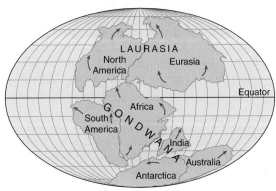

Laurasia and Gondwana, 210 MYA

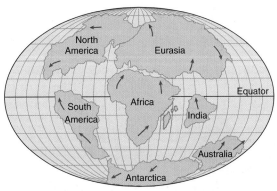

Most modern continents had formed by 65 MYA

FIGURE 14.5 Pangaea, an ancient supercontinent of 200 million years ago, combined all the world's continents in a single landmass. Continents have combined and separated repeatedly.

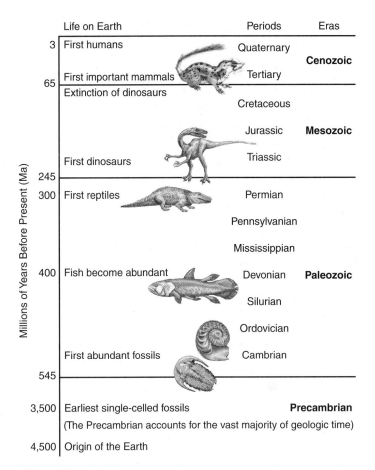

FIGURE 14.6 Periods and eras in geological time, and major life-forms that mark some periods.

of millions of years (fig. 14.5). The redistribution of continents has profound effects on the earth's climate and may help explain the periodic mass extinctions of organisms that mark the divisions between many major geological periods (fig. 14.6).

14.2 Rocks and Minerals

A **mineral** is a naturally occurring, inorganic, solid element or compound with a definite chemical composition and a regular internal crystal structure. "Naturally occurring" means not created by humans (or synthetic). Organic materials, such as coal, produced by living organisms or biological processes are generally not minerals. The two fundamental characteristics of a mineral that distinguish it from all other minerals are its chemical composition and its crystal structure. No two minerals are identical in both respects. Once purified, metals such as iron, aluminum, or copper lack a crystal structure, and thus are not minerals. The ores from which they are extracted, however, are minerals and make up an important part of economic mineralogy.

A **rock** is a solid, cohesive, aggregate of one or more minerals. Within the rock, individual mineral crystals (or grains) are mixed together and held firmly in a solid mass (fig. 14.7). The grains may be large or small, depending on how the rock was formed, but each

FIGURE 14.7 Crystals of different minerals create beautifully colored patterns in a rock sample seen in a polarizing microscope.

grain retains its own unique mineral qualities. Each rock type has a characteristic mixture of minerals (and therefore of different chemical elements), grain sizes, and ways in which the grains are mixed and held together. Granite, for example, is a mixture of quartz, feldspar, and mica crystals. Different kinds of granite have distinct percentages of these minerals and particular grain sizes, depending on how quickly the rock solidified. These minerals, in turn, are made up of a few elements such as silicon, oxygen, potassium, and aluminum.

The rock cycle creates and recycles rocks

What could be harder and more permanent than rocks? Like the continents they create, rocks are also part of a relentless cycle of formation and destruction. They are made and then torn apart, cemented together by chemical and physical forces, crushed, folded, melted, and recrystallized by dynamic processes related to those that shape the large-scale features of the crust. We call this cycle of creation, destruction, and metamorphosis the **rock cycle** (fig. 14.8). Understanding something of how this cycle works helps explain the origin and characteristics of different types of rocks, as well as how they are shaped, worn away, transported, deposited, and altered by geological forces.

There are three major rock classifications: igneous, sedimentary, and metamorphic. In this section we will look at how they are made and some of their properties.

Igneous Rocks

The most common rock-type in the earth's crust is solidified from magma, welling up from the earth's interior. These rocks are classed as **igneous rocks** (from *igni*, the Latin word for fire). Magma extruded to the surface from volcanic vents cools quickly to make basalt, rhyolite, andesite, and other fine-grained rocks. Magma that cools slowly in subsurface chambers or is intruded between overlying layers makes granite, gabbro, or other coarse-grained crystalline rocks, depending on its specific chemical composition.

Metamorphic Rocks

Preexisting rocks can be modified by heat, pressure, and chemical agents to create new forms called **metamorphic rock.** Deeply buried strata of igneous, sedimentary, and metamorphic rocks are subjected to great heat and pressure by deposition of overlying sediments or while they are being squeezed and folded by tectonic processes. Chemical reactions can alter both the composition and the structure of the rocks as they are metamorphosed. Some common metamorphic rocks are marble (from limestone), quartzite (from sandstone), and slate (from mudstone and shale). Metamorphic rocks are often the host rock for economically important minerals such as talc, graphite, and gemstones.

Weathering and sedimentation wear down rocks

Most of these crystalline rocks are extremely hard and durable, but exposure to air, water, changing temperatures, and reactive chemical agents slowly breaks them down in a process called **weathering** (fig. 14.9). *Mechanical weathering* is the physical breakup of rocks into smaller particles without a change in chemical composition of the constituent minerals. You have probably seen mountain valleys scraped by glaciers or river and shoreline pebbles that are rounded from being rubbed against one another as they are tumbled by waves and currents. *Chemical weathering* is the selective removal or alteration of specific components that leads to weakening and disintegration of rock. Among the more important chemical weathering processes are oxidation (combination of oxygen with an element to form an oxide or hydroxide mineral) and hydrolysis (hydrogen atoms from water molecules combine with other chemicals to form acids). The products of

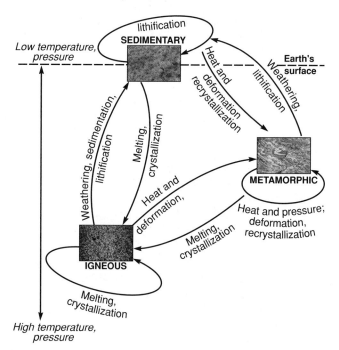

FIGURE 14.8 The rock cycle includes a variety of geological processes that can transform any rock.

FIGURE 14.9 Weathering slowly reduces an igneous rock to loose sediment. Here, exposure to moisture expands minerals in the rock, and frost may also force the rock apart.

FIGURE 14.10 Different colors of soft sedimentary rocks deposited in ancient seas during the Tertiary period 63 to 40 million years ago have been carved by erosion into the fluted spires and hoodoos of the Pink Cliffs of Bryce Canyon National Park.

these reactions are more susceptible to both mechanical weathering and dissolution in water. For instance, when carbonic acid (formed when CO_2 and H_2O combine) percolates through porous limestone layers in the ground, it dissolves the calcium carbonate (limestone) and creates caves.

Particles of rock are transported by wind, water, ice, and gravity until they come to rest again in a new location. The deposition of these materials is called **sedimentation.** Waterborne particles from sediments cover ocean continental shelves and fill valleys and plains. Most of the American Midwest, for instance, is covered with a layer of sedimentary material hundreds of meters thick in the form of glacierborne till (rock debris deposited by glacial ice), windborne loess (fine dust deposits), riverborne sand and gravel, and ocean deposits of sand, silt, and clay. Deposited

material that remains in place long enough, or is covered with enough material to compact it, may once again become rock. Some examples of **sedimentary rock** are shale (compacted mud), sandstone (cemented sand), tuff (volcanic ash), and conglomerates (aggregates of gravel, sand, silt, and clay).

Sedimentary rocks are also formed from crystals that precipitate out of, or grow from, a solution. An example is rock salt, made of the mineral halite, which is the name for ordinary table salt (sodium chloride). Salt deposits often form when a body of salt water dries up and salt crystals are left behind.

Many sedimentary formations have distinctive layers that show different conditions when they were laid down. Sedimentary rocks such as sandstone and limestone can be shaped by erosion into striking features (fig. 14.10). Geomorphology is the study of the processes that shape the earth's surface and the structures they create.

Humans have become a major force in shaping landscapes. Geomorphologist Roger Hooke, of the University of Maine, looking only at housing excavations, road building, and mineral production, estimates that we move about 30 to 35 gigatons (billion tons or Gt) per year worldwide. When combined with the 10 Gt each year that we add to river sediments through erosion, our earth-moving prowess is comparable to, or greater than, any other single geomorphic agent except plate tectonics.

14.3 ECONOMIC GEOLOGY AND MINERALOGY

The earth is unusually rich in mineral variety. Mineralogists have identified some 4,400 different mineral species—far more, we believe, than on any of our neighboring planets. What makes the difference? The processes of plate tectonics and the rock cycle on this planet have gradually concentrated uncommon elements and allowed them to crystallize into new minerals. But this accounts for only about one-third of our geological legacy. The biggest distinction is life. Most of our minerals are oxides, but there was little free oxygen in the atmosphere until it was released by photosynthetic organisms, thus triggering evolution of our great variety of minerals.

Economic mineralogy is the study of resources that are valuable for manufacturing and are, therefore, an important part of domestic and international commerce. Most economic minerals are metal-bearing ores, minerals with unusually high concentrations of metals. Lead, for example, generally comes from the mineral galena (PbS), and copper comes from sulfide ores, such as bornite (Cu5FeS4). Nonmetallic geological resources include graphite, feldspar, quartz crystals, diamonds, and other crystals that are valued for their usefulness or beauty. Metals have been so important in human affairs that major epochs of human history are commonly known by their dominant materials and the technology involved in using those materials (Stone Age, Bronze Age, Iron Age, etc.). The mining, processing, and distribution of these materials have broad implications for both our culture and our environment (fig. 14.11). We still are strongly dependent on the unique lightness, strength, and malleability of

FIGURE 14.11 The availability of metals and the ways we extract and use them have profound effects on our society and environment.

Table 14.2 Primary Uses of Some Major Metals Consumed in the United States

Metal	Use
Aluminum	Packaging foods and beverages (38%), transportation, electronics
Chromium	High-strength steel alloys
Copper	Building construction, electric and electronic industries
Iron	Heavy machinery, steel production
Lead	Leaded gasoline, car batteries, paints, ammunition
Manganese	High-strength, heat-resistant steel alloys
Nickel	Chemical industry, steel alloys
Platinum-group	Automobile catalytic converters, electronics, medical uses
Gold	Medical, aerospace, electronic uses; accumulation as monetary standard
Silver	Photography, electronics, jewelry

metals. Most economically valuable crustal resources exist everywhere in small amounts; the important thing is to find them concentrated in economically recoverable levels (table 14.2).

Metals are essential to our economy

The metals consumed in greatest quantity by world industry include iron (740 million metric tons annually), aluminum (40 million metric tons), manganese (22.4 million metric tons), copper and chromium (8 million metric tons each), and nickel (0.7 million metric tons). Most of these metals are consumed in the United States, Japan, and Europe, in that order. The largest sources are China, Australia, Russia, Canada, and the United States. To some extent the abundance of ores in these countries is simply a matter of land area, but Africa, which is roughly as large as North America, has relatively little metal ore.

The rapid growth of green technologies, such as renewable energy and electric vehicles, has made a group of rare earth metals especially important. Worries about impending shortages of these minerals complicate future developments in this sector (see Exploring Science, p. 303).

Nonmetal minerals include gravel, clay, sand, and gemstones

Nonmetal minerals are a broad class that covers resources from silicate minerals (gemstones, mica, talc, and asbestos) to sand, gravel, salts, limestone, and soils. Durable, highly valuable, and easily portable, gemstones and precious metals have long been a

way to store and transport wealth. Unfortunately these valuable materials also have bankrolled despots, criminal gangs, and terrorism in many countries. In recent years, brutal civil wars in Africa have been financed—and often motivated by—gold, diamonds, tantalum ore, and other high-priced commodities. Much of this illegal trade ends up in the $100-billion-per-year global jewelry trade, two-thirds of which sells in the United States. Many people who treasure a diamond ring or a gold wedding band as a symbol of love and devotion are unaware that it may have been obtained through inhumane labor conditions and environmentally destructive mining and processing methods. Civil rights organizations are campaigning to require better documentation of the origins of gems and precious metals to prevent their use as financing for crimes against humanity (see "Conflict Diamonds" at www.mhhe.com/environmentalscience).

In 2004 a group of Nobel Peace laureates called on the World Bank to overhaul its policies on lending for resource extractive industries. "War, poverty, climate change, and ongoing violations of human rights—all of these scourges are all too often linked to the oil and mining industries" wrote Archbishop Desmond Tutu, winner of the 1984 Nobel Peace Prize for helping to eliminate apartheid in South Africa. In response, the World Bank appointed an Extractive Industries Review headed by former Indonesian environment minister, Emil Salim. In its final report the committee recommended that some areas of exceptionally high biodiversity value should be "no-go" zones for extractive industries, and that the rights of those affected by extractive projects need better protection.

You might be surprised to learn that sand and gravel production comprise by far the greatest volume and dollar value of all nonmetal mineral resources and a far greater volume than all metal ores. Sand and gravel are used mainly in brick and concrete

Could shortages of a group of obscure minerals limit the growth of alternative energy supplies and green technology? A recent decision by China to limit exports of rare earth metals is seen by some experts as a serious threat to the global clean tech industry.

"Rare earth" metals are a collection of metallic elements, including scandium, yttrium, and fifteen lanthanides, such as neodymium, dysprosium, and gadolinium, that are essential in modern electronics. These metals are used in cell phones, high-efficiency lights, hybrid cars, superconductors, high-strength magnets, lightweight batteries, lasers, energy-conserving lamps, and a variety of medical devices. Because of their unusual properties, small amounts of these metals can make motors 90 percent lighter and lights 80 percent more efficient. Without these materials, MP3 players, hybrid vehicles, high-capacity wind turbines, and much other high-tech equipment would be impossible. A Toyota Prius, for example, uses about a kilogram of neodymium and dysprosium for its electric motor and as much as 15 kg of lanthanum for its battery pack.

Despite their name, these elements occur relatively widely in the earth's crust, but commercially viable concentrations are found in only a few locations. Just three nations, China, Russia, and the United States, posses more than 68 percent of the world's known reserves of these materials.

Currently China produces about 97 percent of all rare earth metals, an increase from about 30 percent two decades ago. China's dominance in mining these metals results partly from

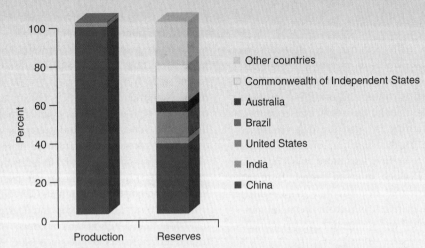

China controls a little more than one-third of known rare earth metals, but currently produces 97 percent of these important materials.

Source: USGS, 2010.

China's use of these materials in electronics production, partly from its low labor costs, and partly from the government's willingness to overlook the environmental damage in extracting and processing these metals. About half of all Chinese production of rare earth metals occurs in a single mine in Baotou in Inner Mongolia; most of the rest come from small, often unlicensed mines in southern China.

Like gold, silver, and other precious metals, rare earth elements are often separated from ore by crushing ore-bearing rocks and washing the ore in strong acids. These acids extract metals from the ore, but when the metals are later separated from the acid slurry, tremendous amounts of toxic wastewater are produced. Often acids are pumped directly into a borehole drilled in the ground, and metals are dissolved from ores in place. The resulting slurry is then pumped to the surface for processing. Acidic wastewater is frequently stored behind earthen dams, which can leak into surface and ground water. Processing also releases the sulfur and radioactive uranium and thorium that frequently occur with rare earth elements.

Having a near-monopoly on rare earth metals production is one reason China has emerged as the world leader in renewable energy (see chapter 1). In 2010, China cut its exports of these essential metals by nearly half. Officials said this was necessary to ensure that domestic electronic needs were met, but it also gave Chinese industry a tremendous advantage in the global market. Establishing better control on illegal mines (and collecting taxes on what is produced) is another reason for China's interest in controlling exports. Other nations are concerned about supplies for strategic needs (such as military guidance systems), as well as ordinary consumer applications. Many firms are simply moving to China. The division of General Motors that deals with miniaturized magnet research, for example, shut down its U.S. office and moved its entire staff to China in 2006. The Danish wind turbine company Vestas also moved much of its production to China in 2009.

In response to expected shortages and rising prices, several companies are working to reopen mines in North America and Australia. Molycorp Minerals expects to have its mine in Mountain Pass, California, back in production by 2012, meeting perhaps 10 percent of global demand, and Avalon Rare Metals of Toronto is working on a mine in Canada's Northwest Territories. Greenland is also jumping into this new gold rush, with hopes to produce up to 25 percent of rare earth metals from recently discovered ore bodies. It remains to be seen whether new environmental controls will be in place for this coming expansion.

construction and paving, as loose road filler, and for sandblasting. High-purity silica sand is our source of glass. These materials usually are retrieved from surface pit mines and quarries, where they were deposited by glaciers, winds, or ancient oceans.

14.4 ENVIRONMENTAL EFFECTS OF RESOURCE EXTRACTION

Each of us depends daily on geological resources mined from sites around the world. We use scores of metals and minerals, many of which we've never even heard of, in our lights, computers, watches, fertilizers, and cars. Mining and purifying all these resources can have severe environmental and social consequences. The most obvious effect of mining is often the disturbance or removal of the land surface. Farther-reaching effects, though, include air and water pollution. The EPA lists more than 100 toxic air pollutants, from acetone to xylene, released from U.S. mines every year. Nearly 80,000 metric tons of particulate matter (dust) and 11,000 tons of sulfur dioxide are released from nonmetal mining alone. Pollution from chemical and sediment runoff is a major problem in many local watersheds.

Mining can affect water quality in several ways. Gold and other metals are often found in sulfide ores. When these minerals are exposed to air and water, they produce sulfuric acid, which is highly mobile and strongly acidic. In addition, metal elements often occur in very low concentrations—10 to 20 parts per billion may be economically extractable for gold, platinum, and other metals. Consequently, vast quantities of ore must be crushed and washed to extract metals. A great deal of water is used in cyanide heap-leaching and other washing techniques. The USGS estimates that in arid Nevada, mining consumes about 230,000 m³ (60 million gal) of fresh water per day. After use in ore processing, much of this water contains sulfuric acid, arsenic, heavy metals, and other contaminants. Mine runoff leaking into lakes and streams damages or destroys aquatic ecosystems.

Nevertheless, public policy in the United States has encouraged mining on public lands as a way of boosting the economy and utilizing natural resources. Today many people think these laws seem outmoded and in need of reform (What Do You Think? p. 305).

Mining can have serious environmental impacts

There are many techniques for extracting geological materials. The most common methods are open-pit mining, strip mining, and underground mining. An ancient method of accumulating gold, diamonds, and coal is placer mining, in which pure nuggets are washed from stream sediments. Since the California gold rush of 1849, placer miners have used water cannons to blast away hillsides. This method, which chokes stream ecosystems with sediment, is still used in Alaska, Canada, and many other regions. Another ancient, and much more dangerous, method is underground mining. Mine tunnels occasionally collapse, and natural gas in coal mines can explode. Water seeping into mine shafts also dissolves toxic minerals. Contaminated

water seeps into groundwater; it is also pumped to the surface, where it enters streams and lakes.

In underground coal mines, another major environmental risk is fires. Hundreds of coal mines smolder in the United States, China, Russia, India, South Africa, and Europe. The inaccessibility and size of these fires make many impossible to extinguish or control. One mine fire in Centralia, Pennsylvania, has been burning since 1962; control efforts have cost at least $40 million, but the fire continues to expand. China, which depends on coal for much of its heating and electricity, has hundreds of smoldering mine fires; one has been burning for 400 years. According to a recent study from the International Institute for Aerospace Survey in the Netherlands, these fires consume up to 200 million tons of coal every year and emit as much carbon dioxide as all the cars in the United States. Toxic fumes, explosive methane, and other hazardous emissions are also released from these fires.

Open-pit mines are used to extract massive beds of metal ores and other minerals. The size of modern open pits can be hard to comprehend. The Bingham Canyon mine, near Salt Lake City, Utah, is 800 m (2,640 ft) deep and nearly 4 km (2.5 mi) wide at the top. More than 5 billion tons of copper ore and waste material have been removed from the hole since 1906. A chief environmental challenge of open-pit mining is that groundwater accumulates in the pit. In metal mines, a toxic soup results. No one yet knows how to detoxify these lakes, which endanger wildlife and nearby watersheds.

Half the coal used in the United States comes from surface or strip mines (fig. 14.12). Because coal is often found in expansive, horizontal beds, the entire land surface can be

FIGURE 14.12 Some giant mining machines stand as tall as a 20-story building and can scoop up thousands of cubic meters of rock per hour.

What Do You Think?

Should We Revise Mining Laws?

In 1872 the U.S. Congress passed the General Mining Law intended to encourage prospectors to open up the public domain and promote commerce. This law, which has been in effect more than a century, says, "All valuable mineral deposits in lands belonging to the United States are hereby declared to be free and open to exploration and purchase . . . by citizens of the United States." Claim holders can "patent" (buy) the land for $2.50 to $5 per acre (0.4 hectares), depending on the type of claim. Once the patent fee is paid, the owners can do anything they want with the land, just like any other private property. Although $2.50 per acre may have been a fair market value in 1872, many people regard it as ridiculously low today, amounting to a scandalous give-away of public property.

In Nevada, for example, a Canadian mining company paid $9,000 for federal land that contains an estimated $20 billion worth of precious metals. Similarly, Colorado investors bought about 7,000 ha (17,000 acres) of rich oil-shale land in 1986 for $42,000 and sold it a month later for $37 million. You don't actually have to find any minerals to patent a claim. A Colorado company paid a total of $400 for 65 ha (160 acres) it claimed would be a gold mine. Almost 20 years later, no mining has been done, but the property—which just happens to border the Keystone Ski Area—is being subdivided for condos and vacation homes.

According to the Bureau of Land Management (BLM), some $4 billion in minerals are mined each year on U.S. public lands. Under the 1872 law, mining companies don't pay a penny for the ores they take. Furthermore, they can deduct a depletion allowance from taxes on mineral profits. Former Senator Dale Bumpers of Arkansas, who calls the 1872 mining law "a license to steal," has estimated that the government could derive $320 million per year by charging an 8 percent royalty on all minerals and probably could save an equal amount by requiring larger bonds to be posted to clean up after mining is finished. The Meridian Gold Company, for example, has posted a $2 million bond for cleaning up the Beartrack Gold Mine in Idaho. Reclamation, however, is expected to cost 15 times that amount. Chapter 13 has more information on how reclamation and restoration can return damaged sites to beneficial uses.

On the other hand, mining companies argue they would be forced to close down if they had to pay royalties or post larger bonds. Many people would lose jobs and the economies of western mining towns would collapse if mining becomes uneconomic. We provide subsidies and economic incentives to many industries to stimulate economic growth. Why not support mining for metals essential for our industrial economy? Mining is a risky and expensive business. Without subsidies, mines would close down and we would be completely dependent on unstable foreign supplies.

Mining critics respond that other resource-based industries have been forced to pay royalties on materials they extract from public lands. Coal, oil, and gas companies pay 12.5 percent royalties on fossil fuels obtained from public lands. Timber companies—although they don't pay the full costs of the trees they take—have to bid on logging sales and clean up when they are finished. Even gravel companies pay for digging up the public domain. Ironically, we charge for digging up gravel, but give gold away free.

Over the past decade, numerous bills have been introduced in Congress to revise the mining law. Those supported by environmental groups generally would require companies mining on federal lands to pay a higher royalty on their production. They also would eliminate the patenting process, impose stricter reclamation requirements, and give federal managers authority to deny inappropriate permits. In contrast, bills offered by western legislators, and enthusiastically backed by mining supporters, would leave most provisions of the 1872 bill in place.

Currently Rosemont Coppera—a subsidiary of a Canadian company—wants to open a large, open-pit copper mine on 3,670 acres (1,485 ha) of public land in the Santa Rita Mountains south of Tucson, Arizona. Three-quarters of this project would be on a national forest. The company says the mine will boost the local economy, but opponents argue that it will use precious groundwater and threaten tourism, biodiversity, and archeological sites. The Forest Service has said it may deny necessary permits. If so, it will be the first serious attempt in more than a century to deny mining on federal land. What do you think we should do about this mining law? Does it remain necessary and justifiable or should it be revoked or revised??

Rosemont Copper, a subsidiary of Canada-based Augusta Resource Corp., wants to dig a large open-pit mine on 3,670 acres of land in the Santa Rita Mountains south of Tucson, Arizona. If blocked by the U.S. Forest Service, this would be the first serious attempt in a century to deny a mine on public land.

stripped away to cheaply and quickly expose the coal. The overburden, or surface material, is placed back into the mine, but usually in long ridges called spoil banks. Spoil banks are very susceptible to erosion and chemical weathering. Because the spoil banks have no topsoil (the complex organic mixture that supports vegetation—see chapter 9), revegetation occurs very slowly.

The 1977 federal Surface Mining Control and Reclamation Act (SMCRA) requires better restoration of strip-mined lands, especially where mines replaced prime farmland. Since then, the record of strip-mine reclamation has improved substantially. Complete mine restoration is expensive, often more than $10,000 per hectare. Restoration is also difficult because the developing soil is usually acidic and compacted by the heavy machinery used to reshape the land surface.

Bitter controversy has grown recently over mountaintop removal, a coal mining method mainly practiced in Appalachia. Long, sinuous ridge-tops are removed by giant mining machines to expose horizontal beds of coal (fig. 14.13). Up to 215 m (700 ft) of ridge-top is pulverized and dumped into adjacent river valleys. The debris can be laden with selenium, arsenic, coal, and other toxic substances. At least 900 km (560 mi) of streams have been buried in West Virginia alone. Environmental lawyers have sued to stop the destruction of streams, arguing that it violates the Clean Water Act (chapter 24). In 2011 the EPA stopped one of the largest proposed mountaintop mines in U.S. history by revoking permits for the Spruce No. 1 in West Virginia to dispose of mining waste in local waterways. Environmentalists cheered this decision, while mining proponents threatened a lawsuit to overturn the decision.

The Mineral Policy Center in Washington, D.C., estimates that 19,000 km (12,000 mi) of rivers and streams in the United States are contaminated by mine drainage. The EPA estimates that cleaning up impaired streams, along with 550,000 abandoned mines, in the United States may cost $70 billion. Worldwide, mine closing and rehabilitation costs are estimated in the trillions of dollars. Because of the volatile prices of metals and coal, many mining companies have gone bankrupt before restoring mine sites, leaving the public responsible for cleanup.

Meanwhile, some mining companies have acknowledged that in the future they will increasingly be held liable for environmental damages. They say they're seeking ways to improve the industry's social and environmental record. Mine executives also recognize that big cleanup bills will increasingly cut into company values and stock prices. Finding creative ways to keep mines cleaner from the start will make good economic sense, even if it's not easy to do.

Processing ores also has negative effects

Metals are extracted from ores by heating or with chemical solvents. Both processes release large quantities of toxic materials that can be even more environmentally hazardous than mining. **Smelting**—roasting ore to release metals—is a major source of air pollution. One of the most notorious examples of ecological devastation from smelting is a wasteland near Ducktown, Tennessee. In the mid-1800s, mining companies began excavating the rich copper deposits in the area. To extract copper from the ore, they built huge, open-air wood fires, using timber from the surrounding forest. Dense clouds of sulfur dioxide released from sulfide ores poisoned the vegetation and acidified the soil over a 50 mi^2 (13,000 ha) area. Rains washed the soil off the denuded land, creating a barren moonscape.

Sulfur emissions from Ducktown smelters were reduced in 1907 after Georgia sued Tennessee over air pollution. In the 1930s the Tennessee Valley Authority (TVA) began treating the soil and replanting trees to cut down on erosion. Recently, upward of $250,000 per year has been spent on this effort. Although the trees and other plants are still spindly and feeble, more than two-thirds of the area is considered "adequately" covered with vegetation. Similarly, smelting of copper-nickel ore in Sudbury, Ontario, a century ago caused widespread ecological destruction that is slowly being repaired following pollution-control measures (see fig. 16.19).

Heap-leach extraction, which is often used to get metals from low-grade ore, has a high potential for environmental contamination. A cyanide solution sprayed on a large pile of ore (fig. 14.14) to dissolve gold can leak into surface or ground water. A case

FIGURE 14.13 Mountaintop removal mining is a highly destructive, and deeply controversial, method of extracting Appalachian coal.

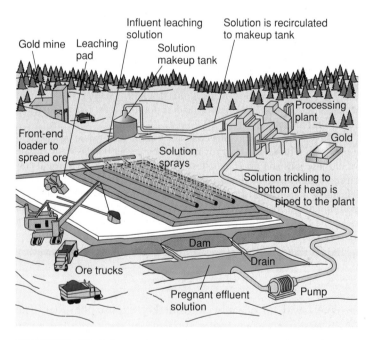

FIGURE 14.14 In a heap-leach operation, huge piles of low-grade ore are heaped on an impervious pad and sprayed with a cyanide-solution. As the leaching solution trickles through the crushed ore, it extracts gold and other precious metals. This technique is highly profitable but carries large environmental risks.

in point is the Beartrack Mine in east-central Idaho. After extracting $220 million in gold, the mine was closed in 2000, leaving about 15 million metric tons of mine waste and huge ponds of cyanide. Meridian Gold, the company that owned the mine, has posted a $2 million bond for reclamation, but it may well cost 30 times that much to restore the area. Meanwhile the acids, metals, and cyanide in the heap threaten the nearby Salmon River and its $2 billion annual tourism and fishing industry.

Every other year the Blacksmith Institute compiles a list of the world's worst pollution problems. For 2010 they analyzed the top six toxic threats to human health. Five of the six are metals or metaloids (lead, mercury, chromium, radionuclides, and arsenic), and the largest pollution problems are generally associated with mining and processing. Altogether they estimate about 60 million people are jeopardized by these pollution sources. These problems are especially disastrous in the developing world and in the former Soviet Union, where funds and political will aren't available to deal with pollution or help people suffering from terrible health effects of pollution.

14.5 CONSERVING GEOLOGICAL RESOURCES

Conservation offers great potential for extending our supplies of economic minerals and reducing the effects of mining and processing. The advantages of conservation are also significant: less waste to dispose of, less land lost to mining, and less consumption of money, energy, and water resources.

Recycling saves energy as well as materials

Some waste products already are being exploited, especially for scarce or valuable metals. Aluminum, for instance, must be extracted from bauxite ore by electrolysis, an expensive, energy-intensive process. Recycling waste aluminum, such as beverage cans, on the other hand, consumes one-twentieth of the energy of extracting new aluminum. Today nearly two-thirds of all aluminum beverage cans in the United States are recycled, up from only 15 percent 20 years ago. The high value of aluminum scrap ($650 a ton versus $60 for steel, $200 for plastic, $50 for glass, and $30 for paperboard) gives consumers plenty of incentive to deliver their cans for collection. Recycling is so rapid and effective that half of all the aluminum cans now on a grocer's shelf will be made into another can within two months. The energy cost of extracting other materials is shown in table 14.3.

Platinum, the catalyst in automobile catalytic exhaust converters, is valuable enough to be regularly retrieved and recycled from used cars (fig. 14.15). Other metals commonly recycled are gold, silver, copper, lead, iron, and steel. The latter four are readily available in a pure and massive form, including copper pipes, lead batteries, and steel and iron auto parts. Gold and silver are valuable enough to warrant recovery, even through more difficult means. See chapter 21 for further discussion of this topic.

Although total U.S. steel production has fallen in recent decades—largely because of inexpensive supplies from new and efficient Japanese steel mills—a new type of mill subsisting

Table 14.3 Energy Requirements in Producing Various Materials from Ore and Raw Source Materials

Product	Energy Requirement (Mj/Kg)	
	New	From Scrap
Glass	25	25
Steel	50	26
Plastics	162	n.a.
Aluminum	250	8
Titanium	400	n.a.
Copper	60	7
Paper	24	15

Source: E. T. Hayes, *Implications of Materials Processing*, 1997.

FIGURE 14.15 The richest metal source we have—our mountains of scrapped cars—offers a rich, inexpensive, and ecologically beneficial resource that can be "mined" for a number of metals.

(a) A meteor impact crater in Arizona

(b) Volcanic eruption

(c) Glacier in Alaska

FIGURE 14.16 Geological events such as meteor or asteroid impacts (a), massive volcanic eruptions (b), or climate change (c) are thought to trigger mass extinctions that mark major eras in the earth's history.

entirely on a readily available supply of scrap/waste steel and iron is a growing industry. Minimills, which remelt and reshape scrap iron and steel, are smaller and cheaper to operate than traditional integrated mills that perform every process from preparing raw ore to finishing iron and steel products. Minimills produce steel at $225 to $480 per metric ton, whereas steel from integrated mills averages $1,425 to $2,250 per metric ton. The energy cost is likewise lower in minimills: 5.3 million BTU/ton of steel compared to 16.08 million BTU/ton in integrated mill furnaces. Minimills now account for about half of all of U.S. steel production. Recycling is slowly increasing as raw materials become more scarce and wastes become more plentiful.

New materials can replace mined resources

Mineral and metal consumption can be reduced by new materials or new technologies developed to replace traditional uses. This is a long-standing tradition; for example, bronze replaced stone technology and iron replaced bronze. More recently, the introduction of plastic pipe has decreased our consumption of copper, lead, and steel pipes. In the same way, the development of fiber-optic technology and satellite communication reduces the need for copper telephone wires.

Iron and steel have been the backbone of heavy industry, but we are now moving toward other materials. One of our primary uses for iron and steel has been machinery and vehicle parts. In automobile production, steel is being replaced by polymers (long-chain organic molecules similar to plastics), aluminum, ceramics, and new, high-technology alloys. All of these reduce vehicle weight and cost while increasing fuel efficiency. Some of the newer alloys that combine steel with titanium, vanadium, or other metals wear much better than traditional steel. Ceramic engine parts provide heat insulation around pistons, bearings, and cylinders, keeping the rest of the engine cool and operating efficiently. Plastics and polymers reinforced with glass fiber are used in body parts and some engine components.

Electronics and communications (telephone) technology, once major consumers of copper and aluminum, now use ultra-high-purity glass cables to transmit pulses of light, instead of metal wires carrying electron pulses. Once again, this technology has been developed for its greater efficiency and lower cost, but it also affects consumption of our most basic metals.

14.6 GEOLOGICAL HAZARDS

Earthquakes, volcanic eruptions, floods, and landslides are among the geological forces that have shaped the world around us (fig. 14.16). Catastrophic events, such as the impact of a giant asteroid 65 million years ago off the coast of what is now Yucatán, or volcanic eruptions, such as those that covered 2 million km^2 of Siberia with basalt up to 2 km deep 250 million years ago, are thought to have triggered the mass extinctions that mark transitions between major historic eras. The asteroid impact 65 million years ago that ended the age of the dinosaurs is calculated to have created a tsunami hundreds of meters high that could have swept around the world several times before subsiding. This impact also ejected so much dust into the air that sunlight was blocked for years and a global winter decimated much of the life on the earth.

Fortunately such massive events are rare. Still, geological hazards are a huge threat. Among direct natural disasters, floods take the largest number of human lives, while windstorms (hurricanes, cyclones, tornadoes) cause the greatest property damage.

Earthquakes are frequent and deadly hazards

The 2010 earthquake in Haiti wasn't the world's worst geological disaster. A far larger toll is thought to have been caused by a 1976 earthquake in Tangshan, China. Government officials reported 655,000 deaths, although some geologists doubt it was that high.

Earthquakes are sudden movements in the earth's crust that occur along faults (planes of weakness) where one rock mass slides past another one, as was the case along the Enriquillo–Plantain Garden Fault in Haiti. The San Andreas

---

Think About It

What are the most dangerous geological hazards where you live? Are you aware of emergency preparedness plans? How would you evacuate your area if it becomes necessary?

---

fault in California is one of the most notorious and highly visible faults in the world (fig. 14.17). It runs about 810 miles (1,300 km) from the Salton Sea in southern California to Point Reyes, just north of San Francisco, where it veers offshore to follow the coast north nearly to Oregon. It marks the boundary where the Pacific plate is moving slowly to the northwest while the North American plate, to the east, is moving in the opposite direction. A number of major quakes have originated along this fault, including those that struck San Francisco in 1906 and 1989. When movement along faults occurs gradually and relatively smoothly, it is called creep or seismic slip and may be undetectable to the casual observer. When friction prevents rocks from slipping easily, stress builds up until it is finally released with a sudden jerk, which can cause the ground to shake like jelly. The point on a fault at which the first movement occurs during an earthquake is called the epicenter.

Table 14.4 Worldwide Frequency of Earthquakes

Richter Scale Magnitude*	Description	Average Number per Year
2–2.9	Unnoticeable	300,000
3–3.9	Smallest felt	49,000
4–4.9	Minor earthquake	6,200
5–5.9	Damaging earthquake	800
6–6.9	Destructive earthquake	120
7–7.9	Major earthquake	18
>8	Great earthquake	1 or 2

*For every unit increase in the Richter scale, ground displacement increases by a factor of 10, while energy release increases by a factor of 30. There is no upper limit to the scale, but the largest earthquake recorded was 9.5, in Chile in 1960.

Source: B. Gutenberg in *Earth* by F. Press and R. Siever, 1978, W. H. Freeman & Company.

Earthquakes have always seemed mysterious, sudden, and violent, coming without warning and leaving in their wake ruined cities and dislocated landscapes (table 14.4). Cities like San Francisco or Port-au-Prince, parts of which are built on soft landfill or poorly consolidated soil, usually suffer the greatest damage from earthquakes. Water-saturated soil can liquefy when shaken. Buildings sometimes sink out of sight or fall down like a row of dominoes under these conditions.

The worst death toll usually occurs in cities with poorly constructed buildings (fig. 14.18). Today contractors in earthquake zones are attempting to prevent damage and casualties by constructing buildings that can withstand tremors. The primary methods used are heavily reinforced structures, strategically placed weak spots in the building that can absorb vibration from the rest of the building, and pads or floats beneath the building on which it can shift harmlessly with ground motion.

FIGURE 14.17 The Elkhorn Scarp shows where the San Andreas fault runs across the Carrizo Plain northwest of Los Angeles.

FIGURE 14.18 Earthquakes are most devastating where building methods can't withstand shaking. The 2010 earthquake in Haiti is thought to have killed 230,000 people. Collapsed buildings, food and water shortages, infectious diseases, and exposure to the elements all contributed to the death toll.

As you can see in figure 14.19, the most seismically active region in the lower 48 U.S. states is along the West Coast, where tectonic plates are colliding. Nearly every state has had significant tremors, however. The largest North American earthquake in recorded history occurred, surprisingly, in New Madrid, Missouri, in 1811. The tremor, which had an estimated magnitude of 8.8, made bells ring as far away as Boston and caused the Mississippi River to run backward for several hours. The region was sparsely inhabited, so there were few deaths. If a similar quake occurred there today, a large part of Memphis would be destroyed.

There's evidence that human activities can trigger earthquakes. The raising and lowering of water levels in reservoirs behind large dams often correlate with increased seismic activity. Chinese geologists, for example, suspect that the 2008 earthquake that killed 69,000 people in Sichuan Province may have been triggered by the recent construction of the Zipingpu Dam on the Min River only a few kilometers from the earthquake epicenter. Similarly, injection of fluids into deep wells also correlates with increased earth tremors. In 2009 two geothermal deep-well projects—one in Switzerland and the other in California—were abruptly canceled when earthquakes in the immediate vicinity suddenly intensified.

Tsunamis can be more damaging than the earthquakes that trigger them

Tsunamis are giant sea waves triggered by earthquakes or landslides. The name is derived from the Japanese for "harbor wave," because the waves often are noticed only when they approach shore. Tsunamis can be more damaging than the earthquakes that create them. The tsunami that struck the coast of Japan on March 11, 2011, for example, was triggered by a

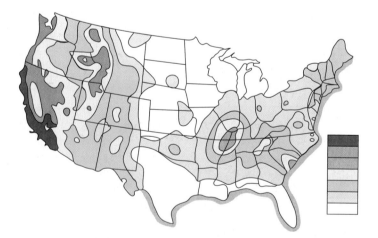

FIGURE 14.19 A seismic map of the lower 48 states shows the risk of earthquakes. Although the highest risk is along the Pacific coast and the Rocky Mountains, you may be surprised to see that an area along the Mississippi River around New Madrid, Missouri, also has a high potential for seismic activity.

FIGURE 14.20 In 2011 a magnitude 9.0 earthquake just off the coast of Japan created a massive tsunami that destroyed homes, killed at least 25,000 people, and damaged four nuclear power plants.

magnitude 9.0 underwater earthquake about 72 km (45 mi) out to sea. Although the earthquake shook buildings, it did relatively little damage because Japan has high construction standards. The waves that followed, however, were up to 38 m (124 ft) high. In some low-lying areas, the waves traveled up to 10 km (6 mi) inland, washing away buildings, boats, cars, and even whole trains. As the wall of water smashed through towns and villages, it carried a thick slurry of vehicles, building materials, mud, and debris that was much more destructive than the water alone. At least 25,000 people were listed as dead or missing, hundreds of thousands of buildings were damaged or destroyed (many by fires resulting from broken gas lines; fig 14.20) and millions of residents were without electricity or water.

An effect of the tsunami that will be very long-lasting is the damage it did to nuclear power plants along Japan's northeast coast. The reactors shut down after the earthquake, as they were designed to do, but the tsunami destroyed the cooling systems (as well as the backup generators) needed to keep the reactors under control. Three of the six reactors at the Fukushima Dai-ichi complex were damaged by fires and hydrogen gas explosions after partial meltdown of the reactor cores, and a fourth was damaged when water levels dropped in the waste fuel storage pool, exposing fuel rods to the air.

The Japanese government ranked the disaster as a seven—the same as the 1986 Chernobyl catastrophe—on the international scale for nuclear accidents. Residents were evacuated from all areas within 20 km (12 mi) of the damaged reactors, and thousands who live just outside that area were also advised to consider leaving. Elevated radiation levels have been detected in milk, vegetables, seafood, and some water supplies from regions surrounding the damaged nuclear plants. Officials have banned the sale of some food items and advised residents of a village northeast of the plant not to drink tap water. The government has estimated economic losses of as much as $300 billion, not including costs such as the cleanup of the nuclear reactors, several of which will probably have to be entombed in concrete forever.

Volcanoes eject gas and ash, as well as lava

Volcanoes and undersea magma vents produce much of the earth's crust. Over hundreds of millions of years, gaseous emissions from these sources formed the earth's earliest oceans and atmosphere. Many of the world's fertile soils are weathered volcanic materials. Volcanoes have also been an ever-present threat to human populations. One of the most famous historic volcanic eruptions was that of Mount Vesuvius in southern Italy, which buried the cities of Herculaneum and Pompeii in A.D. 79. The mountain had been giving signs of activity before it erupted, but many citizens chose to stay and take a chance on survival. On August 24, the mountain buried the two towns in ash. Thousands were killed by the dense, hot, toxic gases that accompanied the ash flowing down from the volcano.

Today more than 500 million people live in the danger zones around volcanoes. Many assume that because the volcano hasn't erupted for years, it never will again. They don't realize that the time between eruptions can be centuries, but the next big one could come at any time without warning.

Nuées ardentes (French for "glowing clouds") are deadly, denser-than-air mixtures of hot gases and ash like those that inundated Pompeii and Herculaneum. These pyroclastic clouds can have temperatures exceeding 1,000°C, and they move at more than 100 km/hour (60 mph). In November 2010, pyroclastic clouds rolled down the slopes of Mt. Merapi just outside Yogyakarta, Indonesia. At least 325 people were killed, and more than 300,000 were forced from their homes (fig. 14.21).

Mudslides are also disasters sometimes associated with volcanoes. The 1985 eruption of Nevado del Ruiz, 130 km (85 mi) northwest of Bogotá, Colombia, caused mudslides that buried most of the town of Armero and devastated the town of Chinchina. An estimated 25,000 people were killed. Volcanic eruptions can release enough ash and dust into the air to change global climate, at least temporarily. In 1815, Mt. Tambora, on the Indonesian island of Sumatra, expelled 175 km^2 of dust and ash. These dust clouds circled the globe and caused such a reduction in sunlight and air temperatures that 1815 was known as the year without a summer.

It is not just a volcano's dust that blocks sunlight. Sulfur emissions from volcanic eruptions combine with rain and atmospheric moisture to produce sulfuric acid (H_2SO_4). The resulting droplets of H_2SO_4 interfere with solar radiation and can significantly cool the world climate. In 1991, Mt. Pinatubo in the Philippines emitted 20 million tons of sulfur dioxide, which combined with water to form tiny droplets of sulfuric acid. This acid aerosol reached the stratosphere, where it circled the globe for two years. This thin haze cooled the entire earth by 1°C and postponed global warming for several years. It also caused a 10 to 15 percent reduction in stratospheric ozone, allowing increased ultraviolet light to reach the earth's surface.

Landslides are examples of mass wasting

Gravity constantly pulls downward on every material everywhere on earth, causing a variety of phenomena collectively termed **mass wasting** or mass movement, in which geological materials are moved downslope from one place to another. The resulting movement is often slow and subtle, but some slope processes such as rockslides, avalanches, and land slumping can be swift, dangerous, and very obvious. *Landslide* is a general term for rapid downslope movement of soil or rock. In the United States alone, over $1 billion in property damage is done every year by landslides and related mass wasting.

In some areas active steps are taken to control landslides or to limit the damage they cause. On the other hand, many human activities such as road construction, forest clearing, agricultural cultivation, and building houses on steep, unstable slopes increase both the frequency and the damage done by landslides. In some cases people are unaware of the risks they face by locating on or under unstable hillsides. In other cases they simply deny clear and obvious danger. Southern California, where people build expensive houses on steep hills of relatively unconsolidated soil, is often the site of large economic losses from landslides. Chapparal fires expose the soil to heavy winter rains. Resulting mudslides carry away whole neighborhoods and bury downslope areas in debris flows (fig. 14.22). Often there's warning that slopes are becoming unstable. In Brazil, however, more than 1,000 people were killed in 2011 when landslides washed away favelas (shanty towns), on steep slopes above Rio de Janeiro.

Floods are the greatest geological hazard

Like earthquakes and volcanoes, floods are normal events that cause damage when people get in the way. As rivers carve and shape the landscape, they build broad floodplains, level expanses

FIGURE 14.21 Volcanic ash covers damaged houses and dead vegetation after the November 2010 eruption of Mt. Merapi *(background)* in central Java. More than 300,000 residents were displaced and at least 325 were killed by this multiday eruption.

FIGURE 14.22 Damaged homes sit on a hillside after a landslide in Laguna Beach, Calif.

FIGURE 14.23 In 2004, unusually strong monsoon rains caused rivers to overflow their banks, and about one-quarter of Bangladesh was under water. About 25 million people were displaced.

that are periodically inundated. Many cities have been built on these flat, fertile plains, which are both good for agriculture and convenient to the river. When floods occur irregularly, people develop a false sense of security. But eventually most floodplains do flood.

Among direct natural disasters, floods take the largest number of human lives and cause the most property damage. A flood on the Yangtze River in China in 1931 killed 3.7 million people, making it the most deadly natural disaster in recorded history. In another flood, on China's Yellow River in 1959, about 2 million people died, mostly due to resultant famine and disease.

Some more recent flooding disasters include Bangladesh in 2004, where 25 million people were displaced when torrential monsoon rains flooded about one-fourth of the low-lying country (fig. 14.23). Similarly, in 2008 cyclone Nargis hit the Irrawaddy

Delta of Burma (officially known as Myanmar). The government made the situation worse by refusing international assistance. It took weeks for aid to reach some areas. An estimated 140,000 people died both directly from the flood and from starvation and disease that followed. And in 2011, after seven years of the worst drought in recorded history, Australia had torrential rains that flooded an area the size of South Africa. Australia has vastly better public services and relief efforts than Burma or Bangladesh, so the death toll in Australia was relatively low, but the economic losses ran into the billions.

Are these recent disasters related to global climate change? Many climate scientists predict that global warming will cause more extreme weather events, including both severe droughts in some places and more intense rainfall in others. In addition, many other human activities increase the severity and frequency of floods. Covering the land with hardened surfaces, such as roads, parking lots, and building roofs, reduces water infiltration into the soil and speeds the rate of runoff into streams and lakes. Clearing forests for agriculture and destroying natural wetlands also increases both the volume and the rate of water discharge after a storm.

In an effort to control floods, many communities build levees and floodwalls to keep water within riverbanks, and river channels are dredged and deepened to allow water to recede faster. Every flood-control structure simply transfers the problem downstream, however. The water has to go somewhere. If it doesn't soak into the ground upstream, it will simply exacerbate floods somewhere downstream.

Rather than spend money on levees and floodwalls, many people think it would be better to restore wetlands, replace groundcover on water courses, build check dams on small streams, move buildings off the floodplain, and undertake other nonstructural ways of reducing flood danger. According to this view, floodplains should be used for wildlife habitat, parks, recreation areas, and other uses not susceptible to flood damage.

The National Flood Insurance Program administered by the Federal Emergency Management Agency (FEMA) was intended to aid people who cannot buy insurance at reasonable rates, but its effects have been to encourage building on the floodplains by making people feel that, whatever happens, the government will take care of them. Many people would like to relocate homes and businesses out of harm's way after the recent floods or to improve them so they will be less susceptible to flooding, but owners of damaged property can collect only if they rebuild in the same place and in the same way as before. This perpetuates problems rather than solves them.

Beaches are vulnerable

Beach erosion occurs on all sandy shorelines because the motion of the waves is constantly redistributing sand and other sediments. One of the world's longest and most spectacular sand beaches runs down the Atlantic Coast of North America from New England to Florida and around the Gulf of Mexico. Much of this beach lies on some 350 long, thin barrier islands that stand between the mainland and the open sea.

Early inhabitants recognized that exposed, sandy shores were hazardous places to live, and they settled on the bay side of barrier islands or as far upstream on coastal rivers as was practical. Modern residents, however, place a high value on living where they have an ocean view and ready access to the beach. Construction directly on beaches and barrier islands can cause irreparable damage to the whole ecosystem. Under normal circumstances, fragile vegetative cover holds the shifting sand in place. Damaging this vegetation and breaching dunes with roads can destabilize barrier islands. Storms then wash away beaches or even whole islands. Hurricane Katrina in 2005 caused $100 billion in property damage along the Gulf Coast of the United States, mostly from the storm washing

FIGURE 14.24 The aftermath of Hurricane Katrina on Dauphin Island, Alabama. Since 1970 this barrier island at the mouth of Mobile Bay has been overwashed at least five times. More than 20 million yd^3 (15 million m^3) of sand has been dredged or trucked in to restore the island. Some beach houses have been rebuilt, mostly at public expense, five times. Does it make sense to keep rebuilding in such an exposed place?

over barrier islands and coastlines (fig. 14.24). In the future, intensified storms and rising sea levels caused by global warming will make barrier islands and low-lying coastal areas even riskier places to live.

Cities and individual property owners often spend millions of dollars to protect beaches from erosion and repair damage after storms. Sand is dredged from the ocean floor or hauled in by the truckload, only to wash away again in the next storm. Building artificial barriers, such as groins or jetties, can trap migrating sand and build beaches in one area, but they often starve downstream beaches and make erosion there even worse.

As is the case for inland floodplains, many government policies encourage people to build where they probably shouldn't. Subsidies for road building and bridges, support for water and sewer projects, tax exemptions for second homes, flood insurance, and disaster relief are all good for the real estate and construction businesses but invite people to build in risky places. In some areas, beach houses have been rebuilt, mostly at public expense, four or five times in the past 20 years. Does it make sense to keep rebuilding in these places?

The Coastal Barrier Resources Act of 1982 prohibited federal support, including flood insurance, for development on sensitive islands and beaches. In 1992, however, the U.S. Supreme Court ruled that ordinances forbidding floodplain development amount to an unconstitutional "taking," or confiscation, of private property.

CONCLUSION

We need materials from the earth to sustain our modern lifestyle, but many of the methods we use to get those materials have severe environmental consequences. Still, there are ways that we can extend resources through recycling and the development of new materials and more efficient ways of using them. We can do much to repair the damage caused by resource extraction, although open-pit mines and mountains with their tops removed will never be returned to their original, pristine condition. Water contamination is one of the main environmental costs of mining. Pollutants include acids, cyanide, mercury, heavy metals, and sediment. Air pollution from smelting also can cause widespread damage. We'll discuss this risk further in chapter 16.

We also should be aware of geological hazards, such as floods, earthquakes, volcanoes, and landslides. Because these hazards often occur on a geological time scale, residents who haven't experienced one of these catastrophic events assume that they never will. People move into highly risky places without considering what the consequences may be of ignoring nature. The earth may seem extremely stable to us, but it's really a highly dynamic system. We remain ignorant of these forces at our own peril.

REVIEWING LEARNING OUTCOMES

By now you should be able to explain the following points:

14.1 Summarize the processes that shape the earth and its resources.

- Earth is a dynamic planet.
- Tectonic processes reshape continents and cause earthquakes.

14.2 Explain how rocks and minerals are formed.

- The rock cycle creates and recycles rocks.
- Weathering and sedimentation wear down rocks.

14.3 Think critically about economic geology and mineralogy.

- Metals are essential to our economy.
- Nonmetal minerals include gravel, clay, sand, and gemstones.

14.4 Critique the environmental effects of resource extraction.

- Mining can have serious environmental impacts.
- Processing ores also has negative consequences.

14.5 Discuss ways we could conserve geological resources.

- Recycling saves energy as well as materials.
- New materials can replace mined resources.

14.6 Describe geological hazards.

- Earthquakes are frequent and deadly hazards.
- Tsunamis can be deadlier than the earthquakes that trigger them.
- Volcanoes eject gas and ash, as well as lava.
- Landslides are examples of mass wasting.
- Floods are the greatest geological hazard.

PRACTICE QUIZ

1. Describe the layered structure of the earth.
2. Define *mineral* and *rock*.
3. What are tectonic plates and why are they important to us?
4. Why are there so many volcanoes, earthquakes, and tsunamis along the "ring of fire" that rims the Pacific Ocean?
5. Describe the rock cycle and name the three main rock types that it produces.
6. Why are rare earth metals important and what concerns do we have about supplies?
7. Give some examples of nonmetal mineral resources and describe how they are used.
8. Describe some ways metals and other mineral resources can be recycled.
9. What are some environmental hazards associated with mineral extraction?
10. Describe some of the leading geological hazards and their effects.

CRITICAL THINKING AND DISCUSSION QUESTIONS

1. Look at the walls, floors, appliances, interior, and exterior of the building around you. How many earth materials were used in their construction?

2. Is your local bedrock igneous, metamorphic, or sedimentary? If you don't know, who might be able to tell you?

3. Suppose you live in a small, mineral-rich country, and a large, foreign mining company proposes to mine and market your minerals. How should revenue be divided between the company, your government, and citizens of the country? Who bears the greatest costs? How should displaced people be compensated? Who will make sure that compensation is fairly distributed?

4. Geological hazards affect a minority of the population who build houses on unstable hillsides, in flood-prone areas, or on faults. What should society do to ensure safety for these people and their property?

5. A persistent question in this chapter is how to reconcile our responsibility as consumers with the damage from mineral extraction and processing. How responsible are you? What are some steps you could take to reduce this damage?

6. If gold jewelry is responsible for environmental and social devastation, should we stop wearing it? Should we worry about the economy in producing areas if we stop buying gold and diamonds? What further information would you need to answer these questions?

 **Data Analysis:** **Mapping Geological Hazards**

Volcanoes

Look at the animated map of volcanoes in the animated National Atlas at nationalatlas.gov/dynamic.html. (You will need Shockwave to display this movie; you can download Shockwave at the link on this page.) Where are most of the volcanoes in the lower 48 states? Roll over the red dots to see photos and elevations of the volcanoes. As you move down the chain of mountains from north to south, is there a trend in elevations? Look at the Alaska map. How are the volcanoes arranged here? Based on your readings in this chapter, can you explain the pattern?

Where Are Recent Earthquakes?

You can find out about recent earthquakes, where they were, and how big they were, by looking at the USGS earthquake information page, earthquake.usgs.gov/. Click on the map to investigate recent events.

1. Describe where most recent earthquakes have occurred, either by dominant states or by mountain ranges/coastlines or other geographic features. Using what you know from this chapter about causes of earth movement and earthquakes, can you explain any of the patterns you see? Find at least one concentration of earthquakes, and try to explain the geological processes that cause them. You might do this with a partner so that you can discuss likely explanations.

2. Now choose a map (conterminous U.S., Alaska, Hawaii, or Puerto Rico), and look at the magnitude of recent earthquakes: compare the size of squares to the magnitude numbers in the legend. What size class is most dominant? What proportion of earthquakes on your chosen map are in the largest size class?

3. Alaska is often a hot spot of earthquake activity. Find the Alaska map (you may have to back up a page or two), and zoom in on a concentration of earthquakes. How many earthquakes are off shore? How many are on shore?

4. Click on an earthquake in a location that is somewhat familiar to you. After you zoom in a few times, you should be redirected to some text information about that earthquake, and you may be able to find still more detailed maps. As you get closer, can you recognize features around the earthquake site? All these earthquakes occurred within the last week or less. Have you heard about these recent earthquakes in the news? Why or why not?

Figures 14.3 and 14.4 show the margins of tectonic plates, or the sections of crust that make up the earth's surface. The movements—and collisions—of these plates are responsible for many of the geological resources and events we see around us. Because these movements are so important, the map is reproduced here. Examine it to answer the following questions:

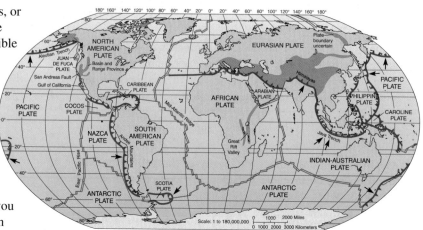

Tectonic plate margins around the globe.

1. What kind of plate movement goes on at the red, spiky lines (for example, on the western coast of South America)? What do the points indicate? Examine the red line that runs across the north side of the Mediterranean Sea. Which direction do the spikes point? What does that tell you about the movement of Africa and Eurasia? Explain the relationship of the red line in northern India and the Himalayan mountains.

2. What kind of plate movement is going on at the double blue line that runs through the middle of the Atlantic and through Iceland? Why does Iceland have abundant geothermal energy?

3. Look at figure 14.3. Then explain why the double blue line through the Atlantic is a ridge. Describe the locations of a spreading, or rifting, zone in or near Africa, and one in or near North America.

4. Examine the red line of subduction that runs north along the eastern side of the Philippine plate and near Japan.

This zone has the world's deepest ocean trench, the Marianas Trench, which was declared a marine preserve at the end of George W. Bush's presidency. Explain how a trench forms, and explain why volcanoes often occur near trenches. Based on your explanation, where in Alaska would you expect to find the most active volcanoes?

For Additional Help in Studying This Chapter, please visit our website at www.mhhe.com/cunningham12e. You will find additional practice quizzes and case studies, flashcards, regional examples, placemarkers for Google Earth™ mapping, and an extensive reading list, all of which will help you learn environmental science.

CHAPTER 15

Arctic sea ice is disappearing at an accelerating rate. The 2008 summer extent shown here was the lowest on record at the time.

Air, Weather, and Climate

Learning Outcomes

After studying this chapter, you should be able to:

15.1 Describe the general composition and structure of the atmosphere.

15.2 Explain why weather events follow general patterns.

15.3 Outline some factors in natural climate variability.

15.4 Explain how we know recent climate change is human-caused.

15.5 List some effects of climate change.

15.6 Identify some solutions being developed to slow climate change.

"I was born in 1992. You have been negotiating all my life. You cannot tell me you need more time."

~ Christina Ora, youth delegate from the Solomon Islands addressing the plenary at COP15, 2009

Case Study

When Wedges Do More Than Silver Bullets

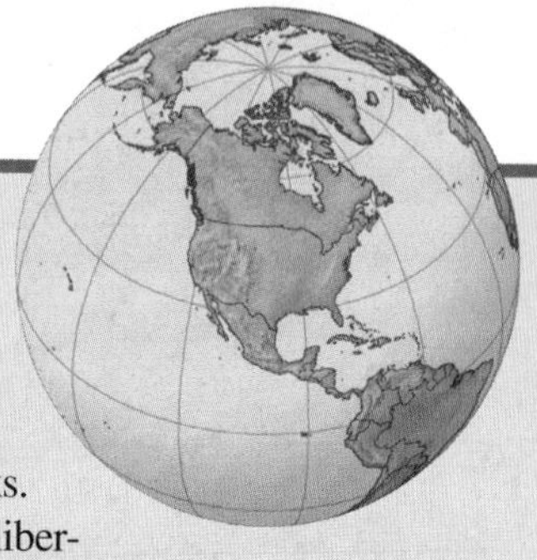

In the summer of 2010, Russians suffered the worst heat wave in 1,000 years of recorded Russian history. In Moscow, temperatures exceeded 100°F (40°C) for the first time in history, and the death rate doubled, from heatstroke and lung ailments caused by smoke from burning forests and peat swamps. Heat and smoke were blamed for 11,000 extra deaths that August. More than seven weeks of extreme temperatures also destroyed one-third of Russian grain crops, and wheat prices doubled worldwide. Extreme conditions can happen in a complex climate system, but these were consistent with the increasingly volatile weather expected by climate scientists, as increasing concentrations of "greenhouse gases" (fig. 15.1) retain more and more energy in our atmosphere.

Global average temperatures are about 1°C (about 2°F) higher than they've been in centuries This difference might seem slight, but the difference between the last glacial maximum and today is only about 5°C. A change of 1°C allows new crop pests and weeds to survive winters farther north. Even moderate warming can dry soil enough to force farmers to irrigate crops more, where irrigation is possible, or to abandon farms in poor countries and move to already-overstressed cities.

In California and other parts of the western United States, cities rely on snowmelt in the mountains for water. Here the specter of declining snowpack is sobering up voters and politicians alike. But still we have a hard time getting around to finding policies to reduce greenhouse gas emissions.

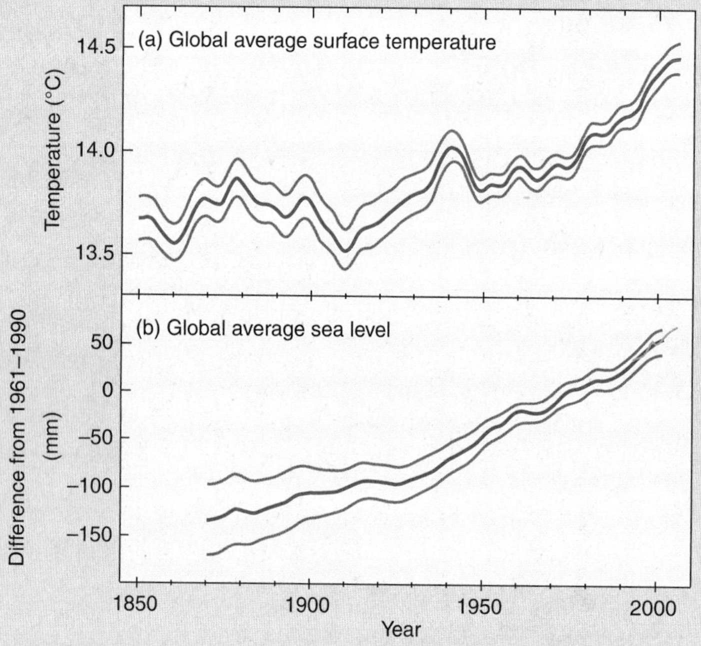

FIGURE 15.1 Observed temperatures have increased in recent decades. Blue lines show uncertainty (range of possible values) for global averages (red lines). **Source:** IPCC, 2007.

Meanwhile, data indicate that if we don't reduce our carbon output in the next few years, we will permanently lose the ice caps and permafrost, which help moderate the global climate. Soon we will be on a path for irreversible and unavoidable increases of 5–7°C within the coming century, with sea-level rises of 1 m or more by 2100.

Among climate scientists who study the data, there is no longer any debate about whether our carbon emissions are triggering climate change or whether that change is likely to be extraordinarily costly, in both human and economic terms. Remaining debates are only about details: how fast sea levels are likely to rise, or where drought will be worst, or about fine-tuning of climate models.

Among policymakers, it's another matter. Politicians are responsible for establishing new rules to reduce our carbon output, but many

still have a hard time understanding the connection between climate change and the increasing incidence of forest fires, drought, water shortages, heat waves, and pest outbreaks. Energy-industry money also pays liberally to sow doubt in the minds of policymakers and the public. Climate changes are gradual, proceeding over decades, so it's hard to get the public focused on remedies today.

Many politicians have hoped for a silver bullet—a technology that will fix the problem all at once—perhaps nuclear fusion, or space-based solar energy, or giant mirrors that would reflect solar energy away from the earth's surface. While these are intriguing ideas, none are workable now, and climate scientists are warning us that immediate action is critical to avoid tipping points such as the loss of polar ice, ancient arctic permafrost, and glaciers.

Wedges Can Work Now

To help us out of this quagmire of indecision, a Princeton ecologist and an engineer have proposed a completely different approach to imagining alternatives. Their approach has come to be called **wedge analysis,** or breaking down a large problem into smaller, bite-size pieces. By calculating the contribution of each wedge, we can add them up, see the magnitude of their collective effect, and decide that it's worth trying to move forward.

Stephen Pacala and Robert Socolow, of Princeton University's Climate Mitigation Initiative, introduced the wedge idea in a 2004 article in the journal *Science*. Their core idea was that currently available technologies—efficient vehicles, buildings, power plants, alternative fuels—could solve our problems today, if we just take them seriously. Future technologies, no matter how brilliant, can do nothing for us right now. Pacala and Socolow have further honed their ideas in subsequent papers, and others have picked up the wedge idea to envision strategies for problems such as reducing transportation energy use or water consumption. The *Science* paper focuses on CO_2 production, but the authors point out that similar analysis could be done for other greenhouse gases.

Pacala and Socolow described three possible trajectories in our carbon emissions. The "business as usual" scenario follows

Case Study continued

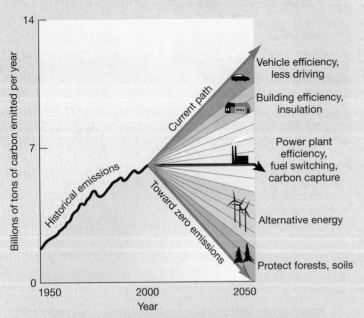

FIGURE 15.2 We could stabilize or even reduce carbon emissions now if we focus on multiple modest strategies.

buildings would equal another wedge. Increased efficiency in our coal power plants would equal another wedge.

These steps add up to 4/7 of the stabilization triangle, using currently available technologies. The remaining 3/7 can be accomplished by capturing and storing carbon at power plants, by changing the way power plants operate, and by reducing reliance on coal power. Another set of seven wedges, including alternative energy, preventing deforestation, and reducing soil loss, could put us on a trajectory to reduce our CO_2 emissions and prevent disastrous rates of climate change. Further details on the wedges are given later in this chapter.

These strategies also offer economic advantages. Improved efficiency and reduced energy consumption mean long-term cost savings. Efficient cars will cut household expenses. Sustainable infrastructure can provide long-term employment stability, rather than the boom-and-bust cycles of coal and oil extraction. We continually replace buildings, roads, and vehicles; if we start now to build them better, we could drastically cut our costs in the near future.

Cleaner power sources will also reduce asthma and other respiratory illnesses, saving health care costs and improving quality of life. Less reliance on coal will reduce toxic mercury in our food, since coal power plants are the main source of airborne mercury, which enters our food chain through aquatic ecosystems and the fish we eat. Better land management can preserve food, water, and wood resources for the future.

Perhaps it's not surprising, then, that thousands of local communities are stepping up to lead the way on these initiatives, even while national governments dither. You don't have to care about climate change to agree about saving money and reducing smog. If you do care about climate change, it feels good to stop fretting and start acting.

In this chapter we'll examine the composition and behavior of our atmosphere and the factors that make it change over time. For related resources, including Google Earth™ placemarks that show locations where these issues can be seen, visit EnvironmentalScience-Cunningham.blogspot.com.

Further Reading:

Pacala, S., and Socolow, R. 2004. Stabilization wedges: Solving the climate problem for the next 50 years with current technologies. *Science*, 305 (5686): 968–72.

the current pattern of constantly increasing CO_2 output. This trajectory heads toward at tripling of CO_2 by 2100, accompanied by temperature increases of around 5°C (9°F) and a sea-level rise of 0.5–1 m (fig. 15.2).

A second trajectory is a "stabilization scenario." In this scenario, we prevent further increases in CO_2 emissions, and we nearly double CO_2 in the atmosphere by 2100. Temperatures increase by about 2–3°C, and sea level rises by about 29–50 cm.

A third trajectory, declining CO_2 emissions, could result from new energy sources and better land management.

To achieve stabilization, we need to reduce our annual carbon emissions by about 7 billion tons (or 7 gigatons, GT) per year within 50 years (fig. 15.2). This 7 GT can be subdivided into seven wedges, each representing 1 GT of carbon we need to cut.

Cutting one of those gigatons could be accomplished by increasing fuel economy in our cars from 30 to 60 mpg. Another gigaton could be eliminated if we reduced reliance on cars (with more public transit or less suburban sprawl, for example) and cut driving from an average 10,000 miles to 5,000 miles per year. Better insulation and efficient appliances in our houses and office

15.1 WHAT IS THE ATMOSPHERE?

Of all the planets in our solar system, only the earth has an atmosphere that makes life possible. The atmosphere retains solar heat, protects us from deadly radiation in space, and distributes the water

that makes up most of your body. The atmosphere consists of gas molecules, held near the earth's surface by gravity and extending upward about 500 km. All the weather we see is in just the lowest 10–12 km, in a constantly circulating and swirling layer known as the troposphere. **Weather** is a term for the short-lived and local patterns of temperature and moisture that result from this circulation. In contrast, **climate** is long-term patterns of temperature and precipitation. Understanding the difference between short-term variations and long-term patterns is important in understanding our climate.

The earth's earliest atmosphere probably consisted mainly of lightweight hydrogen and helium. Over billions of years, most of that hydrogen and helium diffused into space. Volcanic emissions added carbon, nitrogen, oxygen, sulfur, and other elements to the atmosphere. Virtually all the molecular oxygen (O_2) that we breathe was probably produced by photosynthesis in blue-green bacteria, algae, and green plants.

Clean, dry air is mostly nitrogen and oxygen (table 15.1). Water vapor concentrations vary from near zero to 4 percent, depending on air temperature and available moisture. Minute particles and liquid droplets—collectively called **aerosols**—also are suspended in the air (fig. 15.3). Atmospheric aerosols are important in capturing, distributing, or reflecting energy.

The atmosphere has four distinct zones of contrasting temperature, which result from differences in absorption of solar energy (fig. 15.4). The layer of air immediately adjacent to the earth's surface is called the **troposphere** (*tropein* means "to turn or change" in Greek). Within the troposphere, air absorbs energy from the sun-warmed earth's surface, and from moisture evaporating from oceans. Warmed air circulates in great vertical and horizontal **convection currents,** which occur when warm, low-density air rises above a cooler, denser layer. (You can observe a similar process in a pot of simmering water on the stove: water heated at the hot bottom of the pot rises up above the cooler layers at the top, creating convective circulation patterns.) Convection constantly redistributes heat and moisture around the globe. The depth of the troposphere ranges from about 18 km (11 mi) over the equator, where heating and convection are intense, to about 8 km (5 mi) over the poles, where air is cold and dense. The troposphere consists mainly of relatively large, heavy molecules, held close to the

FIGURE 15.3 The atmospheric processes that purify and redistribute water, moderate temperatures, and balance the chemical composition of the air are essential in making life possible. To a large extent, living organisms have created, and help to maintain, the atmosphere on which we all depend.

Table 15.1	Present Composition of the Lower Atmosphere*	
Gas	**Symbol or Formula**	**Percent by Volume**
Nitrogen	N_2	78.08
Oxygen	O_2	20.94
Argon	Ar	0.934
Carbon dioxide	CO_2	0.035
Neon	Ne	0.00182
Helium	He	0.00052
Methane	CH_4	0.00015
Krypton	Kr	0.00011
Hydrogen	H_2	0.00005
Nitrous oxide	N_2O	0.00005
Xenon	Xe	0.000009

*Average composition of dry, clean air.

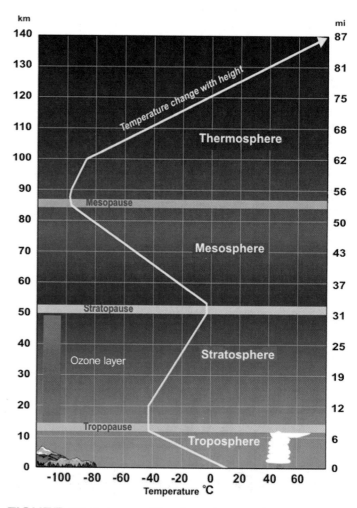

FIGURE 15.4 Layers of the atmosphere vary in temperature and composition. Most weather happens in the troposphere. Stratospheric ozone is important for blocking ultraviolet solar energy.

earth's surface by gravity. Consequently, the troposphere contains about 75 percent of the total mass of the atmosphere. Within the troposphere, temperatures drop rapidly with increasing distance from the earth, reaching about $-60°C$ ($-76°F$). At this point air is no longer warmer than its surroundings, and it ceases to rise. We call this boundary, where mixing ends, the tropopause.

The **stratosphere** extends from the tropopause up to about 50 km (31 mi). This layer is vastly more dilute than the troposphere, but it has similar composition—except that it has almost no water vapor and nearly 1,000 times more **ozone** (O_3). Near the earth's surface ozone is a pollutant, but in the stratosphere it serves a very important function: Ozone absorbs certain wavelengths of ultraviolet solar radiation, known as UV-B (290–330 nm; see fig. 3.10). This absorbed energy warms the stratosphere, and temperature increases with elevation. Stratospheric UV absorption also protects life on the earth's surface, because UV radiation damages living tissues, causing skin cancer, genetic mutations, and crop failures. A number of air pollutants, including Freon, once used in refrigerators, and bromine compounds, used as pesticides, deplete stratospheric ozone, especially over Antarctica. This has allowed increased amounts of UV radiation to reach the earth's surface (see fig. 16.14).

Unlike the troposphere, the stratosphere is relatively calm, because warm layers lie above colder layers. There is so little mixing that when volcanic ash or human-caused contaminants reach the stratosphere, they can remain in suspension there for years.

Above the stratosphere, the temperature diminishes again in the mesosphere, or middle layer. The thermosphere (heated layer) begins at about 80 km. This is a region of highly ionized (electrically charged) gases, heated by a steady flow of high-energy solar and cosmic radiation. In the lower part of the thermosphere, intense pulses of high-energy radiation cause electrically charged particles (ions) to glow. We call this phenomenon the *aurora borealis* and *aurora australis,* or northern and southern lights.

No sharp boundary marks the end of the atmosphere. The density of gas molecules decreases with distance from the earth until it becomes indistinguishable from the near-vacuum of interstellar space.

Absorbed solar energy warms our world

The sun supplies the earth with an enormous amount of energy, but that energy is not evenly distributed over the globe. Incoming solar radiation (insolation) is much stronger near the equator than at high latitudes. Of the solar energy that reaches the outer atmosphere, about one-quarter is reflected by clouds and atmospheric gases, and another quarter is absorbed by carbon dioxide, water vapor, ozone, methane, and a few other gases (fig. 15.5). This energy absorption

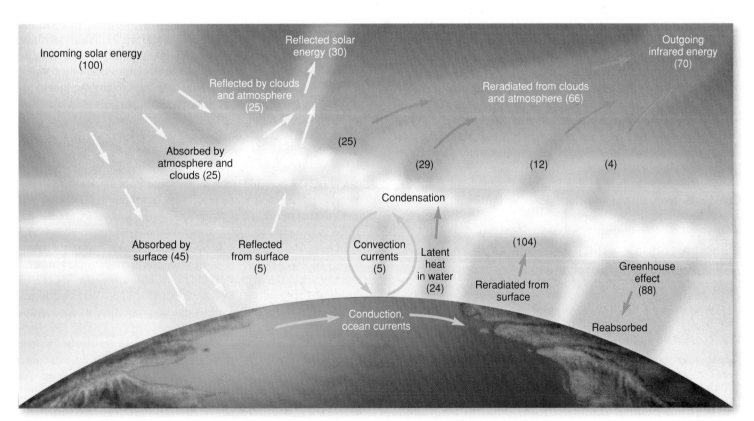

FIGURE 15.5 Energy balance between incoming and outgoing radiation. The atmosphere absorbs or reflects about half of the solar energy reaching the earth. Most of the energy reemitted from the earth's surface is long-wave, infrared energy. Most of this infrared energy is absorbed by aerosols and gases in the atmosphere and is re-radiated toward the planet, keeping the surface much warmer than it would otherwise be. This is known as the greenhouse effect.

Table 15.2 Albedo (Reflectivity) of Surfaces	
Surface	Albedo (%)
Fresh snow	80–85
Dense clouds	70–90
Water (low sun)	50–80
Sand	20–30
Water (sun overhead)	5
Forest	5–10
Black soil	3
Earth/atmosphere average	30

warms the atmosphere. About half of incoming energy reaches the earth's surface. Some of this energy is reflected by bright surfaces, such as snow and ice. The rest is absorbed by the earth's surface and by water. Surfaces that *reflect* energy have a high **albedo** (reflectivity). Most of these surfaces appear bright to us because they reflect light as well as other forms of radiative energy. Surfaces that *absorb* energy have a low albedo and generally appear dark. Black soil, pavement, and open water, for example, have low albedos (table 15.2).

Absorbed energy heats the absorbing surface (such as an asphalt parking lot in summer), evaporates water, or provides the energy for photosynthesis in plants. Following the second law of thermodynamics, absorbed energy is gradually reemitted as lower-quality heat energy. A brick building, for example, absorbs energy in the form of light and reemits that energy in the form of heat.

Water is extremely efficient at absorbing and storing energy. This is why increasing open water at the poles (shown in this chapter's opening photo) worries climatologists. For hundreds of thousands of years, the Arctic has been mostly white, reflecting most energy that reached the icy surface. Now open water increasingly captures and stores that energy, further accelerating ice melting and atmospheric warming. This is a good example of a **positive feedback loop,** in which melting leads to further melting, with probably dramatic consequences.

The greenhouse effect is energy capture by gases in the atmosphere

The change in energy quality shown in fig. 15.5 is important because the atmosphere selectively absorbs longer wavelengths. Most solar energy comes in the form of intense, high-energy light or near-infrared wavelengths (see fig. 3.10), which pass relatively easily through the atmosphere to reach the earth's surface. Energy re-released from the earth's warmed surface ("terrestrial energy") is lower-intensity, longer-wavelength radiation in the far-infrared part of the spectrum. Atmospheric gases, especially carbon dioxide and water vapor, absorb much of this long-wavelength energy and re-release it in the lower atmosphere. This long-wave terrestrial energy provides most of the heat in the lower atmosphere (see red shading in fig. 15.5). If the atmosphere were as transparent to

infrared radiation as it is to visible light, the earth's average surface temperature would be about 18°C (33°F) colder than it is now.

This phenomenon is called the **greenhouse effect** because the atmosphere, loosely comparable to the glass of a greenhouse, transmits sunlight while trapping heat inside. The greenhouse effect is a natural atmospheric process that is necessary for life as we know it. However, too strong a greenhouse effect, caused by burning of fossil fuels and deforestation, can destabilize the environment we're used to. We will discuss this issue later in this chapter.

Greenhouse gases is a general term for gases that are especially effective at capturing the long-wavelength energy from the earth's surface. Water vapor (H_2O) is the most abundant greenhouse gas, and it is always present in the atmosphere. Carbon dioxide (CO_2) is the most abundant human-caused greenhouse gas, followed by methane (CH_4), nitrous oxide (N_2O), and dozens of other gases. These are discussed later in this chapter.

Evaporated water stores energy, and winds redistribute it

Much of the incoming solar energy is used to evaporate water. In fact, every gram of evaporating water absorbs 580 calories of energy as it transforms from liquid to gas. Globally, water vapor contains a huge amount of stored energy, known as **latent heat.** When water vapor condenses, returning from a gas to a liquid form, the 580 calories of heat energy are released. Imagine the sun shining on the Gulf of Mexico in the winter. Warm sunshine and plenty of water allow continuous evaporation that converts an immense amount of solar (light) energy into latent heat stored in evaporated water. Now imagine a wind blowing the humid air north from the Gulf toward Canada. The air cools as it rises and moves north. Cooling causes the water vapor to condense. Rain (or snow) falls as a consequence.

Note that not only water has moved from the Gulf to the Midwest: 580 calories of heat have also moved with every gram of moisture. The heat and water have moved from a place with strong incoming solar energy to a place with much less solar energy and much less water. The redistribution of heat and water around the globe is essential to life on earth. Without oceans to absorb and store heat, and wind currents to redistribute that heat in the latent energy of water vapor, the earth would undergo extreme temperature fluctuations like those of the moon, where it is 100°C (212°F) during the day and −130°C (−200°F) at night. Water performs this vital function because of its unique properties in heat absorption and energy of vaporization (chapter 3).

Uneven heating, with warm air close to the equator and colder air at high latitudes, also produces pressure differences that cause wind, rain, storms, and everything else we know as weather. As noted earlier, air circulation occurs as the sun warms the earth's surface, and air nearest the surface warms and expands, becoming less dense than the air above it. Rising warm air produces vertical convection currents. These convection currents can be as small and as localized as a narrow column of hot air rising over a sun-heated rock, or they can cover huge regions of the earth,

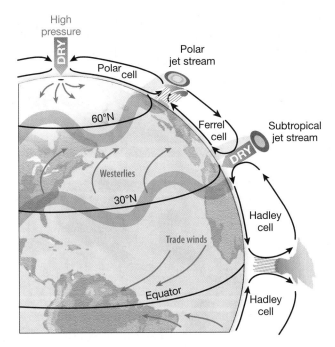

FIGURE 15.6 Convection cells circulate air, moisture, and heat around the globe. Jet streams develop where cells meet, and surface winds result from convection. Convection cells expand and shift seasonally.

circulating air from warm latitudes to cool latitudes and back. At the largest scale, the convection cells are described by a simplified model known as Hadley cells, which redistribute heat globally (fig. 15.6).

Where air rises in convection currents, air pressure at the surface is low. Where air is sinking, or subsiding, air pressure is high. On a weather map these high and low pressure centers, or rising and sinking currents of air, move across continents. In most of North America, they generally move from west to east. Rising air tends to cool with altitude, releasing latent heat, which causes further rising. Very warm and humid air can rise very vigorously, especially if it is rising over a mass of very cold air. As water vapor carried aloft cools and condenses, it releases energy that fuels violent storms, which we will discuss later.

Pressure differences are an important cause of wind. There is always someplace with high pressure (sinking) air and someplace with low pressure (rising) air. Air moves from high-pressure centers toward low-pressure areas, and we call this movement wind.

15.2 Weather Has Regional Patterns

Weather involves the physical conditions in the atmosphere (humidity, temperature, air pressure, wind, and precipitation) over short time scales, usually days or weeks. In this section we'll examine why those patterns occur. In general, most major weather patterns result from uneven solar heating, which causes areas of high and low pressure, together with spinning of the earth.

Why does it rain?

To understand why it rains, remember two things: Water condenses as air cools, and air cools as it rises. Any time air is rising, clouds, rain, or snow might form. Cooling occurs because of changes in pressure with altitude: Air cools as it rises (as pressure decreases); air warms as it sinks (as pressure increases). Air rises in convection currents where solar heating is intense, such as over the equator. Moving masses of air also rise over each other and cool. Air also rises when it encounters mountains. If the air is moist (if it has recently come from over an ocean or an evaporating forest region, for example), condensation and rainfall are likely as the air is lifted (fig. 15.7). Regions with intense solar heating, frequent colliding air masses, or mountains tend to receive a great deal of precipitation.

Where air is sinking, on the other hand, it tends to warm because of increasing pressure. As it warms, available moisture evaporates. Rainfall occurs relatively rarely in areas of high pressure. High pressure and clear, dry conditions occur where convection currents are sinking. High pressure also occurs where air sinks after flowing over mountains. Figure 15.6 shows sinking, dry air at about 30° north and south latitudes. If you look at a world map, you will see a band of deserts at approximately these latitudes.

Another ingredient is usually necessary to initiate condensation of water vapor: condensation nuclei. Tiny particles of smoke, dust, sea salts, spores, and volcanic ash all act as condensation nuclei. These particles form a surface on which water molecules can begin to coalesce. Without them even supercooled vapor can remain in gaseous form. Even apparently clear air can contain large numbers of these particles, which are generally too small to be seen by the naked eye.

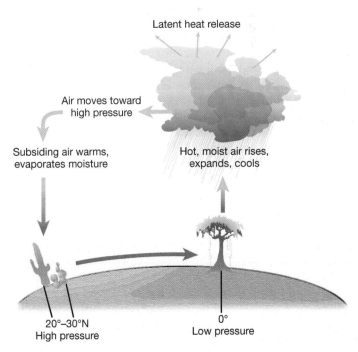

FIGURE 15.7 Convection currents distribute latent energy (heat in evaporated water) around the globe.

The Coriolis effect explains why winds seem to curve

In the Northern Hemisphere, winds generally appear to bend clockwise (right), and in the Southern Hemisphere they appear to bend counterclockwise (left). Examples include the trade winds that brought Columbus to the Americas and the midlatitude Westerlies that bring hurricanes north from Florida to North Carolina (see fig. 15.6). Ocean currents similarly curve clockwise in the Northern Hemisphere (the Gulf Stream) and counterclockwise in the south (the Humboldt Current near Peru). This curving pattern results from the fact that the earth rotates in an eastward direction as the winds move above the surface. The apparent curvature of the winds is known as the **Coriolis effect.** On a global scale, this effect produces predictable wind patterns and currents. On a regional scale, the Coriolis effect produces cyclonic winds, or wind movements controlled by the earth's spin. Cyclonic winds spiral clockwise out of an area of high pressure in the Northern Hemisphere and counterclockwise into a low-pressure zone. If you look at a weather map in the newspaper, you can probably find this spiral pattern.

Why does this curving or spiraling motion occur? Imagine you were looking down on the North Pole of the rotating earth. Now imagine that the earth was a merry-go-round in a playground, with the North Pole at its center and the equator around the edge. As it spins counterclockwise (eastward), the spinning edge moves very fast (a full rotation, 39,800 km, every 24 hours for the real earth, or more than 1,600 km/hour). Near the center, though, there is very little eastward velocity, because the distance around a circle near the pole is relatively short. If you were standing on the edge of the merry-go-round and threw a ball toward the center, the ball would be traveling eastward very fast, at the speed of the spinning edge, as well as toward the center. To someone standing on the merry-go-round, the ball would appear to be traveling east as well as north, making a curve toward the right. If you threw the ball from the center toward the edge, it would start out with no eastward velocity, but the surface below it would spin eastward, making the ball end up, to a person on the merry-go-round, west of its starting point. If you were looking down at the South Pole, you would see the earth spinning clockwise, and winds—or thrown balls—would appear to bend left.

Winds move above the earth's surface much as the ball does. However, it's a myth that bathtubs and sinks spiral in opposite directions in the Northern and Southern Hemispheres. Those movements are far too small to be affected by the spinning of the earth.

At the top of the troposphere are **jet streams,** hurricane-force winds that circle the earth. These powerful winds follow an undulating path approximately where the vertical convection currents known as the Hadley and Ferrell cells meet. The approximate path of one jet stream over the Northern Hemisphere is shown in figure 15.8. Although we can't perceive jet streams on the ground, they are important to us because they greatly affect weather patterns. Sometimes jet streams dip down near the top of the world's highest mountains, exposing mountain climbers to violent, brutally cold winds.

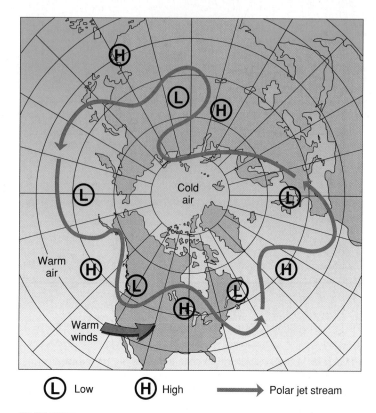

(L) Low (H) High ——→ Polar jet stream

FIGURE 15.8 A typical pattern of the arctic circumpolar vortex. This large, circulating mass of cold air sends "fingers," or lobes, across North America and Eurasia, spreading storms in their path. If the vortex becomes stalled, weather patterns stabilize, causing droughts in some areas and excess rain elsewhere.

Ocean currents modify our weather

Warm and cold ocean currents strongly influence climate conditions on land. Surface ocean currents result from wind pushing on the ocean surface, as well as from the Coriolis effect. As surface water moves, deep water wells up to replace it, creating deeper ocean currents. Differences in water density—depending on the temperature and saltiness of the water—also drive ocean circulation. Huge cycling currents called gyres carry water north and south, redistributing heat from low latitudes to high latitudes (see fig. 17.4). The Alaska current, flowing from Alaska southward to California, keeps San Francisco cool and foggy during the summer.

The Gulf Stream, one of the best-known currents, carries warm Caribbean water north past Canada's maritime provinces to northern Europe. This current is immense, some 800 times the volume of the Amazon, the world's largest river. The heat transported from the Gulf keeps Europe much warmer than it should be for its latitude. As the warm Gulf Stream passes Scandinavia and swirls around Iceland, the water cools and evaporates, becomes dense and salty, and plunges downward, creating a strong, deep, southward current. Oceanographer Wallace Broecker calls this the ocean conveyor system (see fig. 17.4).

Ocean circulation patterns were long thought to be unchanging, but now oceanographers believe that currents can shift abruptly. About 11,000 years ago, for example, as the earth was gradually warming at the end of the Pleistocene ice age, a huge body of meltwater, called Lake Agassiz, collected along the south margin of the North American ice sheet. At its peak it contained more water than all the current freshwater lakes in the world. Drainage of this lake to the east was blocked by ice covering what is now the Great Lakes. When that ice dam suddenly gave way, it's estimated that some 163,000 km³ of fresh water roared down the St. Lawrence Seaway and out into the North Atlantic, where it layered on top of the ocean and prevented the sinking of deep, cold, dense seawater. This, in turn, apparently stopped the oceanic conveyor and plunged the whole planet back into an ice age (called the Younger Dryas after a small tundra flower that became more common in colder conditions) that lasted for another 1,300 years.

Could this happen again? Meltwater from Greenland glaciers is now flooding into the North Atlantic just where the Gulf Stream sinks and creates the deep south-flowing current. Already, evidence shows that the deep return flow has weakened by about 30 percent. Even minor changes in the strength or path of the Gulf Stream might give northern Europe a climate more like that of Siberia—an ironic consequence of polar warming.

Think About It

Find London and Stockholm on a globe. Then find cities in North America at a similar latitude. Temperatures in London and Stockholm rarely get much below freezing. How do you think their climate compares with the cities you've identified in North America? Explain the difference.

Much of humanity relies on seasonal rain

Large parts of the world, especially near the tropics, receive seasonal rains that sustain both ecosystems and human life. Seasonal rains give life, but when they fail to arrive, crop failures and famine can result. Seasonal rains can also cause disastrous flooding, as in the 2010 floods in Pakistan, which left 2 million homeless, or the 2003 floods in China, which forced 100 million people from their homes.

The most regular seasonal rains are known as **monsoons.** In India and Bangladesh, monsoon rains come when seasonal winds blow hot, humid air from the Indian Ocean (fig. 15.9). The hot land surface produces strong convection currents that lift this air, causing heavy rain across the subcontinent. When the rising air reaches the Himalayas, it rises even further, creating some of the heaviest rainfall in the world. During the five-month rainy season of 1970, a weather station in the foothills of the Himalayas recorded 25 m (82 ft) of rain!

Tropical and subtropical regions around the world have seasonal rainy and dry seasons (see the discussion of tropical biomes, chapter 5). The main reason for this variable climate is that the region of most intense solar heating and evaporation

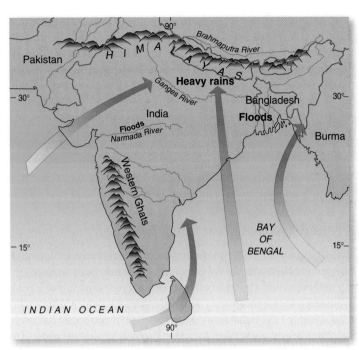

FIGURE 15.9 Summer monsoon air flows over the Indian subcontinent. Warming air rises over the plains of central India in the summer, creating a low-pressure cell that draws in warm, wet oceanic air. As this moist air rises over the Western Ghats or the Himalayas, it cools and heavy rains result. These monsoon rains flood the great rivers bringing water for agriculture, but also causing much suffering.

FIGURE 15.10 Failure of monsoon rains brings drought, starvation, and death to both livestock and people in the Sahel desert margin of Africa. Although drought is a fact of life in Africa, many governments fail to plan for it, and human suffering is much worse than it needs to be.

shifts through the year. Remember that the earth's axis of rotation is at an angle. In December and January the sun is most intense just south of the equator; in June and July the sun is most intense just north of the equator. Wherever the sun shines most directly, evaporation and convection currents—and rainfall and thunderstorms—are very strong. As the earth orbits the sun, the tilt of its axis creates seasons with varying amounts of wind, rain, and heat or cold. Seasonal rains support seasonal tropical forests, and they fill some of the world's greatest rivers, including the Ganges and the Amazon. As the year shifts from summer to winter, solar heating weakens, the rainy season ends, and little rain may fall for months.

Frontal systems create local weather

The boundary between two air masses of different temperature and density is called a front. When cooler air pushes away warmer air, we call the moving boundary a **cold front.** Cold, dense air of a cold front tends to hug the ground and push under the lighter, warmer air as it advances. As warm air is forced upward, it cools, and its cargo of water vapor condenses to water droplets or ice crystals. Air masses near the ground move slowly because of friction and turbulence near the ground surface, so upper layers of a moving air mass often move ahead of the lower layers (fig. 15.11, *below*). Notice that the region of cloud formation and precipitation is relatively narrow. Cold fronts can generate strong convective currents as they push warmer air rapidly upward. Violent winds and thunderstorms can result, with towering thunderheads. The weather after the cold front passes is usually clear, dry, and invigorating.

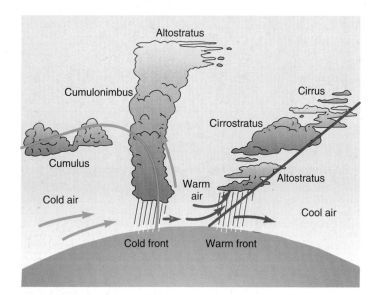

FIGURE 15.11 A cold front assumes a bulbous, "bull-nose" appearance because ground drag retards forward movement of surface air. As warm air is lifted up over the advancing cold front, it cools, producing precipitation. When warm air advances, it slides up over cooler air in front and produces a long, wedge-shaped zone of clouds and precipitation. The high cirrus clouds that mark the advancing edge of the warm air mass may be 1,000 km and 48 hours ahead of the front at ground level.

In a **warm front**, the advancing air mass is warmer than surrounding air. Because warm air is less dense than cool air, an advancing warm front will slide up over cooler air masses, creating a long, wedge-shaped profile with a broad band of clouds and precipitation (fig. 15.11, *right*). Gradual lifting and cooling in a warm front lacks the violent updrafts and strong convection currents that accompany a cold front. A warm front can have many layers of clouds at different heights. The highest layers are often wispy cirrus (mare's tail) clouds, composed mainly of ice crystals, which can extend 1,000 km (600 mi) and 2 days ahead of the front we detect at the ground level. A moist warm front can bring days of drizzle and cloudy skies.

Cyclonic storms can cause extensive damage

Severe cyclonic storms are powerful and dangerous natural forces, especially those spawned by rising, low-pressure air over warm tropical oceans. Winds swirl into this low-pressure area, turning counterclockwise in the Northern Hemisphere due to the Coriolis effect. If water vapor is abundant, as over a warm sea, the latent heat released by condensation intensifies the convection currents, which draw up more warm air and water vapor, further intensifying the wind and rain.

Called **hurricanes** in the Americas, or typhoons in the western Pacific, these storms can be hundreds of kilometers across, with winds of 320 km/hr (200 mph). Equally dangerous are the walls of water (storm surges) they push far inland (fig. 15.12a). In July 1931, torrential rains spawned by a typhoon caused massive flooding on China's Yangtze River that killed an estimated 3.7 million people, the largest known storm death toll in human history. A similar storm in July 1959 caused flooding of the Yellow River that killed 2 million people.

Hurricane Katrina, which devastated coastal Louisiana and the Gulf Coast in 2005, caused most of its damage with storm surges. The category 4 storm, with 232 km/hr (145 mph) winds, pushed a storm surge up to 9 m (29 ft) high onto coastal areas. Aided by shipping and oil-drilling canals, the surge destroyed large parts of New Orleans and many other cities, many of which still have not recovered (fig. 15.12b).

Tornadoes, swirling funnel clouds that form over land, also are considered cyclonic storms. Though never as large or powerful as hurricanes, tornadoes can be just as destructive in the limited areas where they touch down (fig. 15.12c). Tornadoes are generated on the American Great Plains by giant "supercell" frontal systems where strong, dry-air cold fronts from Canada collide with warm, humid air moving north from the Gulf of Mexico. Greater air temperature differences cause more powerful storms. This is why most tornadoes occur in the spring, when arctic cold fronts penetrate far south over the warming plains. As warm air rises rapidly over dense, cold air, intense vertical convection currents generate towering thunderheads with anvil-shaped leading edges and domed tops up to 20,000 m (65,000 ft) high. Water vapor cools and condenses as it rises, releasing latent heat and accelerating updrafts within the supercell. Sometimes penetrating into the stratosphere, the tops of these clouds can encounter jet streams, which help create even stronger convection currents.

(a) Hurricane Floyd, 1999

(b) Gulf Shores, Alabama, 2005

(c) A tornado touches down

FIGURE 15.12 (a) Hurricane Floyd was hundreds of kilometers wide as it approached Florida in 1999. Note the hole, or eye, in the center of the storm. (b) Destruction caused by Hurricane Katrina in 2005. More than 230,000 km² (90,000 mi²) of coastal areas were devastated by this massive storm, and many cities were almost completely demolished. (c) Tornadoes are much smaller than hurricanes, but can have stronger local winds.

15.3 NATURAL CLIMATE VARIABILITY

Until recently, most of us considered climate as relatively constant. Geologists and climatologists, though, have long understood that climates shift on scales of decades, centuries, and millennia. Teasing apart the simultaneous effects of multiple factors is a complex process, but expanding evidence is helping us discern the patterns. Ice cores are among our key sources of data.

Ice cores tell us about climate history

Every time it snows, small amounts of air are trapped in the snow layers. In Greenland, Antarctica, and other places where cold is persistent, yearly snows slowly accumulate over the centuries. New layers compress lower layers into ice, but still tiny air bubbles remain, even thousands of meters deep into glacial ice. Each bubble is a tiny sample of the atmosphere at the time that snow fell.

Climatologists have discovered that by drilling deep into an ice sheet, they can extract ice cores, from which they can collect airbubble samples. Samples taken every few centimeters show how the atmosphere has changed over time. Ice core records have revolutionized our understanding of climate history (see fig. 15.13). We can now see how concentrations of atmospheric CO_2 have varied. We can detect ash layers and spikes in sulfate concentrations that record volcanic eruptions. Most important, we can look at isotopes of oxygen. In cold years, water molecules with slightly lighter oxygen atoms evaporate more easily than water with slightly heavier isotopes. Consequently, by looking at the proportions of heavier and lighter oxygen atoms (isotopes), climatologists can reconstruct temperatures over time, and plot temperature changes against concentrations of CO_2 and other atmospheric gases.

The first very long record was from the Vostok ice core, which reached 3,100 m into the Antarctic ice and gives us a record of temperatures and atmospheric CO_2 over the past 420,000 years. A team of Russian scientists worked for 37 years at the Vostok site, about 1,000 km from the South Pole, to extract this ice core. A similar core has been drilled from the Greenland ice sheet.

FIGURE 15.13 Dr. Mark Twickler, of the University of New Hampshire, holds a section of the 3,000 m Greenland ice sheet core, which records 250,000 years of climate history.

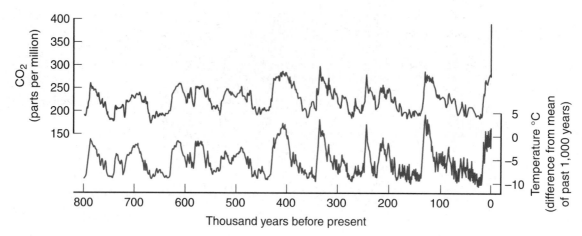

FIGURE 15.14 Atmospheric CO_2 concentrations *(red line)* map very closely to temperatures *(blue, derived from oxygen isotopes)* in air bubbles from the Antarctic Vostok ice core. Temperatures lag behind the recent jump in CO_2 possibly because the ocean has been absorbing heat. In the 800,000-year EPICA ice core there is no evidence of temperatures or CO_2 higher than that anticipated within the coming century.

Sources: UN Environment Programme; J.Bouzel et al., 2007. *EPICA Dorne C Ice Core 800KYr Deuterurm Data and Temperature Estunares.*

More recently the European Project for Ice Coring in Antarctica (EPICA) has produced a record reaching back over 800,000 years (see fig. 15.14). All these cores show that climate has varied dramatically over time but that there is a close correlation between atmospheric temperatures and CO_2 concentrations. From these ice cores, we know that CO_2 concentrations have varied between 180 to 300 ppm (parts per million) in the past 800,000 years. Therefore, we know that today's concentrations of approximately 390 ppm are about one-third higher than the earth has seen in nearly a million years. We also know that present temperatures are nearly as warm as any in the ice core record. Further warming in the coming decades is likely to exceed anything in the ice core records.

Ice core data also show that the climate is warmer now than it has been since the development of civilization, agriculture, and urbanization as we know them. We know from historical accounts that slight climate shifts can be destabilizing for human communities. During the "little ice age" that began in the 1400s, a cooling climate caused crops to fail repeatedly in agricultural regions of northern Europe. Scandinavian settlements in Greenland founded during the warmer period around A.D. 1000 lost contact with Iceland and Europe as ice blocked shipping lanes. It became too cold to grow crops, and fish that once migrated along the coast stayed farther south. The Greenland settlers died out, perhaps in battles with Inuit people who were driven south from the high Arctic by colder weather.

Evidence from ice cores drilled in the Greenland ice cap suggests that world climate can change abruptly. It appears that during the last major interglacial period, 135,000 to 115,000 years ago, temperatures flipped suddenly from warm to cold or vice versa over a period of decades rather than centuries.

Earth's movement explains some cycles

You may notice that figure 15.14 shows repeated peaks and low points. Climatologists have studied many data series like these and observed simultaneous repeating patterns of warming and cooling. The longest-period cycles are known as **Milankovitch cycles,** after Serbian scientist Milutin Milankovitch, who first described them in the 1920s. These cycles are periodic shifts in the earth's orbit and tilt (fig. 15.15). The earth's elliptical orbit stretches and shortens in a 100,000-year cycle, while the axis of rotation changes its angle of tilt in a 40,000-year cycle. Furthermore, over a 26,000-year period, the axis wobbles like an out-of-balance spinning top. These variations change the distribution and intensity of sunlight reaching the earth's surface and, consequently, global climate. Bands of sedimentary rock laid in

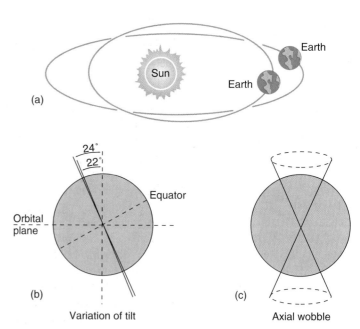

FIGURE 15.15 Milankovitch cycles, which may affect long-term climate conditions: (a) changes in the elliptical shape of the earth's orbit, (b) shifting tilt of the axis, and (c) wobble of the earth.

the oceans seem to match both these Milankovitch cycles and the periodic cold spells associated with worldwide expansion of glaciers every 100,000 years or so.

El Niño is an ocean–atmosphere cycle

On another time scale, there are decades-long oscillations in the oceans and atmosphere. Both the ocean and the atmosphere have regular patterns of flow, or currents, but these shift from time to time. As ocean currents shift, like water swirling in a bathtub, areas of warm water slosh back and forth. Sloshing in the ocean influences low-pressure areas in the atmosphere—and winds and rain change as a consequence. One important example is known as El Niño/Southern Oscillation, or ENSO.

The core of the ENSO system is a huge pool of warm surface water in the Pacific Ocean that sloshes slowly back and forth between Indonesia and South America. Most years this pool is held in the western Pacific by steady equatorial trade winds pushing ocean surface currents westward (fig. 15.16). These surface winds are strengthened by a huge low-pressure area in the warm, western Pacific. Upwelling convection currents of moist tropical air draw in winds from across the Pacific. Towering thunderheads created by rising air bring torrential summer rains to the tropical rainforests of northern Australia and Southeast Asia. Winds high in the troposphere carry a return flow back to the eastern Pacific

where dry subsiding currents create deserts from Chile to southern California. Surface waters driven westward by the trade winds are replaced by the upwelling of cold, nutrient-rich, deep waters off the west coast of South America that support dense schools of anchovies and other fish.

Every three to five years, for reasons we don't fully understand, the Indonesian low collapses and the mass of warm surface water surges back east across the Pacific. One theory is that the high cirrus clouds reduce heating and weaken atmospheric circulation. Another theory is that eastward-flowing deep currents called baroclinic waves periodically interfere with coastal upwelling, warming the sea surface off South America and eliminating the temperature gradient across the Pacific. At any rate, the shift in position of the tropical low-pressure area has repercussions in weather systems across North and South America and perhaps around the world.

Peruvian fishermen were the first to name these irregular cycles, as weakened upwelling currents and warming water resulted in disappearance of the anchovy schools on which they depended. They named these events **El Niño** (Spanish for "the Christ child") because they were observed around Christmastime. Increased attention to these patterns has shown that sometimes, between El Niño events, coastal waters become extremely cool, and these extremes have come to be called **La Niña** (or "the little girl"). Together this cycle is called the El Niño Southern Oscillation (ENSO).

How does the ENSO cycle affect you? During an El Niño year, the northern jet stream—which normally is over Canada—splits and is drawn south over the United States. This pulls moist air inland from the Pacific and Gulf of Mexico, bringing intense storms and heavy rains from California across the Midwestern states. La Niña years bring extreme hot, dry weather to these same areas. El Niño events have brought historic floods to the Mississippi River basin, but Oregon, Washington, and British Columbia tend to be warm and dry in El Niño years. Droughts in Australia and Indonesia during El Niño episodes cause disastrous crop failures and forest fires.

ENSO-related droughts and floods are expected to intensify and become more irregular with global climate change, in part because the pool of warm water is warming and expanding. High sea surface temperatures spawn larger and more violent storms such as hurricanes. On the other hand, increased cloud cover would raise the albedo while upwelling convection currents generated by these storms could pump heat into the stratosphere. This might have an overall cooling effect, or a negative feedback in the warming climate system.

Climatologists have observed many decade-scale oscillations. The Pacific Decadal Oscillation (PDO), for example, involves a vast pool of warm water that moves back and forth across the North Pacific every 30 years or so. From about 1977 to 1997, surface water temperatures in the middle and western parts of the North Pacific Ocean were cooler than average, while waters off the western United States were warmer. During this time, salmon runs in Alaska were bountiful, while those in Washington and Oregon were greatly diminished. In 1997, however, ocean surface

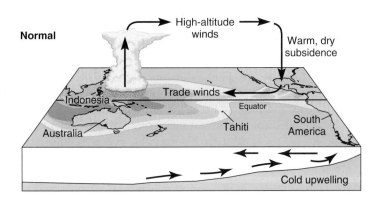

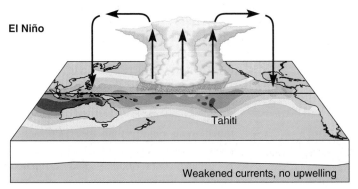

FIGURE 15.16 Normally surface trade winds drive currents from South America toward Indonesia, and cold, deep water wells up near Peru. During El Niño years, winds and currents weaken, and warm, low-pressure conditions shift eastward, bringing storms to the Americas.

temperatures along the coast of western North America turned significantly cooler, perhaps marking a return to conditions that prevailed between 1947 and 1977. Under this cooler regime, Alaskan salmon runs declined while those in Washington and Oregon improved somewhat. A similar North Atlantic Oscillation (NAO), occurs between Canada and Europe.

15.4 ANTHROPOGENIC CLIMATE CHANGE

Many scientists regard anthropogenic (human-caused) global climate change to be the most important environmental issue of our times. The idea that humans might alter world climate is not new. In 1895 Svante Arrhenius, who subsequently received a Nobel Prize for his work in chemistry, predicted that CO_2 released by coal burning could cause global warming. At the time this idea seemed theoretical, though, and real impacts seemed unlikely.

The first data showing human impacts on atmospheric CO_2 came from an observatory on top of the Mauna Loa volcano in Hawaii. The observatory, established in 1957 as part of an International Geophysical Year, was intended to provide data on air chemistry in a remote, pristine environment. Surprisingly, measurements showed CO_2 levels increasing about 0.5 percent per year. Concentrations have risen from 315 ppm in 1958 to 392 ppm in 2011 (fig. 15.17). This graph, first produced by David Keeling at the Mauna Loa observatory, is one of the first and most important pieces of evidence that demonstrates Svante Arrhenius's prediction.

Keeling's graph has some distinctive patterns. One is the annual variation in CO_2 concentrations: every May, CO_2 levels drop slightly as plant growth on the vast northern continents capture CO_2 in photosynthesis. During the northern winter, levels rise again as respiration releases CO_2. Another pattern is that CO_2 levels are rising at an accelerating rate, currently more than 2 ppm each year. We are on track to double the preindustrial concentration of CO_2, which was 280 ppm, in about a century.

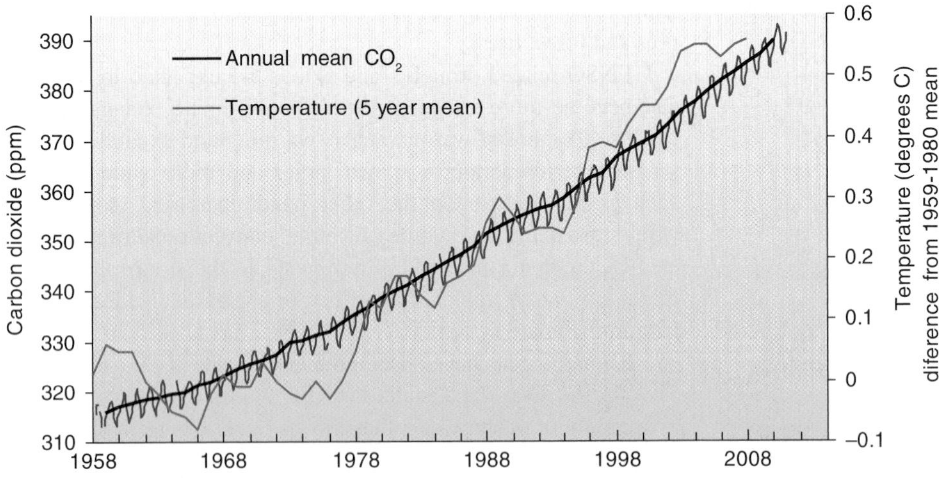

The IPCC assesses data for policymakers

The climate system is complex, confusing, and important, so a great deal of effort has been invested in carefully and thoroughly analyzing observations like those from Mauna Loa. Since 1988 the **Intergovernmental Panel on Climate Change (IPCC)** has brought together scientists and government representatives from 130 countries to review scientific evidence on the causes and likely effects of human-caused climate change. The group's fourth Assessment Report (known as AR4) was published in 2007, representing six years of work by 2,500 scientists, in four volumes. This report stated that there is a 90 percent probability (it is "very likely") that recently observed climate changes result from human activities, and some changes were reported to be "virtually certain," or having a 99 percent probability of being anthropogenic (human-caused).

The wording is cautious, but it represents a remarkable unanimity for scientists, who tend to disagree and to view evidence with skepticism. Among climate scientists who consider trends in the data, there is no disagreement about whether human activities are causing current rapid climate changes. The AR4 report projected warming of about 1–6°C by 2100, depending on what policies we follow to curb climate change. The IPCC's "best estimate" for the most likely scenario was 2–4°C (3–8°F). To put that in perspective, the average global temperature change between now and the middle of the last glacial period is about 5°C. Droughts, heat stress, and increasing hurricane frequency (caused by warming oceans) could have disastrous human and economic costs. Melting ice on the Arctic Ocean, Greenland, and Antarctica was expected to contribute up to 0.6 m (about 1.5 ft) of sea-level rise.

Evidence gathered since the last IPCC report indicates that IPCC estimates were too optimistic. Increases in carbon emissions since the AR4 exceed even the worst business-as-usual model scenario published by the IPCC. Arctic ice is shrinking much more rapidly than the IPCC anticipated, and the impact on energy retention is greater than models had estimated. Revised estimates project a sea-level increase of about 1–2 m (3-6 ft) by 2100. This increase would flood populous coastal regions, including low-lying cities such as Miami, New Orleans (fig. 15.18), Boston, New York, London, and Mumbai.

FIGURE 15.17 Measurements of atmospheric CO_2 taken at the top of Mauna Loa, Hawaii, show an increase of 1.5–2.5 percent each year in recent years. For carbon dioxide, monthly mean (red) and annual mean (black) carbon dioxide are shown. Temperature represents 5-year mean variation from the 1950–1980 mean temperature.
Source: Data from NOAA Earth System Research Laboratory, 2011.

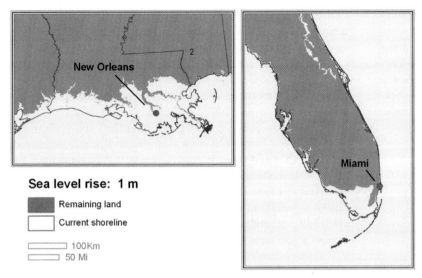

FIGURE 15.18 Approximate change in land surface with the 1 m (3 ft) sea-level rise that the IPCC says is possible by the year 2100. Some analysts expect a 2 m (6 ft) rise if no action is taken.

How does climate change work?

As noted at the beginning of this chapter, the "greenhouse effect" describes the fact that gases in our atmosphere prevent long-wavelength (terrestrial) energy leaving the earth's surface from escaping to space (fig. 15.5). The energy retained in our atmosphere keeps our earth warm enough for life as we know it. Certain gases such as water (H_2O) are especially effective at blocking or absorbing this long-wavelength energy. Human activity is not drastically altering the overall concentrations of water in the atmosphere, however. Industry, forest-clearing, and agriculture have multiplied concentrations of several other greenhouse gases, however (fig. 15.18). Concentrations of these gases are low, less than 0.1 percent of the atmosphere, but multiplying even these low concentrations has considerably increased energy storage, raising both temperatures and storm activity (see section 15.1) in the atmosphere.

Carbon dioxide is the most important greenhouse gas because it is produced abundantly, it lasts decades to centuries in the atmosphere, and it is very effective at capturing long-wave energy. Emissions of CO_2 doubled in the 40 years from 1970 to 2010, from about 14 Gt/yr to more than 30 Gt/yr (fig. 15.19). Carbon dioxide contributes over three-quarters (76.6 percent) of human-caused climate impacts. Burning of fossil fuels is by far the greatest source of CO_2. Deforestation and other land-use changes are the second biggest factor. Deforestation releases carbon stored in standing trees. Organic material in the exposed soil oxidizes and decays, producing still more CO_2 and CH_4. Cement production is also an important contributor, and cement for construction has recently pushed China into the lead for global CO_2 emissions (fig. 15.20).

Methane (CH_4) from agriculture and other sources is the second most important greenhouse gas, accounting for 14 percent of our greenhouse output. Methane absorbs 23 times as much energy per gram as CO_2 does, and it is accumulating at a faster rate than CO_2.

Methane is produced when plant matter decays in oxygen-free conditions, as in the bottom of a wetland. (Where oxygen is abundant, decay produces mainly CO_2). Methane is also released from natural gas wells. Rice paddies are a rich source of CH_4, as are ruminant animals, such as cattle. In a cow's stomach, which has little oxygen, digestion produces CH_4, which cows then burp into the atmosphere. A single cow can't produce much CH_4, but the global population of nearly 1 billion cattle produces enough methane to double the concentration naturally present in the atmosphere.

Nitrous oxide (N_2O), our third most important greenhouse gas, accounts for 8 percent of greenhouse gases. This gas is also released from agricultural processes, plant decay, vehicle engines, denitrification of soils, and other sources. Even though we don't produce as much N_2O as we do other greenhouse gases, this is an important gas because it is especially effective at capturing heat. Many other gases, including chlorofluorocarbons, sulfur hexafluoride, and other fluorine gases, make smaller contributions. Like N_2O and CH_4, these are emitted in relatively small amounts, but their ability to absorb specific energy wavelengths gives them a disproportionate effect.

One way to compare the importance of these various sources is to convert them all to equivalents of our most important greenhouse gas, CO_2. The units used on the Y-axis in fig. 15.19, gigatons of CO_2-equivalent per year (Gt CO_2 eq/yr), let us compare the effects of these sources. All four have increased, but fossil fuel burning rose the most between 1970 and 2010. This is a reason

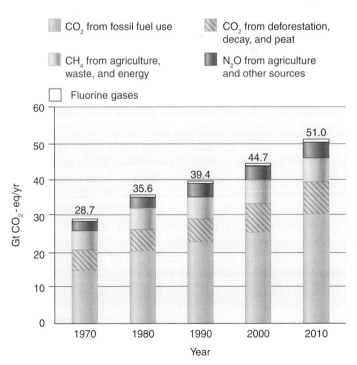

FIGURE 15.19 Contributions to climate change by different gases.
Source: IPCC, 2007.

transportation and coal-burning power plants are two of the key sectors addressed in efforts to slow climate change.

The large orange bars in fig. 15.19 show that burning fossil fuels is also our most abundant source of greenhouse gases. Electricity production, transportation, heating, and industrial activities that depend on fossil fuels together produce 50 percent of our greenhouse gases. Deforestation and agriculture account for another 30 percent. The remaining 20 percent is produced by industry.

Positive feedbacks accelerate change

As noted earlier in this chapter, the melting of polar ice is a concern because it will increase energy absorption (because water has a lower albedo than ice) and enhance warming globally. These and other feedbacks, and some tipping points at which sudden change occurs, are critical factors in climate change.

Another important feedback is the CO_2 release from warming and drying peat. Peat is soggy, semidecayed plant matter accumulated over thousands of years across the vast expanses of tundra in Canada and Siberia. As this peat thaws and dries, it oxidizes and decays, releasing more CO_2 and CH_4. A more ominous consequence of melting in the expanses of frozen arctic lands may be the release of vast stores of frozen, compressed CH_4 (methane hydrate) now locked in permafrost and ocean sediments. Release of these two carbon stores could add as much CO_2 to the atmosphere as all the fossil fuels ever burned.

Negative feedbacks are also possible: increased ocean evaporation could intensify snowfall at high latitudes, restoring some of the high-albedo snow surfaces.

How do we know recent change is human-caused?

The IPCC's third assessment report of 2001 noted that the only way to absolutely prove a human cause for climate change is to do a controlled experiment. In a controlled experiment, you keep all factors unchanging except the one you're testing, and you set aside a group of individuals—a control group—that you can later compare to the group you manipulated (see chapter 2 for more discussion of designing experiments). In the current climate manipulation experiment, however, we have only one earth to work with. So we have no controls, and we cannot keep other factors constant. What we are doing is an uncontrolled experiment—injecting carbon dioxide, methane, and other gases into the atmosphere, and observing changes that follow.

In an uncontrolled experiment, a model is usually the best way to prove cause and effect. You build a computer model, a complex set of equations, that includes variables for all the known natural fluctuations (such as the Milankovitch cycles). You also include

variables for all the known human-caused inputs (CO_2, methane, aerosols, soot, and so on). Then you run the model and see if it can re-create past changes in temperatures.

If you can accurately "predict" past changes, then your model is a good description of how the system works—how the atmosphere responds to more CO_2, how oceans absorb heat, how reduced snow cover contributes feedbacks, and so on.

If you can create a model that represents the system quite well, then you can re-run the model, but this time you leave out the extra CO_2 and other factors we know that humans have contributed. If the model *without* human inputs is *inconsistent* with observed changes in temperature, and if the model *with* human inputs is *consistent* with observations, then you can be extremely confident, beyond the shadow of a reasonable doubt, that the human inputs have made the difference.

Testing detailed climate models against observed temperature trends is exactly what the IPCC and thousands of climate scientists have done in the past 20 years or so. The IPCC provided a comparison of models with and without human inputs (fig. 15.21). In all regions, the models without human inputs (blue) were significantly lower than observed climate records. Models with human-caused changes (pink) are the only way to explain recently observed increases in air temperatures, in ocean temperatures, in declining snow and ice cover, and so on. Different models in the IPCC analysis might vary in the regional severity of changes, or they might disagree on the speed of change, but the direction of change is no longer in doubt.

Scientists are generally cautious about making absolute statements. For a climate scientist, any claims of absolute proof are suspect and probably untrue. Any public statement without measures of uncertainty (how much do you really know, compared to what you don't?) is probably irresponsible. This habit of conservatism makes statements in the Fourth Assessment Report especially emphatic—for a climate scientist. When the report says that "Most of the observed increase in global average temperatures since the mid-twentieth century is *very likely* due to the observed increase in anthropogenic GHG [greenhouse gas] concentrations," it might not look like strong language to you. But this is about as vehement and unanimous a group of scientists as you're likely to find.

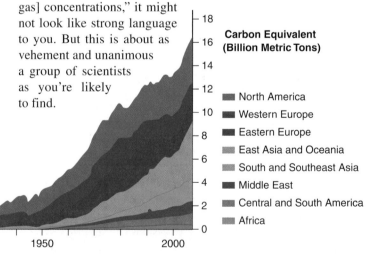

FIGURE 15.20 Carbon emissions by region since 1800. The two largest emitters, China (24%) and the United States (21%), produce nearly half of all emissions.

Data Source: Boden, T.A., G. Marland, and R.J. Andres. 2010. Carbon Dioxide Information Analysis Center, Oak Ridge National Laboratory, U.S. Department of Energy.

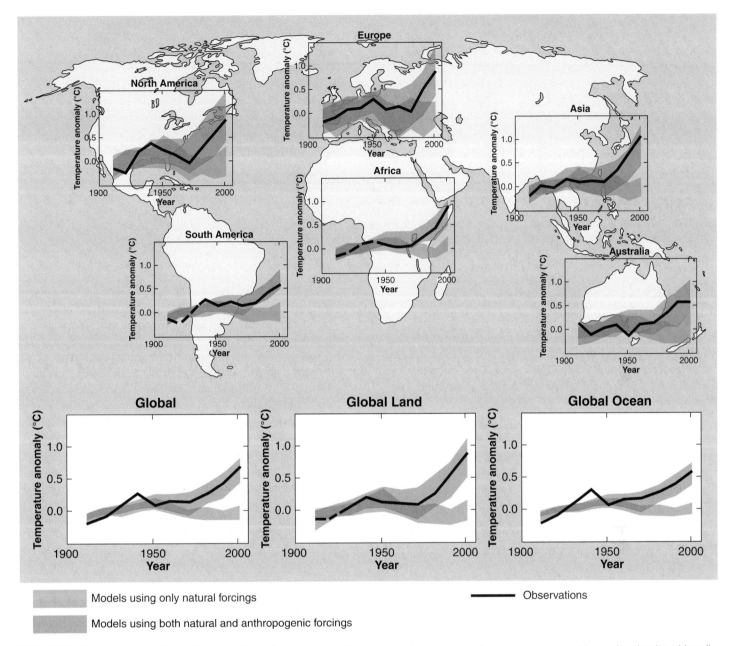

Models using only natural forcings

Models using both natural and anthropogenic forcings

Observations

FIGURE 15.21 Comparison of observed continental- and global-scale changes in surface temperature with results simulated by climate models using either natural or both natural and anthropogenic forcings. Decadal averages of observations are shown for the period 1906–2005 (black line) plotted against the center of the decade and relative to the corresponding average for the period 1901–1950. Lines are dashed where spatial coverage is less than 50 percent. Blue shaded bands show the 5 to 95 percent range for 19 simulations from five climate models using only the natural forcings due to solar activity and volcanoes. Pink shaded bands show the 5 to 95 percent range for 58 simulations from 14 climate models using both natural and anthropogenic forcings.
Source: IPCC 2007, SPM4.

15.5 WHAT EFFECTS ARE WE SEEING?

The American Geophysical Union, one of the nation's largest and most respected scientific organizations, says, "As best as can be determined, the world is now warmer than it has been at any point in the last two millennia, and, if current trends continue, by the end of the century it will likely be hotter than at any point in the last two million years."

Fortunately, as shown by Socolow and Pacala (opening case study) and others, we do have options, if we chose to use them. Mitigating climate change doesn't mean reverting to the Stone Age; it mostly means investing our resources in different kinds of energy. In this section we'll examine some of the consequences of recent climate changes and some of the reasons so many scientists urge us to take action soon. Following this, we'll consider some of the many steps we can take as individuals and as a society to work for a better future.

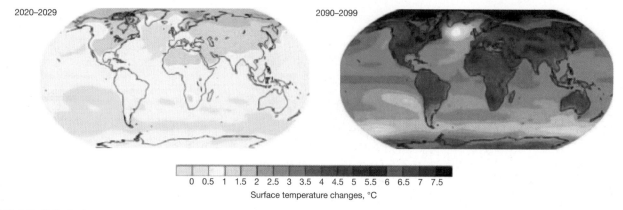

0 0.5 1 1.5 2 2.5 3 3.5 4 4.5 5 5.5 6 6.5 7 7.5
Surface temperature changes, °C

FIGURE 15.22 Surface temperature projections from IPCC scenario B1. This scenario assumes that global population peaks in midcentury and declines thereafter. It also infers rapid introduction of new, cleaner, and more-efficient technologies, but without additional climate initiatives.

Effects include warming, drying, and habitat change

Over the last century the average global temperature has climbed about 0.6°C (1°F). Nineteen of the 20 warmest years in the past 150 have occurred since 1980. New records for hot years are observed with increasing frequency. Here are some effects that have been observed:

- Polar regions have warmed much faster than the rest of the world. In Alaska, western Canada, and eastern Russia, average temperatures have increased as much as 4°C (7°F) over the past 50 years. These extremes are consistent with climate models (fig. 15.22). Permafrost is melting; houses, roads, pipelines, sewage systems, and transmission lines are being damaged as the ground sinks beneath them. Trees are tipping over, and beetle infestations (made possible by warmer winters) are killing millions of hectares of pine and spruce forest from Alaska to Colorado.

- Arctic sea ice is only half as thick now as it was 30 years ago, and the area covered by sea ice has decreased by more than 1 million km^2 (an area larger than Texas and Oklahoma combined) in just three decades. By 2040 the Arctic Ocean could be totally ice-free in the summer. This is bad news for polar bears, which depend on the ice to hunt seals. An aerial survey in 2005 found bears swimming across as much as 260 km (160 mi) of open water to reach the pack ice. When the survey was repeated after a major storm, dozens of bears were missing or spotted dead in the water. The United States has put polar bears on the endangered species list because of loss of Arctic sea ice (fig. 15.23). Loss of sea ice is also devastating for Inuit people, whose traditional lifestyle depends on ice for travel and hunting.

- Ice shelves on the Antarctic Peninsula are breaking up and disappearing rapidly, and Emperor and Adélie penguin populations have declined by half over the past 50 years as the ice shelves on which they depend for feeding and breeding disappear. Ninety percent of the glaciers on the Antarctic Peninsula are now retreating an average of 50 m per year. The Greenland ice cap also is melting twice as fast as it did a few years ago. Because ice shelves are floating, they don't affect sea level when they melt. Greenland's massive ice cap, however, holds enough water to raise sea level by about 7 m (about 23 ft) if it all melts. Melting glaciers and ice caps are contributing about 1 mm per year to sea level rise.

- Half of the world's small glaciers will disappear by 2100, according to a study of 120,000 such glaciers. Mt. Kilimanjaro has lost nearly all its famous ice cap since 1915. In 1972, Venezuela had six glaciers; now it has only two. When Montana's Glacier National Park was created in 1910, it held 150 glaciers. Now fewer than 30 remnants of glaciers remain (fig. 15.24). If current trends continue, all will have melted by 2030.

FIGURE 15.23 Diminishing Arctic sea ice prevents polar bears from hunting seals, their main food source.

FIGURE 15.24 Alpine glaciers everywhere are retreating rapidly. These images show the Grinnell Glacier in 1914 and 1998. By 2030, if present melting continues, there will be no glaciers in Glacier National Park.

- The oceans have apparently been buffering the effects of our greenhouse emissions both by absorbing CO_2 and by storing heat. Deep-diving sensors show that the oceans are absorbing 0.85 watts per m^2 more than is radiated back to space. This absorption slows current warming, but it also means that even if we reduce our greenhouse gas emissions today, it will take centuries to dissipate that stored heat. Absorbed CO_2 is also acidifying the oceans. Because shells of mollusks and corals dissolve at low pH, ocean acidity is likely to alter marine communities.

- Sea level has risen worldwide approximately 15–20 cm (6–8 in.) in the past century. About one-quarter of this increase is ascribed to melting glaciers; roughly half is due to thermal expansion of seawater. If all of Antarctica were to melt, the resulting rise in sea level could be several hundred meters.

- Droughts are becoming more frequent and widespread. In Africa, for example, droughts have increased about 30 percent since 1970. In North America, recent wet winters and hot, dry summers are consistent with climate models (fig. 15.25).

- Extreme droughts in the Amazon rainforest occurred in 2010 and 2005, the two warmest years on record thus far. These droughts, associated with high temperatures in the Atlantic, killed billions more trees than in a normal year, releasing an estimated 8 billion metric tons of CO_2 (more than China produced in 2009). A 2°C temperature rise (the best-case scenario) will destroy 20–40 percent of the Amazon forest, turning it from a carbon sink to a carbon source.

- Biologists report that many animals are breeding earlier or extending their range into new territory as the climate changes. In Europe and North America, for example, 57 butterfly species have either died out at the southern end of their range, or extended the northern limits, or both. Plants also are moving into new territories. Given enough time and a route for migration, many species may adapt to new conditions, but we now are forcing many of them to move much faster than they moved at the end of the last ice age (fig. 15.26). Insect pests and diseases have also expanded their range as hard winters have retreated northward.

- Coral reefs worldwide are "bleaching," losing their photosynthetic algae, as water temperatures rise above 30°C (85°F). With reefs nearly everywhere threatened by pollution, overfishing, and other stressors, biologists worry that rapid climate change could be the final blow for many species in these complex, biologically rich ecosystems.

- Storms are becoming stronger and more damaging. The 2005 Atlantic storm season was the most severe on record, with 26 named tropical storms, twice as many as the average over the past 30 years. This increased frequency and intensity could have other causes, but it is consistent with the expected consequences of rising sea surface temperatures.

Global warming will be costly; preventing it might not be

In 2006 Sir Nicholas Stern, former chief economist of the World Bank, issued a study on behalf of the British government on the costs of global climate change. It was one of the most strongly worded warnings to date from a government report. He said, "Scientific evidence is now overwhelming: climate change is a serious global threat, and it demands an urgent global response." Stern estimated that if we don't act soon, immediate costs of climate

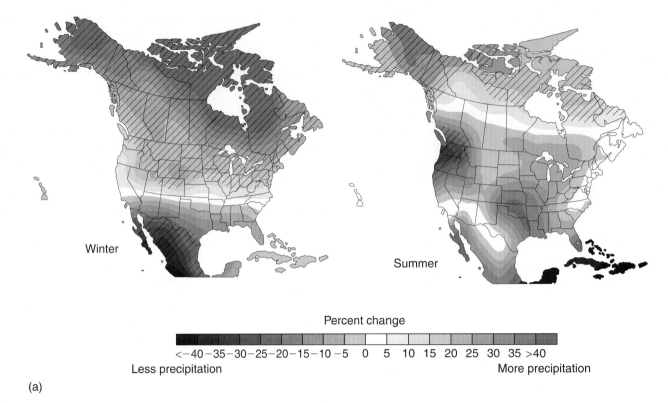

Winter

Summer

Percent change

<-40 -35 -30 -25 -20 -15 -10 -5   0   5 10 15 20 25 30 35 >40

Less precipitation   More precipitation

(a)

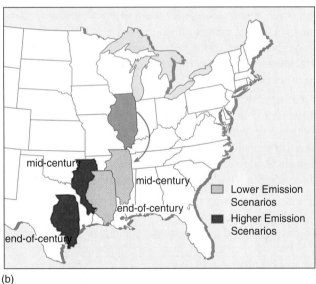

mid-century

mid-century

end-of-century

end-of-century

Lower Emission Scenarios

Higher Emission Scenarios

(b)

FIGURE 15.25 Models predict warmer, wetter winters and drier summers by 2100, compared to recent averages (a). Hatching marks areas with highest confidence in model projections. Midwestern farm states, the core of our food economy, will have summer climate similar to current summers in Louisiana or Texas (b).
Source: U.S. Global Change Research Program, 2009: www.globalchange.gov/usimpacts.

change will be at least 5 percent of the global GDP each year. If a wider range of risks is taken into account, the damage could equal 20 percent of the annual global economy. That would disrupt our economy and society on a scale similar to the great wars and economic depression of the first half of the twentieth century. The fourth IPCC report, meanwhile, estimated that preventing CO_2 doubling and stabilizing the world climate would cost only 0.12 percent of annual global GDP per year.

The Stern report, updated in 2009, estimates that reducing greenhouse gas emissions now to avoid the worst impacts of climate change would cost only about 1 percent of the annual

global GDP. That means that $1 invested now could save us $20 later in this century. "We can't wait the five years it took to negotiate Kyoto," Sir Nicholas says. "We simply don't have time." The actions we take—or fail to take—in the next 10 to 20 years will have a profound effect on those living in the second half of this century and in the next. Energy production, Stern suggests, will have to be at least 80 percent decarbonized by 2050 to stabilize our global climate.

Those of us in the richer countries will likely have resources to blunt problems caused by climate change, but residents of poorer countries will have fewer options. The Stern report says that without action, at least 200 million people could become refugees as their homes are hit by drought or floods. Furthermore, there's a question of intergenerational equity. What kind of world are we leaving to our children and grandchildren? What price will they pay if we fail to act?

The Stern review recommends four key elements for combating climate change. They are: (1) *emissions trading* to promote cost-effective emissions reductions, (2) *technology sharing* that would double research investment in clean-energy technology and accelerate the spread of that technology to developing countries, (3) *reduction of deforestation,* which is a quick and highly

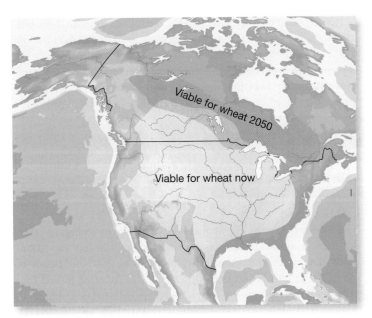

FIGURE 15.26 Most of the central United States is suitable for growing wheat now, but if current trends continue, the climatic conditions for wheat could be in central Canada in 2050.

cost-effective way to reduce emissions, and (4) *helping poorer countries* by honoring pledges for development assistance to adapt to climate change.

Sea-level change will eliminate many cities

About one-third of the world's population now live in areas that would be flooded if all of Greenland's ice were to melt. Even the 75 cm (30 in.) sea-level rise expected by 2050 will flood much of south Florida, Bangladesh, Pakistan, and many other low-lying coastal areas. Most of the world's largest urban areas are on coastlines. Wealthy cities such as New York or London can probably afford to build dikes to keep out rising seas, but poorer cities such as Jakarta, Kolkata, or Manila might simply be abandoned as residents flee to higher ground. Small island countries such as the Maldives, the Bahamas, Kiribati, and the Marshall Islands could become uninhabitable if sea levels rise a meter or more. The South Pacific nation of Tuvalu has already announced that it is abandoning its island homeland. All 11,000 residents will move to New Zealand, perhaps the first of many climate-change refugees.

Insurance companies worry that the $2 trillion in insured property along U.S. coastlines is at increased risk from a combination of high seas and catastrophic storms. At least 87,000 homes in the United States within 150 m (500 ft) of a shoreline are in danger of coastal erosion or flooding in the next 50 years. Accountants warn that loss of land and structures to flooding and coastal erosion together with damage to fishing stocks, agriculture, and water supplies could raise worldwide insurance claims from about $50 billion, which they were in the 1970s, to more than $150 billion per year in 2010. Some of this increase in insurance claims is

due to the fact that more people are living in dangerous places, but extra-severe storms only exacerbate this problem.

Why are there disputes over climate evidence?

Scientific studies have long been unanimous about the direction of climate trends, but commentators on television, newspapers, and radio continue to fiercely dispute the evidence. Why is this? One reason may be that change is threatening, and many of us would rather ignore it or dispute it than acknowledge it. Another reason may be a lack of information. Another is that while scientists tend to look at trends in data, the public might be more impressed by one or two recent events, such as an especially snowy winter in their local area. And on radio and TV, colorful opinions capture more attention than data and graphs. Climate scientists offer the following responses to some of the claims in the popular media:

Reducing climate change requires abandoning our current way of life. Reducing climate change requires that we use different energy. By replacing coal-powered electricity with wind, solar, natural gas, and improved efficiency, we can drastically cut our emissions but keep our computers, TVs, cars, and other conveniences. Reducing coal dependence will also reduce financial costs of pollution damage to health and vegetation.

There is no alternative to current energy systems. As long as we invest only in fossil fuels, this will be true, but Chinese and European energy companies are creating new markets and jobs in energy and improved efficiency. Fossil fuels rely heavily on abundant public subsidies; shifting subsidies toward solar, wind, and other technologies would make these alternatives economical, and more profitable than the energy and transportation technologies of the 1940s.

A comfortable lifestyle requires high CO_2 output. Most northern Europeans produce half the CO_2 of North Americans, per capita. Yet they have higher standards of living (in terms of education, health care, life span, vacation time, financial security) than North Americans. Residents of San Francisco consume about one-sixth as much energy as residents of Kansas City, yet quality of life is not necessarily six times greater in Kansas City than in San Francisco.

Natural changes such as solar variation can explain observed warming. Solar input fluctuates, but the changes are slight and do not coincide with the direction of changes in temperatures. Milankovitch cycles also cannot explain the rapid changes in the past few decades. Increased observed temperature and sea-level changes, however, do correspond closely with GHG emissions (fig. 15.19).

The climate has changed before, so this is nothing new. Today's CO_2 level of roughly 392 ppm exceeds by at least 30 percent anything the earth has seen for nearly a million years, and perhaps as long as 15 million years. Antarctic ice cores indicate that CO_2 concentrations for the past 800,000 years have varied from 180 to 300 ppm (see fig. 15.14). This natural variation in CO_2 appears to be a feedback in glacial cycles, resulting from changes in biotic activity in warm periods. Because temperature closely tracks CO_2, temperatures by 2100 are likely to exceed anything in the past million years. The rate of change is probably also unprecedented. Changes that took 1,000 to 5,000 years at the end of ice ages are now occurring on the scale of a human lifetime.

Temperature changes are leveling off. Over short time frames, temperature trends vary (fig. 15.1), but over decades the trends in surface air temperatures and in sea level continue to rise.

We had cool temperatures and snowstorms last year, not heat and drought. Regional differences in temperature and precipitation trends are predicted by climate models. Most of the United States is expected to see wetter, warmer winters and drier, hotter summer (fig. 15.25).

Climate scientists don't know everything, and they have made errors in the past. The gaps and uncertainties in climate data are minute compared to the evident trends. There are many unknowns—details of precipitation change or interaction of long-term cycles such as El Niño—but the trends are unequivocal. Climatologist James Hansen has noted that while most people make occasional honest mistakes, fraud in data collection is almost unheard of. This is because transparency in the scientific process ensures public visibility of errors. There is much less public accountability in popular media, however, where climate scientists are regularly subjected to personal attacks from climate-change deniers.

15.6 ENVISIONING SOLUTIONS

Dire warnings of climate change are intimidating, but in response, individuals and communities around have been working on countless promising and exciting strategies to mitigate these changes. All of these efforts, at all scales, are valuable. In this section we'll look at some of these strategies. Curbing climate change is a daunting task, but it is also full of opportunity.

We can establish new rules and standards

In 1997 a meeting in Kyoto, Japan, called together climate scientists and government representatives from around the globe. This meeting was a follow-up to the 1992 Earth Summit in Rio de Janeiro, Brazil, at which most nations had agreed in principle that sustainable development—equitable growth that doesn't destroy opportunities for future generations—was a good idea. The **Kyoto Protocol** (agreement) followed this general principle and called on nations to roll back emissions of CO_2, CH_4, and N_2O. The goal was to reach 5 percent below 1990 levels by 2012. Poorer nations, such as India and China, were exempt, allowing them to expand their economies and improve standards of living.

The Kyoto protocol went into effect in 2005. Each country signing the agreement is now responsible for following through and reducing emissions. Considerable progress has occurred, especially in western Europe, but most countries are behind their Kyoto targets. Among developed nations, only the United States and Australia still declined to sign the protocol. The United States has persistently claimed that reducing carbon emissions would be too costly and that we must "put the interests of our own country first and foremost."

Many other smaller agreements have followed Kyoto. New policy strategies have also been implemented, including carbon-trading markets in Europe, Asia, and North America. In carbon trading, or cap-and-trade systems, legal limits are set on emissions, and countries with lower emissions can sell their emissions credits, or their "right to pollute," to someone else. The Kyoto Protocol promoted this approach.

The global market for trading carbon emission credits has grown rapidly. In 2006 about 700 million tons of carbon equivalent credits were exchanged, with a value of some $3.5 billion. By 2010, trade had grown to 7 billion tons, despite economic slowdowns that reduced carbon emissions, and thus prices for carbon credits, in 2009.

Business groups are understandably wary of changing rules, but increasingly they are saying that they can accept new standards if they are clear and fairly applied. In 2007 the heads of ten of the largest business conglomerates in America joined four environmental groups to call for strong national legislation to achieve significant reductions in greenhouse gases. The corporations included Alcoa, BP America, Caterpillar, DuPont, General Electric, and others. The nongovernmental organizations were Environmental Defense, the Pew Center, Natural Resources Defense Council, and World Resources Institute. That initiative was expanded in 2009 by the group Business for Innovative Climate & Energy Policy (BICEP), which has asked the Obama administration to reduce greenhouse gases by 80 percent below 1990 levels by 2050. Members of this group, including Gap, Inc., eBay, and others have received support from EPA administrator Lisa Jackson.

These companies want the U.S. economy to remain competitive as international policies about greenhouse gases change. They also prefer a single national standard rather than a jumble of conflicting local and state rules. This complex landscape of differing rules is a very real possibility, as many states and cities are beginning to lead the way in curbing their own emissions (What Do You Think? p. 339). Knowing that climate controls are inevitable, businesses want to know now how they'll have to adapt, rather than wait until a crisis causes us to demand sudden, radical changes.

Stabilization wedges could work now

As discussed in the opening case study, the idea of stabilization wedges is that they can work just by expanding currently available technologies. To stabilize carbon emissions, we would need to cut about 7 GT in 50 years; to reduce CO_2, as called for in the Kyoto Protocol, we could add another seven wedges (fig. 15.2).

Because most of our CO_2 emissions come from fossil fuel combustion, energy conservation and a switch to renewable fuels are important. Doubling vehicle efficiency and halving the miles we drive would add up to 1.5 of the 1-GT wedges. Installing efficient lighting and appliances, and insulating buildings, could add up to another 2 GT. Capturing and storing carbon released by power plants, gas wells, and other sources could save another gigaton.

States Take the Lead on Climate Change

In 2006, California passed a groundbreaking law that places a cap on emissions of carbon dioxide and other global warming gases from utilities, refineries, and manufacturing plants. The law aims to roll back the state's greenhouse gas releases to 1990 levels (a reduction of 15 percent) by 2020, and to 80 percent below 1990 levels by 2050. Reductions involve enforceable caps on emissions, monitored through regular industry emissions reports. Companies that cut emissions below their maximum allowance can profit by selling credits to other companies that have not met their caps. Putting a price on carbon emissions is creating incentives for innovation, which can now be cheaper than polluting. At the same time the cost of implementing the plan is low, and industries can meet standards in any way they choose. The legislation addresses a wide range of carbon sources, including agriculture, cement production, electricity generation, and suburban sprawl. Utilities and corporations are also prohibited from buying power from out-of-state suppliers whose sources don't meet California's emission standards. All these can be seen online at the "California Climate Change Portal."

This rule is the most aggressive climate-change effort of any state, but California voters strongly support it. When the energy industry challenged it in a ballot initiative in 2010, claiming it would cost the state jobs, 62 percent of voters still voted to keep the law. California has often led the way in improving air quality. In 2004 the state passed revolutionary legislation that required automakers to cut tailpipe emissions of carbon dioxide from cars and trucks, which has since been picked up by New York and other states. When car manufacturers failed to comply, California sued the six largest automakers in 2006, charging that they were costing the state billions of dollars in health and environmental damages.

What inspires such revolutionary steps? One factor is that California's economy relies almost entirely on declining winter snowpack for both urban water use and farm irrigation. Recent years of severe droughts have affected much of the state and worried cities and counties. Californians also have gotten tired of waiting for action in Washington, where the dominant view has been that climate controls will cost jobs. Contrary to this argument, California has seen rapid job growth in clean energy. Between 1995 and 2008, clean-energy businesses grew by 45 percent, 10 times the state's average growth rate. Clean energy employs over 500,000 people and has brought in over \$9 billion in venture capital, or 60 percent of all clean-energy investments nationwide.

Following this lead, most U.S. states and more than 500 cities have taken steps to promote renewable energy and reduce greenhouse gas emissions. Massachusetts announced in 2010 that, like California, it will cut greenhouse gases by 25 percent by 2020. Strategies the states are taking include efficient building standards, support for alternative energy, more-efficient distribution grids, land-use planning standards, support for retrofitting old houses, and auto insurance incentives for efficient vehicles.

Carbon trading has also caught on, with 27 states and four Canadian provinces participating in three regional carbon-trading compacts—the Midwestern Greenhouse Gas Reduction Accord, the Western Climate Initiative, and the northeastern Regional Greenhouse Gas Initiative. The northeastern compact (RGGI) began trading carbon credits for 233 plants in 2008. By 2010, carbon credit auctions produced more than \$700 million in revenue to support conservation and alternative energy initiatives in participating states.

Carbon trading is not perfect: carbon prices are often too low to provide real incentives for some industries; many question whether a "right to pollute" is the best strategy; and carbon revenues risk being diverted to states' general funds, as happened in New York in 2010. However, these compacts are widely considered successful—and palatable—approaches to reducing emissions.

New rules are a challenge to industry, but they can also lead to greater efficiency in operations, and changes are generally manageable if they are predictable and evenly applied. Still, these rules have been difficult to establish in Washington. If you were in Congress, what evidence would you want to see in order to buy into some of these state-led innovations?

Pacala and Socolow's original 14 wedges are paraphrased in table 15.3. As the authors note, nobody will agree that all the wedges are a good idea, and all have some technological limitations, but none are as far off as revolutionary technologies such as nuclear fusion. Some analysts have subsequently proposed still additional wedges, and technologies that make these wedges possible, or that point to new ones, are changing rapidly.

Alternative practices can be important

Carbon capture and storage, one of the important stabilization wedges, is beginning to be widely practiced. Norway's state oil company, Statoil, which extracts oil and gas from beneath the North Sea, has been pumping more than 1 million metric tons of CO_2 per year into an aquifer 1,000 m below the seafloor at one of its North Sea gas wells. Injecting CO_2 increases pressure on oil reservoirs and enhances oil recovery. It also saves money because the company would have to pay a \$50 per ton carbon tax on its emissions. Around the world, deep saltwater aquifers could store a century's worth of CO_2 at current fossil fuel consumption rates.

Carbon capture and injection is widely practiced for improving oil and gas recovery, so the technology is available (fig. 15.27). There are concerns about leaking from deep storage, but the main concern is that there have been few compelling economic arguments. Carbon taxes, or carbon trading, could be strategies to justify carbon capture.

Most attention is focused on CO_2 because it is our most abundant greenhouse gas, but methane is also important because, although we produce less of it, methane is a much more powerful absorber of infrared energy. Some atmospheric scientists think the best short-term strategy might be to focus on methane.

Methane from landfills, oil wells, and coal mines is now being collected in some places for fuel. Rice paddies are another major methane source. Changing flooding schedules and fertilization techniques can reduce some of these emissions. Reducing gas pipeline leaks would conserve this resource as well as reduce warming. Finally, ruminant animals (such as cows, camels, sheep)

Table 15.3	Actions to Reduce Global CO_2 Emissions by 1 Billion Tons over 50 Years
1.	Double the fuel economy for 2 billion cars from 30 to 60 mpg.
2.	Cut average annual travel per car from 10,000 to 5,000 miles.
3.	Improve efficiency in heating, cooling, lighting, and appliances by 25 percent.
4.	Update all building insulation, windows, and weather stripping to modern standards.
5.	Boost efficiency of all coal-fired power plants from 32 percent today to 60 percent (through co-generation of steam and electricity).
6.	Replace 800 large coal-fired power plants with an equal amount of gas-fired power (four times current capacity).
7.	Capture CO_2 from 800 large coal-fired or 1,600 gas-fired, power plants and store it securely.
8.	Replace 800 large coal-fired power plants with an equal amount of nuclear power (twice the current level).
9.	Add 2 million 1 MW windmills (50 times current capacity).
10.	Generate enough hydrogen from wind to fuel a billion cars (4 million 1 MW windmills).
11.	Install 2,000 GW of photovoltaic energy (700 times current capacity).
12.	Expand ethanol production to 2 trillion liters per year (50 times current levels).
13.	Stop all tropical deforestation and replant 300 million ha of forest.
14.	Apply conservation tillage to all cropland (10 times current levels).

Source: Data from Pacala and Socolow, 2004.

are a major source of methane. Modifying human diets, including less beef consumption, could reduce methane significantly.

There are many regional initiatives

Many countries are working to reduce greenhouse emissions. The United Kingdom, for example, had already rolled CO_2 emissions back to 1990 levels by 2000 and vowed to reduce them 60 percent by 2050. Britain already has started to substitute natural gas for coal, promote energy efficiency in homes and industry, and raise its already high gasoline tax. Plans are to "decarbonize" British society and to decouple GNP growth from CO_2 emissions. A revenue-neutral carbon levy is expected to lower CO_2 releases and trigger a transition to renewable energy over the next five decades. In 2007, New Zealand's prime minister, Helen Clark, pledged that her country would be **carbon neutral** by 2025, through a combination of wind and geothermal energy, carbon capture on farms, and other strategies.

Germany also has reduced its CO_2 emissions at least 10 percent by switching from coal to gas and by encouraging energy efficiency throughout society. Atmospheric scientist Steve Schneider calls this a "no regrets" policy—even if we don't need to stabilize our climate, many of these steps save money, conserve resources, and have other environmental benefits. Nuclear power also is being promoted as an energy alternative that produces no greenhouse gases directly and that provides high-volume, centralized power production. It remains an imperfect option because greenhouse gases and other pollutants are produced in mining, processing, and transporting nuclear fuel. There are also security worries and unresolved problems of how to store wastes safely. Still this is an option favored by many states and utility companies.

Many people believe renewable energy sources offer the best solution to climate problems. Chapter 20 discusses options for conserving energy and switching to renewable sources, such

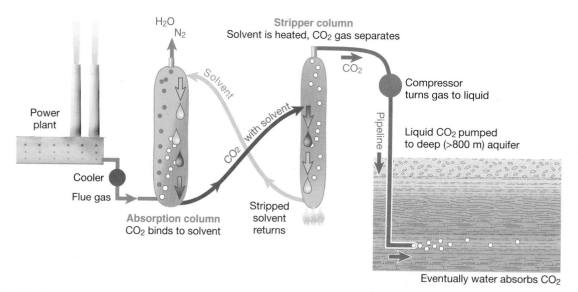

FIGURE 15.27 One method of carbon capture and storage uses a liquid solvent, such as ammonia, to capture CO_2. Steam and nitrogen are released, and the CO_2 is compressed and pumped to deep aquifers for permanent storage.

Reducing Carbon Dioxide Emissions

Individual actions can have tremendous impacts on climate change, because our actions are multiplied by the millions of others who make similar decisions. Many of existing options save money in the long run and have other benefits such as reducing pollution and resource consumption.

The most obvious strategies involve domestic transportation, heating, and lighting, which together make up roughly 40 percent of our national production of CO_2. You can drive less, walk, bike, take public transportation, carpool, or buy a vehicle that gets at least 30 mpg (12.6 km/l). Average annual CO_2 reductions are about 20 lbs (9 kg) for each gallon of gasoline saved. Replacing standard incandescent light bulbs with compact fluorescents or other efficient bulbs is another easy and money-saving fix. Average annual CO_2 reductions are about 500 lbs (0.23 metric tons) per bulb, or 10,000 lbs (4.6 metric tons) for every 20 bulbs in your household.

A recent study of behavior and household options found that we could reduce U.S. emissions by 233 metric tons of carbon with these simple changes. These strategies—another example of wedge analysis at the household level—could reduce total emissions by 7.4 percent in 10 years without any new regulations, technology, or reductions in well-being. Transportation efficiency would make the most rapid difference (fig. 1).

To read more, see T. Dietz et al., 2009. Household actions can provide a behavioral wedge to rapidly reduce U.S. carbon emissions. *Proceedings of the National Academy of Sciences* 106(44): 18452–56

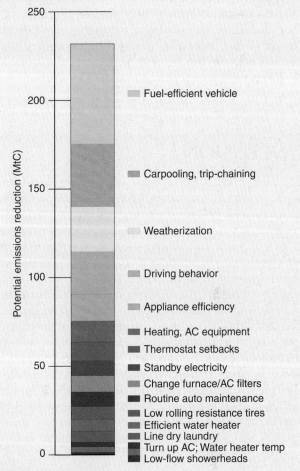

FIGURE 1 Potential impact on emissions in 10 years if available strategies were widely adopted.
Adapted from Dietz at., 2009.

as solar, wind, geothermal, biomass, and fuel cells. Denmark, the world's leader in wind power, now gets 20 percent of its electricity from windmills. Plans are to generate half of the nation's electricity from offshore wind farms by 2030.

Countless individual cities and states have announced their own plans to combat global warming. Among the first of these were Toronto, Copenhagen, and Helsinki, which pledged to reduce CO_2 emissions 20 percent from 1990 levels by 2010. Some corporations are following suit. British Petroleum set a goal of cutting CO_2 releases from all its facilities by 10 percent before 2010. Each of us can make a contribution in this effort. As Professor Socolow and his colleagues point out, simply driving less and buying high-mpg vehicles could save about 1.5 billion tons of carbon emissions by 2054 (fig. 15.28; What Can You Do? see above).

FIGURE 15.28 Burning fossil fuels produces about half our greenhouse gas emission, and transportation accounts for about half of our fossil fuel consumption. Driving less, choosing efficient vehicles, carpooling, and other conservation measures are among our most important personal choices in the effort to control global warming.

In the midst of all the debate about how serious the consequences of global climate change may or may not be, we need to remember that many of the proposed solutions are advantageous in their own right. Even if climate change turns out not to be as much of a threat as we think now, they have other positive benefits. Moving from fossil fuels to renewable energy sources such as solar or wind power, for example, would free us from dependence on foreign oil and improve air quality. Planting trees makes cities pleasant places to live and provides habitat for wildlife. Making buildings more energy efficient and buying efficient vehicles saves money now and in the long run. Walking, biking, and climbing stairs are good for your health, and they help reduce traffic congestion and energy consumption. Reducing waste, recycling, and other forms of sustainable living improve our environment in many ways in addition to helping fight climate change. It's important to focus on these positive effects rather than to look only at the gloom-and-doom scenarios for global climate catastrophes. As the Irish statesman and philosopher Edmund Burke said, "Nobody made a greater mistake than he who did nothing because he could do only a little."

CONCLUSION

Climate change may be the most far-reaching issue in environmental science today. Although the challenge is almost inconceivably large, solutions are possible if we choose to act, as individuals and as a society. Temperatures are now higher than they have been in thousands of years, and climate scientists say that if we don't reduce greenhouse gas emissions soon, drought, flooding of cities, and conflict may be inevitable. The "stabilization wedge" proposal is a list of immediate and relatively modest steps that could be taken to accomplish needed reductions in greenhouse gases.

Understanding the climate system is essential to understanding the ways in which changing composition of the atmosphere (more carbon dioxide, methane, and nitrous oxide, in particular) matters to us. Basic concepts to remember about the climate system include how the earth's surfaces absorb solar heat, how atmospheric convection transfers heat, and that different gases in the atmosphere absorb and store heat that is reemitted from the earth. Increasing heat storage in the lower atmosphere can cause increasingly vigorous convection, more extreme storms and droughts, melting ice caps, and rising sea levels. Changing patterns of monsoons, cyclonic storms, frontal weather, and other precipitation patterns could have extreme consequences for humans and ecosystems.

Despite the importance of natural climate variation, observed trends in temperature and sea level are more rapid and extreme than other changes in the climate record. Exhaustive modeling and data analysis by climate scientists show that these changes can be explained only by human activity. Increasing use of fossil fuels is our most important effect, but forest clearing, decomposition of agricultural soils, and increased methane production are also extremely important.

International organizations, national governments, and local communities have all begun trying to reverse these changes. Individual actions and commitment are also essential if we are to avoid dramatic and costly changes in our own lifetimes.

REVIEWING LEARNING OUTCOMES

By now you should be able to explain the following points:

15.1 Describe the general composition and structure of the atmosphere.

- Absorbed solar energy warms our world.
- The greenhouse effect is energy captured by gases in the atmosphere.
- Evaporated water stores energy, and winds redistribute it.

15.2 Explain why weather events follow general patterns.

- Why does it rain?
- The Coriolis effect explains why winds seem to curve.
- Ocean currents modify our weather.
- Much of humanity relies on seasonal rain.

- Frontal systems create local weather.
- Cyclonic storms cause extensive damage.

15.3 Outline some factors in natural climate variability.

- Ice cores tell us about climate history.
- The earth's movement explains some cycles.
- El Niño is an ocean–atmosphere cycle.

15.4 Explain the nature of anthropogenic climate change.

- The IPCC assesses data for policymakers.
- How does climate change work?
- Positive feedbacks accelerate change.
- How do we know recent change is human-caused?

15.5 What effects are we seeing?

- Effects include warming, drying, and habitat change.
- Global warming will be costly; preventing it might not be.
- Sea-level change will eliminate many cities.
- Why are there disputes over climate evidence?

15.6 Identify some solutions to slow climate change.

- We can establish new rules and standards
- Stabilization wedges could work now.
- Alternative practices can be important.
- There are many regional initiatives.

PRACTICE QUIZ

1. What are the dominant gases that make up clean, dry air?
2. Name and describe four layers of the atmosphere.
3. What is the greenhouse effect? What is a greenhouse gas?
4. What are some factors that influence natural climate variation?
5. Explain the following: Hadley cells, jet streams, Coriolis effect.
6. What is a monsoon, and why is it seasonal?
7. What is a cyclonic storm?
8. Identify 5 to 10 actions we take to increase greenhouse gases in the atmosphere.
9. What is the IPCC, and what is its function?
10. What method has the IPCC used to demonstrate a human cause for recent climate changes? Why can't we do a proper manipulative study to prove a human cause?
11. List 5 to 10 effects of changing climate.
12. What is a climate stabilization wedge? Why is it an important concept?
13. What is the Kyoto Protocol?
14. List several actions cities, states, or countries have taken to unilaterally reduce greenhouse gas emissions.

CRITICAL THINKING AND DISCUSSION QUESTIONS

1. Weather patterns change constantly over time. From your own memory, what weather events can you recall? Can you find evidence in your own experience of climate change? What does your ability to recall climate changes tell you about the importance of data collection?
2. Many people don't believe that climate change is going on, even though climate scientists have amassed a great deal of data to demonstrate it. What factors do you think influence the degree to which a person believes or doesn't believe climatologists' reports?
3. How does the decades-long, global-scale nature of climate change make it hard for new policies to be enacted? What factors might be influential in people's perception of the severity of the problem?
4. What forces influence climate most in your region? in neighboring regions? Why?
5. Of the climate wedges shown in table 15.3, which would you find most palatable? least tolerable? Why? Can you think of any additional wedges that should be included?
6. Would you favor building more nuclear power plants to reduce CO_2 emissions? Why or why not?

The Intergovernmental Panel on Climate Change (IPCC) has a rich repository of figures and data, and because these data are likely to influence some policy actions in your future, it's worthwhile taking a few minutes to look at the IPCC reports.

The most brief and to the point is the Summary for Policy Makers (SPM) that accompanies the fourth Assessment Report. You can find the summary at www.ipcc.ch/ipccreports/ar4-syr .htm. If you have time, the full report is also available at this site.

Open the SPM and look at the first page of text, then look at the first figure, SPM1 (reproduced here). Look at this figure carefully and answer the following questions:

1. What is the subject of each graph? Why are all three shown together?

2. Carefully read the caption. What does the area between the blue lines represent? Why are the blue lines shown in this report?

3. The left axis for all three graphs shows the difference between each year's observations and an average value. What values are averaged?

4. What do the blue lines represent? In the third graph, what is the value of the blue line, in million km^2, for the most recent year shown? Approximately what year had the lowest value shown? What does a decline in this graph represent on the ground?

5. Why is the trend in the snow cover graph less steep than the trends in the other two graphs?

6. Nearly every page of the IPCC report has graphs that show quite interesting details when you take the time to look at them. Choose two other graphs in the SPM document and explain the main messages they give. See if you can explain them clearly enough to communicate the main idea to a friend or family member. Have different students select different graphs and explain them to the class.

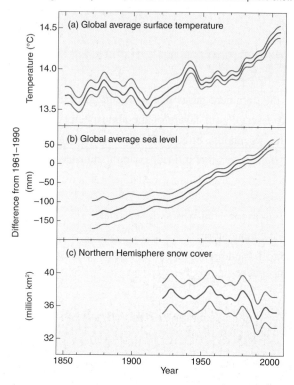

Changes in temperature, sea level and Northern Hemisphere snow cover

See the evidence: view the IPCC report at www.ipcc.ch/graphics/graphics/syr/ spml.jpg.

For Additional Help in Studying This Chapter, please visit our website at www.mhhe.com/cunningham12e. You will find additional practice quizzes and case studies, flashcards, regional examples, placemarks for Google Earth™ mapping, and an extensive reading list, all of which will help you learn environmental science.

The Great London Smog of 1952 killed thousands and helped change the way we see air pollution.

Air Pollution

"The only thing we have to fear is fear itself."

~ Franklin D. Roosevelt

Learning Outcomes

After studying this chapter, you should be able to:

16.1 Describe the air around us.

16.2 Identify natural sources of air pollution.

16.3 Discuss human-caused air pollution.

16.4 Explain how climate topography and atmospheric processes affect air quality.

16.5 Compare the effects of air pollution.

16.6 Evaluate air pollution control.

16.7 Summarize current conditions and future prospects.

Case Study The Great London Smog

London was once legendary for its pea-soup fogs. In the days of Charles Dickens and Sherlock Holmes, darkened skies and blackened buildings, saturated with soot from hundreds of thousands of coal-burning fireplaces, were a fact of life. Londoners had been accustomed to filthy air since the beginning of the industrial revolution, but over a period of four days in 1952, just over 60 years ago, they experienced the worst air pollution disaster on record. Smoke, soot, and acidic droplets of fog made the air opaque. Thousands died, thousands more became ill, and our view of air pollution changed forever.

In early December 1952 a dense blanket of coal smoke and fog settled on the city. Under normal conditions, winds keep polluted air moving, away from the city and out over the countryside. These winds occur, broadly speaking, because air near the earth's surface is usually warmed by the sun-heated ground, while air aloft is cool. Turbulence develops as the warmed air rises and cool air sinks, and the turbulence tends to circulate pollutants away from their sources. But from time to time an inversion develops. As the name implies, an inversion occurs when layers of air are out of order: the still, cold air settles near the ground, trapped by warmer layers above. On a cold, dark December day in London, the cold air can settle in to stay.

Inversions can be unpleasantly chilly and damp, but in a city where coal was the primary fuel, burned in countless low-efficiency stoves and furnaces, the stable inversion conditions also trapped smoke, particulates of coal dust, and tiny droplets of sulfuric acid (from sulfur in coal) in the city.

The "killer smog" of 1952 came on suddenly, on Friday, December 5. Home heaters and industrial furnaces were working in full force, pumping out smoke on the cold winter day. During the afternoon, visibility plummeted, and traffic came to a halt as drivers were blinded by the smoke and fog. Hundreds of cattle at a cattle market were the first to go. With lungs blackened by soot, they suffocated while standing in their pens. People could cover their faces and go indoors, but the soot soon reached inside buildings, as well. Concerts were canceled because of blackened air in the halls, and books in the British Museum were tainted with soot. Visibility fell to a foot in some places by the third day of the inversion.

The ill and elderly, especially those with lung or heart ailments and the heavy smokers, were the next to go. Hospitals filled with victims of bronchitis, pneumonia, lung inflammations, and heart failure. Like the cattle in the market, victims' lungs were clogged by smoke and microscopic soot particles, their lips turned blue, and they asphyxiated due to lack of oxygen. Healthy people tried to stay indoors and keep quiet, and children were kept home from school—as much to prevent them from getting lost in the dark as because of the air quality.

Four days later, a change in the weather brought fresh winds into London, and the inversion dissipated. Studies showed that at least 4,700 deaths were attributable to air pollution during and immediately after the inversion. More-recent epidemiological studies have found that lingering ailments killed perhaps another 8,000 in the months that followed, bringing the total death toll to over 12,000.

Air pollution wasn't generally treated as a problem at the time. The weekend of December 5 just had worse smoky fog than usual. Everyone understood that coal smoke was unhealthy, of course, but grimy air and illness were a cost of living in the city. Controls on smoke had been proposed for centuries, since at least 1300, but pollution was too normal, and too pervasive, to change. We rarely consider normal conditions a problem, or imagine alternatives, until a crisis makes us start to question the costs of customary ways of doing business.

The smog of 1952 turned out to be one such crisis. Alarming death counts caught the attention of politicians and the public alike, and gradually led to changes in expectations and practices. New government policies gradually began to phase out coal fireplaces, replacing them with oil burners and other forms of heat. New efforts were made to monitor air quality and to put limits on industrial pollutants. These changes were solidified in the United Kingdom's Clean Air Act of 1956, which established health standards and helped homeowners convert to other heat sources. A decade later, the 1968 Clean Air Act expanded these rules to address industrial emissions. In the United States a similar Clean Air Act was adopted in 1963, with major amendments in 1970.

Although air quality in cities (and often in the countryside) is frequently worse than we would like, extreme conditions like the killer smog are mainly historical curiosities today. We now have higher expectations for air pollution control, and we no longer find it acceptable—at least in principle—for private citizens or industries to emit pollutants that cause illness or death. Air quality standards exist for smoke, particulate matter, sulfuric acid, heavy metals, and other contaminants, all of which can now be captured before they leave the smokestack. New practices and rules keep our environment cleaner, reduce health care costs, and protect buildings, forests, and farms from the effects of air pollution.

Coal burning remains one of our greatest challenges in air pollution control, though the source is now electricity-producing power plants. But there are many other sources. Every year millions of additional cars create more nitrogen oxides and carbon dioxide. Industry produces new classes of hazardous organic air pollutants, airborne metals are emitted from coal burning and mining, and dust remains a serious problem. Asthma and cardiovascular conditions remain elevated in cities with poor air quality. But things are nowhere near as bad as they were in London in 1952. In this chapter we'll examine major types and sources of air pollutants. We'll also consider policies and technology that help ensure that events like the smog of London doesn't happen in your town.

For related resources, including Google Earth™ placemarks that show locations discussed in this chapter, visit Environmental-Science-Cunningham.blogspot.com.

16.1 THE AIR AROUND US

How does the air taste, feel, smell, and look in your neighborhood? Chances are that wherever you live, the air is contaminated to some degree. Smoke, haze, dust, odors, corrosive gases, noise, and toxic compounds are among our most widespread pollutants. According to the Environmental Protection Agency (EPA), human activities release some 147 million metric tons of air pollutants (not counting carbon dioxide or windblown soil) into the atmosphere each year in the United States alone. Worldwide emissions are around 2 billion metric tons per year (table 16.1).

Especially in the burgeoning megacities of rapidly industrializing countries (chapter 22), air pollution often exceeds World Health Organization standards. In many Chinese cities, for example, airborne dust, smoke, and soot often are ten times higher than levels considered safe for human health (fig. 16.1). Currently, 16 of the 20 smoggiest cities in the world are in China. Worldwide we continue to have low-level, chronic exposure to pollutants. When millions of people are exposed over many years to these risks, the cumulative number of injuries and deaths may actually be greater than from notable events like those in London in 1952.

Table 16.1 Estimated Fluxes of Pollutants and Trace Gases to the Atmosphere

Species	Sources	Approximate Annual Flux (Millions of Metric Tons/Yr)	
		Natural	Anthropogenic
CO_2 (carbon dioxide)	Respiration, fossil fuel burning, land clearing, industrial processes	370,000*	29,600
CH_4 (methane)	Rice paddies and wetlands, gas drilling, landfills, animals, termites	155	350
CO (carbon monoxide)	Incomplete combustion, CH_4 oxidation, plant metabolism	1,580	930
Non-methane hydrocarbons	Fossil fuel burning, industrial uses, volatile compounds from plants	860	92
NO_x (nitrogen oxides)	Fossil fuel burning, lightning, biomass burning, soil microbes	90	140
SO_x (sulfur oxides)	Fossil fuel burning, industry, biomass burning, volcanoes, oceans	35	79
SPM (suspended particulate materials)	Biomass burning, dust, sea salt, biogenic aerosols	583	362

*Natural flux to atmosphere is balanced over time by capture, deposition, or decomposition of gases or SPM.

FIGURE 16.1 On a smoggy day in Shanghai (*left*) visibility is less than 1 km. Twenty-four hours later, after a rainfall (*right*), the air has cleared dramatically.

FIGURE 16.2 Natural pollution sources, such as volcanoes, can be important health hazards.

Despite these challenges, most developed countries no longer have acute air pollution episodes like London's killer smog. Many people are surprised to learn that a generation ago most American cities were much dirtier than they are today. We've cleaned up many of the worst pollution sources, especially those that are large, centralized, and easy to monitor, and we have better standards and technology for many smaller sources. The many improvements in air quality demonstrate that dramatic progress can be made in solving environmental problems. Continuing industry challenges to clean air rules also indicate that we can't be complacent. Public attention is always needed to protect the safeguards we now rely on.

There are many natural air pollutants

It is difficult to give a simple, comprehensive definition of pollution. The word comes from the Latin *pollutus*, which means "made foul, unclean, or dirty." Some authors use the term only for damaging materials that are released into the environment by human activities. There are, however, many natural sources of air quality degradation. Volcanoes spew out ash, acid mists, hydrogen sulfide, and other toxic gases (fig. 16.2). Sea spray and decaying vegetation are major sources of reactive sulfur compounds in the air. Trees and bushes emit millions of tons of volatile organic compounds (terpenes and isoprenes). These compunds create, for example, the blue haze that gave the Blue Ridge Mountains their name. Storms in arid regions raise dust clouds that transport millions of tons of soil. Bacterial metabolism of decaying vegetation in swamps and of cellulose in the guts of termites and ruminant animals is responsible for as much as two-thirds of the methane in the air.

For these compounds, the difference between natural and human-caused sources is mainly in concentrations, as in cities, and in our production of amounts greater than natural systems

can absorb. Many substances are innocuous at naturally occurring levels, but high concentrations in cities or industrial areas can exceed our physical ability to tolerate them. In many cities and agricultural regions, for example, more than 90 percent of the airborne particulate matter is anthropogenic (human-caused). Effects include asthma, allergies, and heart and lung ailments.

16.2 Major Types of Pollutants

Throughout history, countless ordinances have prohibited emission of objectionable smoke, odors, and noise. Air pollution traditionally has been treated as a local problem. The Clean Air Act of 1963 was the first national legislation in the United States aimed at air pollution control. The act provided federal grants to states to combat pollution but was careful to preserve states' rights to set and enforce air quality regulations. But it soon became obvious that some pollution problems could not be solved on a local basis.

Criteria pollutants were addressed first

Amendments to the law in 1970 essentially rewrote the U.S. Clean Air Act. Congress designated new standards, to be applied evenly across the country, for six major pollutants: sulfur dioxide, nitrogen oxides, carbon monoxide, ozone, lead, and particulate matter. These standards were set according to health criteria and environmental quality. National ambient air quality standards (NAAQS) identify maximum allowable limits for these (**ambient air** is the air around us). These six **conventional** or **criteria pollutants** were addressed first because they contributed the largest volume of air quality degradation and also are considered the most serious threat of all air pollutants to human health and welfare. Primary standards (table 16.2) are intended to protect human health. Secondary standards are also set to protect crops, materials, climate, visibility, and personal comfort.

Table 16.2	National Ambient Air Quality Standards (NAAQS)	
Pollutant	**Primary (Health-Based) Averaging Time**	**Standard Concentration**
TSP[a]	Annual geometric mean[b]	50 µg/m^3
	24 hours	150 µg/m^3
SO$_2$	Annual arithmetic mean[c]	80 µg/m^3 (0.03 ppm)
	24 hours	120 µg/m^3 (0.14 ppm)
CO	8 hours	10 mg/m^3 (9 ppm)
	1 hour	40 mg/m^3 (35 ppm)
NO$_2$	Annual arithmetic mean	80 µg/m^3 (0.05 ppm)
O$_3$	Daily max 8 hour avg.	157 µg/m^3 (0.08 ppm)
Lead	Maximum quarterly avg.	1.5 µg/m^3

[a]Total suspended particulate material, PM2.5 and PM10.

[b]The geometric mean is obtained by taking the nth root of the product of n numbers. This tends to reduce the impact of a few very large numbers in a set.

[c]An arithmetic mean is the average determined by dividing the sum of a group of data points by the number of points.

We also distinguish pollutants according to how they are produced. **Primary pollutants** are those released directly from the source into the air in a harmful form (fig. 16.3). **Secondary pollutants** are converted to a hazardous form after they enter the air or are formed by chemical reactions as components of the air mix and interact. Solar radiation often provides the energy for these reactions. Photochemical oxidants and atmospheric acids formed by these mechanisms are among our most important pollutants in terms of health and ecosystem damage.

Fugitive emissions are those that do not go through a smokestack. By far the most massive example of this category is dust from soil erosion, strip mining, rock crushing, and building construction (and destruction). Fugitive industrial emissions are hard to monitor, but they are extremely important sources of air pollution. Leaks around valves and pipe joints, and evaporation of volatile compounds from oil-processing facilities, contribute as much as 90 percent of the hydrocarbons and volatile organic chemicals emitted from oil refineries and chemical plants.

Transportation and power plants are the dominant sources of criteria pollutants (fig. 16.4). We'll examine each of these, then we'll look at additional pollutants that are also monitored under the Clean Air Act.

Sulfur Dioxide (SO₂)

Natural sources of sulfur in the atmosphere include evaporation of sea spray, erosion of sulfate-containing dust from arid soils, fumes from volcanoes and hot springs, and biogenic emissions of hydrogen sulfide (H_2S) and organic sulfur-containing compounds. Total yearly emissions of sulfur from all sources amount to some

FIGURE 16.3 Primary pollutants are released directly from a source into the air. Coal-burning power plants like this one produce about two-thirds of the sulfur oxides, one-third of the nitrogen oxides, and one-half of the mercury emitted in the United States each year.

114 million metric tons. Worldwide, anthropogenic sources represent about two-thirds of the all airborne sulfur, but in most urban areas they contribute as much as 90 percent of the sulfur in the air. The predominant form of anthropogenic sulfur is **sulfur dioxide** (SO_2) from combustion of sulfur-containing fuel (coal and oil), purification of sour (sulfur-containing) natural gas or oil, and industrial processes, such as smelting of sulfide ores. China and the United States are the largest sources of anthropogenic sulfur, primarily from coal burning and smelting.

Sulfur dioxide was a major contaminant, along with particulate matter, responsible for illness and death in London's smog of 1952 (opening case study). This colorless corrosive gas is directly damaging

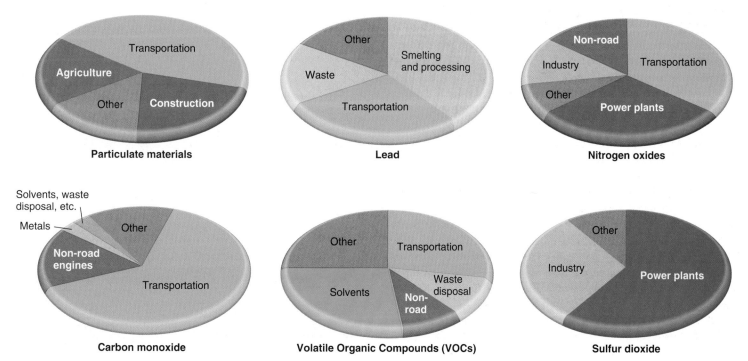

FIGURE 16.4 Anthropogenic sources of primary "criteria" pollutants in the United States. Volatile organic compounds are an important precursor of ozone, one of the 6 criteria pollutants.

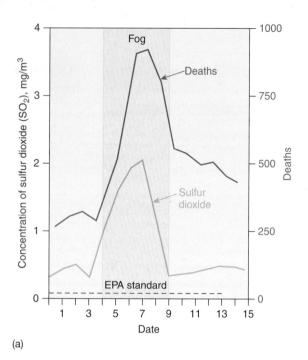

(a)

(b)

FIGURE 16.5 Sulfur dioxide concentrations and deaths during the London smog of December 1952. The EPA standard limit is 0.08 mg/m³ (dashed line, (a). The soybean leaf at right (b) was exposed to 2.1 mg/m³ sulfur dioxide for 24 hours. White patches show where chlorophyll has been destroyed.

to both plants and animals (fig. 16.5). Once in the atmosphere, it can be further oxidized to sulfur trioxide (SO_3), which reacts with water vapor or dissolves in water droplets to form sulfuric acid (H_2SO_4), a major component of acid rain. Very small solid particles or liquid droplets can transport the acidic sulfate ion (SO_4^{-2}) long distances through the air or deep into the lungs where it is very damaging. Sulfur dioxide and sulfate ions are probably second only to smoking as causes of air-pollution-related health damage. Sulfate particles and droplets reduce visibility in the United States as much as 80 percent. Some of the smelliest and most obnoxious air pollutants are sulfur compounds, such as hydrogen sulfide from pig manure lagoons or mercaptans (organosulfur thiols) from paper mills (fig. 16.6).

Nitrogen Oxides (NO$_x$)

Nitrogen oxides are highly reactive gases formed when nitrogen in fuel or in air is heated (during combustion) to temperatures above 650°C (1,200°F) in the presence of oxygen. Bacteria can also form NO as they oxidize nitrogen-containing compounds in soil or water. The initial product, **nitric oxide** (NO), oxidizes further in the atmosphere to **nitrogen dioxide** (NO_2), a reddish-brown gas that gives photochemical smog its distinctive color. In addition, **nitrous oxide** (N_2O) is an intermediate form that results from soil denitrification. Nitrous oxide absorbs ultraviolet light and is an important greenhouse gas (chapter 15). Because nitrogen readily changes from one of these forms to another by gaining or losing O atoms, the general term NO_x is used to describe these gases. Nitrogen oxides combine with water to make nitric acid (HNO_3), a major component of acid rain.

Anthropogenic sources account for 60 percent of the global emissions of about 230 million metric tons of reactive nitrogen compounds each year (see table 16.1). About 95 percent of all human-caused NO_x in the United States is produced by fuel combustion in transportation and electric power generation. Because

FIGURE 16.6 The most annoying pollutants from this paper mill are pungent organosulfur thiols and sulfides. Chlorine bleaching can also produce extremely dangerous organochlorines, such as dioxins.

we continue to drive more miles every year, and to consume abundant electricity, we have had less success in controlling NO_x than other pollutants.

Excess nitrogen from agricultural fertilizer use and production is also an important, but little understood, contributor to airborne NO_x. Fertilizers washing from farmlands also cause excess fertilization and eutrophication of inland waters and coastal seas. Environmental dispersal of nitrogen from fertilizers also may be adversely affecting terrestrial plants by fertilizing weedy and invasive plants.

Carbon Monoxide (CO)

Carbon monoxide (CO) is a colorless, odorless, nonirritating, but highly toxic gas. CO is produced mainly by incomplete combustion of fuel (coal, oil, charcoal, or gas), as in furnaces, incinerators, engines, or fires, as well as in decomposition of organic matter. CO blocks oxygen uptake in blood by binding irreversibly to hemoglobin (the protein that carries oxygen in our blood), making hemoglobin unable to hold oxygen and deliver it to cells. Human activities produce about half of the 1 billion metric tons of CO released to the atmosphere each year. In the United States, two-thirds of the CO emissions are created by internal combustion engines in transportation. Land-clearing fires and cooking fires also are major sources. About 90 percent of the CO in the air is converted to CO_2 in photochemical reactions that produce ozone. Catalytic converters on vehicles are one of the important methods to reduce CO production by ensuring complete oxidation of carbon to carbon dioxide (CO_2).

Carbon dioxide is the predominant form of carbon in the air. Growing recognition of the health and environmental risks associated with climate change (chapter 15) have led to recent regulations on CO_2, which are discussed below.

Ozone (O_3) and Photochemical Oxidants

Ozone (O_3) high in the stratosphere provides a valuable shield for the biosphere by absorbing incoming ultraviolet radiation. But at ground level, O_3 is a strong oxidizing reagent that damages vegetation, building materials (such as paint, rubber, and plastics), and sensitive tissues (such as eyes and lungs). Ozone has an acrid, biting odor that is a distinctive characteristic of photochemical smog. Ground-level O_3 is a product of photochemical reactions (reactions initiated by sunlight) between other pollutants, such as NO_x or volatile organic compounds. A general term for products of these reactions is **photochemical oxidants**. One of the most important of these reactions involves splitting nitrogen dioxide (NO_2) into nitrous oxide (NO) and oxygen (O). This single O atom is then available to combine with a molecule of O_2 to make **ozone** (O_3).

Hydrocarbons in the air contribute to the accumulation of ozone by combining with NO to form new compounds, leaving single O atoms free to form O_3 (fig. 16.7). Many of the NO compounds are damaging photochemical oxidants. A general term for organic chemicals that evaporate easily or exist as gases in the air is **volatile organic compounds (VOCs)**. Plants are the largest source of VOCs, releasing an estimated 350 million tons of isoprene (C_5H_8) and 450 million tons of terpenes ($C_{10}H_{15}$) each year.

Atmospheric oxidant production:

1. $NO + VOC \longrightarrow NO_2$ (nitrogen dioxide)
2. $NO_2 + UV \longrightarrow NO + O$ (nitric oxide + atomic oxygen)
3. $O + O_2 \longrightarrow O_3$ (ozone)
4. $NO_2 + VOC \longrightarrow PAN$, etc. (peroxyacetyl nitrate)

Net results:

$NO + VOC + O_2 + UV \longrightarrow O_3$, PAN, and other oxidants

FIGURE 16.7 Secondary production of urban smog oxidants by photochemical reactions in the atmosphere.

About 400 million tons of methane (CH_4) are produced by natural wetlands and rice paddies and by bacteria in the guts of termites and ruminant animals. These volatile hydrocarbons are generally oxidized to CO and CO_2 in the atmosphere.

In addition to these natural VOCs, a large number of other synthetic organic chemicals, such as benzene, toluene, formaldehyde, vinyl chloride, phenols, chloroform, and trichloroethylene, are released into the air by human activities. About 28 million tons of these compounds are emitted each year in the United States, mainly unburned or partially burned hydrocarbons from transportation, power plants, chemical plants, and petroleum refineries. These chemicals play an important role in the formation of photochemical oxidants.

Lead

Our most abundantly produced metal air pollutant, lead is toxic to our nervous systems and other critical functions. Lead binds to enzymes and to components of our cells, such as brain cells, which then cannot function normally. Airborne lead is produced by a wide range of industrial and mining processes. The main sources are smelting of metal ores, mining, and burning of coal and municipal waste, in which lead is a trace element, and burning of gasoline to which lead has been added. Until recently, leaded gasoline was the main source of lead in the United States, but leaded gas was phased out in the 1980s. Since 1986, when the ban was enforced, children's average blood lead levels have dropped 90 percent and average IQs have risen three points. Banning leaded gasoline in the United States was one of the most successful pollution-control measures in American history. Now, 50 nations have renounced leaded gasoline. The global economic benefit of this step is estimated to be more than $200 billion per year.

Worldwide atmospheric lead emissions amount to about 2 million metric tons per year, or two-thirds of all metallic air pollution. Globally, most of this lead is still from leaded gasoline, as well as metal ore smelting and coal burning.

Particulate Matter

Particulate matter includes solid particles or liquid droplets suspended in a gaseous medium. Very fine solid or liquid particulates suspended in the atmosphere are **aerosols**. This includes dust, ash, soot, lint, smoke, pollen, spores, algal cells, and many other suspended materials. Particulates often are the most obvious form of air pollution because they reduce visibility and leave dirty deposits on windows, painted surfaces, and textiles.

Particulates small enough to breathe are monitored under the Clean Air Act. Particles smaller than 2.5 micrometers in diameter, such as those found in smoke and haze, and produced by fires, power plants, or vehicle exhaust, are among the most dangerous particulates because they can be drawn into the lungs, where they damage respiratory tissues. Asbestos fibers and cigarette smoke are among these dangerous fine particles. This fine particulate matter is referred to as PM2.5, in reference to its size. Reducing sulfur in coal and diesel fuel, which produces aerosol droplets of sulfuric acid, is one important strategy for controlling PM2.5 particulates.

Coarse inhalable particles are larger than 2.5 micrometers but less than 10 micrometers in diameter. These are known as PM10, and they are typically found near roads or other visible dust sources. The "dust bowl" of the 1930s involved this kind of particulates. At that time farmland soils were often left bare, especially during severe drought, and billions of tons of topsoil blew away from farmlands. Soil conservation on farmlands is one strategy for reducing PM10; another strategy is better management of dust at construction sites.

Dust storms can travel remarkable distances. Dust from Africa's Sahara desert regularly crosses the Atlantic and raises particulate levels above federal health standards in Miami and San Juan, Puerto Rico (fig. 16.8). Amazon rainforests receive mineral nutrients carried in dust from Africa; more than half the 50 million tons of dust transported to South America each year has been traced to the bed of the former Lake Chad in Africa. In China, vast dust storms blow out of the Gobi desert every spring, choking Beijing and closing airports and schools in Japan and Korea. The dust plume follows the jet stream across the Pacific to Hawaii and then to the west coast of North America, where it sometimes makes up as much as half the particulate air pollution in Seattle, Washington. Some Asian dust storms have polluted the U.S. skies as far east as Georgia and Maine.

Epidemiological studies have shown that cities with chronically high levels of particulates have higher death rates, mostly from heart and lung disease. Emergency-room visits and death rates rise in days following a dust storm. Some of this health risk comes from the particles themselves, which clog tiny airways and make breathing difficult. The dust also carries pollen, bacteria, viruses, fungi, herbicides, acids, radioactive isotopes, and heavy metals between continents.

Airborne dust is considered the primary source of allergies worldwide. Saharan dust storms are suspected of raising asthma rates in Trinidad and Barbados, where cases have increased 17-fold in 30 years. *Aspergillus sydowii*, a soil fungus from Africa, has been shown to be causing death of corals and sea fans in remote reefs in the Caribbean. Europe also receives airborne pathogens via dust storms. Outbreaks of foot-and-mouth disease in Britain have been traced to dust storms from North Africa.

Mercury and other metals are also regulated

In addition to criteria pollutants or conventional pollutants, many other pollutants are regulated to protect public health and our environment. Standards for these pollutants continue to evolve, as do definitions of which pollutants require regulation. These changes

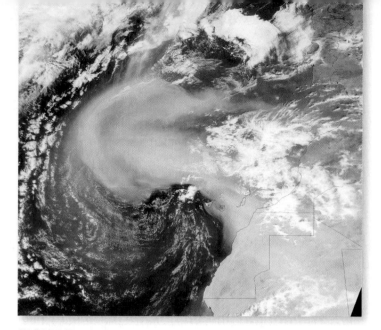

FIGURE 16.8 A massive dust storm extends more than 1,600 km (1,000 mi) from the coast of western Sahara and Morocco. Storms such as this can easily reach the Americas, and they have been linked both to the decline of coral reefs in the Caribbean and to the frequency and intensity of hurricanes formed in the eastern Atlantic Ocean.

reflect increases in certain pollutants, such as airborne mercury; the introduction of new pollutants, such as newly developed organic compounds; and increasing recognition of risks, as in the case of carbon dioxide.

Many toxic metals are released into the air by burning coal and oil, mining, smelting of metal ores, or manufacturing. Lead, mercury, cadmium, nickel, arsenic (a highly toxic metalloid), and others are released in the form of metal fumes or suspended particulates by fuel combustion, ore smelting, and disposal of wastes. Among these, lead and mercury are the most abundantly produced toxic metals.

Mercury has become regulated relatively recently. Like lead, mercury is toxic in minute doses, causing nerve damage and other impairments, especially in young children and developing fetuses. Volcanoes and rock weathering can produce mercury, but 70 percent of airborne mercury derives from coal-burning power plants, metal processing (smelting), waste incineration, and other industrial combustion.

About 75 percent of human exposure to mercury comes from eating fish. This is because aquatic bacteria are mainly responsible for converting airborne mercury into methyl mercury, a form that accumulates in living animal tissues. Once methyl mercury enters the food web, it bioaccumulates in predators. As a consequence, large, long-lived, predatory fish contain especially high levels of mercury in their tissues. Contaminated tuna fish alone is responsible for about 40 percent of all U.S. exposure to mercury (fig. 16.9). Swordfish, shrimp, and other seafood are also important mercury sources in our diet.

Freshwater fish also carry risks. Mercury contamination is the most common cause of impairment of U.S. rivers and lakes, and 45 states have issued warnings against frequent consumption of fresh-caught fish. A 2007 study tested more than 2,700 fish from 636 rivers and streams in 12 western states, and mercury was found in every one of them.

Global air circulation also deposits airborne mercury on land. Half or more of the mercury that falls on North America may come from abroad, much of it from Asian coal-burning power plants. Similarly, North American mercury travels to Europe. A 2009 report by the U.S. Geological Survey found that mercury levels in Pacific Ocean tuna have risen 30 percent in the past 20 years, with another 50 percent rise projected by 2050. Increased coal burning in China, which is building two new coal-burning power plants every week, is understood to be the main cause of growing mercury emissions in the Pacific.

Much of our understanding of mercury poisoning comes from a disastrous case in Minamata, Japan, in the 1950s, where a chemical factory regularly discharged mercury-laden waste into Minamata Bay. Babies whose mothers ate mercury-contaminated fish suffered profound neurological disabilities, including deafness, blindness, mental retardation, and cerebral palsy. In adults, mercury poisoning caused numbness, loss of muscle control, and dementia. The connection between "Minamata disease" and mercury was established in the 1950s, but waste dumping didn't end for another ten years.

The U.S. National Institutes of Health (NIH) estimates that 1 in 12 American women has more mercury in her blood than the 5.8 µg/l considered safe by the EPA. Between 300,000 and 600,000 of the 4 million children born each year in the United States are exposed in the womb to mercury levels that could cause diminished intelligence or developmental impairments. According to the NIH, elevated mercury levels cost the U.S. economy $8.7 billion each year in higher medical and educational costs and in lost workforce productivity.

Mercury emissions in the United States have declined since the Clean Air Act began regulating mercury emissions, and many states have instituted rules for capturing mercury before it leaves the smokestack. In 2009 the EPA took another step in controlling mercury emissions when it issued new rules controlling emissions from cement plants, one of the largest sources of the toxin. Health advocates continue to lobby for international standards on emissions, especially from coal-burning power plants (What Do You Think? p. 354).

FIGURE 16.9 Airborne mercury bioaccumulates in seafood, especially in top predators such as tuna. Mercury contamination is also the most common cause of fish consumption advisories in U.S. lakes and rivers.

Carbon dioxide and halogens are key greenhouse gases

Some 370 billion tons of CO_2 are emitted each year from respiration (oxidation of organic compounds by plant and animal cells; table 16.1). These releases are usually balanced by an equal uptake by photosynthesis in green plants. At normal concentrations, CO_2 is nontoxic and innocuous, but atmospheric levels are steadily increasing (about 0.5 percent per year) due to human activities and are now causing global climate change, with serious implications for both human and natural communities (chapter 15).

Regulating CO_2 has been a subject of intense debate since the 1990s. On the one hand, policymakers have widely acknowledged that climate change is likely to have disastrous effects. On the other hand, CO_2 is difficult to consider limiting because we produce abundant quantities, reductions involve changes to both technology and behavior, and CO_2 production historically has been closely tied to our economic productivity. Although future economic growth is likely to depend on efficiencies and new technologies, these concerns remain an important part of the debate.

Since the midterm elections of 2010, many members of Congress have been intent on eliminating this and other pollution regulation, arguing that it is too costly for industry and the economy (see further discussion in section 16.5). Energy companies and their representatives, in particular, have lobbied to prevent legal limits on greenhouse gases. The 2011 congressional budget proposed to slash EPA funding by one-third, in part to reduce pollution monitoring and regulation.

The question of whether the EPA should regulate greenhouse gases was so contentious that it went to the Supreme Court in 2007. The Court ruled that it was the EPA's responsibility to limit these gases, on the grounds that greenhouse gases endanger public health and welfare within the meaning of the Clean Air Act. The Court, and subsequent EPA documents, noted that these risks include increased drought, more frequent and intense heat waves and wildfires, sea-level rise, and harm to water resources, agriculture, wildlife, and ecosystems. In addition to these risks, the U.S. military has cited climate change as a security threat. A coalition of generals and admirals signed a report from the Center for Naval Analyses stating that climate change "presents significant national security challenges" including violence resulting from scarcity of water, and migration due to sea-level rise and crop failure.

Since the Supreme Court ruling, the EPA is charged with regulating six greenhouse gases: carbon dioxide, methane, nitrous oxide, hydrofluorocarbons, perfluorocarbons, and sulfur hexafluoride. These are gases whose emissions have grown dramatically in recent decades.

Three of these six greenhouse gases contain halogens, a group of lightweight, highly reactive elements (fluorine, chlorine, bromine, and iodine). Because they are generally toxic in their elemental form, they are commonly used as fumigants and

What Do You Think?

Cap and Trade for Mercury Pollution?

Often referred to as quicksilver, mercury is used in a host of products, including paints, batteries, fluorescent lightbulbs, electrical switches, pesticides, skin creams, antifungal agents, and old thermometers. Mercury also is a powerful neurotoxin that destroys the brain and central nervous system at high doses. Minute amounts can cause nerve damage and developmental defects in children. Exposure results mainly from burning garbage, coal, or other mercury-laden materials—the mercury falls to the ground and washes into lakes and wetlands, where it enters the food web. In a survey of freshwater fish from 260 lakes across the United States, the EPA found that every fish sampled contained some level of mercury.

In 1994 the EPA declared mercury a hazardous pollutant regulated under the Clean Air Act. Municipal and medical incinerators were required to reduce their mercury emissions by 90 percent. Industrial and mining operations also agreed to cut emissions. However, the law did not address the 1,032 coal-burning power plants, which produce nearly half of total annual U.S. emissions, some 48 tons per year.

Finally in 2000 the EPA declared mercury from power plants, like that from other sources, a public health risk. The agency could have applied existing air-toxin regulations and required power plants to reduce their emissions by 90 percent in 5 years with existing control technology. But the EPA in 2000 opted instead for a "cap and trade" market mechanism, which should reduce mercury releases 70 percent in about 30 years.

Cap-and-trade approaches set limits (caps) and allow utilities to buy and sell unused pollution credits. This strategy is widely supported because it uses a profit motive rather than rules, and it allows industries to make their own decisions about emission controls. It also allows continued emissions if credits are cheaper than emission controls, and traders have the opportunity to make money on the exchanges.

On the other hand, public health advocates argue that although cap-and-trade systems work well for some pollutants, they are inappropriate for a substance that is toxic at very low levels, and they object that utilities are allowed to continue emitting mercury for years longer than necessary. Many eastern states are especially concerned because they suffer from high mercury pollution generated in the Midwest and blown east by prevailing winds (fig. 1).

Meanwhile, in the Allegheny Mountains of West Virginia, a huge coal-fired power plant is adding fuel to the mercury debate. The enormous

1,600-megawatt Mount Storm plant ranked second in the nation in mercury emissions just a few years ago. When Mount Storm installed new controls to capture sulfur and nitrogen oxides from its stack, this equipment also caught 95 percent of its mercury emissions, at no extra cost. This is excellent news, but it also raises a policy question: If existing technology can cut mercury economically, why wait 30 years to impose similarly cost-effective limits on other power plants?

This case illustrates the complexity of regulating air pollution. Highly mobile, widely dispersed, produced by a variety of sources, and having diverse impacts, air pollutants can be challenging to regulate. Often air quality controversies—such as mercury control—pit a diffuse public interest (improving general health levels or child development) against a very specific private interest (utilities that must pay millions of dollars per year to control pollutants). How would you set the rules if you were in charge? Would you impose rules or allow for trading of mercury emission permits? Why? How would you negotiate the responsibility for controlling pollutants?

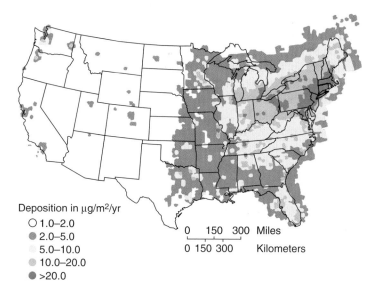

Deposition in µg/m²/yr
○ 1.0–2.0
● 2.0–5.0
  5.0–10.0
  10.0–20.0
● >20.0

0   150   300   Miles
0 150 300   Kilometers

FIGURE 1 Atmospheric mercury deposition in the United States. Due to prevailing westerly winds, and high levels of industrialization, eastern states have high mercury deposition.

Source: EPA, 1998.

disinfectants, but they also have hundreds of uses in industrial and commercial products. Chlorofluorocarbons (CFCs) have been banned for most uses in industrialized countries, but about 600 million tons of these compounds are used annually worldwide in spray propellants and refrigeration compressors and for foam blowing. They diffuse into the stratosphere, where they release chlorine and fluorine atoms that destroy ozone molecules that protect the earth from ultraviolet radiation (see section 16.3).

Halogen compounds are also powerful greenhouse gases: they trap more energy per molecule than does CO_2, and they persist in the atmosphere for decades to centuries. Perfluorocarbons will persist in the atmosphere for thousands of years. The global warming potential (per molecule, over time) of some types of CFCs is 10,900 times that of CO_2 (table 16.3).

Table 16.3	Global Warming Potential (GWP) of Several Greenhouse Gases	
GAS	**Global warming potential[1]**	**Atmospheric lifetime (years)[2]**
Carbon dioxide (CO_2)	1	~ 100
Methane (CH_4)	25	124
Nitrous oxide (N_2O)	298	1144
CFC-12 (CCl_2F_2)	10,900	100
HCFC-142b(CH_3CClF_2)	2,310	18
Sulfur hexafluoride (SF_6)	22,800	3200

[1]A measure of radiative effects, integrated over a 100-yr time horizon, relative to an equal mass of CO_2 emissions. CO_2 is set as 1 for comparison.

[2]Average residence times shown; actual range for CO_2 is decades to centuries.

Source: Carbon Dioxide Information Analysis Center, 2011.

Developing rules and standards for greenhouse gases will take time and considerable debate. Many strategies have been proposed, including subsidies for alternative energy, reducing tax breaks and other subsidies for fossil fuels, imposing a tax on coal, oil, and gas, and cap-and-trade systems, including carbon-trading markets. The last of these options has been the most acceptable, and carbon trading is now worth billions of dollars every year. Data remain inconclusive regarding whether this has produced an overall decline in emissions.

Hazardous air pollutants (HAPs) can cause cancer and nerve damage

Although most air contaminants are regulated because of their potential adverse effects on human health or environmental quality, a special category of toxins is monitored by the U.S. EPA because they are particularly dangerous. Called **hazardous air pollutants (HAPs),** these chemicals include carcinogens, neurotoxins, mutagens, teratogens, endocrine system disrupters, and other highly toxic compounds (chapter 8). Twenty of the most "persistent bioaccumulative toxic chemicals" (see table 8.2) require special reporting and management because they remain in ecosystems for long periods of time and accumulate in animal and human tissues. Most of these chemicals are either metal compounds, chlorinated hydrocarbons, or volatile organic compounds. Gasoline vapors, solvents, and components of plastics are all HAPs that you may encounter on a daily basis.

Only about 50 locations in the United States regularly measure concentrations of HAPs in ambient air. Often the best source of information about these chemicals is the **Toxic Release Inventory (TRI)** collected by the EPA as part of the community right-to-know program. Established by Congress in 1986, the TRI requires 23,000 factories, refineries, hard rock mines, power plants, and chemical manufacturers to report on toxin releases (above certain minimum amounts) and waste management methods for 667 toxic chemicals. Although this total is less than 1 percent of all chemicals registered for use, and represents a limited range of sources, the TRI is widely considered the most comprehensive source of information about toxic pollution in the United States (fig. 16.10).

Most HAP releases are decreasing, but discharges of mercury and dioxins—both of which are bioaccumulative and toxic at extremely low levels—have increased in recent years. Dioxins are created mainly by burning plastics and medical waste containing chlorine. The EPA reports that 100 million Americans live in areas where the cancer rate from HAPs exceeds 10 in 1 million, or ten times the normally accepted standard for action. Benzene, formaldehyde, acetaldehyde, and 1,3 butadiene are responsible for most of this HAP cancer risk. Furthermore, twice that many Americans (70 percent of the U.S. population) live in areas where the risk of death from causes other than cancer exceeds 1 in 1 million. To help residents track local air quality levels, the EPA recently estimated the concentration of HAPs in localities across the continental United States (over 60,000 census tracts). You can access this information on the Environmental Defense Fund web page at www.scorecard.org/env-releases/hap/.

FIGURE 16.10 Harmful air toxics from large industrial sources, such as chemical plants, petroleum refineries, and paper mills, have been reduced by nearly 70 percent since the EPA began regulating them. Many smaller sources remain unregulated.

Aesthetic degradation also results from pollution

Aesthetic degradation is any undesirable change in the physical characteristics or chemistry of the atmosphere, such as noise, odors, and light pollution. These factors rarely threaten life or health directly, but they can strongly impact our quality of life. They also increase stress, which affects health. We are often especially susceptible to noises and odors. Often the most sensitive device for odor detection is the human nose. We can smell styrene, for example, at 44 parts per billion (ppb). Trained panels of odor testers often are used to evaluate air samples. Factories that emit noxious chemicals sometimes spray "odor maskants" or perfumes into smokestacks to cover up objectionable odors. Light pollution also is a concern in most urban areas, where ambient light confuses birds and hides the stars.

Indoor air can be worse than outdoor air

We have spent a considerable amount of effort and money to control the major outdoor air pollutants, but we have only recently begun to address indoor air pollutants. The EPA has found that indoor concentrations of toxic air pollutants are often higher than outdoors. Furthermore, people generally spend more time inside than out, so they are exposed to higher doses of these pollutants.

In some cases, indoor air in homes has concentrations of chemicals that would be illegal outside or in the workplace. The

FIGURE 16.11 Smoky cooking and heating fires may cause more ill health effects than any other source of indoor air pollution except tobacco smoking. Some 2.5 billion people, mainly women and children, spend hours each day in poorly ventilated kitchens and living spaces where carbon monoxide, particulates, and cancer-causing hydrocarbons often reach dangerous levels.

EPA has found that concentrations of such compounds as chloroform, benzene, carbon tetrachloride, formaldehyde, and styrene can be seventy times higher in indoor air than in outdoor air, as plastics, carpets, paints, and other common materials off-gas these compounds. Finding less-toxic paints and fabrics can make indoor spaces both healthier and more pleasant.

In the less-developed countries of Africa, Asia, and Latin America, where such organic fuels as firewood, charcoal, dried dung, and agricultural wastes provide the majority of household energy, smoky, poorly ventilated heating and cooking fires are the greatest source of indoor air pollution (fig. 16.11). The World Health Organization (WHO) estimates that 2.5 billion people—over a third of the world's population—are adversely affected by pollution from this source. Women and small children spend long hours each day around open fires or unventilated stoves in enclosed spaces. Levels of carbon monoxide, particulates, aldehydes, and other toxic chemicals can be 100 times higher than would be legal for outdoor ambient concentrations in the United States. Designing and building cheap, efficient, nonpolluting energy sources for the developing countries would not only save shrinking forests but would make a major impact on health as well.

16.3 ATMOSPHERIC PROCESSES

Topography, climate, and physical processes in the atmosphere play an important role in the transport, concentration, dispersal, and removal of many air pollutants. Cities concentrate dust and pollutants in urban "dust domes"; winds cause mixing between air layers, precipitation, and atmospheric chemistry. All these factors determine whether pollutants will remain in the locality where they are produced or go elsewhere. In this next section we will survey some environmental factors that affect air pollution levels.

Temperature inversions trap pollutants

As in London's smog of 1952, **temperature inversions** can greatly concentrate air pollutants. Inversions occur when a stable layer of warmer air lies above cooler air. The normal conditions, where temperatures decline with increasing height, are inverted, and these stable conditions prevent convection currents from dispersing pollutants. Often these conditions occur when cold air settles in a valley that is surrounded by hills or mountains. When a cold front slides under an adjacent warmer air mass, or when cool air subsides down a mountain slope to displace warmer air in the valley below, the cold air becomes trapped, as in a bowl. Inversions might last from a few hours to a few days.

The most stable inversion conditions are usually created by rapid nighttime cooling in a valley or basin where air movement is restricted. Los Angeles is a classic example, with conditions that create both temperature inversions and photochemical smog (fig. 16.12). The city is surrounded by mountains on three sides and the climate is dry, with abundant sunshine for photochemical oxidation and ozone production. Millions of automobiles and trucks create high pollution levels. Skies are generally clear at night, allowing heat to radiate from the ground. The ground and the lower

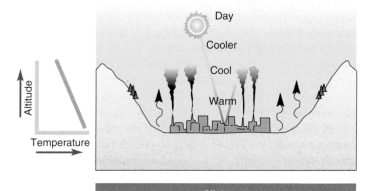

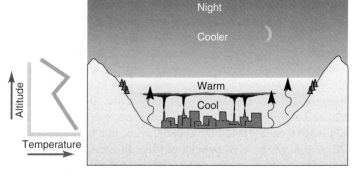

FIGURE 16.12 Atmospheric temperature inversions occur where ground-level air cools more quickly than upper levels. This temperature differential prevents mixing and traps pollutants close to the ground.

layers of air cool quickly at night, while upper air layers remain relatively warm. During the night, cool, humid, onshore breezes also slide in under the contaminated air, which is trapped by a wall of mountains to the east and by the cap of warmer air above.

Morning sunlight is absorbed by the concentrated aerosols and gaseous chemicals caught near the ground by the inversion. This complex mixture quickly cooks up a toxic brew of hazardous compounds. As the ground warms later in the day, convection currents break up the temperature gradient and pollutants are carried back down to the surface, where more contaminants are added. Nitric oxide (NO) from automobile exhaust is oxidized to a brownish haze of nitrogen dioxide (NO_2). As nitrogen oxides are used up in reactions with unburned hydrocarbons, the ozone level begins to rise. By early afternoon an acrid brown haze fills the air, making eyes water and throats burn. In the 1970s, before pollution controls were enforced, the Los Angeles basin often would reach 0.34 ppm or more by late afternoon and the pollution index could be 300, the stage considered a health hazard.

Wind currents carry pollutants worldwide

Dust and contaminants can be carried great distances by the wind. Areas downwind from industrial complexes often suffer serious contamination, even if they have no pollution sources of their own (fig. 16.13). Pollution from the industrial belt between the Great Lakes and the Ohio River Valley, for example, regularly contaminates the Canadian Maritime Provinces, and sometimes can be traced as far as Ireland. As noted earlier, long-range transport is a major source of Asian mercury in North America.

Studies of air pollutants over southern Asia reveal a 3 km thick toxic cloud of ash, acids, aerosols, dust, and photochemical reactants that regularly covers the entire Indian subcontinent and can last for much of the year. Nobel laureate Paul Crutzen estimates that up to 2 million people in India alone die each year from atmospheric pollution. Produced by forest fires, the burning of agricultural wastes, and dramatic increases in the use of fossil fuels, the Asian smog layer cuts by up to 15 percent the amount of solar energy reaching the earth's surface beneath it. Meteorologists suggest that the cloud—80 percent of which is human-made—could disrupt monsoon weather patterns and may be disturbing rainfall and reducing rice harvests over much of South Asia. As UN Environment Programme executive director Klaus Töpfer said, "There are global implications because a pollution parcel like this, which stretches three km high, can travel half way round the globe in a week."

An increase in monitoring activity has revealed industrial contaminants in places usually considered among the cleanest in the world. Samoa, Greenland, Antarctica, and the North Pole all have heavy metals, pesticides, and radioactive elements in their air. Since the 1950s, pilots flying in the high Arctic have reported dense layers of reddish-brown haze clouding the arctic atmosphere. Aerosols of sulfates, soot, dust, and toxic heavy metals,

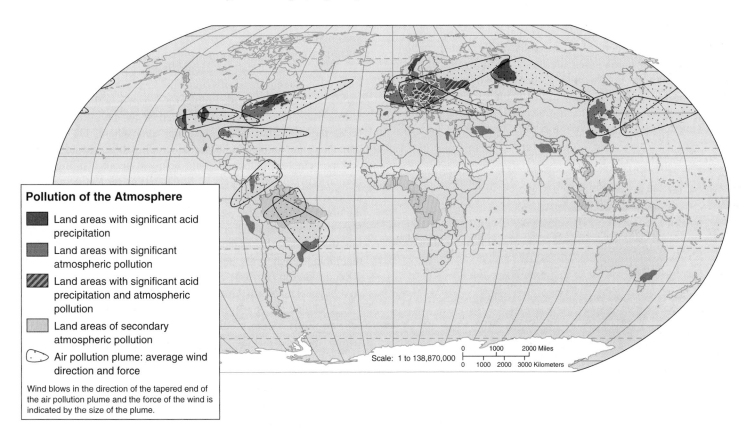

Pollution of the Atmosphere

- Land areas with significant acid precipitation
- Land areas with significant atmospheric pollution
- Land areas with significant acid precipitation and atmospheric pollution
- Land areas of secondary atmospheric pollution
- Air pollution plume: average wind direction and force

Wind blows in the direction of the tapered end of the air pollution plume and the force of the wind is indicated by the size of the plume.

Scale: 1 to 138,870,000

0   1000   2000 Miles
0   1000   2000   3000 Kilometers

FIGURE 16.13 Long-range transport carries air pollution from source regions thousands of kilometers away into formerly pristine areas. Secondary air pollutants can be formed by photochemical reactions far from primary emissions sources.

such as vanadium, manganese, and lead, travel to the pole from the industrialized parts of Europe and Russia.

A process called "grasshopper" transport, or atmosphere distillation, helps deliver contaminants to the poles. Volatile compounds evaporate from warm areas, travel through the atmosphere, then condense and precipitate in cooler regions (fig. 16.14). Over several years, contaminants accumulate in the coldest places, generally at high latitudes where they bioaccumulate in food chains. Whales, polar bears, sharks, and other top carnivores in polar regions have been shown to have dangerously high levels of pesticides, metals, and other HAPs in their bodies. The Inuit people of Broughton Island, well above the Arctic Circle, have higher levels of polychlorinated biphenyls (PCBs) in their blood than any other known population, except victims of industrial accidents. Far from any source of this industrial by-product, these people accumulate PCBs from the flesh of fish, caribou, and other animals they eat. This exacerbates the cultural crisis caused by climate change.

Stratospheric ozone is destroyed by chlorine

In 1985 the British Antarctic Atmospheric Survey announced a startling and disturbing discovery: **stratospheric ozone** concentrations over the South Pole were dropping precipitously during September and October every year as the sun reappears at the end of the long polar winter (fig. 16.15). This ozone depletion has been occurring at least since the 1960s but was not recognized because earlier researchers programmed their instruments to ignore changes in ozone levels that were presumed to be erroneous.

Chlorine-based aerosols, especially **chlorofluorocarbons (CFCs)** and other halon gases, are the principal agents of ozone depletion. Nontoxic, nonflammable, chemically inert, and cheaply produced, CFCs were extremely useful as industrial gases and in refrigerators, air conditioners, Styrofoam inflation, and aerosol spray cans for many years. From the 1930s until the 1980s, CFCs

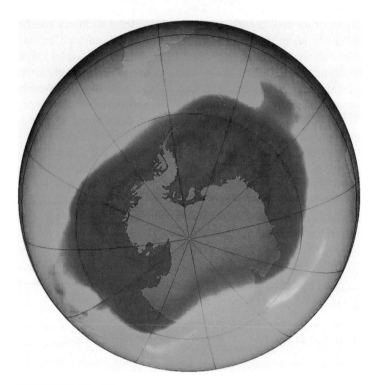

FIGURE 16.15 The region of stratospheric ozone depletion grew steadily to an area of nearly 30 million km² in 2006 (shown here). This ozone "hole" has shown signs of decline since the Montreal Protocol went into effect.

were used all over the world and widely dispersed through the atmosphere.

What we often call an ozone "hole" is really a vast area of reduced concentrations of ozone in the stratosphere. Although ozone is a pollutant in the ambient air, ozone in the stratosphere is important because it absorbs much of the harmful ultraviolet (UV) radiation that enters the outer atmosphere. UV radiation damages plant and animal tissues, including the eyes and the skin. A 1 percent loss of ozone could result in about a million extra human skin cancers per year worldwide, if no protective measures are taken. Excessive UV exposure could reduce agricultural production and disrupt ecosystems. Scientists worry that, for example, high UV levels in Antarctica could reduce populations of plankton, the tiny floating organisms that form the base of a food chain that includes fish, seals, penguins, and whales in Antarctic seas. In 2006 the region of ozone depletion covered 29.5 million km² (an area larger than North America).

Antarctica's exceptionally cold winter temperatures (–85 to –90°C) help break down ozone. During the long, dark winter months, strong winds known as the circumpolar vortex isolate Antarctic air and allow stratospheric temperatures to drop low enough to create ice crystals at high altitudes—something that rarely happens elsewhere in the world. Ozone and chlorine-containing molecules are absorbed on the surfaces of these ice particles. When the sun returns in the spring, it provides energy to liberate chlorine ions, which readily bond with ozone, breaking it down to molecular oxygen (table 16.4).

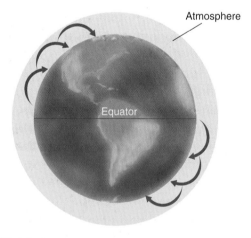

FIGURE 16.14 Air pollutants evaporate from warmer areas and then condense and precipitate in cooler regions. Eventually this "grasshopper" redistribution leads to accumulation in the Arctic and Antarctic.

Table 16.4	Stratospheric Ozone Destruction by Chlorine Atoms and UV Radiation	
Step	**Products**
1. $CFCl_3$ (chlorofluorocarbon) + UV energy	$CFCl_2$ + Cl
2. Cl + O_3	ClO + O_2
3. O_2 + UV energy	20
4. ClO + 20	O_2 + Cl
5. Return to step 2	

It is only during the Antarctic spring (September through December) that conditions are ideal for rapid ozone destruction. During that season, temperatures are still cold enough for high-altitude ice crystals, but the sun gradually becomes strong enough to drive photochemical reactions.

As the Antarctic summer arrives, temperatures moderate somewhat, the circumpolar vortex breaks down, and air from warmer latitudes mixes with Antarctic air, replenishing ozone concentrations in the ozone hole. Slight decreases worldwide result from this mixing, however. Ozone re-forms naturally, but not nearly as fast as it is destroyed. Because the chlorine atoms are not themselves consumed in reactions with ozone, they continue to destroy ozone for years. Eventually they can precipitate out, but this process happens very slowly in the stable stratosphere.

About 10 percent of all stratospheric ozone worldwide has been destroyed in recent years, and levels over the Arctic have averaged 40 percent below normal. Ozone depletion has been observed over the North Pole as well, although it is not as concentrated as that in the south.

The Montreal Protocol is a resounding success

The discovery of stratospheric ozone losses brought about a remarkably quick international response. In 1987 an international meeting in Montreal, Canada, produced the Montreal Protocol, the first of several major international agreements on phasing out most use of CFCs by 2000. As evidence accumulated, showing that losses were larger and more widespread than previously thought, the deadline for the elimination of all CFCs (halons, carbon tetrachloride, and methyl chloroform) was moved up to 1996, and a $500 million fund was established to assist poorer countries in switching to non-CFC technologies. Fortunately, alternatives to CFCs for most uses already exist. The first substitutes are hydrochlorofluorocarbons (HCFCs), which release much less chlorine per molecule. These HCFCs are also being phased out, as newer halogen-free alternatives are developed.

The Montreal Protocol is often cited as the most effective international environmental agreement ever established. Global CFC production has been cut by more than 95 percent since 1988 (fig. 16.16). Some of that has been replaced by HCFCs, which release chlorine, but not as much as CFCs. The amount of chlorine entering the atmosphere already has begun to decrease.

The size of the O_3 "hole" increased steadily from its discovery until the mid-1990s, when the Montreal Protocol began having an effect. Since then it has varied from year to year, but the trend has been to stabilize or decrease in recent years. In one of the world's most remarkable success stories, stratospheric O_3 levels should be back to normal by about 2049. There is variation in this trend, however. The 2006 O_3 hole was the largest ever. Ironically, climate warming in the lower atmosphere has contributed to cooling in the stratosphere. This cooling increases ice crystal formation over the Antarctic and results in more O_3 depletion.

The Montreal Protocol had an added benefit in the fact that CFCs and other ozone-destroying gases are also powerful, persistent greenhouse gases. Reductions in emissions of these gases under the Montreal Protocol amount to one-quarter of all greenhouse gas emissions worldwide. This reduction is having a greater impact on climate-changing gases than the Kyoto Protocol has yet had. Thus the agreements in the Montreal Protocol are having extended, and very encouraging, positive effects.

There's another interesting connection to climate change. Under the Montreal Protocol, China, India, Korea, and Argentina were allowed to continue to produce 72,000 tons (combined) of CFCs per year until 2010. Most of the funds appropriated through the Montreal Protocol are going to these countries to help them phase out CFC production and destroy their existing stocks. Because CFCs are potent greenhouse gases, this phase-out also makes these countries eligible for credits in the climate trading market. In 2006 nearly two-thirds of the greenhouse gas emissions credits traded internationally were for HFC-23 elimination, and almost half of all payments went to China. Some critics think this is double-dipping, but if it eliminates a dangerous risk to all of us, isn't it worth it?

In 1995 chemists Sherwood Rowland, Mario Molina, and Paul Crutzen shared the Nobel Prize in Chemistry for their work on atmospheric chemistry and stratospheric ozone. This was the first Nobel Prize for an environmental issue.

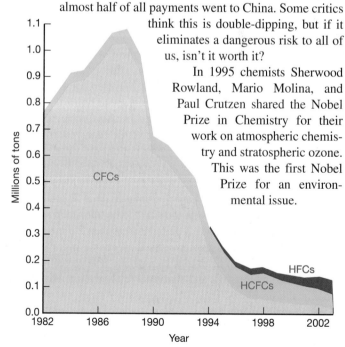

FIGURE 16.16 The Montreal Protocol has been remarkably successful in eliminating CFC production. The remaining HFC and HCFC use is primarily in developing countries, such as China and India.

16.4 EFFECTS OF AIR POLLUTION

Air pollution is a problem of widespread interest because it affects so many parts of our lives. The most obvious effects are on our health. Damage to infrastructure, vegetation, and aesthetic quality—especially visibility—are also important considerations.

Polluted air damages lungs

The World Health Organization estimates that some 5 to 6 million people die prematurely every year from illnesses related to air pollution. Heart attacks, respiratory diseases, and lung cancer all are significantly higher in people who breathe dirty air, compared to matching groups in cleaner environments. Residents of the most polluted cities in the United States, for example, are 15 to 17 percent more likely to die of these illnesses than those in cities with the cleanest air. This can mean as much as a five- to ten-year decrease in life expectancy for those who live in the worst parts of Los Angeles or Baltimore, compared to a place with clean air. Of course, the likelihood of suffering ill health from air pollutants depends on the intensity and duration of exposure as well as age and prior health status. The very young, the very old, and those already suffering from respiratory or cardiovascular disease are much more at risk. Some people are supersensitive because of genetics or prior exposure. And those doing vigorous physical work or exercise are more likely to succumb than more sedentary folks.

The United Nations estimates that at least 1.3 billion people around the world live in areas where outdoor air is dangerously polluted. Mexico City is among the world's most polluted cities, largely because of vehicle exhaust and dust. In Madrid, Spain, smog is estimated to shave one-half year off the life of each resident. This adds up to more than 50,000 years lost annually for the whole city. In China, city dwellers are four to six times more likely than country folk to die of lung cancer. As noted earlier, the greatest air quality problem is often in poorly ventilated homes in poorer countries where smoky fires are used for cooking and heating. Billions of women and children spend hours each day in these unhealthy conditions. The World Health Organization estimates that 2 million children under age 5 die each year from acute respiratory diseases exacerbated by air pollution.

In industrialized countries, one of the biggest health threats from air pollution is from soot or fine particulate material. We once thought that particles smaller than 10 micrometers (10 millionths of a meter) were too small to be trapped in the lungs. Now we know that fine PM2.5 particles (less than 2.5 micrometers in diameter) pose even greater risks than coarse particles. They have been linked with heart attacks, asthma, bronchitis, lung cancer, immune suppression, and abnormal fetal development, among other health problems. Fine particulates have many sources. Until recently power plants were the largest source, but clean air rules will require power plants to install filters and precipitators to remove at least 70 percent of their particulate emissions.

Diesel engines have long been a major source of both soot and SO_2 in the United States (fig. 16.17). Under a new rule announced in 2006, new engines in trucks and buses, in combination with

FIGURE 16.17 Soot and fine particulate material from diesel engines, wood stoves, power plants, and other combustion sources have been linked to asthma, heart attacks, and a variety of other diseases.

low-sulfur diesel fuel that is now required nationwide, will reduce particulate emissions by up to 98 percent when the rule is fully implemented in 2012. These standards will also be applied to off-road vehicles, such as tractors, bulldozers, locomotives, and barges, whose engines previously emitted more soot than all the nation's cars, trucks, and buses together. The sulfur content of diesel fuel is now 500 parts per million (ppm) compared to an average of 3,400 ppm before the regulations were imposed. By 2012 only 15 ppm of sulfur will be allowed in diesel fuel.

The U.S. EPA estimates that at least 160 million Americans—more than half the population—live in areas with unhealthy concentrations of fine particulate matter. PM2.5 levels have decreased about 30 percent over the past 25 years, but health conditions will improve if we can make further reductions.

How does pollution make us sick?

The most common route of exposure to air pollutants is by inhalation, but direct absorption through the skin or contamination of food and water also are important pathways. Because they are strong oxidizing agents, sulfates, SO_2, NO_x, and O_3 act as irritants that damage delicate tissues in the eyes and respiratory passages. Fine particulates, irritants in their own right, penetrate deep into the lungs and carry metals and other HAPs on their surfaces. Inflammatory responses set in motion by these irritants impair lung function and trigger cardiovascular problems as the heart tries to compensate for lack of oxygen by pumping faster and harder. If the irritation is really severe, so much fluid seeps into the lungs through damaged tissues that the victim actually drowns.

Carbon monoxide binds to hemoglobin and decreases the ability of red blood cells to carry oxygen. Asphyxiants such as this cause headaches, dizziness, and heart stress, and can be lethal if concentrations are high enough. Lead also binds to hemoglobin, reducing its oxygen-carrying capacity at high levels. At lower levels, lead causes long-term damage to critical neurons in the brain that results in mental and physical impairment and developmental retardation.

Some important chronic health effects of air pollutants include bronchitis and emphysema. **Bronchitis** is a persistent inflammation of bronchi and bronchioles (large and small airways in the lung) that causes mucus buildup, a painful cough, and involuntary muscle spasms that constrict airways. Severe bronchitis can lead to emphysema, an irreversible **chronic obstructive lung disease** in which airways become permanently constricted and alveoli are damaged or even destroyed. Stagnant air trapped in blocked airways swells the tiny air sacs in the lung (alveoli), blocking blood circulation. As cells die from lack of oxygen and nutrients, the walls of the alveoli break down, creating large empty spaces incapable of gas exchange (fig. 16.18). Thickened walls of the bronchioles lose elasticity, and breathing becomes more difficult. Victims of emphysema make a characteristic whistling sound when they breathe. Often they need supplementary oxygen to make up for reduced respiratory capacity.

Irritants in the air are so widespread that about half of all lungs examined at autopsy in the United States have some degree of alveolar deterioration. The Office of Technology Assessment (OTA) estimates that 250,000 people suffer from pollution-related bronchitis and emphysema in the United States, and some 50,000 excess deaths each year are attributable to complications of these diseases, which are probably second only to heart attack as a cause of death.

Smoking is undoubtedly the largest cause of obstructive lung disease and preventable death in the world. The World Health Organization says that tobacco kills some 3 million people each year. This ranks it with AIDS as one of the world's leading killers. Because of cardiovascular stress caused by carbon monoxide in smoke and chronic bronchitis and emphysema, about twice as many people die of heart failure as die from lung cancer associated with smoking. The Surgeon General estimates that more than 400,000 people die each year in the United States from emphysema, heart attacks, strokes, lung cancer, or other diseases caused by smoking. These diseases are responsible for 20 percent of all mortality in the United States, or four times as much as infectious agents. Lung cancer has now surpassed breast cancer as the leading cause of cancer deaths for U.S. women. Advertising aimed at making smoking appear stylish and liberating has resulted in a 600 percent increase in lung cancer among women since 1950. Total costs for early deaths and smoking-related illnesses in the United States are estimated to be $100 billion per year.

Plants suffer cell damage and lost productivity

Uncontrolled industrial fumes from furnaces, smelters, refineries, and chemical plants destroy vegetation and created desolate, barren landscapes around mining and manufacturing centers. The copper-nickel smelter at Sudbury, Ontario, is a spectacular and notorious example of air pollution effects on vegetation and ecosystems. In 1886 the corporate ancestor of the International Nickel Company (INCO) began open-bed roasting of sulfide ores at Sudbury. Sulfur dioxide and sulfuric acid released by this process caused massive destruction of the plant community within about 30 km of the smelter. Rains washed away the exposed soil, leaving a barren moonscape of blackened bedrock (fig. 16.19a). Super-tall, 400 m smokestacks were installed in the 1950s, and sulfur scrubbers were added 20 years later. Emissions were reduced by 90 percent and the surrounding ecosystem is beginning to recover (fig. 16.19b). Similar destruction occurred at many other sites during the nineteenth century. Copperhill, Tennessee, Butte, Montana, and the Ruhr Valley in Germany are some well-known examples, but these areas also are showing signs of recovery since corrective measures were taken. Norilsk, Russia, is a copper-smelting town that continues to have these extremely barren conditions. Norilsk's far northern latitude puts struggling vegetation at a further disadvantage, and its remote location minimizes public oversight, making conditions even more persistent than in many other smelting areas.

There are two probable ways that air pollutants damage plants. They can be directly toxic, damaging sensitive cell membranes much as irritants do in human lungs. Within a few days of exposure to toxic levels of oxidants, mottling (discoloration) occurs in leaves due to chlorosis (bleaching of chlorophyll), and then necrotic (dead) spots develop (fig. 16.5). If injury is severe, the whole plant may be killed. Sometimes these symptoms are so distinctive that positive identification of the source of damage is possible. Often, however, the symptoms are vague and difficult to separate from diseases or insect damage.

Certain combinations of environmental factors have **synergistic effects** in which the injury caused by exposure to two factors together is more than the sum of exposure to each factor individually. For instance, when white pine seedlings are exposed to subthreshold concentrations of ozone and sulfur dioxide individually, no visible injury occurs. If the same concentrations of pollutants are given together, however, visible damage occurs. In alfalfa, however, SO_2 and O_3 together cause less damage than

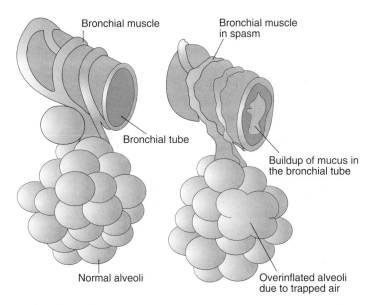

FIGURE 16.18 Bronchitis and emphysema can result in constriction of airways and permanent damage to tiny, sensitive air sacs called alveoli, where oxygen diffuses into blood vessels.

Labels in figure:
Bronchial muscle
Bronchial muscle in spasm
Bronchial tube
Buildup of mucus in the bronchial tube
Normal alveoli
Overinflated alveoli due to trapped air

(a) 1975

(b) 2005

FIGURE 16.19 In 1975, acid precipitation from the copper-nickel smelters (tall stacks in background) had killed all the vegetation and charred the pink granite bedrock black for a large area around Sudbury, Ontario (a). By 2005, forest cover was growing again, although the rock surfaces remain burned black (b).

either one alone. These complex interactions point out the unpredictability of future effects of pollutants. Outcomes might be either more or less severe than previous experience indicates.

Pollutant levels too low to produce visible symptoms of damage may still have important effects. Field studies using open-top chambers (fig. 16.20) and charcoal-filtered air show that yields in some sensitive crops, such as soybeans, may be reduced as much as 50 percent by currently existing levels of oxidants in ambient air. Some plant pathologists suggest that ozone and photochemical oxidants are responsible for as much as 90 percent of agricultural, ornamental, and forest losses from air pollution. The total costs of this damage may be as much as $10 billion per year in North America alone.

Acid deposition has many negative effects

Most people in the United States became aware of problems associated with **acid precipitation** (the deposition of wet acidic solutions or dry acidic particles from the air) within the last decade

FIGURE 16.20 An open-top chamber tests air pollution effects on plants under normal conditions for rain, sun, field soil, and pest exposure.

or so, but English scientist Robert Angus Smith coined the term *acid rain* in his studies of air chemistry in Manchester, England, in the 1850s. By the 1940s it was known that pollutants, including atmospheric acids, could be transported long distances by wind currents. This was thought to be only an academic curiosity until it was shown that precipitation of these acids can have far-reaching ecological effects.

We describe acidity in terms of pH (see figure 3.4). Values below 7 are acidic, while those above 7 are alkaline. Normal, unpolluted rain generally has a pH of about 5.6 due to carbonic acid created by CO_2 in air. Sulfur, chlorine, and other elements also form acidic compounds as they are released in sea spray, volcanic emissions, and biological decomposition. These sources can lower the pH of rain well below 5.6. Other factors, such as alkaline dust can raise it above 7. In industrialized areas, anthropogenic acids in the air usually far outweigh those from natural sources. Acid rain is only one form in which acid deposition occurs. Fog, snow, mist, and dew also trap and deposit atmospheric contaminants. Furthermore, fallout of dry sulfate, nitrate, and chloride particles can account for as much as half of the acidic deposition in some areas.

Aquatic Effects

Lakes and streams can be especially sensitive to acid deposition, especially where vegetation or bedrock makes them naturally acidic to start with. This problem was first publicized in Scandinavia, which receives industrial and automobile emissions —principally H_2SO_4 and HNO_3—generated in northwestern Europe. The thin, acidic soils and oligotrophic lakes and streams in the mountains of southern Norway and Sweden have been severely affected by this acid deposition. Some 18,000 lakes in Sweden are now so acidic that they will no longer support game fish or other sensitive aquatic organisms.

Generally, reproduction is the most sensitive stage in fish life cycles. Eggs and fry of many species are killed when the pH drops to about 5.0. This level of acidification also can disrupt the food chain by killing aquatic plants, insects, and invertebrates on which fish depend for food. At pH levels below 5.0, adult fish die as well. Trout, salmon, and other game fish are usually the most sensitive. Carp, gar, suckers, and other less desirable fish are more resistant.

In the early 1970s, evidence began to accumulate suggesting that air pollutants are acidifying many lakes in North America. Studies in the Adirondack Mountains of New York revealed that about half of the high-altitude lakes (above 1,000 m or 3,300 ft) were acidified and had no fish. Areas showing lake damage correlate closely with average pH levels in precipitation (fig. 16.21). Some 48,000 lakes in Ontario are endangered, and nearly all of Quebec's surface waters, including about 1 million lakes, are believed to be highly sensitive to acid deposition.

Sulfates account for about two-thirds of the acid deposition in eastern North America and most of Europe, while nitrates contribute most of the remaining one-third. In urban areas, where transportation is the major source of pollution, nitric acid is equal to or slightly greater than sulfuric acids in the air. A vigorous program of pollution control has been undertaken by both Canada and the United States, and SO_2 and NO_x emissions have decreased dramatically over the past three decades over much of North America.

Forest Damage

In the early 1980s, disturbing reports appeared of rapid forest declines in both Europe and North America. One of the earliest was a detailed ecosystem inventory on Camel's Hump Mountain in Vermont. A 1980 survey showed that seedling production, tree density, and viability of spruce-fir forests at high elevations had declined about 50 percent in 15 years. A similar situation was found on Mount Mitchell in North Carolina, where almost all red spruce and Fraser fir above 2,000 m (6,000 ft) are in a severe decline. Nearly all the trees are losing needles and about half of them are dead (fig. 16.22). The stress of acid rain and fog, other air pollutants, and attacks by an invasive insect called the woody aldegid are killing the trees.

Many European countries reported catastrophic forest destruction in the 1980s. It still isn't clear what caused this injury. In the longest-running forest-ecosystem monitoring record in North America, researchers at the Hubbard Brook Experimental Forest in New Hampshire have shown that forest soils have become depleted of natural buffering reserves of basic cations such as calcium and magnesium through years of exposure to acid rain. Replacement of these cations by hydrogen and aluminum ions seems to be one of the main causes of plant mortality.

Buildings and Monuments

In cities throughout the world, some of the oldest and most glorious buildings and works of art are being destroyed by air pollution. Smoke and soot coat buildings, paintings, and textiles. Limestone and marble are destroyed by atmospheric acids at an alarming rate. The Parthenon in Athens, the Taj Mahal in Agra, the Colosseum in Rome, frescoes and statues in Florence, medieval cathedrals in Europe (fig. 16.23), and the Lincoln Memorial and Washington Monument in Washington, D.C., are slowly dissolving and flaking away because of acidic fumes in the air. Medieval stained glass windows in Cologne's gothic cathedral are so porous from etching by atmospheric acids that pigments disappear and the glass literally crumbles away. Restoration costs for this one building alone are estimated at 1.5 to 3 billion euros (U.S. $1.8 billion).

On a more mundane level, air pollution also damages ordinary buildings and structures. Corroding steel in reinforced concrete weakens buildings, roads, and bridges. Paint and rubber deteriorate due to oxidation. Limestone, marble, and some kinds of sandstone flake and crumble. The Council on Environmental Quality estimates that U.S. economic losses from architectural damage caused by air pollution amount to about $4.8 billion in direct costs and $5.2 billion in property value losses each year.

Smog and haze reduce visibility

We have realized only recently that pollution affects rural areas as well as cities. Even supposedly pristine places like our national parks are suffering from air pollution. Grand Canyon National Park, where maximum visibility used to be 300 km, is now so smoggy on some winter days that visitors can't see the opposite rim only 20 km across the canyon. Mining operations, smelters, and power plants (some of which were moved to the desert to

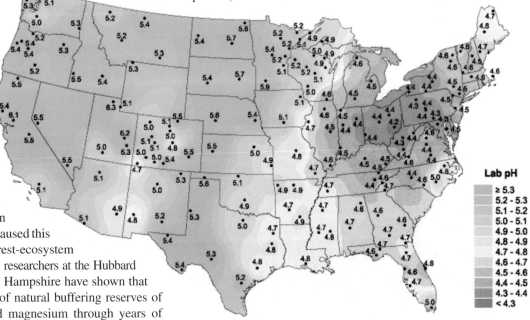

FIGURE 16.21 Acid precipitation over the United States.
Source: National Atmospheric Deposition Program/National Trends Network, 2000. http://nadp.sws.uiuc.edu.

Lab pH
- ≥ 5.3
- 5.2 – 5.3
- 5.1 – 5.2
- 5.0 – 5.1
- 4.9 – 5.0
- 4.8 – 4.9
- 4.7 – 4.8
- 4.6 – 4.7
- 4.5 – 4.6
- 4.4 – 4.5
- 4.3 – 4.4
- < 4.3

FIGURE 16.22 A Fraser fir forest on Mount Mitchell, North Carolina, killed by acid rain, insect pests, and other stressors.

improve air quality in cities like Los Angeles) are the main culprits. Similarly, the vistas from Shenandoah National Park just outside Washington, D.C., are so hazy that summer visibility is often less than 1.6 km because of smog drifting in from nearby urban areas.

Historical records show that over the past four or five decades human-caused air pollution has spread over much of the United States. Researchers report that a gigantic "haze blob" as much as 3,000 km across covers much of the eastern United States in the summer, cutting visibility as much as 80 percent. Smog and haze are so prevalent that it's hard for people to believe that the air once was clear. Studies indicate, however, that if all human-made sources of air pollution were shut down, the air would clear up in a few days and there would be about 150 km visibility nearly everywhere rather than the 15 km to which we have become accustomed.

FIGURE 16.23 Atmospheric acids, especially sulfuric and nitric acids, have almost completely eaten away the face of this medieval statue. Each year the total loss from air pollution damage to buildings and materials amounts to billions of dollars.

16.5 AIR POLLUTION CONTROL

"Dilution is the solution to pollution" was one of the early approaches to air pollution control. Tall smokestacks were built to send emissions far from the source, where they became unidentifiable and largely untraceable. But dispersed and diluted pollutants are now the source of some of our most serious pollution problems. We are finding that there is no "away" to which we can throw our waste products. While most of the discussion in this section focuses on industrial solutions, each of us can make important personal contributions to this effort (What Can You Do? p. 365).

Because most air pollution in the developed world is associated with transportation and energy production, the most effective strategy would be conservation: Reducing electricity consumption, insulating homes and offices, and developing better public transportation could all greatly reduce air pollution in the United States, Canada, and Europe. Alternative energy sources, such as wind and solar power, produce energy with little or no pollution, and these and other technologies are becoming economically competitive (chapter 20). In addition to conservation, pollution can be controlled by technological innovation.

Substances can be captured after combustion

Particulate removal involves filtering air emissions. Filters trap particulates in a mesh of cotton cloth, spun glass fibers, or asbestos-cellulose. Industrial air filters are generally giant bags 10 to 15 m long and 2 to 3 m wide. Effluent gas is blown through the bag, much like the bag on a vacuum cleaner. Every few days or weeks, the bags are opened to remove the dust cake. Electrostatic precipitators are the most common particulate controls in power plants. Ash particles pick up an electrostatic surface charge as they pass between large electrodes in the effluent stream (fig. 16.24).

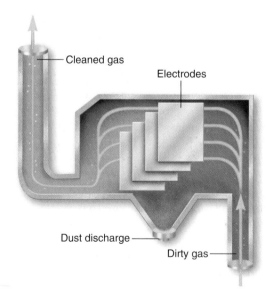

FIGURE 16.24 An electrostatic precipitator traps particulate material on electrically charged plates as effluent makes its way to the smokestack.

Charged particles then collect on an oppositely charged collecting plate. These precipitators consume a large amount of electricity, but maintenance is relatively simple, and collection efficiency can be as high as 99 percent. The ash collected by both of these techniques is a solid waste (often hazardous due to the heavy metals and other trace components of coal or other ash source) and must be buried in landfills or other solid-waste disposal sites.

Sulfur removal is important because sulfur oxides are among the most damaging of all air pollutants in terms of human health and ecosystem viability. Switching from soft coal with a high sulfur content to low-sulfur coal is the surest way to reduce sulfur emissions. High-sulfur coal is frequently politically or economically expedient, however. In the United States, Appalachia, a region of chronic economic depression, produces most high-sulfur coal. In China, much domestic coal is rich in sulfur. Switching to cleaner oil or gas would eliminate metal effluents as well as sulfur. Cleaning fuels is an alternative to switching. Coal can be crushed, washed, and gasified to remove sulfur and metals before combustion. This improves heat content and firing properties, but

may replace air pollution with solid-waste and water pollution problems; furthermore, these steps are expensive.

Sulfur can also be removed to yield a usable product instead of simply a waste disposal problem. Elemental sulfur, sulfuric acid, and ammonium sulfate can all be produced using catalytic converters to oxidize or reduce sulfur. Markets have to be reasonably close and fly ash contamination must be reduced as much as possible for this procedure to be economically feasible.

Nitrogen oxides (NO_x) can be reduced in both internal combustion engines and industrial boilers by as much as 50 percent by carefully controlling the flow of air and fuel. Staged burners, for example, control burning temperatures and oxygen flow to prevent formation of NO_x. The catalytic converter on your car uses platinum-palladium and rhodium catalysts to remove up to 90 percent of NO_x, hydrocarbons, and carbon monoxide at the same time.

Hydrocarbon controls mainly involve complete combustion or controlling evaporation. Hydrocarbons and volatile organic compounds are produced by incomplete combustion of fuels or by solvent evaporation from chemical factories, paints, dry cleaning, plastic manufacturing, printing, and other industrial processes. Closed systems that prevent escape of fugitive gases can reduce many of these emissions. In automobiles, for instance, positive crankcase ventilation (PCV) systems collect oil that escapes from around the pistons and unburned fuel and channels them back to the engine for combustion. Controls on fugitive losses from industrial valves, pipes, and storage tanks can have a significant impact on air quality. Afterburners are often the best method for destroying volatile organic chemicals in industrial exhaust stacks.

Fuel switching and fuel cleaning cut emissions

Switching from soft coal with a high sulfur content to low-sulfur coal can greatly reduce sulfur emissions. In the United States most high-sulfur coal comes from Appalachia, while low-sulfur coal comes mainly from Wyoming, Montana, and other western states. Because Appalachian economies have been heavily dependent on coal mining for generations, discussions of switching fuel sources can be highly political. Changing to another fuel, such as natural gas or nuclear energy, can eliminate all sulfur emissions as well as those of particulates and heavy metals. Natural gas is more expensive and more difficult to ship and store than coal, however, and many people prefer the known risks of coal pollution to the uncertain dangers and costs of nuclear power (chapter 19).

Alternative energy sources, such as wind and solar power, are a more complete form of fuel switching. Alternatives are becoming economically competitive in many areas (chapter 20).

Clean air legislation remains controversial

Since 1970 the Clean Air Act has been modified, updated, and amended many times. Amendments have involved acrimonious debate. As in the case of CO_2 restrictions, discussed earlier, victims of air pollution demand more protection, while industry and energy groups insist that controls are too expensive. Bills have sometimes languished in Congress for years because of disputes over burdens of responsibility, cost, and definitions of risk. A 2002 report concluded that simply by enforcing existing clean air legislation, the United States could prevent at least 6,000 deaths and 140,000 asthma attacks every year.

The most significant amendments were in the 1990 update, which addressed a variety of issues, including acid rain, urban air pollution, and toxic air emissions. These amendments also restricted ozone-depleting chemicals in accordance with the Montreal Protocol.

One of the most contested aspects of the act has been the "new source review," which was established in 1977. This provision was adopted because industry argued that it would be intolerably expensive to install new pollution-control equipment on old power plants and factories that were about to close down anyway. Congress agreed to "grandfather" existing equipment, or exempt it from new pollution limits, with the stipulation that when they were upgraded or replaced, more stringent rules would apply. The result was that owners have kept old facilities operating precisely because they were exempted from pollution control. In fact, corporations poured millions into aging power plants and factories, expanding their capacity, to avoid having to build new ones. Thirty years later, most of those grandfathered plants are still going strong and continue to be among the biggest contributors to smog and acid rain.

Clean air legislation has been very successful

Despite these disputes, the Clean Air Act has been extremely successful in saving money and lives. The EPA estimates that between 1970 and 2010, lead fell 99 percent, SO_2 declined 39 percent, and CO shrank 32 percent (fig. 16.25). Filters, scrubbers, and precipitators on power plants and other large stationary sources are responsible for most of the particulate and SO_2 reductions. Catalytic converters on cars are responsible for most of the CO and O_3 reductions. For 23 of the largest U.S. cities, air quality now reaches hazardous levels 93 percent less frequently than a decade ago. Forty of the 97 metropolitan areas that failed to meet clean air standards in the 1980s are now in compliance, many for the first time in a generation.

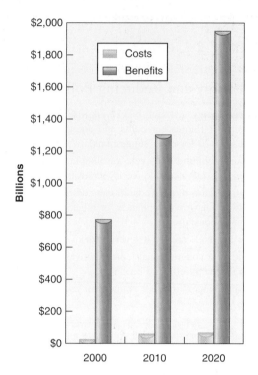

FIGURE 16.26 Direct costs and benefits of Clean Air Act provisions by 2000, 2010, and 2020, in billions of 2006 dollars.
Source: EPA 2011, Clean Air Impacts Summary Report.

The only conventional, "criteria" pollutants that have not dropped significantly are particulates and NO_x. Because automobiles are the main source of NO_x, cities, such as Nashville, Tennessee, and Atlanta, Georgia, where pollution comes largely from traffic, still have serious air quality problems. Rigorous pollution controls are having a positive effect on Southern California air quality. Los Angeles, which had the dirtiest air in the nation for decades, wasn't even in the top 20 polluted cities in 2010.

In a 2011 study of the economic costs and benefits of the 1990 Clean Air Act, the EPA found that the direct benefits of air quality protection by 2020 will be $2 trillion, while the direct costs of implementing those protections was about 1/30th of that, or $65 billion (fig. 16.26). The direct benefits were mainly in prevented costs of premature illness, death, and work losses (table 16.5). About half of the direct costs were improvements in cars and trucks, which now burn cleaner and more efficiently than they did in the past. This cost has been distributed to vehicle owners, who also benefit from lower expenditures on fuel. A quarter of costs involved cleaner furnaces and pollutant capture at electricity-generating power plants and other industrial facilities. The remaining costs involved pollution reductions at smaller businesses, municipal facilities, construction sites, and other sources. Overall, emission controls have not dampened economic productivity, despite widespread fears to the contrary. Emissions of criteria pollutants have declined in recent decades, whereas economic indicators have grown (fig. 16.27).

In addition to these savings, the Clean Air Act has created thousands of jobs in developing, installing, and maintaining technology and in monitoring. At a time when many industries are providing fewer jobs, owing to greater mechanization, jobs have been

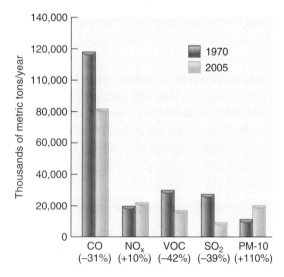

FIGURE 16.25 Air pollution trends in the United States, 1970 to 1998. Although population and economic activity increased during this period, emissions of all criteria air pollutants, except for nitrogen oxides and particulate matter, decreased significantly.
Source: Environmental Protection Agency, 2011.

Table 16.5 Reductions of Health Impairments Resulting from Ozone and Particulate Reductions Since 1990

Health Effect Reductions (PM2.5 & Ozone Only)	Year 2010 (in cases)	Year 2020 (in cases)
Adult Mortality-particles	160,000	230,000
Infant Mortality-particles	230	280
Mortality-ozone	4300	7100
Chronic Bronchitis	54,000	75,000
Heart Disease	130,000	200,000
Asthma Exacerbation	1,700,000	2,400,000
Emergency Room Visits	86,000	120,000
School Loss Days	3,200,000	5,400,000
Lost Work Days	13,000,000	17,000,000

Source: EPA, 2011.

growing in clean technologies and pollution control and monitoring. At the same time, reductions in acid rain have decreased losses to forest resources and building infrastructure.

Market mechanisms have been part of the solution, especially for sulfur dioxide, which is widely considered to have benefited from a cap-and-trade approach. This strategy sets maximum limits for each facility and then lets facilities sell pollution credits if they can cut emissions, or facilities can buy credits if they are cheaper than installing pollution-control equipment. When trading began in 1990, economists estimated that eliminating 10 million tons of sulfur dioxide would cost $15 billion per year. Left to find the most economical ways to reduce emissions, however, utilities have been able to reach clean air goals for one-tenth that price. A serious shortcoming of this approach is that while trading has resulted in overall pollution reduction, some local "hot spots" remain where owners have found it cheaper to pay someone else to reduce pollution than to do it themselves.

Particulate matter (mostly dust and soot) is produced by agriculture, fuel combustion, metal smelting, concrete manufacturing, and other activities. Industrial cities, such as Baltimore, Maryland, and Baton Rouge, Louisiana, also have continuing problems. Eighty-five other urban areas are still considered nonattainment regions. In spite of these local failures, however, 80 percent of the United States now meets the National Ambient Air Quality Standards (fig. 16.28). This improvement in air quality is perhaps the greatest environmental success story in our history.

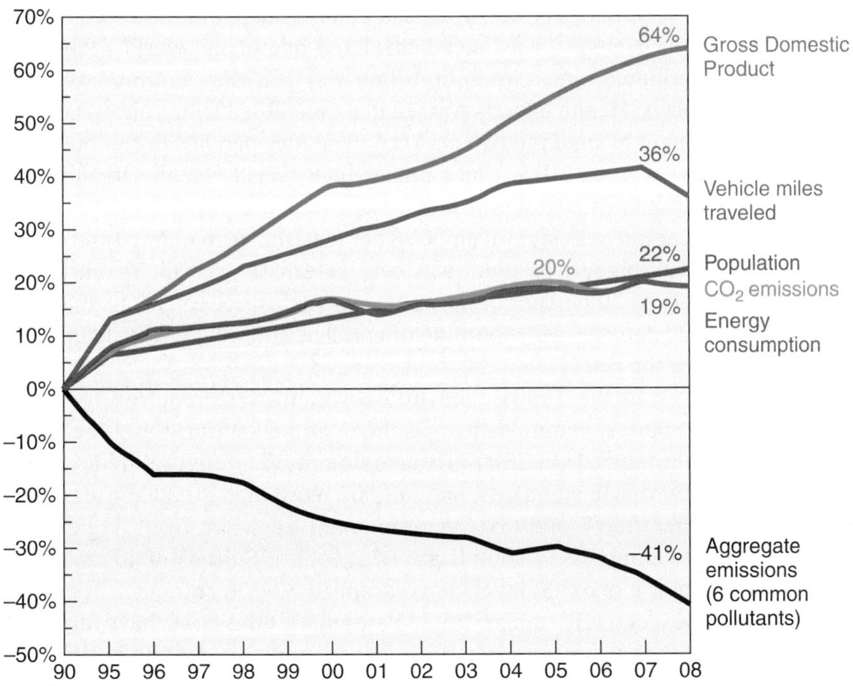

16.6 GLOBAL PROSPECTS

The outlook is not so encouraging in many parts of the world. The major metropolitan areas of many developing countries are growing at explosive rates to incredible sizes (chapter 22), and environmental quality is abysmal in many of them. In Mexico City, notorious for bad air, pollution levels exceed WHO health standards 350 days per year, and more than half of all city children have lead levels in their blood high enough to lower intelligence and retard development. Mexico City's 131,000 industries and 2.5 million vehicles spew out more than 5,500 tons of air pollutants daily. In Santiago, Chile, suspended particulates exceed WHO standards of 90 mg/m^3 about 299 days per year.

Rapid industrialization and urban growth outpace pollution controls

Rapid growth and industrialization in China, India, and many other parts of the developing world are producing emissions much faster than pollution-control agencies can manage. Because China's growth is so rapid, its air quality is increasingly poor. Many of China's 400,000 factories have no air pollution controls. Experts estimate that home coal burners and factories emit 10 million tons of soot and 15 million tons of sulfur dioxide annually and that emissions have increased rapidly over the past 20 years. Sixteen of the 20 cities in the world with the worst air quality are in China. Shenyang, an industrial city in northern China, is thought to have the world's worst continuing particulate problem, with peak winter concentrations over 700 mg/m^3 (nine times U.S. maximum standards). Airborne particulates in Shenyang exceed WHO standards on 347 days per year. It's estimated that air pollution is responsible for 400,000 premature deaths every year in China. Beijing, Xi'an, and Guangzhou also have severe air pollution problems. The high incidence of cancer in Shanghai is thought to be linked to air pollution (see fig. 16.1).

FIGURE 16.27 Comparison of growth measures and emissions of criteria air pollutants, 1990–2008.
Source: EPA, 2011.

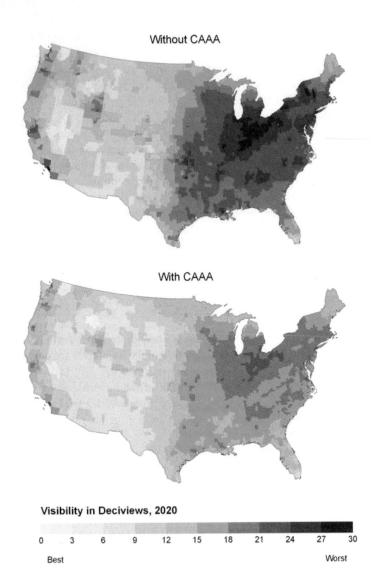

Visibility in Deciviews, 2020

| | | | | | | | | | | |
|0|3|6|9|12|15|18|21|24|27|30|

Best   Worst

FIGURE 16.28 Projected visibility impairments, shown with dark colors, would be considerably worse in 2020 without the 1990 Clean Air Act amendments (CAAA, *top*) than they will be with the amendments *(bottom)*. Units are deciviews, a measure of perceptible change in visibility.
Source: EPA 2011, Clean Air Impacts Summary Report.

Every year the Blacksmith Institute compiles a list of the world's worst-polluted places. Globally, smelters, mining operations, petrochemical industries—which release hazardous organic compounds to the air and water—and chemical manufacturing are frequently the worst sources of pollutants. Often these are in impoverished and developing areas of Africa, Asia, or the Americas, where government intervention is weak and regulations are nonexistent or poorly enforced. Funds and political will are usually unavailable to deal with pollution, much of which is involved with materials going to wealthier countries or waste that is received from developed countries (see chapter 21).You can learn more about these places at www.blacksmithinstitute.org.

Norilsk, Russia (one site highlighted on Blacksmith Institute's list of worst places), is a notorious example of toxic air pollution. Founded in 1935 as a slave labor camp, this Siberian city is considered one of the most polluted places on earth. Norilsk houses the world's largest nickel mine and heavy metals smelting complex, which discharge over 4 million tons of cadmium, copper, lead, nickel, arsenic, selenium, and zinc into the air every year. The snow turns black as quickly as it falls, the air tastes of sulfur, and the average life expectancy for factory workers is ten years below the Russian average (which already is the lowest of any industrialized country). Difficult pregnancies and premature births are much more common in Norilsk than elsewhere in Russia. Children living near the nickel plant are ill twice as much as Russia's average, and birth defects are reported to affect as much as 10 percent of the population. Why do people stay in such a place? Many were attracted by high wages and hardship pay, and now that they're sick, they can't afford to move.

There are also signs of progress

Despite global expansion of chemical industries and other sources of air pollution, there have been some spectacular successes in air pollution control. Sweden and West Germany (countries affected by forest losses due to acid precipitation) cut their sulfur emissions by two-thirds between 1970 and 1985. Austria and Switzerland have gone even farther, regulating even motorcycle emissions. The Global Environmental Monitoring System (GEMS) reports declines in particulate levels in 26 of 37 cities worldwide. Sulfur dioxide and sulfate particles, which cause acid rain and respiratory disease, have declined in 20 of these cities.

Even poor countries can control air pollution. Delhi, India, for example, was once considered one of the world's ten most polluted cities. Visibility often was less than 1 km on smoggy days. Health experts warned that breathing Delhi's air was equivalent to smoking two packs of cigarettes per day. Pollution levels were nearly five times higher than World Health Organization standards. Respiratory diseases were widespread, and the cancer rate was significantly higher than for surrounding rural areas. The biggest problem was vehicle emissions, which contributed about 70 percent of air pollutants (industrial emissions made up 20 percent, while burning of garbage and firewood made up most of the rest).

In the 1990s catalytic converters were required for automobiles, and unleaded gasoline and low-sulfur diesel fuel were introduced. In 2000 private automobiles were required to meet European standards, and in 2002 more than 80,000 buses, auto-rickshaws, and taxis were required to switch from liquid fuels to compressed natural gas (fig. 16.29). Sulfur dioxide and carbon monoxide levels have dropped 80 percent and 70 percent, respectively, since 1997. Particulate emissions have dropped by about 50 percent. Residents report that the air is dramatically clearer and more healthy. Unfortunately, rising prosperity,

FIGURE 16.29 Air quality in Delhi, India, has improved dramatically since buses, auto-rickshaws, and taxis were required to switch from liquid fuels to compressed natural gas. This is one of the most encouraging success stories in controlling pollution in the developing world.

FIGURE 16.30 Cubatao, Brazil, was once considered one of the most polluted cities in the world. Better environmental regulations and enforcement along with massive investments in pollution-control equipment have improved air quality significantly.

driven by globalization of information management, has doubled the number of vehicles on the roads, threatening this progress. Still, the gains made in Delhi are encouraging for people everywhere.

Twenty years ago, Cubatao, Brazil, was described as the "Valley of Death," one of the most dangerously polluted places in the world. Every year a steel plant, a huge oil refinery, and fertilizer and chemical factories churned out thousands of tons of air pollutants that were trapped between onshore winds and the uplifted plateau on which São Paulo sits (fig. 16.30). Trees died on the surrounding hills. Birth defects and respiratory diseases were alarmingly high. Since then, however, the citizens of Cubatao

have made remarkable progress in cleaning up their environment. The end of military rule and restoration of democracy allowed residents to publicize their complaints. The environment became an important political issue. The state of São Paulo invested about $100 million and the private sector spent twice as much to clean up most pollution sources in the valley. Particulate pollution was reduced 75 percent, ammonia emissions were reduced 97 percent, hydrocarbons that cause ozone and smog were cut 86 percent, and sulfur dioxide production fell 84 percent. Fish are returning to the rivers, and forests are regrowing on the mountains. Progress is possible! We hope that similar success stories will be obtainable elsewhere.

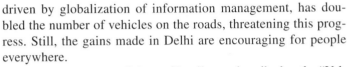

CONCLUSION

Air pollution is often the most obvious and widespread type of pollution. Everywhere on earth, from the most remote island in the Pacific, to the highest peak in the Himalayas, to the frigid ice cap over the North Pole, there are traces of human-made contaminants, remnants of the 2 billion metric tons of pollutants released into the air worldwide every year by human activities.

Adverse effects of air pollution include respiratory diseases, birth defects, heart attacks, developmental disabilities in children, and cancer. Environmental impacts include destruction of stratospheric ozone, poisoning of forests and waters by acid rain, and corrosion of building materials.

We have made encouraging progress in controlling air pollution, progress that has economic benefits as well as health benefits. Many students aren't aware of how much worse air quality was in the industrial centers of North America and Europe a century or two ago compared to today. Cities such as London, Pittsburgh, Chicago, Baltimore, and New York had air quality as bad as or worse

than most megacities of the developing world now. The progress in reducing air pollution in these cities gives us hope that residents can do so elsewhere as well.

The success of the Montreal Protocol in eliminating CFCs is a landmark in international cooperation on an environmental problem. Growth of the stratospheric ozone hole has slowed, and we expect the ozone depletion to end in about 50 years. This is one of the few global environmental threats that has had such a rapid and successful resolution. Let's hope that others will follow.

Developing areas face severe challenges in air quality. Most of the worst air pollution in the world occurs in large cities of developing countries. However, there are dramatic cases of pollution in developing countries. Problems that once seemed overwhelming can be overcome. In some cases this requires lifestyle changes or different ways of doing things to bring about progress, but as the Chinese philosopher Lao Tsu wrote, "A journey of a thousand miles must begin with a single step."

REVIEWING LEARNING OUTCOMES

By now you should be able to explain the following points:

16.1 Describe the air around us.

- What are some natural pollutants?

16.2 Discuss human-caused air pollution.

- Criteria pollutants were addressed first.
- Mercury and other metals are also regulated.
- Carbon dioxide and halogens are greenhouse gases.
- Hazardous air pollutants cause cancer and nerve damage.
- Aesthetic degradation also results from pollution.
- Indoor air is more dangerous for most of us than outdoor air.

16.3 Explain how atmospheric processes affect air quality.

- Temperature inversions trap pollutants.
- Wind currents carry pollutants worldwide.
- Stratospheric ozone is destroyed by chlorine.
- The Montreal Protocol is a resounding success.

16.4 Describe the effects of air pollution.

- Polluted air damages lungs.
- How does pollution make us sick?
- Plants suffer cell damage and lost productivity.
- Acid deposition has many negative effects.
- Smog and haze reduce visibility.

16.5 Outline methods for air pollution control.

- Substances can be captured after combustion.
- Fuel switching and fuel cleaning also are effective.
- Clean air legislation remains controversial.
- Clean air legislation has been very successful.

16.6 Summarize some global prospects.

- Industrialization and urban growth outpace pollution controls.
- There are also signs of progress.

PRACTICE QUIZ

1. Define *primary air pollutants* and *secondary air pollutants.*
2. What are the six criteria pollutants in the original Clean Air Act? Why were they chosen? What are some additional hazardous air toxins that have been added?
3. What pollutants in indoor air may be hazardous to your health? What is the greatest indoor air problem globally?
4. What is acid deposition? What causes it?
5. What is an atmospheric inversion and how does it trap air pollutants?
6. What is the difference between ambient and stratospheric ozone? What is destroying stratospheric ozone?
7. What is long-range air pollution transport? Give two examples.
8. What is the ratio of direct costs and benefits of the Clean Air Act? What costs are mainly saved?
9. Which of the conventional pollutants has decreased most in the recent past and which has decreased least?
10. Give one example of current air quality problems and one success in controlling pollution in a developing country.

CRITICAL THINKING AND DISCUSSION QUESTIONS

1. What might be done to improve indoor air quality? Should the government mandate such changes? What values or worldviews are represented by different sides of this debate?
2. Debate the following proposition: Our air pollution blows onto someone else; therefore, installing pollution controls will not bring any direct economic benefit to those of us who have to pay for them.
3. Utility managers once claimed that it would cost $1,000 per fish to control acid precipitation in the Adirondack lakes and that it would be cheaper to buy fish for anglers than to put scrubbers on power plants. Suppose that is true. Does it justify continuing pollution?
4. Developing nations claim that richer countries created global warming and stratospheric ozone depletion and therefore should bear responsibility for fixing these problems. How would you respond?
5. If there are thresholds for pollution effects, is it reasonable or wise to depend on environmental processes to disperse, assimilate, or inactivate waste products?
6. How would you choose between government "command and control" regulations versus market-based trading programs for air pollution control? Are there situations where one approach would work better than the other?

Data Analysis: Graphing Air Pollution Control

Reduction of acid-forming air pollutants in Europe is an inspiring success story. The first evidence of ecological damage from acid rain came from disappearance of fish from Scandinavian lakes and rivers in the 1960s. By the 1970s, evidence of air pollution damage to forests in northern and central Europe alarmed many people. International agreements reached since the mid-1980s have been highly successful in reducing emissions of SO_2 and NO_x as well as photochemical oxidants, such as O_3. The graph on this page shows reductions in SO_2 emissions in Europe between 1990 and 2002. The light blue area shows actual SO_2 emissions. Blue represents changes due to increased nuclear and renewable energy. Orange shows reductions due to energy conservation. Green shows improvement from switching to low-sulfur fuels. Purple shows declines due to increased abatement measures (flue gas scrubbers). The upper boundary of each area indicates what emissions would have been without pollution control.

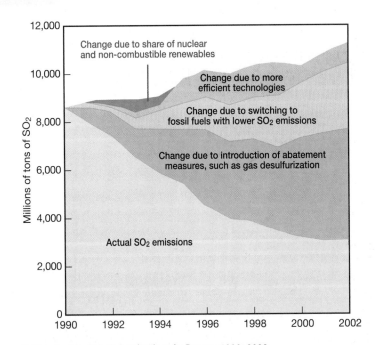

Sulfur dioxide emission reductions in Europe, 1990–2002.

1. How much have actual SO_2 emissions declined since 1990?

2. How much lower were SO_2 emissions in 2002 than they would have been without pollution control (either in percentage or actual amount)?

3. What percentage of this reduction was due to abatement measures, such as flue gas scrubbers?

4. What percent was gained by switching to low-sulfur fuels?

5. How much did energy conservation contribute?

6. What happened to nuclear power?

Water Use and Management

Between 2000 and 2010, the surface level of Lake Mead, the largest reservoir on the Colorado River, fell more than 100 ft (30.5 m) during the worst drought in recorded history. If water levels fall another 100 ft, the reservoir will reach "dead pool" levels at which it can provide neither the water nor the electrical power on which millions of people depend.

Learning Outcomes

After studying this chapter, you should be able to:

17.1 Summarize why water is a precious resource and why shortages occur.

17.2 Compare major water compartments.

17.3 Summarize water availability and use.

17.4 Investigate freshwater shortages.

17.5 Illustrate the benefits and problems of dams and diversions.

17.6 Appreciate how we might get by with less water.

17.7 Understand how we might increase water supplies.

"I tell you gentlemen; you are piling up a heritage of conflict and litigation of water rights, for there is not sufficient water to supply the land."

~ John Wesley Powell

Case Study   When Will Lake Mead Go Dry?

The Colorado River is the life-blood of the American Southwest. More than 30 million people and a $1.2 trillion regional economy in cities such as Los Angeles, Phoenix, Las Vegas, and Denver depend on its water. But the reliability of this essential resource is in doubt. Drought, climate change, and rapid urban growth are creating worries about the sustainability of the water supply for the entire watershed.

In 2008 Tim Barnett and David Pierce from the Scripps Institute in California published a provocative article suggesting that within a decade or so, both Lake Mead and Lake Powell could reach levels at which neither would be able to either produce power or provide water for urban or agricultural use, if no changes are made in current water allocations. For these huge lakes, which constitute more than 85 percent of the water storage for the entire Colorado system, to reach "dead pool" levels would be a catastrophe for the whole region. This warning is based on both historical records and climate models that suggest a 10 to 30 percent runoff reduction in the area over the next 50 years.

The roots of this problem can be traced to the Colorado Compact of 1922, in which state water rights were allocated. The previous decade had been the wettest in more than a thousand years. The estimated annual river flow of 18 million acre-feet (22 billion m^3) that negotiators thought they could allocate was about 20 percent higher than the twentieth-century average. The error didn't matter much at the time, because none of the states were able to withdraw their share of water from the river.

As cities have grown, however, and agriculture has expanded over the past century, competing claims for water have repeatedly caused tensions and disputes. Cumulatively, massive water diversion projects, such as the Colorado River Aqueduct, which provides water for Los Angeles, the All-American Canal, which irrigates California's Imperial Valley, or the Central Arizona Project, which transports water over the mountains and across the desert to Phoenix and Tucson, are capable of diverting as much water as the entire Colorado River flow. In 1944 the United States agreed to provide 1.5 million acre-feet to Mexico so there would be at least a little water (although of dubious quality) in the river when it crossed the border.

To make matters worse, climate change is expected to decrease western river flows by 10 to 30 percent over the next 50 years. The Southwest is currently in its eighth year of drought, which may be the first hint of that change. The maximum water level in Lake Mead (an elevation of 1,220 feet or 372 m) was last reached in 2000. Since then the lake level has been dropping about 12 feet (3.6 m) per year, reaching 1,097 feet in 2010. The minimum power level (the height at which electricity can be produced) is 1,050 feet (320 m). The minimum level at which water can be drawn off by gravity is 900 feet (274 m). Barnett and Pierce estimate that without changes in current management plans, there's a 50 percent chance that minimum power pool levels in both Lakes Mead and Powell will be reached by 2017 and there's an equal chance that live storage in both lakes will be gone by about 2021.

Already, we're at or beyond the sustainable limits of the river. Currently Lake Powell holds a little over half its maximum volume, and Lake Mead is only 43 percent full (fig. 17.1). The shores

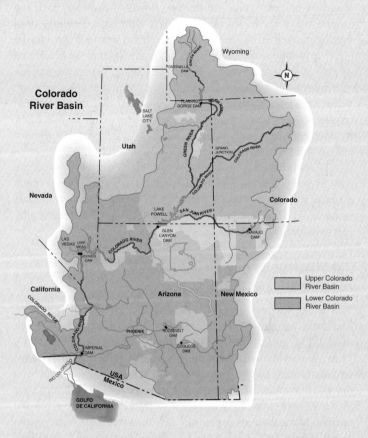

FIGURE 17.1 The Colorado River flows 2,330 km (1,450 mi) through seven western states. Its water supports 30 million people and a $1.2 trillion regional economy, but drought, climate change, and rapid urban growth threaten the sustainability of this resource.

of both lakes now display a wide "bathtub ring" of deposited minerals left by the receding water. One suggestion has been to drain Lake Powell in order to ensure a water supply for Lake Mead. This solution is strenuously opposed by many of the 3 million people per year who recreate in its red rock canyons and sparkling blue water. On the other hand, think of the cost and disruption if Los Angeles, Phoenix, Las Vegas, and other major metropolitan areas of the Southwest were to run out of water and power.

The American Southwest isn't alone in facing this problem. The United Nations warns that water supplies are likely to become one of the most pressing environmental issues of the twenty-first century. By 2025 two-thirds of all humans could be living in places where water resources are inadequate. In this chapter we'll look at the sources of our fresh water, what we do with it, and how we might protect its quality and extend its usefulness.

For further reading, see:

Barnett, T. P., and D. W. Pierce. 2008. When will Lake Mead go dry? *Journal of Water Resources Research*, vol. 44, W03201.

Powell, James L. 2009. *Dead Pool: Lake Powell, Global Warming, and the Future of Water in the West*. University of California Press.

For related resources, including Google Earth™ placemarks that show locations where these issues can be seen, visit EnvironmentalScience-Cunningham.blogspot.com.

17.1 WATER RESOURCES

Water is a marvelous substance—flowing, rippling, swirling around obstacles in its path, seeping, dripping, trickling, constantly moving from sea to land and back again. Water can be clear, crystalline, icy green in a mountain stream, or black and opaque in a cypress swamp. Water bugs skitter across the surface of a quiet lake; a stream cascades down a stairstep ledge of rock; waves roll endlessly up a sand beach, crash in a welter of foam, and recede. Rain falls in a gentle mist, refreshing plants and animals. A violent thunderstorm floods a meadow, washing away streambanks. Water is a most beautiful and precious resource.

Water is also a great source of conflict. Some 2 billion people now live in countries with insufficient fresh water. Some experts estimate this number could double in 25 years. To understand this resource, let's first ask, where does our water come from, and why is it so unevenly distributed?

The hydrologic cycle constantly redistributes water

The water we use cycles endlessly through the environment. The total amount of water on our planet is immense—more than 1,404 million km^3 (370 billion billion gal) (table 17.1). This water evaporates from moist surfaces, falls as rain or snow, passes through living organisms, and returns to the ocean in a process known as the **hydrologic cycle** (see fig. 3.19). Every year about 500,000 km^3, or a layer 1.4 m thick, evaporates from the oceans. More than 90 percent of that moisture falls back on the ocean. The 47,000 km^3 carried onshore joins some 72,000 km^3 that evaporate from lakes, rivers, soil, and plants to become our annual, renewable freshwater supply. Plants play a major role in the hydrologic cycle, absorbing groundwater and pumping it into the atmosphere by transpiration (transport plus evaporation). In tropical forests, as much as 75 percent of annual precipitation is returned to the atmosphere by plants.

Solar energy drives the hydrologic cycle by evaporating surface water, which becomes rain and snow. Because water and sunlight are unevenly distributed around the globe, water resources are very uneven. At Iquique in the Chilean desert, for instance, no rain has fallen in recorded history. At the other end

of the scale, 22 m (72 ft) of rain were recorded in a single year at Cherrapunji in India. Figure 17.2 shows broad patterns of precipitation around the world. Most of the world's rainiest regions are tropical, where heavy rainy seasons occur, or in coastal mountain regions. Deserts occur on every continent just outside the tropics (the Sahara, the Namib, the Gobi, the Sonoran, and many others). Rainfall is also slight at very high latitudes, another high-pressure region.

Water supplies are unevenly distributed

Rain falls unevenly over the planet (fig. 17.2). Some places get almost no precipitation, while others receive heavy rain almost daily. Three principal factors control these global water deficits and surpluses. First, global atmospheric circulation creates regions of persistent high air pressure and low rainfall about 20° to 40° north and south of the equator (chapter 15). These same circulation patterns produce frequent rainfall near the equator and between about 40° and 60° north and south latitude. Second, proximity to water sources influences precipitation. Where prevailing winds come over oceans, they bring moisture to land. Areas far from oceans—in a windward direction—are usually relatively dry.

A third factor in water distribution is topography. Mountains act as both cloud formers and rain catchers. As air sweeps up the windward side of a mountain, air pressure decreases and air cools. As the air cools, it reaches the saturation point, and moisture condenses as either rain or snow. Thus the windward side of a mountain range, as in the Pacific Northwest, is usually wet much of the year. Precipitation leaves the air drier than it was on its way up the mountain. As the air passes the mountaintop and descends the other side, air pressure rises, and the already-dry air warms, increasing its ability to hold moisture. Descending, warming air rarely produces any rain or snow. Places in the **rain shadow**, the dry, leeward side of a mountain range, receive little precipitation. A striking example of the rain shadow effect is found on Mount Waialeale, on the island of Kauai, Hawaii (fig. 17.3). The windward side of the island receives nearly 12 m of rain per year, while the leeward side, just a few kilometers away, receives just 46 cm.

Usually a combination of factors affects precipitation. In Cherrapunji, India, atmospheric circulation sweeps moisture from the warm Indian Ocean toward the high ridges of the Himalayas. Iquique, Chile, lies in the rain shadow of the Andes and in a high-pressure desert zone. Prevailing winds are from the east, so even though Iquique lies near the ocean, it is far from the winds' moisture source—the Atlantic. In the American Southwest, Australia, and the Sahara, high-pressure atmospheric conditions tend to keep the air and land dry. The global map of precipitation represents a complex combination of these forces of atmospheric circulation, prevailing winds, and topography.

Human activity also explains some regions of water deficit. As noted earlier, plant transpiration recycles moisture and produces rain. When forests are cleared, falling rain quickly

---

Table 17.1 Some Units of Water Measurement

One cubic kilometer (km^3) equals 1 billion cubic meters (m^3), 1 trillion liters, or 264 billion gallons.

One acre-foot is the amount of water required to cover an acre of ground 1 foot deep. This is equivalent to 325,851 gallons, or 1.2 million liters, or 1,234 m^3, about the amount consumed annually by a family of four in the United States.

One cubic foot per second of river flow equals 28.3 liters per second or 449 gallons per minute.

See the table at the end of the book for conversion factors.

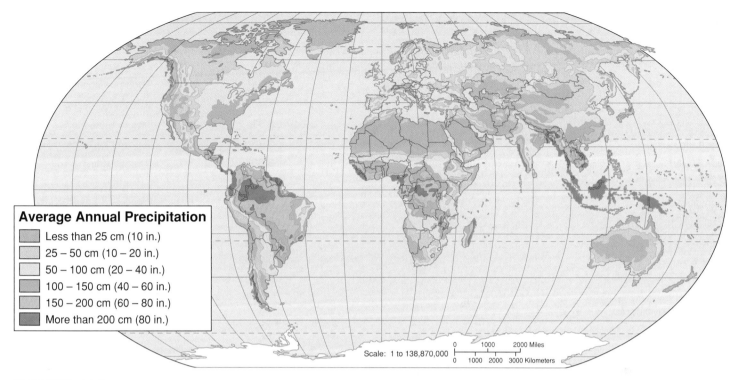

FIGURE 17.2 Average annual precipitation. Note wet areas that support tropical rainforests occur along the equator, while the major world deserts occur in zones of dry, descending air between 20° and 40° north and south.

Average Annual Precipitation

- Less than 25 cm (10 in.)
- 25 – 50 cm (10 – 20 in.)
- 50 – 100 cm (20 – 40 in.)
- 100 – 150 cm (40 – 60 in.)
- 150 – 200 cm (60 – 80 in.)
- More than 200 cm (80 in.)

Scale: 1 to 138,870,000

enters streams and returns to the ocean. In Greece, Lebanon, parts of Africa, the Caribbean, South Asia, and elsewhere, desertlike conditions have developed since the original forests were destroyed.

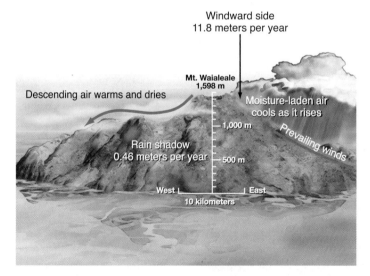

FIGURE 17.3 Rainfall on the east side of Mount Waialeale in Hawaii is more than 20 times as much as on the west side. Prevailing trade winds bring moisture-laden sea air onshore. The air cools as it rises up the flanks of the mountain and the water it carries precipitates as rain—11.8 m (38 ft) per year!

Think About It

We have noted three important natural causes of water surpluses and deficits. Which of these might be important where you live?

Does water availability affect your lifestyle? Should it?

17.2 MAJOR WATER COMPARTMENTS

The distribution of water often is described in terms of interacting compartments in which water resides, sometimes briefly and sometimes for eons (table 17.2). The length of time water typically stays in a compartment is its **residence time**. On average, a water molecule stays in the ocean for about 3,000 years, for example, before it evaporates and starts through the hydrologic cycle again.

Oceans hold 97 percent of all water on earth

Together the oceans contain more than 97 percent of all the *liquid* water in the world. (The water of crystallization in rocks is far larger than the amount of liquid water.) Oceans are too salty for most human uses, but they contain 90 percent of the world's living biomass. Although the ocean basins really form a continuous

Table 17.2   Earth's Water Compartments

Compartment	Volume (1,000 km³)	Percent of Total Water	Average Residence Time
Total	1,386,000	100	2,800 years
Oceans	1,338,000	96.5	3,000 to 30,000 years*
Ice and snow	24,364	1.76	1 to 100,000 years*
Saline groundwater	12,870	0.93	Days to thousands of years*
Fresh groundwater	10,530	0.76	Days to thousands of years*
Fresh lakes	91	0.007	1 to 500 years*
Saline lakes	85	0.006	1 to 1,000 years*
Soil moisture	16.5	0.001	2 weeks to 1 year*
Atmosphere	12.9	0.001	1 week
Marshes, wetlands	11.5	0.001	Months to years
Rivers, streams	2.12	0.0002	1 week to 1 month
Living organisms	1.12	0.0001	1 week

*Depends on depth and other factors.

Source: Data from UNEP, 2002.

reservoir, shallows and narrows between them reduce water exchange, so they have different compositions, climatic effects, and even different surface elevations.

Oceans play a crucial role in moderating the earth's temperature (fig. 17.4). Vast river-like currents transport warm water from the equator to higher latitudes, and cold water flows from the poles to the tropics (fig. 17.5). The Gulf Stream, which flows northeast from the coast of North America toward northern Europe, flows at a steady rate of 10–12 km per hour (6–7.5 mph) and carries more than 100 times more water than all rivers on earth put together.

In tropical seas, surface waters are warmed by the sun, diluted by rainwater and runoff from the land, and aerated by wave action. In higher latitudes, surface waters are cold and much more dense. This dense water subsides or sinks to the bottom of deep ocean basins and flows toward the equator. Warm surface water

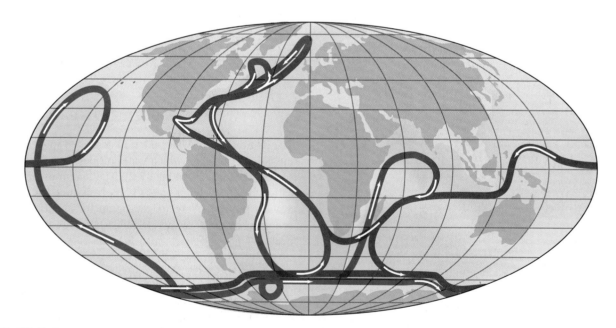

FIGURE 17.4 Ocean currents act as a global conveyor system, redistributing warm (red) and cold (blue) currents around the globe. These currents moderate our climate. For example, the Gulf Stream keeps northern Europe much warmer than northern Canada. Ocean colors show salinity variation from low (*blue*) to high (*yellow*).
Source: NASA.

FIGURE 17.5 Ocean currents, such as the warm Gulf Stream, redistribute heat as they flow around the globe. Here, orange and yellow indicate warm water temperatures (25–30°C); blue and green are cold (0–5°C).

of the tropics stratifies or floats on top of this cold, dense water as currents carry warm water to high latitudes. Sharp boundaries form between different water densities, different salinities, and different temperatures, retarding mixing between these layers.

Glaciers, ice, and snow contain most surface fresh water

Of the 2.4 percent of all water that is fresh, nearly 90 percent is tied up in glaciers, ice caps, and snowfields (fig. 17.6). Although most of this ice is located in Antarctica, Greenland, and the floating ice cap in the Arctic, alpine glaciers and snowfields supply water to billions of people. The winter snowpack on the western slope of the Rocky Mountains, for example, provides 75 percent of the flow in the Colorado River described in the opening case study of this chapter. Drought conditions already have reduced snowfall (and runoff) in the western United States, and global warming is projected to cause even further declines.

As chapter 15 discusses, climate change is shrinking glaciers and snowfields nearly everywhere (fig. 17.7). In Asia, the Tibetan glaciers that are the source of six of the world's largest rivers and supply drinking water for three billion people are shrinking rapidly. There are warnings that these glaciers could vanish in a few decades, which would bring enormous suffering and economic loss in many places.

Groundwater stores large resources

After glaciers, the next largest reservoir of fresh water is held in the ground as **groundwater**. Precipitation that does not evaporate back into the air or run off over the surface percolates through the soil and into fractures and spaces of permeable rocks in a process called **infiltration** (fig. 17.8). Upper soil layers that hold both air and water make up the **zone of aeration**. Moisture for plant growth comes primarily from these layers. Depending on rainfall amount, soil type, and surface topography, the zone of aeration may be very shallow or quite deep. Lower soil layers where all spaces are filled with water make up the **zone of saturation**. The top of this zone is the **water table**. The water table is not flat, but undulates according to the surface topography and subsurface structure. Water tables also rise and fall seasonally, depending on precipitation and infiltration rates.

Porous layers of sand, gravel, or rock lying below the water table are called **aquifers**. Aquifers are always underlain by relatively impermeable layers of rock or clay that keep water from seeping out at the bottom (fig. 17.9).

Folding and tilting of the earth's crust by geological processes can create shapes that generate water pressure in confined aquifers (those trapped between two impervious, confining rock layers). When a pressurized aquifer intersects the surface, or if it is penetrated by a pipe or conduit, the result is an **artesian well** or spring, from which water gushes without being pumped.

Areas where water infiltrates into an aquifer are called **recharge zones**. The rate at which most aquifers are refilled is very slow, however, and groundwater presently is being removed faster than it can be replenished in many areas. Urbanization, road building, and other development often block recharge zones and prevent replenishment of important aquifers. Contamination of surface

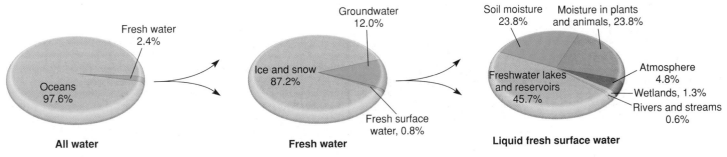

FIGURE 17.6 Less than 1 percent of fresh water, and less than 0.02 percent of all water, is fresh, liquid surface water on which terrestrial life depends.
Source: U.S. Geological Survey.

FIGURE 17.7 Glaciers and snowfields provide much of the water on which billions of people rely. The snowpack in the western Rocky Mountains, for example, supplies about 75 percent of the annual flow of the Colorado River. Global climate change is shrinking glaciers and causing snowmelt to come earlier in the year, disrupting this vital water source.

FIGURE 17.8 Precipitation that does not evaporate or run off over the surface percolates through the soil in a process called infiltration. The upper layers of soil hold droplets of moisture between air-filled spaces. Lower layers, where all spaces are filled with water, make up the zone of saturation, or groundwater.

water in recharge zones and seepage of pollutants into abandoned wells have polluted aquifers in many places, making them unfit for most uses (chapter 18). Many cities protect aquifer recharge zones from pollution or development, both as a way to drain off rainwater and as a way to replenish the aquifer with pure water.

Some aquifers contain very large volumes of water. The groundwater within 1 km of the surface in the United States is more than 30 times the volume of all the freshwater lakes, rivers, and reservoirs on the surface. Although water can flow through limestone caverns in underground rivers, most movement in aquifers is a dispersed and almost imperceptible trickle through tiny fractures and spaces. Depending on geology, it can take anywhere from a few hours to several years for contaminants to move a few hundred meters through an aquifer.

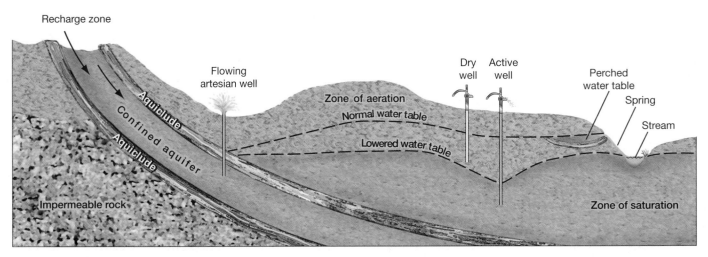

FIGURE 17.9 An aquifer is a porous or cracked layer of rock. Impervious rock layers (aquicludes) keep water within a confined aquifer. Pressure from uphill makes an artesian well flow freely. Pumping can create a cone of depression, which leaves shallower wells dry.

Rivers, lakes, and wetlands cycle quickly

Precipitation that does not evaporate or infiltrate into the ground runs off over the surface, drawn by the force of gravity back toward the sea. Rivulets accumulate to form streams, and streams join to form rivers. Although the total amount of water contained at any one time in rivers and streams is small compared to the other water reservoirs of the world (see table 17.2), these surface waters are vitally important to humans and most other organisms. Most rivers, if they were not constantly replenished by precipitation, meltwater from snow and ice, or seepage from groundwater, would begin to diminish in a few weeks.

We measure the size of a river in terms of its **discharge**, the amount of water that passes a fixed point in a given amount of time. This is usually expressed as liters or cubic feet of water per second. The 16 largest rivers in the world carry nearly half of all surface runoff on earth. The Amazon is by far the largest river in the world (table 17.3), carrying roughly ten times the volume of the Mississippi. Several Amazonian tributaries, such as the Madeira, Rio Negro, and Ucayali, would be among the world's top rivers in their own right.

Ponds are generally considered to be small temporary or permanent bodies of water shallow enough for rooted plants to grow over most of the bottom. Lakes are inland depressions that hold standing fresh water year-round. Maximum lake depths range from a few meters to over 1,600 m (1 mi) in Lake Baikal in Siberia. Surface areas vary in size from less than one-half hectare (one acre) to large inland seas, such as Lake Superior or the Caspian Sea, covering hundreds of thousands of square kilometers. Both ponds and lakes are relatively temporary features on the landscape because they eventually fill with silt or are emptied by cutting of an outlet stream through the barrier that creates them.

While lakes contain nearly 100 times as much water as all rivers and streams combined, they are still a minor component of total world water supply. Their water is much more accessible than groundwater or glaciers, however, and they are important in many ways for humans and other organisms.

Wetlands play a vital and often unappreciated role in the hydrologic cycle. Their lush plant growth stabilizes soil and holds back surface runoff, allowing time for infiltration into aquifers and producing even, year-long stream flow. In the United States, about 20 percent of the 1 billion ha of land area was once wetland. In the past 200 years, more than one-half of those wetlands have been drained, filled, or degraded. Agricultural drainage accounts for the bulk of the losses.

When wetlands are disturbed, their natural water-absorbing capacity is reduced and surface waters run off quickly, resulting in floods and erosion during the rainy season and dry, or nearly dry, streambeds the rest of the year. This has a disastrous effect on biological diversity and productivity, as well as on human affairs.

The atmosphere is among the smallest of compartments

The atmosphere is among the smallest of the major water reservoirs of the earth in terms of water volume, containing less than 0.001 percent of the total water supply. It also has the most rapid turnover rate. An individual water molecule resides in the atmosphere for about ten days, on average. While water vapor makes up only a small amount (4 percent maximum at normal temperatures) of the total volume of the air, movement of water through the atmosphere provides the mechanism for distributing fresh water over the landmasses and replenishing terrestrial reservoirs.

Think About It

Locate the ten rivers in table 17.3 on the physiographic map in the back of your book. Also, check their approximate locations in figure 17.2. How many of these rivers are tropical? in rainy regions? in populous regions? How might some of these rivers affect their surrounding environment or populations?

17.3 WATER AVAILABILITY AND USE

Clean, fresh water is essential for nearly every human endeavor. Perhaps more than any other environmental factor, the availability of water determines the location and activities of humans on earth (fig. 17.10). **Renewable water supplies** are made up, in general, of surface runoff plus the infiltration into accessible freshwater aquifers. About two-thirds of the water carried in rivers and streams every year occurs in seasonal floods that are too large or violent to be stored or trapped effectively for human uses. Stable runoff is the dependable, renewable, year-round supply of surface water. Much of this occurs, however, in sparsely inhabited regions

Table 17.3	Major Rivers of the World	
River	**Countries in River Basin**	**Average Annual Discharge at (m³/sec)**
Amazon	Brazil, Peru	175,000
Orinoco	Venezuela, Colombia	45,300
Congo	Congo	39,200
Yangtze	Tibet, China	28,000
Bramaputra	Tibet, India, Bangladesh	19,000
Mississippi	United States	18,400
Mekong	China, Laos, Burma, Thailand, Cambodia, Vietnam	18,300
Parana	Paraguay, Argentina	18,000
Yenisey	Russia	17,200
Lena	Russia	16,000

$1 m^3$ = 264 gallons.

Source: World Resources Institute.

FIGURE 17.10 Water has always been the key to survival. Who has access to this precious resource and who doesn't has long been a source of tension and conflict.

Periodic droughts create severe regional water shortages. Droughts are most common and often most severe in semiarid zones where moisture availability is the critical factor in determining plant and animal distribution. Undisturbed ecosystems often survive extended droughts with little damage, but introduction of domestic animals and agriculture disrupts native vegetation and undermines natural adaptations to low moisture levels.

Land-use practices often exacerbate drought effects. The worst drought in American history occurred in the 1930s. Poor soil conservation practices and a series of dry years in the Great Plains combined to create the "dust bowl." Wind stripped topsoil from millions of hectares of land, and billowing dust clouds turned day into night. Thousands of families were forced to leave farms and migrate to cities.

As the opening case study shows, much of the western United States has been exceptionally dry over the past decade. Many places are experiencing water crises (fig. 17.11). Is this just a temporary cycle or the beginning of a new climatic regime?

or where technology, finances, or other factors make it difficult to use it productively. Still, the readily accessible, renewable water supplies are very large, amounting to some 1,500 km³ (about 400,000 gal) per person per year worldwide.

Many countries suffer water scarcity and water stress

The United Nations considers 1,000 m³ (264,000 gal) of water per person per year to be the minimum necessary to meet basic human needs. **Water scarcity** occurs when the demand for water exceeds the available amount or when poor quality restricts its use. **Water stress** occurs when renewable water supplies are inadequate to satisfy essential human or ecosystem needs, bringing about increased competition among potential demands. Water stress is most likely to occur in poor countries where the per capita renewable water supply is low.

As you can see in figure 17.2, South America, West Central Africa, and South and Southeast Asia all have areas of very high rainfall. The highest per capita water supplies generally occur in countries with wet climates and low population densities. Iceland, for example, has about 160 million gallons per person per year. In contrast, Bahrain, where temperatures are extremely high and rain almost never falls, has essentially no natural fresh water. Almost all of Bahrain's water comes from imports and desalinized seawater. Egypt, in spite of the fact that the Nile River flows through it, has only about 11,000 gallons of water annually per capita, or about 15,000 times less than Iceland.

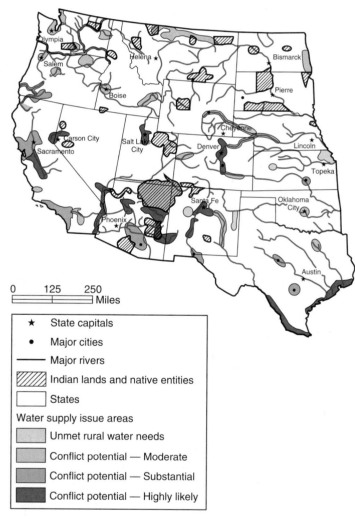

FIGURE 17.11 Rapidly growing populations in arid regions are straining available water supplies. By 2025, the Department of the Interior warns, shortages could cause conflicts in many areas.
Source: Data from U.S. Department of Interior.

The United States government projects that 36 states will have water deficits by 2012. Increased temperatures, disturbed weather patterns, population growth, urban sprawl, waste, and wasteful uses all will contribute to these shortages. The effects on water supplies may well be the most serious consequences of global climate change.

If the government had listened to Major John Wesley Powell, settlement patterns in the western United States would be very different from what we see there today. Powell, who led the first expedition down the Colorado River, went on to be the first head of the U.S. Geological Survey. In that capacity he did a survey of the agricultural and settlement potential of the western desert. His conclusion, quoted at the beginning of this chapter, was that there isn't enough water there to support a large human population.

Powell recommended that the political organization of the West be based on watersheds so that everyone in a given jurisdiction would be bound together by the available water. He thought that farms should be limited to local surface water supplies, and that cities should be small oasis settlements. Instead we've built huge metropolitan areas, such as Los Angeles, Phoenix, Las Vegas, and Denver, in places where there is little or no natural water supply. Will those cities survive impending shortages?

Water use is increasing

Human water use has been increasing about twice as fast as population growth over the past century (fig. 17.12). Water use is stabilizing in industrialized countries, but demand will increase in developing countries where supplies are available. The average amount of water withdrawn worldwide is about 646 m³ (170,544 gal) per person per year. This overall average hides great discrepancies in the proportion of annual runoff withdrawn in different areas. Some countries with a plentiful water supply withdraw a very small percentage of the water available to them. Canada, Brazil,

and the Congo, for instance, withdraw less than 1 percent of their annual renewable supply.

By contrast, in countries such as Libya, Yemen, and Israel, where water is one of the most crucial environmental resources, groundwater and surface water withdrawal together amount to more than 100 percent of their renewable supply. They are essentially "mining" water—extracting groundwater faster than it is being replenished. Obviously, this isn't sustainable in the long run.

The total annual renewable water supply in the United States amounts to an average of about 9,000 m³ (nearly 2.4 million gal) per person per year. We now withdraw about one-fifth of that amount, or some 5,000 liters (1,300 gal) per person per day, including industrial and agricultural water. By comparison, the average water use in Haiti is less than 30 liters (8 gal) per person per day.

In contrast to energy resources, which usually are consumed when used, water can be used over and over if it is not too badly contaminated. Water **withdrawal** is the total amount of water taken from a water body. Much of this water could be returned to circulation in a reusable form. Water **consumption**, on the other hand, is loss of water due to evaporation, absorption, or contamination.

Agriculture is the greatest water consumer worldwide

We can divide water use into three major sectors: agricultural, domestic, and industrial. Of these, agriculture accounts for by far the greatest use and consumption. Worldwide, crop irrigation is responsible for two-thirds of water withdrawal and 85 percent of consumption. Evaporation and seepage from unlined irrigation canals are the principal consumptive water losses. Agricultural water use varies greatly, of course. Over 90 percent of water used in India is agricultural; in Kuwait, where water is especially precious, only 4 percent is used for crops. In the United States, which has both a large industrial sector and a highly urbanized population, about half of all water withdrawal, and about 80 percent of consumption, is agricultural.

A tragic case of water overconsumption is the Aral Sea, which lies in Kazakhstan and Uzbekistan (see map at the end of this book). Once the fourth largest inland water body in the world, this giant saline lake lost 75 percent of its surface area and 90 percent of its volume between 1975 and 2009 (fig. 17.13) when, under the former Soviet Union, 90 percent of the natural flow of the Amu Dar'ya and Syr Dar'ya rivers was diverted to irrigate rice and cotton. Towns that once were

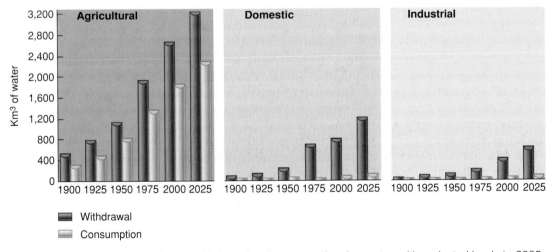

FIGURE 17.12 Growth of water withdrawal and consumption, by sector, with projected levels to 2025.
Source: UNEP, 2002.

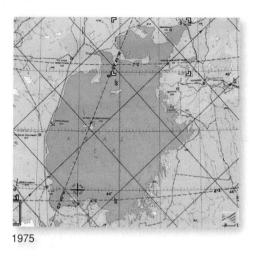

1975

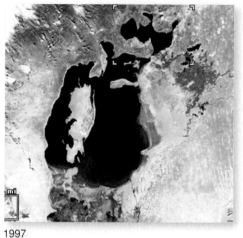

1997

2005

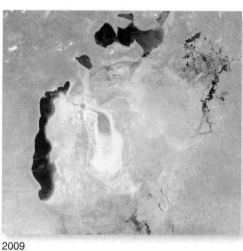

2009

FIGURE 17.13 For 30 years, rivers feeding the Aral Sea were diverted to irrigate cotton and rice fields. The Sea has lost more than 90 percent of its water. The "Small Aral" (upper right lobe) has separated from the main lake, and is now being refilled.

prosperous fish-processing and shipping ports now lie 100 km from the lake shore.

Vozrojdenie Island, which was used for biological weapons productions in the Soviet era, has become connected to the mainland, causing concern about the security of materials stored there. The salt concentration in the remaining water doubled, and fishing, which once produced 20,000 tons per year, ceased completely. Today more than 200,000 tons of salt, sand, and toxic chemicals are blown every day from the dried lake bottom. This polluted cloud is destroying pastures, poisoning farm fields, and damaging the health of residents who remain in the area.

As water levels dropped, the lake split into two lobes. The "Small Aral" in Kazakhstan is now being reclaimed. Some of the river flow has been restored (mainly because Soviet-style rice and cotton farming have been abandoned), and a dam has been built to separate this small lobe from the larger one in Uzbekistan. Water levels in the small, northern lake have risen

more than 8 m and surface area has expanded by 30 percent. With cleaner water pouring into the Small Aral, native fish are being reintroduced, and it's hoped that commercial fishing might one day be resumed. The fate of the larger lake remains clouded. There may never be enough water to refill it, and if there were, the toxins left in the lake bed could make it unusable anyway.

A similar catastrophe has befallen Lake Chad in northern Africa. Sixty thousand years ago, during the last ice age, this area was a verdant savanna sprinkled with freshwater lakes and occupied by crocodiles, hippopotamuses, elephants, and gazelles. At that time Lake Chad was about the present size of the Caspian Sea (400,000 km^2). Climate change has turned the Sahara into a desert, and by the mid-1960s Lake Chad had shrunk to 25,000 km^2 (as large as the United States' Lake Erie). With a maximum depth of 7 m, the lake is highly sensitive to climate, and it expands and contracts dramatically. Persistent drought coupled with increased demand from massive irrigation projects in the 1970s and 1980s has reduced Lake Chad to less than 1,000 km^2. The silty sand left on the dry lake bed is whipped aloft by strong winds funneled between adjacent mountain ranges. In the winter the former lake bed, known as the Bodélé Depression, produces an average of 700,000 tons of dust every day. About 40 million tons of this dust are transported annually from Africa to South America, where it is thought to be the main source of mineral nutrients for the Amazon rainforest (chapter 16).

Irrigation can be very inefficient. Traditionally the main method has been flood or furrow irrigation, in which water floods a field (fig. 17.14*a*). As much as half of this water can be lost through evaporation. Much of the rest runs off before it is used by plants. In arid lands, flood irrigation is needed to help remove toxic salts from soil, but these salts contaminate streams, lakes, and wetlands downstream. Repeated flood irrigation also waterlogs the soil, reducing crop growth. Sprinkler systems can also be inefficient (fig. 17.14*b*). Water spraying high in the air quickly evaporates, rather than watering crops. In recent years, growing pressure on water resources has led to more efficient sprinkler systems that hang low over crops to reduce evaporation (see fig. 10.13).

(a) Flood irrigation

(b) Rolling sprinklers

(c) Drip irrigation

FIGURE 17.14 Agricultural irrigation consumes more water than any other use. Methods vary from flood and furrow (a), which uses extravagant amounts of water but also flushes salts from soils, to sprinklers (b), to highly efficient drip systems (c).

Drip irrigation (fig. 17.14c) is a promising technology for reducing irrigation water use. These systems release carefully regulated amounts of water just above plant roots, so that nearly all water is used by plants. Only about 1 percent of the world's croplands currently use these systems, however.

Irrigation infrastructure, such as dams, canals, pumps, and reservoirs, is expensive. Irrigation is also the economic foundation of many regions. In the United States, the federal government has taken responsibility for providing irrigation for nearly a century. The argument for doing so is that irrigated agriculture is a public good that cannot be provided by individual farmers. A consequence of this policy has frequently been heavily subsidized crops whose costs, in water and in dollars, far outweigh their value.

Domestic and industrial water use is greatest in wealthy countries

Worldwide, domestic water use accounts for only about 6 percent of water withdrawals. Because little of this water evaporates or seeps into the ground, consumptive water use is slight, about 10 percent on average. Where sewage treatment is unavailable, however, water can be badly degraded by urban uses. In wealthy countries, each person uses about 500 to 800 l per day (180,000 to 280,000 l per year), far more than in developing countries (30 to 150 l per day). In North America the largest single use of domestic water is toilet flushing (fig. 17.15). On average, each person in the United States uses about 50,000 l (13,000 gal) of drinking-quality water annually to flush toilets. Bathing accounts for nearly a third of water use, followed by laundry and washing. In western cities such as Palm Desert and Phoenix, lawn watering is also a major water user.

Urban and domestic water use have grown approximately in proportion with urban populations, about 50 percent between 1960 and 2000. Although individual water use seems slight on the scale of world water withdrawals, the cumulative effect of inefficient appliances, long showers, liberal lawn-watering, and other uses is enormous. California has established increasingly stringent standards for washing machines, toilets, and other appliances, in order to reduce urban water demands. Many other cities and states are following this lead to reduce domestic water use.

Industry accounts for 20 percent of global freshwater withdrawals. Industrial use rates range from 70 percent in industrialized parts of Europe to less than 5 percent in countries with little industry. Power production, including hydroelectric, nuclear, and thermoelectric power, make up 50 to 70 percent of industrial uses, and industrial processes make up the remainder. As with domestic water, little of this water is made unavailable after use, but it is often degraded by defouling agents, chlorine, or heat when it is released to the environment. The greatest industrial producer of degraded water is mining. Ores must be washed and treated with chemicals such as mercury and cyanide (chapter 14). As much as 80 percent of water used in mining and processing is released with only minimal treatment. In developed countries, industries have greatly improved their performance in recent decades, however. Water withdrawal and consumption have both fallen relative to industrial production.

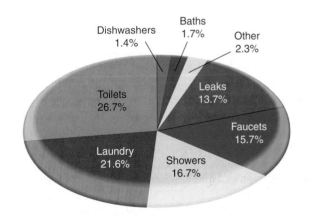

FIGURE 17.15 Typical household water use in the United States.
Source: Data from the American Water Works Association, 2010.

17.4 FRESHWATER SHORTAGES

Clean drinking water and basic sanitation are necessary to prevent communicable diseases and to maintain a healthy life. For many of the world's poorest people, one of the greatest environmental threats to health remains the continued use of polluted water. The United Nations estimates that at least a billion people lack access to safe drinking water and 2.5 billion don't have adequate sanitation. These deficiencies result in hundreds of millions of cases of water-related illness and more than 5 million deaths every year. As populations grow, more people move into cities, and agriculture and industry compete for increasingly scarce water supplies, water shortages are expected to become even worse.

By 2025 two-thirds of the world's people may be living in countries that are water-stressed—defined by the United Nations as consumption of more than 10 percent of renewable freshwater resources. One of the United Nations Millennium goals is to reduce by one-half the proportion of people without reliable access to clean water and improved sanitation.

There have been many attempts to enhance local supplies and redistribute water. Towing icebergs from Antarctica has been proposed, and creating rain in dry regions has been accomplished, with mixed success, by cloud seeding—distributing condensation nuclei in humid air to help form raindrops. Desalination is locally important: in the arid Middle East, where energy and money are available but water is scarce, desalination is sometimes the principal source of water. Some American cities, such as Tampa, Florida, and San Diego, California, also depend partly on energy-intensive desalination (Exploring Science, p. 389).

Many people lack access to clean water

The World Health Organization considers an average of 1,000 m³ (264,000 gal) per person per year to be a necessary amount of water for modern domestic, industrial, and agricultural uses. Some 45 countries, most of them in Africa or the Middle East, cannot meet the minimum essential water needs of all their citizens. In some countries the problem is access to clean water. In Mali, for example, 88 percent of the population lacks clean water; in Ethiopia, it is 94 percent. Rural people often have less access to clean water than do city dwellers. Causes of water shortages include natural deficits, overconsumption by agriculture or industry, and inadequate funds for purifying and delivering good water.

More than two-thirds of the world's households have to fetch water from outside the home (fig. 17.16). This is heavy work, done mainly by women and children and sometimes taking several hours a day. Improved public systems bring many benefits to these poor families.

Availability doesn't always mean affordability. A typical poor family in Lima, Peru, for instance, uses one-sixth as much water as a middle-class American family but pays three times as much for it. If they followed government recommendations to boil all water to prevent cholera, up to one-third of the poor family's income could be used just in acquiring and purifying water.

FIGURE 17.16 Village water supplies in Ghana.

Investments in rural development have brought significant improvements in recent years. Since 1990 nearly 800 million people—about 13 percent of the world's population—have gained access to clean water. The percentage of rural families with safe drinking water has risen from less than 10 percent to nearly 75 percent.

Groundwater is being depleted

Groundwater is the source of nearly 40 percent of the fresh water for agricultural and domestic use in the United States. Nearly half of all Americans and about 95 percent of the rural population depend on groundwater for drinking and other domestic purposes. Overuse of these supplies causes several kinds of problems, including drying of wells and natural springs, and disappearance of surface water features such as wetlands, rivers, and lakes.

In many areas of the United States, groundwater is being withdrawn from aquifers faster than natural recharge can replace it. The Ogallala Aquifer, for example, underlies eight states in the arid high plains between Texas and North Dakota (fig. 17.17). As deep as 400 m (1,200 ft) in its center, this porous bed of sand, gravel, and sandstone once held more water than all the freshwater lakes, streams, and rivers on earth. Excessive pumping for irrigation and other uses has removed so much water that wells have dried up in many places, and farms, ranches, even whole towns are being abandoned.

On a local level, this causes a cone of depression in the water table, as is shown in figure 17.18. A heavily pumped well can lower the local water table so that shallower wells go dry. On a broader scale, heavy pumping can deplete a whole aquifer. Many aquifers have slow recharge rates, so it will take thousands of years to refill them once they are emptied. Much of the groundwater we now are using probably was left there by the glaciers thousands of years ago. It is fossil water, in a sense. It will never be replaced in our lifetimes, and is, essentially, a nonrenewable resource. Covering aquifer recharge zones with urban development or diverting runoff that once replenished reservoirs ensures that they will not refill.

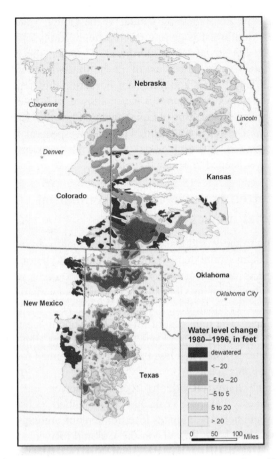

FIGURE 17.17 The Ogallala/High Plains regional aquifer supports a multimillion-dollar agricultural economy, but withdrawal far exceeds recharge. Some areas are down to less than 3 m of saturated thickness.

Map legend:
Water level change
1980–1996, in feet
- dewatered
- < −20
- −5 to −20
- −5 to 5
- 5 to 20
- > 20

0   50   100 Miles

Withdrawal of large amounts of groundwater causes porous formations to collapse, resulting in **subsidence** or settling of the surface above. The U.S. Geological Survey estimates that the San Joaquin Valley in California, for example, has sunk more than 10 m in the last 50 years because of excessive groundwater pumping. Around the world, many cities are experiencing subsidence. Many are coastal cities, built on river deltas or other unconsolidated sediments. Flooding is frequently a problem as these coastal areas sink below sea level. Some inland areas also are affected by severe subsidence. Mexico City is one of the worst examples. Built on an old lake bed, it has probably been sinking since Aztec times. In recent years, however, rapid population growth and urbanization (chapter 22) have caused groundwater overdrafts. Some areas of the city have sunk as much as 8.5 m (25.5 ft). The Shrine of Guadalupe, the cathedral, and many other historic monuments are sinking at odd and perilous angles.

A widespread consequence of aquifer depletion is **saltwater intrusion**. Along coastlines and in areas where saltwater deposits are left from ancient oceans, overuse of freshwater reservoirs often allows saltwater to intrude into aquifers used for domestic and agricultural purposes (fig. 17.18).

Diversion projects redistribute water

Dams and canals are a foundation of civilization because they store and redistribute water for farms and cities. Many great civilizations, including ancient empires of Sumeria, Egypt, and India, have been organized around large-scale canal systems. As modern dams and water diversion projects have grown in scale and number, though, their environmental costs have raised serious questions about efficiency, costs, and the loss of river ecosystems.

More than half of the world's 227 largest rivers have been dammed or diverted (fig. 17.19). Of the 50,000 large dams in the world, 90 percent were built in the twentieth century. Half of those are in China, and China continues to build and plan dams on its remaining rivers. Dams are justified in terms of flood control, water storage, and electricity production. However, the costs of relocating villages, lost fishing and farming, and water losses to evaporation are enormous. Economically speaking, at least one-third of the world's large dams should never have been built.

As the chapter-opening case study shows, many southwestern cities are facing a crisis with the drying of Lake Mead. Las Vegas, Nevada, which gets 40 percent of its water from the lake, has started a $3.5 billion, 525 km (326 m) pipeline to tap aquifers in the northeastern part of the state. Local ranchers fear that groundwater pumping will decimate the range, destroy native vegetation, and cause massive dust storms. They point to the Owens Valley in California, where a similar water grab by Los Angeles in 1913 dried up the river and destroyed both natural vegetation and the ranching economy. Las Vegas also has suggested that if local water supplies fail, they may ask states east of the Mississippi to share some of their water.

Las Vegas is also digging a $3.5 billion tunnel that will burrow into Lake Mead, 100 m (300 ft) below the normal outlet (fig. 10.10). Even if the lake reaches the "dead pool" level as warned in the beginning of this chapter, the city will still be

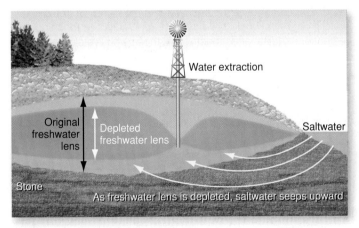

FIGURE 17.18 Saltwater intrusion into a coastal aquifer as the result of groundwater depletion. Many coastal regions of the United States are losing freshwater sources due to saltwater intrusion.

FIGURE 17.19 Hoover Dam powers Las Vegas, Nevada. Lake Mead, behind the dam, loses about 1.3 billion m^3 per year to evaporation.

able to draw off water. Of course, this might prevent refilling the reservoir to provide water and power to downstream users. If you lived downstream, how would you feel about this outcome?

Similarly, China faces a massive water crisis. Northern and western parts of the country are dry and getting drier. The Gobi desert is moving eastward; its leading edge is now only 100 km (60 mi) from Beijing. Of 600 major Chinese cities, more than two-thirds have water shortages, and 100 of those cities have acute water problems. Without a new water source, planners warn, the entire national capital, Beijing, might have to be moved to a new location. To combat this crisis, the Chinese government has embarked on a massive water distribution scheme called the South-Water-North project (What Do You Think? p. 387).

FIGURE 17.20 The five North American Great Lakes contain about 20 percent of all the freshwater in the world. But diverting water from them would have massive repercussions for shipping, recreation, wildlife, and industry. Will those who live along their shores be willing to share?

Might we do something similar in the United States? While the desert Southwest is getting drier, a huge supply of water sits temptingly not so far to the northeast. Together, the five Great Lakes contain about 20 percent of all the freshwater in the world (fig. 17.20). The states and provinces bordering this treasure have signed a compact promising to never allow large-scale transfers out of the watershed. But if Los Angeles, Phoenix, Tucson, Las Vegas, and Denver were faced with the prospect of becoming ghost towns, will this resolve hold? Can we tell western states they should have listened to Major Powell?

Dams often have severe environmental and social impacts

According to the World Dam Commission, there were only about 250 high dams (more than 15 m tall) in the world before 1900. In the twentieth century, however, at least 45,000 dams were built, about half of them in China. Other countries with many dams include Turkey, Japan, Iran, India, Russia, Brazil, Canada, and the United States. The total cost of this building boom is estimated to have been $2 trillion. At least one-third of the dams aren't justified on economic grounds, and less than half have planned for social or environmental impacts.

Though dams provide hydroelectric power and water to distant cities, they also can have unintended consequences. Reservoirs in hot, dry climates lose tremendous amounts of water to evaporation. Lake Powell, on the Colorado River, loses more than 1 billion m^3 of water to evaporation and seepage every year. As the water level drops, it leaves a bathtub ring of salt deposits on the canyon walls (photo, p. 372).

International Rivers, an environmental and human rights organization, reports that dam projects have forced more than 23 million people from their homes and land, and many are still suffering the impacts of dislocation years after it occurred. Often the people being displaced are ethnic minorities. On India's Narmada River, for example, a proposed series of 30 dams have displaced about 1 million villagers and tribal people. Many who have been relocated have never successfully integrated either socially or economically. Protests around this project have raged for 20 years or more.

There's increasing concern that big dams in seismically active areas can trigger earthquakes. In more than 70 cases worldwide, large dams have been linked with increased seismic activity. Geologists suggest that filling the reservoir behind the nearby Zipingpu Dam on the Min River caused the devastating 7.9-magnitude Sichuan earthquake that killed an estimated 90,000 people in 2008. If true, it would be the world's deadliest dam-induced earthquake. But it pales in comparison to the potential catastrophe if the Three Gorges Dam on the Yangtze were to collapse. As one engineer says, "It would be a flood of Biblical proportions for the 100 million people who live downstream."

Dams are also lethal for migratory fish, such as salmon. Adult fish are blocked from migrating to upstream spawning areas. And juvenile fish die if they go through hydroelectric turbines. The slack water in reservoirs behind dams is also a serious problem. Juvenile salmon evolved to ride the surge of spring runoff downstream to the ocean in two or three weeks. Reservoirs slow this journey to as

What Do You Think?

China's South-Water-North Diversion

Water is inequitably distributed in China. In the south, torrential monsoon rains cause terrible floods. A 1931 flood on the Yangtze displaced 56 million people and killed 3.7 million (the worst natural disaster in recorded history). Northern and western China, on the other hand, are too dry, and getting drier. At least 200 million Chinese live in areas without sufficient fresh water. The government has warned that unless new water sources are found soon, many of those people (including the capital Beijing, with roughly 20 million residents) will have to be moved. But where could they go? Southern China has water, but doesn't need more people.

The solution, according to the government, is to transfer some of the extra water from south to north. A gargantuan project is now under way to do just that. Work has begun to build three major canals to carry water from the Yangtze River to northern China. Ultimately it's planned to move 45 billion m³ per year (more than twice the annual flow of the Colorado River through the U.S. Grand Canyon) 1,600 km (1,000 mi) north. The initial cost estimate of this scheme is about 400 billion yuan (roughly U.S. $62 billion), but it could easily be twice that much.

The eastern route uses the Grand Canal, built by Zhou and Sui emperors 1,500 years ago across the coastal plain between Shanghai and Beijing. This project is already operational. It's relatively easy to pump water through the existing waterways, but they're so polluted by sewage

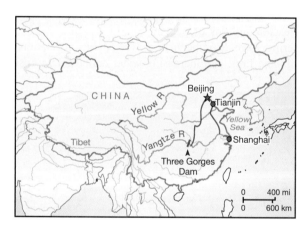

With the Yellow River nearly depleted by overuse, northern China now plans canals (*red*) to deliver Yangtze water to Beijing.

and industrial waste that northern cities—even though they're desperately dry—are reluctant to use this water.

The central route will draw water from the reservoir behind the recently completed Three Gorges Dam on the Yangtze. Part of the motivation for building this controversial dam and flooding the historic Three Gorges was to provide energy and raise the river level for the South-to-North project. This middle canal will cross several major mountain ranges and dozens of rivers, including the Han and Yellow rivers. Already at least 1.5 million people have been displaced by this project, but planners say it's justified by the benefits to a much greater number of people in northern China.

It's hoped this segment will be finished by 2020. The western route is the most difficult and expensive. It would tunnel through rugged mountains, across aqueducts, and over deep canyons for more than 250 km (160 mi), from the upper Yangtze to the Yellow River where they both spill off the Tibetan Plateau. This phase won't be finished until at least 2050. If global warming melts all Tibet's glaciers, however, it may not be feasible anyway.

Planners have waited a lifetime to see this project move forward. Revolutionary leader Mao Zedong proposed it 50 years ago. Environmental scientists worry, however, that drawing down the Yangtze will worsen pollution problems (already exacerbated by the Three Gorges Dam), dry up downstream wetlands, and possibly even alter ocean circulation and climate along China's eastern coast. Although southern China has too much water during the rainy season, southern cities face water shortages because of rapidly growing populations and severe pollution. At least half of all major Chinese rivers are too polluted for human consumption. Drawing water away from the rivers on which millions rely only makes pollution problems worse.

What do you think? Are there other ways that China could adapt to uneven water distribution? If you were advising the Chinese government, what safeguards might you recommend to avoid unexpected consequences from the gargantuan project? And might we do something similar in the United States to relieve water shortages in the desert Southwest? It's about the same distance from Phoenix to Chicago as from Yichang to Beijing.

But residents in the U.S. Great Lakes states are worried about the possible effects of global climate change. They may need their water in the future. Furthermore, they may be reluctant to help pay the cost for people who continue to live in the desert. Over the past 50 years, cities in the industrialized "Rust Belt" have lost millions of jobs and residents to the "Sunbelt." Rather than pay billions of dollars to move water, they might prefer to say, "Come back to Detroit and Toledo and Cleveland, and bring your jobs with you." Is it inconceivable that we'd allow desert cities to become ghost towns? Great cities have risen, flourished, and crumbled to dust in the past. What do you think—could it happen again?

---

much as three months, throwing off the time-sensitive physiological changes that allow the fish to survive in salt water when they reach the ocean. Reservoirs expose young salmon to predators, and warm water in reservoirs increases disease in both young and older fish.

Some dams have fish ladders—a cascading series of pools and troughs—that allow fish to bypass the dam. Another option is to move both adults and juveniles by barge or truck. This can result in the strange prospect of barges of wheat moving downstream while passing barges of fish moving the opposite direction. Both these options are expensive and only partially effective in restoring blocked salmon runs.

The tide may be turning against dams. In 1998 the Army Corps of Engineers announced that it would no longer be building large dams and diversion projects. In the few remaining sites where dams might be built, public opposition is so great that getting approval for projects is unlikely. Instead, the new focus may be on removing existing dams and restoring natural habitats. Former interior secretary Bruce Babbitt said, "Of the 75,000 large dams in the United States, most were built a long time ago and are now obsolete, expensive, and unsafe. They were built with no consideration of the environmental costs. As operating licenses come up for renewal, removal and restoration to original stream flows will be one of the options."

FIGURE 17.21 This dam is now useless because its reservoir has filled with silt and sediment.

Sedimentation limits reservoir life

Rivers with high sediment loads can fill reservoirs quickly (fig. 17.21). In 1957 the Chinese government began building the Sanmenxia Dam on the Huang He (Yellow River) in Shaanxi Province. From the beginning, engineers warned that the river carried so much sediment that the reservoir would have a very limited useful life. Dissent was crushed, however, and by 1960 the dam began filling the river valley and inundating fertile riparian fields that once had been part of China's traditional granaries.

Within two years, sediment accumulation behind the dam had become a serious problem. It blocked the confluence of the Wei and Yellow rivers and backed up the Wei so it threatened to flood the historic city of Xi'an. By 1962 the reservoir was almost completely filled with sediment, and hydropower production dropped by 80 percent. The increased elevation of the riverbed raised the underground water table and caused salinization of wells and farm fields. By 1991 the riverbed was 4.6 m above the surrounding landscape. The river is kept in check only by earthen dams that frequently fail and flood the surrounding countryside. By the time the project was complete, more than 400,000 people had been relocated, far more than planners expected.

Problems are similar, although not so severe, in some American rivers. As the muddy Colorado River slows behind the Glen Canyon and Boulder dams, it drops its load of suspended sand and silt. More than 10 million metric tons of sediment collect every year behind these dams. Even if there is enough water in the future to fill these reservoirs, within about a century they'll be full of mud and useless for either water storage or hydroelectric generation.

Elimination of normal spring floods—and the sediment they would usually drop to replenish beaches—has changed the riverside environment in the Grand Canyon. Invasive species crowd out native riparian plants. Beaches that campers use have disappeared. Boulders dumped in the canyon by side streams fill the riverbed. On several occasions, dam managers have released large surges of water from the Glen Canyon Dam to try to replicate normal spring floods. The results have been gratifying, but they don't last long. The canyon needs regular floods to maintain its character.

Climate change threatens water supplies

The Intergovernmental Panel on Climate Change (IPCC) warns us that climate change threatens to exacerbate water shortages caused by population growth, urban sprawl, wasteful practices, and pollution. The IPCC *Fourth Assessment Report* predicted with "very high confidence" that reduced precipitation and higher evaporation rates caused by higher temperatures will result in a 10 to 30 percent runoff reduction over the next 50 years in some dry regions at midlatitudes (see chapter 15).

Figure 17.22 shows a summary of predictions from several climate models for changes in precipitation in 2090–2099 compared to 1980–1999. White areas are where less than two-thirds of the models agree on likely outcomes: colored areas are where more than 90 percent of the models agree. How does this map compare to figure 17.2? Which areas do you think are most likely to suffer from water shortages by the end of this century? Which areas may benefit from climate change? Where will the largest number of people be affected?

In many parts of the world, severe droughts are already resulting in depleted rivers, empty reservoirs, and severe water shortages for millions of people. South Australia, for example, is suffering from extreme heat waves, dying vegetation, massive wildland fires, and increasing water deficits. The Australian

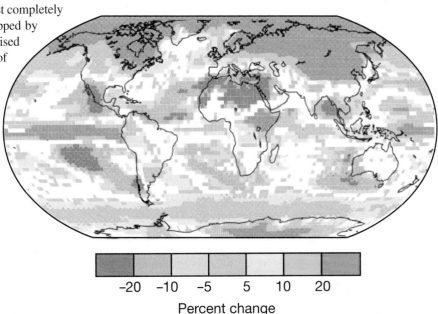

Percent change

FIGURE 17.22 Relative changes in precipitation (in percentage) for the period 2090–2099 compared to 1980–1999, predicted by the Intergovernmental Panel on Climate Change.
Source: IPCC, 2007.

As you've learned in this chapter, the ocean has a vast amount of water. Unfortunately, it's too salty for most human uses. There are ways to remove salt and other minerals from water, but they tend to be energy-intensive and expensive. The most common form of desalination is distillation: water is boiled, and the steam is collected and condensed to make fresh water. This works well. Glass-distilled water remains the standard for purity in most science labs. The world's largest desalination plant is the Jebel Ali Plant in the United Arab Emirates. It uses multistage flash distillation and is capable of producing 300 million cubic meters of water per year. Distillation produces over 85 percent of all desalinated water in the world.

In a multistage distillation facility, water evaporates in a series of spaces called stages containing heat exchangers and condensate collectors. Each compartment has a partial vacuum, which causes water to boil at a much lower temperature than 100°C (212°F). As water passes from one stage to another, temperatures and pressures are adjusted to match the boiling point as salt concentration increases. The total evaporation in all the stages is about 15 percent of the input water. The main limitations of this design are the corrosion caused by the warm brine as well as the energy required to heat and pump water and to create a vacuum. In oil-rich countries, such as Saudi Arabia, where water costs more than pumping oil out of the ground, energy costs aren't important, but in other places it becomes a concern. However, coupling a desalination facility with a power plant can cut costs by as much as two-thirds. Waste heat from the power plant is used to preheat seawater, which simultaneously provides cooling for the power plant. Disposal of warm, salty brine can have serious adverse impacts on local coastal areas.

The other principal method of desalination is reverse osmosis. This filtration process removes large molecules from solutions by applying pressure to the solution on one side of a selectively permeable membrane. Every cell in your body is enclosed by a selectively permeable membrane. It's one of the things that make life possible. The plasma membrane around each cell has tiny pores that allow small molecules, such as water, to pass through but that exclude large molecules, such as salt. Ordinarily osmosis causes small molecules to move from an area of high concentration (pure water) to areas of lower concentration (a salt solution, which has fewer water molecules per unit area because some space is occupied by the salt). You may have observed osmosis in a biology lab. If you put an amoeba in a high salt solution, it shrivels (water is drawn out). If you put it in pure water, it bloats and explodes (the interior is saltier than the water).

Reverse osmosis drives this process backward by applying a pressure to the high-salt side of a semipermeable membrane to filter the water. In practice the membranes are packed in concentric coils inside a tube. A large facility may have tens of thousands of these tubes. Pore sizes can vary from 0.1 nanometers (3.9×10^{-9} in) to 5,000 nanometers (0.0002 in) depending on the solution to be filtered. Reverse osmosis systems can range from industrial-size facilities capable of purifying hundreds of thousands of gallons per day, to a pen-size straw that you can carry in your pocket to sip water from a contaminated source.

Although there are many more reverse osmosis facilities than thermal desalination plants, they produce a relatively small percentage of all desalinated water. Still, they can be more mobile and easier to operate than a distillation plant. And if the pores are less than 1 nanometer, the water produced can be cleaner than distillation (although the production rate is very low with such small pore sizes).

---

government has declared that this drought is most likely the result of global climate change. Although the country has recently agreed to reduce carbon emissions to combat climate change, the only short-term solution, leaders admit, is to try to adapt to these new conditions.

In Yemen, the national capital, Sanaa, may have to be abandoned because it has no water to support its rapidly growing population of more than 2 million. Yemen depends entirely on groundwater, which is rapidly being depleted. Some wells in Sanaa are now 800 to 1,000 meters deep and many are no longer usable because of the sinking water table. Millions of people may have to abandon Sanaa and other mountain cities for the coastal plain. A shrinking resource base played a role in the civil unrest in Yemen in 2011.

A more dire situation is taking place in Somalia where a severe drought threatens 10 million people. Hardest hit is the southwestern region where Somalia, Ethiopia, and Kenya meet. The U.N. refugee agency called this the "worst humanitarian disaster" in the world. The agency estimates that 2 million children are severely malnourished and in need of lifesaving action. Hundreds of thousands of refugees are surging into temporary camps in search of food and water. Political instability in the region makes aid delivery difficult.

Would you fight for water?

Many environmental scientists warn that declining water supplies could lead to wars between nations. *Fortune* magazine wrote, "Water will be to the 21st century what oil was to the 20th." For its 2009 World Water Day, the United Nations is focusing on transboundary water supplies. Nearly 40 percent of the world's population lives in river and lake basins shared by two or more countries. These 263 watersheds include the territory of 145 countries and cover nearly half the earth's land surface. Figure 17.23 shows five of the world's major rivers and the countries they cross. Great reservoirs of fresh water also cross borders. There are more than 270 known transboundary aquifers.

Already we've seen skirmishes—if not outright warfare—over water access. An underlying factor in hostilities between Israel and its neighbors has been control of aquifers and withdrawals from the Jordan River. India, Pakistan, and Bangladesh also

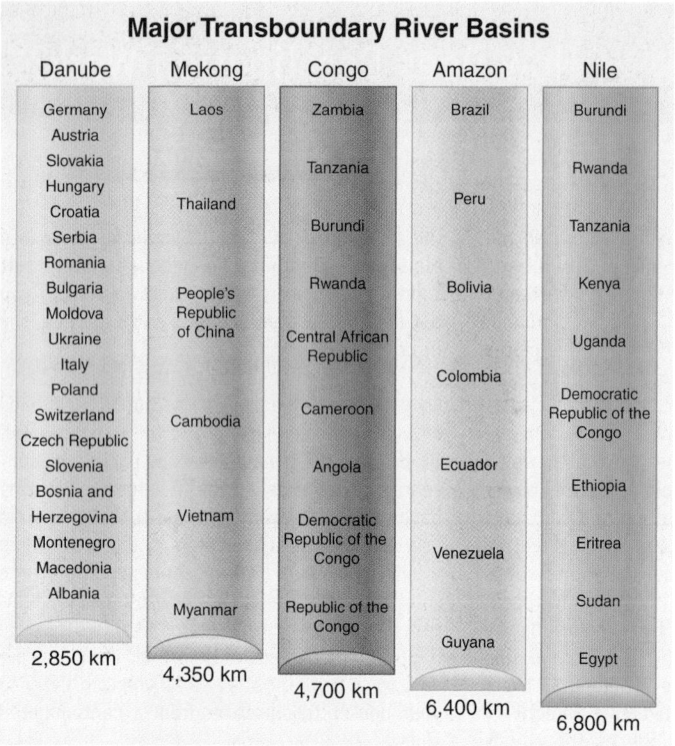

Major Transboundary River Basins

Danube	Mekong	Congo	Amazon	Nile
Germany	Laos	Zambia	Brazil	Burundi
Austria				
Slovakia		Tanzania		Rwanda
Hungary			Peru	
Croatia	Thailand	Burundi		Tanzania
Serbia				
Romania		Rwanda	Bolivia	Kenya
Bulgaria	People's			
Moldova	Republic			
Ukraine	of China	Central African		Uganda
Italy		Republic	Colombia	
Poland				Democratic
Switzerland	Cambodia	Cameroon		Republic of the
Czech Republic				Congo
Slovenia		Angola	Ecuador	
Bosnia and				Ethiopia
Herzegovina	Vietnam	Democratic		
Montenegro		Republic of the	Venezuela	Eritrea
Macedonia		Congo		
Albania				Sudan
	Myanmar	Republic of the		
		Congo	Guyana	
2,850 km				Egypt
	4,350 km	4,700 km	6,400 km	6,800 km

FIGURE 17.23 Together, these five rivers cross 54 countries. Already, skirmishes and sabre-rattling have occurred as neighbors squabble over scarce supplies. **Source:** Data World Water Day, 2009.

have confronted each other over water rights; and Turkey and Iraq threatened to send armies to protect access to water in the Tigris and Euphrates rivers. As chapter 13 reports, Saddam Hussein cut off water flow into the massive Iraq marshes as a way of punishing his enemies among the Marsh Arabs. Drying the marshes drove 140,000 people from their homes and destroyed a unique way of life. It also caused severe ecological damage to what is regarded by some as the biblical Garden of Eden.

In Kenya, nomadic tribes have fought over dwindling water resources. An underlying cause of the ongoing genocide in the Darfur region of Sudan is water scarcity. When rain was plentiful, Arab pastoralists and African farmers coexisted peacefully. Drought—perhaps caused by global warming—has upset that truce. The hundreds of thousands who have fled to Chad could be considered climate refugees as well as war victims. How many more tragedies, such as these, might we see in the future as people struggle for declining water resources?

17.5 GETTING BY WITH LESS WATER

In many cases we may simply have to adapt to less water. An example is a breakthrough agreement for the Klamath River in California. To keep enough water in the river to rebuild fish populations sufficient for sustainable tribal, recreational, and

commercial fisheries, farmers have to reduce their withdrawals. A key provision for farmers is to have a reliable and certain water allocation. When the irrigation gates were closed in 2001, most farmers had already planted their crops. Most of their expenses for the year were already invested. To cut off water to their crops at that point meant financial ruin.

A major feature of the settlement is that farmers agree to a 10 to 25 percent reduction in their historic water use in exchange for a one-time payment to help finance conservation measures. The benefit to the farmers is a greatly reduced threat of having their water completely shut off again to protect fish. In most years, farmers will have to get by on less water than usual. In really dry years, they'll either have to pump groundwater or fallow—temporarily dry up—some cropland. With clearer rules in place, farmers can shift in dry years from planting low-value crops, such as alfalfa, to using just part of their land to grow higher-value crops, thus keeping their income up while still using less water and fertilizer.

A description often used for such a plan is a "land bank." Other Californian water districts are using this same approach. Los Angeles, for example, is paying farmers to agree to fallow land in dry years. Farmers get enough income to cover their fixed costs—buildings, equipment, mortgages, and taxes—while still staying in business until better years come. The city can ensure a supply of water in bad years at a much lower cost than other alternatives.

Farmers in the Klamath basin have also agreed to a similar approach for wetlands. They'll take turns flooding fields on a rotating basis so that waterfowl have a place to rest and feed. Some call this a "walking" wetlands program. No one loses his or her fields permanently, but there's a guarantee of more habitat for birds (fig. 17.24). Money to pay for both wetland mitigation and crop reductions will come from a $1 billion budget provided mainly for endangered species protection. This plan also contains guarantees for stabilized power costs for family farms, ranches, and the two Klamath wildlife refuges.

FIGURE 17.24 A flock of snow geese rises from the Lower Klamath National Wildlife Refuge. Millions of migrating birds use these wetlands for feeding and resting.

A model for the Klamath restoration project comes from conservation progress in the Deschutes River in central Oregon. A century ago, much of the water in the Deschutes was dammed and diverted to irrigate farms. As was the case in the Klamath, Native American tribes on the Warm Springs Reservation downstream from this diversion sued over the destruction of their traditional fishing rights. As part of their settlement, irrigation districts upstream have lined canals to prevent seepage, and switched from flood irrigation (which often loses as much as half of its water to evaporation) to more efficient sprinkler systems. This allows farmers to use less water while still getting the same crop yield. Now salmon are once again making their way upstream from the Columbia River into the Warm Springs Reservation.

17.6 INCREASING WATER SUPPLIES

Where do present and impending freshwater shortages leave us now? On a human time scale, the amount of water on the earth is fixed, for all practical purposes, and there is little we can do to make more water. There are, however, several ways to increase local supplies.

In the dry prairie states of the 1800s and early 1900s, desperate farmers paid self-proclaimed "rainmakers" in efforts to save their withering crops. Centuries earlier, Native Americans danced and prayed to rain gods. We still pursue ways to make rain. Seeding clouds with dry ice or potassium iodide particles has been tested for many years with mixed results. Recently researchers have been having more success using hygroscopic salts, which seem to significantly increase rainfall amounts. This technique is being tested in Mexico, South Africa, and the western United States. There is a concern, however, that rain induced to fall in one area decreases the precipitation somewhere else. Furthermore, there are worries about possible contamination from the salts used to seed clouds.

A technology that might have great potential for increasing freshwater supplies is desalination of ocean water or brackish saline lakes and lagoons. Worldwide, 13,080 desalination plants produce more than 12 billion gallons (45 billion liters) of water a day (see Exploring Science, p. 389). This is expected to grow to about 100 million m^3 (26 billion gal) per day by 2015. Middle Eastern oil-rich states produce about 60 percent of desalinated water. Saudi Arabia is the largest single producer, at about one-third of the world total. The United States is second, at 20 percent. Although desalination is still three to four times more expensive than most other sources of fresh water, it provides a welcome water supply in such places as Oman and Bahrain, where there is no other access to fresh water. If a cheap, inexhaustible source of energy were available, however, the oceans could supply all the water we would ever need.

Domestic conservation can save water

We could probably save as much as half of the water we now use for domestic purposes without great sacrifice or serious changes in our lifestyles. Simple steps, such as taking shorter showers, stopping leaks, and washing cars, dishes, and clothes as efficiently as possible, can go a long way toward forestalling the water shortages that many authorities predict. Isn't it better to adapt to more conservative uses now, when we have a choice, than to be forced to do it by scarcity in the future?

The use of conserving appliances, such as low-volume showerheads and efficient dishwashers and washing machines, can reduce water consumption greatly (What Can You Do? p. 393). If you live in an arid part of the country, you might consider whether you really need a lush green lawn that requires constant watering, feeding, and care. Planting native ground cover in a "natural lawn" or developing a rock garden or landscape in harmony with the surrounding ecosystem can be both ecologically sound and aesthetically pleasing (fig. 17.25). There are about 30 million ha (75 million acres) of cultivated lawns, golf courses, and parks in the United States. They receive more water, fertilizer, and pesticides per hectare than any other kind of land.

The largest U.S. domestic water use is toilet flushing (see fig. 17.15). There are now several types of waterless or low-volume toilets. Waterless composting systems can digest both human and kitchen wastes by aerobic bacterial action, producing a rich, nonoffensive compost that can be used as garden fertilizer. There are also low-volume toilets that use recirculating oil or aqueous chemicals to carry wastes to a holding tank, from which they are periodically taken to a treatment plant. Anaerobic digesters use bacterial or chemical processes to produce usable methane gas from domestic wastes. These systems provide valuable energy and save water but are more difficult to operate than conventional toilets. Few cities are ready to mandate waterless toilets, but a number of cities (including Los Angeles, Orlando, Austin, and Phoenix) have ordered that water-saving toilets,

FIGURE 17.25 By using native plants in a natural setting, residents of Phoenix save water and fit into the surrounding landscape.

showers, and faucets be installed in all new buildings. The motivation was twofold: to relieve overburdened sewer systems and to conserve water.

Significant amounts of water can be reclaimed and recycled. In California, water recovered from treated sewage constitutes the fastest-growing water supply, growing about 30 percent per year. Despite public squeamishness, purified sewage effluent is being used for everything from agricultural irrigation to flushing toilets (fig. 17.26). In a statewide first, San Diego is currently piping water from the local sewage plant directly into a drinking-water reservoir. Residents of Singapore and Queensland, Australia, also are now drinking purified sewage effluent. "Don't rule out desalination because it's expensive, or recycling because it sounds yucky," says Morris Iemma, premier of New South Wales. "We're not getting rain; we have no choice."

Recycling can reduce consumption

In many developing countries as much as 70 percent of all the agricultural water used is lost to leaks in irrigation canals, application to areas where plants don't grow, runoff, and evaporation. Better farming techniques, such as leaving crop residue on fields and ground cover on drainage ways, intercropping, use of mulches, and low-volume irrigation, could reduce these water losses dramatically.

Nearly half of all industrial water use is for cooling of electric power plants and other industrial facilities. Some of this water use could be avoided by installing dry cooling systems

FIGURE 17.26 Recycled water is being used in California and Arizona for everything from agriculture, to landscaping, to industry. Some cities even use treated sewage effluent for human drinking-water supplies.

similar to the radiator of your car. In many cases, cooling water could be reused for irrigation or other purposes in which water does not have to be drinking quality. The waste heat carried by this water could be a valuable resource if techniques were developed for using it.

Prices and policies have often discouraged conservation

Through most of U.S. history, water policies have generally worked against conservation. In the well-watered eastern United States, water policy was based on riparian usufructuary (use) rights—those who lived along a river bank had the right to use as much water as they liked as long as they didn't interfere with its quality or availability to neighbors downstream. It was assumed that the supply would always be endless and that water had no value until it was used. In the drier western regions where water often is a limiting resource, water law is based primarily on the Spanish system of prior appropriation rights, or "first in time are first in right." Even if the prior appropriators are downstream, they can legally block upstream users from taking or using water flowing over their property. But the appropriated water had to be put to "beneficial" use by being consumed. This creates a policy of "use it or lose it." Water left in a stream, even if essential for recreation, aesthetic enjoyment, or to sustain ecological communities, is not being appropriated or put to "beneficial" (that is, economic) use. Under this system, water rights can be bought and sold, but water owners frequently are reluctant to conserve water for fear of losing their rights.

In most federal "reclamation" projects, customers were charged only for the immediate costs of water delivery. The costs of building dams and distribution systems was subsidized, and the potential value of competing uses was routinely ignored. Farmers in California's Central Valley, for instance, for many years paid only about one-tenth of what it cost the government to supply water to them. This didn't encourage conservation. Subsidies created by underpriced water amounted to as much as $500,000 per farm per year in some areas.

Growing recognition that water is a precious and finite resource has changed policies and encouraged conservation across the United States. Despite a growing population, the United States is now saving some 144 million liters (38 million gal) per day—or enough water to fill Lake Erie in a decade—compared to per capita consumption rates of 20 years ago. With 37 million more people in the United Sates now than there were in 1980, we get by with 10 percent less water. New requirements for water-efficient fixtures and low-flush toilets in many cities help to conserve water on the home front. More efficient irrigation methods on farms also are a major reason for the downward trend.

Charging a higher proportion of real costs to users of public water projects has helped encourage conservation, and so have water marketing policies that allow prospective users to bid on water rights. Both the United States and Australia have had effective water pricing and allocation policies that encourage the most socially beneficial uses and discourage wasteful water uses. Market mechanisms for water allotment can be sensitive, however,

What Can You Do?

Saving Water and Preventing Pollution

Each of us can conserve much of the water we use and avoid water pollution in many simple ways.

- Don't flush every time you use the toilet. Take shorter showers; don't wash your car so often.

- Don't let the faucet run while washing hands, dishes, food, or brushing your teeth. Draw a basin of water for washing and another for rinsing dishes. Don't run the dishwasher when half full.

- Dispose of used motor oil, household hazardous waste, batteries, and so on, responsibly. Don't dump anything down a storm sewer that you wouldn't want to drink.

- Avoid using toxic or hazardous chemicals for simple cleaning or plumbing jobs. A plunger or plumber's snake will often unclog a drain just as well as caustic acids or lye. Hot water and soap will clean brushes more safely than organic solvents.

- If you have a lawn, use water sparingly. Water your grass and garden at night, not in the middle of the day. Consider planting native plants, low-maintenance ground cover, a rock garden, or some other xeriphytic landscaping.

- Use water-conserving appliances: low-flow showerheads, low-flush toilets, and aerated faucets.

- Use recycled (gray) water for lawns, house plants, and car washing.

- Check your toilet for leaks. A leaky toilet can waste 50 gallons per day. Add a few drops of dark food coloring to the tank and wait 15 minutes. If the tank is leaking, the water in the bowl will have changed color.

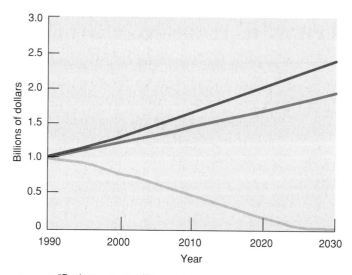

"Business as usual" scenario
Scenario with accelerated investment in water supply and sanitation services
Scenario with accelerated investment and efficiency reforms

FIGURE 17.27 Three scenarios for government investments on clean water and sanitation services, 1990 to 2030.

Source: World Bank estimates based on research paper by Dennis Anderson and William Cavendish, "Efficiency and Substitution in Pollution Abatement: Simulation Studies in Three Sectors."

in developing countries where farmers and low-income urban residents could be outbid for irreplaceable water supplies.

It will be important, as water markets develop, to be sure that environmental, recreational, and wildlife values are not sacrificed to the lure of high-bidding industrial and domestic uses. Given prices based on real costs of using water and reasonable investments in public water supplies, pollution control, and sanitation, the World Bank estimates that everyone in the world could have an adequate supply of clean water by the year 2030 (fig. 17.27). We will discuss the causes, effects, and solutions for water pollution in chapter 18.

CONCLUSION

Water is a precious resource. As human populations grow and climate change affects rainfall patterns, water is likely to become even more scarce in the future. Already about 2 billion people live in water-stressed countries (where water supplies are inadequate to meet all demands), and at least half those people don't have access to clean drinking water. Depending on population growth rates and climate change, by 2050 there could be 7 billion people (about 60 percent of the world population) living in areas with water stress or scarcity. Conflicts over water rights are becoming more common between groups within countries and between neighboring countries that share water resources. This is made more likely by the fact that most major rivers cross two or more countries before reaching the sea. Many experts agree with *Fortune* magazine that "water will be to the 21st century what oil was to the 20th."

There are many ways to make more water available. Huge projects, such as the Chinese scheme to ship water from the well-watered south to the dry north, are already under way. Would we want to do something similar in the United States? Building dams and shipping water between watersheds can have severe ecological and social effects. Perhaps a better way is to practice conservation and water recycling. These efforts, also, are under way in many places, and show great promise for meeting our needs for this irreplaceable resource. There are things you can do as an individual to save water and prevent pollution. Even if you don't have water shortages now where you live, it may be wise to learn how to live in a water-limited world.

REVIEWING LEARNING OUTCOMES

By now you should be able to explain the following points:

17.1 Summarize why water is a precious resource and why shortages occur.
- The hydrologic cycle constantly redistributes water.
- Water supplies are unevenly distributed.

17.2 Compare major water compartments.
- Oceans hold 97 percent of all water on earth.
- Glaciers, ice, and snow contain most surface fresh water.
- Groundwater stores large resources.
- Rivers, lakes, and wetlands cycle quickly.
- The atmosphere is among the smallest of compartments.

17.3 Summarize water availability and use.
- Many countries suffer water scarcity and water stress.
- Water use is increasing.
- Agriculture is the greatest water consumer worldwide.
- Domestic and industrial water use are greatest in wealthy countries.

17.4 Investigate freshwater shortages.
- Many people lack access to clean water.
- Groundwater is being depleted.
- Diversion projects redistribute water.
- Dams often have severe environmental and social impacts.
- Sedimentation limits reservoir life.
- Climate change threatens water supplies.
- Would you fight for water?

17.5 Appreciate how we might get by with less water.

17.6 Understand how we might increase water supplies.
- Domestic conservation can save water.
- Recycling can reduce consumption.
- Prices and policies have often discouraged conservation.

PRACTICE QUIZ

1. What is the difference between withdrawal, consumption, and degradation of water?
2. Explain how water can enter and leave an aquifer (see fig. 17.9).
3. Describe the changes in water withdrawal and consumption by sector shown in figure 17.12.
4. Describe some problems associated with dam building and water diversion projects.
5. Describe the path a molecule of water might follow through the hydrologic cycle from the ocean to land and back again.
6. Where are the five largest rivers in the world (table 17.3)?
7. How do mountains affect rainfall distribution? Does this affect your part of the country?
8. Identify and explain three consequences of overpumping aquifers.
9. How much water is fresh (as opposed to saline) and where is it?
10. Explain how saltwater intrusion happens (fig. 17.18).

CRITICAL THINKING AND DISCUSSION QUESTIONS

1. What changes might occur in the hydrologic cycle if our climate were to warm or cool significantly?
2. Why does it take so long for the deep ocean waters to circulate through the hydrologic cycle? What happens to substances that contaminate deep ocean water or deep aquifers in the ground?
3. Are there ways you could use less water in your own personal life? What obstacles prevent you from taking these steps?
4. Should we use up underground water supplies now or save them for some future time?
5. How should we compare the values of free-flowing rivers and natural ecosystems with the benefits of flood control, water diversion projects, hydroelectric power, and dammed reservoirs?
6. Would it be feasible to change from flush toilets and using water as a medium for waste disposal to some other system? What might be the best way to accomplish this?

One definition of water stress is when annual water supplies drop below 1,700 m^3 per person. Water scarcity is defined as annual water supplies below 1,000 m^3 per person. More than 2.8 billion people in 48 countries will face either water stress or scarcity conditions by 2025. Of these countries, 40 are expected to be in West Asia or Africa. By 2050, far more people could be facing water shortages, depending both on population projections and scenarios for water supplies based on global warming and consumption patterns. The graph shows an estimate for water stress and scarcity in 1995 together with three possible scenarios (high, medium, and low population projections) for 2050. You'll remember from chapter 7 that according to the 2004 UN population revision, the low projection for 2050 is about 7.6 billion, the medium projection is 8.9 billion, and the high projection is 10.6 billion.

1. What combined numbers of people could experience water stress and scarcity under the low, medium, and high scenarios in 2050?

2. What proportion (percentage) of 7.6 billion, 8.9 billion, and 10.6 billion would this be?

3. How does the percentage of the population in these two categories vary in the three estimates?

4. Why is the proportion of people in the scarce category so much larger in the high projection?

5. How many liters are in 1,000 m^3? How many gallons?

6. How does 1,000 m^3 compare to the annual consumption by the average family of four in the United States? (Hint: Look at table 17.1 and the table of units of measurement conversions at the end of this book).

7. Why isn't the United States (as a whole) considered to be water-stressed?

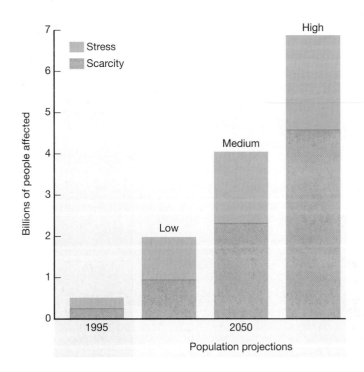

Global water stress and scarcity

Oil and industrial debris burn on the Cuyahoga River.
Events such as this led to the Clean Water Act of 1972.

Water Pollution

Learning Outcomes
After studying this chapter, you should be able to:

18.1 Define *water pollution*.

18.2 Describe the types and effects of water pollutants.

18.3 Investigate water quality today.

18.4 Explain water pollution control.

18.5 Summarize water legislation.

"Water, water everywhere; nor any drop to drink."

~ Samuel Taylor Coleridge

Case Study
Protecting Our Nation's Water

Burn on, big river, burn on

Now the Lord can make you tumble

And the Lord can make you turn

And the Lord can make you overflow

But the Lord can't make you burn

– Randy Newman

Singer-songwriter Randy Newman wrote his ode to Ohio's Cuyahoga River, "Burn On," after one of that river's legendary fires in 1969. It might be hard to imagine that a river could burn, but this river—like many others in 1969—was so choked with oil, tires, and other industrial waste that it caught fire repeatedly, in more than one case burning Cleveland's bridges. Lake Erie, into which the Cuyahoga empties in Cleveland, was also essentially "dead" in 1969, with extremely low oxygen levels and an ecosystem that had nearly collapsed. Cleveland residents aren't necessarily proud of this part of their historical legacy, but the rest of us can be thankful to them for taking the spotlight in 1969 and riveting national attention to the problem of uncontrolled contamination of our water resources.

Today American cities have some of the cleanest tap water in the world. Although many people are skeptical about municipal water, it's actually carefully monitored according to national safety standards, which is why water-related epidemics are extremely rare here despite concentrated urban populations. To appreciate the importance of this, consider that rivers have always provided much of our water, but just half a century ago the main disposal methods for industrial effluent, municipal wastewater, and sewage was to dump it into the nearest river.

Another disaster in 1969 also caught the public eye: an oil well blew out near the coast of Santa Barbara, California, flooding popular beaches with sticky, black oil. Television footage of volunteer crews struggling to clean their beach helped galvanize public opinion. Starting the following year, President Richard Nixon signed into law several of our cornerstone environmental protections—laws we now rely on so completely that most people don't even know they exist.

The Clean Water Act was first introduced to Congress in 1969. For three years the bill was passed back and forth between the House and Senate, for amendments, public comment, and lobbying, before it finally reached the president's desk for a signature. This was not the first U.S. law to address industrial dumping, but it was the first to establish health-related goals. The act called for cities and states, with financial help from the federal government, to make all surface waters safe at least for fishing and swimming. The Clean Water Act also established rules for regulating pollutants that cities and industries were allowed to discharge into public waters. Today there are still discharges into public waters, but all are supposed to be within limits agreed to in a permit from the EPA, and seriously toxic discharges are outlawed.

Through the Clean Water Act, the EPA now monitors water quality in all U.S. cities. Conditions aren't perfect, but water quality is far better than a few decades ago. The upper Cuyahoga River is now part of a scenic national park, and EPA assessments have found steelhead trout, northern pike, and other clean-water fish in the river (fig. 18.1). Lake Erie has largely recovered because of improved wastewater treatment, and the lake now has a robust sport fishery.

Success stories like this are commonplace in the United States these days. The upper Mississippi River, which industry wanted to have designated an open sewer half a century ago, is now clean enough to support mayflies and walleye pike. Chattanooga Creek, which enters the Tennessee River in the middle of Chattanooga, was so polluted from toxic dumping by coke foundries and chemical factories that in 1994 the EPA proposed 2.5 miles of the creek as a Superfund site. It has now been cleaned up, and the city riverfront has become a highly desirable residential and recreational area.

It's hard to overstate the importance of regulating and monitoring environmental quality. Many other countries today still have uncontrolled dumping in public waters, and billions cannot safely drink their tap water. We take our water for granted, but it's only because of the hard work of millions of activists and thousands of elected officials, and the efforts of regulatory staff, that we're able to stay healthy and appreciate a relatively clean environment.

In this chapter we'll look at the causes and effects of water pollution as well as our options for controlling or treating water contaminants. For related resources, including Google Earth™ placemarks that show locations where these issues can be seen, visit EnvironmentalScience-Cunningham.blogspot.com.

FIGURE 18.1 The Cuyahoga River near Cleveland. Since the Clean Water Act passed, its water quality has improved greatly.

18.1 WATER POLLUTION

Most students today are too young to appreciate that water in most industrialized countries was once far more polluted and dangerous than it is now. The Cuyahoga was a particularly egregious case, but forty years ago factories and cities routinely dumped untreated chemicals, metals, oil, solvents, and sewage into rivers and lakes. Toxic solvents and organic chemicals were commonly dumped on the ground, poisoning groundwater that we're now paying billions to clean up. In 1972, President Nixon signed the Clean Water Act, which has been called the United States' most successful and popular environmental legislation. This act established a goal that all the nation's waters should be "fishable and swimmable." While this goal is far from being achieved, the Clean Water Act remains popular because it protects public health (thus saving taxpayer dollars), as well as reducing environmental damage. In addition, water has an aesthetic appeal: The view of a clean lake, river, or seashore makes people happy, and water provides for recreation, so many people feel their quality of life has improved as water quality has been restored.

Clean water is a national, as well as global, priority. Recent polls have found repeatedly that 90 percent of Americans believe we should invest more in clean water and 70 percent would support establishing a trust fund to help communities repair water facilities.

We still have a long way to go in improving water quality. Although pollution from factory pipes has been vastly reduced in the past four decades, erosion from farm fields, construction sites, and streets has, in many areas, gotten worse since 1972. Airborne mercury, sulfur, and other substances are increasingly contaminating lakes and wetlands. Concentrated livestock production and agricultural runoff threaten underground water as well as surface water systems. Increasing industrialization in developing countries has led to widespread water pollution in impoverished regions with little environmental regulation.

Water pollution is anything that degrades water quality

Any physical, biological, or chemical change in water quality that adversely affects living organisms or makes water unsuitable for desired uses might be considered pollution. There are natural sources of water contamination, such as poison springs, oil seeps, and sedimentation from erosion, but in this chapter we will focus primarily on human-caused changes that affect water quality or usability.

Pollution-control standards and regulations usually distinguish between point and nonpoint pollution sources. Factories, power plants, sewage treatment plants, underground coal mines, and oil wells are classified as **point sources** because they discharge pollution from specific locations, such as drain pipes, ditches, or sewer outfalls (fig. 18.2). These sources are discrete and identifiable, so they are relatively easy to monitor and regulate. It is generally possible to divert effluent from the waste streams of these sources and treat it before it enters the environment.

In contrast, **nonpoint sources** of water pollution are scattered or diffuse, having no specific location where they discharge into a particular body of water. Nonpoint sources include runoff from

FIGURE 18.2 Sewer outfalls, industrial effluent pipes, acid draining out of abandoned mines, and other point sources of pollution are generally easy to recognize.

farm fields and feedlots (fig. 18.3), golf courses, lawns and gardens, construction sites, logging areas, roads, streets, and parking lots. Whereas point sources may be fairly uniform and predictable throughout the year, nonpoint sources are often highly episodic. The first heavy rainfall after a dry period may flush high concentrations of gasoline, lead, oil, and rubber residues off city streets, for instance, while subsequent runoff may have lower levels of these pollutants. Spring snowmelt carries high levels of atmospheric acid deposition into streams and lakes in some areas. The irregular timing of these events, as well as their multiple sources and scattered location, makes them much more difficult to monitor, regulate, and treat than point sources.

Perhaps the ultimate in diffuse, nonpoint pollution is **atmospheric deposition** of contaminants carried by air currents and precipitated into watersheds or directly onto surface waters as rain, snow, or dry particles. The Great Lakes, for example, have been found to be accumulating industrial chemicals such as PCBs and dioxins, as well as agricultural toxins such as the insecticide toxaphene that cannot be accounted for by local sources alone. The nearest sources for many of these chemicals are sometimes thousands of kilometers away (chapter 16).

FIGURE 18.3 This scene looks peaceful and idyllic, but allowing cows to trample streambanks is a major cause of bank erosion and water pollution. Nonpoint sources such as this have become the leading unresolved cause of stream and lake pollution in the United States.

Amounts of these pollutants can be quite large. It is estimated that there are 600,000 kg of the herbicide atrazine in the Great Lakes, most of which is thought to have been deposited from the atmosphere. Concentration of persistent chemicals up the food chain can produce high levels in top predators. Several studies have indicated health problems among people who regularly eat fish from the Great Lakes.

Ironically, lakes can be pollution sources as well as recipients. In the past 12 years, about 26,000 metric tons of PCBs have "disappeared" from Lake Superior. Apparently these compounds evaporate from the lake surface and are carried by air currents to other areas where they are redeposited.

18.2 TYPES AND EFFECTS OF WATER POLLUTANTS

Although the types, sources, and effects of water pollutants are often interrelated, it is convenient to divide them into major categories for discussion (table 18.1). Let's look more closely at some of the important sources and effects of each type of pollutant.

Infectious agents remain an important threat to human health

The most serious water pollutants in terms of human health worldwide are pathogenic organisms (chapter 8). Among the most important waterborne diseases are typhoid, cholera, bacterial and amoebic dysentery, enteritis, polio, infectious hepatitis, and schistosomiasis. Malaria, yellow fever, and filariasis are transmitted by insects that have aquatic larvae. Altogether, at least 25 million deaths each year are blamed on these water-related diseases. Nearly two-thirds of the mortalities of children under 5 years old are associated with waterborne diseases.

The main source of these pathogens is untreated or improperly treated human wastes. Animal wastes from feedlots or fields near waterways and food-processing factories with inadequate waste treatment facilities also are sources of disease-causing organisms.

In wealthier countries, sewage treatment plants and other pollution-control techniques have reduced or eliminated most of the worst sources of pathogens in inland surface waters, and drinking water is generally disinfected by chlorination, so epidemics of waterborne diseases are rare in these countries. The United Nations estimates that 90 percent of the people in developed countries have adequate (safe) sewage disposal, and 95 percent have clean drinking water.

The situation is quite different in poor countries. The United Nations estimates that at least 2.5 billion people in these countries lack adequate sanitation, and that about half these people also lack access to clean drinking water. Conditions are especially bad in remote rural areas where sewage treatment is usually primitive or nonexistent and purified water is either unavailable or too expensive to obtain (fig. 18.4). The World Health Organization estimates that 80 percent of all sickness and disease in less-developed countries can be attributed to waterborne infectious agents and inadequate sanitation.

The World Bank estimates that if everyone had pure water and satisfactory sanitation, 200 million fewer episodes of diarrheal illness would occur each year, and 2 million childhood deaths would be avoided. Furthermore, 450 million people would be spared debilitating roundworm or fluke infections. Surely these are goals worth pursuing.

Detecting specific pathogens in water is difficult, time-consuming, and costly; thus, water quality control personnel usually analyze water for the presence of **coliform bacteria**, any of the many types that live in the colon or intestines of humans and other animals. The most common of these is *Escherichia coli* (or *E. coli*). Many strains of bacteria are normal symbionts in mammals, but some, such as *Shigella, Salmonella,* or *Listeria,* can cause fatal diseases. It is usually assumed that if any coliform bacteria are present in a water sample, infectious pathogens are present also.

To test for coliform bacteria, a water sample (or a filter through which a measured water sample has passed) is placed in a dish containing a nutrient medium that supports bacterial growth. After 24 hours in an incubator, living cells will have produced small colonies. If *any* colonies are found in drinking water

Table 18.1 Major Categories of Water Pollutants

Category	Examples	Sources
A. Causes Health Problems		
1. Infectious agents	Bacteria, viruses, parasites	Human and animal excreta
2. Organic chemicals	Pesticides, plastics, detergents, oil, and gasoline	Industrial, household, and farm use
3. Inorganic chemicals	Acids, caustics, salts, metals	Industrial effluents, household cleansers, surface runoff
4. Radioactive materials production, natural sources	Uranium, thorium, cesium, iodine, radon	Mining and processing of ores, power plants, weapons
B. Causes Ecosystem Disruption		
1. Sediment	Soil, silt	Land erosion
2. Plant nutrients	Nitrates, phosphates, ammonium	Agricultural and urban fertilizers, sewage, manure
3. Oxygen-demanding wastes	Animal manure and plant residues	Sewage, agricultural runoff, paper mills, food processing
4. Thermal	Heat	Power plants, industrial cooling

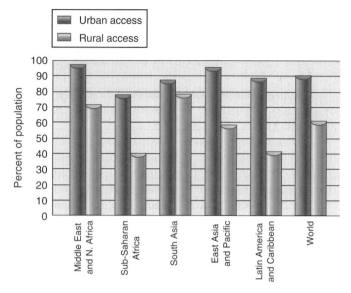

FIGURE 18.4 Proportion of people in developing regions with access to safe drinking water.
Source: © UNESCO, 2002.

samples, the U.S. Environmental Protection Agency considers the water unsafe and requiring disinfection. The EPA-recommended maximum coliform count for swimming water is 200 colonies per 100 ml, but some cities and states allow higher levels. If the limit is exceeded, the contaminated pool, river, or lake usually is closed to swimming (fig. 18.5).

Bacteria are detected by measuring oxygen levels

The amount of oxygen dissolved in water is a good indicator of water quality and of the kinds of life it will support. Water with an oxygen content above 6 parts per million (ppm) will support game

FIGURE 18.5 The national goal of making all surface waters in the United States "fishable and swimmable" has not been fully met, but scenes like this have been reduced by pollution-control efforts.

fish and other desirable forms of aquatic life. Water with less than 2 ppm oxygen will support mainly worms, bacteria, fungi, and other detritus feeders and decomposers. Oxygen is added to water by diffusion from the air, especially when turbulence and mixing rates are high, and by photosynthesis of green plants, algae, and cyanobacteria. Oxygen is removed from water by respiration and chemical processes that consume oxygen.

Organic waste such as sewage, paper pulp, or food waste is rich in nutrients, especially nitrogen and phosphorus. These nutrients stimulate the growth of oxygen-demanding decomposing bacteria. **Biochemical oxygen demand (BOD)** is thus a useful test for the presence of organic waste in water. Most BOD tests involve incubating a water sample for five days, then comparing oxygen levels in the water before and after incubation. An alternative method, called the chemical oxygen demand (COD), uses a strong oxidizing agent (dichromate ion in 50 percent sulfuric acid) to completely break down all organic matter in a water sample. This method is much faster than the BOD test, but it records inactive organic matter as well as bacteria, so it is less useful. A third method of assaying pollution levels is to measure **dissolved oxygen (DO) content** directly, using an oxygen electrode. The DO content of water depends on factors other than pollution (for example, temperature and aeration), so it is best for indicating the health of the aquatic system.

The effects of oxygen-demanding wastes on rivers depends to a great extent on the volume, flow, and temperature of the river water. Aeration occurs readily in turbulent, rapidly flowing rivers, which are therefore often able to recover quickly from oxygen-depleting processes. Downstream from a point source, such as a municipal sewage plant discharge, a characteristic decline and restoration of water quality can be detected either by measuring dissolved oxygen content or by observing the flora and fauna that live in successive sections of the river.

The oxygen decline downstream is called the **oxygen sag** (fig. 18.6). Upstream from the pollution source, oxygen levels support normal populations of clean-water organisms. Immediately below the source of pollution, oxygen levels begin to fall as decomposers metabolize waste materials. Rough fish, such as carp, bullheads, and gar, are able to survive in this oxygen-poor environment where they eat both decomposer organisms and the waste itself. Further downstream, the water may become so oxygen-depleted that only the most resistant microorganisms and invertebrates can survive. We often call this a "dead zone." Eventually most of the nutrients are used up, decomposer populations are smaller, and the water becomes oxygenated once again. Depending on the volumes and flow rates of the effluent plume and the river receiving it, normal communities may not appear for several miles downstream.

Nutrient enrichment leads to cultural eutrophication

Water clarity (transparency) is affected by sediments, chemicals, and the abundance of plankton organisms, and is a useful measure of water quality and water pollution. Rivers and lakes that have clear water and low biological productivity are said to be

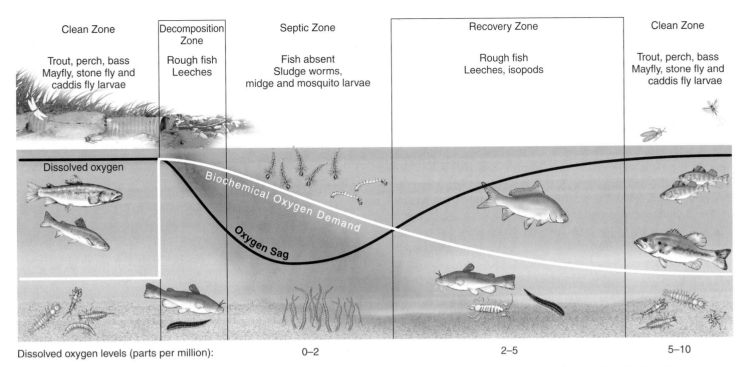

Clean Zone	Decomposition Zone	Septic Zone	Recovery Zone	Clean Zone
Trout, perch, bass Mayfly, stone fly and caddis fly larvae	Rough fish Leeches	Fish absent Sludge worms, midge and mosquito larvae	Rough fish Leeches, isopods	Trout, perch, bass Mayfly, stone fly and caddis fly larvae

Dissolved oxygen

Biochemical Oxygen Demand

Oxygen Sag

Dissolved oxygen levels (parts per million):　　0–2　　　　2–5　　　　5–10

FIGURE 18.6 Oxygen sag downstream of an organic source. A great deal of time and distance may be required for the stream and its inhabitants to recover.

oligotrophic (*oligo* = little + *trophic* = nutrition). By contrast, **eutrophic** (*eu* + *trophic* = truly nourished) waters are rich in organisms and organic materials. Eutrophication is an increase in nutrient levels and biological productivity. Some amount of eutrophication is a normal part of successional changes in most lakes. Tributary streams bring in sediments and nutrients that stimulate plant growth. Over time, ponds or lakes may fill in, eventually becoming marshes. The rate of eutrophication and succession depends on water chemistry and depth, volume of inflow, mineral content of the surrounding watershed, and the biota of the lake itself.

As with BOD, nutrient enrichment sewage, fertilizer runoff, even decomposing leaves in street gutters can produce a human-caused increase in biological productivity called **cultural eutrophication**. Cultural eutrophication can also result from higher temperatures, more sunlight reaching the water surface, or a number of other changes. Increased productivity in an aquatic system sometimes can be beneficial. Fish and other desirable species may grow faster, providing a welcome food source.

Often, however, eutrophication has undesirable results. Elevated phosphorus and nitrogen levels stimulate "blooms" of algae or thick growths of aquatic plants (fig. 18.7). Bacterial populations also increase, fed by larger amounts of organic matter. The water often becomes cloudy or turbid and has unpleasant tastes and odors. In extreme cases, plants and algae die and decomposers deplete oxygen in the water. Collapse of the aquatic ecosystem can result.

Eutrophication can cause toxic tides and "dead zones"

According to the Bible, the first plague to afflict the Egyptians when they wouldn't free Moses and the Israelites was that the water in the Nile turned into blood. All the fish died and the people were unable to drink the water, a terrible calamity in a

FIGURE 18.7 Eutrophic lake. Nutrients from lawn fertilizers and other urban runoff have stimulated growth of algal mats that reduce water quality, alter species composition, and lower the lake's recreational and aesthetic values.

desert country. Some modern scientists believe this may be the first recorded history of a **red tide** or a bloom of deadly aquatic microorganisms. Red tides—and tides of other colors, depending on the species involved—have become increasingly common in slow-moving rivers, brackish lagoons, estuaries, and bays, as well as nearshore ocean waters where nutrients and wastes wash down our rivers.

Eutrophication in marine ecosystems occurs in nearshore waters and partially enclosed bays or estuaries. Some areas, such as the Gulf of Mexico, the Caspian Sea, the Baltic, and Bohai Bay in the Yellow Sea, tend to be in especially critical condition. During the tourist season, the coastal population of the Mediterranean, for example, swells to 200 million people. Eighty-five percent of the effluents from large cities go untreated into the sea. Beach pollution, fish kills, and contaminated shellfish result. Extensive "dead zones" often form where rivers dump oxygen-depleting nutrients into estuaries and shallow seas. The second largest hypoxic (oxygen-depleted) zone in the world occurs during summer months in the Gulf of Mexico at the mouth of the Mississippi River (Exploring Science, p. 403). Studies indicate that as human populations, cities, and agriculture expand, these hypoxic zones will become increasingly common.

It appears that fish and other marine species die in these polluted zones not only because oxygen is depleted but also because of high concentrations of harmful organisms including toxic algae, pathogenic fungi, parasitic protists, and other predators. One of the most notorious and controversial of these is *Pfiesteria piscicida*, a dinoflagellate (a single-celled organism that swims with two whiplike flagella). *Pfiesteria* cells are reported to secrete nerve-damaging toxins, allowing amoeba forms to attack wounded fish. *Pfiesteria* toxins have also been blamed for human health problems including nerve damage. Other studies, however, have raised doubts about *Pfiesteria's* life cycle and its role in both fish mortality and human illnesses.

Inorganic pollutants include metals, salts, acids, and bases

Some toxic inorganic chemicals are released from rocks by weathering, are carried by runoff into lakes or rivers, or percolate into groundwater aquifers. This pattern is part of natural mineral cycles (chapter 3). Humans often accelerate the transfer rates in these cycles thousands of times above natural background levels through the mining, processing, using, and discarding of minerals.

In many areas, toxic inorganic chemicals introduced into water as a result of human activities have become the most serious form of water pollution. Among the chemicals of greatest concern are heavy metals, such as mercury, lead, tin, and cadmium. Super-toxic elements, such as selenium and arsenic, also have reached hazardous levels in some waters. Other inorganic materials, such as acids, salts, nitrates, and chlorine, that normally are not toxic at low concentrations may become concentrated enough to lower water quality or adversely affect biological communities.

Metals

Many metals, such as mercury, lead, cadmium, tin, and nickel, are highly toxic in minute concentrations. Because metals are highly persistent, they can accumulate in food webs and have a cumulative effect in top predators—including humans.

Currently the most widespread toxic metal contamination problem in North America is mercury released from coal-burning power plants. As chapter 16 mentions, an EPA survey of 2,500 fish from 260 lakes across the United States found at least low levels of mercury in every fish sampled. More than half the fish contained mercury levels unsafe for women of childbearing age, and three-quarters exceed the safe limit for young children.

Fifty states have issued warnings about eating freshwater or ocean fish; mercury contamination is by far the most common reason for these advisories (fig. 18.8). Top marine predators, such as shark, swordfish, marlin, king mackerel, and blue-fin tuna, should be avoided completely. You should check local advisories about the safety of fish caught in your local lakes, rivers, and coastal areas. If no advice is available, eat no more than one meal of such fish per week. Public health officials estimate that 600,000 American children now have mercury levels in their bodies high enough to cause mental and developmental problems, and that one in six U.S. women have blood-mercury concentrations that would endanger a fetus.

Mine drainage and leaching of mining wastes are serious sources of metal pollution in water. A survey of water quality in eastern Tennessee—where there has been a great deal of surface mining—found contamination by acids and metals from mine drainage in 43 percent of all surface streams and lakes and more than half of all groundwater used for drinking supplies.

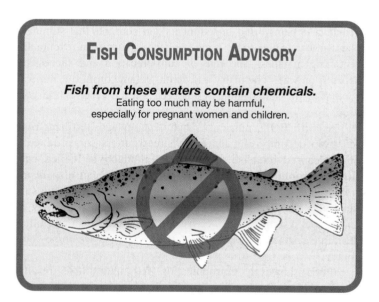

FIGURE 18.8 Mercury contamination is the second most common cause of impairment of U.S. rivers and lakes. Forty-five states have issued warnings about eating locally caught freshwater fish. Long-lived, top predators are especially likely to bioaccumulate toxic concentrations of mercury. The largest source of this highly dangerous toxin is coal-fired power plants.

In the 1980s shrimp boat crews noticed that certain locations off the Gulf Coast of Louisiana were emptied of all aquatic life. Because the region supports shrimp, fish, and oyster fisheries worth $250 to $450 million per year, these "dead zones" were important to the economy as well as to the Gulf's ecological systems. In 1985 marine scientist Nancy Rabelais began mapping areas of low oxygen concentrations in the Gulf waters. Her results, published in 1991, showed that vast areas just above the floor of the Gulf have a summer oxygen concentration of less than 2 parts per million (ppm), a level that eliminated all animal life except microorganisms and primitive worms. Healthy aquatic systems usually have about 10 ppm dissolved oxygen. What caused this hypoxic (oxygen-starved) area to develop?

Rabelais and her team tracked the phenomenon for several years, and it became clear that the dead zone was growing larger over time, that poor shrimp harvests coincided with years when the zone was large, and that the size of the dead zone, which ranges from 5,000 to 20,000 km² (about the size of New Jersey), depended on rainfall and runoff rates from the Mississippi River. Excessive nutrients, mainly nitrogen, from farms and cities far upstream were the suspected culprit.

How did Rabelais and her team know that nutrients were the problem? They noticed that each year, 7 to 10 days after large spring rains in the agricultural parts of the upper Mississippi watershed, oxygen concentrations in the Gulf drop from 5 ppm to below 2 ppm. These rains are known to wash soil, organic debris, and last year's nitrogen-rich fertilizers from farm fields. Scientists also knew that saltwater ecosystems normally have little available nitrogen, a key nutrient for algae and plant growth. Pulses of agricultural runoff were followed by a profuse growth of algae and phytoplankton (tiny floating plants). This burst of biological activity produces an excess of dead plant cells and fecal matter that drifts to the seafloor. Shrimp, clams, oysters, and other filter feeders normally consume this debris, but they can't keep up with the sudden flood of material. Instead, decomposing bacteria in the sediment break down the debris, and consume most of the available

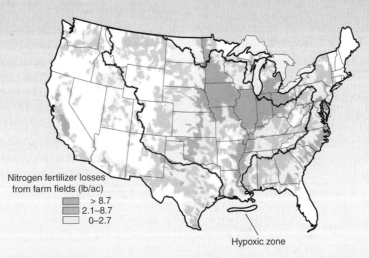

Nitrogen fertilizer losses from farm fields (lb/ac)
- > 8.7
- 2.1–8.7
- 0–2.7

Hypoxic zone

The Mississippi River drains 40 percent of the conterminous United States, including the most heavily farmed states. Nitrogen fertilizer produces a summer "dead zone" in the Gulf of Mexico.

dissolved oxygen as well. Putrefying sediments also produce hydrogen sulfide, which further poisons the water near the seafloor.

In well-mixed water bodies, such as the open ocean, oxygen from upper water layers is frequently mixed into lower layers. Warm, protected water bodies are often stratified, however, as abundant sunlight keeps the upper layers warmer, and less dense, than lower layers. Denser lower layers can't mix with upper layers unless strong currents or winds stir the water.

Many enclosed coastal waters, including Chesapeake Bay, Long Island Sound, the Mediterranean Sea, and the Black Sea, tend to be stratified and suffer hypoxic conditions that destroy bottom and near-bottom communities. There are about 200 dead zones around the world, and the number has doubled each decade since dead zones were first observed in the 1970s. The Gulf of Mexico is second in size behind a 100,000 km² dead zone in the Baltic Sea.

Can dead zones recover? Yes. Water is a forgiving medium, and organisms use nitrogen quickly. In 1996 in the Black Sea region, farmers in collapsing communist economies cut their nitrogen applications by half out of economic necessity; the Black Sea dead zone disappeared, while farmers saw no drop in their crop yields. In the Mississippi watershed, farmers can afford abundant fertilizer, and they fear they can't afford to risk underfertilizing. Because of the great geographic distance between the farm states and the

Gulf, Midwestern states have been slow to develop an interest in the dead zone. At the same time, concentrated feedlot production of beef and pork is rapidly increasing, and feedlot runoff is the fastest growing, and least regulated, source of nutrient enrichment in rivers.

In 2001, federal, state, and tribal governments forged an agreement to cut nitrogen inputs by 30 percent and reduce the size of the dead zone to 5,000 km². This agreement represented remarkably quick political response to scientific results, but it doesn't appear to be enough. Computer models suggest that it would take a 40 to 45 percent reduction in nitrogen to achieve the 5,000 km² goal.

Human activities have increased the flow of nitrogen reaching U.S. coastal waters by four to eight times since the 1950s. Phosphorus, another key nutrient, has tripled. This case study shows how water pollution can connect far-distant places, such as Midwestern farmers and Louisiana shrimpers.

The explosion and fire on the BP Deepwater Horizon rig in 2010 added more pollution to the Gulf. Nearly 5 million barrels (nearly 800 million liters) of oil gushed into the Gulf, but it isn't clear what has happened to most of it. About 1 million barrels were collected at the well, skimmed off the surface, or burned. Another 1.5 million barrels are thought to have evaporated or dissolved. The rest is unaccounted for. Microbes may have eaten a lot of it. The Gulf has many natural oil seeps. It's estimated that around half a million barrels of oil leak into the Gulf each year from natural sources. There's a thriving microbial population adapted to living on this oil, so that much of spilled crude from the Deepwater Horizon was probably metabolized fairly quickly. Still, that metabolism requires oxygen, so the oil will very likely contribute to the dead zone. Another concern is that BP sprayed about 1.3 million gallons (about 5 million liters) of potentially toxic chemicals to break up and disperse the oil slicks. This prevented much of the contamination of beaches and marshes that would otherwise have occurred. But it's unknown what effects these chemicals are having on marine life in the Gulf.

In some cases, metal levels were 200 times higher than what is considered safe for drinking water.

Nonmetallic Salts

Some soils contain high concentrations of soluble salts, including toxic selenium and arsenic. You have probably heard of poison springs and seeps in the desert, where percolating groundwater brings these materials to the surface. Irrigation and drainage of desert soils can mobilize these materials on a larger scale and result in serious pollution problems, as in Kesterson Marsh in California, where selenium poisoning killed thousands of migratory birds in the 1980s.

Salts, such as sodium chloride (table salt), that are nontoxic at low concentrations also can be mobilized by irrigation and concentrated by evaporation, reaching levels that are toxic for many plants and animals. Globally, 20 percent of the world's irrigated farmland is estimated to be affected by salinization, and half that land has enough salt buildup to decrease yields significantly. In the northern United States, at least 25 million tons of sodium chloride and calcium chloride are used every year to melt road ice. Leaching of these road salts into surface waters is having adverse effects on many aquatic ecosystems.

Perhaps the largest human population threatened by naturally occurring arsenic in groundwater is in West Bengal, India, and adjacent areas of Bangladesh (fig. 18.9). Arsenic occurs naturally in the sediments that make up the Ganges River delta. Rapid population growth, industrialization, and intensification of irrigated agriculture have consumed or polluted limited surface water supplies. In an effort to provide clean drinking water for local residents, thousands of deep tube wells were sunk in the 1960s throughout the area. Much of this humanitarian effort was financed by loans from the World Bank.

By the 1980s, health workers became aware of widespread signs of chronic arsenic poisoning among Bengali villagers. Symptoms include watery and inflamed eyes, gastrointestinal cramps, gradual loss of strength, scaly skin and skin tumors, anemia, confusion, and eventually death. Some villages have had wells for centuries without a problem; why is arsenic poisoning appearing now? One theory is that excessive withdrawals now lower the water table during the dry season, exposing arsenic-bearing

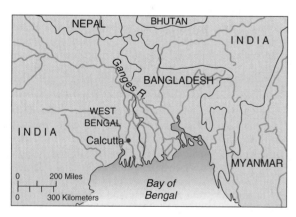

FIGURE 18.9 West Bengal and adjoining areas of Bangladesh have hundreds of millions of people who may be exposed to dangerous arsenic levels in well water.

minerals to oxidation, which converts normally insoluble salts to soluble oxides. When aquifers refill during the next monsoon season, dissolved arsenic can be pumped out. Health workers estimate that the total number of potential victims in India and Bangladesh may exceed 200 million people.

Acids and Bases

Acids are released as by-products of industrial processes, such as leather tanning, metal smelting and plating, petroleum distillation, and organic chemical synthesis. Coal mining is an especially important source of acid water pollution. Sulfur compounds in coal react with oxygen and water to make sulfuric acid. Thousands of kilometers of streams in the United States have been acidified by acid mine drainage, some so severely that they are essentially lifeless.

Coal and oil combustion also leads to formation of atmospheric sulfuric and nitric acids (chapter 16), which are disseminated by long-range transport processes and deposited via precipitation (acidic rain, snow, fog, or dry deposition) in surface waters. Where soils are rich in such alkaline material as limestone, these atmospheric acids have little effect because they are neutralized. In high mountain areas or recently glaciated regions where crystalline bedrock is close to the surface and lakes are oligotrophic, however, there is little buffering capacity (ability to neutralize acids), and aquatic ecosystems can be severely disrupted. These effects were first recognized in the mountains of northern England and Scandinavia about 30 years ago.

Aquatic damage due to acid precipitation has been reported in about 200 lakes in the Adirondack Mountains of New York State and in several thousand lakes in eastern Quebec, Canada. Game fish, amphibians, and sensitive aquatic insects are generally the first to be killed by increased acid levels in the water. If acidification is severe enough, aquatic life is limited to a few resistant species of mosses and fungi. Increased acidity may result in leaching of toxic metals, especially aluminum, from soil and rocks, making water unfit for drinking or irrigation, as well.

Organic pollutants include drugs, pesticides, and other industrial substances

Thousands of different natural and synthetic organic chemicals are used in the chemical industry to make pesticides, plastics, pharmaceuticals, pigments, and other products that we use in everyday life. Many of these chemicals are highly toxic (chapter 8). Exposure to very low concentrations (perhaps even parts per quadrillion in the case of dioxins) can cause birth defects, genetic disorders, and cancer. Some can persist in the environment because they are resistant to degradation and toxic to organisms that ingest them. Contamination of surface waters and groundwater by these chemicals is a serious threat to human health.

The two most important sources of toxic organic chemicals in water are improper disposal of industrial and household wastes and runoff of pesticides from farm fields, forests, roadsides, golf courses, and other places where they are used in large quantities. The U.S. EPA estimates that about 500,000 metric tons of pesticides are used in the United States each year. Much of this material washes into the nearest waterway, where it passes through ecosystems and may accumulate in high levels in nontarget organisms. The bioaccumulation of DDT in aquatic ecosystems was one of the first of these pathways to be understood. Dioxins and other chlorinated hydrocarbons (hydrocarbon molecules that contain chlorine atoms) have been shown to accumulate to dangerous levels in the fat of salmon, fish-eating birds, and humans and to cause health problems similar to those resulting from toxic metal compounds (fig. 18.10).

As chapter 8 reports, atrazine, the most widely used herbicide in America, has been shown to disrupt normal sexual development in frogs at concentrations as low as 0.1 ppb. This level is found regularly wherever farming occurs. Could this be a problem for us as well?

Hundreds of millions of tons of hazardous organic wastes are thought to be stored in dumps, landfills, lagoons, and underground tanks in the United States (chapter 21). Many, perhaps most, of these sites have leaked toxic chemicals into surface waters or groundwater or both. The EPA estimates that about 26,000 hazardous waste sites will require cleanup because they pose an imminent threat to public health, mostly through water pollution.

Countless other organic compounds also enter our water. How do they get there? In some cases, people simply dump unwanted food, medicines, and health supplements down the toilet or sink. More often we consume more than our bodies can absorb, and we excrete the excess, which passes through sewage treatment facilities relatively unchanged. Numerous studies have found quite high levels of caffeine, birth-control hormones, antibiotics, recreational drugs, and other compounds downstream from major cities. This often results in developmental and behavioral changes in fish and other aquatic organisms.

In 2002 the USGS released the first-ever study of pharmaceuticals and hormones in streams. Scientists sampled 130 streams, looking for 95 contaminants, including antibiotics, natural and synthetic

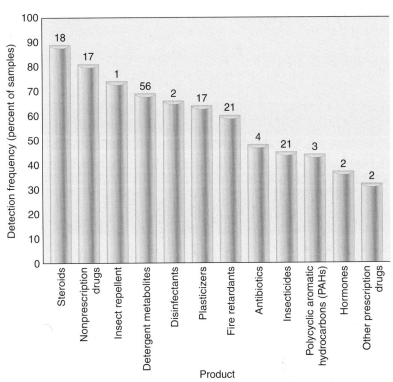

FIGURE 18.11 Detection frequency of organic, wastewater contaminants in a recent USGS survey. Maximum concentrations in water samples are shown above the bars in micrograms per liter. Dominant substances included DEET insect repellent, caffeine, and triclosan, which comes from antibacterial soaps.

hormones, detergents, plasticizers, insecticides, and fire retardants (fig. 18.11). All these substances were found, usually in low concentrations. One stream had 38 of the compounds tested. Drinking water standards exist for only 14 of the 95 substances. A similar study found the same substances in groundwater, which is much harder to clean than surface waters. What are the effects of these widely used chemicals, on our environment or on people consuming the water? Nobody knows. This study is a first step toward filling huge gaps in our knowledge about their distribution, though.

Sediment also degrades water quality

Rivers have always carried sediment to the oceans, but erosion rates in many areas have been greatly accelerated by human activities. Some rivers carry astounding loads of sediment. Erosion and runoff from croplands contribute about 25 billion metric tons of soil, sediment, and suspended solids to world surface waters each year. Forests, grazing lands, urban construction sites, and other sources of erosion and runoff add at least 50 billion additional tons. This sediment fills lakes and reservoirs, obstructs shipping channels, clogs hydroelectric turbines, and makes purification of drinking water more costly. Sediments smother gravel beds in which insects take refuge and fish lay their eggs. Sunlight is blocked so that plants cannot carry out photosynthesis, and oxygen levels decline. Murky, cloudy water also is less attractive for swimming, boating, fishing, and other recreational uses (fig. 18.12).

FIGURE 18.10 The deformed beak of this young robin is thought to be due to dioxins, DDT, and other toxins in its mother's diet.

FIGURE 18.12 A plume of sediment and industrial waste flows from this drainage canal into Lake Erie.

Sediment also can be beneficial. Mud carried by rivers nourishes floodplain farm fields. Sediment deposited in the ocean at river mouths creates valuable deltas and islands. The Ganges River, for instance, builds up islands in the Bay of Bengal that are eagerly colonized by land-hungry people of Bangladesh. Louisiana's coastal wetlands require constant additions of sediment from the muddy Mississippi to counteract coastal erosion. These wetlands are now disappearing at a disastrous rate: Levees now channel the river and its load out into the Gulf of Mexico, where sediments are dumped beyond the continental shelf (see chapter 13).

Thermal pollution is dangerous for organisms

Raising or lowering water temperatures from normal levels can adversely affect water quality and aquatic life. Water temperatures are usually much more stable than air temperatures, so aquatic organisms tend to be poorly adapted to rapid temperature changes. Lowering the temperature of tropical oceans by even one degree can be lethal to some corals and other reef species. Raising water temperatures can have similar devastating effects on sensitive organisms. Oxygen solubility in water decreases as temperatures increase, so species requiring high oxygen levels are adversely affected by warming water.

Humans cause thermal pollution by altering vegetation cover and runoff patterns, as well as by discharging heated water directly into rivers and lakes.

The cheapest way to remove heat from an industrial facility is to draw cool water from an ocean, river, lake, or aquifer, run it through a heat-exchanger to extract excess heat, and then dump the heated water back into the original source. A **thermal plume** of heated water is often discharged into rivers and lakes, where raised temperatures can disrupt many processes in natural ecosystems and drive out sensitive organisms. Nearly half the water we withdraw is used for industrial cooling. Electric power plants, metal smelters, petroleum refineries, paper mills, food-processing factories, and chemical manufacturing plants all use and release large amounts of cooling water.

To minimize thermal pollution, power plants frequently are required to construct artificial cooling ponds or cooling towers in which heat is released into the atmosphere and water is cooled before being released into natural water bodies.

Some species find thermal pollution attractive. Warm water plumes from power plants often attract fish, birds, and marine mammals that find food and refuge there, especially in cold weather. This artificial environment can be a fatal trap, however. Organisms dependent on the warmth may die if they leave the plume or if the flow of warm water is interrupted by a plant shutdown. Endangered manatees in Florida, for example, are attracted to the abundant food and warm water in power plant thermal plumes and are enticed into spending the winter much farther north than they normally would. On several occasions a midwinter power plant breakdown has exposed a dozen or more of these rare animals to a sudden thermal shock that they could not survive.

18.3 WATER QUALITY TODAY

Surface-water pollution is often both highly visible and one of the most common threats to environmental quality. In more developed countries, reducing water pollution has been a high priority over the past few decades. Billions of dollars have been spent on control programs, and considerable progress has been made. Still much remains to be done. In developed countries, poor water quality often remains a serious problem. In this section we will look at progress as well as continuing obstacles in this important area.

The Clean Water Act protects our water

As the opening case study for this chapter shows, the United States and Canada, like most developed countries, have made encouraging progress in protecting and restoring water quality in rivers and lakes over the past 40 years. In 1948, only about one-third of Americans were served by municipal sewage systems, and most of those systems discharged sewage without any treatment or with only primary treatment (the bigger lumps of waste are removed). Most people depended on cesspools and septic systems to dispose of domestic wastes.

The 1972 Clean Water Act established a National Pollution Discharge Elimination System (NPDES), which requires an easily revoked permit for any industry, municipality, or other entity dumping wastes in surface waters. The permit requires disclosure of what is being dumped and gives regulators valuable data and evidence for litigation. As a consequence, only about 10 percent of our water pollution now comes from industrial or municipal point sources. One of the biggest improvements has been in sewage treatment.

Since the Clean Water Act was passed in 1972, the United States has spent more than $180 billion in public funds and perhaps ten times as much in private investments on water pollution control. Most of that effort has been aimed at point sources, especially to build or upgrade thousands of municipal sewage treatment plants. As a result, nearly everyone in urban areas is now served by municipal sewage systems and no major city discharges raw sewage into a river or lake except as overflow during heavy rainstorms.

This campaign has led to significant improvements in surface-water quality in many places. Fish and aquatic insects have returned to waters that formerly were depleted of life-giving oxygen. Swimming and other water-contact sports are again permitted in rivers, lakes, and at ocean beaches that once were closed by health officials.

The Clean Water Act goal of making all U.S. surface waters "fishable and swimmable" hasn't been fully met, but in 1999 the EPA reported that 91.4 percent of all monitored river miles and 87.5 percent of all assessed lake acres are suitable for their designated uses (fig. 18.13). This sounds good, but remember that not all water bodies are monitored. Furthermore, the designated goal for some rivers and lakes is merely to be "boatable." Water quality doesn't have to be very high to be able to put a boat in it. Even in "fishable" rivers and lakes, there isn't a guarantee that you can catch anything other than rough fish like carp or bullheads, nor can you be sure that what you catch is safe to eat. Even with billions of dollars of investment in sewage treatment plants, elimination of much of the industrial dumping and other gross sources of pollutants, and a general improvement in water quality, the EPA reports that 21,000 water bodies still do not meet their designated uses. According to the EPA, an overwhelming majority of the American people—almost 218 million—live within 16 km (10 mi) of an impaired water body.

In 1998 a new regulatory approach to water quality assurance was instituted by the EPA. Rather than issue standards on a river-by-river approach or factory-by-factory permit discharge, the focus is being changed to watershed-level monitoring and protection. Some 4,000 watersheds are monitored for water quality. You can find information about your watershed at www.epa.gov/owow/tmdl/. The intention of this program is to give the public more and better information about the health of their watersheds. In addition, states will have greater flexibility as they identify impaired water bodies and set priorities, and new tools will be used to achieve goals. States are required to identify waters not meeting water quality goals and to develop **total maximum daily loads (TMDL)** for each pollutant and each listed water body. A TMDL is the amount of a particular pollutant that a water body can receive from both point and nonpoint sources. It considers seasonal variation and includes a margin of safety.

By 1999 all 56 states and territories had submitted TMDL lists, and the EPA had approved most of them. Of the 3.5 million mi (5.6 million km) of rivers monitored, only 300,000 mi (480,000 km) currently fail to meet their clean-water goals. Similarly, of 40 million lake acres (99 million ha), only 12.5 percent (in about 20,000 lakes) failed to meet their goal. To give states more flexibility in planning, the EPA has proposed new rules that include allowances for reasonably foreseeable increases in pollutant loadings to encourage "Smart Growth." In the future, TMDLs also will include load allocations from all nonpoint sources, including air deposition and natural background levels.

An encouraging example of improved water quality is seen in Lake Erie. Although widely regarded as "dead" in the 1960s, the lake today is promoted as the "walleye capital of the world." Bacteria counts and algae blooms have decreased more than 90 percent since 1962. Water that once was murky brown is now clear. Ironically, the improved water quality is partly due to immense numbers of invasive zebra mussels, which filter the lake water very efficiently. Swimming is now officially safe along 96 percent of the lake's shoreline. Nearly 40,000 breeding pairs of double-crested cormorants nest in the Great Lakes region, up from only about 100 in the 1970s. Anglers now complain that the cormorants eat too many fish. In 1998 wildlife agents found 800 cormorants shot to death in a rookery on Galloo Island at the east end of Lake Ontario.

The importance of a single word

When the Clean Water Act was passed in 1972, it protected "navigable" waterways. For 30 years the EPA interpreted that to include the tributary streams, wetlands, ponds, and other water sources of navigable rivers. A Michigan shopping-center developer challenged this interpretation, however, when he filled in a wetland without getting a federal permit. The case went to the U.S. Supreme Court, which ruled in 2006 that the law protected only water bodies with a "significant nexus" to navigable streams. The EPA dropped enforcement actions on at least 1,500 water pollution cases (about half of which involved oil spills). And it announced that the act no longer covered millions of acres of wetlands, ponds, tributary streams, or intermittent desert rivers. About 117 million Americans get their drinking water from sources fed by waters that are vulnerable to exclusion from the Clean Water Act, according to EPA reports. And about half of all water pollution may be beyond regulatory reach following this ruling. Attempts have been made to remove the word "navigable" from the Clean Water Act, but passage looks doubtful with Republicans in control of the House of Representatives.

Water quality problems remain

The greatest impediments to achieving national goals in water quality in both the United States and Canada are sediment, nutrients, and pathogens (fig. 18.14), especially from nonpoint discharges of pollutants. These sources are harder to identify and to reduce

FIGURE 18.13 Not all rivers and lakes are "fishable or swimmable," but we've made substantial progress since the Clean Water Act was passed in 1972.

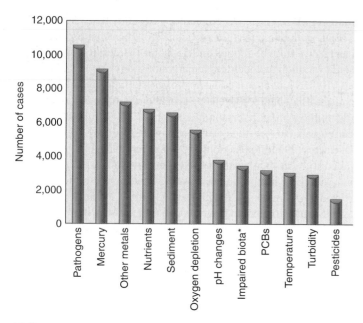

FIGURE 18.14 Twelve leading causes of surface-water impairment in the United States. *Undetermined causes.
Source: Data EPA, 2009.

or treat than are specific point sources. About three-fourths of the water pollution in the United States comes from soil erosion, fallout of air pollutants, and surface runoff from urban areas, farm fields, and feedlots. In the United States, as much as 25 percent of the 46,800,000 metric tons (52 million tons) of fertilizer spread on farmland each year is carried away by runoff.

Cattle in feedlots produce some 129,600,000 metric tons (144 million tons) of manure each year, and the runoff from these sites is rich in viruses, bacteria, nitrates, phosphates, and other contaminants. A single cow produces about 30 kg (66 lb) of manure per day, or about as much as that produced by ten people. Some feedlots have 100,000 animals with little provision for capturing or treating runoff water. Imagine drawing your drinking water downstream from such a facility. Pets also can be a problem. It is estimated that the wastes from about a half million dogs in New York City are disposed of primarily through storm sewers, and therefore do not go through sewage treatment.

Loading of both nitrates and phosphates in surface water has decreased from point sources but has increased about fourfold since 1972 from nonpoint sources. Fossil fuel combustion has become a major source of nitrates, sulfates, arsenic, cadmium, mercury, and other toxic pollutants that find their way into water. Carried to remote areas by atmospheric transport, these combustion products now are found nearly everywhere in the world. Toxic organic compounds, such as DDT, PCBs, and dioxins, also are transported long distances by wind currents.

Other countries also have serious water pollution

Japan, Australia, and most of western Europe also have improved surface-water quality in recent years. Sewage treatment in the wealthier countries of Europe generally equals or surpasses that in the United States. Sweden, for instance, serves 98 percent of its population with at least secondary sewage treatment (compared with 70 percent in the United States), and the other 2 percent have primary treatment. Poorer countries have much less to spend on sanitation. Spain serves only 18 percent of its population with even primary sewage treatment. In Ireland, it is only 11 percent, and in Greece, less than 1 percent of the people have even primary treatment. Most of the sewage, both domestic and industrial, is dumped directly into the ocean.

The fall of the "iron curtain" in 1989 revealed appalling environmental conditions in much of the former Soviet Union and its satellite states in eastern and central Europe. The countries closest geographically and socially to western Europe, the Czech Republic, Hungary, East Germany, and Poland, have made massive investments and encouraging progress toward cleaning up environmental problems. Parts of Russia itself, however, along with former socialist states in the Balkans and Central Asia, remain some of the most polluted places on earth. In Russia, for example, only about half the tap water is fit to drink. In cities like St. Petersburg, even boiling and filtering isn't enough to make municipal water safe. As we saw in chapter 17, at least 200 million Chinese live in areas without sufficient fresh water. Sadly, pollution makes much of the limited water unusable (fig. 18.15). It's estimated that 70 percent of China's surface water is unsafe for human consumption, and that the water in half the country's major rivers is so contaminated that it's unsuited for any use, even agriculture.

Economic growth has been pursued in recent decades at the expense of environmental quality. According to the Chinese Environmental Protection Agency, the country's ten worst polluted cities are all in Shanxi Province. Factories have been allowed to exceed pollution discharges with impunity. For example, 3 million tons of wastewater are produced every day in the province, with two-thirds of it discharged directly into local rivers without any treatment. Locals complain that the rivers, which once were clean and fresh, now run black with industrial waste. Among the

FIGURE 18.15 Half the water in China's major rivers is too polluted to be suitable for any human use. Although the government has spent billions of yuan in recent years, dumping of industrial and domestic waste continues at dangerous levels.

26 rivers in the province, 80 percent were rated Grade V (unfit for any human use) or higher in 2006. More than half the wells in Shanxi are reported to have dangerously high arsenic levels. Many of the 85,000 reported public protests in China in 2006 involved complaints about air and water pollution.

There are also some encouraging pollution-control stories. In 1997, Japan's Minamata Bay, long synonymous with mercury poisoning, was declared officially clean again. Another important success is found in Europe, where one of its most important rivers has been cleaned up significantly through international cooperation. The Rhine, which starts in the rugged Swiss Alps and winds 1,320 km through five countries before emptying through a Dutch delta into the North Sea, has long been a major commercial artery into the heart of Europe. More than 50 million people live in its catchment basin, and nearly 20 million get their drinking water from the river or its tributaries. By the 1970s the Rhine had become so polluted that dozens of fish species disappeared and swimming was discouraged along most of its length.

Efforts to clean up this historic and economically important waterway began in the 1950s, but a disastrous fire at a chemical warehouse near Basel, Switzerland, in 1986 provided the impetus for major changes. Through a long and sometimes painful series of international conventions and compromises, land-use practices, waste disposal, urban runoff, and industrial dumping have been changed and water quality has significantly improved. Oxygen concentrations have gone up fivefold since 1970 (from less than 2 mg/l to nearly 10 mg/l or about 90 percent of saturation) in long stretches of the river. Chemical oxygen demand has fallen fivefold during this same period, and organochlorine levels have decreased as much as tenfold. Many species of fish and aquatic invertebrates have returned to the river. In 1992, for the first time in decades, mature salmon were caught in the Rhine.

Most of the poorest countries of South America, Africa, and Asia have disastrous water quality (fig. 18.16). In most areas sewage treatment is either totally lacking or woefully inadequate. In some urban areas, 95 percent of all sewage is discharged untreated into rivers, lakes, or the ocean. Low technological capabilities and little money for pollution control are made even worse by burgeoning populations, rapid urbanization, and the shift of much heavy industry (especially the dirtier ones) from developed countries where pollution laws are strict to less-developed countries where regulations are more lenient.

Two-thirds of India's surface water, for example, is so contaminated that even coming into contact with it is considered dangerous to human health. Hundreds of millions of people drink and bathe in this water. Consider the Yamuna River, which flows through New Delhi and past the magnificent Taj Mahal in Agra. About 57 million people depend on the Yamuna for agriculture, domestic, and industrial use. Much of the runoff from these activities goes back into the river either untreated or only partially cleaned. In New Delhi, for example, only about half the 15 million residents have access to the municipal wastewater treatment system, which doesn't have enough capacity to treat the sewage it does collect. During the dry season the Yamuna's flow as it leaves the city is reduced to a trickle, 80 percent of which is sewage and

FIGURE 18.16 Ditches in this Haitian slum serve as open sewers into which all manner of refuse and waste are dumped. The health risks of living under these conditions are severe.

industrial effluent. Coliform bacterial counts can be millions of times the level considered safe for drinking or bathing. Although the Indian government has spent more than (U.S.) $500 million in recent years to upgrade the sewage system, the problem has been made worse rather than better by urban sprawl, a rapidly growing economy, and ineffective administration.

For a decade Indian environmental scientists have urged the government to take a new approach to reducing pollution in both the Yamuna and the sacred Ganges, into which it flows. Rather than build more mechanical sewage treatment plants that depend on often-unreliable electrical supplies, they suggest living systems in which sewage flows by gravity into local, low-cost, artificial wetlands. Treating wastes with aquatic plants and solar oxidation, they argue, will be both cheaper and more effective than current Western-style processes. In 2008 the Varanasi city government agreed to start a pilot project to test this biological system.

Similarly, China has announced plans to spend at least (U.S.) $125 billion over the next five years to reduce water pollution and bring clean drinking water to everyone in the country. Already there are indications of success (fig. 18.17).

Groundwater is hard to monitor and clean

About half the people in the United States, including 95 percent of those in rural areas, depend on underground aquifers for their drinking water. This vital resource is threatened in many areas by overuse and pollution and by a wide variety of industrial, agricultural, and domestic contaminants. For decades it was widely assumed that groundwater was impervious to pollution because soil would bind chemicals and cleanse water as it percolated through. Springwater or artesian well water was considered to be the definitive standard of water purity, but that is no longer true in many areas.

One of the serious sources of groundwater pollution throughout the United States is MTBE (methyl tertiary butyl ether), a suspected carcinogen. MTBE is a gasoline additive that has been used since the 1970s to reduce the amount of carbon monoxide and unburned

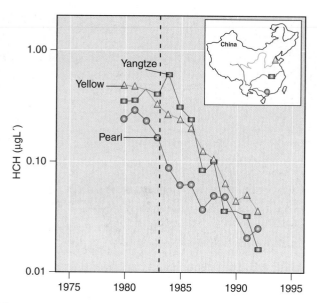

FIGURE 18.17 Annual average concentrations of hexachlorocyclohexane (HCH) in three Chinese rivers after its use was banned in 1983. This chemical is the main ingredient of the insecticide Lindane.

hydrocarbons in vehicle exhaust. By the time the health dangers of MTBE were confirmed in the late 1990s, aquifers across the country had been contaminated—mainly from leaking underground storage tanks at gas stations. Nationwide about 250,000 of these tanks are leaking MTBE into groundwater. In one U.S. Geological Survey (USGS) study, 27 percent of tested shallow urban wells contained MTBE. The additive is being phased out, but plumes of tainted water will continue to move through aquifers for decades to come. (Surface waters have also been contaminated, especially by two-stroke engines, such as those on personal watercraft.)

Treating MTBE-laced aquifers is expensive but not impossible. Douglas MacKay of the University of Waterloo in Ontario suggests that if oxygen could be pumped into aquifers, then naturally occurring bacteria could metabolize (digest) the compound. It could take decades or even centuries for natural bacteria to eliminate MTBE from a water supply, however. Water can also be pumped out of aquifers, reducing the flow and spread of contamination. Thus far little funding has been invested in finding cost-effective remedies, however.

The U.S. EPA estimates that every day some 4.5 trillion liters (1.2 trillion gal) of contaminated water seep into the ground in the United States from septic tanks, cesspools, municipal and industrial landfills and waste disposal sites, surface impoundments, agricultural fields, forests, and wells (fig. 18.18). The most toxic of these are probably waste disposal sites. Agricultural chemicals and wastes are responsible for the largest total volume of pollutants and area affected. Because deep underground aquifers often have residence times of thousands of years, many contaminants are extremely stable once underground. It is possible, but expensive, to pump water out of aquifers, clean it, and then pump it back.

In farm country, especially in the Midwest's corn belt, fertilizers and pesticides commonly contaminate aquifers and wells. Herbicides such as atrazine and alachlor are widely used on corn and soybeans and show up in about half of all wells in Iowa, for example. Nitrates from fertilizers often exceed safety standards in rural drinking water. These high nitrate levels are dangerous to infants (nitrate combines with hemoglobin in the blood and results in "blue-baby"

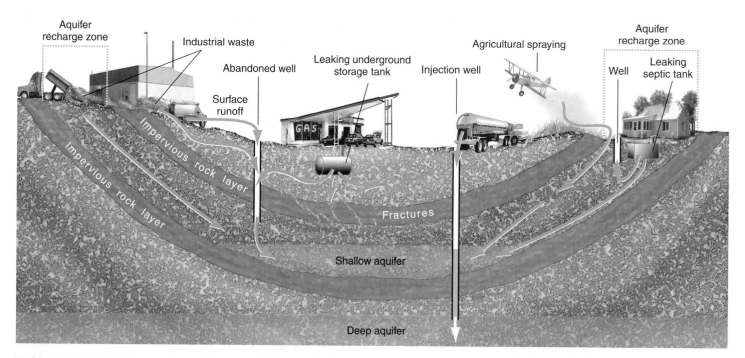

FIGURE 18.18 Sources of groundwater pollution. Septic systems, landfills, and industrial activities on aquifer recharge zones leach contaminants into aquifers. Wells provide a direct route for injection of pollutants into aquifers.

syndrome). They also are transformed into cancer-causing nitrosamines in the human gut. In Florida, 1,000 drinking water wells were shut down by state authorities because of excessive levels of toxic chemicals, mostly ethylene dibromide (EDB), a pesticide used to kill nematodes (roundworms) that damage plant roots.

Although most of the leaky, single-walled underground storage tanks once common at filling stations and factories have now been removed and replaced by more modern ones, a great deal of soil in American cities remains contaminated by previous careless storage and disposal of petroleum products. Considering that a single gallon (3.8 l) of gasoline can make a million gallons of water undrinkable, soil contamination remains a serious problem.

In addition to groundwater pollution problems, contaminated surface waters and inadequate treatment make drinking water unsafe in many areas. Data collected by the EPA in 2008 show that about 30 million people get water from community systems that don't meet all health-based drinking water standards. Most of these systems are small, serving fewer than 3,000 customers. Problems often arise because small systems can't afford modern purification and distribution equipment, regular testing, and trained operators to bring water quality up to acceptable standards.

Every year epidemiologists estimate that around 1.5 million Americans fall ill from infections caused by fecal contamination. In 1993, for instance, the pathogen cryptosporidium got into the Milwaukee public water system, making 400,000 people sick and killing at least 100 people. The total costs of these diseases amount to billions of dollars per year. Preventive measures such as protecting water sources and aquifer recharge zones, providing basic treatment for all systems, installing modern technology and distribution networks, consolidating small systems, and strengthening the Clean Water Act and the Safe Drinking Water Act would cost far less. Unfortunately, in the present climate of budget cutting and anti-regulation, these steps seem unlikely.

There are few controls on ocean pollution

Coastal zones, especially bays, estuaries, shoals, and reefs near large cities or the mouths of major rivers, often are overwhelmed by human-caused contamination. Suffocating and sometimes poisonous blooms of algae regularly deplete ocean waters of oxygen and kill enormous numbers of fish and other marine life. High levels of toxic chemicals, heavy metals, disease-causing organisms, oil, sediment, and plastic refuse are adversely affecting some of the most attractive and productive ocean regions. The potential losses caused by this pollution amount to billions of dollars each year.

Discarded plastic flotsam and jetsam are lightweight and non-biodegradable. They are carried thousands of miles on ocean currents and last for years (fig. 18.19). Even the most remote beaches of distant islands are likely to have bits of polystyrene foam containers or polyethylene packing material that were discarded half a world away. It has been estimated that some 6 million metric tons of plastic bottles, packaging material, and other litter are tossed from ships every year into the ocean, where they ensnare and choke seabirds, mammals (fig. 18.20), and even fish. For further discussion of ocean pollution and the Great

FIGURE 18.19 Beach pollution, including garbage, sewage, and contaminated runoff, is a growing problem associated with ocean pollution.

Pacific garbage patch, see chapter 21. In one day, volunteers in Texas gathered more than 300 tons of plastic refuse from Gulf Coast beaches.

Few coastlines in the world remain uncontaminated by oil or oil products. Oceanographers estimate that 3 to 6 million metric tons of oil are discharged into the world's oceans each year from both land- and sea-based operations. About half of this amount is due to maritime transport. Most oil spills result not from catastrophic, headliner accidents but from routine open-sea bilge pumping and tank cleaning. These procedures are illegal but are easily carried out once ships are beyond sight of land. Much of the rest comes from land-based municipal and industrial runoff or from atmospheric deposition of residues from refining and combustion of fuels.

Our addiction to oil creates increasing risks for further major oil spills. Increased transport from distant locations combined with ultra-deep ocean drilling (see chapter 19 for further discussion), particularly in risky locations, such as the Arctic Ocean, make it likely that more oil spills will occur. Plans to drill for oil along the

FIGURE 18.20 A deadly necklace. Marine biologists estimate that castoff nets, plastic beverage yokes, and other packing residue kill hundreds of thousands of birds, mammals, and fish each year.

seismically active California and Alaska coasts have been controversial because of the damage that oil spills could cause to these biologically rich coastal ecosystems.

18.4 WATER POLLUTION CONTROL

Appropriate land-use practices and careful disposal of industrial, domestic, and agricultural wastes are essential for control of water pollution.

Source reduction is often the cheapest and best way to reduce pollution

The cheapest and most effective way to reduce pollution is usually to avoid producing it or releasing it to the environment in the first place. Elimination of lead from gasoline has resulted in a widespread and significant decrease in the amount of lead in surface waters in the United States. Studies have shown that as much as 90 percent less road-deicing salt can be used in many areas without significantly affecting the safety of winter roads. Careful handling of oil and petroleum products can greatly reduce the amount of water pollution caused by these materials. Although we still have problems with persistent chlorinated hydrocarbons spread widely in the environment, the banning of DDT and PCBs in the 1970s has resulted in significant reductions in levels in wildlife.

Modified agricultural practices in headwater streams in the Chesapeake Bay watershed and the Catskill Mountains of New York have had positive and cost-effective impacts on downstream water quality (What Do You Think? p. 413).

Industry can reduce pollution by recycling or reclaiming materials that otherwise might be discarded in the waste stream. Both of these approaches usually have economic as well as environmental benefits. It turns out that a variety of valuable metals can be recovered from industrial wastes and reused or sold for other purposes. The company benefits by having a product to sell, and the municipal sewage treatment plant benefits by not having to deal with highly toxic materials mixed in with millions of gallons of other types of wastes.

Controlling nonpoint sources requires land management

Among the greatest remaining challenges in water pollution control are diffuse, nonpoint pollution sources. Unlike point sources, such as sewer outfalls or industrial discharge pipes, which represent both specific locations and relatively continuous emissions, nonpoint sources have many origins and numerous routes by which contaminants enter ground and surface waters. It is difficult to identify—let alone monitor and control—all these sources and routes. Some main causes of nonpoint pollution are:

- *Agriculture:* The EPA estimates that 60 percent of all impaired or threatened surface waters are affected by sediment from eroded fields and overgrazed pastures; fertilizers, pesticides, and nutrients from croplands; and animal wastes from feedlots.

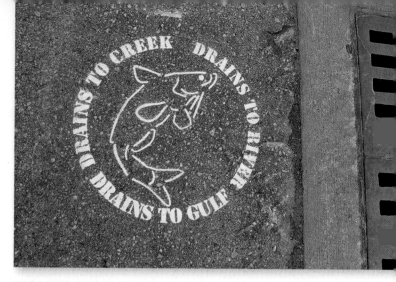

FIGURE 18.21 People often dump waste oil and other pollutants into street drains without thinking about where their wastes go. Painting reminders, such as this one, is a good project for students and youth groups.

- *Urban runoff:* Pollutants carried by runoff from streets, parking lots, and industrial sites contain salts, oily residues, rubber, metals, and many industrial toxins (fig. 18.21). Yards, golf courses, parklands, and urban gardens often are treated with far more fertilizers and pesticides per unit area than farmlands. Excess chemicals are carried by storm runoff into waterways.
- *Construction sites:* New buildings and land development projects such as highway construction affect relatively small areas but produce vast amounts of sediment, typically 10 to 20 times as much per unit area as farming.
- *Land disposal:* When done carefully, land disposal of certain kinds of industrial waste, sewage sludge, and biodegradable garbage can be a good way to dispose of unwanted materials. Some poorly run land disposal sites, abandoned dumps, and leaking septic systems, however, contaminate local waters.

Generally soil conservation methods (chapter 9) also help protect water quality. Applying precisely determined amounts of fertilizer, irrigation water, and pesticides saves money and reduces contaminants entering the water. Preserving wetlands that act as natural processing facilities for removing sediment and contaminants helps protect surface and groundwaters.

In urban areas, reducing materials carried away by storm runoff is helpful. Citizens can be encouraged to recycle waste oil and to minimize use of fertilizers and pesticides. Regular street sweeping greatly reduces contaminants. Runoff can be diverted away from streams and lakes. Many cities are separating storm sewers and municipal sewage lines to avoid overflow during storms.

A good example of watershed management is seen in Chesapeake Bay, the United States' largest estuary. Once fabled for its abundant oysters, crabs, shad, striped bass, and other valuable fisheries, the Bay had deteriorated seriously by the early 1970s. Citizens' groups, local communities, state legislatures, and the federal government together established an innovative pollution-control program that made the bay the first estuary in America targeted for protection and restoration (see chapter 3).

What Do You Think?

Watershed Protection in the Catskills

New York City has long been proud of its excellent municipal drinking water. Drawn from the rugged Catskill Mountains 100 km (60 mi) north of the city, stored in hard-rock reservoirs, and transported through underground tunnels, the city water is outstanding for so large an urban area. Yielding 450,000 m³ (1.2 billion gal) per day, and serving more than 9 million people, this is the largest surface-water storage and supply complex in the world (see map). As the metropolitan agglomeration has expanded, however, people have moved into the area around the Catskill Forest Preserve, and water quality is not as high as it was a century ago.

When the 1986 U.S. Safe Drinking Water Act mandated filtration of all public surface-water systems, the city was faced with building an $8 billion water treatment plant that would cost up to $500 million per year to operate. In 1989, however, the EPA ruled that the city could avoid filtration if it could meet certain minimum standards for microbial contaminants such as bacteria, viruses, and protozoan parasites. In an attempt to avoid the enormous cost of filtration, the city proposed land-use regulations for the five counties (Green, Ulster, Sullivan, Schoharie, and Delaware) in the Catskill/Delaware watershed from which it draws most of its water.

With a population of 50,000 people, the private land within the 520 km² (200 mi²) watershed is mostly devoted to forestry, small farms, housing, and recreational activities. Among the changes the city called for was elimination of storm water runoff from barnyards, feedlots, or grazing areas into watersheds. In addition, farmers would be required to reduce erosion and surface runoff from crop fields and logging operations. Property owners objected strenuously to what they regarded as onerous burdens that would cost enough to put many of them out of business. They also bristled at having the huge megalopolis impose rules on them. It looked like a long and bitter battle would be fought through the courts and the state legislature.

To avoid confrontation, a joint urban/rural task force was set up to see if a compromise could be reached, and to propose alternative solutions to protect both the water supply and the long-term viability of agriculture in the region. The task force agreed that agriculture is the "preferred land use" on private land, and that agriculture has "significant present and future environmental benefits." In addition, the task force proposed a voluntary, locally developed and administered program of "whole farm planning and best management approaches" very similar to ecosystem-based, adaptive management (chapter 9).

This grassroots program, financed mainly by the city but administered by local farmers themselves, attempts to educate landowners and provides alternative marketing opportunities that help protect the watershed. Economic incentives are offered to encourage farmers and foresters to protect the water supply. Collecting feedlot and barnyard runoff in infiltration ponds together with solid conservation practices such as terracing, contour plowing, strip farming, leaving crop residue on fields, ground cover on waterways, and cultivation of perennial crops such as orchards and sugarbush have significantly improved watershed water quality. As of 1999 about 400 farmers—close to the 85 percent participation goal—have signed up for the program. The cost, so far, to the city has been about $50 million—or less than 1 percent of constructing a treatment plant.

Although landowners often object to any restrictions on development, many in the Catskills have found that land-use rules also protect rural lifestyles. Protection of the forests and waters has also helped the area retain a recreational economy and regional identity. Watershed management saved New York billions of dollars; it can also save traditional land uses and livelihoods.

What do you think? Are land-use restrictions a reasonable approach for saving on water treatment? How much should cities pay for watershed protection?

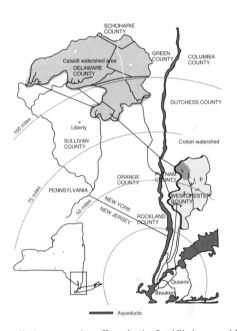

Investment in pollution prevention efforts in the Catskills has saved New York City billions of dollars in water filtration costs.

One of the principal objectives of this plan is reducing nutrient loading through land-use regulations in the six watershed states to control agricultural and urban runoff. Pollution prevention measures such as banning phosphate detergents also are important, as are upgrading wastewater treatment plants and improving compliance with discharge and filling permits. Efforts are under way to replant thousands of hectares of seagrasses and to restore wetlands that filter out pollutants. Since the 1980s, annual phosphorous discharges into the bay dropped 40 percent. Nitrogen levels, however, have remained constant or have even risen in some tributaries. Although progress has been made, the goals of reducing both nitrogen and phosphate levels by 40 percent and restoring viable fish and shellfish populations are still decades away. Still, as former EPA administrator Carol Browner says, it demonstrates the "power of cooperation" in environmental protection.

Human waste disposal occurs naturally when concentrations are low

As we have already seen, human and animal wastes usually create the most serious health-related water pollution problems. More than 500 types of disease-causing (pathogenic) bacteria, viruses, and parasites can travel from human or animal excrement through water. In this section we will look at how to prevent the spread of these diseases.

Natural Processes

In the poorer countries of the world, most rural people simply go out into the fields and forests to relieve themselves as they have always done. Where population densities are low, natural processes eliminate wastes quickly, making this a feasible method of sanitation. The high population densities of cities make this practice unworkable, however. Even major cities of many less-developed countries are often littered with human waste that has been left for rains to wash away or for pigs, dogs, flies, beetles, or other scavengers to consume. This is a major cause of disease, as well as being extremely unpleasant. Studies have shown that a significant portion of the airborne dust in Mexico City is actually dried, pulverized human feces.

Where intensive agriculture is practiced—especially in wet rice paddy farming in Asia—it has long been customary to collect "night soil" (human and animal waste) to be spread on the fields as fertilizer. This waste is a valuable source of plant nutrients, but it is also a source of disease-causing pathogens in the food supply. It is the main reason that travelers in less-developed countries must be careful to surface sterilize or cook any fruits and vegetables they eat. Collecting night soil for use on farm fields was common in Europe and America until about 100 years ago, when the association between pathogens and disease was recognized.

Until about 70 years ago most rural American families and quite a few residents of towns and small cities depended on a pit toilet or "outhouse" for waste disposal. Untreated wastes tended to seep into the ground, however, and pathogens sometimes contaminated drinking water supplies. The development of septic tanks and properly constructed drain fields considerably improved public health (fig. 18.22). In a typical septic system, wastewater is first drained into a septic tank. Grease and oils rise to the top and solids settle to the bottom, where they are subject to bacterial decomposition. The clarified effluent from the septic tank is channeled out through a drainfield of small perforated pipes embedded in gravel just below the surface of the soil. The rate of aeration is high in this drainfield so that pathogens (most of which are anaerobic) will be killed, and soil microorganisms can metabolize any nutrients carried by the water. Excess water percolates up through the gravel and evaporates. Periodically the solids in the septic tank are pumped out into a tank truck and taken to a treatment plant for disposal.

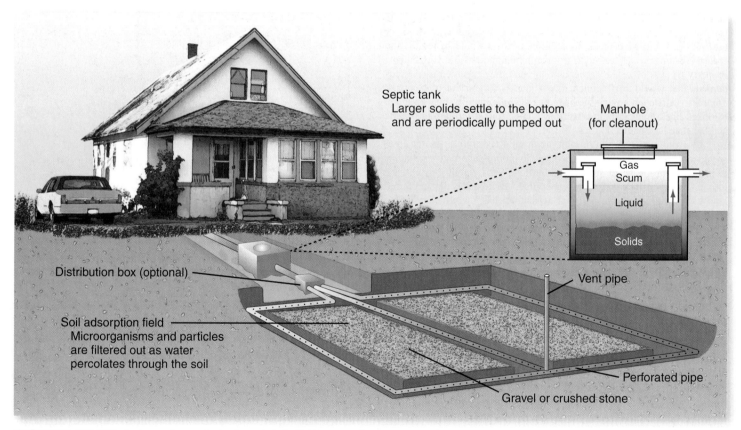

FIGURE 18.22 A domestic septic tank and drain field system for sewage and wastewater disposal. To work properly, a septic tank must have healthy microorganisms, which digest toilet paper and feces. For this reason, antimicrobial cleaners and chlorine bleach should never be allowed down the drain.

Where land is available and population densities are not too high, this can be an effective method of waste disposal. It is widely used in rural areas, but aging, leaky septic systems can be a huge cumulative problem. As chapter 13 points out, the Chesapeake Bay watershed has 420,000 individual septic systems, which constitute a major source of nutrients. Maryland alone plans to spend $7.5 million annually to upgrade failing septic systems.

Municipal Sewage Treatment

Over the past 100 years, sanitary engineers have developed ingenious and effective municipal wastewater treatment systems to protect human health, ecosystem stability, and water quality. This topic is an important part of pollution control, and is a central focus of every municipal government; therefore, let's look more closely at how a typical municipal sewage treatment facility works.

Primary treatment is the first step in municipal waste treatment. It physically separates large solids from the waste stream. As raw sewage enters the treatment plant, it passes through a metal grating that removes large debris. A moving screen then filters out smaller items. Brief residence in a grit tank allows sand and gravel to settle. The waste stream then moves to the primary sedimentation tank, where about half the suspended organic solids settle to the bottom as sludge. Many pathogens remain in the effluent, which is not yet safe to discharge into waterways or onto the ground.

Secondary treatment consists of biological degradation of the dissolved organic compounds. The effluent from primary treatment flows into a trickling filter bed, an aeration tank, or a sewage lagoon. The trickling filter is simply a bed of stones or corrugated plastic sheets through which water drips from a system of perforated pipes or a sweeping overhead sprayer. Bacteria and other microorganisms in the bed catch organic material as it trickles past and aerobically decompose it.

Aeration tank digestion is also called the activated sludge process. Effluent from primary treatment is pumped into the tank and mixed with a bacteria-rich slurry. Air pumped through the mixture encourages bacterial growth and decomposition of the organic material (fig. 18.23). Water flows from the top of the tank, and sludge is removed from the bottom. Some of the sludge is used as an inoculum for incoming primary effluent. The remainder would be valuable fertilizer if it were not contaminated by metals, toxic chemicals, and pathogenic organisms. The toxic content of most sewer sludge necessitates disposal by burial in a landfill or incineration. Sludge disposal is a major cost in most municipal sewer budgets (fig. 18.24). In some communities this is accomplished by land farming, composting, or anaerobic digestion, but these methods don't inactivate metals and some other toxic materials.

Where space is available for sewage lagoons, the exposure to sunlight, algae, aquatic organisms, and air does the same job more slowly but with less energy cost. Effluent from secondary treatment processes is usually disinfected with chlorine, UV light, or ozone to kill harmful bacteria before it is released to a nearby waterway.

Tertiary treatment removes plant nutrients, especially nitrates and phosphates, from the secondary effluent. Although wastewater

(a)

(b)

FIGURE 18.23 In conventional sewage treatment, aerobic bacteria digest organic materials in trickling filter beds in high-pressure aeration tanks. This is described as secondary treatment.

is usually free of pathogens and organic material after secondary treatment, it still contains high levels of inorganic nutrients, such as nitrates and phosphates. When discharged into surface waters, these nutrients stimulate algal blooms and eutrophication. To preserve water quality, these nutrients also must be removed. Passage through a wetland or lagoon can accomplish this. Alternatively, chemicals often are used to bind and precipitate nutrients.

In many American cities, sanitary sewers are connected to storm sewers, which carry runoff from streets and parking lots. Storm sewers are routed to the treatment plant rather than discharged into surface waters, because runoff from streets, yards, and industrial sites generally contains a variety of refuse, fertilizers, pesticides, oils, rubber, tars, lead (from gasoline), and other undesirable chemicals. During dry weather this plan works well.

FIGURE 18.24 "Well, if *you* can't use it, do you know anyone who *can* use 3,000 tons of sludge every day?"
Source: ScienceCartoonsplus.com.

Heavy storms often overload the system, however, causing bypass dumping of large volumes of raw sewage and toxic surface runoff directly into receiving waters. To prevent this overflow, cities are spending hundreds of millions of dollars to separate storm and sanitary sewers. These are huge, disruptive projects. When they are finished, surface runoff will be diverted into a river or lake and cause another pollution problem.

Low-Cost Waste Treatment

The municipal sewage systems used in developed countries are often too expensive to build and operate in the developing world where low-cost, low-tech alternatives for treating wastes are needed. One option is **effluent sewerage**, a hybrid between a traditional septic tank and a full sewer system. A tank near each dwelling collects and digests solid waste just like a septic system. Rather than using a drainfield, however, to dispose of liquids—an impossibility in crowded urban areas—effluents are pumped to a central treatment plant. The tank must be emptied once a year or so, but because only liquids are treated by the central facility, pipes, pumps, and treatment beds can be downsized and the whole system is much cheaper to build and run than a conventional operation.

Another alternative is to use natural or artificial wetlands to dispose of wastes. Constructed wetlands can cut secondary treatment costs to one-third of mechanical treatment costs, or less. Variations on this design are now operating in many places (fig. 18.25). Effluent from these operations can be used to irrigate crops or even raise fish for human consumption if care is taken to first destroy pathogens. Usually 20 to 30 days of exposure to sun, air, and aquatic plants is enough to make the water safe. These systems also can make an important contribution to human food

supplies. A 2,500-ha (6,000-acre) waste-fed aquaculture facility in Calcutta, for example, supplies about 7,000 metric tons of fish annually to local markets. The World Bank estimates that some 3 billion people will be without sanitation services by the middle of the next century under a business-as-usual scenario (fig. 18.26). With investments in innovative programs, however, sanitation could be provided to about half those people and a great deal of misery and suffering could be avoided.

Water remediation may involve containment, extraction, or phytoremediation

Remediation means finding remedies for problems. Just as there are many sources for water contamination, there are many ways to clean it up. New developments in environmental engineering are providing promising solutions to many water pollution problems.

Containment methods confine or restrain dirty water or liquid wastes *in situ* (in place) or cap the surface with an impermeable layer to divert surface water or groundwater away from the site and to prevent further pollution. Where pollutants are buried too deeply to be contained mechanically, materials sometimes can be injected to precipitate, immobilize, chelate, or solidify them. Bentonite slurries, for instance, can effectively stabilize liquids in porous substrates. Similarly, straw or other absorbent material is spread on surface spills to soak up contaminants.

Extraction techniques pump out polluted water so it can be treated. Many pollutants can be destroyed or detoxified by chemical reactions that oxidize, reduce, neutralize, hydrolyze, precipitate, or otherwise change their chemical composition. Where chemical techniques are ineffective, physical methods may work. Solvents and other volatile organic compounds, for instance, can be stripped from solution by aeration and then burned in an incinerator. Some contaminants can be removed by semipermeable membranes or

FIGURE 18.25 This constructed wetland purifies water and provides an attractive landscape at the El Monte Sagrado Resort in Taos, NM.

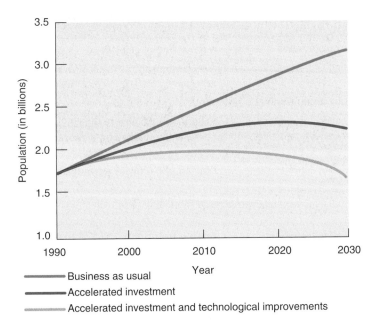

FIGURE 18.26 World population without adequate sanitation—three scenarios in the year 2030. If business as usual continues, more than 3 billion people will lack safe sanitation. Accelerated investment in sanitation services could lower this number. Higher investment, coupled with technological development, could keep the number of people without adequate sanitation from growing even though the total population increases.

Source: World Bank estimates based on research paper by Dennis Anderson and William Cavendish, "Efficiency and Substitution in Pollution Abatement: Simulation Studies in Three Sectors."

resin filter beds that bind selectively to specific materials. Some of the same techniques used to stabilize liquids *in situ* can also be used *in vitro* (in a reaction vessel). Metals, for instance, can be chelated or precipitated in insoluble, inactive forms.

Often, living organisms can be used effectively and inexpensively to clean contaminated water. We call this bioremediation (chapter 21). Restored wetlands, for instance, along streambanks or lake margins can be very effective in filtering out sediment and removing pollutants. They generally cost far less than mechanical water treatment facilities, and they provide wildlife habitat as well.

Lowly duckweed (*Lemna* sp.), the green scum you often see covering the surface of eutrophic ponds, grows fast and can remove large amounts of organic nutrients from water. Under optimal conditions, a few square centimeters of these tiny plants can grow to cover nearly a hectare (about 2.5 acres) in four months. Large duckweed lagoons are being used as inexpensive, low-tech sewage treatment plants in developing countries. Where conventional wastewater purification typically costs $300 to $600 per person served, a duckweed system can cost one-tenth as much. The duckweed can be harvested and used as feed, fuel, or fertilizer. Up to 35 percent of its dry mass is protein—about twice as much as alfalfa, a popular animal feed.

Where space for open lagoons is unavailable, bioremediation can be carried out in tanks or troughs. This has the advantage of controlling conditions more precisely and doesn't release organisms into the environment. Some of the most complex, holistic

systems for water purification are designed by Ocean Arks International (OAI) in Falmouth, Massachusetts. Their "living machines" combine living organisms—chosen to perform specific functions—in contained environments. In a typical living machine, water flows through a series of containers, each with a distinct ecological community designed for a particular function. Wastes generated by the inhabitants of one vessel become the food for inhabitants of another. Sunlight provides the primary source of energy.

OAI has created or is in the process of building water treatment plants in a dozen states and foreign countries. Designs range from remediating toxic wastes from Superfund sites to simply treating domestic wastes. Starting with microorganisms in aerobic and anaerobic environments where different kinds of wastes are metabolized or broken down, water moves through a series of containers containing hundreds of different kinds of plants and animals, including algae, rooted aquatic plants, clams, snails, and fish, each chosen to provide a particular service. Technically the finished water is drinkable, although few people feel comfortable drinking it. More often the final effluent is used to flush toilets or for irrigation. Called ecological engineering, this novel approach can save resources and money as well as help clean up our environment (fig. 18.27).

18.5 WATER LEGISLATION

As the opening case study for this chapter shows, water pollution control has been among the most broadly popular and effective of all environmental legislation in the United States. It has not been without controversy, however. In this section we will look

FIGURE 18.27 Bioreactors, such as these "living machines" from Ocean Arks International use living communities that mimic natural ecosystems to treat water. Polluted water (right flask) can be purified to drinking water quality (left flask).

Table 18.2 Some Important U.S. and International Water Quality Legislation

1. *Federal Water Pollution Control Act* (1972). Established uniform nationwide controls for each category of major polluting industries.

2. *Marine Protection Research and Sanctuaries Act* (1972). Regulates ocean dumping and established sanctuaries for protection of endangered marine species.

3. *Ports and Waterways Safety Act* (1972). Regulates oil transport and the operation of oil handling facilities.

4. *Safe Drinking Water Act* (1974). Requires minimum safety standards for every community water supply. Among the contaminants regulated are bacteria, nitrates, arsenic, barium, cadmium, chromium, fluoride, lead, mercury, silver, pesticides; radioactivity and turbidity also are regulated. This act also contains provisions to protect groundwater aquifers.

5. *Resource Conservation and Recovery Act* (RCRA) (1976). Regulates the storage, shipping, processing, and disposal of hazardous wastes and sets limits on the sewering of toxic chemicals.

6. *Toxic Substances Control Act* (TOSCA) (1976). Categorizes toxic and hazardous substances, establishes a research program, and regulates the use and disposal of poisonous chemicals.

7. *Comprehensive Environmental Response, Compensation, and Liability Act* (CERCLA) (1980) and *Superfund Amendments and Reauthorization Act* (SARA) (1984). Provide for sealing, excavation, or remediation of toxic and hazardous waste dumps.

8. *Clean Water Act* (1985) (amending the 1972 Water Pollution Control Act). Sets as a national goal the attainment of "fishable and swimmable" quality for all surface waters in the United States.

9. *London Dumping Convention* (1990). Calls for an end to all ocean dumping of industrial wastes, tank washing effluents, and plastic trash. The United States is a signatory to this international convention.

at some of the major issues concerning water quality laws and their provisions (table 18.2).

The Clean Water Act was ambitious, bipartisan, and largely successful

Passage of the U.S. Clean Water Act of 1972 was a bold, bipartisan step determined to "restore and maintain the chemical, physical, and biological integrity of the Nation's waters" that made clean water a national priority. Along with the Endangered Species Act and the Clean Air Act, this is one of the most significant and effective pieces of environmental legislation ever passed by the U.S. Congress. It also is an immense and complex law, with more than 500 sections regulating everything from urban runoff, industrial discharges, and municipal sewage treatment to land-use practices and wetland drainage.

The ambitious goal of the Clean Water Act was to return all U.S. surface waters to "fishable and swimmable" conditions. For specific "point" sources of pollution, such as industrial discharge pipes or sewage outfalls, the act requires discharge permits and **best practicable control technology (BPT)**. It sets national goals of **best available, economically achievable technology (BAT)** for toxic substances and zero discharge for 126 priority toxic pollutants. As we discussed earlier in this chapter, these regulations have had a positive effect on water quality. Although surface waters are not yet swimmable or fishable everywhere, surface-water quality in the United States has significantly improved on average over the past quarter century. Perhaps the most important result of the act has been investment of $54 billion in federal funds and more than $128 billion in state and local funds for municipal sewage treatment facilities.

Not everyone, however, is completely happy with the Clean Water Act. Industries, state and local governments, farmers, land developers, and others who have been forced to change their

What Can You Do?

Steps You Can Take to Improve Water Quality

Individual actions have important effects on water quality. Here are some steps you can take to make a difference.

- Compost your yard waste and pet waste. Nutrients from decayed leaves, grass, and waste are a major urban water pollutant. Many communities have public compost sites available.

- Don't fertilize your lawn or apply lawn chemicals. Untreated grass can be just as healthy, and it won't poison your pets or children.

- Make sure your car doesn't leak fluids, oil, or solvents on streets and parking lots, from which contaminants wash straight into rivers and lakes. Recycle motor oil at a gas station or oil-change shop.

- Create a "rain garden" to capture and filter surface runoff. This helps recharge groundwater aquifers and keeps nutrients and toxins out of rivers and lakes (fig. 18.28).

- Don't buy lawnmowers, personal watercraft, or other vehicles with two-cycle engines, which release abundant fuel and oil into air and water. Instead, buy more efficient four-stroke engines.

- Visit your local sewage treatment plant. Often public tours are available or group tours can be arranged, and these sites can be fascinating.

- Keep informed about water policy debates at local and federal levels. Policies change often, and public input is important.

FIGURE 18.28 A rain garden is a shallow depression situated to collect runoff from streets or parking lots. It's planted with species that can survive in saturated soils. This vegetation helps evaporate and cleanse runoff, while temporary storage in the basin allows groundwater recharge. You might build a rain garden in your yard, or on your campus, or elsewhere in your city.

operations or spend money on water protection often feel imposed upon. One of the most controversial provisions of the act has been Section 404, which regulates draining or filling of wetlands. Although the original bill mentions wetlands only briefly, this section has evolved through judicial interpretation and regulatory policy to become one of the principal federal tools for wetland protection. Many people applaud the protection granted to these ecologically important areas that were being filled in or drained at a rate of about half a million hectares per year before the passage of the Clean Water Act. Farmers, land developers, and others who are prevented from converting wetlands to other uses often are outraged by what they consider "taking" of private lands.

Another sore point for opponents of the Clean Water Act are what are called "unfunded mandates," or requirements for state or local governments to spend money that is not repaid by Congress. You will notice that the $128 billion already spent by cities to install sewage treatment and stormwater diversion to meet federal standards far exceeds the $54 billion in congressional assistance for these projects. Estimates are that local units of government could be required to spend another $130 billion to finish the job without any further federal funding. Small cities that couldn't afford or chose not to participate in earlier water quality programs, in which the federal government paid up to 90 percent of the costs, are especially hard hit by requirements that they upgrade municipal sewer and water systems. They now are faced with carrying out those same projects entirely on their own funds.

Clean water reauthorization remains contentious

Opponents of federal regulation have tried repeatedly to weaken or eliminate the Clean Water Act. They regard restriction of their "right" to dump toxic chemicals and waste into wetlands and waterways to be an undue loss of freedom. They resent being forced to clean up municipal water supplies, and call for cost/benefit analysis that places greater weight on economic interests in all environmental planning. Most of all they view any limitation on use of private property to be a "taking" for which they should be fully compensated.

Even those who support the Clean Water Act in principle would like to see it changed and strengthened. Among these proposals are a shift from "end-of-the-pipe" focus on removing specific pollutants from effluents to more attention to changing industrial processes so toxic substances won't be produced in the first place. Another important issue is nonpoint pollution from agricultural runoff and urban areas, which has become the largest source of surface-water degradation in the United States. Regulating these sources remains a difficult problem.

Environmentalists also would like to see stricter enforcement of existing regulations, mandatory minimum penalties for violations, more effective community right-to-know provisions, and increased powers for citizen lawsuits against polluters. Studies have found that, in practice, polluters are given infrequent and light fines for polluting. Under the current law, using data that polluters themselves are required to submit, groups such as the Natural Resources Defense Council and the Citizens for a Better Environment have won million-dollar settlements in civil lawsuits (the proceeds generally are applied to cleanup projects) and some transgressors have even been sent to jail. Not surprisingly, environmentalists want these powers expanded, while polluters find them very disagreeable.

Other important legislation also protects water quality

In addition to the Clean Water Act, several other laws help to regulate water quality in the United States and abroad. Among these is the Safe Drinking Water Act, which regulates water quality in commercial and municipal systems. Critics complain that standards and enforcement policies are too lax, especially for rural water districts and small towns. Some researchers report pesticides, herbicides, and lead in drinking water at levels they say should be of concern. Atrazine, for instance, was detected in 96 percent of all surface-water samples in one study of 374 communities across 12 states. Remember, however, that simply detecting a toxic compound is not the same as showing dangerous levels.

The Superfund program for remediation of toxic waste sites was created in 1980 by the Comprehensive Environmental Response, Compensation, and Liability Act (CERCLA) and was amended by the Superfund Amendments and Reauthorization Act (SARA) of 1984. This program is designed to provide immediate response to emergency situations and to provide permanent remedies for abandoned or inactive sites. These programs provide many jobs for environmental science majors in monitoring and removal of toxic wastes and landscape restoration. A variety of methods have been developed for remediation of problem sites.

CONCLUSION

Half a century ago, rivers in the United States were so polluted that some caught fire while others ran red, black, orange, or other unnatural colors with toxic industrial wastes. Many cities still dumped raw sewage into local rivers and lakes, so that warnings had to be posted to avoid any bodily contact. We've made huge progress since that time. Not all rivers and lakes are "fishable or swimmable," but federal, state, and local pollution controls have greatly improved our water quality in most places.

In rapidly developing countries, such as China and India, water pollution remains a serious threat to human health and ecosystem well-being. Billions of people don't have access to clean drinking water or adequate sanitation. It will take a massive investment to correct this growing problem. But there are relatively low-cost solutions to many pollution issues. Constructed wetlands for ecological sewage treatment show us that we can find low-tech, inexpensive ways to reduce pollution. Living machines for water treatment in individual buildings or communities also offer hope for better ways to treat our wastes. Perhaps you can use the information you've learned by studying environmental science to make constructive suggestions for your own community.

REVIEWING LEARNING OUTCOMES

By now you should be able to explain the following points:

18.1 Define *water pollution*.

- Water pollution is anything that degrades water quality.

18.2 Describe the types and effects of water pollutants.

- Infectious agents remain an important threat to human health.
- Bacteria are detected by measuring oxygen levels.
- Nutrient enrichment leads to cultural eutrophication.
- Eutrophication can cause toxic tides and "dead zones."
- Inorganic pollutants include metals, salts, acids, and bases.
- Organic pollutants include drugs, pesticides, and other industrial substances.
- Sediment also degrades water quality.
- Thermal pollution is dangerous for organisms.

18.3 Investigate water quality today.

- The Clean Water Act protects our water.
- The importance of a single word.
- Water quality problems remain.

- Other countries also have serious water pollution.
- Groundwater is hard to monitor and clean.
- There are few controls on ocean pollution.

18.4 Explain water pollution control.

- Source reduction is often the cheapest and best way to reduce pollution.
- Controlling nonpoint sources requires land management.
- Human waste disposal occurs naturally when concentrations are low.
- Water remediation may involve containment, extraction, or phytoremediation.

18.5 Summarize water legislation.

- The Clean Water Act was ambitious, bipartisan, and largely successful.
- Clean water reauthorization remains contentious.
- Other important legislation also protects water quality.

PRACTICE QUIZ

1. Define *water pollution*.
2. List eight major categories of water pollutants and give an example for each category.
3. Describe eight major sources of water pollution in the United States. What pollution problems are associated with each source?
4. What are red tides, and why are they dangerous?
5. What is eutrophication? What causes it?
6. What is an oxygen sag? How much dissolved oxygen, in ppm, is present at each stage?
7. What are the origins and effects of siltation?
8. Describe primary, secondary, and tertiary processes for sewage treatment. What is the quality of the effluent from each of these processes?
9. Why do combined storm and sanitary sewers cause water quality problems? Why does separating them also cause problems?
10. What pollutants are regulated by the Clean Water Act? What goals does this act set for abatement technology?
11. What is MTBE? Why is it so widespread and hard to control?
12. Describe remediation techniques and how they work.

CRITICAL THINKING AND DISCUSSION QUESTIONS

1. Cost is the greatest obstacle to improving water quality. How would you decide how much of the cost of pollution control should go to private companies, government, or individuals?

2. How would you define *adequate sanitation*? Think of some situations in which people might have different definitions for this term.

3. What sorts of information would you need to make a judgment about whether water quality in your area is getting better or worse? How would you weigh different sources, types, and effects of water pollution?

4. Imagine yourself in a developing country with a severe shortage of clean water. What would you miss most if your water supply were suddenly cut by 90 percent?

5. Proponents of deep well injection of hazardous wastes argue that it will probably never be economically feasible to pump water out of aquifers more than 1 kilometer below the surface. Therefore, they say, we might as well use those aquifers for hazardous waste storage. Do you agree? Why or why not?

6. Arsenic contamination in Bangladesh results from geological conditions, World Bank and U.S. aid, poverty, government failures, and other causes. Who do you think is responsible for finding a solution? Why? Would you answer differently if you were a poor villager in Bangladesh?

 **Data Analysis:** Examining Pollution Sources

Understanding the origins of pollution is the first step toward considering policies for reducing it. The chapter you have just read includes several graphs displaying pollution data. The following questions ask you to think more about the sources of this pollution:

1. In figure 18.13, which of the causes of stream or lake impairment do you think are mainly from point or non-point sources? Most of these contaminants have multiple sources, but try to imagine the most common origin.

2. Figure 18.10 shows a group of less common—but still significant—organic contaminants in surface waters. What do you think are the most likely sources of these chemicals?

3. Based on what you've learned in this chapter, which of the pollutants in these two graphs (figs. 18.11 and 18.14) do you think are most likely to come from the following sources?

   Agriculture _____

   Sewage treatment _____

   Dams, diversion projects _____

   Urban runoff _____

   Mining, smelting _____

   Power plants _____

   Other industry _____

   Forestry _____

   Removal of streamside vegetation _____

4. How would you design a sampling strategy to assess water pollution on your school campus?

5. Figure 18.11 shows some of the new developments in water pollution assessment. Conventional treatment systems were not designed to remove thousands of newly invented chemical compounds, or increasingly widespread compounds, including those shown in the figure. Explain the units used on the Y-axis. What are the numbers above the bars?

6. How many of the pollutants shown do *you* use? Don't forget to include caffeine (which is classed as a nonprescription drug) and antibacterial soaps (disinfectants). Steroids include cholesterol, which occurs naturally in foods.

7. Try to think of additional substances that you might contribute to wastewater.

8. The graph in figure 18.11 results from a reconnaissance study done by the U.S. Geological Survey. The researchers wanted to assess whether a list of 95 contaminants could be detected at all in public waterways. If you wanted to design a study like this, what sorts of sites would you select for sampling? How might your results and your conclusions differ if you did a random sample?

CHAPTER 19

In 2010 the drilling rig Deepwater Horizon exploded and sank, spilling 5 million barrels (800 million liters) of crude oil into the Gulf of Mexico.

Conventional Energy

Learning Outcomes

After studying this chapter, you should be able to:

19.1 Define *energy, work,* and how our energy use has varied over time.

19.2 Describe the benefits and disadvantages of using coal.

19.3 Explain the consequences and rewards of exploiting oil.

19.4 Illustrate the advantages and disadvantages of natural gas.

19.5 Summarize the potential and risk of nuclear power.

19.6 Evaluate the problems of radioactive wastes.

19.7 Discuss the changing fortunes of nuclear power.

19.8 Identify the promise and peril of nuclear fusion.

"The pessimist complains about the wind; the optimist expects it to change; the realist adjusts the sails."

~ *William Arthur Ward*

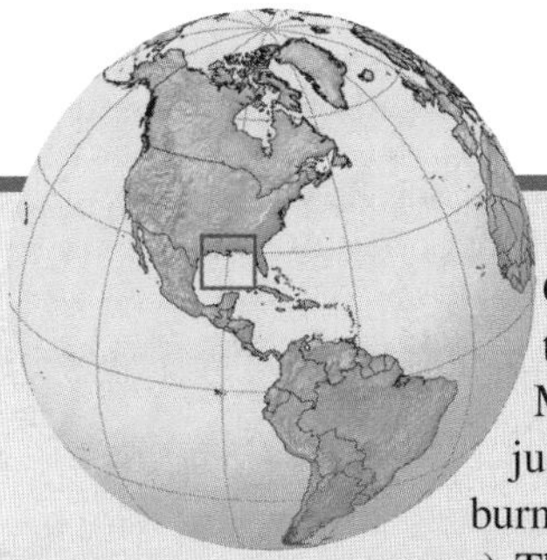

On April 20, 2010, the Deepwater Horizon, which was drilling the Macondo well in the Gulf of Mexico just off the Louisiana coast, exploded, burned, and sank (see photo, opposite page). The well was in nearly a mile (1,600 m) of water, and had reached oil at a depth of 13,360 ft (4,100 m) below the seafloor when a bubble of methane gas shot up through the drill pipe, expanding quickly as it rose and bursting into flame when it reached the surface. The seals and safety barriers designed to prevent escaping gas failed. After burning for about a day, the $560 million drill rig capsized and sank. Eleven workers were killed and 17 others were injured.

Crude oil gushed out of the ruptured drill pipe. BP, the company that owned the well, claimed the spill was about 5,000 barrels per day, but others said that it was at least ten times as much. Despite a number of efforts by BP to cap the well or to inject heavy drilling mud to stop the flow, oil continued to pour into the Gulf. Finally, after four months of drilling, a relief well intersected the damaged borehole just above the spot at which it entered the oil reservoir. This allowed engineers to pump cement into the bottom of the well and seal it permanently.

Altogether, it's estimated that about 5 million barrels (800 million l) of oil were released into the Gulf, making this the largest accidental marine oil spill in world history. It was about 20 times as much oil as was spilled in the wreck of the *Exxon Valdez* in Alaska in 1989. We don't yet know the total impact of this disaster, but the effects on marine life, fishing, and tourism (which bring in about $34 billion annually to Gulf States) could be tragic and long-lasting. Six months after the spill, the Fish and Wildlife Service reported that 6,100 birds, 610 sea turtles, and 100 dolphins died from oil contamination (fig. 19.1). It's estimated that 20 percent of the juvenile bluefin tuna in the Gulf were killed by oil pollution in one of the species' most important spawning areas. This could be a serious blow to an already endangered population.

On the other hand, several factors may reduce ecological damage from the spill. The oil was released in deep water so it had a much greater chance to disperse than if it were in shallow, coastal water. Furthermore, the Gulf water is warm, which speeds up evaporation and metabolism by microorganisms. The Gulf has natural seeps that release about 1,000 barrels of oil per day, so many

FIGURE 19.1 Thousands of birds, turtles, and marine mammals were contaminated by crude oil released from the Macondo well in the Gulf of Mexico.

microbe species were already adapted to metabolize oil. And a government study concluded that nearly three-quarters of the oil was either recovered at the wellhead, dispersed, dissolved, evaporated, broken up by chemical treatments, burned, or skimmed from the ocean surface.

The use of chemical dispersants remains controversial. Altogether, BP sprayed about 1.8 million gallons (6.8 million liters) of a chemical mixture called Corexit either on the ocean surface or next to the gushing wellhead. This solvent mixture is known to be toxic, although no one knows what the effects of such a large amount in the ocean will be. The dispersant was successful in preventing much of the oil from reaching the shore, where it would have been a public relations nightmare for BP, but it also created huge plumes of tiny oil droplets deep under water where it may be more toxic to fish and other sea life than if it had been on the surface.

Immediately after the spill, the companies involved began blaming each other. Although BP owned the well, they had subcontracted with Transocean, which owned the Deepwater Horizon, to do the drilling, while Halliburton supplied the drilling mud, cement, and other supplies. Both subcontractors claimed that BP pressured them to take shortcuts and avoid safety warnings to cut costs. One of the most critical errors was to use a defective blowout preventer on the well. This apparatus is a giant valve that's supposed to be the last line of defense, with huge shears that can cut off the well and prevent a gusher. But the one BP chose had a dead battery, debilitating hydraulic-system leaks, and shears that weren't strong enough to seal the well. It had failed several crucial safety tests, but was used anyway.

A government investigative committee concluded that all three firms made bad choices: "Whether purposeful or not, many of the decisions that BP, Halliburton, and Transocean made that increased the risk of the Macondo blowout clearly saved those companies significant time (and money)." Discovery of offshore deposits has substantially increased our oil and gas supplies, but our addiction to fossil fuel is forcing us to look in ever more dangerous and expensive places for energy.

In this chapter we'll examine the fossil fuels and nuclear sources that now provide 90 percent of our energy. In chapter 20 we'll look at some renewable energy alternatives. For related resources, including Google Earth™ placemarks that show locations discussed in this chapter, visit EnvironmentalScience-Cunningham.blogspot.com.

19.1 ENERGY RESOURCES AND USES

Energy drives our economy today, and many of our most important questions in environmental science have some link to energy resources—from air pollution, climate change, and mining impacts, to technological innovations in alternative energy sources.

Fire was probably the first external energy source used by humans. Charcoal from fires has been found at sites occupied by our early ancestors 1 million years ago. Muscle power provided by domestic animals has been important at least since the dawn of the Neolithic Age 10,000 years ago. Wind and water power have been used nearly as long. Firewood was by far the largest source of energy for cooking and heating in the United States from colonial days until the mid-nineteenth century. The invention of the steam engine, together with diminishing supplies of wood, caused a switch to coal as the major energy source during the industrial revolution (fig. 19.2). Coal, in turn, was replaced by oil in the twentieth century due to the ease of shipping and burning liquid fuels. As easily accessible petroleum supplies have been depleted, however, we look increasingly to remote, dangerous, or politically unstable places for the oil on which we have become dependent.

Our dependence on—some would say addiction to—oil creates serious critical geopolitical and economic problems. The United States, for example, spends about $400 billion every year on imported oil, not counting the costs of maintaining armed forces to ensure access to those resources. And the huge amounts of carbon-based fuels we now burn to support our lifestyles are now causing unsustainable environmental impacts, ranging from oceanic oil spills, such as the Deepwater Horizon disaster in 2010, to mountaintop removal for coal extraction, water pollution from extracting tar sands, or air pollution and global climate change from burning all those fuels.

Table 19.1	Some Energy Units
1 joule (J) = the force exerted by a current of 1 amp per second flowing through a resistance of 1 ohm	
1 watt (W) = 1 joule (J) per second	
1 kilowatt-hour (kWh) = 1 thousand (10^3) watts exerted for 1 hour	
1 megawatt (MW) = 1 million (10^6) watts	
1 gigawatt (GW) = 1 billion (10^9) watts	
1 petajoule (PJ) = 1 quadrillion (10^{15}) joules	
1 PJ = 947 billion BTU, or 0.278 billion kWh	
1 British thermal unit (BTU) = energy to heat 1 lb of water 1°F	
1 standard barrel (bbl) of oil = 42 gal (160 l) or 5.8 million BTU	
1 metric ton of standard coal = 27.8 million BTU or 4.8 bbl oil	

How do we measure energy?

To understand the magnitude of energy use, it is helpful to know the units used to measure it. **Work** is the application of force over distance, and we measure work in **joules** (table 19.1). **Energy** is the capacity to do work. **Power** is the rate of energy flow or the rate of work done: for example, one **watt** (W) is one joule per second. If you use a 100-watt lightbulb for 10 hours, you have used 1,000 watt-hours, or one kilowatt-hour (kWh). Most American households use about 11,000 kWh per year (table 19.2).

Fossil fuels supply most of the world's energy

Currently **fossil fuels** (petroleum, natural gas, and coal) supply about 88 percent of world commercial energy needs (fig. 19.3). Oil makes up roughly 35 percent of that total, while natural gas (24 percent) and coal (29 percent) follow close behind. Renewable sources—solar, wind, geothermal, and hydroelectricity— make up about 7 percent of U.S. commercial power (but hydro and biomass account for most of that). Although growing rapidly, solar, wind, and geothermal power still make up less than 1 percent of the world energy supply. One reason for this disparity is the subsidies we provide for conventional fuels. In 2009, global subsidies for

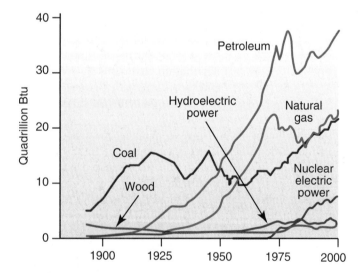

FIGURE 19.2 Although the relative importance of various fuels has shifted over the past century, fossil fuels supply about 84 percent of all energy used currently in the United States, and petroleum makes up the largest share of that total. As you can see, total energy use has surged about tenfold over the past century.

Table 19.2	Energy Uses
Uses	**kWh/year***
Computer	100
Television	125
100 W light bulb	250
15 W fluorescent bulb	40
Dehumidifier	400
Dishwasher	600
Electric stove/oven	650
Clothes dryer	900
Refrigerator	1100

* Averages shown; actual rates vary greatly.
Source: U. S. Department of Energy.

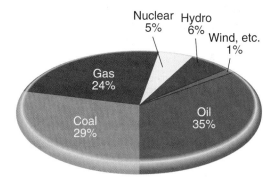

FIGURE 19.3 Global commercial energy sources. This does not include energy collected for personal use or traded in informal markets.
Source: Data from British Petroleum, 2010.

fossil fuels amounted to about $300 billion, or about ten times the support for wind, solar, and other alternative energy sources.

Nuclear power provides less U.S. energy than renewables (about 5 percent of all U.S. energy), but it generates about 20 percent of all electricity. There's enough nuclear fuel to produce power for a long time, but as we discuss later in this chapter, safety concerns and waste storage problems make this option unacceptable to most people.

World energy consumption rose slightly more than 1 percent annually between 1970 and 2000, but between 2003 and 2007 the rate of growth jumped to nearly 5 percent per year. Rapidly expanding economies in developing countries—especially in China—are responsible for most of that growth. For many years the richer countries, with about 20 percent of the world population, consumed roughly 80 percent of all commercial energy, while the other 80 percent of the world had only 20 percent of the total supply. That situation is changing now. By 2035, energy experts predict, emerging economies such as China and India will be consuming about 60 percent of all commercial energy.

In 2008, oil prices surged to $147 per barrel. Americans saw gasoline prices over $4 per gallon. This brought record profits to oil companies (together, ExxonMobil and Chevron made more than $69 billion that year), but pained many consumers. Oil imports cost the United States more than $400 billion every year, not counting the costs of trying to maintain peace in the Middle East.

President Barack Obama said, "Every year, we become more, not less, addicted to oil—a nineteenth-century fossil fuel that is dirty, dwindling, and dangerously expensive." As a centerpiece of his economic recovery plan, Obama vowed to encourage conservation together with clean, renewable energy sources that will create jobs, reduce pollution, and help stop global warming. Because they make up so much of our current energy supply, however, this chapter will focus on fossil fuels and nuclear power. Chapter 20 looks at some options for conservation and sustainable energy.

How do we use energy?

The largest share of the energy used in the United States is consumed by industry (fig. 19.4). Mining, milling, smelting, and forging of primary metals consume about one-quarter of that industrial energy share. The chemical industry is the second largest industrial user of fossil fuels, but only half of its use is for energy generation. The remainder is raw material for plastics, fertilizers, solvents, lubricants, and hundreds of thousands of organic chemicals in commercial use. The manufacture of cement, glass, bricks, tile, paper, and processed foods also consumes large amounts of energy. Although coal provides about one-quarter of our total energy in the United States, it supplies about half our electricity.

Residential and commercial customers use roughly 41 percent of the primary energy consumed in the United States, mostly for space heating, air conditioning, lighting, and water heating. Transportation requires about 28 percent of all energy used in the United States each year. About 98 percent of that energy comes from petroleum products refined into gasoline and diesel fuel, and the remaining 2 percent is provided by natural gas and electricity.

Almost three-quarters of all transport energy is used by motor vehicles. Nearly 3 trillion passenger miles and 600 billion ton miles of freight are carried annually by motor vehicles in the United States. About 75 percent of all freight traffic in the United States is carried by trains, barges, ships, and pipelines, but because they are very efficient, they use only 12 percent of all transportation fuel.

Think About It

Years ago, Europe decided to discourage private automobiles and encourage mass transit by making gasoline expensive (about $5 per gal, on average). What changes would America have to make to achieve the same result?

Producing and transporting energy also consumes and wastes energy. About half of all the energy in primary fuels is lost during conversion to more useful forms, while being shipped to the site of end use, or during use. Electricity is generally promoted as a clean, efficient source of energy because, when it is used to run a resistance heater or an electrical appliance, almost 100 percent of its energy is converted to useful work and no pollution is given off.

What happens, however, before electricity reaches us? Coal-fired power plants supply about half our electrical energy, and large amounts of pollution are released during mining and burning of that coal. Furthermore, nearly two-thirds of the energy in the coal was lost in thermal conversion in the power plant. About 10 percent more is lost during transmission and stepping down to household voltages.

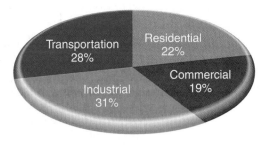

FIGURE 19.4 U.S. energy consumption by sector in 2010.

How much energy do you use every year? Most of us don't think about it much, but maintaining the luxuries we enjoy usually requires an enormous energy input. On average, each person in the United States or Canada uses more than 300 gigajoules (GJ) (the equivalent of about 60 standard barrels or 8 metric tons of oil) per year. By contrast, in the poorest countries of the world, such as Bangladesh, Yemen, and Ethiopia, each person, on average, consumes less than one GJ per year. Put another way, each of us in the richer countries consumes nearly as much energy in a single day as the poorest people in the world consume in a year. In general, income and standards of living rise with increasing energy availability, but the correlation isn't absolute (see Data Analysis, p. 444). Some energy-rich countries, such as Qatar, use vast amounts of energy, although their level of human development isn't correspondingly high. Perhaps more important is that some countries, such as Norway, Denmark, and Japan, have a much higher standard of living by almost any measure than the United States, while using about half as much energy. This suggests abundant opportunities for energy conservation without great sacrifices.

Clearly, energy consumption is linked to the comfort and convenience of our lives. Those of us in the richer countries enjoy many amenities not available to most people in the world. The link isn't absolute, however. Several European countries, including Sweden, Denmark, and Finland, have higher standards of living than does the United States by almost any measure but use about half as much energy.

19.2 COAL

Coal is fossilized plant material preserved by burial in sediments and altered by geological forces that compact and condense it into a carbon-rich fuel. Coal is found in every geologic system since the Silurian Age 400 million years ago, but graphite deposits in very old rocks suggest that coal formation may date back to Precambrian times. Most coal was laid down during the Carboniferous period (286 million to 360 million years ago) when the earth's climate was warmer and wetter than it is now. Because coal takes so long to form, it is essentially a nonrenewable resource.

Coal resources are vast

World coal deposits are ten times greater than conventional oil and gas resources combined. Coal seams can be 100 m thick and can extend across tens of thousands of square kilometers that were vast swampy forests in prehistoric times. The total resource is estimated to be 10 trillion metric tons. If all this coal could be extracted, and if coal consumption continued at present levels, this would amount to several thousand years' supply. At present rates of consumption, these proven-in-place reserves—those explored and mapped but not necessarily economically recoverable—will last about 200 years. Note that "known reserves" have been identified but not thoroughly mapped. **Proven reserves** have been mapped, measured, and shown to be economically recoverable. Ultimate reserves include unknown as well as known resources (fig. 19.5).

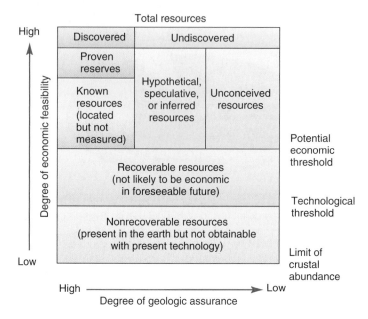

FIGURE 19.5 Categories of natural resource according to economic and technology feasibility, as well as geologic assurance.

Where are these coal deposits located? They're not evenly distributed throughout the world. North America, Europe, and Asia contain more than 90 percent of the world's coal, and five nations (United States, Russia, China, India, and Australia) account for three-quarters of that amount (fig. 19.6). In part, countries with large land areas are more likely to have coal deposits, but this resource is very rare in Africa, the Middle East, or Central and South America (fig. 19.7). Antarctica is thought to have large coal deposits, but they would be difficult, expensive, and ecologically damaging to mine.

It would seem that the abundance of coal deposits is a favorable situation. But do we really want to use all of the coal? In the next section we will look at some of the disadvantages and dangers of mining and burning coal using conventional techniques.

Coal mining is a dirty, dangerous business

Underground mines are subject to cave-ins, fires, accidents, and accumulation of poisonous or explosive gases (carbon monoxide, carbon dioxide, methane, hydrogen sulfide). Between 1870 and 1950, more than 30,000 American coal miners died of accidents and injuries in Pennsylvania alone, equivalent to one person per day for 80 years. Untold thousands have died of respiratory diseases. In some mines, nearly every miner who did not die early from some other cause was eventually disabled by **black lung disease**—inflammation and fibrosis caused by accumulation of coal dust in the lungs or airways (chapter 8). Few of these miners or their families were compensated for their illnesses by the companies for which they worked.

China is reported to have the world's most dangerous coal mines currently. In 2009 the *China Daily* reported that 91,172 workers

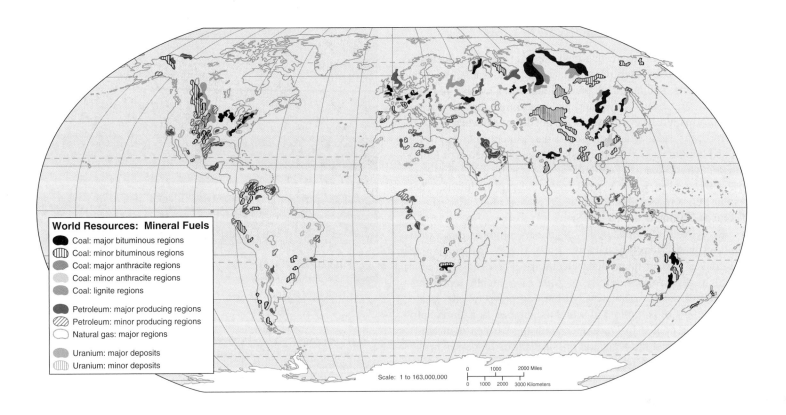

FIGURE 19.6 Where are fossil fuels and uranium located? North America, Europe, the Middle East, and parts of Asia are richly endowed. Africa, most of South America, and island states, like Japan, generally lack these fuels, greatly limiting their economic development.

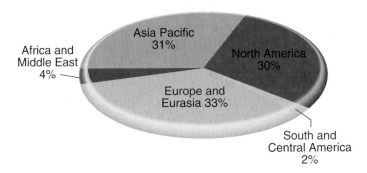

FIGURE 19.7 Proven-in-place coal reserves by region, 2008.
Source: British Petroleum, 2010.

died in 413,700 mine accidents in the previous year. This was the first time since 1995, the paper said, that the annual death toll had fallen below 100,000. Government officials quickly disputed these numbers, but acknowledged that 80 percent of China's 16,000 coal mines operate illegally.

Strip mining or surface mining is cheaper and safer than underground mining but often makes the land unfit for any other use. Mine reclamation is now mandated in the United States, but land is rarely restored to its original contour or biological community. Coal mining also contributes to water pollution. Sulfur and other water-soluble minerals make mine drainage and run-off from coal piles and mine tailings acidic and highly toxic.

Thousands of miles of streams in the United States have been poisoned by coal-mining operations.

Perhaps the most egregious type of strip mining is "mountaintop removal," practiced mainly in Appalachia, where the tops of mountain ridges are scraped off and dumped into valleys below to get at coal seams (fig. 19.8). Streams, farms, even whole towns are buried under hundreds of meters of toxic rubble by this practice.

FIGURE 19.8 One of the most environmentally destructive methods of coal mining is mountaintop removal. Up to 100 m of the mountain is scraped off and pushed into the valley below, burying forests, streams, farms, cemeteries, and sometimes houses.

Burning coal releases many pollutants

Many people aren't aware that coal burning releases radioactivity and many toxic metals. Uranium, arsenic, lead, cadmium, mercury, rubidium, thallium, and zinc—along with a number of other elements—are absorbed by plants and concentrated in the process of coal formation. These elements are not destroyed when the coal is burned; instead they are released as gases or concentrated in fly ash and bottom slag. You are likely to get a higher dose of radiation living next door to a coal-burning power plant than living next to a nuclear plant under normal (nonaccident) conditions. Coal combustion is responsible for about 25 percent of all atmospheric mercury pollution in the United States.

Every year some 82,000 U.S. miners produce more than 1 billion metric tons of coal (fig. 19.9). Eighty-five percent of that coal is burned to produce electricity. The electricity consumed by the average family of four each month represents about 1,140 pounds of coal. Coal-burning power plants create huge amounts of ash, most of which is pumped as a slurry into open storage ponds. In December 2008 the earthen dam holding an ash pond at the Kingston Fossil Plant in Tennessee broke and released at least 5.4 million cubic yards of toxic sludge into the Emory and Clinch rivers. Cleanup costs for this toxic flood, which contained arsenic, chromium, lead, nickel, selenium, and thallium, are estimated to be over $825 million. The EPA revealed that at least 140 sites in the United States were at least as large and dangerous as the Kingston facility. In addition to a risk of catastrophic spills, these waste disposal ponds can leach toxins into local water supplies.

Coal also contains up to 10 percent sulfur (by weight). Unless this sulfur is removed by washing or flue-gas scrubbing, it is released during burning and oxidizes to sulfur dioxide (SO_2) or sulfate (SO_4) in the atmosphere. The high temperatures and rich air mixtures ordinarily used in coal-fired burners also oxidize nitrogen compounds (mostly from the air) to nitrogen monoxide, dioxide, and trioxide. Every year the coal burned in the United States releases some 18 million metric tons of SO_2, 5 million metric tons of nitrogen oxides (NO_x), 4 million metric tons of airborne particulates, 600,000 metric tons of hydrocarbons and carbon monoxide, and about 2 trillion metric tons of CO_2. This is about three-quarters of the SO_2, one-third of the NO_x, and about half of the industrial CO_2 released in the United States each year. Coal burning is the largest single source of greenhouse gases and acid rain in many areas.

These air pollutants have many deleterious effects, including human health costs, injury to domestic and wild plants and animals, and damage to buildings and property (chapters 15 and 16). Total losses from air pollution are estimated at $5 to $10 billion per year in the United States alone. By some accounts, at least 5,000 excess human deaths per year can be attributed to coal production and burning.

Sulfur can be removed from coal before it is burned, or sulfur compounds can be removed from the flue gas after combustion. Formation of nitrogen oxides during combustion also can be minimized. Perhaps the ultimate limit to our use of coal as a fuel

FIGURE 19.9 Every year, about 1 billion tons of coal are mined in the United States. At least 85 percent is burned to produce electricity.

will be the release of carbon dioxide into the atmosphere. As we discussed in chapter 15, carbon traps heat in the atmosphere and is a major contributor to global warming. Since 2001, according to the Sierra Club, plans to build more than 100 coal-burning power plants in the United States have been abandoned. Utilities expect they'll be charged for carbon emissions eventually and don't want to be locked into an obsolete technology.

Clean coal technology could be helpful

Carbon dioxide emissions are the most important limit to our use of coal in conventional boilers. As we discussed in chapter 15, greenhouse gases are now changing our global climate in ways that could have catastrophic consequences. **Carbon sequestration** appears to be a good option for reducing emissions. The United Nations estimates that at least half the CO_2 we release every year could be pumped into deep geologic formations. This can enhance gas and oil recovery. Norway's Statoil already is doing this. Since 1996 the company has injected more than 1 million tons of CO_2 into an oil reservoir beneath the North Sea because otherwise it would have to pay a carbon tax on its emissions. Alternatively, CO_2 can be stored in depleted oil or gas wells, forced into tight sandstone formations, injected into deep briny aquifers, or compressed and pumped to the bottom of the ocean.

New technologies, such as integrated gasification combined cycle (IGCC) or ultra supercritical boilers, could solve many of the problems currently caused by coal combustion (fig. 19.10). Though sulfur-removal from flue gases in conventional power plants has been effective in reducing acid rain in much of the United States, it might be done much better before the coal is burned. Similarly, mercury can be removed from flue gases, but it is captured more cheaply after gasification. And NO_x formation is said to be much lower in an IGCC than in most coal-fired boilers.

China currently leads in supercritical boiler technology, with about 25 units in operation. We can only hope they'll continue to move in this direction. Coal consumption is rising rapidly in

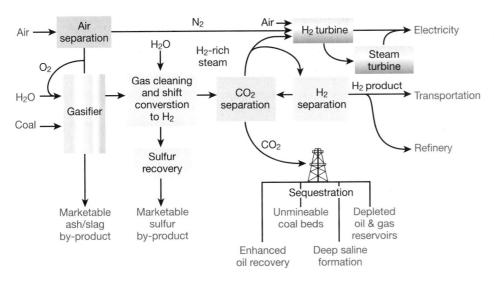

FIGURE 19.10 Clean coal technology could contribute to energy independence, while also reducing our greenhouse gas emissions.

China, and they've become the world's leading importer of coal (see fig. 1.15). China is building roughly one new coal-fired power plant per week, and they've become the world's leading source of CO_2. There are reports that China can capture CO_2 for about $30 per ton, which is about what carbon-trading permits are selling for in Europe.

Some utilities are burning coal together with biomass crops in their power plants. This produces less CO_2 than coal alone, and also improves the combustion characteristics of biomass alone. "Flex fuel" boilers could be a bridge solution until more truly renewable energy sources become available. Another proposal is to convert coal to either liquid or gas. Both fuel types would be more convenient to transport and burn than solid coal, but both are very expensive and produce huge amounts of carbon dioxide (often twice as much as simply burning the coal directly) and polluted water. In much of the western United States, where coal deposits are located, water shortages make synfuel schemes unlikely.

19.3 OIL

Like coal, petroleum is derived from organic molecules created by living organisms millions of years ago and buried in sediments where high pressures and temperatures concentrated and transformed them into energy-rich compounds. Depending on its age and history, a petroleum deposit will have varying mixtures of oil, gas, and solid tarlike materials. Some very large deposits of heavy oils and tars are trapped in porous shales, sandstone, and sand deposits in the western areas of Canada and the United States.

Liquid and gaseous hydrocarbons can migrate out of the sediments in which they formed through cracks and pores in surrounding rock layers. Oil and gas deposits often accumulate under layers of shale or other impermeable sediments, especially where folding and deformation of systems create pockets that will trap

upward-moving hydrocarbons. Contrary to the image implied by its name, an oil pool is not usually a reservoir of liquid in an open cavern but instead individual droplets or a thin film of liquid permeating spaces in a porous sandstone or limestone, much like water saturating a sponge.

As oil exploration techniques improve, we are finding deposits more effectively and in places once thought to be either devoid of oil or impossible to drill. Oil companies are now drilling in increasingly remote and risky places (What Do You Think? p. 430). This could extend our usable oil supplies but with severe environmental and economic risks (fig. 19.11). Energy companies planning to drill in the Arctic National Wildlife Refuge in Alaska claim that directional drilling will allow them to impact only 2 percent of the land surface while seeking out oil-bearing strata. Critics doubt that damage to the land will be so limited.

Pumping oil out of a reservoir is much like sucking liquid out of a sponge. The first fraction comes out easily, but removing subsequent fractions requires increasing effort. We never recover all the oil in a formation; in fact, a 30 to 40 percent yield is about average. There are ways of forcing steam or CO_2 into the oil-bearing formations to "strip" out more of the oil, but at least half the total deposit usually remains in the ground at the point at which it is uneconomical to continue pumping. Methods for squeezing more oil from a reservoir are called secondary recovery techniques.

Have we passed peak oil?

In the 1940s Dr. M. King Hubbert, a Shell Oil geophysicist, predicted that oil production in the United States would peak in the 1970s, based on estimates of U.S. reserves at the time. Hubbert's predicted peak was correct, and subsequent calculations have estimated a similar peak in global oil production in about 2005–2010 (fig. 19.12). Although global production has not yet

FIGURE 19.11 In our search for new supplies of oil, we increasingly turn to places like the deep oceans or the high Arctic, but the social, environmental, and economic costs of our dependence on these energy sources can be high.

Ultradeep Drilling

The Deepwater Horizon, a floating drill rig that sank and spilled about 200 million gallons (nearly 800 million liters) of crude oil into the Gulf of Mexico, wasn't by any means working on the deepest or most remote well in the world. For the Gulf, the record is currently held by the Perdido Spar rig, which is drilling wells in 9,627 ft (nearly 3,000 m) of water about 320 km east of Brownsville, Texas. The Perdido (which means "lost," "missing," or "damned" in Spanish) is a technological marvel. It's a spar platform, meaning that the drill rig sits on top of a huge, hollow cylinder that's nearly as tall as Paris's Eiffel Tower. The spar is tethered to the ocean floor by nine thick cables, and supposedly will withstand hurricane winds, fierce ocean currents, and giant waves. The rig can support 35 wells that go down as much as 6 km below the ocean floor and radiate out horizontally up to 16 km from the well head. It's expected to produce about 130,000 barrels of oil plus 200 million ft³ (about 6 million m³) of natural gas per day.

A spar platform is one of several designs for underwater drilling. In shallow water, up to about 500 m, a fixed platform with very long legs that sit on the sea floor can be used. Beyond 500 m, floating or semi-submersible rigs are used. The Deepwater Horizon, which could drill in up to 2,000 m of water, was built on giant pontoons partially filled with seawater for ballast. All the drilling equipment and living quarters for hundreds of workers sat on decks held up by the pontoons. Like spar rigs, floating platforms are anchored for stability. Even greater depths are accessible from dynamically positioned drilling ships. For example, the Discover Clear Leader, owned and operated by Transocean, is capable of drilling in water nearly 4,000 m deep, and then punching down another 10,000 m through the seabed to ultradeep oil deposits.

Conditions at these depths are extreme. The oil can be 200°C, while water temperatures at the seafloor are just above freezing. Temperature shocks can rupture drill pipe. Oil deposits often accumulate beneath splintery shale or thick layers of taffy-like salt. Corrosion from the salt or sulfur in sediments erodes metal while strong ocean currents sweep equipment away. These depths are too great for human divers to do repairs, so drill operators have to depend on remotely operated robots to do all their work. We think of the seafloor at great depths to be a feature-less, lifeless, mud-covered abyssal plain, but in fact it's often a jumble of deep canyons, sharp ridges, and huge piles of jumbled rocks with a rich, if largely unknown, community of life. All this makes drilling extremely difficult. Even under normal conditions, operating a drill rig, such as the Perdido, costs about $500,000 per day.

In spite of the disaster at the Deepwater Horizon, many countries are rushing to drill in harsh frontier environments. Before 1995 only about 10 percent of oil from the Gulf of Mexico came from deep water

(more than 2,000 m), but now, as the shallow fields are being exhausted, about 70 percent does. The economic rewards of hitting a big find are enormous. The Bureau of Ocean Energy Management, Regulation and Enforcement (the successor to the disgraced Minerals Management Service) estimates that the U.S. outer continental shelf holds about 86 billion barrels of oil and 420 trillion cubic feet (12 trillion m³) of gas. This represents about 60 percent of the oil and 40 percent of the natural gas resources for the United States. Brazil has recently begun tapping an ultradeep oil field that could hold between 50 to 100 billion barrels of oil about 300 km offshore in the Atlantic Ocean. This find could be worth $10 trillion and make Brazil a major player in international oil.

And even after seeing crude oil hemorrhaging into the Gulf of Mexico, fish and seabirds wallowing in black sludge, and BP responsible for billions in damages, other nations are rushing to do their own ultradeep drilling. Ghana, Nigeria, Angola, Congo, Libya, Egypt, and Australia are among the countries offering deep-sea oil leases in their oceanic territories. If the agencies in the United States that are supposed to regulate offshore drilling are rife with cronyism, corruption, and incompetence, think what the oversight may be in some of these other places. America, Canada, and Russia also are exploring drilling in the Arctic Ocean (remember the Titanic?). All this is fueled, of course, by our insatiable appetite for oil. What do you think? What are the limits to the risks we are willing to take for the oil to which we've become accustomed?

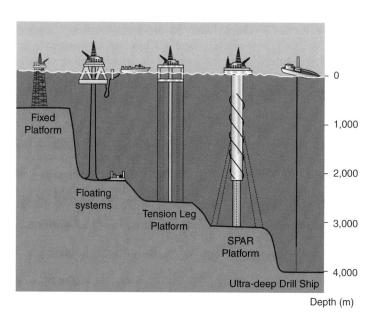

Types of drilling rigs. Note that the rigs aren't drawn to scale. The cylindrical spar, for example is about 200 m tall, while the drill rig below it reaches down as much as 10,000 m.

---

slowed significantly, many oil experts expect that we will pass this peak in the next few years.

About half of the world's original 4 trillion bbl (600 billion metric tons) of liquid oil are thought to be ultimately recoverable. (The rest is too diffuse, too tightly bound in rock formations, or too deep to be extracted.) Of the 2 trillion recoverable barrels, roughly 1.26 trillion bbl are in proven reserves. We have already used more than 0.5 trillion bbl—almost half of proven reserves—and the remainder is expected to last 41 years at current consumption

rates of 30.7 billion bbl per year. Middle Eastern countries have more than half of world supplies (fig. 19.13).

By far the largest supply of proven-in-place oil is in Saudi Arabia, which claims more than 400 billion barrels, at least one-fifth of the total proven world reserve. Just a dozen countries, seven of them in what the Organization for Economic Cooperation and Development (OECD) consider the "Greater Middle East," contain 91 percent of the proven, economically recoverable oil. With our insatiable appetite for (some would say addiction to) oil, it is

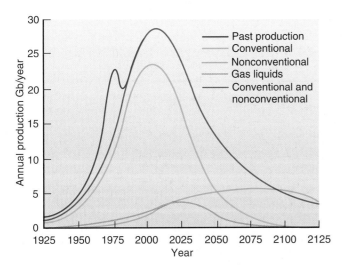

FIGURE 19.12 Worldwide production of crude oil with predicted Hubbert production. Gb = billion barrels.
Source: Jean Laherrère, www.hubbertpeak.org.

not difficult to see why this volatile region plays such an important role in world affairs. Although they didn't change the flow of oil very much, civil unrest and democratic revolutions across North Africa and the Middle East made oil markets uneasy in 2011. The price of crude oil spiked to over $100 per barrel, and reminded us of how dependent we are on imported energy.

Note that we have been discussing *proven* reserves. Oil companies estimate that reservoirs on outer continental shelves and in the deep ocean may hold several hundred billion barrels. Altogether, the United States has already used more than half of its original recoverable petroleum resource. Of the 120 billion barrels thought to remain, about 21 billion barrels are proven-in-place. If we stopped importing oil and depended exclusively on indigenous supplies, our proven reserves would last only four years at current rates of consumption.

Consumption rates continue to climb, however, both in developed countries and in the fast-growing economies such as China, India, and Brazil. China's energy demands have more

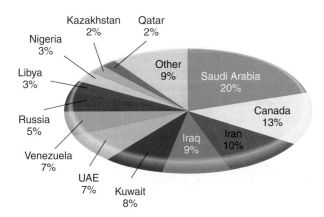

FIGURE 19.13 Proven oil reserves. Twelve countries (eight of them in the Greater Middle East) account for 91 percent of all known, economically recoverable oil.
Source: Data from U.S. DOE, 2008.

than tripled in the past 35 years (much of this energy is used to produce goods for the U.S. and European markets), and China anticipates another doubling of energy demands in the next 15 years. If that occurs, China will be using more energy than the United States. Although renewables are supplying a growing share of China's energy, it's clear that competition is growing for global oil and gas supplies.

The instability of global oil prices makes it difficult to establish conservation practices and renewable technologies. When oil is expensive, people are more likely to invest in alternative energy sources or change their lifestyle choices. When oil prices come tumbling back down, however, most people return to wasteful ways. China has an advantage in developing renewable resources in that the government can simply order homeowners and businesses to install solar collectors or to buy wind power. Currently the Chinese government charges a renewable energy fee to all electricity users. This fee raises billions of yuan annually to help subsidize alternative energy supplies.

Like other fossil fuels, oil has negative impacts

Oil extraction isn't as destructive to landscapes as strip-mining coal, but oil wells can be dirty and disruptive, especially in pristine landscapes. The largest remaining untapped land-based oil field in the United States is thought to be in the Arctic National Wildlife Refuge (ANWR) in northeastern Alaska. Drilling on the coastal plain could disrupt wildlife and wilderness in what has been called "North America's Serengeti."

Refining oil—at least as it's currently done in the United States—releases high levels of air pollution. Some of the worst air quality in America is found near the heavy concentrations of petrochemical industries in Texas, Louisiana, and New Jersey. BP, which is blamed for most of the bad decisions that led to the 2010 Gulf oil spill, has a terrible record of refinery operations. In the past three years, according to the Center for Public Integrity, BP accounted for "97 percent of all flagrant violations found in the refining industry." In 2005 an explosion at a BP Texas City refinery killed 15 workers and injured 170 others.

Like other fossil fuels, burning petroleum produces CO_2 emissions and contributes to global climate change. Sulfur is generally removed from gasoline, so that automobiles don't contribute very much to SO_2 emissions. Until recently, however, diesel fuel in the United States had high sulfur levels and also produced high amounts of very unhealthy particulate emissions. Internal combustion engines also produce large amounts of nitrogen oxides (NO_x). In addition to the danger of spills from wells and drilling platforms, shipping represents a major environmental threat. Every year about 1.5 billion tons (90 billion barrels) of oil are shipped in ocean tankers. On average, about 1 percent of the cargo (9 billion barrels, or about twice as much as the Deepwater Horizon accident) is spilled or discharged annually. Most of the oil enters the ocean from bilge washing; accidents represent only about one-quarter of the oil lost from tankers. Still, the effects can be catastrophic. On March 16, 1978, for example, the *Amoco Cadiz* ran aground 2 km off the coast of Brittany, France, when its steering failed during a storm (fig. 19.14). The entire cargo of 1.6 million barrels (68.7 million gallons) spilled into the sea. The oil contaminated approximately 350 km of the Brittany coastline, including the beaches of 76 communities. Fishing and tourism were devastated. Total economic losses were claimed to be about (U.S.) $1 billion. This was the largest tanker spill in history, more than six times as large as the wreck of the *Exxon Valdez* in Alaska in 1989. International tankers are supposed to be double hulled now to reduce the chances of such a disastrous spill, but many ship owners have ignored this requirement.

Oil shales and tar sands contain huge amounts of petroleum

Estimates of our recoverable oil supplies usually don't account for the very large potential from unconventional resources. The World Energy Council estimates that oil shales, tar sands, and other unconventional deposits contain ten times as much oil as liquid petroleum reserves. **Tar sands** are composed of sand and shale particles coated with bitumen, a viscous mixture of long-chain hydrocarbons. Shallow tar sands are excavated and mixed with hot water and steam to extract the bitumen, then fractionated to make useful products. For deeper deposits, superheated steam is injected to melt the bitumen, which can then be pumped to the surface like liquid crude. Once the oil has been retrieved, it still must be cleaned and refined to be useful.

Canada and Venezuela have the world's largest and most accessible tar sand resources. Canadian deposits in northern Alberta are estimated to be equivalent to 1.7 trillion bbl of oil, only about 10 percent of which is currently economically recoverable, and Venezuela has nearly as much. Together these deposits are three times as large as all conventional liquid oil reserves. By 2010 Alberta hopes to increase its flow to 2 million bbl per day, or twice the maximum projected output of the Arctic National Wildlife Refuge (ANWR). Furthermore, because Athabascan tar sand beds are 40 times larger and much closer to the surface than ANWR oil, the Canadian resource will last longer and may be cheaper to extract. Canada is already the largest supplier of oil to the United States, having surpassed Saudi Arabia in 2000.

FIGURE 19.14 The *Amoco Cadiz* ran aground off the coast of Brittany, France, on March 16, 1978, spilling 1.6 million barrels of oil and contaminating more than 350 km of coastline. The risk of similar spills is one cost of depending on imported oil.

FIGURE 19.15 Alberta tar sands are now the largest single source of oil for the United States, but there are severe environmental and social costs of extracting this oil.

There are severe environmental costs, however, in producing this oil (fig. 19.15). A typical facility producing 125,000 bbl of oil per day creates about 15 million m^3 of toxic sludge, releases 5,000 tons of greenhouse gases, and consumes or contaminates billions of liters of water each year. Surface mining in Canada could destroy millions of hectares of boreal forest. Native Cree, Chipewyan, and Metis people worry about effects on traditional ways of life if forests are destroyed and wildlife and water are contaminated. There are worries about the safety of oil pipelines, and environmentalists argue that investing billions of dollars to extract this resource simply makes us more dependent on fossil fuels.

Similarly, vast deposits of oil shale occur in the western United States. **Oil shale** actually is neither oil nor shale but a fine-grained sedimentary rock rich in solid organic material called kerogen. When heated to about 480°C (900°F), the kerogen liquefies and can be extracted from the stone. Oil shale beds up to 600 m (1,800 ft) thick occur in the Green River Formation in Colorado, Utah, and Wyoming, and lower-grade deposits are found over large areas of the eastern United States. If these deposits could be extracted at a reasonable price and with acceptable environmental impacts, they might yield the equivalent of several trillion barrels of oil.

Mining and extracting shale oil also creates many problems. It is expensive; it uses vast quantities of water, a scarce resource in the arid West; it has a high potential for air and water pollution; and it produces enormous quantities of waste. In the early 1980s, when the search for domestic oil supplies was at fever pitch, serious discussions occurred about filling whole canyons, rim to rim, with oil shale waste. One experimental mine used a nuclear explosion to break up the oil shale. All the oil shale projects dried up when oil prices fell in the mid-1980s, however. In 2009, Interior Secretary Ken Salazar announced that he would open leases for oil shale research and development in Colorado and Utah. "We need to push forward aggressively . . . to see if we can find a safe and economically viable way to unlock these resources on a commercial scale," he said. What do you think? How much environmental damage is acceptable to extend our oil supplies?

19.4 NATURAL GAS

Natural gas (mostly methane) is the world's third-largest commercial fuel, making up 24 percent of global energy consumption. Gas burns more cleanly than either coal or oil, and it produces only half as much CO_2 as an equivalent amount of coal. Substituting gas for coal could help reduce global warming. Many communities are now switching to natural gas for their energy supply.

Most of the world's known natural gas is in a few countries

Two-thirds of all proven natural gas reserves are in the Middle East and the former Soviet Union. The republics of the former Soviet Union have nearly 31 percent of known natural gas reserves (mostly in Siberia and the Central Asian republics) and account for about 40 percent of all production. Both eastern and western Europe buy substantial quantities of gas from these wells.

The total ultimately recoverable natural gas resources are estimated to be 10,000 trillion ft^3, corresponding to about 80 percent as much energy as the recoverable reserves of crude oil. The proven world reserves of natural gas are 6,200 trillion ft^3 (176 million metric tons). Because gas consumption rates are only about half of those for oil, current gas reserves represent roughly a 60-year supply at present usage rates. Proven reserves in North America are about 185 trillion ft^3, or 3 percent of the world total. This is a ten-year supply at current rates of consumption. Known reserves are more than twice as large. Figure 19.16 shows the distribution of proven natural gas reserves in the world.

As it breaks down, coal is slowly transformed into methane. Accumulation of this explosive gas is one of the things that makes coal mining so dangerous. In many places where mining coal seams isn't economically feasible, it is relatively cheap and easy to extract the methane.

New methane sources could be vast

As you learned from our discussion of offshore drilling, vast new natural gas deposits have recently been discovered in oceanic sediments. In addition, large amounts of methane are associated coal deposits on land. Geologists estimate that at least 346 trillion ft^3 of "technically recoverable" natural gas and 62 billion barrels of petroleum liquids occur in five intermountain basins stretching from Montana to New Mexico. This could be 10 percent of the total world methane supply. And shale deposits in the East could be even larger.

The Marcellus and Devonian Shales underlie parts of ten eastern states ranging from Georgia to New York (fig. 19.17). It has long been recognized that methane can be extracted from these formations, but estimates of recoverable amounts were relatively small. New developments in horizontal drilling and hydraulic fracturing

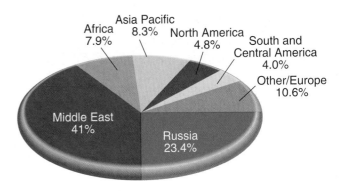

FIGURE 19.16 Proven natural gas reserves by region, 2008. **Source:** Data from British Petroleum, 2010.

along with increased exploratory drilling have now made this deposit a potentially "super-giant gas field." The U.S. Geological Survey now estimates that the Marcellus/Devonian formation may contain 500 trillion ft^3 (13 trillion m^3) of methane. If all of it were recoverable, it would make a 100-year supply for the United States at current consumption rates. And the Utica Shale, which lies below the Marcellus, could have even more methane. But the same issues, concerning a multitude of wells, water pollution, and threats to water supplies on which millions of people depend, raise thorny problems.

Much of this methane is found in relatively shallow formations, which makes it vastly cheaper to extract than most other gas supplies. Drilling a typical offshore gas well costs tens to hundreds of millions of dollars, while a deep conventional gas well costs about the same amount, but a coal-bed or shale-bed gas well is generally less than $100,000. The total value of the methane in the

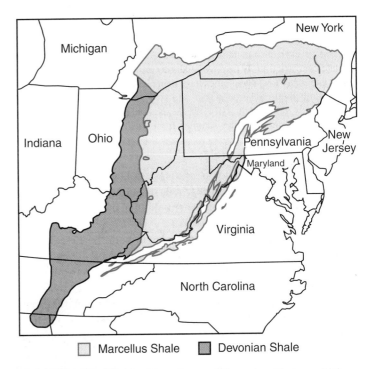

FIGURE 19.17 The Marcellus and Devonian Shales, which underlie much of the Appalachian Mountain chain, contains a "super-giant" gas field.

What Do You Think?

Coal-Bed Methane

Vast amounts of methane may lie in relatively shallow sediments under large areas of North America. This gas is a highly desirable fuel. It burns more cleanly than coal or oil, is easier to ship, and produces less CO_2. But often the resource is held in "tight" formations through which gas doesn't flow easily. To boost well output, mining companies rely on hydraulic fracturing (or "fracking"). A mixture of water, sand, and potentially toxic chemicals is pumped into the ground and rock formations at extremely high pressure. The pressurized fluid cracks sediments and releases the gas. Fracturing rock formations often disrupts aquifers, however, and contaminates water wells.

For years coal-bed methane extraction was a problem only in western states, but this controversial technology is now moving to the East Coast as well. Questions raised in the West concerning a multitude of wells, water pollution, and threats to drinking water supplies and public health are now arising in eastern states. While well drilling in the West is declining, it's being replaced by intense activity in the East. In the past two years, tens of thousands of wells have been drilled into the Marcellus shale in Pennsylvania, West Virginia, New York, and Ohio (see fig. 19.17). Much of this area lies in the environmentally sensitive Chesapeake Bay watershed or the Delaware River basin. Contamination of groundwater or surface runoff in these watersheds is of special concern (see chapter 3).

Drilling companies often refuse to reveal the chemical composition of the fluids they use in fracking. They claim it's a proprietary secret, but it's well known that a number of petroleum distillates, such as diesel fuel, benzene, toluene, xylene, polycyclic aromatic hydrocarbons, glycol ethers, as well as hydrochloric acid or sodium hydroxide, may be used. Many of these chemicals are known to be toxic to humans and wildlife. The U.S. EPA recently forced mining companies to reveal the contents, but not specific fractional composition, of their fracking fluids. Because hydraulic fracturing has a special exemption from the federal Safe Drinking Water Act (in an amendment known as the Halliburton clause), it's up to states and local units of government to protect public health. Several states and many towns and cities have demanded details on the chemicals being pumped into the ground.

A study released in 2011 by the National Academy of Sciences reported that drinking water samples from shallow wells near methane

There could be huge deposits of natural gas in North America, but the costs of extracting this gas could be unacceptable in many places.

drilling sites in Pennsylvania and New York had 17 times as much methane as those from sites far from drilling. And a study by researchers at Dartmouth concluded that 3.6 to 7.9 percent of the methane from shale-gas wells escapes to the atmosphere in leaks and venting over the life of the well. These methane emissions are up to twice those from conventional gas wells. Compared to coal, the footprint of shale gas is at least 20 percent greater for a comparable amount of energy, and may be twice as much over 100 years.

There also have been cases in which methane released by drilling has infiltrated into homes, some of which have exploded. In 2010 a well blowout in Pennsylvania sent more than 100,000 liters of fracking fluid into the air and onto the surrounding forest. Campers in a nearby campground were evacuated, although no serious injuries were reported. Wells in the Marcellus Shale often release radioactivity from uranium in sediments. A number of cities have passed ordinances prohibiting drilling within city limits or near schools or hospitals.

What do you think? Does having access to cleaner fuels justify the social and environmental costs of their extraction? If you were voting on this issue, what restrictions would you impose on the companies drilling wells in your hometown?

Marcellus formation could be trillions of dollars. But the prospect of drilling thousands of methane wells in densely populated places has raised new protests about water pollution, drinking water supplies, and public health.

The methane-containing sediments in both the East and the West are generally considered "tight formations"—that is, gas doesn't migrate easily through tiny pores in the rock. It often takes many closely spaced wells and directional drilling to extract methane from them. In Wyoming's Powder River basin, for example, 140,000 wells have been proposed for methane extraction. Together with the vast network of roads, pipelines, pumping stations, and service facilities, this industry is having serious impacts on ranching, wildlife, and recreation in formerly remote areas. For example, in the Upper Green River Basin 50,000 pronghorn antelope and 10,000 elk migrate through a narrow corridor every year

on their way between summer and winter ranges. The Jonah Gas Fields (fig. 19.18) lies across this migration route, and biologists worry that the noise, traffic, polluted waste water pits, and activity around the wells may interrupt the migration and doom the herd.

Water consumption and pollution are also huge problems in the arid West. It takes large amounts of water to drill the wells, and once in production, each well can produce up to 75,000 liters of salty water per day. Dumping this toxic waste into streams poisons wildlife and domestic livestock. In several western states, ranchers, hunters, anglers, conservationists, water users, and renewable energy activists have banded together in an unlikely coalition to fight against coal-bed gas extraction, calling on Congress to protect private property rights, preserve water quality, and conserve sensitive public lands. "It may be a clean fuel," says one rancher, "but it's a dirty business."

FIGURE 19.18 An aerial view of the Jonah Field in the Upper Green River Basin.

To boost well output, mining companies rely on hydraulic fracturing (or "fracking"). A mixture of water, sand, and potentially toxic chemicals is pumped into the ground and rock formations at extremely high pressure. The pressurized fluid cracks sediments and releases the gas. This often disrupts aquifers, however, and contaminates wells (What Do You Think? p. 434).

Gas can be shipped to market

In many places, gas and oil are found together in sediments, and both can be recovered at the same time. In remote areas, however, where no shipping facilities exist for the gas, it often is simply flared (burned) off—a terrible waste of a valuable resource (fig. 19.11). The World Bank estimates that 100 billion m^3 of gas, or 1.5 times the amount used annually in Africa, are flared every year. Increasingly, however, these "stranded" gas deposits are being captured and shipped to market.

World consumption of natural gas is growing by about 2.2 percent per year, considerably faster than either coal or oil. Much of this increase is in the developing world, where concerns about urban air pollution encourage the switch to cleaner fuel. Gas can be shipped easily and economically through buried pipelines. The United States has been fortunate to have abundant gas resources accessible by an extensive pipeline system. Until 2001, Canada was the primary source of natural gas for the United States, providing about 105 billion m^3 per year. Over the next 20 years Canadian exports are expected to decrease as more of its gas supply is used to heat and extract tar sands. Liquefied natural gas (LNG) imports, on the other hand, are expected to increase to about one-fourth of the 600 billion m^3 of natural gas consumed in the United States each year.

In other places gas lines have been subject to political or economic pressures. Russia, for example, has cut off gas supplies to Ukraine and Belarus in a dispute over prices and has threatened shipments to northern Europe over policy differences. Recent political unrest in North Africa made southern Europe nervous about its access to gas.

Intercontinental gas shipping can be difficult and dangerous. To make the process economical, gas is compressed and liquefied. At 160°C (260°F) the liquid takes up about one-six-hundredth the volume of gas. Special refrigerated ships transport LNG (fig. 19.19). Finding sites for terminals to load and unload these ships is difficult. Many cities are unwilling to accept the risk of an explosion of the volatile cargo. A fully loaded LNG ship contains about as much energy as a medium-size atomic bomb. Furthermore, huge amounts of seawater are used to warm and re-gasify the LNG. This can have deleterious effects on coastal ecology. To override local objections, the federal government has assumed jurisdiction over LNG terminal siting.

Other unconventional gas sources

Natural gas resources have been less extensively investigated than petroleum reserves. There may be extensive "unconventional" sources of gas in unexpected places. Prime examples are recently discovered methane hydrate deposits in arctic permafrost and beneath deep ocean sediments. **Methane hydrate** is composed of small bubbles or individual molecules of natural gas trapped in a crystalline matrix of frozen water. At least 50 oceanic deposits and a dozen land deposits are known. Altogether they are thought to hold some 10,000 gigatons (10^{13} tons) of carbon, or twice as much as the combined amount of all coal, oil, and conventional natural gas. This could be a valuable energy source but would be difficult to extract, store, and ship. If climate change causes melting of these deposits, it could trigger a catastrophic spiral of global warming because methane is 20 times as powerful a greenhouse gas as CO_2. Japan plans exploratory extraction of methane hydrate in the next few years, first on land near Prudhoe Bay, Alaska, and then in Japanese waters.

Methane also can be produced by digesting garbage or manure. Some U.S. cities collect methane from landfills and sewage sludge digestion. Because methane is so much more potent than CO_2 as a greenhouse gas, stopping leaks from pipelines and other sources is

FIGURE 19.19 As domestic supplies of natural gas dwindle, the United States is turning increasingly to shipments of liquefied gas in specialized ships, such as this one at an Australian terminal. An explosion of one of these ships would release about as much energy as a medium-size atomic bomb.

important in preventing global warming. In developing countries, small-scale manure digesters provide a valuable, renewable source of gas for heating, lighting, and cooking (chapter 20).

19.5 NUCLEAR POWER

In 1953 President Dwight Eisenhower presented his "Atoms for Peace" speech to the United Nations. He announced that the United States would build nuclear-powered electrical generators to provide clean, abundant energy. He predicted that nuclear energy would fill the deficit caused by predicted shortages of oil and natural gas. It would provide power "too cheap to meter" for continued industrial expansion of both the developed and the developing world. It would be a supreme example of "beating swords into plowshares." Technology and engineering would tame the evil genie of atomic energy and use its enormous power to do useful work.

Glowing predictions about the future of nuclear energy continued into the early 1970s. Between 1970 and 1974, American utilities ordered 140 new reactors for power plants (fig. 19.20). Some advocates predicted that by the end of the century there would be 1,500 reactors in the United States alone. In 1970 the International Atomic Energy Agency (IAEA) projected worldwide nuclear power generation of at least 4.5 million megawatts (MW) by the year 2000, 18 times more than our current nuclear capacity and twice as much as present world electrical capacity from all sources.

Rapidly increasing construction costs, declining demand for electric power, and safety fears have made nuclear energy much less attractive than promoters expected. Electricity from nuclear plants was about half the price of coal in 1970 but twice as much by 1990. Wind energy is already cheaper than new

FIGURE 19.20 In light of the 2011 disaster in Japan, officials in many countries are reexamining their commitment to nuclear power. According to the U.S. Nuclear Regulatory Commission, the riskiest facility in the United States is this one, the Indian Point Generating Station just 24 miles (38 km) up the Hudson River from New York City. Providing about one-third of the electricity to the city, this plant is estimated to have a 1 in 10,000 chance of core damage from an earthquake. What would the costs be if New York City had to be evacuated?

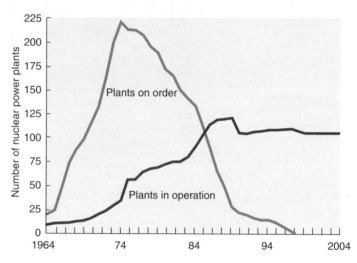

FIGURE 19.21 The changing fortunes of nuclear power in the United States are evident in this graph showing the number of nuclear plants on order and plants in operation.

nuclear plants in many areas, and solar power is becoming competitive as well (chapter 20).

After 1975 only 13 orders were placed for new nuclear reactors in the United States, and all of those orders subsequently were canceled (fig. 19.21). In fact, 100 of the 140 reactors on order in 1975 were canceled. It began to look as if the much-acclaimed nuclear power industry might have been a very expensive wild goose chase that would never produce enough energy to compensate for the amount invested in research, development, mining, fuel preparation, and waste storage.

How do nuclear reactors work?

The most commonly used fuel in nuclear power plants is U^{235}, a naturally occurring radioactive isotope of uranium. Ordinarily U^{235} makes up only about 0.7 percent of uranium ore, too little to sustain a chain reaction in most reactors. It must be purified and concentrated by mechanical or chemical procedures (fig. 19.22). Mining and processing uranium to create nuclear fuel is even more dirty and dangerous than coal mining. In some uranium mines 70 percent of the workers—most of whom were Native Americans—died from lung cancer caused by high radon and dust levels. In addition, mountains of radioactive tailings and debris have been left around fuel preparation plants.

When the U^{235} concentration reaches about 3 percent, the uranium is formed into cylindrical pellets slightly thicker than a pencil and about 1.5 cm long. These small pellets pack an amazing amount of energy. Each 8.5-gram pellet is equivalent to a ton of coal or four barrels of crude oil.

The pellets are stacked in hollow metal rods approximately 4 m long. About 100 of these rods are bundled together to make a **fuel assembly.** Thousands of fuel assemblies containing 100 tons of uranium are bundled together in a heavy steel vessel called the reactor core. Radioactive uranium atoms are unstable—that is, when struck by a high-energy subatomic particle called a

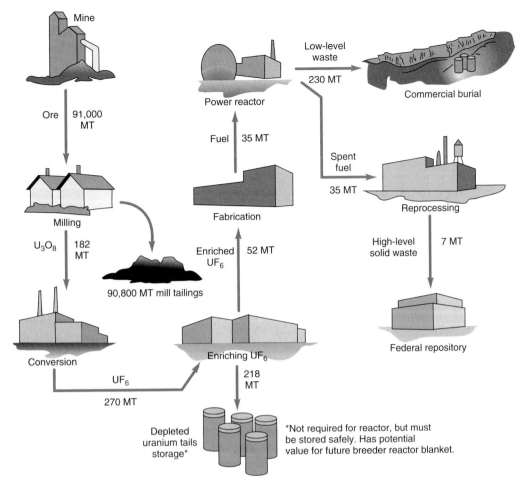

FIGURE 19.22 The nuclear fuel cycle. Quantities represent the average annual fuel requirements for a typical 1,000 MW light water reactor (MT = metric tons). About 35 MT, or one-third of the reactor fuel, is replaced every year. Reprocessing is not currently done in the United States.

*Not required for reactor, but must be stored safely. Has potential value for future breeder reactor blanket.

There are many different reactor designs

Seventy percent of the nuclear plants in the United States and in the world are pressurized water reactors (PWR) (fig. 19.24). Water is circulated through the core, absorbing heat as it cools the fuel rods. This primary cooling water is heated to 317°C (600°F) and reaches a pressure of 2,235 psi. It then is pumped to a steam generator where it heats a secondary water-cooling loop. Steam from the secondary loop drives a high-speed turbine generator that produces electricity. Both the reactor vessel and the steam generator are contained in a thick-walled concrete and steel containment building that prevents radiation from escaping and is designed to withstand high pressures and temperatures in case of accidents. Engineers operate the plant from a complex, sophisticated control room containing many gauges and meters to tell them how the plant is running.

Overlapping layers of safety mechanisms are designed to prevent accidents, but these fail-safe controls make reactors very expensive and very complex. A typical nuclear power plant has 40,000 valves, compared to only 4,000 in a fossil fuel-fired plant of similar size. In some cases the controls are so complex that they confuse operators and cause accidents rather than prevent them. Under normal operating conditions a PWR releases very little radioactivity and is probably less dangerous for nearby residents than a coal-fired power plant.

A simpler but dirtier and more dangerous reactor design is the boiling water reactor (BWR). In this model, water from the reactor core boils to make steam, which directly drives the turbine-generators. This means that highly radioactive water and steam leave the containment structure. Controlling leaks is difficult, and the chances of releasing radiation in an accident are very high, as was the case in the BWR in Fukushima Daiichi, Japan.

In Britain, France, and the former Soviet Union, a common reactor design uses graphite, both as a moderator and as the structural material for the reactor core. In the British MAGNOX design (named after the magnesium alloy used for its fuel rods), gaseous carbon dioxide is blown through the core to cool the fuel assemblies and carry heat to the steam generators. In the Soviet design, called RBMK (the Russian initials for a graphite-moderated, water-cooled reactor), low-pressure cooling water circulates through the core in thousands of small metal tubes.

neutron, they undergo **nuclear fission** (splitting), releasing energy and more neutrons. When uranium is packed tightly in the reactor core, the neutrons released by one atom will trigger the fission of another uranium atom and the release of still more neutrons (fig. 19.23). Thus, a self-sustaining **chain reaction** is set in motion and vast amounts of energy are released.

The chain reaction is moderated (slowed) in a power plant by a neutron-absorbing cooling solution that circulates between the fuel rods. In addition, **control rods** of neutron-absorbing material, such as cadmium or boron, are inserted into spaces between fuel assemblies to shut down the fission reaction or are withdrawn to allow it to proceed. Water or some other coolant is circulated between the fuel rods to remove excess heat.

The greatest danger in one of these complex machines is a cooling system failure. If the pumps fail or pipes break during operation, the nuclear fuel quickly overheats and a "meltdown" can result that releases deadly radioactive material. Although nuclear power plants cannot explode like a nuclear bomb, the radioactive releases from a worst-case disaster like the 2011 meltdown at Japan's Fukushima Daiichi nuclear complex can be just as devastating as a bomb.

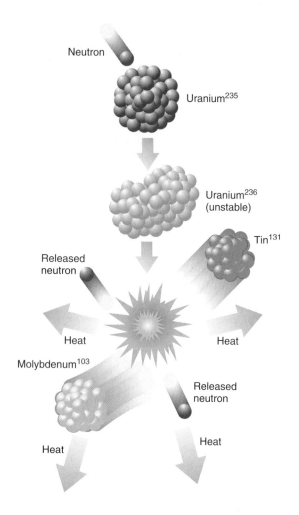

FIGURE 19.23 The process of nuclear fission is carried out in the core of a nuclear reactor. In the sequence shown here, the unstable isotope, uranium-235, absorbs a neutron and splits to form tin-131 and molybdenum-103. Two or three neutrons are released per fission event and continue the chain reaction. The total mass of the reaction product is slightly less than the starting material. The residual mass is converted to energy (mostly heat).

These designs were originally thought to be very safe because graphite has high capacity for both capturing neutrons and dissipating heat. Designers claimed that these reactors could not possibly run out of control; unfortunately, they were proven wrong. The small cooling tubes are quickly blocked by steam if the cooling system fails and the graphite core burns when exposed to air. Two of the most disastrous reactor accidents in the world involved fires in graphite cores that allowed the nuclear fuel to melt and escape into the environment. A 1956 fire at the Windscale Plutonium Reactor in England contaminated hundreds of square kilometers of countryside. Similarly, burning graphite in the Chernobyl nuclear plant in Ukraine made the fire much more difficult to control than it might have been in another reactor design.

The most serious accident at a North American commercial reactor occurred in 1979 when the Three Mile Island nuclear plant near Harrisburg, Pennsylvania, suffered a partial meltdown of the reactor core. The containment vessel held in most radioactive material. No deaths or serious injuries were verified, but the accident was a serious blow to future nuclear development.

Some alternative reactor designs may be safer

Several other reactor designs are inherently safer than the ones we now use. Among these is the modular high-temperature, gas-cooled reactor (HTGCR), which is sometimes called a "pebble-bed reactor." Uranium is encased in tiny ceramic-coated pellets; gaseous helium blown around these pellets is the coolant. If the reactor core is kept small, it cannot generate enough heat to melt the ceramic coating, even if all coolant is lost; thus, a meltdown is impossible and operators could walk away during an accident without risk of a fire or radioactive release. Fuel pellets are loaded into the reactor from the top, shuffle through the core as the uranium is consumed, and emerge from the bottom as spent fuel. This type of reactor can be reloaded during operation. Because the reactors are small, they can be added to a system a few at a time, avoiding the costs, construction time, and long-range commitment of large reactors. Only two of these reactors have been tried in the United States: the Brown's Ferry reactor in Alabama and the Fort St. Vrain reactor near Loveland, Colorado. Both were continually plagued with problems (including fires in control buildings and turbine-generators), and both were closed without producing much power.

A much more successful design has been built in Europe by General Atomic. In West German tests, an HTGCR was subjected to total coolant loss while running at full power. Temperatures remained

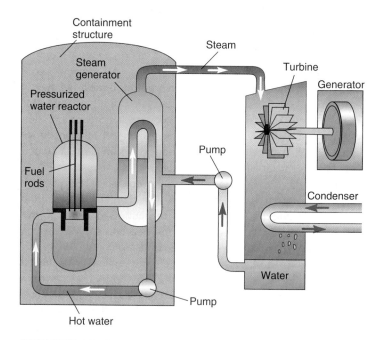

FIGURE 19.24 Pressurized water nuclear reactor. Water is superheated and pressurized as it flows through the reactor core. Heat is transferred to nonpressurized water in the steam generator. The steam drives the turbogenerator to produce electricity.

well below the melting point of fuel pellets and no damage or radiation releases occurred. These reactors might be built without expensive containment buildings, emergency cooling systems, or complex controls. They would be both cheaper and safer than current designs.

Some engineers argue that downsized, simplified reactors could be useful. Micro-reactors, with a standard design about the size of a minivan, might be mass-produced to bring down prices. They might produce only a few megawatts rather than the 1,000 MW typical of conventional nuclear plants, and perhaps could be sealed like a big battery and buried underground for decades, so terrorists couldn't get into them and radioactive waste couldn't get out. At the end of their useful life, they could be picked up and trucked to a reprocessing facility, avoiding the costly and time-consuming process of deconstructing and shipping a commercial reactor. Proponents of this new design claim that it could produce electricity at a price competitive with fossil fuels, and would be useful in remote areas where electricity isn't available.

Small reactors have already been used in a variety of applications, including submarines and satellites, but most of those examples have been military installations that don't have the same safety or security concerns as ordinary civilian use. How safe would it be if there were hundreds or thousands of micro-reactors spread across the country?

Breeder reactors might extend the life of our nuclear fuel

For more than 30 years, nuclear engineers have been proposing high-density, high-pressure **breeder reactors** that produce fuel rather than consume it. These reactors create fissionable plutonium and thorium isotopes from the abundant, but stable, forms of uranium (fig. 19.25). The starting material for this reaction is plutonium reclaimed from spent fuel from conventional fission reactors. After about ten years of operation, a breeder reactor would produce enough plutonium to start another reactor. Sufficient uranium currently is stockpiled in the United States to produce electricity for 100 years at present rates of consumption, if breeder reactors can be made to work safely and dependably.

Several problems have held back the breeder reactor program in the United States. One problem is the concern about safety. The reactor core of the breeder must be at a very high density for the breeding reaction to occur. Water does not have enough heat capacity to carry away the high heat flux in the core, so liquid sodium generally is used as a coolant. Liquid sodium is very corrosive and difficult to handle. It burns with an intense flame if exposed to oxygen, and it explodes if it comes into contact with water. Because of its intense heat, a breeder reactor will melt down and self-destruct within a few seconds if the primary coolant is lost, as opposed to a few minutes for a normal fission reactor.

Another very serious concern about breeder reactors is that they produce excess plutonium that can be used for bombs. It is essential to have a spent-fuel reprocessing industry if breeders are used, but the existence of large amounts of weapons-grade

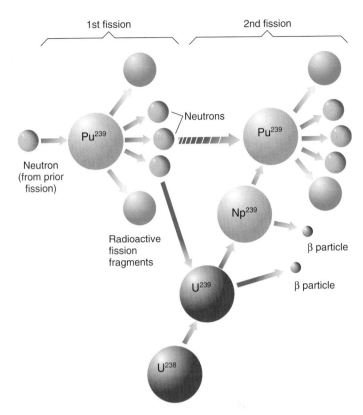

FIGURE 19.25 Reactions in a "breeder" fission process. Neutrons from a plutonium fission change U^{238} to U^{239} and then to Pu^{239} so that the reactor creates more fuel than it uses.

plutonium in the world would surely be a dangerous and destabilizing development. The chances of some of that material falling into the hands of terrorists or other troublemakers are very high. Japan planned to purchase 30 tons of this dangerous material from France and ship it half way around the world through some of the most dangerous and congested shipping lanes on the planet to fuel a breeder program. In 1995 a serious accident at Japan's Moju breeder reactor caused reevaluation of the whole program.

A proposed $1.7 billion breeder-demonstration project in Clinch River, Tennessee, was on and off for 15 years. At last estimate it would cost up to five times the original price if it is ever completed. In 1986 France put into operation a full-sized commercial breeder reactor, the SuperPhénix, near Lyons. It cost three times the original estimate to build and produces electricity at twice the cost per kilowatt of conventional nuclear power. After only a year of operation, a large crack was discovered in the inner containment vessel of the SuperPhénix, and in 1997 it was shut down permanently.

19.6 RADIOACTIVE WASTE MANAGEMENT

One of the most difficult problems associated with nuclear power is the disposal of wastes produced during mining, fuel production, and reactor operation. How these wastes are managed may ultimately be the overriding obstacle to nuclear power.

We lack safe storage for radioactive wastes

Enormous piles of radioactive mine wastes and abandoned mill tailings are the first disposal problem in the nuclear fuel cycle (see fig. 19.22). Production of 1,000 tons of uranium fuel typically generates 100,000 tons of tailings and 3.5 million liters of liquid waste. There now are approximately 200 million tons of radioactive waste in piles around mines and processing plants in the United States. This material is carried by the wind or washes into streams, contaminating areas far from its original source.

In addition to the leftovers from fuel production, the United States has about 100,000 tons of low-level waste (contaminated tools, clothing, building materials, and so on) and about 77,000 tons of high-level (very radioactive) wastes. The high-level wastes consist mainly of spent fuel rods from commercial nuclear power plants and assorted wastes from nuclear weapons production. While they're still intensely radioactive, spent fuel assemblies are stored in deep, water-filled pools at the power plants. These pools were originally intended only as temporary storage until the wastes were shipped to reprocessing centers or permanent disposal sites.

In 1987, after a years-long search in which all the states with possible sites said, "Not in my back yard (NIMBY)," the U.S. Department of Energy announced plans to build the first high-level waste repository on a barren desert ridge under Yucca Mountain in Nevada. Waste was to be buried deep in the ground, where it was hoped it would remain unexposed to groundwater and earthquakes for the thousands of years required for the radioactive materials to decay to a safe level. But continuing worries about the stability of the site led the Obama administration to cut off funding for the project in 2009 after 20 years of research and $100 billion in exploratory drilling and development.

For the foreseeable future, the high-level wastes that were to go to Yucca Mountain will be held in storage pools and dry casks located at 131 sites in 39 states (fig. 19.26). But local residents living near these sites fear casks will leak. Most nuclear power plants are built near rivers, lakes, or seacoasts. Radioactive materials could spread quickly over large areas if leaks occur. A hydrogen gas explosion and fire in 1997 in a dry storage cask at Wisconsin's Point Beach nuclear plant intensified opponents' suspicions about this form of waste storage.

However, to say that we haven't yet decided how to store nuclear waste doesn't prove there is no safe way to store it. My neighbor and I can't agree on where the property line is between our yards. But that doesn't mean there is no line—just that we can't decide. Still, if the owners of nuclear facilities had to pay the full cost for fuel, waste storage, and insurance against catastrophic accidents, no one would be interested in this energy source. Rather than being too cheap to meter, it would be too expensive to matter.

Russia has offered to store nuclear waste from other countries. Plans are to transport wastes to the Mayak in the Ural Mountains. The storage site is near Chelyabinsk, where an explosion at a waste facility in 1957 contaminated about 24,000 km^2 (9,200 mi^2). The region is now considered the most radioactive place on earth, so the Russians feel it can't get much worse. They expect that storing 20,000 tons of nuclear waste should pay about $20 billion.

Some nuclear experts believe that monitored, retrievable storage would be a much better way to handle wastes. This method involves holding wastes in underground mines or secure surface facilities where they can be watched. If canisters begin to leak, they could be removed for repacking. Safeguarding the wastes would be expensive, and the sites might be susceptible to wars or terrorist attacks. We might need a perpetual priesthood of nuclear guardians to ensure that the wastes are never released into the environment.

Shipping nuclear waste to a storage site worries many people—especially those whose cities will be on the shipping route. The Energy Department has performed crash tests on the shipping containers and assures us they are safe (fig. 19.27). Some people still worry about accidents or terrorist attacks. How would you feel if these trains were coming through your city? Or would it be better to keep the waste where it is, at 100 separate power plants?

FIGURE 19.26 Spent fuel is being stored temporarily in large, aboveground "dry casks" at many nuclear power plants.

FIGURE 19.27 A test railcar carrying a spent nuclear fuel shipping cask slams into a concrete wall at 130 km/hr (81 mph). The cask survived without injury. Even so, many people don't want nuclear waste shipped through their city. Would you?

Decommissioning old nuclear plants is expensive

Old power plants themselves eventually become waste when they have outlived their useful lives. Most plants are designed for a life of only 30 years. After that, pipes become brittle and untrustworthy because of the corrosive materials and high radioactivity to which they are subjected. Plants built in the 1950s and early 1960s already are reaching the ends of their lives. You don't just lock the door and walk away from a nuclear power plant; it is much too dangerous. It must be taken apart, and the most radioactive pieces have to be stored just like other wastes. This includes not only the reactor and pipes but also the meter-thick, steel-reinforced concrete containment building. The pieces are cut apart by remote-control robots because it's too dangerous to work directly on them.

Think About It

Is the energy from nuclear power worth the costs? Should we build new reactors and allow existing ones to continue to operate in order to reduce our dependence on fossil fuels? How would you evaluate the risks and benefits of this technology?

Altogether, the U.S. reactors now in operation might cost somewhere between $200 billion and $1 trillion to decommission. No one knows how much it will cost to store the debris for thousands of years or how it will be done. However, we would face this problem, to some degree, even without nuclear electric power plants. Plutonium production plants and nuclear submarines also have to be decommissioned. Originally the Navy proposed to just tow old submarines out to sea and sink them. The risk that the nuclear reactors would corrode away and release their radioactivity into the ocean makes this method of disposal unacceptable, however.

19.7 CHANGING FORTUNES OF NUCLEAR POWER

Although promoted originally as a new wonder of technology that could open the door to wealth and abundance, nuclear power has long been highly controversial. Public opinion about nuclear power has fluctuated over the years. Before the Three Mile Island accident in 1978, two-thirds of Americans supported nuclear power. After Chernobyl exploded in 1986, less than one-third of Americans favored this power source. But in recent years as memories of these disasters faded, public support for nuclear energy has been rising.

With oil and natural gas prices soaring and worries about global warming causing concern about coal usage, nuclear advocates—and even some prominent conservationists—have been promoting nuclear reactors as clean and environmentally friendly because they don't emit greenhouse gases. There's been talk about a "nuclear renaissance," not only in the United States but in other countries, as well. In 2010 the Obama administration pledged a conditional $8.3 billion loan to support construction of two nuclear reactors in Georgia, which, if built, would be the first new U.S. nuclear plants in more than three decades. The administration's fiscal 2011 budget requested an additional $36 billion to guarantee loans for seven to ten additional power plants. And in Japan, where 102 nuclear reactors provided about one-third of the country's electricity, there were plans to expand nuclear energy to about half the nation's power supply.

All that changed, however, with the magnitude-9 earthquake and huge tsunami that hit the northeast coast of Japan on March 11, 2011 (see fig. 14.20). Reactors at the Fukushima Daiichi nuclear power complex shut down, as they were designed to do, when the earthquake hit. But that cut the electrical supply needed to pump cooling water through the reactor core. Backup generators and connections to the regional power grid that would have provided emergency power were destroyed by the tsunami. The reactors quickly overheated, and hydrogen gas explosions damaged several containment buildings. Much worse was that the fuel rods began to melt inside three of the six reactors in the complex, and when water boiled away from nuclear waste storage pools, fuel rods there also began to melt and burn. Even several months after the disaster, the site was so radioactive that it was impossible to inspect the damage to assess how bad it was. Some nuclear experts warned that molten blobs of radioactive metal could be burning through the bottoms of the reactor vessels and into the ground below. The Japanese government rated the disaster a 7—the highest rating on an international scale—equaling Chernobyl. Some observers warned that Fukushima Daiichi could be much worse than Chernobyl because more radioactive material is involved.

Residents were ordered to evacuate a large area around Fukushima, and elevated radiation levels were detected in milk, vegetables, seafood, and some water supplies in the region. The Daiichi reactors were 40 years old and past their designed operating life. It turns out that many reactors in the United States and other countries are the same age and share similar risk factors. A number of governments have changed their policies about nuclear power. Japan announced that it would abandon plans to expand nuclear energy and would reassess the safety of all its aging reactors. Perhaps most extreme, Germany announced that it would phase out all 17 of its reactors and switch to renewable energy sources, such as solar, wind, and geothermal power. Perhaps rather than having a renaissance, this catastrophe may be the death knell of nuclear power.

Fusion energy is an alternative to nuclear fission that could have virtually limitless potential. **Nuclear fusion** energy is released when two smaller atomic nuclei fuse into one larger nucleus. Nuclear fusion reactions, the energy source for the sun and for hydrogen bombs, have not yet been harnessed by humans to produce useful net energy. The fuels for these reactions are deuterium and tritium, two heavy isotopes of hydrogen.

It has been known for 50 years that if temperatures in an appropriate fuel mixture are raised to 100 million degrees Celsius and pressures of several billion atmospheres are obtained, fusion of deuterium and tritium will occur. Under these conditions, the electrons are stripped away from atoms and the forces that normally keep nuclei apart are overcome. As nuclei fuse, some of their mass is converted into energy, some of which is in the form of heat. There are two main schemes for creating these conditions: magnetic confinement and inertial confinement.

Inertial confinement involves a small pellet (or a series of small pellets) bombarded from all sides at once with extremely high-intensity laser light (fig. 19.28a). The sudden absorption of energy causes an implosion (an inward collapse of the material) that will increase densities by 1,000 to 2,000 times and raise temperatures above the critical minimum. So far, no lasers powerful enough to create fusion conditions have been built.

Magnetic confinement involves the containment and condensation of plasma, a hot, electrically neutral gas of ions and free electrons, in a powerful magnetic field inside a vacuum chamber. Compression of the plasma by the magnetic field should raise temperatures and pressures enough for fusion to occur. The most promising example of this approach, so far, has been a Russian design called *tokomak* (after the Russian initials for "torodial magnetic chamber"), in which the vacuum chamber is shaped like a large donut (fig. 19.28b).

In both of these cases, high-energy neutrons escape from the reaction and are absorbed by molten lithium circulating in the walls of the reactor vessel. The lithium absorbs the neutrons and transfers heat to water via a heat exchanger, making steam that drives a turbine generator, as in any steam power plant. The advantages of fusion reactions, if they are ever feasible, include production of fewer radioactive wastes, the elimination of fissionable products that could be made into bombs, and a fuel supply that is much larger and less hazardous than uranium.

Despite 50 years of research and a $25 billion investment, fusion reactors never have reached the break-even point at which they produce more energy than they consume. A major setback occurred in 1997 when Princeton University's Tokomak Fusion Test Reactor was shut down. Three years earlier this reactor had set a world's record by generating 10.7 million watts for one second, but researchers conceded that the technology was still decades away from self-sustaining power generation. In 2006, China, South Korea, Russia, Japan, the United States, and the European Union announced a new (U.S.) $13 billion fusion reactor to be built jointly near Marseilles, France. Opponents view this project as just another expensive wild-goose chase and predict that it will never generate enough energy to pay back the fortune spent on its development. A standard joke among workers in this field is that we're only 20 years from achieving nuclear fusion—and always will be.

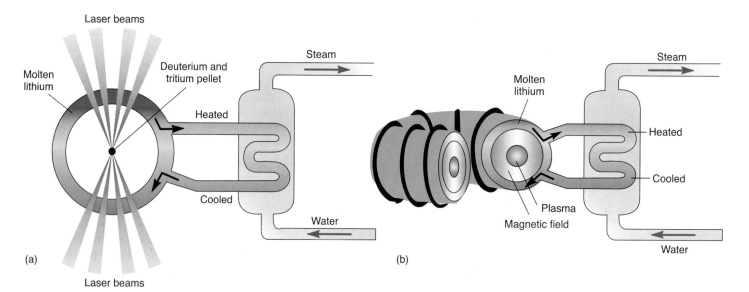

FIGURE 19.28 Nuclear fusion devices. (a) Inertial confinement is created by laser beams that bombard and ignite fuel pellets. Molten lithium transfers heat to a steam generator. (b) In the tokomak design, a powerful magnetic field confines the plasma and compresses it so that critical temperatures and pressures are reached.

CONCLUSION

Our energy future is far from certain. We have probably used half of the easily accessible liquid petroleum reserves in the world. This provided a lifestyle of luxury and convenience for those of us lucky enough to live in the industrialized countries of the world, but it has created titanic environmental problems—including acid rain, strip-mined landscapes, massive oil spills, huge payments to unstable countries, and, perhaps most importantly, global climate change. There are still very large supplies of unconventional fossil fuels, including tar sands, oil shale, coalbed methane, and methane hydrates, but the environmental costs of extracting those resources may preclude their use.

What, then, should we do? Some people hold out the promise of technological solutions to this dilemma. They point to IGCC, nuclear power, and possibly fusion reactors as attractive energy sources. Others, however, point to nuclear disasters and unacceptable waste storage options. Many argue that we ought to move immediately toward conservation and renewable energy, such as solar, wind, biofuels, small-scale hydro, and geothermal power. Even if we do this, however, it will probably take decades to replace our dependence on fossil fuels. Therefore, it's important to understand the relative benefits and disadvantages of each of our conventional energy sources.

As consumers, each of us needs to examine our energy use and its environmental impacts. In chapter 20 we'll investigate conservation and renewable energy options.

REVIEWING LEARNING OUTCOMES

By now you should be able to explain the following points:

19.1 Define *energy, work,* and how our energy use has varied over time.
- How do we measure energy?
- Fossil fuels supply most of the world's energy.
- How do we use energy?

19.2 Describe the benefits and disadvantages of using coal.
- Coal resources are vast.
- Coal mining is a dirty, dangerous business.
- Burning coal releases many pollutants.
- Clean coal technology could be helpful.

19.3 Explain the consequences and rewards of exploiting oil.
- Have we passed peak oil?
- Like other fossil fuels, oil has negative impacts.
- Oil shales and tar sands contain huge amounts of petroleum.

19.4 Illustrate the advantages and disadvantages of natural gas.
- Most of the world's known natural gas is in a few countries.
- New methane sources could be vast.
- Gas can be shipped to market.
- Other unconventional sources.

19.5 Summarize the potential and risk of nuclear power.
- How do nuclear reactors work?
- There are many different reactor designs.
- Some alternative reactor designs may be safer.
- Breeder reactors might extend the life of our nuclear fuel.

19.6 Evaluate the problems of radioactive wastes.
- We lack safe storage for radioactive wastes.
- Decommissioning old nuclear plants is expensive.

19.7 Discuss the changing fortunes of nuclear power.

19.8 Identify the promise and peril of nuclear fusion.

PRACTICE QUIZ

1. What is energy? What is power?
2. What are the major sources of commercial energy worldwide and in the United States? Why are data usually presented in terms of commercial energy?
3. How does energy use in the United States compare with that in other countries?
4. How much coal, oil, and natural gas are in proven reserves worldwide? Where are those reserves located?
5. What is coal-bed methane, and why is it controversial?
6. What are the most important health and environmental consequences of our use of fossil fuels?
7. Describe how a nuclear reactor works and why reactors can be dangerous.
8. What are the four most common reactor designs? How do they differ from each other?
9. What are the advantages and disadvantages of the breeder reactor?
10. Describe methods proposed for storing and disposing of nuclear wastes.

CRITICAL THINKING AND DISCUSSION QUESTIONS

1. We have discussed a number of different energy sources and energy technologies in this chapter. Each has advantages and disadvantages. If you were an energy policy analyst, how would you compare such different problems as the risk of a nuclear accident versus air pollution effects from burning coal?

2. If your local utility company were going to build a new power plant in your community, what kind would you prefer? Why?

3. The nuclear industry is placing ads in popular magazines and newspapers claiming that nuclear power is environmentally friendly because it doesn't contribute to the greenhouse effect. How do you respond to that claim?

4. Our energy policy effectively treats some strip-mine and well-drilling areas as national sacrifice areas, knowing they will never be restored to their original state when extraction is finished. How do we decide who wins and who loses in this transaction?

5. Storing nuclear wastes in dry casks outside nuclear power plants is highly controversial. Opponents claim the casks will inevitably leak. Proponents claim they can be designed to be safe. What evidence would you consider adequate or necessary to choose between these two positions?

6. The policy of the United States has always been to make energy as cheap and freely available as possible. Most European countries charge three to four times as much for gasoline as we do. Who benefits and who or what loses in these different approaches? How have our policies shaped our lives? What does existing policy tell you about how governments work?

 **Data Analysis:** Comparing Energy Use and Standards of Living

In general, income and standard of living increase with energy availability. This makes sense because cheap energy makes it possible to heat and air condition our homes, travel easily and frequently, obtain fresh foods out of season, have a wide variety of entertainment, work, and educational opportunities, and use machines to extend our productivity. However, energy use per capita isn't strictly tied to quality of life. Some countries use energy extravagantly without corresponding increases in income or standard of living. Look at the graph on this page and answer the following questions:

1. What country in this graph has the highest energy use?

2. How much energy does it use, and how much per capita income does it have?

3. What do you think might explain these values?

4. What do you know about the standard of living in this country?

5. How much energy per person do the United States and Denmark use annually?

6. How do you think the standard of living in the United States and Denmark compare?

7. How would you characterize energy use and income in Malaysia and Poland compared to Luxembourg?

8. In which of these countries would you rather live?

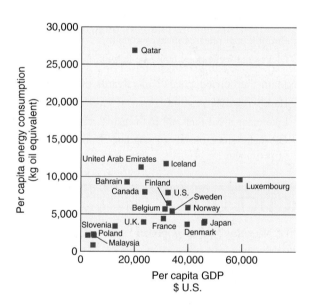

Per capita energy consumption and GDP.

Concentrating Solar Power (CSP) plants in desert regions could supply more electricity than all current world use.

Sustainable Energy

Learning Outcomes

After studying this chapter, you should be able to:

20.1 Describe renewable energy resources.

20.2 Explain how we could tap solar energy.

20.3 Grasp the potential of fuel cells.

20.4 Explain how we get energy from biomass.

20.5 Summarize the prospects for hydropower.

20.6 Report on the applications for wind power.

20.7 Visualize the uses of waves, tides, and geothermal energy.

20.8 Discuss our energy future.

"We know the country that harnesses the power of clean, renewable energy will lead the twenty-first century."

~ *Barack Obama*

445

Desertech: A Partnership for Renewable Energy

Northern Europe has a problem. They'd like to be environmentally responsible and wean themselves away from fossil fuels. Coastal regions generally have good wind power resources, and Great Britain, Germany, the Netherlands, and Scandinavia lead the world in offshore wind farms. But the most abundant renewable energy supply—solar—is often sorely lacking in the notoriously dark, cloudy, northern regions. Look at the location of northern Europe on a globe. Stockholm, Oslo, and Helsinki, for example, are all at about the same latitude as Anchorage, Alaska.

A great solar resource exists, however, just across the Mediterranean Sea in the Sahara Desert, where the skies are cloudless and the sun shines fiercely nearly every day. An area about 125 × 125 km—or about 0.3 percent of North Africa—receives enough sunlight to supply all the current electrical consumption in Europe. And high-voltage, direct-current (HVDC) transmission lines have advanced, so it's economically and technically feasible to ship electrical current from Africa to Europe. Transmission losses are only 3 percent per 1,000 km and add just 1–2 cents per kilowatt-hour, an insignificant amount when you consider that the fuel is free.

A consortium led by the German Aerospace Center has been studying this issue for a decade. Operating under the name Desertech, about a dozen German banks and energy companies, together with other interested parties in more than 20 countries, have begun building a giant network of renewable energy facilities and a HVDC Supergrid they hope will eventually link Europe, the Middle East, and North Africa (EU-MENA) to make a significant contribution both to regional development and to combating global climate change.

Some three dozen concentrating solar power (CSP) plants, spread across North Africa and the Middle East, together with about 20 offshore wind farms, a dozen hydroelectric dams, and a few biomass or geothermal facilities (fig. 20.1) linked together by HVDC "electric highways" form the heart of this ambitious plan. We'll discuss details of CSP later in this chapter, but basically it captures solar energy to generate steam that produces electricity. This technology is already competitive with fossil fuels. In fact, in

2008, when oil hit $140 per barrel, CSP was less than half the price of an equivalent amount of oil energy.

Why would oil-rich Arab countries want to help Europe kick their fossil fuel habit? As we saw in chapter 19, the world is approaching—or may have already passed—peak oil production. And remaining supplies are becoming increasingly expensive and difficult to reach. Many formerly oil-rich countries are facing the prospect of life without oil. Why not sell an endless supply of solar power, and save your remaining oil for your own use or to sell for higher prices at a later date?

Wouldn't this just mean trading dependency on unstable Middle Eastern countries for oil to dependency for their solar electricity? Perhaps. But if Desertech leads to local economic development (it's expected that building and operating all those power plants will add 80,000 well-paying local jobs), and if local economies become dependent on the power and water from renewable energy, mutual benefits from the system may help make it safe from political threats and civil unrest.

The first steps in Desertech implementation are now taking place. In 2011, contracts were issued for 65 km of HVDC to connect Spain and France—the first link in the Supergrid. And at the same time, Morocco, which has been selected for the first CSP plant, announced it had chosen both the site for the facility and four consortia partners to design, finance, build, and operate it.

Many other parts of the world are following this development with interest. China, Australia, South Africa, and western North America also have vast solar potential. The Desertech Consortium points out that within 6 hours, world deserts receive more energy from the sun than humankind consumes in an entire year. Perhaps many others of us could benefit from a similar system.

In this chapter we'll look at our options for finding environmentally and socially sustainable ways to meet our energy needs. For related resources, including Google Earth™ placemarks that show locations where these issues can be seen, visit EnvironmentalScience-Cunningham.blogspot.com.

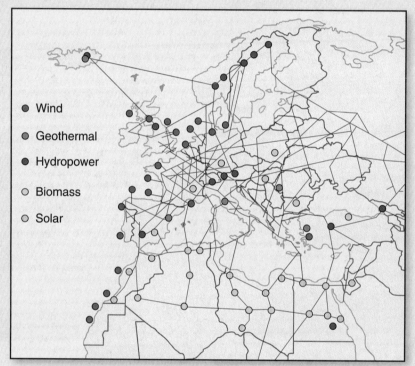

FIGURE 20.1 A supergrid of HVDC transmission lines may link a network of renewable energy facilities in Europe, North Africa, and the Middle East and provide both a substantial percentage of electricity for the region as well as drinking water for desert nations.
Source: German Aerospace Center, 2010.

Legend:
- Wind
- Geothermal
- Hydropower
- Biomass
- Solar

20.1 RENEWABLE ENERGY

In his 2011 State of the Union speech, President Barack Obama said, "To truly transform our economy, protect our security, and save our planet from the ravages of climate change, we need to ultimately make clean, renewable energy the profitable kind of energy ... So tonight, I challenge you to join me in setting a new goal: By 2035, 80 percent of America's electricity will come from clean energy sources." He called this our "Sputnik moment." For those who don't remember Sputnik, it was the first space satellite. Launched in 1957 by the Soviet Union, Sputnik shocked the United States by suggesting that the United States lagged in rocket technology. The U.S. Congress responded with a massive mobilization of money and resources in science and technology. Many benefits eventually came from this investment, including computer chips, photovoltaic energy, GPS, and the Internet. Many of us hope that investment in clean energy today can similarly provide millions of jobs, freedom from dependence on foreign oil, a boost to our economy, and a cleaner environment.

The good news is that using currently available technology and only those sites where energy facilities are socially, economically, and politically acceptable, there's more than enough power from the sun, wind, geothermal, biomass, and other sources to meet all our present energy needs (fig. 20.2). We'll look at each of those sources in this chapter.

One of the easiest ways to avoid energy shortages and to relieve environmental and health effects of our current energy technologies is simply to use less. Per capita energy consumption in the United States rose sharply in the 1960s, but price shocks in the 1970s led to dramatic improvements in household

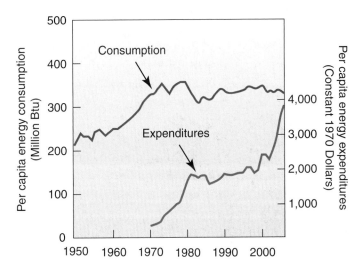

FIGURE 20.3 Per capita energy consumption in the United States rose rapidly in the 1960s. Price shocks in the 1970s encouraged conservation. Although GDP continued to grow in the 1980s, higher efficiency kept per capita consumption relatively constant. Expenditures per person, however have risen sharply. **Source:** Data from U.S. Department of Energy, 2010.

and industrial energy use (fig. 20.3). Although population and GDP have continued to grow since then, energy intensity, or the amount of energy needed to provide goods and services, has declined, and total energy use has remained relatively stable. Energy prices have grown rapidly, however, as we use up readily available supplies and search for energy in ever more distant and dangerous places. Even those who don't care about environmental or social issues may be interested in reducing energy consumption for economic reasons.

There are many ways to save energy

Much of the energy we consume is wasted. This statement isn't a simple admonishment to turn off lights and turn down furnace thermostats in winter; it's a technological challenge. Our ways of using energy are so inefficient that most potential energy in fuel is lost as waste heat, becoming a form of environmental pollution. Of the energy we do extract from primary resources, however, much is used for frankly trivial or extravagant purposes. As chapter 19 shows, several European countries have higher standards of living than the United States, and yet use 30 to 50 percent less energy.

Many conservation techniques are relatively simple and highly cost-effective. Compact fluorescent bulbs, for example, produce four times as much light as an incandescent bulb of the same wattage, and last up to ten times as long. Although they cost more initially, total lifetime savings can be $30 to $50 per fluorescent bulb.

Light-emitting diodes (LEDs) are even more efficient, consuming 90 percent less energy and lasting hundreds of times as long as ordinary lightbulbs. They can produce

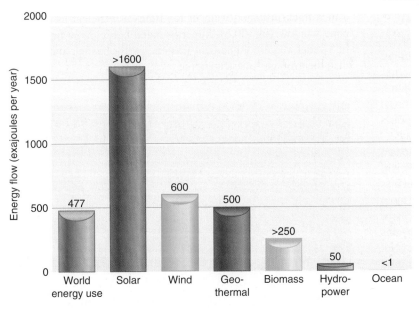

FIGURE 20.2 Potential energy available from renewable resources using currently available technology in presently accessible sites. Together, these sources could supply more than six times current world energy use. **source:** Adapted from UNDP and International Energy Agency.

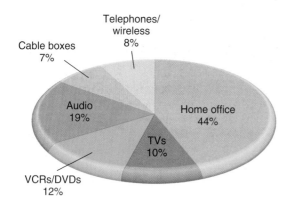

Telephones/
wireless
8%

Cable boxes
7%

Audio
19%

Home office
44%

TVs
10%

VCRs/DVDs
12%

FIGURE 20.4 Typical standby energy consumption by household electrical appliances.
Source: U.S. Department of Energy.

Table 20.1	Typical Net Efficiencies of Some Power Sources	
Electric Power Plants		**Yield (Percent)**
Hydroelectric (best case)		90
Cogeneration		80
Fuel cell (hydrogen)		80
IGCC		45
Coal-fired generator		38
Oil-burning generator		38
Nuclear generator		30
Photovoltaic solar		15

Source: U.S. Department of Energy.

millions of colors and be adjusted in brightness to suit ambient conditions. They are being used now in everything from flashlights and Christmas lights, to advertising signs, brake lights, exit signs, and street lights. New York city has replaced 11,000 traffic lights with LEDs. It also replaced 180,000 old refrigerators with new energy-saving models. Ann Arbor, Michigan, replaced 1,000 streetlights with LED models. These lights saved the city over $80,000 in the first year, and paid for themselves in just over two years.

Few of us realize how much electricity is used by appliances in standby mode. You may think you've turned off your TV, DVD player, cable box, or printer, but they're really continuing to draw power in an "instant-on" mode (fig. 20.4). For the average home, these "vampire currents" can represent up to a quarter of the monthly electric bill. Putting your computer to sleep saves about 90 percent of the energy it uses when fully on, but turning it completely off is even better. Plugging appliances into an inexpensive power strip allows you to switch them off when not in use.

Industrial energy savings are another important part of our national energy budget. More efficient electric motors and pumps, new sensors and control devices, advanced heat-recovery systems, and material recycling have reduced industrial energy requirements significantly. In the early 1980s, U.S. businesses saved $160 billion per year through conservation. When oil prices collapsed, however, many businesses returned to wasteful ways.

Energy efficiency is a measure of energy produced compared to energy consumed. Table 20.1 shows the typical energy efficiencies of some power sources. Thermal-conversion machines, such as steam turbines in coal-fired or nuclear power plants, can turn no more than 40 percent of the energy in their primary fuel into electricity or mechanical power because of the need to reject waste heat. Does this mean that we can never increase the efficiency of fossil fuel use? No. In cogeneration technology, waste heat can be recaptured and used for space heating, raising the net yield to nearly 80 percent. In fuel cells the chemical energy of a fuel is converted directly into electricity without combustion. Because this process is not limited by waste heat elimination, its efficiencies also can approach 80 percent with such fuel as hydrogen gas or methane. We'll discuss the special case of biofuel efficiency later in this chapter.

Green buildings can cut energy costs by half

Innovations in "green" building have been stirring interest in both commercial and household construction. Much of the innovation has occurred in large commercial structures, which have larger budgets—and more to save through efficiency—than most homeowners have. Energy audits can help show you where energy losses are occurring (fig. 20.5). Sealing leaks with caulk or weather stripping is one of the quickest and most cost-effective things you can do to save energy.

New houses can also be built with extra-thick, superinsulated walls and roofs. Windows can be oriented to let in sunlight, and eaves can provide shade. Double-glazed windows that have internal reflective coatings and that are filled with an inert gas (argon or xenon) have an insulation factor of R11, the same as a standard 4-inch-thick insulated wall, or ten times as efficient as a single-pane window (fig. 20.6). Superinsulated houses now being built in Sweden require 90 percent less energy for heating and cooling than the average American home.

FIGURE 20.5 Infrared photography shows heat loss in a building.

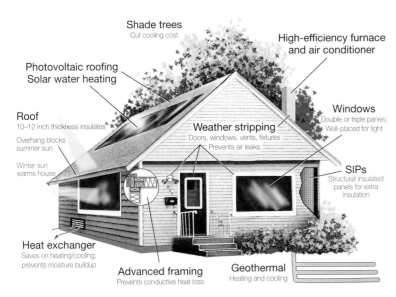

Shade trees
Cut cooling cost

High-efficiency furnace
and air conditioner

Photovoltaic roofing
Solar water heating

Roof
10–12 inch thickness insulates

Overhang blocks
summer sun

Winter sun
warms house

Weather stripping
Doors, windows, vents, fixtures
Prevents air leaks

Windows
Double or triple panes;
Well-placed for light

SIPs
Structural insulated
panels for extra
insulation

Heat exchanger
Saves on heating/cooling;
prevents moisture buildup

Advanced framing
Prevents conductive heat loss

Geothermal
Heating and cooling

FIGURE 20.6 Energy-efficient buildings can lower energy costs dramatically. Many features can be added to older structures. New buildings that start with energy-saving features (such as SIPs or advanced framing) can save even more money.

Transportation could be far more efficient

One of the areas in which most of us can accomplish the greatest energy conservation is in our transportation choices. You may not be able to build an energy-efficient house or persuade your utility company to switch from coal or nuclear to solar energy, but you can decide every day how you travel to school, to work, or for shopping or entertainment. Automobiles and light trucks account for 40 percent of the U.S. oil consumption and produce one-fifth of its carbon dioxide emissions. According the U.S. EPA, raising the average fuel efficiency of the passenger fleet by 3 miles per gallon (approx. 1.4 l/100 km) will save American consumers about $25 billion a year in fuel costs, reduce carbon dioxide emissions by 140 million metric tons per year, and save more oil than the maximum expected production from Alaska's Arctic National Wildlife Refuge.

The Bureau of Transportation Statistics reports that there are now more vehicles in the United States (214 million) than licensed drivers (190 million). More importantly, those vehicles are used for an average of 1 billion trips per day. Many of us drive now for errands or short shopping trips that previously might have been made on foot. Some of that is due to the design of our cities (chapter 22). Suburban subdivisions have replaced compact downtown centers in most cities. Shopping areas are surrounded by busy streets and vast parking lots that are highly pedestrian-unfriendly. But sometimes we use fuel inefficiently simply because we haven't thought about alternatives. The Census Bureau reports that three-quarters of all workers commute alone in private vehicles. Less than 5 percent use public transportation or carpool, and a mere 0.38 percent walk or travel by bicycle.

In response to the 1970s oil price shocks, automobile gas-mileage averages in the United States more than doubled from 13.3 mpg in 1973 to 25.9 mpg in 1988. Unfortunately, falling fuel prices of the 1990s discouraged further conservation, and mileage hardly improved for nearly 20 years. President Obama has called for a minimum of 39 mpg for cars in the United States and 30 mpg for light trucks by 2016. This will add about $1,300 to the sale price of each vehicle, but drivers should recoup this cost in about three years through lower fuel expenses.

What can you do if you want to be environmentally responsible? The cheapest, least environmentally damaging, and healthiest alternative for short trips is walking. You need to get some exercise every day, why not make walking part of it? Next, in terms of minimal expense and environmental impact, is an ordinary bicycle. For trips less than 2 km, it's often quicker to go by bicycle than to find a parking space for your car (fig. 20.7). While many cities have downgraded their mass transit systems, you might be surprised at the places you can go with this option.

You probably already know that **hybrid gasoline-electric engines** offer the best fuel economy and lowest emissions of any currently available vehicles. During most city driving, they depend mainly on quiet, emission-free, battery-powered electric motors. A small gasoline engine kicks in to help accelerate or when the batteries need recharging. This extends their range compared to pure electric vehicles. In 2011 the Toyota Prius had the highest mileage rating of any mass-produced automobile sold in America: 60 mpg (25 km/l) in city driving and 51 mpg (22 km/l)

FIGURE 20.7 Bicycles can be an efficient source of travel in bicycle-friendly cities.

on the highway. However, in 2011, U.S. automakers agreed with the Obama administration that by 2025 the corporate fuel average will double to 54.5 miles per gallon (23 km/l). Some of this advance will be accomplished by adding hybrid vehicles, but much will be credits for better air conditioning systems, low emission paint, etc.

An even greater savings might be achieved by **plug-in hybrids**. Recharging the batteries from ordinary household current at night can allow these vehicles to travel up to 100 km (60 mi) on the electric motor alone. Because most Americans only drive about 30 miles per day, they'd rarely have to buy any gasoline. In most places electricity costs the equivalent of about 50 cents per gallon. This means that we'll be generating more electricity, but it's easier to capture pollutants and greenhouse gases at a single, stationary power plant than from thousands of individual, mobile vehicles. And if the electricity comes from renewable sources, such as the Desertech project described in the case study for this chapter, most of your transportation could be relatively pollution-free.

Diesels already make up about half the autos sold in Europe because of their superior efficiency. LightningHybrids, a Colorado startup, has designed a biodiesel-powered hydraulic hybrid four-seat sports sedan that gets 100 mpg (42.1 km/l) on the highway. Most Americans think of diesels as noisy, smoke-belching, truck engines, but recent advances have made them much cleaner and quieter than they were a generation ago. Ultra low-sulfur diesel fuel and effective tailpipe emission controls now make these engines nearly as clean and energy-efficient as hybrids. Perhaps best of all would be to have flex-fuel or diesel plug-in hybrids that could burn ethanol or biodiesel when they need fuel. That could make us entirely independent from imported oil.

Both the United States and the European Union have spent billions of dollars on research and development of hydrogen fuel-cell-powered vehicles. Using hydrogen gas for fuel, these vehicles would produce water as their only waste product. We'll discuss how fuel cells work in more detail later in this chapter. Although prototype fuel cell vehicles are already being tested in several places, even the most optimistic predictions are that it will take at least 20 years for this technology to be mass produced at a reasonable cost. Although hydrogen fuel could be produced with electricity from remote wind or solar facilities, providing a convenient and inexpensive way to get surplus energy to market, most hydrogen currently is created from natural gas, making it no cleaner or more efficient than simply burning the gas directly. While not calling for an end to fuel cell research, conservation groups are urging the government not to abandon other useful technologies, such as hybrid engines and conventional pollution control, while waiting for fuel cells.

Cogeneration produces both electricity and heat

One of the fastest-growing sources of new energy is **cogeneration**, the simultaneous production of both electricity and steam or hot water in the same plant. By producing two kinds of useful energy in the same facility, the net energy yield from the primary fuel is increased from 30–35 percent to 80–90 percent. In 1900, half the electricity generated in the United States came from plants that also provided industrial steam or district heating. As power plants became larger, dirtier, and less acceptable to neighbors, they were forced to move away from their customers. Waste heat from the turbine generators became an unwanted pollutant to be disposed of in the environment. Furthermore, long transmission lines, which are unsightly and lose up to 20 percent of the electricity they carry, became necessary.

Think About It

What barriers do you see to walking, biking, or mass transit in your hometown? How could cities become more friendly to sustainable transportation? Why not write a letter to your city leaders or the editor of your newspaper describing your ideas?

By the 1970s cogeneration had fallen to less than 5 percent of our power supplies, but interest in this technology is being renewed. The capacity for cogeneration more than doubled in the 1980s to about 30,000 megawatts (MW). District heating systems are being rejuvenated, and plants that burn municipal wastes are being studied. Small neighborhood- or apartment-building-size power-generating units are being built that burn methane (from biomass digestion), natural gas, diesel fuel, or coal (fig. 20.8). The Fiat Motor Company makes a small generator for about $10,000 that produces enough electricity and heat for four or five energy-efficient houses. These units are especially valuable for facilities like hospitals or computer centers that can't afford power outages.

FIGURE 20.8 A technician adjusts a gas microturbine that produces on-site heat and electricity for businesses, industry, or multiple housing units.

Although you may not be buying a new house or car for a few years, and you probably don't have much influence over industrial policy or utility operation, there are things that all of us can do to save energy every day (What Can You Do? at right).

20.2 SOLAR ENERGY

The sun serves as a giant nuclear furnace in space, constantly bathing our planet with a free energy supply. Solar heat drives winds and the hydrologic cycle. All biomass, as well as fossil fuels and our food (both of which are derived from biomass), results from conversion of light energy (photons) into chemical bond energy by photosynthetic bacteria, algae, and plants. The average amount of solar energy arriving at the top of the atmosphere is 1,330 watts per square meter. About half of this energy is absorbed or reflected by the atmosphere (more at high latitudes than at the equator), but the amount reaching the earth's surface is some 10,000 times all the commercial energy used each year. However, this tremendous infusion of energy comes in a form that, until recently, has been too diffuse and low in intensity to be used except for environmental heating and photosynthesis. But as the opening case study for this chapter shows, there may be ways to use this vast power source, so we might never again have to burn fossil fuels. Figure 20.9 shows world solar energy potential.

Solar collectors can be passive or active

Our simplest and oldest use of solar energy is **passive heat absorption**, using natural materials or absorptive structures with no moving parts to simply gather and hold heat. For thousands of years people have built thick-walled stone and adobe dwellings that slowly collect heat during the day and gradually release that heat at night (fig. 20.10). After cooling at night, these massive building materials maintain a comfortable daytime temperature within the house, even as they absorb external warmth.

What Can You Do?

Some Things You Can Do to Save Energy

1. Drive less: make fewer trips, use telecommunications and mail instead of going places in person.
2. Use public transportation, walk, or ride a bicycle.
3. Use stairs instead of elevators.
4. Join a car pool or drive a smaller, more efficient car; reduce speeds.
5. Insulate your house or add more insulation to the existing amount.
6. Turn thermostats down in the winter and up in the summer.
7. Weather-strip and caulk around windows and doors.
8. Add storm windows or plastic sheets over windows.
9. Create a windbreak on the north side of your house; plant deciduous trees or vines on the south side.
10. During the winter, close windows and drapes at night; during summer days, close windows and drapes if using air conditioning.
11. Turn off lights, television sets, and computers when not in use.
12. Stop faucet leaks, especially hot water.
13. Take shorter, cooler showers; install water-saving faucets and showerheads.
14. Recycle glass, metals, and paper; compost organic wastes.
15. Eat locally grown food in season.
16. Buy locally made, long-lasting materials.

A modern adaptation of this principle is a glass-walled "sunspace" or greenhouse on the south side of a building (fig. 20.11). Incorporating massive energy-storing materials, such as brick walls, stone floors, or barrels of heat-absorbing water into buildings also collects heat to be released slowly at night. An interior, heat-absorbing wall called a Trombe wall is an effective passive heat collector. Some Trombe walls are built of glass blocks enclosing a water-filled space or water-filled circulation tubes, so heat from solar rays can be absorbed and stored while light passes through to inside rooms.

Active solar systems generally pump a heat-absorbing, fluid medium (air, water, or an antifreeze solution) through a relatively small collector, rather than passively collecting heat in a stationary medium like masonry. Active collectors can be located adjacent to or on top of buildings rather than being built into the structure. Because they are relatively small and structurally independent, active systems can be retrofitted to existing buildings.

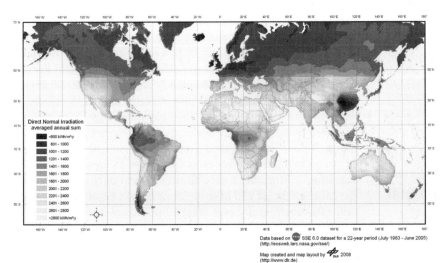

Direct Normal Irradiation
averaged annual sum

<800 kWh/m²/y
801 - 1000
1001 - 1200
1201 - 1400
1401 - 1600
1601 - 1800
1801 - 2000
2001 - 2200
2201 - 2400
2401 - 2600
2601 - 2800
>2800 kWh/m²/y

Data based on SSE 6.0 dataset for a 22-year period (July 1983 - June 2005)
(http://eosweb.larc.nasa.gov/sse/)

Map created and map layout by DLR 2008
(http://www.dlr.de)

FIGURE 20.9 Cumulative average annual solar radiation. Within 6 hours, deserts receive more energy from the sun than humans consume in a year.
Source: German Aerospace Center, 2008.

FIGURE 20.10 Taos Pueblo in northern New Mexico uses adobe construction to keep warm at night and cool during the day.

FIGURE 20.12 Solar water heaters can be scaled up to provide hot water and space heating for whole cities.

A flat black surface sealed with a double layer of glass makes a good solar collector. Water pumped through the collector picks up heat for space heating or to provide hot water. A collector with about 5 m² of surface can reach 95°C (200°F) and can provide enough hot water for an average family. China currently produces about 80 percent of the world's solar water heaters, which cost less than $200 each. At least 30 million Chinese homes get hot water and/or space heat from solar energy (see chapter 1). In Europe, municipal solar systems provide district heating for whole cities (fig. 20.12).

In a symbolic act to illustrate his commitment to solar energy, President Obama restored to the White House roof the solar electric

FIGURE 20.11 The Adam Joseph Lewis Center for Environmental Studies at Oberlin College is designed to be self-sustaining even in northern Ohio's cool, cloudy climate. Large, south-facing windows let in sunlight, while 370 m² of solar panels on the roof generate electricity. A constructed wetland outside and a living machine inside (see fig. 18.27) purify wastewater.

panels and a solar water heater that were removed 30 years earlier by the Reagan administration.

High-temperature solar energy

High-temperature solar thermal plants are suitable for industrial-size facilities. The solar farms being built in North Africa for the Desertech project, for example, are concentrating solar power (CSP) systems. They use long trough-shaped parabolic mirrors to reflect and concentrate sunlight on a central tube containing a heat-absorbing fluid (fig. 20.13a). Reaching much higher temperatures than possible in a basic flat panel collector, the fluid passes through a heat exchanger, where it generates steam to turn a turbine to produce electricity. Research by the German Aerospace Center suggests that CSP plants in North Africa and the Middle East should be able to provide 470,000 MW by 2050, or about 17 percent of the power used by the European Union. Costs, they estimate, should be equal to or lower than nuclear or fossil fuel power.

There are several advantages for a CSP plant besides fuel cost. Heat from the transfer fluid can be stored in a medium, such as molten salt, for later use. This allows the system to continue to generate electricity on cloudy days or at night. Desertech expects to be able to produce power nearly around the clock. In addition, those plants located near coastlines (fig. 20.1) can use seawater to cool the power cycle (necessary to keep turbines operating). But the heat absorbed from turbines isn't all wasted. Much of it can be used to flash-evaporate water to create pure drinking water—something sorely lacking in most of North Africa and the Middle East. A 250 MW collector field is expected to provide 200 MW of electricity plus 100,000 m³ (about 26 million gal) of distilled water per day.

But wouldn't highly polished mirrors in a CSP plant be damaged by desert sand storms? The parabolic troughs follow the sun to maximize solar energy absorption. On days when storms are forecast, the mirrors can be rotated into a protective position.

Solar-thermal power plants in California's Mojave Desert have been operating for over 20 years, and have withstood hailstorms, sandstorms, and gale-force winds. Wouldn't it take huge areas of land to capture solar energy? According to the German Aerospace Center, supplying 17 percent of Europe's energy requirements will take 2,500 km², or less than 0.3 percent of the Sahara desert.

This doesn't necessarily mollify critics, however. While some people regard deserts as useless, barren wastelands, others view them as beautiful, biologically rich, and captivating (fig. 20.13b). In 2010, state regulators approved 13 large solar thermal facilities and wind farms for California's Mojave Desert. Subsequently, however, most of these projects were canceled or delayed by protests over land use. Some people argued that these plants would harm rare or endangered species, such as the desert tortoise or the fringe-toed lizard. Native American groups protested that some areas were sacred cultural sites, while others simply love the solitude and mystery of the desert and believe that large, industrial facilities are an unwelcome intrusion. In response to this challenge, millions of hectares of desert have been added to new or existing protected areas to forestall energy development.

Another high-temperature system uses thousands of smaller mirrors arranged in concentric rings around a tall central tower (fig. 20.13c). The mirrors, driven by electric motors, track the sun and focus light on a heat absorber at the top of the "power tower" where a transfer medium is heated to temperatures as high as 500°C (1,000°F), which then drives a steam-turbine electric generator. Under optimum conditions, a 50 ha (130 acre) mirror array should be able to generate 100 MW of clean, renewable power. Southern California Edison's Solar II plant in the Mojave Desert east of Los Angeles is an example. Its 2,000 mirrors focused on a 100 m (300 ft) tall tower generate 10 MW, or enough electricity for 5,000 homes at an operating cost far below that of nuclear power or oil. Because all the mirrors are focused on a single point, the heat transfer medium has to be capable of absorbing much higher energy levels than in solar troughs. So far most of these plants use liquid sodium or molten nitrate salt for heat absorption. These materials are much more corrosive and difficult to handle than the lower temperature fluids suitable for a solar trough.

The Worldwatch Institute estimates that U.S. deserts could provide more than 7,000 GW of solar energy—nearly seven times

(a)

(b)

(c)

FIGURE 20.13 Concentrating solar power (CSP) uses either mirrored troughs (a) or movable mirrors focused on a central tower (c) to generate steam that turns a turbine to generate electricity. Often these facilities are sited in deserts where sunshine is intense and land is unoccupied. Large industrial facilities can intrude, however, on desert wildlife, solitude, and beauty (b).

current U.S. electrical capacity from all sources, while solar water heaters could easily provide half the hot water we use.

Public policy can promote renewable energy

Energy policies in some states include measures to encourage conservation and alternative energy sources. Among these are: (1) "distributional surcharges" in which a small per kWh charge is levied on all utility customers to help renewable energy finance research and development, (2) "renewables portfolio" standards to require power suppliers to obtain a minimum percentage of their energy from sustainable sources, and (3) **green pricing** that allows utilities to profit from conservation programs and charge premium prices for energy from renewable sources. Perhaps your state has some or all of these in place.

Iowa, for example, has a Revolving Loan Fund supported by a surcharge on investor-owned gas and electric utilities. This fund provides low-interest loans for renewable energy and conservation. Many utilities now offer green energy options. You agree to pay a couple of dollars extra on your monthly bill, and they promise to use the money to build or buy renewable energy. Buying a 100 kW "block" of wind power provides the same environmental benefits as planting a half acre of trees or not driving an automobile 4,000 km (2,500 mi) per year.

California is currently the U.S. leader in its renewable portfolio. In 2009 Governor Arnold Schwarzenegger signed an executive order requiring 33 percent of electricity sold in the state to be from renewable energy by 2020.

Photovoltaic cells generate electricity directly

Photovoltaic cells capture solar energy and convert it directly to electrical current by separating electrons from their parent atoms and accelerating them across a one-way electrostatic barrier formed

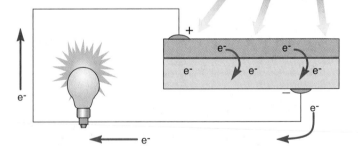

FIGURE 20.14 The operation of a photovoltaic cell. Boron impurities incorporated into the upper silicon crystal layers cause electrons (e⁻) to be released when solar radiation hits the cell. The released electrons move into the lower layer of the cell, thus creating a shortage of electrons, or a positive charge, in the upper layer and an oversupply of electrons, or negative charge, in the lower layer. The difference in charge creates an electric current in a wire connecting the two layers.

by the junction between two different types of semiconductor material (fig. 20.14). The first photovoltaic cells were made by slicing thin wafers from giant crystals of extremely pure silicon.

Over the past 25 years the efficiency of energy captured by photovoltaic cells has increased from less than 1 percent of incident light to more than 15 percent under field conditions and over 75 percent in the laboratory. Promising experiments are under way using exotic metal alloys, such as gallium arsenide, and semiconducting polymers of polyvinyl alcohol, which are more efficient in energy conversion than silicon crystals.

One of the most promising developments in photovoltaic cell technology in recent years is the invention of amorphous, thin-film silicon collectors. First described in 1968 by Stanford Ovshinky, these noncrystalline silicon semiconductors can be made into lightweight, paper-thin films that require much less material than conventional crystalline silicon cells. They also are much cheaper to manufacture and can be made in a variety of shapes and sizes, permitting ingenious applications. Roof tiles with amorphous silicon collectors layered on their surface already are available (fig. 20.15a). Photovoltaic cells already are providing power to places where conventional power is unavailable, such as lighthouses, mountaintop microwave repeater stations, and remote islands. In developing countries, solar power could make electricity available to some of the nearly 2 billion people who aren't served by a power grid (fig. 20.15b).

In 2010, thin-film photovoltaic cells finally broke the $1-per-watt barrier, a price that begins to make them competitive with fossil fuels and nuclear power in many situations. As further research improves their efficiency and lifespan, industry experts believe they could produce electricity for less than 10¢ per kilowatt-hour by 2020. This makes solar competitive with fossil fuels in many places for utility-scale baseload power arrays (fig. 20.15c). The largest photovoltaic plant under construction in the world in 2011 was a 2,000 MW giant (the size of two large nuclear plants) being built in China using thin-film technology from Arizona's First Solar Company.

A photovoltaic array of about 30–40 m² will generate enough electricity for an efficient house. There's a huge potential for rooftop solar energy. In the United States, it's estimated that more than 1,000 mi² (2,590 km²) of roofs suitable for photovoltaic systems could generate about three-quarters of present electrical consumption. In 2010, Southern California Edison started construction of photovoltaic arrays on roofs of warehouses and big-box retail stores (fig. 20.15d). Over the next five years, the utility expects to install a total of 250 MW of solar voltaic power. Overall, California's 1 million solar roofs project aims to build 3,000 MW of photovoltaic energy on homes and apartments by 2016. More than $2.8 billion in incentives are available to homeowners to cover costs.

Initial cost of this technology is a barrier for most homeowners. A 3 kw system still costs about $30,000 installed. Most of us can't pay that up front. Innovative financing programs, however, are helping make energy independence a reality. First introduced in Berkeley, California, Property Assessed Clean Energy (PACE) uses city bonds to pay for renewable energy and conservation

(a) Flexible, thin-film solar tiles

(b) Electricity for developing countries

(c) Base-load solar power facility

(d) Roof-top solar

FIGURE 20.15 Solar photovoltaic energy is highly versatile and can be used in a variety of dispersed settings. (a) Thin flim PV collectors can be printed on flexible backing and used like ordinary roof tiles. (b) Developing countries can install solar panels where a utility grid isn't available. (c) Utility-scale PV arrays can provide base-load power. (d) Millions of square meters of commercial rooftops could be fitted with solar panels.

Think About It

The 2005 U.S. Energy Bill had more than $12 billion in subsidies for the oil, coal, gas, and nuclear industries, but only one-sixth that much for renewable energy. Where might we be if that ratio had been reversed?

expenses. The bonds are paid off through a 20-year assessment on property taxes. Decreased utility bills can offset tax increases, so that switching to renewable energy is relatively painless for the property owner.

Feed-in tariffs (requiring utilities to buy surplus power from small producers at a fair price) are generally essential to individual solar installations. Rather than pay for large battery arrays to store solar energy, your meter simply spins backward during the day when you sell surplus electricity to your local utility. At night, when your solar panels aren't making any electricity, you buy some back from your power company. Depending on how big your system is and how frugal you are with electrical use, you might end up making money. Utilities usually resist such requirements. They complain the power from homeowners isn't "clean" (that is uniform, stable current). Solar proponents claim that utilities just don't like competition.

Smart metering can save money

When you install a meter smart enough to measure whether you're producing or consuming energy, you'll have an opportunity to make many more energy choices. For example, your meter could tell you where the power you buy comes from and when the cheapest prices are available. In the middle of the night, when consumption is low, utility companies often pay customers to take unwanted power off their hands. With a smart house system, you could instruct your appliances (dishwasher, water heater, etc.) to run only when a certain price is available, or only when the electricity comes from wind or solar sources.

An exciting potential of plug-in hybrids and electric vehicles is that they could serve as an enormous, distributed battery array. Automobiles in particular could be programmed to charge only at night when electric prices are at their lowest. If we had millions of vehicles connected to a smart grid, they'd collectively have enough storage capacity to smooth out peaks and valleys in the energy supply. Wouldn't all those vehicles require building new power plants and burning lots of dirty coal? Not necessarily. Utilities report that they could provide power for 4 million plug-in vehicles just with their existing surplus power supplies. And if the batteries were recharged from solar or wind facilities, our fossil fuel consumption would be sharply reduced.

The Pacific Gas and Electric utility in California is in the process of installing 9 million smart meters for its customers, while Europe is projected to have 80 million smart meters installed by 2013. Many customers love saving money and knowing exactly how much power costs for each appliance. Others, however, fear this new technology. Some worry about the radio signals that convey information between the utility and their home, scared by the electromagnetic radiation involved. But it's not ionizing radiation, nor all that different from the radiation emitted by their cell phone, microwave, cordless phone, wireless router, or a whole host of remote controls. It's ironic that people who accept all these other radiation sources reject smart meters. Others worry that this is an example of Big Brother spying on them. Who knows what information will be gathered with this system? But does it really matter if the utility knows what time your dishwasher ran last night?

20.3 FUEL CELLS

Fuel cells are devices that use ongoing electrochemical reactions to produce an electric current. They are very similar to batteries except that rather than recharging them with an electrical current, you add more fuel for the chemical reaction.

Fuel cells are not new; the basic concept was recognized in 1839 by William Grove, who was studying the electrolysis of water. He suggested that rather than use electricity to break apart water and produce hydrogen and oxygen gases, it should be possible to reverse the process by joining oxygen and hydrogen to produce water and electricity. The term *fuel cell* was coined in 1889 by Ludwig Mond and Charles Langer, who built the first practical device using a platinum catalyst to produce electricity from air and coal gas. The concept languished in obscurity until the 1950s when the U.S. National Aeronautics and Space Administration (NASA) was searching for a power source for spacecraft. Research funded by NASA eventually led to development of fuel cells that now provide both electricity and drinkable water on every space shuttle flight. The characteristics that make fuel cells ideal for space exploration—small size, high efficiency, low emissions, net water production, no moving parts, and high reliability—also make them attractive for a number of other applications.

All fuel cells have similar components

All fuel cells consist of a positive electrode (the cathode) and a negative electrode (the anode) separated by an electrolyte, a material that allows the passage of charged atoms, called ions, but is impermeable to electrons (fig. 20.16). In the most common systems, hydrogen or a hydrogen-containing fuel is passed over the anode while oxygen is passed over the cathode. At the anode, a reactive catalyst, such as platinum, strips an electron from each hydrogen atom, creating a positively charged hydrogen ion (a proton). The hydrogen ion can migrate through the electrolyte to the cathode, but the electron is excluded. Electrons pass through an external circuit, and the electrical current generated by their passage can be used to do useful work. At the cathode, the electrons and protons are reunited and combined with oxygen to make water.

The fuel cell provides direct-current electricity as long as it is supplied with hydrogen and oxygen. For most uses, oxygen is provided by ambient air. Hydrogen can be supplied as a pure gas, but storing hydrogen gas is difficult and dangerous because of its volume and explosive nature. Liquid hydrogen takes far less space

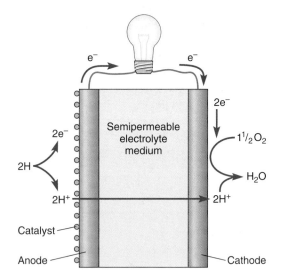

FIGURE 20.16 Fuel cell operation. Electrons are removed from hydrogen atoms at the anode to produce hydrogen ions (protons) that migrate through a semipermeable electrolyte medium to the cathode, where they reunite with electrons from an external circuit and oxygen atoms to make water. Electrons flowing through the circuit connecting the electrodes create useful electrical current.

FIGURE 20.17 The Long Island Power Authority has installed 75 stationary fuel cells to provide reliable backup power.

than the gas, but must be kept below −250°C (−400°F), not a trivial task for most mobile applications. The alternative is a device called a **reformer** or converter that strips hydrogen from fuels such as natural gas, methanol, ammonia, gasoline, ethanol, or even vegetable oil. Many of these fuels can be derived from sustainable biomass crops. Even methane effluents from landfills and wastewater treatment plants could be used as a fuel source. Where a fuel cell can be hooked permanently to a gas line, hydrogen could be provided by solar, wind, or geothermal facilities that use electricity to hydrolyze water.

A fuel cell run on pure oxygen and hydrogen produces no waste products except drinkable water and radiant heat. When a reformer is coupled to the fuel cell, some pollutants are released (most commonly carbon dioxide), but the levels are typically far less than conventional fossil fuel combustion in a power plant or automobile engine. Although the theoretical efficiency of electrical generation of a fuel cell can be as high as 70 percent, the actual yield is closer to 40 or 45 percent. This is about the same as an integrated gasification combined cycle (IGCC) plant (chapter 19). On the other hand, the quiet, clean operation and variable size of fuel cells make them useful in buildings where waste heat can be captured for water heating or space heating. A new 45-story office building at 4 Times Square, for example, has two 200-kilowatt fuel cells on its fourth floor that provide both electricity and heat. This same building has photovoltaic panels on its façade, natural lighting, fresh air intakes to reduce air conditioning, and a number of other energy conservation features.

U.S. automakers have focused a great deal of attention on fuel cells in recent years. While this would reduce pollution, eliminate our dependence on imported oil, and make a good use for wind or solar energy, critics claim that it will take 20 years or more to develop automotive fuel cells and build the necessary infrastructure. We should concentrate, instead, they say, on plug-in hybrids or all-electric vehicles. Iceland, with no fossil fuels but abundant geothermal energy, is determined to be the world's first hydrogen-based economy. They have one hydrogen filling station and a fleet of fuel cell buses.

The current from a fuel cell is proportional to the size (area) of the electrodes, while the voltage is limited to about 1.23 volts per cell. A number of cells can be stacked together until the desired power level is reached. A fuel cell stack that provides almost all of the electricity needed by a typical home (along with hot water and space heating) would be about the size of a refrigerator. A 200-kilowatt unit fills a medium-size room and provides enough energy for 20 houses or a small factory (fig. 20.17). Tiny fuel cells running on methanol may soon be used in cell phones, pagers, toys, computers, videocameras, and other appliances now run by batteries. Rather than buy new batteries or spend hours recharging spent ones, you might just add an eyedropper of methanol every few weeks to keep your gadgets operating.

20.4 BIOMASS ENERGY

Photosynthetic organisms have been collecting and storing the sun's energy for more than 2 billion years. Plants capture about 0.1 percent of all solar energy that reaches the earth's surface. That kinetic energy is transformed, via photosynthesis, into chemical bonds in organic molecules (chapter 3). A little more than half of the energy that plants collect is spent in such metabolic activities as pumping water and ions, mechanical movement, maintenance of cells and tissues, and reproduction; the rest is stored in biomass.

The magnitude of this resource is difficult to measure. Most experts estimate useful biomass production at 15 to 20 times the amount we currently get from all commercial energy sources. It would be ridiculous to consider consuming all green plants as fuel, but biomass has the potential to become a prime source of energy. It has many advantages over nuclear and fossil fuels because of its renewability and easy accessibility. Biomass resources used as fuel include wood, wood chips, bark, branches, leaves, starchy roots, and other plant and animal materials.

We can burn biomass

Wood fires have been a primary source of heating and cooking for thousands of years. As recently as 1850, wood supplied 90 percent of the fuel used in the United States. Wood now provides less than 1 percent of the energy in advanced economies, but 2 billion people—about 30 percent of the world population—depend on firewood and charcoal as their primary energy source.

In many countries firewood gathering is a major source of forest destruction and habitat degradation. Furthermore, it can be a social burden. Poor people often spend a high proportion of their income on cooking and heating fuel. In rural families, women and children may spend hours every day searching for fuel. Development agencies are working to design and distribute highly efficient stoves, both as a way to improve the lives of poor people and to reduce forest degradation.

Even in rich countries, fuelwood and other biomass sources are becoming increasingly popular in the face of rising oil prices. Many homeowners have installed wood-burning stoves or outdoor boilers to replace fossil fuels. This can be good for your pocketbook, but the smoke can be noxious for your neighbors. In Oregon's Willamette Valley or in the Colorado Rockies, where woodstoves are popular and topography concentrates contaminants, as much as 80 percent of air pollution on winter days is attributable to wood fires. Some cities and towns have banned installation of new fireplaces or stoves.

Still, biomass can make a significant contribution to renewable energy supplies. In Denmark, the energy-independent islands of Samsø and Ærø get about half their space heating from biomass, both from agricultural wastes (such as straw) and biomass crops, such as reeds and elephant grass growing on land unsuitable for crops (all the rest comes from solar and wind). Burning these crops in an industrial boiler for district heating makes it easier to install and maintain pollution-control equipment than in individual stoves. Most plant material has low sulfur, so it doesn't contribute to acid rain. And because it burns at a lower temperature than coal, it doesn't create as much nitrogen oxides. Of course, these crops are carbon neutral—that is, they absorb as much CO_2 in growing as they emit when burned.

Some utilities are installing "flex-fuel" boilers in their power plants that can burn wood chips, agricultural waste, or other biomass fuels (fig. 20.18). As chapter 19 points out, co-combustion of coal together with biomass can have benefits over burning either alone. Including biomass in the mix reduces greenhouse gas emissions, while also improving combustion properties. Even higher efficiencies can be achieved by capturing waste heat for beneficial use. A district heating plant in St. Paul, Minnesota, for example, uses 275,000 tons of wood per year (mostly from urban trees killed by storms and disease) to provide heating, air conditioning, and electricity to 25 million square feet of offices and living space in 75 percent of all downtown buildings. Although the efficiency of electrical generation in this plant is less than 40 percent (as it is in most power plants), the net yield is about 80 percent because waste heat is used rather than discarded.

Methane from biomass is clean and efficient

Where wood and other fuels are in short supply, people often dry and burn animal manure. This may seem like a reasonable use of waste biomass, but it can intensify food shortages in poorer countries. Not putting this manure back on the land as fertilizer reduces crop production and food supplies. In India, for example, where fuelwood supplies have been chronically short for many years, a limited manure supply must fertilize crops and provide household fuel. Cows in India produce more than 800 million tons of dung per year, more than half of which is dried and burned in cooking fires. If that dung were applied to fields as fertilizer, it could boost crop production of edible grains by 20 million tons per year, enough to feed about 40 million people.

When cow dung is burned in open fires, however, more than 90 percent of the potential heat and most of the nutrients are lost. Compare that to the efficiency of using dung to produce methane gas, an excellent fuel. In the 1950s, simple, cheap methane digesters were designed for villages and homes, but they were not widely used. Now, however, 35 million Chinese households use biogas for cooking and lighting, and that number is expected to double in 20 years. Two large municipal facilities in Nanyang will soon provide fuel for more than 20,000 families. Perhaps other countries will follow China's lead.

Methane gas is the main component of natural gas. It is produced by anaerobic decomposition (digestion by anaerobic bacteria) of any moist organic material. Many people are familiar with the fact that swamp gas is explosive. Swamps are simply large methane digesters, basins of wet plant and animal wastes sealed from the air by a layer of water. Under these conditions, organic materials are decomposed by anaerobic (oxygen-free) rather than aerobic (oxygen-using) bacteria, producing flammable gases instead of carbon dioxide. This same process may be reproduced artificially by placing organic wastes in a container and providing warmth and water (fig. 20.19). Bacteria are ubiquitous enough to start the culture spontaneously.

Burning methane produced from manure provides more heat than burning the dung itself, and the sludge left over from bacterial digestion is a rich fertilizer, containing healthy bacteria as well as most of the nutrients originally in the dung. Whether the manure is of livestock or human origin, airtight digestion also eliminates some health hazards associated with direct use of dung, such as exposure to fecal pathogens and parasites.

How feasible is methane—from manure or from municipal sewage—as a fuel resource in developed countries? Methane is a clean fuel that burns efficiently. Any kind of organic waste material: livestock manure, kitchen and garden scraps, and even municipal garbage and sewage can be used to generate gas. In fact,

FIGURE 20.18 This Michigan power plant uses wood chips to fuel its boilers. Where wood supplies are nearby, this is a good choice both economically and environmentally.

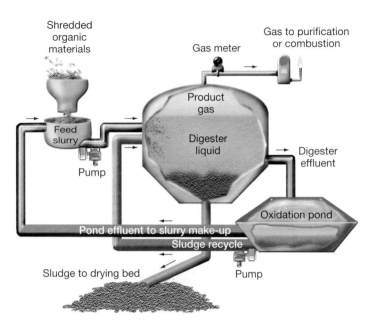

FIGURE 20.19 Continuous unit for converting organic material to methane by anaerobic fermentation. One kilogram of dry organic matter will produce 1–1.5 m³ of methane, or 2,500–3,600 million calories per metric ton.

municipal landfills are active sites of methane production, contributing as much as 20 percent of the annual output of methane to the atmosphere. This is a waste of a valuable resource and a threat to the environment because methane absorbs infrared radiation and contributes to the greenhouse effect (chapter 15). About 300 landfills in the United States currently burn methane and generate enough electricity together for a million homes. Another 600 landfills have been identified as potential sources for methane development. Hydrologists worry, however, that water will be pumped into landfills to stimulate fermentation, thus increasing the potential for groundwater contamination.

Cattle feedlots and chicken farms in the United States are a tremendous potential fuel source. Collectible crop residues and feedlot wastes each year contain 4.8 billion gigajoules (4.6 quadrillion Btus) of energy, more than all the nation's farmers use. The Haubenschild farm in central Minnesota, for instance, uses manure from 850 Holsteins to generate all the power needed for their dairy operation and still have enough excess electricity for an additional 80 homes. In January 2001 the farm saved 35 tons of coal and, 1,200 gallons of propane, and made $4,380 from electric sales.

A number of colleges around the United States are weaning themselves off fossil fuels. In Vermont, Middlebury College feeds locally harvested wood chips into a gasification plant that provides both heat and electricity to the campus. It will reduce the school's carbon footprint about 40 percent. And the University of New Hampshire in Durham plans to provide 80 percent of its heating and electrical needs by buying and burning methane gas given off by a landfill a few miles away. At the University of Minnesota's Morris campus, a gasification plant uses about 1,700 tons of corn stover (stalks and cobs) and other local agricultural waste

every year to provide as much as 80 percent of the school's heating and cooling needs, while a wind turbine provides most of its electricity. Together these alternative energy systems replace at least $1 million per year in fossil fuels (mostly natural gas) and could make the campus not only carbon neutral but even carbon negative in a few years. These are just a few of the efforts across the country of campuses to "walk the walk" by not only teaching about environmental issues but also by changing the way they operate. Altogether, 614 colleges and universities representing about one-third of the students in the United States have made a commitment to reduce their carbon footprint. Could you convince the administration at your school to join this movement?

Ethanol and biodiesel can contribute to fuel supplies

Biofuels, such as ethanol and biodiesel, are by far the biggest recent news in biomass energy. Globally, production of these two fuels is booming, from Brazil, which gets about 40 percent of its transportation energy from ethanol generated from sugarcane, to Southeast Asia, where millions of hectares of tropical forest have been cleared for palm oil plantations, to the United States, where about one-fifth of the corn (maize) crop currently is used to make ethanol (fig. 20.20). In 2009 President Obama proposed spending $150 billion over 10 years to develop renewable fuels and create 5 million "green collar" jobs. He supported a plan to increase ethanol production in the United States from 9 billion to 36 billion gallons per year (30 billion to 136 billion liters) by 2022. However, it would take the entire U.S. corn crop to produce that much ethanol from corn. We need to find other ways to create biofuels.

Crops with a high oil content, such as soybeans, sunflower seed, rape seed (usually called canola in America), and palm oil fruits are relatively easy to make into biodiesel (fig. 20.21). In

FIGURE 20.20 No room was available in elevators to store the bumper corn harvest in 2010. Millions of bushes sat in the open, much of which was slated to be made into ethanol.

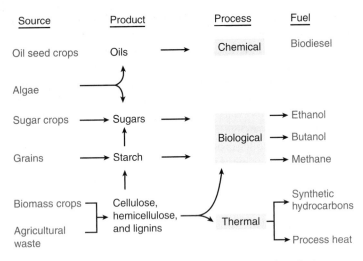

FIGURE 20.21 Methods for turning biomass into fuel.

Source	Product	Process	Fuel
Oil seed crops	Oils →	Chemical	Biodiesel
Algae			
Sugar crops →	Sugars →		→ Ethanol
Grains →	Starch →	Biological	→ Butanol
			→ Methane
Biomass crops	Cellulose, hemicellulose, and lignins	Thermal	Synthetic hydrocarbons
Agricultural waste			Process heat

some cases the oil needs only minimal cleaning to be used in a standard diesel engine. Yields per hectare for many of these crops are low, however. It would take a very large land area to meet our transportation needs with soy or sunflowers, for example. Furthermore, diversion of these oils for vehicles deprives humans of an important source of edible oils.

Oil palms are eight to ten times more productive per unit area than soy or sunflower (although palm fruit is more expensive to harvest and transport). Currently millions of hectares of species-rich forests in Southeast Asia are being destroyed to create palm oil plantations. Indonesia already has 6 million ha of palm oil plantations and Malaysia has nearly as much. Together these two countries produce nearly 90 percent of the world's palm oil. Burning of Indonesian forests and the peat lands on which they stand currently releases some 1.8 billion tons of carbon dioxide every year. Indonesia is currently third in the world—behind the United States and China—in human-made greenhouse gas emissions. At least 100 species, including orangutans, Sumatran tigers, and the Asian rhinoceros, are threatened by habitat loss linked to palm oil expansion. In 2011, however, Indonesia signed an agreement with Norway to protect its forests in exchange for $1 billion in development aid (see chapter 12).

A shrubby tree called *Jatropha curcas* has recently been promoted as a good alternative to both oil palm and soybean oil. This native of Mexico and the Caribbean, has nuts with a high (but toxic) oil content that can be easily converted to diesel fuel. Indian scientists have bred prolific new varieties they claim grow well on marginal soil with few inputs. Field tests suggested that *Jatropha* might produce as much as 15,000 liters of oil per hectare (1,600 gal per acre or about three times as much as palm oil). India has set aside 50 million ha (123 million acres) of land for *Jatropha* that it hopes will provide 20 percent of its diesel fuel. In 2008, Air New Zealand flew a Boeing 747 using a 50–50 blend of *Jatropha* oil and aviation fuel. But test plantings in other areas suggest that *Jatropha* needs more water and fertilizer, is more sensitive to pests and diseases, and has lower yield than promised. As so often happens, what's touted as a miracle solution may not be so great when we look closer.

Cellulosic ethanol may offer hope for the future

Crops such as sugarcane and sugar beets have a high sugar content that can be fermented into ethanol, but sugar is expensive and the yields from these crops are generally low, especially in temperate climates. Starches in grains, such as corn, have higher yields and can be converted into sugars that can be turned by yeast into ethanol (this is the same process used to make drinking alcohol), butanol (which burns in engines much like gasoline), or methanol (fig. 20.22). The idea of burning ethanol in vehicles isn't new. Henry Ford designed his 1908 Model T to run on ethanol.

The need to move away from imported oil has created boom times for corn-based ethanol production in America. Since 1980

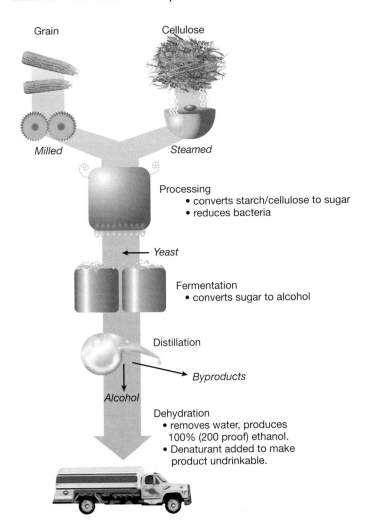

FIGURE 20.22 Ethanol (or ethyl alcohol) can be produced from a wide variety of sources. Maize (corn) and other starchy grains are milled (ground) and then processed to convert starch to sugar, which can be fermented by yeast into alcohol. Distillation removes contaminants and yields pure alcohol. Cellulosic crops, such as wood or grasses, can also be converted into sugars, but the process is more difficult. Steam blasting, alkaline hydrolysis, enzymatic conditioning, and acid pretreatment are a few of the methods for breaking up woody material. Once sugars are released, the processes are similar.

FIGURE 20.23 More than 100 ethanol plants distill biofuels from corn and other grains in the United States.

more than 100 new refineries have been built, and U.S. ethanol production has grown from about 500 million liters to 30 billion liters per year (fig. 20.23). The United States and Brazil now produce about 95 percent of all the ethanol in the world. A sudden collapse in oil prices in 2008, however, bankrupted many U.S. ethanol producers and casts doubt on continued growth in this area. There has been a great deal of debate about the net energy yield and environmental costs of ethanol from corn (see Exploring Science, p. 463), but everyone agrees that cellulosic ethanol—if we can find ways to produce it economically—would have considerable environmental, social, and economic advantages over using edible grains or sugar crops for transportation fuel.

Most plants put the bulk of the energy they capture from the sun into cellulose and a related polymer, hemicelluloses, which are made of long chains of simple sugars. Woody plants add a sticky glue, called lignin, to hold cells together. If we can find ways to economically release those simple sugars so they can be fermented into ethanol or other useful liquid fuels, we could greatly increase the net energy yield from all sorts of crops. But it's difficult. If it were easy for microbes to dismantle woody material, we probably wouldn't have forests or prairies—the landscape would simply be covered with green slime. A number of techniques have been proposed for extracting sugars from cellulosic materials. Most involve mechanical chopping or shredding followed by treatment with bacteria or fungi (or enzymes derived from them).

So far, there are no commercial-scale cellulosic ethanol factories operating in North America, but the Department of Energy has provided $385 million in grants for six cellulosic biorefinery plants. These pilot projects will test a wide variety of feedstocks including rice and wheat straw, milo stubble, switchgrass hay, almond hulls, corn stover, and wood chips.

Switchgrass (*Panicum virgatum*), a tall grass native to the Great Plains, has been one recent focus of attention. Switchgrass is a perennial, with deep roots that store carbon (and thus capture atmospheric greenhouse gases). Long-lasting perennial roots

hold soil in place, unlike annual corn crops that require tillage for planting and weed control. Fewer trips through the field with a tractor or cultivator require much less fuel and improve net energy yield (fig. 20.24).

An even better biofuel crop may be *Miscanthus x giganteus*, a perennial grass from Asia. Often called elephant grass (although this name is also used for other species), *Miscanthus* is a sterile, hybrid grass that grows 3 or 4 meters in a single season (fig. 20.25). Europeans have been experimenting with this species for several decades, but it has only recently been introduced to the United States. *Miscanthus* can produce at least five times as much dry biomass per hectare as corn. Part of the reason for this is that *Miscanthus* starts growing four to six weeks earlier in the spring than corn, and stays green a month or so longer in the fall. This longer growing season, coupled with the nutrients and energy stored in underground rhizomes, gives this giant grass a huge advantage compared to annual crops, such as corn. Its perennial growth and long-lasting canopy also protect the soil from erosion and require much less fuel for cultivation.

Where using corn or switchgrass to produce enough ethanol to replace 20 percent of U.S. gasoline consumption would take about one-quarter of all current U.S. cropland out of food production, *Miscanthus* could produce the same amount on less than half that much area. And it wouldn't need to be prime farm fields. *Miscanthus* can grow on marginal soil with far less fertilizer than corn needs. In the fall, *Miscanthus* moves nutrients into underground rhizomes. This means that the standing stalks are almost entirely cellulose and next year's crop needs very little fertilizer.

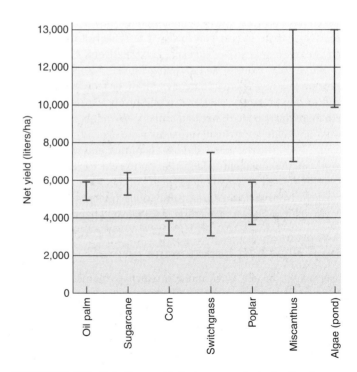

FIGURE 20.24 Proven biofuel sources include oil palms, sugarcane, and corn grain (maize). Other experimental sources may produce better yields, however.
Source: Data from E. Marris, 2006. *Nature* 444:670–678.

FIGURE 20.25 *Miscanthus x giganteus* is a perennial grass that can grow 3 or 4 meters in a single season. It thrives on marginal land with little fertilizer or water and can produce five times as much biomass as corn.

And because it stores far more carbon in the soil than other crops, *Miscanthus* may be eligible for climate offset credits.

Currently, there are no known diseases or pests for *Miscanthus*. However, Professor S. Raghu and his colleagues at the University of Illinois point out that the characteristics that make it an attractive energy crop—rapid growth, highly efficient photosynthesis, low need for nutrients, no known pests, high water-use efficiency—also make it a good candidate to become an invasive pest. The fact that the variety being tested for growth in the United States is a sterile hybrid may make it less likely to spread, but there are cases of invasives that spread vegetatively.

Harvesting, storing, and shipping biomass crops remains a problem. The low energy content of straw or wood chips, compared to oils or sugars, makes it prohibitively expensive to ship them more than about 50 km to a refinery. We might need to have a very large number of small refineries if we depend on cellulosic ethanol. Interestingly, some authors claim that you could drive a hybrid automobile about twice as far on the electricity generated by burning a ton of dry biomass than you could on the ethanol fermented from that same ton. So, burning biomass may sill be a better solution than fermentation if we move to hybrid engines.

Could algae be a hope for the future?

Algae might be an even more productive biofuel crop than any we've discussed so far. While *Miscanthus* can yield up to 13,000 liters (3,500 gal) of ethanol per hectare, some algal species growing in a photobioreactor might theoretically produce 30 times as much high-quality oil. This is partly because single-celled algae can grow 30 times as fast as higher plants. Furthermore, some algae store up to half their total mass as oil. Photobioreactors are much more expensive to build and operate than planting crops, but they could be placed on land unsuitable for agriculture and they could use recycled water. Open ponds are much cheaper than photobioreactors, but they also produce far less biomass per unit area. So far, the actual yield from algal growth facilities is actually about the same as *Miscanthus*.

One of the most intriguing benefits of algal growth facilities is that they could be placed next to conventional power plants, where CO_2 from burning either fossil fuels or biomass could be captured and used for algal growth. Thus, they'd actually be carbon negative: providing a net reduction in atmospheric carbon while also creating useful fuel.

An algal bioreactor started producing biodiesel in South Africa in 2006, and one in Brazil aims to start trapping CO_2 from a coal-fired power plant soon. A number of U.S. companies, including Solix Biofuels, Sapphire Energy, OriginOil, PetroAlgae, and Shell Oil, are exploring algal biofuels. In 2009 Japan Airlines made a test flight using a combination of jet fuel and algal oils. Another tantalizing fact is that some algae produce hydrogen gas as a photosynthetic by-product. If fuel cells ever become economically feasible, algae might provide them with a good energy source that doesn't depend on fossil fuels.

20.5 HYDROPOWER

The winds, waves, tides, ocean thermal gradients, and geothermal areas are renewable energy sources. Although available only in selected locations, these sources could make valuable contributions to our total energy supply.

Falling water has been used as an energy source since ancient times

The invention of water turbines in the nineteenth century greatly increased the efficiency of hydropower dams. By 1925, falling water generated 40 percent of the world's electric power. Since then, hydroelectric production capacity has grown 15-fold, but fossil fuel use has risen so rapidly that water power is now only 20 percent of total electrical generation. Still, many countries produce most of their electricity from falling water (fig. 20.26). Norway, for instance, depends on hydropower for 99 percent of its electricity. Currently, total world hydroelectric production is about 3,000 terrawatt hours (10^{12} Whr). Six countries—Canada, Brazil, the United States, China, Russia, and Norway—account for more than half that total. In fact, of the approximately 50,000 dams in the world taller than 15 m (45 ft), roughly half are in China. Untapped hydro resources are still abundant in Latin America, Central Africa, India, and China.

Much of the hydropower development in recent years has been in enormous dams. There is a certain efficiency of scale in giant dams, and they bring pride and prestige to the countries that build them, but, as we discussed in chapter 17, they can have unwanted social and environmental effects. The largest hydroelectric dam in the world at present is the Three Gorges Dam on China's Yangtze River, which spans 2 km and is 185 m (600 ft) tall. Designed to generate 25,000 MW of power, this dam produces as much energy as 25 large nuclear power plants. The reservoir behind the dam displaced at least 1.5 million people and submerged 5,000 archaeological sites.

Biofuels (alcohol refined from plant material or diesel fuel made from vegetable oils or animal fats) might be the answer to both our farm crisis and our fuel needs. But do these crops represent a net energy gain? Or does it take more fossil fuel energy to grow, harvest, and process crop-based biofuels than you get back in the finished product? The answer depends on the assumptions you make in calculating net energy yields. How much energy does it take to grow crops, and what crop harvest can you expect? What return do you assume for fermentation, and what credit do you assign for recycled heat or useful by-products?

For years David Pimental from Cornell University has published calculations showing a net energy loss in biofuels. In 2005 he was joined by Tad Patzek from the University of California–Berkeley in claims that it takes 29 percent more energy to refine ethanol from corn than it yields. Soy-based biodiesel is equally inefficient, these authors maintain, and cellulose-based biofuels are even worse. Switchgrass and woodchips take at least 50 percent more energy than they produce as ethanol, according to their calculations.

A counterargument came from Bruce Dale of Michigan State University and John Sheehan from the U.S. National Renewable Energy Laboratory, who maintain that biofuels produced by modern techniques represent a positive energy return. They say that Pimental and Patzek used outdated data and unreasonably pessimistic assumptions in making their estimates. Dramatic improvements in farming productivity, coupled with much greater efficiency in ethanol fermentation, now yield about 35 percent more energy in ethanol from corn than is consumed in growing and harvesting the crop. A major difference between the outcomes is how far back you assign costs. Is the energy used to manufacture farm equipment included, or only that which is needed for fermentation and purification? Pimental and Patzek assume this energy comes from fossil fuels, but Dale and Sheehan argue that fermentation waste can be burned to make manufacturing more efficient, as is already being done in Brazil.

A valuable addition to this debate comes from the work of ecologist David Tilman and his colleagues at the Cedar Creek Natural History Area in Minnesota. This group studies the effects of biodiversity on ecosystem resilience. They have shown that plots with high species diversity have greater net productivity than species grown in a monoculture. Tilman suggests that diverse mixtures of native perennial plants could grow on marginal land with far lower inputs of water, nutrients, and less fossil fuels for planting, cultivation, and weed-control than grains such as soy or corn. Tilman and his colleagues calculate a corn-based ethanol net energy ratio of 1.2, while cellulosic ethanol from prairie species can yield about 5.4 times as much energy as it takes to grow, harvest, and process the crop.

In 2009 Tilman joined with economist Steven Polasky and others to compare a broader set of environmental and health considerations for different biofuels. They calculated the climate-change and health costs of different crops. An important assumption in this study is that diversion of food crops, such as soy and corn, to biofuel production results in prairie and forest destruction when food shortages and rising prices force people in developing countries to seek new land for agriculture. This land conversion creates a carbon debt that can take centuries to balance against the higher efficiency of the biofuel.

The greatest debt, according to these authors, is from palm oil grown on tropical peatlands, which would take 423 years to repay. Corn ethanol, in these calculations, would take 93 years to repay if its cultivation results in conversion of existing grasslands. Prairie grasses grown on marginal land with minimal inputs, according to this study, would have no carbon debt. When Tilman and his colleagues add up health costs (from fine particulate materials released during processing) and climate costs (from release of greenhouse gases), they calculate that a billion-gallon increase in fuel consumption (about the U.S. growth between 2006 and 2007) would cost $469 million for gasoline, between $472 million and $952 million for corn ethanol (depending on biorefinery technology and heat source), but only $123 million to $208 million for cellulosic ethanol.

These conclusions were immediately challenged by Adam Liska and his colleagues from the University of Nebraska, who claim that Tilman and his colleagues used outdated data for their net energy yields. Modern refineries, this group claims, produce 1.8 times as much energy in corn ethanol as the crop inputs. They didn't address other health or environmental effects, however.

Obviously, there are many assumptions in all these studies. If you were asked to calculate the yields and effects of various biofuels, where would you start?

Biomass Fuel Efficiency			
Fuel	Inputs (GJ/ha)	Outputs (GJ/ha)	Net Energy Ratio
Corn ethanol	75.0	93.8	1.2
Soy ethanol	15.0	28.9	1.9
Cellulosic electricity	4.0	22.0	5.5
Cellulosic ethanol	4.0	21.8	5.4
Cellulosic synfuel	4.0	32.4	8.1

Source: Tilman et al. 2006. *Science* 314:1598.

There are other problems with big dams, besides human displacement, ecosystem destruction, and wildlife losses. Dam failure can cause catastrophic floods and thousands of deaths. Sedimentation often fills reservoirs rapidly and reduces the usefulness of the dam for either irrigation or hydropower. In China, the Sanmenxia Reservoir silted up in only two years, and the Laoying Reservoir filled with sediment before the dam was even finished. Schistosomiasis, caused by parasitic flatworms called blood flukes (chapter 8), is transmitted to humans by snails that thrive in slow-moving, weedy tropical waters behind these

FIGURE 20.26 Hydropower dams produce clean renewable energy but can be socially and ecologically damaging.

FIGURE 20.27 Solar collectors capture power only when the sun shines, but hydropower is available 24 hours a day. Small turbines such as this one can generate enough power for a single-family house with only 15 m (50 ft) of head and 200 l (50 g) per minute flow. The turbine can have up to four nozzles to handle greater water flow and generate more power.

dams. It is thought that 14 million Brazilians suffer from this debilitating disease.

Rotting vegetation in artificial impoundments can have disastrous effects on water quality. When Lake Brokopondo in Suriname flooded a large region of uncut rainforest, underwater decomposition of the submerged vegetation produced hydrogen sulfide that killed fish and drove out villagers over a wide area. Acidified water from this reservoir ruined the turbine blades, making the dam useless for power generation. A recent study of one reservoir in Brazil suggested that decaying vegetation produced more greenhouse gases (carbon dioxide and methane) than would have come from generating an equal amount of energy by burning fossil fuels.

In warm climates, large reservoirs often suffer enormous water losses. Lake Nasser, formed by the Aswan High Dam in Egypt, loses 15 billion m³ each year to evaporation and seepage. Unlined canals lose another 1.5 billion m³. Together these losses represent one-half of the Nile River flow, or enough water to irrigate 2 million ha of land. The silt trapped by the Aswan Dam formerly fertilized farmland during seasonal flooding and provided nutrients that supported a rich fishery in the Delta region. Farmers now must buy expensive chemical fertilizers, and the fish catch has dropped almost to zero. As in South America, schistosomiasis is an increasingly serious problem.

If big dams—our traditional approach to hydropower—have so many problems, how can we continue to exploit the great potential of hydropower? Fortunately, there is an alternative to gigantic dams and destructive impoundment reservoirs. Small-scale, **low-head hydropower** technology can extract energy from small headwater dams that cause much less damage than larger projects. Some modern, high-efficiency turbines can even operate on **run-of-the-river flow**. Submerged directly in the stream and small enough not to impede navigation in most cases, these turbines don't require a dam or diversion structure and can generate useful power with a current of only a few kilometers per hour. They also cause minimal environmental damage and don't interfere with fish movements, including spawning migration. **Micro-hydro generators** operate on similar principles but are small enough to provide economical power for a single home. If you live close to a small stream or river that runs year-round and you have sufficient water pressure and flow, hydropower is probably a cheaper source of electricity for you than solar or wind power (fig. 20.27).

However, small-scale hydropower systems also can cause abuses of water resources. The Public Utility Regulatory Policies Act of 1978 included economic incentives to encourage small-scale energy projects. As a result, thousands of applications were made to dam or divert small streams in the United States. Many of these projects have little merit. All too often, fish populations, aquatic habitat, recreational opportunities, and the scenic beauty of free-flowing streams and rivers are destroyed primarily to provide tax benefits for wealthy investors.

20.6 WIND

Wind power played a crucial role in the settling of the American West, much of which has abundant underground aquifers, but little surface water. The strong, steady winds blowing across the prairies provided the energy to pump water that allowed ranchers and farmers to settle the land. By the end of the nineteenth century, nearly every farm or ranch west of the Mississippi River had at least one windmill, and the manufacture, installation, and repair of windmills was a major industry. The Rural Electrification Act of 1935 brought many benefits to rural America, but it effectively killed wind power development, and shifted electrical generation to large dams and fossil fuel-burning power plants. It's interesting to speculate what the course of history might have been if we had not spent trillions of dollars on fossil fuels and nuclear power, but instead had invested that money on small-scale, renewable energy systems.

The oil price shocks of the 1970s spurred a renewed interest in wind power. In the 1980s, the United States was a world leader in wind technology, and California hosted 90 percent of all wind power generators in the world. Poor management, technical flaws, and overdependence on subsidies, however, led to bankruptcy of many of the most important companies of that era, including Kenetech, once the world's largest manufacturer of wind generators. Now China dominates both the global solar and wind power markets.

Modern wind machines are far different from those employed a generation ago. The largest wind turbines now being built have towers up to 150 m tall with 62 m long blades that reach as high as a 45-story building. Each can generate 5 MW of electricity, or enough for 2,500 typical American homes. Out of commission for maintenance only about three days per year, many can produce power 90 percent of the time. Theoretically up to 60 percent efficient, modern windmills typically produce about 35 percent of peak capacity under field conditions. Currently wind farms are the cheapest source of *new* power generation, costing as little as 3 cents/kWh compared to 4 to 5 cents/kWh for coal and five times that much for nuclear fuel. If the carbon "cap and trade" program proposed by President Obama becomes law, wind energy could be cheaper in many places than fossil fuels.

As table 20.2 shows, when the land consumed by mining is taken into account, wind power takes about one-third as much area and creates about five times as many jobs to create the same amount of electrical energy as coal.

Wind could meet all our energy needs

Wind power offers an enormous potential for renewable energy. The World Meteorological Organization estimates that 80 million MW of wind power could be developed economically worldwide. This would be five times the total current global electrical generating capacity. Wind has a number of advantages over most other power sources. Wind farms have much shorter planning and construction times than fossil fuel or nuclear power plants. Wind farms are modular (more turbines can be added if loads grow) and they have no fuel costs or air emissions (fig. 20.28).

FIGURE 20.28 Wind is the fastest growing power source in the United States and represented more than 40 percent of new installed capacity in 2010.

In the past decade, total wind generating capacity has increased nearly 20-fold making it the fastest growing energy source in the world. With 190,000 MW of global installed capacity in 2011, wind power produced about 400 TWhr of electricity. The Wind Energy Association predicts that 1.5 million MW of capacity could be possible by 2020. In 2009–2010, wind turbine prices fell by about 25 percent. In many areas the cost of wind-generated energy is already equal to that of coal power.

Wind does have limitations, however. Like solar energy, it is an intermittent source. Furthermore, not every place has strong enough or steady enough wind to make this an economical resource. Although modern windmills are more efficient than those of a few years ago, it takes a wind velocity between 7 m per second (16 mph) and 27 m per second (60 mph) to generate useful amounts of electricity.

In 2009 China passed Denmark, Germany, and Spain to become the world's largest producer of wind turbines. Clean technology provides more than 1 million Chinese jobs building equipment, much of which is exported. In 2010, for the first time, more than half the new wind energy was added outside Europe and North America. This was mainly driven by China, which installed 16.5 GW.

Although Denmark still provides a greater share of its electricity (20 percent) from wind than any other nation, the United States, with 35 GW of wind farms, has the largest installed capacity—for now. Steady, powerful winds give the United States a greater potential for this energy source than any other industrialized country. American wind farms averaged 35 percent of their theoretical potential in 2010. That compares very favorably with Germany and China at about 16 percent, and India at only 10 percent. This means that a U.S. wind farm produces about twice as much electricity per year, on average, as the same size installation in Germany or China.

As the opening case study for this chapter shows, Europe is focusing for indigenous renewable energy production on offshore wind farms. Although the United States does have good wind potential on the mid-Atlantic continental shelf, the bulk of North America's wind potential is situated on land (fig. 20.29). Compared to offshore installations, which are costly because of the

Table 20.2	Jobs and Land Required for Alternative Energy Sources	
Technology	**Land Use (m² per Gigawatt-Hour for 30 Years)**	**Jobs (per Terawatt-Hour per Year)**
Coal	3,642	116
Photovoltaic	3,237	175
Solar thermal	3,561	248
Wind	1,335	542

Source: Lester R. Brown, 1991.

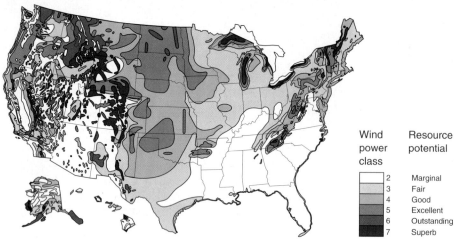

FIGURE 20.29 United States wind resource map. Mountain ranges and areas of the High Plains have the highest wind potential, but much of the country has a fair to good wind supply.
Source: Data from U.S. Department of Energy.

Wind power class	Resource potential
2	Marginal
3	Fair
4	Good
5	Excellent
6	Outstanding
7	Superb

Some people object to the sight of large machines looming on the horizon, and there's controversy about how close to houses and schools wind turbines should be allowed. Some people claim that the low-frequency sound waves from moving blades cause headaches, insomnia, digestive problems, panic attacks, and other health issues. Others dismiss these symptoms as too vague and general to assign to a specific source. It's difficult to study the issue scientifically because the low-frequency sounds attributed to wind towers are indistinguishable from wind and other ambient noises. To some people, on the other hand, windmills offer a welcome alternative to nuclear or fossil fuel-burning plants.

We need a supergrid

Many of the places with the greatest potential for both solar and wind development are far from the urban centers where power is needed. This means we'll need a vastly increased network of power lines if we're going to depend on wind or solar for a much greater proportion of our energy. In introducing his plans to double the amount of renewable energy over the next three years, President Obama said, "Today, the electricity we use is carried along a grid of lines and wires that dates back to Thomas Edison—a grid that can't support the demands of clean energy." He designated $4.5 billion to modernize and expand the transmission grid as part of the $86 billion in clean-energy investments in the economic recovery bill (fig. 20.30).

Fortunately, as we've seen earlier in this chapter, high-voltage direct current lines make it possible to transmit electricity over long distances with relatively minor loses. Interestingly, studies in California show that integration of renewable resources can smooth out daily variations (fig. 20.31). The wind blows more

need to operate in deep water and withstand storms and waves, wind tower construction on land is relatively simple and cheap. There also is growing demand for wind projects from farmers, ranchers, and rural communities because of the economic benefits that wind energy brings. One thousand megawatts of wind power (equivalent to one large nuclear or fossil fuel plant) can create more than 3,000 permanent jobs, while paying about $4 million in rent to landowners and $3.6 million in tax payments to local governments. Texas is currently the wind leader for the United States with 10,085 MW of installed capacity. Iowa is second with 3,675 MW, and 14 other states have at least 1,000 MW.

With each tower taking only about a 0.1 ha (0.25 acre) of cropland, farmers find that they can continue to cultivate 90 percent of their land while getting $2,000 or more in annual rent for each wind machine. An even better return results if the landowner builds and operates the wind generator, selling the electricity to the local utility. Annual profits can be as much as $100,000 per turbine, a wonderful bonus for use of 10 percent of your land. Cooperatives are springing up to help landowners finance, build, and operate their own wind generators. About 20 Native American tribes, for example, have formed a coalition to study wind power. Together their reservations (which were sited in the windiest, least productive parts of the Great Plains) could generate at least 350,000 MW of electrical power, equivalent to about half of the current total U.S. installed capacity.

There are problems with wind energy. In some places, high bird mortality has been reported around wind farms. This seems to be particularly true in California, where rows of generators were placed at the summit of mountain passes where wind velocities are high but where migrating birds and bats are likely to fly into rotating blades. New generator designs and more careful tower placement seems to have reduced this problem in most areas. Although national polls in the United States show that 82 percent of the public supports additional wind power, the rate of support is often considerably less among people who live close to the towers.

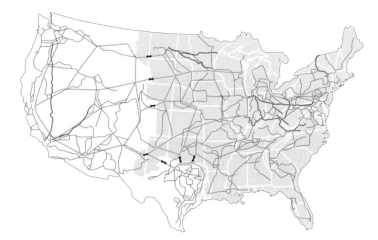

FIGURE 20.30 New high-voltage power lines will be needed if the United States is to make effective use of its renewable energy potential. The pink area served by the Eastern electrical grid needs to be connected to the west by interlinks (black dots) for maximum efficiency.

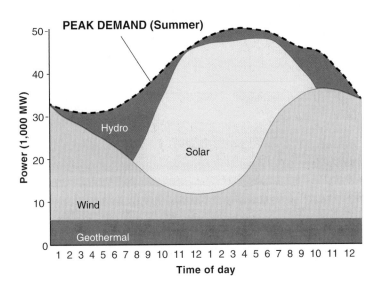

FIGURE 20.31 The wind does't blow all the time, nor is sunshine always available, but a mix of renewable resources could supply all the energy we need, especially if distant facilities are linked together. This graph shows hypothetical energy supplies for a typical July day in California.

strongly at night, and the sun shines (obviously) during the day. And because hydropower can start up quickly, it easily fills in gaps. Even though the wind doesn't blow every day in most locations, linking together wind farms even a few hundred kilometers apart can give a more steady electrical supply than does a single site.

A super grid, such as the one proposed for Desertech, could make our entire energy supply more robust, reliable, and sustainable.

20.7 OTHER ENERGY SOURCES

The earth's internal temperature can provide a useful source of energy in many cases. High-pressure, high-temperature energy exists just below the earth's surface. Around the edges of continental plates or where the earth's crust overlays magma (molten rock) pools close to the surface, this **geothermal energy** is experienced in the form of hot springs, geysers, and fumaroles. Yellowstone National Park is the largest geothermal region in the United States. Iceland, Japan, and New Zealand also have high concentrations of geothermal springs and vents. Depending on the shape, heat content, and access to groundwater, these sources produce wet steam, dry steam, or hot water.

Although few places have geothermal steam, the earth's warmth can help reduce energy costs nearly everywhere. Pumping water through buried pipes can extract enough heat so that a heat pump will operate more efficiently. Similarly, the relatively uniform temperature of the ground can be used to augment air conditioning in the summer (fig. 20.32). This can cut home heating costs by half in many areas, and pay for itself in five years.

Engineers are now exploring deep wells for community geothermal systems. Drilling 2,000 m (6,000 ft) in the American West gets you into rocks above 100°C. Fracturing them to expose more surface area, and pumping water in can produce enough

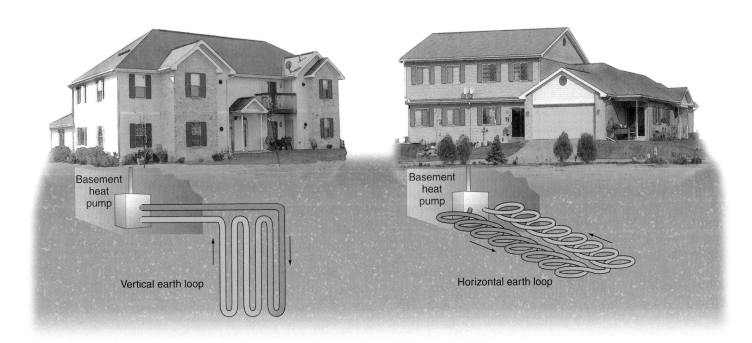

FIGURE 20.32 Geothermal energy can cut heating and cooling costs by half in many areas. In summer (shown here), warm water is pumped through buried tubing (earth loops) where it is cooled by constant underground temperatures. In winter, the system reverses and the relatively warm soil helps heat the house. Where space is limited (*left*), earth loops can be vertical. If more space is available (*right*) the tubing can be laid in shallow horizontal trenches. A heat exchanger concentrates heat, so water entering the house is far warmer than the ground temperature.

steam to run an electrical generator at a cost significantly lower than conventional fossil fuel or nuclear power. The well is no more expensive than most oil wells, and the resource is essentially perpetual. Currently about 60 new geothermal energy projects are being developed in the United States. However, there are cautions about this technology. As we saw in chapter 19, fracturing (or fracking) can contaminate ground water aquifers. And in 2010 two large geothermal projects (one in California and another in Switzerland) were abruptly canceled over concerns that they seemed to be triggering earthquakes.

Tides and waves contain significant energy

Ocean tides and waves contain enormous amounts of energy that can be harnessed to do useful work. A tidal station works like a hydropower dam, with its turbines spinning as the tide flows through them. A high-tide/low-tide differential of several meters is required to spin the turbines. Unfortunately, variable tidal periods often cause problems in integrating this energy source into the electric utility grid. Nevertheless, demand has kept some plants running for many decades.

Ocean wave energy can easily be seen and felt on any seashore. The energy that waves expend as millions of tons of water are picked up and hurled against the land, over and over, day after day, can far exceed the combined energy budget for both insolation (solar energy) and wind power in localized areas. Captured and turned into useful forms, that energy could make a substantial contribution to meeting local energy needs.

Dutch researchers estimate that 20,000 km of ocean coastline are suitable for harnessing wave power. Among the best places in the world for doing this are the west coasts of Scotland, Canada, the United States (including Hawaii), South Africa, and Australia. Wave energy specialists rate these areas at 40 to 70 kW per meter of shoreline. Altogether, it's calculated, if the technologies being studied today become widely used, wave power could amount to as much as 16 percent of the world's current electrical output.

Some of the designs being explored include oscillating water columns that push or pull air through a turbine, and a variety of floating buoys, barges, and cylinders that bob up and down as waves pass, using a generator to convert mechanical motion into electricity. It's difficult to design a mechanism that can survive the worst storms.

An interesting new development in this field is the Pelamis wave-power generator developed by the Scottish start-up company Ocean Power Delivery (fig. 20.33). The first application of this technology is now being built 5 km off the coast of Portugal. It will use three units capable of producing some 2.25 MW of electricity, or enough to supply 1,500 Portuguese households. If preliminary trials go well, plans are to add 40 more units in a year or two. Each of the units consists of four cylindrical steel sections linked by hinged joints. Anchored to the seafloor at its nose, the snakelike machine points into the waves and undulates up and down and side to side as swells move along its 125 m length. This motion pumps fluid to hydraulic motors that drive

FIGURE 20.33 The Pelamis wave converter (named after a sea snake) is a 125 m long and 3.5 m diameter tube-hinged, so it undulates as ocean swells pass along it. This motion drives pistons that turn electrical generators. Energy experts calculate that capturing just 1 to 2 percent of global wave power could supply at least 16 percent of the world's electrical demand.

electrical generators to produce power, which is carried to shore by underwater cables.

Pelamis's inventor, Richard Yemm, says that survivability is the most important feature of a wave-power device. Being offshore, the Pelamis isn't exposed to the pounding breakers that destroy shore-based wave-power devices. If waves get too steep, the Pelamis simply dives under them, much as a surfer dives under a breaker. These wave converters lie flat in the water and are positioned far offshore, so they are unlikely to stir up as much opposition as do the tall towers of wind generators.

Think About It

Some people object to the sight of giant windmills. They think it's an intrusion on the land and spoils the view. Yet those same people don't object to other forms of modern technology. Is this resistance just because wind power is new, or is there something truly different about it?

Ocean thermal electric conversion might be useful

Temperature differentials between upper and lower layers of the ocean's water also are a potential source of renewable energy. In a closed-cycle **ocean thermal electric conversion (OTEC)** system, heat from sun-warmed upper ocean layers is used to evaporate a working fluid, such as ammonia or Freon, which has a low boiling point. The pressure of the gas produced is high enough to spin turbines to generate electricity. Cold water then is pumped from the ocean depths to condense the gas.

As long as a temperature difference of about 20°C (36°F) exists between the warm upper layers and cooling water, useful amounts of net power can, in principle, be generated with one of these systems. This differential corresponds, generally, to a depth of about 1,000 m in tropical seas. The places where this much temperature difference is likely to be found close to shore are islands that are the tops of volcanic seamounts, such as Hawaii, or the edges of continental plates along subduction zones (chapter 14) where deep trenches lie just offshore. The west coast of Africa, the south coast of Java, and a number of South Pacific islands, such as Tahiti, have usable temperature differentials for OTEC power.

Although their temperature differentials aren't as great as the ocean, deep lakes can have very cold bottom water. Ithaca, New York, has recently built a system to pump cold water out of Lake Cayuga to provide natural air conditioning during the summer. Cold water discharge from a Hawaiian OTEC system has been used to cool the soil used to grow cool-weather crops such as strawberries.

20.8 WHAT'S OUR ENERGY FUTURE?

Former vice president Al Gore has issued a bold and inspiring challenge to the United States. Currently, he said, "We're borrowing money from China to buy oil from the Persian Gulf to burn in ways that destroy the planet." He urged America to repower itself with 100 percent carbon-free electricity within a decade. Doing so, he proposed, would solve the three biggest crises we face—environmental, economic, and security—simultaneously. This ambitious project could create millions of jobs, spur economic development, and eliminate our addiction to imported fossil fuels.

But could we realistically get all our electricity from renewable, environmentally friendly sources in such a short time? Mark Jacobson from Stanford University and Mark Delucchi from the University of California-Davis believe we can. Moreover, they calculate that currently available wind, water, and solar technologies could supply 100 percent of the world's energy by 2030 and completely eliminate all our use of fossil fuels. They calculate that it would take 3.8 million large wind turbines (each rated at 5 MW), 1.7 billion rooftop photovoltaic systems, 720,000 wave converters, half a million tidal turbines, 89,000 concentrated solar power plants and industrial-sized photovoltaic arrays, 5,350 geothermal plants, and 900 hydroelectric plants, worldwide.

Wouldn't it be an overwhelming job to build and install all that technology? It would be a huge effort, but it's not impossible. Jacobson and Delucchi point out that society has achieved massive transformations before. In 1956 the United Sates began building the Interstate Highway System, which now extends 47,000 mi (75,600 km) and has changed commerce, landscapes, and society. And every year roughly 60 million new cars and trucks are added to the world's highways.

Is there enough clean energy to meet our needs? Yes, there is. As we've already seen, the readily available wind, solar, and water power sources are at least 100 times larger than our current power consumption. Even allowing for growth as people in the developing world improve their standard of living, there's more than enough environmentally friendly energy for everyone.

The World Energy Council projects that renewables could provide about 60 percent of world cumulative energy consumption in 2030 assuming that political leaders take global warming seriously and pass taxes to encourage conservation and protect the environment (fig. 20.34). This idealized "ecological scenario" also envisions measures to shift wealth from the north to south, and to enhance economic equity. By the end of the twenty-first century, renewable sources could provide all our energy needs if we take the necessary steps to make this happen. After the meltdown of four nuclear reactors in Japan in 2011, Germany, Japan, Switzerland, Sweden and several other countries announced intentions to move away from both nuclear and fossil fuels and to emphasize renewable energy sources in the future.

Interestingly, it would take about 30 percent less total energy to meet our needs with sun, wind, and water than to continue using fossil fuels. That's because electricity is a more efficient way to use energy than burning dead plants and animals. For example, only about 20 percent of the energy in gasoline is used to move a vehicle (the rest is wasted as heat). An electric vehicle, on the other hand, uses about three-quarters of the energy in electricity for motion. Furthermore, much of the energy from renewable sources could often be produced closer to where it's used, so there are fewer losses in transmission and processing.

Won't it be expensive to install so much new technology? Yes it will be, but the costs of continuing our current dependence on fossil fuels would be much higher. It's estimated that investing $700 billion per year now in clean energy will avoid twenty times that much in a few decades from the damages of climate change.

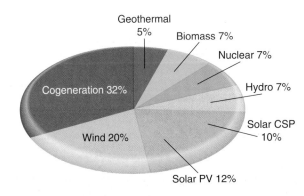

FIGURE 20.34 A renewable energy scenario for 2030. Co-generation would mostly burn natural gas to generate both electricity and space heating.
Source: 2008 Worldwatch Report, p. 178.

CONCLUSION

None of the renewable energy sources discussed in this chapter are likely to completely replace fossil fuels and nuclear power in the near future. They could, however, make a substantial collective contribution toward providing us with the conveniences we crave in a sustainable, environmentally friendly manner. They could also make us energy independent and balance our international payment deficit.

Accidents, such as the Gulf Oil Spill, together with rising fuel prices and civil unrest the Middle East have prompted a turn in U.S. energy policy. Our dependence on imported oil is getting a second look. When he took office, President Obama called for at least $86 billion in incentives and grants for conservation and renewable energy. He set a goal of 10 percent of the nation's electricity from renewables by 2012 and 25 percent by 2025.

Some people think we could do better—and that we need to if we're to stabilize the national budget, create green jobs, and reduce global climate change. We could get all the energy we need from renewable sources. And many of the steps needed to meet this goal would save money, improve our environment, and have social benefits. The question remains, however, whether we'll have the courage, foresight, and resolve to do so. What do you think? How should we move toward a sustainable energy future?

REVIEWING LEARNING OUTCOMES

By now you should be able to explain the following points:

20.1 Describe how renewables can help us meet our energy needs.
- There are many ways to save energy.
- Green buildings can cut energy costs by half.
- Transportation could be far more efficient.
- Cogeneration produces both electricity and heat.

20.2 Explain how we could tap solar energy.
- Solar collectors can be passive or active.
- CSP is an example of high-temperature solar.
- Public policy can promote renewable energy.
- Photovoltaics generate electricity directly.
- Smart metering can save money and energy.

20.3 Visualize how fuel cells work.
- All fuel cells share common components.

20.4 Grasp the potential of biomass.
- We can burn biomass.
- Methane from biomass is clean and efficient.

- Ethanol and biodiesel can contribute to fuel supplies.
- Cellulosic ethanol seems to offer hope for the future.
- Could algae be a hope for the future?

20.5 Explain the benefits and drawbacks of hydropower.
- Falling water has been used as an energy source since ancient times.

20.6 Investigate wind energy.
- Wind could meet all our energy needs.
- We need a supergrid.

20.7 Understand the prospects for other energy sources.
- Geothermal heat, tides, and waves could be valuable resources.
- Ocean thermal electric conversion might be useful.

20.8 Discuss our energy future.

PRACTICE QUIZ

1. Describe five ways we could conserve energy individually or collectively.
2. Explain the principle of net energy yield. Give some examples.
3. What is the difference between active and passive solar energy?
4. How do photovoltaic cells generate electricity?
5. What is a fuel cell and how does it work?
6. Why might *Jatropha* be a good source of biodiesel?
7. Why might *Miscanthus* be a good source of ethanol?
8. What are some advantages and disadvantages of large hydroelectric dams?
9. How can geothermal energy be used for home heating?
10. Describe how tidal power or ocean wave power generate electricity.

CRITICAL THINKING AND DISCUSSION QUESTIONS

1. What alternative energy sources are most useful in your region and climate? Why?

2. What can you do to conserve energy where you live? In personal habits? In your home, dormitory, or workplace?

3. Do you think building wind farms in remote places, parks, or scenic wilderness areas would be damaging or unsightly?

4. If you were the energy czar of your state, where would you invest your budget?

5. What could (or should) we do to help developing countries move toward energy conservation and renewable energy sources? How can we ask them to conserve when we live so wastefully?

 **Data Analysis:** **Energy Calculations**

Most college students either already own or are likely to buy an automobile and a computer sometime soon. How do these items compare in energy usage? Suppose that you were debating between a high-mileage car, such as the Honda Insight, or a sport utility vehicle, such as a Ford Excursion. How do the energy requirements of these two purchases measure up? To put it another way, how long could you run a computer on the energy you would save by buying an Insight rather than an Excursion?

Here are some numbers you need to know. The Insight gets about 75 mpg, while the Excursion gets about 12 mpg. A typical American drives about 15,000 mi per year. A gallon of regular, unleaded gasoline contains about 115,000 Btu on average. Most computers use about 100 watts of electricity. One kilowatt-hour (kWh) = 3,413 Btu.

1. How much energy does the computer use if it is left on continuously? (You really should turn it off at night or when it isn't in use, but we'll simplify the calculations.)
   100 watt/h + 24 h/day + 365 days/yr = _____ kWh/yr

2. How much gasoline would you save in an Insight, compared with an Excursion?
   a. Excursion:
      15,000 mi/yr ÷ 12 mpg = _____ gal/yr
   b. Insight:
      15,000 mi/yr ÷ 75 mpg = _____ gal/yr
   c. Gasoline savings (a − b) = _____ gal/yr
   d. Energy savings:
      (gal + 115,000 Btu) = _____ Btu/yr
   e. Converting Btu to kWh:
      (Btu + 0.00029 Btu/kWh) = _____ kWh/yr saved

3. How long would the energy saved run your computer?
   kWh/yr saved by Insight ÷ kWh/yr consumed by computer = _____

For Additional Help in Studying This Chapter, please visit our website at www.mhhe.com/cunningham12e. You will find additional practice quizzes and case studies, flashcards, regional examples, placemarks for Google Earth™ mapping, and an extensive reading list, all of which will help you learn environmental science.

CHAPTER 21

An endangered Hawaiian monk seal is disentangled
from abandoned fishing nets in the
Papahānaumokuākea Marine National Monument.

Solid, Toxic,
and Hazardous Waste

Learning Outcomes

After studying this chapter, you should be able to:

21.1 Identify the components of solid waste.

21.2 Describe how wastes have been—and are
being—disposed of or treated.

21.3 Identify how we might shrink the waste stream.

21.4 Investigate hazardous and toxic wastes.

*"We have no knowledge, so we have
stuff; but stuff without knowledge is
never enough."*

~ *Greg Brown*

The Papahānaumokuākea Marine National Monument, the largest conservation area in the United States, was established by President George W. Bush in 2006. With the designation of this sanctuary, the president protected a chain of islands, atolls, and reefs extending across 140,000 mi², northwest of the larger inhabited islands of Hawaii. The monument protects some of the most pristine and diverse deep coral reefs and over 7,000 marine species, including rare and endangered species such as the Laysan albatross and the Hawaiian monk seal. The string of isolated islets and coral atolls make up the world's largest tropical seabird rookery, supporting 14 million nesting seabirds. The preserve is also home to a wealth of cultural and historic heritage sites, including ship wrecks and World Heritage cultural sites for native Hawaiians.

Despite its remote location, Papahānaumokuākea,* also known as the Northwestern Hawaiian Islands Marine National Monument, remains vulnerable to the flotsam and jetsam of modern life. The islands and reefs lie within the vast circulating currents known as the Pacific gyre. These swirling currents, driven by winds and the Coriolis effect (chapter 15) concentrate nutrients, organic debris, and, in recent decades, an ocean of plastic trash. Often called the Great Pacific Garbage Patch, or the Pacific Garbage Gyre, this region of floating plastic debris is really a drifting cloud of plastic particles, soda bottles and caps, disposable shopping bags, packaging, discarded fishing nets, and other debris. Much of it consists of tiny fragments floating just below the surface, but some pieces are large and recognizable, and some float 20–30 m deep. The greatest concentrations of plastic debris occur in the eastern Pacific, between California and Hawaii, and in the western Pacific near Japan. But the trash field extends across the ocean, with lesser aggregations near the Papahānaumokuākea preserve. Similar garbage patches have been identified in the Atlantic and elsewhere in the world's oceans, but the Pacific cases are the best studied.

The Pacific garbage gyre is thought to contain more than 100 million tons of plastic. In some areas this debris outweighs the living biomass. Fish have been found with stomachs full of plastic fragments. Seabirds gulp down plastic fragments, then regurgitate them for their chicks. With stomachs blocked by indigestible bottle caps, disposable lighters, and other items, chicks starve to death. In one study of Laysan albatrosses, 90 percent of the carcasses of dead albatross chicks contained plastic fragments (fig. 21.1). Seals, turtles, porpoises, and seabirds become ensnared in ghost fishing nets and drown, or they die from ingesting indigestible materials. Oceanographers worry that this debris is slowly starving ocean ecosystems.

Surveys at sea and on beaches indicate that 50–80 percent of the floating material originates onshore. The rest is discarded or lost at sea. Stray shopping bags, drink containers, fast-food boxes, and other refuse fall from dumpsters, wash away from landfills, or are discarded on the street, then wash into storm sewers and streams. Eventually these items travel to the sea, where they gradually break into smaller pieces as they join the great global masses of ocean plastic.

The problem has been extraordinarily difficult to address because it is widespread, diffuse, abundant, and constantly replenished by careless or incomplete disposal of waste onshore and at sea. But growing awareness is starting to make a difference. Cleanup cruises in Papahānaumokuākea have col-

FIGURE 21.1 A Laysan albatross chick, which died after being fed plastic debris rather than fish. Starvation after plastic ingestion is a leading cause of death for these albatross chicks.

lected more than 700 metric tons of discarded fishing gear that had clogged reefs.

In Papahānaumokuākea and elsewhere, marine debris has also caught the public's attention, and widespread beach cleanups are having an effect. According to the EPA, beach cleanups involved 183,000 people across the United States, collecting nearly 2,000 tons of debris from 9,000 miles of coastline in 2008. Increasing awareness is also encouraging many fishing boats to reduce disposal of plastic garbage at sea. Because all this material fouls fishing gear, costing time and money, it is in their interest to bring in the garbage they produce or collect in their nets.

You can help, too: the next time you see plastic debris that's about to wash into a storm sewer, remember that everything ends up eventually in the ocean. Pick it up if you can, and try to prevent

*Pronounced Pa-pa-ha-nao-Mo-kua-kea; To hear the pronunciation, visit the monument's website, www.papahanaumokuakea.gov.

Case Study continued

your own plastic bottles, caps, and packaging from escaping into waterways. You can also try to reduce the amount of disposable containers, bottles, and packaging you buy. Containing and minimizing loose garbage is one of the best ways to reduce marine debris.

The remote atolls of northwestern Hawaii show us that no place is too remote to be affected by our waste production and disposal. The materials we buy and the ways we manage our garbage can have dramatic impacts on living systems at home and far away. At the same time, responses to the problem have shown that

people everywhere have an interest in taking care of the land and oceans, and in keeping them beautiful. Often the obstacles, and the volumes of waste, seem insurmountable. But clean-up efforts in Hawaii shows that progress can be real if we keep at it. In this chapter we'll examine the waste we produce, our methods to dispose of it, and strategies to reduce, reuse, and recycle it.

For related resources, including Google Earth™ placemarks that show locations discussed in this chapter, visit Environmental-Science-Cunningham.blogspot.com.

21.1 SOLID WASTE

Waste is everyone's business. We all produce wastes in nearly everything we do. According to the Environmental Protection Agency, the United States produces 11 billion tons of solid waste each year. About half of that amount consists of agricultural waste, such as crop residues and animal manure, which are generally recycled into the soil on the farms where they are produced. They represent a valuable resource as ground cover to reduce erosion and fertilizer to nourish new crops, but they also constitute the single largest source of nonpoint air and water pollution in the country. More than one-third of all solid wastes are mine tailings, overburden from strip mines, smelter slag, and other residues produced by mining and primary metal processing. Road and building construction debris is another major component of solid waste. Much of this material is stored in or near its source of production and isn't mixed with other kinds of wastes. Improper disposal practices, however, can result in serious and widespread pollution.

Industrial waste—other than mining and mineral production—amounts to some 400 million metric tons per year in the United States. Most of this material is recycled, converted to other forms, destroyed, or disposed of in private landfills or deep injection wells. About 60 million metric tons of industrial waste falls in a special category of hazardous and toxic waste, which we will discuss later in this chapter.

Municipal waste—a combination of household and commercial refuse—amounts to more than 200 million metric tons per year in the United States (fig. 21.2). That's approximately two-thirds of a ton for each man, woman, and child every year—twice as much per capita as Europe or Japan, and five to ten times as much as most developing countries.

The waste stream is everything we throw away

Think for a moment about how much we discard every year. There are organic materials, such as yard and garden wastes, food wastes, and sewage sludge from treatment plants; junked cars; worn out furniture; and consumer products of all types. Newspapers,

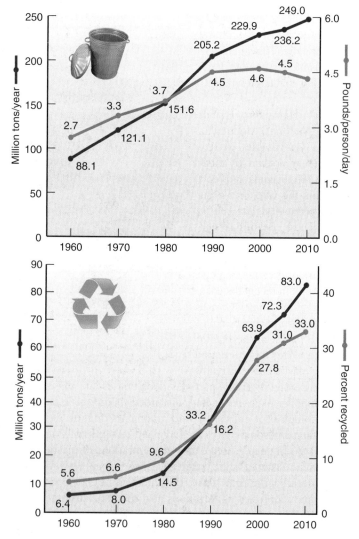

FIGURE 21.2 Bad news and good news in solid waste production. Per capita waste has risen steadily to more than 2 kg per person per day. Recycling rates are also rising, however. Recycling data include composting.

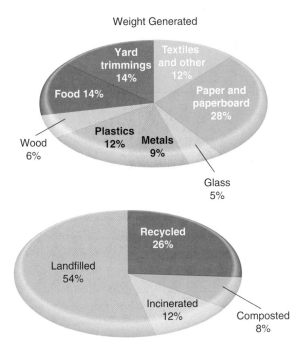

Weight Generated

FIGURE 21.3 Composition of municipal solid waste in the United States by weight, before recycling, and disposal methods.
Source: Data from U.S. Environmental Protection Agency. Office of Solid Waste Management, 2010.

Think About It

Figure 21.2 shows a continuing increase in waste production per capita. What is the percentage increase per capita from 1960 to 2005? (*Hint:* calculate (4.5 − 2.7) ÷ 2.7.) What might account for this increase? Is there a relationship between waste production and our quality of life?

21.2 WASTE DISPOSAL METHODS

Where are our wastes going now? In this section, we will examine some historic methods of waste disposal as well as some future options. We'll begin with the least desirable but most common methods, then proceed to some preferable options. Keep in mind as you read this that modern waste management reverses this order and stresses the "three R's" of reduction, reuse, and recycling before destruction or, finally, secure storage of wastes.

Open dumps release hazardous materials into air and water

For many people, the way to dispose of waste is to simply drop it someplace. Open, unregulated dumps are still the dominant method of waste disposal in most developing countries (fig. 21.4). The giant developing-world megacities have enormous garbage problems. Mexico City, one of the largest cities in the world, generates some 10,000 tons of trash *each day*. Until recently, most of this torrent of waste was left in giant piles, exposed to the wind and rain, as well as rats, flies, and other vermin. Manila, in the Philippines, generates a similar amount of waste, half of which goes to a giant, constantly smoldering dump called "Smoky Mountain." Over 20,000 people live and work on this mountain of refuse, scavenging for recyclable items or edible food scraps. In July 2000, torrential rains spawned by

magazines, catalogs, and office refuse make paper one of our major wastes (fig. 21.3). In spite of recent progress in recycling, many of the 200 *billion* metal, glass, and plastic food and beverage containers used every year in the United States end up in the trash. Wood, concrete, bricks, and glass come from construction and demolition sites, dust and rubble from landscaping and road building. All of this varied and voluminous waste has to arrive at a final resting place somewhere.

The **waste stream** is a term that describes the steady flow of varied wastes that we all produce, from domestic garbage and yard wastes to industrial, commercial, and construction refuse. Many of the materials in our waste stream would be valuable resources if they were not mixed with other garbage. Unfortunately, our collecting and dumping processes mix and crush everything together, making separation an expensive and sometimes impossible task. In a dump or incinerator, much of the value of recyclable materials is lost.

Another problem with refuse mixing is that hazardous materials in the waste stream get dispersed through thousands of tons of miscellaneous garbage. This mixing makes the disposal or burning of what might have been rather innocuous stuff a difficult, expensive, and risky business. Spray paint cans, pesticides, batteries (zinc, lead, or mercury), cleaning solvents, smoke detectors containing radioactive material, and plastics that produce dioxins and PCBs when burned are mixed willy-nilly with paper, table scraps, and other nontoxic materials. The best thing to do with household toxic and hazardous materials is to separate them for safe disposal or recycling, as we will see later in this chapter.

FIGURE 21.4 Trash disposal has become a crisis in the developing world, where people have adopted cheap plastic goods and packaging but lack good recycling or disposal options.

(a)

(b)

FIGURE 21.5 Plastic trash dumped on land and at sea ends up on remote beaches (a) and kills unknown numbers of marine organisms (b). This sea turtle is tangled in abandoned fishing nets.

Typhoon "Kai Tak" caused part of the mountain to collapse, burying at least 215 people. The government would like to close these dumps, but finding another disposal method has been a challenge.

Most developed countries forbid open dumping, at least in metropolitan areas, but illegal dumping is still a problem. You have undoubtedly seen trash accumulating along roadsides and in vacant, weedy lots in the poorer sections of cities. As noted in the opening case study, this is not just a question of aesthetics. Plastic and other waste products are long-lasting and sometimes dangerous. Discarded oil and solvents, from cars, paints, and household chemicals are also toxic. An estimated 200 million liters of waste motor oil are poured into the sewers or allowed to soak into the ground every year in the United States. This is about five times as much as was spilled by the *Exxon Valdez* in Alaska in 1989! No one knows the volume of solvents and other chemicals disposed of by similar methods.

Ocean dumping is nearly uncontrollable

We have long treated the oceans as a universal dumping ground. An estimated 20 million tons of plastic debris ends up in the ocean each year. This includes some 25,000 metric tons (55 million lbs) of packaging, including half a million bottles, cans, and plastic containers, which are dumped at sea. Beaches, even in remote regions, are littered with the nondegradable flotsam and jetsam of industrial society (fig. 21.5a). About 150,000 tons (330 million lbs)

of fishing gear—including more than 1,000 km (660 mi) of nets—are lost or discarded at sea each year (fig. 21.5b). Wildlife advocates estimate that 50,000 northern fur seals are entangled in this refuse and drown or starve to death every year in the North Pacific alone.

We often export waste to countries ill-equipped to handle it

The United States disposes of about 47 million computers and 1 million cell phones every year, each containing a complex mix of often-toxic metals and plastics. Since 1989 it has been illegal to export this **electronic waste**, or **e-waste**, to developing countries, but we continue to do so. About 80 percent of our e-waste is shipped overseas, mostly to China and other developing countries in Asia and Africa. There, villagers, including young children, break it apart to retrieve valuable metals. Often, this scrap recovery is done under primitive conditions where workers have little or no protective gear (fig. 21.6a) and residue goes into open dumps. Health risks in this work are severe, especially for growing children. Soil, groundwater, and surface water contamination at these sites has been found to be as much as 200 times the World Health Organization's standards. An estimated 100,000 workers handle e-waste in China alone. With increasing regulation in China, however, the trade is shifting to India, Ghana, and other impoverished areas.

E-waste generation is increasing, and soon developing countries themselves will be the leading producers of these toxic materials (fig. 21.6b). Outdated electronic devices are one of the greatest sources of toxic material currently going to developing countries. There are at least 2 billion television sets and personal computers in use globally. Televisions often are discarded after only about five years, while computers, play-stations, cell phones, and other electronics become obsolete even faster. As many as 600 million computers are in use in the United States (twice as many as there are residents), and most will be discarded in the next few years. Only about 10 percent of the components are currently recycled. These computers contain at least 2.5 billion kg of lead (as well as mercury, gallium, germanium, nickel, palladium, beryllium, selenium, arsenic), and valuable metals, such as gold, silver, copper, and steel.

Toxic waste exportation is a chronic problem even though it is banned in most countries. In 2006, for example, 400 tons of toxic waste were illegally dumped at 14 open dumps in Abidjan, the capital of the Ivory Coast. The black sludge—petroleum wastes containing hydrogen sulfide and volatile hydrocarbons—killed ten people and injured many others. At least 100,000 city residents sought medical treatment for vomiting, stomach pains, nausea, breathing difficulties, nosebleeds, and headaches. The sludge—which had been refused entry at European ports—was transported by an Amsterdam-based multinational company on a Panamanian-registered ship and handed over to an Ivorian firm (thought to be connected to corrupt government officials) to be dumped in the Ivory Coast. The Dutch company agreed to clean up the waste and pay the equivalent of (U.S.) $198 million to settle claims.

Most of the world's obsolete ships are now dismantled and recycled in poor countries. The work is dangerous, and old ships often are full of toxic and hazardous materials, such as oil, diesel fuel, asbestos, and heavy metals. On India's Anlang Beach, for example, more than 40,000 workers tear apart outdated vessels using crowbars, cutting torches, and even their bare hands. Metal is dragged away and sold for recycling. Organic waste is often simply burned on the beach, where ashes and oily residue wash back into the water.

Landfills receive most of our waste

Over the past 50 years most American and European cities have recognized the health and environmental hazards of open dumps. Instead we have **sanitary landfills**, where solid waste is contained more effectively. To decrease smells and litter and to discourage insect and rodent populations, landfill operators are required to compact the refuse and cover it every day with a layer of dirt. This method helps control pollution, but the dirt fill also takes up as much as 20 percent of landfill space. Since 1994, all operating landfills in the United States have been required to control such hazardous substances as oil, chemical compounds, toxic metals, and contaminated rainwater that seeps through piles of waste. An impermeable clay and/or plastic lining underlies and encloses the storage area (fig. 21.7). Drainage systems are installed in and around the liner to catch drainage and to help monitor chemicals that may be leaking. Modern municipal solid-waste landfills now have many of the safeguards of hazardous waste repositories described later in this chapter.

Think About It

Ocean dumping of both solid waste and hazardous waste is a chronic problem. Suppose you were a captain or a sailor on an ocean-going ship. What factors might influence your decision to dump waste oil, garbage, or occasional litter overboard? (Money? Time? Legal considerations about your cargo or waste?) Whose responsibility is ocean dumping? What steps could the international community take to reduce it?

(a)

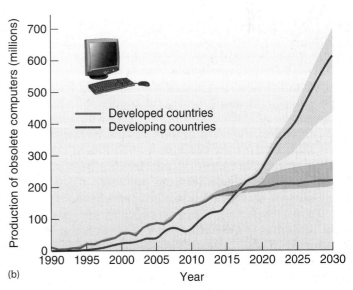

(b)

FIGURE 21.6 A Chinese woman smashes a computer monitor to remove valuable metals, releasing a host of health risks (a). Increasingly, this industry will serve e-waste producers in developing areas (b).

FIGURE 21.7 A plastic liner being installed in a sanitary landfill. This liner and a bentonite clay layer below it prevent leakage to groundwater. Trash is also compacted and covered with earth fill every day.

Landfill space near population centers is becoming scarce and expensive. Just 25 years ago the United States had 8,000 landfills; today we have fewer than 2,000. Fresh Kills Landfill on Staten Island, New York, was the world's largest until it closed in 2001. New York now sends its garbage to Pennsylvania and Ohio. Many other cities are running out of local landfill space and must export trash, at enormous expense, to neighboring states. More than half the solid waste from New Jersey goes out of state, some of it up to 800 km (500 mi) away.

More careful attention is now paid to the siting of new landfills. Sites located on highly permeable or faulted rock formations are passed over in favor of sites with less leaky geologic foundations. Landfills are being built away from rivers, lakes, floodplains, and aquifer recharge zones rather than near them, as was often done in the past. More care is being given to a landfill's long-term effects so that costly cleanups and rehabilitation can be avoided.

Historically, landfills have been a convenient and relatively inexpensive waste-disposal option in most places, but this situation is changing rapidly. Rising land prices and shipping costs, as well as increasingly demanding landfill construction and maintenance requirements, are making this a more expensive disposal method. The cost of disposing a ton of solid waste in Philadelphia went from $20 in 1980 to more than $100 in 2010. Union County, New York, experienced an even steeper price rise. In 1987, it paid $70 to get rid of a ton of waste; a year later, that same ton cost $420, or about $10 for a typical garbage bag. In the past decades, costs have continued to rise steadily, though not as sharply. The United States now spends over $10 billion per year to dispose of trash.

Suitable places for waste disposal are becoming scarce in many areas. Other uses compete for open space. Communities have become more concerned and vocal about health hazards, as well as aesthetics. It is difficult to find a neighborhood willing to accept a new landfill. Since 1984, when stricter financial and environmental protection requirements for landfills took effect, thousands of landfills have closed.

A positive trend in landfill management is methane recovery. Methane, or natural gas, is a natural product of decomposing garbage deep in a landfill. Methane is also a potent greenhouse gas. Normally methane seeps up to the landfill surface and escapes. At 300 U.S. landfills, the methane is being collected and burned. Cumulatively, these landfills could provide enough electricity for a city of a million people. Three times as many landfills could be recovering methane. Tax incentives could be developed to encourage this kind of resource recovery.

Incineration produces energy but causes pollution

Landfilling is still the disposal method for the majority of municipal waste in the United States (fig. 21.8). Faced with growing piles of garbage and a lack of available landfills at any price, however, public officials are investigating other disposal methods. The method to which they frequently turn is burning. Another term commonly used for this technology is **energy recovery**, or waste-to-energy, because the heat derived from incinerated refuse is a useful resource. Burning garbage can produce steam used directly for heating buildings or generating electricity. Internationally, well over 1,000 waste-to-energy plants in Brazil, Japan, and western Europe generate much-needed energy while also reducing the amount that needs to be landfilled. In the United States, more than 110 waste incinerators burn 45,000 tons of garbage daily. Some of these are simple incinerators; others produce steam and/or electricity.

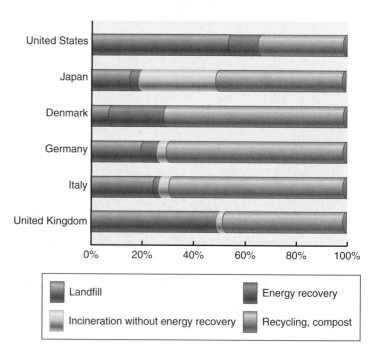

FIGURE 21.8 Disposal methods for municipal solid waste in several developed countries.

Types of Incinerators

Municipal incinerators are specially designed burning plants capable of burning thousands of tons of waste per day. In some plants, refuse is sorted as it comes in to remove unburnable or recyclable materials before combustion. This is called **refuse-derived fuel** because the enriched burnable fraction has a higher energy content than the raw trash. Another approach, called **mass burn**, is to dump everything smaller than sofas and refrigerators into a giant furnace and burn as much as possible (fig. 21.9). This technique avoids the expensive and unpleasant job of sorting through the garbage for nonburnable materials, but it often causes greater problems with air pollution and corrosion of burner grates and chimneys.

In either case, residual ash and unburnable residues representing 10 to 20 percent of the original volume are usually taken to a landfill for disposal. Because the volume of burned garbage is reduced by 80 to 90 percent, disposal is a smaller task. However, the residual ash usually contains a variety of toxic components that make it an environmental hazard if not disposed of properly. Ironically, one worry about incinerators is whether enough garbage will be available to feed them. Some communities in which recycling has been really successful have had to buy garbage from neighbors to meet contractual obligations to waste-to-energy facilities. In other places, fears that this might happen have discouraged recycling efforts.

Incinerator Cost and Safety

The cost-effectiveness of garbage incinerators is the subject of heated debates. Initial construction costs are high—usually between $100 million and $300 million for a typical municipal facility. Tipping fees at an incinerator, the fee charged to haulers for each ton of garbage dumped, are often much higher than those at a landfill. As landfill space near metropolitan areas becomes more scarce and more expensive, however, landfill rates are certain to rise. It may pay in the long run to incinerate refuse so that the lifetime of existing landfills will be extended.

Environmental safety of incinerators is another point of concern. The EPA has found alarmingly high levels of dioxins, furans, lead, and cadmium in incinerator ash. These toxic materials were more concentrated in the fly ash (lighter, airborne particles capable of penetrating deep into the lungs) than in heavy

bottom ash. Dioxin levels can be as high as 780 parts per billion. One part per billion of TCDD, the most toxic dioxin, is considered a health concern. All of the incinerators studied exceeded cadmium standards, and 80 percent exceeded lead standards. Proponents of incineration argue that if they are run properly and equipped with appropriate pollution-control devices, incinerators are safe to the general public. Opponents counter that neither public officials nor pollution-control equipment can be trusted to keep the air clean. They argue that recycling and source reduction efforts are better ways to deal with waste problems.

The EPA, which generally supports incineration, acknowledges the health threat of incinerator emissions but holds that the danger is very slight. The EPA estimates that dioxin emissions from a typical municipal incinerator may cause one death per million people in 70 years of operation. Critics of incineration claim that a more accurate estimate is 250 deaths per million in 70 years.

One way to reduce these dangerous emissions is to remove batteries containing heavy metals and plastics containing chlorine before wastes are burned. Bremen, West Germany, is one of several European cities now trying to control dioxin emissions by keeping all plastics out of incinerator waste. Bremen is requiring households to separate plastics from other garbage. This is expected to eliminate nearly all dioxins and other combustion by-products and prevent the expense of installing costly pollution-control equipment that otherwise would be necessary to keep the burners operating. Several cities have initiated a recycling program for the small "button" batteries used in hearing aids, watches, and calculators in an attempt to lower mercury emissions from its incinerator.

21.3 SHRINKING THE WASTE STREAM

Having less waste to discard is obviously better than struggling with disposal methods, all of which have disadvantages and drawbacks. In this section we will explore some of our options for recycling, reuse, and reduction of the wastes we produce.

Recycling captures resources from garbage

The term *recycling* has two meanings in common usage. Sometimes we say we are *recycling* when we really are *reusing* something, such as refillable beverage containers. In terms of solid waste management, however, **recycling** is the reprocessing of discarded materials into new, useful products (fig. 21.10). Some recycling processes reuse materials for the same purposes; for instance, old aluminum cans and glass bottles are usually melted and recast into new cans and bottles. Other recycling processes turn old materials into entirely new products. Old tires, for instance, are shredded and turned

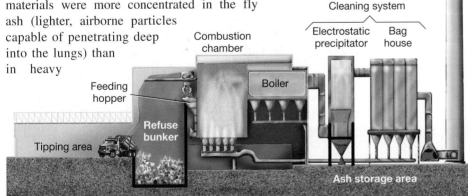

FIGURE 21.9 A diagram of a municipal "mass burn" garbage incinerator. Steam produced in the boiler can be used to generate electricity or to heat nearby buildings.

FIGURE 21.10 Creating a stable, economically viable market for recycled products is essential for recycling success. Consumers can help by buying recycled products.

into rubberized road surfacing. Newspapers become cellulose insulation, kitchen wastes become a valuable soil amendment, and steel cans become new automobiles and construction materials.

The high value of aluminum scrap has spurred a large percentage of aluminum recycling nearly everywhere (fig. 21.11). About two-thirds of all aluminum beverage cans are now recycled; up from only 15 percent in 1970. Aluminum is also valuable, lightweight, and easy to recycle rapidly: half of the cans now on grocery shelves will be made into another can within two months. Even so, we

throw away staggering amounts of materials. Every three months Americans throw away enough aluminum drink cans to rebuild the entire commercial airline fleet.

Recycling saves money, materials, and energy

Recycling is usually a better alternative to either dumping or burning wastes. It saves money, energy, raw materials, and landfill space, while also reducing pollution. Recycling also encourages individual awareness and responsibility for the refuse produced. Household sorting is the bedrock of many recycling systems (fig. 21.12). But many cities now have recycling facilities with mechanical sorting machines so that homeowners don't have to separate recyclables into different categories. Everything can be placed in a single container.

Curbside pickup of recyclables costs around $35 per ton, as opposed to the $80 paid to dispose of them at an average metropolitan landfill. Many recycling programs cover their own expenses with materials sales and may even bring revenue to the community. Landfills continue to dominate American waste disposal but recycling (including composting) has quadrupled since 1980 (fig. 21.13).

Another benefit of recycling is that it could cut our waste volumes drastically. Philadelphia is investing in neighborhood collection centers that will recycle 600 tons a day, enough to eliminate the need for a previously planned, high-priced incinerator. With its Fresh Kills landfill now closed, New York exports its daily 11,000 tons of waste by truck, train, and barge, to New Jersey, Pennsylvania, Virginia, South Carolina, and Ohio. New York has set ambitious recycling goals of 50 percent waste reduction, but still the city recycles less than 30 percent of its household and office waste. In contrast, Minneapolis and Seattle recycle nearly

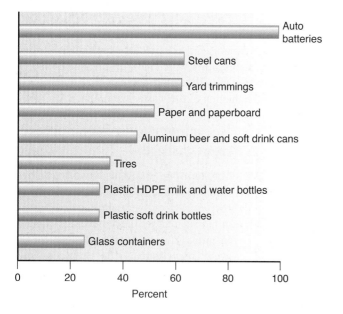

FIGURE 21.11 Recycling rates for selected materials in the United States. Battery recycling, which is required by law, is very successful. Other materials, even though valuable for reuse, have mixed recycling success.
Source: Data from Environmental Protection Agency, 2010.

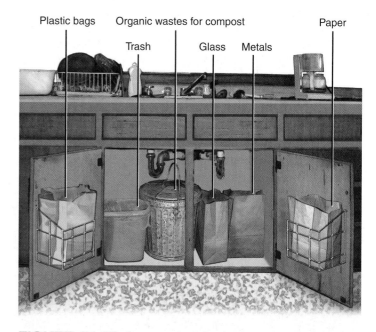

FIGURE 21.12 Source separation in the kitchen—the first step in a strong recycling program. One benefit of recycling is that it reminds us of our responsibility for waste management.

Environmental Justice

Who do you suppose lives closest to toxic waste dumps, Superfund sites, or other polluted areas in your city or county? If you answered poor people and minorities, you are probably right. Everyday experiences tell us that minority neighborhoods are much more likely to have high pollution levels and unpopular industrial facilities such as toxic waste dumps, landfills, smelters, refineries, and incinerators than are middle- or upper-class, white neighborhoods.

One of the first systematic studies showing this inequitable distribution of environmental hazards based on race in the United States was conducted by Robert D. Bullard in 1978. Asked for help by a predominantly black community in Houston that was slated for a waste incinerator, Bullard discovered that all five of the city's existing landfills and six of eight incinerators were located in African-American neighborhoods. In a book entitled *Dumping on Dixie,* Bullard showed that this pattern of risk exposure in minority communities is common throughout the United States (fig. 1).

In 1987, the Commission for Racial Justice of the United Church of Christ published an extensive study of environmental racism. Its conclusion was that race is the most significant variable in determining the location of toxic waste sites in the United States. Among the findings of this study are:

- Three of the five largest commercial hazardous waste landfills accounting for about 40 percent of all hazardous waste disposal in the United States are located in predominantly black or Hispanic communities.

- 60 percent of African Americans and Latinos and nearly half of all Asians, Pacific Islanders, and Native Americans live in communities with uncontrolled toxic waste sites.

- The average percentage of the population made up by minorities in communities without a hazardous waste facility is 12 percent. By contrast, communities with one hazardous waste facility have, on average, twice as high (24 percent) a minority population, while those with two or more such facilities average three times as high a minority population (38 percent) as those without one.

- The "dirtiest" or most polluted zip codes in California are in riot-torn South Central Los Angeles where the population is predominantly African American or Latino. Three-quarters of all blacks and half of all Hispanics in Los Angeles live in these polluted areas, while only one-third of all whites live there.

Race is claimed to be the strongest determinant of who is exposed to environmental hazards. Where whites can often "vote with their feet" and move out of polluted and dangerous neighborhoods, minorities are restricted by color barriers and prejudice to less desirable locations. In some areas, though, class or income also are associated with environmental hazards. The difference between *environmental racism* and other kinds of *environmental injustice* can be hard to define. Economic opportunity is often closely tied to race and cultural background in the United States.

Racial inequities also are revealed in the way the government cleans up toxic waste sites and punishes polluters (fig. 2). White communities see faster responses and get better results once toxic wastes are discovered than do minority communities. Penalties assessed against polluters of white communities average six times higher than those against polluters of minority communities. Cleanup is more thorough in white communities as well. Most toxic wastes in white communities are removed or destroyed. By contrast, waste sites in minority neighborhoods are generally only "contained" by putting a cap over them, leaving contaminants in place to potentially resurface or leak into groundwater at a later date. The growing environmental justice movement works to combine civil rights and social justice with environmental concerns to call for a decent, livable environment and equal environmental protection for everyone.

Ethical Considerations

What are the ethical considerations in waste disposal? Does everyone have a right to live in a clean environment or only a right to buy one if they can afford it? What would be a fair way to distribute the risks of toxic wastes? If you had to choose between an incinerator, a secure landfill, or a composting facility for your neighborhood, which would you take?

FIGURE 1 Native Americans protest toxic waste dumping on tribal lands.

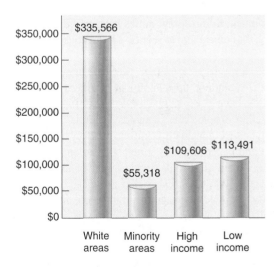

FIGURE 2 Hazardous waste law enforcement. The average fines or penalties per site for violation of the Resource Conservation and Recovery Act vary dramatically with racial composition of the communities where waste was dumped.

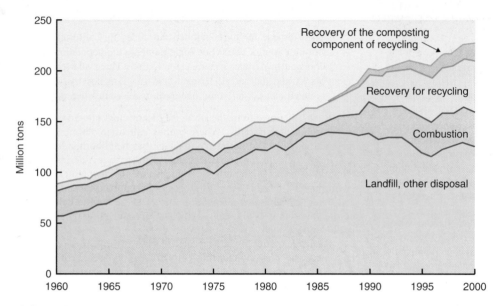

FIGURE 21.13 Disposal of municipal solid waste from 1960 to 2000. Landfills remain the dominant destination, but recycling and composting are increasing.
Source: Environmental Protection Agency.

60 percent of domestic waste, Los Angeles and Chicago over 40 percent. In 2002, New York Mayor Michael Bloomberg raised a national outcry by canceling most of the city's recycling program. He argued that the program didn't pay for itself and the money should be spent to balance the city's budget. A year later, Bloomberg relented after realizing that it cost more to ship garbage to Ohio than to recycle. Recycling was reinstated for nearly all recyclable materials.

Japan is probably the world's leader in recycling (see fig. 21.8). Short of land for landfills, Japan recycles about half its municipal waste and incinerates about 30 percent. The country has begun a push to increase recycling, because incineration costs almost as much. Some communities have raised recycling rates to 80 percent, and others aim to reduce waste altogether by 2020. This level of recycling is most successful when waste is well sorted. In Yokohama, a city of 3.5 million, there are now 10 categories of recyclables, including used clothing and sorted plastics. Some communities have 30 or 40 categories for sorting recyclables.

Recycling lowers our demands for raw resources. In the United States, we cut down 2 million trees every day to produce newsprint and paper products, a heavy drain on our forests. Recycling the print run of a single Sunday issue of the *New York Times* would spare 75,000 trees. Every piece of plastic we make reduces the reserves supply of petroleum and makes us more dependent on foreign oil. Recycling 1 ton of aluminum saves 4 tons of bauxite (aluminum ore) and 700 kg (1,540 lb) of petroleum coke and pitch, as well as keeping 35 kg (77 lb) of aluminum fluoride out of the air.

Recycling also reduces energy consumption and air pollution. Plastic bottle recycling can save 50 to 60 percent of the energy needed to make new ones. Making new steel from old scrap offers up to 75 percent energy savings. Producing aluminum from scrap

instead of bauxite ore cuts energy use by 95 percent, yet we still throw away more than a million tons of aluminum every year. If aluminum recovery were doubled worldwide, more than a million tons of air pollutants would be eliminated every year.

Recycling plastic is especially difficult

Much of the plastic ocean debris (opening case study) results from carelessness, but another part of the problem is that plastic is tricky to reuse and recycle. Contamination is a major reason for this difficulty. Most of the 24 billion plastic soft drink bottles sold every year in the United States are made of PET (polyethylene terephthalate), which can be melted and remanufactured into carpet, fleece clothing, plastic strapping, and nonfood packaging. However, even a trace of vinyl—a single PVC (polyvinyl chloride) bottle in a truckload, for example—can make PET useless.

Although most bottles are now marked with a recycling number, it's hard for consumers to remember which is which. Another obstacle is that many soft drink bottles are sold and consumed on the go, and never make it into recycling bins. As a consequence Americans have an extremely low recovery rate for plastics (fig. 21.14).

Reducing litter is an important benefit of recycling. Ever since disposable paper, glass, metal, foam, and plastic packaging began to accompany nearly everything we buy, these discarded wrappings have collected on our roadsides and in our lakes, rivers, and oceans. Without incentives to properly dispose of beverage cans, bottles, and papers, it often seems easier to just toss them aside when we have finished using them. Litter is a costly as well as unsightly problem. The United States pays an estimated 32 cents for each piece of litter picked up by crews along state highways, which adds up to $500 million every

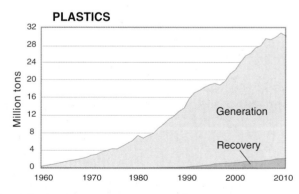

FIGURE 21.14 Recycling of plastics in the United States is improving but remains extremely low. This helps explain why so much plastic ends up in the ocean.

year. "Bottle-bills" requiring deposits on bottles and cans have reduced littering in many states.

Our present public policies often tend to favor extraction of new raw materials. Energy, water, and raw materials are often sold to industries below their real cost, in order to create jobs and stimulate the economy. A pound of recycled clear PET, the material in most soft drink bottles, is worth about 40¢, while a pound of virgin PET costs about 25¢. Setting the prices of natural resources at their real cost would tend to encourage efficiency and recycling.

Price fluctuations are a constant challenge for businesses trying to make an income on recycling. As with any primary materials, prices can vary dramatically. Some years copper has been so valuable that copper pipes, wires, and flashing are stolen from houses; other years prices and demand are low. States and cities have often helped to stabilize markets by requiring government agencies to purchase a minimum amount of recycled materials. Each of us can play a role in creating markets, as well. If we buy things made from recycled materials—or ask for them if they aren't available—we will help make it possible for recycling programs to succeed. Growing world demand, with expanding consumer economies around the world, is also likely to increase the value of recycled materials.

Commercial-scale recycling and composting are areas of innovation

Recycling household waste is the bedrock of recycling programs, but large-scale recycling is growing rapidly. The most common large-scale recycling is **composting** municipal yard waste and tree trimmings. Composting allows natural aerobic (oxygen-rich) decomposition to reduce organic debris to a nutrient-rich soil amendment. Many people compost yard and garden waste in their backyards. Increasingly, cities and towns are providing compost facilities in order to save landfill space. Organic debris such as yard waste makes up 13 percent of the waste we generate (see fig. 21.3). Almost two-thirds of our yard waste is composted.

While compost is a useful material, its market value is low. Many new and exciting technologies are emerging that create still more marketable products, especially energy, from garbage. The Swiss company Kompogas, for example, ferments organic waste in giant tanks, producing natural gas (methane), compost, and fertilizer. The company makes money on both ends, by collecting waste and selling energy and fertilizer. Increasingly, German and Swiss cities provide biogas generation and composting for municipal organic waste and household scraps. Cities save money in waste disposal and make money selling gas and compost.

Demolition and construction debris is another major source of waste. Every year thousands of tons of debris from building sites heads to landfills, but recycling facilities are beginning to collect, sort, and resell increasing portions of this debris. Taylor Recycling, in Newburgh, New York, recycles and sells 97 percent of the mixed demolition debris it receives, well above the industry average of 30 to 50 percent. Trees are ground up and converted to mulch for landscaping. Dirt from stumps is screened and sold as clean garden soil. Mixed materials are sorted into recyclable glass, metals, and plastics. Construction debris is sorted and ground: broken drywall is ground to fresh gypsum, which is sold to drywall producers; wood is composted or burned; bricks are crushed for fill and construction material. Organic waste that can't be separated, such as food-soaked paper, is sent to a gasifier. The gasifier is like an enclosed, oxygen-free pressure cooker, which converts biomass to natural gas. The gas runs electric generators for the plant, and any extra gas can be sold. Waste heat warms the recycling facility. The 3 percent of incoming waste that doesn't get recycled is mainly mixed plastics, which currently are landfilled.

Demanufacturing is necessary for appliances and e-waste

Demanufacturing is the disassembly and recycling of obsolete products, such as TV sets, computers, refrigerators, and air conditioners. As we mentioned earlier, electronics and appliances are among the fastest-growing components of the global waste stream. Americans throw away about 54 million household appliances, such as stoves and refrigerators, 12 million computers, and uncounted cell phones each year. Most office computers are used only 3 years; televisions last 5 years or so; refrigerators last longer, an average of 12 years. In the United States, an estimated 300 million computers await disposal in storage rooms and garages.

Demanufacturing is key to reducing the environmental costs of e-waste and appliances. A single personal computer can contain 700 different chemical compounds, including toxic materials (mercury, lead, and gallium), and valuable metals (gold, silver, copper), as well as brominated fire retardants and plastics. A typical personal computer has about $6 worth of gold, $5 worth of copper, and $1 of silver. Approximately 40 percent of lead entering U.S. landfills, and 70 percent of heavy metals, comes from e-waste. Batteries and switches in toys and electronics make up another 10 to 20 percent of heavy metals in our waste stream. These contaminants can enter groundwater if computers are landfilled, or the air if they are incinerated. When collected, these materials can become a valuable resource—and an alternative to newly mined materials.

To reduce these environmental hazards, the European Union now requires cradle-to-grave responsibility for electronic products. Manufacturers now have to accept used products or fund independent collectors. An extra $20 (less than one percent of the price of most computers) is added to the purchase price to pay for collection and demanufacturing. Manufacturers selling computers, televisions, refrigerators, and other appliances in Europe must also phase out many of the toxic compounds used in production. Japan is rapidly adopting European environmental standards, and some U.S. companies are following suit, in order to maintain their international markets. In the United States, at least 29 states have passed, or are considering, legislation to control disposal of appliances and computers, in order to protect groundwater and air quality.

Reuse is even more efficient than recycling

Even better than recycling or composting is cleaning and reusing materials in their present form, thus saving the cost and energy of remaking them into something else. We do this already with some specialized items. Auto parts are regularly sold from scrap yards, especially for older car models. In some areas, stained glass windows, brass fittings, fine woodwork, and bricks salvaged from old houses bring high prices. Some communities sort and reuse a variety of materials received in their dumps (fig. 21.15).

In many cities, glass and plastic bottles are routinely returned to beverage producers for washing and refilling. The reusable, refillable bottle is the most efficient beverage container we have. This is better for the environment than remelting and more profitable for local communities. A reusable glass container makes an average of 15 round-trips between factory and customer before it becomes so scratched and chipped that it has to be recycled. Reusable containers also favor local bottling companies and help preserve regional differences.

Since the advent of cheap, lightweight, disposable food and beverage containers, many small, local breweries, canneries, and bottling companies have been forced out of business by huge national conglomerates. These big companies can afford to ship food and beverages great distances as long as it is a one-way trip. If they had to collect their containers and reuse them, canning and bottling factories serving large regions would be uneconomical. Consequently, the national companies favor recycling rather than refilling because they prefer fewer, larger plants and don't want to be responsible for collecting and reusing containers. In some circumstances, life-cycle assessment shows that washing and decontaminating containers takes as much energy and produces as much air and water pollution as manufacturing new ones.

In less affluent nations, reuse of manufactured goods is an established tradition. Where most manufactured products are expensive and labor is cheap, it pays to salvage, clean, and repair products.

Reducing waste is often the cheapest option

Excess packaging of food and consumer products is one of our greatest sources of unnecessary waste. Paper, plastic, glass, and metal packaging material make up 50 percent of our domestic trash by volume. Much of that packaging is primarily for marketing and has little to do with product protection (fig. 21.16). Manufacturers and retailers might be persuaded to reduce these wasteful practices if consumers ask for products without excess packaging. Canada's National Packaging Protocol (NPP) recommends that packaging minimize depletion of virgin resources and production of toxins in manufacturing. The preferred hierarchy is (1) no packaging, (2) minimal packaging, (3) reusable packaging, and (4) recyclable packaging.

Where disposable packaging is necessary, we still can reduce the volume of waste in our landfills by using materials that are compostable or degradable. **Photodegradable plastics** break down when exposed to ultraviolet radiation. **Biodegradable plastics** incorporate such materials as cornstarch that can be decomposed by microorganisms. These degradable plastics often don't decompose completely; they only break down to small particles that remain in

FIGURE 21.15 Reusing discarded products is a creative and efficient way to reduce wastes. This recycling center in Berkeley, California, is a valuable source of used building supplies and a money saver for the whole community.

FIGURE 21.16 How much more do we need? Where will we put what we already have?

FIGURE 21.17 According to the U.S. Environmental Protection Agency, industries produce about one ton of hazardous waste per year for every person in the United States. Responsible handling and disposal is essential.

the environment. In doing so, they can release toxic chemicals into the environment. And in modern, lined landfills they don't decompose at all. Furthermore, they make recycling less feasible and may lead people to believe that littering is okay.

Most of our attention in waste management focuses on recycling. But slowing the consumption of throw-away products is by far the most effective way to save energy, materials, and money. The 3R waste hierarchy—reduce, reuse, recycle—lists the most important strategy first. Industries are increasingly finding that reducing saves money. Soft drink makers use less aluminum per can than they did 20 years ago, and plastic bottles use less plastic. 3M has saved over $500 million in the past 30 years by reducing its use of raw materials, reusing waste products, and increasing efficiency. Individual action is essential too (What Can You Do? above).

In 2007 the European Union adopted new regulations that aim to reduce both landfills and waste incineration. For the first time, the waste hierarchy—prevention, reuse, recycling, then disposal only as a last resort—is formalized in law. By 2020, half of all EU municipal solid waste and 70 percent of all construction waste is expected to be reused or recycled as a result of this law. No recyclable waste will be allowed in landfills. This law also establishes the "polluter pays" principle (those who create pollution should pay for it), and the "proximity principle," which says that waste should be treated in the nearest appropriate facility to the site at which it was produced. Mixing of toxic waste is also forbidden, making reuse and reprocessing easier.

21.4 HAZARDOUS AND TOXIC WASTES

The most dangerous aspect of the waste stream we have described is that it often contains highly toxic and hazardous materials that are injurious to both human health and environmental quality. We now produce and use a vast array of flammable, explosive, caustic, acidic, and highly toxic chemical substances for industrial, agricultural, and domestic purposes (fig. 21.17). According to the EPA, industries in the United States generate about 265 million metric tons of *officially* classified hazardous wastes each year, slightly more than 1 ton for each person in the country. In addition, considerably more toxic and hazardous waste material is generated by industries or processes not regulated by the EPA. Shockingly, at least 40 million metric tons (22 billion lbs) of toxic and hazardous wastes are released into the air, water, and land in the United States each year. The biggest source of these toxins are the chemical and petroleum industries (fig. 21.18).

Hazardous waste must be recycled, contained, or detoxified

Legally, a **hazardous waste** is any discarded material, liquid or solid, that contains substances known to be (1) fatal to humans or laboratory animals in low doses, (2) toxic, carcinogenic, mutagenic, or teratogenic to humans or other life-forms, (3) ignitable with a flash point less than 60°C, (4) corrosive, or (5) explosive or highly reactive (undergoes violent chemical reactions either by itself or when mixed with other materials). Notice that this definition includes both toxic and hazardous materials as defined in chapter 8. Certain compounds are exempt from regulation as hazardous waste if they are accumulated in less than 1 kg (2.2 lb) of commercial chemicals or 100 kg of contaminated soil, water, or debris. Even larger amounts (up to 1,000 kg) are exempt when stored at an approved waste treatment facility for the purpose of being beneficially used, recycled, reclaimed, detoxified, or destroyed.

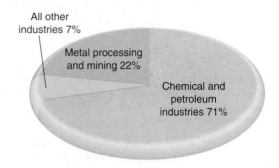

FIGURE 21.18 Producers of hazardous wastes in the United States.
Source: Data from the U.S. EPA, 2002.

Most hazardous waste is recycled, converted to nonhazardous forms, stored, or otherwise disposed of on-site by the generators—chemical companies, petroleum refiners, and other large industrial facilities—so that it doesn't become a public problem. Still, the hazardous waste that does enter the waste stream or the environment represents a serious environmental problem. And orphan wastes left behind by abandoned industries remain a serious threat to both environmental quality and human health. For years, little attention was paid to this material. Wastes stored on private property, buried, or allowed to soak into the ground were considered of little concern to the public. An estimated 5 billion metric tons of highly poisonous chemicals were improperly disposed of in the United States between 1950 and 1975 before regulatory controls became more stringent.

Federal Legislation

Two important federal laws regulate hazardous waste management and disposal in the United States. The Resource Conservation and Recovery Act (RCRA, pronounced "rickra") of 1976 is a comprehensive program that requires rigorous testing and management of toxic and hazardous substances. A complex set of rules require generators, shippers, users, and disposers of these materials to keep meticulous account of everything they handle and what happens to it from generation (cradle) to ultimate disposal (grave) (fig. 21.19).

The Comprehensive Environmental Response, Compensation and Liability Act (CERCLA or Superfund Act), passed in 1980 and modified in 1984 by the Superfund Amendments and Reauthorization Act (SARA), is aimed at rapid containment, cleanup, or remediation of abandoned toxic waste sites. This statute authorizes the Environmental Protection Agency to undertake emergency actions when a threat exists that toxic material will leak into the environment. The agency is empowered to bring suit for the recovery of its costs from potentially responsible parties such as site owners, operators, waste generators, or transporters.

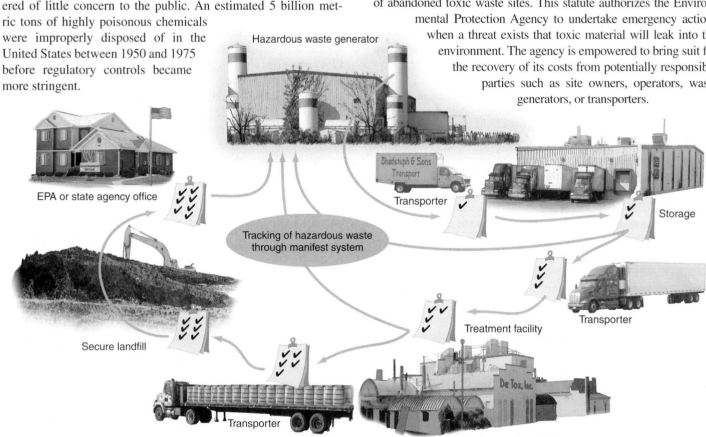

FIGURE 21.19 Toxic and hazardous wastes must be tracked from "cradle to grave" by detailed shipping manifests.

SARA also established that communities have a right to know about toxic substances that are produced or used nearby, which could be released into air or water. To give the public access to this information, SARA established a reporting system, the **Toxic Release Inventory**. More than 20,000 manufacturing facilities are required to report annually on the use, release, or transfer of toxic substances. This inventory is not always maintained completely, but it is the best available source of public information on exposure risk. You can find it on the EPA web site and see what potential sources are in your neighborhood.

The government does not have to prove that anyone violated a law or what role they played in a Superfund site. Rather, liability under CERCLA is "strict, joint, and several," meaning that anyone associated with a site can be held responsible for the entire cost of cleaning it up no matter how much of the mess they made. In some cases, property owners have been assessed millions of dollars for removal of wastes left there years earlier by previous owners. This strict liability has been a headache for the real estate and insurance business, but it also allows for protection of public health.

Superfund sites are those listed for federal cleanup

The EPA estimates that there are at least 36,000 seriously contaminated sites in the United States. The General Accounting Office (GAO) places the number much higher, perhaps more than 400,000 when all are identified. By 2007, some 1,680 sites had been placed on the National Priority List (NPL) for cleanup with financing from the federal Superfund program. That number declined to 1,280 by 2011, as sites were cleaned up or deleted from the list.

The **Superfund** is a revolving pool designed to (1) provide an immediate response to emergency situations that pose imminent hazards, and (2) to clean up or remediate abandoned or inactive sites. Without this fund, sites would languish for years or decades while the courts decided who was responsible to pay for the cleanup. Originally a $1.6 billion pool, the fund peaked at $3.6 billion. From its inception, the fund was financed by taxes on producers of toxic and hazardous wastes. Industries opposed this "polluter pays" tax, because current manufacturers are often not the ones responsible for the original contamination. In 1995, Congress agreed to let the tax expire. Since then the Superfund has dwindled, and the public has picked up an increasing share of the bill. In the 1980s the public covered less than 20 percent of the Superfund; now public tax dollars from the general fund cover nearly the entire cost of toxic waste cleanup (fig. 21.20).

Reliance on general funds makes cleanup progress vulnerable to political winds in Congress. In some years Congress is able and willing to cover the costs needed to protect the public from exposure to hazardous substances; in other years it can't or won't provide sufficient funding. EPA estimates suggest that the Superfund is likely to have just half of what it needs for the years 2010–2014.

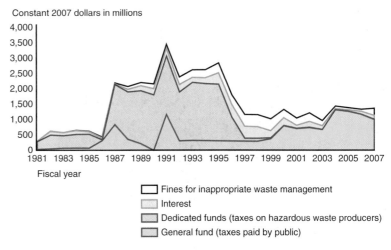

Constant 2007 dollars in millions

☐ Fines for inappropriate waste management
☐ Interest
☐ Dedicated funds (taxes on hazardous waste producers)
☐ General fund (taxes paid by public)

FIGURE 21.20 Sources of money for the Superfund Trust Fund, used to clean up toxic and hazardous waste sites, 1981–2007.

Total costs for hazardous waste cleanup in the United States are estimated between $370 billion and $1.7 trillion, depending on how clean sites must be and what methods are used. For years, Superfund money was spent mostly on lawyers and consultants, and cleanup efforts were often bogged down in disputes over liability and best cleanup methods. During the 1990s, however, progress improved substantially, with a combination of rule adjustments and administrative commitment to cleanup. From 1993 to 2000, the number of completed NPL cleanups jumped from 155 to 757, almost half the 1,680 sites on the list at that time. Since 2000, progress has slowed again, due to underfunding and a lower priority in the federal government.

What qualifies a site for the NPL? These sites are considered to be especially hazardous to human health and environmental quality because they are known to be leaking or have a potential for leaking supertoxic, carcinogenic, teratogenic, or mutagenic materials (chapter 8). The ten substances of greatest concern or most commonly detected at Superfund sites are lead, trichloroethylene, toluene, benzene, PCBs, chloroform, phenol, arsenic, cadmium, and chromium. These and other hazardous or toxic materials are known to have contaminated groundwater at 75 percent of the sites now on the NPL. In addition, 56 percent of these sites have contaminated surface waters, and airborne materials are found at 20 percent of the sites.

Where are these thousands of hazardous waste sites, and how did they get contaminated? Old industrial facilities such as smelters, mills, petroleum refineries, and chemical manufacturing plants are highly likely to have been sources of toxic wastes. Regions of the country with high concentrations of aging factories such as the "rust belt" around the Great Lakes or the Gulf Coast petrochemical centers have large numbers of Superfund sites (fig. 21.21). Mining districts also are prime sources of toxic and hazardous waste. Within cities, factories and places such as railroad yards, bus repair barns, and filling stations where solvents, gasoline, oil, and other petrochemicals were spilled or dumped on the ground often are highly contaminated.

• Hazardous waste site
▢ Aquifers

FIGURE 21.21 Some of the hazardous waste sites on the EPA priority cleanup list. Sites located on aquifer recharge zones represent an especially serious threat. Once groundwater is contaminated, cleanup is difficult and expensive. In some cases, it may not be possible.

Some of the most infamous toxic waste sites were old dumps where many different materials were mixed together indiscriminately. For instance, Love Canal in Niagara Falls, New York, was an open dump used by both the city and nearby chemical factories as a disposal site. More than 20,000 tons of toxic chemical waste was buried under what later became a housing development. Another infamous example occurred in Hardeman County, Tennessee, where about a quarter of a million barrels of chemical waste were buried in shallow pits that leaked toxins into the groundwater. In other sites, liquid wastes were pumped into open lagoons or abandoned in warehouses.

Studies of populations living closest to Superfund and toxic release inventory sites reveal that minorities often are overrepresented in neighborhoods near waste sites. Charges of environmental racism have been made, but they are difficult to prove conclusively (see What Do You Think? p. 481).

Brownfields present both liability and opportunity

Among the biggest problems in cleaning up hazardous waste sites are questions of liability and the degree of purity required. In many cities, these problems have created large areas of contaminated properties known as **brownfields** that have been abandoned or under-utilized because of real or suspected pollution. Up to one-third of all commercial and industrial sites in the urban core of many big cities fall in this category. In heavy industrial corridors the percentage typically is higher.

For years, no one was interested in redeveloping brownfields because of liability risks. Who would buy a property knowing that they might be forced to spend years in litigation and negotiations and be forced to pay millions of dollars for pollution they didn't

create? Even if a site has been cleaned to current standards, there is a worry that additional pollution might be found in the future or that more stringent standards might be applied.

In many cases, property owners complain that unreasonably high levels of purity are demanded in remediation programs. Consider the case of Columbia, Mississippi. For many years a 35 ha (81 acre) site in Columbia was used for turpentine and pine tar manufacturing. Soil tests showed concentrations of phenols and other toxic organic compounds exceeding federal safety standards. The site was added to the Superfund NPL and remediation was ordered. Some experts recommended that the best solution was to simply cover the surface with clean soil and enclose the property with a fence to keep people out. The total costs would have been about $1 million.

Instead, the EPA ordered Reichhold Chemical, the last known property owner, to excavate more than 12,500 tons of soil and haul it to a commercial hazardous waste dump in Louisiana at a cost of some $4 million. The intention was to make the site safe enough to be used for any purpose, and to remove risk from exposure to anybody, even children, who might be exposed to soil on the site.

Similarly, in places where contaminants have seeped into groundwater, the EPA generally demands that cleanup be carried to drinking water standards. Many critics believe that these pristine standards are unreasonable. Former Congressman Jim Florio, a principal author of the original Superfund Act, says, "It doesn't make any sense to clean up a rail yard in downtown Newark so it can be used as a drinking water reservoir." Depending on where the site is, what else is around it, and what its intended uses are, much less stringent standards may be perfectly acceptable.

Recognizing that reusing contaminated properties can play a significant role in rebuilding old cities, creating jobs, increasing the tax base, and preventing needless destruction of open space at urban margins, programs have been established at both federal and state levels to encourage brownfield recycling. Adjusting purity standards according to planned uses and providing liability protection for nonresponsible parties gives developers and future purchasers confidence that they won't be unpleasantly surprised in the future with further cleanup costs. In some communities, former brownfields are being turned into "eco-industrial parks" that feature environmentally friendly businesses and bring in much needed jobs to inner-city neighborhoods.

Hazardous waste storage must be safe

What shall we do with toxic and hazardous wastes? In our homes, we can reduce waste generation and choose less toxic materials. Buy only what you need for the job at hand, and use up what you buy. Consider whether you can replace some toxic substances with safer and cheaper alternatives (What Can You Do? p. 491). Dispose of unneeded materials responsibly (table 21.1). In general there are several strategies for addressing the problem of waste management.

Produce Less Waste

As with other wastes, the safest and least expensive way to avoid hazardous waste problems is to avoid creating the wastes in the first place. Manufacturing processes can be modified to

Science

Phytoremediation: Cleaning Up Toxic Waste with Plants

Getting contaminants out of soil and groundwater is one of the most widespread and persistent problems in waste cleanup. Once leaked into the ground, solvents, metals, radioactive elements, and other contaminants are dispersed and difficult to collect and treat. The main method of cleaning up contaminated soil is to dig it up, then decontaminate it or haul it away and store it in a landfill in perpetuity. At a single site, thousands of tons of tainted dirt and rock may require incineration or other treatment. Cleaning up contaminated groundwater usually entails pumping vast amounts of water out of the ground—hopefully extracting the contaminated water faster than it can spread through the water table or aquifer. In the United States alone, there are tens of thousands of contaminated sites on factories, farms, gas stations, military facilities, sewage treatment plants, landfills, chemical warehouses, and other types of facilities. Cleaning up these sites is expected to cost at least $700 billion.

Recently, a number of promising alternatives have been developed using plants, fungi, and bacteria to clean up our messes. *Phytoremediation* (remediation, or cleanup, using plants) can include a variety of strategies for absorbing, extracting, or neutralizing toxic compounds. Certain types of mustards and sunflowers can extract lead, arsenic, zinc, and other metals (*phytoextraction*). Poplar trees can absorb and break down toxic organic chemicals (*phytodegradation*). Reeds and other water-loving plants can filter water tainted with sewage, metals, or other contaminants. Natural bacteria in groundwater, when provided with plenty of oxygen, can neutralize contaminants in aquifers, minimizing or even eliminating the need to extract and treat water deep in the ground. Radioactive strontium and cesium have been extracted from soil near the Chernobyl nuclear power plant using common sunflowers.

How do the plants, bacteria, and fungi do all this? Many of the biophysical details are poorly understood, but in general, plant roots are designed to efficiently extract nutrients, water, and minerals from soil and groundwater. The mechanisms involved may aid extraction of metallic and organic contaminants. Some plants also use toxic elements as a defense against herbivores—locoweed, for example, selectively absorbs elements such as selenium, concentrating toxic levels in its leaves. Absorption can be extremely effective. Braken fern growing in Florida was found to contain arsenic at concentrations more than 200 times higher than the soil in which it was growing.

Genetically modified plants are also being developed to process toxins. Poplars have been grown with a gene borrowed from bacteria that transform a toxic compound of mercury into a safer form. In another experiment, a gene for producing mammalian liver enzymes, which specialize in breaking down toxic organic compounds, was inserted into tobacco plants. The plants succeeded in producing the liver enzymes and breaking down toxins absorbed through their roots.

These remediation methods are not without risks. As plants take up toxins, insects could consume leaves, allowing contaminants to enter the food web. Some absorbed contaminants are volatilized, or emitted in gaseous form, through pores in plant leaves. Once toxic contaminants are absorbed into plants, the plants themselves are usually toxic and must be landfilled. But the cost of phytoremediation can be less than half the cost of landfilling or treating toxic soil, and the volume of plant material requiring secure storage ends up being a fraction of a percent of the volume of the contaminated dirt.

Cleaning up hazardous and toxic waste sites will be a big business for the foreseeable future, both in the United States and around the world. Innovations such as phytoremediation offer promising prospects for business growth as well as for environmental health and saving taxpayers' money.

Plants can absorb, concentrate, and even decompose toxic contaminants in soil and groundwater.

reduce or eliminate waste production. In Minnesota, the 3M Company reformulated products and redesigned manufacturing processes to eliminate more than 140,000 metric tons of solid and hazardous wastes, 4 billion l (1 billion gal) of wastewater, and 80,000 metric tons of air pollution each year. They frequently found that these new processes not only spared the environment but also saved money by using less energy and fewer raw materials.

Recycling and reusing materials also eliminates hazardous wastes and pollution. Many waste products of one process or industry are valuable commodities in another. Already, about 10 percent of the wastes that would otherwise enter the waste stream in the United States are sent to surplus material exchanges where they are sold as raw materials for use by other industries. This figure could probably be raised substantially with better waste management. In Europe, at least one-third of all industrial wastes are exchanged through clearinghouses where beneficial uses are found. This represents a double savings: The generator doesn't have to pay for disposal, and the recipient pays little, if anything, for raw materials.

Table 21.1	How Should You Dispose of Household Hazardous Waste?
Flush to sewer system (drain or toilet)	Cleaning agents with ammonia or bleach, disinfectants, glass cleaner, toilet cleaner
Put dried solids in household trash	Cosmetics, putty, grout, caulking, empty solvent containers, water-based glue, fertilizer (without weed killer)
Save and deliver to a waste collection center	*Solvents:* cleaning agents (drain cleaner, floor wax-stripper, furniture polish, metal cleaner, oven cleaner), paint thinner and other solvents, glue with solvents, varnish, nail polish remover
	Metals: mercury thermometers, button batteries, NiCad batteries, auto batteries, paints with lead or mercury, fluorescent light bulbs/tubes/ballasts, electronics and appliances
	Poisons: bug spray, pesticides, weed killers, rat poison, insect poison, mothballs
	Other chemicals: antifreeze, gasoline, fuel oil, brake fluid, transmission fluid, paint, rust remover, hairspray, photo chemicals

Source: EPA, 2005.

Convert Substances to Less Hazardous Forms

Several processes are available to make hazardous materials less toxic. *Physical treatments* tie up or isolate substances. Charcoal or resin filters absorb toxins. Distillation separates hazardous components from aqueous solutions. Precipitation and immobilization in ceramics, glass, or cement isolate toxins from the environment so that they become essentially nonhazardous. One of the few ways to dispose of metals and radioactive substances is to fuse them in silica at high temperatures to make a stable, impermeable glass that is suitable for long-term storage.

Incineration is a quick way to dispose of many kinds of hazardous waste. Incineration is not necessarily cheap—nor always clean—unless it is done correctly. Wastes must be heated to over 1,000°C (2,000°F) for a sufficient period of time to complete destruction. The ash resulting from thorough incineration is reduced in volume up to 90 percent and often is safer to store in a landfill or other disposal site than the original wastes. Nevertheless, incineration remains a highly controversial topic (fig. 21.22).

Several sophisticated features of modern incinerators improve their effectiveness. Liquid injection nozzles atomize liquids and mix air into the wastes so they burn thoroughly. Fluidized bed burners pump air from the bottom up through burning solid waste as it travels on a metal chain grate through the furnace. The air velocity is sufficient to keep the burning waste partially suspended. Plenty of oxygen is available, and burning is quick and complete. Afterburners add to the completeness of burning by igniting gaseous hydrocarbons not consumed in the incinerator. Scrubbers and precipitators remove minerals, particulates, and other pollutants from the stack gases.

Chemical processing can transform materials so they become nontoxic. Included in this category are neutralization, removal of metals or halogens (chlorine, bromine, etc.), and oxidation. The Sunohio Corporation of Canton, Ohio, for instance, has developed a process called PCBx in which chlorine in such molecules as PCBs is replaced with other ions that render the compounds less toxic. A portable unit can be moved to the location of the hazardous waste, eliminating the need for shipping them.

Biological waste treatment or **bioremediation** taps the great capacity of microorganisms to absorb, accumulate, and detoxify a variety of toxic compounds. Bacteria in activated sludge basins, aquatic plants (such as water hyacinths or cattails), soil microorganisms, and other species remove toxic materials and purify effluents. Recent experiments have produced bacteria that can decontaminate organic waste metals by converting them to harmless substances. After the Deepwater Horizon oil spill in the Gulf of Mexico in 2007, most of the waste remediation was performed by naturally occurring bacteria, which were adapted to metabolizing oil compounds. Bioremediation holds exciting possibilities for addressing many organic pollutants, including oils, PCBs, and other toxic compounds.

Store Permanently

Inevitably, there will be some materials that we can't destroy, make into something else, or otherwise cause to vanish. We will have to store them out of harm's way. There are differing opinions about how best to do this.

Retrievable Storage Dumping wastes in the ocean or burying them in the ground generally means that we have lost control of them. If we learn later that our disposal technique was a mistake, it is difficult, if not impossible, to go back and recover the wastes. For many supertoxic materials, the best way to store them may be in **permanent retrievable storage**. This means placing waste storage containers in a secure building, salt mine, or bedrock cavern where they can be inspected periodically and retrieved, if necessary, for repacking or for transfer if a better means of disposal is developed. This technique is more expensive than burial in a landfill because the storage area must be guarded and monitored

FIGURE 21.22 Actor Martin Sheen joins local activists in a protest in East Liverpool, Ohio, site of the largest hazardous waste incinerator in the United States. About 1,000 people marched to the plant to pray, sing, and express their opposition. Involving celebrities draws attention to your cause. A peaceful, well-planned rally builds support and acceptance in the broader community.

What Can You Do?

Alternatives to Hazardous Household Chemicals

Chrome cleaner: Use vinegar and nonmetallic scouring pad.

Copper cleaner: Rub with lemon juice and salt mixture.

Floor cleaner: Mop linoleum floors with 1 cup vinegar mixed with 2 gallons of water. Polish with club soda.

Brass polish: Use Worcestershire sauce.

Silver polish: Rub with toothpaste on a soft cloth.

Furniture polish: Rub in olive, almond, or lemon oil.

Ceramic tile cleaner: Mix 1/4 cup baking soda, 1/2 cup white vinegar, and 1 cup ammonia in 1 gallon warm water (good general purpose cleaner).

Drain opener: Use plunger or plumber's snake, pour boiling water down drain.

Upholstery cleaner: Clean stains with club soda.

Carpet shampoo: Mix 1/2 cup liquid detergent in 1 pint hot water. Whip into stiff foam with mixer. Apply to carpet with damp sponge. Rinse with 1 cup vinegar in 1 gal water. Don't soak carpet—it may mildew.

Window cleaner: Mix 1/3 cup ammonia, 1/4 cup white vinegar in 1 quart warm water. Spray on window. Wipe with soft cloth.

Spot remover: For butter, coffee, gravy, or chocolate stains: Sponge up or scrape off as much as possible immediately. Dab with cloth dampened with a solution of 1 teaspoon white vinegar in 1 quart cold water.

Toilet cleaner: Pour 1/2 cup liquid chlorine bleach into toilet bowl. Let stand for 30 minutes, scrub with brush, flush.

Pest control: Spray plants with soap-and-water solution (3 tablespoons soap per gallon water) for aphids, mealybugs, mites, and whiteflies. Interplant with pest repellent plants such as marigolds, coriander, thyme, yarrow, rue, and tansy. Introduce natural predators such as ladybugs or lacewings.

Indoor pests: Grind or blend 1 garlic clove and 1 onion. Add 1 tablespoon cayenne pepper and 1 quart water. Add 1 tablespoon liquid soap.

Moths: Use cedar chips or bay leaves.

Ants: Find where they are entering house, spread cream of tartar, cinnamon, red chili pepper, or perfume to block trail.

Fleas: Vacuum area, mix brewer's yeast with pet food.

Mosquitoes: Brewer's yeast tablets taken daily repel mosquitoes.

Note: test cleaners in small, inconspicuous area before using.

---

continuously to prevent leakage, vandalism, or other dispersal of toxic materials. Remedial measures are much cheaper with this technique, however, and it may be the best system in the long run.

Secure Landfills One of the most popular solutions for hazardous waste disposal has been landfilling. Although, as we saw earlier in this chapter, many such landfills have been environmental disasters, newer techniques make it possible to create safe, modern **secure landfills** that are acceptable for disposing of many hazardous wastes. The first line of defense in a secure landfill is a thick

---

bottom cushion of compacted clay that surrounds the pit like a bathtub (fig. 21.23). Moist clay is flexible and resists cracking if the ground shifts. It is impermeable to groundwater and will safely contain wastes. A layer of gravel is spread over the clay liner and perforated drain pipes are laid in a grid to collect any seepage that escapes from the stored material. A thick polyethylene liner, protected from punctures by soft padding materials, covers the gravel bed. A layer of soil or absorbent sand cushions the inner liner and the wastes are packed in drums, which then are placed into the pit, separated into small units by thick berms of soil or packing material.

When the landfill has reached its maximum capacity, a cover much like the bottom sandwich of clay, plastic, and soil—in that order—caps the site. Vegetation stabilizes the surface and improves its appearance. Sump pumps collect any liquids that filter through the landfill, either from rainwater or leaking drums. This leachate is treated and purified before being released. Monitoring wells check groundwater around the site to ensure that no toxins have escaped.

Most landfills are buried below ground level to be less conspicuous; however, in areas where the groundwater table is close to the surface, it is safer to build above-ground storage. The same protective construction techniques are used as in a buried pit. An advantage to such a facility is that leakage is easier to monitor because the bottom is at ground level.

Transportation of hazardous wastes to disposal sites is of concern because of the risk of accidents. Emergency preparedness officials conclude that the greatest risk in most urban areas is not nuclear war or natural disaster but crashes involving trucks or trains carrying hazardous chemicals through densely packed urban corridors. Another worry is who will bear financial responsibility for abandoned waste sites. The material remains toxic long after the businesses that created it are gone. As is the case with nuclear wastes (chapter 19), we may need new institutions to ensure perpetual care of these wastes.

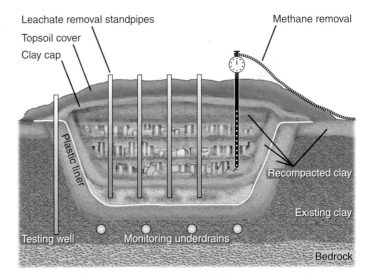

FIGURE 21.23 A secure landfill for toxic waste. A thick plastic liner and two or more layers of impervious compacted clay enclose the landfill. A gravel bed between the clay layers collects any leachate, which can then be pumped out and treated. Well samples are tested for escaping contaminants and methane is collected for combustion.

CONCLUSION

Waste is a global problem. Each year we consume more materials and produce more waste. Finding ways to dispose of all that garbage and hazardous substances is a constant challenge. In the United States, recycling rates are improving, but we still landfill more than half our municipal waste. Many other countries, especially those short on landfill space, recycle over half their waste. Modern landfills seek to keep trash from contaminating air and groundwater. These sites are a great improvement over the past, but they are often remote from major cities, which must transport garbage long distances for disposal. Incineration is a costly but widely used alternative to landfilling.

Waste disposal is expensive, but our policies are often better set up for landfilling or incinerating waste than for recycling. Government policies and economies of scale make it cheaper and more convenient to extract virgin raw materials than to reuse or recycle. But the increasing toxicity of modern products, including

e-waste, makes it more urgent that we reduce, reuse, and recycle materials worldwide. Strategies and opportunities for recycling are expanding, including reuse markets, bioenergy generation, deconstructing demolition debris, and composting yard waste. Creating gold from garbage—making money from biogas, compost, and efficient waste disposal—is a growing industry, especially in Europe.

Hazardous and toxic waste remains a serious health threat and environmental risk. Many abandoned and derelict sites must be cleaned up by the Superfund, although that fund has dwindled since it lost its main source of support, industry contributions, in 1995. These sites threaten public health, especially for minority groups, and are a serious problem. At the same time we are producing new hazardous substances that require safe disposal. We can all help by thinking carefully about what we buy, use, and dispose of in our own communities.

REVIEWING LEARNING OUTCOMES

By now you should be able to explain the following points:

21.1 Identify the components of solid waste.

- The waste stream is everything we throw away.

21.2 Describe how wastes have been—and are being—disposed of or treated.

- Open dumps release hazardous materials into air and water.
- Ocean dumping is nearly uncontrollable.
- We often export waste to countries ill-equipped to handle it.
- Landfills receive most of our waste.
- Incineration produces energy but causes pollution.

21.3 Identify how we might shrink the waste stream.

- Recycling captures resources from garbage.
- Recycling saves money, materials, energy, and space.

- Commercial-scale recycling and composting is an area of innovation.
- Demanufacturing is necessary for appliances and e-waste.
- Reuse is even more efficient than recycling.
- Reducing waste is often the cheapest option.

21.4 Investigate hazardous and toxic wastes.

- Hazardous waste must be recycled, contained, or detoxified.
- Superfund sites are those listed for federal cleanup.
- Brownfields present both liability and opportunity.
- Hazardous waste storage must be safe.

PRACTICE QUIZ

1. What are solid wastes and hazardous wastes? What is the difference between them?

2. Describe the difference between an open dump, a sanitary landfill, and a modern, secure, hazardous waste disposal site.

3. Why are landfill sites becoming limited around most major urban centers in the United States? What steps are being taken to solve this problem?

4. Describe some concerns about waste incineration.

5. List some benefits and drawbacks of recycling wastes. What are the major types of materials recycled from municipal waste and how are they used?

6. What is composting, and how does it fit into solid waste disposal?

7. Describe some ways that we can reduce the waste stream to avoid or reduce disposal problems.

8. List ten toxic substances in your home and how you would dispose of them.

9. What are brownfields and why do cities want to redevelop them?

10. What societal problems are associated with waste disposal? Why do people object to waste handling in their neighborhoods?

CRITICAL THINKING AND DISCUSSION QUESTIONS

1. A toxic waste disposal site has been proposed for the Pine Ridge Indian Reservation in South Dakota. Many tribal members oppose this plan, but some favor it because of the jobs and income it will bring to an area with 70 percent unemployment. If local people choose immediate survival over long-term health, should we object or intervene?

2. There is often a tension between getting your personal life in order and working for larger structural changes in society. Evaluate the trade-offs between spending time and energy sorting recyclables at home compared to working in the public arena on a bill to ban excess packaging.

3. Should industry officials be held responsible for dumping chemicals that were legal when they did it but are now known to be extremely dangerous? At what point can we argue that they *should* have known about the hazards involved?

4. Look at the discussion of recycling or incineration presented in this chapter. List the premises (implicit or explicit) that underlie the presentation as well as the conclusions (stated or not) that seem to be drawn from them. Do the conclusions necessarily follow from these premises?

5. The Netherlands incinerates much of its toxic waste at sea by a shipborne incinerator. Would you support this as a way to dispose of our wastes as well? What are the critical considerations for or against this approach?

Data Analysis: How Much Do You Know About Recycling?

As people become aware of waste disposal problems in their communities, more people are recycling more materials. Some things are easy to recycle, such as newsprint, office paper, or aluminum drink cans. Other things are harder to classify. Most of us give up pretty quickly and throw things in the trash if we have to think too hard about how to recycle them.

1. Take a poll to find out how many people in your class know how to recycle the items in the table shown here. Once you have taken your poll, convert the numbers to percentages: divide the number who know how to recycle each item by the number of students in your class, and then multiply by 100.

2. Now find someone on your campus who works on waste management. This might be someone in your university/college administration, or it might be someone who actually empties trash containers. (You might get more interesting and straightforward answers from the latter.) Ask the following questions: (1) Can this person fill in the items your class didn't know about? (2) Is there a college/university policy about recycling? What are some of the points on that policy? (3) How much does the college spend each year on waste disposal? How many tuition payments does that total? (4) What are the biggest parts of the waste stream? (5) Does the school have a plan for reducing that largest component?

Item	Percentage Who Know How to Recycle
Newspapers	
Paperboard (cereal boxes)	
Cardboard boxes	
Cardboard boxes with tape	
Plastic drink bottles	
Other plastic bottles	
Styrofoam food containers	
Food waste	
Plastic shopping bags	
Plastic packaging materials	
Furniture	
Last year's course books	
Left-over paint	

For Additional Help in Studying This Chapter, please visit our website at www.mhhe.com/cunningham12e. You will find additional practice quizzes and case studies, flashcards, regional examples, placemarks for Google Earth™ mapping, and an extensive reading list, all of which will help you learn environmental science.

Car-free roads provide a cleaner, safer, healthier environment for residents of Vauban, Germany.

Urbanization and Sustainable Cities

Learning Outcomes

After studying this chapter, you should be able to:

22.1   Define *urbanization*.
22.2   Describe why cities grow.
22.3   Understand urban challenges in the developing world.
22.4   Identify urban challenges in the developed world.
22.5   Explain smart growth.

"What kind of world do you want to live in? Demand that your teachers teach you what you need to know to build it."

~ Peter Kropotkin

Case Study   Vauban: A Car-free Suburb

What would it be like to live in a city without automobiles? Residents of Vauban, a district in the city of Freiburg, Germany, have a lifestyle that suggests it might be both enjoyable and economical. In Vauban, it's so easy to get around by tram, bicycle, and on foot that there is little need to depend on cars. The community is designed using "smart growth" principles with stores, banks, schools, and restaurants within easy walking distance of homes. Jobs and office space are available nearby, and trams to the city center run every few minutes through the center and around the edges of Vauban. Residential streets are narrow and vehicle-free, making a great place for bicycles and playing children. Cars must be parked in a large municipal ramp at the edge of town, and buying a space there costs $40,000. Consequently, nearly three-quarters of Vauban's families don't own a car, and more than half sold their car to move there.

Fewer vehicles means less air pollution and greater safety for pedestrians, but most families moved to Vauban not for environmental reasons but because they believe a car-free lifestyle is healthier for children. Schools, child-care services, playgrounds, and sports facilities are a short bike ride from all houses. Children can play outside and can walk or bike to school without having to cross busy streets.

In most American cities, one third of the land area is dedicated to cars, mainly for parking and roads. In Vauban, reducing car dependence has saved so much space that neighborhoods have abundant green space, gardens, and play areas while still being small enough for easy walking.

A highly successful and growing car-sharing program makes it still easier to live without cars. Starting in about 1992, the city's car-sharing program has grown to some 2,500 members, who save money and parking space by using shared cars. The cars are available all around town, and they can be reserved online or by mobile phone. In addition, a single monthly bus ticket covers all regional trains and buses, making it especially easy to get around by public transportation.

A car-free lifestyle makes economic sense. Owning and operating a vehicle in Germany is even more expensive than it is in the

FIGURE 22.1 Narrow streets in Vauban are designed for children and bicycles first, with limited car use.

U.S., where the average car costs about $9,000 per year. Residents of areas like Vauban can put that money to other uses.

Vauban's comfortable row houses, with balconies and private gardens, are designed to conserve energy but maximize quality of life. Clever use of space, beautiful woodwork, large balconies, and large, superinsulated windows make the homes feel spacious while maintaining a small footprint. Just having shared walls minimizes energy losses. Many houses are so efficient that they don't need a heating system at all. In addition, a highly efficient wood-burning co-generation plant provides much of the space heating and electricity for the district, and rooftop solar collectors and photovoltaic panels provide hot water and power for individual homes. While not entirely carbon neutral, Vauban is highly sustainable. Many of the houses produce more energy than they consume.

Similar projects are being built across Europe and even in some developing countries, such as China. On Dongtan Island in the mouth of the Yangtze River near Shanghai, the Chinese government is planning an eco-city for 50,000 people that is expected to be energy, water, and food self-sufficient. In the United States, the Environmental Protection Agency is promoting "car reduced" communities. In California, for example, developers are planning a Vauban-like community called Quarry Village on the outskirts of Oakland, accessible to the Bay Area Rapid Transit system and to the California State University campus in Hayward.

Decades of advertisements and government policies in the United States have persuaded most people that the dream home is a single-family residence on a spacious lot in the suburbs, where a car—regardless of the costs in energy use, insurance, accidents, or land consumption—is essential for every trip, no matter how short the distance. Whether we can break those patterns remains to be seen.

Vauban illustrates a number of ways urban design can help us live sustainably with our environment and our neighbors. In this chapter we'll look at other aspects of city planning and urban environments as well as some principles of ecological economics that help us understand the nature of resources and the choices we face both as individuals and communities.

22.1 URBANIZATION

For most of human history, the vast majority of people have lived in rural areas where they engaged in hunting and gathering, farming, fishing, or other natural-resource based occupations. Since the beginning of the Industrial Revolution about 300 years ago, however, cities have grown rapidly in both size and power (fig. 22.2). Now, for the first time ever, more people live in urban areas than in the country. In 1950, only 38 percent of the world population lived in cities (table 22.1). By 2030, that percentage is expected to nearly double. This means that over the next three decades about 3 billion people will crowd into cities. Some areas—Europe, North America, and Latin America—are already highly urbanized. Only Africa and Asia are below 45 percent urbanized.

Demographers predict that 90 percent of the human population growth in this century will occur in developing countries, and that almost all of that growth will occur in cities (fig. 22.3). Already huge **urban agglomerations** (mergers of multiple

Table 22.1	Urban Share of Total Population (Percentage)		
	1950	**2000**	**2030***
Africa	18.4	40.6	57.0
Asia	19.3	43.8	59.3
Europe	56.0	75.0	81.5
Latin America	40.0	70.3	79.7
North America	63.9	77.4	84.5
Oceania	32.0	49.5	60.7
World	38.3	59.4	70.5

*Projected

Source: United Nations Population Division, 2010.

municipalities) are forming throughout the world. Some of these **megacities** (urban areas with populations over 10 million people) are already truly enormous, some claiming up to 30 million people. Can cities this size—especially in poorer countries—supply all the public services necessary to sustain a civilized life?

If we are to learn to live sustainably—that is, to depend on renewable resources while also protecting environmental quality, biodiversity, and the ecological services on which all life depends—that challenge will have to be met primarily in the cities of the world, where most people will live. Many of us dream of moving to a secluded hideaway in the country, where we could grow our own food, chop wood, and carry water. But it probably isn't possible for 6.5 billion people (let alone the 8 to 9 billion expected by the end of this century) to live rural, subsistence lifestyles. Learning to live together in cities is probably the only way we'll survive.

Cities can be engines of economic progress and social reform. Some of the greatest promise for innovation comes from cities like Vauban, where innovative leaders can focus knowledge and resources on common problems. Cities can be efficient places to live, where mass transportation can move people around and goods and services are more readily available than in the

FIGURE 22.2 In less than 20 years, Shanghai, China, has built Pudong, a new city of 1.5 million residents and 500 skyscrapers on former marshy farmland across the Huang Pu River from the historic city center. This kind of rapid urban growth is occurring in many developing countries.

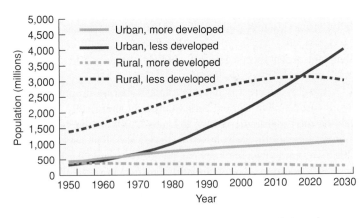

FIGURE 22.3 Growth of urban and rural populations in more-developed regions and in less-developed regions.
Source: United Nations Population Division. *World Urbanization Prospects*, 2004.

FIGURE 22.4 Since their earliest origins, cities have been centers of education, religion, commerce, politics, and culture. Unfortunately, they have also been sources of pollution, crowding, disease, and misery.

FIGURE 22.5 This village in Chiapas, Mexico, is closely tied to the land through culture, economics, and family relationships. While the timeless pattern of life here gives a great sense of identity, it can also be stifling and repressive.

country. Concentrating people in urban areas leaves open space available for farming and biodiversity. But cities can also be dumping grounds for poverty, pollution, and unwanted members of society. Providing food, housing, transportation, jobs, clean water, and sanitation to the 2 or 3 billion new urban residents expected to crowd into cities—especially those in the developing world—in this century may be one of the preeminent challenges of this century.

As the case study of Vauban shows, there is much we can do to make our cities more livable. Of course, Vauban is in Germany, a very wealthy country that can afford major transformations to its cities and transporation systems. What are the prospects for such transformations in countries that are very poor? What hope is there for them?

Cities have specialized functions as well as large populations

Since their earliest origins, cities have been centers of education, religion, commerce, record keeping, communication, and political power. As cradles of civilization, cities have influenced culture and society far beyond their proportion of the total population (fig. 22.4). Until about 1900, only a small percentage of the world's people lived permanently in urban areas, and even the greatest cities of antiquity were small by modern standards. The vast majority of humanity has always lived in rural areas where they subsisted on natural resources—farming, fishing, hunting, timber harvesting, animal herding, or mining.

Just what makes up an urban area or a city? Definitions differ. The U.S. Census Bureau considers any incorporated community to be a city, regardless of size, and defines any city with more than 2,500 residents as urban. More meaningful definitions are based on *functions*. In a **rural area**, most residents depend on agriculture or other ways of harvesting natural resources for their livelihood. In an **urban area**, by contrast, a majority of the people are not directly dependent on natural resource-based occupations.

A **village** is a collection of rural households linked by culture, custom, family ties, and association with the land (fig. 22.5). A **city**, by contrast, is a differentiated community with a population

and resource base large enough to allow residents to specialize in arts, crafts, services, or professions rather than natural resource-based occupations. While the rural village often has a sense of security and connection, it also can be stifling. A city offers more freedom to experiment, to be upwardly mobile, and to break from restrictive traditions, but it can be harsh and impersonal.

Beyond about 10 million inhabitants, an urban area is considered a supercity or **megacity**. Megacities in many parts of the world have grown to enormous size. Chongqing, China, having annexed a large part of Sichuan province and about 30 million people, claims to be the biggest city in the world. In the United States, urban areas between Boston and Washington, D.C., have merged into a nearly continuous megacity (sometimes called Bos-Wash) containing about 35 million people. The Tokyo-Yokohama-Osaka-Kobe corridor contains nearly 50 million people. Because these agglomerations have expanded beyond what we normally think of as a city, some geographers prefer to think of them as urbanized **core regions** that dominate the social, political, and economic life of most countries (fig. 22.6).

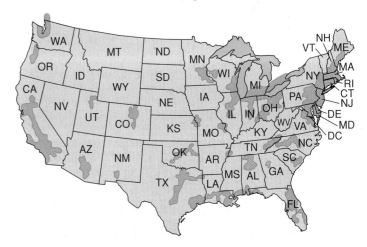

FIGURE 22.6 Urban core agglomerations (*lavender areas*) are forming megalopolises in many areas. While open space remains in these areas, the flow of information, capital, labor, goods, and services links each into an interacting system.
Source: U.S. Census Bureau.

Large cities are expanding rapidly

You can already see the dramatic shift in size and location of big cities. In 1900 only 13 cities in the world had populations over 1 million (table 22.2). All of those cities, except Tokyo and Peking were in Europe or North America. London was the only city in the world with more than 5 million residents. By 2007, there were at least 300 cities—100 of them in China alone—with more than 1 million residents. Of the 13 largest of these metropolitan areas, none are in Europe. Only New York City and Los Angeles are in a developed country. By 2025, it's expected that at least 93 cities will have populations over 5 million, and three-fourths of those cities will be in developing countries (fig. 22.7). In just the next 25 years, Mumbai, India; Delhi, India; Karachi, Pakistan; Manila, Philippines; and Jakarta, Indonesia all are expected to grow by at least 50 percent.

China represents the largest demographic shift in human history. Since the end of Chinese collectivized farming and factory work in 1986, around 250 million people have moved from rural areas to cities. And in the next 25 years an equal number is expected to join this vast exodus. In addition to expanding existing cities, China plans to build 400 new urban centers with populations of at least 500,000 over the next 20 years. Already at least half of the concrete and one-third of the steel used in construction around the world each year is consumed in China.

Table 22.2 The World's Largest Urban Areas (Populations in Millions)

1900		2015**	
London, England	6.6	Tokyo, Japan	31.0
New York, USA	4.2	New York, USA	29.9
Paris, France	3.3	Mexico City	21.0
Berlin, Germany	2.4	Seoul, Korea	19.8
Chicago, USA	1.7	São Paulo, Brazil	18.5
Vienna, Austria	1.6	Osaka, Japan	17.6
Tokyo, Japan	1.5	Jakarta, Indonesia	17.4
St. Petersburg, Russia	1.4	Delhi, India	16.7
Philadelphia, USA	1.4	Los Angeles, USA	16.6
Manchester, England	1.3	Beijing, China	16.0
Birmingham, England	1.2	Cairo, Egypt	15.5
Moscow, Russia	1.1	Manila, Philippines	13.5
Peking, China*	1.1	Buenos Aires, Brazil	12.9

*Now spelled Beijing.
**Projected.

Source: T. Chandler, Three Thousand Years of Urban Growth, 1974, Academic Press and World Gazetter, 2003.

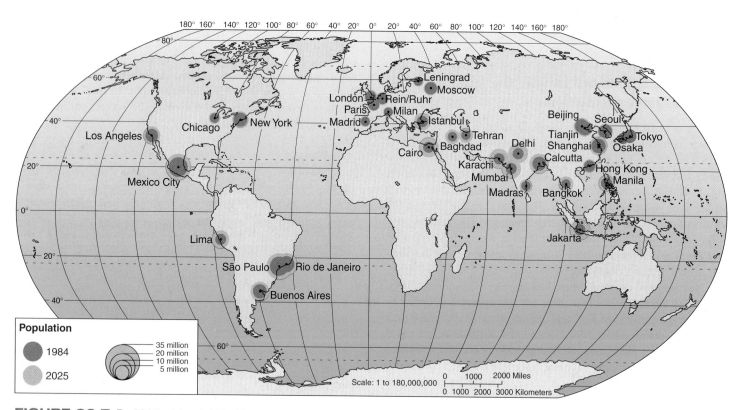

FIGURE 22.7 By 2025, at least 400 cities will have populations of 1 million or more, and 93 supercities will have populations above 5 million. Three-fourths of the world's largest cities will be in developing countries that already have trouble housing, feeding, and employing their people.

Consider Shanghai, for example. In 1985, the city had a population of about 10 million. It's now about 19 million—including at least 4 million migrant laborers. In the past decade, Shanghai has built 4,000 skyscrapers (buildings with more than 25 floors). The city already has twice as many tall buildings as Manhattan, and proposals have been made for 1,000 more. The problem is that most of this growth has taken place in a swampy area called Pudong, across the Huang Pu River from the historic city center (see fig. 22.2). Pudong is now sinking about 1.5 cm per year due to groundwater drainage and the weight of so many buildings.

Other Chinese cities have plans for similar massive building projects to revitalize blighted urban areas. Harbin, an urban complex of about 9 million people and the capital of Heilongjiang Province, for example, recently announced plans to relocate the entire city across the Songhua River on 740 km^2 (285 mi^2) of former farmland. Residents hope these new towns will be both more livable for their residents and more ecologically sustainable than the old cities they're replacing. In 2005, the Chinese government signed a long-term contract with a British engineering firm to build at least five "eco-cities," each with about half a million residents. Plans call for these cities to be self-sufficient in energy, water, and most food products, with the aim of zero emissions of greenhouse gases from transportation.

22.2 WHY DO CITIES GROW?

Urban populations grow in two ways: by natural increase (more births than deaths) and by immigration. Natural increase is fueled by improved food supplies, better sanitation, and advances in medical care that reduce death rates and cause populations to grow both within cities and in the rural areas around them (chapter 7). In Latin America and East Asia, natural increase is responsible for two-thirds of urban population growth. In Africa and West Asia, immigration is the largest source of urban growth. Immigration to cities can be caused both by **push factors** that force people out of the country and by **pull factors** that draw them into the city.

Immigration is driven by push and pull factors

People migrate to cities for many reasons. In some areas, the countryside is overpopulated and simply can't support more people. The "surplus" population is forced to migrate to cities in search of jobs, food, and housing. Not all rural-to-urban shifts are caused by overcrowding in the country, however. In some places, economic forces or political, racial, or religious conflicts drive people out of their homes. The countryside may actually be depopulated by such demographic shifts. The United Nations estimated that in 2002 at least 19.8 million people fled their native country and that about another 20 million were internal refugees within their own country, displaced by political, economic, or social instability. Many of these refugees end up in the already overcrowded megacities of the developing world.

Land tenure patterns and changes in agriculture also play a role in pushing people into cities. The same pattern of agricultural mechanization that made farm labor largely obsolete in the United States early in this century is spreading now to developing countries. Furthermore, where land ownership is concentrated in the hands of a wealthy elite, subsistence farmers are often forced off the land so it can be converted to grazing lands or monoculture cash crops. Speculators and absentee landlords let good farmland sit idle that otherwise might house and feed rural families.

Even in the largest and most hectic cities, many people are there by choice, attracted by the excitement, vitality, and opportunity to meet others like themselves. Cities offer jobs, housing, entertainment, and freedom from the constraints of village traditions. Possibilities exist in the city for upward social mobility, prestige, and power not ordinarily available in the country. Cities support specialization in arts, crafts, and professions for which markets don't exist elsewhere.

Modern communications also draw people to cities by broadcasting images of luxury and opportunity. An estimated 90 percent of the people in Egypt, for instance, have access to a television set. The immediacy of television makes city life seem more familiar and attainable than ever before. We generally assume that beggars and homeless people on the streets of developing nations' teeming cities have no other choice of where to live, but many of these people want to be in the city. In spite of what appears to be dismal conditions, living in the city may be preferable to what the country had to offer.

Government policies can drive urban growth

Government policies often favor urban over rural areas in ways that both push and pull people into the cities. Developing countries commonly spend most of their budgets on improving urban areas (especially around the capital city where leaders live), even though only a small percentage of the population lives there or benefits directly from the investment. This gives the major cities a virtual monopoly on new jobs, housing, education, and opportunities, all of which bring in rural people searching for a better life. In Peru, for example, Lima accounts for 20 percent of the country's population, but has 50 percent of the national wealth, 60 percent of the manufacturing, 65 percent of the retail trade, 73 percent of the industrial wages, and 90 percent of all banking in the country. Similar statistics pertain to São Paulo, Mexico City, Manila, Cairo, Lagos, Bogotá, and a host of other cities.

Governments often manipulate exchange rates and food prices for the benefit of more politically powerful urban populations but at the expense of rural people. Importing lower-priced food pleases city residents, but local farmers then find it uneconomical to grow crops. As a result, an increased number of people leave rural areas to become part of a large urban workforce, keeping wages down and industrial production high. Zambia, for instance, sets maize prices below the cost of local production to discourage farming and to maintain a large pool of workers for the mines. Keeping the currency exchange rate high stimulates export trade but makes it difficult for small farmers to buy the fuels, machinery, fertilizers, and seeds that they need. This depresses rural employment and rural income while stimulating the urban economy. The effect is to transfer wealth from the country to the city.

22.3 URBAN CHALLENGES IN THE DEVELOPING WORLD

Large cities in both developed and developing countries face similar challenges in accommodating the needs and by-products of dense populations. The problems are most intense, however, in rapidly growing cities of developing nations.

Cities in developing nations are also where most population growth will occur in coming decades (fig. 22.7). These cities already struggle to supply food, water, housing, jobs, and basic services for their residents. The unplanned and uncontrollable growth of those cities causes tragic urban environmental problems. What responsibilities might we in richer countries have to help people in these developing areas?

Think About It

How many of the large cities shown in figure 22.7 are in developing countries? What are some differences between large cities in wealthy countries and those in less-wealthy countries? If you were a farmer in India or China, what would encourage you to move to one of these cities?

Traffic congestion and air quality are growing problems

A first-time visitor to a supercity—particularly in a less-developed country—is often overwhelmed by the immense crush of pedestrians and vehicles of all sorts that clog the streets. The noise, congestion, and confusion of traffic make it seem suicidal to venture onto the street. Jakarta, for instance, is one of the most congested cities in the world (fig. 22.8). Traffic is chaotic almost all the time. People commonly spend three or four hours each way commuting to work from outlying areas. Bangkok also has monumental traffic problems. The average resident spends the equivalent of 44 days a year sitting in traffic jams. About 20 percent of all fuel is consumed by vehicles standing still. Hours of work lost each year are worth at least $3 billion.

Traffic congestion is expected to worsen in many developing countries as the number of vehicles increases but road construction fails to keep pace (fig. 22.9). All this traffic, much of it involving old, poorly maintained vehicles, combines with smoky factories, and use of wood or coal fires for cooking and heating to create a thick pall of air pollution in the world's supercities. Lenient pollution laws, corrupt officials, inadequate testing equipment, ignorance about the sources and effects of

FIGURE 22.8 Motorized rickshaws, motor scooters, bicycles, street vendors, and pedestrians all vie for space on the crowded streets of Jakarta. But in spite of the difficulties of living here people work hard and have hope for the future.

pollution, and lack of funds to correct dangerous situations usually exacerbate the problem.

What is its human toll? An estimated 60 percent of Kolkata's residents are thought to suffer from respiratory diseases linked to air pollution. Lung cancer mortality in Shanghai is reported to be four to seven times higher than rates in the countryside. There have been some encouraging success stories, however. As we saw in chapter 16, air pollution in Delhi decreased dramatically after vehicles were required to install emission controls and use cleaner fuels.

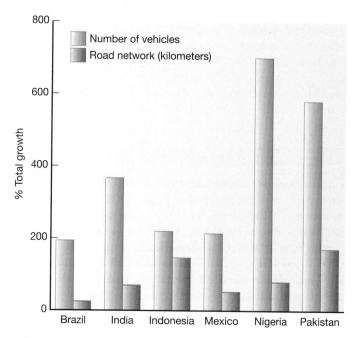

FIGURE 22.9 Transport growth in selected developing countries, 1980–2000.

Source: Earth Trends, 2006.

Insufficient sewage treatment causes water pollution

Few cities in developing countries can afford to build modern waste treatment systems for their rapidly growing populations. The World Bank estimates that only 35 percent of urban residents in developing countries have satisfactory sanitation services. The situation is especially desperate in Latin America, where only 2 percent of urban sewage receives any treatment. In Egypt, Cairo's sewer system was built about 50 years ago to serve a population of 2 million people. It is now being overwhelmed by more than 10 million people. Less than one-tenth of India's 3,000 towns and cities have even partial sewage systems and water treatment facilities. Some 150 million of India's urban residents lack access to sanitary sewer systems.

Figure 22.10 shows one of many tidal canals that crisscross Jakarta, and serve as the sewage disposal system for many of the 10 million city residents. In 2007, unusually heavy rain backed up these canals and flooded about half the city. Health officials braced for disease epidemics.

Some 400 million people, or about one-third of the population, in developing world cities do not have safe drinking water, according to the World Bank. Although city dwellers are somewhat more likely than rural people to have clean water, this still represents a large problem. Where people have to buy water from merchants, it often costs 100 times as much as piped city water and may not be any safer to drink. Many rivers and streams in developing countries are little more than open sewers, and yet they are all that poor people have for washing clothes, bathing, cooking, and—in the worst cases—for drinking. Diarrhea, dysentery, typhoid, and cholera are widespread diseases in these countries, and infant mortality is tragically high (chapter 8).

Many cities lack adequate housing

The United Nations estimates that at least 1 billion people—15 percent of the world's population—live in crowded, unsanitary slums of the central cities and in the vast shantytowns and squatter settlements that ring the outskirts of most developing world cities. Around 100 million people have no home at all. In Mumbai, India, for example, it is thought that half a million people sleep on the streets, sidewalks, and traffic circles because they can find no other place to live. In Brazil, perhaps 1 million "street kids" who have run away from home or been abandoned by their parents live however and wherever they can. This is surely a symptom of a tragic failure of social systems.

Slums are generally legal but inadequate multifamily tenements or rooming houses, either custom built to rent to poor people or converted from some other use. The chals of Mumbai, India, for example, are high-rise tenements built in the 1950s to house immigrant workers. Never very safe or sturdy, these dingy, airless buildings are already crumbling and often collapse without warning. Eighty-four percent of the families in these tenements live in a single room; half of those families consist of six or more people. Typically, they have less than 2 square meters of

FIGURE 22.10 This tidal canal in Jakarta serves as an open sewer. By some estimates, about half of the 10 million residents of this city have no access to modern sanitation systems.

floor space per person and only one or two beds for the whole family. They may share kitchen and bathroom facilities down the hall with 50 to 75 other people. Even more crowded are the rooming houses for mill workers where up to 25 men sleep in a single room only 7 meters square. Because of this crowding, household accidents are a common cause of injuries and deaths in developing world cities, especially to children. Charcoal braziers or kerosene stoves used in crowded homes are a routine source of fires and injuries. With no place to store dangerous objects beyond the reach of children, accidental poisonings and other mishaps are a constant hazard.

Shantytowns are settlements created when people move onto undeveloped lands and build their own houses. Shacks are built of corrugated metal, discarded packing crates, brush, plastic sheets, or whatever building materials people can scavenge. Some shantytowns are simply illegal subdivisions where the landowner rents land without city approval. Others are spontaneous or popular settlements or **squatter towns** where people occupy land without the owner's permission. Sometimes this occupation involves thousands of people who move onto unused land in a highly organized, overnight land invasion, building huts and laying out streets, markets, and schools before authorities can root them out. In other cases, shantytowns just gradually "happen."

FIGURE 22.11 Homeless people have built shacks along this busy railroad track in Jakarta. It's a dangerous place to live, with many trains per day using the tracks, but for the urban poor, there are few other choices.

FIGURE 22.12 Plastic waste has created mountains of garbage, especially in developing countries with insufficient waste management systems.

Called *barriads, barrios, favelas,* or *turgios* in Latin America, *bidonvillas* in Africa, or *bustees* in India, shantytowns surround every megacity in the developing world (fig. 22.11). They are not an exclusive feature of poor countries, however. Some 250,000 immigrants and impoverished citizens live in the *colonias* along the southern Rio Grande in Texas. Only 2 percent have access to adequate sanitation. Many live in conditions as awful as you would see in any developing world city.

About three-quarters of the residents of Addis Ababa, Ethiopia, or Luanda, Angola, live in squalid refugee camps. Two-thirds of the population of Calcutta live in unplanned squatter settlements and nearly half of the 20 million people in Mexico City live in uncontrolled, unauthorized shantytowns. Many governments try to clean out illegal settlements by bulldozing the huts and sending riot police to drive out the settlers, but the people either move back in or relocate to another shantytown.

These populous but unauthorized settlements usually lack sewers, clean water supplies, electricity, and roads. Often the land on which they are built was not previously used because it is unsafe or unsuitable for habitation. In Bhopal, India, and Mexico City, for example, squatter settlements were built next to deadly industrial sites. In Rio de Janeiro, La Paz (Bolivia), Guatemala City, and Caracas (Venezuela), they are perched on landslide-prone hills. In Lima (Peru), Khartoum (Sudan), and Nouakchott (Mauritania), shantytowns have spread onto sandy deserts. In Manila, thousands of people live in huts built on towering mounds of garbage and burning industrial waste in city dumps (fig. 22.12).

As desperate and inhumane as conditions are in these slums and shantytowns, many people do more than merely survive there. They keep themselves clean, raise families, educate their children, find jobs, and save a little money to send home to their parents. They learn to live in a dangerous, confusing, and rapidly changing world and have hope for the future. The people have parties; they sing and laugh and cry. They are amazingly adaptable and resilient. In many ways, their lives are no worse than those in the early industrial cities of Europe and America a century ago. Perhaps continuing development will bring better conditions to cities of the developing world as it has for many in the developed world.

22.4 Urban Challenges in the Developed World

For the most part, the rapid growth of central cities that accompanied industrialization in nineteenth- and early twentieth-century Europe and North America has now slowed or even reversed. London, for instance, once the most populous city in the world, has lost nearly 2 million people, dropping from its high of 8.6 million in 1939 to about 6.7 million now. While the greater metropolitan area surrounding London has been expanding to about 10 million inhabitants, the city itself is now only the twelfth largest city in the world.

Many of the worst urban environmental problems of the more developed countries have been substantially reduced in recent years. Minority groups in inner cities, however, remain vulnerable to legacies of environmental degradation in industrial cities (What Do You Think? p. 503).

In most developed countries, improved sanitation and medical care have reduced or totally eliminated many of the communicable diseases that once afflicted urban residents. Air and water quality have improved dramatically as heavy industry such as steel smelting and chemical manufacturing have moved to developing countries. In consumer and information economies, workers no longer need to be concentrated in central cities. They can

What Do You Think?

People for Community Recovery

The Lake Calumet Industrial District on Chicago's far South Side is an environmental disaster area. A heavily industrialized center of steel mills, oil refineries, railroad yards, coke ovens, factories, and waste disposal facilities, much of the site is now a marshy wasteland of landfills, toxic waste lagoons, and slag dumps, around a system of artificial ship channels.

At the southwest corner of this degraded district sits Altgeld Gardens, a low-income public housing project built in the late 1940s by the Chicago Housing Authority. The 2,000 units of "The Gardens" or "The Projects," as they are called by the largely minority residents, are low-rise row houses, many of which are vacant or in poor repair. But residents of Altgeld Gardens are doing something about their neighborhood. People for Community Recovery (PCR) is a grassroots citizen's group organized to work for a clean environment, better schools, decent housing, and job opportunities for the Lake Calumet neighborhood.

PCR was founded in 1982 by Mrs. Hazel Johnson, an Altgeld Gardens resident whose husband died from cancer that may have been pollution-related. PCR has worked to clean up more than two dozen waste sites and contaminated properties in their immediate vicinity. Often this means challenging authorities to follow established rules and enforce existing statutes. Public protests, leafleting, and community meetings have been effective in public education about the dangers of toxic wastes and have helped gain public support for cleanup projects.

PCR's efforts successfully blocked construction of new garbage and hazardous waste landfills, transfer stations, and incinerators in the Lake Calumet district. Pollution prevention programs have been established at plants still in operation. And PCR helped set up a community monitoring program to stop illegal dumping and to review toxic inventory data from local companies.

Education is an important priority for PCR. An environmental education center administered by community members organizes workshops, seminars, fact sheets, and outreach for citizens and local businesses. A public health education and screening program has been set up to improve community health. Partnerships have been established with nearby Chicago State University to provide technical assistance and training in environmental issues.

PCR also works on economic development. Environmentally responsible products and services are now available to residents. Jobs that are being created as Green businesses are brought into the community. Wherever possible, local people and minority contractors from the area are hired to clean up waste sites and restore abandoned buildings. Job training for youth and adults as well as retraining for displaced workers is a high priority.

In the 1980s, a young community organizer named Barack Obama worked with PCR on jobs creation, housing issues, and education. He credits the lessons he learned there for much of his subsequent political successes. In his best-selling memoir *Dreams from My Father* Obama devotes more than 100 pages to his formative experiences at Altgeld Gardens and other nearby neighborhoods.

PCR and Mrs. Johnson have received many awards for their fight against environmental racism and despair. In 1992, PCR was the recipient of the President's Environmental and Conservation Challenge Award. PCR is the only African-American grassroots organization in the country to receive this prestigious award.

Although Altgeld Gardens is far from clean, much progress has been made. Perhaps the most important accomplishment is community education and empowerment. Residents have learned how and why they need to work together to improve their living conditions. Could these same lessons be useful in your city or community? What could you do to help improve urban environments where you live?

The Calumet industrial district in South Chicago.

live and work in dispersed sites. Automobiles now make it possible for much of the working class to enjoy amenities such as single-family homes, yards, and access to recreation that once were available only to the elite.

In the United States, old, dense manufacturing cities such as Philadelphia and Detroit have lost population as industry has moved to developing countries. In a major demographic shift, both businesses and workers have moved west and south. Some of the most rapidly growing metropolitan areas like Phoenix, Arizona; Boulder, Colorado; Austin, Texas; and San Jose, California, are centers for high-technology companies located in landscaped suburban office parks. These cities often lack a recognizable downtown, being organized instead around low-density housing developments, national-chain shopping malls, and extensive freeway networks. For many high-tech companies, being located near industrial centers and shipping is less important than a good climate, ready access to air travel, and amenities such as natural beauty and open space.

Urban sprawl consumes land and resources

While the move to suburbs and rural areas has brought many benefits to the average citizen, it also has caused numerous urban problems. Cities that once were compact now spread over the landscape, consuming open space and wasting resources. This pattern of urban growth is known as **sprawl.** While there is no universally accepted definition of the term, sprawl generally includes the characteristics outlined in table 22.3. As former Maryland Governor Parris N. Glendening said, "In its path, sprawl consumes thousands of acres of forests and farmland, woodlands and wetlands. It requires government to spend millions extra to build new schools, streets, and water and sewer lines." And Christine Todd Whitman, former New Jersey governor and head of the Environmental Protection Agency, said, "Sprawl eats up our open space. It creates traffic jams that boggle the mind and pollute the air. Sprawl can make one feel downright claustrophobic about our future."

In most American metropolitan areas, the bulk of new housing is in large, tract developments that leapfrog out beyond the edge of the city in a search for inexpensive rural land with few restrictions on land use or building practices (fig. 22.13). The U.S. Department of Housing and Urban Development estimates that urban sprawl consumes some 200,000 ha (roughly 500,000 acres) of farmland each year. Because cities often are located in fertile river valleys or shorelines, much of that land would be especially valuable for producing crops for local consumption. But with

FIGURE 22.13 Huge houses on sprawling lots consume land, alienate us from our neighbors, and make us ever more dependent on automobiles. They also require a lot of lawn mowing!

planning authority divided among many small, local jurisdictions, metropolitan areas have no way to regulate growth or provide for rational, efficient resource use. Small towns and township or county officials generally welcome this growth because it profits local landowners and business people. Although the initial price of tract homes often is less than comparable urban property, there are external costs in the form of new roads, sewers, water mains, power lines, schools, and shopping centers and other extra infrastructure required by this low-density development.

Landowners, builders, real estate agents, and others who profit from this crazy-quilt development pattern generally claim that growth benefits the suburbs in which it occurs. They promise that adding additional residents will lower the average taxes for everyone, but in fact, the opposite often is true. In a study titled *Better Not Bigger,* author Eben Fodor analyzed the costs of medium-density and low-density housing. In suburban Washington, D.C., for instance, each new house on a quarter acre (0.1 ha) lot cost $700 more than it paid in taxes. A typical new house on a 5 acre (2 ha) lot, however, cost $2,200 more than it paid in taxes because of higher expenses for infrastructure and services. Ironically, people who move out to rural areas to escape from urban problems such as congestion, crime, and pollution often find that they have simply brought those problems with them. A neighborhood that seemed tranquil and remote when they first moved in, soon becomes just as crowded, noisy, and difficult as the city they left behind as more people join them in their rural retreat.

In a study of 58 large American urban areas, author and former mayor of Albuquerque, David Rusk, found that between 1950 and 1990, populations grew 80 percent, while land area grew 305 percent. In Atlanta, Georgia, the population grew 32 percent between 1990 and 2000, while the total metropolitan area increased by 300 percent. The city is now more than 175 km across. Atlanta loses an estimated $6 million to traffic delays every day. By far the fastest growing metropolitan region in the United States is Las Vegas, Nevada, which doubled its population but quadrupled its size in the 1990s (fig. 22.14*a* and *b*).

Table 22.3	**Characteristics of Urban Sprawl**

1. Unlimited outward extension.
2. Low-density residential and commercial development.
3. Leapfrog development that consumes farmland and natural areas.
4. Fragmentation of power among many small units of government.
5. Dominance of freeways and private automobiles.
6. No centralized planning or control of land uses.
7. Widespread strip malls and "big-box" shopping centers.
8. Great fiscal disparities among localities.
9. Reliance on deteriorating older neighborhoods for low-income housing.
10. Decaying city centers as new development occurs in previously rural areas.

Source: Excerpt from a speech by Anthony Downs at the CTS Transportation Research Conference, as appeared on Website by Planners Web, Burlington, VT, 2001.

(a) 1972

(b) 2002

FIGURE 22.14 Satellite images of Las Vegas, Nevada, in 1972 (a) and 2002 (b). The metropolitan area quadrupled in three decades.

Transportation is crucial in city development

Getting people around within a large urban area has become one of the most difficult problems that many city officials face. A century ago most American cities were organized around transportation corridors. First horse-drawn carriages, then electric streetcars provided a way for people to get to work, school, and shops. Everyone, rich or poor, wanted to live as close to the city center as possible.

The introduction of automobiles allowed people to move to suburbs, and cities began to spread over the landscape. The U.S. Interstate Highway System was the largest construction project in human history. Originally justified as necessary for national defense, it was really a huge subsidy for the oil, rubber, automobile, and construction industries. Its 72,000 km (45,000 mi) of freeways probably did more than anything to encourage sprawl and change America into an auto-centered society.

Because many Americans now live far from where they work, shop, and recreate, most consider it essential to own a private automobile. The average U.S. driver spends about 443 hours per year behind a steering wheel. This means that for most people, the equivalent of one full 8-hour day per week is spent sitting in an automobile. Of the 5.8 billion barrels of oil consumed each year in the United States (60 percent of which is imported), about two-thirds is burned in cars and trucks. As chapter 16 shows, about two-thirds of all carbon monoxide, one-third of all nitrogen oxides, and one-quarter of all volatile organic compounds emitted each year from human-caused sources in the United States are released by automobiles, trucks, and buses.

Building the roads, parking lots, filling stations, and other facilities needed for an automobile-centered society takes a vast amount of space and resources (fig. 22.15). In some metropolitan areas it is estimated that one-third of all land is devoted to the automobile. To make it easier for suburban residents to get from their homes to jobs and shopping, we provide an amazing network of freeways and highways. At a cost of several trillion dollars to build, the interstate highway system was designed to allow us to drive at high speeds from source to destination without ever

having to stop. As more and more drivers clog the highways, however, the reality is far different. In Los Angeles, for example, which has the worst congestion in the United States, the average speed in 1982 was 58 mph (93 km/hr), and the average driver

FIGURE 22.15 Freeways give us the illusion of speed and privacy, but they consume land, encourage sprawl, and create congestion as people move farther from the city to get away from traffic and then have to drive to get anywhere.

spent less than 4 hours per year in traffic jams. In 2000, the average speed in Los Angeles was only 35.6 mph (57.3 km/hr), and the average driver spent 82 hours per year waiting for traffic. Although new automobiles are much more efficient and cleaner operating than those of a few decades ago, the fact that we drive so much farther today and spend so much more time idling in stalled traffic means that we burn more fuel and produce more pollution than ever before.

Altogether, it is estimated that traffic congestion costs the United States $78 billion per year in wasted time and fuel. Some people argue that the existence of traffic jams in cities shows that more freeways are needed. Often, however, building more traffic lanes simply encourages more people to drive farther than before. Rather than ease congestion and save fuel, more freeways can exacerbate the problem.

Sprawl impoverishes central cities from which residents and businesses have fled. With a reduced tax base and fewer civic leaders living or working in downtown areas, the city is unable to maintain its infrastructure. Streets, parks, schools, and civic buildings fall into disrepair at the same time that these facilities are being built at great expense in new suburbs. The poor who are left behind when the upper and middle classes abandon the city center often can't find jobs where they live and have no way to commute to the suburbs where jobs are now located. About one-third of Americans are too young, too old, or too poor to drive. For these people, car-oriented development causes isolation and makes daily tasks like grocery shopping very difficult. Parents, especially mothers, spend long hours transporting young children. Teenagers and aging grandparents are forced to drive, often presenting a hazard on public roads.

Sprawl also is bad for your health. By encouraging driving and discouraging walking, sprawl promotes a sedentary lifestyle that contributes to heart attacks and diabetes, among other problems. In Atlanta, for example, the lowest-density suburbs tend to have significantly higher rates of overweight residents than the highest-density neighborhoods.

Think About It

Who benefits most from urban sprawl, and who benefits least? In what ways do you benefit and suffer from sprawl? Do home buyers initiate the process of urban expansion, or do developers? What conditions help make this process so persistent?

Finally, sprawl fosters uniformity and alienation from local history and natural environment. Housing developments often are based on only a few standard housing styles, while shopping centers and strip malls everywhere feature the same national chains. You could drive off the freeway in the outskirts of almost any big city in America and see exactly the same brands of fast-food restaurants, motels, stores, filling stations, and big-box shopping centers.

Mass transit could make our cities more livable

Many American cities are now rebuilding the public transportation systems that were abandoned in the 1950s (fig. 22.16). Consider how different your life might be if you lived an automobile-free life in a city with good mass transit.

A famous example of successful mass transit is found in Curitiba, Brazil. High-speed, bi-articulated buses, each of which can carry 270 passengers, travel on dedicated roadways closed to all other vehicles. These bus-trains are linked to 340 feeder routes extending throughout the city. Everyone in the city is within walking distance of a bus stop that has frequent, convenient, affordable service. Curitiba's buses carry some 1.9 million passengers per day, or about three-quarters of all personal trips within the city. Working with existing roadways for the most part, the city was able to construct this system for one-tenth the cost of a light rail system or freeway system, and one-hundredth the cost of a subway.

But many developing countries are adopting the automobile-centered model of the United States rather than Curitiba's model. Traffic accidents have become the third-largest cause of years of lost life worldwide. For example, the number of vehicles increased eightfold in Nigeria and sixfold in Pakistan between 1980 and 2000, while the road networks in those countries expanded by only 10 to 20 percent in the same time.

The recent introduction of the Tata Nano in India raises nightmares for both urban planners and energy experts. Costing less than $2,000 brand new, these tiny vehicles put car ownership within reach for millions who could never afford it before. But they will probably increase gasoline consumption greatly and result in huge traffic jams as inexperienced drivers take to the road for the first time.

FIGURE 22.16 Many American cities are now rebuilding light rail systems that were abandoned in the 1950s when freeways were built. Light rail is energy-efficient and popular, but it can cost up to $100 million per mile ($60 million per kilometer).

In 2010 more than 18 million motor vehicles were sold in China, making it both the world's largest manufacturer and the largest market for automobiles. The number of cars, buses, vans, and trucks on the road in China is expected to surpass the number in the United States by 2050. How those vehicles will be powered is of vital importance to our global ecosystem. Already Chinese efficiency standards are higher than those in the United States, and Chinese companies are making rapid progress in developing hybrid and all-electric vehicles.

22.5 SMART GROWTH

Smart growth is a term that describes such strategies for well-planned developments that make efficient and effective use of land resources and existing infrastructure. An alternative to haphazard, poorly planned sprawling developments, smart growth involves thinking ahead to develop pleasant neighborhoods while minimizing the wasteful use of space and tax dollars for new roads and extended sewer and water lines.

Smart growth aims to make land-use planning democratic. Public discussions allow communities to guide planners. Mixing land uses, rather than zoning exclusive residential areas far separate from commercial areas, makes living in neighborhoods more enjoyable. By planning a range of housing styles and costs, smart growth allows people of all income levels, including young families and aging grandparents, to find housing they can afford. Open communication between planners and the community helps make urban expansion fair, predictable, and cost-effective.

Smart growth approaches acknowledge that urban growth is inevitable; the aim is to direct growth, to make pleasant spaces for us to live, and to preserve some accessible, natural spaces for all to enjoy (table 22.4). It strives to promote the safety, livability, and revitalization of existing urban and rural communities.

Smart growth protects environmental quality. It attempts to reduce traffic and to conserve farmlands, wetlands, and open space. This may mean restricting land use, but it also means finding economically sound ways to reuse polluted industrial areas within the city (fig. 22.17). As cities grow and transportation and communications enable communities to interact more, the need for regional planning becomes both more possible and more pressing. Community and business leaders need to make decisions based on a clear understanding of regional growth needs and how infrastructure can be built most efficiently and for the greatest good.

Table 22.4 Goals for Smart Growth
1. Create a positive self-image for the community.
2. Make the downtown vital and livable.
3. Alleviate substandard housing.
4. Solve problems with air, water, toxic waste, and noise pollution.
5. Improve communication between groups.
6. Improve community member access to the arts.

Source: Vision 2000, Chattanooga, TN.

FIGURE 22.17 Many cities have large amounts of unused open space that could be used to grow food. Residents often need help decontaminating soil and gaining access to the land.

One of the best examples of successful urban land-use planning in the United States is Portland, Oregon, which has rigorously enforced a boundary on its outward expansion, requiring, instead, that development be focused on in-filling unused space within the city limits. Because of its many urban amenities, Portland is considered one of the most livable cities in America. Between 1970 and 1990, the Portland population grew by 50 percent but its total land area grew only 2 percent. During this time, Portland property taxes decreased 29 percent and vehicle miles traveled increased only 2 percent. By contrast, Atlanta, which had similar population growth, experienced an explosion of urban sprawl that increased its land area three-fold, drove up property taxes 22 percent, and increased traffic miles by 17 percent. A result of this expanding traffic and increasing congestion was that Atlanta's air pollution increased by 5 percent, while Portland's, which has one of the best public transit systems in the nation, decreased by 86 percent.

Garden cities and new towns were early examples of smart growth

The twentieth century saw numerous experiments in building **new towns** for society at large that try to combine the best features of the rural village and the modern city. One of the most influential of all urban planners was Ebenezer Howard (1850–1929), who not only wrote about ideal urban environments but also built real cities to test his theories. In *Garden Cities of Tomorrow,* written in 1898, Howard proposed that the congestion of London could be relieved by moving whole neighborhoods to **garden cities** separated from the central city by a greenbelt of forests and fields.

In the early 1900s, Howard worked with architect Raymond Unwin to build two garden cities outside of London, Letchworth and Welwyn Garden. Interurban rail transportation provided access to these cities. Houses were clustered in "superblocks" surrounded by parks, gardens, and sports grounds. Streets were curved. Safe and convenient walking paths and overpasses protected pedestrians from traffic. Businesses and industries were screened from housing areas by vegetation. Each city was limited to about 30,000 people to facilitate social interaction. Housing and jobs were designed to create a mix of different kinds of people and to integrate work, social activities, and civic life. Trees and natural amenities were carefully preserved and the towns were laid out to maximize social interactions and healthful living. Care was taken to meet residents' psychological needs for security, identity, and stimulation.

Letchworth and Welwyn Garden each have 70 to 100 people per acre. This is a true urban density, about the same as New York City in the early 1800s and five times as many people as most suburbs today. By planning the ultimate size in advance and choosing the optimum locations for housing, shopping centers, industry, transportation, and recreation, Howard believed he could create a hospitable and satisfying urban setting while protecting open space and the natural environment. He intended to create parklike surroundings that would preserve small-town values and encourage community spirit in neighborhoods.

Planned communities also have been built in the United States following the theories of Ebenezer Howard, but most plans have been based on personal automobiles rather than public transit. Radburn, New Jersey, was designed in the 1920s, and two highly regarded new towns of the 1960s are Reston, Virginia, and Columbia, Maryland. More recent examples, such as Seaside in northern Florida, represent a modern movement in new towns known as "new urbanism."

New urbanism advanced the ideas of smart growth

New towns and garden cities included many important ideas, but they still left cities behind. Rather than abandon the cultural history and infrastructure investment in existing cities, a group of architects and urban planners is attempting to redesign metropolitan areas to make them more appealing, efficient, and livable. In the United States, Andres Duany, Peter Calthorpe, and others have led this movement and promoted the term "new urbanism" to describe it. Sometimes called a neo-traditionalist approach, their designs attempt to recapture a small-town neighborhood feel in new developments. The goal of new urbanism has been to rekindle Americans' enthusiasm for cities. New urbanist architects do this by building charming, integrated, walkable developments. Sidewalks, porches, and small front yards encourage people to get outside and be sociable. A mix of apartments, townhouses, and detached houses in a variety of price ranges ensures that neighborhoods will include a diversity of ages and income levels. Some design principles of this movement include:

- Limit city size or organize them in modules of 30,000 to 50,000 people, large enough to be a complete city but small enough to be a community. A greenbelt of agricultural and recreational land around the city limits growth while promoting efficient land use. By careful planning and cooperation with neighboring regions, a city of 50,000 people can have real urban amenities such as museums, performing arts centers, schools, hospitals, etc.

- Determine in advance where development will take place. Such planning protects property values and prevents chaotic development in which the lowest uses drive out the better ones. It also recognizes historical and cultural values, agricultural resources, and such ecological factors as impact on wetlands, soil types, groundwater replenishment and protection, and preservation of aesthetically and ecologically valuable sites.

- Locate everyday shopping and services so people can meet daily needs with greater convenience, less stress, less automobile dependency, and less use of time and energy. Provide accessible, sociable public spaces (fig. 22.18).

- Increase jobs in the community by locating offices, light industry, and commercial centers in or near suburbs, or by enabling work at home via computer terminals. These alternatives save commuting time and energy and provide local jobs.

- Encourage walking or the use of small, low-speed, energy-efficient vehicles (microcars, motorized tricycles, bicycles, etc.) for many local trips now performed by full-size automobiles.

- Promote more diverse, flexible housing as alternatives to conventional, detached single-family houses. "In-fill" building between existing houses saves energy, reduces land costs, and might help provide a variety of living arrangements. Allowing owners to turn unused rooms into rental units provides space for those who can't afford a house and brings income to retired people who don't need a whole house themselves.

- Create housing "superblocks" that use space more efficiently and foster a sense of security and community. Widen peripheral arterial streets and provide pedestrian overpasses so traffic flows smoothly around residential areas; narrow streets within blocks, to slow traffic so children can play more safely. The land released from streets can be used for gardens, linear parks, playgrounds, and other public areas that will foster community spirit and encourage people to get out and walk.

FIGURE 22.18 This walking street in Queenstown, New Zealand, provides opportunities for shopping, dining, and socializing in a pleasant outdoor setting.

Green urbanism promotes sustainable cities

While new urbanism has promoted livable neighborhoods and raised interest in cities, critics point out that green urbanist developments, like garden cities and new towns, have often been **greenfield developments**, projects built on previously undeveloped farmlands or forests on the outskirts of large cities. In addition to contributing to sprawl, developments built on greenfields still require most residents to commute to work by private car, which undermines efforts to reduce car dependence. Goals of mixed-income neighborhoods also fall short, because the architect-designed houses rarely fall into middle- or low-income price ranges.

A new vision is emerging for "Smart Cities" with minimal environmental impacts. Rooftop solar panels and wind turbines will capture most or all of the energy needed by the city. Plug-in hybrid cars will serve as a massive dispersed electrical storage system. When excess energy is available, it will be stored in car batteries and then released back into the grid as demand rises. Food will be grown on rooftops and in empty lots and sold or bartered in local markets. Mass transportation will move residents around the city quickly and inexpensively. Rainwater will be collected, filtered, and reused. Metal and glass will be collected and recycled; organic waste will be composted to produce biogas for energy.

"Green urbanism" is another term that describes many strategies to redevelop existing cities to promote ecologically sound practices. Many green urbanist ideas are demonstrated in the BedZED project in London, England (What Do You Think? p. 510).

European cities have been especially innovative in green planning. Stockholm, Sweden, has expanded by building small satellite suburbs linked to the central city by commuter rails and by bicycle routes that pass through a network of green spaces that reach far into the city. Copenhagen, Denmark, has rebuilt most of its transportation infrastructure since the 1960s, including more than 300 km of well-marked bike lanes and separated bike trails. Thirty percent of all trips through central Copenhagen are made using public transportation, and 14 percent of the trips are made by bicycle.

Green building strategies are encouraged in many European cities. Many German cities now require that half of all new development must be vegetated. An increasingly popular strategy to meet this rule is "green roofs"—growing grass or other vegetation (fig. 22.19). Green roofs absorb up to 70 percent of rain water,

FIGURE 22.19 This award-winning green roof on the Chicago City Hall is functional as well as beautiful. It reduces rain runoff by about 50 percent, and keeps the surface as much as 30°F cooler than a conventional roof on hot summer days.

provide bird and butterfly habitat, insulate homes, and, contrary to old mythology, are structurally sturdy and long-lasting.

These are some common principles of green urbanist planning:

- When building new structures, focus on in-fill development—filling in the inner city so as to help preserve green space in and around cities. Where possible, focus on **brownfield developments**, building on abandoned, reclaimed industrial sites. Brownfields have been eyesores and environmental liabilities in cities for decades, but as urban growth proceeds, they are becoming an increasingly valuable land resource.

- Build high-density, attractive, low-rise, mixed-income housing near the center of cities or near public transportation routes (fig. 22.20a). Densely packed housing saves energy as well as reducing infrastructure costs per person.

- Provide incentives for alternative transportation, such as reserved parking for shared cars (fig. 22.20b) or bicycle routes and bicycle parking spaces. Figure 22.20c shows an 8,000-bicycle parking garage at the train station in Leiden,

(a)

(b)

(c)

FIGURE 22.20 Green urbanism includes (a) concentrated, low-rise housing, (b) car-sharing clubs that receive special parking allowances, and (c) alternative transportation methods. These examples are from Amsterdam and Leiden, the Netherlands.

The Architecture of Hope

How sustainable and self-sufficient can urban areas be? An exciting experiment in minimal impact in London gives us an image of what our future may be. BedZED, short for the Beddington Zero Energy Development, is an integrated urban project built on the grounds of an old sewage plant in South London. BedZED's green strategies begin with recycling the ground on which it stands. Designed by architect Bill Dunster and his colleagues, the complex demonstrates dozens of energy-saving and water-saving ideas. BedZED has been occupied, and winning awards, since it was completed in 2003.

Like Vauban (opening case study), most of BedZED's innovations involve simple, even conventional ideas. Expansive, south-facing, triple-glazed windows provide abundant light, minimize the use of electric lamps, and provide passive solar heat in the winter. Thick, superinsulated walls keep interiors warm in winter and cool in summer. Rotating "wind cowls" on roofs turn to catch fresh breezes, which cool spaces in summer. In winter, heat exchangers warm incoming fresh air with the heat from stale, outgoing air. Energy used in space heating is nearly eliminated. Building materials are recycled, reclaimed, or renewable, which reduces the "embodied energy" invested in producing and transporting them.

BedZED does use energy, but the complex generates its own heat and electricity with a small, on-site, superefficient plant that uses local tree trimmings for fuel. Thus BedZED uses no *fossil* fuels, and it is "carbon-neutral" because the carbon dioxide released by burning wood was recently captured from the air by trees. In addition, photovoltaic cells on roofs provide enough free energy to power 40 solar cars. Fuel bills for BedZED residents can be as little as 10 percent of what other Londoners pay for similar-sized homes.

Water-efficient appliances and toilets reduce water use. Rainwater collection systems provide "green water" for watering gardens, flushing toilets, and other nonconsumptive uses. Reed-bed filtration systems purify used water without chemicals. Water meters allow residents to see how much water they use. Just knowing about consumption rates helps encourage conservation. Residents use about half as much water per person as other Londoners.

BedZED residents can save money and time by not using, or even owning a car. Office space is available on-site, so some residents can work where they live, and the commuter rail station is just a ten-minute walk away. The site is also linked to bicycle trails that facilitate bicycle commuting. Car pools and rent-by-the-hour auto memberships allow many residents to avoid owning (and parking) a vehicle altogether.

Building interiors are flooded with natural light, ceilings are high, and most residences have rooftop gardens. Community events and common

South-facing windows heat homes, and colorful, rotating "wind cowls" ventilate rooms at BedZED, an ecological housing complex in South London, U.K.

spaces encourage humane, healthy lifestyles and community ties. Child-care services, shops, entertainment, and sports facilities are built into the project. The approximately 100 housing units are designed for a range of income levels ensuring a racially, ethnically, and age-diverse community. Prices are lower than many similar-sized London homes, and few in this price range or inner-city location have abundant sunlight or gardens.

Similar projects are being built across Europe and even in some developing countries, such as China. Architect Dunster says that BedZED-like developments on cleaned-up brownfields could provide all the 3 million homes that the U.K. expects to need in the next decade with no sacrifice of open space. And as green building techniques, designs, and materials become standard, he argues, they will cost no more than conventional, energy-wasting structures.

What do you think? Would you enjoy living in a dense, urban setting such as BedZED? Would it involve a lower or higher standard of living than you now have? How much would it be worth to avoid spending 8 to 10 hours per week not fighting bumper-to-bumper traffic while commuting to school or work? The average cost of owning and driving a car in the United States is about $9,000 per year. What might you do with that money if owning a vehicle were unnecessary? Try to imagine what urban life might be like if most private automobiles were to vanish. If you live in a typical American city, how much time do you have to enjoy the open space that the suburbs and freeways were supposed to provide? Perhaps, most importantly, how will we provide enough water, energy, and space for the 3 billion people expected to crowd into cities worldwide over the next few decades if we don't adopt some of the sustainable practices and approaches represented by BedZED or Vauban?

the Netherlands. An 8,000-car parking garage at the station would cut out the heart of the city. Discourage car use by minimizing the amount of space devoted to driving and parking cars, or by charging for parking space, once realistic alternatives are available.

- Encourage ecological building techniques, including green roofs, passive solar energy use, water conservation systems, solar water heating, wind turbines, and appliances that conserve water and electricity.

- Encourage co-housing—groups of households clustered around a common green space that share child care, gardening,

maintenance, and other activities. Co-housing can reduce consumption of space, resources, and time while supporting a sense of community.

- Provide facilities for recycling organic waste, building materials, appliances, and plastics, as well as metals, glass, and paper.

- Invite public participation in decision making. Emphasize local history, culture, and environment to create a sense of community and identity. Coordinate regional planning through metropolitan boards that cooperate with but do not supplant local governments.

Open space design preserves landscapes

Traditional suburban development typically divides land into a checkerboard layout of nearly identical 1 to 5 ha parcels with no designated open space (fig. 22.21, *top*). The result is a sterile landscape consisting entirely of house lots and streets. This style of development, which is permitted—or even required—by local zoning and ordinances, consumes agricultural land and fragments wildlife habitat. Many of the characteristics that people move to the country to find—space, opportunities for outdoor recreation, access to wild nature, a rural ambience—are destroyed by dividing every acre into lots that are "too large to mow but too small to plow."

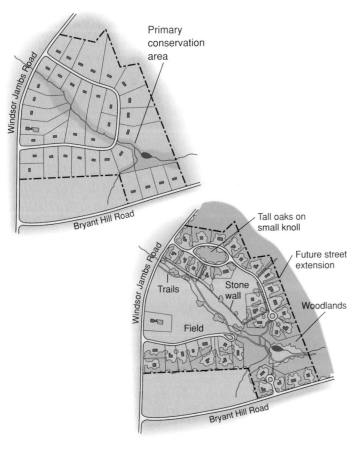

FIGURE 22.21 Conventional subdivision (*top*) and an open space plan (*bottom*). Although both plans provide 36 home sites, the conventional development allows for no public space. Cluster housing on smaller lots in the open space design preserves at least half the area as woods, prairie, wetlands, farms, or other conservation lands, while providing residents with more attractive vistas and recreational opportunities than a checkerboard development.

FIGURE 22.22 Jackson Meadows, an award-winning cluster development near Stillwater, Minnesota, groups houses at sociable distances and preserves surrounding open space for walking, gardening, and scenic views from all houses.

An interesting alternative known as **conservation development**, cluster housing, or open space zoning preserves at least half of a subdivision as natural areas, farmland, or other forms of open space. Among the leaders in this design movement are landscape architects Ian McHarg, Frederick Steiner, and Randall Arendt. They have shown that people who move to the country don't necessarily want to own a vast acreage or to live miles from the nearest neighbor; what they most desire is long views across an interesting landscape, an opportunity to see wildlife, and access to walking paths through woods or across wildflower meadows.

By carefully clustering houses on smaller lots, a conservation subdivision can provide the same number of buildable lots as a conventional subdivision and still preserve 50 to 70 percent of the land as open space (fig. 22.21, *bottom*). This not only reduces development costs (less distance to build roads, lay telephone lines, sewers, power cables, etc.) but also helps foster a greater sense of community among new residents. Walking paths and recreation areas get people out of their houses to meet their neighbors. Home owners have smaller lots to care for and yet everyone has an attractive vista and a feeling of spaciousness.

An award-winning example of cluster development is Jackson Meadow, near Stillwater, Minnesota (fig. 22.22). The 64 single-family, custom-designed houses are gathered on just one-third of the project's 336 acres. Two hundred acres are reserved for recreation and scenery. Because the houses are clustered, the developer was able to share one central well and pump house between them, instead of drilling 64 separate wells. In most remote developments of this size, wastewater from these homes would be treated in 64 separate, underground septic systems and leach fields. Here, wastewater is collected and drained into a constructed wetland septic system, where natural bacteria treat water. Water treatment in constructed wetlands is clean, safe, chemical-free, and odorless when it is designed correctly.

Urban habitat can make a significant contribution toward saving biodiversity. In a ground-breaking series of habitat conservation plans triggered by the need to protect the endangered California gnatcatcher, some 85,000 ha (210,000 acres) of coastal scrub near San Diego was protected as open space within the rapidly expanding urban area. This is an area larger than Yosemite Valley, and will benefit many other species as well as humans.

CONCLUSION

What can be done to improve conditions in cities? Vauban, Germany, is an outstanding example of green design to improve transportation, protect central cities, and create a sense of civic pride. Other cities have far to go, however, before they reach this standard. Among the immediate needs are housing, clean water, sanitation, food, education, health care, and basic transportation for their residents. The World Bank estimates that interventions to improve living conditions in urban households in the developing world could average the annual loss of almost 80 million "disability-free" years of life. This is about twice the feasible benefit estimated from all other environmental programs studied by the World Bank.

Many planners argue that social justice and sustainable economic development are answers to the urban problems we have discussed in this chapter. If people have the opportunity and money to buy better housing, adequate food, clean water, sanitation, and other things they need for a decent life, they will do so. Democracy, security, and improved economic conditions help in slowing population growth and reducing rural-to-city movement. An even more important measure of progress may be institution of a social welfare safety net guaranteeing that old or sick people will not be abandoned and alone.

Some countries have accomplished these goals even without industrialization and high incomes. Sri Lanka, for instance, has lessened the disparity between the core and periphery of the country. Giving all people equal access to food, shelter, education, and health care eliminates many incentives for interregional migration. Both population growth and city growth have been stabilized, even though the per capita income is only $800 per year. What do you think; could we help other countries do something similar?

REVIEWING LEARNING OUTCOMES

By now you should be able to explain the following points:

22.1 Define *urbanization*.
- Cities have specialized functions as well as large populations.
- Large cities are expanding rapidly.

22.2 Describe why cities grow.
- Immigration is driven by push and pull factors.
- Government policies can drive urban growth.

22.3 Understand urban challenges in the developing world.
- Traffic congestion and air quality are growing problems.
- Insufficient sewage treatment causes water pollution.
- Many cities lack adequate housing.

22.4 Identify urban challenges in the developed world.
- Urban sprawl consumes land and resources.
- Transportation is crucial in city development.
- Mass transit could make our cities more livable.

22.5 Explain smart growth.
- Garden cities and new towns were early examples of smart growth.
- New urbanism advanced the ideas of smart growth.
- Green urbanism promotes sustainable cities.
- Open space design preserves landscapes.

PRACTICE QUIZ

1. What is the difference between a city and a village and between rural and urban?
2. How many people now live in cities, and how many live in rural areas worldwide?
3. What changes in urbanization are predicted to occur in the next 30 years, and where will that change occur?
4. From memory, list five of the world's largest cities. Check your list against table 22.2. How many were among the largest in 1900?
5. Describe the current conditions in a typical megacity of the developing world. What forces contribute to its growth?
6. Describe the difference between slums and shantytowns.
7. Why are urban areas in U.S. cities decaying?
8. How has transportation affected the development of cities? What have been the benefits and disadvantages of freeways?
9. Describe some ways that American cities and suburbs could be redesigned to be more ecologically sound, socially just, and culturally amenable.
10. Explain the difference between greenfield and brownfield development. Why is brownfield development becoming popular?

CRITICAL THINKING AND DISCUSSION QUESTIONS

1. Picture yourself living in a rural village or a developing world city. What aspects of life there would you enjoy? What would be the most difficult for you to accept?
2. Why would people move to one of the megacities of the developing world if conditions are so difficult there?

3. A city could be considered an ecosystem. Using what you learned in chapters 3 and 4, describe the structure and function of a city in ecological terms.

4. Look at the major urban area(s) in your state. Why were they built where they are? Are those features now a benefit or drawback?

5. Weigh the costs and benefits of automobiles in modern American life. Is there a way to have the freedom and convenience of a private automobile without its negative aspects?

6. Boulder, Colorado, has been a leader in controlling urban growth. One consequence is that the city has stayed small and charming, so housing prices have skyrocketed and poor people have been driven out. If you lived in Boulder, what solutions might you suggest? What do you think is an optimum city size?

Data Analysis: Using a Logarithmic Scale

We've often used very large numbers in this book. Millions of people suffer from common diseases. Hundreds of millions are moving from the country to the city. Billions of people will probably be added to the world population in the next half century. Cities that didn't exist a few decades ago now have millions of residents. How can we plot such rapid growth and such huge numbers? If you use ordinary graph paper, making a scale that goes to millions or billions will run off the edge of the page unless you make the units very large.

Figure 1, for example, shows the growth of Mumbai, India, over the past 150 years plotted with an **arithmetic scale** (showing constant intervals) for the Y-axis. It looks as if there is very little growth in the first third of this series and then explosive growth during the last few decades, yet we know that the *rate* of growth was actually greater at the beginning than at the end of this time. How could we display this differently? One way to make the graph easier to interpret is to use a **logarithmic scale.** A logarithmic scale, or "log scale," progresses by factor of 10. So the Y-axis would be numbered 0, 1, 10, 100, 1,000. . . . The effect on a graph is to spread out the smaller values and compress the larger values. In figure 2, the

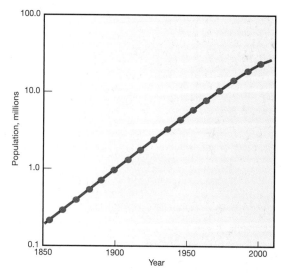

FIGURE 2 The growth of Mumbai.

same data are plotted using a log scale for the Y-axis, which makes it much easier to see what happened throughout this time period.

1. Do these two graphing techniques give you a different impression of what's happening in Mumbai?

2. How might researchers use one or the other of these scales to convey a particular message or illustrate details in a specific part of the growth curve?

3. Approximately how many people lived in Mumbai in 1850?

4. How many lived there in 2000?

5. When did growth of Mumbai begin to slow?

6. What percentage did the population increase between 1850 and 2000?

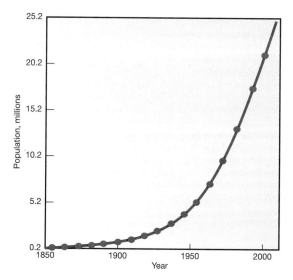

FIGURE 1 The growth of Mumbai.

A small amount of seed money would allow this young mother to expand her business and help provide for her family.

Ecological Economics

"Unleashing the energy and creativity in each human being is the answer to poverty."

~ *Muhammad Yunus*

Learning Outcomes

After studying this introduction, you should be able to:

23.1 Identify some assumptions of classical and neoclassical economics.

23.2 Explain key ideas of environmental economics.

23.3 Describe relationships among population, technology, and scarcity.

23.4 Understand ways we measure growth.

23.5 Summarize how market mechanisms can reduce pollution.

23.6 Discuss the importance of trade, development, and jobs.

23.7 Evaluate the aims of green business.

Case Study   Loans That Change Lives

Ni Made is a young mother of two children who lives in a small Indonesian village. Her husband is a day laborer who makes only a few dollars per day—when he can find work. To supplement their income, Made goes to the village market every morning to sell a drink she makes out of boiled pandanus leaves, coconut milk, and pink tapioca (opening photograph). A small loan would allow her to rent a covered stall during the rainy season and to offer other foods for sale. The extra money she could make could change her life. But traditional banks consider Made too risky to lend to, and the amounts she needs too small to bother with.

Around the world, billions of poor people find themselves in the same position as Made; they're eager to work to build a better life for themselves and their families, but lack resources to succeed. Now, however, a financial revolution is sweeping around the world. Small loans are becoming available to the poorest of the poor. This new approach was invented by Dr. Muhammad Yunus, professor of rural economics at Chittagong University in Bangladesh. Talking to a woman who wove bamboo mats in a village near his university, Dr. Yunus learned that she had to borrow the few taka she needed each day to buy bamboo and twine. The interest rate charged by the village moneylenders consumed nearly all her profits. Always living on the edge, this woman, and many others like her, couldn't climb out of poverty (fig. 23.1).

To break this predatory cycle, Dr. Yunus gave the woman and several of her neighbors small loans totaling about 1,000 taka (about $20). To his surprise, the money was paid back quickly and in full. So he offered similar amounts to other villagers with similar results. In 1983, Dr. Yunus started the Grameen (village) Bank to show that "given the support of financial capital, however small, the poor are fully capable of improving their lives." His experiment has been tremendously successful. By 2009, the Grameen Bank had nearly 2 billion customers, 97 percent of them women. It had loaned more than $8 billion with 98 percent repayment, nearly twice the collection rate of commercial Bangladesh banks.

The Grameen Bank provides credit to poor people in rural Bangladesh without the need for collateral. It depends, instead, on mutual trust, accountability, participation, and creativity of the

FIGURE 23.1 For the poorest people in developing countries, a small business loan can be the most sustainable strategy for development.

borrowers themselves. Microcredit is now being offered by hundreds of organizations in 43 other countries. Institutions from the World Bank to religious charities make small loans to worthy entrepreneurs. Wouldn't you like to be part of this movement? Well, now you can. You don't have to own a bank to help someone in need.

A brilliant way to connect entrepreneurs in developing countries with lenders in wealthy countries is offered by Kiva, a San Francisco–based technology startup. The idea for Kiva, which means "unity" or "cooperation" in Swahili, came from Matt and Jessica Flannery. Jessica had worked in East Africa with the Village Enterprise Fund, a California nonprofit that provides training, capital, and mentoring to small businesses in developing countries. Jessica and Matt wanted to help some of the people she had met, but they weren't wealthy enough to get into microfinancing on their own. Joining with four other young people with technology experience, they created Kiva, which uses the power of the Internet to help the poor.

Kiva partners with about a dozen development nonprofits with staff in developing countries. The partners identify hardworking entrepreneurs who deserve help. They then post a photo and brief introduction to each one on the Kiva web page. You can browse the collection to find someone whose story touches you. The minimum loan is generally $25. Your loan is bundled with that of others until it reaches the amount needed by the borrower. You make your loan using your credit card (through PayPal, so it's safe and easy). The loan is generally repaid within 12 to 18 months (although without interest). At that point, you can either withdraw the money, or use it to make another loan.

The in-country staff keeps track of the people you're supporting and monitors their progress, so you can be confident that your money will be well used. Loan requests often are on their web page for only a few minutes before being filled. It's easy to take part in this innovative human development project. Check out Kiva.org.

In this chapter we'll look further at both microlending and conventional financing for human development. We'll also look at the role of natural resources in economies, and how ecological economics is bringing ecological insights into economic analysis. We'll examine cost–benefit analysis as well as other measures of human well-being and genuine progress. Finally, we'll look at how market mechanisms can help us solve environmental problems, and how businesses can contribute to sustainability.

23.1 PERSPECTIVES ON THE ECONOMY

Economy is the management of resources, ideally to meet our needs as efficiently as possible. The terms *ecology* and *economy* share a common root, *oikos* (ecos), the Greek word for "household." Economics is the *nomos*, or counting, of the household resources. Ecology is the *logos*, or logic, of how the household works.

Much of our economy involves using natural resources, such as oil, wood, or iron, to produce goods. Some resources are renewable, others are not. Ideas and actions also generate economic activity. Musicians, for example, support an industry based largely on ideas, culture, and knowledge. Economics involves choices and trade-offs, because we don't have unlimited abilities to produce goods. Understanding the balance of costs and benefits of these choices is a concern for economists (fig. 23.2). Investing money in a Kiva loan, for example, has a low cost (just a few dollars), and a small financial benefit (interest on a few dollars). The potential benefits to society are tremendous, however, making it easy for many people to decide to make a Kiva loan.

Can development be sustainable?

Environmental economics, like environmental science, tends to ask questions about long-term resource use: Are we using resources efficiently? Are the costs of our resource use reflected in the prices we pay for goods? Are there alternative strategies that

FIGURE 23.2 Bread or bullets? What are the costs and benefits of each? And what are the trade-offs between them?

Table 23.1	Goals for Sustainable Natural Resource Use

- Harvest rates for renewable resources (those like organisms that regrow or those like fresh water that are replenished by natural processes) should not exceed regeneration rates.
- Waste emissions should not exceed the ability of nature to assimilate or recycle those wastes.
- Nonrenewable resources (such as minerals) may be exploited by humans, but only at rates equal to the creation of renewable substitutes.

could help us produce goods and services with fewer resources? Does our use of resources limit the opportunities of others—either future generations or people in other regions—to lead healthy and productive lives?

One of the most important questions in environmental science is how we can continue to improve human welfare within the limits of the earth's natural resources and biological systems. *Development* means improving people's lives, usually through increased access to goods (such as food) or services (such as education). *Sustainability* means living on the earth's renewable resources without damaging the ecological processes that support us all (table 23.1). **Sustainable development** is an effort to marry these two ideas. A definition developed by the World Commission on Environment and Development in 1987 is that "sustainable development is development that meets the needs of the present without compromising the ability of future generations to meet their own needs."

Is this possible? Not at our present population and rates of consumption. Some observers insist that there is no way that more people can live at a high standard of living without irreversibly degrading our environment. Others say that as natural resources become scarce, we will simply find alternatives. Still others argue that there's enough for everyone if we can just share equitably and consume less. Much of this debate depends on how we define resources and economic growth.

Resources can be renewable or nonrenewable

A **resource** is anything with potential use in creating wealth or giving satisfaction. Natural resources can be either renewable or nonrenewable. In general, **nonrenewable resources** are materials present in fixed amounts in the environment, especially earth resources such as minerals, metals, and fossil fuels (fig. 23.3). Many of these resources are renewed or recycled over geological time, as are oil and coal, but on a human time scale they are not renewable. Predictions abound that we are in imminent danger of running out of one or another of these exhaustible resources. Supplies of metals and other commodities, however, have frequently been extended by more efficient use, recycling, substitution of one material for another, or new technologies that can extract resources from dilute or remote sources.

FIGURE 23.3 Nonrenewable resources, such as the oil from this forest of derricks in Huntington Beach, California, are irreplaceable. Once they're exhausted (as this oil field was half a century ago) they will never be restored on a human time scale.

FIGURE 23.4 Nature provides essential ecological services, such as the biological productivity, water storage and purification, and biodiversity protection in this freshwater marsh and its surrounding forest. Ironically, while biological resources are infinitely renewable, if they're damaged by our actions they may be lost forever.

Renewable resources are things that can be replenished or replaced. These include living organisms, fresh water from rain and snow, and sunlight—our ultimate energy source. These systems also provide essential ecological services on which we depend, although most of us don't think of these resources very often (fig. 23.4). We discuss these ideas further in section 23.2.

Because biological organisms and ecological processes are self-renewing, we often can harvest surplus organisms or take advantage of ecological services without diminishing future availability, if we do so carefully. Unfortunately, our stewardship of these resources often is less than ideal. Even once vast biological populations such as passenger pigeons, American bison, and Atlantic cod, for instance, were exhausted by overharvesting in only a few years. Similarly, we are now reducing renewable water resources (from rainfall) in many regions by modifying the climate system. This modification of a renewable resource is leading to drought and reduced crop production in dry regions (chapter 15). Mismanagement of renewable resources, then, often makes them more ephemeral and limited than nonrenewable resources.

We also depend on **intangible resources**, such as open space, beauty, serenity, wisdom, and diversity (fig. 23.5). Paradoxically, these resources can be both infinite *and* exhaustible. There is no upper limit to the amount of beauty, knowledge, or compassion that can exist in the world, yet they can be easily destroyed. A single piece of trash can ruin a beautiful vista, or a single cruel remark can spoil an otherwise perfect day. On the other hand, unlike tangible resources that usually are reduced by use or sharing, intangible resources often are increased by use and multiplied by being shared. Nonmaterial assets can be important economically. Information management and tourism—both based on intangible resources—have become two of the largest and most powerful industries in the world.

Another term used to describe resources is **capital**, or wealth that can be used to produce more wealth. Usually capital refers to something that has been built up or accumulated over time. There can be many forms of capital. Microlending, as described in the opening case study, provides *financial capital* (money) that small businesses need to start or grow. Economists also consider *manufactured* or *built capital* (tools, infrastructure, and technology), *natural capital* (goods and services provided by nature), *human* or *cultural capital* (knowledge, experience, and ideas about how to make or do things), and even *social capital* (shared values, trust, cooperative spirit, and community organization). All these kinds of capital may be needed to produce marketable goods and services.

FIGURE 23.5 Scenic beauty, solitude, and relatively untouched nature, as in this Colorado wilderness, are treasured by many people but are hard to evaluate in economic terms.

FIGURE 23.6 Informal markets such as this one in Bali, Indonesia, may be the purest example of willing sellers and buyers setting prices based on supply and demand.

Classical economics examines supply and demand

Classical economics originally was a branch of moral philosophy concerned with how individual interest and values intersect with larger social goals. We trace many of our ideas about economy to Adam Smith (1723–1790), a moral philosopher concerned with individual freedom of choice. Smith's landmark book *Inquiry into the Nature and Causes of the Wealth of Nations,* published in 1776, argued:

> Every individual endeavors to employ his capital so that its produce may be of the greatest value. He generally neither intends to promote the public interest, nor knows how much he is promoting it. He intends only his own security, only his own gain. And he is in this led by an *invisible hand* to promote an end which was no part of his intention. By pursuing his own interests he frequently promotes that of society more effectually than when he really intends to.

This statement often is taken as justification for the capitalist system, in which willing sellers and fully informed buyers agree on a fair price for goods in the market. Smith proposed that this agreement would bring about the greatest efficiency of resource use, because efficiency is necessary to produce goods at an acceptably low price. Assuming that all buyers and sellers are free to make any choice, this system also ensures individual liberty (fig. 23.6). The British economist John Maynard Keynes summarized faith in free markets and pricing this way: "Capitalism is the astounding belief that the most wickedest of men will do the most wickedest of things for the greatest good of everyone."

In a real market, producing goods at a low cost often requires that some costs are **externalized**, or passed off to someone else. Environmental costs and social costs are often supported by communities. For example, producing electricity at a power plant requires a stable and educated work force to run the plant, and the cost of educating workers normally is borne by society in general. Transportation networks, mining and oil drilling, and insurance costs are often supported by the public. Tax breaks also represent a subsidy because they transfer costs from a company to other sectors of society. Producing electricity also usually involves some pollution, and a power company allows the public to absorb the cost of that pollution (such as health care or reduced crop production). All these costs are involved in producing and selling electricity, but they are external to the company's cost calculations. Many economists, both conservative and liberal, argue that subsidies are a problem because they mask the real costs of production and lead to inefficiency in our economy.

David Ricardo (1772–1823), a contemporary of Adam Smith, introduced a description of the relationship between supply and demand in economics. **Demand** is the amount of a product or service that consumers are willing and able to buy at various possible prices, assuming they are free to express their preferences. **Supply** is the quantity of that product being offered for sale at various prices, other things being equal. Classical economics proposes that there is a direct, inverse relationship between supply and demand (fig. 23.7). With increasing quantity of production, supply increases and prices fall. With decreasing quantity, prices rise and demand falls. The difference between the cost of production and the price buyers are willing to pay, Ricardo called "rent." Today we call it profit.

In a free market of independent and rational buyers and sellers, an optimal price is achieved at the intersection of the supply and demand curves (fig. 23.7). This intersection is known as the **market equilibrium**. In real life, prices are not determined strictly by total supply and demand as much as by what economists called **marginal costs and benefits**. Sellers ask themselves, "What would it cost to produce one more unit of this product or service? Suppose I add one more worker or buy an extra supply of raw materials,

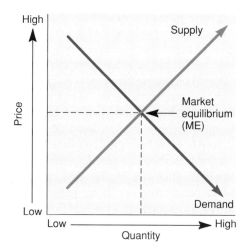

FIGURE 23.7 Classic supply/demand curves. When price is low, supply is low and demand is high. As prices rise, supply increases but demand falls. Market equilibrium is the price at which supply and demand are equal.

how much more profit could I make?" Buyers ask themselves similar questions, "How much would I benefit and what would it cost if I bought one more widget?" If both buyer and seller find the marginal costs and benefits attractive, a sale is made.

There are exceptions to this theory of supply and demand. Consumers will buy some things regardless of cost. Raising the price of cigarettes, for instance, doesn't necessarily reduce demand. We call this price inelasticity. Other items have **price elasticity:** they follow supply/demand curves exactly. When price goes up, demand falls and vice versa.

Neoclassical economics emphasizes growth

Toward the end of the nineteenth century, the field of economics divided into two broad camps. **Political economy** continued the tradition of moral philosophy and concerned itself with social structures and relationships among the classes. This group included reformers such as Karl Marx and later E. F. Schumacher, who argued that unfettered capital accumulation inevitably leads to inequity, which leads to instability in society. The other camp, called **neoclassical economics,** strove to adapt methods of modern science, and to be mathematically rigorous, noncontextual, abstract, and predictive. Neoclassical economists claim to be objective and value-free, leaving social concerns to other disciplines. Like their classical predecessors, they retain an emphasis on scarcity and the interaction of supply and demand in determining prices and resource allocation (fig. 23.8).

Constant economic growth is considered necessary and desirable, in the neoclassical view. Growth keeps people happy by always offering more income or goods than people had last year. In a growing population, economic growth is seen as the only way to maintain full employment and avoid class conflict arising from inequitable distribution of wealth. Growth is also essential because businesses borrow resources to operate and grow. Few lenders are willing to share their money without a promise of greater returns later. Thus businesses must continue to expand in order to increase profits and maintain the confidence of shareholders, whose money they are using to run their operations.

John Stuart Mill (1806–1873), a classical economist and philosopher, argued that perpetual growth in material well-being is neither possible nor desirable. Economies naturally mature to a steady state, he proposed, leaving people free to pursue nonmaterialistic goals. He didn't regard this equilibrium state to be necessarily one of stagnation or poverty. Instead, he wrote that in a stable economy, "There would be as much scope as ever for all kinds of mental culture, and moral and social progress; as much room for improving the art of living, and much more likelihood of its being improved when minds cease to be engrossed by the art of getting on."

Some neoclassical economists point out that not all growth involves increased resource consumption and pollution. Growth based on education, entertainment, and nonconsumptive activities, as suggested by Mill, can still contribute to economic expansion.

Neoclassical economics tends to view natural resources as interchangeable. As one resource becomes scarce, neoclassical economists believe, substitutes will be found. Labor is also substitutable. Because materials and labor are substitutable, they are not considered indispensable. Debates about the nature of growth, consumption, and resource scarcity run through recent developments in economics, including environmental economics, the concept of a steady-state economy, and sustainable development.

23.2 ECOLOGICAL ECONOMICS

Classical and neoclassical economics shape most of our economic activities, but the externalized costs and social inequity of these approaches have led to an interest in new ways to evaluate economic progress. **Ecological economics** has emerged as a way to understand the relationship between our economy and the ecological systems that support it. These economists point out that even Adam Smith wrote that the profit motive is good at guiding decision making in the short term, but it is ill-suited to long-term decision making. Ecological economists argue that we need to improve our long-term decision making, if we intend to be here for the long term.

Ecological economists have followed two main approaches to resolving the short view of conventional economics: One approach is to question the necessity of constant growth. Another is to identify externalities and calculate their costs. If we know these costs, we can make the price reflect the total costs of production, or we can reduce those costs by changing the ways we make things.

Ecological economics draws on ecological concepts such as systems (chapter 2), thermodynamics, and material cycles (chapter 3). In a system, all components are interdependent. Disrupting one component (such as climate conditions) risks destabilizing other components (such as agricultural production) in unpredictable and possibly catastrophic ways. Thermodynamics and material cycles teach that energy and materials are continually reused. One organism's waste

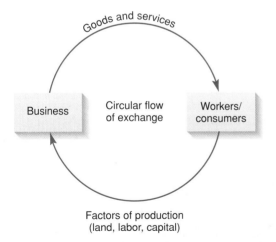

FIGURE 23.8 The neoclassical model of the economy focuses on the flow of goods, services, and factors of production (land, labor, capital) between business and individual workers and consumers. The social and environmental consequences of these relationships are irrelevant in this view.

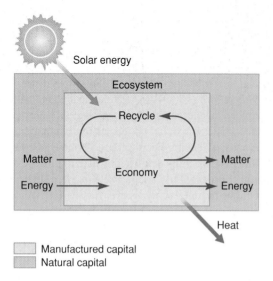

FIGURE 23.9 Ecological economics includes services such as recycling and resource provision and economic accounting. An effort is made to internalize, rather than externalize, natural resources and services.

is another's nutrient or energy source. This perspective allows us to consider relationships among the many parts of our environment, our economy, and our own well-being (fig. 23.9).

Systems analysis also proposes that the earth has a limited carrying capacity for human populations. A limited carrying capacity means that unlimited economic growth is not possible. Many ecological economists, such as Herman Daly, argue for a **steady-state economy,** characterized by low human birth rates and death rates, the use of renewable energy sources, material recycling, and an emphasis on durability, efficiency, and stability. **Throughput,** the volume of materials and energy consumed and of waste produced, should be minimized. Most neoclassical growth models have assumed constantly increasing consumption and waste production, with resources being substituted as they run out. The steady-state idea, in contrast, reflects the notion of an ecosystem in equilibrium, or a population below its carrying capacity, where overall conditions remain generally stable over time.

Ecological economics assigns cost to ecosystems

Ecological economics seeks to draw natural systems and ecological processes into economic accounting. Classical and neoclassical economics usually have focused on the value of human capital and built capital and resources, such as buildings, roads, or labor. Natural systems are considered free of cost—in part because in fact they have been free, abundant, or unregulated. Rivers absorb wastewater, bacteria decompose

waste, and winds carry away smoke, but until recently there were few mechanisms for monitoring these processes or assigning them a value.

Consider energy production again, in which producers weigh their profits against the costs of coal, labor, buildings, and furnaces. These are **internal costs,** part of the normal expenses of doing business. Costs such as the climate's absorption of carbon dioxide, sulfur dioxide, or mercury are external costs, outside the accounting system. Similarly, if populations downwind suffer from air pollution, incurring lost workdays or illness, traditional economics has argued that society should bear responsibility.

Can we internalize these costs? Identifying and calculating values takes considerable effort because it is a new process, but a study by Paul Epstein and colleagues in 2011 produced an unusually complete estimate of the externalized costs of coal power. Epstein and his colleagues examined the costs of land disturbance, methane emissions from mines, fatalities during coal transportation, climate impacts, downwind costs to health, child development, cancers, direct federal subsidies to coal miners and energy producers, the costs of restoring abandoned mines, and other costs. Their best estimate of externalized costs was approximately 18 ¢/kilowatt-hour (kWh). The average cost

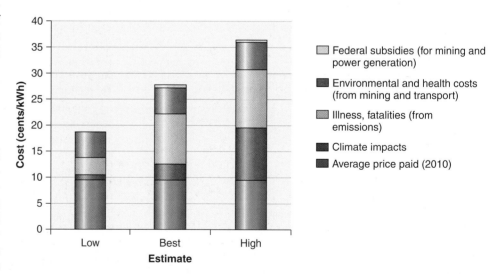

FIGURE 23.10 Estimates of externalized costs of coal power, in ¢/kilowatt-hour (kWh). Externalized costs (red colors) are shown added to the average price paid in 2010 (blue).

of electric power in 2010 was about 9.5 ¢/kWh, or just half of the externalized costs. In other words, the real cost was about triple the price paid for electricity (fig. 23.10).

High and low estimates were also calculated in this study, to account for uncertainties in the data. These suggested that the public absorbs about 9¢ to 28¢ for every kWh of coal-based electricity. This study focused on coal because it is the world's dominant source of electric power, but similar accounting could be done for any power source or economic activity.

Accounting for all costs should make production more efficient, because an accurate price can help the public make more informed decisions. In general, the cost of cleaning up a power plant usually is lower than the cost of health care and lost productivity. In economic terms, the extra costs of illness and environmental damage are "market inefficiencies": they represent inefficient overall use of resources (money, time, energy, materials) because of incomplete accounting of costs and benefits.

Ecosystem services include provisioning, regulating, and aesthetic values

Ecosystem services is a general term for the resources provided and waste absorbed by our environment. These services are often grouped into four general classes (table 23.2): regulation (of climate, water supplies, and other factors), provision (of foods and other resources), supporting or preserving (of crop pollinators, nutrient cycling), and aesthetic or cultural benefits (fig. 23.11).

Although many ecological processes have no direct market value, we can estimate replacement costs, contingent values, shadow prices, and other methods of indirect assessment to determine a rough value. For instance, we now dispose of much of our wastes by letting nature detoxify them. How much would it cost if we had to do this ourselves?

Estimates of the annual value of all ecological goods and services provided by nature range from $16 trillion to $54 trillion, with a median worth of $33 trillion, or about three-fourths the combined annual GNP of all countries in the world (table 23.3). These estimates are lower than the real value because they omit ecosystem services from several biomes, such as deserts and tundra, that are poorly understood in terms of their economic contributions.

FIGURE 23.11 We rely on ecosystem services to provide resources; they also regulate our environment and support essential biogeochemical processes that support life.

Accounting for ecosystem services is a focus of several global initiatives on sustainable development. A UN program called The Economics of Ecosystems and Biodiversity (TEEB) has been working to improve estimates of the value of ecosystem services. TEEB studies have shown that preserving ecosystems is far more cost-effective than using up their resources. Even restoring already-damaged ecosystems has enormous paybacks (fig. 23.12). Calculating a price for carbon storage in natural ecosystems has been the aim of REDD (Reducing greenhouse gas Emissions through Deforestation and Degradation) programs. These efforts

---

Table 23.2 Important Ecological Services

1. *Regulate* global energy balance; chemical composition of the atmosphere and oceans; local and global climate; water catchment and groundwater recharge; production, storage, and recycling of organic and inorganic materials; maintenance of biological diversity.

2. *Provide* space and suitable substrates for human habitation, crop cultivation, energy conversion, recreation, and nature protection.

3. *Produce* oxygen, fresh water, food, medicine, fuel, fodder, fertilizer, building materials, and industrial inputs.

4. *Supply* aesthetic, spiritual, historic, cultural, artistic, scientific, and educational opportunities and information.

Source: R. S. de Groot, *Investing in Natural Capital,* 1994.

Table 23.3 Estimated Annual Value of Ecological Services

Ecosystem Services	Value (Trillion $U.S.)
Soil formation	17.1
Recreation	3.0
Nutrient cycling	2.3
Water regulation and supply	2.3
Climate regulation (temperature and precipitation)	1.8
Habitat	1.4
Flood and storm protection	1.1
Food and raw materials production	0.8
Genetic resources	0.8
Atmospheric gas balance	0.7
Pollination	0.4
All other services	1.6
Total value of ecosystem services	**33.3**

Source: Adapted from R. Costanza et al., "The Value of the World's Ecosystem Services and Natural Capital," *Nature,* Vol. 387 (1997).

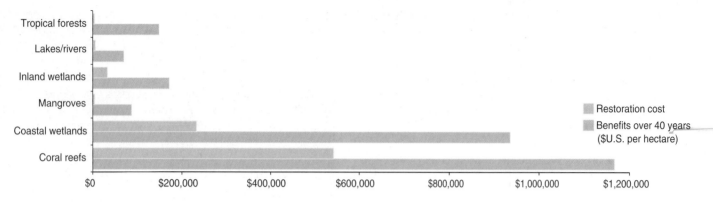

FIGURE 23.12 Can we afford to restore biodiversity? It's harder to find money to restore ecosystems than to destroy them, but the benefits over time greatly exceed the average cost of restoration.

are discussed in chapter 5. A 2003 economic study from Cambridge University (U.K.) estimated that protecting a series of nature reserves representing samples of all major biomes would cost (U.S.) $45 billion per year, but would preserve ecological services worth 100 times that cost—$4.4 trillion to $5.2 trillion annually.

23.3 POPULATION, TECHNOLOGY, AND SCARCITY

Despite changing perspectives on resources, many analysts continue to ask the question that worried Adam Smith and Thomas Malthus (chapter 1): Are we about to run out of essential natural resources? It stands to reason that if we consume a fixed supply of nonrenewable resources at a constant rate, eventually we'll use up all the economically recoverable reserves. There have been many warnings that our extravagant depletion of nonrenewable resources sooner or later will result in catastrophe. The dismal prospects of Malthusian diminishing returns and a life of misery, starvation, and social decay inspire many people to call for immediate changes to lower consumption rates. Historic observations have often followed projections of resource depletion. The **Hubbert curve**, for example, developed by Stanley Hubbert, has fairly accurately described the peak and decline of U.S. oil supplies (fig. 23.13).

On the other hand, many economists contend that resource supplies and demand are not rigidly fixed. Human ingenuity and enterprise often allow us to respond to scarcity in ways that postpone or contradict dire warnings of collapse. In this section we consider some of the arguments for and against limits to growth of the global economy.

Communal property resources are a classic problem in ecological economics

In 1968 biologist Garret Hardin wrote an influential article entitled "The Tragedy of the Commons." He argued that any commonly held resource inevitably is degraded or destroyed because the narrow self-interest of individuals tends to outweigh public

interest. Hardin offered as a metaphor the common woodlands and pastures held by most colonial New England villages. In deciding how many cattle to put on the commons, Hardin explained, each villager would attempt to maximize his or her own personal gain. Adding one more cow to the commons could mean a substantially increased income for an individual farmer. The damage done by overgrazing, however, would be shared among all the farmers (fig. 23.14). This is known as the "free rider" problem, where individuals take more than their fair share of commonly held resources. Hardin concluded that the only solutions would be either to give coercive power to the government or to privatize the resource.

Hardin intended this parable to warn about human overpopulation and overexploitation of resources. Other authors have used his metaphor to explain such diverse problems as famines, air pollution, or collapsing fisheries.

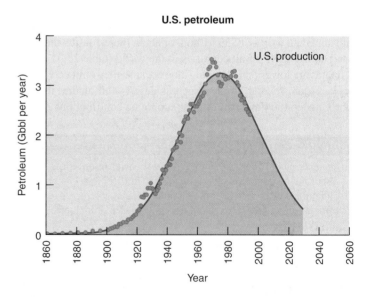

FIGURE 23.13 United States petroleum production shows the Hubbert curve. Dots indicate actual production. The shaded area under the curve represents the estimated amount of economically recoverable oil, 220 Gbbl (Gbbl = Gigabarrels or billions of barrels).

FIGURE 23.14 Adding more cattle to the Brazilian Cerrado (savanna) increases profits for individual ranchers, but is bad for biodiversity and environmental quality.

What Hardin was really describing, however, was an **open access system** in which there are no rules to manage resource use. In fact, many communal resources have been successfully managed for centuries by cooperative arrangements among users. Some examples include Native American management of wild rice beds and hunting grounds; Swiss village-owned mountain forests and pastures; Maine lobster fisheries; communal irrigation systems in Spain, Bali, and Laos; and nearshore fisheries almost everywhere in the world.

A large body of literature in economics and social sciences describes how these cooperative systems work. Among the features shared by **communal resource management systems** are: (1) community members have lived on the land or used the resource for a long time and anticipate that their children and grandchildren will as well, thus giving them a strong interest in sustaining the resource and maintaining bonds with their neighbors; (2) the resource has clearly defined boundaries; (3) the community group size is known and enforced; (4) the resource is relatively scarce and highly variable so that the community is forced to be interdependent; (5) management strategies appropriate for local conditions have evolved over time and are collectively enforced; that is, those affected by the rules have a say in them; (6) the resource and its use are actively monitored, discouraging anyone from cheating or taking too much; (7) conflict resolution mechanisms reduce discord; and (8) incentives encourage compliance with rules, while sanctions for noncompliance keep community members in line.

In some cases privatization leads to degradation of common pool resources. Where small villages have owned and managed jointly held forests or fishing grounds for generations, privatization has led to shortsighted decision making, leading to rapid destruction of both society and ecosystems. A tragic example of this was the forced privatization of Indian reservations in the United States. Where communal systems once enforced restraint over harvesting, privatization encouraged narrow self-interest.

With individuals making decisions for personal, near-term benefit, many people chose to sell their resources to outsiders, who could easily take advantage of the weakest members of the community. Failing to recognize or value local knowledge and forcing local people to participate in a market economy allowed outsiders to disenfranchise native people and resulted in disastrous resource exploitation. Distinguishing between open-access systems and communal property regimes is important in understanding how best to manage natural resources.

Scarcity can lead to innovation

In a pioneer or frontier economy, methods for harvesting resources and turning them into useful goods and services tend to be inefficient and wasteful. The history of logging in the United States, for example, is a classic case of inefficient resource exploitation. Between about 1860 and 1930, the supply of American forests was vast and unregulated. Logging companies cleared eastern states, then swarmed across the Great Lakes forests. As prime timber in the northern forests was depleted, the companies simply shifted to the Rocky Mountains and the Pacific Northwest. At each stage, logging wasted a vast amount of wood, but this inefficiency seemed unimportant because the supply of trees was so great. Labor, financial capital, and transportation to market were the scarce resources.

Today our use of forest products is considerably more efficient. We have smaller supplies and greater demand, and we have developed better technology and methods that allow us to create the same amount of goods and services using fewer resources. Instead of using giant old-growth timbers for building, we use laminated beams, chipboard, and other products that can be produced from what once was scrap wood. We also require that logging companies replant new forests after logging, because they can no longer simply move on to a new region.

Scarcity often is a catalyst for innovation and change (fig. 23.15). As materials become more expensive and difficult to obtain, it becomes cost-effective to discover new supplies or to use available ones more efficiently. Several important factors play a role in this cycle of technological development:

- Technical inventions can increase efficiency of extraction, processing, use, and recovery of materials.

- Substitution of new materials or commodities for scarce ones can extend existing supplies or create new ones.

- Trade makes remote supplies of resources available and may also bring unintended benefits in information exchange and cultural awakening.

- Discovery of new reserves through better exploration techniques, more investment, and looking in new areas becomes rewarding as supplies become limited and prices rise.

- Recycling becomes feasible and accepted as resources become more valuable. Recycling now provides about 37 percent of the iron and lead, 20 percent of the copper, 10 percent of the aluminum, and 60 percent of the antimony that is consumed each year in the United States.

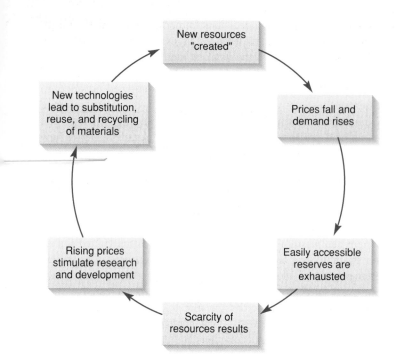

FIGURE 23.15 Scarcity/development cycle. Paradoxically, resource use and depletion of reserves can stimulate research and development, the substitution of new materials, and the effective creation of new resources.

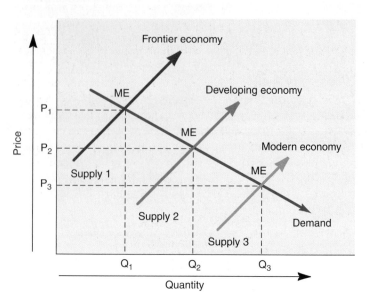

FIGURE 23.16 Supply and demand curves at three different stages of economic development. At each stage there is a market equilibrium point at which supply and demand are in balance. As the economy becomes more efficient, the equilibrium shifts so there is a larger quantity available at a lower price than before. (P = price, Q = quantity, ME = market equilibrium)

Increasing technological efficiency can dramatically shift supply and demand relationships. As technology makes goods and services cheaper to produce, the quantity available at a given price can increase greatly. The market equilibrium, or the point at which supply and demand equilibrate, will shift to lower prices and higher quantities as a market matures (fig. 23.16).

Carrying capacity is not necessarily fixed

Despite repeated warnings that rapidly growing populations and increasing affluence are bound to exhaust natural resources and result in rapid price increases, technological developments have resulted in price decreases for most raw materials over the last hundred years. Consider copper for example. Twenty years ago worries about impending shortages led the United States to buy copper and store it in strategic stockpiles. Estimated demand for this important metal, essential for electric motors, telephone lines, transoceanic cables, and other uses, far exceeded known reserves. It looked as if severe shortages and astronomical price increases were inevitable. But then aluminum power lines, satellites, fiber optic cables, integrated circuits, microwave transmission, and other inventions greatly diminished the need for copper. Although prices are highly variable, the general trend for most materials was downward in the twentieth century.

Recent reports have warned that increasing demand for consumer goods and infrastructure in China will raise demand, and prices, to previously unimagined levels. Will this increase lead to shortages and crisis? Or will it lead to innovation and resource substitution? Economists generally believe that substitutability and technological development will help us avoid catastrophe. Ecologists generally argue that there are bound to be limits to how much we can consume.

A noted example of this debate occurred in 1980. Ecologist Paul Ehrlich bet economist Julian Simon that increasing human populations and growing levels of material consumption would inevitably lead to price increases for natural resources. They chose a package of five metals—chrome, copper, nickel, tin, and tungsten—priced at the time at $1,000. If, in ten years, the combined price (corrected for inflation) was higher than $1,000, Simon would pay the difference. If the combined price had fallen, Ehrlich would pay. In 1990 Ehrlich sent Simon a check for $576.07; the price for these five metals had fallen 47.6 percent.

Does this prove that resource abundance will continue indefinitely? Hardly. Ehrlich argued that the timing and set of commodities chosen simply were the wrong ones. The fact that we haven't yet run out of raw materials doesn't mean that it will never happen. Many ecological economists now believe that some nonmarket resources such as ecological processes may be more irreplaceable than tangible commodities like metals. What do you think? Are we approaching limits to consumption? Which resources, if any, do you think are most likely to be limiting in the future?

Economic models compare growth scenarios

In the early 1970s, an influential study of resource limitations was funded by the Club of Rome, an organization of wealthy business owners and influential politicians. The study was carried out by a team of scientists from the Massachusetts Institute of Technology headed by the late Donnela Meadows. The results of this study were published in the 1972 book *Limits to Growth*. A complex computer model of the world economy was used to examine various scenarios of different resource depletion rates, growing population, pollution, and industrial output.

Given the Malthusian assumptions built into this model, catastrophic social and environmental collapse seemed inescapable (fig. 23.17a). Food supplies and industrial output rise as population grows and resources are consumed. Once past the carrying capacity of the environment, however, a crash occurs as population, food production, and industrial output all decline precipitously. Pollution continues to grow as society decays and people die, but, eventually, it also falls. Notice the similarity between this set of curves and the "boom and bust" population cycles described in chapter 6.

Many economists criticized these results because they discount technological development and factors that might mitigate the effects of scarcity. In 1992, the Meadows group published updated computer models in *Beyond the Limits* that include technological progress, pollution abatement, population stabilization, and new public policies that work for a sustainable future. If we adopt these changes sooner rather than later, all factors in the model stabilize sometime in this century at an improved standard of living for everyone (fig. 23.17b). Of course, neither of these computer models shows what will happen, only what some possible outcomes *might* be, depending on the choices we make.

23.4 MEASURING GROWTH

How do we monitor our resource consumption and its effects? In order to know if conditions in general are getting better or worse, economists have developed indices that countries or regions can use to monitor change over time. These indices track a variety of activities and values, to produce an overall picture of the economy. Which factors we choose to monitor, though, reflect judgments about what is important in society, and those judgments can vary substantially.

GNP is our dominant growth measure

The most common way to measure a nation's output is **gross national product (GNP).** GNP can be calculated in two ways. One is the money flow from households to businesses in the form of goods and services purchased. The other is to add up all the costs of production in the form of wages, rent, interest, taxes, and profit. In either case, a subtraction is made for capital depreciation, the wear and tear on machines, vehicles, and buildings used in production. Some economists prefer **gross domestic product (GDP),** which includes only the economic activity within national boundaries. Thus the vehicles made and sold by Ford in Europe don't count in GDP.

Both GNP and GDP have been criticized as measures of real progress or well-being because they don't attempt to distinguish between beneficial activities and harmful activities. A huge oil spill that pollutes beaches and kills wildlife, for example, shows up as a positive addition to GNP because it generates economic activity in the costs of cleanup.

Ecological economists also argue that GNP doesn't account for natural resources used up or ecosystems damaged by economic activities. Robert Repeto of the World Resources Institute estimates that soil erosion in Indonesia, for instance, reduces the value of crop production about 40 percent per year. If natural capital were taken into account, Indonesian GNP would be reduced by at least 20 percent annually.

Similarly, Costa Rica experienced impressive increases in timber, beef, and banana production between 1970 and 1990. But decreased natural capital during this period represented by soil erosion, forest destruction, biodiversity losses, and accelerated water runoff add up to at least $4 billion or about 25 percent of annual GNP.

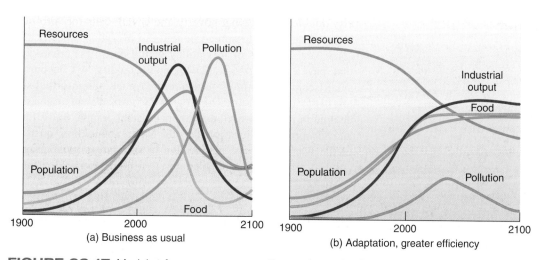

FIGURE 23.17 Models of resource consumption and scarcity. Running the model with assumptions of Malthusian limits and high consumption causes food, productivity, and populations to crash, while pollution increases *(left)*. Running the same model with assumptions of slowing population growth and consumption, with better technologies produces stable output and population *(right)*. Which of these models will we follow?

Alternate measures account for well-being

A number of systems have been proposed as alternatives to GNP that reflect genuine progress and social welfare. In their 1989 book, Herman Daly and John Cobb proposed a genuine progress index (GPI) that takes into account real per capita income, quality of life, distributional equity, natural resource depletion, environmental damage, and the value of unpaid labor. They point out that while per capita GDP in the United States nearly doubled between 1970 and 2000, per capita GPI increased only 4 percent (fig. 23.18). Some social service organizations would add to this index the costs of social breakdown and crime, which would decrease real progress even further over this time span.

A newer measure is the **Environmental Performance Index (EPI)** created by researchers at Yale and Columbia Universities to evaluate national sustainability and progress toward achievement of the United Nations Millennium Development Goals. The EPI is based on sixteen indicators tracked in six categories: environmental health, air quality, water resources, productive natural resources, biodiversity and habitat, and sustainable energy. The top-ranked countries—New Zealand, Sweden, Finland, the Czech Republic, and the United Kingdom—all commit significant resources and effort to environmental protection. In 2006, the United States ranked 28th in the EPI, or lower than Malaysia, Costa Rica, Columbia, and Chile, all of which have between 6 and 15 times lower GDP than the United States. See Data Analysis (p. 537) for a graph of human development index (HDI) versus EPI.

The United Nations Development Programme (UNDP) uses a benchmark called the human development index (HDI) to track social progress. HDI incorporates life expectancy, educational attainment, and standard of living as critical measures of development. Gender issues are accounted for in the gender development index (GDI), which is simply HDI adjusted or discounted for inequality or achievement between men and women.

In its annual Human Development Report, the UNDP compares country-by-country progress. As you might expect, the highest development levels are generally found in North America, Europe, and Japan. In 2006, Norway ranked first in the world in both HDI and GDI. The United States ranked eighth while Canada was sixth. The 25 countries with the lowest HDI in 2006 were all in Africa. Haiti ranks the lowest in the Western Hemisphere.

Although poverty remains widespread in many places, encouraging news also can be found in development statistics. Poverty has fallen more in the past 50 years, the UNDP reports, than in the previous 500 years. Child death rates in developing countries as a whole have been more than halved. Average life expectancy has increased by 30 percent while malnutrition rates have declined by almost a third. The proportion of children who lack primary school has fallen from more than half to less than a quarter. And the share of rural families without access to safe water has fallen from nine-tenths to about one-quarter.

Some of the greatest progress has been made in Asia. China and a dozen other countries with populations that add up to more than 1.6 billion, have decreased the proportion of their people living below the poverty line by half. Still, in the 1990s the number of people with incomes less than $1 per day increased by almost 100 million to 1.3 billion—and the number appears to be growing in every region except Southeast Asia and the Pacific. Even in industrial countries, more than 100 million people live below the poverty line and 37 million are chronically unemployed.

Economic growth can be a powerful means of reducing poverty, but its benefits can be mixed. The GNP of Honduras, for instance, grew 2 percent per year in the 1980s and yet poverty doubled. To combat poverty, the UNDP calls for "pro-poor growth" designed to spread benefits to everyone. Specifically, some key elements of this policy would be to: (1) create jobs that pay a living wage, (2) lessen inequality, (3) encourage small-scale agriculture, microenterprises, and the informal sector, (4) foster technological progress, (5) reverse environmental decline in marginal regions, (6) speed demographic transitions, and (7) provide education for all. Since about three-quarters of the world's poorest people live in rural areas, raising agricultural productivity and incomes is a high priority for these actions (fig. 23.19).

Cost–benefit analysis aims to optimize benefits

One way to evaluate public projects is to analyze the costs and benefits they generate in a **cost–benefit analysis (CBA).** This process attempts to assign values to resources as well as to social and environmental effects of carrying out or not carrying out a given undertaking. It tries to find the optimal efficiency point at which the marginal cost of pollution control equals the marginal benefits (fig. 23.20).

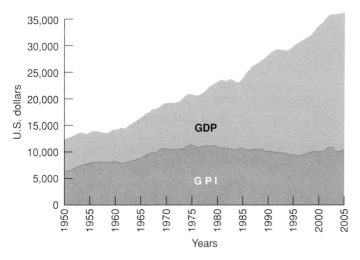

FIGURE 23.18 Although per capita GDP in the United States nearly doubled between 1970 and 2000 in inflation-adjusted dollars, a genuine progress index that takes into account natural resource depletion, environmental damage, and options for future generations hardly increased at all.

FIGURE 23.19 Raising agricultural productivity and rural incomes are high priorities of the UN Millennium Development Goals.

FIGURE 23.21 What is the value of solitude or beauty? How would you assign costs and benefits to a scene such as this?

CBA is one of the main conceptual frameworks of resource economics and is used by decision makers around the world as a way of justifying the building of dams, roads, and airports, as well as in considering what to do about biodiversity loss, air pollution, and global climate change. Deeply entrenched in bureaucratic practice and administrative culture, this technique has become much more widespread in American public affairs since the Reagan administration's executive orders in the 1980s calling for the application of CBA to all regulatory decisions and legislative proposals. Many conservatives see CBA as a way of eliminating what they consider to be unnecessary and burdensome requirements to protect clean air, clear water, human health, or biodiversity. They would like to add a requirement that all regulations be shown to be cost-effective.

The first step in CBA is to identify who or what might be affected by a particular plan. What are the potential outcomes and results? What alternative actions might be considered? After identifying and quantifying all effects, an attempt is made to assign monetary costs and benefits to each one. Usually the direct

expenses of a project are easy to ascertain. How much will you have to pay for land, materials, and labor? The monetary value of lost opportunities—to swim or fish in a river, or to see birds in a forest—is much harder to appraise (fig. 23.21). How would you put a price on good health or a long life? It's also important to ask who will bear the costs and who will reap the benefits of any proposal. Are there external costs that should be accounted for? Eventually the decision maker compares all the costs and benefits to see whether the project is justified or whether some alternative action might bring more benefits at less cost.

Because of the difficulty of assigning monetary prices to intangible or public resources we value, many people object to CBA. In analyzing the costs and benefits of a hydroelectric dam, for example, economists often cannot assign suitable values to land, forests, streams, fisheries and livelihoods, and community. Ordinary people often cannot answer questions about how much money they would pay to save a wilderness or how much they would accept to allow it to be destroyed.

A study by the Economic Policy Institute of Washington, D.C., found that costs for complying with environmental regulations are almost always less than industry and even governments estimate they will be. For example, electric utilities in the United States claimed that it would cost $4 to $5 billion to meet the 1990 Clean Air Act. But by 1996, utilities were actually saving $150 million per year. Similarly, when CFCs were banned, automobile manufacturers protested it would add $1,200 to the cost of each new car. The actual cost was about $40.

Cost–benefit analysis is also criticized for its absence of standards and inadequate attention to alternatives. Who judges how costs and benefits will be estimated? How can we compare things as different as the economic gain from cheap power with loss of biodiversity or the beauty of a free-flowing river? Critics claim that placing monetary values on everything could lead to a belief that only money and profits count and that any behavior is acceptable as long as you can pay for it. Sometimes speculative or even hypothetical results are given specific numerical values in CBA and then treated as if they are hard facts. Risk-assessment techniques (see chapter 8) may be more appropriate for comparing uncertainties.

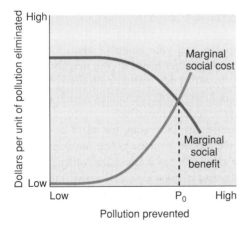

FIGURE 23.20 To achieve maximum economic efficiency, regulations should require pollution prevention up to the optimum point (P_0) at which the costs of eliminating pollution just equal the social benefits of doing so.

23.5 MARKET MECHANISMS CAN REDUCE POLLUTION

We are becoming increasingly aware that our environment and economy are mutually interconnected. Natural resources and ecological services are essential for a healthy economy, and a vigorous economy can provide the means to solve environmental problems. In this section, we'll explore some of these links.

Using market forces

Most scientists regard global climate change as the most serious environmental problem we face. In 2006, the business world got a harsh warning about this problem from British economist, Sir Nicolas Stern. Commissioned by the British treasury department to assess the threat of global warming, Sir Nicolas, who formerly was chief economist at the World Bank, issued a 700-page study that concluded that if we don't act to control greenhouse gases, the damage caused by climate change could be equivalent to losing as much as 20 percent of the global GDP every year. This could have an impact on our lives and environment greater than the worldwide depression or the great wars of the twentieth century.

We have many options for combating climate change, but many economists believe **market forces** can reduce pollution more efficiently than rules and regulations. Assessing a tax, for example, on each ton of carbon emitted could have the desired effect of reducing greenhouse gases and controlling climate change, but could still allow industry to search for the most cost-effective ways to achieve these goals. It also creates a continuing incentive to search for better ways to reduce emissions. The more you reduce your discharges, the more you save.

The cost of climate change will be far greater than steps we could take now to reduce climate change. Stern calculates that it will take about $500 billion per year (1 percent of global GDP) to avoid the worst impacts of climate change if we act now. That is a lot of money, but it's a bargain compared to his estimates of $10 trillion in annual losses and costs of climate change in 50 years if we don't change our practices. And the longer we wait, the more expensive carbon reduction and adaptation are going to be.

On the other hand, reducing greenhouse gas emissions and adapting to climate change will create significant business opportunities, as new markets are created in low-carbon energy technologies and services (fig. 23.22). These markets could create millions of jobs and be worth hundreds of billions of dollars every year. Already, Europe has more than 5 million jobs in renewable energy, and the annual savings from solar, wind, and hydro power are saving the European Union about $10 billion per year in avoided oil and natural gas imports. Being leaders in the fields of renewable energy and carbon reduction gives pioneering countries a tremendous business advantage in the global marketplace. Markets for low-carbon energy could be worth $500 billion per year by 2050, according to the Stern report.

FIGURE 23.22 Markets for low-carbon energy could be worth $500 billion per year by 2050, and could create millions of high-paying jobs.

Is emissions trading the answer?

The Kyoto Protocol, which was negotiated in 1997, and has been ratified by every industrialized nation in the world except the United States and Australia, sets up a mechanism called **emissions trading** to control greenhouse gases. This is also called a **cap-and-trade** approach. The first step is to mandate upper limits (the cap), on how much each country, sector, or specific industry is allowed to emit. Companies that can cut pollution by more than they're required can sell the **credit** to other companies that have more difficulty meeting their mandated levels.

Suppose you've just built a state-of-the-art power plant that allows you to capture and store CO_2 for about $20 per ton, and that allows you to cut your CO_2 emissions far below the amount you are permitted to produce. Suppose, further, that your neighboring utility has a dirty, old coal-fired power plant for which it would cost $60 per ton to reduce CO_2 emissions. You might strike a deal with your neighbor. You reduce your CO_2 emissions, and he pays you $40 for each ton you reduce, so he doesn't have to reduce. You make $20 per ton, and your neighbor saves $20 per ton. Both of you benefit. On the other hand, if your neighbor can find an even cheaper way to **offset** his carbon emissions, he's free to do so. This creates an incentive to continually search for ever more cost-effective ways to reduce emissions.

Opportunities are increasing for all of us to buy carbon offsets. When you buy an airplane ticket, for example, some airlines offer you the chance to pay a few extra dollars, which will be used to pay for planting trees, which will absorb carbon. You can also buy carbon offsets if you have an old, inefficient car. For about $20 per ton (or about $100 per year for the average American car), they'll plant trees, build a windmill, or provide solar lights to a village in a developing country to compensate for your emissions. You can take pride in being **carbon-neutral** at a far lower price than buying a new automobile.

Sulfur trading offers a good model

The 1990 U.S. Clean Air Act created one of the first market-based systems for reducing air pollution. It mandated a decrease in acid-rain-causing sulfur dioxide (SO_2) from power plants and other industrial facilities. An SO_2 targeted reduction was set at 10 million tons per year, leaving it to industry to find the most efficient way to do this. The government expected that meeting this goal would cost companies up to $15 billion per year, but the actual cost has been less than one-tenth of that. Prices on the sulfur exchange have varied from $60 to $800 per ton depending on the availability and price of new technology, but most observers agree that the market has found much more cost-effective ways to achieve the desired goal than rigid rules would have created.

This program is regarded as a shining example of the benefits of market-based approaches. There are complaints, however, that while nationwide emissions have come down, "hot spots" remain where local utilities have paid for credits rather than install pollution abatement equipment. If you're living in one of these hot spots and continuing to breathe polluted air, it's not much comfort to know that nationwide average air quality has improved. Currently, credits and allowances of more than 30 different air pollutants are traded in international markets.

Carbon trading is already at work

Climate change is revolutionizing global economics. In 2006, approximately (U.S.) $28 billion worth of climate credits, equivalent to 1 billion tons of CO_2, traded hands on international markets. The market grew more than four-fold by 2010, to $120 billion. By far the most active market currently is the Amsterdam-based European Climate Exchange. The United States has a climate market in Chicago, and regional agreements are developing nationwide (chapter 15), but thus far participation is only voluntary because the United States doesn't have mandatory emissions limits, and carbon credits are selling for only about one-tenth the price they are in Europe.

In 2010 more than 80 percent of the international emissions payments went to just four countries, and nearly two-thirds of those payments were for reductions of the refrigerant HFC-23 (fig. 23.23). Most entrepreneurs are uninterested in deals less than about $250,000. Smaller projects just aren't worth the time and expense of setting them up. In one of the biggest deals so far, a consortium of British bankers signed a contract to finance an incinerator on a large chemical factory in Quzhou, in China's Zhejiang Province. The incinerator will destroy hydrofluorocarbon (HFC-23) that previously had been vented into the air. This has a double benefit: HFC-23 destroys stratospheric ozone, and it also is a potent greenhouse gas (approximately 11,700 times as powerful as CO_2). The $500 million deal will remove the climate-changing equivalent of the CO_2 emitted by 1 million typical American cars each driven 20,000 km per year. But the incinerator will cost only $5 million—a windfall profit to be split between the bankers and the factory owners.

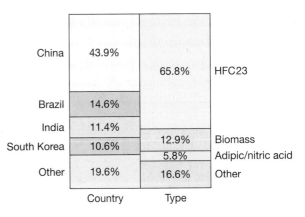

FIGURE 23.23 Worldwide emissions reductions payments by country and type. Currently, four countries are collecting 80 percent of all proceeds from emissions trading, and two-thirds of those payments are going for relatively cheap HFC23 incineration. Is this fair?
Source: United Nations, 2010.

There's a paradox in this deal. HFC production in China and India is soaring because a growing middle class fuels a demand for refrigerators and air conditioners. The huge payments flowing into these countries under the Kyoto Protocol are helping their economies grow and increasing middle-class affluence, and thus creating more demand for refrigerators and air conditioners. Furthermore, air conditioners using this refrigerant are much less energy efficient than newer models, so their increasing numbers are driving the demand for electricity, which currently is mostly provided by coal-fired power plants.

Critics of our current emissions markets point out that this mechanism was originally intended to encourage the spread of renewable energy and nonpolluting technology to developing countries in places such as sub-Saharan Africa. It was envisioned as a way to spread solar panels, windmills, tree farms, and other technologies that would provide climate control and also speed development of the poorest people. Instead, marketing emission credits, so far, is benefiting primarily bankers, consultants, and factory owners and is leading to short-term fixes rather than fundamental, long-term solutions.

23.6 TRADE, DEVELOPMENT, AND JOBS

Trade can be a powerful tool in making resources available and raising standards of living. Think of the things you now enjoy that might not be available if you had to live exclusively on the resources available in your immediate neighborhood. Too often, the poorest, least powerful people suffer in this global marketplace. To balance out these inequities, nations can deliberately invest in economic development projects. In this section, we'll look at some aspects of trade, development, business, and jobs that have impacts on our environment and welfare.

International trade brings benefits but also intensifies inequities

The banking and trading systems that regulate credit, currency exchange, shipping rates, and commodity prices were set up by the richer and more powerful nations in their own self-interest. The General Agreement on Tariffs and Trade (GATT) and World Trade Organization (WTO) agreements, for example, negotiated primarily between the largest industrial nations, regulate 90 percent of all international trade.

These systems tend to keep the less-developed countries in a perpetual role of resource suppliers to the more-developed countries. The producers of raw materials, such as mineral ores or agricultural products, get very little of the income generated from international trade (fig. 23.24).

Policies of the WTO and the IMF have provoked criticism and resistance in many countries. As a prerequisite for international development loans, the IMF frequently requires debtor nations to adopt harsh "structural adjustment" plans that slash welfare programs and impose cruel hardships on poor people. The WTO has issued numerous rulings that favor international trade over pollution prevention or protection of endangered species. Trade conventions such as the North American Free Trade Agreement (NAFTA) have been accused of encouraging a "race to the bottom" in which companies can play one country against another and move across borders to find the most lax labor and environmental protection standards.

No single institution has more influence on financing and policies of developing countries than the World Bank. Of some $25 billion loaned each year for development projects by international agencies, about two-thirds comes from the World Bank. Founded in 1945 to fund reconstruction of Europe and Japan, the World Bank shifted its emphasis to aid developing countries in the 1950s. Many of its projects have had adverse environmental and social effects, however. Its loans often go to corrupt governments and fund ventures such as nuclear power plants, huge dams, and giant water diversion schemes. Former U.S. treasury secretary Paul O'Neill said that these loans have driven poor countries "into a ditch" by loading them with unpayable debt. He said that funds should not be loans, but rather grants to fight poverty.

Microlending helps the poorest of the poor

Global aid from the WTO usually aids banks and industries more than it helps the impoverished populations who most need assistance. Often structural adjustment leads the poorest to pay back loans negotiated by their governments and industries. These concerns led Dr. Muhammad Yunus of Bangladesh to initiate the micro-loan plan of the Grameen Bank (opening case study).

Microlending programs have assisted billions of people—most of them low-status women who have no other way to borrow money at reasonable interest rates. This model is now being used by hundreds of other development agencies around the world (fig. 23.25). Even in the United States, organizations assist microenterprises with loans, grants, and training. The Women's Self-Employment Project in Chicago, for instance, teaches job skills to single mothers in housing projects. Similarly, "tribal circle" banks on Native American reservations successfully finance microscale economic development ventures. Kiva.org, mentioned in the opening case study for this chapter, raised $71 million in just four years to help 171,000 entrepreneurs in developing countries.

One of the most important innovations of the Grameen Bank is that borrowers take out loans in small groups. Everyone in the

FIGURE 23.25 With a loan of only a few dollars, this Chinese coal deliverer could buy his own cart and more than double his daily income. If you could make a tiny loan that would change his life, wouldn't you do it?

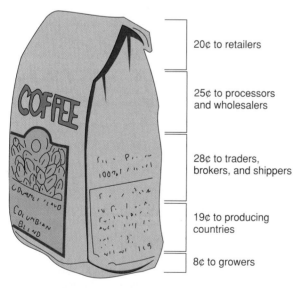

20¢ to retailers

25¢ to processors and wholesalers

28¢ to traders, brokers, and shippers

19¢ to producing countries

8¢ to growers

FIGURE 23.24 What do we really pay for when we purchase a dollar's worth of coffee?

group is responsible for each other's performance. The group not only guarantees loan repayment, it helps businesses succeed by offering support, encouragement, and advice. Where banks depend on the threat of foreclosure and a low credit rating to ensure debt repayment, the Grameen Bank has something at least as powerful for poor villagers—the threat of letting down your neighbors and relatives. Becoming a member of a Grameen group also requires participation in a savings program that fosters self-reliance and fiscal management.

The process of running a successful business and repaying the loan transforms many individuals. Women who previously had little economic power, influence, or self-esteem are empowered with a sense of pride and accomplishment. Dr. Yunus also discovered that money going to families through women helped the families much more than the same amount of money in men's salaries. Women were more likely to spend money on children's food or education, producing generational benefits with the increased income.

The most recent venture for the Grameen Bank is providing mobile phone service to rural villages. Supplying mobile phones to poor women not only allows them to communicate, it provides another business opportunity. They rent out their phone to neighbors, giving the owner additional income, and linking the whole village to the outside world. Suddenly, people who had no access to communication can talk with their relatives, order supplies from the city, check on prices at the regional market, and decide when and where to sell their goods and services. This is a great example of "bottom-up development." Founded in 1996, Grameen Phone now has 2.5 million subscribers and is Bangladesh's largest mobile phone company. In 2006 Dr. Yunus received the Nobel Peace Prize for his groundbreaking work in helping the poor.

23.7 GREEN BUSINESS

Business leaders are increasingly discovering that they can save money and protect our environment by greening up their business practices. They can save substantial amounts of money through fuel efficiency and reducing electricity consumption. These steps also cut greenhouse gases. Recycling waste, and minimizing use of hazardous materials, saves on disposal costs. In addition these companies win public praise and new customers by demonstrating an interest in our shared environment. By conserving resources, they also help ensure the long-term survival at their own corporations.

Known by a variety of names, including eco-efficiency, clean production, pollution prevention, industrial ecology, natural capitalism, restorative technology, the natural step, environmentally preferable products, design for the environment, and the next industrial revolution, this movement has had some remarkable successes and presents an encouraging pathway for how we might achieve both environmental protection and social welfare. Some of the leaders in this new approach to business include Paul Hawken, William McDonough, Ray Anderson, Amory Lovins, David Crockett, and John and Nancy Todd.

Operating in a socially responsible manner consistent with the principles of sustainable development and environmental protection, they have shown, can be good for employee morale, public relations, and the bottom line simultaneously. Environmentally conscious or "green" companies such as the Body Shop, Patagonia, Aveda, Malden Mills, Johnson and Johnson, and Interface, Inc. (What Do You Think? p. 532) consistently earn high marks from community and environmental groups. Conserving resources, reducing pollution, and treating employees and customers fairly may cost a little more initially, but can save money and build a loyal following in the long run.

New business models follow concepts of ecology

Paul Hawken's 1993 book, *The Ecology of Commerce,* was a seminal influence in convincing many people to reexamine the role of business and economics in environmental and social welfare. Basing his model for a new industrial revolution on the principles of ecology, Hawken points out that almost nothing is discarded or unused in nature. The wastes from one organism become the food of another. Industrial processes, he argues, should be designed on a similar principle (table 23.4). Rather than a linear pattern in which we try to maximize the throughput of material and minimize labor, products and processes should be designed to

- be energy efficient;
- use renewable materials;
- be durable and reusable or easily dismantled for repair and remanufacture, nonpolluting throughout their entire life cycle;
- provide meaningful and sustainable livelihoods for as many people as possible;
- protect biological and social diversity;
- use minimum and appropriate packaging made of reusable or recyclable materials.

Table 23.4 Goals for an Eco-Efficient Economy

- Introduce no hazardous materials into the air, water, or soil.
- Measure prosperity by how much natural capital we can accrue in productive ways.
- Measure productivity by how many people are gainfully and meaningfully employed.
- Measure progress by how many buildings have no smokestacks or dangerous effluents.
- Make the thousands of complex governmental rules unnecessary that now regulate toxic or hazardous materials.
- Produce nothing that will require constant vigilance from future generations.
- Celebrate the abundance of biological and cultural diversity.
- Live on renewable solar income rather than fossil fuels.

What Do You Think?

Eco-Efficient Business Practices

In 1994, in response to customers' concerns about health problems caused by chemical fumes from new carpeting, wall coverings, and other building materials, Ray Anderson, founder and CEO of Interface, Inc., a billion-dollar-a-year interior furnishing company, decided to review the company environmental policy. What he found was that the company really didn't have an environmental vision other than to obey all relevant laws and comply with regulations. He also learned that carpeting—of which Interface was the world's third-largest manufacturer—is one of the highest volume and longest lasting components in landfills. A typical carpet is made of nylon embedded in fiberglass and polyvinyl chloride. After a useful life of about 12 years, most carpeting is ripped up and discarded. Every year, more than 770 million m^2 (920 million yd^2) of carpet weighing 1.6 billion kg (3.5 billion lbs) ends up in U.S. landfills. The only recycling that most manufacturers do is to shave off some of the nylon for remanufacture. Everything else is buried in the ground where it will last at least 20,000 years.

At about the same time that Interface was undergoing its environmental audit, Anderson was given a copy of Paul Hawken's book *The Ecology of Commerce*. Reading it, he said, was like "a spear through the chest." He vowed to turn his company around, to make its goal sustainability instead of simply maximizing profits. Rather than sell materials, Interface would focus on selling service. The key is what Anderson calls an "evergreen lease." First of all, the carpet is designed to be completely recyclable. Where most flooring companies merely sell carpet, Interface offers to lease carpets to customers. As carpet tiles wear out, old ones are removed and replaced as part of the lease. The customer pays no installation or removal charges, only a monthly fee for constantly fresh-looking and functional carpeting. Everything in old carpet is used to make new product. Only after many reincarnations as carpet, are materials finally sent to the landfill.

Dramatic changes have been made at Interface's 26 factories. Toxic air emissions have been nearly eliminated by changing manufacturing processes and substituting nontoxic materials for more dangerous ones. Solar power and methane from a landfill are replacing fossil fuel use. Interface may be the first carbon-neutral manufacturing company in America. Less waste is produced as more material is recycled and products are designed for eco-efficiency. The total savings from pollution prevention and recycling in 2007 was $150 million.

Not only has Interface continued to be an industry leader, it was named one of the "100 Best Companies to Work for in America" by *Fortune* magazine. Ray Anderson became a popular speaker on the topic of

Ray Anderson.
Source: Courtesy Ray Anderson, Interface, Inc.

eco-efficiency and clean production. He became co-chair the President's Council on Sustainable Development, was named Entrepreneur of the Year by Ernest & Young, and was the Georgia Conservancy's Conservationist of the Year in 1998. Anderson's book, *Mid-Course Correction: Toward a Sustainable Enterprise*, published in 1999 by Chelsea Green, won critical acclaim.

Transforming an industry as large as interior furnishing required persistence. "Like aircraft carriers," Anderson said, "big businesses don't turn on a dime." Still, he showed that the principles of sustainability and financial success can coexist and can lead to a new prosperity that includes both environmental and human dividends. His motto, that we should "put back more than we take and do good to the Earth, not just no harm," has become a vision for a new industrial revolution that now is reaching many companies beyond his own.

Ethical Considerations

What responsibilities do businesses have to protect the environment or save resources beyond the legal liabilities spelled out in the law? None whatever, according to conservative economist Milton Friedman. In fact, Friedman argues, it would be unethical for corporate leaders to consider anything other than maximizing profits. To spend time or resources doing anything other than making profits and increasing the value of the company is a betrayal of their duty. What do you think? Should social justice, sustainability, or environmental protection be issues of concern to corporations?

We can do all this and at the same time increase profits, reduce taxes, shrink government, increase social spending, and restore our environment, Hawken claims. Recently, Hawken has served as chairperson for The Natural Step in America, a movement started in Sweden by Dr. K. H. Robert, a physician concerned about the increase in environmentally related cancers. Through a consensus process, a group of 50 leading scientists endorsed a description of the living systems on which our economy and lives depend. More than 60 major European corporations and 55 municipalities have incorporated sustainability principles (table 23.5) into their operations.

Another approach to corporate responsibility is called the **triple bottom line.** Rather than reporting only net profits as a

Table 23.5 The Natural Step: System Conditions for Sustainability
1. Minerals and metals from the earth's crust must not systematically increase in nature.
2. Materials produced by human society must not systematically increase in nature.
3. The physical basis for biological productivity must not be systematically diminished.
4. The use of resources must be efficient and just with respect to meeting human needs.

measure of success, ethically sensitive corporations include environmental effects and social justice programs as indications of genuine progress.

Corporations committed to eco-efficiency and clean production include such big names as Monsanto, 3M, DuPont, Duracell, and Johnson and Johnson. Following the famous three Rs—reduce, reuse, recycle—these firms have saved money and gotten welcome publicity. Savings can be substantial. Slashing energy use and redesigning production to use less raw material and to produce less waste is reported to have saved DuPont $3 billion over the past decade, while also reducing greenhouse emissions 72 percent.

Think About It

Most designs for environmental efficiency involve relatively simple rethinking of production or materials. Many of us might be able to save money, time, or other resources in our own lives just by thinking ahead. Think about your own daily life: Could you use new strategies to reduce consumption or waste in recreational activities, cooking, or shopping? In transportation? In housing choices?

Efficiency starts with product design

Our current manufacturing system often is incredibly wasteful. On average, for every truckload of products delivered in the United States, 32 truckloads of waste are produced along the way. The automobile is a typical example. Industrial ecologist, Amory Lovins, calculates that for every 100 gallons (380 l) of gasoline burned in your car engine, only one percent (1 gal or 3.8 l) actually moves passengers. All the rest is used to move the vehicle itself. The wastes produced—carbon dioxide, nitrogen oxides, unburned hydrocarbons, rubber dust, heat—are spread through the environment where they pollute air, water, and soil.

Architect William McDonough urges us to rethink design approaches (table 23.6). In the first place, he says, we should question whether the product is really needed. Could we provide the same service in a more eco-efficient manner? According to McDonough, products should be divided into three categories:

1. *Consumables* are products like food, natural fabrics, or paper that can harmlessly go back to the soil as compost.

2. *Service products* are durables such as cars, TVs, and refrigerators. These products should be leased to the customer to provide their intended service, but would always belong to the manufacturer. Eventually they would be returned to the maker, who would be responsible for recycling or remanufacturing the product. Knowing that they will have to dismantle the product at the end of its life will encourage manufacturers to design for easy disassembly and repair.

3. *Unmarketables* are compounds like radioactive isotopes, persistent toxins, and bioaccumulative chemicals. Ideally, no one would make or use these products. But because eliminating their use will take time, McDonough suggests that in the mean time these materials should belong to the manufacturer

---

Table 23.6 McDonough Design Principles

Inspired by the way living systems actually work, Bill McDonough offers three simple principles for redesigning processes and products:

1. *Waste equals food.* This principle encourages elimination of the concept of waste in industrial design. Every process should be designed so that the products themselves, as well as leftover chemicals, materials, and effluents, can become "food" for other processes.

2. *Rely on current solar income.* This principle has two benefits: First, it diminishes, and may eventually eliminate, our reliance on hydrocarbon fuels. Second, it means designing systems that sip energy rather than gulping it down.

3. *Respect diversity.* Evaluate every design for its impact on plant, animal, and human life. What effects do products and processes have on identity, independence, and integrity of humans and natural systems? Every project should respect the regional, cultural, and material uniqueness of its particular place.

---

and be molecularly tagged with the maker's mark. If they are discovered to be discarded illegally, the manufacturer would be held liable.

Following these principles, McDonough Bungart Design Chemistry has created nontoxic, easily recyclable materials to use in buildings and for consumer goods. Among some important and innovative "green office" projects designed by the McDonough and Partners architectural firm are the Environmental Defense Fund headquarters in New York City, the Environmental Studies Center at Oberlin College in Ohio (see fig. 20.9), the European Headquarters for Nike in Hilversum, the Netherlands, and the Gap Corporate Offices in San Bruno, California (fig. 23.26). Intended to promote employee well-being and productivity as well as eco-efficiency, the Gap building has high ceilings, abundant skylights, windows that open, a full-service fitness center (including pool), and a landscaped atrium for each office bay that brings the outside in. The roof is covered with native grasses.

FIGURE 23.26 The award-winning Gap, Inc. corporate offices in San Bruno, California, demonstrate some of the best features of environmental design. A roof covered with native grasses provides insulation and reduces runoff. Natural lighting, an open design, and careful relation to its surroundings all make this a pleasant place to work.

Warm interior tones and natural wood surfaces (all wood used in the building was harvested by certified sustainable methods) give a friendly feeling. Paints, adhesives, and floor coverings are low toxicity and the building is one-third more energy efficient than strict California laws require. A pleasant place to work, the offices help recruit top employees and improve both effectiveness and retention. As for the bottom line, Gap, Inc. estimates that the increased energy and operational efficiency will have a four- to eight-year payback.

Green consumerism gives the public a voice

Consumer choice can play an important role in persuading businesses to produce eco-friendly goods and services (What Can You Do? at right). Increasing interest in environmental and social sustainability has caused an explosive growth of green products. You can find ecotravel agencies, telephone companies that donate profits to environmental groups, entrepreneurs selling organic foods, shade-grown coffee, efficient houses, paint thinner made from orange peels, sandals made from recycled auto tires, earthworms for composting, sustainable clothing, shoes, rugs, balm, shampoo, and insect repellent. Although these eco-entrepreneurs represent a tiny sliver of the $7-trillion-per-year U.S. economy, they often serve as pioneers in developing new technologies and offering innovative services.

In some industries eco-entrepreneurs have found profitable niches within a larger market. In other cases, once a consumer demand has built up, major companies add green products or services to their inventory. Natural foods, for instance, have grown from the domain of a few funky, local co-ops to a $7 billion market segment. Most supermarket chains now carry some organic food choices. Similarly, natural-care health and beauty products are now more than 10 percent of a $33 billion industry. By supporting these products, you can ensure that they will continue to be available and, perhaps, even help expand their penetration into the market.

Walmart, the large-volume discount price chain, has established a name as the world's largest seller of organic products and compact fluorescent lightbulbs, among other green products. The incentive for Walmart has been that green products and green production processes are often less wasteful, and thus cheaper (when produced in volume), than other products and processes. Does this mean that Walmart has internalized all its costs and produced sustainable relationships with suppliers? Some critics say no, and that Walmart has diluted the idea of sustainable and organic production. But supporters point out that the chain has also helped to legitimate these ideas for the public. Evidently Walmart shoppers also are enthusiastic about contributing to environmental solutions while they shop, because they eagerly buy the organic and energy-efficient products offered.

Environmental protection creates jobs

For years business leaders and politicians have portrayed environmental protection and jobs as mutually exclusive. Pollution control, protection of natural areas and endangered species, and limits on use of nonrenewable resources, they claim, will strangle the

What Can You Do?

Personally Responsible Economy

There are many things that each of us can do to lower our ecological impacts and support green businesses through responsible consumerism and ecological economics.

- Practice living simply. Ask yourself if you really need more material goods to make your life happy and fulfilled.
- Minimize consumption of resources, to save personal and global costs of electricity, gas, metals, plastics, and other resources. Recycle or reuse products, and avoid excessive packaging.
- Look at the amount of your garbage on trash day. Is that the amount of throughput you would like to produce?
- Support environmentally friendly businesses. Consider spending a little more for high-quality, fairly produced goods, at least some of the time. Ask companies what they are doing about environmental protection and human rights.
- Buy green products. Look for efficient, high quality materials that will last and that are produced in the most environmentally friendly manner possible. Subscribe to clean-energy programs if they are available in your area.
- Think about the total life-cycle costs of the things you buy, including environmental impacts, service charges, energy use, and disposal costs as well as initial purchase price.
- Invest in socially and environmentally responsible mutual funds or green businesses when you have money for investment.
- Try making a Kiva or similar micro-loan. You may find that it's fun and educational, and it can feel good to help others.
- Vote thoughtfully. Think carefully about the long-term vs. the short-term social and environmental impacts of economic policies, and work with others in your community to push elected representatives to act in ways that safeguard resources for the generations to come.

economy and throw people out of work. Ecological economists dispute this claim, however. Their studies show that only 0.1 percent of all large-scale layoffs in the United States in recent years were due to government regulations (fig. 23.27). Environmental protection, they argue, is not only necessary for a healthy economic system, it actually creates jobs and stimulates business.

Recycling, for instance, makes more new jobs than extracting virgin raw materials. This doesn't necessarily mean that recycled goods are more expensive than those from virgin resources. We're simply substituting labor in the recycling center for energy and huge machines used to extract new materials in remote places.

Japan, already a leader in efficiency and environmental technology, has recognized the multibillion dollar economic potential of green business. The Japanese government is investing (U.S.) $4 billion per year on research and development that targets seven areas, ranging from utilitarian projects such as biodegradable plastics and heat-pump refrigerants to exotic schemes such as carbon-dioxide-fixing algae and hydrogen-producing microbes.

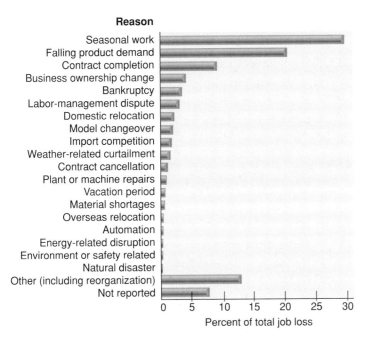

FIGURE 23.27 Although opponents of environmental regulation often claim that protecting the environment costs jobs, studies by economist E. S. Goodstein show that only 0.1 percent of all large-scale layoffs in the United States were the result of environmental laws.
Source: E. S. Goodstein, Economic Policy Institute, Washington, D.C.

Increasingly, people argue that the United States needs a new Apollo Project (like the one that sent men to the moon, but this time focusing on saving planet Earth) to develop renewable energy, break our dependence on fossil fuels, create green jobs, and reinvigorate the economy. The global recession of 2008–2009 strengthened this idea. In 2009 President Barack Obama signed an economic-recovery bill with at least $62 billion in direct spending on green initiatives and $20 billion in green tax incentives. Among the provisions in this bill are $19 billion for renewable energy and upgrading the electrical transmission grid; $20 billion for energy conservation, including weatherizing building and providing efficient appliances; $17 billion for mass transit and advanced automobiles; and $500 million for green jobs programs. More than a million "green collar" jobs could result from these investments. Check out the Apollo Alliance for current news about a new green economy.

Economists report the renewable energy sector already employs more than 2 million workers worldwide. If we were to get half our energy from sustainable sources, it would probably sustain nearly 10 million jobs. Even more people could be employed in energy conservation, ecosystem restoration, and climate remediation programs. Morgan Stanley, a global financial services firm, estimates that global sales from clean energy alone could grow to as much as (U.S.) $1 trillion per year by 2030. Already, authors are rushing books to publication giving advice on how to make a fortune investing in green corporations and renewable energy technologies. For students contemplating career choices, clean energy and conservation could be good areas to explore.

CONCLUSION

At the 1972 Stockholm Conference on the Human Environment, Indira Gandhi, then prime minister of India, stated that "Poverty is the greatest polluter of them all." She meant that the world's poorest people are too often both the victims and the agents of environmental degradation. They are forced to meet short-term survival needs at the cost of long-term sustainability. But "charity is not an answer to poverty," according to Dr. Muhammad Yunus of the Grameen Bank, "It only helps poverty to continue. It creates dependency and takes away individual's initiative ... Poverty isn't created by the poor, it's created by the institutions and policies that surround them ... All we need to do is to make appropriate changes in the institutions and policies, and/or create new ones." The microcredit revolution he started may be the key for breaking the cycle of poverty and changing the lives of the poor.

Economics has given us many tools to understand development, trade, and strategies to improve human well-being. Ecological economics is increasingly finding ways to include natural services, including regulation, provisioning, and aesthetic and cultural accounting, in that accounting. Because the poorest populations often depend directly on environmental services, this new approach could provide real assistance to needy people in developing regions. Emissions trading, green business, fair trade, and other strategies are also being used to aid poor countries.

These strategies also promise to aid wealthier countries by improving efficiency, lowering externalized costs to society, and encouraging the spread of renewable energy and nonpolluting technologies worldwide. Although economists remain divided about the necessity of constant growth, steady-state economies, and the degree to which resources are fixed or flexible, it is increasingly accepted that greater internalization of environmental and social costs are important in any effort toward sustainable development.

REVIEWING LEARNING OUTCOMES

By now you should be able to explain the following points:

23.1 Explain classical and neoclassical perspectives on economy.
- Can development be sustainable?
- Resources can be renewable or nonrenewable.
- Classical economics examines supply and demand.
- Neoclassical economics emphasizes growth.

23.2 Describe ideas of ecological economics.
- Ecological economics assigns cost to ecosystems.
- Ecosystem services include provisioning, regulating, and aesthetic values.

23.3 Describe relationships among population, technology, and scarcity.
- Communal property resources are a classic problem in ecological economics.
- Scarcity can lead to innovation.
- Carrying capacity is not necessarily fixed.
- Economic models compare growth scenarios.

23.4 Evaluate measures of growth.
- GNP is our dominant growth measure.
- Alternate measures account for well-being.
- Cost–benefit analysis aims to optimize resource use.

23.5 Summarize how market mechanisms can reduce pollution.
- Using market forces.
- Is emissions trading the answer?
- Sulfur trading offers a good model.
- Carbon trading is already at work.

23.6 Explain the importance of trade, development, and jobs.
- International trade brings benefits but also intensifies inequities.
- Aid often doesn't help the people who need it.
- Microlending helps the poorest of the poor.

23.7 Evaluate green business.
- New business models follow concepts of ecology.
- Efficiency starts with design of products and processes.
- Green consumerism gives the public a voice.
- Environmental protection creates jobs.

PRACTICE QUIZ

1. Define *economics* and distinguish the emphasis of classical, neoclassical, and ecological economics.
2. Define *resources* and give some examples of renewable, non-renewable, and intangible resources.
3. Describe the relationship between supply and demand.
4. What do we mean by "externalizing costs"? Give several examples.
5. Identify some important ecological services on which our economy depends.
6. Describe how cost–benefit ratios are determined and how they are used in natural resource management.
7. Explain how scarcity and technological progress can extend resource availability and extend the carrying capacity of the environment.
8. Describe how GNP is calculated and explain why this may fail to adequately measure human welfare and environmental quality. Discuss some alternative measures of progress.
9. What is microlending, and what are its benefits?
10. List some of the characteristics of an eco-efficient economic system.

CRITICAL THINKING AND DISCUSSION QUESTIONS

1. When the ecologist warns that we are using up irreplaceable natural resources and the economist rejoins that ingenuity and enterprise will find substitutes for most resources, what underlying premises and definitions shape their arguments?

2. How can intangible resources be infinite and exhaustible at the same time? Isn't this a contradiction in terms? Can you find other similar paradoxes in this chapter?

3. What would be the effect on the developing countries of the world if we were to change to a steady-state economic system? How could we achieve a just distribution of resource benefits while still protecting environmental quality and future resource use?

4. Resource use policies bring up questions of intergenerational justice. Suppose you were asked: "What has posterity ever done for me?" How would you answer?

5. If you were doing a cost–benefit study, how would you assign a value to the opportunity for good health or the existence of rare and endangered species in faraway places? Is there a danger or cost in simply saying some things are immeasurable and priceless and therefore off limits to discussion?

6. If natural capitalism or eco-efficiency has been so good for some entrepreneurs, why haven't all businesses moved in this direction?

 **Data Analysis:** Evaluating Human Development

The human development index (HDI) is a measure created by the United Nations Development Programme to track social progress. HDI incorporates life expectancy, adult literacy, children's education, and standard of living indicators to measure human development. The 2006 report draws on statistics from 175 countries. While there has been encouraging progress in most world regions, the index shows that widening inequality is taking a toll on global human development.

The graph shows trends in the HDI by world region. Study this graph carefully, and answer the following questions: (*Note:* you may have to search online to find some answers.)

1. Which region has the highest HDI rating?

2. What does OECD stand for?

3. Which region has made the greatest progress over the past 30 years, and how much has its HDI increased?

4. Which region has shown the least progress in human development?

5. What historic events could explain the reduction in Europe and the CIS between 1990 and 1995?

6. How much lower is the HDI ranking of sub-Saharan Africa from the OECD?

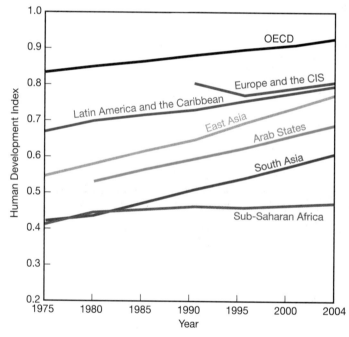

Trends in human development, 1975–2004.
Source: United Nations Development Programme, 2006.

These stockpiled tusks are stored because international agreements make most sales illegal.

Environmental Policy, Law, and Planning

Learning Outcomes

After studying this chapter, you should be able to:

24.1 List several basic concepts in policy.

24.2 Describe some major environmental laws.

24.3 Identify ways that executive, judicial, and legislative bodies shape policy.

24.4 Explain the purposes of international treaties and conventions.

24.5 Outline dispute resolution and planning.

"The power to command frequently causes failure to think."

~ *Barbara Tuchman*

Case Study

Can Policy Protect Elephants?

Elephants are among the most charismatic and fascinating of African animals. Powerful national symbols in many parts of Africa, they also draw millions of tourists and support essential tourism economies. Ecologists prize the elephant as a keystone species, a critical part of its ecosystem. African elephants are also considered an umbrella species: habitat saved for elephants also supports countless other species.

Elephants are also valued for their ivory. Ivory is a luxury commodity that has been traded globally for centuries, but the trade has transformed dramatically with modern guns, growing trade networks, and increasing global wealth. The 1960s–1980s saw unprecedented slaughter of Africa's elephants, mainly for ivory but increasingly for bushmeat, skin, feet, and other parts, as well. In the face of this lucrative global trade, individual countries have been powerless to protect elephant populations. Already by the 1960s, this symbolic species was dwindling and disappearing in many parts of the continent.

African elephants are just one of many species falling to increasingly efficient poaching in the late twentieth century. Indian elephants, South American parrots, and Asian turtles have diminished sharply. Legal but uncontrolled slaughter threatened most of the world's whale species, as well as sea turtles and many species of fish. By the 1950s, people around the world were questioning whether new policies were needed to prevent extinctions. It was clear that international trade was the primary incentive for most wildlife collectors and poachers. So international trade became the focus of efforts to save these species.

The world's first widely adopted biodiversity policy, the Convention on International Trade in Endangered Species (CITES), was drafted at a 1963 meeting of the International Union for Conservation of Nature (IUCN). Because this was a convention (agreement) among countries, representatives of each participating state had to debate the policies and decide whether to ratify (agree to and sign) the convention. The convention was finally adopted, with 80 countries signing on, at a 1973 IUCN meeting, a full decade after its initial drafting.

What exactly does CITES do? The convention marries science and policy to protect biodiversity. It establishes the principle

FIGURE 24.1 African elephants (*Elphas maximus* and *Loxodonta africana*) are among the many species CITES is designed to protect.

that exports and imports of endangered species and their parts should be restricted, that scientific evidence should be used in deciding if a species is endangered, that participating states should establish scientific agencies with the power to monitor threats to species, and that an export permit can be granted only if the scientific agency certifies that the export will not further endanger the species. Export of an individual animal for zoos, for example, is often permitted because just one or two individuals can bring a large sum to an impoverished country. Participating countries also agree to penalize anyone found breaking the rules, and to confiscate illegally traded species or body parts (fig. 24.1).

CITES also establishes a list of species that require monitoring. There are three levels of protection. Species of urgent concern can be traded only in exceptional circumstances and require a permit from both importing and exporting countries (the Appendix 1 species list). Others require export permits but can be traded relatively freely (Appendix II). A few require permits in only a few countries (Appendix III). These lists make it possible to monitor all these species and to follow standard procedures for protecting them.

Establishing and enforcing a new policy is a contentious process. Conflicting economic interests pit ivory marketers against wildlife conservationists and the general public, who tend to be enthusiastic about protecting wildlife. Details require years of bitter negotiation. Establishing independent monitoring and policing agencies supports national pride, but it's also expensive. Enforcing new rules is difficult where established tradition is an unfettered free-for-all.

Despite these inevitable challenges, CITES has been extremely popular. It protects the charismatic and symbolically powerful species that people cherish worldwide, such as India's tigers and China's pandas, as well as Africa's elephants. Most of these species are worth far more as tourist attractions than as body parts, so local support for protection is often strong. There are now 175 signatory nations, all but 20 of the world's independent states. Enforcement varies tremendously, and countries dispute whether or not a species should be added to the endangered list, but CITES now provides some protection to about 30,000 species. Some 900 of these are on the high-concern Appendix I list.

Case Study continued

The African elephant was put on the Appendix II list in 1976. That listing proved insufficient to restrict poaching: ivory is too valuable, and importing countries bore no legal responsibility to help restrict trade. The population plummeted by half between 1979 and 1989, from about 1.2 million to 600,000. Finally in 1989 the African elephant was upgraded to Appendix I. International trade in ivory is allowed only in exceptional circumstances. The hope is that if ivory cannot be sold, it will no longer be profitable to kill elephants for their tusks. Perhaps most important, it has raised international awareness of the precarious state of Africa's elephant populations. Protected populations in many areas have stabilized. In some national parks in southern Africa, populations have rebounded to the extent that parks must periodically cull some animals, which cannot migrate outside park boundaries and which threaten to exceed local carrying capacity.

As with any policy that threatens lucrative trade, CITES's protection for elephants has been controversial. Debates have been long and bitter. Special exemptions to the Appendix I listing have been granted to four African countries (Botswana, Namibia, South Africa, and Zimbabwe) that insist their populations are stable and that they should be permitted to sell some of the ivory stockpiled during culling operations. Nonetheless CITES is an important and largely effective agreement. CITES also set global precedents: among the first broadly accepted environmental policies, the convention predates the U.S. Endangered Species Act by a decade, and many other countries have adopted internal laws following the principles set by CITES.

In this chapter we will examine how environmental policies are formed. We will examine the structures involved in shaping, refining, and enforcing laws, at national, local, and international scales.

For related resources, including Google Earth™ placemarks that show locations where these issues can be seen, visit EnvironmentalScience-Cunningham.blogspot.com.

24.1 BASIC CONCEPTS IN POLICY

The basic idea of **policy** is a statement of intentions and rules that outline acceptable behaviors or accomplish some end. In the case of CITES, these rules were formally stated and voted on, but policy can also be less formally defined. You might have an informal policy always to do your homework; a church might have an open-door policy for visitors; most countries have policies to ensure the welfare of citizens. Laws, such as the U.S. Clean Water Act (chapter 18), are formal statements of national policy. Like the rules that define how you play a board game, policies define agreed-upon limits of behavior. Because it sets the rules we live by, policy making is a contentious and extremely important process.

We will take **environmental policy** to mean official rules and regulations concerning the environment, as well as public opinion that helps shape environmental policy. Ideally, environmental policy serves the needs of human health, economic stability, and ecosystem health. Often these interests have been seen as contradictory, leading to bitter disputes in policy making. Increasingly, however, the public and policy makers acknowledge that these interests overlap. Protecting species, for example, supports the economy and is often a source of pride for communities. Controlling pollution protects human health, saves money in health care, and preserves essential resources.

The drafting of CITES in 1963 was part of a wave of environmental protections that started in many countries in the 1960s and 1970s. Increasing evidence of the damage caused by unconstrained resource use and pollution, following the industrial expansions of World War II, helped foster widespread interest in pollution control, species protection, health and human safety, and other safeguards. Since then global interest in environmental quality has remained high and has even strengthened in recent years, as more people become aware of our dependence on a healthy environment. In a 2007 BBC poll of 22,000 residents of 21 countries, 70 percent said they were personally ready to make sacrifices to protect the environment (fig. 24.2). Overall, 83 percent agreed that individuals would definitely or probably have to make lifestyle changes to reduce the amount of climate-changing gases they

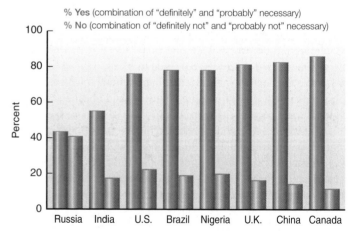

FIGURE 24.2 About 70 percent of the 22,000 people in 21 countries polled by the BBC in 2007 agreed with the statement, "I am ready to make significant changes to the way I live to help prevent global warming or climate change."

produce. Concern about environmental quality varied by country: just over 40 percent of Russians polled were willing to change their lifestyle to prevent global warming, compared to nearly 90 percent of Canadians. The Chinese were the most enthusiastic about energy taxes to prevent climate change. Eighty-five percent of the Chinese polled agreed that such taxes are necessary.

Basic principles guide environmental policy

Protection of fundamental rights is usually one aim of policy making. For many groups these rights include a right to a safe, sustainable environment (fig. 24.3). The 1987 World Commission on Environment and Development, for example, stated, "All human beings have the fundamental right to an environment adequate for their health and well-being." Of the 195 independent nations in the world, 109 now have constitutional provisions for protection of the environment and natural resources. Often such policies are not legally enforceable, but they provide a statement of intentions that can guide subsequent laws.

In theory, democratic societies try to establish policies that are fair to everyone, at least in principle, and that defend everyone's basic needs. Ideally the majority interests are served, but interests of the minority are also defended. Fairness in policy making requires that rules are transparent and decided by public debate and input from many groups in society.

Cost–benefit analysis is often involved in policy making. The aim is to assign standard values, such as dollar value, to competing concerns, then compare the costs and benefits of a plan. This way we can decide if benefits outweigh costs of a policy. In principle this is a utilitarian approach to ensuring an objective, methodical decision process. In practice there are many difficulties in implementing rational cost–benefit analysis:

- Many conflicting values and needs cannot be compared: they are fundamentally different or we lack complete information.

FIGURE 24.3 How do we ensure a safe and healthy environment for everyone? Policies of individuals, communities, and governing institutions are all responsible.

- We often lack agreed-upon broad societal goals; benefits to different groups and individuals often conflict.
- Policy makers may be motivated less by societal goals than by their own interests in power, status, money, or reelection.
- Past investments in existing programs and policies create "path dependence" (a tendency to follow familiar ways) and hidden costs that prevent policy makers from considering good alternatives foreclosed by previous decisions.
- Policy makers often lack the information or models to calculate costs and benefits accurately. Some values, such as health, freedom, or scenery, are very hard to price in dollars.

Another consideration is the **precautionary principle**: when an activity threatens to harm health or the environment, we should fully understand risks before initiating that activity. According to this principle, for example, we shouldn't mass-market new chemicals, new cars, or new children's toys until we're sure they are safe. These are four widely accepted tenets of this principle:

- People have a duty to take steps to prevent harm. If you suspect something bad might happen, you have an obligation to try to stop it.
- The burden of proof of carelessness of a new technology, process, activity, or chemical lies with the proponents, not with the general public.
- Before using a new technology, process, or chemical, or starting a new activity, people have an obligation to examine a full range of alternatives, including the alternative of not using it.
- Decisions using the precautionary principle must be open and democratic and must include the affected parties.

The European Union has adopted this precautionary principle as the basis of its environmental policy. In the United States, opponents of this approach claim that it threatens productivity and innovation. However, many American firms that do business in Europe—including virtually all of the largest corporations—are having to change their manufacturing processes to adapt to more careful E.U. standards. For example, lead, mercury, and other hazardous materials must be eliminated from electronics, toys, cosmetics, clothing, and a variety of other consumer products. A proposal being debated by the E.U. would require testing of thousands of chemicals, cost industry billions of dollars, and lead to many more products and compounds being banned as they are shown to be unsafe to the public. What would you do about this? Is this proposal just common sense, or is it an invitation to decision paralysis?

Corporate money influences policy

Politics is often seen as a struggle for power among competing interest groups, which strive to shape public policy to suit their own agendas. Usually money is a key to political power, so the issue of money in politics is a matter of perennial and contentious debate. Even in a democratic society, money is an important factor in what rules are made, or not made. Often those with the most money can advertise to promote their point of view, draw voters to their side, and win the most legislative seats. Wealthy individuals and corporations can buy influence and friends in government agencies. Large donations are now essential to success in a political campaign, so wealthy donors can influence policy makers with generous funding, or by threatening to withhold funds.

Debates about power and money in politics in the United States have increased since 2010, when the U.S. Supreme Court decided that corporations can spend unlimited amounts of money in political advertising (see section 24.3, below). Because this decision opens elections to potentially vast new infusions of money, many observers expect this decision to shape politics for decades to come. What do you think? Does it matter if an oil company can spend millions on political advertising? Would it matter if the company were a foreign-owned corporation working to shape U.S. laws?

Public awareness and action shape policy

Although power and money are important forces, they can't account for our many policies that serve the public interest. Public participation by scientists and citizens is also an essential force in policy formation. Take the example of CITES, an almost universally adopted set of rules protecting species, mostly for noneconomic purposes. Concerned scientists and communities, as well as policy makers with a strong conscience who objected to excessive poaching, led to the establishment of this convention. Within the United States, civic action has led to many of our strongest environmental and social protections, such as the Clean Water Act, the Clean Air Act, and the Voting Rights Act, which defend the interests of all citizens. Many people consider public citizenship the most important force in a democratic government.

Policy changes often start with protests over environmental contamination, pollution, or resource waste. Media attention helps publicize these protests and the problems they are challenging. One of the most lauded U.S. environmental rules, the Clean Water Act, followed shocking and widely broadcast images of burning of the Cuyahoga River (chapter 18) and the 1969 Santa Barbara oil spill. This oil spill resulted from an oil well blowout near the California coast. An estimated 100,000 barrels (16,000 m^3) of black, gooey crude oil was spilled, and much of it washed up onto beautiful Southern California beaches. Dead birds, dolphins, and sea lions caught public attention, and cleanup crews rallied to try to protect the beaches (fig. 24.4). For weeks, images were broadcast nationwide on TV news. Daily updates were broadcast nationwide, using live satellite feeds, for which the technology had just become available.

These images helped to build national concern for environmental protections. Images of young volunteers, smudged with oil, trying vainly to sweep gooey oil off a beautiful beach, were ideal for TV. Like the Cuyahoga fires, the Santa Barbara oil spill

FIGURE 24.4 Beach cleanup efforts after the Santa Barbara oil spill in 1969 made excellent media material and had an important role in U.S. environmental policy.

played an important role in mobilizing public opinion and was a major factor in passage of the 1972 U.S. Clean Water Act.

Opportunities for public awareness and participation have increased in recent years. There is a growing diversity of print media outlets and online news. YouTube, blogs, and email lists let anyone post images and news. In addition, there are many groups committed to encouraging public involvement in environmental policy. One of these is the Environmental Working Group, which monitors policy and budget issues in Congress. Another is the League of Conservation Voters, which publishes a scorecard for all members of Congress, recording how they voted on environmental laws. You can see how your own representatives are doing by looking at their website: www.lcv.org/scorecard/.

24.2 MAJOR ENVIRONMENTAL LAWS

We depend on many different laws to protect resources such as clean water, clean air, safe food, and biodiversity (table 24.1). Here we'll review a few of these laws, and in the sections that follow we'll examine how laws are created. We will focus mainly on U.S. laws in this section. Many other countries have followed the early lead of the United States in environmental policy formation. In recent years other countries and the international community have increasingly led the way, as global concern for environmental policy has expanded.

Most of the laws we take for granted now are relatively recent. For most of its history, U.S. policy has had a laissez-faire or hands-off attitude toward business and private property. Pollution and environmental degradation were regarded as the unfortunate but necessary costs of doing business. There were some early laws

Table 24.1   Major U.S. Environmental Laws

Legislation	Provisions
Wilderness Act of 1964	Established the national wilderness preservation system.
National Environmental Policy Act of 1969	Declared national environmental policy, required Environmental Impact Statements, created Council on Environmental Quality.
Clean Air Act of 1970	Established national primary and secondary air quality standards. Required states to develop implementation plans. Major amendments in 1977 and 1990.
Clean Water Act of 1972	Set national water quality goals and created pollutant discharge permits. Major amendments in 1977 and 1996.
Federal Pesticides Control Act of 1972	Required registration of all pesticides in U.S. commerce. Major modifications in 1996.
Marine Protection Act of 1972	Regulated dumping of waste into oceans and coastal waters.
Coastal Zone Management Act of 1972	Provided funds for state planning and management of coastal areas.
Endangered Species Act of 1973	Protected threatened and endangered species, directed FWS to prepare recovery plans.
Safe Drinking Water Act of 1974	Set standards for safety of public drinking-water supplies and to safeguard groundwater. Major changes made in 1986 and 1996.
Toxic Substances Control Act of 1976	Authorized EPA to ban or regulate chemicals deemed a risk to health or the environment.
Federal Land Policy and Management Act of 1976	Charged the BLM with long-term management of public lands. Ended homesteading and most sales of public lands.
Resource Conservation and Recovery Act of 1976	Regulated hazardous waste storage, treatment, transportation, and disposal. Major amendments in 1984.
National Forest Management Act of 1976	Gave statutory permanence to national forests. Directed USFS to manage forests for "multiple use."
Surface Mining Control and Reclamation Act of 1977	Limited strip mining on farmland and steep slopes. Required restoration of land to original contours.
Alaska National Interest Lands Act of 1980	Protected 40 million ha (100 million acres) of parks, wilderness, and wildlife refuges.
Comprehensive Environmental Response, Compensation and Liability Act of 1980	Created $1.6 billion "Superfund" for emergency response, spill prevention, and site remediation for toxic wastes. Established liability for cleanup costs.
Superfund Amendments and Reauthorization Act of 1994	Increased Superfund to $8.5 billion. Shares responsibility for cleanup among potentially responsible parties. Emphasizes remediation and public "right to know."

Source: N. Vig and M. Kraft, *Environmental Policy in the 1990s,* 3rd Congressional Quarterly Press.

forbidding gross interference with another person's property or rights—the Rivers and Harbors Act of 1899, for example, made it illegal to dump so much refuse in waterways that navigation was blocked. But in general there were few rules regarding actions on private property, even when those actions impaired the health or resources of neighbors.

Many of these attitudes and rules changed in the 1960s and 1970s, which marked a dramatic turning point in our understanding of the dangerous consequences of pollution. Rachel Carson's *Silent Spring* (1962) and Barry Commoner's *Closing Circle* (1971) alerted the public to the ecological and health risks of pesticides, hazardous wastes, and toxic industrial effluents. Public activism in the civil rights movement and protests against the war in Vietnam carried over to environmental protests and demands for environmental protection.

Rising public concern and activism about environmental issues, such as DDT-poisoned birds (chapter 8), water pollution, and rising smog levels, led to more than 27 major federal laws for environmental protection and hundreds of administrative regulations established in the environmental decade of the 1970s. Among the most important were the establishment of the Environmental Protection Agency (EPA) and the National Environmental Policy Act (NEPA), which requires environmental impact statements for all major federal projects. Because of their power, the EPA and NEPA have both been the targets of repeated attacks by groups that dislike regulations imposed on their pollution emissions or resource uses.

In the initial phase of this environmental revolution, the main focus was on direct regulation and lawsuits to force violators to obey the law. In recent years, attention has shifted to pollution prevention and collaborative methods that can provide win/win solutions for all stakeholders. At the same time, many corporations now recognize that unregulated pollution is unacceptable, and cleaner, more efficient practices are widely understood to be good for profits as well as for the environment.

Environmental, health, and public safety laws, like other rules, impose burdens for some people and provide protections for others. The Clean Water Act, for example, requires that industries take responsibility for treating waste, rather than discharging it into public waters for free. Cities have had to build and maintain expensive sewage treatment plants, instead of discharging sewage into lakes and rivers. These steps internalize costs that previously had been left to the public to deal with. These national laws are intended to protect public health and shared resources for all areas and all citizens. In fact, enforcement varies, but the existence of national, legally enforceable laws allows some recourse for victims when environmental laws are broken.

You can see the text of these laws, together with some explanation, on the EPA's website: www.epa.gov/lawsregs/laws/index.html#env. If you have never examined the text of a law, you should take a look at these.

NEPA (1969) establishes public oversight

Signed into law by President Nixon in 1970, the **National Environmental Policy Act (NEPA)** is the cornerstone of U.S. environmental policy.

NEPA does three important things: (1) it authorizes the Council on Environmental Quality (CEQ), the oversight board for general environmental conditions; (2) it directs federal agencies to take environmental consequences into account in decision making; and (3) it requires an **environmental impact statement (EIS)** be published for every major federal project likely to have an important impact on environmental quality (fig. 24.5). NEPA doesn't forbid environmentally destructive activities if they comply otherwise with relevant laws, but it demands that agencies admit publicly what they plan to do. Once embarrassing information is

FIGURE 24.5 Every major federal project in the United States must be preceded by an Environmental Impact Statement.

revealed, however, few agencies will bulldoze ahead, ignoring public opinion. And an EIS can provide valuable information about government actions to public interest groups, which wouldn't otherwise have access to this information.

What kinds of projects require an EIS? The activity must be federal and it must be major, with a significant environmental impact. Evaluations are always subjective as to whether specific activities meet these characteristics. Each case is unique and depends on context, geography, the balance of beneficial versus harmful effects, and whether any areas of special cultural, scientific, or historical importance might be affected. A complete EIS for a project is usually time-consuming and costly. The final document is often hundreds of pages long and generally takes six to nine months to prepare. Sometimes just requesting an EIS is enough to sideline a questionable project. In other cases, the EIS process gives adversaries time to rally public opposition and information with which to criticize what's being proposed. If agencies don't agree to prepare an EIS voluntarily, citizens can petition the courts to force them to do so.

Every EIS must contain the following elements: (1) purpose and need for the project, (2) alternatives to the proposed action (including taking no action), (3) a statement of positive and negative environmental impacts of the proposed activities. In addition, an EIS should make clear the relationship between short-term resources and long-term productivity, as well as any irreversible commitment of resources resulting from project implementation.

Many lawmakers in recent years have tried to ignore or limit NEPA in forest policy, energy exploration, and marine wildlife protection. The "Healthy Forest Initiative," for example, eliminated public oversight of many logging projects by bypassing EIS reviews and prohibiting citizen appeals of forest management plans (chapter 12). Similarly, when the Bureau of Land Management proposed 77,000 coal-bed methane wells in Wyoming and Montana, promoters claimed that water pollution and aquifer depletion associated with this technology didn't require environmental review (chapter 19). And in the 2005 Energy Bill, Congress inserted a clause that exempts energy companies from NEPA requirements in a number of situations, with the aim of speeding energy development on federal land.

The Clean Air Act (1970) regulates air emissions

The first major environmental legislation to follow NEPA was the Clean Air Act (CAA) of 1970. Air quality has been a public concern at least since the beginning of the industrial revolution, when coal smoke, airborne sulfuric acid, and airborne metals such as mercury became common in urban and industrial areas around the world (fig. 24.6). Sometimes these conditions produced public health crises: one infamous event was the 1952 Great Smog of London, several days of cold, still weather that trapped coal smoke in the city and killed some 4,000 people from infections and asphyxiation. Another 8,000 died from respiratory illnesses in the months that followed (chapter 16).

FIGURE 24.6 Severely polluted air was once normal in cities. The Clean Air Act has greatly reduced the health and economic losses associated with air pollution.

Although crises of this magnitude have been rare, chronic exposure to bad air has long been a leading cause of illness in many areas. The Clean Air Act provided the first nationally standardized rules in the United States to identify, monitor, and reduce air contaminants. The core of the act is an identification and regulation of seven major "criteria pollutants," also known as "conventional pollutants." These seven include sulfur oxides, lead, carbon monoxide, nitrogen oxides (NO_x), particulates (dust), volatile organic compounds, and metals and halogens (such as mercury and bromine compounds). Recent revisions to the Clean Air Act require the EPA to monitor carbon dioxide, our dominant greenhouse gas, which endangers the public through heat stress, drought, and other mechanisms (chapters 15, 16).

Most of these pollutants have declined dramatically since 1970. An exception is NO_x, which derives from internal combustion engines such as those in our cars. Further details on these pollutants are given in chapter 16.

The Clean Water Act (1972) protects surface water

Water protection has been a goal with wide public support, in part because clean water is both healthy and an aesthetic amenity. The act aimed to make the nation's waters "fishable and swimmable," that is, healthy enough to support propagation of fish and shellfish that could be consumed by humans, and low in contaminants so that they were safe for children to swim and play in them.

The first goal of the Clean Water Act (CWA) was to identify and control point source pollutants, end-of-the-pipe discharges from factories, municipal sewage treatment plants, and other sources. Discharges are not eliminated, but water at pipe outfalls must be tested, and permits are issued that allow moderate discharges of low-risk contaminants such as nutrients or salts. Metals, solvents, oil, high counts of fecal bacteria, and other more serious contaminants must be captured before water is discharged from a plant.

By the late 1980s, point sources were increasingly under control, and the CWA was used to address nonpoint sources, such as runoff from urban storm sewers. The act has also been used to promote watershed-based planning, in which communities and agencies collaborate to reduce contaminants in their surface waters. As with the CAA, the CWA provides funding to aid pollution-control projects. Those funds have declined in recent years, however, leaving many municipalities struggling to pay for aging and deteriorating sewage treatment facilities. For more details on the CWA and water pollution control, see chapter 18.

The Endangered Species Act (1973) protects wildlife

While CITES (opening case study) aims to provide international protections for species, the Endangered Species Act (ESA) regulates species within the United States. The ESA provides a structure for identifying and listing species that are vulnerable, threatened, or endangered. Once a species is listed as endangered, the ESA provides rules for protecting it and its habitat, ideally to make recovery possible (fig. 24.7). Listing of a species can be a controversial process because habitat conservation can get in the way of land development. Many ESA controversies arise when developers want to put new housing developments in scenic areas where the last remnants of a species occur. In many cases, however, disputes have been resolved by negotiation and more creative design of developments, which can sometimes allow both for development and for species protection.

There is considerable collaboration between CITES and the ESA in listing and monitoring endangered species. The ESA maintains a worldwide list of endangered species, as well as a U.S. list. The ESA also provides for grants and programs to help land owners protect species. The responsibility for studying and attempting to restore threatened and endangered species lies mainly with the Fish and Wildlife Service and the National Oceanic and Atmospheric Administration. You can read more about endangered species, biodiversity, and the ESA in chapter 11.

FIGURE 24.7 The Endangered Species Act is charged with protecting species and their habitat. The gray wolf has recovered in most of its range because of ESA protection.

The Superfund Act (1980) lists hazardous sites

Most people know this law as the Superfund Act because it created a giant fund to help remediate abandoned toxic sites. The proper name of this law is informative, though: the Comprehensive Environmental Response, Compensation, and Liability Act (CERCLA). The act aims to be comprehensive, addressing abandoned sites, emergency spills, or uncontrolled contamination, and it allows the EPA to try to establish liability, so that polluters help to pay for cleanup. It's much cheaper to make toxic waste than to clean it up, so we have thousands of chemical plants, gas stations, and other sites that have been abandoned because they were too expensive to clean properly. The EPA is responsible for finding a contractor to do cleanup, and the Superfund was established to cover the costs, which can be in the billions of dollars. Until recently, the fund was financed mainly by contributions from industrial producers of hazardous wastes. In the 1990s, however, Congress voted to end that source, and the Superfund was allowed to dwindle to negligible levels. Site cleanup is now funded by taxpayer dollars.

According to the EPA, one in four Americans lives within 3 miles of a hazardous waste site. The Superfund program has identified more than 47,000 sites that may require cleanup. The most serious of these (or the most serious for which proponents have been sufficiently vigorous) have been put on a National Priorities List. About 1,600 sites have been put on this list, and about 700 cleanups have been completed. The total cost of remediation is thought to be something between $370 billion and $1.7 trillion. To read more, see chapter 21.

24.3 HOW ARE POLICIES MADE?

The general process by which policies are developed is often called the *policy cycle*, because rules are developed, enacted, and revised repeatedly (fig. 24.8). The cycle starts when a problem is

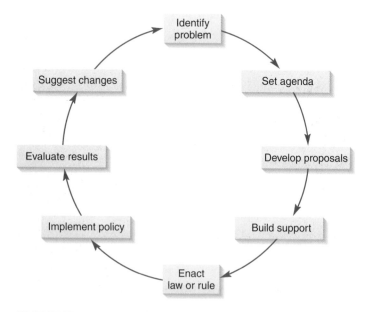

FIGURE 24.8 The policy cycle.

FIGURE 24.9 Making a ruckus on behalf of environmental protection can attract attention to your cause.

identified as a priority. In the case of endangered species protections (both CITES and the ESA), for example, the public became concerned about accelerating damage to species and ecosystems. Citizen groups and wildlife advocates initiated public debates on the issue. Communities, students, and environmental groups worked to organize stakeholders, choose tactics, and aggregate related issues into a case for species preservation that was of interest to a broad range of groups and communities.

Citizens often draw attention to their cause by organizing protests, marches, street theater, or other kinds of public events (fig. 24.9). Getting involved in local election campaigns, writing letters, or making telephone calls to legislators also influences the decision process. You'd be surprised at how few letters or calls legislators receive from voters, even on important national issues. Your voice can have an important impact (see chapter 25).

Once an issue is defined and support has been gathered, the next stage in the policy cycle is to propose a new law or rule. The rule is debated and negotiated. Media campaigns, public education, and personal lobbying of decision makers are needed to build support for a new policy. Getting a proposal enacted as a law or a rule takes persistence, negotiation, and usually years of effort.

After a rule or law is enacted, continued public oversight and monitoring are usually needed to ensure that government agencies faithfully carry out policies. Measuring impacts and results are also essential, so that amendments can be proposed, if necessary, to make the policy fairer or more effective.

Each branch of government plays a role in establishing laws. **Statute law** consists of documents or laws ("statutes") voted on and enacted by the legislative branch of government. **Case law** is derived from law suits (legal cases) in court. **Administrative law** rises from executive orders, administrative rules and regulations. Because every country has different legislative and legal processes, this discussion focuses on the U.S. system, which shares many general similarities with other democratic systems.

Congress and legislatures vote on statutory laws

Elected legislative bodies, such as Congress, state legislatures, or town councils, debate and vote on policies that become legally enforceable laws. Federal laws (statutes) are enacted by Congress and signed by the president. They originate as legislative proposals called "bills", which are usually drafted by the congressional staff, often in consultation with representatives of various interest groups.

Thousands of bills are introduced every year in Congress. Some are very narrow, providing funds to build a specific section of road or to help a particular person, for instance. Others are extremely broad, perhaps overhauling the social security system or changing the entire tax code.

Bills Move Slowly Through Congress

After a bill is introduced, it goes to a committee, where it is discussed and debated. Most hearings take place in Washington, but if the bill is controversial or legislators want to attract publicity for themselves or the issue, they may conduct field hearings closer to the site of the controversy. The public often has an opportunity to give testimony at field hearings (fig. 24.10). Elected officials may be swayed by public opinion, and they need public support in policy making.

The language of a bill is debated, revised, and negotiated in a committee until it is considered widely acceptable enough to send to the full House of Representatives or Senate. Compromises are necessary to make the bill acceptable to different parties. The House has one version of the bill, which is debated on the floor of the House. The Senate has another version that it debates. Often there are further amendments at these stages.

By the time an issue has passed through both the House and Senate, the versions approved by the two bodies are likely to be different. They go then to conference committee to iron out any differences between them. After going back to the House and Senate for confirmation, the final bill goes to the president, who may either sign it into law or veto it. If the president vetoes the bill, it may still become law if two-thirds of the House and Senate vote to override the veto. If the president takes no action within ten days of receiving a bill from Congress, the bill becomes law without his signature.

Each step of this convoluted process is published in print and online in the *Congressional Quarterly Weekly,* which you can access at any time through the official congressional website, thomas.gov. It can take a little practice to find the legislation you want, but it's a rich repository of public records.

Legislative Riders

There are two types of legislation: authorizing bills become laws, while appropriation bills provide the money for federal agencies and programs. Eliminating funding for an agency in an appropriations bill is often an effective way to prevent laws from being implemented. Appropriation bills are not supposed to make policy, but merely to fund existing plans and projects. Legislators who can't muster enough votes to pass pet projects through regular channels often will try to add authorizing amendments called **riders** into completely unrelated funding bills. Even if they oppose the riders, other members of Congress have a difficult time voting against an appropriation package for disaster relief or to fund programs that benefit their districts. Often this happens in conference committee because when the conference report goes back to the House and Senate, the vote is either to accept or reject with no opportunity to debate or amend further.

Starting with the 104th Congress (in 1995) industry groups began using this tactic to roll back environmental protections and gain access to natural resources. For instance, riders that put a moratorium on listing additional species under the Endangered Species Act were attached to 1996 spending bills that exempted "salvage" (postfire) logging on public lands from environmental laws. Subsequently, riders have been attached to many appropriation bills. The 2004 Omnibus spending bill, for example, included numerous amendments to prevent administrative appeals and judicial reviews of environmentally destructive government policies, allow increased logging and road building in Alaska's Tongass National Forest, cut funding for land conservation, weakened national organic labeling standards, and expanded forest-thinning projects. Generally riders are tacked onto completely unrelated bills that legislators will have difficulty voting against. A rider to eliminate critical habitat for endangered species, for example, was hung on a veteran's health care bill.

Lobbying Influences Government

Groups or individuals with an interest in pending legislation can often have a great deal of influence through lobbying, or visiting congressional offices, talking directly with representatives, and using personal contacts to persuade elected representatives to vote in their favor. The term *lobbying* derives from the habit of people waiting in hallways and lobbies of Congress to catch the elbow of a passing legislator and plead their case.

FIGURE 24.10 Citizens line up to testify at a legislative hearing. By getting involved in the legislative process, you can be informed and have an impact on governmental policy.

Citizens often make trips to Washington—or to state capitals, county seats, or city halls—to try to personally persuade elected officials on upcoming votes. This direct contact is a basic part of the democratic process, but it can sometimes work unevenly because most people can't abandon work or school and fly to Washington to lobby.

Not everyone can go to Washington, but many people join organizations that can collectively send representatives, or hire professional lobbyists, to make sure their message is heard. Most major organizations now have lobbyists in Washington. The biggest single citizen lobbying group is probably the American Association of Retired Persons (AARP), which actively lobbies for issues considered of interest to senior citizens. The National Rifle Association (NRA), Union of Concerned Scientists, and many other groups participate in lobbying. Environmental organizations such as the Natural Resources Defense Council, Audubon, and the Sierra Club lobby on many environmental bills. Lobbyists and volunteer activists attend hearings, draft proposed legislation, and meet with officials. The range of interests involved in lobbying is astounding. Business organizations, workers, property owners, religious and ethnic groups, are all there. Walking the halls of Congress, you see an amazing mixture of people attempting to be heard.

One group of professional lobbyists who have been in the news are the K Street lobbyists. These lobbying firms hire powerful lawyers, former Senators, military officers, and others, and their offices are concentrated on Washington D.C.'s K Street. (Much of Washington's street grid is named by letters, such as A, B, C, and so on.) Former Congress members and military personnel are valuable because they have personal ties that can be a great aid in catching the ear of voting members of Congress. In the lobbying world, K street has garnered special attention because it is where the big industry groups set up shop. These groups have especially large rewards to reap through lobbying. The *Washington Post* reported on a case in which a group of corporations invested $1.6 million lobbying for a special low overseas tax rate, and the effort saved them over $100 billion in tax payments. In another case, the Carmen Group, a lobbying firm, charged $500,000 to lobby for insurance claims following the September 11, 2001, attacks on the World Trade Center, and as a result the government agreed to cover $1 billion in insurance premiums for its clients.

Often lobbyists write the bills that a legislator introduces. This ensures favorable terms for their clients. Consequently, the gains to be won through lobbying have grown into billions of dollars per year. The number of lobbyists registered in Washington more than doubled between 2000 and 2005, from 16,000 to almost 35,000. The biggest industry lobbying firms, such as The Federalist Group, can charge $20,000 to $40,000 per month for their services.

Lobbying is, by its nature, about tipping the tables in the favor of an interest group. But lobbying is also something that many people see as necessary, as part of getting voices heard in a democratic process. What do you think? Is corporate lobbying important? Is it necessary? How would you distinguish the actions of an oil industry's lobbyists from those of an environmental or community group's lobbyists?

Judges decide case law

Often environmental policies are established when groups bring complaints to the courts, involving damage to property or health, failure to enforce existing laws, or infringement of rights. Judges or juries decide these cases by determining whether written law (statutes) or customary law (common law) has been violated. The body of legal opinions built up by many court cases is called **case law.**

The United States is divided into 96 federal court districts, each of which has at least one trial court. Disputes about procedural issues and interpretations of the law in district courts are sent on to one of 12 regional appeals courts. Cases might involve criminal prosecutions, claims against the federal government, or complaints in which parties come from multiple states. Each state has its own courts that generally parallel the federal system. These courts decide cases involving state laws.

Legislation is often written in vague and general terms, which help make a bill widely enough accepted to gain passage. Congress often leaves it to the courts to "fill in the gaps," especially in environmental laws. As one senator said when Congress was about to pass the Superfund legislation, "All we know is that the American people want these hazardous waste sites cleaned up … Let the courts worry about the details."

The Supreme Court Decides Major Cases

Lawsuits with very far-reaching implications are decided by the United States Supreme Court, a group of nine justices whose job is to judge whether a law is consistent with the U.S. Constitution, or whether a policy is consistent with a law as written by Congress. States also have Supreme Courts for deciding cases at the state level. The Supreme Court can only study and judge a few cases a year, and for many people the Court and its actions are little known and poorly understood. But this is a body that makes pivotal and far-reaching decisions, many of them affecting our national environmental policies (fig. 24.11).

Perhaps the most sweeping rule change the Supreme Court has made in recent years was the 2010 decision in *Citizens United v. Federal Election Commission*. For nearly a century, anticorruption laws had limited corporate and union spending on political campaigns, generally on the grounds that corporations have more money than individuals, which gives them unfair influence in elections. In a hotly debated, 5–4 split vote, the Court decided that these laws limited free speech (in the form of political advertising), which is protected under the Constitution. The five-person majority argued that because U.S. law gives corporations the same rights and protections as individual people, corporate advertising could not be limited. Within months of the decision, a coalition of oil, gas, and coal corporations announced that they would substantially increase their campaign contributions in the next congressional races.

Many commentators, including dissenting members of the Supreme Court, argue that the *Citizens United* decision protects corporations from a wide range of public oversight and legal restrictions. Among other concerns, this decision opens the door to new levels of corporate influence in justice, as well as politics.

FIGURE 24.11 The Supreme Court decides pivotal cases, many of them bearing on natural resources or environmental health.

For example, a coal company could easily fund a judge's election campaign. If a community filed a suit against the coal company for polluting streams, or for unsafe working conditions, could they be sure to receive a fair judgment?

A dissenting opinion from the four justices who voted against the *Citizens United* decision stated that "corporations have no consciences, no beliefs, no feelings, no thoughts, no desires … and their 'personhood' often serves as a useful legal fiction. But they are not themselves members of 'We the People' by whom and for whom our Constitution was established." The dissent states further, "The fact that corporations are different from human beings might seem to need no elaboration, except that the majority opinion almost completely elides it."

In another 2010 case, *Monsanto v. Geertson Seed Farms*, the Court overturned a ban on planting genetically modified, herbicide-tolerant alfalfa. Because alfalfa is the fourth-largest crop in the United States, and because genetically modified plants can readily contaminate fields of organic alfalfa used to feed organic dairy cattle, this decision is likely to have far-reaching impacts on farming, organic foods, and other environmental issues.

Legal Standing

Before a trial can start, the litigants must establish that they have **standing,** or a right to stand before the bar and be heard. The main criteria for standing is a valid interest in the case. Plaintiffs must show that they are materially affected by the situation they petition the court to redress. This is an important point in environmental cases. Groups or individuals often want to sue a person or corporation for degrading the environment. But unless they can show that they personally suffer from the degradation, courts are likely to deny standing.

In a landmark 1969 case, *Sierra Club v. Morten*, the Sierra Club challenged a decision of the Forest Service and the Department of the Interior to lease public land in California to Walt Disney Enterprises for a ski resort. The land in question was a beautiful valley that cut into the southern boundary of Sequoia National Park. Building a road into the valley would have necessitated cutting down a grove of giant redwood trees within the park. The Sierra Club argued that it should be granted standing in the case to represent the trees, animals, rocks, and mountains that couldn't defend their own interests in court. After all, the club pointed out corporations—such as Disney Enterprises—are treated as persons and represented by attorneys in the courts. Why not grant trees the same rights?

The case went all the way to the Supreme Court, which ruled that the Sierra Club failed to show that its members would be materially affected by the development. However, the Court established a key precedent in stating that "aesthetic and environmental wellbeing, like economic wellbeing, are important ingredients of the quality of life" and are "deserving of legal protection."

Criminal Law Prosecutes Lawbreakers

Violation of many environmental statutes constitutes criminal offenses. In 1975 the U.S. Supreme Court ruled that corporate officers can be held criminally liable for violations of environmental laws if they were grossly negligent, or the illegal actions can be considered willful and knowing violations. In 1982 the EPA created an Office for Criminal Investigation. Under the Clinton administration, prosecutions for environmental crimes rose to nearly 600 per year. They fell by 75 percent under George W. Bush, however. The Obama administration has again increased new prosecutions to about 400 per year.

Civil law regulates relations between individuals or between individuals and corporations (which have the rights of individuals under U.S. law). Property rights and personal dignity and freedom are protected by civil law. Sometimes legislative statutes, such as the Civil Rights Act, establish specific aspects of civil law. Custom and previous court decisions, collectively called **common law,** can also establish precedents that constitute a working definition of individual rights and responsibilities.

Criminal offenses can lead to jail, while civil cases lead only to fines. Civil judgments can be costly, however. A group of Alaskan fishermen won $5 billion from the Exxon oil company for damages caused by the 1989 *Exxon Valdez* oil spill. In 2000 the Koch oil company, one of the largest pipeline and refinery operators in the United States, agreed to pay $35 million in fines and penalties to state and federal authorities for negligence in more than 300 oil spills in Texas, Oklahoma, Kansas, Alabama, Louisiana, and Missouri between 1990 and 1997. Koch also agreed to spend more than $1 billion on cleanup and improved operations.

Sometimes the purpose of a civil suit is to prevent harmful actions. You might ask the courts, for example, to order the government to cease and desist from activities that are in violation of either the spirit or the letter of the law. Public interest groups have often asked courts to stop logging and mining operations, to enforce implementation of laws regarding endangered species, public health, air and water pollution laws.

Lawsuits can also be used to stop public interest groups from challenges to industry. Because defending a lawsuit is so expensive, the mere threat of litigation can be a chilling deterrent. Increasingly environmental activists are being harassed with

strategic lawsuits against public participation (SLAPP). Citizens who criticize businesses that pollute or government agencies that neglect their responsibility to protect the public are often sued in retaliation.

Most of these preemptive strikes are groundless and ultimately dismissed, but defending yourself against them can be cripplingly expensive and they can halt progress on the issue in question. Public interest groups and individual activists—many of whom have little money to defend themselves—often are intimidated from taking on polluters. For example, when a West Virginia farmer wrote an article about a coal company's pollution of the Buckhannon River, the company sued him for $200,000 for defamation. Similarly, citizen groups fighting a proposed incinerator in upstate New York were sued for $1.5 million by their own county governments. A Texas woman called a nearby landfill a dump—and her husband was named in a $5 million suit for failing to "control his wife."

Executive agencies make rules and enforce laws

More than 100 federal agencies and thousands of state and local boards and commissions oversee environmental policies. They set rules, adjudicate disputes, and investigate misconduct.

In the federal government, the president's cabinet (a group of department heads in charge of various tasks) includes the heads of major departments such as Agriculture, Interior, or Justice (fig. 24.12). Rules that these agencies make and enforce decide many of our most important environmental, resource, and health issues.

The public can influence agency rule-making by giving comments on a proposed rule. Comments can be made at regional hearings, which agencies are required to offer, and by mail or email during a "public comment period," usually a few months, that precedes the acceptance of a new rule or action. Often thousands of comments are collected, and these can substantially influence whether or not a proposed rule or plan is enacted.

Executive orders also can be powerful agents for change. In 1994, for instance, President Bill Clinton issued Executive Order 12898 requiring all federal agencies to collect data on effects of pollution on minorities, and to develop strategies to promote environmental justice. The availability of this information allows the public to see and debate matters that are otherwise hidden from view. During his two terms in office, Clinton used the Antiquities Act to establish 22 new national monuments (fig. 24.13). In addition, he expanded dozens of existing

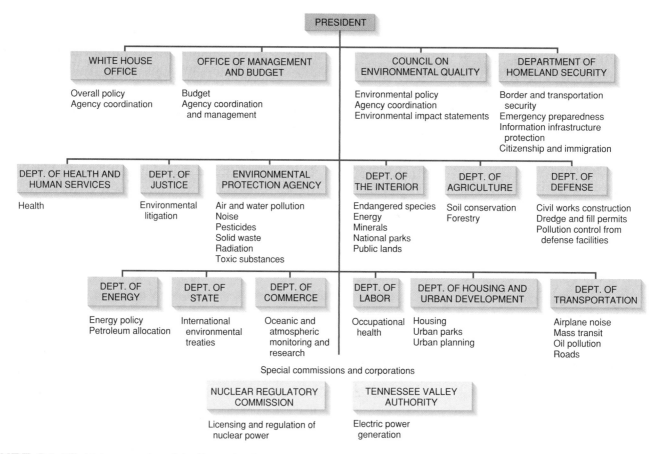

FIGURE 24.12 Major agencies of the Executive Branch of the U.S. federal government with responsibility for resource management and environmental protection.
Source: U.S. General Accounting Office.

FIGURE 24.13 Many national monuments, including the U.S. Virgin Islands Coral Reef Monument shown here, are created by executive rules.

national parks and wildlife refuges. Altogether, Clinton ordered protection for about 36 million ha (90 million acres) of nature preserves, the largest of which was the Pacific Ocean reserve composed of 34 million ha of ocean and coral reefs northwest of Hawaii.

Rules and policies made by executive decree in one administration can be quickly undone in the next one. In his first day in office, President George W. Bush ordered all federal agencies to suspend or ignore more than 60 rules and regulations from the Clinton administration. In addition, President Bush called for a sweeping overhaul of environmental laws to ease restrictions on businesses and to speed decisions on development projects, and he prevented implementation of many other environmental rules. Because most of this agenda was pursued through agency regulations and executive orders, most Americans, distracted by fears of terrorism and lingering wars in Afghanistan and Iraq, were unaware of the magnitude and implications of this abrupt policy shift. Barack Obama, in turn, reversed many of President Bush's executive rules, restoring environmental and social protections.

An important administrative rule that has been reversed repeatedly by these presidents is the moratorium (ban) on building new logging roads in nearly 24 million ha of roadless public land. This rule, also established under Clinton, defends wild areas that have no other legal protection (*de facto* wilderness). The pro-industry Bush administration reversed Clinton's rules; the pro-conservation Obama administration then reversed Bush's reversals.

Regulatory Agencies

The EPA is the primary agency with responsibility for protecting environmental quality. It was created in 1970, at the same time as NEPA, and its head is appointed by the president as part of his cabinet, the small body of top administrators who work directly with the president. Like other cabinet positions, the EPA is strongly influenced by political winds. EPA funding, enforcement activity, and rule-making strongly reflect the views of the president's political advisers. Under the Nixon and Carter administrations, the EPA grew rapidly and enforced air and water quality standards vigorously. EPA activity and funding declined sharply during the Reagan administration, recovered under Bill Clinton, declined again during the Bush administrations, and rose again under Obama.

The Departments of the Interior and Agriculture manage natural resources. The National Park Service, which is responsible for more than 376 national parks, monuments, historic sites, and recreational areas, is part of the Interior Department. Other Interior agencies are the Bureau of Land Management (BLM), which administers some 140 million ha of land, mostly in the western United States, and the Fish and Wildlife Service, which operates more than 500 national wildlife refuges and administers endangered species protection.

The Department of Agriculture is home to the U.S. Forest Service, which manages about 175 national forests and grasslands, totaling some 78 million ha (fig. 24.14). With 39,000 employees, the Forest Service is nearly twice as large as the EPA. The Department of Labor houses the Occupational Safety and Health Agency (OSHA), which oversees workplace safety, including exposure to toxic substances. In addition, several independent agencies that are not tied to any specific department also play a role in environmental protection and public health. The Consumer Products Safety Commission passes and enforces regulations to protect consumers, and the Food and Drug Administration is responsible for the purity and wholesomeness of food and drugs.

All of these agencies have a tendency to be "captured" by the industries they are supposed to be regulating. Many of the people with expertise to regulate specific areas came from the industry or sector of society that their agency oversees. Furthermore, the

FIGURE 24.14 Smokey Bear symbolizes the Forest Service's role in extinguishing forest fires.

FIGURE 24.15 When a record-breaking spate of tornadoes destroyed communities in several states in the spring of 2011, it was clear that government agencies were needed to aid recovery.

people they work most closely with and often develop friendships with are those they are supposed to watch. And when they leave the agency to return to private life—as many do when the administration changes—they are likely to go back to the same industry or sector where their experience and expertise lies. The effect is often what's called a "revolving door," where workers move back and forth between industry and government. As a result, regulators often become overly sympathetic with and protective of the industry they should be overseeing.

How much government do we want?

In his 1981 inaugural address, President Ronald Reagan famously said, "Government is not the solution to our problem; government is the problem." In this, he invoked a perennial debate in American politics: Is government a power that undermines personal liberties? Or is government a representative of the people and a defender of personal liberties against bullies?

The answer sometimes depends on when you ask. During political campaigns, many of us decry the size and cost of government agencies. But in a crisis most of us assume the government will be there to help us out, as it did after a record-breaking spate of tornadoes swept from Mississippi to Georgia in 2011. Within hours, affected states welcomed federal rescue teams, emergency aid, and federal funding for reconstruction (fig. 24.15).

Debates about the proper size and role of government are common. We value self-reliance and rugged individualism. Yet we also want someone to protect us from contaminated food and drugs, to educate our children, and to provide roads, bridges, and safe drinking water. President Reagan was among those who favor "free market" capitalism, with businesses unfettered by rules such as the Clean Air Act or the Clean Water Act. Political strategist Grover Norquist, president of the Americans for Tax Reform, famously said he'd like to "shrink the government down to the size where we can drown it in the bathtub." Other observers note that while Reagan's language focused on freedom for individuals

and small business owners, the elimination of public health and safety regulations tends to undermine the interests of individuals and small businesses. Advantages are often given to the biggest players in a rule-less game.

Part of the reason these disputes persist may be that both views are partially correct. Regulations, such as those imposing expensive pollution abatement technologies for polluters, require private businesses to bear the cost of protecting public resources. Businesses are squeezed between shareholders' demands for ever-higher profits and agency demands for safer, sometimes costly, operating standards. Viewed another way, regulations require businesses to clean up their own messes. Opponents of agency regulations point out that corporations produce the economic vitality on which our prosperity depends. Proponents of regulations point out that corporations couldn't prosper without subsidies, tax breaks, transportation infrastructure, and a healthy and educated workforce. These costs are necessary to doing business but are normally external to the accounting of business costs and profits.

Since about 1981 the small-government philosophy has dismantled much of the regulatory structure set up during President Nixon's term in office. Often this has been done by agency heads, who have been appointed despite openly opposing the existence of those agencies and their laws. For example, President George W. Bush appointed Christopher Cox, a proponent of bank deregulation, to chair the Securities and Exchange Commission, which oversees Wall Street trading. Subsequent dismantling of trading rules led to risky behavior by banks, which culminated in the Wall Street collapse in 2007–2008. Business failures and high unemployment spread nationwide and have lasted for years. President Bush also oversaw dramatic reductions in USDA food safety inspections, on the grounds that they represented unnecessary interference in the private business of the food industry. Increasingly frequent food contamination scares have made many Americans rethink the importance of government inspectors in the food system.

These debates probably will always be with us. Which view is most correct depends on many factors: your interest group, life experience, philosophical perspective, and economic position, the time frame you analyze, and other priorities. What factors influence your view on these issues? Do you think there is room for compromise? If not, why not? If yes, where?

24.4 INTERNATIONAL CONVENTIONS

Growing interconnections in our global environment and economy have made nations increasingly interested in signing on to international agreements (conventions) for environmental protection. A principal motivation for participating in these treaties is the recognition that countries can no longer act alone to protect their resources and interests. Water resources, the atmosphere, trade in endangered species, and many other concerns cross international boundaries. Over time the number of parties taking part in negotiations has grown sharply (fig. 24.16). The speed at which agreements take force has also increased. The Convention on

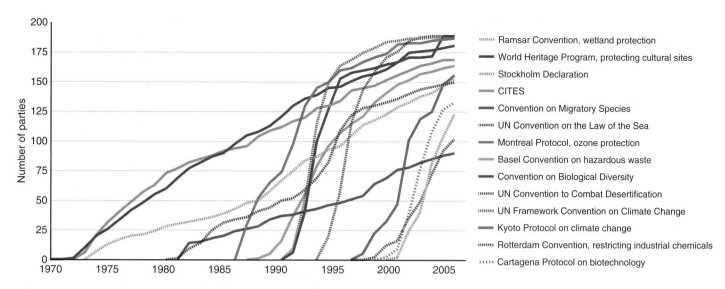

FIGURE 24.16 Major international environmental agreements, listed in order of ratification dates.

International Trade in Endangered Species (CITES), for example, was not enforced until 14 years after its ratification in 1973, but the Convention on Biological Diversity (1991) was enforceable after just one year, with 160 nations signing the agreement just four years after its introduction.

Over the past 25 years, more than 170 treaties and conventions have been negotiated to protect our global environment. These agreements have focused on concerns ranging from intercontinental shipping of hazardous waste, to deforestation, overfishing, trade in endangered species, global warming, and wetlands protection.

Major International Agreements

International accords and conventions have emerged slowly but fairly steadily from meetings such as those in Stockholm and Rio (table 24.2). A few of the important benchmark agreements are discussed here.

The **Convention on International Trade in Endangered Species** (CITES, 1973) declared that wild flora and fauna are valuable, irreplaceable, and threatened by human activities. To protect disappearing species, CITES maintains a list of threatened and endangered species that may be affected by trade. As with most international agreements, this one takes no position on movement or loss of species within national boundaries, but it establishes rules to restrict unauthorized or illegal trade across boundaries. In particular, an export permit is required specifying that a state expert declares an export is legal, that it is not cruel, and that it will not threaten a wild population.

The **Montreal Protocol** (1987) protects stratospheric ozone. This treaty committed signatories to phase out the production and use of several chemicals that break down ozone in the atmosphere. The ozone "hole," a declining concentration of ozone (O_3) molecules over the South Pole, threatened living things: ozone high in the atmosphere blocks cancer-causing ultraviolet radiation, keeping it from reaching the earth's surface. The stable chlorine- and

Table 24.2 Some Important International Treaties
CBD: Convention on Biological Diversity 1992 (1993)
CITES: Convention on International Trade on Endangered Species of Wild Fauna and Flora 1973 (1987)
CMS: Convention on the Conservation of Migratory Species of Wild Animals 1979 (1983)
Basel: Basel Convention on the Transboundary Movements of Hazardous Wastes and their Disposal 1989 (1992)
Ozone: Vienna Convention for the Protection of the Ozone Layer and Montreal Protocol on Substances that Deplete the Ozone Layer 1985 (1988)
UNFCCC: United Nations Framework Convention on Climate Change 1992 (1994)
CCD: United Nations Convention to Combat Desertification in those Countries Experiencing Serious Drought and/or Desertification, Particularly in Africa 1994 (1996)
Ramsar: Convention on Wetlands of International Importance especially as Waterfowl Habitat 1971 (1975)
Heritage: Convention Concerning the Protection of the World Cultural and Natural Heritage 1972 (1975)
UNCLOS: United Nations Convention on the Law of the Sea 1982 (1994)

fluorine-based chemicals at fault for reducing ozone are used mainly as refrigerants. Alternative refrigerants have since been developed, and the use of chlorofluorocarbons (CFCs) and related molecules has plummeted. Although the ozone "hole" has not disappeared, it has declined as predicted by atmospheric scientists since the phase-out of CFCs. The Montreal Protocol is often held up as an example of a highly successful and effective international environmental agreement.

The Montreal Protocol was effective because it bound signatory nations not to purchase CFCs or products made using them from countries that refused to ratify the treaty. This trade restriction put substantial pressure on producing countries. Initially the protocol called for only a 50 percent reduction in CFC production, but subsequent research showed that ozone was being depleted faster than previously thought (chapter 9). The protocol was strengthened to an outright ban on CFC production, in spite of the objection of a few countries.

The **Basel Convention** (1992) restricts shipment of hazardous waste across boundaries. The aim of this convention, which has 172 signatories, is to protect health and the environment, especially in developing areas, by stating that hazardous substances should be disposed of in the states that generated them. Signatories are required to prohibit the export of hazardous wastes unless the receiving state gives prior informed consent, in writing, that a shipment is allowable. Parties are also required to minimize production of hazardous materials and to ensure that there are safe disposal facilities within their own boundaries. This convention establishes that it is the responsibility of states to make sure that their own corporations comply with international laws. The Basel Convention was enhanced by the Rotterdam Convention (1997), which places similar restrictions on unauthorized transboundary shipment of industrial chemicals and pesticides.

The 1994 **UN Framework Convention on Climate Change (UNFCC)** directs governments to share data on climate change, to develop national plans for controlling greenhouse gases, and to cooperate in planning for adaptation to climate change. Where the UNFCC encouraged reduction in greenhouse gas (GHG) emissions, the Kyoto Protocol (1997) set binding targets for signatories to reduce greenhouse gas emissions to less than 1990 levels by 2012. While the idea of binding targets is strong, and some countries (such as Sweden) are likely to achieve their goals, most countries are still falling short of their target. The protocol has been controversial because it sets tighter restrictions on industrialized countries, which are responsible for roughly 90 percent of GHG emissions up to the present, than for developing countries. Signatories are required to report their GHG emissions in order to document changes in their production. The protocol went into force in 2005, when 198 states and the European Union had signed the agreement. These signatories contribute almost 64 percent of global GHG emissions. The United States and China, the largest GHG emitters, have not signed the Kyoto Protocol, out of concern for their economic growth. Subsequent meetings in Copenhagen (2009) and Cancun (2010) have added incremental progress, although real reductions remain modest (chapter 15).

Enforcement often depends on national pride

Enforcement of international agreements often depends on the idea that countries care about their international reputation. Except in extreme cases such as genocide, the global community is unwilling to send an external police force into a country, because states are wary of interfering with the internal sovereignty of other states. However, most countries are reluctant to appear irresponsible or immoral in the eyes of the international community, so moral persuasion and public embarrassment can be effective enforcement strategies. Shining a spotlight on transgressions will often push a country to comply with international agreements.

Often international negotiators aim for unanimous agreement to ensure strong acceptance of international policies. Though this approach makes a strong agreement, a single recalcitrant nation can have veto power over the wishes of the vast majority. For instance, more than 100 countries at the UN Conference on Environment and Development (UNCED), held in Rio de Janeiro in 1992, agreed to restrictions on the release of greenhouse gases. At the insistence of U.S. negotiators, however, the climate convention was reworded so that it only urged—but did not require—nations to stabilize their emissions. Similarly, in 2010 negotiations over CITES species protection, Japan almost single-handedly derailed global protections for bluefin tuna, a huge, long-lived fish whose populations have dropped below 15 percent of historic levels (see chapter 6).

When a consensus cannot be reached, negotiators may seek an agreement acceptable to a majority of countries. This approach was used in negotiating the Kyoto Protocol on climate change, which sought, and eventually achieved, agreement from a majority of countries. Only signing countries are bound by such a treaty, but nonsigning countries may comply anyway, to avoid international embarrassment (fig. 24.17).

FIGURE 24.17 Global awareness of environmental issues can push countries to comply with treaties. Here a youth group from the Maldives, an island nation threatened by rising sea levels, stages a protest as part of the global 350.org movement for controlling climate change.

More recently the 2009 Copenhagen climate summit sought and failed to find a widely acceptable compromise with enforceable strategies for greenhouse gas emissions. Disagreements between the world's two largest greenhouse gas producers, China and the United States, effectively derailed any binding agreements. The meeting did produce a nonbinding agreement of principles, however. Among these principles were the statements that "climate change is one of the greatest challenges of our time," and that "deep cuts in global emissions are required … to hold the increase in global temperature below 2 degrees Celsius."

This nonbinding statement was quickly accepted by over 110 countries. While these statements don't commit countries to meaningful action—the primary aim of the Copenhagen meeting—they do commit countries to acknowledging the principle. Perhaps establishing agreement on the idea will be palatable and make some progress, where wrangling over penalties has thus far failed.

When strong accords with meaningful sanctions cannot be passed, sometimes the pressure of world opinion generated by revealing the sources of pollution can be effective. Activists can use this information to expose violators. For example, the environmental group Greenpeace discovered monitoring data in 1990 showing that Britain was disposing of coal ash in the North Sea. Although not explicitly forbidden by the Oslo Convention on ocean dumping, this evidence proved to be an embarrassment, and the practice was halted.

Trade sanctions can be an effective tool to compel compliance with international treaties. The Montreal Protocol used the threat of trade sanctions very effectively to cut CFC production dramatically. On the other hand, trade agreements also can work against environmental protection. The World Trade Organization (WTO) was established to promote free international trade and to encourage economic development. The WTO's emphasis on unfettered trade, however, has led to weakening of local environmental rules. In 1990 the United States banned the import of tuna caught using methods that kill thousands of dolphins each year. Shrimp caught with nets that kill endangered sea turtles were also banned. Mexico filed a complaint with the WTO, contending that dolphin-safe tuna laws represented an illegal barrier to trade. Thailand, Malaysia, India, and Pakistan filed a similar suit against turtle-friendly shrimp laws. The WTO ordered the United States to allow the import of both tuna and shrimp from countries that allow fisheries to kill dolphins and turtles. Environmental advocates point out that the WTO has never ruled against a corporation because it is composed of industry leaders. As such, the WTO mainly defends the interests of the business community, not the broader public interest.

24.5 NEW APPROACHES TO POLICY

As we gain experience with environmental governance, new policy strategies are being developed. These approaches are growing in part because environmental protection has remained a high priority for the public (fig. 24.18). One of the key changes has been to seek win/win compromises in environmental debates. Dispute resolution and mediation are strategies for reaching

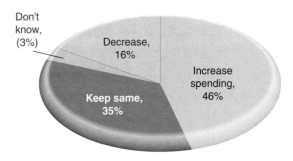

FIGURE 24.18 Americans persistently favor maintaining or increasing spending on environmental protections.

agreements without the mutual suspicions and hostility inherent in a lawsuit. Dispute resolution can avoid the time, expense, and winner-take-all confrontation inherent in lawsuits, these techniques encourage compromise and workable solutions with which everyone can live.

Arbitration is a formal process of dispute resolution somewhat like a trial. There are stringent rules of evidence, cross-examination of witnesses, and the process results in a legally binding decision. The arbitrator can actively work to find creative resolutions to the dispute. **Mediation** is generally less formal. Disputants are encouraged to sit down and talk to see if they can come up with a solution. Often in face-to-face meetings people are more willing to see their opponent's viewpoint and seek solutions.

Less rigid strategies for rule making are also being developed. One of these is **adaptive management,** or "learning by doing." This approach proposes that management should be experimental. Environmental policies should be designed from the outset to test clearly formulated hypotheses about the ecological, social, and economic impacts of the actions being undertaken (fig. 24.19). What initially seemed to be the best policy may not always be best, so we need to carefully monitor how conditions are changing. And we need to be able and willing to revise plans if our initial assumptions don't hold up over time.

Ecological principles also suggest that policy makers should plan for resilience—that is, for changes and recovery in a system (table 24.3). This means, for example, that protected forests must be large enough to allow for disturbance (such as fire or pest outbreaks) and recovery, or that policy makers should anticipate the possibility climate or species abundance may change over time.

Community-based planning uses local knowledge

Over the past several decades, natural resource managers have come to recognize the value of holistic planning that acknowledges multiple users and perspectives. Involving all stakeholders and interest groups early in the planning process can help avoid the "train wrecks" in which adversaries become entrenched in non-negotiable positions. Working with local communities can tap into traditional knowledge and gain acceptance for management

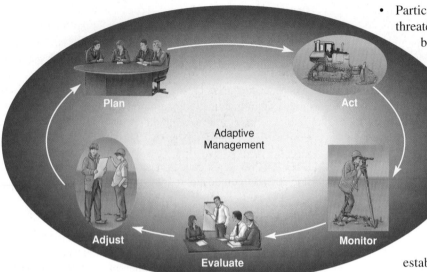

FIGURE 24.19 Adaptive management recognizes that we need to treat management plans for ecosystem as a scientific experiment in which we monitor, evaluate, and adjust our policies to fit changing conditions and knowledge.

plans that finally emerge from policy planning. There are many reasons to use collaborative approaches:

- Incorporating a variety of perspectives early in the process is more likely to lead to the development of acceptable solutions in the end. Public buy-in to an idea is likely to be better if many people have a voice from the start.

- Two heads are better than one. Involving multiple stakeholders and multiple sources of information enriches the process.

- Community-based planning provides access to situation-specific information and experience that can often only be obtained by active involvement of local residents.

Table 24.3	Planning for Resilience

1. Interdisciplinary, integrated modes of inquiry are needed for adaptive management of wicked problems.

2. We must recognize that these problems are fundamentally nonlinear and that we need nonlinear approaches to them.

3. We must attend to interactions between long-term processes, such as climate change or soil erosion in the American Corn Belt, and rapid events, such as the collapse of Antarctic ice sheets or the appearance of a dead zone in the Gulf of Mexico.

4. The spatial and temporal scales of our concerns are widening. We must consider global interconnections in our planning.

5. We need adaptive management policies that focus on building resilience and the capacity of renewal both in ecosystems and in human institutions.

- Participation is an important management tool. Project-threatening resistance on the part of certain stakeholders can be minimized by inviting active cooperation of all stakeholders throughout the planning process.

  - The knowledge and understanding needed by those who will carry out subsequent phases of a project can only be gained through active participation.

Community-based planning can be seen in the Atlantic Coastal Action Programme (ACAP) in eastern Canada. The purpose of this project is to develop blueprints for the restoration and maintenance of environmentally degraded harbors and estuaries in ways that are both biologically and socially sustainable. Officially established under Canada's Green Plan and supported by Environment Canada, this program created 13 community groups, some rural and some urban, with membership in each dominated by local residents. Federal and provincial government agencies are represented primarily as nonvoting observers and resource people. Each community group is provided with core funding for full-time staff who operate an office in the community and facilitate meetings.

Four of the 13 ACAP sites are in the Bay of Fundy, an important and unique estuary lying between New Brunswick and Nova Scotia. Approximately 270 km long, and with an area of more than 12,000 km^2, the bay, together with the nearby Georges Bank and the Gulf of Maine once formed one of the richest fisheries in the world (fig. 24.20a). With the world's highest recorded tidal range (up to 16 m at maximum spring tide), the bay still sustains a great variety of fishery and wildlife resources, and provides habitat for a number of rare or endangered species. Now home to more than 1 million people, the coastal region is an important agricultural, lumbering, and paper-producing region. Pollution and sediment damage harbors and biological communities. Overfishing and introduction of exotic species have resulted in endemic species declines. The collapse of cod, halibut, and haddock fishing has had devastating economic effects on the regional economy and the livelihoods of local residents. To cope with these complex, intertwined social and biological problems, ACAP is bringing together different stakeholders from around the bay to create comprehensive plans for ecological, economic, and social sustainability, including citizen monitoring and adaptive management.

Green plans outline goals for sustainability

Several national governments have undertaken integrated environmental planning that incorporates community round-tables for vision development. Canada, New Zealand, Sweden, and Denmark all have so-called **green plans** or comprehensive, long-range national environmental strategies. The best of these plans weave together complex systems, such as water, air, soil, and energy, and

(a)

(b)

(c)

FIGURE 24.20 Innovations in environmental policy and planning can be found worldwide. Examples include community planning in Canada's Bay of Fundy (a), the Dutch Green plan that restores ecosystems (b), and the Law of Mother Earth, protecting rights of all Bolivians to a healthy environment (c).

plan contains 223 policy changes aimed at reducing pollution and establishing economic stability. Three important mechanisms have been adopted for achieving these goals: integrated life-cycle management, energy conservation, and improved product quality. These measures should make consumer goods last longer and be more easily recycled or safely disposed of when no longer needed. For example, auto manufacturers are now required to design cars so they can be repaired or recycled rather than being discarded.

Among the guiding principles of the Dutch green plan are: (1) the "stand-still" principle that says environmental quality will not deteriorate, (2) abatement at the source rather than cleaning up afterward, (3) the "polluter pays" principle that says users of a resource pay for negative effects of that use, (4) prevention of unnecessary pollution, (5) application of the best practicable means for pollution control, (6) carefully controlled waste disposal, and (7) motivating people to behave responsibly.

The Netherlands have invested billions of euros in implementing this comprehensive plan. Some striking successes already have been accomplished. Between 1980 and 1995, emissions of sulfur dioxide, nitrogen oxides, ammonia, and volatile organic compounds were reduced more than 30 percent; pesticide use had been reduced 25 percent; chlorofluorocarbon use had been virtually eliminated; and industrial wastewater discharge into the Rhine River was down 70 percent. Some 250,000 ha (more than 600,000 acres) of former wetlands that had been drained for agriculture are being restored as nature preserves and 40,000 ha (99,000 acres) of forest are being replanted. This is remarkably generous and foresighted for Europe's most densely populated country.

Bolivia's Law of Mother Earth

The small, impoverished country of Bolivia has taken remarkably strong stands on many environmental issues since the election of President Evo Morales in 2005. Most of Bolivia's population lives in poor farming communities, directly dependent on natural resources, including water, healthy soil, and natural biodiversity (fig. 24.20c).

In 2011 Bolivia set a world precedent by proposing the "Law of Mother Earth." Following indigenous Andean traditions of considering Mother Earth, or *Pachamama*, to be a living being, the new law explicitly aims to protect life and biodiversity. It grants all people equal rights to a clean environment, including safe water, protection of biodiversity, clean air, and essential ecological functions. Specific terms of the law include requiring the government to transition toward renewable energy, to develop new economic indicators that account for environmental costs of economic activities, to focus on food sovereignty, and to invest in energy efficiency.

None of these steps will be easy, but like all polices, the first step is to identify a goal. The law also sets a standard by which later policy decisions can be judged. Figuring out how to reach that goal may take years. But without a statement of policy intent, progress might never happen. Can other nations do the same thing? What factors might support or discourage other places from developing their own Mother Earth Laws?

mesh them with human factors such as economics, health, and carrying capacity. Perhaps the most thorough and well-thought-out green plan in the world is that of the Netherlands (fig. 24.20b).

Developed in the 1980s through a complex process involving the public, industry, and government, the 400-page Dutch

CONCLUSION

Otto von Bismark once said, "Laws are like sausages, it is better not to see them being made." Still, if you hope to improve your environmental quality, it's helpful to understand how policies and laws are made and enforced. Laws, such as the Clean Water Act have been among the most effective tools that conservationists have had to protect biodiversity and habitat. But there is a constant struggle between those who want to strengthen environmental laws and those who want to reduce or eliminate them.

The legislative, administrative, and judicial branches of government all contribute to stages in the policy cycle. Though ordinary individuals might feel powerless in these various stages, you might be surprised at how much impact you can have if you get involved. Probably the best way to participate in environmental policy formation or passage of environmental laws is to join a group that shares your concerns. Being part of a group amplifies your influence. But even as an individual, you can make an impression. Write to or call your legislator. They do pay attention to constituents.

Global cooperation has also emerged as a key part of environmental protection. Dozens of laws protect resources, biodiversity, and environmental quality. Mechanisms for enforcement are not as obvious as they are within a single country, but creative strategies are evolving, and most nations see it to be in their interest to cooperate with their neighbors, most of the time.

On a smaller scale, community planning and community knowledge have also become key parts of policy formation. Understanding all these aspects of policy is a first step toward empowering yourself to influence the health of the environment in which you live.

REVIEWING LEARNING OUTCOMES

By now you should be able to explain the following points:

24.1 List several basic concepts in policy.

- Basic principles guide policy.
- Corporate money influences policy.
- Public awareness and action shape policy.

24.2 Describe some major environmental laws.

- NEPA (1969) establishes public oversight.
- The Clean Air Act (1970) regulates air emissions.
- The Clean Water Act (1972) protects surface water.
- The Endangered Species Act (1973) protects wildlife.
- The Superfund Act (1980) lists hazardous sites.

24.3 Explain how policies are made.

- Congress and legislatures vote on statutory laws.
- Judges decide case law.

- Executive agencies make rules and enforce laws.
- How much government do we want?

24.4 Explain the purposes of international conventions.

- Major international agreements.
- Enforcement often depends on national pride.

24.5 Outline some new approaches to policy.

- Community-based planning uses local knowledge.
- Green plans outline goals for sustainability.
- Bolivia's Law of Mother Earth.

PRACTICE QUIZ

1. What is the policy cycle, and how does it work?
2. Describe the path of a bill through Congress. When are riders and amendments attached?
3. What are the differences and similarities between statutory law and administrative law?
4. List some of the major U.S. environmental laws of the past 30 years.
5. Why have some international environmental treaties and conventions been effective while most have not? Describe two such treaties.
6. Why are international treaties increasingly common?
7. Explain some concerns about the influence of money on policy.
8. What is resilience? Why is it important?
9. What is collaborative, community-based planning?
10. What is the idea of a green plan?

CRITICAL THINKING AND DISCUSSION QUESTIONS

1. In your opinion, how much environmental protection is too much? Think of a practical example in which some stakeholders may feel oppressed by government regulations. How would you justify or criticize these regulations?

2. Among the steps in the policy cycle, where would you put your efforts if you wanted influence in establishing policy?

3. Do you believe that trees, wild animals, rocks, or mountains should have legal rights and standing in the courts? Why or why not? Are there other forms of protection you would favor for nature?

4. It's sometimes difficult to determine whether a lawsuit is retaliatory or based on valid reason. How would you define a SLAPP suit, and differentiate it from a legitimate case?

5. Create a list of arguments for and against an international body with power to enforce global environmental laws. Can you see a way to create a body that could satisfy both reasons for and against this power?

6. Identify a current environmental problem, and outline some policy approaches that could be used to address it. What strengths and weaknesses would different approaches have?

 ## Data Analysis: Examine Your Environmental Laws

The federal government publicizes the text of laws in multiple locations on the Internet. Reading about these laws is a good way to get a sense of the structures of environmental regulation, and to understand some of the compromises and the complexity of making rules that apply to thousands of different cases across the country. The primary way to access government rules and laws is through thomas.gov. A more direct source for environmental legislation is to go to the EPA website: www.epa.gov/lawsregs/laws/index.html#env.

Go to this website, and select one bill that bears on an issue you find interesting. Links are provided to the text of the law, usually in PDF format. Open the text of the law you have chosen, and look through the table of contents to see what sections ("titles") are covered in the bill.

1. What are the topics listed in the table of contents?

2. Definitions of terms come next. What terms are defined?

3. Choose a short section, perhaps 1–3 pages long. Read it carefully. Explain the content of those pages to your class. Also try to explain what the context of the bill might be: Why were those words written? By whom? As a result of what kind of problem?

The web address listed above gives you direct access to federal laws that define how the U.S. environment and resources are managed.

For Additional Help in Studying This Chapter, please visit our website at www.mhhe.com/cunningham12e. You will find additional practice quizzes and case studies, flashcards, regional examples, placemarks for Google Earth™ mapping, and an extensive reading list, all of which will help you learn environmental science.

Hundreds of Indian students created an elephant threatened by climate change. Intended to be visible from space, these EARTH art projects are grassroots efforts to create a new sense of urgency and of possibility for our planet.

What Then Shall We Do?

Learning Outcomes

After studying this chapter, you should be able to:

25.1 Explain how we can make a difference.

25.2 Summarize environmental education.

25.3 Evaluate what individuals can do.

25.4 Review how we can work together.

25.5 Investigate campus greening.

25.6 Define the challenge of sustainability.

"When spiders unite, their web can tie down an elephant."

~ African proverb

Case Study 350.org: Making a Change

Could a small group of students at a liberal arts college in Vermont mobilize an international campaign to tackle the most important environmental challenge we face today? Yes, they did. And it's an example from which all of us could learn something. It all started several years ago when author Bill McKibben, who is writer in residence at Middlebury College, organized a group of students to do something positive about the problem of global climate change. They realized they didn't have the money or resources to put together a national rally of the sort that has been so successful at attracting attention for civil rights and environmental causes in the past. Furthermore it seemed oxymoronic to encourage thousands of people to fly or drive long distances to protest excessive burning of fossil fuels.

They decided, instead, to use the power of the Internet and social networks to mobilize activists to do something meaningful and newsworthy in their own neighborhoods. By acting together on the same day and publishing photos and news releases to show the interconnections between actions, they could create a meta-event with greater power to influence local citizens and decision makers than a mammoth march in a single place might have.

The event they created was called Step It Up. With a minuscule budget and little previous experience, they inspired tens of thousands of citizens in 2007 to participate in more than 1,400 events in iconic places in all 50 of the United States. These creative actions—from skiers descending a melting glacier to protest global warming, or planting endangered chestnut trees to absorb CO_2, or flying thousands of handmade kites with environmental messages—were designed to both attract attention and to educate the public about the need to cut carbon emissions 80 percent by 2050.

Building on this success, the group decided to broaden their campaign to the international stage. Renaming themselves 350.org, they expanded their team to include young people from all over the world. They chose 350 because it's the number (in parts per million) that climate scientists say is the safe upper limit for carbon dioxide in the atmosphere. We're already at about 390 ppm, and politicians are debating whether we might hold emissions to 450 ppm, so the students have chosen an ambitious goal. But why not dream big?

On October 24, 2009, the 350 team mobilized 5,200 events in 181 countries. CNN called it "the most widespread day of political action in the planet's history." The team took their message of global concern to the big UN climate conference in Copenhagen, Denmark, in December 2009, but they were disappointed in the inability of government leaders to forge a meaningful agreement for combating climate change. Undaunted, 350.org planned even more protests. On October 10, 2020 (10/10/10), they organized a "global work party" with more than 7,000 projects in 187 countries. People put up solar panels, dug community gardens, planted trees, and did other actions to help reduce carbon emissions. Working together on pragmatic local projects empowers people, gives them hope, and helps build grassroots networks.

One of the most creative projects the team has launched is the 350 EARTH ART. So far, more than 15 major art installations have been created involving thousands of people. Each is designed to be visible from space (fig. 25.1). Not only did the art pieces turn out beautifully, they captured media attention and demonstrated to political leaders our widespread desire for environmental protection. They also unleashed creativity, got people motivated, and offered a hopeful way to express opinions about the future of our world. Wouldn't you like to get involved? Contact 350.org to get suggestions for how to plan an event, create a press release, invite elected officials and media, follow up, and get other useful resources.

In this chapter we'll look at how individuals and other groups are working to protect the earth and build a sustainable future. For related resources, including Google Earth™ placemarks that show locations where these issues can be explored via satellite images, visit EnvironmentalScience-Cunningham.blogspot.com.

FIGURE 25.1 On 10/10/10, over a thousand New Mexicans of all ages flooded the Santa Fe River's dry riverbed with blue-painted cardboard and other blue materials to show where the River should be flowing.

25.1 MAKING A DIFFERENCE

Throughout this book you have read about environmental problems, from climate change to biodiversity to energy policy debates. Biodiversity is disappearing at the fastest rate ever known; major ocean fisheries have collapsed; within 50 years, it is expected that two-thirds of countries will experience water shortages, and 3 billion people may live in slums. You have also seen that, as we have come to understand these problems, many exciting innovations have been developed to deal with them. New irrigation methods reduce agricultural water use; bioremediation provides inexpensive methods to treat hazardous waste; new energy sources, including wind, solar, geothermal, and even pressure-cooked garbage, offer strategies for weening our society from its dependence on oil and gas. Growth of green consumerism has developed markets for recycled materials, low-energy appliances, and organic foods. Population growth continues, but its rate has plummeted from a generation ago.

Stewardship for our shared resources is increasingly understood to be everybody's business. The environmental justice movement (chapter 21) has shown that minority groups and the poor frequently suffer more from pollution than wealthy or white people. African Americans, Latinos, Native Americans, and other minority groups have a clear interest in pursuing environmental solutions. Religious groups are voicing new concerns about preserving our environment (chapter 1). Farmers are seeking ways to save soil and water resources (chapter 10). Loggers are learning about sustainable harvest methods (chapter 12). Business leaders are discovering new ways to do well by doing good work for society and the environment (chapter 23). These changes are exciting, but many challenges remain.

Whatever your skills and interests, you can contribute to understanding and protecting our common environment. If you enjoy science, many disciplines contribute to this cause. As you know by now, biology, chemistry, geology, ecology, climatology, geography, demography, and other sciences all provide ideas and data that are essential to understanding our environment. Environmental scientists usually focus on one of these disciplines, but their work also serves the others. An environmental chemist, for example, might study contaminants in a stream system, and this work might help an aquatic ecologist understand changes in a stream's food web.

You can also help seek environmental solutions if you prefer writing, art, working with children (fig. 25.2), history, politics, economics, or other areas of study. As you have read, environmental science depends on communication, education, good policies, and economics as well as on science.

In this chapter, we will discuss some of the steps you can take to help find solutions to environmental dilemmas. You have already taken the most important step, educating yourself. When you understand how environmental systems function—from nutrient cycles and energy flows to ecosystems, climate systems, population dynamics, agriculture, and economies—you can develop well-informed opinions and help find useful answers.

FIGURE 25.2 Helping children develop a sense of wonder is a first step in protecting nature. As the Senegalese poet Baba Dioum said, "In the end, we conserve only what we love. We will love only what we understand. We will understand only what we are taught."

25.2 ENVIRONMENTAL EDUCATION

In 1990 Congress recognized the importance of environmental education by passing the National Environmental Education Act. The act established two broad goals: (1) to improve understanding among the general public of the natural and built environment and the relationships between humans and their environment, including global aspects of environmental problems, and (2) to encourage postsecondary students to pursue careers related to the environment. Specific objectives proposed to meet these goals include developing an awareness and appreciation of our natural and social/cultural environment, knowledge of basic ecological concepts, acquaintance with a broad range of current environmental issues, and experience in using investigative, critical-thinking, and problem-solving skills in solving environmental problems (fig. 25.3). Several states, including Arizona, Florida, Maryland, Minnesota, Pennsylvania, and Wisconsin, have successfully incorporated these goals and objectives into their curricula (table 25.1).

A number of organizations have been established to teach ecology and environmental ethics to elementary and secondary school students, as well as to get them involved in active projects to clean up their local community. Groups such as Kids Saving the Earth or Eco-Kids Corps are an important way to reach this vital audience. Family education results from these efforts as well. In a World Wildlife Fund survey, 63 percent of young people said they "lobby" their parents about recycling and buying environmentally responsible products.

Environmental literacy means understanding our environment

Speaking in support of the National Environmental Education Act, former Environmental Protection Agency administrator William K. Reilly called for broad **environmental literacy** in which every citizen is fluent in the principles of ecology and has a "working

FIGURE 25.3 Environmental education helps develop awareness and appreciation of ecological systems and how they work.

Table 25.1 Outcomes from Environmental Education

The natural context: An environmentally educated person understands the scientific concepts and facts that underlie environmental issues and the interrelationships that shape nature.

The social context: An environmentally educated person understands how human society is influencing the environment, as well as the economic, legal, and political mechanisms that provide avenues for addressing issues and situations.

The valuing context: An environmentally educated person explores his or her values in relation to environmental issues; from an understanding of the natural and social contexts, the person decides whether to keep or change those values.

The action context: An environmentally educated person becomes involved in activities to improve, maintain, or restore natural resources and environmental quality for all.

Source: *A Greenprint for Minnesota,* Minnesota Office of Environmental Education, 1993.

knowledge of the basic grammar and underlying syntax of environmental wisdom." Environmental literacy, according to Reilly can help establish a stewardship ethic—a sense of duty to care for and manage wisely our natural endowment and our productive resources for the long haul. "Environmental education," he says, "boils down to one profoundly important imperative: preparing ourselves for life in the next century. When the twenty-first century rolls around, it will not be enough for a few specialists to know what is going on while the rest of us wander about in ignorance."

You have made a great start toward learning about your environment by reading this book and taking a class in environmental science. Pursuing your own environmental literacy is a life-long process.

Table 25.2 The Environmental Scientist's Bookshelf

What are some of the most influential and popular environmental books? In a survey of environmental experts and leaders around the world, the top 12 best books on nature and the environment were:

A Sand County Almanac by Aldo Leopold (100)

Silent Spring by Rachel Carson (81)

State of the World by Lester Brown and the Worldwatch Institute (31)

The Population Bomb by Paul Ehrlich (28)

Walden by Henry David Thoreau (28)

Wilderness and the American Mind by Roderick Nash (21)

Small Is Beautiful: Economics as if People Mattered by E. F. Schumacher (21)

Desert Solitaire: A Season in the Wilderness by Edward Abbey (20)

The Closing Circle: Nature, Man, and Technology by Barry Commoner (18)

The Limits to Growth: A Report for the Club of Rome's Project on the Predicament of Mankind by Donella H. Meadows et al. (17)

The Unsettling of America: Culture and Agriculture by Wendell Berry (16)

Man and Nature by George Perkins Marsh (16)

[1]Indicates number of votes for each book. Because the preponderance of respondents were from the United States (82 percent), American books are probably overrepresented.

Source: From Robert Merideth, *The Environmentalist's Bookshelf: A Guide to the Best Books,* 1993, by G. K. Hall, an imprint of Macmillan, Inc. Reprinted by permission.

Some of the most influential environmental books of all time examine environmental problems and suggest solutions (table 25.2). To this list we'd add some personal favorites: *The Singing Wilderness* by Sigurd F. Olson, *My First Summer in the Sierra* by John Muir, and *Encounters with the Archdruid* by John McPhee.

In addition to reading about your environment, you can also learn about it directly by getting outdoors and experiencing the beauty and wonder of the natural world. As author Edward Abbey wrote,

> It is not enough to fight for the land; it is even more important to enjoy it. While it is still there. So get out there and mess around with your friends, ramble out yonder and explore the forests, encounter the grizz, climb the mountains. Run the rivers, breathe deep of that yet sweet and lucid air, sit quietly for a while and contemplate the precious stillness, that lovely mysterious and awesome space.

Citizen science encourages everyone to participate

While university classes often tend to be theoretical and abstract, many students are discovering they can make authentic contributions to scientific knowledge through active learning and undergraduate research programs. Internships in agencies or environmental organizations are one way of doing this. Another is to get involved in organized **citizen science** projects in which ordinary people join with established scientists to answer real scientific questions. Community-based research was pioneered in the Netherlands, where several dozen research centers now study environmental issues ranging from water quality in the Rhine River, cancer rates by geographic area, and substitutes for harmful organic solvents. In each project, students and neighborhood groups team with scientists and university personnel to collect data. Their results have been incorporated into official government policies.

Similar research opportunities exist in the United States and Canada. The Audubon Christmas Bird Count is a good example (Exploring Science, p. 565). Earthwatch offers a much smaller but more intense opportunity to take part in research. Every year hundreds of Earthwatch projects each field a team of a dozen or so volunteers who spend a week or two working on issues ranging from loon nesting behavior to archaeological digs. The American River Watch organizes teams of students to measure water quality. You might be able to get academic credit as well as helpful practical experience in one of these research experiences.

FIGURE 25.4 Many interesting, well-paying jobs are opening up in environmental fields. Here an environmental technician takes a sample from a monitoring well for chemical analysis.

Environmental careers range from engineering to education

The need for both environmental educators and environmental professionals opens up many job opportunities in environmental fields. The World Wildlife Fund estimates, for example, that 750,000 new jobs will be created over the next decade in the renewable energy field alone. Scientists are needed to understand the natural world and the effects of human activity on the environment. Lawyers and other specialists are needed to develop government and industry policy, laws, and regulations to protect the environment. Engineers are needed to develop technologies and products to clean up pollution and to prevent its production in the first place. Economists, geographers, and social scientists are needed to evaluate the costs of pollution and resource depletion and to develop solutions that are socially, culturally, politically, and economically appropriate for different parts of the world. In addition, business will be looking for a new class of environmentally literate and responsible leaders who appreciate how products sold and services rendered affect our environment.

Trained people are essential in these professions at every level, from technical and clerical support staff to top managers. Perhaps the biggest national demand over the next few years will be for environmental educators to help train an environmentally literate populace. We urgently need many more teachers at every level who are trained in environmental education. Outdoor activities and natural sciences are important components of this mission, but environmental topics such as responsible consumerism, waste disposal, and respect for nature can and should be incorporated into reading, writing, arithmetic, and every other part of education.

Green business and technology are growing fast

Can environmental protection and resource conservation—a so-called green perspective—be a strategic advantage in business? Many companies think so. An increasing number are jumping on the environmental bandwagon, and most large corporations now have an environmental department. A few are beginning to explore integrated programs to design products and manufacturing processes to minimize environmental impacts. Often called "design for the environment," this approach is intended to avoid problems at the beginning rather than deal with them later on a case-by-case basis. In the long run, executives believe this will save money and make their businesses more competitive in future markets. The alternative is to face increasing pollution control and waste disposal costs—now estimated to be more than $100 billion per year for all American businesses—as well as to be tied up in expensive litigation and administrative proceedings.

The market for pollution-control technology and know-how is also expected to be huge. Many companies are positioning themselves to cash in on this enormous market. Germany and Japan appear to be ahead of America in the pollution-control field because they have had more stringent laws for many years, giving them more experience in reducing effluents.

The rush to "green up" business is good news for those looking for jobs in environmentally related fields, which are predicted to be among the fastest growing areas of employment during the next few years. The federal government alone projects a need to hire some 10,000 people per year in a variety of environmental disciplines (fig. 25.4). How can you prepare yourself to enter this market? The best bet is to get some technical training: Environmental engineering, analytical chemistry, microbiology, ecology, limnology, groundwater hydrology, or computer science all have great potential. Currently, a chemical engineer with a graduate degree and some experience in an environmental field can practically name his or her salary. Some other very good possibilities are environmental law and business administration, both rapidly expanding fields.

For those who aren't inclined toward technical fields, there are many opportunities for environmental careers. A good liberal arts education will help you develop skills such as

Exploring Science

Every Christmas since 1900, dedicated volunteers have counted and recorded all the birds they can find within their team's designated study site (fig. 1). This effort has become the largest, longest-running, citizen-science project in the world. For the 100th count, nearly 50,000 participants in about 1,800 teams observed 58 million birds belonging to 2,309 species. Although about 70 percent of the counts in 2000 were made in the United States or Canada, 650 teams in the Caribbean, Pacific Islands, and Central and South America also participated. Participants enter their bird counts on standardized data sheets, or submit their observations over the Internet. Compiled data can be viewed and investigated online, almost as soon as they are submitted.

Frank Chapman, the editor of *Bird-Lore* magazine and an officer in the newly formed Audubon Society, started the Christmas Bird Count in 1900. For years, hunters had gathered on Christmas Day for a competitive hunt, often killing hundreds of birds and mammals as teams tried to outshoot each other. Chapman suggested an alternative contest: to see which team could observe and identify the most birds, and the most species, in a day. The competition has grown and spread. In the 100th annual count, the winning team was in Monte Verde, Costa Rica, with an amazing 343 species tallied in a single day.

The tens of thousands of bird-watchers participating in the count gather vastly more information about the abundance and distribution of birds than biologists could gather alone. These data provide important information for scientific research on bird migrations, populations, and habitat change. Now that the entire record for a century of bird data is available on the Bird-Source website (www.birdsource.org), both professional ornithologists and

FIGURE 1 Citizen-science projects, such as the Christmas Bird Count, encourage people to help study their local environment.

amateur bird-watchers can study the geographical distribution of a single species over time, or they can examine how all species vary at a single site through the years. Those concerned about changing climate can look for variation in long-term distribution of species. Climatologists can analyze the effects of weather patterns such as El Niño or La Niña

on where birds occur. One of the most intriguing phenomena revealed by this continent-wide data collection is irruptive behavior: that is, appearance of massive numbers of a particular species in a given area in one year, and then their move to other places in subsequent years following weather patterns, food availability, and other factors.

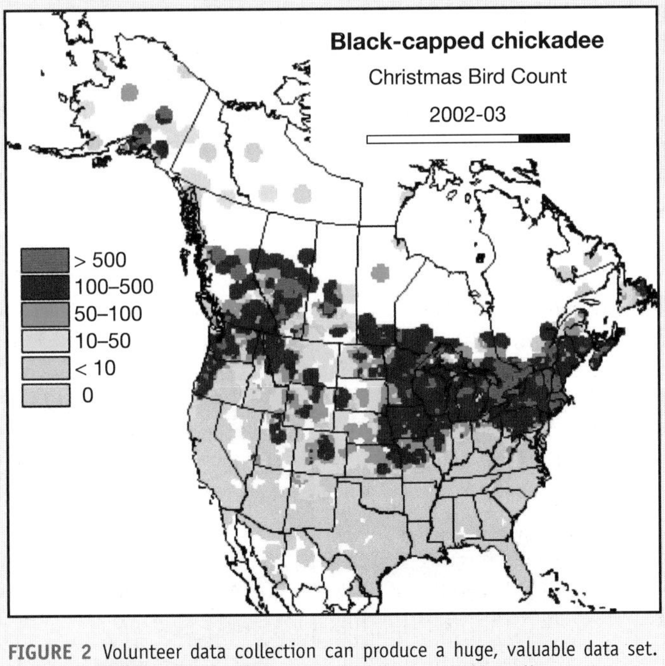

In 2010, the 110th Christmas Bird Count included 60,753 observers who collected data on 55 million birds in 2,160 locations across North America. This citizen-science effort has produced a rich, geographically broad data set far larger than any that could be produced by professional scientists (fig. 2). Following the success of the Christmas Bird Count, other citizen-science projects have been initiated. Project Feeder Watch, which began in the 1970s, has more than 15,000 participants, from schoolchildren and backyard bird-watchers to dedicated birders. The Great Backyard Bird Count of 2010 collected records on over 600 species and more than 6 million individual birds. In other areas, farmers have been enlisted to monitor pasture and stream health; volunteers monitor water quality in local streams and rivers; and nature reserves solicit volunteers to help gather ecological data. You can learn more about your local environment, and contribute to scientific research, by participating in a citizen-science project. Contact your local Audubon chapter or your state's department of natural resources to find out what you can do.

How does counting birds contribute to sustainability? Citizen-science projects are one way individuals can learn more about the scientific process, become familiar with their local environment, and become more interested in community issues. In this chapter, we'll look at other ways individuals and groups can help protect nature and move toward a sustainable society.

Black-capped chickadee

Christmas Bird Count

2002-03

> 500
100–500
50–100
10–50
< 10
0

FIGURE 2 Volunteer data collection can produce a huge, valuable data set. Christmas Bird Count data, such as this map, are available online.
Source: Data from Audubon Society.

communication, critical thinking, balance, vision, flexibility, and caring that should serve you well. Large companies need a wide variety of people; small companies need a few people who can do many things well. There are many opportunities for planners (chapter 22), health professionals (chapter 8), writers, teachers, and policy makers.

25.3 WHAT CAN INDIVIDUALS DO?

Some prime reasons for our destructive impacts on the earth are our consumption of resources and disposal of wastes. Technology has made consumer goods and services cheap and readily available in the richer countries of the world. As you already know, we in the industrialized world use resources at a rate out of proportion to our percentage of the population. If everyone in the world were to attempt to live at our level of consumption, given current methods of production, the results would surely be disastrous. In this section we will look at some options for consuming less and reducing our environmental impacts. Perhaps no other issue in this book represents so clear an ethical question as the topic of responsible consumerism.

How much is enough?

Our consumption of resources and disposal of wastes often have destructive impacts on the earth. Air and water quality, biodiversity, open space, and global climate all suffer because of the lifestyle choices we make. But do we really need all the stuff we consume or accumulate? Might our lives be simpler and more fulfilling if we could learn to live more sustainably? While billions of people don't have enough to survive, many of us suffer from too much.

A century ago, economist and social critic, Thorstein Veblen, in his book, *The Theory of the Leisure Class,* coined the term **conspicuous consumption** to describe buying things we don't want or need just to impress others. How much more shocked he would be to see current trends. The average American now consumes twice as many goods and services as in 1950. The average house is now more than twice as big as it was 50 years ago, even though the typical family has half as many people. We need more space to hold all the stuff we buy. Shopping has become the way many people define themselves. As Marx predicted, everything has become commodified; getting and spending have eclipsed family, ethnicity, even religion as the defining matrix of our lives. But the futility and irrelevance of much American consumerism leaves a psychological void. Once we possess things, we find they don't make us young, beautiful, smart, and interesting as they promised. With so much attention on earning and spending money, we don't have time to have real friends, to cook real food, to have creative hobbies, or to do work that makes us feel we have accomplished something with our lives. Some social critics call this drive to possess stuff **"affluenza."**

A growing number of people find themselves stuck in a vicious circle: They work frantically at a job they hate, to buy things they don't need, so they can save time to work even longer hours (fig. 25.5). Seeking a measure of balance in their lives, some opt out of the rat race and adopt simpler, less-consumptive lifestyles. As Thoreau wrote in *Walden,* "Our life is frittered away by detail . . . simplify, simplify."

The United Nations Environment Programme (UNEP) has held workshops on sustainable consumption in Paris and Tokyo. Recognizing that making people feel guilty about their lifestyles and purchasing habits isn't working, UNEP is attempting to find ways to make sustainable living something consumers will adopt willingly. The goal is economically, socially, and environmentally viable solutions that allow people to enjoy a good quality of life while consuming fewer natural resources and polluting less. A good example of this approach is a British automaker that provides a mountain bike with every car it sells, urging buyers to use the bike for short journeys. Another example cited by UNEP is European detergent makers who encourage customers to switch to low-temperature washing liquids and powders, not just to save energy but because it's good for their clothes.

FIGURE 25.5 Is this our highest purpose?
Source: © 1990 Bruce von Alten.

We can choose to reduce our environmental impacts

Often, seemingly small steps can have significant environmental effects. Would you be surprised to learn that for most of us, switching from a red-meat based diet to a vegetarian one can reduce our greenhouse footprint as much as trading in a normal size sedan for a hybrid? That's the conclusion of a study by researchers at the University of Chicago. Raising beef takes a lot more energy than growing the equivalent amount of vegetables, grains, and fruit. Where it takes about 2 calories of fossil fuel energy to grow most produce, the ratio can be as high as 80 to 1 for cattle raised in confined feeding operations. Furthermore, those cattle eat mostly grain (350 million tons of it in 2010), and much of the fertilizer used to grow that grain, together with the millions of tons of manure from feedlots, washes down the Mississippi River to create a massive dead zone in the Gulf of Mexico. And the cattle emit methane, which is 22 times as strong a greenhouse gas as CO_2.

Switching to a vegetarian diet can also be good for your health. Many studies show that consuming less fat reduces cardiovascular problems. If you really like meat, on the other hand, there are some alternatives to strict vegetarianism that can be good for you and the planet. Growing chickens or farm-raised fish (preferably vegetarian ones, such as tilapia or catfish) takes about one-tenth as much energy as beef, and has far less fat. Or eating only locally grown, grass-fed beef has far less environmental impact than those from confined feeding operations.

Collectively, the choices we make can be important. The What Can You Do? list on this page has some other suggestions for lowering your environmental impacts.

"Green washing" can mislead consumers

Although many people report they prefer to buy products and packaging that are socially and ecologically sustainable, there is a wide gap between what consumers say in surveys about purchasing habits and the actual sales data. Part of the problem is accessibility and affordability. In many areas, green products either aren't available or are so expensive that those on limited incomes (as many living in voluntary simplicity are) can't afford them. Although businesses are beginning to recognize the size and importance of the market for "green" merchandise, the variety of choices and the economies of scale haven't yet made environmentally friendly products as accessible as we would like.

Another problem is that businesses, eager to cash in on this premium market, offer a welter of confusing and often misleading claims about the sustainability of their offerings. Consumers must be wary to avoid "green scams" that sound great but are actually only overpriced standard items. Many terms used in advertising are vague and have little meaning. For example:

- "Nontoxic" suggests that a product has no harmful effects on humans. Since there is no legal definition of the term, however, it can have many meanings. How nontoxic is the product? And to whom? Substances not poisonous to humans can be harmful to other organisms.

- "Biodegradable," "recyclable," "reusable," or "compostable" may be technically correct but not signify much. Almost everything will biodegrade *eventually*, but it may take thousands of years. Similarly, almost anything is potentially recyclable or reusable; the real question is whether there are programs to do so in your community. If the only recycling or composting program for a particular material is half a continent away, this claim has little value.

What Can You Do?

Reducing Your Impact

Purchase Less

Ask yourself whether you really need more stuff.

Avoid buying things you don't need or won't use.

Use items as long as possible (and don't replace them just because a new product becomes available).

Use the library instead of purchasing books you read.

Make gifts from materials already on hand, or give nonmaterial gifts.

Reduce Excess Packaging

Carry reusable bags when shopping and refuse bags for small purchases.

Buy items in bulk or with minimal packaging; avoid single-serving foods.

Choose packaging that can be recycled or reused.

Avoid Disposable Items

Use cloth napkins, handkerchiefs, and towels.

Bring a washable cup to meetings; use washable plates and utensils rather than single-use items.

Buy pens, razors, flashlights, and cameras with replaceable parts.

Choose items built to last and have them repaired; you will save materials and energy while providing jobs in your community.

Conserve Energy

Walk, bicycle, or use public transportation.

Turn off (or avoid turning on) lights, water, heat, and air conditioning when possible.

Put up clotheslines or racks in the backyard, carport, or basement to avoid using a clothes dryer.

Carpool and combine trips to reduce car mileage.

Save Water

Water lawns and gardens only when necessary.

Use water-saving devices and fewer flushes with toilets.

Don't leave water running when washing hands, food, dishes, and teeth.

Based on material by Karen Oberhauser, Bell Museum Imprint, University of Minnesota, 1992. Used by permission.

- "Natural" is another vague and often misused term. Many natural ingredients—lead or arsenic, for instance—are highly toxic. Synthetic materials are not necessarily more dangerous or environmentally damaging than those created by nature.

- "Organic" can connote different things in different places. There are loopholes in standards so that many synthetic chemicals can be included in "organics." On items such as shampoos and skin-care products, "organic" may have no significance at all. Most detergents and oils are organic chemicals, whether they are synthesized in a laboratory or found in nature. Few of these products are likely to have pesticide residues anyway. Some cigarette brands advertise that they're organic, but they're still toxic.

- "Environmentally friendly," "environmentally safe," and "won't harm the ozone layer" are often empty claims. Since there are no standards to define these terms, anyone can use them. How much energy and nonrenewable material are used in manufacture, shipping, or use of the product? How much waste is generated, and how will the item be disposed of when it is no longer functional? One product may well be more environmentally benign than another, but be careful who makes this claim.

Certification identifies low-impact products

Products that claim to be environmentally friendly are being introduced at 20 times the normal rate for consumer goods. To help consumers make informed choices, several national programs have been set up to independently and scientifically analyze environmental impacts of major products. Germany's Blue Angel, begun in 1978, is the oldest of these programs. Endorsement is highly sought after by producers since environmentally conscious shoppers have shown that they are willing to pay more for products they know have minimum environmental impacts. To date, more than 2,000 products display the Blue Angel symbol. They range from recycled paper products, energy-efficient appliances, and phosphate-free detergents to refillable dispensers.

Similar programs are being proposed in every Western European country as well as in Japan and North America. Some are autonomous, nongovernmental efforts like the United States' Green Seal program (managed by the Alliance for Social Responsibility in New York). Others are quasi-governmental institutions such as the Canadian Environmental Choice programs.

The best of these organizations attempt "cradle-to-grave" **life-cycle analysis** (fig. 25.6) that evaluates material and energy inputs and outputs at each stage of manufacture, use, and disposal of the product. While you need to consider your own situation in making choices, the information supplied by these independent agencies is generally more reliable than self-made claims from merchandisers.

Green consumerism has limits

To quote Kermit the Frog, "It's not easy being green." Even with the help of endorsement programs, doing the right thing from an environmental perspective may not be obvious. Often we are

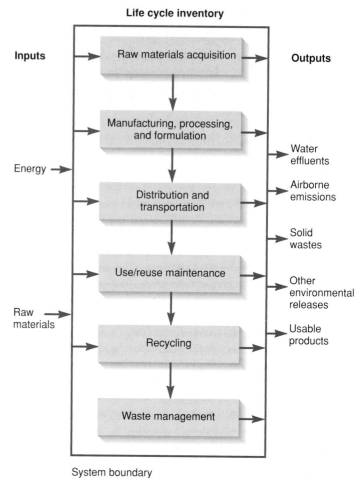

FIGURE 25.6 At each stage in its life cycle, a product receives inputs of materials and energy, produces outputs of materials or energy that move to subsequent phases, and releases wastes into the environment.

faced with complicated choices. Do the social benefits of buying rainforest nuts justify the energy expended in transporting them here, or would it be better to eat only locally grown products? In switching from Freon propellants to hydrocarbons, we spare the stratospheric ozone but increase hydrocarbon-caused smog. By choosing reusable diapers over disposable ones, we decrease the amount of material going to the landfill, but we also increase water pollution, energy consumption, and pesticide use (cotton is one of the most pesticide-intensive crops grown in the United States).

When the grocery store clerk asks you, "Paper or plastic?" you probably choose paper and feel environmentally virtuous, right? Everyone knows that plastic is made by synthetic chemical processes from nonrenewable petroleum or natural gas. Paper from naturally growing trees is a better environmental choice, isn't it? Well, not necessarily. In the first place, paper making consumes water and causes much more water pollution than does plastic manufacturing. Paper mills also release air pollutants, including foul-smelling sulfides and captans as well as highly toxic dioxins.

Furthermore, the brown paper bags used in most supermarkets are made primarily from virgin paper. Recycled fibers aren't strong enough for the weight they must carry. Growing, harvesting, and transporting logs from agroforestry plantations can be as environmentally disruptive as oil production. It takes a great deal of energy to pulp wood and dry newly made paper. Paper is also heavier and bulkier to ship than plastic. Although the polyethylene used to make a plastic bag contains many calories, in the end, paper bags are generally more energy-intensive to produce and market than plastic ones.

If both paper and plastic go to a landfill in your community, the plastic bag takes up less space. It doesn't decompose in the landfill, but neither does the paper in an air-tight, water-tight landfill. If paper is recycled but plastic is not, then the paper bag may be the better choice. If you are lucky enough to have both paper and plastic recycling, the plastic bag is probably a better choice since it recycles more easily and produces less pollution in the process. The best choice of all is to bring your own reusable cloth bag.

Complicated, isn't it? We often must make decisions without complete information, but it's important to make the best choices we can. Don't assume that your neighbors are wrong if they reach conclusions different from yours. They may have valid considerations of which you are unaware. The truth is that simple black and white answers often don't exist.

Taking personal responsibility for your environmental impact can have many benefits. Recycling, buying green products, and other environmental actions not only set good examples for your friends and neighbors, they also strengthen your sense of involvement and commitment in valuable ways. There are limits, however, to how much we can do individually through our buying habits and personal actions to bring about the fundamental changes needed to save the earth. Green consumerism generally can do little about larger issues of global equity, chronic poverty, oppression, and the suffering of millions of people in the developing world. There is a danger that exclusive focus on such problems as whether to choose paper or plastic bags, or to sort recyclables for which there are no markets, will divert our attention from the greater need to change basic institutions.

25.4 How Can We Work Together?

While some people can be effective working alone to bring about change, most of us find it more productive and more satisfying to work with others.

Collective action, such as that mobilized by 350.org, multiplies individual power (fig. 25.7). You get encouragement and useful information from meeting regularly with others who share your interests. It's easy to get discouraged by the slow pace of change; having a support group helps maintain your enthusiasm. You should realize, however, that there is a broad spectrum of environmental and social action groups. Some will suit your particular interests, preferences, or beliefs more than others. In this section, we will look at some environmental organizations as well as options for getting involved.

As the opening case study for this chapter shows, collective action can help change public and governmental perceptions. 350.org's theory of change is simple: if an international grassroots movement holds our leaders accountable to the latest climate science, we can start the global transformation we so desperately need. By carrying out local projects linked in an international movement, they hope to encourage their neighbors to take positive steps to protect our environment, while also giving decision makers a sense of urgency and possibility for our planet.

National organizations are influential but sometimes complacent

Among the oldest, largest, and most influential environmental groups in the United States are the National Wildlife Federation, the World Wildlife Fund, the Audubon Society, the Sierra Club, the Izaak Walton League, Friends of the Earth, Greenpeace, Ducks Unlimited, the Natural Resources Defense Council, and The Wilderness Society. Each of these groups arose in response to different challenges. The Audubon Society organized to protect egrets and other birds, which were being slaughtered for their plumes to decorate ladies' hats. The Sierra Club was organized to protect the giant redwood trees of California, and the valleys where they grew, which were rapidly being logged for timber. Ducks Unlimited formed during the Dust Bowl years to protect ducks and their habitat, in a time when farmers were draining wetlands as fast as they could, and converting them to crop fields. The Wilderness Society organized about the same time to protect open space for all citizens, not just the richest, who could afford private retreats in the mountains.

Sometimes known as the "group of 10," these large organizations have become more established and lost many of their radical roots. They have professional staff that work with lawmakers in state and federal governments. Membership in these groups frequently rises in response to political trends, as in the 1980s (fig. 25.8), but most members are passive and only occasionally involved in organizational activities. Many operations now are run by professional staffs, rather than citizen volunteers.

FIGURE 25.7 Working together with others can give you energy, inspiration, and a sense of accomplishment.

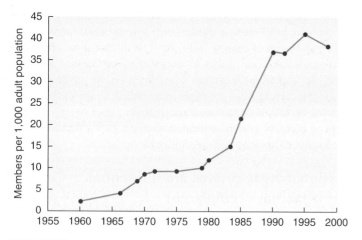

FIGURE 25.8 Growth of national environmental organizations in the United States.

FIGURE 25.9 Street theater can be a humorous, yet effective, way to convey a point in a nonthreatening way. Seeing that you're committed to a cause may encourage your neighbors to get involved as well.

Still, the established groups are powerful and important forces in environmental protection. Their mass membership, large professional staffs, and long history give them a degree of respectability and influence not found in newer, smaller groups. The Sierra Club, for instance, with about half a million members and chapters in almost every state, has a national staff of about 400, an annual budget over $20 million, and 20 full-time professional lobbyists in Washington, D.C. These national groups have become a potent force in Congress, especially when they band together to pass specific legislation, such as the Alaska National Interest Lands Act or the Clean Air Act.

In a survey that asked congressional staff and officials of government agencies to rate the effectiveness of groups that attempt to influence federal policy on pollution control, the top five were national environmental organizations. In spite of their large budgets and important connections, the American Petroleum Institute, the Chemical Manufacturers Association, and the Edison Electric Institute ranked far behind these environmental groups in terms of influence.

Although much of the focus of the big environmental groups is in Washington, Audubon, Sierra Club, and Izaak Walton have local chapters, outings, and publicity campaigns (fig. 25.9). This can be a good way to get involved. Go to some meetings, volunteer, offer to help. You may have to start out stuffing envelopes or some other unglamorous job, but if you persevere, you may have a chance to do something important and fun. It's a good way to learn and meet people.

Some environmental groups, such as the Environmental Defense Fund (EDF), The Nature Conservancy (TNC), the National Resources Defense Council (NRDC), and the Wilderness Society (WS), have limited contact with ordinary members except through their publications. They depend on a professional staff to carry out the goals of the organization through litigation (EDF and NRDC), land acquisition (TNC), or lobbying (WS). Although not often in the public eye, these groups can be very effective because of their unique focus. TNC buys land of high ecological value that is threatened by development. With more than 3,200 employees and assets around $3 billion, TNC manages 7 million acres in

FIGURE 25.10 The Nature Conservancy buys land with high biodiversity or unique natural values to protect it from misuse and development.

what it describes as the world's largest private sanctuary system (fig. 25.10). Still, the Conservancy is controversial for some of its management decisions, such as gas and oil drilling in some reserves, and including executives from some questionable companies on its governing board and advisory council. The Conservancy replies that it is trying to work with these companies to bring about change rather than just criticize them.

New players bring energy to environmental policy

There's a new energy today in environmental groups. In 2004, Michael Shellenberger and Ted Nordhaus, consultants for the Environmental Grant-making Foundation, proclaimed the "Death of Environmentalism." The major environmental organizations, they claimed, had become so embedded in Washington politics and concerned about their own jobs that they had become largely irrelevant. The greatest evidence of this impotence, they charged, was the failure to influence policy on global climate change despite years of work and hundreds of millions of dollars spent on lobbying. Even political candidates, such as John Kerry and Al Gore, who had stellar records in Congress as environmental advocates, barely mentioned it when running for office.

But just when the prospects for environmental progress seemed darkest, we seemed to reach a tipping point. The evidence of global climate change became impossible to ignore. But several emerging leaders deserve credit and thanks as well. Al Gore, who lost the U.S. presidency by just one vote in the Supreme Court, reinvented himself as an environmental champion. In a single year, he won a Grammy, an Emmy, and an Oscar, wrote a best-selling book, and shared the Nobel Peace Prize with the IPCC. Once reviled and rejected, he became a global environmental hero.

Now a broader, diverse, savvy, and passionate movement is taking shape. The 350.org network has inspired tens of thousands of citizens to work to stop climate change. One of their more inspiring campaigns was to restore Carter-era solar panels to the White House. Installed in 1979 by then-president Jimmy Carter, the solar water heating panels were removed when Ronald Reagan took office. Unity College in Maine rescued the panels and installed them on a student dining hall, where they provided hot water for 30 years. In 2010 a 350.org delegation delivered one of the panels to Washington, D.C., to be reinstalled on the White House roof. During their long—and well publicized—road trip from Maine to Washington, they made many stops to demonstrate to the press and other interested parties how well the panel still works after 36 years.

New leaders, including Majora Carter and Van Jones, have brought poverty, jobs, and justice groups into the conversation. Business and religious leaders also joined in. More than 450 colleges pledged to adopt renewable energy and other measures to become climate-neutral. A new coalition called the Climate Action Network was organized to lobby for climate policy, and a group called 1Sky sought to mobilize ordinary Americans to work for green jobs, transform energy policy, and freeze climate pollution levels. Together these groups hope to bring about a major change in human history. Wouldn't you like to be part of this effort?

International nongovernmental organizations mobilize many people

International **nongovernmental organizations (NGOs)** can be vital in the struggle to protect areas of outstanding biological value. Without this help, local groups could never mobilize the public interest or financial support for projects, such as saving Laguna San Ignacio.

The rise in international NGOs in recent years has been phenomenal. At the Stockholm Conference in 1972, only a handful of environmental groups attended, almost all from fully developed countries. Twenty years later, at the Rio Earth Summit, more than 30,000 individuals representing several thousand environmental groups, many from developing countries, held a global Ecoforum to debate issues and form alliances for a better world.

Some NGOs are located primarily in the more highly developed countries of the north and work mainly on local issues. Others are headquartered in the north but focus their attention on the problems of developing countries in the south. Still others are truly global, with active groups in many different countries. A few are highly professional, combining private individuals with representatives of government agencies on quasi-government boards or standing committees with considerable power. Others are on the fringes of society, sometimes literally voices crying in the wilderness. Many work for political change, more specialize in gathering and disseminating information, and some undertake direct action to protect a specific resource.

Public education and consciousness-raising using protest marches, demonstrations, civil disobedience, and other participatory public actions and media events are generally important tactics for these groups. Greenpeace, for instance, carries out well-publicized confrontations with whalers, seal hunters, toxic waste dumpers, and others who threaten very specific and visible resources. Greenpeace may well be the largest environmental organization in the world, claiming some 2.5 million contributing members.

In contrast to these highly visible groups, others choose to work behind the scenes, but their impact may be equally important. Conservation International has been a leader in debt-for-nature swaps to protect areas particularly rich in biodiversity. It also has some interesting initiatives in economic development, seeking products made by local people that will provide livelihoods along with environmental protection (fig. 25.11).

FIGURE 25.11 International conservation groups often initiate economic development projects that provide a local alternative to natural resource destruction.

25.5 CAMPUS GREENING

Colleges and universities can be powerful catalysts for change. Across North America, and around the world, students and faculty are studying sustainability and carrying out practical experiments in sustainable living and ecological restoration.

Organizations for secondary and college students often are among our most active and effective groups for environmental change. The largest student environmental group in North America is the **Student Environmental Action Coalition (SEAC).** Formed in 1988 by students at the University of North Carolina at Chapel Hill, SEAC has grown rapidly to more than 30,000 members in some 500 campus environmental groups. SEAC is both an umbrella organization and grassroots network that functions as an information clearinghouse and a training center for student leaders. Member groups undertake a diverse spectrum of activities ranging from politically neutral recycling promotion to confrontational protests of government or industrial projects. National conferences bring together thousands of activists who share tactics and inspiration while also having fun. If there isn't a group on your campus, why not look into organizing one?

Another important student organizing group is the network of Public Interest Research Groups active on most campuses in the United States. While not focused exclusively on the environment, the PIRGs usually include environmental issues in their priorities for research. By becoming active, you could probably introduce environmental concerns to your local group if they are not already working on problems of importance to you. Remember that you are not alone. Others share your concerns and want to work with you to bring about change; you just have to find them. There is power in working together.

Electronic communication is changing the world

One of the most important skills you are likely to learn in SEAC or other groups committed to social change is how to organize. This is a dynamic process in which you must adapt the realities of your circumstances and the goals of your group, but there are some basic principles that apply to most situations (table 25.3). Using communications media to get your message out is an important part of the modern environmental movement. Table 25.4 suggests some important considerations in planning a media campaign.

It's probably not a surprise to anyone that the Internet is changing our world. The power of social media to organize mass demonstrations and change political systems—sometimes practically overnight—was demonstrated across North Africa and the Middle East in 2011. In many places, youth with few resources other than cell phones, blogs, and a social network were able to have a powerful impact.

You may not want to overthrow your government, but you can influence public opinion and political decision making in ways never possible a generation ago. Using electronic communications, it's now possible to link many local systems into a virtual network. Where the media was once controlled by a tiny number of corporations and publishing moguls, now all of us have the power to communicate widely. You can organize events to educate and mobilize your neighbors or

Table 25.3	**Organizing an Environmental Campaign**
1.	What do you want to change? Are your goals realistic, given the time and resources you have available?
2.	What and who will be needed to get the job done? What resources do you have now and how can you get more?
3.	Who are the stakeholders in this issue? Who are your allies and constituents? How can you make contact with them?
4.	How will your group make decisions and set priorities? Will you operate by consensus, majority vote, or informal agreement?
5.	Have others already worked on this issue? What successes or failures did they have? Can you learn from their experience?
6.	Who has the power to give you what you want or to solve the problem? Which individuals, organizations, corporations, or elected officials should be targeted by your campaign?
7.	What tactics will be effective? Using the wrong tactics can alienate people and be worse than taking no action at all.
8.	Are there social, cultural, or economic factors that should be recognized in this situation? Will the way you dress, talk, or behave offend or alienate your intended audience? Is it important to change your appearance or tactics to gain support?
9.	How will you know when you have succeeded? How will you evaluate the possible outcomes?
10.	What will you do when the battle is over? Is yours a single-issue organization, or will you want to maintain the interest, momentum, and network you have established?

Source: Based on material from "Grassroots Organizing for Everyone" by Claire Greensfelder and Mike Roselle from *Call to Action,* 1990 Sierra Book Club Books.

fellow students (fig. 25.12). And, collectively, you can let your political representatives know what you think about what they're doing.

Schools can be environmental leaders

Colleges and universities can be sources of information and experimentation in sustainable living. They have the knowledge and expertise to figure out how to do new things, and they have students who have the energy and enthusiasm to do much of the research, and for whom that discovery will be a valuable learning experience. At many colleges and universities, students have undertaken campus audits to examine water and energy use, waste production and disposal, paper consumption, recycling, buying locally produced food, and many other examples of sustainable resource consumption. At more than 100 universities and colleges across America, graduating students have taken a pledge that reads:

> I pledge to explore and take into account the social and environmental consequences of any job I consider and will try to improve these aspects of any organization for which I work.

Could you introduce something similar at your school?

Campuses often have building projects that can be models for sustainability research and development. More than 110 colleges have built, or are building structures certified by the U.S. Green Building Council. Some recent examples of prize-winning

Table 25.4 Using the Media to Influence Public Opinion

Shaping opinion, reaching consensus, electing public officials, and mobilizing action are accomplished primarily through the use of the communications media. To have an impact in public affairs, it is essential to know how to use these resources. Here are some suggestions:

1. *Assemble a press list.* Learn to write a good press release by studying books from your public library on press relations techniques. Get to know reporters from your local newspaper and TV stations.

2. *Appear on local radio and TV talk shows.* Get experts from local universities and organizations to appear.

3. *Write letters to the editor, feature stories, and news releases.* You may include black and white photographs. Submit them to local newspapers and magazines. Don't overlook weekly community shoppers and other "freebie" newspapers, which usually are looking for newsworthy material.

4. *Try to get editorial support from local newspapers, radio, and TV stations.* Ask them to take a stand supporting your viewpoint. If you are successful, send a copy to your legislator and to other media managers.

5. *Put together a public service announcement and ask local radio and TV stations to run it* (preferably not at 2 A.M.). Your library or community college may well have audiovisual equipment that you can use. Cable TV stations usually have a public access channel and will help with production.

6. *If there are public figures in your area who have useful expertise, ask them to give a speech or make a statement.* A press conference, especially in a dramatic setting, often is a very effective way of attracting attention.

7. *Find celebrities or media personalities to support your position.* Ask them to give a concert or performance, both to raise money for your organization and to attract attention to the issue. They might like to be associated with your cause.

8. *Hold a media event that is photogenic and newsworthy.* Clean up your local river and invite photographers to accompany you. Picket the corporate offices of a polluter, wearing eye-catching costumes and carrying humorous signs. Don't be violent, abusive, or obnoxious; it will backfire on you. Good humor usually will go farther than threats.

9. *If you hear negative remarks about your issue on TV or radio, ask for free time under the Fairness Doctrine to respond.* Stations have to do a certain amount of public service to justify relicensing and may be happy to accommodate you.

10. *Ask your local TV or newspaper to do a documentary or feature story about your issue or about your organization and what it is trying to do.* You will not only get valuable free publicity, but you may inspire others to follow your example.

FIGURE 25.12 Protests, marches, and public demonstrations can be an effective way to get your message out and to influence legislators.

for Sustainable Buildings, a booklet that covers everything from energy-efficient lighting to native landscaping. With 275 photovoltaic panels to catch sunlight, there should be no need to buy electricity for the building. In fact, it's expected that surplus energy will be sold back to local utility companies to help pay for building operation.

Oberlin's Environmental Studies Center, designed by architect Bill McDonough, features 370 m² of photovoltaic panels on its roof, a geothermal well to help heat and cool the building, large south-facing windows for passive solar gain, and a "living machine" for water treatment, including plant-filled tanks in an indoor solarium and a constructed wetland outside (see fig. 20.11).

UCSB's Bren School of Environmental Science and Management looks deceptively institutional but claims to be the most environmentally state-of-the-art structure of its kind in the United States (fig. 25.13). It wasn't originally intended to be a particularly green building, but planners found that some simple features like

FIGURE 25.13 The University of California at Santa Barbara claims its new Bren School of Environmental Science and Management is the most environmentally friendly building of its kind in the United States.

sustainable designs can be found at Stanford University, Oberlin College in Ohio, and the University of California at Santa Barbara. Stanford's Jasper Ridge building will provide classroom, laboratory, and office space for its biological research station. Stanford students worked with the administration to develop *Guidelines*

having large windows that harvest natural light and open to let ocean breezes cool the interior make the building both more functional and more appealing. Motion detectors control light levels and sensors monitor and refresh the air when there is too much CO_2 putting students to sleep. More than 30 percent of interior materials are recycled. Solar panels supply 10 percent of the electricity, and the building exceeds federal efficiency standards by 30 percent. "The overriding and very powerful message is it really doesn't cost any more to do these things," says Dennis Aigner, dean of Bren School.

These facilities can become important educational experiences. At Carnegie Mellon University in Pittsburgh, students helped design a green roof for Hamershlag Hall. They now monitor how the living roof is reducing storm water drainage and improving water quality. A kiosk inside the dorm shows daily energy use and compares it to long-term averages. Classrooms within the dorm offer environmental science classes in which students can see sustainability in action. Green dorms are popular with students. They appreciate natural lighting, clean air, lack of allergens in building materials, and other features of LEED-certified buildings. One of the largest green dorms in the country is at the University of South Carolina, where more than 100 students are on a waiting list for a room.

A recent study by the Sustainable Endowments Institute evaluated more than 100 of the leading colleges and universities in the United States on their green building policies, food and recycling programs, climate change impacts, and energy consumption. The report card ranked Dartmouth, Harvard, Stanford, and Williams as the top of the "A list" of 23 greenest campuses. Berea College in Kentucky got special commendation as a small school with a strong commitment to sustainability. It's "ecovillage" has a student-designed house that produces its own electricity and treats waste water in a living system. The college has a full-time sustainability coordinator to provide support to campus programs, community outreach, and teaching. Some other campuses with academic programs in sustainability include Arizona State in Tempe, and Northern Arizona University in Flagstaff.

Your campus can reduce energy consumption

The Campus Climate Challenge, recently launched by a coalition of nonprofit groups, seeks to engage students, faculty, and staff at 500 college campuses in the United States and Canada in a long-term campaign to eliminate global warming pollution. Many campuses have invested in clean energy, set strict green building standards for new construction, purchased fuel-efficient vehicles, and adopted other policies to save energy and reduce their greenhouse gas emissions. Some examples include Concordia University in Austin, Texas, the first college or university in the country to purchase all of its energy from renewable sources. The 5.5 million kilowatt-hours of "green power" it uses each year will eliminate about 8 million pounds of CO_2 emissions annually, the equivalent of planting 1,000 acres of trees or taking 700 cars off the roads. Emory University in Atlanta, Georgia, is a leader in green building standards, with 11 buildings that are or could become LEED certified.

Emory's Whitehead Biomedical Research Building was the first facility in the Southeast to be LEED certified. Like a number of other colleges, Carleton College in Northfield, Minnesota, has built its own windmill, which is expected to provide about 40 percent of the school's electrical needs. The $1.8 million wind turbine is expected to pay for itself in about ten years. The Campus Climate Challenge website at www.energyaction.net contains valuable resources, including strategies and case studies, an energy action packet, a campus organizing guide, and more.

At many schools, students have persuaded the administration to buy locally produced food and to provide organic, vegetarian, and fair trade options in campus cafeterias. This not only benefits your health and the environment, but can also serve as a powerful teaching tool and everyday reminder that individuals can make a difference. Could you do something similar at your school? See the Data Analysis box at the end of this chapter for other suggestions.

25.6 SUSTAINABILITY IS A GLOBAL CHALLENGE

As the developing countries of the world become more affluent, they are adopting many of the wasteful and destructive lifestyle patterns of the richer countries. Automobile production in China, for example, is increasing at about 19 percent per year, or doubling every 3.7 years. By 2030 there could be nearly as many automobiles in China as in the United States. What will be the effect on air quality, world fossil fuel supplies, and global climate if that growth rate continues? Already, two-thirds of the children in Shenzhen, China's wealthiest province, suffer from lead poisoning, probably caused by use of leaded gasoline. And, as chapter 8 points out, diseases associated with affluent lifestyles—such as obesity, diabetes, heart attacks, depression, and traffic accidents—are becoming the leading causes of morbidity and mortality worldwide.

On the other hand, there appears to be a dramatic worldwide shift in public attitudes toward environmental protection. In a BBC poll of 22,000 residents of 21 countries, 83 percent agreed that individuals would definitely or probably have to make lifestyle changes to reduce the amount of climate-changing gases they produce. Overall, 70 percent said they were personally ready to make sacrifices to protect the environment. Similarly, in 2011 the Greenberg Quinlan Rosner Research company conducted a bipartisan survey of public opinion about clean air standards. Despite claims from some conservatives that the public opposes EPA pollution standards, and greenhouse gas regulation in particular, this poll found overwhelming support for even stronger EPA rules among likely 2012 American voters. Seventy-seven percent of respondents in this survey supported stricter CO_2 standards, and 69 percent said that EPA scientists rather than Congress should set pollution standards.

We would all benefit by helping developing countries access more efficient, less-polluting technologies. Education, democracy, and access to information are essential for sustainability. It is in our best interest to help finance protection of our common

FIGURE 25.14 Sustainable development to ensure a healthy environment for all the world's people is the aim of the millenium development goals.

future in some equitable way. Maurice Strong, chair of the Earth Charter Council, estimates that development aid from the richer countries should be some $150 billion per year, while internal investments in environmental protection by developing countries will need to be about twice that amount. Many scholars and social activists believe that poverty is at the core of many of the world's most serious human problems: hunger, child deaths, migrations, insurrections, and environmental degradation (fig. 25.14). One way to alleviate poverty is to foster economic growth so there can be a bigger share for everyone.

Strong economic growth already is occurring in many places. The World Bank projects that if current trends continue, economic output in developing countries will rise by 4 to 5 percent per year in the next 40 years. Economies of industrialized countries are expected to grow more slowly but could still triple over that period. Altogether, the total world output could be quadruple what it is today.

That growth could provide funds to clean up environmental damage caused by earlier, wasteful technologies and misguided environmental policies. It is estimated to cost $350 billion per year to control population growth, develop renewable energy sources, stop soil erosion, protect ecosystems, and provide a decent standard of living for the world's poor. This is a great deal of money, but it is small compared to over $1 trillion per year spent on wars and military equipment.

While growth simply implies an increase in size, number, or rate of something, development, in economic terms, means a real increase in average welfare or well-being. **Sustainable development** based on the use of renewable resources in harmony with ecological systems is an attractive compromise to the extremes of no growth versus unlimited growth (fig. 25.15). Perhaps the best definition of this goal is that of the World Commission on Environment and Development,

which defined sustainable development in *Our Common Future* as "meeting the needs of the present without compromising the ability of future generations to meet their own needs." Some goals of sustainable development include:

- A demographic transition to a stable world population of low birth and death rates.
- An energy transition to high efficiency in production and use, coupled with increasing reliance on renewable resources.
- A resource transition to reliance on nature's "income" without depleting its "capital."
- An economic transition to sustainable development and a broader sharing of its benefits.
- A political transition to global negotiation grounded in complementary interests between North and South, East and West.
- An ethical or spiritual transition to attitudes that do not separate us from nature or each other.

Notice that these goals don't apply just to developing countries. It's equally important that those of us in the richer countries adopt these targets as well. Supporting our current lifestyles is much more resource intensive and has a much greater impact on our environment than the billions of people in poorer countries. Many environmental scientists prefer to simply use the term **sustainability** to describe the search for ways of living more lightly on the earth because it can include residents of both the developed and developing world.

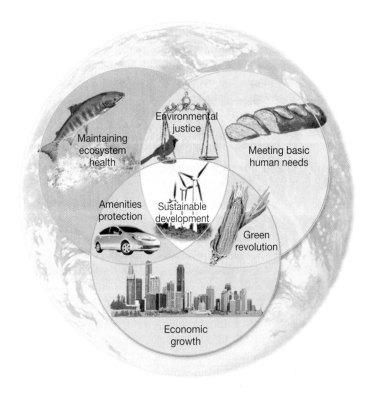

FIGURE 25.15 A model for integrating ecosystem health, human needs, and sustainable economic growth.

Table 25.5 Millennium Development Goals

Goals	Specific Objectives
1. Eradicate extreme poverty and hunger.	1a. Reduce by half the proportion of people living on less than a dollar a day.
	1b. Reduce by half the proportion of people who suffer from hunger.
2. Achieve universal primary education.	2a. Ensure that all boys and girls complete a full course of primary schooling.
3. Promote gender equality and empower women.	3a. Eliminate gender disparity in primary and secondary education by 2015.
4. Reduce child mortality.	4a. Reduce by two-thirds the mortality rate among children under five.
5. Improve maternal health.	5a. Reduce by three-quarters the maternal mortality ratio.
6. Combat HIV/AIDS, malaria, and other diseases.	6a. Halt and begin to reverse the spread of HIV/AIDS.
	6b. Halt and begin to reverse the spread of malaria and other major diseases.
7. Ensure environmental sustainability.	7a. Integrate the principles of sustainable development into policies and programs; reverse the loss of environmental resources.
	7b. Reduce by half the proportion of people without sustainable access to safe drinking water.
	7c. Achieve significant improvement, in the lives of 100 million slum dwellers by 2020.
8. Develop a global partnership for development.	8a. Develop further an open trading and financial system that is rule-based, predictable, and nondiscriminatory, including a commitment to good governance, development, and poverty reduction.
	8b. Address the least-developed countries' special needs. This includes tariff-and quota-free access for their exports; enhanced debt relief for heavily indebted poor countries.

In 2000, United Nations Secretary-General Kofi Annan called for a **millennium assessment** of the consequences of ecosystem change on human well-being as well as the scientific basis for actions to enhance the conservation and sustainable use of those systems. More than 1,360 experts from around the world worked on technical reports about the conditions and trends of ecosystems, scenarios for the future, and possible responses.

The findings from the millennium assessment goals serve as a good summary for this book. Among the key conclusions are:

- All of us depend on nature and ecosystem services to provide the conditions for a decent, healthy, and secure life.

- We have made unprecedented changes to ecosystems in recent decades to meet growing demands for food, fresh water, fiber, and energy.

- These changes have helped improve the lives of billions, but at the same time they weakened nature's ability to deliver other key services, such as purification of air and water, protection from disasters, and the provision of medicine.

- Among the outstanding problems we face are the dire state of many of the world's fish stocks, the intense vulnerability of the 2 billion people living in dry regions, and the growing threat to ecosystems from climate change and pollution.

- Human actions have taken the planet to the edge of a massive wave of species extinctions, further threatening our own well-being.

- The loss of services derived from ecosystems is a significant barrier to reducing poverty, hunger, and disease.

- The pressures on ecosystems will increase globally unless human attitudes and actions change.

- Measures to conserve natural resources are more likely to succeed if local communities are given ownership of them, share the benefits, and are involved in decisions.

- Even today's technology and knowledge can reduce considerably the human impact on ecosystems. They are unlikely to be deployed fully, however, until ecosystem services cease to be perceived as free and limitless.

- Better protection of natural assets will require coordinated efforts across all sections of governments, businesses, and international institutions.

As a result of this assessment, the United Nations has developed a set of goals and objectives for sustainable development (table 25.5). From what you've learned in this book, how do you think we could work—individually and collectively—to accomplish these goals?

CONCLUSION

All through this book you've seen evidence of environmental degradation and resource depletion, but there are also many cases in which individuals and organizations are finding ways to stop pollution, use renewable rather than irreplaceable resources, and even restore biodiversity and habitat. Sometimes all it takes is the catalyst of a pilot project to show people how things can be done differently to change attitudes and habits. In this chapter, you've learned some practical approaches to living more lightly on the world individually as well as working collectively to create a better world.

Public attention to issues in the United States seems to run in cycles. Concern builds about some set of problems, and people are willing to take action to find solutions, but then interest wanes and other topics come to the forefront. For the past decade, the American public has consistently said that the environment is very important, and that government should pay more attention to environmental quality. Nevertheless, people haven't shown this concern for the environment to be a very high priority, either in personal behavior or in how they vote.

Recently, however, the whole world seems to have reached a tipping point. Countries, cities, companies, and campuses all are vying to be the most green. This may be a very good time to work on social change and sustainable living. We hope that you'll find the information in this chapter helpful. As the famous anthropologist Margaret Mead said, "Never doubt that a small group of thoughtful, committed people can change the world. Indeed, it is the only thing that ever has."

REVIEWING LEARNING OUTCOMES

By now you should be able to explain the following points:

25.1 Explain how we can make a difference.

25.2 Summarize environmental education.
- Environmental literacy means understanding our environment.
- Citizen science encourages everyone to participate.
- Environmental careers range from engineering to education.
- Green business and technology are growing fast.

25.3 Evaluate what individuals can do.
- How much is enough?
- We can choose to reduce our environmental impacts.
- "Green washing" can mislead consumers.
- Certification identifies low-impact products.
- Green consumerism has limits.

25.4 Review how we can work together.
- National organizations are influential but sometimes complacent.
- New players bring energy to environmental policy.
- International nongovernmental organizations mobilize many people.

25.5 Investigate campus greening.
- Electronic communication is changing the world.
- Schools can be environmental leaders.
- Your campus can reduce energy consumption.

25.6 Define the challenge of sustainability.

PRACTICE QUIZ

1. Describe four major contexts for outcomes from environmental education.
2. Define *conspicuous consumption.*
3. Explain why vegetarianism can have a lower climate change impact than a beef-based diet.
4. Give two examples of green washing.
5. List five things that you can do to reduce your environmental impact.
6. List six stages in the Life Cycle Inventory at which we can analyze material and energy balances of products.
7. Identify the ten biggest environmental organizations.
8. List six goals of sustainable development.
9. Identify two key messages from the UN millennium assessment that you believe are most important for environmental science.
10. Identify two goals or objectives from the UN millennium goals that you believe are most important for environmental science.

CRITICAL THINKING AND DISCUSSION QUESTIONS

1. What lessons do you derive from the case study about 350.org? If you were interested in bringing about change in your neighborhood or in the wider world, which of the tactics used in this effort might you use for your campaign?

2. Reflect on how you learned about environmental issues. What have been the most important formative experiences or persuasive arguments in shaping your own attitudes. If you were designing an environmental education program for youth, what elements would you include?

3. How might it change your life if you were to minimize your consumption of materials and resources? Which aspects could you give up, and what is absolutely essential to your happiness and well-being. Does your list differ from that of your friends and classmates?

4. Have you ever been involved in charitable or environmental work? What were the best and worst aspects of that experience? If you haven't yet done anything of this sort, what activities seem appealing and worthwhile to you?

5. What green activities are now occurring at your school? How might you get involved?

6. In the practice quiz, we asked you to identify two key messages from the millennium assessment and two goals and objectives that you believe are most important for environmental science. Why did you choose these messages and goals? How might we accomplish them?

 Data Analysis: **Campus Environmental Audit**

How sustainable is your school? What could you, your fellow students, the faculty, staff, and administration do to make your campus more environmentally friendly? Perhaps you and your classmates could carry out an environmental audit of your school. Some of the following items are things you could observe for yourself; other information you'd need to get from the campus administrators.

1. *Energy.* How much total energy does your campus use each year? Is any of it from renewable sources? How does your school energy use compare to that of a city with the same population? Could you switch to renewable sources? How much would that cost? How long would the payback time be for various renewable sources? Is there a campus policy about energy conservation? What would it take to launch a campaign for using resources efficiently?

2. *Buildings.* Are any campus buildings now LEED certified? Do any campus buildings now have compact fluorescent bulbs, high-efficiency fans, or other energy-saving devices? Do you have single-pane or double-pane windows? Are lights turned off when rooms aren't in use? At what temperatures (winter and summer) are classrooms, offices, and dorms maintained? Who makes this decision? Could you open a window in hot weather? Are new buildings being planned? Will they be LEED certified? If not, why?

3. *Transportation.* Does your school own any fuel-efficient vehicles (hybrids or other high-mileage models)? If you were making a presentation to an administrator to encourage him or her to purchase efficient vehicles, what arguments would you use? How many students commute to campus? Are they encouraged to carpool or use public transportation? How might you promote efficient transportation? How much total

space on your campus is devoted to parking? What's the cost per vehicle to build and maintain parking? How else might that money be spent to facilitate efficient transportation? Where does runoff from parking lots and streets go? What are the environmental impacts of this storm runoff?

4. *Water use.* What's the source of your drinking water? How much does your campus use? Where does wastewater go? How many toilets are on the campus? How much water does each use for every flush? How much would it cost to change to low-flow appliances? How much would it save in terms of water use and cost?

5. *Food.* What's the source of food served in campus dining rooms? Is any of it locally grown or organic? How much junk food is consumed annually? What are the barriers to buying locally grown, fair-trade, organic, free-range food? Does the campus grow any of its own food? Would that be possible?

6. *Ecosystem restoration.* Are there opportunities for reforestation, stream restoration, wetland improvements, or other ecological repair projects on your campus. What percentage of the vegetation on campus is native? What might be the benefits of replacing non-native species with indigenous varieties? Have gardeners considered planting species that provide food and shelter for wildlife?

What other aspects of your campus life could you study to improve sustainability? How could you organize a group project to promote beneficial changes in your school's environmental impacts?

Glossary

A

abiotic Nonliving.

abundance The number or amount of something.

acid precipitation Acidic rain, snow, or dry particles deposited from the air due to an increased release of acids by anthropogenic or natural resources.

acids Substances that release hydrogen ions (protons) in water.

active learner Someone who understands and remembers best by doing things physically.

active solar system A mechanical system that actively collects, concentrates, and stores solar energy.

acute effects Sudden, severe effects.

acute poverty Insufficient income or access to resources needed to provide the basic necessities for life, such as food, shelter, sanitation, clean water, medical care, and education.

adaptation The acquisition of traits that allow a species to survive and thrive in its environment.

adaptive management A management plan designed from the outset to "learn by doing" and to actively test hypotheses and adjust treatments as new information becomes available.

administrative law Executive orders, administrative rules and regulations, and enforcement decisions by administrative agencies and special administrative courts.

aerobic Living or occurring only in the presence of oxygen.

aerosols Minute particles or liquid droplets suspended in the air.

aesthetic degradation Changes in environmental quality that offend our aesthetic senses.

affluenza An addiction to spending and consuming beyond one's needs.

albedo A description of a surface's reflective properties.

allergens Substances that activate the immune system.

allopatric speciation Speciation deriving from a common ancestor due to geographic barriers that cause reproductive isolation.

ambient air The air immediately around us.

amino acid An organic compound containing an amino group and a carboxyl group; amino acids are the units or building blocks that make peptide and protein molecules.

amorphous silicon collectors Photovoltaic cells that collect solar energy and convert it to electricity using noncrystalline (randomly arranged) thin films of silicon.

anaerobic respiration The incomplete intracellular breakdown of sugar or other organic compounds in the absence of oxygen that releases some energy and produces organic acids and/or alcohol.

analytical thinking Thinking that breaks a problem down into its constituent parts.

anemia Low levels of hemoglobin due to iron deficiency or lack of red blood cells.

annual A plant that lives for a single growing season.

anthropocentrism The belief that humans hold a special place in nature; being centered primarily on humans and human affairs.

antigens Chemical compounds to which antibodies bind.

appropriate technology Technology that can be made at an affordable price by ordinary people using local materials to do useful work in ways that do the least possible harm to both human society and the environment.

aquaculture Growing aquatic species in net pens or tanks.

aquifers Porous, water-bearing layers of sand, gravel, and rock below the earth's surface; reservoirs for groundwater.

arbitration A formal process of dispute resolution in which there are stringent rules of evidence, cross-examination of witnesses, and a legally binding decision made by the arbitrator that all parties must obey.

arithmetic scale A scale that uses ordinary numbers as units in a linear sequence.

artesian well A pressurized aquifer from which water gushes without being pumped, due to the aquifer's intersecting the surface or being penetrated by a pipe or conduit; *also called* a spring.

asthma A distressing disease characterized by shortness of breath, wheezing, and bronchial muscle spasms.

atmospheric deposition Sedimentation of solids, liquids, or gaseous materials from the air.

atom The smallest unit of matter that has the characteristics of an element; consists of three main types of subatomic particles: protons, neutrons, and electrons.

atomic number The characteristic number of protons per atom of an element; used as an identifying attribute.

autotroph An organism that synthesizes food molecules from inorganic molecules by using an external energy source, such as light energy.

B

barrier islands Low, narrow, sandy islands that form offshore from a coastline.

bases Substances that bond readily with hydrogen ions.

BAT *See* best available, economically achievable technology.

Batesian mimicry Evolution by one species to resemble the coloration, body shape, or behavior of another species that is protected from predators by a venomous stinger, bad taste, or some other defensive adaptation.

benthic The bottom of a sea or lake.

best available, economically achievable technology (BAT) The best pollution control available.

best practicable control technology (BPT) The best technology for pollution control available at reasonable cost and operable under normal conditions.

beta particles High-energy electrons released by radioactive decay.

bill A piece of legislation introduced in Congress and intended to become law.

binomials Two-part names (genus and species, usually in Latin) invented by Carl Linneaus to show taxonomic relationships.

bioaccumulation The selective absorption and concentration of molecules by cells.

biocentric preservation A philosophy that emphasizes the fundamental right of living organisms to exist and to pursue their own goods.

biocentrism The belief that all creatures have rights and values; being centered on nature rather than humans.

biochemical oxygen demand A standard test of water pollution, measured by the amount of dissolved oxygen consumed by aquatic organisms over a given period.

biocide A broad-spectrum poison that kills a wide range of organisms.

biodegradable plastics Plastics that can be decomposed by microorganisms.

biodiversity The genetic, species, and ecological diversity of the organisms in a given area.

biodiversity hot spots Areas with exceptionally high numbers of endemic species.

biofuel Fuels such as ethanol, methanol, or vegetable oils from crops.

biogeochemical cycles Movement of matter within or between ecosystems; caused by living organisms, geological forces, or chemical reactions. The cycling of nitrogen, carbon, sulfur, oxygen, phosphorus, and water are examples.

biogeographical area An entire self-contained natural ecosystem and its associated land, water, air, and wildlife resources.

biological community The populations of plants, animals, and microorganisms living and interacting in a certain area at a given time.

biological controls Use of natural predators, pathogens, or competitors to regulate pest populations.

biological or biotic factors Organisms and products of organisms that are part of the environment and potentially affect the life of other organisms.

biological oxygen demand (BOD) A standard test for measuring the amount of dissolved oxygen utilized by aquatic microorganisms.

579

biological pests Organisms that reduce the availability, quality, or value of resources useful to humans.

biological resources The earth's organisms.

biomagnification An increase in the concentration of certain stable chemicals (for example, heavy metals or fat-soluble pesticides) in successively higher trophic levels of a food chain or web.

biomass The total mass or weight of all the living organisms in a given population or area.

biomass fuel Organic material produced by plants, animals, or microorganisms that can be burned directly as a heat source or converted into a gaseous or liquid fuel.

biomass pyramid A metaphor or diagram that explains the relationship between the amounts of biomass at different trophic levels.

biome A broad, regional type of ecosystem characterized by distinctive climate and soil conditions and a distinctive kind of biological community adapted to those conditions.

bioremediation The use of biological organisms to remove or detoxify pollutants from a contaminated area.

biosphere The zone of air, land, and water at the surface of the earth that is occupied by organisms.

biosphere reserves World heritage sites identified by the IUCN as worthy of national park or wildlife refuge status because of high biological diversity or unique ecological features.

biota All organisms in a given area.

biotic Pertaining to life; environmental factors created by living organisms.

biotic potential The maximum reproductive rate of an organism, given unlimited resources and ideal environmental conditions. *Compare* environmental resistance.

birth control Any method used to reduce births, including abstinence, delayed marriage, contraception, devices or medications that prevent implantation of fertilized zygotes, and induced abortions.

black lung disease Inflammation and fibrosis caused by accumulation of coal dust in the lungs or airways. *See* respiratory fibrotic agents.

blind experiments Experiments in which those carrying out the experiment don't know, until after the gathering and analysis of data, which was the experimental treatment and which was the control.

blue revolution New techniques of fish farming that may contribute as much to human nutrition as miracle cereal grains but also may create social and environmental problems.

body burden The sum total of all persistent toxins in our body that we accumulate from our air, water, diet, and surroundings.

bog An area of waterlogged soil that tends to be peaty; bogs are fed mainly by precipitation and have low productivity, and some are acidic.

boom-and-bust cycles Population cycles characterized by repeated overshoot of the carrying capacity of the environment followed by population crashes.

boreal forest A broad band of mixed coniferous and deciduous trees that stretches across northern North America (and also Europe and Asia); its northernmost edge, the taiga, intergrades with the arctic tundra.

BPT *See* best practical control technology.

breeder reactor A nuclear reactor that produces fuel by bombarding isotopes of uranium and thorium with high-energy neutrons that convert inert atoms to fissionable ones.

bronchitis A persistent inflammation of bronchi and bronchioles (large and small airways in the lungs).

brownfield development Building on abandoned or reclaimed polluted industrial sites.

brownfields Abandoned or underused urban areas in which redevelopment is blocked by liability or financing issues related to toxic contamination.

buffalo commons A large open area proposed for the Great Plains in which wildlife and native people could live as they once did without interference by industrialized society.

C

cancer Invasive, out-of-control cell growth that results in malignant tumors.

cap-and-trade An approach to controlling pollution by mandating upper limits (the cap) on how much each country, sector, or specific industry is allowed to emit. Companies that can cut pollution by more than they're required to can sell the credit to other companies that have more difficulty meeting their mandated levels.

capital Any form of wealth, resources, or knowledge available for use in the production of more wealth.

captive breeding Raising plants or animals in zoos or other controlled conditions to produce stock for subsequent release into the wild.

carbamates Urethanes, such as carbaryl and aldicarb, that are used as pesticides.

carbohydrate An organic compound consisting of a ring or chain of carbon atoms with hydrogen and oxygen attached; examples are sugars, starches, cellulose, and glycogen.

carbon cycle The circulation and reutilization of carbon atoms, especially via the processes of photosynthesis and respiration.

carbon monoxide (CO) A colorless, odorless, non-irritating, but highly toxic gas produced by incomplete combustion of fuel, incineration of biomass or solid waste, or partially anaerobic decomposition of organic material.

carbon neutral A system or process that doesn't release more carbon to the atmosphere than it consumes.

carbon sequestration Storing carbon (usually in the form of CO_2) in geological formations or at the bottom of the ocean.

carbon sink Places of carbon accumulation, such as in large forests (organic compounds) or ocean sediments (calcium carbonate); carbon is thus removed from the carbon cycle for moderately long to very long periods of time.

carbon source The originating point of carbon that reenters the carbon cycle; sources include cellular respiration and combustion.

carcinogens Substances that cause cancer.

carnivores Organisms that prey mainly upon animals.

carrying capacity The maximum number of individuals of any species that can be supported by a particular ecosystem on a long-term basis.

case law Precedents from both civil and criminal court cases.

cash crops Crops that are sold rather than consumed or bartered.

catastrophic systems Dynamic systems that jump abruptly from one seemingly steady state to another without any intermediate stages.

cell Minute biological compartments within which the processes of life are carried out.

cellular respiration The process in which a cell breaks down sugar or other organic compounds to release energy used for cellular work; may be anaerobic or aerobic, depending on the availability of oxygen.

cellulosic Composed primarily of cellulose.

chain reaction A self-sustaining reaction in which the fission of nuclei produces subatomic particles that cause the fission of other nuclei.

chaotic systems Systems that exhibit variability, which may not necessarily be random, yet whose complex patterns are not discernible over a normal human time scale.

chaparral Thick, dense, thorny, evergreen scrub found in Mediterranean climates.

chemical bond The force that holds atoms together in molecules and compounds.

chemical energy Potential energy stored in chemical bonds of molecules.

chemosynthesis The process in which inorganic chemicals, such as hydrogen sulfide (HS) or hydrogen gas (H^2), serve as an energy source for synthesis of organic molecules.

chlorinated hydrocarbons Hydrocarbon molecules to which chlorine atoms are attached.

chlorofluorocarbons (CFCs) Chemical compounds with a carbon skeleton and one or more attached chlorine and fluorine atoms. Commonly used as refrigerants, solvents, fire retardants, and blowing agents.

chloroplasts Chlorophyll-containing organelles in eukaryotic organisms; sites of photosynthesis.

chronic effects Long-lasting results of exposure to a toxin; can be caused by a single, acute exposure or a continuous, low-level exposure.

chronic food shortages Long-term undernutrition and malnutrition; usually caused by people's lack of money to buy food or lack of opportunity to grow it themselves.

chronic obstructive lung disease Irreversible damage to the linings of the lungs caused by irritants.

chronically undernourished Those people whose diet doesn't provide the 2,200 kcal per day, on average, considered necessary for a healthy, productive life.

citizen science Projects in which trained volunteers work with scientific researchers to answer real-world questions.

city A differentiated community with a sufficient population and resource base to allow residents to specialize in arts, crafts, services, and professional occupations.

civil law A body of laws regulating relations between individuals, or between individuals and corporations, concerning property rights, personal dignity and freedom, and personal injury.

classical economics Modern, Western economic theories of the effects of resource scarcity, monetary policy, and competition on supply of and demand for goods and services in the marketplace. This is the basis for the capitalist market system.

clear-cut Cutting every tree in a given area, regardless of species or size; an appropriate harvest method for some species; can be destructive if not carefully controlled.

climate A description of the long-term pattern of weather in a particular area.

climax community A relatively stable, long-lasting community reached in a successional series; usually determined by climate and soil type.

closed canopy A forest where tree crowns spread over 20 percent of the ground; has the potential for commercial timber harvests.

cloud forests High mountain forests where temperatures are uniformly cool and fog or mist keeps vegetation wet all the time.

coal gasification The heating and partial combustion of coal to release volatile gases, such as methane and carbon monoxide; after pollutants are washed out, these gases become efficient, clean-burning fuel.

coal-to-liquid (CTL) technology Technology that turns coal into liquid fuel.

coal washing Coal technology that involves crushing coal and washing out soluble sulfur compounds with water or other solvents.

Coastal Zone Management Act Legislation of 1972 that gave federal money to 30 seacoast and Great Lakes states for development and restoration projects.

co-composting Microbial decomposition of organic materials in solid waste into useful soil additives and fertilizer. Often, extra organic materials, such as sewer sludge, animal manure, leaves, and grass clippings, are added to solid waste to speed the process and make the product more useful.

coevolution The process in which species exert selective pressure on each other and gradually evolve new features or behaviors as a result of those pressures.

cogeneration The simultaneous production of electricity and steam or hot water in the same plant.

cold front A moving boundary of cooler air displacing warmer air.

coliform bacteria Bacteria that live in the intestines (including the colon) of humans and other animals; used as a measure of the presence of feces in water or soil.

commensalism A symbiotic relationship in which one member is benefited and the second is neither harmed nor benefited.

common law The body of court decisions that constitute a working definition of individual rights and responsibilities where no formal statutes define these issues.

communal resource management systems Resources managed by a community for long-term sustainability.

community-supported agriculture (CSA) A program in which you make an annual contribution to a local farm in return for weekly deliveries of a "share" of whatever the farm produces.

competitive exclusion A theory that no two populations of different species will occupy the same niche and compete for exactly the same resources in the same habitat for very long.

complexity (ecological) The number of species at each trophic level and the number of trophic levels in a community.

composting The biological degradation of organic material under aerobic (oxygen-rich) conditions to produce compost, a nutrient-rich soil amendment and conditioner.

compound A molecule made up of two or more kinds of atoms held together by chemical bonds.

conclusion A statement that follows logically from a set of premises.

condensation The aggregation of water molecules from vapor to liquid or solid when the saturation concentration is exceeded.

confidence limits A statistical measure of the quality of data that tells you how close the sample's average probably is to the average for the entire population of that species.

confined animal feeding operations Facilities in which large numbers of animals spend most or all of their life in confinement.

conifers Needle-bearing trees that produce seeds in cones.

conservation development Consideration of landscape history, human culture, topography, and ecological values in subdivision design. Using cluster housing, zoning, covenants, and other design features, at least half of a subdivision can be preserved as open space, farmland, or natural areas.

conservation medicine A medical field that attempts to understand how environmental changes threaten our own health as well as that of the natural communities on which we depend for ecological services.

conservation of matter In any chemical reaction, matter changes form; it is neither created nor destroyed.

conspicuous consumption A term coined by economist and social critic Thorstein Veblen to describe buying things we don't want or need in order to impress others.

consumer An organism that obtains energy and nutrients by feeding on other organisms or their remains. *See also* heterotroph.

consumption The fraction of withdrawn water that is lost in transmission or that is evaporated, absorbed, chemically transformed, or otherwise made unavailable for other purposes as a result of human use.

contour plowing Plowing along hill contours; reduces erosion.

control rods Neutron-absorbing material inserted into spaces between fuel assemblies in nuclear reactors to regulate fission reactions.

controlled studies Studies in which comparisons are made between experimental and control populations that are (as far as possible) identical in every factor except the one variable being studied.

convection currents Rising or sinking air currents that stir the atmosphere and transport heat from one area to another. Convection currents also occur in water; *see* spring overturn.

conventional pollutants The seven major pollutants (sulfur dioxide, carbon monoxide, particulates, hydrocarbons, nitrogen oxides, photochemical oxidants, and lead) identified and regulated by the U.S. Clean Air Act.

cool deserts Deserts such as the American Great Basin characterized by cold winters and sagebrush.

coral bleaching Whitening of corals caused by expulsion of symbiotic algae—often resulting from high water temperatures, pollution, or disease.

coral reefs Prominent oceanic features composed of hard, limy skeletons produced by coral animals; usually formed along edges of shallow, submerged ocean banks or along shelves in warm, shallow, tropical seas.

core (earth's) The dense, intensely hot mass of molten metal, mostly iron and nickel, thousands of kilometers in diameter at the earth's center.

core habitat Essential habitat for a species.

core region The primary industrial region of a country; usually located around the capital or largest port; has both the greatest population density and the greatest economic activity of the country.

Coriolis effect The influence of friction and drag on air layers near the earth; deflects air currents to the direction of the earth's rotation.

cornucopian fallacy The belief that nature is limitless in its abundance and that perpetual growth is not only possible but essential.

corridor A strip of natural habitat that connects two adjacent nature preserves to allow migration of organisms from one place to another.

cost–benefit analysis (CBA) An evaluation of large-scale public projects by comparing the costs and benefits that accrue from them.

cover crops Plants, such as rye, alfalfa, or clover, that can be planted immediately after harvest to hold and protect the soil.

creative thinking Asks, how could I do this differently?

credit An amount of pollution a company is allowed to sell when they reduce emissions below their allowed cap. *See* cap-and-trade.

criminal law A body of court decisions based on federal and state statutes concerning wrongs against persons or society.

criteria pollutants *See* conventional pollutants.

critical factor The single environmental factor closest to a tolerance limit for a given species at a given time. *See* limiting factors.

critical thinking An ability to evaluate information and opinions in a systematic, purposeful, efficient manner.

croplands Lands used to grow crops.

crude birth rate The number of births in a year divided by the midyear population.

crude death rate The number of deaths per thousand persons in a given year; *also called* crude mortality rate.

crust The cool, lightweight, outermost layer of the earth's surface that floats on the soft, pliable, underlying layers; similar to the "skin" on a bowl of warm pudding.

cultural eutrophication An increase in biological productivity and ecosystem succession caused by human activities.

D

debt-for-nature swap Forgiveness of international debt in exchange for nature protection in developing countries.

deciduous Trees and shrubs that shed their leaves at the end of the growing season.

decline spiral A catastrophic deterioration of a species, community, or whole ecosystem; accelerates as functions are disrupted or lost in a downward cascade.

decomposers Fungi and bacteria that break down complex organic material into smaller molecules.

deductive reasoning Deriving testable predictions about specific cases from general principles.

deep ecology A philosophy that calls for a profound shift in our attitudes and behavior toward nature.

degradation (of a water resource) Deterioration in water quality due to contamination or pollution; makes water unsuitable for other desirable purposes.

Delaney clause A controversial amendment to the Federal Food, Drug, and Cosmetic Act, added in 1958, prohibiting the addition of any known cancer-causing agent to processed foods, drugs, or cosmetics.

delta Fan-shaped sediment deposit found at the mouth of a river.

demand The amount of a product that consumers are willing and able to buy at various possible prices, assuming they are free to express their preferences.

demanufacturing Disassembly of products so components can be reused or recycled.

demographic bottleneck A population founded when just a few members of a species survive a catastrophic event or colonize new habitat geographically isolated from other members of the same species.

demographic transition A pattern of falling death rates and birthrates in response to improved living conditions; could be reversed in deteriorating conditions.

demography Vital statistics about people: births, marriages, deaths, etc.; the statistical study of human populations relating to growth rate, age structure, geographic distribution, etc., and their effects on social, economic, and environmental conditions.

denitrifying bacteria Free-living soil bacteria that convert nitrates to gaseous nitrogen and nitrous oxide.

density-dependent Factors affecting population growth that change as population size changes.

dependency ratio The number of nonworking members compared to working members for a given population.

dependent (response) variable A variable that is affected by the condition being altered in a manipulative experiment.

desalinization (or desalination) Removal of salt from water by distillation, freezing, or ultrafiltration.

desert A type of biome characterized by low moisture levels and infrequent and unpredictable precipitation. Daily and seasonal temperatures fluctuate widely.

desertification Conversion of productive lands to desert.

detritivore Organisms that consume organic litter, debris, and dung.

dew point The temperature at which condensation occurs for a given concentration of water vapor in the air.

dieback A sudden population decline; *also called* population crash.

diminishing returns A condition in which unrestrained population growth causes the standard of living to decrease to a subsistence level where poverty, misery, vice, and starvation make life permanently drab and miserable. This dreary prophecy of diminishing returns has led economics to be called "the dismal science."

disability-adjusted life years (DALY) A measure of premature deaths and losses due to illnesses and disabilities in a population.

discharge The amount of water that passes a fixed point in a given amount of time; usually expressed as liters or cubic feet of water per second.

disclimax community *See* equilibrium community.

discount rates The difference between present value and future value of a resource; generally equivalent to an interest rate.

disease A deleterious change in the body's condition in response to destabilizing factors such as nutrition, chemicals, or biological agents.

dissolved oxygen (DO) content The amount of oxygen dissolved in a given volume of water at a given temperature and atmospheric pressure; usually expressed in parts per million (ppm).

disturbance-adapted species Species that depend on disturbances to succeed.

disturbances Periodic destructive events such as fires or floods; changes in an ecosystem that affect (positively or negatively) the organisms living there.

diversity (species diversity, biological diversity) The number of species present in a community (species richness), as well as the relative abundance of each species.

DNA (deoxyribonucleic acid) A giant molecule composed of millions or billions of nucleotides (sugars and bases called purines and pyramidines held together by phosphate bonds) that form a double helix and store genetic information in all living cells.

dominant plants Those plant species in a community that provide a food base for most of the community; they usually take up the most space and have the largest biomass.

double-blind design A design in which neither the experimenter nor the subjects know, until after the gathering and analysis of data, which was the experimental treatment and which was the control.

drip irrigation The use of pipe or tubing perforated with very small holes to deliver water one drop at a time directly to the soil around each plant.

dry alkali injection Spraying dry sodium bicarbonate into flue gas to absorb and neutralize acidic sulfur compounds.

E

Earth Charter A set of principles for sustainable development, environmental protection, and social justice developed by a council appointed by the United Nations.

earthquakes Sudden, violent movement of the earth's crust.

ecocentric (ecologically centered) A philosophy that claims moral values and rights for both organisms and ecological systems and processes.

ecofeminism A pluralistic, nonhierarchical, relationship-oriented philosophy that suggests how humans could reconceive themselves and their relationships to nature in nondominating ways as an alternative to patriarchal systems of domination.

ecojustice Justice in the social order and integrity in the natural order.

ecological development A gradual process of environmental modification by organisms.

ecological diseases Emergent diseases (new or rarely seen diseases) that cause devastating epidemics among wildlife and domestic animals.

ecological economics A relatively new field that brings the insights of ecology to economic analysis.

ecological equivalents Different species that occupy similar ecological niches in similar ecosystems in different parts of the world.

ecological footprint A measure that computes the demands placed on nature by individuals and nations.

ecological niche The functional role and position of a species (population) within a community or an ecosystem, including what resources it uses, how and when it uses the resources, and how it interacts with other populations.

ecological succession The process by which organisms occupy a site and gradually change environmental conditions so that other species can replace the original inhabitants.

ecology The scientific study of relationships between organisms and their environment. It is concerned with the life histories, distribution, and behavior of individual species as well as the structure and function of natural systems at the level of populations, communities, and ecosystems.

economic development A rise in real income *per person;* usually associated with new technology that increases productivity or resources.

economic growth An increase in the total wealth of a nation; if population grows faster than the economy, there may be real economic growth, but the share per person may decline.

economic thresholds In pest management, the point at which the cost of pest damage exceeds the costs of pest control.

ecosystem A specific biological community and its physical environment interacting in an exchange of matter and energy.

ecosystem management An integration of ecological, economic, and social goals in a unified systems approach to resource management.

ecosystem restoration Reinstating an entire community of organisms to as near its natural condition as possible.

ecotone A boundary between two types of ecological communities.

ecotourism A combination of adventure travel, cultural exploration, and nature appreciation in wild settings.

edge effects A change in species composition, physical conditions, or other ecological factors at the boundary between two ecosystems.

effluent sewerage A low-cost alternative sewage treatment for cities in poor countries that combines some features of septic systems and centralized municipal treatment systems.

electron A negatively charged subatomic particle that orbits around the nucleus of an atom.

electronic waste *See* e-waste.

electrostatic precipitators The most common particulate controls in power plants; fly ash particles pick up an electrostatic surface charge as they pass between large electrodes in the effluent stream, causing particles to migrate to the oppositely charged plate.

element A molecule composed of one kind of atom; cannot be broken into simpler units by chemical reactions.

El Niño A climatic change marked by the shifting of a large warm water pool from the western Pacific Ocean toward the east. Wind direction and precipitation patterns are changed over much of the Pacific and perhaps around the world.

emergent diseases A new disease or one that has been absent for at least 20 years.

emergent properties Characteristics of whole, functioning systems that are quantitatively or qualitatively greater than the sum of the system's parts.

emigration The outward movement of members from a population.

emissions standards Regulations for restricting the amounts of air pollutants that can be released from specific point sources.

emissions trading Programs in which companies that have cut pollution more than they're required to can sell "credits" to other companies that still exceed allowed levels.

endangered species A species considered to be in imminent danger of extinction.

endemism A state in which species are restricted to a single region.

endocrine disrupters Chemicals that disrupt normal hormone functions.

energy The capacity to do work (that is, to change the physical state or motion of an object).

energy crops Crops that can be used to make ethanol or diesel fuel.

energy efficiency A measure of energy produced compared to energy consumed.

energy pyramid A representation of the loss of useful energy at each step in a food chain.

energy recovery Incineration of solid waste to produce useful energy.

entropy Disorder in a system.

environment The circumstances or conditions that surround an organism or group of organisms as well as the complex of social or cultural conditions that affect an individual or community.

environmental economics *See* ecological economics.

environmental ethics A search for moral values and ethical principles in human relations with the natural world.

environmental governance Rules and regulations that govern our impacts on the environment and natural resources.

environmental health The science of external factors that cause disease, including elements of the natural, social, cultural, and technological worlds in which we live.

environmental hormones Chemical pollutants that substitute for, or interfere with, naturally occurring hormones in our bodies; these chemicals may trigger reproductive failure, developmental abnormalities, or tumor promotion.

environmental impact statement (EIS) An analysis, required by provisions in the National Environmental Policy Act of 1970, of the effects of any major program a federal agency plans to undertake.

environmental indicators Organisms or physical factors that serve as a gauge for environmental changes. More specifically, organisms with these characteristics are called bioindicators.

Environmental Performance Index (EPI) A measure that evaluates national sustainability and progress toward achievement of the United Nations Millennium Development Goals.

environmentalism Active participation in attempts to solve environmental pollution and resource problems.

environmental justice A recognition that access to a clean, healthy environment is a fundamental right of all human beings.

environmental law The special body of official rules, decisions, and actions concerning environmental quality, natural resources, and ecological sustainability.

environmental literacy Fluency in the principles of ecology that gives us a working knowledge of the basic grammar and underlying syntax of environmental wisdom.

environmental policy The official rules or regulations concerning the environment adopted, implemented, and enforced by some governmental agency.

environmental racism Decisions that restrict certain people or groups of people to polluted or degraded environments on the basis of race.

environmental resistance All the limiting factors that tend to reduce population growth rates and set the maximum allowable population size or carrying capacity of an ecosystem.

environmental resources Anything an organism needs that can be taken from the environment.

environmental science The systematic, scientific study of our environment as well as our role in it.

enzymes Molecules, usually proteins or nucleic acids, that act as catalysts in biochemical reactions.

epidemiology The study of the distribution and causes of disease and injuries in human populations.

epiphyte A plant that grows on a substrate other than the soil, such as the surface of another organism.

equilibrium community A community subject to periodic disruptions, usually by fire, that prevent it from reaching a climax stage; *also called disclimax community.*

estuary A bay or drowned valley where a river empties into the sea.

ethics A branch of philosophy concerned with right and wrong.

eukaryotic cell A cell containing a membrane-bounded nucleus and membrane-bounded organelles.

eutrophic Rivers and lakes rich in organisms and organic material (*eu* = truly; *trophic* = nutritious).

evolution A theory that explains how random changes in genetic material and competition for scarce resources cause species to change gradually.

e-waste Discarded electronic equipment, such as computers, cell phones, and television sets.

exhaustible resources Generally considered the earth's geologic endowment: minerals, nonmineral resources, fossil fuels, and other materials present in fixed amounts in the environment.

existence value The importance we place on just knowing that a particular species or a specific organism exists.

exotic organisms Alien species introduced by human agency into biological communities where they would not naturally occur.

exponential growth Growth at a constant rate of increase per unit of time; can be expressed as a constant fraction or exponent. *See* geometric growth.

external costs Expenses, monetary or otherwise, borne by someone other than the individuals or groups who use a resource.

extinction The irrevocable elimination of species; can be a normal process of the natural world as species outcompete or kill off others or as environmental conditions change.

extirpate To destroy totally; extinction caused by direct human action, such as hunting or trapping.

extreme poverty Living on less than (U.S.) $1 per day.

F

family planning Controlling reproduction; planning the timing of birth and having the number of babies that are wanted and can be supported.

famines Acute food shortages characterized by large-scale loss of life, social disruption, and economic chaos.

fauna All of the animals present in a given region.

fecundity The physical ability to reproduce.

fen An area of waterlogged soil that tends to be peaty; fed mainly by upwelling water; low productivity.

feral A domestic animal that has taken up a wild existence.

fermentation (alcoholic) A type of anaerobic respiration that yields carbon dioxide and alcohol.

fertility A measurement of the actual number of offspring produced through sexual reproduction; usually described in terms of number of offspring of females, because paternity can be difficult to determine.

fetal alcohol syndrome A tragic set of permanent physical, mental, and behavioral birth defects that result from the mother's drinking alcohol during pregnancy.

fibrosis The general name for accumulation of scar tissue in the lung.

filters A porous mesh of cotton cloth, spun glass fibers, or asbestos-cellulose that allows air or liquid to pass through but holds back solid particles.

fire-climax community An equilibrium community maintained by periodic fires; examples include grasslands, chaparral shrubland, and some pine forests.

flex-fuel vehicles Vehicles that can burn variable mixtures of gasoline and ethanol.

floodplains Low lands along riverbanks, lakes, and coastlines subjected to periodic inundation.

flora All of the plants present in a given region.

flue-gas scrubbing Treating combustion exhaust gases with chemical agents to remove pollutants. Spraying crushed limestone and water into the exhaust gas stream to remove sulfur is a common scrubbing technique.

food aid Financial assistance intended to boost less-developed countries' standards of living.

food chain A linked feeding series; in an ecosystem, the sequence of organisms through which energy and materials are transferred, in the form of food, from one trophic level to another.

food security The ability of individuals to obtain sufficient food on a day-to-day basis.

food surpluses Excess food supplies.

food web A complex, interlocking series of individual food chains in an ecosystem.

forest Any area where trees cover more than 10 percent of the land.

forest management Scientific planning and administration of forest resources for sustainable harvest, multiple use, regeneration, and maintenance of a healthy biological community.

fossil fuels Petroleum, natural gas, and coal created by geological forces from organic wastes and dead bodies of formerly living biological organisms.

founder effect The effect on a population founded when just a few members of a species survive a catastrophic event or colonize new habitat geographically isolated from other members of the same species.

freezing condensation A process that occurs in the clouds when ice crystals trap water vapor. As the ice crystals become larger and heavier, they begin to fall as rain or snow.

fresh water Water other than seawater; covers only about 2 percent of earth's surface; includes streams, rivers, lakes, ponds, and water associated with several kinds of wetlands.

freshwater ecosystems Ecosystems in which the fresh (nonsalty) water of streams, rivers, ponds, or lakes plays a defining role.

front The boundary between two air masses of different temperature and density.

fuel assembly A bundle of hollow metal rods containing uranium oxide pellets; used to fuel a nuclear reactor.

fuel cells Mechanical devices that use hydrogen or hydrogen-containing fuels such as methane to produce an electric current. Fuel cells are clean, quiet, and highly efficient sources of electricity.

fuel-switching A change from one fuel to another.

fuelwood Branches, twigs, logs, wood chips, and other wood products harvested for use as fuel.

fugitive emissions Substances that enter the air without going through a smokestack, such as dust from soil erosion, strip mining, rock crushing, construction, and building demolition.

fumigants Toxic gases such as methyl bromine that are used to kill pests.

fungi One of the five kingdom classifications. Fungi are nonphotosynthetic, eukaryotic organisms with cell walls, filamentous bodies, and absorptive nutrition.

fungicide A chemical that kills fungi.

G

Gaia hypothesis A theory that the living organisms of the biosphere form a single, complex interacting system that creates and maintains a habitable Earth; named after Gaia, the Greek "Earth mother" goddess.

gamma rays Very short-wavelength forms of the electromagnetic spectrum.

gap analysis A biogeographical technique of mapping biological diversity and endemic species to find gaps between protected areas that leave endangered habitats vulnerable to disruption.

garden city A new town with special emphasis on landscaping and rural ambience.

gasohol A mixture of gasoline and ethanol.

gene A unit of heredity; a segment of DNA that contains information for the synthesis of a specific protein, such as an enzyme.

gene banks Storage for seed varieties for future breeding experiments.

general fertility rate Crude birthrate multiplied by the percentage of reproductive-age women.

genetic assimilation The disappearance of a species as its genes are diluted through crossbreeding with a closely related species.

genetic drift The gradual changes in gene frequencies in a population due to random events.

genetic engineering Laboratory manipulation of genetic material using molecular biology techniques to create desired characteristics in organisms.

genetically modified organisms (GMOs) Organisms whose genetic code has been altered by artificial means such as interspecies gene transfer.

genuine progress index (GPI) An alternative to GNP or GDP for economic accounting that measures real progress in quality of life and sustainability.

geographic information systems (GIS) A system of spatial data, such as boundaries or road networks, and computer software to display and analyze those data.

geographic isolation *See* allopatric speciation.

geometric growth Growth that follows a geometric pattern of increase, such as 2, 4, 8, 16, …. *See* exponential growth.

geothermal energy Energy drawn from the internal heat of the earth, either through geysers, fumaroles, hot springs, or other natural geothermal features, or through deep wells that pump heated groundwater.

germ plasm Genetic material (plant seeds or parts or animal eggs, sperm, and embryos) that may be preserved for future agricultural, commercial, and ecological purposes.

global environmentalism A concern for, and action to help solve, global environmental problems.

globalization The revolution in communications, transportation, finances and commerce that has brought about increasing interdependence of national economies.

grasslands A biome dominated by grasses and associated herbaceous plants.

Great Pacific Garbage Patch A huge expanse of the Pacific Ocean, stretching from about 500 nautical miles off the coast of California almost to Japan, in which floating refuse and trash is accumulated and concentrated by ocean currents. It's estimated that this swirling garbage vortex contains at least 100 million tons of flotsam and jetsam, much of it plastic that has been ground up into tiny particles.

greenfield development Housing projects built on previously undeveloped farmlands or forests on the outskirts of large cities.

greenhouse effect Gases in the atmosphere are transparent to visible light but absorb infrared (heat) waves that are re-radiated from the earth's surface.

greenhouse gases Chemical compounds that trap heat in the atmosphere. The principal anthropogenic greenhouse gases are carbon dioxide, methane, chlorofluorocarbons, nitrous oxide, and sulfur hexafluoride.

green plans Integrated national environmental plans for reducing pollution and resource consumption while achieving sustainable development and environmental restoration.

green political parties Political organizations based on environmental protection, participatory democracy, grassroots organization, and sustainable development.

green pricing Setting prices to encourage conservation or renewable energy; plans that invite customers to pay a premium for energy from renewable sources.

green revolution Dramatically increased agricultural production brought about by "miracle" strains of grain; usually requires high inputs of water, plant nutrients, and pesticides.

gross domestic product (GDP) The total economic activity within national boundaries.

gross national product (GNP) The sum total of all goods and services produced in a national economy. Gross domestic product (GDP) is used to distinguish economic activity within a country from that of offshore corporations.

groundwater Water held in gravel deposits or porous rock below the earth's surface; does not include water or crystallization held by chemical bonds in rocks or moisture in upper soil layers.

gully erosion Removal of layers of soil, creating channels or ravines too large to be removed by normal tillage operations.

H

habitat The place or set of environmental conditions in which a particular organism lives.

habitat conservation plans Agreements under which property owners are allowed to harvest resources or develop land as long as habitat is conserved or replaced in ways that benefit threatened endangered or threatened species in the long run. Some incidental "taking" or loss of endangered species is generally allowed in such plans.

Hadley cells Circulation patterns of atmospheric convection currents as they sink and rise in several intermediate bands.

hazardous The term used for chemicals that are dangerous, including flammables, explosives, irritants, sensitizers, acids, and caustics; these chemicals may be relatively harmless in diluted concentrations.

hazardous air pollutants (HAPs) Especially dangerous air pollutants, including carcinogens, neurotoxins, mutagens, teratogens, endocrine system disrupters, and other highly toxic compounds.

hazardous waste Any discarded material containing substances known to be toxic, mutagenic, carcinogenic, or teratogenic to humans or other life-forms; or ignitable, corrosive, explosive, or highly reactive alone or with other materials.

health A state of physical and emotional well-being; the absence of disease or ailment.

heap-leach extraction A technique for separating gold from extremely low-grade ores. Crushed ore is piled in huge heaps and sprayed with a dilute alkaline-cyanide solution, which percolates through the pile to extract the gold, which is separated from the effluent in a processing plant. This process has a high potential for water pollution.

heat A form of energy transferred from one body to another because of a difference in temperatures.

heat capacity The amount of heat energy that must be added or subtracted to change the temperature of a body; water has a high heat capacity.

heat of vaporization The amount of heat energy required to convert water from a liquid to a gas.

herbicide A chemical that kills plants.

herbivore An organism that eats only plants.

heterotroph An organism that is incapable of synthesizing its own food and, therefore, must feed upon organic compounds produced by other organisms.

high-level waste repository A place where intensely radioactive wastes can be buried and remain unexposed to groundwater and earthquakes for tens of thousands of years.

high-quality energy Intense, concentrated, and high-temperature energy that is especially useful in carrying out work.

HIPPO Habitat destruction, Invasive species, Pollution, Population (human), and Overharvesting—the leading causes of extinction.

holistic science The study of entire, integrated systems rather than isolated parts; often takes a descriptive or interpretive approach.

homeostasis The maintenance of a dynamic, steady state in a living system through opposing, compensating adjustments.

Homestead Act Legislation passed in 1862 allowing any U.S. citizen or applicant for citizenship over 21 years old and head of a family to acquire 160 acres of public land by living on it and cultivating it for five years.

host organism An organism that provides lodging for a parasite.

hot desert Deserts of the American Southwest and Mexico; characterized by extreme summer heat and cacti.

Hubbert curve A curve describing a peak and decline in production of natural resources, especially oil production, defined by M. King Hubbert in 1956.

human development index (HDI) A measure of quality of life using life expectancy, child survival, adult literacy, childhood education, gender equity, and access to clan water and sanitation as well as income.

human ecology The study of the interactions of humans with the environment.

human resources Human wisdom, experience, skill, labor, and enterprise.

humus Sticky, brown, insoluble residue from the bodies of dead plants and animals; gives soil its structure, coating mineral particles and holding them together; serves as a major source of plant nutrients.

hurricanes Large cyclonic oceanic storms with heavy rain and winds exceeding 119 km/hr (74 mph).

hybrid gasoline-electric engine A small gasoline engine that generates electricity that is stored in batteries and powers an electric motor that drives a vehicle's wheels.

hybrid gasoline-electric vehicles Automobiles that run on electric power and a small gasoline or diesel engine.

hydrologic cycle The natural process by which water is purified and made fresh through evaporation and precipitation. This cycle provides all the freshwater available for biological life.

hypothesis A provisional explanation that can be tested scientifically.

I

igneous rocks Crystalline minerals solidified from molten magma from deep in the earth's interior; basalt, rhyolite, andesite, lava, and granite are examples.

inbreeding depression In a small population, an accumulation of harmful genetic traits (through random mutations and natural selection) that lowers the viability and reproductive success of enough individuals to affect the whole population.

independent variable A factor not affected by the condition being altered in a manipulative experiment.

index value A value to which other values are adjusted so that they are all on the same scale or magnitude.

indicator species Species whose critical tolerance limits can be used to judge environmental conditions.

inductive reasoning Inferring general principles from specific examples.

industrial revolution Advances in science and technology that have given us power to understand and change our world.

industrial timber Trees used for lumber, plywood, veneer, particleboard, chipboard, and paper; *also called* roundwood.

inertial confinement A nuclear fusion process in which a small pellet of nuclear fuel is bombarded with extremely high-intensity laser light.

infiltration The process of water percolation into the soil and the pores and hollows of permeable rocks.

informal economy Small-scale family businesses in temporary locations outside the control of normal regulatory agencies.

inherent value Ethical values or rights that exist as an intrinsic or essential characteristic of a particular thing or class of things simply by the fact of their existence.

inholdings Private lands within public parks, forests, or wildlife refuges.

inorganic pesticides Inorganic chemicals such as metals, acids, or bases used as pesticides.

insecticide A chemical that kills insects.

insolation Incoming solar radiation.

instrumental value The value or worth of objects that satisfy the needs and wants of moral agents. Objects that can be used as a means to some desirable end.

intangible resources Factors such as open space, beauty, serenity, wisdom, diversity, and satisfaction that cannot be grasped or contained. Ironically, these resources can be both infinite and exhaustible.

integrated gasification combined cycle (IGCC) A power plant that heats fuel (usually coal, but could be biomass or other sources) to high temperatures and pressures in the presence of 96 percent oxygen. Hydrogen is separated from hydrocarbons and separated from CO_2 and other contaminants. The hydrogen is burned in a gas turbine and surplus heat drives a steam turbine, both of which generate electricity.

integrated pest management (IPM) An ecologically based pest-control strategy that relies on natural mortality factors, such as natural enemies, weather, cultural control methods, and carefully applied doses of pesticides.

Intergovernmental Panel on Climate Change (IPCC) An international organization formed to assess global climate change and its impacts. The IPCC is concerned with social, economic, and environmental impacts of climate change, and it was established by the United Nations Environment Programme and the World Meteorological Organization.

internal costs The expenses, monetary or otherwise, borne by those who use a resource.

interplanting The system of planting two or more crops, either mixed together or in alternating rows, in the same field; protects the soil and makes more efficient use of the land.

interpretive science Explanation based on observation and description of entire objects or systems rather than isolated parts.

interspecific competition In a community, competition for resources between members of *different* species.

intraspecific competition In a community, competition for resources among members of the *same* species.

invasive species Organisms that thrive in new territory where they are free of predators, diseases, or resource limitations that may have controlled their population in their native habitat.

ionizing radiation High-energy electromagnetic radiation or energetic subatomic particles released by nuclear decay.

ionosphere The lower part of the thermosphere.

ions Electrically charged atoms that have gained or lost electrons.

I-PAT formula A formula that says our environmental impacts (I) are the product of our population size (P) times our affluence (A) and the technology (T) used to produce the goods and services we consume.

island biogeography The study of rates of colonization and extinction of species on islands or other isolated areas, based on size, shape, and distance from other inhabited regions.

isotopes Forms of a single element that differ in atomic mass due to having a different number of neutrons in the nucleus.

J

J curve A J-shaped growth curve that depicts exponential growth.

jet streams Powerful winds or currents of air that circulate in shifting flows; similar to oceanic currents in their extent and effect on climate.

joule A unit of energy. One joule is the energy expended in 1 second by a current of 1 amp flowing through a resistance of 1 ohm.

K

k-selected species Species that reproduce more slowly, occupy higher trophic levels, have fewer offspring, longer life spans, and greater intrinsic control of population growth than *r*-selected species.

keystone species A species whose impacts on its community or ecosystem are much larger and more influential than would be expected from mere abundance.

kinetic energy Energy contained in moving objects, such as a rock rolling down a hill, the wind blowing through the trees, or water flowing over a dam.

known resources Resources that have been located, have not been completely mapped, but nevertheless are likely to become economical in the foreseeable future.

kwashiorkor A widespread human protein-deficiency disease resulting from a starchy diet low in protein and essential amino acids.

Kyoto Protocol An international agreement to reduce greenhouse gas emissions.

L

La Niña The part of a large-scale oscillation in the Pacific (and, perhaps, other oceans) in which trade winds hold warm surface waters in the western part of the basin and cause upwelling of cold, nutrient-rich, deep water in the eastern part of the ocean.

landfills Land disposal sites for solid waste. Refuse is compacted and covered with a layer of dirt to minimize rodent and insect infestation, wind-blown debris, and leaching by rain.

land reform Democratic redistribution of landownership to recognize that those who work the land have a right to a fair share of the products of their labor.

landscape ecology The study of the reciprocal effects of spatial pattern on ecological processes. The study of how landscape history shapes the features of the land and the organisms that inhabit it as well as our reaction to, and interpretation of, the land.

landslide The sudden fall of rock and earth from a hill or cliff. Often triggered by an earthquake or heavy rain.

latent heat Stored energy in a form that is not sensible (cannot be detected by ordinary senses).

LD50 A chemical dose lethal to 50 percent of a test population.

less-developed countries (LDC) Nonindustrialized nations characterized by low per capita income, high birthrates and death rates, high population growth rates, and low levels of technological development.

life-cycle analysis The evaluation of material and energy inputs and outputs at each stage of manufacture, use, and disposal of a product.

life expectancy The average age that individuals born in a particular time and place can be expected to attain.

life span The longest duration of life reached by a type of organism.

limiting factors Chemical or physical factors that limit the existence, growth, abundance, or distribution of an organism.

lipid A nonpolar organic compound that is insoluble in water but soluble in solvents, such as alcohol and ether; includes fats, oils, steroids, phospholipids, and carotenoids.

liquid metal fast breeder A nuclear power plant that converts uranium 238 to plutonium 239; thus, it creates more nuclear fuel than it consumes. Because of the extreme heat and density of its core, the breeder uses liquid sodium as a coolant.

lobbying Using personal contacts, public pressure, or political action to persuade legislators to vote in a particular manner.

locavore Someone who eats locally grown, seasonal food.

logarithmic scale A scale that uses logarithms as units in a sequence that progresses by a factor of 10. That is, each subsequent increment on the scale is 10 times the one that precedes it.

logical learner Someone who understands and remembers best by thinking through a topic and finding logical reasons for statements.

logical thinking Asks, can the rules of logic help understand this?

logistic growth Growth rates regulated by internal and external factors that establish an equilibrium with environmental resources.

longevity The length or duration of life; *compare* survivorship.

low-head hydropower Small-scale hydro technology that can extract energy from small headwater dams; causes much less ecological damage.

low-input high-diversity biofuels Fuels derived from mixed polycultures of perennial native species that require only minimal amounts of cultivation, fertilizer, irrigation, or pesticides when grown as energy crops.

low-quality energy Diffuse, dispersed, low-temperature energy that is difficult to gather and use for productive purposes.

LULUs *Locally Unwanted Land Uses*, such as toxic waste dumps, incinerators, smelters, airports, freeways, and other sources of environmental, economic, or social degradation.

M

magma Molten rock from deep in the earth's interior; called lava when it spews from volcanic vents.

magnetic confinement A technique for enclosing a nuclear fusion reaction in a powerful magnetic field inside a vacuum chamber.

malignant tumor A mass of cancerous cells that have left their site of origin, migrated through the body, and invaded normal tissues, and are growing out of control.

malnourishment A nutritional imbalance caused by a lack of specific dietary components or an inability to absorb or utilize essential nutrients.

Man and Biosphere (MAB) program A design for nature preserves that divides protected areas into zones with different purposes. A highly protected core is surrounded by a buffer zone and peripheral regions in which multiple-use resource harvesting is permitted.

mangroves Trees from a number of genera that live in salt water.

manipulative experiment An experiment in which some conditions are deliberately altered while others are held constant to study cause-and-effect relationships.

mantle A hot, pliable layer of rock that surrounds the earth's core and underlies the cool, outer crust.

marasmus A widespread human protein-deficiency disease caused by a diet low in calories and protein or imbalanced in essential amino acids.

marginal costs and benefits The costs and benefits of producing one additional unit of a good or service.

marine Living in or pertaining to the sea.

market equilibrium The dynamic balance between supply and demand under a given set of conditions in a "free" market (one with no monopolies or government interventions).

market forces Dynamic relationships in "free" market systems that capitalist systems rely upon to achieve national goals.

marsh Wetland without trees; in North America, this type of land is characterized by cattails and rushes.

mass burn Incineration of unsorted solid waste.

mass wasting Mass movement of geologic materials downhill caused by rockslides, avalanches, or simple slumping.

matter Anything that takes up space and has mass.

mean An average.

mediation An informal dispute resolution process in which parties are encouraged to discuss issues openly but in which all decisions are reached by consensus and any participant can withdraw at any time.

Mediterranean climate areas Specialized landscapes with warm, dry summers; cool, wet winters; many unique plant and animal adaptations; and many levels of endemism.

megacity *See* megalopolis.

megalopolis Also known as a megacity or supercity; an urban area with more than 10 million inhabitants.

megawatt (MW) Unit of electrical power equal to 1,000 kilowatts or 1 million watts.

mesosphere The atmospheric layer above the stratosphere and below the thermosphere; the middle layer. Temperatures here are usually very low.

metabolism All the energy and matter exchanges that occur within a living cell or organism; collectively, the life processes.

metamorphic rock Igneous and sedimentary rocks modified by heat, pressure, and chemical reactions.

metapopulation A collection of populations that have regular or intermittent gene flow between geographically separate units.

methane hydrate Small bubbles or individual molecules of methane (natural gas) trapped in a crystalline matrix of frozen water.

microbial agents Beneficial microbes (bacteria, fungi) that can be used to suppress or control pests; *also called* biological controls.

micro-hydro generators Small power generators that can be used in low-level rivers to provide economical power for four to six homes, freeing them from dependence on large utilities and foreign energy supplies.

mid-ocean ridges Mountain ranges on the ocean floor created where molten magma is forced up through cracks in the planet's crust.

Milankovitch cycles Periodic variations in tilt, eccentricity, and wobble in the earth's orbit; Milutin Milankovitch suggested that these are responsible for cyclic weather changes.

millennium assessment A set of ambitious environmental and human development goals established by the United Nations in 2000.

milpa agriculture An ancient farming system in which small patches of tropical forest are cleared for a period of perennial polyculture agriculture, and then the land is left fallow for many years to restore the soil; *also called* swidden agriculture.

mineral A naturally occurring, inorganic, crystalline solid with definite chemical composition and characteristic physical properties.

minimum viable population size The number of individuals needed for long-term survival of rare and endangered species.

mitigation Repairing or rehabilitating a damaged ecosystem or compensating for damage by providing a substitute or replacement area.

mixed perennial polyculture Growing a mixture of different perennial crop species (where the same plant persists for more than one year) together in the same plot.

models Simple representations of more complex systems.

molecule A combination of two or more atoms.

monitored, retrievable storage Holding wastes in underground mines or secure surface facilities, such as dry casks, where they can be watched and repackaged if necessary.

monoculture agroforestry Intensive planting of a single species; an efficient wood production approach, but one that encourages pests and disease infestations and conflicts with wildlife habitat or recreational uses.

monsoon A seasonal reversal of wind patterns caused by the different heating and cooling rates of the oceans and continents.

montane coniferous forests Coniferous forests of the mountains, consisting of belts of different forest communities along an altitudinal gradient.

moral agents Beings capable of making distinctions between right and wrong and acting accordingly. Those whom we hold responsible for their actions.

moral extensionism Expansion of our understanding of inherent value or rights to persons, organisms, or things that might not be considered worthy of value or rights under some ethical philosophies.

morals A set of ethical principles that guide our actions and relationships.

moral subjects Beings that are not capable of distinguishing between right or wrong or that are not able to act on moral principles and yet are capable of being wronged by others.

moral value The value or worth of something based on moral principles.

morbidity Illness or disease.

more-developed countries (MDC) Industrialized nations characterized by high per capita incomes, low birth and death rates, low population growth rates, and high levels of industrialization and urbanization.

mortality Death rate in a population; the probability of dying.

Muellerian mimicry Evolution of two species, both of which are unpalatable and have poisonous stingers or some other defense mechanism, to resemble each other.

mulch Protective ground covers, including both natural products and synthetic materials, that protect the soil, save water, and prevent weed growth.

multiple use Many uses that occur simultaneously; a component of forest management; limited to mutually compatible uses.

mutagens Agents, such as chemicals or radiation, that damage or alter genetic material (DNA) in cells.

mutation A change, either spontaneous or by external factors, in the genetic material of a cell. Mutations in the gametes (sex cells) can be inherited by future generations of organisms.

mutualism A symbiotic relationship between individuals of two different species in which both species benefit from the association.

mycorrhizal symbiosis An association between the roots of most plant species and certain fungi. The plant provides organic compounds to the fungus, while the fungus provides water and nutrients to the plant.

N

NAAQS *National Ambient Air Quality Standards*; federal standards specifying the maximum allowable levels (averaged over specific time periods) for regulated pollutants in ambient (outdoor) air.

natality The production of new individuals by birth, hatching, germination, or cloning.

natural experiment A study of events that have already happened.

natural history The study of where and how organisms carry out their life cycles.

natural increase Crude death rate subtracted from crude birthrate.

natural organic pesticides "Botanicals" or organic compounds naturally occurring in plants, animals, or microbes that serve as pesticides.

natural resources Goods and services supplied by the environment.

natural selection The mechanism for evolutionary change in which environmental pressures cause certain genetic combinations in a population to become more abundant. Genetic combinations best adapted for present environmental conditions tend to become predominant.

negative feedback loop A situation in which a factor or condition causes changes that reduce that factor or condition.

neoclassical economics A branch of economics that attempts to apply the principles of modern science to economic analysis in a mathematically rigorous, noncontextual, abstract, predictive manner.

neo-Luddites People who reject technology, believing that it is the cause of environmental degradation and social disruption. Named after the followers of Ned Ludd, who tried to turn back the Industrial Revolution in England.

neo-Malthusians People who see the world as characterized by scarcity and competition, with too many people fighting for too few resources. Named for Thomas Malthus, who predicted a dismal cycle of misery, vice, and starvation as a result of human overpopulation.

NEPA National Environmental Policy Act, the cornerstone of U.S. environmental policy. Authorizes the Council on Environmental Quality, directs federal agencies to take environmental consequences into account when making decisions, and requires an environmental impact statement for every major federal project likely to have adverse environmental effects.

net energy yield Total useful energy produced during the lifetime of an entire energy system minus the energy used, lost, or wasted in making useful energy available.

neurotoxins Toxic substances, such as lead or mercury, that specifically poison nerve cells.

neutron A subatomic particle, found in the nucleus of the atom, that has no electromagnetic charge.

new towns Experimental urban environments that seek to combine the best features of the rural village and the modern city.

nihilists Those who believe the world has no meaning or purpose other than a dark, cruel, unceasing struggle for power and existence.

NIMBY *Not In My BackYard*: the rallying cry of those opposed to LULUs.

nitrate-forming bacteria Bacteria that convert nitrites into compounds that can be used by green plants to build proteins.

nitrite-forming bacteria Bacteria that combine ammonia with oxygen to form nitrites.

nitrogen cycle The circulation and reutilization of nitrogen in both inorganic and organic phases.

nitrogen-fixing bacteria Bacteria that convert nitrogen from the atmosphere or a soil solution into ammonia that can then be converted to plant nutrients by nitrite- and nitrate-forming bacteria.

nitrogen oxides Highly reactive gases formed when nitrogen in fuel or combustion air is heated to over 650°C (1,200°F) in the presence of oxygen, or when bacteria in soil or water oxidize nitrogen-containing compounds.

noncriteria pollutants *See* unconventional air pollutants.

nongovernmental organizations (NGOs) A term referring collectively to pressure and research groups, advisory agencies, political parties, professional societies, and other groups concerned about environmental quality, resource use, and many other issues.

nonpoint sources Scattered, diffuse sources of pollutants, such as runoff from farm fields, golf courses, and construction sites.

nonrenewable resources Minerals, fossil fuels, and other materials present in essentially fixed amounts (within human time scales) in our environment.

nuclear fission The radioactive decay process in which isotopes split apart to create two smaller atoms.

nuclear fusion A process in which two smaller atomic nuclei fuse into one larger nucleus and release energy; the source of power in a hydrogen bomb.

nucleic acids Large organic molecules made of nucleotides that function in the transmission of hereditary traits, in protein synthesis, and in control of cellular activities.

nucleus The center of the atom; occupied by protons and neutrons. In cells, the organelle that contains the chromosomes (DNA).

nuées ardentes Deadly, denser-than-air mixtures of hot gases and ash ejected from volcanoes.

numbers pyramid A diagram showing the relative population sizes at each trophic level in an ecosystem; usually corresponds to the biomass pyramid.

O

obese Generally considered to be a body mass greater than 30 kg/m², or roughly 30 pounds above normal for an average person.

oceanic islands Islands in the ocean; formed by breaking away from a continental landmass, volcanic action, coral formation, or a combination of sources; support distinctive communities.

ocean shorelines Rocky coasts and sandy beaches along the oceans, which support rich, stratified communities.

ocean thermal electric conversion (OTEC) Energy derived from temperature differentials between warm ocean surface waters and cold deep waters. This differential can be used to drive turbines attached to electric generators.

offset allowances A controversial component of air quality regulations that allows a polluter to avoid installation of control equipment on one source with an "offsetting" pollution reduction at another source.

oil shale A fine-grained sedimentary rock rich in solid organic material called kerogen. When heated, the kerogen liquefies to produce a fluid petroleum fuel.

oligotrophic The condition of rivers and lakes that have clear water and low biological productivity (*oligo* = little; *trophic* = nutrition); usually clear, cold, infertile headwater lakes and streams.

omnivore An organism that eats both plants and animals.

open access system A commonly held resource for which there are no management rules.

open canopy A forest where tree crowns cover less than 20 percent of the ground; *also called* woodland.

open range Unfenced, natural grazing lands; includes woodland as well as grassland.

open system A system that exchanges energy and matter with its environment.

optimum The most favorable condition in regard to an environmental factor.

orbital The space or path in which an electron orbits the nucleus of an atom.

organic compounds Complex molecules organized around skeletons of carbon atoms arranged in rings or chains; includes biomolecules, molecules synthesized by living organisms.

organophosphates Organic molecules to which one or more phosphate groups are attached.

overburden Overlying layers of noncommercial sediments that must be removed to reach a mineral or coal deposit.

overgrazing Allowing livestock to eat so much forage that the ecological health of the habitat is damaged.

overharvesting Harvesting so much of a resource that its existence is threatened.

overnutrition Receiving too many calories.

overshoot The extent to which a population exceeds the carrying capacity of its environment.

oxygen cycle The circulation and reutilization of oxygen in the biosphere.

oxygen sag Oxygen decline downstream from a pollution source that introduces materials with high biological oxygen demands.

ozone A highly reactive molecule containing three oxygen atoms; a dangerous pollutant in ambient air. In the stratosphere, however, ozone forms an ultraviolet absorbing shield that protects us from mutagenic radiation.

P

Pacific decadal oscillation (PDO) A large pool of warm water that moves north and south in the Pacific Ocean every 30 years or so and has large effects on North America's climate.

parabolic mirrors Curved mirrors that focus light from a large area onto a single, central point, thereby concentrating solar energy and producing high temperatures.

paradigm A model that provides a framework for interpreting observations.

parasite An organism that lives in or on another organism, deriving nourishment at the expense of its host, usually without killing it.

parsimony Choosing the simpler of two equally plausible explanations.

particulate material Atmospheric aerosols, such as dust, ash, soot, lint, smoke, pollen, spores, algal cells, and other suspended materials. The term originally was applied only to solid particles but now is extended to droplets of liquid.

parts per billion (ppb) Number of parts of a chemical found in 1 billion parts of a particular gas, liquid, or solid mixture.

parts per million (ppm) Number of parts of a chemical found in 1 million parts of a particular gas, liquid, or solid mixture.

parts per trillion (ppt) Number of parts of a chemical found in 1 trillion (10^{12}) parts of a particular gas, liquid, or solid mixture.

passive heat absorption The use of natural materials or absorptive structures without moving parts to gather and hold heat; the simplest and oldest use of solar energy.

pastoralists People who make a living by herding domestic livestock.

pasture Grazing lands suitable for domestic livestock.

patchiness Within a larger ecosystem, the presence of smaller areas that differ in some physical conditions and thus support somewhat different communities; a diversity-promoting phenomenon.

pathogen An organism that produces disease in a host organism, disease being an alteration of one or more metabolic functions in response to the presence of the organism.

peat Deposits of moist, acidic, semidecayed organic matter.

pelagic Zones in the vertical water column of a water body.

pellagra Lassitude, torpor, dermatitis, diarrhea, dementia, and death brought about by a diet deficient in tryptophan and niacin.

peptides Two or more amino acids linked by a peptide bond.

perennial species Plants that grow for more than two years.

permafrost A permanently frozen layer of soil that underlies the arctic tundra.

permanent retrievable storage Placing waste storage containers in a secure building, salt mine, or bedrock cavern where they can be inspected periodically and retrieved if necessary.

persistent organic pollutants (POPs) Chemical compounds that persist in the environment and retain biological activity for long times.

pest Any organism that reduces the availability, quality, or value of a useful resource.

pesticide Any chemical that kills, controls, drives away, or modifies the behavior of a pest.

pesticide treadmill A need for constantly increasing doses or new pesticides to prevent pest resurgence.

pest resurgence Rebound of pest populations due to acquired resistance to chemicals and nonspecific destruction of natural predators and competitors by broadscale pesticides.

pH A value that indicates the acidity or alkalinity of a solution on a scale of 0 to 14, based on the proportion of H^+ ions present.

phosphorus cycle The movement of phosphorus atoms from rocks through the biosphere and hydrosphere and back to rocks.

photochemical oxidants Products of secondary atmospheric reactions. *See* smog.

photodegradable plastics Plastics that break down when exposed to sunlight or to a specific wavelength of light.

photosynthesis The biochemical process by which green plants and some bacteria capture light energy and use it to produce chemical bonds. Carbon dioxide and water are consumed while oxygen and simple sugars are produced.

photosynthetic efficiency The percentage of available light captured by plants and used to make useful products.

photovoltaic cell An energy-conversion device that captures solar energy and directly converts it to electrical current.

physical or abiotic factors Nonliving factors, such as temperature, light, water, minerals, and climate, that influence an organism.

phytoplankton Microscopic, free-floating, autotrophic organisms that function as producers in aquatic ecosystems.

pioneer species In primary succession on a terrestrial site, the plants, lichens, and microbes that first colonize the site.

plankton Primarily microscopic organisms that occupy the upper water layers in both freshwater and marine ecosystems.

plasma A hot, electrically neutral gas of ions and free electrons.

plug-in hybrids Vehicles with hybrid gasoline-electric engines adapted with a larger battery array (enough to propel the vehicle for 50 km or so on the batteries alone) and a plug-in to recharge the batteries from a standard electric outlet.

poachers Persons who hunt wildlife illegally.

point sources Specific locations of highly concentrated pollution discharge, such as factories, power plants, sewage treatment plants, underground coal mines, and oil wells.

policy A societal plan or statement of intentions intended to accomplish some social good.

policy cycle The process by which problems are identified and acted upon in the public arena.

political economy The branch of economics concerned with modes of production, distribution of benefits, social institutions, and class relationships.

pollution Anything that makes the environment foul, unclean, dirty; any physical, chemical, or biological change that adversely affects the health, survival, or activities of living organisms or that alters the environment in undesirable ways.

pollution charges Fees assessed per unit of pollution, based on the "polluter pays" principle.

polycentric complex A city with several urban cores surrounding a once-dominant central core.

population A group of individuals of the same species occupying a given area.

population crash A sudden population decline caused by predation, waste accumulation, or resource depletion; *also called* a dieback.

population explosion Growth of a population at exponential rates to a size that exceeds environmental carrying capacity; usually followed by a population crash.

population momentum A potential for increased population growth as young members reach reproductive age.

positive feedback loop A situation in which a factor or condition causes changes that further enhance that factor or condition.

postmaterialist values A philosophy that emphasizes quality of life over acquisition of material goods.

postmodernism A philosophy that rejects the optimism and universal claims of modern positivism.

potential energy Stored energy that is latent but available for use; for example, a rock poised at the top of a hill, or water stored behind a dam.

power The rate of energy delivery; measured in horsepower or watts.

precautionary principle The decision to leave a margin of safety for unexpected developments.

precycling Making environmentally sound decisions at the store and reducing waste before we buy.

predation A predator's act of feeding.

predator An organism that feeds directly on other organisms in order to survive; live-feeders, such as herbivores and carnivores.

predator-mediated competition A situation in which predation reduces prey populations and gives an advantage to competitors that might not otherwise be successful.

premises Introductory statements that set up or define a problem. Those things taken as given.

prevention of significant deterioration A clause of the Clean Air Act that prevents degradation of existing clean air; opposed by industry as an unnecessary barrier to development.

price elasticity A situation in which supply and demand of a commodity respond to price.

primary forest Forest composed primarily of native species, where there are no clearly visible indications of human activity and ecological processes are not significantly disturbed.

primary pollutants Chemicals released directly into the air in a harmful form.

primary productivity Synthesis of organic materials (biomass) by green plants using the energy captured in photosynthesis.

primary standards Regulations of the 1970 Clean Air Act; intended to protect human health.

primary succession An ecological succession that begins in an area where no biotic community previously existed.

primary treatment A process that removes solids from sewage before it is discharged or treated further.

principle of competitive exclusion A result of natural selection whereby two similar species in a community occupy different ecological niches, thereby reducing competition for food.

producer An organism that synthesizes food molecules from inorganic compounds by using an external energy source; most producers are photosynthetic.

production frontier The maximum output of two competing commodities at different levels of production.

productivity The synthesis of new organic material; synthesis done by green plants using solar energy is called primary productivity.

prokaryotic Cells that do not have a membrane-bounded nucleus or membrane-bounded organelles.

promoters Agents that are not carcinogenic but that assist in the progression and spread of tumors; sometimes called cocarcinogens.

pronatalist pressures Influences that encourage people to have children.

proteins Chains of amino acids linked by peptide bonds.

proton A positively charged subatomic particle found in the nucleus of an atom.

proven reserves *See* proven resources.

proven resources Resources that have been thoroughly mapped and are economical to recover at current prices with available technology.

public trust A doctrine obligating the government to maintain public lands in a natural state as guardians of the public interest.

pull factors (in urbanization) Conditions that draw people from the country into the city.

push factors (in urbanization) Conditions that force people out of the country and into the city.

R

r-selected species Species that tend to have rapid reproduction and high offspring mortality. They frequently overshoot the carrying capacity of their environment and display boom-and-bust life cycles. They lack intrinsic population controls and tend to occupy lower trophic levels in food webs than *k-selected species*.

radiative evolution Divergence from a common ancestor into two or more new species.

radioactive An unstable isotope that decays spontaneously and releases subatomic particles or units of energy.

radioactive decay A change in the nuclei of radioactive isotopes that spontaneously emit high-energy electromagnetic radiation and/or subatomic particles while gradually changing into another isotope or different element.

radionuclides Isotopes that exhibit radioactive decay.

rainforest A forest with high humidity, constant temperature, and abundant rainfall (generally over 380 cm [150 in.] per year); can be tropical or temperate.

rain shadow A dry area on the downwind side of a mountain.

rangeland Grasslands and open woodlands suitable for livestock grazing.

rational choice Public decision making based on reason, logic, and science-based management.

recharge zone An area where water infiltrates an aquifer.

reclamation Chemical, biological, or physical cleanup and reconstruction of severely contaminated or degraded sites to return them to something like their original topography and vegetation.

recoverable resources Resources accessible with current technology but not economical under current conditions.

re-creation (in ecology) Construction of an entirely new biological community to replace one that has been destroyed on that or another site.

recycling Reprocessing of discarded materials into new, useful products; not the same as reuse of materials for their original purpose, but the terms are often used interchangeably.

red tide A population explosion or bloom of minute, single-celled marine organisms called dinoflagellates. Billions of these cells can accumulate in protected bays, where the toxins they contain can poison other marine life.

reduced tillage systems Systems, such as minimum till, conserve-till, and no-till, that preserve soil, save energy and water, and increase crop yields.

reflective thinking Asks, what does this all mean?

reformer A device that strips hydrogen from fuels such as natural gas, methanol, ammonia, gasoline, or vegetable oil so they can be used in a fuel cell.

refuse-derived fuel Processing of solid waste to remove metal, glass, and other unburnable materials; organic residue is shredded, formed into pellets, and dried to make fuel for power plants.

regenerative farming Farming techniques and land stewardship that restore the health and productivity of the soil by rotating crops, planting ground cover, protecting the surface with crop residue, and reducing synthetic chemical inputs and mechanical compaction.

regulations Rules established by administrative agencies. Regulations can be more important than statutory law in the day-to-day management of resources.

rehabilitate To rebuild elements of structure or function in an ecological system without necessarily achieving complete restoration to its original condition.

rehabilitation of land A utilitarian program to make an area useful to humans.

relativists Those who believe that moral principles are always dependent on the particular situation.

remediation Cleaning up chemical contaminants from a polluted area.

renewable resources Resources normally replaced or replenished by natural processes; resources not depleted by moderate use; examples include solar energy, biological resources such as forests and fisheries, biological organisms, and some biogeochemical cycles.

renewable water supplies Annual freshwater surface runoff plus annual infiltration into underground freshwater aquifers that are accessible for human use.

replication Repeating studies or tests to verify reliability.

reproducibility The capacity for a particular result to be observed or obtained more than once.

reproductive isolation Barriers (geographical, behavioral, or biological) that prevent gene flow between members of a species.

residence time The length of time a component, such as an individual water molecule, spends in a particular compartment or location before it moves on through a particular process or cycle.

resilience The ability of a community or ecosystem to recover from disturbances.

resistance (inertia) The ability of a community to resist being changed by potentially disruptive events.

resource In economic terms, anything with potential use in creating wealth or giving satisfaction.

resource partitioning In a biological community, various populations sharing environmental resources through specialization, thereby reducing direct competition. *See also* ecological niche.

resource scarcity A shortage or deficit in some resource.

restoration Bringing something back to a former condition. Ecological restoration involves active manipulation of nature to re-create conditions that existed before human disturbance.

restoration ecology An area of ecology that seeks to repair or reconstruct ecosystems damaged by human actions.

riders Amendments attached to bills in conference committee, often completely unrelated to the bill to which they are added.

rill erosion The removal of thin layers of soil by little rivulets of running water that gather and cut small channels in the soil.

risk The probability that something undesirable will happen as a consequence of exposure to a hazard.

risk assessment Evaluation of the short-term and long-term risks associated with a particular activity or hazard; usually compared to benefits in a cost–benefit analysis.

RNA Ribonucleic acid; nucleic acid used for transcription and translation of the genetic code found on DNA molecules.

Roadless rule A Clinton-era ban on logging, road building, and other development on the lands identified as deserving of wilderness protection in the Roadless Area Review and Evaluations (RARE).

rock A solid, cohesive aggregate of one or more crystalline minerals.

rock cycle The process whereby rocks are broken down by chemical and physical forces; sediments are moved by wind, water, and gravity, sedimented and reformed into rock, and then crushed, folded, melted, and recrystallized into new forms.

rotational grazing Confining animals to a small area for a short time (often only a day or two) before shifting them to a new location.

ruminant animals Cud-chewing animals, such as cattle, sheep, goats, and buffalo, with multi-chambered stomachs in which cellulose is digested with the aid of bacteria.

runoff The excess of precipitation over evaporation; the main source of surface water and, in broad terms, the water available for human use.

run-of-the-river flow Ordinary river flow not accelerated by dams, flumes, etc. Some small, modern, high-efficiency turbines can generate useful power with run-of-the-river flow or with a current of only a few kilometers per hour.

rural area An area in which most residents depend on agriculture or the harvesting of natural resources for their livelihood.

S

S curve An S-shaped curve that depicts logistic growth.

salinity The amount of dissolved salts (especially sodium chloride) in a given volume of water.

salinization A process in which mineral salts accumulate in the soil, killing plants; occurs when soils in dry climates are irrigated profusely.

salt marsh Shallow wetlands along coastlines that are flooded regularly or occasionally with seawater.

saltwater intrusion Movement of saltwater into freshwater aquifers in coastal areas where groundwater is withdrawn faster than it is replenished.

salvage logging Harvesting timber killed by fire, disease, or windthrow.

sample To analyze a small but representative portion of a population to estimate the characteristics of the entire class.

sanitary landfills A landfill in which garbage and municipal waste are buried every day under enough soil or fill to eliminate odors, vermin, and litter.

savannas Open grasslands with sparse tree cover.

scavenger An organism that feeds on the dead bodies of other organisms.

science A process for producing knowledge methodically and logically.

scientific consensus A general agreement among informed scholars.

scientific method A systematic, precise, objective way to study a problem. Generally this requires observation, hypothesis development and testing, data gathering, and interpretation.

scientific theory An explanation supported by many tests and accepted by a general consensus of scientists.

secondary pollutants Chemicals that acquire a hazardous form after entering the air or that are formed by chemical reactions as components of the air interact.

secondary recovery technique Pumping pressurized gas, steam, or chemical-containing water into a well to squeeze more oil from a reservoir.

secondary standards Regulations of the 1972 Clean Air Act intended to protect materials, crops, visibility, climate, and personal comfort.

secondary succession Succession on a site where an existing community has been disrupted.

secondary treatment Bacterial decomposition of suspended particulates and dissolved organic compounds that remain after primary sewage treatment.

secure landfill A solid-waste disposal site lined and capped with an impermeable barrier to prevent leakage or leaching. Drain tiles, sampling wells, and vent systems provide monitoring and pollution control.

sedimentary rock Deposited material that remains in place long enough, or is covered with enough material, to compact into stone; examples include shale, sandstone, breccia, and conglomerates.

sedimentation The deposition of organic materials or minerals by chemical, physical, or biological processes.

selection pressures Factors in the environment that favor the successful reproduction of individuals possessing certain heritable traits and that reduce the viability and fertility of individuals that do not possess those traits.

seriously undernourished Persons who receive less than 80 percent of their minimum daily caloric requirements.

shantytowns Settlements created when people move onto undeveloped lands and build their own shelters with cheap or discarded materials. Some are simply illegal subdivisions where a landowner rents land without city approval; others are land invasions.

sheet erosion The peeling off of thin layers of soil from the land surface; accomplished primarily by wind and water.

sick building syndrome Headaches, allergies, chronic fatigue, and other symptoms caused by poorly vented indoor air contaminated by pathogens or toxins.

significant numbers Meaningful numbers whose accuracy can be verified.

sinkholes A large surface crater caused by the collapse of an underground channel or cavern; often triggered by groundwater withdrawal.

sludge A semisolid mixture of organic and inorganic materials that settles out of wastewater at a sewage treatment plant.

slums Legal but inadequate multifamily tenements or rooming houses; some are custom built for rent to poor people, others are converted from some other use.

smart growth Efficient use of land resources and existing urban infrastructure.

smelting Heating ores to extract metals.

smog The term used to describe the combination of smoke and fog in the stagnant air of London; now often applied to photochemical pollution products or urban air pollution of any kind.

social justice Equitable access to resources and the benefits derived from them; a system that recognizes inalienable rights and adheres to what is fair, honest, and moral.

soil A complex mixture of weathered mineral materials from rocks, partially decomposed organic molecules, and a host of living organisms.

soil horizons Horizontal layers that reveal a soil's history, characteristics, and usefulness.

soil profile All the vertical layers or horizons that make up a soil in a particular place.

southern pine forest A U.S. coniferous forest ecosystem characterized by a warm, moist climate.

speciation The generation of new species.

species A population of morphologically similar organisms that can reproduce sexually among themselves but that cannot produce fertile offspring when mated with other organisms.

species diversity The number and relative abundance of species present in a community.

species recovery plan A plan for restoration of an endangered species through protection, habitat management, captive breeding, disease control, or other techniques that increase populations and encourage survival.

sprawl Unlimited outward extension of city boundaries that lowers population density, consumes open space, generates freeway congestion, and causes decay in central cities.

spring overturn Springtime lake phenomenon that occurs when the surface ice melts and the surface water temperature warms to its greatest density at 4°C and then sinks, creating a convection current that displaces nutrient-rich bottom waters.

squatter towns Shantytowns that occupy land without owner's permission. Some are highly organized movements in defiance of authorities; others grow gradually.

stability In ecological terms, a dynamic equilibrium among the physical and biological factors in an ecosystem or a community; relative homeostasis.

stable runoff The fraction of water available year-round; usually more important than total runoff when determining human uses.

Standard Metropolitan Statistical Area (SMSA) An urbanized region with at least 100,000 inhabitants with strong economic and social ties to a central city of at least 50,000 people.

standing The right to take part in legal proceedings.

state shift A permanent or long-lasting change in a system to a new set of conditions and relations in response to a disturbance.

statistics Numbers that let you evaluate and compare things.

statute law Formal documents or decrees enacted by the legislative branch of government.

statutory law Rules passed by a state or national legislature.

steady-state economy Characterized by low birth and death rates, use of renewable energy sources, recycling of materials, and emphasis on durability, efficiency, and stability.

stewardship A philosophy that holds that humans have a unique responsibility to manage, care for, and improve nature.

strategic lawsuits against public participation (SLAPPs) Lawsuits that have no merit but are brought merely to intimidate and harass private citizens who act in the public interest.

strategic metals and minerals Materials a country cannot produce itself but that it uses for essential materials or processes.

stratosphere The zone in the atmosphere extending from the tropopause to about 50 km (30 mi) above the earth's surface; temperatures are stable or rise slightly with altitude; has very little water vapor but is rich in ozone.

stratospheric ozone The ozone (O_3) occurring in the stratosphere 10 to 50 km above the earth's surface.

stress-related diseases Diseases caused or accentuated by social stresses such as crowding.

strip farming Planting different kinds of crops in alternating strips along land contours. When one crop is harvested, the other crop remains to protect the soil and prevent water from running straight down a hill.

strip mining Removing surface layers over coal seams using giant earth-moving equipment to create a huge open pit; coal is scooped up by enormous surface-operated machines and transported by trucks; an alternative to deep mines.

structure (in ecological terms) Patterns of organization, both spatial and functional, in a community.

Student Environmental Action Coalition (SEAC) A student-based environmental organization that is both an umbrella organization and a grassroots network to facilitate environmental action and education on college campuses.

subduction The process by which one tectonic plate is pushed down below another as plates crash into each other.

subsidence A settling of the ground surface caused by the collapse of porous formations that result from a withdrawal of large amounts of groundwater, oil, or other underground materials.

subsoil A layer of soil beneath the topsoil that has lower organic content and higher concentrations of fine mineral particles; often contains soluble compounds and clay particles carried down by percolating water.

sulfur cycle The chemical and physical reactions by which sulfur moves into or out of storage and through the environment.

sulfur dioxide A colorless, corrosive gas directly damaging to both plants and animals.

Superfund A fund established by Congress to pay for containment, cleanup, or remediation of abandoned toxic waste sites. The fund is financed by fees paid by toxic waste generators and by cost recovery from cleanup projects.

supply The quantity of a product being offered for sale at various prices, other things being equal.

surface mining The mining of minerals from surface pits. *See* strip mining.

surface soil The first true layer of soil; the layer in which organic material is mixed with mineral particles; thickness ranges from a meter or more under virgin prairie to zero in some deserts.

surface tension A condition in which the water surface meets the air and acts like an elastic skin.

survivorship The percentage of a population reaching a given age, or the proportion of the maximum life span of the species reached by any individual.

sustainability Living within the bounds of nature based on renewable resources used in ways that don't deplete nonrenewable resources, harm essential ecological services, or limit the ability of future generations to meet their own needs.

sustainable agriculture An ecologically sound, economically viable, socially just, and humane agricultural system. Stewardship, soil conservation, and integrated pest management are essential for sustainability.

sustainable development A real increase in well-being and standard of life for the average person that can be maintained over the long-term without degrading the environment or compromising the ability of future generations to meet their own needs.

sustained yield Utilization of a renewable resource at a rate that does not prevent the resource from being fully renewed on a long-term basis.

swamp A wetland with trees, such as the extensive swamp forests of the southern United States.

swidden agriculture *See* milpa agriculture.

symbiosis The intimate living together of members of two different species; includes mutualism, commensalism, and, in some classifications, parasitism.

sympatric speciation Species that arise from a common ancestor due to biological or behavioral barriers that cause reproductive isolation even though the organisms live in the same place.

synergism An interaction in which one substance exacerbates the effects of another. The sum of the interaction is greater than the parts.

synergistic effects When an injury caused by exposure to two environmental factors together is greater than the sum of exposure to each factor individually.

systemic A condition or process that affects the whole body. Many metabolic poisons are systemic.

systems Networks of interactions among many interdependent factors.

T

taiga The northernmost edge of the boreal forest, including species-poor woodland and peat deposits; intergrading with the arctic tundra.

tailings Mining waste left after mechanical or chemical separation of minerals from crushed ore.

taking Unconstitutional confiscation of private property.

tar sands Sand deposits containing petroleum or tar.

technological optimists Those who believe that technology and human enterprise will find cures for all our problems. *Also called* Promethean environmentalism.

tectonic plates Huge blocks of the earth's crust that slide around slowly, pulling apart to open new ocean basins or crashing ponderously into each other to create new, larger landmasses.

temperate rainforest The cool, dense, rainy forest of the northern Pacific coast; enshrouded in fog much of the time; dominated by large conifers.

temperature A measure of the speed of motion of a typical atom or molecule in a substance.

temperature inversion A stable layer of warm air overlying cooler air, trapping pollutants near ground level.

teratogens Chemicals or other factors that specifically cause abnormalities during embryonic growth and development.

terracing Shaping the land to create level shelves of earth to hold water and soil; requires extensive hand labor or expensive machinery, but enables farmers to farm very steep hillsides.

territoriality An intense form of intraspecific competition in which organisms define an area surrounding their home site or nesting site and defend it, primarily against other members of their own species.

tertiary treatment The removal of inorganic minerals and plant nutrients after primary and secondary treatment of sewage.

thermal plume A plume of hot water discharged into a stream or lake by a heat source, such as a power plant.

thermocline In water, a distinctive temperature transition zone that separates an upper layer that is mixed by the wind (the epilimnion) and a colder, deep layer that is not mixed (the hypolimnion).

thermodynamics A branch of physics that deals with transfers and conversions of energy.

thermodynamics, first law Energy can be transformed and transferred, but cannot be destroyed or created; i.e., energy is *conserved*.

thermodynamics, second law With each successive energy transfer or transformation, less energy is available to do work.

thermosphere The highest atmospheric zone; a region of hot, dilute gases above the mesosphere extending out to about 1,600 km (1,000 mi) from the earth's surface.

Third World Less-developed countries that are not either capitalistic and industrialized (First World) or centrally planned socialist economies (Second World); not intended to be derogatory.

threatened species A species that is still abundant in parts of its territorial range but that has declined significantly in total numbers and may be on the verge of extinction in certain regions or localities.

throughput The flow of energy and matter into, through, and out of a system.

tidal station A dam built across a narrow bay or estuary that traps tide water flowing both in and out of the bay. Water flowing through the dam spins turbines attached to electric generators.

tide pool Depressions in a rocky shoreline that are flooded at high tide but cut off from the ocean at low tide.

timberline In mountains, the highest-altitude edge of forest that marks the beginning of the treeless alpine tundra.

tolerance limits *See* limiting factors.

tornado A violent storm characterized by strong swirling winds and updrafts. Tornadoes form when a strong cold front pushes under a warm, moist air mass over the land.

tort law Court cases that seek compensation for damages.

total fertility rate The number of children born to an average woman in a population during her entire reproductive life.

total growth rate The net rate of population growth resulting from births, deaths, immigration, and emigration.

total maximum daily loads (TMDLs) The amount of particular pollutants that a water body can receive from both point and nonpoint sources and still meet water quality standards.

toxic colonialism Shipping toxic wastes to a weaker or poorer nation.

Toxic Release Inventory A program created by the Superfund Amendments and Reauthorization Act of 1984 that requires manufacturing facilities and waste-handling and disposal sites to report annually on releases of more than 300 toxic materials.

toxins Poisonous chemicals that react with specific cellular components to kill cells or to alter growth or development in undesirable ways; often harmful, even in dilute concentrations.

tradable permits Pollution quotas or variances that can be bought or sold.

tragedy of the commons An inexorable process of degradation of communal resources due to the selfishness of "free riders" who use or destroy more than their fair share of common property. *See* open access system.

transitional zone A zone in which populations from two or more adjacent communities meet and overlap.

triple bottom line Corporate accounting that reports social and environmental costs and benefits as well as merely economic ones.

trophic level A step in the movement of energy through an ecosystem; an organism's feeding status in an ecosystem.

tropical rainforests Forests in which rainfall is abundant—more than 200 cm (80 in.) per year—and temperatures are warm to hot year-round.

tropical seasonal forest Semi-evergreen or partly deciduous forests tending toward open woodlands and grassy savannas dotted with scattered, drought-resistant tree species; distinct wet and dry seasons, hot year-round.

tropopause The boundary between the troposphere and the stratosphere.

troposphere The layer of air nearest to the earth's surface; both temperature and pressure usually decrease with increasing altitude.

tsunami Giant seismic sea swells that move rapidly from the center of a submarine earthquake; can be 10 to 20 meters high when they reach shorelines hundreds or even thousands of kilometers from their source.

tundra Treeless arctic or alpine biome characterized by cold, harsh winters, a short growing season, and a potential for frost any month of the year; vegetation includes low-growing perennial plants, mosses, and lichens.

U

unconventional air pollutants Toxic or hazardous substances, such as asbestos, benzene, beryllium, mercury, polychlorinated biphenyls, and vinyl chloride, not listed in the original Clean Air Act because at that time they were not released in large quantities; *also called* noncriteria pollutants.

unconventional oil Resources such as shale oil and tar sands that can be liquefied and used like oil.

undernourished Persons who receive less than 90 percent of the minimum dietary intake over a long time period; they lack energy for an active, productive life and are more susceptible to infectious diseases.

undiscovered resources Speculative or inferred resources or those that we haven't even thought about.

universalists Those who believe that some fundamental ethical principles are universal and unchanging. In this vision, these principles are valid regardless of the context or situation.

upwelling Convection currents within a body of water that carry nutrients from bottom sediments toward the surface.

urban agglomerations An aggregation of many cities into a large metropolitan area.

urban area An area in which a majority of the people are not directly dependent on natural-resource-based occupations.

urbanization An increasing concentration of the population in cities and a transformation of land use to an urban pattern of organization.

utilitarian conservation A philosophy that resources should be used for the greatest good for the greatest number for the longest time.

utilitarianism *See* utilitarian conservation.

utilitarians Those who hold that an action is right if it produces the greatest good for the greatest number of people.

V

values An estimation of the worth of things; sets of ethical beliefs and preferences that determine people's sense of right and wrong.

verbal learner Someone who understands and remembers best by listening to the spoken word.

vertical stratification The vertical distribution of specific subcommunities within a community.

vertical zonation Terrestrial vegetation zones determined by altitude.

village A collection of rural households linked by culture, custom, and association with the land.

visible light A portion of the electromagnetic spectrum that includes the wavelengths used for photosynthesis.

visual learner Someone who understands and remembers best by reading, or looking at pictures and diagrams.

vitamins Organic molecules essential for life that we cannot make for ourselves; must be obtained from our diet; act as enzyme cofactors.

volatile organic compounds (VOCs) Organic chemicals that evaporate readily and exist as gases in the air.

volcanoes Vents in the earth's surface through which gases, ash, or molten lava are ejected. Also a mountain formed by this ejecta.

voluntary simplicity Deliberately choosing to live at a lower level of consumption as a matter of personal and environmental health.

vulnerable species Naturally rare organisms or species whose numbers have been so reduced by human activities that they are susceptible to actions that could push them into threatened or endangered status.

W

warm front A long, wedge-shaped boundary caused when a warmer advancing air mass slides over neighboring cooler air parcels.

waste stream The steady flow of varied wastes, from domestic garbage and yard wastes to industrial, commercial, and construction refuse.

water cycle The recycling and reutilization of water on earth, including atmospheric, surface, and underground phases and biological and nonbiological components.

water droplet coalescence A mechanism of condensation that occurs in clouds too warm for ice crystal formation.

waterlogging Water saturation of soil that fills all air spaces and causes plant roots to die from lack of oxygen; a result of overirrigation.

water scarcity Annual available freshwater supplies less than 1,000 m³ per person.

watershed The land surface and groundwater aquifers drained by a particular river system.

water stress A situation when residents of a country don't have enough accessible, high-quality water to meet their everyday needs.

water table The top layer of the zone of saturation; undulates according to the surface topography and subsurface structure.

watt (W) The force exerted by 1 joule, or the equivalent of a current of 1 amp per second flowing through a resistance of 1 ohm.

weather Description of the physical conditions of the atmosphere (moisture, temperature, pressure, and wind).

weathering Changes in rocks brought about by exposure to air, water, changing temperatures, and reactive chemical agents.

wedge analysis Policy options proposed by R. Socolow and S. Pacala for reducing greenhouse gas emissions using existing technologies. Each wedge represents a cumulative reduction of the equivalent of 1 billion tons of carbon over the next 50 years.

wetland mitigation Replacing a wetland damaged by development (roads, buildings, etc.) with a new or refurbished wetland.

wetlands Ecosystems of several types in which rooted vegetation is surrounded by standing water during part of the year. *See also* swamp, marsh, bog, fen.

wicked problems Problems with no simple right or wrong answer where there is no single, generally agreed-upon definition of or solution for the particular issue.

wilderness An area of undeveloped land affected primarily by the forces of nature; an area where humans are visitors who do not remain.

Wilderness Act Legislation of 1964 recognizing that leaving land in its natural state may be the highest and best use of some areas.

wildlife Plants, animals, and microbes that live independently of humans; plants, animals, and microbes that are not domesticated.

wildlife refuges Areas set aside to shelter, feed, and protect wildlife; due to political and economic pressures, refuges often allow hunting, trapping, mineral exploitation, and other activities that threaten wildlife.

windbreak Rows of trees or shrubs planted to block wind flow, reduce soil erosion, and protect sensitive crops from high winds.

wind farms Large numbers of windmills concentrated in a single area; usually owned by a utility or large-scale energy producer.

wise use groups A coalition of ranchers, loggers, miners, industrialists, hunters, off-road vehicle users, land developers, and others who call for unrestricted access to natural resources and public lands.

withdrawal A description of the total amount of water taken from a lake, river, or aquifer.

woodland A forest where tree crowns cover less than 20 percent of the ground; *also called* open canopy.

work The application of force through a distance; requires energy input.

world conservation strategy A proposal for maintaining essential ecological processes, preserving genetic diversity, and ensuring that the utilization of species and ecosystems is sustainable.

World Trade Organization (WTO) An association of 153 nations that meet to regulate international trade.

worldviews Sets of basic beliefs, images, and understandings that shape how we see the world around us.

X

X ray Radiation of very short wavelength in the electromagnetic spectrum; can penetrate soft tissue; although useful in medical diagnosis, also damages tissue and causes mutations.

Y

yellowcake The concentrate of 70 to 90 percent uranium oxide extracted from crushed ore.

Z

zero population growth (ZPG) The number of births at which people are just replacing themselves; *also called* the replacement level of fertility.

zone of aeration Upper soil layers that hold both air and water.

zone of leaching The layer of soil just beneath the topsoil where water percolates, removing soluble nutrients that accumulate in the subsoil.

zone of saturation Lower soil layers where all spaces are filled with water.

Credits

PHOTOGRAPHS

Design Elements

Subject Index: © imagebroker/Alamy RF; About the authors: © Glow Images/SuperStock RF; Brief TOC, detailed TOC: © Digital Vision/PunchStock RF; Preface: © Comstock Images/PictureQuest RF; Glossary: © Purestock/Getty RF; Credits: © Digital Vision/PunchStock RF

Learning to Learn

Page 1: © William P. Cunningham; p. 2: © The McGraw-Hill Companies, Inc./Barry Barker, photographer; p. 5: © BananaStock/JupiterImages/PictureQuest RF; p. 6: © PhotoDisc RF

Chapter 1

Opener: © Liu Jin/AFP/Getty; 1.1: Courtesy Jean Ku/DOE/NREL; 1.3: © Vol. 6/Corbis RF; 1.5a: NOAA Geophysical Fluid Dynamics Laboratory; 1.5b: © Norbert Schiller/The Image Works; 1.5c: © Roger A. Clark/Photo Researchers; 1.6a: © Dimas Ardian/Getty; 1.6b: © Christopher S. Collins, Pepperdine University; 1.6c: © William P. Cunningham; 1.7: © The Bridgeman Art Library/Getty; 1.8a: Courtesy of the Bancroft Library. University of California, Berkeley; 1.8b: Courtesy of Grey Towers National Historic Landmark; 1.8c: © Bettmann/Corbis; 1.8d: © AP/Wide World Photos; 1.9: © William. P. Cunningham; 1.10a: © AP/Wide World Photos; 1.10b: Tom Turney/The Brower Fund/Earth Island Institute; 1.10c: With permission of the University Archives, Columbia University in the City of New York; 1.10d: © AP/Wide World Photos; 1.11: © Stockgrek/age fotostock RF; 1.12: © William P. Cunningham; 1.14: © Justin Guariglia/Corbis; 1.17: © The McGraw-Hill Companies, Inc./Barry Barker, photographer; 1.18: © Check Six/Getty; 1.19: © The McGraw-Hill Companies, Inc./Barry Barker, photographer; 1.22: © Carr Clifton; 1.23: © William P. Cunningham; 1.24 © Sam Kittner; 1.25 © Basel Action Network 2006

Chapter 2

Opener and 2.2: © Artur Stefanski; 2.4 2.5: David L. Hansen, University of Minnesota Agricultural Experiment Station; 2.6: © Artur Stefanski; 2.10, 2.12: © William P. Cunningham; 2.13: © The McGraw-Hill Companies, Inc./John A. Karachewski, photographer

Chapter 3

Opener: © iStockphoto; p. 55: © G.I. Bernard/Animals Animals Earth Scenes; 3.8: © William P. Cunningham; 3.9: NOAA; Fig. 2 p. 68: The SeaWIFS Project, NASA/Goddard Space Flight Center and ORBIMAGE; 3.22: © Nigel Cattlin/Visuals Unlimited

Chapter 4

Opener: © Galen Rowell/Corbis; 4.1: Portrait of Charles Darwin, 1840 by George Richmond (1809-96). Down House, Downe, Kent, UK/Bridgeman Art Library; 4.2: © Vol. 6/Corbis RF; 4.3: © William P. Cunningham; 4.5: © visual safari/Alamy RF; 4.6, 4.13: © William P. Cunningham; 4.14: © Corbis RF; 4.15: © D. P. Wilson/Photo Researchers; 4.16: © Creatas/PunchStock RF; 4.17a,b: © Edward S. Ross; 4.18: © Tom Finkle; 4.19a: © William P. Cunningham; 4.19b: PhotoDisc RF; 4.19c: © William P. Cunningham; 4.20 © Vol. 6 PhotoDisc/Getty RF; 4.22a: © Corbis RF; 4.22b: © Eric and David Hosking/Corbis; 4.22c: © image100/

PunchStock RF; 4.24: © Wm. P. Cunningham; p. 91 (left & right): © Paul Ojanen; 4.27 & 4.28: © William P. Cunningham; 4.29: © Gerals Lacz/Peter Arnold/Photolibrary

Chapter 5

Opener: © iStockphoto; p. 99: © AP/Wide World Photos; 5.6: © Vol. 60/PhotoDisc RF; 5.7: © Vol. 6/Corbis RF; 5.8: © William P. Cunningham; 5.9: © Mary Ann Cunningham; 5.10: © William P. Cunningham; 5.11: © Vol. 90/Corbis RF; 5.12: © William P. Cunningham; 5.13: © Vol. 74/PhotoDisc RF; 5.14: Courtesy of SeaWiFS/NASA; 5.16: NOAA; 5.17a: © Vol. 89/PhotoDisc RF; 5.17b: © Robert Garvey/Corbis; 5.17c: © Andrew Martinez/Photo Researchers; 5.17d: © Pat O'Hara/Corbis; 5.18: USGS; 5.19: © Stephen Rose; 5.22a,b: © Vol. 16/PhotoDisc RF; 5.22c: © William P. Cunningham

Chapter 6

Opener: Image courtesy of the Food and Agriculture Organization; p. 117: © PunchStock RF; 6.2: Courtesy of National Museum of Natural History, Smithsonian Institution; p. 122: © Vol. 72/Corbis RF; 6.8a: © Digital Vision/Getty RF; 6.8b: © Stockbyte RF; 6.8c: © Corbis RF; 6.10: © Art Wolfe/Stone/Getty; Exploring Science Fig 1 p. 127: NOAA; 6.14: © The McGraw-Hill Companies, Inc./Barry Barker, photographer

Chapter 7

Opener: © Pierre Tremblay/Masterfile; 7.1: © Apichart Weerawong/AP/Wide World Photos; 7.2: Courtesy of John Cunningham; 7.5: © William P. Cunningham; 7.6: © Frans Lemmens/Getty; p. 140: © Alain Le Grasmeur/Corbis; 7.15a: © William P. Cunningham; 7.15b: © Vol. 17/PhotoDisc/Getty RF

Chapter 8

Opener: © Brand X Pictures/Jupiterimages RF; 8.1: © Getty RF; 8.4: © William P. Cunningham; 8.5a: © Vol. 40/Corbis RF; 8.5b: © Image Source/Getty RF; 8.5c: Courtesy of Stanley Erlandsen, University of Minnesota; 8.7: © Digital Vision/Getty RF; p. 171: © Dana Dolinoy, University of Michigan; 8.17: © The McGraw-Hill Companies, Inc./Sharon Farmer, photographer

Chapter 9

Opener: © Digital Vision/Getty RF; 9.1: © Simon Craven, 2005, Vassar College; 9.4: © Norbert Schiller/The Image Works; 9.8a: © Scott Daniel Peterson; 9.8b: © Lester Bergman/Corbis; 9.10: © William P. Cunningham; 9.12: Courtesy U.S. Department of Agriculture Conservation Services; 9.13: © FAO Photo/R.Faidutti; 9.14: Photo by Jeff Vanuga, courtesy of USDA Natural Resources Conservation Center; 9.15: USDA/NRCS; p. 187, 9.16, 9.17: © William P. Cunningham

Chapter 10

Opener: © John Maier/The New York Times/Redux Pictures; 10.4a,b: Courtesy USDA/NSCC; 10.7: © FAO Photo/R.Faidutti; 10.9a,b: Courtesy Natural Resource Conservation Service; 10.11a: Photo by Lynn Betts, courtesy of USDA Natural Resources Conservation Service; 10.11b: Photo by Jeff Vanuga, courtesy of USDA Natural Resources Conservation Center; 10.11c: © Corbis RF; 10.12: © William P. Cunningham; 10.13: © Vol. 13/Corbis RF; 10.14: © William P. Cunningham; 10.15: © Bettmann/

Corbis; 10.19: © Corbis RF; 10.21: © William P. Cunningham; 10.22: © Joe Munroe/Photo Researchers; Box Fig. 1 p.209: © The Capuchin Soup Kitchen-Earthworks Urban Farm, Detroit, MI; 10.24: Photo by Tim McCabe, courtesy of USDA Natural Resources Conservation Service; 10.25, 10.26,10.27: © William P. Cunningham; 10.28: © FAO Photo/W.Ciesla; 10.30: © W.C. Wood/Tanimura and AntLe, Inc.; 10.32a: Photo by Lynn Betts, courtesy of USDA Natural Resources Conservation Service; 10.32b & 10.33: © William P. Cunningham; p. 218 (left & right): Courtesy Dr. Bruno Glaser, University Bayreuth, Deutschland; 10.34: © Tom Sweeney/Minneapolis Star Tribune

Chapter 11

Opener: © William P. Cunningham; 11.1: Courtesy U.S. Fish & Wildlife Service/J & K Hollingworth; 11.2: © Vol. 112/PhotoDisc RF; 11.3: © Vol. 244/PhotoDisc RF; 11.5, 11.6, 11.7: © William P. Cunningham; 11.10: Courtesy USGS; 11.11: Courtesy of Bell Museum, University of MN. Photo taken by Mary Ann Cunningham; 11.13a: Courtesy of Dr. Mitchell Eaton; 11.13b: © William P. Cunningham; 11.13c: © Lynn Funkhouser/Peter Arnold/Photolibrary; 11.14: © Corbis RF; 11.15: © 2009 J.T. Oris; 11.16 Courtesy U.S. Fish and Wildlife Service, photographer Dave Menke; p. 239: © Getty RF; 11.19: © William P. Cunningham; 11.20: © Punchstock RF; 11.21: Courtesy of Dr. Ronald Tilson, Minnesota Zoological Garden

Chapter 12

Opener: © Jacques Jangous/Getty; 12.1: © Digital Vision/PunchStock RF; 12.4: © Digital Vision/Getty RF; 12.5,12.6: © William P. Cunningham; 12.8a-c: Courtesy United Nations Environment Programme; 12.10,12.11: © William P. Cunningham; 12.12: © Gary Braasch/Stone/Getty; 12.13: © Creatas/PunchStock RF; p. 257: © Ian Mcallister/Getty; 12.14: © William P. Cunningham; 12.16: © Mary Ann Cunningham; 12.17, 12.18,12.20: William P. Cunningham; 12.22: © Yellowstone National Park Wildlife Photo File, American Heritage Center, University of Wyoming; 12.23: © William P. Cunningham; 12.24: © Digital Vision/Getty RF; 12.25: © Corbis/RF; 12.26: © William P. Cunningham; 12.30: Courtesy of R.O. Bierregaard

Chapter 13

Opener: © Mary Ann Cunningham; 13.3, 13.4, 13.5b, 13.5c,13.6: © William P. Cunningham; 13.7: © Mark Edwards/Peter Arnold/Photolibrary; 13.8: © Vol. 5/PhotoDisc/Getty RF; 13.9, 13.10, 13.12: © William P. Cunningham; 13.14a: Courtesy of Wisconsin-Madison Archives; 13.14b: Courtesy University of Wisconsin Arboretum; 13.15: Courtesy Minnesota Department of Natural Resources; 13.16,13.17 © William P. Cunningham; 13.19: © Joel Sartore/Getty; 13.20: Courtesy U.S. Army Corps of Engineers; 13.21: © Chris Harris; 13.22: Courtesy U.S. Army Corps of Engineers; 13.23: © Jeff Greenberg/Photo Researchers; 13.25: Courtesy of South Florida Water Management District; p. 287: © Mary Ann Cunningham; 13.26: Courtesy NOAA; 13.27: Photo by Don Poggensee, USDA Natural Resources Conservation Service; 13.28: © William P. Cunningham; 13.29: Courtesy Federal Interagency Stream Corridor Restoration Handbook, NRCS, USDA; 13.31a-c: © William P. Cunningham; 13.33: Courtesy FAO/9717/I. de Borhegyi; 13.34: © William P. Cunningham

Figure 9.6: Data source: UN FAO; Figure 9.9: Data from USDA and UN FAO, 2008; Figure 9.20: USDA Economics Research Service, 2011.

Chapter 10

What can you do?: Source: Citizen's Guide to Pest Control and Pesticide Safety, EPA 730-K-95-001; Table 10.1: Source: Environmental Working Group, 2002; Table 10.2: Source: Based on 14 years' data from Missouri Experiment Station, Columbia, MO; Figure 10.8: USDA Natural Resource Conservation Service; Figure 10.9: Natural Resource Conservation Service; Figure 10.10: Source: ISRIC Global Assessment of Human-induced Soil Degradation, 2008; Figure 10.16: Source: US EPA, 2000; Figure 10.17: Data from the UN Food and Agriculture Organization, 2009; Figure 10.18: Source: USDA, 2009; Figure 10.20: Source: USDA; Figure 10.23: Worldwatch Institute, 2003; Figure 10.31: Source: Tolba, et al., World Environment, 1972–1992, p. 307, Chapman & Hall, 1992 United Nations Environment Programme; TA 10.2: Data source: USDA National Agricultural Statistics Service (NASS).

Chapter 11

Figure 11.4: Source: Conservation International, 2005; Figure 11.8: Source: U.S. Fish and Wildlife Service, 2008; Figure 11.12: Source: SeaWeb; Figure 11.17: Source: Copyright 1990 by Herb Block in The Washington Post. Reprinted by permission of The Herb Block Foundation; Table 11.1: Source: IUCN Red List 2008; Table 11.3: Source: W.W. Gibbs, 2001.

Chapter 12

Table 12.1: Source: Data from USFS, 2002; Table 12.2: Source: Data from World Conservation Union, 1990; Table 12.3: Source: World Commission on Protected Areas, 2007; Figure 12.2: Source: UN Food and Agriculture Organization (FAO), 2002; Figure 12.3: Source: UN Food and Agricultural Organization (FAO); Figure 12.7: Source: Data from FAO 2008; Figure 12.9: Source: Data from FAO, 2007; TA12.1: Source: Wildlife Conservation Society; TA 12.2: Source: Wildlife Conservation Society; Figure 12.15: Source: World Resource Institute, 2004; Figure 12.19: Source: UN World Commission on Protected Areas; Figure 12.21: Data from J. Hoekstra, et al, 2005.

Chapter 13

Figure 13.1: Source: USGS; Figure 13.2: Data Source: Walker and Moral, 2003; Figure 13.11: Data from S. Packard and J. Balaban, 1994; TA 13.1: Data from Walker and Moral, 2003.

Chapter 14

Table 14.3: Source: Blacksmith Institute 2006; Table 14.4: Source: E.T. Hayes, Implications of Materials Processing, 1997; Table 14.5: Source: The Disaster Center 2005; Table 14.6: Source: B. Gutenberg in Earth by F. Pres and R. Siever, 1978, W.H. Freeman & Company.

Chapter 15

Table 15.3: Source: Data from Pacala and Socolow, 2004; Figure 15.1: Source: IPCC 2007; Figure 15.4: Source: National Weather Service: http://www.srh.noaa. gov/jetstream/atmos/atmprofile.htm; Figure 15.14, p. 32: Source: Data from United Nations Environment Programme; Figure 15.17: Source:Data from NOAA Earth System Research Laboratory; Figure 15.19: Source: Climate Change 2007: Synthesis Report. Contribution of Working Groups I, II and III to the Fourth Assessment Report of the Intergovernmental Panel on Climate Change.

Figure SPM.1. IPCC, Geneva, Switzerland; Figure 15.20: Source: Data from IPCC, 2007; Figure 15.21: Source: Data from IPCC, 2007; TA 15.2 Climate Change 2007: Synthesis Report. Contribution of Working Groups I, II and III to the Fourth Assessment Report of the Intergovernmental Panel on Climate Change. Figure SPM.4. IPCC, Geneva, Switzerland.

Chapter 16

Table 16.1: Source: UNEP, 1999; Figure 16.1: Source: EPA, 1998; Figure 16.5: Source: UNEP, 1999; Figure 16.6: Source: UNEP, 1999; Figure 16.8: Source: UNEP, 1999; Figure 16.11: Source: UNEP, 1999; Figure 16.13: Source: Environmental Defense Fund based on EPA data, 2003; Figure 16.26: Source: National Atmospheric Deposition Program/National Trends Network, 2000. http://nadp.sws.uiuc.edu; Figure 16.31: Source: Environmental Protection Agency, 2002.

Chapter 17

Table 17.2: Data from UNEP, 2002; Table 17.3: Source: World Resources Institute; Figure 17.6: Source: U.S. Geological Survey; Figure 17.11: Source: Data from U.S. Department of Interior; Figure 17.12: Source: UNEP, 2002; Figure 17.15: Source: EPA, 2004; Figure 17.19: Source: Adapted from Climate Change 2007: Synthesis Report. Contribution of Working Groups I, II and III to the Fourth Assessment Report of the Intergovernmental Panel on Climate Change; Figure 3.3. IPCC, Geneva, Switzerland.; Figure 17.21: Data World Water Day, 2009; Figure 17.27: Source: World Bank estimates based on research paper by Dennis Anderson and William Cavendish, "Efficiency and Substitu- tion in Pollution Abatement: Simulation Studies in Three Sectors"; TA 17.3: Freshwater Stress and Scarcity in Africa by 2025, VITAL WATER GRAPHICS, reproduced with kind permission from the United Nations Environment Programme (www.unep.org/ vitalwater/management.htm).

Chapter 18

Figure 18.4: Source: UNESCO, 2002; Figure 18.13: Source: Data EPA 2009; Figure 18.18: Source: EPA Safe Drinking Water Information System 2001; Figure 18.24: © 2001 Sidney Harris, ScienceCartoonsPlus.com. Reprinted by permission; Figure 18.26: Source: World Bank estimates based on research paper by Dennis Anderson and William Cavendish, "Efficiency and Substitution in Pollution Abatement: Simulation Studies in Three Sectors."

Chapter 19

Table 19.2: Source: U.S. Department of Energy; Figure 19.3: Source: Data from British Petroleum, 2006; Figure 19.11: Source: Jean Laherrere, www.hubbertpeak.org; Figure 19.12: Source: Data from DOE, 2008; Figure 19.17: Source: British Petroleum, 2006.

Chapter 20

Table 20.1: Source: U.S. Department of Energy; Table 20.2: Source: Alan C. Lloyd, 1999; Untable: Tilman, et al., 2006. Science 314: 1598; Table 20.3: Source: Lester R. Brown, 1991; Figure 20.2: Source: U.S. Department of Energy; Figure 20.3: Source: U.S. Department of Energy; Figure 20.7: Source: National Weather Bureau, U.S. Department of Commerce; Figure 20.23: Source: Data from E. Marris, 2006. Nature 444: 670–678; Figure 20.28: Source: Data from U.S. Department of Energy; Figure 20.32: Source: World Energy Council, 2002.

Chapter 21

What can you do?: Source: Minnesota Pollution Control Agency; Table 21.1: Source: EPA, 2005; Figure 21.1:

Data Source: U.S. EPA; Figure 21.2: Source: Data from Environmental Protection Agency, 2006; Figure 21.3: Source: Data from U.S. Environmental Protection Agency Office of Solid Waste Management, 2006; Figure 21.8: Source: Eurostat, UNEP, 2003; TA 21.2: Source: M. Lavelle and M. Coyle, The National Law Journal, Vol. 15:52–56, No. 3, September 21, 1992; Figure 21.11: Source; Environmental Protection Agency, 2003; Figure 21.13: Source: Environmental Protection Agency, 2003; Figure 21.16: Source: Reprinted with special permission of Universal Press Syndicate; Figure 21.18: Source: Data from the U.S. EPA, 2002; Figure 21.20: Source: Environmental Protection Agency, 2004; Figure 21.21: Source: Environmental Protection Agency.

Chapter 22

Table 22.1: Source: United Nations Population Division, 2004; Table 22.2: Source: T. Chandler, Three Thousand Years of Urban Growth, 1974, Academic Press and World Gazetter, 2003; Table 22.3: Source: Excerpt from speech by Anthony Downs at the CTS Transportation Research Conference, as appeared on Website by Planners Web, Burlington, VT 2001; Table 22.4: Source: Vision 2000, Chattanooga, TN; Figure 22.3: Source: United Nations Population Division, World Urbanization Prospects, 2004; Figure 22.6: Source: U.S. Census Bureau; Figure 22.9: Source: EarthTrends, 2006.

Chapter 23

Table 23.2: Source: R. S. de Groot, Investing in Natural Capital, 1994; Table 23.3: Source: Adapted from R. Costanza et al, "The Value of the World's Ecosystem Services and Natural Capital," Nature, Vol. 387 (1997); Figure 23.9: Source: T. Prugh, National Capital and Human Economic Survival 1995, ISEE; Figure 23.10: Source: Her- man Daly in A. M. Jansson, et al, Investing in Natural Capi- tal, ISEE; Figure 23.17: Source: Redefining Progress, 2006; Figure 23.22: Source: United Nations, 2007; Figure 23.26: Source: E.S. Goodstein, Economic Policy Institute, Washing- ton, DC; TA23.3: Source: United Nations Development Programme, 2006.

Chapter 24

Table 24.1: Source: N. Vig and M. Kraft, Environmental Policy in the 1990s, 3rd Congressional Quarterly Press; Table 24.3: Kai N. Lee, Compass and Gyroscope, 1993, Copyright © 1993 Island Press, Reprinted by permission of Alexander Hoyt & Associates; Table 24.4: Source: L. Gunderson, C. Holling and S. Light, Barriers and Bridges to the Renewal of Ecosystems and Institutions, 1995, Columbia University Press; Figure 24.13: Source: U.S. General Accounting Office; Figure 24.16: Source: United Nations Environment Programme from Global Environment Outlook 2000.

Chapter 25

What Can You Do?: Based on material by Karen Oberhauser, Bell Museum Imprint, University of Minnesota, 1992. Used by permission; Table 25.1: Source: A Greenprint for Minne- sota, Minnesota Office of Environmental Education, 1993; Table 25.2: From Robert Merideth, The Environmentalist's Bookshelf: A Guide to the Best Books, 1993 by G. K. Hall, an imprint of Macmillan, Inc. Reprinted by permission; What can you do?: Based on material by Karen Oberhauser, Bell Museum Imprint, University of Minnesota, 1992. Used by permission; Table 25.3: Based on material from "Grassroots Organizing for Everyone" by Claire Greensfelder and Mike Roselle from Call to Action, 1990 Sierra Book Club Books; TA25.3: Reprinted by permission of Audubon; Figure 25.5: © 1990 Bruce von Alten; Figure 25.15: Source: UN, 2002.

Index

public citizenship, in environmental policy, 546
Public Interest Research Groups (PIRGs), 572
Public Utility Regulatory Policies Act (1978), 464
Pueblo to People, 27
pull factors, immigration, 499
pupfish, desert, temperature tolerance limits and, 77
purple loosestrife, 274
push factors, immigration, 499

Q

quality of life indicators, 25

R

r-selected species, 121
Rabelais, Nancy, 403
race, environmental health hazards and, 481
racism, environmental, 33
radioactive waste. *See* nuclear waste
Raghu, S., 462
rain shadow, 374
rainfall. *See also* precipitation
    acid precipitation, 362–363
    ingredients for, 323
    monsoons, 325–326
rainforests, 257, 265
rangelands
    grazing fees, 259
    overgrazing, 258–259
Reagan, Ronald, 552
recharge zones, 377–378
reclamation
    defined, 273
    restoration and, 291–292
    water policies and, 393
recycling, 569
    benefits, 479–480, 482
    commercial scale, 483
    composting, 483
    defined, 479
    geologic resources, 307–308
    hazardous wastes, 485–486
    jobs creation and, 534
    plastics, 482–483
    water, 392
red tides, 401–402
Redefining Progress, 19
reduced tillage systems, 217
reduction, of atoms, 54
reefs. *See* coral reefs
reflective thinking, 8
reformers, in fuel cells, 457
refuse-derived fuel, 479
regolith, 200
regulatory agencies, 551–552
rehabilitation, 273
Reichhold Chemical, 488
Reilly, William K., 562–563
reintroduction, 273
religion, and family planning, 150
remediation
    defined, 273
    water, 416–417
remote sensing, 68
renewable energy, 447–451, 454
renewable energy islands, 458
renewable resources, 517
Repeto, Robert, 525
replication, in science, 39

reproducibility, science and, 39
reproductive isolation, 224
reservoirs, sedimentation levels in, 388
residence time, 375
resilience
    in biological communities, 89–90
    in economic policy, 555, 556
    in systems, 46
Resource Conservation and Recovery Act (1976), 486
resource extraction, 304–307
resource partitioning, 78, 79
resources
    conservation (*see* conservation)
    defined, 516
    in economics, 516–517
    intangible, 517
    nonrenewable, 516, 517
    partitioning, 78, 79
    renewable, 517
    waste of, in start of environmental movement, 22
respiration, cellular, 61, 67
restoration ecology
    benefits of, 277–280
    common terms used in, 273
    components of restoration, 273–274
    defined, 272–273
    early conservationists, 274–275
    letting nature heal itself, 275
    origins of, 274–277
    pragmatic side of, 273
    protecting, 275
    reintroduction of native species, 275–277
    restoring prairies, 280–284
    restoring wetlands and streams, 284–292
rhinos, white and Javanese, 242, 243
ribonucleic acid (RNA), 57
Ricardo, David, 518
rice
    genetically modified, 190
    as major food crop, 183
ricin, 169
riders, legislative, 547
rill erosion, 202
risk, 170
    acceptance, 172
    assessment, 170
    defined, 170
    management, 173–174
    perception, 170, 172
rivers
    restoration, 289–291
    as water compartment, 379
Rivers and Harbors Act (1899), 543
Roadless Rule, 255
Robert, K. H., 532
Robinson, Frances, 5
rock cycle, 300
rock salt, 301
rocks
    defined, 299–300
    igneous, 300
    metamorphic, 300
    sedimentary, 301
    weathering and sedimentation, 300–301
Rogers, Will, 177
Roosevelt, Franklin D., 345
Roosevelt, Theodore, 21, 23, 275
rotational grazing, 259
Rowland, Sherwood, 359
run-of-the-river flow, 464

rural areas
    defined, 497
    mobile phone service to, 531
    Rural Electrification Act (1935), 464
Rusk, David, 504
Russia. *See also* Soviet Union (former)
    Chernobyl nuclear accident (1985), 291, 438, 441
    ice core drilling of Greenland ice sheet, 327
    nuclear waste site explosion near Chelyabinsk, 440
    population growth rate, 137
    toxic air pollution in Norilsk, 368
    wood and paper pulp production in, 249

S

S population curve, 119–120
saccharin, 169
Sachs, Jeffrey, 28, 161
Safe Drinking Water Act (1986), 411, 419, 434
Safe Harbor Policy, 238
Sagan, Carl, 48
saguaro cactus, 76–77
Sahara Desert (Africa), overgrazing, 103
Salazar, Ken, 432–433
salinization, 204
salmon
    dams and lethal effects on, 386–387
    endangered, 238
    genetically modified, 190, 191
salt marshes, 108, 109
saltwater intrusion, 385
sample, in statistics, 42
Sand County Almanac, A (Leopold), 22
sanitary landfills, 477
Sargasso Sea, 107
SARS (severe acute respiratory syndrome), 157
satellites, earth-imaging, 68
Saudi Arabia
    desalination in, 391
    oil production, 430
    parks and preserves in, 261
savannas, 103, 248, 278–279
Savory, Allan, 259
scarcity, 523–524
scavenger organisms, 63, 83–84
Schaller, George, 243
schistosomiasis, 463–464
Schneider, Steve, 340
Schumacher, E. F., 519
Science, 30
science
    accuracy and precision, 39–40
    basic principles, 39
    consensus and conflict in, 47–48
    deductive and inductive reasoning in, 40
    defined, 39
    experimental design and, 41–43
    hypotheses and theories, 40–41
    models, 43–44
    probability in, 41
    pseudoscience, detecting, 47–48
    skepticism and accuracy, dependence on, 39–40
    statistics, 42–43
    systems, 44–46
scientific consensus, 47
scientific theory, 40–41
Scott, J. Michael, 240–241
seafood, dietary, 185–186
seasonal winds, 324
SeaWIFS (satellite), 68
secondary air pollutants, 349

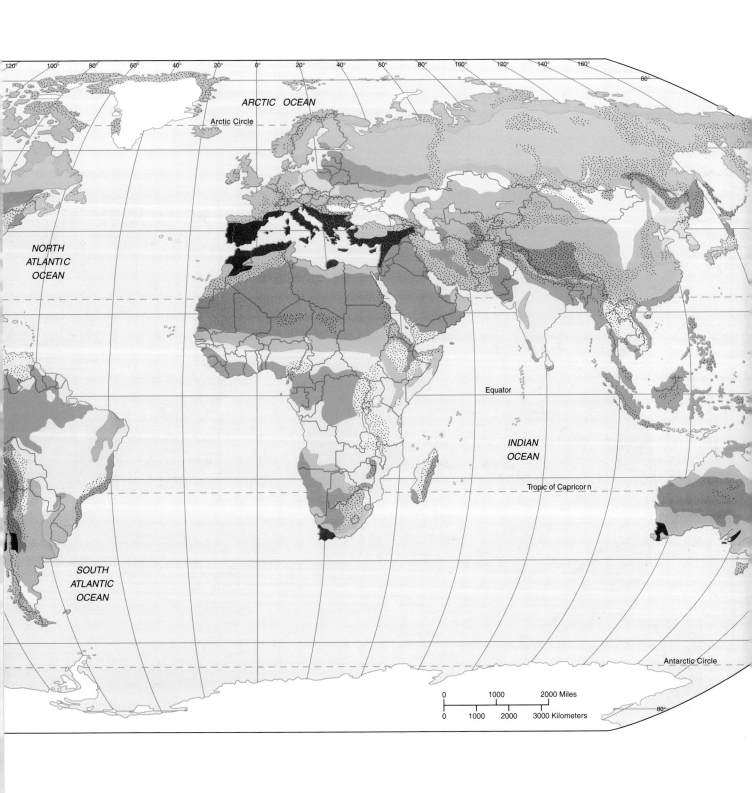

ARCTIC OCEAN

Arctic Circle

NORTH
ATLANTIC
OCEAN

SOUTH
ATLANTIC
OCEAN

Equator

INDIAN
OCEAN

Tropic of Capricorn

Antarctic Circle

| 0 | | 1000 | | 2000 Miles |
| 0 | 1000 | 2000 | 3000 Kilometers |

NORTH
PACIFIC
OCEAN

Tropic of Cancer

60°
40°
20°
0°
20°
40°
60°